Engineering Mathematics

Volume 1

Engineering Mathematics

Volume 1

BABU RAM

Formerly Dean, Faculty of Physical Sciences,
Maharshi Dayanand University, Rohtak

Pearson

Acquisitions Editor: Anita Yadav
Production Editor: Vamanan Namboodiri

ISBN: 9788131784709

First Impression, 2012
Sixth Impression, 2024
Seventh Impression, 2025

Published by Pearson India Education Services Pvt. Ltd, CIN: U72200TN2005PTC057128

Head Office: 1st Floor, Berger Tower, Plot No. C-001A/2, Sector 16B, Noida – 201 301,Uttar Pradesh, India.

Registered Office: Featherlite, 'The Address' 5th Floor, Survey No 203/10B, 200 Ft MMRD Road, Zamin Pallavaram, Chennai - 600044
Website: in .pearson.com; Email:companysecretary. india@pearson.com

Digitally printed in India by Trinity Academy for Corporate Training Ltd, New Delhi in the year of 2025.

★ ★ ★

In the Memory of My Parents

Smt. Manohari Devi and Sri. Makhan Lal

★ ★ ★

Contents

Preface to the Revised Edition

The book Engineering Maths I caters to the requirements of the revised syllabi of various Indian universities. Eight more chapters have been incorporated in this book to enable the students cover all the topics in the syllabus. Accordingly, the contents of the present book has also been divided into two volumes. Volume I of the book consists of six new chapters, namely, Tangents and Normals, Rectification and Quadrature, Centre of Gravity and Moments of Inertia, Logic, Elements of Fuzzy Logic, and Graphs in addition to the eleven chapters of Part I of first edition of the book. Volume II of the book consists of two new chapters entitled Calculus of Variation and Dynamics in addition to the fifteen chapters of Part II of the first edition of the book.

The contents of some of the previous chapters have been rearranged along with additions and refinements. A number of examples, selected generally from various university question papers, have been added in almost all chapters of this book.

I would like to thank Suresh Kumar Godara for the effort he has taken in preparing the figures of the book.

My wife Meena Kumari and daughter-in-law Ritu provided the moral support during the revision of this book. My son Aman Kumar, working with Goldman-Sach, Bangalore, offered, as usual, valid comments on the contents of some chapters.

Special thanks are due to Thomas Mathew Rajesh, Anita Yadav and Vamanan Namboodiri at Pearson Education for their constructive support.

Suggestions and feedback on the contents of the book will be gratefully acknowledged.

Babu Ram

Symbols and Basic Formulae

1 Greek Letters

α	alpha	ϕ	phi
β	beta	Φ	capital phi
γ	gamma	ψ	psi
Γ	capita gamma	Ψ	capital psi
δ	delta	ξ	xi
Δ	capital delta	η	eta
ε	epsilon	ζ	zeta
ι	iota	χ	chi
θ	theta	π	pi
λ	lambda	σ	sigma
μ	mu	Σ	capital sigma
ν	nu	τ	tau
ω	omega	ρ	rho
Ω	capital omega	κ	kapha

2 Algebraic Formulae

(i) Arithmetic Progression $a, a + d, a + 2d, \ldots,$ nth term $T_n = a + (n-1)d$;
Sum of n terms = $\frac{n}{2}$ $[2a + (n-1)d]$.

(ii) Geometrical Progression: $a, ar, ar^2, \ldots,$ nth term $T_n = ar^{n-1}$;

Sum of n terms = $\dfrac{a(1-r^n)}{1-r}$.

(iii) Arithmetic Mean of two number a and b is $\frac{1}{2}(a+b)$

(iv) Geometric Mean of two numbers a and b is $\sqrt{ab}.$

(v) Harmonic Mean of two numbers a and b is $\frac{2ab}{a+b}$.

(vi) If $ax^2 + bx + c = 0$ is quadratic, then

(a) its roots are given by $\dfrac{-b \pm \sqrt{b^2 - 4ac}}{2a}$,

(b) the sum of the roots is equal to $-\dfrac{b}{a}$,

(c) product of the roots is equal to $\dfrac{c}{a}$,

(d) $b^2 - 4ac = 0 \Rightarrow$ the roots are equal,

(e) $b^2 - 4ac > 0 \Rightarrow$ the roots are real and distinct,

(f) $b^2 - 4ac < 0 \Rightarrow$ the roots are complex,

(g) if $b^2 - 4ac$ is a perfect square, the roots are rational.

3 Properties of Logarithm

(i) $\log_a 1 = 0, \log_a 0 = -\infty$ for $a > 1, \log_a a = 1.$
$\log_e 2 = 0.6931, \log_e 10 = 2.306, \log_{10} e = 0.4343.$

(ii) $\log_a P + \log_a q = \log_a pq.$

(iii) $\log_a p - \log_a q = \log_a \dfrac{p}{q}.$

(iv) $\log_a p^q = q \log_a p.$

(v) $\log_a n = \log_b b . \log_b n = \dfrac{\log_b n}{\log_b a}.$

4 Angles Relations

(i) 1 radian = $\dfrac{180^0}{\pi}$

(ii) $1^0 = 0.0174$ radian

5 Algebraic Signs of Trigonometrical Ratios:

(a) First quadrant: All trig. Ratios are positive.

(b) Second quadrant: sin θ and cosec θ are positive, all others negative.

(c) Third quadrant: tan θ and cot θ are positive, all others negative.

(d) Fourth quadrant: cos θ and sec θ are positive, all others negative.

6 Commonly Used Values of Triganometrical Ratios:

$\sin\frac{\pi}{2}=1,\ \cos\frac{\pi}{2}=0,\ \tan\frac{\pi}{2}=\infty,$

$\operatorname{cosec}\frac{\pi}{2}=1,\ \sec\frac{\pi}{2}=\infty,\ \cos\frac{\pi}{2}=0,$

$\sin\frac{\pi}{6}=\frac{1}{2},\ \cos\frac{\pi}{6}=\frac{\sqrt{3}}{2},\ \tan\frac{\pi}{6}=\frac{1}{\sqrt{3}},$

$\operatorname{cosec}\frac{\pi}{6}=2,\ \sec\frac{\pi}{6}=\frac{2}{\sqrt{3}},\ \cot\frac{\pi}{6}=\sqrt{3},$

$\sin\frac{\pi}{3}=\frac{\sqrt{3}}{2},\ \cos\frac{\pi}{3}=\frac{1}{2},\ \tan\frac{\pi}{3}=\sqrt{3},$

$\operatorname{cosec}\frac{\pi}{3}=\frac{2}{\sqrt{3}},\ \sec\frac{\pi}{3}=2,\ \cot\frac{\pi}{3}=\frac{1}{\sqrt{3}},$

$\sin\frac{\pi}{4}=\frac{1}{\sqrt{2}},\ \cos\frac{\pi}{4}=\frac{1}{\sqrt{2}},\ \tan\frac{\pi}{4}=1,$

$\operatorname{cosec}\frac{\pi}{4}=\sqrt{2},\ \sec\frac{\pi}{4}=\sqrt{2},\ \cot\frac{\pi}{4}=1.$

7 Trig. Ratios of Allied Angles

(a) $\sin(-\theta)=-\sin\theta$, $\cos(-\theta)=\cos\theta$,
$\tan(-\theta)=-\tan\theta$,
$\operatorname{Cosec}(-\theta)=-\operatorname{cosec}\theta$, $\sec(-\theta)=\sec\theta$,
$\cot(-\theta)=-\cot\theta$.

(b) Any trig ratio of
$(n.\ 90+\theta)=$

$$\begin{cases}\pm\ \text{same trig. ratio } \theta \text{ when } n \text{ is even}\\ \pm\ \text{co-ratio of } \theta \text{ when } n \text{ is odd}\end{cases}$$

For example: $\sin(4620)=\sin[90^0(52)-60^0]$

$=\sin(-60^0)=-\sin 60^0=-\frac{\sqrt{3}}{2}.$

Similarly, $\operatorname{cosec}(270^0-\theta)=\operatorname{cosec}(90^0(3)-\theta)=-\sec\theta$.

8 Transformations of Products and Sums

(a) $\sin(A+B)=\sin A\cos B+\cos A\sin B$

(b) $\sin(A-B)=\sin A\cos B-\cos A\sin B$

(c) $\cos(A+B)=\cos A\cos B-\sin A\sin B$

(d) $\cos(A-B)=\cos A\cos B+\sin A\sin B$

(e) $\tan(A+B)=\dfrac{\tan A+\tan B}{1-\tan A\tan B}$

(f) $\tan(A-B)=\dfrac{\tan A-\tan B}{1+\tan A\tan B}$

(g) $\sin 2A=2\sin A\cos A=\dfrac{2\tan A}{1+\tan^2 A}$

(h) $\cos 2A=\cos^2 A-\sin^2 A=1-2\sin^2 A$
$=2\cos^2 A-1=\dfrac{1-\tan^2 A}{1+\tan^2 A}$

(i) $\tan 2A=\dfrac{\sin 2A}{\cos 2A}=\dfrac{2\tan A}{1-\tan^2 A}$

(j) $\sin 3A=3\sin A-4\sin^3 A$

(k) $\cos 3A=4\cos^3 A-3\cos A$

(l) $\tan 3A=\dfrac{3\tan A-\tan^3 A}{1-3\tan^2 A}$

(m) $\sin A+\sin B=2\sin\dfrac{A+B}{2}\cos\dfrac{A-B}{2}$

(n) $\sin A-\sin B=2\cos\dfrac{A+B}{2}\sin\dfrac{A-B}{2}$

(o) $\cos A+\cos B=2\cos\dfrac{A+B}{2}\cos\dfrac{A-B}{2}$

(p) $\cos A-\cos B=2\sin\dfrac{A+B}{2}\sin\dfrac{B-A}{2}$

(q) $\sin A\cos B=\frac{1}{2}[\sin(A+B)+\sin(A-B)]$

(r) $\cos A\sin B=\frac{1}{2}[\sin(A+B)-\sin(A-B)]$

(s) $\cos A\cos B=\frac{1}{2}[\cos(A+B)-\cos(A-B)]$

(t) $\sin A\sin B=\frac{1}{2}[\cos(A-B)-\cos(A+B)]$

9 Expressions for $\sin\frac{A}{2}$, $\cos\frac{A}{2}$ and $\tan\frac{A}{2}$

(a) $\sin\frac{A}{2}=\pm\sqrt{\dfrac{1-\cos A}{2}}$

(b) $\cos\frac{A}{2} = \pm\sqrt{\frac{1+\cos A}{2}}$

(c) $\tan\frac{A}{2} = \pm\sqrt{\frac{1-\cos A}{1+\cos A}}$

(d) $\sin\frac{A}{2} + \cos\frac{A}{2} = \pm\sqrt{1+\sin A}$

(e) $\sin\frac{A}{2} - \cos\frac{A}{2} = \pm\sqrt{1-\sin A}$

10 Relations Between Sides and Angles of a Triangle

(a) $\frac{a}{\sin A} = \frac{b}{\sin B} = \frac{c}{\sin C}$ (sine formulae)

(b) $$\left.\begin{aligned} \cos A &= \frac{b^2+c^2-a^2}{2bc} \\ \cos B &= \frac{c^2+a^2-b^2}{2ca} \\ \cos C &= \frac{a^2+b^2-c^2}{2cb} \end{aligned}\right\}\text{ cosine formulae}$$

(c) $$\left.\begin{aligned} a &= b\cos C + c\cos B \\ b &= c\cos A + a\cos C \\ c &= a\cos B + b\cos A \end{aligned}\right\}\text{ Projection fromulae.}$$

11 Permutations and Combinations Formulae

$${}^nP_r = \frac{n!}{(n-r)!},$$

$${}^nC_r = \frac{n!}{r!(n-r)!} = {}^nC_{n-r},$$

$${}^nC_0 = {}^nC_n = 1$$

12 Differentiations Formulae

(a) $\frac{d}{dx}(\sin x) = \cos x$ (b) $\frac{d}{dx}(\cos x) = -\sin x$

(c) $\frac{d}{dx}(\tan x) = \sec^2 x$ (d) $\frac{d}{dx}(\cot x) = -\operatorname{cosec}^2 x$

(e) $\frac{d}{dx}(\sec x) = \sec x\tan x$ (f) $\frac{d}{dx}(\operatorname{cosec} x) = -\operatorname{cosec} x\cot x$

(g) $\frac{d}{dx}(e^x) = e^x$ (h) $\frac{d}{dx}(a^x) = a^x\log_e a$

(i) $\frac{d}{dx}(\log_a x) = \frac{1}{x\log a}$ (j) $\frac{d}{dx}(\log_a x) = \frac{1}{x}$

(k) $\frac{d}{dx}(ax+b)^n = na(ax+b)^{n-1}$

(l) $\frac{d^n}{dx^n}(ax+b)^m = m(m-1)(m-2)\ldots(m-n+1)(ax+b)^{m-n}$

(m) $\frac{d}{dx}(\sin^{-1}x) = \frac{1}{\sqrt{1-x^2}}$ (n) $\frac{d}{dx}(\cos^{-1}x) = -\frac{1}{\sqrt{1-x^2}}$

(o) $\frac{d}{dx}(\tan^{-1}x) = \frac{1}{1+x^2}$ (p) $\frac{d}{dx}(\cot^{-1}x) = \frac{1}{1+x^2}$

(q) $\frac{d}{dx}(\sec^{-1}x) = \frac{1}{x\sqrt{x^2-1}}$ (r) $\frac{d}{dx}(\operatorname{cosec}^{-1}x) = -\frac{1}{x\sqrt{x^2-1}}$

(s) $\frac{d}{dx}(\sinh x) = \cosh x$ (t) $\frac{d}{dx}(\cosh x) = \sinh x$

(u) $D^n(uv) = D^n u + nc_1 D^{n-1}uDv + nc_2 D^{n-2}uD^2v + \ldots + {}^nC_r D^{n-r}uD^r v + \ldots + {}^nC_n uD^n v$

(Leibnitz's Formulae)

13 Integration Formulae

(a) $\int \sin x\,dx = -\cos x$ (b) $\int \cos x\,dx = \sin x$

(c) $\int \tan x\,dx = -\log\cos x$ (d) $\int \cot x\,dx = \log\sin x$

(e) $\int \sec x\,dx = \log(\sec x + \tan x)$

(f) $\int \operatorname{cosec} x\,dx = \log(\operatorname{cosec} x - \cot x)$

(g) $\int \sec^2 x\,dx = \tan x$ (h) $\int \operatorname{cosec}^2 x\,dx = -\cot x$

(i) $\int e^x\,dx = e^x$ (j) $\int a^x\,dx = \frac{a^x}{\log_e a}$

(k) $\int \frac{1}{x} dx = \log_e x$ (l) $\int x^n dx = \frac{x^{n+1}}{n+1}, n \neq -1$

(m) $\int \frac{dx}{a^2 - x^2} = \frac{1}{a}\tan^{-1}\frac{x}{a}$ (n) $\int \frac{dx}{a^2 - x^2} = \frac{1}{2a} \frac{a+x}{\log \hat{e}\, a - 2x}$

(o) $\int \frac{dx}{x^2 - a^2} = \frac{1}{2a}\log \hat{e} \frac{x-a}{x+a}$ (p) $\int \frac{dx}{\sqrt{a^2 - x^2}} = \sin^{-1}\frac{x}{a}$

(q) $\int \frac{dx}{\sqrt{a^2 + x^2}} = \sinh^{-1}\frac{x}{a}$ (r) $\int \frac{dx}{\sqrt{x^2 - a^2}} = \cosh^{-1}\frac{x}{a}$

(s) $\int \sqrt{a^2 + x^2}\, dx = \frac{x\sqrt{a^2 + x^2}}{2} + \frac{a^2}{2}\sinh^{-1}\frac{x}{a}$

(t) $\int \sqrt{x^2 - a^2}\, dx = \frac{x\sqrt{x^2 - a^2}}{2} + \frac{a^2}{2}\cosh^{-1}\frac{x}{a}$

(u) $\int \sqrt{a^2 - x^2}\, dx = \frac{x\sqrt{a^2 - x^2}}{2} + \frac{a^2}{2}\sin^{-1}\frac{x}{a}$

(v) $\int e^{ax} \sin bx\, dx = \frac{e^{ax}}{a^2 + b^2}(a \sin bx - b \cos bx)$

(w) $\int e^{ax} \cos bx\, dx = \frac{e^{ax}}{a^2 + b^2}(a \cos bx - b \sin bx)$

(x) $\int_0^{\frac{\pi}{2}} \sin^n x\, dx = \int_0^{\frac{\pi}{2}} COS^n x\, dx$

$$= \begin{cases} \frac{(n-1)(n-3)(n-5)\ldots}{n(n-2)(n-4)\ldots} & \text{if } n \text{ is odd} \\ \frac{(n-1)(n-3)(n-5)\ldots\pi}{n(n-2)(n-4)\ldots\ 2} & \text{if } n \text{ is even} \end{cases}$$

(y) $\int_0^{\frac{\pi}{2}} \sin^m x \cos^n x\, dx$

$$= \begin{cases} \frac{(m-1)(m-3)\ldots(n-1)(n-3)\ldots}{(m+n)(m+n-2)(m+n-4)\ldots} & \text{if } m \text{ and } n \text{ are not simultaneously even} \\ \frac{(m-1)(m-3)(n-5)\ldots\pi}{(m+n)(m+n-2)(m+n-4)\ldots 2} & \text{if both } m \text{ and } n \text{ are even} \end{cases}$$

14 Beta and Gamma Functions

(a) $\beta(m, n) = \int_0^1 x^{m-1}(1-x)^{n-1} dx$ converges for $m, n > 0$.

(b) $\Gamma(n) = \int_0^\infty e^{-x} x^{n-1} dx$ converges for $n > 0$

(c) $\Gamma(n+1) = n\Gamma(n)$ and $\Gamma(n+1) = n!$ if n is positive integer

(d) $\Gamma(1) = 1 = \Gamma(2)$ and $\Gamma(\frac{1}{2}) = \sqrt{\pi}$

(e) $\beta(m, n) = \frac{\Gamma(m)\Gamma(n)}{\Gamma(m+n)}$

(f) $\int_0^{\frac{\pi}{2}} \sin^p x \cos^q x\, dx = \frac{1}{2}\beta\left(\frac{p+1}{2}, \frac{q+1}{2}\right)$

$$= \frac{\Gamma\left(\frac{p+1}{2}\right)\Gamma\left(\frac{q+1}{2}\right)}{2\Gamma\left(\frac{p+q+1}{2}\right)}$$

(g) $\int_0^{\frac{\pi}{2}} \sqrt{\tan\theta}\, d\theta = \int_0^{\frac{\pi}{2}} \sin^{\frac{1}{2}}\theta \cos^{-\frac{1}{2}}\theta\, d\theta$

$$= \frac{\Gamma\left(\frac{3}{4}\right)\Gamma\left(\frac{1}{4}\right)}{2\Gamma(1)} = \tfrac{1}{2}\Gamma\left(\tfrac{1}{4}\right)\Gamma\left(\tfrac{3}{4}\right).$$

1 Sequences and Series

The intuitive concept of sequences of numbers involves not only a set of numbers but also an order in which these numbers have been placed. This suggests that for each positive integer, there is a number associated in the sequence. Thus, real sequence is a function whose domain is the set N of natural numbers and range a set of real numbers. Our aim in this chapter is to study the convergence of real sequences and then to apply the results to study the convergence behavior of various infinite series.

1.1 SEQUENCES

A function $f: N \to R$ whose domain is the set N of all natural numbers and range a set of real numbers is called a *sequence of real number* or simply a *real sequence.*

If $n \in N$, then $f(n)$ is generally denoted by x_n, a_n, or u_n and is called the nth *term of the sequence* $\{x_n\}$, $\{a_n\}$ or $\{u_n\}$. The set of all distinct terms of a sequence is called its range.

For example, consider the sequence

$$\{x_n\} = \{(-1)^n\} = \{-1,1,-1,1,-1\ldots\}.$$

Its nth term is $x_n = (-1)^n$ and the range is $\{-1,1\}$ Similarly, for the sequence $\{x_n\} = \left\{\frac{1}{n}\right\}_{n\in N}$, the nth term is $x_n = \frac{1}{n}$. All the elements in this sequence are distinct. Thus, the range for this sequence is an infinite set.

A sequence $\{x_n\}$ defined by $x_n = c$, where c is a fixed real number, for all $n \in N$, is called a *constant sequence.*

A sequence x_n is said to be *bounded above* if there exists a real number k, such that $x_n \le K$ for all $n \in N$. The number K is called an *upper bound* of the sequence $\{x_n\}$.

A sequence $\{x_n\}$ is said to be *bounded below* if there exists a real number k, such that $x_n \ge k$ for all $n \in N$. The number k is called a *lower bound* of the sequence $\{x_n\}$.

A sequence which is bounded both above and below is called a *bounded sequence.*

Thus a sequence $\{x_n\}$ is said to be bounded if there exist two real numbers k and K such that $k \le x_n \le K$ for all $n \in N$.

If we choose $M = \max\{|k|, |K|\}$, then $\{x_n\}$ is *bounded* if $|x_n| \le M$ for all $n \in N$.

If there exists no real number M such that $|x_n| \le M$ for all $n \in N$, then the sequence $\{x_n\}$ is said to be *unbounded.*

For example, consider the sequence $\left\{\frac{1}{n}\right\}_{n\in N}$. Since $0 < x_n \le 1$ for all $n \in$ N, the sequence $\left\{\frac{1}{n}\right\}$ is bounded. Here 1 is an upper bound, while 0 is a lower bound. We note that no element of the sequence is actually equal to zero, while one element is equal to the upper bound 1. Thus for this sequence, the upper bound is attained while the lower bound is not attained.

But if we consider the sequence $\{x_n\} = \{2^{n-1}\}$, then $\{x_n\} = \{1, 2^2, 2^3, \ldots\}$.

For this sequence $1 \le a_n$, $n \in$ N. But there no real number K such that $x_n \le K$ for all $n \in$ N. Therefore the given sequence is *not bounded above.*

1.2 CONVERGENCE OF SEQUENCES

A sequence $\{x_n\}$ is said to *converge to the limit* λ, if for every $\varepsilon > 0$ there exists a positive integer $n_0(\varepsilon)$ such that

$$|x_n - \lambda| < \varepsilon \text{ for all } n \ge n_0(\varepsilon).$$

If the sequence $\{x_n\}$ converges to the limit λ, then we write $\lim_{n\to\infty} x_n = \lambda$.

A sequence $\{x_n\}$ is said to be *divergent* if $\lim_{n\to\infty} x_n$ is not finite, that is, if $\lim_{n\to\infty} x_n$ is $+\infty$ or $-\infty$.

For example, consider the sequence $\{x_n\} = \left\{\frac{1}{2^n}\right\}_{n\in N}$. Then $\lim_{n\to\infty} x_n = \lim_{n\to\infty} \frac{1}{2^n} = 0$ (finite). Hence the sequence $\left\{\frac{1}{2^n}\right\}$ is convergent.

On the other hand, if we consider the sequence $\{x_n\} = \{n^2\}$, then $\lim_{n\to\infty} n^2 = \infty$. Hence the sequence $\{n^2\}$ is divergent.

A sequence $\{x_n\}$ which neither converges to a finite number nor diverges to ∞ or $-\infty$, is called an *oscillatory sequence.*

For example, consider the sequence $\{x_n\} = \{(-1)^n\}$. Here, the even elements are all +1 and so $x_{2n} \to 1$, whereas the odd elements are –1 and so $x_{2n-1} \to -1$. Hence the sequence $\{(-1)^n\}$ *oscillates finitely* between −1 and +1.

On the other hand, if $\{x_n\} = \{n\,(-1)^n\}$, then the sequence *oscillates infinitely* between $-\infty$ and $+\infty$.

Remark 1.1.

The positive integer n_0 in the definition of convergence of the sequence, in general, depends on the value of ε. For example, consider the sequence $\{a_n\} = 1, \frac{1}{2}, \frac{1}{3}, \ldots$ for which $a_n = \frac{1}{n} (n = 1, 2, \ldots)$. We would naturally guess that the sequence has the limit 0. To prove that 0 is the limit, we choose $\varepsilon > 0$. Then for the convergence we must find n_0 such that

$$|a_n - 0| < \varepsilon \text{ for all } n \geq n_0,$$

that is, such that

$$\left|\tfrac{1}{n} - 0\right| < \varepsilon \text{ for all } n \geq n_0 \qquad (1)$$

or such that

$$\frac{1}{n} < \varepsilon \text{ for all } n \geq n_0.$$

Thus, if we choose n_0 such that $\frac{1}{n_0} < \varepsilon$, then since $\frac{1}{n} < \frac{1}{n_0}$ if $n \geq n_0$, (1) will hold. But $\frac{1}{n_0} < \varepsilon$ if and only if $n_0 > \frac{1}{\varepsilon}$. Thus the condition (1) is satisfied if we choose the integer n_0 greater than $\frac{1}{\varepsilon}$. Thus, the sequence $\left\{\frac{1}{n}\right\}$ converges to 0 if $n_0 > \frac{1}{\varepsilon}$. Obviously, n_0 depends upon ε in this case.

If we consider the sequence $\{a_n\}$, where $a_n = 1 (n = 1, 2, \ldots)$, then the given sequence has the limit $\lambda = 1$. To prove this, we note that

$$a_n - \lambda = 1 - 1 = 0$$

so that for any $\varepsilon > 0$,

$$|a_n - \lambda| < \varepsilon \text{ for all } n \geq 1.$$

Thus for any $\varepsilon > 0$, we can choose $n_0 = 1$. *This is one of the rare cases where n_0 does not depend on ε.*

EXAMPLE 1.1

Show that the sequence $\{a_n\} = \{n\}$ does not have a limit.

Solution. Assume, on the contrary, that the sequence $\{n\}$ has the limit λ. Then for a given $\varepsilon > 0$, there must exist a positive integer n_0 such that

$$|a_n - \lambda| < \varepsilon \text{ for all } n \geq n_0,$$

that is,

$$|n - \lambda| < \varepsilon \text{ for all } n \geq n_0,$$

Choose $\varepsilon = 1$. Then, we have

$$|n - \lambda| < 1 \text{ for all } n \geq n_0,$$

or

$$\lambda - 1 < n < \lambda + 1 \text{ for all } n \geq n_0.$$

Thus all values of n satisfying $n \geq n_0$ must lie between $\lambda - 1$ and $\lambda + 1$. This is absurd. Thus, we arrive at a contradiction and so the given sequence has no limit.

EXAMPLE 1.2

If $a_n = 2 + \frac{(-1)}{n^2}$, find n_0 such that

$$\text{(i)}\ |a_n - 2| < \frac{1}{10^4} \text{ and (ii)}\ |a_n - 2| < \varepsilon$$

for all $n > n_0$. What conclusion can you draw regarding the convergence of $\{a_n\}$?

Solution. (i) We have

$$|a_n - 2| = \left|\frac{(-1)^n}{n^2}\right| = \frac{1}{n^2}.$$

Hence, $|a_n - 2| < \frac{1}{10^4}$ implies

$$\left|\frac{1}{n^2}\right| < \frac{1}{10^4} \text{ or } n^2 > 10^4 \text{ or } n > 10^2.$$

Therefore, if we choose $n_0 = 10^2 = 100$, then

$$|a_n - 2| < \frac{1}{10^4} \text{ for all } n > n_0.$$

(ii) The expression $|a_n - 2| < \varepsilon$ implies

$$\frac{1}{n^2} < \varepsilon \text{ or } n^2 > \frac{1}{\varepsilon} \text{ or } n > \frac{1}{\sqrt{\varepsilon}}.$$

Choose a positive integer $n_0 > \frac{1}{\sqrt{\varepsilon}}$. Hence for this n_0, we have

$$|a_n - 2| < \varepsilon \text{ for all } n < n_0,$$

Therefore, we conclude that the sequence $\{a_n\}$ converges to 2.

EXAMPLE 1.3

If $a_n = \frac{2n+1}{3n+5}$, show that $\lim_{n\to\infty} a_n = \frac{2}{3}$.

Solution. Let $\varepsilon > 0$. Then

$$\left|a_n - \frac{2}{3}\right| = \left|\frac{2n+1}{3n+5} - \frac{2}{3}\right|$$

$$= \left|\frac{-7}{3(3n+5)}\right|$$

$$= \frac{7}{3(3n+5)} < \varepsilon$$

implies

$$3n + 5 > \frac{7}{3\varepsilon}$$

or

$$n > \frac{7}{9\varepsilon} - \frac{5}{3}.$$

Let n_0 be a positive integer greater than $\frac{7}{9\varepsilon} - \frac{5}{3}$. Then,

$$\left|a_n - \frac{2}{3}\right| < \varepsilon \text{ for all } n > n_0.$$

Hence, $\lim_{n\to\infty} a_n = \frac{2}{3}$.

Theorem 1.1.

Every convergent sequence has a unique limit.

Proof: Suppose, on the contrary, that a sequence $\{x_n\}$ converges to two distinct real numbers λ and μ. Then for $\varepsilon > 0$, there exists positive integers m_1 and m_2 such that

$$|x_n - \lambda| < \frac{\varepsilon}{2} \text{ for all } n \geq m_1 \text{ and}$$

$$|x_n - \mu| < \frac{\varepsilon}{2} \text{ for all } n \geq m_2.$$

Therefore, for $n \geq \max(m_1, m_2)$, we have

$$|\lambda - \mu| = |\lambda - x_n + x_n - \mu| \leq |x_n - \lambda| + |x_n - \mu|$$

$$\leq \tfrac{\varepsilon}{2} + \tfrac{\varepsilon}{2} \text{ for all } n \geq \max(m_1, m_2)$$

$$= \varepsilon \text{ for all } n \geq \max(m_1, m_2).$$

If we select $\varepsilon = \frac{1}{2}|\lambda - \mu|$, then

$$|\lambda - \mu| < \tfrac{1}{2}|\lambda - \mu|. \text{ which is absurd.}$$

Hence, the sequence $\{x_n\}$ cannot converge to two different limits.

Theorem 1.2.

Every convergent sequence is bounded.

Proof: Let the sequence $\{x_n\}$ converge to the limit λ. Then, by the definition of convergence, to each $\varepsilon > 0$, there exists a positive integer m such that

$$|x_n - \lambda| < \varepsilon \text{ for all } n \geq m,$$

or

$$\lambda - \varepsilon < x_n < \lambda + \varepsilon \text{ for all } n \geq m.$$

If k and K are the least and greatest of $x_1, x_2, \ldots, x_{m-1}, \lambda - \varepsilon, \lambda + \varepsilon$, then

$$k \leq x_n \leq K \text{ for all } n \in N.$$

Hence the sequence $\{x_n\}$ is bounded.

Remark 1.2.

The converse of Theorem 1.2 may not be true. There are bounded sequences which do not converge. For example, the sequence $\{(-1)^n\}$ is bounded but it does not converge. In fact, it does not have a unique limit.

1.3 THE UPPER AND LOWER LIMITS OF A SEQUENCE

A number M is said to be the least upper bound of a sequence if

(i) no element of the sequence $\{x_n\}$ exceeds M

(ii) at least one element of the sequence $\{x_n\}$ exceeds $M - \varepsilon$, where ε is any positive number.

Similarly, a number m is called the *greatest lower bound* of a sequence $\{x_n\}$ if

(i) no element of the sequence is less than m

(ii) at least one element of the sequence is less than $m + \varepsilon$, where ε is a positive number.

For example, for the sequence $\{x_n\} = \{\frac{1}{n}\} = \{1, \frac{1}{2}, \ldots, \frac{1}{n}, \ldots\}$, the least upper bound is 1, whereas the greatest lower bound is 0. Since one element in the sequence is equal to the least upper bound, we say that the least upper bound is attained by this sequence. Since the greatest lower bound is not attained, there are infinite number of elements of the sequence that are less than $m + \varepsilon$.

A *superior number* for a given sequence is a number such that no element or only a finite number of elements of the sequence exceed this number.

An *inferior number* for a given sequence is a number such that only a finite number of elements or no element of the sequence is less than this number.

Thus,

(i) if a sequence is bounded above, then superior numbers always exists
(ii) if a sequence is bounded below, then inferior number always exists.

The greatest lower bound of the set of superior numbers of a sequence is called the *upper limit* or *limit superior* of the sequence. It is denoted by U or $\overline{\lim}\, a_n$.

The least upper bound of the set of inferior numbers of a sequence $\{a_n\}$ is called the *lower limit* or *limit inferior* of the sequence. We denote it by L or $\underline{\lim}\, a_n$.

For example,

(i) In the sequence $\{\frac{1}{n}\}$, the element $1, \frac{1}{2}, \frac{1}{3}, \ldots$ are all superior numbers. If we take $\frac{1}{985}$, only 984 number of the sequence are greater than $\frac{1}{985}$. The greatest lower bound of the superior number is 0. Hence, the upper limit of the sequence is 0. On the other hand, the inferior numbers of the given sequence are 0 and all negative numbers. Hence the lower limit is zero.

(ii) The superior numbers of the sequence $\{\frac{n}{n+1}\}$ consist of 1 and all numbers greater than 1. The greatest lower bound of the set of superior number is 1. Hence, the upper limit U for this sequence is 1. Also $\frac{n}{n+1}$ is an inferior number for all value of n. The least upper bound of this set of inferior numbers is 1. Thus $L = U = 1$. Also, upper bound $M = 1$ and lower bound of the sequence is $m = \frac{1}{2}$. The lower bound is attained by the sequence, but not the upper bound.

(iii) The superior numbers of the sequence $(-1)^n$ consists of 1 and all numbers greater than 1. The greatest lower bound of the set of superior numbers is 1. Hence, the upper limit $U = 1$. The inferior numbers of this sequence are $-1, -2, \ldots$. The least upper bound of this set is -1. Thus $L = -1$.

(iv) Consider the sequence $\{(-1)^n(1+\frac{1}{n})\} = \{-2, \frac{3}{2}, -\frac{4}{3}, \frac{5}{4}, -\frac{6}{5}, \ldots\}$. Clearly, $M = \frac{3}{2}$ and $m = 2$. Further, $\frac{3}{2}, \frac{5}{4}, \frac{7}{6}, \ldots$ are superior numbers and $-2, \frac{4}{3}, -\frac{6}{5}, \ldots$ are inferior numbers.

Therefore, $U = 1$ and $L = -1$.

Theorem 1.3.

The necessary and sufficient condition that a sequence should be convergent is that its upper and lower limits are equal.

Proof: Suppose that the sequence $\{x_n\}$ converges to the limit λ. Then to each $\varepsilon > 0$, there exists a positive integer n_0, such that

$$|x_n - \lambda| < \varepsilon \text{ for all } n > n_0$$

or

$$\lambda - \varepsilon < x_n < \lambda + \varepsilon \text{ for all } n > n_0.$$

Hence $\lambda + \varepsilon$ is a superior number. Then U, being the greatest lower bound of the set of superior numbers, we have $U \le \lambda$. But $\lambda - \varepsilon$ is not a superior number. Therefore, U cannot be less than λ. Hence $U = \lambda$.

Similarly, $\lambda - \varepsilon$ is an inferior number for all $\varepsilon > 0$ and the least upper bound of such number is λ. Thus $L = \lambda$. Hence $U = L = \lambda$.

Suppose conversely that $U = L$. By definition, only a finite number of terms of the sequence exceed $U + \varepsilon$ and only a finite number of terms are less than $L - \varepsilon$. Thus, except for finite number of terms, all the other terms lie between $L - \varepsilon$ and $U + \varepsilon$, that is, between $U - \varepsilon$ and $U + \varepsilon$. Hence, $U - \varepsilon < x_n < U + \varepsilon$ after a certain value of n. This implies $\lim_{n\to\infty} x_n = U$ and so the sequence $\{x_n\}$ converges.

1.4 CAUCHY'S PRINCIPLE OF CONVERGENCE

The following fundamental theorem is useful for determining whether a sequence converges or not.

Theorem 1.4.
(Cauchy's Principle of Convergence). A necessary and sufficient condition for the convergence of a sequence $\{x_n\}$ is that for each $\varepsilon > 0$ there exists a positive integer n_0 such that

$$|x_m - x_n| < \varepsilon \text{ for all } m > n > n_0.$$

Proof: *Necessity*: If the sequence $\{x_n\}$ converges to the limit λ, then to each $\varepsilon > 0$ there exists a positive integer n_0 such that

$$|x_n - \lambda| < \frac{\varepsilon}{2} \text{ for all } n > n_0,$$

$$|x_m - \lambda| < \frac{\varepsilon}{2} \text{ for all } m > n_0.$$

Therefore,

$$\begin{aligned}|x_m - x_n| &= |x_m - \lambda - x_n + \lambda|\\ &= |(x_m - \lambda) - (x_n - \lambda)|\\ &\le |(x_m - \lambda)| + |(x_n - \lambda)|\\ &< \frac{\varepsilon}{2} + \frac{\varepsilon}{2} = \varepsilon \text{ for all } m,\ n > n_0.\end{aligned}$$

Sufficiency: Suppose

$$|x_m - x_n| < \varepsilon \text{ for all } m > n > n_0.$$

Therefore,

$$x_m - \varepsilon < x_n < x_m + \varepsilon \text{ for all } m > n > n_0.$$

Thus, $x_m + \varepsilon$ is a superior number whereas $x_m - \varepsilon$ is an inferior number. Therefore, if U and L denotes upper and lower limits, respectively, we have

$$U < x_m + \varepsilon \text{ and } L > x_m - \varepsilon,$$

which yields

$$U - L < 2\varepsilon.$$

Also $U \ge L$, that is, $U - L \ge 0$. Hence,

$$0 \le U - L < 2\varepsilon.$$

Since $\varepsilon > 0$ is arbitrary, it follows that $U = L$. Hence, by Theorem 1.3, the sequence $\{x_n\}$ converges. This completes the proof of Theorem 1.4.

A sequence $\{x_n\}$ is called a *Cauchy Sequence* or a *Fundamental Sequence* if for each $\varepsilon > 0$, there exists a positive integer n_0 such that

$$|x_m - x_n| < \varepsilon \text{ for } m > n \ge n_0.$$

Thus, Theorem 1.4 may be restated in the following way:

Theorem 1.5.
A sequence of real numbers is convergent if and only if it is a Cauchy Sequence.

EXAMPLE 1.4
Show that the sequence

$$\{x_n\} = \left\{\frac{\cos n\pi}{n}\right\}$$

is a Cauchy Sequence.

Solution. We have

$$\begin{aligned}|x_n - x_m| &= \left|\frac{\cos n\pi}{n} - \frac{\cos m\pi}{m}\right|\\ &\le \frac{m\,|\cos n\pi| + n\,|\cos m\pi|}{nm}\\ &= \frac{m+n}{nm}.\end{aligned}$$

If $m > n$, we have

$$|x_n - x_m| \le \frac{n+m}{nm} < \frac{2m}{nm} = \frac{2}{n}.$$

Choose $\varepsilon > 0$ and $n_0 > \frac{2}{\varepsilon}$. Then

$$|x_n - x_m| < \varepsilon \text{ for } n,m > n_0.$$

Hence $\{x_n\}$ is a Cauchy Sequence.

EXAMPLE 1.5
Using Cauchy's Principle of Convergence, show that the sequence $\{x_n\}$, where $x_n = 1 + \frac{1}{2} + \frac{1}{3} + \ldots + \frac{1}{n}$, is not convergent.

Solution. Suppose, on the contrary, that $\{x_n\}$ is convergent. Then taking $\varepsilon = \frac{1}{2}$, the Cauchy's Principal yields

$$|x_{2m} - x_m| < \frac{1}{2} \tag{1}$$

But

$$\begin{aligned}x_{2m} - x_m &= \frac{1}{m+1} + \frac{1}{m+2} + \ldots + \frac{1}{2m}\\ &> \frac{1}{2m} + \frac{1}{2m} + \ldots + \frac{1}{2m} \quad (m \text{ terms})\\ &= \frac{m}{2m} = \frac{1}{2},\end{aligned}$$

that is,

$$|x_{2m} - x_m| > \frac{1}{2},$$

which is a contradiction to (1). Hence, the sequence is not convergent.

1.5 MONOTONIC SEQUENCE

A sequence $\{x_n\}$ is said to be *monotonic increasing or steadily increasing* if $x_{n+1} \geq x_n$ for all n.

A sequence $\{x_n\}$ is said to be *monotonic decreasing* or *steadily decreasing* if $x_{n+1} \leq x_n$ for all n.

A sequence $\{x_n\}$ is said to be *monotonic* if it is either monotonic increasing or monotonic decreasing.

A sequence $\{x_n\}$ is *strictly increasing* if $x_{n+1} > x_n$ for all n and *strictly decreasing* if $x_{n+1} < x_n$ for all n.

For example,

(i) $1+\frac{1}{2}, \frac{1}{3}, \frac{1}{4}, \ldots$ is a monotonic decreasing sequence.

(ii) $\left\{\frac{n}{n+1}\right\}$ is a monotonic increasing sequence.

(iii) $\{n\}$ is a monotonic increasing sequence.

Theorem 1.6.

A monotonic sequence always tends to a limit (finite or infinite). (Thus, a monotonic sequence is either convergent or divergent, it cannot oscillate).

Proof: Consider the monotonic increasing sequence, $\{x_n\}$. Suppose that its upper bound M is finite. Therefore, $x_n > M - \varepsilon$ for some value of n, say n_0 and, hence, for all $n > n_0$. Therefore,

$$M - \varepsilon < x_n \leq M \text{ for all } n > n_0$$

and so $\lim_{n\to\infty} x_n = M$ proving that $\{x_n\}$ is convergent.

If M is infinite, we can find n_0 such that x_n is greater than any positive N for all $n > n_0$. Therefore, $\lim_{n\to\infty} x_n = \infty$. Hence, the theorem is proved for monotonic increasing sequence. Also, we have shown that the *limit of a monotonic increasing sequence is its upper bound.*

Similarly, a monotonic decreasing sequence tends to its lower bound. If the lower bound is finite, say L, the monotonic decreasing sequence converges to L. If the lower bound is not finite, then the sequence converges to $-\infty$.

Theorem 1.7.

A necessary and sufficient condition for the convergence of a monotonic sequence is that it is bounded.

Proof: *Necessity*: We have already proved in Theorem 1.2 that a convergent sequence is bounded.

Sufficiency: Suppose that $\{x_n\}$ is a bounded monotonic increasing sequence. Thus range of $\{x_n\}$ is bounded and so, by completeness property, it has least upper bound, say M. We assert that $\{x_n\}$ converges to M. To prove it, let $\varepsilon > 0$ be a positive number. Since M is supremum, there exists at least one member say x_m such that $x_m > M - \varepsilon$. Since $\{x_n\}$ is monotonic increasing, we have

$$x_n \geq x_m > M - \varepsilon \text{ for all } n \geq m.$$

Also, since M is supremum,

$$x_n \leq M < M + \varepsilon \text{ for all } n.$$

Hence

$$M - \varepsilon < x_n < M + \varepsilon \text{ for all } n \geq m$$

or

$$|x_n - M| < \varepsilon \text{ for all } n \geq m.$$

Hence, $\{x_n\}$ converges to M.

The case of a bounded monotonic decreasing sequence can be considered similarly.

EXAMPLE 1.6

Show that the sequence $\{x_n\}$, where

$$x_n = 1 + \frac{1}{1!} + \frac{1}{2!} + \ldots + \frac{1}{n!}$$

is convergent.

Solution. We have

$$x_{n+1} - x_n = \frac{1}{(n+1)!} > 0 \text{ for all } n.$$

Hence $\{x_n\}$ is monotonic increasing. Further,

$$x_n = 1 + \frac{1}{1!} + \frac{1}{2!} + \ldots + \frac{1}{n!} < 1 + 1 + \frac{1}{2}$$
$$+ \ldots + \frac{1}{2^{n-1}} < 1 + \frac{1}{1-\frac{1}{2}} = 3$$

and so $\{x_n\}$ is bounded. Hence by Theorem 1.6, the given sequence converges to its upper bound, which is less than 3. In fact, $x_n \to e = 2.71\ldots$, the base of the natural logarithm.

EXAMPLE 1.7

Show that the sequence $\{x_n\}$, where

$$x_n = \left(1+\frac{1}{n}\right)^n,$$

is convergent.

Solution. We have

$$x_n = \left(1+\frac{1}{n}\right)^n.$$

By Binomial Theorem for a positive integer,

$$x_n = 1+\frac{n}{n}+\frac{n(n-1)}{2!}\cdot\frac{1}{n^2}+\frac{n(n-1)(n-2)}{3!}\cdot\frac{1}{n^3}$$
$$+\ldots+\frac{n(n-1)\ldots 1}{n!}\cdot\frac{1}{n^n}$$
$$=1+1+\frac{1-\frac{1}{n}}{2!}+\frac{\left(1-\frac{1}{n}\right)\left(1-\frac{2}{n}\right)}{3!}$$
$$+\ldots+\frac{\left(1-\frac{1}{n}\right)\left(1-\frac{2}{n}\right)\ldots\left(1-\frac{n-1}{n}\right)}{n!}.$$

Therefore, changing n to $n+1$, we get

$$x_{n+1} = 1+1+\frac{1-\frac{1}{n+1}}{2}+\frac{\left(1-\frac{1}{n+1}\right)\left(1-\frac{2}{n+1}\right)}{3!}$$
$$+\ldots+\frac{\left(1-\frac{1}{n+1}\right)\left(1-\frac{2}{n+1+1}\right)\ldots\left(1-\frac{n}{n+1}\right)}{(n+1)!}$$

All the factors in the numerators of x_n and x_{n+1} are positive. Further, each factor in the numerator of x_{n+1} is greater than the corresponding factor in the numerator of x_n whereas the denominators of x_n and x_{n+1} are same. Therefore, $x_{n+1} > x_n$ and so $\{x_n\}$ is monotonic increasing. Also,

$$x_n < 1+1+\frac{1}{2!}+\ldots+\frac{1}{n!}$$
$$<1+1+\frac{1}{2}+\frac{1}{2^2}+\frac{1}{2^3}+\ldots+\frac{1}{2^{n-1}}<3$$

and so $\{x_n\}$ is bounded. Hence, $\{x_n\}$ converges to a finite positive limit whose value is less than 3. In fact the $\lim\limits_{n\to\infty} x_n = e = 2.71\ldots$.

EXAMPLE 1.8

Show that $\lim\limits_{x\to 0}(1+x)^{1/x} = e$.

Solution. Substituting $x=\frac{1}{y}$, we get

$$\lim_{y\to\infty}\left(1+\frac{1}{y}\right)^y$$

and so the result follows from Example 1.7.

EXAMPLE 1.9

Show that the sequence whose nth term is

$$x_n = \frac{2n-7}{3n+2},$$

(i) is monotonic increasing, (ii) is bounded above, (iii) is bounded below, and (iv) has a limit.

Solution. For the given sequence,

$$x_n = \frac{2n-7}{3n+2},\quad x_{n+1} = \frac{2(n+1)-7}{3(n+1)+2} = \frac{2n-5}{3n+5}.$$

Then

$$x_{n+1}-x_n = \frac{2n-5}{3n+5}-\frac{2n-7}{3n+2} > 0 \text{ for all } n.$$

Hence, the sequence is monotonic increasing. The sequence is bounded above and below. The upper and lower bounds being $\frac{2}{3}$ and -1. Further,

$$\lim_{n\to\infty} x_n = \lim_{n\to\infty}\frac{2-\frac{7}{n}}{3+\frac{2}{n}} = \frac{2}{3}.$$

EXAMPLE 1.10

Show that $\lim\limits_{n\to\infty}(n)^{1/n} = 1$.

Solution. Let $a_n = (n)^{1/n} = 1 + h_n$, where $h_n > 0$. Then,

$$n = a_n^n = (1+h_n)^n$$
$$= 1+nh_n+\frac{1}{2}n(n-1)h_n^2$$
$$+\ldots+\frac{1}{n!}n(n-1)\ldots h_n^n > \frac{n(n-1)}{2}h_n^2.$$

Thus,

$$h_n^2 < \frac{2}{n-1}$$

and so

$$0 < h_n < \left(\frac{2}{n-1}\right)^{1/2}.$$

Let $\varepsilon > 0$ be given. Then,

$$\left(\frac{2}{n-1}\right)^{1/2} < \varepsilon \text{ if and only if } n > 1+\frac{2}{\varepsilon^2}.$$

If n_0 is a positive integer greater than $1+\frac{2}{\varepsilon^2}$, then

$$-\varepsilon < 0 < h_n < \left(\frac{2}{n-1}\right)^{1/2} < \varepsilon.$$

Hence, $h_n \to 0$ and so $\lim_{n\to\infty} a_n = \lim_{n\to\infty}(1+h_n) = 1$.

1.6 THEOREMS ON LIMITS

We will now discuss some important properties of the limits of sequences.

Theorem 1.8.

If $\{a_n\}$, $\{b_n\}$ are two sequences converging to a and b, respectively, then

(i) $\lim_{n\to\infty}(a_n \pm b_n) = \lim_{n\to\infty} a_n \pm \lim_{n\to\infty} b_n = a \pm b$

(ii) $\lim_{n\to\infty}(a_n b_n) = (\lim_{n\to\infty} a_n)(\lim_{n\to\infty} b_n) = ab$

(iii) $\lim_{n\to\infty}\left(\frac{a_n}{b_n}\right) = \frac{\lim_{n\to\infty} a_n}{\lim_{n\to\infty} b_n} = \frac{a}{b}, b \neq 0, b_n \neq 0$ for all n.

Proof: Let $\varepsilon > 0$. Since, $a_n \to a$, $b_n \to b$ as $n \to \infty$, there exists positive integers m_1 and m_2 such that

$$|a_n - a| < \frac{\varepsilon}{2} \text{ for all } n \geq m_1$$

and

$$|b_n - b| < \frac{\varepsilon}{2} \text{ for all } n \geq m_2.$$

Therefore, for $m = \max(m_1, m_2)$, we have

$$|a_n - a| < \frac{\varepsilon}{2} \text{ for all } n \geq m$$

and

$$|b_n - b| < \frac{\varepsilon}{2} \text{ for all } n \geq m.$$

Therefore,

$$\begin{aligned}|(a_n \pm b_n) - (a \pm b)| &= |(a_n - a) \pm (b_n - b)| \\ &\leq |a_n - a| + |b_n - b| \\ &< \frac{\varepsilon}{2} + \frac{\varepsilon}{2} = \varepsilon \text{ for all } n \geq m.\end{aligned}$$

Hence, by definition of convergence, the sequence $\{(a_n \pm b_n)\}$ converges to $a \pm b$.

(ii) Let $\varepsilon > 0$. Then, as in part (i),

$$|a_n - a| < \frac{\varepsilon}{2}, \ |b_n - b| < \frac{\varepsilon}{2} \text{ for all } n \geq m.$$

Also, since $\{a_n\}$ and $\{b_n\}$ are convergent sequences, they are bounded. Therefore, there exist positive real numbers k and K such that $|a_n| < k$ and $|b_n| \leq k$ for all $n \in N$. Thus, for $n \geq m$, we have

$$\begin{aligned}|(a_n b_n - ab)| &= |a_n(b_n - b) + b(a_n - a)| \\ &\leq |a_n||b_n - b| + |b||a_n - a| \\ &\leq k\frac{\varepsilon}{2} + |b|.\frac{\varepsilon}{2} = \frac{\varepsilon}{2}(k + |b|).\end{aligned}$$

Hence

$$\lim_{n\to\infty}(a_n b_n) = ab.$$

(iii) If $b \neq 0$, we have

$$\left|\frac{1}{b_n} - \frac{1}{b}\right| = \left|\frac{b - b_n}{bb_n}\right|.$$

Choose n_0 such that $|b_n - b| < \varepsilon|b||b_n|$ for all $n > n_0$. Therefore,

$$\left|\frac{1}{b_n} - \frac{1}{b}\right| < \varepsilon \text{ for all } n > n_0.$$

Hence if $b_n \to b$ as $n \to \infty$, then $\frac{1}{b_n} \to \frac{1}{b}$ provided $b \neq 0$. Now $a_n \to a$, $\frac{1}{b_n} \to \frac{1}{b}$, $b \neq 0$

Therefore, by part (ii), $\frac{a_n}{b_n} \to a\left(\frac{1}{b}\right) = \frac{a}{b}$ as $n \to \infty$.

Theorem 1.9.

If $\{b_n\}$ is a monotonic increasing sequence tending to infinity and $b_n > 0$ for all n, then

$$\lim_{n\to\infty}\frac{a_n}{b_n} = \lim_{n\to\infty}\frac{a_{n+1} - a_n}{b_{n+1} - b_n},$$

provided, the latter limit exists.

Proof: Suppose that $\lim_{n\to\infty}\frac{a_{n+1} - a_n}{b_{n+1} - b_n} = \lambda$ (finite). Therefore, to each $\varepsilon > 0$, there exists a positive integer n_0 such that

$$\left|\frac{a_{n+1} - a_n}{b_{n+1} - b_n} - \lambda\right| < \varepsilon \text{ for all } n \geq n_0.$$

Therefore,

$$\lambda - \varepsilon < \frac{a_{n+1} - a_n}{b_{n+1} - b_n} < \lambda + \varepsilon \text{ for all } n \geq n_0.$$

or

$$\begin{aligned}(\lambda - \varepsilon)(b_{n+1} - b_n) &< a_{n+1} - a_n \\ &< (\lambda + \varepsilon)(b_{n+1} - b_n) \text{ for all } n \geq n_0.\end{aligned}$$

Substituting $n = m, m + 1, m + 2, \ldots, n - 1$ and adding, we get

$$(\lambda - \varepsilon)(b_n - b_m) < a_n - a_m < (\lambda + \varepsilon)(b_n - b_m)$$

or

$$(\lambda-\varepsilon)\left(1-\frac{b_m}{b_n}\right)+\frac{a_m}{b_n}<\frac{a_n}{b_n}<(\lambda+\varepsilon)\left(1-\frac{b_m}{b_n}\right)+\frac{a_m}{b_n}.$$

Since $b_n \to \infty$,

$$\lim_{n\to\infty}(\lambda-\varepsilon)\left(1-\frac{b_m}{b_n}\right)+\frac{a_m}{b_n}=\lambda-\varepsilon$$

and

$$\lim_{n\to\infty}(\lambda+\varepsilon)\left(1-\frac{b_m}{b_n}\right)+\frac{a_m}{b_n}=\lambda+\varepsilon.$$

Therefore, we can find a positive integer n_0 such that for $n \geq n_0$, we have

$$(\lambda-\varepsilon)\left(1-\frac{b_m}{b_n}\right)+\frac{a_m}{b_n}>\lambda-\varepsilon-\varepsilon=\lambda-2\varepsilon$$

and

$$(\lambda+\varepsilon)\left(1-\frac{b_m}{b_n}\right)+\frac{a_m}{b_n}<\lambda+\varepsilon+\varepsilon=\lambda+2\varepsilon.$$

Thus,

$$\lambda-2\varepsilon<\frac{a_n}{b_n}<\lambda+2\varepsilon \text{ for all } n\geq n_0.$$

Hence,

$$\lim_{n\to\infty}\frac{a_n}{b_n}=\lambda\cdot$$

Theorem 1.10. (Cauchy).

If a_n is positive for all values of n, then

$$\lim_{n\to\infty} a_n^{1/n}=\lim_{n\to\infty}\frac{a_n+1}{a_n},$$

provided, the latter limit exists.

(While studying convergence of infinite series, we shall observe that this theorem of Cauchy implies that whenever D'Alembert's Ratio Test is applicable, Cauchy's Root Test is also applicable to test the convergence of a series of positive terms).

Proof: Suppose that $\lim_{n\to\infty}\frac{a_{n+1}}{a_n}=\lambda$ (finite). Therefore, for $\varepsilon > 0$, we have

$$\lambda-\varepsilon<\frac{a_{n+1}}{a_n}<\lambda+\varepsilon \text{ for all } n\geq m.$$

Substituting $n = m, m + 1,\ldots n-1$ in succession and multiplying vertically, we get

$$(\lambda-\varepsilon)^{n-m}<\frac{a_n}{a_m}<(\lambda+\varepsilon)^{n-m}$$

or

$$(\lambda-\varepsilon)^{n-m}a_m<a_n<(\lambda+\varepsilon)^{n-m}a_m.$$

Taking nth root, we get

$$(\lambda-\varepsilon)^{1-\frac{m}{n}}a_m^{\frac{1}{n}}<a_n^{\frac{1}{n}}<(\lambda+\varepsilon)^{1-\frac{m}{n}}a_m^{\frac{1}{n}}.$$

Setting $A_n=(\lambda-\varepsilon)^{1-\frac{m}{n}}a_m^{\frac{1}{n}}$ and $B_n=(\lambda+\varepsilon)^{1-\frac{m}{n}}a_m^{\frac{1}{n}}$ and taking into account that a_m is finite, we have

$$\lim_{n\to\infty}A_n=\lambda-\varepsilon \text{ and } \lim_{n\to\infty}B_n=\lambda+\varepsilon.$$

Therefore, we can choose n_0 such that

$$A_n>\lambda-2\varepsilon \text{ and } B_n<\lambda+2\varepsilon \text{ for all } n\geq n_0$$

and so

$$\lambda-2\varepsilon<a_n^{\frac{1}{n}}<\lambda+2\varepsilon \text{ for all } n\geq n_0.$$

Hence

$$\lim_{n\to\infty}a_n^{\frac{1}{n}}=\lambda.$$

Theorem 1.11.

(Cauchy's First Theorem on Limits). If $\lim_{n\to\infty}a_n=\lambda$, then

$$\lim_{n\to\infty}\left(\frac{a_1+a_2+\ldots+a_n}{n}\right)=\lambda.$$

Proof: Write $b_n = a_n - \lambda$. Since $\lim_{n\to\infty}a_n=\lambda$, it follows that $\lim_{n\to\infty}b_n=0$. Also,

$$\frac{a_1+a_2+\ldots+a_n}{n}=\lambda+\frac{b_1+b_2+\ldots+b_n}{n}.$$

Therefore, to prove the theorem, it is sufficient to show that if $\lim_{n\to\infty}b_n=0$, then

$$\lim_{n\to\infty}\frac{b_1+b_2+\ldots+b_n}{n}=0\cdot$$

Since $\{b_n\}$ converges, it is bounded and so there exists $k > 0$ such that $|b_n| \leq K$ for all n. Since $\lim_{n\to\infty} b_n = 0$, to each $\varepsilon > 0$, there exists a positive integer m such that $|b_n| < \varepsilon$ for all $n \geq m$.

Therefore,

$$\left|\frac{b_1+b_2+\ldots+b_n}{n}\right|$$

$$=\left|\frac{b_1+b_2+\ldots b_m}{n}+\frac{b_{m+1}+b_{m+2}+\ldots+b_n}{n}\right|$$

$$\leq\frac{|b_1|+|b_2|+\ldots+|b_m|}{n}$$

$$+\frac{|b_{m+1}|+|b_{m+2}|+\ldots+|b_n|}{n}$$

$$<\frac{mK}{n}+\frac{(n-m)\varepsilon}{n} \text{ for all } n\geq m.$$

$$<\frac{mK}{n}+\frac{\varepsilon}{2}.$$

If m_1 is a positive integer greater than $\frac{2mK}{\varepsilon}$, then $\frac{mK}{n}<\frac{\varepsilon}{2}$ when $n\geq m_1$. Thus for $n>\max(m, m_1)$, we have

$$\left|\frac{b_1+b_2+\ldots+b_n}{n}\right|<\frac{\varepsilon}{2}+\frac{\varepsilon}{2}=\varepsilon$$

and so

$$\lim_{n\to\infty}\frac{b_1+b_2+\ldots+b_n}{n}=0.$$

This proves the theorem.

Remark 1.3. The converse of Theorem 1.11 is not true. For example, consider the sequence $\{a_n\}=\{(-1)^n\}$. Then

$$\frac{a_1+a_2+\ldots+a_n}{n}=\begin{cases}0 & \text{if } n \text{ is even}\\ -\frac{1}{n} & \text{if } n \text{ is odd.}\end{cases}$$

Therefore,

$$\lim_{n\to\infty}\frac{a_1+a_2+\ldots+a_n}{n}=0.$$

But the sequence $\{a_n\}$ is not convergent

Theorem 1.12. (Cesaro). If the sequence $\{a_n\}$ and $\{b_n\}$ tend to definite finite limits A and B, respectively, then

$$\lim_{n\to\infty}\frac{a_1b_n+a_2b_{n-1}+\ldots+a_nb_1}{n}=AB.$$

Proof: Write $a_n = a_n - A$. Since $a_n\to A$ as $n\to\infty$ it follows that a_n and hence $|a_n|\to 0$ as $n\to\infty$. Therefore,

$$+\frac{a_1b_n+a_2b_{n-1}+\ldots+a_nb_1}{n}.$$

$$=\frac{A(b_1+b_2+\ldots+b_n)}{n}$$

$$+\frac{a_1b_n+a_2b_{n-1}+\ldots+a_nb_1}{n}.$$

Since $\lim_{n\to\infty} b_n=B$, by Theorem 1.11 $\frac{b_1+b_2+\ldots+b_n}{n}\to B$ as $n\to\infty$ Since $\{b_n\}$ converges, it is bounded and, therefore, there exists $K>0$ such that $|b_n|\leq K$. Therefore,

$$\left|\frac{a_1b_n+a_2b_{n-1}+\ldots+a_nb_1}{n}\right|$$

$$\leq K\left\{\frac{|a_1|+|a_2|+\ldots+|a_n|}{n}\right\}.$$

Since $|a_n|\to 0$ as $n\to\infty$, by Theorem 1.11, $\frac{|a_1|+|a_2|+\ldots+|a_n|}{n}$ tends to 0 as $n\to\infty$. Hence

$$\lim_{n\to\infty}\frac{a_1b_n+a_2b_{n-1}+\ldots+a_nb_1}{n}=AB.$$

EXAMPLE 1.11

Show that

$$\lim_{n\to\infty}\frac{1}{n}(1+2^{1/2}+3^{1/3}+\ldots+n^{1/n})=1.$$

Solution. Let $a_n = n^{1/n}$. Then $\lim_{\to\infty} a_n=\lim_{n\to\infty} n^{\frac{1}{n}}=1$. Therefore, by Cauchy's First Theorem on limits,

$$\frac{a_1+a_2+\ldots+a_n}{n}\to 1 \text{ as n}\to\infty,$$

that is,

$$\frac{1+2^{\frac{1}{2}}+3^{\frac{1}{3}}+\ldots+n^{\frac{1}{n}}}{n}\to 1 \text{ as } n\to\infty.$$

EXAMPLE 1.12

Show that

$$\lim_{n\to\infty}\frac{1}{n}\left(1+\frac{1}{2}+\frac{1}{3}+\cdots+\frac{1}{n}\right)=0.$$

Solution. Let $a_n = \frac{1}{n}$. Then $\lim_{n\to\infty} a_n = \lim_{n\to\infty} \frac{1}{n} = 0$. Therefore, by Cauchy's First Theorem on limits,

$$\lim_{n\to\infty} \frac{a_1 + a_2 + \cdots + a_n}{n} = 0,$$

that is,

$$\lim_{n\to\infty} \frac{1 + \frac{1}{2} + \cdots + \frac{1}{n}}{n} = 0.$$

EXAMPLE 1.13

Show that

$$\lim_{n\to\infty} \frac{1+2+\cdots+n}{n^2} = \frac{1}{2}.$$

Solution. Let

$$a_n = 1 + 2 + \ldots + n \text{ and } b_n = n^2.$$

Then

$$\frac{a_{n+1} - a_n}{b_{n+1} - b_n} = \frac{1}{n} \frac{(1+\frac{1}{n})}{(1+\frac{1}{n})^2 - 1}$$

$$= \frac{1}{n}\left[\frac{1+\frac{1}{n}}{1+\frac{1}{n^2}+\frac{2}{n}-1}\right]$$

$$= \frac{1+\frac{1}{n}}{2+\frac{1}{n}}$$

and so

$$\lim_{n\to\infty} \frac{a_{n+1} - a_n}{b_{n+1} - b_n} = \frac{1}{2}.$$

Then, by Theorem 1.9, we have

$$\lim_{n\to\infty} \frac{a_n}{b_n} = \lim_{n\to\infty} \frac{a_{n+1} - a_n}{b_{n+1} - b_n} = \frac{1}{2},$$

that is,

$$\lim_{n\to\infty} \frac{1+2+\ldots+n}{n^2} = \frac{1}{2}.$$

1.7 SUBSEQUENCES

Let $\{a_n\}$ be a sequence and let

$$n_1 < n_2 < n_3 < \ldots < n_k < \ldots$$

be infinite subset of the set N of all the positive integers. Then $\{a_n\}$ is called the *subsequence* of the sequence $\{a_n\}$.

Thus, any infinite succession of the terms of $\{a_n\}$, picked out in any way but preserving the original order, is called a subsequence of $\{a_n\}$.

For example, consider the sequence $\{\frac{1}{n}\}$. Then, the sequence $\{\frac{1}{2}, \frac{1}{4}, \frac{1}{6}, \frac{1}{8}, \ldots, \frac{1}{2n}, \ldots\}$ is subsequence of $\{\frac{1}{n}\}$.

If the subsequence converges, its limit is called *a subsequential limit* of $\{a_n\}$.

Theorem 1.13. If a sequence $\{a_n\}$ of real numbers converges to a limit λ, then every subsequence $\{a_{n_k}\}$ of $\{a_n\}$ converges to the same limit.

Proof: Since $\{a_n\}$ converges to λ for each $\varepsilon > 0$, there exists a positive integer n_0 such that

$$|a_n - \lambda| < \varepsilon \text{ for all } n > n_0.$$

Then taking $n \geq n_k \geq n_0$, we have

$$|a_n - \lambda| \leq \varepsilon \text{ for all } n \geq n_0.$$

Hence $\{a_{n_k}\}$ converges to λ.

Remark 1.4. The converse of Theorem 1.13 need not be true. For example, consider the sequence $\{a_n\} = \{(-1)^n\}$. This sequence is not convergent. But the subsequence $\{a_n\} = \{1,1,1,\ldots\}$ whose all terms are 1 converges to 1 as $n \to \infty$. Thus, the subsequence $\{a_{2n}\}$ of $\{a_n\}$ is convergent but the original sequence $\{a_n\}$ is not convergent.

However, if the union of the subsequences is the whole sequence, then it seems plausible that the whole sequence will converge to the same limit. For example, the converse of Theorem 1.13 holds good if we take two subsequences—one formed by odd terms alone and the other by even terms alone—if these two sequences tend to a limit, then the whole sequence also converges to that limit. Thus we have the following theorem.

Theorem 1.14. If for a sequence $\{a_n\}$, the subsequences $\{a_{2n-1}\}$ and $\{a_{2n}\}$ converges to a, then $\{a_n\}$ converges to a.

Proof: Since $a_{2n-1} \to a$ as $n \to \infty$, given $\varepsilon > 0$, there exists a positive integer m_1 such that

$$|a_{2n-1} - a| < \varepsilon \text{ for all } n > m_1. \qquad (1)$$

Since $a_{2n} \to a$ as $n \to \infty$, given $\varepsilon > 0$, there exists a positive integer m_2 such that

$$|a_{2n} - a| < \varepsilon \text{ for all } n > m_2. \quad (2)$$

Let $m = \max(m_1, m_2)$ and let $n > 2m$. If n is even and equal to $2p$, then $p > m \geq m_2$ and so by (2),

$$|a_n - a| < \varepsilon \text{ for all } n > 2m. \quad (3)$$

On the other hand, if n is odd and equal to $2p + 1$, then since $n > 2m, p > m \geq m_1$, we have by (1),

$$|a_n - a| < \varepsilon \text{ for all } n > 2m. \quad (4)$$

Hence, whether n is even or odd, we have from (3) and (4) that

$$|a_n - a| < \varepsilon \text{ for all } n > 2m$$

Therefore, $\{a_n\}$converges to a.

Theorem 1.15. Every bounded sequence $\{a_n\}$ of real numbers contains a convergent subsequence.

Proof: Let A be the range of the bounded sequence $\{a_n\}$. If A is finite, then there is at least one point of A, say b, which appears infinite number of times in the sequence. Hence $\{b,b,b,\ldots\}$ is a subsequence of $\{a_n\}$ which converges to b.

If A is infinite, then it is a bounded infinite subset of real numbers. But, by Weierstrass's Theorem, a bounded infinite subset has a limit point, say a, on the real line. Choose n_1 so that $|a_{n_1} - a| < 1$. Having chosen $n_1, n_2, \ldots, n_{i-1}$, there is an integer $n_i > n_{i-1}$ such that $\left|a_{n_i} - a\right| < \frac{1}{i}$. Hence, the subsequence $\{a_{n_j}\}$ converges to a.

1.8 SERIES

An expression of the form $a_1 + a_2 + \ldots + a_n + \ldots$, denoted by $\sum_{n=1}^{\infty} a_n$, is called an *infinite series*. The term a_n is called the *nth term* of the series. Further,

$$S_n = a_1 + a_2 + \cdots + a_n$$

is called the *partial sum of the series* $\sum_{n=1}^{\infty} a_n$.

The definition of convergence or divergence of the series $\sum_{n=1}^{\infty} a_n$ depends on the convergence or divergence of the sequence $\{S_n\}_{n=1}^{\infty}$ of partial sums.

Let $\sum_{n=1}^{\infty} a_n$ be a series of real number with partial sums $S_n = a_1 + a_2 + \ldots + a_n$. If the sequence $\{S_n\}_{n=1}^{\infty}$ converges to l, we say that the series $\sum_{n=1}^{\infty} a_n$ converges to the sum l and we write $\sum_{n=1}^{\infty} a_n = l$. If $\{S_n\}_{n=1}^{\infty}$ diverges, then the series $\sum_{n=1}^{\infty} a_n$ also diverges.

In case of infinite series, the Cauchy's criteria for the convergence takes the form

$$|S_m - S_n| < \varepsilon, \; m > n > n_0(\varepsilon),$$

that is,

$$|a_{n+1} + a_{n+2} + \cdots + a_m| < \varepsilon, \; m > n > n_0(\varepsilon).$$

If we take $m = n + 1$, then the last expression takes the form

$$|a_{n+1}| < \epsilon, \; n > n_0(\epsilon)$$

or

$$\lim a_n = 0.$$

which is a *necessary condition for the convergence of the series* $\sum_{n=1}^{\infty} a_n$.

However, $\lim a_n = 0$ is not a *sufficient condition for the convergence of the series* $\sum_{n=1}^{\infty} a_n$ of positive terms. For example,

$$\sum_{1}^{\infty} \frac{1}{n}, \sum_{n=1}^{\infty} \frac{1}{\sqrt{n}}, \sum_{n=2}^{\infty} \frac{1}{n \log n}$$

are all divergent series inspite of the fact that $\lim_{n\to\infty} a_n = 0$. But the condition is sometimes useful. Infact, if $\lim a_n \neq 0$, *we can at once say that the series is not convergent.*

Remark 1.5. Two convergent series can be added (or subtracted) term by term to give a convergent series. Thus, if $\Sigma u_n = S$ and $\Sigma v_n = T$, then the sum of n terms of the series obtained by addition is $S_n + T_n$. But $\lim_{n\to\infty}(S_n + T_n) = \lim_{n\to\infty} S_n + \lim_{n\to\infty} T_n = S + T$ by Theorem 1.8. Hence, the sum series converges to $S+T$.

(ii) We note that the *sum of two non-convergent series form a convergent series*. For example, let

$$S = 1 - 1 + 1 - 1 + 1 - 1 + \ldots$$

and

$$T = 0+1-1+1-1+1-\ldots$$

We observe that S and T are both oscillatory, but $S+T=1$.

Theorem 1.16. (Pringsheim). If the terms of the series Σu_n of positive terms steadily (monotonically) decrease, then it is necessary but not a sufficient condition for the convergence of the series that $\lim_{n\to\infty} nu_n = 0$.

Proof: *The condition is necessary*: Let the series Σu_n of positive terms be convergent. Then, by Cauchy's Principle of Convergence, for a given $\varepsilon > 0$ there exists a positive integer m such that

$$|u_{m+1} + u_{m+2} + \ldots + u_n| < \frac{\varepsilon}{2} \text{ for all } n \geq m.$$

Since the terms are monotonically decreasing, $u_n \geq u_{n+1} \geq u_{n+2} \geq \ldots \geq u_{m+1}$. Hence, the above expression reduces to

$$(n-m)u_n < \frac{\varepsilon}{2} \text{ for all } n \geq m$$

Since Σu_n converges, $\lim_{n\to\infty} u_n = 0$ and, therefore, we can choose $\mu > m$ such that $n\,u_n < \frac{\varepsilon}{2}$ for all $n \geq \mu > m$ and so $\lim_{n\to\infty} nu_n = 0$.

To show that the *condition is not sufficient*, consider the series $\sum \frac{1}{n\log n}$. Then, $\lim_{n\to\infty} nu_n = \lim_{n\to\infty} \frac{1}{\log n} = 0$. But, the series $\sum \frac{1}{n\log n}$ is divergent.

Remark 1.6. It follows from the above theorem that if Σu_n series of positive terms is decreasing monotonically and if $\lim_{n\to\infty} nu_n \neq 0$, then the series Σu_n cannot converge.

Theorem 1.17. A positive term series either converges or diverges to $+\infty$.

Proof: Let Σu_n be a positive term series and let S_n be its partial sum. Then,

$$S_{n+1} = u_1 + u_2 + \ldots + u_{n+1},$$

$$S_n = u_1 + u_2 + \ldots + u_n$$

and so

$$S_{n+1} - S_n = u_{n+1} > 0 \text{ for all } n.$$

Thus $S_{n+1} > S_n$ for all n. Hence $\{S_n\}$ is a monotonically increasing sequence.

If $\{S_n\}$ is bounded above, then $\{S_n\}$ is monotonically increasing and bounded above sequence and so, by Theorem 1.7, it is convergent. Hence Σa_n is convergent.

If $\{S_n\}$ is not bounded above, then S_n is monotonically increasing and not bounded above and so it diverges to $+\infty$. Hence, Σu_n diverges to $+\infty$.

Theorem 1.18. Convergence, divergence, or oscillation of a series of positive terms is not affected by the addition or omission of a finite number of its terms.

Proof: Let Σu_n be a series of positive terms. It is sufficient to show that the two series

$$u_1 + u_2 + \ldots + u_m + u_{m+1} + u_{m+2} + \ldots$$

and

$$u_{m+1} + u_{m+2} + \ldots\ldots$$

converge or diverge together. To show it, let S_n and T_n denote the nth partial sums, respectively, of the above two series. Then

$$S_n = u_1 + u_2 + \ldots + u_n$$

and

$$\begin{aligned} T_n &= u_{m+1} + u_{m+2} + \ldots + u_{m+n} \\ &= (u_1 + u_2 + \ldots + u_{m+n}) - (u_1 + u_2 + \ldots + u_m) \\ &= S_{m+n} - S_m. \end{aligned}$$

But S_m, being the sum of a finite number of terms of Σu_n, is a fixed finite quantity. Therefore,

(i) If S_{m+n} tends to a finite limit, then T_n also tends to a finite limit

(ii) If S_{m+n} tends to ∞, then T_n also tends to ∞

(iii) If $\lim_{\infty\to 0} S_{m+n}$ does not exist, the $\lim_{n\to\infty} T_n$ also does not exist.

Thus, the two sequences S_n and T_n converge or diverge together. Hence, the series $u_1 + u_2 + \ldots + u_{m+1} + u_{m+2} + \ldots$ and the series $u_{m+1} + u_{m+2} + \ldots$ converge or diverge together.

EXAMPLE 1.14

Using Cauchy's general principle of convergence, show that the series $\sum_{n=1}^{\infty} \frac{1}{n}$ does not converge.

Solution. We shall prove our result by contradiction. So, suppose that the given series is convergent. We choose $\varepsilon = \frac{1}{2}$. Then, by Cauchy's criterion, we have

$$|a_n - a_m| < \frac{1}{2} \text{ for all } n > m > n_0\left(\frac{1}{2}\right)$$

or

$$\left|\frac{1}{m+1} + \frac{1}{m+2} + \cdots \frac{1}{n}\right| < \frac{1}{2} \text{ for all } n > m > n_0\left(\frac{1}{2}\right)$$

or

$$\frac{1}{m+1} + \frac{1}{m+2} + \cdots + \frac{1}{n} < \frac{1}{2} \text{ for all } n > m > n_0\left(\frac{1}{2}\right).$$

We take $n = 2m$ and so we should have

$$\frac{1}{m+1} + \frac{1}{m+2} + \ldots + \frac{1}{2m} < \frac{1}{2}.$$

But,

$$\frac{1}{m+1} + \frac{1}{m+2} + \cdots + \frac{1}{2m}$$
$$> \frac{1}{2m} + \frac{1}{2m} + \cdots + \frac{1}{2m} \text{ (}m\text{ terms)}$$
$$= \frac{m}{2m} = \frac{1}{2}.$$

Thus, we arrive at a contradiction. Hence, the given series is not convergent.

EXAMPLE 1.15

Test the convergence of the series

$$\sqrt{\frac{1}{4}} + \sqrt{\frac{2}{6}} + \sqrt{\frac{3}{8}} + \ldots + \sqrt{\frac{n}{2(n+1)}} + \ldots$$

Solution. The nth term of the given series is

$$u_n = \sqrt{\frac{n}{2(n+1)}} = \sqrt{\frac{1}{2(1+\frac{1}{n})}}.$$

Therefore,

$$\lim_{n\to\infty} u_n = \frac{1}{\sqrt{2}} \neq 0.$$

Thus necessary condition for convergence is not satisfied. Hence Σu_n is divergent.

EXAMPLE 1.16

Discuss the convergence of the *geometric series*

$$1 + a + a^2 + a^3 + \cdots$$

Solution: The partial sum for the given series is

$$S_n = 1 + a + a^2 + \cdots + a^{n-1}$$
$$= \frac{a^n - 1}{a - 1}, \quad a \neq 1.$$

If $|a| < 1$, then

$$S_n = \frac{1 - a^n}{1 - a} \text{ and so}$$

$\lim_{n\to\infty} S_n = \frac{1}{1-a}$ since, $a^n \to 0$ as $n \to \infty$.

Thus, if $|a| < 1$, then Σa^{n-1} is convergent and its sum is $\frac{1}{1-a}$.

If $a = 1$, then

$$S_n = 1 + 1 \cdots + 1 \text{ (}n\text{ terms)} = n$$

and so $\lim_{n\to\infty} S_n = \infty$. Hence Σa^{n-1} is divergent.

If $a > 1$, then $S_n > n$ and so $\lim_{n\to\infty} S_n = \infty$. Hence the series is divergent.

If $a = -1$, the series becomes

$$1 - 1 + 1 + 1 - 1 + \cdots$$

and then

$$S_{2n} = 0 \text{ and } S_{2n-1} = 1.$$

Hence the series oscillates between 0 and 1.

If $a < -1$, the successive terms increase in magnitude. Then,

$$S_{2n} = \frac{a^{2n} - 1}{a - 1} (+ve)$$

is monotonic increasing while S_{2n+1} is negative and numerically increasing with n. Hence the series oscillates between $+\infty$ and $-\infty$.

EXAMPLE 1.17

Show by direct summation of n terms that the series

$$\frac{1}{1.2}+\frac{1}{2.3}+\frac{1}{3.4}+\cdots$$

is convergent.

Solution. The nth term of the given series is

$$a_n=\frac{1}{n(n+1)}=\frac{1}{n}-\frac{1}{n+1}.$$

Therefore,

$$a_1=1-\frac{1}{2}$$

$$a_2=\frac{1}{2}-\frac{1}{3}$$

$$a_3=\frac{1}{3}-\frac{1}{4}$$

$$a_n=\frac{1}{n}-\frac{1}{n+1}$$

Then the partial sum is

$$S_n=a_1+a_2+\ldots+a_n=1-\frac{1}{n+1}.$$

Therefore,

$$\lim_{n\to\infty}S_n=1-0=1.$$

Hence, the given series converges to 1.

1.9 COMPARISON TESTS

The convergence or divergence behavior of a given series is generally determined by comparing its terms with the terms of another series whose convergence behavior is known. Such comparisons are called *Comparison Tests*.

Comparison Test I(a): If Σu_n and Σv_n are two series of positive terms such that $u_n < v_n$ and Σv_n is convergent, then Σu_n is also convergent.

Proof: Let

$$S_n=u_1+u_2+\ldots+u_n$$

and

$$T_n=v_1+v_2+\ldots+v_n$$

be the partial sums of Σu_n and Σv_n, respectively. Since $u_n < v_n$ for all n, we have $S_n < T_n$ and so

$\lim_{n\to\infty}S_n<\lim_{n\to\infty}T_n$. Since Σv_n is convergent, $\lim_{n\to\infty}T_n$ is finite and so $\lim_{n\to\infty}S_n$ is finite. Hence, Σu_n is convergent.

Comparison Test 1(b): *If* Σu_n and Σv_n are two series of positive terms such that $u_n > v_n$ and Σv_n is divergent, then Σu_n also diverges.

Proof: Let

$$S_n=u_1+u_2+\ldots+u_n$$

and

$$T_n=v_1+v_2+\ldots+v_n$$

be partial sums of Σu_n and Σv_n, respectively. Since $u_n > v_n$, it follows that $S_n > T_n$ and so

$\lim_{n\to\infty}S_n>\lim_{n\to\infty}T_n=\infty$, since Σv_n diverges. Thus $\lim_{n\to\infty}S_n=\infty$ and hence, Σu_n diverges.

Comparison Test II. If Σu_n and Σv_n are two series of positive terms, then

(i) if $\lim_{n\to\infty}\frac{u_n}{v_n}=l$ (finite and non-zero), then Σu_n and Σv_n both converge or diverge together

(ii) if $\lim_{n\to\infty}\frac{u_n}{v_n}=0$ and Σv_n converges, then Σu_n also converges

(iii) if $\lim_{n\to\infty}\frac{u_n}{v_n}=\infty$ and Σv_n diverges, then Σu_n also diverges.

Proof: (i). Since u_n and v_n are positive terms, $\frac{u_n}{v_n}>0$ and so $\lim_{n\to\infty}\frac{u_n}{v_n}\geq 0$. But $\lim_{n\to\infty}\frac{u_n}{v_n}=l\neq 0$. Therefore, $l > 0$. Since $\lim_{n\to\infty}\frac{u_n}{v_n}=l$, for every $\varepsilon > 0$, there exists a positive integer m such that

$$\left|\frac{u_n}{v_n}-l\right|<\varepsilon \text{ for all } n>m,$$

that is,

$$l-\varepsilon<\frac{u_n}{v_n}<l+\varepsilon$$

or

$$(l-\varepsilon)v_n < u_n < (l+\varepsilon)v_n.$$

Thus $u_n < p\ v_n$, where $p > 1$. Hence, by Comparison Test I(a), the series Σu_n converges, if Σv_n converges.

Also $u_n > q\ v_n$, where $q < 1$. If Σv_n diverges then Σu_n also diverges. Similarly, we can show that Σv_n converges or diverges according as Σu_n converges or diverges.

(ii) Since $\lim_{n\to\infty} \frac{u_n}{v_n} = 0$, given $\varepsilon > 0$ there exists a positive integer m such that

$$\left|\frac{u_n}{v_n} - 0\right| < \varepsilon \text{ for all } n > m$$

or

$$-\varepsilon < \frac{u_n}{v_n} < \varepsilon \text{ for all } n > m$$

$$u_n < \varepsilon v_n \text{ for all } n > m.$$

Since Σv_n is convergent, by Comparison Test I(a), the series Σu_n also converges.

(iii) Since $\lim_{n\to\infty} \frac{u_n}{v_n} = \infty$, given $M > 0$, however large, there exists integer m such that $\frac{u_n}{v_n} > M$ for all $n > m$. Hence $u_n > Mv_n$ for all $n > m$. Since Σv_n diverges, it follows by Comparison Test I(b) that Σu_n also diverges.

Comparison Test III: If Σu_n and Σv_n are two series of positive terms, then

(i) if $\frac{u_{n+1}}{u_n} < \frac{v_{n+1}}{v_n}$ and Σv_n converges, then Σu_n also converges.

(ii) if $\frac{u_{n+1}}{u_n} > \frac{v_{n+1}}{v_n}$ and Σv_n diverges, then Σu_n also diverges.

Proof: Let S_n and T_n be, respectively, the partial sums of the series Σu_n and Σv_n. Then

$$S_n = u_1\left(1 + \frac{u_2}{u_1} + \frac{u_3}{u_2}.\frac{u_2}{u_1} + \frac{u_4}{u_3}.\frac{u_3}{u_2}.\frac{u_2}{u_1} + \cdots\right)$$

$$< u_1\left(1 + \frac{v_2}{v_1} + \frac{v_3}{v_2}.\frac{v_2}{v_1} + \frac{v_4}{v_3}.\frac{v_3}{v_2}.\frac{v_2}{v_1} + \cdots\right),$$

$$\text{since } \frac{u_{n+1}}{u_n} < \frac{u_{n+1}}{v_n},$$

$$= \frac{u_1}{v_1}(v_1 + v_2 + \cdots + v_n) = \frac{u_1}{v_1}T_n$$

Hence

$$\lim_{n\to\infty} S_n < \lim_{n\to\infty} \frac{u_1}{v_1} T_n \cdot$$

Since Σv_n converges, $\lim_{n\to\infty} \frac{u_1}{v_1} T_n$ is finite and so is $\lim_{n\to\infty} S_n$ which in turn implies that Σu_n converges.

Part (ii). This can be proved by replacing '<' by '>' in the above proof.

EXAMPLE 1.18

Show that the series $\sum_{n=1}^{\infty} \frac{1}{n^p}$, called the *Harmonic Series*, converges if $p > 1$ and diverges if $p \le 1$.

Solution. Take $p > 1$, first. Group the terms of the series as follows:

$$\frac{1}{1^p} + \left(\frac{1}{2^p} + \frac{1}{3^p}\right) + \left(\frac{1}{4^p} + \frac{1}{5^p} + \frac{1}{6^p} + \frac{1}{7^p}\right)$$
$$+ \left(\frac{1}{8^p} + \cdots + \frac{1}{15^p}\right) + \cdots$$

But

$$\frac{1}{2^p} + \frac{1}{3^p} < \frac{1}{2^p} + \frac{1}{2^p} = \frac{2}{2^p} = \frac{1}{2^{p-1}}$$

$$\frac{1}{4^p} + \frac{1}{5^p} + \frac{1}{6^p} + \frac{1}{7^p} < \frac{1}{4^p} + \frac{1}{4^p} + \frac{1}{4^p} + \frac{1}{4^p}$$
$$= \frac{1}{4^{p-1}} = \frac{1}{(2^{p-1})^2},$$

$$\frac{1}{8^p} + \frac{1}{9^p} + \cdots + \frac{1}{15^p} < \frac{1}{8^{p-1}} = \frac{1}{(2^{p-1})^3},$$

and so on.

Hence,

$$\sum_{n=1}^{\infty} \frac{1}{n^p} < 1 + \frac{1}{2^{p-1}} + \frac{1}{(2^{p-1})^2} + \frac{1}{(2^{p-1})^3} + \cdots.$$

The right hand side is a Geometric progression with common ratio $\frac{1}{2^{p-1}} < 1$, since $p > 1$. Hence, by Comparison Test the series is convergent.

If $p = 1$, then the series is

$$1+\frac{1}{2}+\frac{1}{3}+\frac{1}{4}+\cdots$$

and

$$1+\frac{1}{2}=1+\frac{1}{2},$$

$$\frac{1}{3}+\frac{1}{4}>\frac{1}{4}+\frac{1}{4}=\frac{1}{2},$$

$$\frac{1}{5}+\frac{1}{6}+\frac{1}{7}+\frac{1}{8}>\frac{1}{8}+\frac{1}{8}+\frac{1}{8}+\frac{1}{8}$$

$$=\frac{1}{2}$$

and so on.

Therefore, the given series is less than the series

$$1+\frac{1}{2}+\frac{1}{2}+\frac{1}{2}+\ldots$$

Leaving aside the first term, this series is a Geometric progression whose common ratio is 1. Therefore, this series is divergent and so is the given series.

If $0 < p < 1$, then $n^p < n$, that is, $\frac{1}{n^p} < \frac{1}{n}$. But the series $\sum \frac{1}{n}$ is divergent. Hence, $\sum \frac{1}{n^p}$ is divergent for $p < 1$.

If p = 0, then the series becomes 1 + 1 + 1 + ..., which is divergent.

EXAMPLE 1.19

Test the convergence of the series

$$\frac{1}{1.2}+\frac{1}{2.3}+\frac{1}{3.4}+\ldots$$

Solution. We note that

$$\frac{1}{1.2}+\frac{1}{2.3}+\frac{1}{3.4}+\cdots<\frac{1}{1}+\frac{1}{2^2}+\frac{1}{3^2}+\cdots=\sum\frac{1}{n^2}.$$

But $\sum \frac{1}{n^2}$ is convergent (Harmonic Series for $p = 2$). Hence, the given series is convergent.

EXAMPLE 1.20

Test the convergence of the series

$$\sum\frac{(n+1)(n+2)}{(n^2+1)(n^2+2)}.$$

Solution. The nth term of the given series is

$$u_n=\frac{(n+1)(n+2)}{(n^2+1)(n^2+2)}=\frac{(1+\frac{1}{n})(1+\frac{2}{n})}{n^2(1+\frac{1}{n^2})(1+\frac{2}{n^2})}.$$

Take $v_n = \frac{1}{n^2}$. Then,

$$\lim_{n\to\infty}\frac{u_n}{v_n}=\lim_{n\to\infty}\frac{(1+\frac{1}{n})(1+\frac{2}{n})}{(1+\frac{1}{n^2})(1+\frac{2}{n^2})}$$

$$= 1, \text{ finite and non-zero:}$$

Therefore, Σu_n and Σv_n converge or diverge together. But the series $\sum v_n = \sum \frac{1}{n^2}$ converges. Therefore, Σu_n is convergent.

EXAMPLE 1.21

Examine the convergence of the series

$$\sum[(n^3+1)^{\frac{1}{3}}-n].$$

Solution. The nth term of the series is

$$u_n = (n^3+1)^{\frac{1}{3}} - n$$

$$=\left(n^3\left(1+\frac{1}{n^3}\right)\right)^{\frac{1}{3}}-n=n\left(1+\frac{1}{n^3}\right)^{\frac{1}{3}}-n$$

$$=n\left[\left(1+\tfrac{1}{n^3}\right)^{\frac{1}{3}}-1\right]$$

$$=n\left[1+\frac{1}{3}\cdot\frac{1}{n^3}+\frac{\frac{1}{3}(\frac{1}{3}-1)}{2!}\left(\tfrac{1}{n^3}\right)^2+\cdots-1\right]$$

$$=\frac{n}{n^3}\left[\frac{1}{3}-\frac{1}{9}\cdot\frac{1}{n^3}-\cdots\right]$$

$$=\frac{1}{n^2}\left[\frac{1}{3}-\frac{1}{9}\cdot\frac{1}{n^3}-\cdots\right].$$

Choose $v_n = \frac{1}{n^2}$. Then

$$\lim_{n\to\infty}\frac{u_n}{v_n}=\lim_{n\to\infty}\left[\frac{1}{3}-\frac{1}{9}\cdot\frac{1}{n^3}-\cdots\right]$$

$$=\tfrac{1}{3}, \text{ finite and non-zero.}$$

Hence, Σu_n and Σv_n converge or diverge together. But $\sum v_n = \sum \frac{1}{n^2}$ is convergent. Therefore, Σu_n is also convergent.

EXAMPLE 1.22

Examine the convergence of the series

$$\sum(\sqrt{n+1}-\sqrt{n}).$$

Solution: The nth term of the given series is

$$u_n = \sqrt{n+1}-\sqrt{n}$$
$$= \frac{\sqrt{n+1}-\sqrt{n}}{\sqrt{n+1}+\sqrt{n}}[\sqrt{n+1}+\sqrt{n}]$$
$$= \frac{n+1-n}{\sqrt{n+1}+\sqrt{n}}$$
$$= \frac{1}{\sqrt{n}\left[\sqrt{1+\frac{1}{n}}+1\right]}.$$

Take $v_n = \frac{1}{\sqrt{n}}$. Then

$$\lim_{n\to\infty}\frac{u_n}{v_n} = \lim_{n\to\infty}\frac{1}{\sqrt{1+\frac{1}{n}}+1}$$
$$= \tfrac{1}{2} \text{ finite and non-zero.}$$

Therefore, Σu_n and Σv_n converge or diverge together. But $\sum\frac{1}{\sqrt{n}}$ diverges. Therefore, Σu_n is also divergent.

EXAMPLE 1.23

Examine the convergence of the series

$$\sum_{n=1}^{\infty}\frac{\sqrt{n}}{n^2+1}.$$

Solution. The nth term of the series is

$$u_n = \frac{\sqrt{n}}{n^2+1} = \frac{\sqrt{n}}{n^2(1+\frac{1}{n^2})} = \frac{1}{n^{\frac{3}{2}}(1+\frac{1}{n^2})}.$$

Take $v_n = \frac{1}{n^{\frac{3}{2}}}$.. Then

$$\lim_{n\to\infty}\frac{u_n}{v_n} = \lim_{n\to\infty}\frac{1}{1+\frac{1}{n^2}} = 1, \text{ finite and non-zero.}$$

Therefore, Σu_n and Σv_n converge or diverge together. But the series $\Sigma v_n = \Sigma\frac{1}{n^{\frac{3}{2}}}$ is convergent. Hence, Σu_n is also convergent.

EXAMPLE 1.24

Examine the convergence of the series

(i) $\frac{1^p}{2^q}+\frac{2^p}{3^q}+\frac{3^p}{4^q}+\ldots$

(ii) $\sum\frac{1}{n}\sin\frac{1}{n}$

Solution. (i) The nth term of the series is

$$u_n = \frac{n^p}{(n+1)^q} = \frac{n^p}{n^q\left(1+\frac{1}{n}\right)^q} = \frac{1}{n^{q-p}\left(1+\frac{1}{n}\right)^q}.$$

Take

$$v_n = \frac{1}{n^{q-p}}.$$

Then,

$$\lim_{n\to\infty}\frac{u_n}{v_n} = \lim_{n\to\infty}\frac{1}{\left(1+\frac{1}{n}\right)^q} = 1.$$

Therefore, Σu_n and Σv_n converge or diverge together. But $\Sigma v_n = \Sigma\frac{1}{n^{q-p}}$ converges if $q-p>1$ and diverges, if $q-p\le 1$. Hence, Σu_n converges if $q-p>1$ and diverges if $q-p\le 1$.

(ii) Here

$$u_n = \frac{1}{n}\sin\frac{1}{n}$$

Since $\lim_{n\to\infty}\frac{\sin\frac{1}{n}}{\frac{1}{n}} = 1$, it follows that $\sin\frac{1}{n}\sim\frac{1}{n}$ and so $u_n\sim\frac{1}{n^2}$. We, therefore, take $v_n = \frac{1}{n^2}$, and then

$$\lim_{n\to\infty}\frac{u_n}{v_n} = \lim_{n\to\infty}\frac{n^2}{n}\sin\frac{1}{n} = \lim_{n\to\infty}\frac{\sin\frac{1}{n}}{\frac{1}{n}} = 1.$$

Therefore, Σu_n and Σv_n converge or diverge together. But $\Sigma v_n = \Sigma\frac{1}{n^2}$ converges. Hence, Σu_n is also convergent, by Comparison Test.

EXAMPLE 1.25

If Σu_n is a convergent series of positive terms, show that Σu_n^2 is also convergent.

Give an example to show that the converse need not be true.

Solution. Since Σu_n is convergent, $u_n\to 0$ as $n\to\infty$. Therefore, there exists a positive integer m such that $0\le u_n\le 1$ for all $n\ge m$ and so $u_n^2\le u_n$ for all $n\ge m$. Hence, by Comparison

Test, Σu_n^2 is convergent.

However, the converse need not be true.

For example, if we take $\Sigma u_n = \Sigma \frac{1}{n}$, then

$\Sigma u_n^2 = \Sigma \frac{1}{n^2}$. The series $\Sigma \frac{1}{n^2}$ is convergent, but $\Sigma \frac{1}{n}$ does not converge.

EXAMPLE 1.26

Show that the series

$$\frac{1}{2}+\frac{1}{3^2}+\frac{1}{2^3}+\frac{1}{3^4}+\cdots$$

is convergent.

Solution. Since 2 < 3, each term, of the given series in less than or equal to the corresponding term of the series

$$\frac{1}{2}+\frac{1}{2^2}+\frac{1}{2^3}+\frac{1}{2^4}+\cdots,$$

which is a geometric series with common ratio $\frac{1}{2}$ and so, converges. Hence, by Comparison Test, the given series converges.

EXAMPLE 1.27

Show that the series

$$\frac{1}{1.2.3}+\frac{3}{2.3.4}+\frac{5}{3.4.5}+\cdots$$

is convergent.

Solution. The *n*th term of the given series is

$$u_n = \frac{2n-1}{n(n+1)(n+2)}$$

$$= \frac{2-\frac{1}{n}}{n^2\left(1+\frac{1}{n}\right)\left(1+\frac{2}{n}\right)}.$$

Taking $v_n = \frac{1}{n^2}$, we note that $\lim\limits_{n\to\infty} \frac{u_n}{v_n} = 2$ (finite). Therefore, Σu_n and Σv_n converge or diverge together. But $\Sigma \frac{1}{n^2}$ is convergent. Hence Σu_n is convergent.

EXAMPLE 1.28

Examine the convergence of the series

$$\sum \frac{2n^3+5}{4n^5+1}.$$

Solution. The *n*th term of the given series is

$$u_n = \frac{2n^3+5}{4n^3+1}$$

$$= \frac{n^3\left(2+\frac{5}{n^3}\right)}{n^5\left(4+\frac{1}{n^5}\right)} = \frac{\left(2+\frac{5}{n^3}\right)}{n^2\left(4+\frac{1}{n^5}\right)}.$$

Take $v^n = \frac{1}{n^2}$. Then

$$\lim_{n\to\infty}\frac{u_n}{v_n} = \lim_{n\to\infty}\frac{2+\frac{2}{n^3}}{4+\frac{1}{n^5}} = \frac{1}{2}, \text{ finite and non-zero.}$$

Hence, Σu_n and Σv_n converge or diverge together. But $\Sigma v_n = \Sigma \frac{1}{n^2}$ converges. Therefore, Σu_n converges.

EXAMPLE 1.29

Test the convergence of the series

$$\frac{1}{1.2.3}+\frac{1}{2.3.4}+\frac{1}{3.4.5}+\cdots$$

Solution. The *n*th term of the series is

$$u_n = \frac{1}{n(n+1)(n+2)} = \frac{1}{n^3(1+\frac{1}{n})(1+\frac{2}{n})}$$

Take $v_n = \frac{1}{n^3}$. Then,

$$\lim_{n\to\infty}\frac{u_n}{v_n} = \lim_{n\to\infty}\frac{1}{(1+\frac{1}{n})(1+\frac{2}{n})}$$

$$=1, \text{ finite and non-zero.}$$

Therefore, Σu_n and Σv_n converge or diverge together. But the series $\Sigma v_n = \Sigma \frac{1}{n^3}$ is convergent. Hence, the given series Σu_n is also convergent.

EXAMPLE 1.30

Examine the convergence of the series

$$\sum\left(\sqrt{n^4+1}-\sqrt{n^4-1}\right).$$

Solution. The *n*th term of the series is

$$u_n = \sqrt{n^4+1}-\sqrt{n^4-1}$$

$$= \frac{\sqrt{n^4+1}-\sqrt{n^4-1}}{\sqrt{n^4+1}+\sqrt{n^4-1}}.[\sqrt{n^4+1}+\sqrt{n^4-1}]$$

$$= \frac{2}{\sqrt{n^4+1}+\sqrt{n^4-1}}$$

$$= \frac{2}{n^2\left[\left(1+\frac{1}{n^4}\right)^{\frac{1}{2}}+\left(1-\frac{1}{n^4}\right)^{\frac{1}{2}}\right]}.$$

Take $v_n = \frac{2}{n^2}$. Then,

$$\lim_{n\to\infty}\frac{u_n}{v_n} = \lim_{n\to\infty}\frac{1}{(1+\frac{1}{n^4})^{\frac{1}{2}}+(1-\frac{1}{n^4})^{\frac{1}{2}}}$$

$$= \frac{1}{2}, \text{ finite and non-zero.}$$

Therefore, the series Σu_n and Σv_n converge or diverge together. But the series $\Sigma v_n = \Sigma \frac{2}{n^2}$ is convergent. Hence, the given series is convergent.

1.10 D'ALEMBERI'S RATIO TEST

Let Σu_n be a series of positive terms. Then

(i) Σu_n is convergent if $\lim_{n\to\infty}\frac{u_{n+1}}{u_n} < 1$

(ii) Σu_n is divergent if $\lim_{n\to\infty}\frac{u_{n+1}}{u_n} > 1$

(If $\lim_{n\to\infty}\frac{u_{n+1}}{u_n} = 1$, this test gives no information about the convergence or divergence of the series Σu_n).

Proof: Suppose that

$$\lim_{n\to\infty}\frac{u_{n+1}}{u_n} = l.$$

Then to each $\varepsilon > 0$, there exists a positive integer m such that

$$\left|\frac{u_{n+1}}{u_n} - l\right| < \varepsilon \text{ for all } n \geq m$$

and so

$$l - \varepsilon < \frac{u_{n+1}}{u_n} < l + \varepsilon \text{ for all } n \geq m.$$

If $l < 1$, choose $\varepsilon > 0$ such that $l + \varepsilon = r < 1$. Then,

$$\frac{u_{n+1}}{u_n} < r \text{ for all } n \geq m$$

and so

$$\frac{u_{n+1}}{u_n} < \frac{r^{n+1}}{r^n} \text{ for all } n \geq m.$$

Take $v_r = r^n$. Then

$$\frac{u_{n+1}}{u_n} < \frac{v_{n+1}}{v_n}$$

and $\Sigma v_n = \Sigma r^n$, $r < 1$ is convergent. Therefore, by Comparison Test, Σu_n also converges.

To prove (ii), proceed with the same process as in (i) and obtain

$$\frac{u_{n+1}}{u_n} > r = \frac{r^{n+1}}{r^n}, \ r > 1 \text{ for all } n \geq m.$$

Take $v_n = r^n$. Then,

$$\frac{u_{n+1}}{u_n} > \frac{v_{n+1}}{v_n}$$

and $\Sigma v_n = \Sigma r^n$ diverges. Hence, by Comparison Test, Σu_n also diverges.

Remark 1.7. For comparison with other tests, D'Alembert Ratio Test is generally used in its inverted form:

"*If Σu_n is a series positive terms and $\lim_{n\to\infty}\frac{u_n}{u_{n+1}} > 1$ then Σu_n converges. If $\lim_{n\to\infty}\frac{u_n}{u_{n+1}} < 1$, then Σu_n diverges.*"

EXAMPLE 1.31

Examine the convergence of the series

$$1 + \frac{1^2\cdot 2^2}{1\cdot 3\cdot 5} + \frac{1^2\cdot .2^2.\cdot 3^2}{1\cdot 3\cdot 5\cdot 7\cdot 9} + \cdots.$$

Solution. The nth term of the given series is

$$u_n = \frac{1^2.\cdot 2^2\cdot 3^2\cdots n^2}{1\cdot 3\cdot 5\cdots(4n-5)(4n-3)}$$

Then,

$$u_{n+1} = \frac{1^2\cdot 2^2\cdot 3^2\cdots n^2(n+1)^2}{1\cdot 3\cdot 5\cdots(4n-5)(4n-3)(4n-1)(4n+1)}$$

and so

$$\frac{u_n}{u_{n+1}} = \frac{(4n-1)(4n+1)}{(n+1)^2}$$

$$= \frac{16n^2 - 1}{(n+1)^2}$$

$$= \frac{16 - \frac{1}{n^2}}{\left(1 - \frac{1}{n}\right)^2}.$$

Further,

$$\lim_{n\to\infty} \frac{u_n}{u_{n+1}} = \lim_{n\to\infty} \frac{16 - \frac{1}{n^2}}{\left(1 - \frac{1}{n}\right)^2} = \frac{16}{1} > 1.$$

Hence, by D'Alembert's Ratio Test, the series Σu_n converges.

EXAMPLE 1.32

Examine the convergence of the series

$$1 + \frac{3}{7}x + \frac{3.6}{710}x^2 + \frac{3.6.9}{7.10.13}x^3 + \cdots$$

Solution. Neglecting the first term, we have

$$u_n = \frac{3\cdot 6\cdot 9\cdots 3n}{7\cdot 10\cdot 13\cdots(3n+4)}x^n \text{ and}$$

$$u_{n+1} = \frac{3\cdot 6\cdot 9\cdots(3n)(3n+3)}{7\cdot 10\cdot 13\cdots(3n+4)(3n+7)}x^{n+1}.$$

Therefore,

$$\lim_{n\to\infty} \frac{u_n}{u_{n+1}} = \lim_{n\to\infty} \frac{3n+7}{3n+3}.\frac{1}{x}$$

$$= \lim_{n\to\infty} \frac{n(3+\frac{7}{3})}{n(3+\frac{3}{n})}.\frac{1}{x} = \frac{1}{x}$$

Therefore, by D'Alembert's Ratio Test, the given series converges if $\frac{1}{X} > 1$ or if $x < 1$ and diverges if $\frac{1}{X} < 1$ or if $x > 1$. If $x = 1$, then D'Alembert's Ratio Test gives no information.

EXAMPLE 1.33

Test the convergence of the series

$$\frac{1}{2\sqrt{1}} + \frac{x^2}{3\sqrt{2}} + \frac{x^4}{4\sqrt{3}} + \frac{x^6}{5\sqrt{4}} + \cdots,\ x > 0.$$

Solution. The nth term of the given series is

$$u_n = \frac{x^{2n-2}}{(n+1)\sqrt{n}} \text{ and so } u_{n+1} = \frac{x^{2n}}{(n+2)\sqrt{n+1}}.$$

Therefore,

$$\lim_{n\to\infty} \frac{u_n}{u_{n+1}} = \lim_{n\to\infty} \frac{(n+2)\sqrt{n+1}}{(n+1)\sqrt{n}}\cdot\frac{1}{x^2}$$

$$= \lim_{n\to\infty} \frac{n(1+\frac{2}{n})}{n(1+\frac{1}{4})}.\frac{\sqrt{n}(1+\frac{1}{n})^{\frac{1}{2}}}{\sqrt{n}}.\frac{1}{x^2} = \frac{1}{x^2}.$$

Thus, by D'Alembert's Ratio Test, the given series converges if $\frac{1}{x^2} > 1$, that is if, $x^2 < 1$, and diverges if $\frac{1}{x^2} < 1$, that is, if $x^2 > 1$. When $x^2 = 1$, the D'Alembert's Ratio Test fails and we have

$$u_n = \frac{1}{(n+1)\sqrt{n}} = \frac{1}{n^{\frac{3}{2}}(1+\frac{1}{n})}.$$

Taking $v_n = \frac{1}{n^{\frac{3}{2}}}$, we have

$$\lim_{n\to\infty} \frac{u_n}{v_n} = \lim_{n\to\infty} \frac{1}{1+\frac{1}{n}} = 1,$$ finite and non-zero:

Hence, by Comparison Test, Σu_n and Σv_n converge or diverge together. But $\Sigma v_n = \Sigma \frac{1}{n^{\frac{3}{2}}}$ converges. Hence Σu_n also converges. Therefore, the given series converges for $x^2 \leq 1$ and diverges for $x^2 < 1$.

EXAMPLE 1.34

Test the convergence of the following series:

(i) $\sum \frac{x^n}{n!}$

(ii) $1 + \frac{2^p}{2!} + \frac{3^p}{3!} + \frac{4^p}{4!} + \cdots$

Solution. (i) The nth term of the series is

$$u_n = \frac{x^n}{n!}.$$

Therefore,

$$u_{n+1} = \frac{x^{n+1}}{(n+1)!}$$

and so

$$\lim_{n\to\infty} \frac{u_{n+1}}{u_n} = \lim_{n\to\infty} \frac{x}{n+1}$$

$= 0$, less than 1, for all finite values of x.

Hence, by D'Alembert's Ratio Test, the given series converges for all finite values of x.

(ii) We have

$$u_n = \frac{n^p}{n!}, \quad u_{n+1} = \frac{(n+1)^p}{(n+1)!}$$

and so

$$\lim_{n\to\infty}\frac{u_n}{u_{n+1}} = \lim_{n\to\infty}\frac{n+1}{\left(1+\frac{1}{n}\right)^p}$$

$$= \infty \text{ for all values of } p.$$

Hence, the given series converges for all values of p.

EXAMPLE 1.35

Examine the convergence of the series

$$\sum\frac{\sqrt{n}}{\sqrt{n^2+1}}x^n, \quad x>0$$

Solution. For the given series

$$u_n = \frac{\sqrt{n}}{\sqrt{n^2+1}}x^n, \; u_{n+1} = \frac{\sqrt{n+1}}{\sqrt{(n+1)^2+1}}x^{n+1}$$

and so

$$\lim_{n\to\infty}\frac{u_n}{u_{n+1}} = \lim_{n\to\infty}\sqrt{\frac{n}{n+1}\cdot\frac{n^2+2+2n}{n^2+1}}\cdot\frac{1}{x}$$

$$= \lim_{n\to\infty}\sqrt{\frac{1}{1+\frac{1}{n}}\cdot\frac{1+\frac{2}{n^2}+\frac{2}{n}}{1+\frac{1}{n^2}}}\cdot\frac{1}{x} = \frac{1}{x}$$

Hence, by D'Alembert's Ratio Test, Σu_n converges if $\frac{1}{x}>1$, that is, if $x<1$ and diverges if $\frac{1}{x}<1$, that is, if $x>1$. When $x=1$, this test gives no information. But in that case,

$$u_n = \frac{\sqrt{n}}{\sqrt{n^2+1}} = \sqrt{\frac{n}{n^2(1+\frac{1}{n^2})}} = \frac{1}{\sqrt{n}}\cdot\frac{1}{\sqrt{1+\frac{1}{n^2}}}.$$

Taking $v_n = \frac{1}{\sqrt{n}}$ we note that

$$\lim_{n\to\infty}\frac{u_n}{v_n} = \lim_{n\to\infty}\frac{1}{\sqrt{1+\frac{1}{n^2}}} = 1, \text{ finite and non-zero.}$$

Therefore, Σu_n and Σv_n converge or diverge together. But $\Sigma v_n = \Sigma\frac{1}{\sqrt{n}}$ diverges. Therefore, the given series diverges for $x=1$. Hence, the given series converges if $x<1$ and diverges if $x\geq 1$.

EXAMPLE 1.36

Test for the convergence of the series

(i) $\sum\frac{n^2(n+1)^2}{n!}$

(ii) $\left(\frac{1}{3}\right)^2+\left(\frac{1\cdot 2}{3\cdot 5}\right)^2+\left(\frac{1\cdot 2\cdot 3}{3\cdot 5\cdot 7}\right)^2+\left(\frac{1\cdot 2\cdot 3\cdot 4}{3\cdot 5\cdot 7\cdot 9}\right)^2+\cdots$.

(iii) $\sum\frac{n!}{n^n}$

Solution. (i) For the given series

$$u_n = \frac{n^2(n+1)^2}{n!}, \quad u_{n+1} = \frac{(n+1)^2(n+2)^2}{(n+1)!}$$

and so

$$\lim_{n\to\infty}\frac{u_n}{u_{n+1}} = \lim_{n\to\infty}\frac{n^2(n+1)^2}{n!}\cdot\frac{(n+1)n!}{(n+1)^2(n+2)^2}$$

$$= \lim_{n\to\infty}\frac{n^3(1+\frac{1}{n})}{n^2(1+\frac{2}{n})^2} = \infty.$$

Hence, the given series converges by D'Alembert's Ratio Test.

(ii) For the given series

$$u_n = \left(\frac{1.2.3.4\ldots..n}{3.5.7.9\ldots.(2n+1)}\right)^2,$$

$$u_{n+1} = \left(\frac{1.2.3.4\ldots.(n+1)}{3.5.7.9\ldots.(2n+3)}\right)^2.$$

Therefore,

$$\lim_{n\to\infty}\frac{u_n}{u_{n+1}} = \lim_{n\to\infty}\frac{(2n+3)^2}{(n+1)^2} = \lim_{n\to\infty}\frac{(2+\frac{1}{n})^2}{(1+\frac{1}{n})^2} = 2.$$

Hence, the given series converges by D'Alembert's Ratio Test.

(iii) The nth term of the given series is

$$u_n = \frac{n!}{n^n}.$$

Therefore,

$$u_{n+1} = \frac{(n+1)!}{(n+1)^{n+1}} = \frac{(n+1)n!}{(n+1)^{n+1}} = \frac{n!}{(n+1)^n}$$

and so

$$\lim_{n\to\infty}\frac{u_n}{u_{n+1}} = \lim_{n\to\infty}\frac{(n+1)^n}{n^n} = \left(\frac{n+1}{n}\right)^n$$

$= \left(1+\frac{1}{4}\right)^n = e = 2.7$, greater than 1.

Hence, the given series converges by D'Alembert's Ratio Test.

EXAMPLE 1.37

Test for the convergence of the series

(i) $x + 2x^2 + 3x^3 + 4x^4 + \cdots$

(ii) $\sum \frac{x^n}{(2n)!}$

(iii) $\sum \frac{n}{n^2+1} x^n$, $x > 0$

(iv) $1 + \frac{x}{2} + \frac{x^2}{5} + \frac{x^3}{10} + \cdots + \frac{x4}{n^2+1} + \cdots$

Solution. (i) The given series is Σnx^n. Therefore,

$$u_n = nx^n,\ u_{n+1} = (n+1)x^{n+1}$$

and so

$$\lim_{n\to\infty} \frac{u_n}{u_{n+1}} = \lim_{n\to\infty} \frac{nx^n}{(n+1)x^{n+1}}$$

$$= \lim_{n\to\infty} \frac{1}{(1+\frac{1}{n})x} = \frac{1}{x}.$$

Hence, by D'Alembert's Ratio Test, the series converges if $\frac{1}{x} > 1$, that is, if $x < 1$ and diverges if $\frac{1}{x} < 1$, that is, if $x > 1$. If $x = 1$, the Ratio Test gives no information about convergence. But for $x = 1$, the series becomes Σn, which is divergent. Hence, the given series converges for $x < 1$ and diverges for $x \geq 1$.

(ii) We have

$$u_n = \frac{x^n}{(2n)!},\ u_{n+1} = \frac{x^{n+1}}{(2n+2)!}$$

Therefore,

$$\lim_{n\to\infty} \frac{u_n}{u_{n+1}} = \lim_{n\to\infty} \frac{x^n}{(2n)!} \cdot \frac{(2n+2)(2n+1)(2n)!}{x^{n+1}}$$

$$= \lim_{n\to\infty} \frac{(2n+2)(2n+1)}{x}$$

$$= \lim_{n\to\infty} \frac{n^2(2+\frac{2}{n})(2+\frac{1}{n})}{x} = \infty > 1.$$

Hence, by D'Alembert's Ratio Test, the given series converges.

(iii) The nth term of the series is

$$u_n = \frac{n}{n^2+1} x^n,\ x > 0.$$

Therefore,

$$u_{n+1} = \frac{(n+1)}{(n+1)^2+1} x^{n+1}$$

and so

$$\lim_{n\to\infty} \frac{u_n}{u_{n+1}} = \lim_{n\to\infty} \frac{n}{n^2+1} x^n \cdot \frac{(n+1)^2+1}{(n+1)x^{n+1}}$$

$$= \lim_{n\to\infty} \frac{n(n^2+2+2n)}{(n+1)(n^2+1)} \cdot \frac{1}{x}$$

$$= \lim_{n\to\infty} \frac{n^3\left(1+\frac{2}{n^2}+\frac{2}{n}\right)}{n^3\left(1+\frac{1}{n}\right)\left(1+\frac{1}{n^2}\right)} \frac{1}{x} = \frac{1}{x}.$$

Hence, by D'Alembert's Ratio Test, the series converges if $\frac{1}{x} > 1$ or $x < 1$ and diverges if $\frac{1}{x} < 1$ or $x > 1$. If $x = 1$, the Ratio Test gives no information. But in that case,

$$u_n = \frac{n}{n^2+1} = \frac{n}{n^2\left(1+\frac{1}{n^2}\right)} = \frac{1}{n\left(1+\frac{1}{n^2}\right)}.$$

Take $v_n = \frac{1}{n}$. Then,

$$\lim_{n\to\infty} \frac{u_n}{v_n} = 1, \text{ finite and non-zero:}$$

Hence, Σu^n and Σv^n converges or diverges together. But $\Sigma v_n = \Sigma \frac{1}{n}$ diverges. Therefore Σu^n also diverges. Hence, the given series converges for $x < 1$ and diverges for $x \geq 1$.

(iv) Neglecting the first term, the nth term of the series is

$$u_n = \frac{x^n}{n^2+1},\ u_{n+1} = \frac{x^{n+1}}{(n+1)^2+1}.$$

Therefore,

$$\lim_{n\to\infty} \frac{u_n}{u_{n+1}} = \lim_{n\to\infty} \frac{x^n}{n^2+1} \cdot \frac{n^2+2+2n}{x^{n+1}}$$

$$= \lim_{n\to\infty} \cdot \frac{n^2\left(1+\frac{2}{n^2}+\frac{2}{n}\right)}{n^2\left(1+\frac{1}{n^2}\right)} \cdot \frac{1}{x} = \frac{1}{x}.$$

Therefore, the series converges if $\frac{1}{x} > 1$ or $x < 1$ and diverges for $\frac{1}{x} < 1$ or $x > 1$. For $x = 1$, the D'Alembert's Ratio Test given no information.

But in such a case,

$$u_n = \frac{1}{n^2+1} = \frac{1}{n^2\left(1+\frac{1}{n^2}\right)}.$$

Taking $v_n = \frac{1}{n^2}$, we get

$$\lim_{n\to\infty}\frac{u_n}{v_n} = \lim_{n\to\infty}\frac{1}{\left(1+\frac{1}{n^2}\right)}$$

$$= 1, \text{ finite and non-zero.}$$

Hence, Σu_n and Σv_n converge or diverge together. But $\Sigma v_n = \Sigma\frac{1}{n^2}$ is convergent. Hence, Σu_n converges for $x = 1$. Thus, the given series is convergent for $x \le 1$ and diverges for $x > 1$.

EXAMPLE 1.38

Examine the convergence for the following series of positive term.

(i) $1+\frac{\alpha+1}{\beta+1}+\frac{(\alpha+1)(2\alpha+1)}{(\beta+1)(2\beta+1)}+\frac{(\alpha+1)(2\alpha+1)(3\alpha+1)}{(\beta+1)(2\beta+1)(3\beta+1)}+\ldots$

(ii) $\frac{4}{18}+\frac{4.12}{18.27}+\frac{4.12.20}{18.27.36}+\ldots$

(iii) $1+\frac{1}{2}.\frac{x^2}{4}+\frac{1.3.5}{2.4.6}.\frac{x^4}{8}+\frac{1.3.5.7.9}{2.4.6.8.10}.\frac{x^6}{12}+\ldots$

Solution. (i) Without taking notice of the first term, the nth term of the given series is

$$u_n = \frac{(\alpha+1)(2\alpha+1)\ldots(n\alpha+1)}{(\beta+1)(2\beta+1)\ldots(n\beta+1)}.$$

Therefore,

$$u_{n+1} = \frac{(\alpha+1)(2\alpha+1)\ldots(n\alpha+1)[(n+1)\alpha+1]}{(\beta+1)(2\beta+1)\ldots(n\beta+1)[(n+1)\beta+1]}$$

and so

$$\lim_{n\to\infty}\frac{u_n}{u_{n+1}} = \lim_{n\to\infty}\frac{(n+1)\beta+1}{(n+1)\alpha+1}$$

$$= \lim_{n\to\infty}\frac{(1+\frac{1}{n})\beta+\frac{1}{n}}{(1+\frac{1}{n})\alpha+\frac{1}{n}} = \frac{\beta}{\alpha}.$$

Hence, by D'Alembert's Ratio Test, the series is convergent if $\frac{\beta}{\alpha}>1$ or $\beta>\alpha>0$ and divergent if $\frac{\beta}{\alpha}<1$ or $\alpha>\beta>0$. If $\alpha=\beta$, then $\lim_{n\to\infty}\frac{u_n}{u_{n+1}}=1$ and so the Ratio Test gives no information. But then the series becomes

$$1+1+1+\ldots,$$

which is divergent. Hence, the given series converges if $\beta>\alpha>0$ and diverges if $\alpha\ge\beta>0$.

(ii) The given series is

$$\frac{4}{18}+\frac{4.12}{18.27}+\frac{4.12.20}{18.27.36}+\ldots$$

whose nth term is given by

$$u_n = \frac{4\cdot12\cdot20\ldots(8n-4)}{18\cdot27\cdot36\ldots(9n+9)}.$$

Therefore,

$$u_{n+1} = \frac{4\cdot12\cdot20\ldots(8n-4)(8n+4)}{18\cdot27\cdot36\ldots(9n+9)(9n+18)}$$

and so

$$\lim_{n\to\infty}\frac{u_n}{u_{n+1}} = \lim_{n\to\infty}\frac{9n+18}{8n+4}$$

$$= \lim_{n\to\infty}\frac{n\left(9+\frac{18}{n}\right)}{n\left(8+\frac{4}{n}\right)} = \frac{9}{8}>1.$$

Hence, by D'Alembert's Ratio Test, the given series converges.

(iii) For the series (iii), we have

$$u_n = \frac{1.3.5\ldots(4n-7)}{(4n-6)}.\frac{x^{2n-2}}{(4n-4)},$$

$$u_{n+1} = \frac{1.3.5\ldots(4n-7)(4n-5)(4n-3)}{2.4.6\ldots(4n-6)(4n-4)(4n-2)}.\frac{x^{2n}}{4n}.$$

Therefore,

$$\lim_{n\to\infty}\frac{u_n}{u_{n+1}} = \lim_{n\to\infty}\frac{(4n-2)(4n-4)}{(4n-5)(4n-3)}.\frac{4n}{(4n-4)}.\frac{1}{x^2}$$

$$= \lim_{n\to\infty}\frac{16n^2-8n}{16n^2-32n+15}.\frac{1}{x^2} = \frac{1}{x^2}$$

Hence, by D'Alembert Ratio Test, the series converges if $\frac{1}{x^2}>1$ or $x^2 < 1$ and diverges if $x^2 > 1$. If $x^2 = 1$, the test fails. But if $x^2 = 1$, then

$$u_n = \frac{1\cdot3\cdot5\ldots(4n-7)}{2\cdot4.6\ldots(4n-6)(4n-4)}$$

$$=\frac{1\cdot 3\cdot 5\ldots n(4-\frac{7}{n})}{2\cdot 4\cdot 6\ldots n^2(4-\frac{6}{n})(4-\frac{4}{n})}.$$

Take $v_n = \frac{1\cdot 3\cdot 5\ldots n}{2\cdot 4\cdot 6\ldots n}$. Then $\frac{u_n}{v_n} \to A$ (finite and nonzero). Hence, the series converges for $x^2 = 1$.

Therefore, the series converges when $x^2 \le 1$ and diverges when $x^2 > 1$.

1.11 CAUCHY'S ROOT TEST

If Σu_n is a series of positive terms, then

(i) the series Σu_n converges if $\lim_{n\to\infty}(u_n)^{\frac{1}{n}} < 1$.

(ii) the series Σu_n diverges if $\lim_{n\to\infty}(u_n)^{\frac{1}{n}} > 1$.

(When $\lim_{n\to\infty}(u_n)^{\frac{1}{n}} = 1$, the root test fails to give any information regarding convergence of the series Σu_n).

Proof: Suppose that $\lim_{n\to\infty}(u_n)^{\frac{1}{n}} = r$. Then, by definition, to each $\varepsilon > 0$, there exists a positive integer m such that

$$|(u_n)^{\frac{1}{n}} - r| < \varepsilon \text{ for all } n \ge m$$

or

$$r-\varepsilon < (u_n)^{\frac{1}{n}} < r+\varepsilon \text{ for all } n \ge m$$

or

$$(r-\varepsilon)^n < u^n < (r+\varepsilon)^n \cdot$$

Case (i). If $r < 1$, choose $\varepsilon > 0$ such that $r+\varepsilon < 1$. Taking $r+\varepsilon = t < 1$, we have

$$u_n < t^n \text{ for all } n \ge m.$$

But the geometric series, $\Sigma t^n, t<1$ is convergent. Hence, by Comparison Test, Σu_n converges.

Case (ii). If $r > 1$, choose $\varepsilon > 0$ such that, $r-\varepsilon > 1$. Taking $r - \varepsilon = T > 1$, we have

$$u_n < T^n \text{ for all } n \ge m.$$

But the geometric series $\Sigma T^n, T > 1$ is divergent. Hence, by Comparison Test, the series Σu_n diverges.

Remark 1.8. (i) If u_n is positive for all values of n, then, by Cauchy's Limit Theorem,

$$\lim_{n\to\infty}(u_n)^{\frac{1}{n}} = \lim_{n\to\infty}\frac{u_{n+1}}{u_n},$$

provided the latter limit exists. It follows from this result that if D'Alembert. Ratio Test is applicable to any series, then Cauchy's Root Test is also applicable. But the converse is not true. To show that Cauchy's Root Test is more powerful than D'Alembert's Ratio Test, consider the series with nth term u_n defined by

$$u_n = 2^{-n-(-1)^n}.$$

Then

$$\lim_{n\to\infty}(u^n)^{\frac{1}{n}} = \frac{1}{2}$$

Hence, by Cauchy's Root Test, the series Σu_n converges. On the other hand

$$\frac{u_{n+1}}{u_n} = 2^{-1+(-1)^n-(-1)^{n+1}} = \begin{cases} 2 \text{ if } n \text{ is even} \\ \frac{1}{2^3} \text{ if } n \text{ is odd.}\end{cases}$$

Thus, $\overline{\lim}\frac{u_{n+1}}{u_n} = 2$ and $\underline{\lim}\frac{u_{n+1}}{u_n} = \frac{1}{8}$. Therefore, D'Alembet's Ratio Test yields no definite result.

(ii) The two tests are, of course, completely equivalent if, $\lim_{n\to\infty}\frac{u_{n+1}}{u_n}$ exists.

EXAMPLE 1.39

Examine the convergence of the series

$$\frac{2}{1^2}x+\frac{3^2}{2^3}x^2+\frac{4^3}{2^4}x^3+\ldots$$

Solution. The nth term of the given series is

$$u_n = \frac{(n+1)^n}{n^{n+1}}x^n = (\frac{n+1}{n})^n\cdot\frac{x^n}{n}.$$

Therefore,

$$\lim_{n\to\infty}u_n^{\frac{1}{n}} = \lim_{n\to\infty}\frac{n+1}{n}.\frac{x}{n^{\frac{1}{n}}} = \lim_{n\to\infty}\frac{(1+\frac{1}{n})}{n^{\frac{1}{n}}} = x.$$

Hence, by Cauchy's Root Test, the given series converges if $x < 1$ and diverges if $x > 1$. When $x = 1$, we have

$$u_n = \left(\frac{n+1}{n}\right)^n \cdot \frac{1}{n}.$$

Take $v_n = \frac{1}{n}$. Then,

$$\lim_{n\to\infty} \frac{u_n}{v_n} = \lim_{n\to\infty} \left(1+\frac{1}{n}\right)^n = e \text{ (finite and non-zero).}$$

But $\Sigma v_n = \Sigma \frac{1}{n}$ is divergent. So Σu_n diverges for $x = 1$. Hence, the given series converges if $x < 1$ and diverges if $x \geq 1$.

EXAMPLE 1.40

Examine the convergence of the following series

(i) $\sum \frac{(n+1)^{n^2}}{n^{n^2} \cdot 3^n}$

(ii) $\frac{1}{2} + \frac{2}{3}x + \left(\frac{3}{4}\right)^2 x^2 + \left(\frac{4}{5}\right)^2 x^3 + \left(\frac{5}{6}\right)^4 x^4 + \ldots$ for $x > 0$.

Solution. (i) The nth term for the series is

$$u_n = \frac{(n+.1)^{n^2}}{n^{n^2} 3^n}.$$

Therefore,

$$\lim_{n\to\infty} u_n^{\frac{1}{n}} = \lim_{n\to\infty} \frac{(n+1)^n}{3n^n} = \lim_{n\to\infty} \frac{1}{3}\left(1+\frac{1}{n}\right)^n = \frac{1}{3}e < 1.$$

Hence, by Cauchy's Root Test, the given series is convergent.

(ii) Ignoring the first term, the nth term of the series is

$$u_n = \left(\frac{n+1}{n+2}\right)^n x^n.$$

Therefore,

$$\lim_{n\to\infty} u_n^{\frac{1}{n}} = \lim_{n\to\infty} \left(\frac{n+1}{n+2}\right) x = \lim_{n\to\infty} = \frac{1+\frac{1}{n}}{1+\frac{2}{n}} x = x.$$

Hence, the series converges if $x < 1$ and diverges if $x > 1$. If $x = 1$, then

$$u_n = \left(\frac{n+1}{n+2}\right)^n = \frac{\left(1+\frac{1}{n}\right)^n}{\left(1+\frac{2}{n}\right)^n} = \frac{\left(1+\frac{1}{n}\right)^n}{[(1+\frac{2}{n})^{\frac{n}{2}}]^2}$$

and so $\lim_{n\to\infty} u_n = \frac{e}{e^2} = \frac{1}{e} \neq 0$. Since $\lim_{n\to\infty} u_n \neq 0$, the series diverges for $x = 1$. Thus the given series converges for $x < 1$ and diverges for $x \geq 1$.

EXAMPLE 1.41

Examine the convergence of the series

(i) $1 + \frac{x}{2} + \frac{x^2}{3^2} + \frac{x^3}{4^3} + \ldots (x > 0)$

(ii) $\sum \left(1+\frac{1}{\sqrt{n}}\right)^{-n^{\frac{3}{2}}}$

Solution. (i). Neglecting the first term, the nth term of the given series is

Therefore, $$u_n = \frac{x^n}{(n+1)^n}.$$

$$\lim_{n\to\infty} (u_n)^{\frac{1}{n}} = \lim_{n\to\infty} \left(\frac{x^n}{(n+1)^n}\right)^{\frac{1}{n}}$$

$$= \lim_{n\to\infty} \left(\frac{x^n}{n^n \left(1+\frac{1}{n}\right)^n}\right)^{\frac{1}{n}}$$

$$= \lim_{n\to\infty} \frac{x}{n\left(1+\frac{1}{n}\right)} = 0 < 1.$$

Hence, the given series is convergent by Cauchy's Root Test.

(ii) For this series,

$$u_n = \left(1+\frac{1}{\sqrt{n}}\right)^{-n^{\frac{3}{2}}} = \frac{1}{\left(1+\frac{1}{\sqrt{n}}\right)^{\sqrt{n}.n}}.$$

Therefore,

$$\lim_{n\to\infty} u_n^{\frac{1}{n}} = \lim_{n\to\infty} \frac{1}{\left(1+\frac{1}{\sqrt{n}}\right)^{\sqrt{n}}} = \frac{1}{e} < 1.$$

Hence, the given series converges by Cauchy's Root Test.

EXAMPLE 1.42

Test the convergence of the series

$$\sum \left(\frac{n}{n+1}\right)^{n^2}$$

Solution. For the given series,

$$u_n = \left(\frac{n}{n+1}\right)^{n^2}.$$

Therefore,

$$u_n^{\frac{1}{n}} = \left(\frac{n}{n+1}\right)^n = \left(\frac{1}{1+\frac{1}{n}}\right)^n = \frac{1}{\left(1+\frac{1}{n}\right)^n}$$

and so

$$\lim_{n\to\infty} u_n^{\frac{1}{n}} = \lim_{n\to\infty}\frac{1}{\left(1+\frac{1}{n}\right)^n} = \frac{1}{e} < 1.$$

Hence, by Cauchy's Root Test, the given series converges.

1.12 RAABE'S TEST

Let Σu_n be a series of positive terms, then

(i) Σu_n converges if $\lim_{n\to\infty} n\left(\frac{u_n}{u_{n+1}}-1\right) > 1$

(ii) Σu_n diverges if $\lim_{n\to\infty} n\left(\frac{u_n}{u_{n+1}}-1\right) < 1$.

(This test gives no information if $\lim_{n\to\infty} n\left(\frac{u_n}{u_{n+1}}-1\right) = 1$).

Proof: Let $\lim_{n\to\infty} n\left(\frac{u_n}{u_{n+1}}-1\right) = k$. If $\underline{k} > 1$, choose a number p so that $k > p > 1$. Now compare the series Σu_n with the auxiliary series Σv_n, where $v_n = \frac{1}{n^p}$. Then Σv_n is convergent if $p > 1$. Therefore, Σu_n is convergent if after some particular term

$$\frac{u_n}{u_{n+1}} > \frac{v_n}{v_{n+1}}$$

or if

$$\frac{u_n}{u_{n+1}} > \left(\frac{n+1}{n}\right)^p = \left(1+\frac{1}{n}\right)^p$$

$$= 1 + \frac{p}{n} + \frac{p(p-1)}{2!}\cdot\frac{1}{n^2} + \frac{p(p-1)(p-2)}{3!}\cdot\frac{1}{n^3} + \ldots$$

or if

$$n\left(\frac{u_n}{u_{n+1}}-1\right) > p + \frac{p(p-1)}{2!}\cdot\frac{1}{n} + \frac{p(p-1)(p-2)}{2!}\cdot\frac{1}{n^2} + \ldots$$

or if

$$\lim_{n\to\infty} n\left(\frac{u_n}{u_{n+1}}-1\right) > \lim_{n\to\infty}\left(p + \frac{p(p-1)}{2!}\cdot\frac{1}{n} + \frac{p(p-1)(p-2)}{2!}\cdot\frac{1}{n^2} + \ldots\right)$$

or if

$$\lim_{n\to\infty} n\left(\frac{u_n}{u_{n+1}}-1\right) > p > 1.$$

Thus, Σu_n converges if $\lim_{n\to\infty} n\left(\frac{u_n}{u_{n+1}}-1\right) > p > 1$.

To prove (ii), we use the fact that $\sum v_n = \sum \frac{1}{n^p}$ diverges if $p \le 1$. Then, Σu_n will also diverge if

$$\frac{u_n}{u_{n+1}} < \frac{v_n}{v_{n+1}}$$

or if

$$\frac{u_n}{u_{n+1}} < 1 + \frac{p}{n} + \frac{p(p-1)}{2!}\cdot\frac{1}{n^2} + \frac{p(p-1)(p-2)}{3!}\cdot\frac{1}{n^3} + \ldots$$

or if

$$n\left(\frac{u_n}{u_{n+1}}-1\right) < p + \frac{p(p-1)}{2!}\cdot\frac{1}{n} + \frac{p(p-1)(p-2)}{3!}\cdot\frac{1}{n^2} + \ldots$$

or if

$$\lim_{n\to\infty} n\left(\frac{u_n}{u_{n+1}}-1\right) < p \le 1.$$

Thus, the series Σu_n diverges if $\lim_{n\to\infty} n\left(\frac{u_n}{u_{n+1}}-1\right) < 1$.

EXAMPLE 1.43

Examine the convergence of the series

$$\frac{1}{4} + \frac{1\cdot 3}{4\cdot 6} + \frac{1\cdot 3\cdot 5}{4\cdot 6\cdot 8} + \cdots + \frac{1\cdot 3\cdot 5\ldots(2n-1)}{4\cdot 6\cdot 8\ldots(2n+2)} + \cdots$$

Solution. The nth term of the given series is

$$u_n = \frac{1\cdot 3\cdot 5\ldots(2n-1)}{4\cdot 6\cdot 8\ldots(2n+2)}.$$

Therefore,

$$u_{n+1} = \frac{1\cdot 3\cdot 5\ldots(2n+1)}{4\cdot 6\cdot 8\ldots(2n+4)}$$

and so

$$\lim_{n\to\infty}\frac{u_n}{u_{n+1}}=\lim_{n\to\infty}\frac{(2n+4)}{(2n+1)}=1.$$

Hence, D'Alembert's Ratio Test is not applicable. Further,

$$\frac{u_n}{u_{n+1}}=\frac{(2n+4)}{(2n+1)}$$

implies

$$n\left(\frac{u_n}{u_{n+1}}-1\right)=n\left(\frac{2n+4}{2n+1}-1\right)$$

$$=n\left(\frac{2n+4-(2n+1)}{2n+1}\right)$$

$$=\frac{3n}{2n+1}=\frac{3n}{n\left(2+\frac{1}{n}\right)}.$$

Therefore,

$$\lim_{n\to\infty}n\left(\frac{u_n}{u_{n+1}}-1\right)=\lim_{n\to\infty}\frac{3n}{n\left(2+\frac{1}{n}\right)}=\frac{3}{2}>1.$$

Hence, by Raabe's Test, the given series converges.

EXAMPLE 1.44

Examine the following series for convergence

(i) $x^2(\log 2)^q+x^3(\log 3)^q+x^4(\log 4)^q+\ldots$

(ii) $x+\frac{1}{2}\cdot\frac{x^3}{3}+\frac{1\cdot 3}{2\cdot 4}\cdot\frac{x^5}{5}+\frac{1\cdot 3\cdot 5}{2\cdot 4\cdot 6}\cdot\frac{x^7}{7}+\ldots.$

Solution. (i). The nth term of the given series is

$$u_n=x^{n+1}\left[\log(n+1)\right]^q.$$

Therefore,

$$u_{n+1}=x^{n+2}[\log(n+2)]^q$$

and so

$$\frac{u_n}{u_{n+1}}=\frac{1}{x}\left[\frac{\log(n+1)}{\log(n+2)}\right]^q$$

$$=\frac{1}{x}\left[\frac{\log n\left(1+\frac{1}{n}\right)}{\log n\left(1+\frac{2}{n}\right)}\right]^q$$

$$=\frac{1}{x}\left[\frac{\log n+\log\left(1+\frac{1}{n}\right)}{\log n+\log\left(1+\frac{2}{n}\right)}\right]^q$$

$$=\frac{1}{x}\left[\frac{\log n+\frac{1}{n}-\frac{1}{2n^2}+\ldots}{\log n+\frac{2}{n}-\frac{1}{n}\left(\frac{2}{n}\right)^2+\ldots}\right]^q$$

$$=\frac{1}{x}\left[\frac{1+\frac{1}{n\log n}-\frac{1}{2n^2\log n}+\ldots}{1+\frac{2}{n\log n}-\frac{2}{n^2\log n}+\ldots}\right]^q.$$

Hence $\lim_{n\to\infty}\frac{u_n}{u_{n+1}}=\frac{1}{x}$ and, by D'Alembert's Ratio Test, the series Σu_n converges if $\frac{1}{x}>1$, that is, if $x<1$ and diverges if $\frac{1}{x}<1$, that is, if $x>1$. For $x=1$, the Ratio Test gives no information. But for $x=1$, we have

$$\frac{u_n}{u_{n+1}}=\left[\frac{1+\frac{1}{n\log n}-\frac{1}{2n^2\log n}+\ldots}{1+\frac{2}{n\log n}-\frac{2}{n^2\log n}+\ldots}\right]^q$$

$$=\left[1+\frac{1}{n\log n}-\frac{1}{2n^2\log n}+\ldots\right]^q$$

$$\times\left[1+\frac{2}{n\log n}-\frac{2}{n^2\log n}+\ldots\right]^{-q}$$

$$=\left[1+q\left(\frac{1}{n\log n}-\frac{1}{2n^2\log n}+\ldots\right)+\ldots\right]$$

$$\times\left[1+q\left(\frac{2}{n\log n}-\frac{2}{2n^2\log n}+\ldots\right)+\ldots\right]$$

$$=1+q\left(\frac{1}{n\log n}-\frac{2}{n\log n}+\ldots\right)+\ldots,$$

which yields

$\frac{u_n}{u_{n+1}}-1=-\frac{q}{n\log n}+$ term containing higher powers of n and $\log n$ in the denominator or $n\left(\frac{u_n}{u_{n+1}}-1\right)=\frac{-q}{\log n}+$ term containing powers of n and $\log n$ in the denominator.

Therefore,

$$\lim_{n\to\infty}n\left(\frac{u_n}{u_{n+1}}-1\right)=0<1.$$

Hence, by Raabe's Test, the given series diverges for $x=1$. It follows that the given series converges for $x<1$ and diverges for $x\geq 1$.

(ii) Neglecting the first term, the nth term of the given series is

$$u_n=\frac{1.3.5\ldots(2n-1)}{2.4.6\ldots(2n)}\cdot\frac{x^{2n+1}}{(2n+1)}.$$

Thus

$$u_{n+1}=\frac{1.3.5\ldots(2n-1)(2n+1)}{2.4.6\ldots(2n)(2n+2)}\cdot\frac{x^{2n+3}}{2n+3}$$

and so

$$\frac{u_n}{u_{n+1}}=\frac{(2n+2)(2n+3)}{(2n+1)(2n+1)}\cdot\frac{1}{x^2}$$

$$=\frac{2n\left(1+\frac{1}{n}\right)2n\left(1+\frac{3}{2n}\right)}{2n\left(1+\frac{1}{2n}\right)2n\left(1+\frac{1}{2n}\right)}\cdot\frac{1}{x^2}.$$

Therefore,

$$\lim_{n\to\infty}\frac{u_n}{u_{n+1}}=\lim_{n\to\infty}\frac{\left(1+\frac{1}{n}\right)\left(1+\frac{3}{2n}\right)}{\left(1+\frac{1}{2n}\right)\left(1+\frac{1}{2n}\right)}\cdot\frac{1}{x^2}=\frac{1}{x^2}.$$

Hence, by D'Alembert's Raito Test, the series Σu_n converges if $\frac{1}{x^2}>1$, that is, if $x^2<1$, and diverges if $\frac{1}{x^2}<1$, , that is, if $x^2>1$. If $x^2=1$ the Ratio Test gives no information. But for $x^2 = 1$, we have

$$\frac{u_n}{u_{n+1}}=\frac{(2n+2)(2n+3)}{(2n+1)(2n+1)}=\frac{4n^2+10n+6}{4n^2+4n+1}.$$

Thus

$$n\left(\frac{u_n}{u_{n+1}}-1\right)=n\left(\frac{4n^2+10n+6}{4n^2+4n+1}-1\right)$$

$$=\frac{6n^2+5n}{4n^2+4n+1}$$

and so

$$\lim_{n\to\infty} n\left(\frac{u_n}{u_{n+1}}-1\right)=\lim_{n\to\infty}\frac{6n^2+5n}{4n^2+4n+1}$$

$$=\lim_{n\to\infty}\frac{n^2\left(6+\frac{5}{n}\right)}{n^2 4+\frac{4}{n}+\frac{1}{n^2}}=\frac{3}{2}>1.$$

Therefore, by Raabe's Test, the given series converges for $x^2 = 1$. Hence, Σu_n converges for $x^2 \leq 1$ and diverges for $x^2 > 1$.

EXAMPLE 1.45

Examine the convergence of the series

$$1+\frac{1}{2}x+\frac{1\cdot 3}{2\cdot 4}x^2+\frac{1\cdot 3\cdot 5}{2\cdot 4\cdot 6}x^3+\cdots$$

Solution. The nth term of the series is

$$u_n=\frac{1\cdot 3\cdot 5\ldots(2n-1)}{2\cdot 4\cdot 6\ldots(2n)}x^n.$$

Therefore,

$$u_{n+1}=\frac{1\cdot 3\cdot 5\ldots(2n-1)(2n+1)}{2\cdot 4\cdot 6\ldots(2n)(2n+2)}x^{n+1}$$

and so

$$\lim_{n\to\infty}\frac{u_n}{u_{n+1}}=\lim_{n\to\infty}\frac{2n+2}{2n+1}\cdot\frac{1}{x}=\frac{1}{x}.$$

By D'Alembert's Ratio Test, the series converges if $\frac{1}{x}>1$, that is if $x<1$ and diverges if $\frac{1}{x}<1$, that is if $x>1$. The test fails to give any information if $x=1$. However, for $x=1$, we have

$$\frac{u_n}{u_{n+1}}=\frac{2n+2}{2n+1}$$

or

$$n\left(\frac{u_n}{u_{n+1}}-1\right)=\frac{n}{2n+1}=\frac{1}{2+\frac{1}{n}}.$$

Therefore,

$$\lim_{n\to\infty} n\left(\frac{u_n}{u_{n+1}}-1\right)=\lim_{n\to\infty}\frac{1}{2+\frac{1}{n}}=\frac{1}{2}<1$$

and so, by, Raabe's Test, the given series diverges for $x = 1$. Hence, the given series converges if $x<1$ and diverges for $x\geq 1$.

EXAMPLE 1.46

Use Raabe's Test to examine the convergence of series of Example 1.36 (iii) for the case $x^2=1$.

Solution. The series in Example 1.36 (iii) was

$$1+\frac{1}{2}\cdot\frac{x^2}{4}+\frac{1\cdot 3\cdot 5}{2\cdot 4\cdot 6\cdot 8}x^4+\frac{1\cdot 3\cdot 5\cdot 7\cdot 9}{2\cdot 4\cdot 6\cdot 8\cdot 10}\cdot\frac{x^6}{12}+\cdots$$

The nth term is

$$u_n=\frac{1\cdot 3\cdot 5\cdots(4n-7)}{2\cdot 4\cdot 6\cdots(4n-6)}\cdot\frac{x^{2n-2}}{4n-4}.$$

Then

$$\frac{u_n}{u_{n+1}}=\frac{16n^2-8n}{16n^2-32n+15}\cdot\frac{1}{x^2}.$$

By D'Alembert's Ratio Test, the series converges for $x^2<1$ and diverges for $x^2>1$. For $x^2=1$, the Ratio Test gives no information. In this case, we have

$$u_{n+1}=\frac{16n^2-8n}{16n^2-32n+15}$$

and so

$$n\left(\frac{u_n}{u_{n+1}}-1\right)=n\left(\frac{24n-15}{16n^2-32n+15}\right).$$

Therefore,

$$\lim_{n\to\infty} n\left(\frac{u_n}{u_{n+1}}-1\right)=\frac{3}{2}>1$$

and so the given series converges for $x^2 = 1$. Hence, the series converges if $x^2 \le 1$ and diverges if $x^2 > 1$.

EXAMPLE 1.47

Examine the convergence of the series

$$x+\frac{2^2x^2}{2!}+\frac{3^3x^3}{3!}+\frac{4^4x^4}{4!}+\cdots.$$

Solution. The nth term of the series is

$$u_n=\frac{n^nx^n}{n!}.$$

Therefore,

$$u_{n+1}=\frac{(n+1)^{n+1}x^{n+1}}{(n+1)!}$$

and so

$$\frac{u_n}{u_{n+1}}=\frac{n^nx^n}{n!}\cdot\frac{(n+1).!}{(n+1)^{n+1}x^{n+1}}=\left(\frac{n}{n+1}\right)^n\cdot\frac{1}{x}.$$

Hence,

$$\lim_{n\to\infty}\frac{u_n}{u_{n+1}}=\lim_{n\to\infty}\left(\frac{1}{1+\frac{1}{n}}\right)^n\cdot\frac{1}{x}=\frac{1}{e}\cdot\frac{1}{x}\cdot$$

By D'Alembert's Ratio Test, the series converges if $\frac{1}{e}\cdot\frac{1}{x}>1$, that is, if $x<\frac{1}{e}$, and diverges if $\frac{1}{e}\cdot\frac{1}{x}<1$, that is, if $x>\frac{1}{e}$. Since, this question involves e, we do not apply Raabe's Test. We shall solve it by another test, known as Logarithmic Test (see Example 1.53).

EXAMPLE 1.48

Test the convergence of the series

$$\frac{x}{1\cdot 2}+\frac{x^2}{2\cdot 3}+\frac{x^3}{3\cdot 4}+\frac{x^4}{4.5}+\cdots.$$

Solution. The nth term of the series is

$$u_n=\frac{x^n}{n(n+1)}.$$

Therefore,

$$u_{n+1}=\frac{x^{n+1}}{(n+1)(n+2)}$$

and so

$$\lim_{n\to\infty}\frac{u_n}{u_{n+1}}=\lim_{n\to\infty}\frac{n+2}{n}.\frac{1}{x}=\frac{1}{x}.$$

Therefore, by D'Alembert's Ratio Test, the series converges if $\frac{1}{x}>1$, that is, if $x<1$ and diverges if $\frac{1}{x}<1$, that is, if $x>1$. If $x=1$, then Ratio Test gives no information. However, for $x=1$, we have

$$\frac{u_n}{u_{n+1}}=\frac{n+2}{n}$$

or

$$n\left(\frac{u_n}{u_{n+1}}-1\right)=n\left(\frac{n+2}{n}-1\right)=n\left(\frac{2}{n}\right)=2.$$

Therefore,

$$\lim_{n\to\infty} n\left(\frac{u_n}{u_{n+1}}-1\right)=2$$

and so the series converges by Raabe's Test for $x=1$. Hence, the given series converges if $x\le 1$ and diverges if $x>1$.

EXAMPLE 1.49

Test the convergence of the series

$$1+\frac{\alpha}{\beta}x+\frac{\alpha(\alpha+1)}{\beta(\beta+1)}x^2+\frac{\alpha(\alpha+1)(\alpha+2)}{\beta(\beta+1)(\beta+2)}x^3+\cdots$$

Solution. Neglecting the first term, the nth term of the given series is

$$u_n=\frac{\alpha(\alpha+1)(\alpha+2)\cdot\ldots\cdot(\alpha+(n-1))}{\beta(\beta+1)(\beta+2)(\beta+(n-1))}x^n.$$

Therefore,

$$u_{n+1}=\frac{\alpha(\alpha+1)(\alpha+2)\cdots(\alpha+(n-1))(\alpha+n)}{\beta(\beta+1)(\beta+2)\cdots(\beta+(n-1))(\beta+n)}x^{n+1}$$

and

$$\lim_{n\to\infty}\frac{u_n}{u_{n+1}}=\lim_{n\to\infty}\frac{\beta+n}{\alpha+n}\cdot\frac{1}{x}$$

$$=\lim_{n\to\infty}\frac{n\left(1+\frac{\beta}{n}\right)}{n\left(1+\frac{\alpha}{n}\right)}\cdot\frac{1}{x}=\frac{1}{x}.$$

Thus, by D'Alembert's Ratio Test, the series converges if $\frac{1}{x}>1$, that is, if $x<1$ and diverges if $\frac{1}{x}<1$ that is, if $x>1$. For $x=1$, the Ratio Test gives no infomiation. But for $x=1$, we have

$$\frac{u_n}{u_{n+1}}=\frac{\beta+n}{\alpha+n}.$$

Therefore,

$$n\left(\frac{u_n}{u_{n+1}}-1\right)=n\left(\frac{\beta+1}{\alpha+n}-1\right)$$

$$=n\left(\frac{\beta-\alpha}{\alpha+n}\right)=n\left(\frac{\beta-\alpha}{n\left(1+\frac{\alpha}{n}\right)}\right)$$

and so

$$\lim_{n\to\infty}n\left(\frac{u_n}{u_{n+1}}-1\right)=\beta-\alpha.$$

Thus, by Raabe's Test, the series converges for $\beta-\alpha>1$ and diverges for $\beta-\alpha<1$, when $x=1$. If $\beta-\alpha=1$ then the series becomes

$$1+\frac{\alpha}{1+\alpha}+\frac{\alpha(\alpha+1)}{(1+\alpha)(2+\alpha)}+\frac{\alpha(\alpha+2)}{(1+\alpha)(2+\alpha)(3+\alpha)}+\cdots$$

or

$$1+\frac{\alpha}{1+\alpha}+\frac{\alpha}{2+\alpha}+\frac{\alpha}{3+\alpha}+\cdots,$$

which is divergent. Hence, the given series converges for $x<1$ and diverges for $x>1$. If $x=1$, then the series in question converges for $\beta-\alpha>1$ and diverges for $\beta-\alpha<1$.

EXAMPLE 1.50

Test for convergence of the series

$$1+\frac{3}{7}x+\frac{3.6}{7.10}x^2+\frac{3.6.9}{7.10.13}x^3+\cdots.$$

Solution. Ignoring the first term, the nth term of the given series is

$$u_n=\frac{3.6.9\ldots 3n}{7.10.13\ldots(3n+4)}x^n.$$

Therefore,

$$\lim_{n\to\infty}\frac{u_n}{u_{n+1}}=\lim_{n\to\infty}\frac{3n+7}{3n+4}\cdot\frac{1}{x}=\frac{1}{x}$$

and so, by D'Alembert's Ratio Test, the series converges if $\frac{1}{x}>1$, that is, if $x<1$ and diverges if $\frac{1}{x}<1$, that is, if $x>1$. If $x=1$, the test fails. But for $x=1$, we get

$$\frac{u_n}{u_{n+1}}=\frac{3n+7}{3n+3}$$

and thus

$$\lim_{n\to\infty}n\left(\frac{u_n}{u_{n+1}}-1\right)=\lim_{n\to\infty}n\left(\frac{3n+7}{3n+3}-1\right)$$

$$=\lim_{n\to\infty}n\left(\frac{4}{3n+3}\right)=\frac{4}{3}>1.$$

Therefore, the series converges for $x=1$. Hence, the given series converges for $x\le 1$ and diverges for $x>1$.

1.13 LOGARITHMIC TEST

A series Σu_n of positive terms converges or diverges according as

$$\lim_{n\to\infty}n\log\frac{u_n}{u_{n+1}}>1 \text{ or } <1.$$

(This Test gives no information if $\lim_{n\to\infty}n\log\frac{u_n}{u_{n+1}}=1$).

Proof: Consider the auxiliary series $\Sigma v_n=\Sigma\frac{1}{n^p}$, which converges if $p>1$ and diverges if $p<1$. Then,

$$\frac{v_n}{v_{n+1}}=\left(\frac{n+1}{n}\right)^p=\left(1+\frac{1}{n}\right)^p.$$

If $p>1$, then Σv_n is convergent. The series Σu_n will be convergent if

$$\frac{u_n}{u_{n+1}}>\frac{v_n}{v_{n+1}}$$

or if

$$\frac{u_n}{u_{n+1}}>\left(1+\frac{1}{n}\right)^p$$

or if

$$\log\frac{u_n}{u_{n+1}} > \log\left((1+\frac{1}{n}\right)^p = p\log\left(1+\frac{1}{n}\right)$$

or if

$$\log\frac{u_n}{u_{n+1}} > p\left[\frac{1}{n}-\frac{1}{2n^2}+\frac{1}{3n^3}-\cdots\right]$$

or if

$$n\log\frac{u_n}{u_{n+1}} > p\left[1-\frac{1}{2n}+\frac{1}{3n^2}-\cdots\right]$$

or if

$$\lim_{n\to\infty} n\log\frac{u_n}{u_{n+1}} > p > 1.$$

If $p < 1$, then Σv_n diverges. The series Σu_n will diverge if

$$\frac{u_n}{u_{n+1}} < \frac{v_n}{v_{n+1}}$$

or if

$$\log\frac{u_n}{u_{n+1}} < \log\left(1+\frac{1}{n}\right)^p = p\log\left(1+\frac{1}{n}\right)$$

or if

$$\log\frac{u_n}{u_{n+1}} < p\left[\frac{1}{n}-\frac{1}{2n^2}+\frac{1}{3n^3}-\cdots\right]$$

or if

$$\lim_{n\to\infty} n\log\frac{u_n}{u_{n+1}} < \lim_{n\to\infty} p\left[1-\frac{1}{2n}+\frac{1}{3n^2}-\cdots\right] < p < 1.$$

Thus, the series Σu_n converges if $\lim_{n\to\infty} n\log\frac{u_n}{u_{n+1}} > 1$ and diverges if $\lim_{n\to\infty}\log\frac{u_n}{u_{n+1}} < 1$.

Remark 1.9. The Logarithmic Test should be applied after the failure of D'Alembert's Ratio Test and generally when the ratio $\frac{u_n}{u_{n+1}}$ involves 'e.'

EXAMPLE 1.51

Test for convergence of the series

$$1+\frac{x}{2}+\frac{2!}{3^2}x^2+\frac{3!}{4^3}x^3+\frac{4!}{5^4}x^4+\cdots.$$

Solution. Ignoring the first term, the nth term of the given series is

$$u_n = \frac{n!}{(n+1)^n}x^n.$$

Therefore,

$$u_{n+1} = \frac{(n+1)!}{(n+2)^{n+1}}x^{n+1}$$

so that

$$\frac{u_n}{u_{n+1}} = \frac{n!}{(n+1)^n}x^n\cdot\frac{(n+2)^{n+1}}{(n+1)!x^{n+1}}$$

$$= \frac{(n+2)^{n+1}}{(n+1)^{n+1}}\cdot\frac{1}{x}.$$

$$= \frac{(n+2)^n(n+2)1}{(n+1)^n(n+1)x} = \frac{\left(1+\frac{2}{n}\right)^n\left(1+\frac{2}{n}\right)}{\left(1+\frac{1}{n}\right)^n\left(1+\frac{1}{n}\right)}\cdot\frac{1}{x}.$$

Thus

$$\lim_{n\to\infty}\frac{u_n}{u_{n+1}} = \lim_{n\to\infty}\frac{\left(1+\frac{2}{n}\right)^n}{\left(1+\frac{1}{n}\right)^n}\cdot\frac{\left(1+\frac{2}{n}\right)}{\left(1+\frac{1}{n}\right)}\cdot\frac{1}{x} = \frac{e^2}{e}\cdot\frac{1}{x} = \frac{e}{x}$$

and so the series converges if $\frac{e}{x} > 1$, that is, if $x < e$ and diverges if $\frac{e}{x} < 1$, that is, if $x > e$. If $x = e$, then $\lim_{n\to\infty}\frac{u_n}{u_{n+1}} = 1$ and so the D'Alembert's Ratio Test gives no information about convergence. But when $x = e$, then

$$\frac{u_n}{u_{n+1}} = \frac{\left(1+\frac{2}{n}\right)^{n+1}}{\left(1+\frac{1}{n}\right)^{n+1}}\cdot\frac{1}{e}.$$

Since, the expression for $\frac{u_n}{u_{n+1}}$ involves e, we do no apply Raabe's Test and so Logarithmic Test is preferred. Thus,

$$\log\frac{u_n}{u_{n+1}} = (n+1)\log\left(1+\frac{2}{n}\right)$$

$$-(n+1)\log\left(1+\frac{1}{n}\right) - \log e$$

$$= (n+1)\left[\log\left(1+\frac{2}{n}\right) - \log\left(1+\frac{1}{n}\right)\right] - 1$$

$$= (n+1)\left[\frac{2}{n}-\frac{1}{2}\cdot\frac{4}{n^2}+\frac{1}{3}\cdot\frac{8}{n^3}-\cdots\right.$$

$$\left.-\left(\frac{1}{n}-\frac{1}{2n^2}+\frac{1}{3n^3}+\cdots\right)\right] - 1$$

$$= (n+1)\left[\frac{1}{n}-\frac{3}{2n^2}+\cdots\right] - 1$$

$$= 1-\frac{3}{2n}+\frac{1}{n}-\frac{3}{2n^2}+\cdots-1$$

$$= -\frac{1}{2n} - \frac{3}{2n^2} + \cdots$$

and so

$$\lim_{n\to\infty} n \log \tfrac{u_n}{u_{n+1}} = \lim_{n\to\infty} n\left[-\frac{1}{2n} - \frac{3}{2n^2} + \ldots\right]$$

$$= -\frac{1}{2} < 1.$$

Hence, by Logarithmic Test, the series diverges. It follows, therefore, that the given series converges if $x < e$ and diverges if $x \geq e$.

EXAMPLE 1.52

Test for convergence of the series

$$\frac{1}{(\log 2)^p} + \frac{1}{(\log 3)^p} + \cdots + \frac{1}{(\log n)^p} + \cdots$$

Solution. The nth term of the series is

$$u_n = \frac{1}{(\log n)^p}.$$

Therefore,

$$\frac{u_n}{u_{n+1}} = \left(\frac{\log(n+1)}{\log n}\right)^p$$

$$= \left(\frac{\log n + \log(1+\frac{1}{n})}{\log n}\right)^p$$

$$= \left[\frac{\log n + \frac{1}{n} - \frac{1}{2n^2} + \frac{1}{3n^2} - \cdots}{\log n}\right]^p$$

$$= \left[1 + \frac{1}{n \log n} - \frac{1}{2n^2 \log n} + \cdots\right]^p$$

and so $\lim_{n\to\infty} \frac{u_n}{u_{n+1}} = 1$. Thus the D'Alembert's Ratio Test fails. We have

$$\log \frac{u_n}{u_{n+1}} = p \log\left[1 + \frac{1}{n \log n} - \frac{1}{2n^2 \log n} + \cdots\right]$$

$$= p\left[\frac{1}{n \log n} + \cdots\right]$$

and so

$$\lim_{nn\to\infty} n \log \frac{u_n}{u_{n+1}} = \lim_{n\to\infty} p\left(\frac{1}{\log n} + \cdots\right)$$

$$= 0, \text{ for all values of } p.$$

Hence, by Log Test, the given series diverges for all values of p.

EXAMPLE 1.53

Examine the convergence of the series

$$x + \frac{2^2x^2}{2!} + \frac{3^3x^2}{3!} + \frac{4^4x^4}{4!} + \cdots.$$

Solution. As shown is Example 1.47 the given series converges for $x < \frac{1}{e}$ and diverges for $x > \frac{1}{e}$. If $x = \frac{1}{e}$, then

$$\frac{u_n}{u_{n+1}} = \frac{e}{(1+\frac{1}{n})^n}$$

and so

$$\log \tfrac{u_n}{u_{n+1}} = \log e - n \log\left(1 + \frac{1}{n}\right)$$

$$= 1 - n\left(\frac{1}{n} - \frac{1}{2n^2} + \frac{1}{3n^3} + \cdots\right)$$

$$= \frac{1}{2n} - \frac{1}{3n^2} + \cdots.$$

Thus

$$\lim_{n\to\infty} n \log \frac{u_n}{u_{n+1}} = \lim_{n\to\infty}\left(\frac{1}{2} - \frac{1}{3n} + \cdots\right) = \frac{1}{2} < 1,$$

and so, by Log Test, the given series diverges, for $x = \frac{1}{e}$. Hence, the given series converges if $x < \frac{1}{e}$ and diverges if $x \geq \frac{1}{e}$.

1.14 DE MORGAN–BERIRAND TEST

The series Σu_n of positive terms converges or diverges according to

$$\lim_{n\to\infty}\left[\log n\left\{n\left(\frac{u_n}{u_{n+1}} - 1\right) - 1\right\}\right] > 1 \text{ or } < 1.$$

Proof: Consider the Auxiliary Series $\Sigma v_n = \sum \frac{1}{n(\log n)^p}$, which is convergent when $p > 1$. Then

$$\frac{v_n}{v_{n+1}} = \frac{(n+1)[\log(n+1)]^p}{n(\log n)^p}$$

$$= \left(1 + \frac{1}{n}\right)\left[\frac{\log n + \log(1+\frac{1}{n})}{\log n}\right]^p$$

$$=\left(1+\frac{1}{n}\right)\left[1+\frac{1}{\log n}\left(\frac{1}{n}-\frac{1}{2n^2}+\cdots\right)\right]^p$$

$$=\left(1+\frac{1}{n}\right)\left[1+\frac{p}{\log n}\left(\frac{1}{n}-\frac{1}{2n^2}+\cdots\right)\right]$$

$$=1+\frac{1}{n}+\frac{p}{n\log n}+\cdots.$$

But, by Comparison Test, the series Σu_n will converge if

$$\frac{u_n}{u_{n+1}}>\frac{v_n}{v_{n+1}},$$

that is, if

$$\frac{u_n}{u_{n+1}}>1+\frac{1}{n}+\frac{p}{n\log n}+\cdots$$

or if

$$\frac{u_n}{u_{n+1}}-1>\frac{1}{n}+\frac{p}{n\log n}+\cdots$$

or if

$$n\left(\frac{u_n}{u_{n+1}}-1\right)>1+\frac{p}{\log n}+\cdots$$

or if

$$\log n\left[n\left(\frac{u_n}{u_{n+1}}-1\right)-1\right]>p+\cdots$$

or if

$$\lim_{n\to\infty}\log n\left[n\left(\frac{u_n}{u_{n+1}}-1\right)-1\right]>p>1.$$

Similarly, it could be proved that the series diverges if

$$\lim_{n\to\infty}\log n\left[n\left(\frac{u_n}{u_{n+1}}-1\right)-1\right]<1.$$

EXAMPLE 1.54

Examine the convergence of the series

$$\frac{1^2}{2^2}+\frac{1^2\cdot 3^2}{2^2\cdot 4^2}x+\frac{1^2\cdot 3^2\cdot 5^2}{2^2\cdot 4^2\cdot 6^2}x^2+\cdots$$

Solution. The nth term of the series is

$$u_n=\frac{1^2\cdot 3^2\cdot 5^2\cdots(2n-1)^2}{2^2\cdot 4^2\cdots(2n)^2}x^{n-1}$$

and so

$$\lim_{n\to\infty}\frac{u_n}{u_{n+1}}=\lim_{n\to\infty}\left(\frac{2n+2}{2n+1}\right)^2\cdot\frac{1}{x}=\frac{1}{x}.$$

Thus, the series converges for $\frac{1}{x}<1$, that is, if $x<1$ and diverges when $x>1$. When $x=1$, the D'Alembert's Ratio Test fails. For $x=1$, we have

$$\frac{u_n}{u_{n+1}}=\left(\frac{2n+2}{2n+1}\right)^2$$

or

$$\frac{u_n}{u_{n+1}}-1=\frac{4n^2+8n+4}{4n^2+4n+1}-1=\frac{4n+3}{4n^2+4n+1}$$

or

$$\lim_{n\to\infty}n\left(\frac{u_n}{u_{n+1}}-1\right)=\lim_{n\to\infty}\frac{4n^2+3n}{4n^2+4n+1}=1.$$

Thus Raabe's Test also fails. Now

$$n\left(\frac{u_n}{u_{n+1}}-1\right)-1=\frac{4n^2+3n}{4n^2+4n+1}-1$$

$$=\frac{-n-1}{4n^2+4n+1}.$$

Therefore,

$$\lim_{n\to\infty}\log n\left[n\left(\frac{u_n}{u_{n+1}}-1\right)-1\right]$$

$$=\lim_{n\to\infty}\frac{-n-1}{4n^2+4n+1}\cdot\frac{n\log n}{n}$$

$$=\lim_{n\to\infty}\frac{-n^2-n}{4n^2+4n+1}\cdot\frac{\log n}{n}$$

$$=-\frac{1}{n}(0)=0<1,$$

and so, by de Morgan–Bertrand Test, the series diverges for $x=1$. Hence, the series converges for $x<1$ and diverges for $x\geq 1$.

1.15 GAUSS'S TEST

Let Σu_n be a series of positive terms. If

$$\frac{u_n}{u_{n+1}} = 1 + \frac{\lambda}{n} + O\left(\frac{1}{n^{p+1}}\right), p > 1,$$

then the series Σu_n converges if $\lambda > 1$, and diverges if $\lambda \leq 1$.

Proof: We have

and so $$\frac{u_n}{u_{n+1}} = 1 + \frac{\lambda}{n} + O\left(\frac{1}{n^{p+1}}\right), \; p > 1$$

$$n\left(\frac{u_n}{u_{n+1}} - 1\right) = \lambda + O\left(\frac{1}{n^p}\right), \; p > 1$$

Thus

$$\lim_{n\to\infty} n\left(\frac{u_n}{u_{n+1}} - 1\right) = \lambda$$

and by Raabe's Test, the series Σu_n converges if $\lambda > 1$ and diverges if $\lambda < 1$. When $\lambda = 1$, then

$$\lim_{n\to\infty}\left[n\left(\frac{u_n}{u_{n+1}} - 1\right) - 1\right]\log n$$

$$= \lim_{n\to\infty}\left(\frac{\log n}{n^p}\right) \to 0 \text{ as } n \to \infty,$$

by L'Hospital Rule. Hence, by de Morgan–Bertrand Test, the given series diverges. Thus, the given series converges for $\lambda > 1$ and diverges for $\lambda \leq 1$.

EXAMPLE 1.55

Discuss the convergence of the *Hypergeometric Series*

$$1 + \frac{\alpha \cdot \beta}{1 \cdot \gamma} x + \frac{\alpha(\alpha.+1)\beta(B+1)}{1 \cdot 2 \cdot \gamma(\gamma+1)} x^3$$

$$+ \frac{\alpha(\alpha+.1)(\alpha+2)\beta(\beta+1)(\beta+2)}{1 \cdot 2 \cdot 3 \cdot \gamma(\gamma+1)(\gamma+2)} \cdot x^3 + \cdots.$$

Solution. Since $\alpha + n$ is positive after a certain stage, we can assume that α, β, γ are positive. Then ignoring the first term, the nth term of the series is

$$u_n = \frac{\alpha(\alpha+1)\cdots(\alpha+n-1)\beta(\beta+1)\cdots(\beta+n-1)}{1 \cdot 2 \cdot 3 \cdot n \cdot \gamma(\gamma+1)\cdots(\gamma+n-1)} x^n.$$

Therefore,

$$\frac{u_n}{u_{n+1}} = \frac{(\gamma+n)(n+1)}{(\alpha+n)(\beta+n)} \cdot \frac{1}{x}$$

and so

$$\lim_{n\to\infty} \frac{u_n}{u_{n+1}} = \lim_{n\to\infty} \frac{\left(1+\frac{\gamma}{n}\right)\left(1+\frac{1}{n}\right)}{\left(1+\frac{\alpha}{n}\right)\left(1+\frac{\beta}{n}\right)} \cdot \frac{1}{x} = \frac{1}{x}$$

Hence, by D'Alembert's Ratio Test, the series Σu_n converges if $\frac{1}{x} > 1$, that is, if $x < 1$ and diverges if $\frac{1}{x} < 1$, that is, if $x > 1$. For $x = 1$, we observe that

$$\frac{u_n}{u_{n+1}} = \frac{\left(1+\frac{\gamma}{n}\right)\left(1+\frac{1}{n}\right)}{\left(1+\frac{\alpha}{n}\right)\left(1+\frac{\beta}{n}\right)}$$

$$= \left(1+\frac{\gamma}{n}\right)\left(1+\frac{1}{n}\right)\left(1+\frac{\alpha}{n}\right)^{-1}\left(1+\frac{\beta}{n}\right)^{-1}$$

$$= \left(1 + \frac{1}{n} + \frac{\gamma}{n} + \frac{\gamma}{n^2}\right)$$

$$\times\left(1 - \frac{\alpha}{n} - \frac{\beta}{n} + \frac{\alpha\beta}{n^2} + \cdots\right)$$

$$= 1 + \frac{1}{n}(1 - \gamma - \alpha - \beta) + O\left(\frac{1}{n^2}\right).$$

Therefore, by Gauss's Test, Σu_n converges if $1 - \gamma - \alpha - \beta > 1$, that is, if $\gamma > \alpha + \beta$ and diverges if $\gamma \leq \alpha + \beta$. It follows, therefore, that the Hypergeometric Series converges if $x < 1$ and diverges if $x > 1$. For $x = 1$, the series converges if $\gamma > \alpha + \beta$ and diverges if $\gamma \leq \alpha + \beta$.

EXAMPLE 1.56

Discuss the convergence of the *Harmonic Series* $\Sigma \frac{1}{n^p}$ using Gauss's Test.

Solution. The nth term of the given series is $u_n = \frac{1}{n^p}$. Therefore,

$$\frac{u_n}{u_{n+1}} = \left(\frac{n+1}{n}\right)^p = \left(1 + \frac{1}{n}\right)^p = 1 + \frac{p}{n} + O\left(\frac{1}{n^2}\right)$$

Hence, by Gauss's Test, the Harmonic Series converges if $p > 1$ and diverges if $p \leq 1$.

EXAMPLE 1.57

Discuss the convergence of

(i) $1+\frac{3}{7}+\frac{3\cdot 6}{7\cdot 10}+\frac{3\cdot 6\cdot 9}{7\cdot 10\cdot 13}+\cdots$.

(ii) $\frac{1^2}{2^2}+\frac{1^2\cdot 3^2}{2^2\cdot 4^2}+\frac{1^2\cdot 3^2\cdot 5^2}{2^2\cdot 4^2\cdot 6^2}+\cdots$.

Solution. (i) The nth term is

$$u_n=\frac{3\cdot 6\cdot 9\cdots(3n)}{7\cdot 10\cdot 13\cdots(3n+4)}$$

Therefore,

$$\frac{u_n}{u_{n+1}}=\frac{3n+7}{3n+3}=1+\frac{4}{3n}+O\left(\frac{1}{n^2}\right).$$

Hence, by Gauss's Test, the series Σu_n converges.
(ii) For this series

$$u_n=\frac{1^2.3^2...(2n.-1)^2}{2^2.4^26^2...(2n)^2}.$$

Therefore,

$$\frac{u_n}{u_{n+1}}=\frac{(2n+2)^2}{(2n+1)^2}=\frac{4n^2+8n+4}{4n^2+4n+1}=1+\frac{1}{n}+O\left(\frac{1}{n^2}\right)$$

and so the series diverges by Gauss's Test.

1.16 CAUCHY'S INTEGRAL TEST

Let f be a non-negative monotonic decreasing function of $x \geq 1$ and let $f(n) = u_n$ for all positive integral values of n. Then, the series Σu_n and the integral $\int_1^\infty f(x)\text{dx}$ converge or diverge together.

Proof: Since f is monotonic in any interval $1<x<\infty$, it is integrable in that interval. If x lies between n and $(n + 1)$, we have

$$f(n)>f(x)>f(n+1).$$

Integrating, we have

$$\int_n^{n+1} f(n)\text{dx} > \int_n^{n+1} f(x)\text{dx} > \int_n^{n+1} f(n+1)\text{dx}. \quad (1)$$

Let $I_n=\int_1^n f(x)\text{dx}$. Then

$$I_{n+1}-I_n=\int_n^{n+1} f(x)\text{dx}$$

and so the expression (1) reduces to

$$f(n)>I_{n+1}-I_n>f(n+1).$$

Since $f(n)=u_n, f(n+1)=u_{n+1}$, we get

$$u_n>I_{n+1}-I_n>u_{n+1}. \quad (2)$$

Therefore,

$$u_{n-1}>I_n-I_{n-1}>u_n,$$
$$u_{n-2}>I_{n-1}-I_{n-2}>u_{n-1},$$
$$u_1>I_2-I_1>u_2$$

Adding, we have

$$u_1+u_2+\cdots+u_{n-1}>I_n-I_1$$
$$>u_2+u_3+\cdots+u_n$$

or

$$S_n-u_n>I_n-I_1>S_n-u_1,$$

where $S_n=u_1+u_2+\ldots+u_n$ is the partial sum of the series Σu_n. But,

$$I_1=\int_1^1 f(x)\text{dx}=0.$$

Therefore,

$$S_n-u_n>I_n>S_n-u_1$$

or

$$u_1>S_n-I_n>u_n>0.$$

Thus, the sequence $(S_n - I_n)$ is bounded. Also, $(S_n - I_n) - (S_{n+1} - I_{n+1}) = I_{n+1} - I_n - u_{n+1} > 0$ by (2) Hence, the sequence $(S_n - I_n)$ is monotonically decreasing and bounded and so converges to a definite limit between 0 and u_1. Therefore S_n and I_n converge or diverge together. Consequently, the series Σu_n and the integral $\int_1^\infty f(x)\text{dx}$ converge or diverge together.

Remark. 1.10 (a) It is important to observe that whether S_n is convergent or divergent, the sequence $S_n - I_n$ is always convergent and the limit lies between 0 and u_1.

1.10 (b) Setting $f(n)=\frac{1}{n}$, we have

$$S_n=u_1+u_2+\cdots+u_n=1+\frac{1}{2}+\frac{1}{3}+\cdots+\frac{1}{n},$$

$$I_n = \int_1^n \frac{1}{x} dx = [\log x]_1^n = \log n.$$

Hence, $S_n - I_n = 1 + \frac{1}{2} + \ldots + \frac{1}{n} - \log n$ tends to a finite limit γ, called *Euler's Constant*, whose value 0.55721 (less than $u_1 = 1$) is correct to five places of decimals.

EXAMPLE 1.58

Apply Cauchy's Integral Test to show that the series $\sum \frac{1}{n^p}$ converges if $p > 1$, and diverges if $0 < p \leq 1$.

Solution. The nth term of the series is

$$u_n = \frac{1}{n^p}.$$

As per Cauchy's Integral Test, we have

$$f(n) = u_n = \frac{1}{n^p}.$$

Therefore,

$$f(x) = \frac{1}{x^p}.$$

We observe that f is positive and monotonic decreasing for $x \geq 1$. Therefore, by Cauchy's Integral Test, Σu_n and $\int_1^\infty f(x)\text{dx}$ converge or diverge together.

If $p \neq 1$, then

$$\int_1^\infty f(x)\text{dx} = \int_1^\infty \frac{1}{x^p}\text{dx} = \int_1^\infty x^{-p}\text{dx} = \left[\frac{x^{1-p}}{1-p}\right]_1^\infty.$$

When $p > 1$, then

$$\int_1^\infty f(x)\text{dx} = -\frac{1}{p-1}\left[\frac{1}{x^{p-1}}\right]_1^\infty$$

$$= -\frac{1}{p-1}[o-1]$$

$$= \frac{1}{p-1} \text{ (finite)}.$$

Thus $\int_1^\infty f(x)\text{dx}$ converges and so, by Cauchy's Integral Test, Σu_n also converges for $p > 1$.

When $0 < p < 1$, then

$$\int_1^\infty f(x)\text{dx} = \frac{1}{1-p}[x^{1-p}]_1^\infty = \frac{1}{1-p}[\infty - 1] = \infty$$

Thus $\int_1^\infty f(x)\text{dx}$ diverges and so, by Cauchy's Integral Test, Σu_n diverges for $p < 1$.

If $p = 1$, then

$$\int_1^\infty f(x)\text{dx} = \int_1^\infty \frac{1}{x}\text{dx}$$

$$= [\log x]_1^\infty$$

$$= \infty - \log 1$$

$$= \infty - 0$$

$$= \infty$$

Thus, $\int_1^\infty f(x)\text{dx}$ diverges in this case and so, by Cauchy's Integral Test, Σu_n diverges for $p = 1$ Hence, $\Sigma u_n = \Sigma \frac{1}{n^p}$ converges for p > 1 and diverges for p ≤ 1.

EXAMPLE 1.59

Apply Cauchy's Integral Test to show that the series $\sum_{n=2}^\infty \frac{1}{n(\log n)^p}$ converges if $p > 1$ and diverges if $0 < p \leq 1$.

Solution. For the given series, the nth term is

$$u_n = \frac{1}{n(\log n)^p}.$$

Taking $u_n = f(n)$, one gets

$$f(x) = \frac{1}{x(\log x)^p}.$$

For $x \geq 2$ and positive p, f is positive and monotonically decreasing. Hence, Cauchy's Integral Test is applicable.

When $p \neq 1$,

$$\int_2^\infty f(x)\text{dx} = \int_2^\infty \frac{1}{x}(\log x)^{-p}\text{dx} = \left[\frac{(\log x)^{-p+1}}{-p+1}\right]_2^\infty.$$

Now if $p > 1$, then $p - 1$ is positive and so

$$\int_2^\infty f(x)\text{ dx} = -\frac{1}{p-1}\left[\frac{1}{(\log x)^{p-1}}\right]_2^\infty$$

$$= -\frac{1}{p-1}\left[0 - \frac{1}{(\log 2)^{p-1}}\right]$$

$$= \frac{1}{(p-1)(\log 2)^{p-1}} \text{ (finite).}$$

Thus, $\int_2^{\infty} f(x)$ dx converges and hence, by Cauchy's Integral Test, the given series Σu_n converges for $p > 1$.

If $p < 1$, then $1-p$ is positive and so

$$\int_2^{\infty} f(x)\, dx = \frac{1}{1-p}[(\log x)^{1-p}]_2^{\infty}$$

$$= \frac{1}{p}[\infty - (\log 2)^{1-p}] = \infty$$

Hence, by Cauchy's Integral Test, Σu_n diverges for $p < 1$.

If $p = 1$, then $f(x) = \dfrac{1}{x \log x}$. Therefore,

$$\int_2^{\infty} f(x)\, dx = \int_2^{\infty} \frac{\frac{1}{x}}{\log x} dx$$

$$= [\log \log x]_2^{\infty}$$

$$= \infty - \log \log 2 = \infty.$$

Thus, $\int_2^{\infty} f(x)$ dx diverges and so Σu_n also diverges for $p = 1$.

Hence, the series $\sum_{n=2}^{\infty} \frac{1}{n(\log n)^p}$ converges if $p > 1$ and diverges if $p \le 1$.

EXAMPLE 1.60

Examine the convergence of the series $\sum \frac{1}{\sqrt{n}}$.

Solution. We have

$$u_n = \frac{1}{\sqrt{n}} = f(n) \cdot$$

Therefore, $f(x) = \frac{1}{\sqrt{x}}$ and it is positive and monotonic decreasing. Therefore, Cauchy's Integral Test is applicable. We have

$$\int_1^{\infty} f(x)\, dx = \int_1^{\infty} \frac{1}{\sqrt{x}} dx$$

$$= \left[\frac{1}{2} x^{\frac{1}{2}}\right]_1^{\infty} = \infty.$$

Thus, the integral $\int_1^{\infty} f(x)\, dx$ is divergent. Hence, by Cauchy's Integral Test, the series $\sum \frac{1}{\sqrt{n}}$ is divergent.

1.17 CAUCHY'S CONDENSATION TEST

If f is positive and monotonic decreasing to the limit 0 as $x \to \infty$ and a is any constant greater than 1, then

$$\Sigma f(x) \text{ and } \Sigma a^n f(a^n)$$

converge or diverge together.

Proof: Since $a > 1$, a^n is an increasing function of n. Also f is monotonic decreasing. Therefore,

$$f(a^{n-1}) > f(x) > f(a^n) \text{ for } a^{n-1} < x < a^n$$

and so

$$\int_{a^{n-1}}^{a^n} f(a^{n-1})\, dx > \int_{a^{n-1}}^{a^n} f(x)\, dx > \int_{a^{n-1}}^{a^n} f(a^n) dx.$$

If

$$I_t = \int_1^t f(x)\, dx,$$

then

$$\int_{a^{n-1}}^{a^n} f(x)\, dx = I_{a^n} - I_{a^{n-1}}$$

and the above expression reduces to

$$(a^n - a^{n-1}) f(a^{n-1}) > I_{a^n} - I_{a^{n-1}}$$

$$> (a^n - a^{n-1}) f(a^n)$$

or

$$(a-1)a^{n-1} f(a^{n-1}) > I_{a^n} - I_{a^{n-1}}$$

$$> \left(1 - \frac{1}{a}\right) a^n f(a^n).$$

Summing up for $n = 2, 3, \ldots, n-1, n$, we have

$$(a-1)\sum_{\infty-1}^{n-1} a^r f(a^r) > I_{a^n} - I_a$$

$$> \left(1 - \frac{1}{a}\right) \sum_{r=2}^{n} a^r f(a^r).$$

If follows that $\Sigma a^n f(a^n)$ and I_{a^n} converge or diverge together. But

$$\lim_{n\to\infty} I_{a^n} = \int_1^{\infty} f(x)\,dx.$$

By the Integral Test, this integral and $\Sigma f(x)$ converge or diverge together. Hence, $\Sigma f(n)$ and $\Sigma a^n f(a^n)$ converge or diverge together.

EXAMPLE 1.61

Applying Cauchy's Condensation Test, examine the convergence of the series

$$\sum_{n=2}^{\infty} \frac{1}{n(\log n)^p}$$

Solution. We have

$$f(x) = \frac{1}{x(\log x)^p},$$

which satisfies the conditions of the Cauchy's Condensation Test. So Σu_n and $\sum a^n \frac{1}{a^n(\log a^n)^p}$ converge or diverge together. But

$$\sum a^n \frac{1}{a^n(\log a^n)^p} = \sum \frac{1}{n^p(\log a)^p}$$

Since $(\log a)^p$ is constant, it follows that Σu_n and $\Sigma \frac{1}{n^p}$ converge or diverge together. But, $\Sigma \frac{1}{n^p}$ converges if $p > 1$ and diverges if $p \le 1$. Hence, the given series converges if $p \ge 1$ and diverges if $p \le 1$.

EXAMPLE 1.62

Examine the convergence of the series

(i) $\displaystyle\sum_2^{\infty} \frac{1}{\log n}$

(ii) $\displaystyle\sum_2^{\infty} \frac{1}{(n \log n)^p}$

(iii) $\displaystyle\sum_2^{\infty} \frac{1}{(\log n)^{\log n}}$.

Solution. (i) Here, $u_n = \frac{1}{\log n}$. Therefore, $f(n) = \frac{1}{\log n}$ and

$$a^n f(a^n) = \frac{a^n}{\log a^n} = \frac{a^n}{n \log a} > \frac{1}{n \log a} \text{ since } a > 1.$$

But, the series $\sum \frac{1}{n \log a}$ is divergent due to the divergence of $\Sigma \frac{1}{n}$. Hence, by Cauchy's Condensation Test, the series $\sum \frac{1}{\log n}$ is divergent.

(ii) Here

$$u_n = f(n) = \frac{1}{(n \log n)^p}$$

and

$$a^n f(a^n) = \frac{a^n}{(a^n \log a^n)^p} = \frac{1}{a^{n(p-1)}(n \log a)^p}$$

$$= \frac{1}{a^{n(p-1)} \cdot n^p (\log a)^p}.$$

If $p > 1$, $p - 1$ then is positive. Since $\sum \frac{1}{n^p}$ converges when $p > 1$, the series $\Sigma a^n f(a^n)$ converges and so, by Cauchy's Condensation Test, the series $\sum \frac{1}{(n \log n)^p}$ converges for $p > 1$.

For $p = 1$, we have

$$a^n f(a^n) = \frac{1}{n \log a}$$

Since $\log a$ is constant and $\Sigma \frac{1}{n}$ is divergent $\Sigma a^n f(a^n)$ diverges for $p = 1$. Hence, the given series diverges for $p = 1$.

If $p < 1$, then $a^{n(p-1)} < 1$ since $a > 1$. Therefore,

$$a^n f(a^n) > \frac{1}{n^p(\log a)^p}.$$

But the series $\sum \frac{1}{n^p}$ diverges when $p < 1$, therefore, $\Sigma a^n f(a^n)$ diverges for $p < 1$. Hence, the given series diverges for $p < 1$. It follows, therefore, that the series $\sum \frac{1}{(n \log n)^p}$ converges for $p > 1$ and diverges for $p \le 1$.

(iii) We have

$$f(n) = \frac{1}{(\log n)^{\log n}}.$$

Therefore,

$$\sum a^n f(a^n) = \sum \frac{a^n}{(\log a^n)^{\log a^n}}$$

$$= \sum \frac{a^n}{(n \log a)^{n \log a}}.$$

Calling this series as Σv_n, we have

$$v_n^{\frac{1}{n}} = \frac{a}{(n \log a)^{\log a}} \to 0 \ as \ n \to \infty .$$

Therefore, by Cauchy's Root Test, $\Sigma a^n f(a^n)$ converges. Hence by Cauchy's. Condensation Test, the series $\sum \frac{1}{(\log n)^{\log n}}$ converges.

EXAMPLE 1.63

Discuss the convergence of

$$\sum \frac{1}{n \log n(\log \ \log n)^p} .$$

Solution. Here

$$f(n) = \frac{1}{n \log n(\log \ \log n)^p} .$$

Therefore,

$$a^n f(a^n) = \frac{a^n}{a^n \ \log a^n (\log \ \log a^n)^p}$$

$$= \frac{1}{n \log a(\log(n \log a))^p}$$

$$= \frac{1}{n \log a(\log n + \ \log a)^p}$$

$$= \frac{1}{n \log a(\log n)^p \left[1 + \frac{\log a}{\log n}\right]^p}$$

Therefore, compare the series $\Sigma a^n f(a^n)$ with $\sum v_n = \sum \frac{1}{n(\log n)^p}$, which converges for $p > 1$ and diverges for $p \le 1$. (see Example 1.62). Thus $\Sigma a^n f(a^n)$ converges for $p > 1$ and diverges for $p \le 1$. Hence, by Cauchy's Condensation Test, the given series converges for $p > 1$ and diverges for $p \le 1$.

1.18 KUMMER'S TEST

(a) Let Σu_n be a series of positive terms and let D_n be a positive sequence. If

$$\underline{\lim} \left(D_n \frac{u_n}{u_{n+1}} - D_{n+1} \right) > 0,$$

then the series Σu_n is convergent.

(b) Let Σu_n be a series positive terms and let $\Sigma \frac{1}{D_n}$ be divergent. If

$$\overline{\lim} \left(D_n \frac{u_n}{u_{n+1}} - D_{n+1} \right) < 0.$$

then the series Σu_n is divergent.

Proof: (a) Let $\underline{\lim} \left(D_n \frac{u_n}{u_{n+1}} - D_{n+1} \right) = p$ and let $0 < h < p$. Thus h is an inferior number. Therefore, $D_n \frac{u_n}{u_{n+1}} - D_{n+1} > h$ after a certain value of n, say $n = m$. Thus,

$$D_m u_m - D_{m+1} u_{m+1} > h \ u_{m+1},$$

$$D_{m+1} u_{m+1} - D_{m+2} u_{m+2} > h \ u_{m+2},$$

$$D_{n-1} u_{n-1} - D_n u_n > h u_n .$$

Hence,

$$h(u_{m+1} + u_{m+2} + \ldots + u_n) < D_m u_m - D_n u_n < D_m u_m ,$$

or

$$u_{m+1} + u_{m+2} + \cdots + u_n < \frac{D_m u_m}{h},$$

which is independent of n. Therefore, $\sum_{m+1}^{n} u_n$ remains bounded as $n \to \infty$ and so Σu_n is convergent.

(b) Let $\overline{\lim}(D_n \frac{u_n}{u_{n+1}} - D_{n+1}) < 0$. Then, 0 is a superior number of this sequence. Therefore,

$$D_n u_n - D_{n+1} u_{n+1} < 0 \text{ for some } n \ge m$$

and so

$$D_m u_m < D_{n+1} u_{n+1} < \cdots < D_n u_n ,$$

which yields

$$u_n > \frac{D_m u_m}{D_n} .$$

Since $\Sigma \frac{1}{D_n}$ diverges, therefore, by Comparison Test, Σu_n also diverges.

Deductions:

(i) Let $D_n = 1$ for all n. Then Kummer's Test reduces to "The series Σu_n converges if

$\underline{\lim}(\frac{u_n}{u_{n+1}}-1) > 0$, that is if $\underline{\lim}\frac{u_n}{n+1} > 1$ and diverges if

$\overline{\lim}\left(\frac{u_n}{u_{n+1}}-1\right) < 0$ or if $\overline{\lim}\frac{u_n}{u_{n+1}} < 1$."

Thus D'Alembert's Ratio Test is a particular case of Kummer's Test.

(ii) Let $D_n = n$. Then Kummer's Test reduces to "Σu_n converges if

$$\overline{\lim}(n\frac{u_n}{u_{n+1}}-(n+1)) > 0,$$

that is, if $\underline{\lim}\, n(\frac{u_n}{u_{n+1}}-1) > 1$

and diverges if $\overline{\lim}(n\frac{u_n}{u_{n+1}}-(n+1)) < 0$, that is, if $\overline{\lim}\, n(\frac{u_n}{u_{n+1}}-1) < 1$. Thus Raabe's Test is a particular case of Kummer's Test".

(iii) Let $D_n = n \log n$. The series $\sum\frac{1}{D_n} = \sum\frac{1}{n \log n}$ is divergent. Then

$$n \log n\frac{u_n}{u_{n+1}} - (n+1)\log(n+1)$$

$$= n \log n\frac{u_n}{u_{n+1}} - (n+1)\log n$$

$$-(n+1)\log\left(1+\frac{1}{n}\right)$$

$$= \log n\left[n\left(\frac{u_n}{u_{n+1}}-1\right)-1\right]$$

$$-\log\left(1+\frac{1}{n}\right)^n - \log\left(1+\frac{1}{n}\right).$$

Now $\lim_{n\to\infty}(1+\frac{1}{n})^n = e$ and $\log e = 1$. Hence, Kummar's Test reduces to "Σu_n converges if $\underline{\pm\pm}\pm\pm$ $n\left[\left(\frac{u_n}{u_{n+1}}- \right)- \right]>$ and diverges if

$$\overline{\lim}\log n\left[n\left(\frac{u_n}{u_{n+1}}-1\right)-1\right] < 1$$

Thus de Morgan–Bertrand Test is also a particular case of Kummer's Test.

1.19 ALTERNATING SERIES

A series in which positive and negative terms occur alternately is called an *Alternating Series*. Regarding convergence behavior of an alternating series, we have the following theorem, known as *Leibnitz's Rule*.

Theorem 1.19. (Leibnitz's Rule). If u_n is positive and monotonically decreases to the limit zero, then the alternating series $u_1 - u_2 + u_3 - u_4 + \dots$ is convergent.

Proof: Consider

$$u_n - u_{n+1} + u_{n+2} - \dots + (-1)^p u_{n+p}.$$

Writing this expression as

$$(u_n - u_{n+1}) + (u_{n+2} - u_{n+3}) + \cdots,$$

we see that when p is odd, $-u_{n+p}$ occurs in the last bracket; and when p is even, u_{n+p} is the last term. Since $u_n \geq u_{n+1}$, each bracket is non-negative and so the expression discussed earlier is non-negative.

Now we write the same expression as

$$u_n - (u_{n+1} - u_{n+2}) - (u_{n+3} - u_{n+4}) + \dots,$$

then we observe that when p is even, the term u_{n+p} is included in the last bracket and when p is odd, $-u_{n+p}$ is the last term. Since all the brackets are non-negative, the expression is certainly less then u_n. It follows, therefore, that

$$0 \leq \left|u_n - u_{n+1} + u_{n+2} - \dots + (-1)^p u_{n+p}\right| < u_n.$$

Since, $u_n \to 0$ as $n \to \infty$ for each $\varepsilon > 0$, there exists an integer n_0 such that $u_n < \varepsilon$ for all $n > n_0$. Thus, $|u_n - u_{n+1} + u_{n+2} - \dots + (-1)^p u_{n+p}| < \varepsilon$ for all $n > n_0$. Hence, by Cauchy's Principle of Convergence, the alternating series $u_1 - u_2 + u_3 - \dots$ is convergent.

Corollary: If u_n is positive, monotonically decreasing and tends to the finite limit α, then the series $\Sigma(-1)^{n-1}u_n$ oscillates finitely.

Proof: Let $u_n = \alpha + v_n$. Then $v_n \to 0$. Since α is the lower bound of $\{u_n\}$, v_n is positive for all n.

Also, since u_n is monotonically decreasing, so is v_n. Therefore,

$$S_{2n} = (\alpha + v_1) - (\alpha + v_2) + \ldots - (\alpha + v_{2n})$$
$$= v_1 - v_2 + \ldots - v_{2n},$$
$$S_{2n+1} = \alpha + (v_1 - v_2 + \ldots + v_{2n+1}).$$

Since v_n is positive and monotonically decreasing to zero, the series $\Sigma(-1)^n u_n$ converges by Leibnitz's Rule to the sum V. Then $\Sigma(-1)^{n-1}u_n$ oscillates between V and $V + \alpha$.

Remark 1.11. If the term of a series are alternately positive and negative and the term continuously decreases, we cannot say that the series is convergent unless $\lim_{n\to\infty} u_n = 0$. For example, consider the series

$$2 - \frac{3}{2} + \frac{4}{3} - \frac{5}{4} + \ldots.$$

We note that $\lim_{n\to\infty} u_n = \lim_{n\to\infty}\{(-1)^{n-1}\frac{n+1}{n}\} = 1$ or -1 according as $n \to \infty$ through odd or even values. This series is oscillatory.

EXAMPLE 1.64

Show that the series $1 - \frac{1}{2^k} + \frac{1}{3^k} - \frac{1}{4^k} + \ldots$ is convergent for all positive values of k.

Solution. For the given series, the nth term is $u_n = \frac{1}{n^k}$.

Then $u_n \ge u_{n+1}$ and $\lim_{n\to\infty} \frac{1}{n^k} = 0$. Therefore, by Leibnitze's Rule, the given series converges. *In particular, the series* $1 - \frac{1}{2} + \frac{1}{3} - \frac{1}{4} + \ldots$ *converges.*

EXAMPLE 1.65

Examine the convergence of the series

$$\frac{1}{1.2} - \frac{1}{3.4} + \frac{1}{5.6} - \frac{1}{7.8} + \ldots.$$

Solution. For the given series, we have

$$u_n = \frac{1}{(2n-1)(2n)} = \frac{1}{2n^2(2 - \frac{1}{n})}.$$

Therefore, $\lim_{n\to\infty} u_n = 0$. Further,

$$u_n - u_{n+1} = \frac{1}{(2n-1)(2n)} - \frac{1}{(2n+1)(2n+2)}$$
$$= \frac{8n+2}{(2n-1)(2n)(2n+1)(2n+1)} > 0$$

for all n. Thus $\{u_n\}$ is monotonically decreasing to zero. Therefore, by Leibnitz's Rule, the given series is convergent.

EXAMPLE 1.66

Examine for convergence of the series

(i) $\frac{1}{x+1} - \frac{1}{x+2} + \frac{1}{x+3} - \cdots$, where x is not a negative integer.

(ii) $1 - \frac{1}{3} + \frac{1}{5} - \frac{1}{7} + \cdots$

(iii) $\frac{\log 2}{2^2} - \frac{\log 3}{3^2} + \frac{\log 4}{4^2} - \cdots$

(iv) $\frac{1}{6} - \frac{2}{11} + \frac{3}{16} - \frac{4}{21} + \frac{5}{26} - \cdots$.

Solution. (i) If $x > -1$, the terms are altemating from the beginning. If $x < -1$ (except negative integers), the terms are ultimately altering their signs. Since, the removal of finite number of terms does not affect the convergence of the series, we may assume the series to be alternating in this case also. The nth term of the series is $u_n = \frac{1}{x+n}$ and so $\lim_{n\to\infty} u_n = 0$. Further,

$$u_n - u_{n+1} = \frac{1}{x+n} - \frac{1}{x+n+1}$$
$$= \frac{1}{(x+n)(x+n+1)} > 0$$

and so $\{u_n\}$ is monotonically decreasing sequence. Hence, by Leibnitz's Test, the given series converges.

(ii) Clearly, $\{u_n\}$ is monotonic decreasing and

$$\lim_{n\to\infty} u_n = \lim_{n\to\infty} \frac{1}{2n-1} = 0.$$

Hence, the given alternating series is convergent.

(iii) The nth term of the given alternating series is

$$u_n = \frac{\log(n+1)}{(n+1)^2}.$$

Then,

$$\lim_{n\to\infty} u_n = \lim \frac{\log(n+1)}{(n+1)^2}\left(\frac{\infty}{\infty}\text{ form}\right)$$

$$= \lim_{n\to\infty} \frac{\frac{1}{n+1}}{2(n+1)}, \text{ (L'Hospital Rule)}$$

$$= \lim_{n\to\infty} \frac{1}{2(n+2)^2} = 0.$$

To examine the monotonicity, we make use of a corollary to Mean Value Theorem, according to which "a function is monotonic decreasing if its derivative is negative." So, let us take

$$f(n) = \frac{\log n}{n^2}, \; n > 0.$$

Then

$$f'(n) = \frac{n^2\left(\frac{1}{n}\right) - 2n \log n}{n^4}$$

$$= \frac{1-2\log n}{n^3} < 0 \text{ if } 1—2 \log n < 0,$$

that is, if $\log n > \frac{1}{2}$, that is, if $n > e^{\frac{1}{2}} = 1.65$. But, $n > 1$ and so the condition is satisfied. Thus,

$$f(n+1) > f(n+2) \text{ for all } n \geq 1,$$

which shows that $u_n > u_{n+1}$. Hence, by Leibnitz's Test, the given series is convergent.

(iv) The nth term of the given series is

$$u_n = \frac{n}{5n+1} = \frac{1}{5+\frac{1}{n}} \to \frac{1}{5} \text{ as } n \to \infty.$$

Thus the terms are monotonically decreasing and u_n tends to a finite limit. Hence, by corollary to Leibnitz's Test, the given series oscillates finitely.

EXAMPLE 1.67

Examine the convergence of the series

$$\frac{1}{\log 2} - \frac{1}{\log 3} + \frac{1}{\log 4} - \frac{1}{\log 5} + \ldots$$

Solution. For this series, the nth term is

$$u_n = \frac{1}{\log(n+1)}.$$

Then,

$$\lim_{n\to\infty} u_n = \lim_{n\to\infty} \frac{1}{\log(n+1)} = 0$$

and

$$u_n - u_{n+1} = \frac{1}{\log(n+1)} - \frac{1}{\log(n+2)} > 0.$$

Hence, by Libnitz's Test, the given series converges.

EXAMPLE 1.68

Examine the convergence of the series

$$\frac{1}{\sqrt{2}+1} - \frac{1}{\sqrt{3}-1} + \frac{1}{\sqrt{4}+1} - \frac{1}{\sqrt{5}-1} + \cdots.$$

Solution. The given series is an alternating series. We observe that

$$u_n = \frac{(-1)^n}{\sqrt{n}+(-1)^n} \quad (n \geq 2)$$

and that $u_n \to 0$ as $n \to \infty$. But the series is not convergent. In fact, the Leibnitz's Test is not applicable because the terms do not decrease monotonically. Further,

$$u_n = \frac{(-1)^n}{\sqrt{n}+(-1)^n}$$

$$= \frac{(-1)^n[\sqrt{n}-(-1)^n]}{n-1}$$

$$= \frac{(-1)^n\sqrt{n}}{n-1} - \frac{1}{n-1}.$$

Therefore,

$$\sum u_n = \sum \frac{(-1)^n\sqrt{n}}{n-1} - \sum \frac{1}{n-1}.$$

The first series on the right converges by Leibnitz's Test while the second series diverges. Therefore, Σu_n, whose terms are the difference of the terms of these two series, diverges to $-\infty$.

1.20 ABSOLUTE CONVERGENCE OF A SERIES

A series Σu_n containing both positive and negative terms is said to be *absolutely convergent*, if $\Sigma|u_n|$ is convergent.

Thus, the series which becomes convergent when all its negative terms are made positive is called absolutely convergent series.

For example, the series

$$1-\frac{1}{2}+\frac{1}{2^2}-\frac{1}{2^3}+\cdots$$

is absolutely convergent because the series $1+\frac{1}{2}+\frac{1}{2^2}+\frac{1}{2^3}+\cdots$ is convergent.

If the series Σu_n converges and the series $\Sigma|u_n|$ diverges, then the series Σu_n is said to be *non- absolutely* or *semi-convergent* series.

For example, the series

$$1-\frac{1}{2}+\frac{1}{3}-\frac{1}{4}+\frac{1}{5}-\frac{1}{6}+\cdots$$

converges by Leibnitz's Test but the series

$$1+\frac{1}{2}+\frac{1}{3}+\frac{1}{4}+\frac{1}{5}+\frac{1}{6}+\cdots$$

is divergent. Hence, the series $1-\frac{1}{2}+\frac{1}{3}-\frac{1}{4}+\frac{1}{5}-\frac{1}{6}+\cdots$ is non-absolutely convergent.

A convergent series which is unaffected by any rearrangement of its terms is called an *unconditional convergent series*, whereas the convergent series which is affected by any rearrangement of its terms is called *conditionally convergent series*.

For example, the series

$$1-\frac{1}{2}+\frac{1}{2^2}-\frac{1}{2^3}+\cdots$$

is unconditionally convergent since its value is not affected by rearrangement. On the other hand, the series

$$1-\frac{1}{2}+\frac{1}{3}-\frac{1}{4}+\frac{1}{5}-\cdots$$

is conditionally convergent since its value can be affected by rearrangement.

Remark 1.12. To test for absolute convergence, we have to apply only the test for series with positive terms, as discussed earlier.

Theorem 1.20. An absolutely convergent series is itself convergent.

Proof: Suppose $\Sigma|u_n|$ converges. Therefore, for a given $\varepsilon > 0$, there exists a positive integer n_0 such that

$$|u_{n+1}|+|u_{n+2}|+\cdots+|u_m|<\varepsilon \text{ for } m, n>n_0.$$

Therefore,

$$|u_{n+1}+u_{n+2}+\cdots+|\le|u_{n+1}|+|u_{n+2}|$$
$$+|u_{n+3}|+\cdots+|u_m|$$
$$<\varepsilon \text{ for } m, n>n_0.$$

Hence, by Cauchy's Principle of Convergence, the series Σu_n is convergent.

Remark 1.13. The converse of Theorem 1.20 is not true. For example, the series $1-\frac{1}{2}+\frac{1}{3}-\frac{1}{4}+\frac{1}{5}-\frac{1}{6}+\cdots$ is convergent by Leibnitz's Test, but the series $\sum|u_n|=1+\frac{1}{2}+\frac{1}{3}+\frac{1}{4}+\frac{1}{5}+\frac{1}{6}+\cdots$ is divergent.

Theorem 1.21. In an absolutely convergent series, the series formed by positive terms only is convergent and the series formed by negative terms only is also convergent.

Proof: Let Σu_n be the series and let

$$S_n=u_1+u_2+\cdots+u_n,$$
$$\sigma_n=|u_1|+|u_2|+\cdots+|u_n|.$$

If $P(n)$ and $-Q(n)$ denotes the sum of positive and negative terms, respectively, in S_n, then

$$S_n=P(n)-Q(n),$$
$$\sigma_n=P(n)+Q(n).$$

Therefore,

$$P(n)=\frac{1}{2}(\sigma_n+S_n)$$

and

$$Q(n)=\frac{1}{2}(\sigma_n-S_n).$$

Since the series $\Sigma|u_n|$ converges, both S_n and σ_n tend to a finite limit as $n\to\infty$. Let these limits be s and t, respectively. Then $P(n)\to\frac{1}{2}(s+t)$ and $Q(n)\to\frac{1}{2}(s-t)$, and so the series formed by both the positive and by negative terms are convergent.

EXAMPLE 1.69

Show that the exponential series

$$1+x+\frac{x^2}{2!}+\frac{x^3}{3!}+\cdots+\frac{x^4}{n!}+\cdots$$

is absolutely convergent for all values of x.

Solution. For the given series, we have

$$\left|\frac{u^n}{u_{n+1}}\right|=\left|\frac{x^n}{n!}\frac{(n+1)!}{x^{n+1}}\right|=\frac{n+1}{|x|}$$

and so

$$\lim_{n\to\infty}\left|\frac{u_n}{u_{n+1}}\right|=\infty>1.$$

Hence, the series is absolutely convergent by extended D'Alembert's Test.

EXAMPLE 1.70

Examine the logarithmic series

$$x-\frac{x^2}{2}+\frac{x^3}{3}-\dots+(-1)^{n+1}\frac{x^n}{n}+\dots$$

for absolute convergence.

Solution. It is an alternating series for which

$$\left|\frac{u_n}{u_{n+1}}\right|=\left|\frac{x^n}{n}\frac{n+1}{x^{n+1}}\right|=\frac{n(1+\frac{1}{n})}{n}\cdot\frac{1}{|x|}.$$

Therefore,

$$\lim_{n\to\infty}\left|\frac{u_n}{u_{n+1}}\right|=\frac{1}{|x|}$$

and so, by extended D'Alembert's Ratio Test, the given series converges absolutely if $\frac{1}{|x|}>1$, that is, if $|x| < 1$.

When $x = 1$, the series becomes $1-\frac{1}{2}+\frac{1}{3}-\frac{1}{4}+\cdots$, which converges by Leibnitz's Test. But the series $1+\frac{1}{2}+\frac{1}{3}-\frac{1}{4}+\cdots$ diverges.

When $x = -1$, the series becomes $-(1+\frac{1}{2}+\frac{1}{3}+\frac{1}{4}+\cdots)$ and is hence divergent.

When $x < -1$, the terms are all negative. Removing a common negative sign, the term becomes positive. Since

$$\lim_{n\to\infty}\frac{u_n}{u_{n+1}}=\frac{1}{|x|}<1,$$

the series diverges by D'Alembert's Ratio Test.

When $x > 1$, the nth term does not tend to zero. For if $\frac{x^n}{n}=y$, then

$$\log y = n\log x-\log n$$

$$=n\left(\log x-\frac{\log n}{n}\right)\to\infty,$$

$$\text{since }\frac{\log n}{n}\to 0\text{ as }n\to\infty.$$

Therefore, $\frac{x^n}{n}\to\infty$ as $n\to\infty$. The series, therefore, oscillates infinitely.

EXAMPLE 1.71

Show that the series $\sum\frac{\cos n\theta}{n^p}$ and $\sum\frac{\sin n\theta}{n^p}$ are absolutely convergent for $p > 1$.

Solution. Since,

$$\left|\frac{\cos n\theta}{n^p}\right|\le\frac{1}{n^p}\text{ and }\left|\frac{\sin n\theta}{n^p}\right|\le\frac{1}{n^p}$$

and since $\sum\frac{1}{n^p}$ converges for $p > 1$, it follows that $\sum\frac{\cos n\theta}{n^p}$ and $\sum\frac{\sin n\theta}{n^p}$ converge for $p > 1$.

EXAMPLE 1.72

Examine absolute convergence of the hypergeometric series

$$1+\frac{\alpha.\beta}{1\cdot\gamma}x+\frac{\alpha(\alpha+1)\beta(\beta+1)}{1\cdot2(\gamma+1)}x^2+\cdots.$$

Solution. We have

$$\lim_{n\to\infty}\left|\frac{u_n}{u_{n+1}}\right|=\lim_{n\to\infty}\left|\frac{(n+1)(n+\gamma)}{(\alpha+n)(\beta+n)}\cdot\frac{1}{x}\right|=\frac{1}{|x|}.$$

Therefore, by D'Alembert's Ratio Test, the hyper geometric series is absolutely convergent if $\frac{1}{|x|}>1$, that is, if $|x|<1$.

When, $x = 1$, we have

$$\frac{u_n}{u_{n+1}} = \frac{\left(1+\frac{1}{n}\right)\left(1+\frac{\gamma}{n}\right)}{\left(1+\frac{\gamma}{n}\right)\left(1+\frac{\beta}{n}\right)}$$

$$= 1 + \frac{1+\gamma-\alpha-\beta}{n} + O\left(\frac{1}{n^2}\right) \text{ by binomial expansion.}$$

Therefore, by Gauss's Test, the series is convergent if $1 + \gamma - \alpha - \beta > 1$ or if $\gamma > \alpha + \beta$ and divergent if $1 + \gamma - \alpha - \beta < 1$ or if $\gamma \leq \alpha + \beta$.

When $x > 1, \lim_{n\to\infty} \frac{u_n}{u_{n+1}} = \frac{1}{|x|} < 1$. Therefore, the series diverges whatever α, β, and γ may be.

EXAMPLE 1.73

Discuss convergence of the *Binomial Series*

$$1 + mx + \frac{m(m-1)}{2!}x^2 + \cdots + \frac{m(m-1)\ldots(m-n+1)}{n!}x^n + \cdots$$

Solution. We have

$$\lim_{n\to\infty}\left|\frac{u_n}{u_{n+1}}\right| = \lim_{n\to\infty}\left|\frac{n}{m-n+1}\right|\left|\frac{1}{x}\right|$$

$$= \lim_{n\to\infty}\frac{1}{\frac{m}{n}, -1+\frac{1}{n}}\left|\frac{1}{x}\right| = \frac{1}{|x|}.$$

Hence, the series is absolutely convergent if $|x| < 1$. Further,

$$\lim_{n\to\infty} |u_n|^{1/n} = \lim_{n\to\infty}\left|\frac{u_{n+1}}{u_n}\right| = |x|.$$

Therefore,

$$\lim_{n\to\infty} |u_n| = \lim_{n\to\infty} |x|^n = \infty \; if \; |x| > 1.$$

Hence, the series cannot converge when $|x| > 1$. When $x < 0$, the terms are ultimately of the same sign and hence the series is divergent if $x < -1$. If $x > 1$, the terms are alternately positive and negative after a certain stage and the series oscillates infinitely. When $x = -1$, the series is

$$1 - m + \frac{m(m-1)}{2!} - \ldots + (-1)^n \times \frac{m(m-1)\ldots(m-n-1)}{n!} + \ldots$$

Whatever m may be, the terms are of the same sign after a certain value of n. We have

$$\frac{u_n}{u_{n+1}} = 1 + \frac{m+1}{n} + O\left(\frac{1}{n^2}\right).$$

Hence, by Gauss's Test, the series is convergent if $m + 1 > 1$, that is, if $m > 0$ and divergent if $m < 0$. If $m = 0$, the series reduce to single term 1. If $x = 1$, the series is

$$1 + m + \frac{m(n-1)}{2!} + \cdots + \frac{m(m-1)\ldots(m-n-1)}{n!} + \cdots.$$

The terms are alternately positive and negative after a certain value of n. From above, we have

$$\left|\frac{u_n}{u_{n+1}}\right| = 1 + \frac{m+1}{n} + O\left(\frac{1}{n^2}\right).$$

Hence, by Gauss's Test, the series converges absolutely if $m + 1 > 1$, that is, if $m > 0$.

EXAMPLE 1.74

Examine the following series for convergence/absolute convergence.

(i) $\Sigma \frac{(-1)^{n-1} \sin nx}{n^3}$

(ii) $x - \frac{x^2}{2} + \frac{x^3}{3} - \frac{x^4}{4} + \frac{x^5}{5} - \cdots$

(iii) $x - \frac{x^2}{\sqrt{2}} + \frac{x^3}{\sqrt{3}} - \frac{x^4}{\sqrt{4}} + \cdots$

Solution. (i) The nth term of the series is

$$u_n = \frac{(-1)^{n+1} \sin nx}{n^3}$$

and so

$$|u_n| = \frac{|\sin nx|}{n^3} \leq \frac{1}{n^3}.$$

Taking v_n as $\frac{1}{n^3}$, we have $\sum v_n = \sum \frac{1}{n^3}$, which converges. Therefore, by comparison test, $\Sigma| u_n|$

converges. Hence, the given series is absolutely convergent.

(ii) We have $u_n = \frac{x^n}{n}$ and so

$$\frac{u_n}{u_{n+1}} = \frac{x^n}{n} \cdot \frac{n+1}{x^{n+1}} = \left(\frac{n+1}{n}\right) \cdot \frac{1}{x}.$$

Thus

$$\lim_{n\to\infty} \left|\frac{u_n}{u_{n+1}}\right| = \left(1+\frac{1}{n}\right)\left|\frac{1}{x}\right| = \left|\frac{1}{x}\right|.$$

Hence, by extended D'Alembert's Ratio Test, the given series converges absolutely if $\left|\frac{1}{x}\right| > 1$, that is, if $|x| < 1$. If $x = 1$, the series becomes

$$1 - \frac{1}{2} + \frac{1}{3} - \frac{1}{4} + \frac{1}{5} - \frac{1}{6} + \cdots,$$

which is convergent by Leibnitz's Test. If $x = -1$, then the series becomes $-1 - \frac{1}{2} - \frac{1}{3} - \frac{1}{4} - \frac{1}{5} - \ldots = -(1 + \frac{1}{2} + \frac{1}{3} + \ldots)$, which is divergent. Hence, the given series converges for $-1 < x \le 1$ and converges absolutely for $-1 < x < 1$.

(iii) The nth term of the series is $u_n = \frac{x^n}{\sqrt{n}}$. Therefore,

$$\frac{u_n}{u_{n+1}} = \frac{x^n}{\sqrt{n}} \frac{\sqrt{n+1}}{x^{n+1}} = \sqrt{\frac{n+1}{n}} \cdot \frac{1}{x}$$

and so

$$\lim_{n\to\infty} \left|\frac{u_n}{u_{n+1}}\right| = \lim_{n\to\infty} \left(1+\frac{1}{n}\right)^{\frac{1}{2}} \cdot \left|\frac{1}{x}\right| = \frac{1}{|x|}.$$

Hence, by extended D'Alembert Ratio Test, the series converges absolutely if $\frac{1}{|x|} > 1$, that is, if $|x| < 1$, that is if $-1 < x < 1$. If $x = 1$, then the series becomes

$$1 - \frac{1}{\sqrt{2}} + \frac{1}{\sqrt{3}} - \frac{1}{\sqrt{4}} + \ldots,$$

whose nth term is $u_n = \frac{1}{\sqrt{n}} \to 0$ as $\to \infty$. Also $u_n \ge u_{n+1}$. Hence, by Leibnitz's Test, Σu_n converges. If $x = -1$, then the series becomes

$$1 - \frac{1}{\sqrt{2}} - \frac{1}{\sqrt{3}} - \frac{1}{\sqrt{4}} - \cdots$$

or

$-\left(1 + \frac{1}{\sqrt{2}} + \frac{1}{\sqrt{3}} + \frac{1}{\sqrt{4}} + \cdots\right)$, which is divergent: Hence, the given series converges for $-1 < x \le 1$ and converges absolutely for $-1 < x \le 1$.

EXAMPLE 1.75

Examine the series $\sum_{n=2}^{\infty} \frac{(-1)^n}{n(\log n)^2}$ for convergence and absolute convergence

Solution. The nth term of the given alternating series is $\frac{(-1)^n}{n(\log n)^2}$. So

$$u_n = \frac{1}{n(\log n)^2}.$$

We observe that

$$\lim_{n\to\infty} u_n = \lim_{n\to\infty} \frac{1}{n(\log n)^2} = 0$$

and $u_n > u_{n+1}$, $n > 2$. Therefore, by Leibnitz's Test, the given series is convergent.

To examine for absolute convergence, we have

$$|u_n| = \frac{1}{n(\log n)^2} = f(n), \text{ say.}$$

Therefore,

$$f(x) = \frac{1}{x(\log x)^2},$$

which is positive and monotonic decreasing. Hence Cauchy's Integral Test is applicable. We have

$$\int_2^\infty f(x)dx = \int_2^\infty \frac{dx}{x(\log x)^2} = \int_2^\infty (\log x)^{-2} \cdot \frac{1}{x} \text{dx}$$

$$= \left(\frac{(\log x)^{-1}}{-1}\right)_2^\infty$$

$$= \frac{1}{\log 2} \text{ (finite).}$$

Thus, $\int_2^\infty f(x)dx$ converges and so by Cauchy's Integral Test, the series $\sum_{n=2}^{\infty} |u_n|$ converges. Hence, the given series converges absolutely.

1.21 CONVERGENCE OF THE SERIES OF THE TYPE $\sum_{n=1}^{\infty} u_n v_n$

To study the convergence behavior of the series of the type $\Sigma u_n v_n$, we have two theorems of great importance known as Abel's Test and Dirichlet's Test. The following lemma would be required to establish these tests.

Theorem 1.22. (Abel's Lemma). If $v_1, v_2 \ldots, v_n$ are positive and decreasing, then

$$Bv_1 < u_1v_1 + u_2v_2 + \cdots + u_nv_n < Av_1,$$

where A and B are, respectively, the greatest and the least of the quantities

$$u_1, u_1 + u_2, u_1 + u_2 + u_3, \cdots, u_1 + u_2 + \cdots + u_n.$$

Proof: Let

$$S_n = u_1 + u_2 + \cdots + u_n.$$

Then

$$u_1 = S_1, u_2 = S_2 - S_1, \ldots, u_n = S_n - S_{n-1}.$$

Since A is greatest of $S_1, S_2, \cdots, S_n$, we obtain

$$\begin{aligned}\sum_{m=1}^{n} u_m v_m &= u_1v_1 + u_2v_2 + u_3v_3 + \cdots + u_nv_n \\ &= S_1v_1 + (S_2 - S_1)v_2 + (S_3 - S_2)v_3 \\ &\quad + \cdots + (S_n - S_{n-1})v_n \\ &= S_1(v_1 - v_2) + S_2(v_2 - v_3) \\ &\quad + \cdots + S_{n-1}(v_{n-1} - v_n) + S_nv_n \\ &< A[v_1 - v_2 + v_2 - v_3 \\ &\quad + \cdots + v_{n-1} - v_n + v_n] = Av_1.\end{aligned}$$

Similarly, if A is least of $S_1, S_2, \ldots, S_n$, then we get

$$\sum_{m=1}^{n} u_m v_m > Bv_1.$$

Hence

$$Bv_1 < u_1v_1 + u_2v_2 + \cdots + u_nv_n < Av_1.$$

Theorem 1.23. (Abel's Test). If $\{v_n\}$ is positive and monotonically decreasing and if Σu_n converges, then $\Sigma u_n v_n$ is convergent.

Proof: Let $\{v_n\}$ be positive and monotonically decreasing, whereas u_n may be positive or negative. Then for the convergence of the series $\Sigma u_n v_n$, we must have

$$\left|u_nv_n + u_{n+1}v_{n+1} + \cdots + u_{n+p}v_{n+p}\right| < \varepsilon.$$

By Abel's Lemma,

$$Bv_n < u_nv_n + u_{n+1}v_{n+1} + \cdots + u_{n+p}v_{n+p} < Av_n,$$

where A and B are respectively the greatest and the least of

$$u_n, u_n + u_{n+1}, \ldots u_n + u_{n+1} + \ldots + u_{n+p}.$$

Therefore,

$$|u_nv_n + u_{n+1}v_{n+1} + \cdots + u_{n+p}v_{n+p}| < \alpha v_n,$$

where α is the maximum of $|A|$ and $|B|$. Since $\{v_n\}$ is positive and decreasing, it is bounded, that is, $v_n < m$ for all n. Since Σu_n converges, $\alpha < \frac{\varepsilon}{m}$ for $n >$ some n_0. Thus

$|u_nv_n + u_{n+1}v_{n+1} + \cdots + u_{n+p}v_{n+p}| < \varepsilon$ for $n > n_0$.

Hence, by Cauchy's Principle of Convergence, the series $\Sigma u_n v_n$ is convergent.

Theorem 1.24. (Dirichlet's Test). If $\{v_n\}$ is positive and monotonically decreasing to the limit 0 and if Σu_n is either convergent or oscillates finitely, then $\Sigma u_n v_n$ is convergent.

Proof: By Abel's Lemma, we have

$$Bv_n < u_nv_n + u_{n+1}v_{n+1} + \cdots + u_{n+p}v_{n+p} < Av_n,$$

where A and B are, respectively, the greatest and the least of

$$u_n, u_n + u_{n+1}, \cdots, u_n + u_{n+1} + \cdots + u_{n+p}.$$

Therefore, if $\alpha = \max(|A|, |B|)$, then

$$\left|u_nv_n + u_{n+1}v_{n+1} + \cdots + u_{n+p}v_{n+p}\right| < \alpha v_n.$$

Since Σu_n is convergent or oscillates, α is finite, say less then k for all n. Since v_n tends monotonically to 0, we have $v_n < \frac{\varepsilon}{k}$ for some $n > n_0$. Hence,

$$\left|u_nv_n + u_{n+1}v_{n+1} + \ldots + u_{n+p}v_{n+p}\right| < \varepsilon \text{ for } n > n_0.$$

Hence, the series $\Sigma u_n v_n$ is convergent by Cauchy's Principle of Convergence.

Remark 1.14. If $\Sigma u_n = 1 - 1 + 1 + \ldots$, then it follows that *Leibnitz's Test is a particular case of Dirichlet's Test.*

EXAMPLE 1.76

Test the following series for convergence:

$$0-\frac{1}{2}+\frac{1}{2^2}-\frac{1}{3}+\frac{2}{3^2}-\frac{1}{4}+\frac{3}{4^2}-\ldots.$$

Solution. Consider the series

$$\sum u_n = 1-1+\frac{1}{2}-\frac{1}{2}+\frac{1}{3}-\frac{1}{3}+\frac{1}{4}-\frac{1}{4}+\cdots.$$

This series is convergent since $\lim_{n\to\infty} S_{2n} = 0$, $\lim_{n\to\infty} S_{2n+1} = \lim_{n\to\infty}\frac{1}{n+1} = 0$. Now consider the sequence $\{v_n\}$ with elements

$$0, \frac{1}{2}, \frac{1}{2}, \frac{2}{3}, \frac{2}{3}, \frac{3}{4}, \frac{3}{4}, \ldots$$

This sequence is positive and monotonic decreasing. Then multiplying the terms of the above two series, we get

$$\sum u_n v_n = 0-\frac{1}{2}+\frac{1}{2^2}-\frac{1}{3}+\frac{1}{3^2}-\frac{1}{4}$$
$$+\frac{3}{4^2}-\ldots \text{ (given series)}$$

Hence, by Abel's Test, the given series $\Sigma u_n v_n$ is convergent.

EXAMPLE 1.77

Test the following series for convergence:

(i) $\sum \frac{(n^3+1)^{\frac{1}{3}}-n}{\log n}$

(ii) $1-\frac{1}{3.2^2}+\frac{1}{5.3^2}-\frac{1}{74^2}+\cdots$

Solution. (i) If $u_n = (n^3+1)^{\frac{1}{3}} - n$ and $v_n = \frac{1}{\log n}$, then

$$\Sigma u_n v_n = \Sigma \frac{(n^3+1)^{\frac{1}{3}}-n}{\log n} \text{ (given series).}$$

The series $\sum u_n = (n^3+1)^{\frac{1}{3}} - n$ is convergent and $\{v_n\}$ is a positive monotonically decreasing sequence. Hence, by Abel's Test, the given series $\Sigma u_n v_n$ is convergent.

(ii) If $\sum u_n = \frac{1}{1^2}-\frac{1}{2^2}+\frac{1}{3^2}-\frac{1}{4^2}+\cdots$ and $v_n = \frac{1}{2n-1}$, then the given series can be written as $\Sigma u_n v_n$. But, Σu_n is convergent by Leibnitz's Test and v_n decreases monotonically to zero. Hence, by Dirichlet's Test, the given series converges.

EXAMPLE 1.78

Examine the convergence of the series $\sum \frac{1}{n} \sin n\theta$ and $\sum \frac{1}{n} \cos n\theta$.

Solution. We have

$$\sum_{m=1}^{n} \sin m\theta = \frac{1}{2\sin\frac{\theta}{2}}\left[2\sin\theta \sin\frac{\theta}{2} + 2\sin 2\theta \sin\frac{\theta}{2} + \cdots + 2\sin n\theta \sin\frac{\theta}{2}\right]$$

$$= \frac{1}{2\sin\frac{\theta}{2}}\left[\left(\cos\frac{\theta}{2} - \cos\frac{3\theta}{2}\right) + \left(\cos\frac{3\theta}{2} - \frac{\cos 5\theta}{2}\right) + \cdots + \left(\cos\frac{2n-1}{2}\theta - \cos\frac{2n+1}{2}\theta\right)\right]$$

$$= \frac{1}{2\sin\frac{\theta}{2}}\left[\cos\frac{\theta}{2} - \cos\frac{2n+1}{2}\theta\right]$$

and so

$$\left|\sum_{m=1}^{n} \sin m\theta\right| \le \left|\operatorname{cosec}\frac{\theta}{2}\right|.$$

Hence the series $\Sigma \sin m\theta$ oscillates finitely. Also, the sequence $1, \frac{1}{2}, \frac{1}{3}, \ldots, \frac{1}{n}$ is a decreasing sequence of positive terms which tends to zero as $n \to \infty$. Hence, by Dirichlet's Test, the series $\sum \frac{1}{n} \sin n\theta$ converges if $\sin\frac{\theta}{2} \ne 0$, that is, if $\theta \ne 0$ or not an even multiple of π. If $\theta = 0$ or $2m\pi$ where m is an integer, the series $\sum \frac{1}{n} \sin n\theta$ is merely a series of zeros and is, therefore, convergent.

Similarly,

$$\sum_{m=1}^{n} \cos m\theta = \frac{1}{2\sin\frac{\theta}{2}}\left[\sin\frac{2n+1}{2}\theta - \sin\frac{\theta}{2}\right]$$

and so

$$\left|\sum_{m=1}^{n} \cos m\theta\right| \le \left|\operatorname{cosec}\frac{\theta}{2}\right|.$$

Therefore, as in the case of series $\sum \frac{1}{n}\sin n\theta$ the given series $\sum \frac{1}{n}\cos n\theta$ converges.

If $\theta = 0$ or $2m\pi$ where m is integer, then we get the series $1+\frac{1}{2}+\ldots+\frac{1}{n}+\ldots+$, which is divergent.

1.22 DERANGEMENT OF SERIES

If a series consists of finite number of terms, then its terms can be arranged in any manner what so ever. On the other hand, if terms of an infinite series are deranged, its sum or even the nature of convergence may alter. However, we will notice that derangement does not affect the sum for convergence of the series, if the infinite series is absolutely convergent.

To study the effect of derangement, consider the series

$$1-\frac{1}{2}+\frac{1}{3}-\frac{1}{4}+\frac{1}{5}-\cdots.$$

This series is convergent by Leibnitz's Test. Further, by the definition of *Euler's Constant* γ, we know that

$$1+\frac{1}{2}+\frac{1}{3}+\cdots+\frac{1}{2n} \to \log 2n+\gamma \qquad (1)$$

and

$$1+\frac{1}{2}+\frac{1}{3}+\cdots+\frac{1}{n} \to \log n+\gamma$$

or

$$2\left(\frac{1}{2}+\frac{1}{4}+\frac{1}{6}+\frac{1}{8}+\cdots+\frac{1}{2n}\right) \to \log n+\gamma. \qquad (2)$$

But

$$1-\frac{1}{2}+\frac{1}{3}-\frac{1}{4}+\frac{1}{5}-\frac{1}{6}+\cdots-\frac{1}{2n}$$

$$=\left(1+\frac{1}{2}+\frac{1}{3}+\cdots+\frac{1}{2n}\right)$$

$$-2\left(\frac{1}{2}+\frac{1}{4}+\frac{1}{6}+\cdots+\frac{1}{2n}\right).$$

Therefore, by (1) and (2), we get

$$S_{2n} \to \log(2n)+\gamma - \log n - \gamma = \log\frac{2n}{n} = \log 2.$$

Thus, the sum of the series in question is log 2.

Now consider the following derangement of the above series:

$$1-\frac{1}{2}-\frac{1}{4}+\frac{1}{3}-\frac{1}{6}-\frac{1}{8}+\frac{1}{5}-\frac{1}{10}-\frac{1}{12}+\cdots$$

Then

$$S_{3n} = \left(1-\frac{1}{2}\right)-\frac{1}{4}+\left(\frac{1}{3}-\frac{1}{6}\right)-\frac{1}{8}+\left(\frac{1}{5}-\frac{1}{10}\right)$$

$$-\frac{1}{12}+\cdots+\left(\frac{1}{2n-1}-\frac{1}{4n-2}\right)-\frac{1}{4n}$$

$$=\frac{1}{2}-\frac{1}{4}+\frac{1}{6}-\frac{1}{8}+\cdots+\frac{1}{4n-2}-\frac{1}{4n}$$

$$=\frac{1}{2}\left(1-\frac{1}{2}+\frac{1}{3}-\frac{1}{4}+\cdots+\frac{1}{2n-1}-\frac{1}{2n}\right)$$

$$=\frac{1}{2}S_{2n}$$

and so

$$\lim_{n\to\infty} S_{3n} = \lim_{n\to\infty}\frac{1}{2}S_{2n} = \frac{S}{2} = \frac{1}{2}\log 2.$$

If follows, therefore, that derangement of the series has changed its sum.

We now show that derangement of terms does not affect the sum of an absolutely convergent series.

Theorem 1.25. The sum of an absolutely convergent series is not altered by any derangement.

Proof: Let $\Sigma u_n = u_1 + u_2 + u_3 + \ldots + u_n + \ldots$ be an absolutely convergent series. Let $\Sigma v_n = v_1 + v_2 + \ldots + v_n + \ldots$ be the deranged series of Σu_n. Thus, the terms in Σv_n are the same as the terms in Σu_n but in other order. Let S_n and T_n denote the sum of n terms of the two series Σu_n and Σv_n, respectively. We can find p such that T_p contains all the terms of S_n. Thus taking r an integer greater then p, $T_r - S_n$ contains a number of terms of the first series beyond u_n. Since Σu_n is absolutely convergent, there exists n_0 such that

$$|u_{n+1}|+|u_{n+2}|+|u_{n+2}|+\cdots<\frac{\varepsilon}{2}, n\geq n_0.$$

Since, $|T_r - S_n|$ < the sum of a certain number of terms on the left side of the above expression, it follows that

$$|T_r - S_n|<\frac{\varepsilon}{2}, n\geq n_0, r>p.$$

If S denotes the sum of Σu_n, then

$$|S-S_n|\leq|u_{n+1}|+|u_{n+2}|+\cdots<\frac{\varepsilon}{2}, n\geq n_0.$$

Hence,

$$\begin{aligned}|S-T_r| &= |S-S_n+S_n-T_r|\\ &\leq |S-S_n|+|S_n-T_r|\\ &< \frac{\varepsilon}{2}+\frac{\varepsilon}{2}=\varepsilon, r>p.\end{aligned}$$

Hence, $\Sigma v_n = S = \Sigma u_n$.

1.23 NATURE OF NON-ABSOLUTELY CONVERGENT SERIES

The following theorem provides a deep insight into the nature of non-absolutely (or conditionally) convergent series:

Theorem 1.26. (Riemann). The terms of a non-absolutely convergent series can be deranged in such a manner that the resulting series converges to any given sum or diverges or becomes oscillatory.

Proof: Let $\sum_{n=1}^{\infty} u_n$ be a non – absolutely convergent series. Listing the positive and negative terms of the given series separately in the order in which they occur, we have

$$V \equiv v_1 + v_2 + v_3 + \cdots \quad (1)$$

$$W \equiv -w_1 - w_2 - w_3 - \cdots \quad (2)$$

We claim that V and W are divergent. To prove it, let V_n and W_n denote the sum of n terms of V and W, respectively. If V_n and W_n are both bounded as $n \to \infty$, then the partial sums of $\Sigma|u_n|$ are also bounded and, hence, Σu_n is absolutely convergent, which contradicts the hypothesis of the theorem. On the other hand, if only one of them is bounded, then the partial sum of Σu_n tend to infinity, as a consequence of which Σu_n diverges. This also contradicts the hypothesis, since Σu_n is convergent. Hence, both V and W are divergent.

Let λ be a given number. Since the series (1) diverges, we can choose the first r terms from (1) such that their sum just exceeds λ. Thus,

$$\sum_{k=1}^{r-1} v_k < \lambda < \sum_{k=1}^{r} v_k. \quad (3)$$

Since, the series (2) diverges, we can add the first s terms from (2) to the terms in (3) such that the resulting sum is just less then λ. Thus,

$$\sum_{k=1}^{r} v_k - \sum_{j=1}^{s} w_j < \lambda < \sum_{k=1}^{r} v_k - \sum_{j=1}^{s-1} w_j. \quad (4)$$

Now add t terms from (1) commencing from v_{r+1} such that the resulting sum just exceeds λ. Thus,

$$\sum_{k=1}^{r} v_k - \sum_{p=1}^{s} w_p + \sum_{k=r+1}^{r+t-1} v_k < \lambda < \sum_{k=1}^{r} v_k - \sum_{p=1}^{s} w_p + \sum_{k=r+1}^{r+t} v_k \quad (5)$$

Now add p terms from (2) commencing from $-w_{s+1}$ so that the resulting sum is just less then λ. Continue this process indefinitely. At any stage, we have

$$\sum_{k=1}^{r} v_k - \sum_{j=1}^{S} w_j + \sum_{k=r+1}^{r+t-1} v_k - \cdots + \sum^{n-1} v_k < \lambda$$

$$< \sum_{k=1}^{r} v_k - \sum_{j=1}^{r} w_j + \sum_{k=r+1}^{r+t-1} v_k - \cdots + \sum^{n} v_k \quad (6)$$

or

$$\sum_{k=1}^{r} v_k - \sum_{p=1}^{r} w_p + \sum_{k=r+1}^{s+t-1} v_k - \cdots - \sum^{m} w_j < \lambda$$

$$< \sum_{k=1}^{r} v_k + \sum_{k=t+1}^{r+t-1} v_k - \cdots + \sum^{m-1} w_j \quad (7)$$

If the resulting series be called S and the total number of terms on the left had sides of (6) and (7) be N and M, respectively, then

$$S_N < \lambda < S_N + v_n \text{ and } S_M < \lambda < S_M - w_m.$$

Since the series Σu_n is convergent, $v_n \to 0$ as $n \to \infty$ and $w_m \to 0$ as $m \to \infty$. Therefore, $S_n \to \lambda$ as $n \to \infty$.

Now to get a divergent or oscillatory series, we replace λ in (3), (4), and (5) by the successive terms of a sequence $\lambda_1, \lambda_2, \ldots$, which may by divergent or oscillatory. Thus,

$$S_N < \lambda_p < S_N + v_n \text{ and } S_M < \lambda_q < S_M + w_m.$$

Therefore, $\lim S_N$ or $\lim S_M = \lambda_p$ or λ_p. Thus, if λ_p diverges as $p \to \infty$, then S_N diverges and if λ_p oscillates, then S_N oscillates.

1.24 EFFECT OF DERANGEMENT OF NON-ABSOLUTELY CONVERGENT SERIES

The following theorem of Pringsheim tells us about the change in sum due to derangement of terms of a non-absolutely convergent series.

Theorem 1.27. (Pringsheim). Let f be a positive function monotonically decreasing to the limit zero and let the series $\sum_{n=1}^{\infty}(-1)^{n-1} f(n)$ be deranged so that in the first $p + n$ terms, there are p positive terms and n negative terms. If $\lim_{n\to\infty} nf(n) = g$ and $\lim_{n\to\infty} \frac{p}{n} = k$, then the sum of the series is increased by $\frac{1}{2} g \log k$.

Proof: In the deranged series, there are p positive terms and n negative terms in the first $p + n$ terms. If $p > n$, then the sum of $p + n$ terms of the deranged series is equal to

$$[f(1) - f(2) + f(3) - \cdots - f(2n)] + [f(2n+1) + f(2n+3) + \cdots + f(2p-1)].$$

Thus, the excess over the sum of $2n$ terms of the original series is

$$E = f(2n+1) + f(2n+3) + \cdots + f(2p-1).$$

Since $\lim_{n\to\infty} nf(n) = g$, for $\varepsilon > 0$, there exists positive integer n_0 such that

$$|nf(n) - g| < \varepsilon \text{ for } n > n_0$$

or

$$g - \varepsilon < nf(n) < g + \varepsilon \text{ for } n > n_0$$

or

$$\frac{g-\varepsilon}{n} < f(n) < \frac{g+\varepsilon}{n} \text{ for } n > n_0.$$

Substituting $n = 2n+1, 2n+2, \ldots, 2p-1$ and adding we get

$$(g-\varepsilon)\left[\frac{1}{2n+1} + \frac{1}{2n+3} + \cdots + \frac{1}{2p-1}\right]$$

$$< f(2n+1) + f(2n+3) + \cdots + f(2p-1)$$

$$< (g+\varepsilon)\left[\frac{1}{2n+1} + \frac{1}{2n+3} + \cdots + \frac{1}{2p-1}\right] \quad (1)$$

But

$$\frac{1}{2n+1} + \frac{1}{2n+3} + \cdots + \frac{1}{2p-1}$$

$$= \left(1 + \frac{1}{2} + \frac{1}{3} + \cdots + \frac{1}{2p}\right) - \left(1 + \frac{1}{2} + \cdots + \frac{1}{2n}\right) - \frac{1}{2}\left(\frac{1}{n+1} + \cdots + \frac{1}{p}\right)$$

$$= \log 2p + \gamma_{2p} - (\log 2n + \gamma_{2n}) - \frac{1}{2}(\gamma_p + \log p - \gamma_n - \log n).$$

Therefore,

$$\lim_{p,n\to\infty}\left(\frac{1}{2n+1} + \frac{1}{2n+3} + \cdots + \frac{1}{2p-1}\right)$$

$$= \lim_{p,n\to\infty}\Big[\log 2p + \gamma_{2p} - (\log 2n + \gamma_{2n}) - \frac{1}{2}(\gamma_p + \log p - \gamma_n - \log n)\Big]$$

$$= \lim_{p,n\to\infty} \frac{1}{2} \log\frac{p}{n} = \frac{1}{2} \log k, \text{ by hypothesis.}$$

Hence, by (1), we have

$$E = f(2n+1)+f(2n+3)$$

$$+\cdots+f(2p-1) \to \frac{1}{2} g \log k.$$

Thus the sum of the series is increased by $\frac{1}{2}g \log k$. If $p < n$, the sum of $p + n$ terms of the deranged series is $[f(1) - f(2) + f(3) - \ldots + f(2p-1)] - [f(2p) + f(2p+2) + \cdots + f(2n)]$ and the excess E is now $[f(2p) + \cdots + f(2n)]$. Proceeding as above, the increase in the sum is found to be $\frac{1}{2}g \log k$.

EXAMPLE 1.79

Find the sum of the series

$$1+\frac{1}{3}-\frac{1}{2}+\frac{1}{5}+\frac{1}{7}-\frac{1}{4}+\cdots$$

Solution. The given series is the derangement of the series

$$1-\frac{1}{2}+\frac{1}{3}-\frac{1}{4}+\frac{1}{5}-\frac{1}{6}+\cdots$$

The sum of the latter series is log 2. We observe that in the deranged series, two positive terms are followed by one negative term. Thus $p = 2$, $n = 1$ so that $k = \frac{p}{n} = 2$ and $g = \lim_{n\to\infty} nf(n) = \lim_{n\to\infty} n.\frac{1}{n} = 1$. Therefore, the increase in the sum due to derangement is $\frac{1}{2}g \log k = \frac{1}{2}\log 2$. Hence, the sum of the deranged series is $\log 2 + \frac{1}{2} \log 2 = \frac{3}{2}\log 2$.

Remark 1.15. If we consider the derangement of the series $1-\frac{1}{2}+\frac{1}{3}-\frac{1}{4}+\frac{1}{5}-\frac{1}{6}+\cdots$, then the total sum of the deranged series is $\log 2 + \frac{1}{2} g \log k$. But $g = \lim_{n\to\infty} nf(n) = \lim_{n\to\infty} n.\frac{1}{n} = 1$. Therefore, the total sum of the deranged series is

$$\log 2 + \frac{1}{2} \log k = \log 4^{\frac{1}{2}} + \frac{1}{2} \log k$$

$$= \frac{1}{2} \log 4 + \frac{1}{2} \log k$$

$$= \frac{1}{2}(\log 4 + \log k) = \frac{1}{2} \log 4k.$$

EXAMPLE 1.80

Show that

$$1+\frac{1}{3}-\frac{1}{2}-\frac{1}{4}+\frac{1}{5}+\frac{1}{7}-\frac{1}{6}-\frac{1}{8}+\cdots = \log 2.$$

Solution. The given series is derangement of the series

$$1-\frac{1}{2}+\frac{1}{3}-\frac{1}{4}+\frac{1}{5}-\frac{1}{6}+\cdots$$

Here two positive terms are being followed by two negative terms. Thus,

$$k = \frac{p}{n} = \frac{2}{2} = 1.$$

Hence, the sum of the deranged (given) series is

$$\frac{1}{2} \log 4k = \frac{1}{2} \log 4 = \log 4^{\frac{1}{2}} = \log 2.$$

EXAMPLE 1.81

Find the sum of the series

$$1-\frac{1}{2}-\frac{1}{4}+\frac{1}{3}-\frac{1}{6}-\frac{1}{8}+\frac{1}{5}-\cdots$$

Solution. The given series is derangement of the series

$$1-\frac{1}{2}+\frac{1}{3}-\frac{1}{4}+\frac{1}{5}-\frac{1}{6}+\cdots$$

Here one positive term is followed by two negative terms. Thus $k = \frac{p}{n} = \frac{1}{2}$. Hence, the sum of the deranged series is

$$\frac{1}{2} \log 4k = \frac{1}{2} \log 2.$$

EXAMPLE 1.82

Show that

$$1-\frac{1}{2}-\frac{1}{4}-\frac{1}{6}-\frac{1}{8}+\frac{1}{3}-\frac{1}{10}-\frac{1}{12}-\frac{1}{14}-\frac{1}{16}+\frac{1}{5}-\cdots = 0.$$

Solution. The sum of the given series is $\frac{1}{2} \log 4k$, where $k = \frac{1}{4}$. Hence, the sum is $\frac{1}{2} \log 1 = 0$.

EXAMPLE 1.83

Investigate what derangement of the series $1-\frac{1}{2}+\frac{1}{3}-\frac{1}{4}+\frac{1}{5}-\frac{1}{6}+\cdots$ will reduce its sum to S.

Solution. We know that the sum of the deranged series is $\frac{1}{2}\log 4k$. Thus we have

$$\frac{1}{2}\log 4k = S \text{ or } k=\frac{1}{4}e^{2S}.$$

Thus *to obtain the deranged series with sum S, the ratio of number of positive terms followed by the number of negative terms should be* $\frac{1}{4}e^{2S}$.

EXAMPLE 1.84

Investigate, what derangement of the series

$$1-\frac{1}{2}+\frac{1}{3}-\frac{1}{4}+\frac{1}{5}-\frac{1}{6}+\cdots$$

will reduce its sum to zero.

Solution. It follows from Example (1.83) that $k=\frac{1}{4}e^{2S}$. Here $S=0$, so that $k=\frac{1}{4}e^{0}=\frac{1}{4}$. Thus, the deranged series is

$$1-\frac{1}{2}-\frac{1}{4}-\frac{1}{6}-\frac{1}{8}+\frac{1}{3}-\frac{1}{10}-\frac{1}{12}-\frac{1}{14}-\frac{1}{16}+\frac{1}{5}-\cdots$$

EXAMPLE 1.85

Show that

$$1+\frac{1}{3}+\frac{1}{5}-\frac{1}{2}-\frac{1}{4}+\frac{1}{7}+\frac{1}{9}+\frac{1}{11}-\frac{1}{6}-\frac{1}{8}+\cdots=\frac{1}{2}\log 6.$$

Solution. The given series is the derangement of the series

$$1-\frac{1}{2}+\frac{1}{3}-\frac{1}{4}+\frac{1}{5}-\frac{1}{6}+\cdots.$$

Here three positive terms are being followed by two negative terms. So $k=\frac{p}{n}=\frac{3}{2}$. Hence, the sum of the given series is $\frac{1}{2}\log 4k=\frac{1}{2}\log 6$.

EXAMPLE 1.86

Criticize the following paradox:

(i) $1-\frac{1}{2}+\frac{1}{3}+\frac{1}{4}+\frac{1}{5}-\cdots=(1+\frac{1}{2}+\frac{1}{3}+\frac{1}{4}+\cdots)$
$\qquad -2(\frac{1}{2}+\frac{1}{4}+\frac{1}{6}+\cdots)$

$=(1+\frac{1}{2}+\frac{1}{3}+\frac{1}{4}+\cdots)-(1+\frac{1}{2}+\frac{1}{3}+\cdots)=0.$

(ii) $2(\frac{1}{2}+\frac{1}{4}+\frac{1}{6}+\cdots)=1+\frac{1}{2}+\frac{1}{3}+\frac{1}{4}+\cdots$
$=(1+\frac{1}{3}+\frac{1}{5}+\cdots)+(\frac{1}{2}+\frac{1}{4}+\frac{1}{6}+\cdots).$

Therefore,

$$\frac{1}{2}+\frac{1}{4}+\frac{1}{6}+\cdots=1+\frac{1}{3}+\frac{1}{5}+\cdots$$

which is absurd since each term of the series on the left is less than the corresponding term in the series on the right.

Solution. (i) The series is semi-convergent (conditionally convergent) and so we cannot rearrange its term. The rearrangement of the term changes the sum of the series to zero, which is otherwise log 2. (ii) The series $2(\frac{1}{2}+\frac{1}{4}+\frac{1}{6}+\cdots)$ is a divergent series of positive terms. Therefore, we can derange it in any manner. Thus, the first two steps in the example are correct. However, the third step is wrong. In this step we have taken

$$2\left(\frac{1}{2}+\frac{1}{4}+\frac{1}{6}+\cdots\right)-\left(\frac{1}{2}+\frac{1}{4}+\frac{1}{6}+\cdots\right)$$
$$=\frac{1}{2}+\frac{1}{4}+\frac{1}{6}+\cdots$$

Both the series in the bracket are divergent and tend to ∞ so that we get the indeterminate form $\infty-\infty$. On account of this fallacy, we get the absurd result.

1.25 UNIFORM CONVERGENCE

We now consider sequences or infinite series whose terms are functions of a variable, say x, in some interval $[a,b]$. The convergence of such sequences (or series) in that interval will be called uniform convergence. We define uniform convergence of such sequence $\{s_n(x)\}$ as follows:

A sequence $\{s_n(x)\}$ is said to converge **uniformly to the limit** $s(x)$ in the interval $[a,b]$, if given $\varepsilon>0$, there exists a positive integer m, *independent of x* such that

$|s_n(x)-s(x)|<\varepsilon$ for all $n>m$ and all $x\in[a,b]$.

EXAMPLE 1.87

Show that the sequence $\{s_n(x)\}$, where

$$s_n(x)=\frac{nx}{1+n^2x^2},\ 0\le x\le a$$

is non-uniformly convergent in any interval which includes the origin.

Solution. Let $\varepsilon > 0$. Then $|s_n(x) - 0| < \varepsilon$ implies

$$\frac{nx}{1+n^2x^2} < \varepsilon$$

or

$$1+n^2x^2 - \frac{nx}{\varepsilon} > 0$$

or

$$nx > \frac{1}{2\varepsilon} + \frac{1}{2}\sqrt{\frac{1}{\varepsilon^2} - 4}$$

or

$$n > \frac{1}{x}\left[\frac{1}{2\varepsilon} + \frac{1}{2}\sqrt{\frac{1}{\varepsilon^2} - 4}\right].$$

Thus, we can find an upper bound for n in any interval $0 < a \le x \le b$, but the upper bound is infinite if the interval includes 0. Hence, the given sequence is non-uniformly convergent in any interval which includes the origin.

EXAMPLE 1.88

Show that 1 is the point of non-uniform convergence for the sequence $\{s_n(x)\}$, where $s_n(x) = x_n, x \in [0,1]$.

Solution. We note that

$$s(x) = \lim_{n\to\infty} S_n(x) = \begin{cases} 0 \text{ if } 0 \le x < 1 \\ 1 \text{ if } x = 1. \end{cases}$$

To see whether the convergence is uniform, let $\varepsilon > 0$. Then, $|s_n(x) - 0| < \varepsilon$ implies

$$x^n < \varepsilon$$

or

$$\left(\frac{1}{x}\right)^n > \frac{1}{\varepsilon}$$

or

$$n \log\frac{1}{x} > \log\frac{1}{\varepsilon} \text{ or } n > \frac{\log\frac{1}{\varepsilon}}{\log\frac{1}{x}}.$$

Thus, for uniform convergence we have to take the integer m next higher to $\frac{\log\frac{1}{\varepsilon}}{\log\frac{1}{x}}$. If we take 0 $o \le x \le a < 1$, then the upper bound of m will be $\frac{\log\frac{1}{\varepsilon}}{\log\frac{1}{a}}$. But this upper bound becomes infinite if $a = 1$. Hence for $a = 1$, m cannot be selected Therefore, the sequence converges uniformly in $[0, a]$, where $a < 1$ but it is not uniformly convergent if the interval includes 1. This means that 1 is the point of non-uniform convergence of the sequence $\{s_n(x)\}$.

The following theorem provides characterization for uniform convergence of a sequence $\{s_n(x)\}$.

Theorem 1.28. (Cauchy's General Principle of Uniform Convergence): The necessary and sufficient condition for the sequence $\{s_n(x)\}$ to converge uniformly in the interval $[a,b]$ is that, given $\varepsilon > 0$, there exists positive integer p independent of x, such that

$$|s_n(x) - s_m(x)| < \varepsilon \text{ for all } n > m \ge p \text{ and all } x \in [a,b].$$

The following theorem serves as a test for uniform convergence.

Theorem 1.29. (M_n-Test). Let $\{s_n(x)\}$ be a sequence on $[a,b]$ and let $M_n = \sup\{|s_n(x) - s(x)| : x \in [a,b]\}$. Then $\{s_n(x)\}$ converges uniformly to $s(x)$ if and only if $M_n \to 0$ as $n \to \infty$.

Proof: *Necessity*: Suppose that $\{s_n(x)\}$ converges uniformly to $s(x)$ in $[a,b]$. Then for a given $\varepsilon > 0$, there exists a positive integer m, independent of x, such that

$$|s_n(x) - s(x)| < \varepsilon \text{ for all } n \ge m \text{ and } x \in [a,b].$$

Since M_n is the supremum of $|s_n(x) - s(x)|$ over $[a,b]$, it follows that $M_n \le \varepsilon$ for all $n \ge m$. Hence $M_n \to 0$ as $n \to \infty$.

Sufficiency: Suppose that $M_n = \sup\{|s_n(x) - s(x)|\} \to 0$ as $n \to \infty$. Hence, $M_n < \varepsilon$ for all $n \ge m$ and all $x \in [a,b]$. This implies that

$$\sup\{|s_n(x) - s(x)|\} < \varepsilon \text{ for all } n \ge m \text{ and}$$
all $x \in [a,b]$. Hence

$$\{|s_n(x) - s(x)|\} < \varepsilon \text{ for all } n \ge m \text{ and all}$$
$x \in [a,b]$.

Hence, $\{s_n(x)\}$ converges uniformly to $s(x)$.

1.26 UNIFORM CONVERGENCE OF A SERIES OF FUNCTIONS

The series $\Sigma u_n(x)$ is said to converge uniformly on $[a,b]$ if the sequence $\{s_n(x)\}$ of its partial sums converges uniformly on $[a,b]$.

Theorem 1.30. (Weierstrass's M-Test). A series $\Sigma u_n(x)$ converges uniformly and absolutely on $[a,b]$ if $|u_n(x)| \le M_n$ for all n and all $x \varepsilon [a,b]$, where M_n is independent of x and ΣM_n is convergent.

Proof: Since $|u_n(x)| \le M_n$ and ΣM_n converges, the series $\Sigma |u_n(x)|$ converges by comparison test. To prove uniform convergence, we note that

$$\begin{aligned}|s_n(x) - s_m(x)| &= |u_{n+1}(x) + u_{n+2}(x) \\ &\quad + \cdots + u_m(x)| \\ &\le M_{n+1} + M_{n+2} + \cdots + M_m.\end{aligned}$$

Since ΣM_n converges, $M_{n+1} + M_{n+2} + \cdots + M_m < \varepsilon$ for $n > m \ge p$, where p is independent of x. Hence

$$|s_n(x) - s_m(x)| < \varepsilon \text{ for } n > m \ge p$$

and all $x \varepsilon [a,b]$.

Hence, by Cauchy's General Principle of Uniform Convergence, $\Sigma u_n(x)$ converges uniformly on $[a, b]$.

EXAMPLE 1.89

Show that the series

$$\frac{\cos x}{1^p} + \frac{\cos 2x}{2^p} + \cdots + \frac{\cos nx}{n^p} + \cdots$$

converges uniformly on the real line for $p > 1$.

Solution. We have

$$|u_n| = \left|\frac{\cos nx}{n^p}\right| \le \frac{1}{n^p} \text{ for all } x \in \Re.$$

But the series $\sum \frac{1}{n^p}$ converges for $p > 1$. Hence, by Weierstrass's M-Test, the given series $\Sigma u_n(x)$ converges uniformly on $\Re$ for $p > 1$.

EXAMPLE 1.90

Prove that the series $\sum \frac{\sin(x+nx)}{n(n+1)}$ converges uniformly for all real values of x.

Solution. The result follows taking $M_n = \frac{1}{n(n+1)}$ and applying Weierstrass's M-Test.

EXAMPLE 1.91

Test for uniform convergence of the series

$$1 + a\cos x + a^2 \cos 2x + \cdots + a^n \cos nx + \cdots$$

Solution. We have

$$\sum u_n(x) = \sum_{n=0}^{\infty} a^n \cos nx$$

and $|u_n(x)| = |a^n \cos x| \le |a^n| = a^n$ *if* $a > 0$.

Then, $\sum M_n = \sum a^n = 1 + a + a^2 + \cdots + a^n + \cdots = \frac{1}{1-a}$, $0 < a < 1$. Therefore, the series $\Sigma a_n(x)$ converges uniformly, by Weierstrass's M-Test, if $0 < a < 1$.

EXAMPLE 1.92

If Σa_n converges absolutely, show that the series $\sum \frac{a^n x^n}{1+x^{2n}}$ converges uniformly for all real x.

Solution. We have

$$u_n(x) = \frac{a^n x^n}{1+x^{2n}}.$$

Substituting the first differential of $u_n(x)$ to zero, we get

$$1 + x^{2n} - 2x^{2n} = 0, \text{ which yields } x^n = 1.$$

Also $\frac{d^2 u_n}{dx^2} < 0$ when, $x^n = 1$. Therefore,

$$M_n = \max |u_n(x)| = \max \left|\frac{a_n x^n}{1+x^{2n}}\right| = \frac{|a_n|}{2}$$

Thus, $|u_n(x)| \le M_n$ and $\sum M_n = \frac{1}{2}\sum |a_n| < \infty$ (by hypothesis). Hence, by Weierstrass's M-Test, the given series converges uniformly for all real values of x.

EXAMPLE 1.93

Examine the series

$$\sum_{n=1}^{\infty} u_n(x) = \sum_{n=1}^{\infty} \left[\frac{nx}{1+n^2x^2} - \frac{(n-1)x}{1+(n-1)^2x^2}\right]$$

for uniform convergence.

Solution. We have

$$u_1(x) = \frac{x}{1+x^2} - 0,$$

$$u_2(x) = \frac{2x}{1+2^2x^2} - \frac{x}{1+x^2},$$

$$u_n(x) = \frac{nx}{1+n^2x^2} - \frac{(n-1)x}{1+(n-1)^2x^2}.$$

Therefore, the partial sum $S_n(x)$ is given by

$$S_n(x) = u_1(x) + (x) + \ldots + u_n(x) = \frac{nx}{1+n^2x^2}.$$

The sequence $\left\{\frac{nx}{1+n^2x^2}\right\}$ has 0 as the point of non-uniform convergence (see Example 1.87).

1.27 PROPERTIES OF UNIFORMLY CONVERGENT SERIES

(A) We know that the sum of two continuous functions is also continuous and that this result can be extended to the sum of a finite number of functions. The question arises, "Can we extend this result for infinite number of functions?" The following theorem provides a sufficient condition for the sum function of an infinite series of continuous functions to be continuous.

Theorem 1.31. *If $u_1(x)$, $u_2(x)$,..., $u_n(x)$ are all continuous functions of $x \in [a,b]$ and if $S(x) = \Sigma u_n(x)$ is uniformly convergent in $[a,b]$, then $S(x)$ is continuous in $[a,b]$.*

Remark 1.16. The condition of uniform convergence of $\Sigma u_n(x)$ is sufficient but not necessary for the sum $S(x)$ to be continuous. For example, consider the series,

$$\sum\left[\frac{n^2x}{1+n^3x^2} - \frac{(n-1)^2x}{1+(n-1)^3x^2}\right].$$

For this series, the partial sum $s_n(x)$ is given by

$$s_n(x) = \frac{n^2x}{1+n^3x^2}.$$

We have

$$S(x) = \lim_{n\to\infty} s_n(x) = \lim_{n\to\infty} \frac{n^2x}{1+n^3x^2}\left(\frac{\infty}{\infty} form\right)$$

$$= \lim_{n\to\infty} \frac{x}{\frac{1}{n^2} + nx^2} = 0, x \in [0,1].$$

Hence, $S(x)$ is continuous at 0 on [0, 1] But it can be seen that 0 is a point of non-uniform convergence for the given series.

(B) We know that the sum of two integrable functions is integrable and that the result can be extended to the sum of a finite number of functions. The following theorem provides a sufficient condition to extend this result to an infinite number of functions.

Theorem 1.32. (Term-by-Term Integration):

If $u_1(x), u_2(x), \ldots, u_n(x)$ are all continuous (hence, integrable) functions of x in $[a,b]$ and if $S(x) = \Sigma u_n(x)$ is uniformly convergent in $[a,b]$, then

$$\int_a^b S(x)dx = \int_a^b [\sum_a^b u_n(x)]dx$$

$$= \int_a^b u_1(x)dx + \int_a^b u_2(x)dx$$

$$+ \cdots + \int_a^b u_n(x)dx + \cdots.$$

Remark 1.17. The condition of uniform convergence in the above theorem is sufficient but not necessary. For example, we have

$$1 - x + x^2 - x^3 + \cdots = \frac{1}{1+x}, 0 \le x < 1$$

The result

$$\int_0^1 \frac{dx}{1+x} = \int_0^1 dx - \int_0^1 xdx + \int_0^1 x^2dx - \cdots.$$

is correct because

$$\log 2 = 1 - \frac{1}{2} + \frac{1}{3} - \frac{1}{4} + \cdots.$$

Thus, for this series termby-term integration is valid but the series is not convergent at $x = 1$.

(C) We now extend the result

$$\frac{d}{dx}[f_1(x) + f_2(x) + \cdots f_n(x)]$$

$$= f_1'(x) + f_2'(x) + \cdots + f_n'(x)$$

for the sum of an infinite series

Theorem 1.33. (Term-by-Term Differentiation): *If (i) $\Sigma u_n(x)$ converges to the sum S(x) in [a,b], (ii) $\Sigma u_n'(x)$ converges uniformly to the sum F(x) and (iii) $u_n'(x)$ is a continuous function of x in [a, b] for all n, then $F(x) = S'(x)$.*

Remark 1.18. Again the condition of uniform convergence of $\Sigma u_n'(x)$ is sufficient but not necessary. For example, consider the series for which

$$S_n(x) = \sum_{k=1}^{n} u_k(x) = \frac{1}{2n}\log(1+n^2x^2).$$

Then

$$S(x) = \lim_{n\to\infty} S_n(x) = \lim_{n\to\infty} \frac{\log(1+n^2x^2)}{2n} \left(\frac{\infty}{\infty} \textit{ form}\right)$$

$$= \lim_{n\to\infty} \frac{nx^2}{1+n^2x^2} = 0 \text{ for } 0 \le x \le 1$$

and so $S'(x) = 0$. Also

$$\lim_{n\to\infty} S_n'(x) = \lim_{n\to\infty} \frac{nx}{1+n^2x^2} = 0 \text{ for } 0 \le x \le 1.$$

Therefore,

$$S'(x) = \lim_{n\to\infty} S_n'(x) = \sum_{n=1}^{\infty} u_n'(x).$$

Thus term-by-term differentiation is valid, but the series $\Sigma u_n'(x)$ has 0 as the point of non-uniform convergence.

1.28 POWER SERIES

A series of the form

$$a_0 + a_1x + a_2x^2 + \ldots + a_nx^n + \ldots = \sum_{n=0}^{\infty} a_nx^n \quad (1)$$

is called a *power series*. The numbers a_n are called coefficients of the power series.

If we apply Cauchy's root test for the convergence of the power series, we observe that the series is *absolutely convergent* if

$$|x| < \tfrac{1}{l}, \text{ where } l = \lim |a_n|^{\frac{1}{n}}$$

and divergent if $|x| < \frac{1}{l}$. Thus, if $r = \frac{1}{l}$, it follows that the power series converges absolutely if $| x | < r$ and diverges if $| x | > r$. If *a1, a2,...* are all real and *x* is real, then the interval $-r < x < r$, inside which the power series is convergent, is called the *interval of convergence* of the power series and *r* is called the *radius of convergence* of the power series. Further, if $\lim|a_n|^{\frac{1}{n}} = 0$, then $r = \infty$ and, thus, in that case, the series converges for all finite values of *x*. The function represented by the series in such a case is an *entire function.* For example, the series

$$1 + x + \frac{x^2}{2!} + \frac{x^3}{3!} + \frac{x^4}{4!} + \ldots = \sum \frac{x^n}{n!}$$

has infinite radius of convergence and as such the function e^x, representing this series is an entire function.

Differentiating the power series (1) term by term, we obtain the series

$$a_1 + 2a_2x + 3a_3x^2 + \ldots + na_nx^{n-1} + \ldots a \quad (2)$$

Then the *n*th term of the series is

$$u_n = na_nx^{n-1}.$$

We have

$$\lim |na_n|^{\frac{1}{n}} = \lim n^{\frac{1}{n}}.\lim|a_n|^{\frac{1}{n}} = \lim|a_n|^{\frac{1}{n}}$$

and so the radius of convergence of the differentiated series (2) is the same as that of the series (1). Thus, the differentiated series has same interval of convergence as the original series. In general,

"A real power series can be differentiated any number of times inside the interval of convergence".

Furthermore, a power series Σa_nx^n with radius of convergence *r* converges uniformly on $[-r+\varepsilon, r+\varepsilon]$, $\varepsilon > 0$. Thus, if *f* is the sum of the series Σa_nx^n, then

$$\int_a^x f(x)\,dx = \sum_{n=1}^{\infty} \int_a^x a_nx^n\,dx, \ |x| < r.$$

Hence,

"A real power series can be integrated term by term any number of times inside its interval of convergence".

EXAMPLE 1.94

Examine for convergence the series $\sum_{n=0}^{\infty} \frac{x^n}{n^n}$ for convergence.

Solution. The given series is a power series with $a_n = \frac{1}{n^n}$.

Therefore, its radius of convergence is

$$r = \frac{1}{\lim |a_n|^{\frac{1}{n}}} = \frac{1}{\lim \left|\frac{1}{n}\right|^{\frac{1}{n}}} = \infty.$$

Hence, the series converges absolutely for all values of x.

EXAMPLE 1.95

Examine for convergence the series

$$x - \frac{x^2}{2} + \frac{x^3}{3} - \frac{x^4}{4} + \ldots$$

Solution. The radius of convergence for this series is

$$r = \frac{1}{\lim \left|\frac{1}{n}\right|^{\frac{1}{n}}} = 1.$$

Therefore, it converges absolutely on the interval $-1 < x < 1$. It converges at $x = 1$ by Leibnitz test but diverges at $x = -1$.

EXERCISES

I Sequences

1. Show that the sequence $\{r^n\}$ converges if $-1 < r \leq 1$.
2. Show that $\lim\limits_{n\to\infty} \frac{3+2\sqrt{n}}{\sqrt{n}} = 2$.
3. Give an example of a monotonically increasing sequence that is (i) convergent (ii) not convergent

   Ans. (i) $\{1-\frac{1}{n}\}$, (ii) $\{n\}$.
4. If $a_n = (-1)^n \left(1+\frac{1}{n}\right)$ for all natural number n, show that $\overline{\lim} a_n = 1$ and $\underline{\lim} a_n = -1$

   Hint: The sequence is $\{-2, \frac{3}{2}, \frac{-4}{3}, \frac{5}{4}, \frac{-6}{5}, \ldots\}$. The set inferior numbers and superior numbers are $\{-2, -\frac{4}{3}, -\frac{6}{5}, \ldots\}$ and $\{\frac{3}{2}, \frac{5}{4}, \ldots\}$. Therefore,

   $$\underline{\lim}\, a_n = \text{lub}\left\{-2, -\frac{4}{3}, -\frac{6}{5}\right\} = -1,$$

   $$\overline{\lim}\, a_n = \text{glb}\left\{-\frac{3}{2}, -\frac{5}{4}, \ldots\right\} = 1,$$
5. Show that the sequence $\{a_n\}$, where

   $$a_n = 1 + \frac{1}{4} + \frac{1}{7} + \cdots + \frac{1}{3n-2}$$

   cannot converge
6. Using Cauchy's General Principal of Convergence, show that the sequence $\left\{\frac{n}{n+1}\right\}$ is convergent
7. Show that $\lim\limits_{n\to\infty} [\sqrt{n+1} - \sqrt{n}] = 0$.

II Comparison Test

Examine the convergence of the following series:

8. $\sum\limits_{n=1}^{\infty} (n^2+1)^{\frac{1}{2}} - n$ **Ans.** Divergent.
9. $\sum\limits_{n=2}^{\infty} \frac{1}{\log n}$ **Ans.** Divergent.
10. $\sum\limits_{n=1}^{\infty} \frac{1}{n^p (n+1)^p}$ **Ans.** Converges for $p > \frac{1}{2}$ and diverges for $p \leq \frac{1}{2}$
11. $\sum\limits_{n=1}^{\infty} (n+1)^{\frac{1}{3}} - n^{\frac{1}{3}}$ **Ans.** Divergent.
12. $\frac{1}{1^2} + \frac{1+2}{1^2+2^2} + \frac{1+2+3}{1^2+2^2+3^2} + \ldots$

    Ans. Divergent.
13. $\frac{\sqrt{2}-1}{3^2-1} + \frac{\sqrt{3}-1}{4^3-1} + \frac{\sqrt{4}-1}{5^3-1} + \ldots$

    Ans. Convergent.
14. $\sqrt{\frac{1}{4}} - + \sqrt{\frac{2}{6}} + \frac{3}{8} + \cdots + \sqrt{\frac{n}{2(n+1)}} + \ldots$

    Ans. Divergent.
15. $\frac{1}{1+\sqrt{2}} + \frac{2}{1+2\sqrt{3}} + \frac{3}{1+3\sqrt{4}} + \ldots$

    Ans. Divergent.

16. $\frac{1}{12}+\frac{1}{2.3}+\frac{1}{34}+\ldots$ **Ans.** Convergent.

17. $\sum\frac{(n+1)(n+2)}{(n^2+1)(n^2+2)}$ **Ans.** Convergent.

18. $\sum\sin\frac{1}{n}$

Hint: take $u_n = \sin\frac{1}{n}, v_n = \frac{1}{n}$, then $\lim\frac{u_n}{v_n} = 1$, but $\sum\frac{1}{n}$ diverges.

Ans. Divergent.

III D'Alemberl's Ratio Test

Examine the following series for convergence

19. $1+\frac{2!}{2^2}+\frac{3!}{3^3}+\frac{4!}{4^4}+\ldots$ **Ans.** Convergent.

20. $\sum\frac{(3x+5)^n}{(n+1)!}$

Ans. Convergent for all values of x.

21. $\sum\left(\frac{n^2}{2^n}+\frac{1}{n^2}\right)$ **Ans.** Convergent.

22. $\sum_{n=1}^{\infty}\frac{nx^n}{(n+1)(n+2)}$

Ans. Convergent for $x<1$, divergent for $x\leq 1$.

23. $\sum\frac{n^3-n+1}{n!}$ **Ans.** Convergent.

24. $\frac{x}{2\sqrt{3}}+\frac{x^2}{3\sqrt{4}}+\frac{x^3}{4\sqrt{5}}+\cdots$

Ans. Convergent for $x\geq 1$ and divergent for $x>1$.

25. $\sum\frac{369\cdots 3n}{4.7.10\cdots(3n+1)}\cdot\frac{5n}{3n+2}$

Ans. Convergent.

26. $\sum\frac{n}{1+2^n}$ **Ans.** Convergent.

27. $\sum\frac{2^{n+1}-2}{2^{n+1}+1}x^n$

Ans. Converges for $x<1$ and diverges if $x\geq 1$.

28. $\frac{X}{13}+\frac{x^2}{35}+\frac{x^3}{57}+\cdots$

Ans. Converges for $x\leq 1$ and diverges if $x>1$.

29. $\sum\frac{n^2}{3^n}$ **Ans.** Convergent.

IV Cauchy's Root Test

Examine the following series for convergence

30. $\sum\frac{(n+1)^n}{n^{n+1}}x^n$

Ans. Converges for $x<1$ and diverges if $x\geq 1$.

31. $\sum_{2}^{\infty}\frac{1}{(\log n)^n}$ **Ans.** Convergent.

32. $\sum 5^{-n-(-1)^n}$ **Ans.** Convergent.

33. $\frac{2}{1^2}x+\frac{3^2}{2^3}x^2+\frac{4^3}{3}x^3+\cdots+\frac{(n+1)^n}{n^{n+1}}+\cdots$

Ans. Converges for $x<1$ and diverges if $x\geq 1$.

34. $\sum\left[\left(\frac{n+1}{n}\right)^{n+1}-\frac{n+1}{n}\right]^{-n}$

Ans. Convergent.

35. $\sum\frac{(1+nx)^n}{n^n}$

Ans. converges for $x<1$, diverges for $x\geq 1$.

V Raabe's Test, Logarithmic Test, deMorgon-Bertrand Test, and Gauss Test

Examine for the convergence of the series

36. $\sum \frac{4.7.10\ldots(3n+1)}{1.2.3\ldots n}x^n$

Ans. Converges for $x<\frac{1}{3}$, diverges for $x \geq \frac{1}{3}$.

37. $\sum \frac{1.3.5\ldots(2n-1)}{2.4.6\ldots 2n}$

Ans. Divergent.

38. $\sum \frac{3.6.9\ldots(3n)}{7.10.13\ldots(3n+4)}x^n$

Ans. Converges for $x \leq 1$, diverges for $x > 1$.

39. $\sum \frac{1^2 \cdot 3^2 \ldots (2.n-1)^2}{2^2 \cdot 4^2 \cdot 6^2 \ldots (2n)^2}x^n$

Ans. Converges for $x < 1$, diverges for $x \geq 1$.

Hint: Use Ratio Test and then Gauss's Test as Raabe's Test fails

40. $\sum \frac{(a+nx)^n}{n!}$

Ans. Converges for $x<\frac{1}{e}$,

diverges for $x \geq \frac{1}{e}$.

41. $\sum \frac{x^{n+1}}{(n+1)\log(n+1)}$

Ans. Converges for $x \leq 1$, diverges for $x > 1$.

42. $\sum_{n=1}^{\infty} \frac{(2n)!}{(n!)^2}x^n$

Ans. Converges for $x<\frac{1}{4}$, diverges for $x \geq \frac{1}{4}$.

43. $\frac{2^2}{3^2}+\frac{2^2.4^2}{3^2.5^2}+\frac{2^2.4^2.6^2}{3^2.5^2.7^2}+\cdots$

Hint: $u_n = \frac{2^2.4^2\ldots(2n)^2}{3^2.5^2.7^2\ldots(2n+1)^2}$,

$$\frac{u_n}{u_{n+1}} = \left(1+\frac{3}{2n}\right)^2\left(1+\frac{1}{n}\right)^{-2}$$

$$= \left(1+\frac{3}{n}+\frac{9}{4n^2}\right)\left(1-\frac{2}{n}+\frac{3}{n^2}-\ldots\right)$$

$$= 1+\frac{1}{n}-\frac{3}{4n^2}+\cdots$$

By Gauss's Test, the series diverges.

44. $\frac{x}{1}+\frac{1}{2}.\frac{x^3}{3}+\frac{1.3}{2.4}.\frac{x^5}{5}+\frac{1.3.5}{2.4.6}.\frac{x^7}{7}+\cdots$

Ans. Converges for $x \leq 1$, and diverges for $x > 1$.

45. $\frac{x}{1.2}+\frac{x^2}{3.4}+\frac{x^3}{5.6}+\frac{x^n}{7.8}+\cdots$

Ans. Converges for $x \leq 1$, diverges for $x > 1$.

46. $1+\frac{(1!)}{2!}x^2+\frac{(2!)^2}{4!}x^4+\frac{(3!)^2}{6^1}x^6+\cdots$

Ans. Converges for $x^2 < 4$, diverges for $x^2 \geq 4$.

VI Cauchy's Integral Test, Cauchy's Condensation Test

47. Apply Cauchy's Integral Test to examine the convergence of the series.

(a) $\sum_{n=1}^{\infty} \frac{1}{n^2+1}$ **Ans.** Convergent.

(b) $\sum_{n=1}^{\infty} \frac{1}{n^2+n}$ **Ans.** Convergent.

48. Prove that $\sum \frac{1}{(n \log n \log^2 n \cdots (\log^k n)^p}$, where $\log^2 n$ denotes log log n, etc, converges if $p > 1$, diverges if $p \leq 1$.

VII Alternating Series

49. Examine the convergence of

(a) $\frac{x}{1+x}-\frac{x^2}{1+x^2}+\frac{x^3}{1+x^3}-\frac{x^4}{1+x^4}+\cdots$

Ans. Convergent.

Hint: Terms decrease monotonically, $u^n = \frac{x^n}{1+x^n} = \frac{1}{1+\frac{1}{x^n}} \to 0$ as $n \to \infty$.

Hence, by Leibnitz's Test, the given alternating series converges.

(b) $\sum(-1)^{n-1}\dfrac{n}{n^2+1}$ **Ans.** Convergent.

(c) $\left(\dfrac{1}{2}-\dfrac{1}{\log 2}\right)-\left(\dfrac{1}{2}-\dfrac{1}{\log 3}\right)+\left(\dfrac{1}{2}-\dfrac{1}{\log 4}\right)-\left(\dfrac{1}{2}-\dfrac{1}{\log 5}\right)+\cdots$

Hint: $u_n \to \frac{1}{2}$ (finite) and so the series oscillates finitely.

(d) $1-\dfrac{1}{2\sqrt{2}}+\dfrac{1}{3\sqrt{3}}-\dfrac{1}{4\sqrt{4}}+\cdots$

Ans. Convergent.

VIII Absolute Convergence / Conditional Convergence

50. Show that the series

(a) $\dfrac{1}{2}-\dfrac{1}{2}\cdot\dfrac{1}{2^2}+\dfrac{1}{3}\cdot\dfrac{1}{2^3}-\dfrac{1}{4}\cdot\dfrac{1}{2^4}+\cdots$

(b) $\sum\left(-\dfrac{1}{2}\right)^n\cdot\dfrac{3.6\ldots(3n)}{2.5\ldots(3n-1)}$

(c) $(-1)^{n+1}\frac{n^3}{2^n}$ and

(d) $\sum\dfrac{x^n}{n+1}, 0<x<1$

are absolutely convergent.

51. Show that the series

(a) $\sum\dfrac{(-1)^n}{\sqrt{n}}$

(b) $\sum\dfrac{(-1)^n}{n-\log n}$ and

(c) $\sum\dfrac{(-1)^{n+1}}{\log(n+1)}$

are conditionally convergent.

52. Show that the series

$\dfrac{\cos x}{1^3}-\dfrac{\cos 2x}{2^3}+\dfrac{\cos 3x}{3^3}-\cdots$

converges absolutely.

Hint: $\left|\dfrac{\cos nx}{n^3}\right|\le\dfrac{1}{n^3}$ and $\sum\dfrac{1}{n^3}$ converges.

IX Derangement of Series

53. Show that

$1-\dfrac{1}{2}-\dfrac{1}{4}+\dfrac{1}{3}-\dfrac{1}{6}-\dfrac{1}{8}+\dfrac{1}{5}-\cdots=\dfrac{1}{2}\log 2$

54. Find the sum of the series

$1-\dfrac{1}{2}-\dfrac{1}{4}+\dfrac{1}{6}-\dfrac{1}{3}-\dfrac{1}{8}+\dfrac{1}{10}+\dfrac{1}{5}-\cdots$

Ans. $\frac{1}{2}\log\frac{4}{3}$.

X Uniform Convergence

55. Show that the sequence $\{s_n(x)\}=\left\{\frac{1}{x+n}\right\}$ converges uniformly in $[a, b]$, $a \ge 0$.

56. The partial sum $s_n(x)$ of a series $\Sigma u_n(x)$ is $s_n(x)=\frac{n^2x}{1+n^4x^2}$. Show that its convergence is non-uniform on $[0,1]$.

57. Examine for uniform convergence the series $\sum_{n=0}^{\infty} x.e^{-nx}$ in the interval $[0,1]$

Ans. 0 is point of non-uniform convergence.

58. Show that the series

$\dfrac{2x}{1+x^2}+\dfrac{4x^3}{1+x^4}+\dfrac{8x^7}{1+x^6}+\cdots$

is uniformly convergent in $-1<x<1$.

59. Test for uniform convergence of the series

$\sum\dfrac{x}{(n+x^2)^2}$

Ans. Converges uniformly for all values of x.

60. Show that the sequence $\{s_n(x)\}=\left\{\frac{x}{1+,nx^2}\right\}$ converges uniformly on the real line.

61. Apply Weierstrass's M-Test to show that the series $\sum_{n-1}^{\infty}\frac{1}{1+n^2x}$ converges uniformly on $[1,\infty]$.

Hint: $|u_n(x)|\le\frac{1}{1+n^2x}<\frac{1}{n^2}, x\in[1,\infty]$.

2 Successive Differentiation, Mean Value Theorems and Expansion of Functions

2.1 SUCCESSIVE DIFFERENTIATION

Let f be a real-valued function defined on an interval $[a, b]$. Then, it is said to be derivable at an interior point c if $\lim_{x\to c}\frac{f(x)-f(c)}{x-c}$ exists. This limit, if exists, is called the *derivative* or the *differential coefficient* of the function at $x = c$ and is denoted by $f'(c)$.

The above limit exists, if both the following limits exist and are equal:

(i) $\lim_{x\to c-0}\frac{f(x)-f(c)}{x-c}$, called the *left-hand derivative* and denoted by $f'(c-0)$,

(ii) $\lim_{x\to c+0}\frac{f(x)-f(c)}{x-c}$, called the *right-hand derivative* and denoted by $f'(c+0)$.

The derivative $f'(c)$ exists when $f'(c-0)=f'(c+0)$. If $y=f(x)$ be a derivable (differentiable) function of x, then $f'(x)=\frac{dy}{dx}$ is called the *first differential coefficient* of y with respect to x. If $f'(x)$ is again derivable, then $\frac{d}{dx}\left(\frac{dy}{dx}\right)$ is called the *second differential coefficient* of y with respect to x and denoted by $f''(x)$ *or* $\frac{d^2y}{dx^2}$. If $f''(x)$ is further derivable, then $\frac{d}{dx}\left(\frac{d^2y}{dx^2}\right)$, denoted by $\frac{d^3y}{dx^3}$ is called the *third differential coefficient* of y with respect to x. In general, the nth differential coefficient of y with respect to x is $\frac{d^ny}{dx^n}=\frac{d}{dx}\left(\frac{d^{n-1}y}{dx^{n-1}}\right)$. The differential operator $\frac{d}{dx}$ is also denoted by D. As such, the nth derivative of y is denoted by $D^n y$. For the sake of convenience, the derivatives $Dy, D^2y,\dots,D^ny$ are generally represented by $y_1, y_2, \dots, y_n$.

EXAMPLE 2.1

Find the nth derivative of

(i) $y=e^{ax+b}$, (ii) $y=(ax+b)^m$

(iii) $y=\frac{1}{ax+b}$, (iv) $y=a^x$

(v) $y=\sin(ax+b)$, and $y=\cos(ax+b)$

(vi) $y=e^{ax}\sin(bx+c)$ and $y=e^{ax}\cos(bx+c)$

(vii) $y = \log(ax + b)$.

Solution.

(i) We are given that $y=e^{ax+b}$. Therefore

$$y_1=\frac{dy}{dx}=a\,e^{ax+b},$$
$$y_2=\frac{d^2y}{dx^2}=a^2e^{ax+b},$$
$$y_3=a^3e^{ax+b}$$

and so on. Therefore the nth derivative $D^n y$ is given by $y_n=a^ne^{ax+b}$.

(ii) We have $y=(ax+b)^m$. Therefore

$$y_1=ma(ax+b)^{m-1},$$
$$y_2=m(m-1)a^2(ax+b)^{m-2},$$
$$y_3=m(m-1)(m-2)a^3(ax+b)^{m-3},$$

and so on. Hence, in general,

$$y_n=m(m-1)(m-2)\dots(m-n+1)a^n(ax+b)^{m-n}$$

Further, if m is a positive integer, then

$$y_n=m(m-1)(m-2)\dots(m-n+1)$$
$$x\frac{(m-n)(m-n-1)\dots 2.1}{(m-n)(m-n-1)\dots 2.1}.a^n(ax+b)^{m-n}$$

$$=\frac{m!}{(m-n)!}a^n(ax+b)^{m-n} \qquad (1)$$

From (1), it follows that the mth derivative of the given function is

$$y_m = \frac{m!}{o!} a^m (ax+b)^o = m\,!\,a^m.$$

In particular, taking $a = 1, b = 0$, we get $y_m = m!$.

Thus, if m is a positive integer, the mth differential coefficient of $(ax+b)^m$ is constant.

In case m is negative, then $m = -p$, where p is a positive integer and so

$$y_n = (-p)(-p-1)(-p-2)\ldots(-p-n+1)a^n \times (ax+b)^{-p-n}$$

$$= (-1)^n p(p+1)(p+2)\ldots(p+n-1)a^n \times (ax+b)^{-p-n}$$

$$= (-1)^n \frac{(p+n-1)!}{(p-1)!} a^n (ax+b)^{-p-n}.$$

(iii) We have $y = (ax+b)^{-1}$ and so

$$y_1 = (-1)a(ax+b)^{-2},$$

$$y_2 = (-1)(-2)a^2(ax+b)^{-3},$$

$$y_3 = (-1)(-2)(-3)a^3(ax+b)^{-4}$$

and so on. Hence, in general,

$$y_n = (-1)(-2)(-3)\ldots(-n)a^n(ax+b)^{-(n+1)}$$

$$= (-1)^n n\,!\; a^n (ax+b)^{-(n+1)}.$$

(iv) If is given that $y = a^x$ Therefore

$$y_1 = a^x \log a,$$

$$y_2 = a^x (\log a)^2,$$

$$y_3 = a^x (\log a)^2,$$

and so on. Hence, in general,

$$y_n = a^x (\log a)^n.$$

(v) We have $y = \sin(ax+b)$. Therefore

$$y_1 = a\cos(ax+b) = a\sin\left(ax+b+\tfrac{\pi}{2}\right),$$

$$y_2 = a^2\cos\left(ax+b+\tfrac{\pi}{2}\right)) = a^2\sin(ax+b+\pi),$$

$$y_3 = a^3\cos(ax+b+\pi) = a^3\sin\left(ax+b+\tfrac{3\pi}{2}\right),$$

and so on. Hence, in general,

$$y_n = a^n \sin\left(ax+b+\tfrac{n\pi}{2}\right).$$

In a similar fashion, we can show that if $y = \cos(ax+b)$, then

$$y_n = a^n \cos\left(ax+b+\tfrac{n\pi}{2}\right).$$

(vi) The given function is $y = e^{ax}\sin(bx+c)$. Therefore

$$y_1 = ae^{ax}\sin(bx+c) + b\,e^{ax}\cos(bx+c)$$

$$= e^{ax}[a\sin(bx+c) + b\cos(bx+c)]. \quad (2)$$

Let us choose $a = r\cos\phi$ and $b = r\sin\phi$. Then $r = \sqrt{a^2+b^2}$ and $\phi = \tan^{-1}\left(\tfrac{b}{a}\right)$, and (2) reduces to

$$y_1 = re^{ax}\sin(bx+c+\phi).$$

Similarly, repeating the above argument, we have

$$y_2 = r^2e^{ax}\sin(bx+c+2\phi),$$

$$y_3 = r^3e^{ax}\sin(bx+c+3\phi),$$

and so on. Hence, in general,

$$y_n = r^n e^{ax}\sin(bx+c+n\phi),$$

where $r = \sqrt{a^2+b^2}$ and $\phi = \tan^{-1}\left(\tfrac{b}{a}\right)$.

In a similar fashion, we can show that if $y = e^{ax}\cos(bx+c)$, then

$$y_n = r^n e^{ax}\cos(bx+c+n\phi).$$

(vii) When $y = \log(ax+b)$, then

$$y_1 = \frac{a}{ax+b},$$

$$y_2 = \frac{dy_1}{dx} = (-1)a^2(ax+b)^{-2},$$

$$y_3 = (-1)(-2)a^2(ax+b)^{-3},$$

and so on. Therefore, in general,

$$y_n = (-1)(-2)\ldots(-(n-1))a^n(ax+b)^{-n}$$

$$= \frac{(-1)^{n-1}(n-1)\,!\,a^n}{(ax+b)^n}.$$

EXAMPLE 2.2

Find the nth derivative of

(i) $y = e^{2x}\sin^3 x$, (ii) $\cos^2 x\sin^3 x$

(ii) $y = \cos x\cos 2x\cos 3x$,

(iii) $y = e^{ax}\sin bx\cos cx$

Solution. (i) We have

$$y = e^{2x}\sin^3 x = \tfrac{1}{4}e^{2x}[3\sin x - \sin 3x].$$

$$= \tfrac{3}{4}e^{2x}\sin x - \tfrac{1}{4}e^{2x}\sin 3x.$$

Therefore, using Example 2.1 (vi), we have

$$y_n = \tfrac{3}{4}\left[\sqrt{2^2+1^2}\right]^n e^{2x} \sin\left[x+n\tan^{-1}\tfrac{1}{2}\right]$$
$$-\tfrac{1}{4}\left[\sqrt{2^2+3^2}\right]^n e^{2x} \sin\left[2x+n\tan^{-1}\tfrac{3}{2}\right].$$

(ii) We have $y=\cos^2 x \sin^3 x$

$$=\tfrac{1}{2}(1+\cos 2x)\left[\tfrac{1}{4}(3\sin x-\sin 3x)\right]$$
$$=\tfrac{1}{8}\left[3\sin x-\sin 3x+\tfrac{3}{2}(2\sin x\cos 2x)\right.$$
$$\left.-\tfrac{1}{2}(2\sin 3x\cos 2x)\right]$$
$$=\frac{1}{8}\left[3\sin x-\sin 3x+\frac{3}{2}(\sin 3x\right.$$
$$\left.-\tfrac{1}{2}(2\sin 3x\cos 2x)\right]$$
$$=\tfrac{1}{16}[2\sin x+\sin 3x-\sin 5x].$$

Therefore, using Example 2.1 (v), we have

$$y_n=\frac{1}{16}\left[2\sin\left(x\times\frac{n\pi}{2}\right)+3^n\sin\left(3x+\frac{n\pi}{2}\right)\right.$$
$$\left.-5^n\sin\left(5x+\tfrac{n\pi}{2}\right)\right].$$

(iii) We are given that $y=\cos x\cos 2x\cos 3x$.

$$=\cos x\left[\tfrac{1}{2}(\cos 5x+\cos x)\right]$$
$$=\tfrac{1}{4}\left(2\cos x\cos 5x+2\cos^2 x\right)$$
$$=\tfrac{1}{4}[(\cos 6x+\cos 4x)+(1+\cos 2x)].$$

Therefore, using Example 2.1 (v), we have

$$y_n=\tfrac{1}{4}D^n(1)+\tfrac{1}{4}D^n[\cos 2x+\cos 4x+\cos 6x]$$
$$=0+\frac{1}{4}\left[2^n\cos\left(2x\times\frac{n\pi}{2}\right)+4^n\cos\left(4x+\frac{n\pi}{2}\right)\right.$$
$$\left.+6^n\cos\left(6+\tfrac{n\pi}{2}\right)\right].$$
$$=\tfrac{1}{4}\left[2^n\cos\left(2x+\tfrac{n\pi}{2}\right)+4^n\cos\left(4x+\tfrac{n\pi}{2}\right)\right.$$
$$\left.+6^n\cos\left(6x+\tfrac{n\pi}{2}\right)\right].$$

(iv) We have $y=e^{2x}\sin bx\cos cx$

$$=e^{ax}\left[\tfrac{1}{2}(2\sin bx\cos cx)\right]$$
$$=\tfrac{1}{2}e^{ax}[\sin(bx+cx)+\sin(bx-cx)]$$
$$=\tfrac{1}{2}e^{ax}\sin(b+c)x+\tfrac{1}{2}e^{ax}\sin(b-c)x.$$

Therefore, using Example 2.1 (vi), we have

$$y_n=\tfrac{1}{2}\left(\sqrt{a^2+(b+c)^2}\right)^n e^{ax}$$
$$\times\sin\left\{(b+c)x+n\tan^{-1}\tfrac{b+c}{a}\right\}$$
$$+\tfrac{1}{2}\left(\sqrt{a^2+(b-c)^2}\right)^n e^{ax}$$
$$\times\sin\left\{(b-c)x+n\tan^{-1}\tfrac{b-c}{a}\right\}.$$

EXAMPLE 2.3

Find second derivatives of the following functions with respect to x:

(i) $x=a(t+\sin t),\ y=a(1+\cos t)$

(ii) $x=a\cos^3\theta,\ y=b\sin^3\theta.$

Solution.

(i) For the given equations, we have

$$\tfrac{dx}{dt}=a(1+\cos t)=2a\cos^2\tfrac{t}{2},$$
$$\tfrac{dy}{dt}=-a\sin t=-2a\sin\tfrac{t}{2}\cos\tfrac{t}{2}.$$

Therefore

$$\frac{dy}{dx}=\frac{-2a\sin\frac{t}{2}\cos\frac{t}{2}}{2a\cos^2\frac{t}{2}}=-\tan\tfrac{t}{2}$$

and so

$$\frac{d^2y}{dx^2}=-\frac{1}{2}\sec^2\frac{t}{2}\frac{dt}{dx}$$
$$=-\frac{1}{2}\sec^2\frac{t}{2}\left[\frac{1}{2a\cos^2\frac{t}{2}}\right]=-\frac{1}{4a}\sec^4\frac{t}{2}.$$

(iii) We are given that $x=a\cos^3\theta,\ y=b\sin^3\theta.$

Therefore

$$\frac{dx}{d\theta}=-3a\sin\theta\cos^2\theta,\ \frac{dy}{d\theta}=3b\sin^2\theta\cos\theta,$$

and so

$$\frac{dy}{dx}=\frac{\frac{dy}{d\theta}}{\frac{dx}{d\theta}}=-\frac{b}{a}\tan\theta.$$

Differentiating once more with respect to x, we get

$$\frac{d^2y}{dx^2}=-\frac{b}{a}\sec^2\theta\cdot\frac{d\theta}{dx}$$
$$=-\frac{b}{a}\sec^2\theta\left[\frac{1}{-3a\sin\theta\cos^2\theta}\right]$$
$$=\frac{b}{3a^2}\operatorname{cosec}\theta\sec^4\theta.$$

EXAMPLE 2.4

Find the nth derivative of

(i) $\dfrac{x^2-4x+1}{x^3+2x^2-x-2}$ (ii) $\dfrac{1}{x^2-6x+8}$.

Solution. We have

$$y=\frac{x^2-4x+1}{x^3+2x^2-x-2}=\frac{x^2-4x+1}{(x-1)(x+1)(x+2)}$$

$$=\frac{13}{3(x+2)}-\frac{3}{x+1}-\frac{1}{3(x-1)} \text{ (partial fractions).}$$

Since $D^n\{(ax+b)^{-1}\}=(-1)^n n!\,a^n(ax+b)^{-n-1}$,

we have

$$y_n=(-1)^n n!\left[\frac{13}{3(x+2)^{n+1}}-\frac{3}{(x+1)^{n+1}}-\frac{1}{3(x-1)^{n+1}}\right].$$

$$-\frac{1}{3(x-1)^{n+1}}\Bigg].$$

(ii) We have $y=\dfrac{1}{x^2-6x+8}=\dfrac{1}{(x-4)(x-2)}$

$$=\frac{1}{2(x-4)}-\frac{1}{2(x-2)} \text{ (partial fractions).}$$

Therefore, using

$$D^n\{(ax+b)^{-1}\}=(-1)^n n!\,a^n(ax+b)^{-(n+1)},$$

we have

$$y_n=\frac{(-1)^n n!}{2}\left[\frac{1}{(x-4)^{n+1}}-\frac{1}{(x-2)^{n+1}}\right].$$

EXAMPLE 2.5

If $x=\sin t$, $y=\sin pt$, show that

$$(1-x^2)\frac{d^2y}{dx^2}-x\frac{dy}{dx}+p^2y=0.$$

Solution. We have

$x=\sin t$ and $y=\sin pt$

Therefore

$$\frac{dx}{dt}=\cos t,\quad \frac{dy}{dt}=p\cos pt,$$

and so

$$\frac{dy}{dx}=\frac{\frac{dy}{dt}}{\frac{dx}{dt}}=p\frac{\cos pt}{\cos t}.$$

Differentiating once more with respect to x, we get

$$\frac{d^2y}{dx}=$$

$$=p\left[\frac{\cos t(-p\sin pt)-\cos pt(-\sin t)}{\cos^2 t}\right]\cdot\frac{dt}{dx}$$

$$=\frac{p}{\cos^2 t}[\sin t\cos pt-p\sin pt\cos t]\cdot\frac{1}{\cos t}.$$

Therefore

$$\left(1-x^2\right)\frac{d^2y}{dx^2}-x\frac{dy}{dx}+p^2y$$

$$=\left(1-\sin^2 t\right)\frac{d^2y}{dx^2}-x\frac{p\cos pt}{\cos t}+p^2\sin pt$$

$$=p[\sin t\cos pt-p\sin pt\cos t]\cdot\frac{1}{\cos t}$$

$$-\frac{p\sin t\cos pt}{\cos t}+p^2\sin pt$$

$$=\frac{1}{\cos t}\Big[p\sin t\cos pt-p^2\sin pt\cos t$$

$$-p\sin t\cos pt-p^2\sin pt\cos t\Big]=0.$$

EXAMPLE 2.6

If $y=a\cos(\log x)+b\sin(\log x)$, show that

$$x^2\frac{d^2y}{dx^2}+x\frac{dy}{dx}+y=0.$$

Solution. We have

$$y=a\cos(\log x)+b\sin(\log x).$$

Therefore

$$y_1=\frac{dy}{dx}=-a\sin(\log x)\frac{1}{x}+b\cos(\log x).\frac{1}{x}$$

or

$$xy_1=-a\sin(\log x)+b\cos(\log x).$$

Differentiating again with respect to x, we get

$$xy_2+y_1=-a\cos(\log x)\frac{1}{x}-b\sin(\log x).\frac{1}{x}$$

$$=\frac{1}{x}[a\cos(\log x)+b\sin(\log x)]=-\frac{y}{x}.$$

Hence

$$x^2y_2 + xy_1 + y = 0.$$

EXAMPLE 2.7

If $ax^2 + 2hxy + by^2 = 1$, show that

$$\frac{d^2y}{dx^2} = \frac{h^2 - ab}{(hx+by)^3}.$$

Solution: We have

$$ax^2 + 2hxy + by^2 = 1. \quad (1)$$

Differentiating with respect to x, we get

$$2ax + 2h\left[x\frac{dy}{dx} + y\right] + 2by\frac{dy}{dx} = 0$$

or

$$\frac{dy}{dx} = -\frac{ax+by}{hx+by}.$$

Differentiating again with respect to x, we have

$$\frac{d^2y}{dx^2} = \frac{\left(h^2-ab\right)\left(ax^2+2hxy+by^2\right)}{\left(hx+by\right)^3}$$

$$= \frac{h^2-ab}{\left(hx+by\right)^3}, \text{ using (1).}$$

EXAMPLE 2.8

Find the nth derivative of $\tan^{-1}\left(\frac{x}{a}\right)$.

Solution. We have $y = \tan^{-1}\left(\frac{x}{a}\right)$.

Therefore

$$y_1 = \frac{a}{x^2+a^2} = \frac{a}{(x+ia)(x-ia)}$$

$$= \frac{1}{2i}\left[\frac{1}{x-ia} - \frac{1}{x+ia}\right]. \quad (1)$$

Since

$$D^{n-1}\left\{(x-ia)^{-1}\right\} = (-1)^{n-1}(n-1)\,!\,(x-ia)^{-n},$$

differentiating (n–1) times the expression (1) with respect to x, we get

$$y_n = \frac{(-1)^{n-1}(n-1)!}{2i}\left[(x-ia)^{n} - (x+ia)^{-n}\right].$$

Putting $x = r\cos\phi$ and $a = r\sin\phi$, we get

$$y_n = (-1)^{n-1}(n-1)\,!\,a^{-n}\sin^n\phi\,\sin n\phi,$$

where $\phi = \tan^{-1}\left(\frac{a}{x}\right)$.

2.2 LEIBNITZ'S THEOREM AND ITS APPLICATIONS

This theorem determines the nth differential coefficient of the product of two differentiable functions. The statement of the theorem is as follows:

Theorem 2.1. (Leibnitz's). Let u and v be two functions of x such that nth derivatives $D^n u$ and $D^n v$ exist. Then the nth differential coefficient of their product is given by

$$D^n(uv) = (D^n u)v + {}^nC_1 D^{n-1}u.Dv + {}^nC_2 D^{n-2}uD^2v$$
$$+\ldots + {}^nC_r D^{n-r}u.D^r v + \ldots + {}^nC_n uD^n v.$$

Proof. We shall prove the theorem using mathematical induction. For $n = 1$, we have

$$D(uv) = (Du).v + u.Dv.$$

Therefore, the theorem holds for $n = 1$. Let us assume that the theorem holds for $n = m$. Thus,

$$D^m(uv) = (D^m u).v + {}^mC_1 D^{m-1}u.Dv +$$
$${}^mC_2 D^{m-2}uD^2v + \ldots + {}^mC_r D^{m-r}u.D^r v$$
$$+\ldots + {}^mC_m uD^m v.$$

Differentiating with respect to x, we get

$$D^{m+1}(uv) = \left[\left(D^{m+1}u\right)v + D^m u.Dv\right]$$
$$+{}^mC_1\left(D^m u.Dv + D^{m-1}uD^2v\right)$$
$$+{}^mC_2\left(D^{m-1}u.D^2v + D^{m-2}uD^3v\right) + \ldots$$
$$+{}^mC_r\left(D^{m-r+1}u.D^r v + D^{m-r}u.D^{r+1}v\right)$$
$$+\ldots + \left(Du.D^m v + uD^{m+1}v\right)$$
$$= \left(D^{m+1}u\right)v + \left({}^mC_0 + {}^mC_1\right)D^m u.Dv$$
$$+({}^mC_1 + {}^mC_2)D^{m-1}u.D^2v + \ldots$$
$$+({}^mC_r + {}^mC_{r+1})D^{m-r}u.D^{r+1}v$$
$$+\ldots + uD^{m+1}v$$
$$= \left(D^{m+1}u\right)v + {}^{m+1}C_1 D^m u.Dv$$
$$+{}^{m+1}C_2 D^{m-1}u.D^2v$$
$$+\ldots + {}^{m+1}C_{r+1} D^{m-r}u.D^{r+1}v$$
$$+\ldots + u.D^{m+1}v.$$

Thus, the theorem holds for $n = m + 1$ also. Hence, by mathematical induction, the theorem holds for all positive integral values of n.

EXAMPLE 2.9

Find the nth derivative of $x^{n-1}\log x$.

Solution. We have $y = x^{n-1}\log x$.

Therefore,

$$y_1 = x^{n-1}\left(\frac{1}{x}\right) - (n-1)x^{n-2}\log x$$

or

$$xy_1 = x^{n-1} + (n-1)y.$$

Differentiating both the sides by $(n-1)$ times, we get $D^{n-1}(xy_1) = D^{n-1}(x^{n-1}) + D^{n-1}[(n-1)y]$

or

$$\left(D^{n-1}y_1\right)x + {}^{n-1}C_1 D^{n-2}y_1 Dx = (n-1)! + (n-1)y_{n-1}$$

or

$$xy_n + (n-1)y_{n-1} = (n-1)! + (n-1)y_{n-1}$$

or

$$xy_n = (n-1)!,$$

which yields

$$y_n = \frac{(n-1)!}{x}.$$

EXAMPLE 2.10

If $y = a\cos(\log x) + b\sin(\log x)$, show that $x^2y_2 + x\,y_1 + y = 0$ and $x_2 y_{n+2} + (2n+1)x\ y_{n+1} + \left(n^2+1\right)y_n = 0.$

Solution. We are given that

$$y = a\cos(\log x) + b\sin(\log x).$$

Therefore,

$$y_1 = -\frac{a}{x}\sin(\log x) + \frac{b}{x}\cos(\log x)$$

or

$$xy_1 = -a\sin(\log x) + b\cos(\log x) \quad (1)$$

Differentiating (1) with respect to x, we get

$$xy_2 + y_1 = -\frac{a}{x}\cos(\log x) - \frac{b}{x}\sin(\log x)$$

or

$$x^2y_2 + xy_1 = -y \quad (2)$$

Differentiating (2), with respect to x, n times by using Leibnitz's Theorem, we obtain

$$D^n(x^2y_2) + D^n(xy_1) = -D^n y$$

or

$$(D^n y_2)x^2 + {}^nC_1 D^{n-1}y_2 Dx^2 + {}^nC_2 D^{n-2}y_2 . D^2x^2$$
$$+ \left(D^n y_1\right)x + {}^nC_1 D^{n-1}y_1 Dx + D^n y = 0$$

or

$$x^2y_{n+2} + 2nxy_{n+1} + \frac{2n(n-1)}{2}y_n + xy_{n+1}$$
$$+ ny_n + y_n = 0$$

or

$$x^2y_{n+2} + (2n+1)xy_{n+1} + \left(n^2+1\right)y_n = 0.$$

EXAMPLE 2.11

If $y = e^{a\sin^{-1}x}$, show that $\left(1-x^2\right)y_{n+2} - (2n+1)xy_{n+1} - \left(n^2+a^2\right)y_n = 0.$

Solution. We have $y = e^{a\sin^{-1}x}$. (1)
Therefore,

$$y_1 = e^{a\sin^{-1}x}\left[\frac{a}{\sqrt{1-x^2}}\right]$$

or

$$y_1\sqrt{1-x^2} = ae^{a\sin^{-1}x}$$

or

$$y_1^2\left(1-x^2\right) = a^2y^2, \text{ using (1).}$$

Differentiating with respect to x once more, we get

$$2\,y_1y_2\left(1-x^2\right) + y_1^2(-2x) = 2a^2y\,y_1$$

or

$$y_2(1-x^2) - y_1\,x - a^2y = 0. \quad (2)$$

Using Leibnitz's Theorem, we differentiate (2), n times with respect to x, and get

$$D^n\left[y_2\left(1-x^2\right)\right] - D^n(y_1x) - a^2D^n y = 0$$

or

$$y_{n+2}\left(1-x^2\right) + {}^nC_1 y_{n+1}(-2x) + {}^nC_2 y_n(-2)$$
$$- y_{n+1}(x) + {}^nC_1 y_n - a^2y_n = 0$$

or

$$(1-x^2)y_{n+2}-(2n+1)xy_{n+1}-(n^2+a^2)y_n=0.$$

EXAMPLE 2.12

If $y=e^{\tan^{-1}x}$, show that

$$(1+x^2)y_{n+2}+[2(n+1)x-1]y_{n+1}+n(n+1)y_n=0.$$

Solution. It is given that $y=e^{\tan^{-1}x}$.
Differentiating with respect to x, we get

$$y_1=e^{\tan^{-1}x}\cdot\frac{1}{1+x^2}=\frac{y}{1+x^2}$$

or

$$(1+x^2)y_1=y\cdot$$

Differentiating $(n+1)$ times by using Leibnitz's Theorem, we get

$$D^{n+1}\left[\left(1+x^2\right)y_1\right]=D^{n+1}y$$

or

$$\left(1+x^2\right)y_{n+2}+{}^{n+1}C_1y_{n+1}(2x)+{}^{n+1}C_2y_n.2=y_{n+1}$$

or

$$\left(1+x^2\right)y_{n+2}+[2(n+1)x-1]y_{n+1}+n(n+1)y_n=0.$$

EXAMPLE 2.13

If $\cos^{-1}\left(\frac{y}{b}\right)=\log\left(\frac{x}{n}\right)^n$, show that

$$x^2y_{n+2}+(2n+1)x\,y_{n+1}+2n^2y_n=0.$$

Solution. We have

$$\cos^{-1}\left(\tfrac{y}{b}\right)=\log\left(\tfrac{x}{n}\right)^n=n\log\left(\tfrac{x}{n}\right)$$

$$=n[\log x-\log n].$$

Differentiating with respect to x, we get

$$-\frac{1}{\sqrt{1-\frac{y^2}{b^2}}}\cdot\frac{y_1}{b}=\frac{n}{x}$$

or

$$-\frac{y_1}{\sqrt{b^2-y^2}}=\frac{n}{x}$$

or

$$y_1^2x^2=n^2\left(b^2-y^2\right).$$

Differentiating once more, we get

$$2\,y_1y_2\,x^2+2x\,y_1^2=2n^2y\,y_1$$

or

$$y_2x^2+y_1\,x+n^2y=0.$$

Differentiating n times by Leibnitz's Theorem, we get

$$y_{n+2}x^2+{}^nC_1y_{n+1}(2x)+{}^nC_2y_n2+y_{n+1}x\cdot$$
$$+{}^nC_1\,y_n1+n^2y_n=0$$

or

$$x^2y_{n+2}+(2n+1)x\,y_{n+1}+2n^2\,y_n=0.$$

EXAMPLE 2.14

If $y=\tan^{-1}x$, find $(y_n)_0$.

Solution. Since $y=\tan^{-1}x$, we get $y_1=\frac{1}{1+x^2}$ and so, $\left(1+x^2\right)y_1-1=0$

Differentiating once more, we get

$$\left(1+x^2\right)y_2+2x\,y_1=0.$$

Now differentiating n times using Leibnitz's Theorem, we obtain

$$y_{n+2}\left(1+x^2\right)+ny_{n+1}(2x)+\frac{n(n+1)}{2}y_n.2$$
$$+2x\,y_{n+1}+2ny_n=0$$

or

$$\left(1+x^2\right)y_{n+2}+2(n+1)xy_{n+1}+n(n+1)y_n=0.\quad(1)$$

Substituting $x=0$ in the expression for y, y_1, and y_2, we get

$$(y)_0=0,(y_1)_0=1,(y_2)_0=0.$$

Also substituting $x=0$ in (1), we get

$$(y_{n+2})_0=-[n(n+1)](y_n)_0\cdot\quad(2)$$

Putting $n-4$ and $n-2$ in place of n, respectively, in (2), we get

$$(y_{n-2})_0=-[(n-4)(n-3)](y_{n-4})_0\text{ and}$$

$$(y_n)_0=-[(n-2)(n-1)](y_{n-2})_0$$

$$=-[(n-1)(n-2)]\left[-(n-3)(n-4)](y_{n-4})_0\cdot\right]$$

If n is even, we have

$$(y_n)_0 = [-(n-1)(n-2)]$$

$$\times [-(n-3)(n-4)]\cdots[-(3)(2)](y_2)_0 = 0$$

since $(y_2)_0 = 0$.

If n is odd, then

$$(y_n)_0 = [-(n-1)(n-2)]$$

$$\times[-(n-3)(n-4)]\cdots[-(4)(3)]$$

$$\times[-(2)(1)](y_1)_0$$

$$= (-1)^{\frac{n-1}{2}}(n-1)\,! \text{ since } (y_1)_0 = 1.$$

EXAMPLE 2.15

If $y=\left[x+\sqrt{1+x^2}\right]^m$, find the value of the nth differential coefficient of y for $x = 0$.

Solution. We have

$$y=\left[x+\sqrt{1+x^2}\right]^m. \tag{1}$$

Therefore,

$$y_1 = m\left[\sqrt{1+x^2}\right]^{m-1}.\left[1+\frac{1}{2}\frac{2x}{\sqrt{1+x^2}}\right]$$

$$=\frac{m}{\sqrt{1+x^2}}\left[x+\sqrt{1+x^2}\right]^m=\frac{my}{\sqrt{1+x^2}}. \tag{2}$$

Squaring, we get

$$y_1^2(1+x^2)-m^2y^2=0.$$

Differentiating once again, we get

$$2y_1y_2\left(1+x^2\right)+2xy_1^2-2m^2y\,y_1=0$$

or

$$y_2\left(1+x^2\right)+xy_1-m^2y=0. \tag{3}$$

Differentiating (3) n times by Leibnitz's Theorem, we get

$$y_{n+2}\left(1+x^2\right)+{}^nC_1y_{n+1}(2x)+{}^nC_2y_n(2)$$

$$+xy_{n+1}+{}^nC_1y_n-m^2y_n=0$$

or

$$y_{n+2}\left(1+x^2\right)(2n+1)xy_{n+1}+\left(n^2-m^2\right)y_n=0. \tag{4}$$

Substituting $x = 0$ in (1), (2), (3), and (4), we get

$(y)_0 = 1$, $(y_1)_0 = m$, $(y_2)_0 = m^2$, and

$$(y_{n+2})_0=(m^2-n^2)(y_n)_0. \tag{5}$$

If n is odd, that is, if n = 1, 3, 5, …, then (5) yields

$$(y_3)_0=\left(m^2-1^2\right)(y_1)_0=\left(m^2-1\right)m,$$

$$(y_5)_0=\left(m^2-3^2\right)(y_3)_0=\left(m^2-3^2\right)(m^2-1^2)m,$$

$$(y_7)_0=\left(m^2-5^2\right)(y_1)_0=\left(m^2-5^2\right)\left(m^2-3^2\right)\times\ m^2-$$

$$\times\left(m^2-1^2\right)m,$$

and so on. Hence, when n is odd, we have

$$(y_n)_0=\left[m^2-(n-2)^2\right]\times\left[m^2-(n-4)^2\right]\ldots$$

$$\left(m^2-3^2\right)\left(m^2-1^2\right)m$$

If n is even, putting n = 2, 4, 6, … in (5), we get

$$(y_4)_0=\left(m^2-2^2\right)(y_2)_0=\left(m^2-2^2\right)m^2,$$

$$(y_6)_0=\left(m^2-4^2\right)(y_n)_0=\left(m^2-4^2\right)\left(m^2-2^2\right)m^2,$$

$$(y_8)_0=(m^2-6^2)(y_6)_0=(m^2-6^2)(m^2-4^2)\times$$

$$\times(m^2-2^2)m^2,$$

and so on. Hence, when n is even, we get

$$(y_n)_0=\left[m^2-(n-2)^2\right]$$

$$\times\left[m^2-(n-4)^2\right]\ldots\left(m^2-4^2\right)\left(m^2-2^2\right)m^2.$$

EXAMPLE 2.16

Find the nth derivatives of

(i) $e^{-x}\sin^2 x$. and (ii) $\dfrac{x}{(x-1)(2x+3)}$

Solution. Let $y = e^{-x}\sin 2x$. Then

$$y=\frac{1}{2}e^{-x}(2\sin^2x)=\frac{1}{2}e^{-x}(1-\cos 2x)$$

$$=\frac{1}{2}[e^{-x}-e^{-x}\cos 2x].$$

But $D^ne^{ax} = a^ne^{ax}$ and $D^n\left(e^{ax}\cos(bx+c)\right)$

$$=(a^2+b^2)^{\frac{n}{2}}e^{ax}\cos\left(bx+c-n\tan^{-1}\frac{b}{a}\right)$$

Therefore

$$D^n y = \frac{1}{2}\Big[(-1)^n e^{-x} - (5)^{\frac{1}{2}} e^{-x} \cos(2x+n \times \tan^{-1}(-2))\Big]$$

(ii) Let $y = (ax + b)^{-1}$. Then $y_1 = -a(ax + b)^{-2}$, $y_2 = (-1)(-2)a^2(ax + b)^{-3}$, and in general $y_n = (-1)(-2)\ldots a^n(ax + b)^{-(n+1)}$.

Thus $\quad D^n y = (-1)^n n! a^n (ax + b)^{-n-1}$. (1)

Now

$$f(x) = \frac{x}{(x-2)(2x+3)}$$

$$= \frac{2}{5(x-1)} + \frac{3}{5(2x+3)} \text{ (partial fractions)}.$$

Using (1), we have

$$D^n(x-1)^{-1} = (-1)^n\, n!\, (x - 1)^{-n-1}$$

$$D^n(2x + 3)^{-1} = (-1)^n\, n!\, (2x - 3)^{-n-1}$$

Therefore

$$D^n(f) = \frac{1}{5}(-1)^n n![2(x-1)^{-n-1} + 3\{2^n(2x+3)^{-n-1}\}]$$

EXAMPLE 2.17

Prove that

$$D^n\left[\frac{\log x}{x}\right] = \frac{(-1)^n n!}{x^{n+1}}\left[\log x - 1 - \frac{1}{2} - \frac{1}{3} - \ldots - \frac{1}{n}\right].$$

Solution. By Leibnitz's Theorem, we have,

$$D^n\left[\frac{\log x}{x}\right] = D^n[x^{-1} \log x]$$

$$= \frac{(-1)^n n!}{x^{n+1}} \log x + {}^nC_1 \frac{(-1)^{n-1}(n-1)!}{x^n} \cdot \frac{1}{x}$$

$$+ {}^nC_2 \frac{(-1)^{n-2}(n-2)!}{x^{n-1}}(-1).\frac{1}{x^2}$$

$$+ {}^nC_3 \frac{(-1)^{n-3}(n-3)!}{x^{n-2}}(-1)(-2)\cdot\frac{1}{x^3} + \ldots$$

$$+ \frac{(-1)^{n-1}(n-1)!}{x^n}\cdot\frac{1}{x}$$

$$= \frac{(-1)^n n!}{x^{n+1}}\left[\log x - 1 - \frac{1}{2} - \frac{1}{3} - \ldots - \frac{1}{n}\right]$$

2.3 GENERAL THEOREMS

The theorems applicable to a class of functions are known as *General Theorems*. In this section, we prove the most useful general theorems.

Theorem 2.2. (Rolle's Theorem). If a function f is

(i) continuous in the closed interval $[a, b]$,
(ii) derivable in the open interval (a, b), and
(iii) $f(a) = f(b)$,

then there exists at least one real number $c \in (a, b)$, such that $f'(c) = 0$.

Proof: Since f is continuous in $[a, b]$, it is bounded and attains its bounds. Let the upper and lower bounds of f be M and m, respectively. Then there exists points c and d of $[a, b]$ such that

$$f(c) = m \text{ and } f(d) = M.$$

If $m = M$, then f is a constant over $[a, b]$ and, therefore, $f'(x) = 0$ for all x in $[a, b]$. If $m \neq M$, at least one of these is different from $f(a)$ or $f(b)$. Suppose that $m \neq f(a)$, then

$$f(c) = m \neq f(a) \text{ implies } c \neq a \text{ and}$$

$$f(c) = m \neq f(b) \text{ implies } c \neq b.$$

Therefore, $c \in (a, b)$. We claim that $f'(c) = 0$. In fact, if $f'(c) < 0$, then there exists $\delta > 0$, such that

$$f(x) < f(c) = m \text{ for } x \in (c, c + \delta).$$

This contradicts the fact that m is the infimum. If $f'(c) > 0$, then there exists $\delta > 0$, such that

$$f(x) < f(c) = m \text{ for } x \in (c - \delta, c),$$

which again contradicts the fact that m is the infimum. Hence, $f'(c) = 0$.

A. Geometric Interpretation of Rolle's Theorem

If we draw the curve $y = f(x)$, which is continuous on $[a, b]$ and derivable on (a, b), then Rolle's Theorem states that between two points with equal ordinates on the graph of f, there exists at least one point where the tangent is parallel to the x-axis.

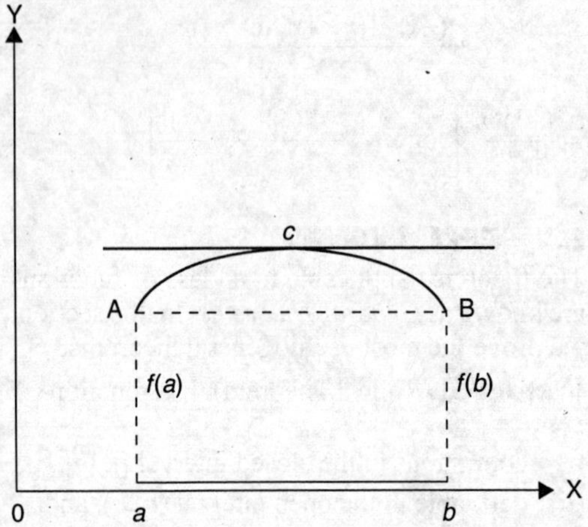

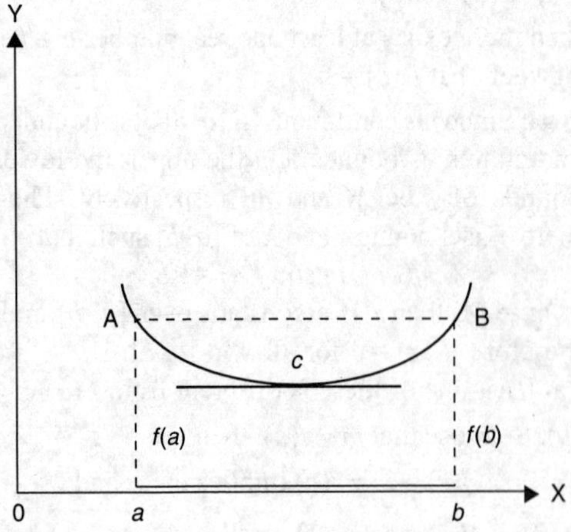

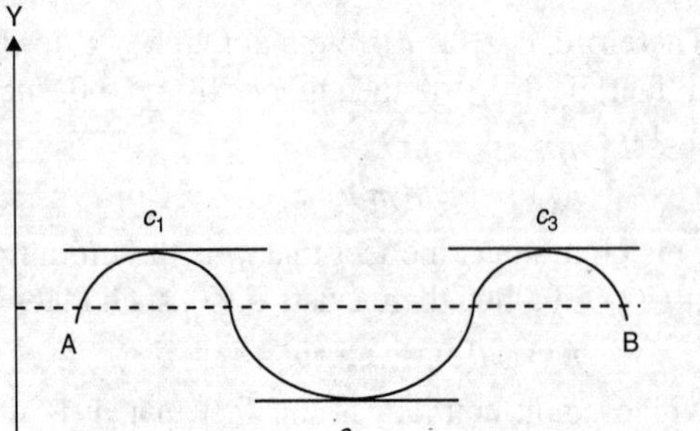

B. Algebraic Interpretation of Rolle's Theorem

In Rolle's Theorem, $f(a) = f(b)$. So if a and b are zeros of f, then Rolle's Theorem says that between two zeros a and b of f, there exists at least one zero of $f'(x)$.

EXAMPLE 2.18

Verify Rolle's Theorem in case of the function $f(x) = 2x^3 + x^2 - 4x-2$.

Solution. The function f is continuous and its derivative exists for any real value of x. If $f(x) = 0$, then $2x^3 + x^2 - 4x -2 = 0$ or $(x^2 - 2)(2x + 1) = 0$, which yields $x = \pm\sqrt{2}, -\frac{1}{2}$. Thus,

$$f\left(\sqrt{2}\right) = f\left(-\sqrt{2}\right) = f\left(-\frac{1}{2}\right) = 0.$$

We choose the interval $\left[-\sqrt{2}, \sqrt{2}\right]$ so that in this interval all the conditions of Rolle's Theorem are satisfied. Now,

$$f'(c) = 6x^2 + 2x - 4 = (3x - 2)(x + 1),$$

which vanishes for $x = -1$ and $x = \frac{2}{3}$. Both the points lie in the interval $\left(-\sqrt{2}, \sqrt{2}\right)$ Hence, the Rolle's Theorem is verified.

EXAMPLE 2.19

Verify Rolle's Theorem when $f(x) = (x-a)m\ (x-b)n$, where m and n are positive integers.

Solution. $f(x) = 0$ yields $x = a$ and $x = b$. So, we consider the interval $[a, b]$. The given function is clearly continuous and differentiable in $[a, b]$ and $f(a) = f(b) = 0$. Also

$$f'(x) = (x - a)^m.n(x-b)^{n-1} + m(x-a)^{m-1}(x-b)^n.$$

Substituting $f'(x) = 0$, we get

$$x = a,\ b, \text{ and } \frac{na + mb}{m + n}.$$

Out of these values $\frac{na+mb}{m+n}$ lies in the open interval (a, b) and hence, the Rolle's Theorem is verified.

EXAMPLE 2.20

Discuss the application of Rolle's Theorem to the function

$$f(x) = 2 + (x-1)^{\frac{2}{3}} \text{ in the interval } [0,\ 2].$$

Solution. We note that

$$f(0) = 2 + (-1)^{\frac{2}{3}} \text{ and } f(2) = 2 + 1 = 3$$

and so, $f(0) \neq f(2)$. Hence, the Rolle's Theorem is not applicable to this function.

EXAMPLE 2.21

If f, ϕ, and ψ are continuous on $[a, b]$ and derivable on (a, b), show that there exists a point $c \in (a, b)$ such that

$$\begin{vmatrix} f(a) & \phi(a) & \psi(a) \\ f(b) & \phi(b) & \psi(b) \\ f'(c) & \phi'(c) & \psi'(c) \end{vmatrix} = 0.$$

Solution. Consider the function F defined by

$$F(x) = \begin{vmatrix} f(a) & \phi(a) & \psi(a) \\ f(b) & \phi(b) & \psi(b) \\ f(x) & \phi(x) & \psi(x) \end{vmatrix}.$$

In $F(a)$ and $F(b)$, two rows in the determinant become identical in each case. Therefore, $F(a) = F(b) = 0$. Since f, ϕ, and ψ are continuous on $[a, b]$ and differentiable on (a, b), F also has these properties. Hence, all the conditions of Rolle's Theorem are satisfied for F in this interval. Therefore, there exists $c \in (a, b)$ such that $F'(c) = 0$. But

$$F'(x) = \begin{vmatrix} f(a) & \phi(a) & \psi(a) \\ f(b) & \phi(b) & \psi(b) \\ f'(x) & \phi'(x) & \psi'(x) \end{vmatrix}.$$

Hence, $F'(c) = 0$ implies

$$\begin{vmatrix} f(a) & \phi(a) & \psi(a) \\ f(b) & \phi(b) & \psi(b) \\ f'(c) & \phi'(c) & \psi'(c) \end{vmatrix} = 0.$$

EXAMPLE 2.22

If $C_0 + \frac{C_1}{2} + \ldots + \frac{C_{n-1}}{n} + \frac{C_n}{n+1} = 0$, where $C_0, C_1, \ldots, C_n$ are real constants, show that the equation

$$C_0 + C_1 x + \ldots + C_{n-1} x^{n-1} + C_n x^n = 0$$

has at least one root between 0 and 1.

Solution. Let

$$f(x) = C_0 + C_1 x + \ldots + C_{n-1} x^{n-1} + C_n x^n.$$

Consider

$$\phi(x) = C_0 x + C_1 \frac{x^2}{2} + C_2 \frac{x^3}{3} + \ldots + C_n \frac{x^{n+1}}{n+1}.$$

Then, $\phi(0) = 0$ and $\phi(1) = C_0 + \frac{C_1}{2} + \ldots + \frac{C_{n-1}}{n} + \frac{C_n}{n+1} = 0$ (given).

Therefore, $\phi(0)=\phi(1)=0$. Also, $\phi(x)$ is continuous in $[0,1]$ and derivable in $(0, 1)$. Thus, $\phi(x)$ satisfies all the conditions of Rolle's Theorem. Hence, there exists $\xi \in (0,1)$ such that $\phi'(\xi) = 0$.

But, $\quad \phi'(x) = f(x)$

and so, $\quad 0 = \phi'(\xi) = f(\xi)$.

This implies that $\xi \in (0,1)$ is a root of $f(x) = 0$, which is the thing we wished to highlight.

EXAMPLE 2.23

If f is continuous in $[a, c]$ and derivable in (a, c), deduce from Rolle's Theorem by considering

$$\phi(x) = f(x) - f(a) - \frac{x-a}{c-a}[f(c) - f(a)],$$

that $f(c) - f(a) = (c-a)f'(\xi)$, $a < \xi < c$.

Solution. We note that

(i) being the sum of continuous functions, $\phi(x)$ is continuous,

(ii) $\phi(x)$ is differentiable for all real values of x, and

(iii) $\phi(a) = 0 = \phi(c)$.

Thus, all conditions of Rolle's Theorem are satisfied. Therefore, there exists a point $\xi \in (a,c)$ such that $\phi'(\xi) = 0$. But,

$$\phi'(x) = f'(x) - \frac{1}{c-a}[f(c) - f(a)]$$

and so $\phi'(\xi) = 0$ implies

$$f(c) - f(a) = (c-a)f'(\xi),\ \xi \in (a,c).$$

Theorem 2.3. (Lagrange's Mean Value Theorem). If a function f defined on $[a, b]$ is

(i) continuous on $[a, b]$ and

(ii) derivable in (a, b),

then there exists at least one real number $c \in (a,c)$ such that

$$\frac{f(b) - f(a)}{b-a} = f'(c).$$

Proof: Consider the function

$$\phi(x) = f(x) + Ax,$$

where A is a constant to be determined such that $\phi(a) = \phi(b)$. Since

$$\phi(a) = f(a) + Aa, \text{ and } \phi(b) = f(b) + Ab,$$

therefore, $\phi(a) = \phi(b)$ yields

$$A = \frac{f(b)-f(a)}{b-a}.$$

The function ϕ is

(i) continuous on $[a, b]$, being the sum of two continuous functions,

(ii) derivable on (a, b), and

(iii) $\phi(a) = \phi(b)$.

Thus, ϕ satisfies the conditions of Rolle's Theorem. Hence, there exists a real number $c \in (a,b)$ such that $\phi'(c) = 0$. But,

$$\phi'(x) = f'(x) + A.$$

Therefore, $\phi'(c) = 0$ implies

$$0 = f'(c) + A = f'(c) - \frac{f(b)-f(a)}{b-a}.$$

Hence,

$$\frac{f(b)-f(a)}{b-a} = f'(c).$$

Remark 2.1

(i) Lagrange's Mean Value Theorem *is also known as First Mean Value Theorem of Differential Calculus.*

(ii) If we put $b = a + h$, then the number between a and b can be written as $a + \theta h$, $0 < \theta < 1$. Thus, Lagrange's Mean Value Theorem takes the form

$$f(a+h) - f(a) = hf'(a+\theta h)$$

or

$$f(a+h) - f(a) = hf'(a+\theta h),\ 0 < \theta < 1.$$

Geometric Interpretation

The Lagrange's Mean Value Theorem states that between two points A and B of the graph of f there exists at least one point where the tangent is parallel to the chord AB.

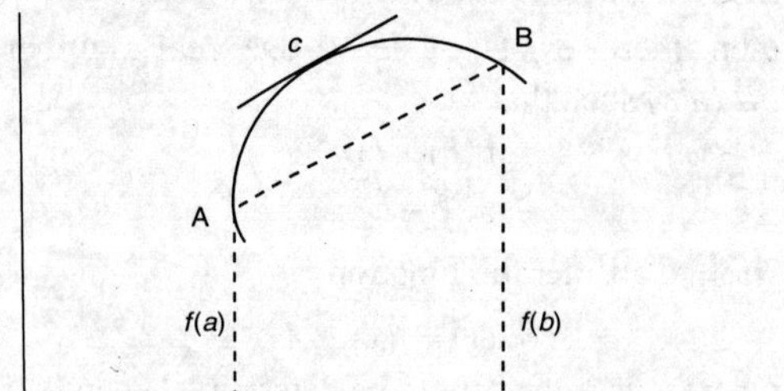

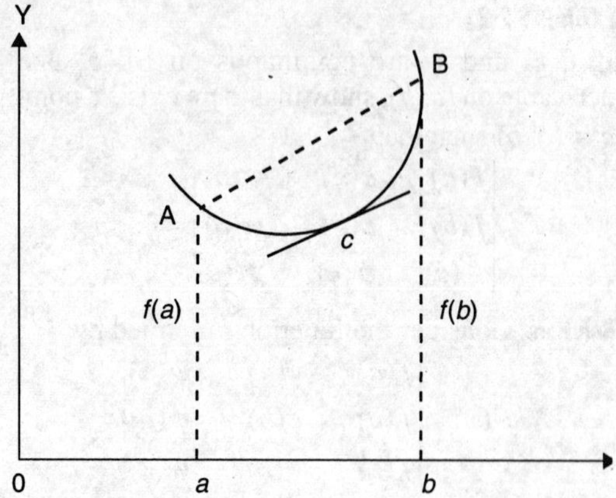

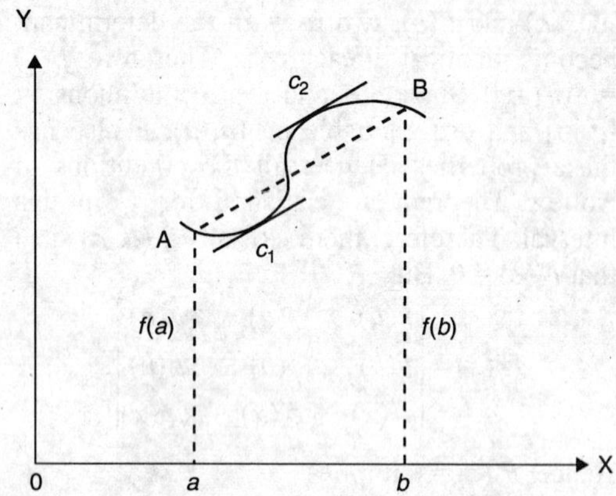

Deductions: Mean Value Theorem asserts that

$$\frac{f(b)-f(a)}{b-a} = f'(c), c \in (a,b).$$

Thus,

(i) if $f'(x) = 0$ for all $x \in [a,b]$, then f is a constant on $[a, b]$. In fact, if x_1 and x_2 are two distinct points of $[a, b]$ and $x_1 < x_2$, then

$$\frac{f(x_2)-f(x_1)}{x_2-x_1} = f'(c) = 0, x_1 < c < x_2.$$

Hence, $f(x_2) = f(x_1)$ and the function f keeps the same value over $[a, b]$. Therefore, f is constant on $[a, b]$.

(ii) If f is continuous on $[a, b]$ and derivable on (a, b) and if $f'(x) > 0$ for all $x \in [a,b]$, then f is strictly increasing on $[a, b]$. In fact, if x_1, $x_2 \in [a,b]$, then by Lagrange's Mean Value Theorem,

$$\frac{f(x_2)-f(x_1)}{x_2-x_1} = f'(c) > 0, x_1 < c < x_2.$$

Hence,

$$f(x_2) - f(x_1) > 0 \text{ or } f(x_2) > f(x_1) \text{ for } x_2 > x_1$$

and so, f is strictly increasing on $[a, b]$. Similarly, if $f'(x) < 0$ for $x \in [a,b]$, then f is strictly decreasing in $[a, b]$.

Theorem 2.4. (Cauchy's Mean Value Theorem). If two functions f and F defined on $[a, b]$ are

(i) continuous on $[a, b]$,

(ii) derivable on (a, b), and

(iii) $F'(x) \neq 0$ for any $x \in (a,b)$,

then there exists a point $c \in (a,b)$ such that

$$\frac{f(b)-f(a)}{F(b)-F(a)} = \frac{f'(c)}{F'(c)}, a < c < b.$$

[Obviously $F(b) \neq F(a)$. Otherwise, F satisfies all the conditions of Rolle's Theorem and as such, $F'(c) = 0$, which contradicts condition (iii) of the statement].

Proof: Consider the function ϕ defined by

$$\phi(x) = f(x) + AF(x),$$

where A is a constant to be determined such that $\phi(a) = \phi(b)$. Since

$$\phi(a) = f(a) + AF(a) \text{ and}$$
$$\phi(b) = f(b) + AF(b),$$

$\phi(a) = \phi(b)$ implies

$$f(a) + AF(a) = f(b) + AF(b).$$

This relation yields

$$A = -\frac{f(b)-f(a)}{F(b)-F(a)}.$$

The function ϕ, being the sum of two continuous and derivable functions is itself

(i) continuous on $[a,b]$,

(ii) derivable on (a,b), and

(iii) $\phi(a) = \phi(b)$.

Hence, by Rolle's Theorem, there exists a point $c \in (a,b)$ such that $\phi'(c) = 0$. But

$$\phi'(x) = f'(x) + AF'(x).$$

Therefore, $\phi'(c) = 0$ implies

$$f'(c) + AF'(c) = 0$$

or

$$f'(c) - \frac{f(b)-f(a)}{F(b)-F(a)}F'(c) = 0$$

or

$$\frac{f(b)-f(a)}{F(b)-F(a)} = \frac{f'(c)}{F'(c)}.$$

Another Form of the Statement: If two functions f and F are continuous in $[a, a+h]$, derivable in $[a, a+h]$, and $F'(x) \neq 0$ for any $x \varepsilon (a, a+h)$, then there exists at least one number θ between 0 and 1 such that

$$\frac{f(a+h)-f(a)}{F(a+h)-F(a)} = \frac{f'(a+\theta h)}{F'(a+\theta h)}.$$

[$f'(x)$ and $F'(x)$ should not vanish for the same value of x].

Deduction. If we define F as $F(x) = x$, then Cauchy's Mean Value Theorem reduces to the Lagrange's Mean Value Theorem. In fact, in that case $F'(x) = 1$, $F(b) = b$, $F(a) = a$ and so,

$$\frac{f(b)-f(a)}{b-a} = f'(c), a < c < b.$$

EXAMPLE 2.24

A twice-differentiable function f is such that $f(a) = f(b) = 0$ and $f(c) > 0$ for $a < c < b$. Show that there is atleast one value ξ between a and b for which $f''(\xi) < 0$.

Solution. Since f'' exists, f and f' both exists and are continuous on $[a,b]$. As $a < c < b$, then by applying Lagrange's Mean Value Theorem to f on the intervals $[a, c]$ and $[c, b]$, we have

$$\frac{f(c)-f(a)}{c-a} = f'(\xi_1), a < \xi_1 < c$$

and

$$\frac{f(b)-f(c)}{b-c} = f'(\xi_2), c < \xi_2 < b.$$

Since $f(a) = f(b) = 0$, these equations reduce to

$$f'(\xi_1) = \frac{f(c)}{c-a}, \quad a < \xi_1 < c$$

and

$$f'(\xi_2) = -\frac{f(c)}{b-c}, \quad c < \xi_2 < b.$$

Since, f' is continuous and derivable on $[\xi_1, \xi_2]$, the use of Lagrange's Mean Value Theorem yields

$$\frac{f'(\xi_2)-f'(\xi_1)}{\xi_2-\xi_1}=f''(\xi),\xi_1<\xi<\xi_2$$

or

$$f''(\xi)=\frac{\frac{-f(c)}{b-c}-\frac{f(c)}{c-a}}{\xi_2-\xi_1}=-\frac{(b-a)f(c)}{(\xi_2-\xi_1)(b-c)(c-a)}<0$$

EXAMPLE 2.25

Show that $\dfrac{\tan x}{x}>\dfrac{x}{\sin x}, 0<x<\dfrac{\pi}{2}.$

Solution. We want to show that

$$\frac{\tan x\sin x-x^2}{x\sin x}>0, 0<x<\frac{\pi}{2}.$$

Since $x\sin x>0$ for $0<x<\frac{\pi}{2}$, it is sufficient to show that

$$\tan x\sin x-x^2>0, 0<x<\frac{\pi}{2}.$$

Let $f(x)=\tan x\sin x-x^2$. Then,

$$f'(x)=\cos x\tan x+\sin x\sec^2 x-2x$$
$$=\sin x+\sin x\sec^2 x-2x.$$

The function f' is continuous and derivable on $\left(0,\frac{\pi}{2}\right)$. Therefore,

$$f''(x)=\cos x+\cos x\sec^2 x$$
$$+2\sin x\sec^2 x\tan x-2$$
$$=\sec x+\cos x-2+2\sin x\tan x\sec^2 x$$
$$=\left(\sqrt{\sec x}-\sqrt{\cos x}\right)^2$$
$$+2\sin x\tan x\sec^2 x>0 \text{ for } 0<x<\frac{\pi}{2}.$$

Since $f''(x)>0$, it follows that f' is strictly increasing in $\left(0,\frac{\pi}{2}\right)$. Also $f'(0)=0$. Therefore, $f'(x)>0$ for $0<x<\frac{\pi}{2}$. This implies that f is strictly increasing in $\left[0,\frac{\pi}{2}\right]$. Also $f(0)=0$ and so, $f(x)>0$ for $0<x<\frac{\pi}{2}$. Hence, $\tan x\sin x-x^2>0$ for $0<x<\frac{\pi}{2}$, which in turn implies that

$$\frac{\tan x}{x}>\frac{x}{\sin x}, 0<x<\frac{\pi}{2}.$$

EXAMPLE 2.26

Show that $|\cos b-\cos a|<|b-a|$.

Solution. Let $f(x)=\cos x$, $a<x<b$. This function satisfies the conditions of Lagrange's Mean Value Theorem. Therefore, using the theorem, we get

$$\frac{f(b)-f(a)}{b-a}=f'(c)=-\sin c,$$

which gives $\cos b-\cos a=(b-a)(-\sin c)$.

Since $|-\sin c|<1$, we have

$$|\cos b-\cos a|=|b-a|\,|-\sin c|<|b-a|.$$

EXAMPLE 2.27

Show that

$$\frac{v-u}{1+v^2}<\tan^{-1}v-\tan^{-1}u<\frac{v-u}{1+u^2}, 0<u<v.$$

Solution. Consider the function f defined by $f(x)=\tan^{-1}x$. We have

$$f'(x)=\frac{1}{1+x^2}.$$

Lagrange's Mean Value Theorem is applicable to f and we have

$$\frac{\tan^{-1}v-\tan^{-1}u}{v-u}=\frac{1}{1+\xi^2}, \ u<\xi<v.$$

Since $\xi<v$, we have

$$\frac{1}{1+\xi^2}>\frac{1}{1+v^2}.$$

Similarly, $\xi>u$, yields

$$\frac{1}{1+\xi^2}<\frac{1}{1+u^2}.$$

Therefore,

$$\frac{1}{1+v^2}<\frac{\tan^{-1}v-\tan^{-1}u}{v-u}<\frac{1}{1+u^2}$$

or

$$\frac{v-u}{1+v^2}<\tan^{-1}v-\tan^{-1}u<\frac{v-u}{1+u^2}, 0<u<v.$$

EXAMPLE 2.28

Show that

$$\frac{\sin\alpha-\sin\beta}{\cos\beta-\cos\alpha}=\cot\theta,\ 0<\alpha<\theta<\beta<\frac{\pi}{2}.$$

Solution. Let $f(x)=\sin x$ and $F(x)=\cos x$, $x\in(\alpha,\beta)$. Since f and F are continuous in $[\alpha,\beta]$, the conditions of the Cauchy's Mean Value Theorem are satisfied. Therefore, there exists $\theta \in (\alpha, \beta)$ such that

$$\frac{f(\beta)-f(\alpha)}{F(\beta)-F(\alpha)}=\frac{f'(\theta)}{F'(\theta)},$$

that is,

$$\frac{\sin\beta-\sin\alpha}{\cos\beta-\cos\alpha}=-\frac{\cos\theta}{\sin\theta},$$

that is,

$$\frac{\sin\alpha-\sin\beta}{\cos\beta-\cos\alpha}=\cot\theta,\ 0<\alpha<\theta<\beta<\frac{\pi}{2}.$$

EXAMPLE 2.29

If a function f is such that its derivatives f' is continuous on $[a, b]$ and derivable on (a, b), show that there exists a number $c \in (a,b)$ such that

$$f(b)=f(a)+(b-a)f'(a)+\frac{1}{2}(b-a)^2 f''(c).$$

Solution. Consider the function $\phi(x) = f(b) - f(x) - (b-x) f'(x) - (b-x)^2 A$, where A is the constant such that $\phi(a) = \phi(b)$. The function ϕ satisfies all the conditions of Rolle's Theorem. Since $\phi(a) = \phi(b)$, we have

$$f(b)-f(a)-(b-a)f'(a)-(b-a)^2A=0. \quad (1)$$

Further,

$$\phi'(x)=f'(x)-f'(x)-(b-x)f''(x)+2(b-x)A$$

$$=-(b-x)f''(x)+2(b-x)A.$$

By Rolle's Theorem, there exists $c \in (a,b)$ such that $\phi'(x) = 0$. Therefore,

$$-(b-c)f''(c)-2(b-c)A=0$$

or

$$(b-c)[2A-f''(c)]=0$$

or

$$A=\frac{1}{2}f''(c),\ b\neq c.$$

Then (1) yields

$$f(b)=f(a)+(b-a)f'(a)+\frac{1}{2}(b-a)^2 f''(c).$$

Theorem 2.5. (Taylor's Theorem). If a function f defined on $[a,a + h]$ is such that

(i) the $(n-1)$th derivative $f^{(n-1)}$ is continuous in $[a, a + h]$ and

(ii) the n th derivative $f^{(n)}$ exists in $(a, a + h)$,

then there exists at least one real number θ between 0 and 1 such that

$$f(a+h)=f(a)+hf'(a)+\frac{h^2}{2!}f''(a)$$
$$+\ldots+\frac{h^{n-1}}{(n-1)!}f^{(n-1)}(a)$$
$$+\frac{h^n(1-\theta)^{n-p}}{p(n-1)!}f^{(n)}(a+\theta h),$$

where p is a positive integer.

Proof: The condition (i) of the statement implies that all the derivatives $f', f'', \ldots, f^{(n-1)}$ exist and are continuous on $[a, a + h]$. Let $a + h = b$. Consider the function

$$F(x)=f(x)+(b-x)f'(x)$$
$$+\ldots+\frac{(b-x)^{n-1}}{(n-1)!}f^{(n-1)}(x)+A(b-x)^p,$$

where A is a constant to be determined, such that $F(a) = F(b)$. But,

$$F(a)=f(a)+(b-a)f'(a)$$
$$+\ldots+\frac{(b-a)^{n-1}}{(n-1)!}f^{(n-1)}(a)+A(b-a)^p,$$

$$F(b) = f(b).$$

Therefore, $F(a) = F(b)$ implies

$$f(b)=f(a)+(b-a)f'(a)+\ldots+\frac{(b-a)^{n-1}}{(n-1)!}$$
$$\times f^{(n-1)}(a)+A(b-a)^p \quad (1)$$

The function F is continuous in $[a, a + h]$, derivable in $(a, a + h)$, and $F(a) = F(a + h)$. Hence, by Rolle's Theorem, there exists at least one real number θ between 0 and 1 such that $F'(a + \theta h) = 0$. But,

$$F'(x)=\frac{(b-x)^{n-1}}{(n-1)!}f^{(n)}(x)-pA(b-x)^{p-1}.$$

Therefore, $F'(a+\theta h)=0$ implies

$$\frac{h^{n-1}}{(n-1)!}f^{(n)}(a+\theta h)-pA(h-\theta h)^{p-1}=0$$

or

$$\frac{h^{n-1}}{(n-1)!}f^{(n)}(a+\theta h)-pAh^{p-1}(1-\theta)^{p-1}=0$$

and so,

$$A=\frac{h^{n-p}(1-\theta)^{n-p}}{p(n-1)!}f^{(n)}(a+\theta h), h\neq 0,\ \theta\neq 1.$$

Substituting this value of A in (1), we get

$$f(b)=f(a)+(b-a)f'(a)+\frac{(b-a)^2}{2!}f''(a)$$

$$+\ldots+\frac{(b-a)^{n-1}}{(n-1)!}f^{(n-1)}(a)$$

$$+\frac{h^{n-1}(1-\theta)^{n-p}}{p(n-1)!}f^{(n)}(a+\theta h) \text{ and}$$

or

$$f(a+h)=f(a)+hf'(a)+\frac{h^2}{2!}f''(a)$$

$$+\ldots+\frac{h^{n-1}}{(n-1)!}f^{(n-1)}(a)$$

$$+\frac{h^n(1-\theta)^{n-p}}{p(n-1)!}f^{(n)}(a+\theta h).$$

The term

$$R_n=\frac{h^n(1-\theta)^{n-p}}{p(n-1)!}f^{(n)}(a+\theta h)$$

is called *Taylor's remainder after n terms* and the theorem with this form of *remainder* is known as *Taylor's Theorem with Schlomilch and Roche form of remainder*.

Substituting $p = 1$, we get

$$R_n=\frac{h^n(1-\theta)^{n-1}}{(n-1)!}f^{(n)}(a+\theta h),$$

which is called *Cauchy's form of the remainder*.

Substituting $p = n$, we get

$$R_n=\frac{h^n}{n!}f^{(n)}(a+\theta h),$$

which is called *Lagrange's form of the remainder*. If we put $a+h=x$, then the Taylor's Theorem reads

$$f(x)=f(a)+(x-a)f'(a)+\frac{(x-a)^2}{2!}f''(a)$$

$$+\ldots+\frac{(x-a)^{n-1}}{(n-1)!}f^{(n-1)}(a)$$

$$+\frac{(x-a)^n}{p(n-1)!}(1-\theta)^{n-p}f^{(n)}(a+\theta(x-a)).$$

Substituting $a = 0$ in this last expression, we have

$$f(x)=f(0)+xf'(0)+\frac{x^2}{2!}f''(0)$$

$$+\ldots+\frac{x^{n-1}}{(n-1)!}f^{(n-1)}(0)$$

$$+\frac{x^n}{p(n-1)!}(1-\theta)^{n-p}f^{(n)}(\theta x),$$

which is known as *Maclaurin's Theorem with Schlomilch and Roche form of remainder*.

Substituting $p = 1$, we get

$$R_n=\frac{x^n}{(n-1)!}(1-\theta)^{n-1}f^{(n)}(\theta x),$$

which is *Cauchy's form of remainder for Maclaurin's Theorem*.

Substituting $p = n$, we get

$$R_n=\frac{x^n}{n!}f^{(n)}(\theta x),$$

which is *Lagrange's form of remainder for Maclaurin's Theorem*.

2.4 TAYLOR'S INFINITE SERIES AND POWER SERIES EXPANSION

The Taylor's Theorem asserts that

$$f(a+h)=f(a)+h\,f'(a)+\frac{h^2}{2!}f''(a)$$

$$+\ldots+\frac{h^{n-1}}{(n-1)!}f^{(n-1)}(a)+R_n,$$

Where R_n is the remainder after n terms. Thus,

$$f(a+h)=S_n+R_n,$$

where

$$S_n = f(a)+h f'(a)+\frac{h^2}{2!}f''(a)+\ldots+\frac{h^{n-1}}{(n-1)!}f^{(n-1)}(a).$$

Thus, if $R_n \to 0$ as $n \to \infty$, then $\lim_{n\to\infty} S_n = \lim_{n\to\infty}\left[f(a+h)-R_n\right] = f(a+h)$,

and so the series

$$f(a)+h f'(a)+\frac{h^2}{2!}f''(a)+\ldots+\frac{h^{n-1}}{(n-1)!}f^{(n-1)}(a)+\ldots,$$

called Taylor's Series, converges to $f(a + h)$. On the other hand, if we put $a + h = x$, then

$$f(x) = f(a)+(x-a)f'(a)+\frac{(x-a)^2}{2!}f''(a)+\ldots,$$

which is the expansion of $f(x)$ in powers of $(x - a)$.

2.5 MACLAURIN'S INFINITE SERIES

If the remainder R_n in the Maclaurin's expansion of a function tends to zero, then we get

$$f(x) = f(0)+xf'(0)+\frac{x^2}{2!}f''(0)+\frac{x^3}{3!}f''(0)+\ldots,$$

which is called *Maclaurin's Infinite Expansion* of f in power of x. The series $f(0)+x f'(0)+\frac{x^2}{2!}f'(0)+\ldots$ is called *Maclaurin's Infinite Series*.

2.6 EXPANSION OF FUNCTIONS

EXAMPLE 2.30

Expand log sin x in powers of $(x - 3)$.

Solution. If $f(x) = \log \sin x$, then

$$f(x) = f(3+(x-3)) = f(3+h),\ h = x-3.$$

Therefore, by Taylor's Series Expansion,

$$f(3+h) = f(3)+hf'(3)+\frac{h^2}{2!}f''(3)+\frac{h3}{3!}f''(3)+\ldots$$

Since $f(x) = \log \sin x$, we have

$$f'(x) = \frac{\cos x}{\sin x} = \cot x,$$

$$f''(x) = -\text{cosec}^2 x,$$

$$f'''(x) = -2 \text{ cosec } x(-\text{cosec } x \cot x) = 2\text{cosec}^2 x \cot x,$$

and so on. Hence,

$$\log \sin x = f(3+h) = \log \sin 3+(x-3)\cot 3 - \frac{(x-3)^2}{2}\text{cosec}^2 3 + \frac{(x-3)^3}{3}\text{cosec}^2 3 \cot 3+\ldots$$

EXAMPLE 2.31

If $0<\theta<1$, show that

$$\log(1+x) = x-\frac{x^2}{2}+\frac{x^3}{3(1+\theta x)^3}$$

and hence,

$$\log(1+x) < x-\frac{x^2}{2}+\frac{x^3}{3},\ x>0.$$

Solution. By Maclaurim's Theorem with remainder R_3, we have

$$f(x) = f(0)+xf'(0)+\frac{x^2}{2}f''(0)+\frac{x^3}{3!}f'''(\theta x),\ 0<\theta<1.$$

It is given that $f(x) = \log(1+x)$, which gives $f(0) = 0$. Therefore,

$$f'(x) = \frac{1}{1+x} \Rightarrow f'(0) = 1,$$

$$f''(x) = -\frac{1}{(1+x)^2} \Rightarrow f''(0) = -1, \text{ and}$$

$$f'''(x) = \frac{2}{(1+x)^3} \Rightarrow f'''(0) = \frac{2}{(1+\theta x)^3}.$$

Hence,

$$f(x) = \log(1+x) = x - \frac{x^2}{2} + \frac{x^3}{3(1+\theta x)^3}.$$

Further, since $x > 0$ and $\theta > 0$, we have $\theta x > 0$ and so,

$$(1+\theta x)^3 > 1 \text{ or } \frac{1}{(1+\theta x)^3} < 1.$$

Therefore,

$$\log(1+x) < x - \frac{x^2}{2} + \frac{x^3}{3}, \text{ for } x > 0.$$

EXAMPLE 2.32

Use Taylor's Series Expansion to compute the value of cos 32°, correct to four decimal places.

Solution. Let $f(x) = \cos x$. Then, $f(x+h) = \cos(x+h)$. By Taylor's Expansion,

$$\cos(x+h) = f(x+h) = f(x) + hf'(x) + \frac{h^2}{2} f''(x)$$

$$+ \frac{h^3}{3!} f'''(x) + \ldots$$

But,

$f'(x) = -\sin x$, $f''(x) = -\cos x$, and

$f'''(x) = \sin x, \ldots$

Substituting $x = 30^o$ and $h = 2^o = \frac{\pi}{90} = 0.03490$, we have

$$\cos(x+h) = \cos x - h\sin x - \frac{h^2}{2}\cos x + \frac{h^3}{6}\sin x + \ldots$$

or

$$\cos 32^o = \cos 30^o - 0.03490 \sin 30^o$$

$$- \frac{(0.03490)^2}{2} \cos 30^o$$

$$+ \frac{(0.03490)^3}{6} \sin 30^o + \ldots$$

$$= \sqrt{\frac{3}{2}} - 0.03490\left(\frac{1}{2}\right)$$

$$- \frac{0.0006}{2}\sqrt{3} + \ldots$$

$$= 0.84857 - 0.01745$$

$$-0.0006\,(0.86602) + \ldots$$

$= 0.8481$, correct to four decimal places.

EXAMPLE 2.33

Expand $\log_e x$ in powers of $(x-1)$ and hence evaluate $\log_e 1.1$, correct to four decimal places.

Solution. We have given that $f(x) = \log_e x$. Therefore, $f(1) = 0$. Further

$$f'(x) = \frac{1}{x} \text{ and so } f'(1) = 1,$$

$$f''(x) = -\frac{1}{x^2} \text{ and so } f''(1) = -1,$$

$$f'''(x) = \frac{2}{x^3} \text{ and so } f'''(1) = 2,$$

$$f^{iv}(x) = -\frac{6}{x^4} \text{ and so } f^{iv}(1) = -6, \text{ and so on.}$$

Substituting these values in the Taylor's Series

$$f(x) = f(a) + (x-a)f'(a) + \frac{(x-a)^2}{2!} f''(a)$$

$$+ \frac{(x-a)^3}{3!} f'''(a) + \ldots,$$

we get

$$\log_e x = f(x) = f(1) + (x-1)f'(1)$$

$$+ \frac{(x-1)^2}{2} f''(1) + \frac{(x-1)^3}{6} f'''(1)$$

$$+ \frac{(x-1)^4}{24} f^{iv}(1) + \ldots$$

$$= (x-1) - \frac{(x-1)^2}{2} + \frac{(x-1)^3}{3}$$

$$- \frac{(x-1)^4}{4} + \ldots,$$

which is the required expansion. Substituting $x = 1.1$ in the expansion, we get

$$\log_e 1.1 = 0.1 - 0.005 + 0.0003 - 0.00002 + \ldots$$

$$= 0.0953.$$

EXAMPLE 2.34

Expand

(i) e^x in power of $(x-1)$ up to four terms.

(ii) $\tan^{-1}x$ in power of $\left(x-\frac{\pi}{4}\right)$.

(iii) $\sin x$ in power of $\left(x-\frac{\pi}{2}\right)$ and determine $\sin 91^\circ$, correct to four decimal places.

Solution. (i) Let $f(x)=e^x$. Then, expanding in Taylor's Series, we have

$$e^x = f(x) = f[1+(x-1)]$$

$$= f(1)+(x-1)f'(1)+\frac{(x-1)^2}{2}f''(1)$$

$$+\frac{(x-1)^3}{3!}f'''(1)+\ldots$$

But,

$$f(x)=e^x \text{ and so } f(1)=e,$$

$$f'(x)=e^x \text{ and so } f'(1)=e,$$

$$f'''(x)=e^x \text{ and so } f''(1)=e,$$

and so on. Hence,

$$e^x = e+(x-1)e+\frac{(x-1)^2}{2}e+\frac{(x-1)^3}{3!}e$$

$$+\frac{(x-1)^4}{4!}e+\ldots$$

$$= e\left[1+(x-1)+\frac{(x-1)^2}{2}+\frac{(x-1)^3}{3!}+\frac{(x-1)^4}{4!}+\ldots\right].$$

(ii) Let $f(x)=\tan^{-1}x$. Then, by Taylor's Series Expansion, we have

$$\tan^{-1}x = f(x) = f\left[\frac{\pi}{4}+\left(x-\frac{\pi}{4}\right)\right]$$

$$= f\left(\frac{\pi}{4}\right)+\left(x-\frac{\pi}{4}\right)f'\left(\frac{\pi}{4}\right)+\frac{\left(x-\frac{\pi}{4}\right)^2}{2!}f''\left(\frac{\pi}{4}\right)+\ldots$$

But,

$$f(x)=\tan^{-1}x \text{ and so } f\left(\frac{\pi}{4}\right)=\tan^{-1}\frac{\pi}{4},$$

$$f'(x)=\frac{1}{1+x^2} \text{ and so } f'\left(\frac{\pi}{4}\right)=\frac{1}{1+\frac{\pi^2}{16}},$$

$$f''(x)=\frac{-2x}{\left(1+x^2\right)^2} \text{ and so } f''\left(\frac{\pi}{4}\right)=\frac{-\pi}{2\left(1+\frac{\pi^2}{16}\right)^2},$$

and so on. Substituting these values in the above expansion, we get

$$\tan^{-1}x = \tan^{-1}\frac{\pi}{4}+\left(x-\frac{\pi}{4}\right).\frac{1}{1+\frac{\pi^2}{16}}-\pi\frac{\left(x-\frac{\pi}{4}\right)^2}{4\left(1+\frac{\pi^2}{16}\right)^2}+\ldots$$

(iii) Let $f(x)=\sin x$. Then, by Taylor's Series Expansion, we get

$$\sin x = f(x) = f\left(\frac{\pi}{2}+\left(x-\frac{\pi}{2}\right)\right)$$

$$= f\left(\frac{\pi}{2}\right)+\left(x-\frac{\pi}{2}\right)f'\left(\frac{\pi}{2}\right)$$

$$+\frac{\left(x-\frac{\pi}{2}\right)^2}{2}f''\left(\frac{\pi}{2}\right)+\frac{\left(x-\frac{\pi}{2}\right)^3}{3!}f'''\left(\frac{\pi}{2}\right)+\ldots$$

But,

$$f(x)=\sin x \text{ and so } f\left(\frac{\pi}{2}\right)=\sin\frac{\pi}{2}=1,$$

$$f'(x)=\cos x \text{ and so } f'\left(\frac{\pi}{2}\right)=\cos\frac{\pi}{2}=0,$$

$$f''(x)=-\sin x \text{ and so } f''\left(\frac{\pi}{2}\right)=-\sin\frac{\pi}{2}=-1,$$

$$f'''(x)=-\cos x \text{ and so } f'''\left(\frac{\pi}{2}\right)=0,$$

and so on. Hence,

$$\sin x = 1+\left(x-\frac{\pi}{2}\right)(0)+\frac{\left(x-\frac{\pi}{2}\right)^2}{2}(-1)$$

$$+\frac{\left(x-\frac{\pi}{2}\right)^3}{3!}(0)+\frac{\left(x-\frac{\pi}{2}\right)^4}{4!}(1)+\ldots$$

$$=1-\frac{\left(x-\frac{\pi}{2}\right)^2}{2}+\frac{\left(x-\frac{\pi}{2}\right)^4}{4!}-\ldots$$

Substituting $x=91^o$, we get

$$x-\frac{\pi}{2}=91^o-90^0=1^o=\frac{\pi}{180}=0.01746$$

Hence,

$$\sin 91^o = 1 - \frac{(0.01746)^2}{2} + \frac{(0.01746)^4}{24} - \ldots$$

$$= 0.9998.$$

EXAMPLE 2.35

Expand $2x^3 + 7x^2 + x - 1$ in power of $(x-2)$.

Solution. Let $f(x) = 2x^3 + 7x^2 + x - 1$. Then, the expansion of $f(x)$, by Taylor's Theorem, in powers of $(x-2)$ is

$$f(x) = f(2+(x-2))$$

$$= f(2) + (x-2)f'(2) + \frac{(x-2)^2}{2!}f''(2) + \frac{(x-3)^2}{3!}f'''(2) + \ldots$$

But,

$f(x) = 2x^3 + 7x^2 + x - 1$ giving $f(2) = 45$,

$f'(x) = 6x^2 + 14x + 1$ giving $f'(2) = 53$,

$f''(x) = 12x + 14$ giving $f''(2) = 38$,

$f'''(x) = 12$ giving $f'''(2) = 12$,

$f^{iv}(x) = 0$ giving $f^{iv}(2) = 0$,

and $f^{(n)}(2) = 0$ for $n \geq 4$. Substituting these values in the above expansion, we get

$$f(x) = 45 + 53(x-2) + 19(x-2)^2 + 2(x-3)^3.$$

EXAMPLE 2.36

Use Taylor's Theorem to show that

(i) $f\left(\frac{x^2}{1+x}\right) = f(x) - \frac{x}{1+x}f'(x) + \frac{x^2}{(1+x)^2}\cdot\frac{1}{2!}f'(x) - \ldots$

(ii) $\tan^{-1}(x+h) = \tan^{-1}x + h\sin\theta \cdot \frac{\sin\theta}{1} - \frac{(h\sin\theta)^2\sin 2\theta}{2} + \frac{(h\sin\theta)^3\sin 3\theta}{3} - \ldots + (-1)^{n-1}(h\sin\theta)^n\frac{\sin n\theta}{n} + \ldots,$

where $\theta = \cot^{-1}x$.

(iii) $f(mx) = f(x) + (m-1)xf'(x) + \frac{(m-1)^2}{2}x^2 f''(x) + \ldots.$

Solution. (i) From the statement it is clear that we wish to expand $f\left(\frac{x^2}{1+x}\right)$ in power of $\left(-\frac{x}{1+x}\right)$. We have, by Taylor's Theorem,

$$f\left(\frac{x^2}{1+x}\right) = f\left(x + \left(-\frac{x}{1+x}\right)\right)$$

$$= f(x) + \left(-\frac{x}{1+x}\right)f'(x) + \frac{1}{2!}\left(-\frac{x}{1+x}\right)^2 f''(x) + \ldots$$

$$= f(x) - \frac{x}{1+x}f'(x) + \frac{x^2}{2(1+x)^2}f''(x) + \ldots.$$

(ii) Let $f(x) = \tan^{-1}x$. Then

$$\tan^{-1}(x+h) = f(x+h)$$

$$= f(x) + hf'(x) + \frac{h^2}{2}f''(x) + \ldots + \frac{h^n}{n!}f^{(n)}(x) + \ldots$$

But, $f^{(n)}(x) = D^n\tan^{-1}x$

$$= (-1)^{n-1}(n-1)!\sin^n\theta\sin n\theta,$$

where $\theta = \cot^{-1}x$. So substituting $n = 1, 2, 3, \ldots$, we get

$$f'(x) = \sin\theta\sin\theta,$$

$$f''(x) = -\sin^2\theta\sin 2\theta,$$

$$f'''(x) = 2!\left(\sin^3\theta\sin 3\theta\right),$$

and so on. Substituting these values in the above expression, we get

$$\tan^{-1}(x+h) = \tan^{-1}x + h\sin\theta\sin\theta - \frac{h^2}{2}\sin^2\theta\sin 2\theta + \frac{h^3}{3!}2!\sin^3\theta\sin 3\theta - \ldots + \frac{h^n}{n!}(-1)^{n-1} \times (n-1)!\sin^n\theta\sin n\theta + \ldots$$

$$= \tan^{-1}x + h\sin\theta\frac{\sin\theta}{1}(h\sin\theta)^2 .\frac{\sin 2\theta}{2} + (h\sin\theta)^3\frac{\sin 3\theta}{3} + \ldots + (-1)^{n-1}(h\sin\theta)^n\frac{\sin n\theta}{n} + \ldots$$

(iii) We have

$$f(mx) = f(x+(m-1)x) = f(x)+(m-1)xf'(x)$$

$$+\frac{(m-1)^2}{2!}x^2 f(x)+\ldots.$$

EXAMPLE 2.37

If $f(x) = x^3 + 8x^2 + 15x - 24$, calculate the value of $f\left(\frac{11}{10}\right)$ by using Taylor's Series.

Solution. We have $f(x)=x^3+8x^2+15x-24$. Then, by Taylor's Series, we have

$$f(x+h)=f(x)+h f'(x)+\frac{h^2}{2!}f''(x)$$

$$+\frac{h^3}{3!}f'''(x)+\frac{h^4}{4!}f^{iv}(x)+-$$

We take $x = 1$ and $h=\frac{1}{10}=0.1$. Also

$$f'(x)=3x^2+16x+15,$$

$$f''(x)=6x+16,$$

$$f'''(x)=6, \text{ and}$$

$$f^{iv}(x)=0\cdot$$

Therefore,

$$f\left(\frac{11}{10}\right)=f(1)+0.1f'(1)+\frac{0.01}{2}f''(1)$$

$$+\frac{0.001}{6}f'''(1)$$

$$=0+3.4+0.11+0.001=3.511\cdot$$

EXAMPLE 2.38

Show that

$$\log(x+h)= \log h+\frac{x}{h}-\frac{x^2}{2h^2}+\frac{x^3}{3h^3}-\ldots$$

Solution. We wish to expand log $(x + h)$ in the ascending power of x. So, let $f(h) = \log(h)$. Then

$$\log(h+x)=f(h+x)$$

$$=f(h)+xf(h)+\frac{x^2}{2!}f''(h)$$

$$+\frac{x^3}{3!}f'''(h)+\ldots.$$

But,

$$f(h)=\log h,\quad f'(h)=\frac{1}{h},$$

$$f''(h)=-\frac{1}{h^2},\ \text{and}\ f'''(h)=\frac{2}{h^3},\ldots.$$

Hence,

$$\log(h+x)= \log h+\frac{x}{h}-\frac{x^2}{2h^2}+\frac{x^3}{3h^3}-\ldots.$$

EXAMPLE 2.39

Expand the following by Maclaurin's Theorem:
(i) tan x. (ii) a^x.
(iii) log sec x.
(iv) log $(1 + x)$, log $(1 - x)$, and $\log\sqrt{\frac{1+x}{1-x}}$.

Solution. Let $f(x) = \tan x$. Then, by Maclaurin's Theorem,

$$f(x)=f(0)+xf'(0)+\frac{x^2}{2!}f''(0)+\frac{x^3}{3!}f'''(0)+\ldots.$$

But,

$y=f(x) = \tan x$ giving $y(0) = \tan 0 = 0$,

$$y_1=f'(x)=\sec^2 x=1+\tan^2 x$$

$$= 1 + y^2 \text{ giving } y_1(0) = 1,$$

$$y_2=f''(x)=2yy_1 \text{ giving}$$

$$y_2(0)=2(y)_0(y_1)_0=0,$$

$$y_3=f'''(x)=2y_1^2+2yy_2 \text{ giving } y_3(0)=2,$$

$$y_4=f^{iv}(x)=4y_1y_2+2y_1y_2+2yy_3$$

$$=6y_1y_2+2yy_3 \text{ giving } y_4(0)=0,$$

$$y_5=f^{(5)}(x)=6y_2^2+8y_1y_3+2yy_4 \text{ giving}$$

$y_5(0) = 16$, and so on. Therefore,

$f(x) = \tan x$

$$=0+x+\frac{x^2}{2!}(0)+\frac{x^3}{3!}(2)+\frac{x^4}{4!}(0)$$

$$+\frac{x^5}{5!}(16)+\ldots$$

$$=x+\frac{x^3}{3}+\frac{2}{15}x^5+\ldots.$$

(ii) Let $f(x) = a^x$. Then, by Maclaurin's Theorem, we have

$$f(x) = f(0) + xf'(0) + \frac{x^2}{2!} f''(0)$$

$$+ \frac{x^3}{3!} f'''(0) + \ldots + \frac{x^4}{n!} f^{(n)}(0) + \ldots.$$

But,
$f(x) = a^x$ so that $f(0) = 1$,
$f'(x) = a^x \log a$ so that $f'(0) = \log a$,
$f''(x) = a^x (\log a)^2$ so that $f''(0) = (\log a)^2$,
$f'''(x) = a^x (\log a)^3$ so that $f'''(0) = (\log a)^3$,
and $f^{(n)}(x) = a^x (\log a)^n$
so that $f^n(0) = (\log a)^n$.
Hence,

$$f(x) = a^x = 1 + x \log a + \frac{x^2}{2!} (\log a)^2$$

$$+ \frac{x^3}{3!} (\log a)^3 + \ldots + \frac{x^n}{n!} (\log a)^n + \ldots.$$

(iii) Let $y = \log \sec x$ so that
$y(0) = \log \sec 0 = \log 1 = 0$.
Then,

$$y_1 = \frac{1}{\sec x}. \sec x \tan x = \tan x$$

so that $y_1(0) = \tan 0 = 0$,

$y_2 = \sec^2 x = 1 + \tan^2 x$

$= 1 + y_1^2$ so that $y_2(0) = 1$,

$y_3 = 2y_1 y_2$ so that $y_3(0) = 2y_1(0) y_2(0) = 0$,

$y_4 = 2y_2^2 + 2y_1 y_3$ so that $y_4(0) = 2 + 0 = 2$,

$y_5 = 4y_2 y_3 + 2y_2 y_3 + 2y_1 y_4$

$= 6y_2 y_3 + 2y_1 y_4$ so that $y_5(0) = 0$,

$y_6 = 6y_3^2 + 8y_2 y_4 + 2y_1 y_5$ so that $y_6(0) = 16$,

and so on. Using Maclaurin's Theorem, we have

$$y = y(0) + xy_1(0) + \frac{x^2}{2!} y_2(0) + \frac{x^3}{3!} y_3(0)$$

$$+ \frac{x^4}{4!} y_4(0) + \frac{x^5}{5!} y_5(0) + \frac{x^6}{6!} y_6(0) + \ldots$$

$$= \frac{x^2}{2} + \frac{x^4}{12} + \frac{x^6}{45} + \ldots$$

(iv) Let

$f(x) = \log(1 + x)$ so that $f(0) = \log 1 = 0$,

$f'(x) = \dfrac{1}{1+x}$ so that $f'(0) = 1$,

$f''(x) = -\dfrac{1}{(1+x)^2}$ so that $f''(0) = -1$,

$f'''(x) = \dfrac{2}{(1+x)^3}$ so that $f'''(0) = 2$,

$f^{iv}(x) = -\dfrac{-6}{(1+x)^4}$ so that $f^{iv}(0) = -6$,

and so on. Therefore, by Maclaurin's Theorem, we have $\log(1 + x) = f(x)$

$$= f(0) + xf'(0) + \frac{x^2}{2!} f''(0) + \frac{x^3}{3!} f'''(0)$$

$$+ \frac{x^4}{4!} f^{iv}(0) + \ldots$$

$$= x - \frac{x^2}{2} + \frac{x^3}{3} - \frac{x^4}{4} + \ldots.$$

Now changing x to $-x$, we get

$$\log(1-x) = -x - \frac{x^2}{2} - \frac{x^3}{3} - \frac{x^4}{4} \ldots.$$

Further,

$$\log\sqrt{\frac{1+x}{1-x}} = \frac{1}{2} \log \frac{1+x}{1-x} = \frac{1}{2} [\log(1+x) - \log(1-x)]$$

$$= \frac{1}{2}\left[x - \frac{x^2}{2} + \frac{x^3}{3} - \frac{x^4}{4} + \ldots\right.$$

$$\left.-\left(-x - \frac{x^2}{2} - \frac{x^3}{3} - \frac{x^4}{4} - \ldots\right)\right]$$

$$= \frac{1}{2}\left[2x + \frac{2x^3}{3} + \frac{2x^5}{5} + \ldots\right]$$

$$= x + \frac{x^3}{3} + \frac{x^5}{5} + \ldots.$$

EXAMPLE 2.40

Show that

(i) $\sin^{-1} x = x + \frac{1}{2} . \frac{x^3}{3} + \frac{1.3}{2.4} . \frac{x^5}{5} + \frac{1.3.5}{2.4.6} . \frac{x^7}{7} + \ldots$

and hence, find the value of π

(ii) $\tan^{-1}x = x - \frac{x^3}{3} + \frac{x^5}{5} - \ldots$

(iii) $\log(1+\sin x) = x - \frac{x^2}{2} + \frac{x^3}{6} - \frac{x^4}{12} + \frac{x^5}{24} + \ldots$.

(iv) $e^x \sec x = 1 + x + \frac{2x^2}{2!} + \frac{4x^3}{3!} + \ldots$

Solution. (i) Let $y = \sin^{-1} x$. (1)

Then, $y_1 = \dfrac{1}{\sqrt{1-x^2}}$ (2)

so that $y_1^2(1-x^2) - 1 = 0$.

Differentiating again, we get

$$(1-x^2)2y_1y_2 - 2xy_1^2 = 0$$

or

$$(1-x^2)y_2 - xy_1 = 0. \quad (3)$$

Differentiating n times, by Leibnitz's Theorem, we have

$$(1-x^2)y_{n+2} - (2n+1)xy_{n+1} - n^2y_n = 0. \quad (4)$$

Substituting $x = 0$ in (1), (2), and (3), we get

$$y(0) = 0, y_1(0) = 1, y_2(0) = 0.$$

Substituting $n = 1$, 2, and 3, in (4), we get

$y_3(0) = y_1(0) = 1$, $y_4(0) = 2^2 y_2(0) = 0$,

$y_5(0) = 3^2 y_3(0) = 9$, $y_6(0) = 4^2 y_4(0) = 0$, and

$y_7(0) = 5^2 y_5(0) = 5^2.3^2.1^2$.

Therefore, by Maclaurin's Theorem, we get

$$\sin^{-1}x = y = y(0) + xy_1(0) + \frac{x^2}{2!}y_2(0) + \frac{x^3}{3!}y_3(0)$$

$$+\frac{x^4}{4!}y_4(0) + \frac{x^5}{5!}y_5(0) + \frac{x^6}{6!}y_6(0)$$

$$+\frac{x^7}{7!}y_7(0) + \cdots$$

$$= x + \frac{1}{2}.\frac{x^3}{3} + \frac{1.3}{2.4}.\frac{x^5}{5} + \frac{1.3.5}{2.4.6}.\frac{x^7}{7} + \cdots$$

Substituting $x = \frac{1}{2}$, we get

$$\sin^{-1}\frac{1}{2} = \frac{1}{2} + \frac{1}{6}\left(\frac{1}{2}\right)^3 + \frac{3}{40}\left(\frac{1}{2}\right)^5 + \frac{5}{112}\left(\frac{1}{2}\right)^7$$

or

$$\frac{\pi}{6} = 0.5 + 0.0208 + 0.0023 + 0.0003 + \ldots \approx 0:5234.$$

Hence, $\pi \approx 3.1404$.

(ii) Let $y = \tan^{-1}x$. (1)

1. Then,

$y_1 = \dfrac{1}{1+x^2}$ or $(1+x^2)y_1 - 1 = 0$. (2)

2. Differentiating once again, we get

$$(1+x^2)y_2 + 2xy_1 = 0. \quad (3)$$

Differentiating (2) n times by Leibnitz's Theorem, we get

$$y_{n+2}(1+x^2) + ny_{n+1}(2x) + \frac{n(n-1)}{2!}y_n(2)$$

$$+2xy_{n+1} + 2ny_n = 0$$

or

$$(1+x^2)y_{n+2} + 2(n+1)xy_{n+1} + n(n+1)y_n = 0. \quad (4)$$

Substituting $x = 0$ in (1), (2), (3), and (4), we get

$y(O) = 0$, $y(0) = 0$, $y_1(0) = 1$, $y_2(0) = 0$, and

$y_{n+2}(0) = -n(n+1)y_n(0)$.

Taking $n = 1,2,3,\ldots$, we get

$y_3(0) = -2, y_4(0) = 0, y_5(0) = 4!$, and so on.

Therefore, by Maclaurin's Theorem, we have

$$\tan^{-1}x = y = y(0) + xy_1(0) + \frac{x^2}{2!}y_2(0) + \frac{x^3}{3!}y_3(0)$$

$$+\frac{x^4}{4!}y_4(0) + \frac{x^5}{5!}y_5(0) + \ldots$$

$$= x - \frac{x^3}{3} + \frac{x^5}{5} - \ldots.$$

(iii) Let $y = \log(1 + \sin x)$ so that $y(0) = \log 1 = 0$.

Then, $y_1 = \dfrac{\cos x}{1+\sin x}$ so that $y_1(0) = 1$,

$$y_2 = \frac{(1+\sin x)(-\sin x) - \cos^2 x}{(1+\sin x)^2}$$

$$= \frac{-(1+\sin x)}{(1+\sin x)^2} = -\frac{1}{1+\sin x} \text{ so that } y_2(0) = -1,$$

$$y_3 = \frac{\cos x}{(1+\sin x)^2} = \frac{\cos x}{1+\sin x}\cdot\frac{1}{1+\sin x}$$

$= -y_1 y_2$ so that $y_3(0) = 1$,

$y_4 = -y_1 y_3 - y_2^2$ so that $y_4(0) = -1-1 = -2$,

$y_5 = -y_1 y_4 - 3y_2 y_3$ so that $y_5(0) = 5$,

and so on. Therefore, Maclaurin's Theorem implies

$$\log(1+\sin x) = y = y(0) + xy_1(0) + \frac{x^2}{2!}y_2(0)$$

$$+\frac{x^3}{3!}y_3(0) + \frac{x^4}{4!}y_4(0) + \frac{x^5}{5!}y_5(0) + \ldots$$

$$= x - \frac{x^2}{2} + \frac{x^3}{6} - \frac{x^4}{12} + \frac{x^5}{24} + \ldots.$$

(iv) Let $y = e^x \sec x$ so that $y(0) = 1$. Then,

$y_1 = e^x \sec x + e^x \sec x \tan x$

$= y + y\tan x$ so that $y_1(0) + 0 = 1$,

$y_2 = y_1 + y_1 \tan x + y\sec^2 x$ so that

$y_2(0) = 1+0+1 = 2$,

$y_3 = y_2 + y_2 \tan x + 2y_1 \sec^2 x + 2y\sec^2 x \tan x$

so that $y_3(0) = 4$,

and so on. Therefore, application of Maclaurin's Theorem yields

$e^x \sec x = y$

$$= y(0) + xy_1(0) + \frac{x^2}{2!}y_2(0) + \frac{x^3}{3!}y_3(0) + \ldots$$

$$= 1 + x + \frac{2x^2}{2!} + \frac{4x^3}{3!} + \ldots.$$

EXAMPLE 2.41

Expand the following functions by Maclaurin's Theorem:

(i) $e^{\sin x}$.

(ii) $\log_e(1+e^x)$.

(iii) $e^x \cos x$.

(iv) $\frac{e^x}{1+e^x}$.

Solution. (i) Let $y = e^{\sin x}$ so that $y(0) = e^0 = 1$. Then,

$y_1 = e^{\sin x}\cos x = y\cos x$ so that

$y_1(0) = y(0)\cos 0 = 1$,

$y_2 = y_1 \cos x - y\sin x$ so that $y_2(0) = 1-0 = 1$,

$y_3 = y_2 \cos x - y_1 \sin x - y_1 \sin x - y\cos x$

$= y_2 \cos x - 2y_1 \sin x - y\cos x$ so that

$y_3(0) = 1-0-1 = 0$,

$y_4 = y_3 \cos x - 3y_2 \sin x - 3y_1 \cos x + y\sin x$

so that $y_4(0) = -3$,

and so on. Therefore, Maclaurin's Theorem yields

$$e^{\sin x} = y = y(0) + xy_1(0) + \frac{x^2}{2!}y_2(0)$$

$$+\frac{x^3}{3!}y_3(0) + \frac{x^4}{4!}y_4(0) + \ldots$$

$$= 1 + x + \frac{x^2}{2!} + \frac{x^4}{4!}(-3) + \ldots$$

$$= 1 + x + \frac{x^2}{2} - \frac{x^4}{8} + \ldots.$$

(ii) We have $y = \log_e(1+e^x)$ so that $y(0) = \log_e 2$. Then,

$y_1 = \frac{e^x}{1+e^x} = 1 - \frac{1}{1+e^x}$ so that $y_1(0) = 1 - \frac{1}{2} = \frac{1}{2}$,

$$y_2 = \frac{e^x}{(1+e^x)^2} = \frac{e^x}{1+e^x}\cdot\frac{1}{1+e^x}$$

$= y_1(1-y_1) = y_1 - y_1^2$ so that $y_2(0) = \frac{1}{4}$,

$y_3 = y_2 - 2y_1 y_2$ so that $y_3(0) = \frac{1}{4} - \frac{1}{4} = 0$,

$y_4 = y_3 - 2\left[y_2^2 + 2y_1 y_3\right]$ so that $y_4(0) = -\frac{1}{8}$,

and so on. Therefore, Maclaurin's Theorem yields

$$\log_e(1+e^x) = y = y(0) + xy_1(0) + \frac{x^2}{2!}y_2(0)$$

$$+\frac{x^3}{3!}y_3(0) + \frac{x^4}{4!}y_4(0) + \ldots$$

$$= \log_e 2 + \frac{x}{2} + \frac{x^2}{8} - \frac{x^4}{192} + \ldots.$$

(iii) Let $y = e^x \cos x$ so that $y(0) = e^0 = 1$. Then,

$$y_1 = e^x \cos x - e^x \sin x$$

$= e^x(\cos x - \sin x)$ so that $y_1(0) = 1$,

$y_2 = e^x(\cos x - \sin x) + e^x(-\sin x - \cos x)$

$= -2e^x \sin x$ so that $y_2(0) = 0$,

$y_3 = -2e^x \sin x - 2e^x \cos x$

$= -2e^x(\sin x + \cos x)$ so that $y_3(0) = -2$,

$y_4 = -2e^x(\sin x + \cos x) - 2e^x(\cos x - \sin x)$

$= -4e^x \cos x = -2^2 y$ so that $y_4(0) = -2^2$,

$y_5 = -2^2 y_1$ so that $y_5(0) = -2^2$,

$y_6 = -2^2 y_2$ so that $y_6(0) = 0$,

$y_7 = -2^2 y_3$ so that $y_7(0) = 2^3$, and

$y_n = 2^{\frac{n}{2}} \cos\left(x + \frac{n\pi}{4}\right)$ so that

$y_n(0) = 2^{\frac{n}{2}} \cos\frac{n\pi}{4}$.

Therefore, by Maclaurin's Theorem, we get

$$e^x \cos x = y = y(0) + xy_1(0) + \frac{x^2}{2!}y_2(0) + \frac{x^3}{3!}y_3(0)$$

$$+\frac{x^4}{4!}y_4(0) + \frac{x^5}{5!}y_5(0) + \frac{x^5}{6!}y_6(0) + \frac{x^7}{7!}y_7(0) + \ldots$$

$$+\frac{x^n}{n!}y_n(0) + \ldots$$

$$= 1 + x - \frac{2x^2}{3!} - \frac{2^2x^4}{4!} - \frac{2^2x^5}{5!} + \frac{2^3x^7}{7!} + \ldots$$

$$+2^{\frac{n}{2}} \cos\frac{n\pi x^n}{4n!} + \ldots.$$

(iv) Let $y = \frac{e^x}{1+e^x} = 1 - \frac{1}{1+e^x}$ so that $y(0) = \frac{1}{2}$. Then,

$$y_1 = \frac{e^x}{(1+e^x)^2} = \frac{e^x}{1+e^x}.\frac{1}{1+e^x} = y(1-y)$$

$= y - y^2$ so that $y_1(0) = \frac{1}{2} - \left(\frac{1}{2}\right)^2 = \frac{1}{4}$,

$y_2 = y_1 - 2yy_1$ so that

$$y_2(0) = \frac{1}{4} - 2\left(\frac{1}{2}\right)\left(\frac{1}{4}\right) = 0,$$

$y_3 = y_2 - 2y_1^2 - 2yy_2$ so that $y_3(0) = -\frac{1}{8}$,

and so on. Hence Maclaurin's Theorem yields

$$\frac{e^x}{1+e^x} = y = y(0) + xy_1(0) + \frac{x^2}{2!}y_2(0) + \frac{x^3}{3!}y_3(0) + \ldots$$

$$= \frac{1}{2} + \frac{1}{4}x - \frac{1}{48}x^3 + \ldots.$$

EXAMPLE 2.42

Use Maclaurin's Theorem to expand

(i) $\sin(m \sin^{-1} x)$ and hence, expand $\sin m\theta$ in powers of $\sin\theta$.

(ii) $e^{a\sin^{-1}x}$ and hence, show that $e^\theta = 1$

$$+ \sin\theta + \frac{1}{2!}\sin^2\theta + \frac{2}{3!}\sin^3\theta + \ldots.$$

Solution. (i) Let

$y = \sin(m \sin^{-1} x)$ so that $y(0) = 0$.

Then,

$$y_1 = \cos(m \sin^{-1} x).\frac{m}{\sqrt{1-x^2}}$$

or

$$(1-x^2)y_1^2 = m^2\cos^2(m \sin^{-1} x) = m^2(1-y^2)$$

or

$(1-x^2)y_1^2 + m^2y^2 = 0$ so that $y_1(0) = m$.

Differentiating again, we get

$$2(1-x^2)y_1y_2 - 2xy_1^2 + 2m^2yy_1 = 0$$

or

$2(1-x^2)y_2 - xy_1 + m^2y = 0$ so that $y_2(0) = 0$

Now differentiating n times by Leibnitz's Theorem, we get

$$(1-x^2)y_{n+2} + n_{C_1}y_{n+1}(-2x) + n_{C_2}y_n(-2) - xy_{n+1} - n_{C_1}y_n + m^2y_n = 0$$

or

$$(1-x^2)y_{n+2} - (2n+1)xy_{n+1} - (n^2 - m^2)y_n = 0.$$

Substituting $x = 0$, we get

$y_{n+2}(0) = (n^2 - m^2)y_n(0)$.

Taking $n = 1,2,3,\ldots$, we get

$$y_3 ±±±(-m^2)y_1 = m(-m^2)$$

$$y_4(0) = (2^2 - m^2)\, y_2(0) = 0,$$

$$y_5(0) = (3^2 - m^2)\, y_3(0) = (3^2 - m^2)(1^2 - m^2)\, m,$$

and so on. Hence, the application of Leibnitz's Theorem yields

$$\sin(m \sin^{-1}x) = y(0) + xy_1(0) + \frac{x^2}{2!}y_2(0) + \frac{x^3}{3!}y_3(0) + \frac{x^4}{4!}y_4(0) + \frac{x^5}{5!}y_5(0) + \ldots$$

$$= mx + \frac{m(1^2 - m^2)}{3!}x^2 + \frac{m(1^2 - m^2)(3^2 - m^2)}{5!}x^5 + \ldots,$$

which is the required expansion. Substituting $x = \sin\theta$, we get

$$\sin m\theta = m \sin\theta + \frac{m(1^2 - m^2)}{3!}\sin^2\theta + \frac{m(1^2 - m^2)(3^2 - m^2)}{5!}\sin^5\theta + \ldots.$$

(ii) Let $y = e^{a\sin^{-1}x}$ so that $y(0) = 1$.
Then,

$$y_1 = e^{a\sin^{-1}x}.\frac{a}{\sqrt{1-x^2}}$$

or

$(1-x^2)y_1^2 - a^2y^2 = 0$ so that $y_1(0) = a$.

Differentiating once more, we get

$(1-x^2)y_2 - xy_1 - a^2y = 0$ so that $y_2(0) = a^2$.

Now differentiating n times by Leibnitz's Theorem, we have

$$(1-x^2)y_{n+2} - (2n+1)xy_{n+1} - (n^2 + a^2)y_n = 0$$

Substituting $x = 0$, we get

$$y_{n+2}(0) = (n^2 + a^2)\, y_n(0).$$

Now taking $n = 1,2,3\ldots$ we get

$$y_3(0) = a(1^2 + a^2),$$

$$y_4(0) = a^2(2^2 + a^2), \text{ and}$$

$$y_5(0) = a(1^2 + a^2)(3^2 + a^2).$$

Hence, using Maclaurin's Theorem, we get

$$e^{a\sin^{-1}x} = y(0) + xy_1(0) + \frac{x^2}{2!}y_2(0) + \frac{x^3}{3!}y_3(0) + \frac{x^4}{4!}y_4(0) + \frac{x^5}{5!}y_5(0) + \ldots$$

$$= 1 + ax + \frac{a^2x^2}{2!} + \frac{a(1^2 + a^2)}{3!}x^3 + \frac{a(2^2 + a^2)}{4!}x^4 + \frac{a(1^2 + a^2)(3^2 + a^2)}{5!}x^5 + \ldots$$

EXAMPLE 2.43

Show that

(i) $\log[x + \sqrt{1+x^2}] = x - \frac{x^3}{3!}1^2 + \frac{x^5}{5!}(3^2.1^2) - \frac{x^7}{7!}(5^2.3^2.1^2) + \ldots$

(ii) $\log(1 + \tan x) = x - \frac{1}{2}x^2 + \frac{2}{3}x^3 - \frac{7}{12}x^4 + \ldots$

Solution. (i) Let

$$y = \log\left[x + \sqrt{1+x^2}\right] \text{ so that } y(0) = 0.$$

Then,

$$y_1 = \frac{1}{x + \sqrt{1+x^2}}\left[1 + \frac{2}{\sqrt{x^2+1}}\right] = \frac{1}{\sqrt{x^2+1}}$$

or

$$(1+x)^2 y_1^2 - 1 = 0 \text{ so that } y_1(0) = 0.$$

Differentiating again, we get

$$(1+x^2)y_2 + xy_1 = 0 \text{ so that } y_2(0) = 0.$$

Differentiating n times by Leibnitz's Theorem, we get

$$(1+x^2)y_{n+2} + (2^{n+1})xy_{n+1} + n^2y_n = 0.$$

Substituting $x = 0$, we get

$$y_{n+2}(0) = -n^2y_n(0).$$

Taking $n = 1,2,3,\ldots$, we get

$$y_3(0) = -1^2,\ y_4(0) = 0,$$

$$y_5(0) = 3^2.1^2,\ y^6(0) = 0,$$

$$y_7(0) = -5^2.3^2.1^2, \text{ and } y_8(0) = 0.$$

Thus we note that $yn(0) = 0$, when n is even and

$$y_n(0) = (-1)^{\frac{n-1}{2}}(n-2)^2(n-4)^2\ldots.5^2.3^2.1^2,$$

when n is odd.

Hence, by Maclaurin's Theorem, we have

$$\log\left[x+\sqrt{1+x^2}\right] = y(0)+xy_1(0)+\frac{x^2}{2!}y_2(0)$$

$$+\frac{x^3}{3!}y_3(0)+\frac{x^4}{4!}y_4(0)$$

$$+\frac{x^5}{5!}y_5(0)+\ldots$$

$$= x-\frac{x^3}{3!}1^2+\frac{x^5}{5!}(3^2.1^2)$$

$$-\frac{x^7}{7!}(5^2.3^2.1^2)+\ldots$$

(ii) Let

$y = \log(1 + \tan x)$ so that $y(0) = \log 1 = 0$.

We have

$$e^y = 1+\tan x.$$

Differentiating with respect to x, we get

$$e^y y_1 = \sec^2 x \text{ so that } y_1(0) = 1.$$

Differentiating once more, we have

$$e^y\left(y_1^2+y_2\right) = 2\sec^2 x\tan x \text{ and so } y_2(0) = -1.$$

Differentiating again, we get

$$e^y\left(y_1^3+3y_1y_2+y_3\right) = 2\sec^4 x+4\sec^2 x\tan^2 x,$$

which yields $y_3(0) = 4$.

If we differentiate once again, we have

$$e^y\left(y_1^4+6y_1^2y_2+4y_1y_3+3y_2^2+y^4\right)\cdot$$

$$= 16\sec^4 x\tan x+8\sec^2 x\tan^2 x$$

= 16 sec^4 x tan x + 8 sec2 x tan2 x

and so, $y_4(0)=-14$. Hence, Maclaurin's Theorem yields

$$\log(1+\tan x) = x-\frac{1}{2}x^2+\frac{2}{3}x^3-\frac{7}{12}x^4+\ldots.$$

2.7 INDETERMINATE FORMS

A function is said to assume an *indeterminate form* if it involves the independent variable in such a way that for a certain assigned value of the independent variable its value cannot be found by simply substituting that value of the variable in the function.

Let f and g be real-valued functions. We know that

$$\lim_{x\to a}\frac{f(x)}{g(x)} = \frac{\lim_{x\to a} f(x)}{\lim_{x\to a} g(x)}, \lim_{x\to a} g(x)\neq 0. \quad (1)$$

But if both $\lim_{x\to a} f(x)$ and $\lim_{x\to a} g(x)$ are zero, then (1) gives no information about the $\lim_{x\to a}\frac{f(x)}{g(x)}$ since the quotient $\frac{\lim_{x\to a} f(x)}{\lim_{x\to a} g(x)}$ in that case takes the form $\frac{0}{0}$, which is meaningless. Thus, $\frac{0}{0}$ is one of the indeterminate forms. Similarly, if $\lim_{x\to a} f(x) = \lim_{x\to a} g(x) = \infty$, then the quotient $\frac{f(x)}{g(x)}$ assumes the indeterminate value $\frac{\infty}{\infty}$. The other indeterminate forms are $\infty-\infty$, $0\times\infty$, $0^0, 1^\infty$, and ∞^0.

Special methods are required to evaluate these indeterminate forms. These methods are called *L'Hospital Rule* after the name of the French mathematician L'Hospital.

Theorem 2.6. Let f and g be functions such that

(i) $\lim_{x\to a} f(x) = \lim_{x\to a} g(x) = 0$.

(ii) $f'(a)$ and $g'(a)$ exist and $g'(a)\neq 0$.

Then,

$$\lim_{x\to a}\frac{f(x)}{g(x)} = \frac{f'(a)}{g'(a)}.$$

Proof: Since, by hypothesis of the theorem, f and g are derivable at $x = a$, they are also continuous at $x = a$ and so,

$$\lim_{x\to a} f(x) = f(a) \text{ and } \lim_{x\to a} g(x) = g(a).$$

Therefore, by condition (i), $f(a) = g(a) = 0$. On the other hand,

$$f'(a) = \lim_{x\to a}\frac{f(x)-f(a)}{x-a} = \lim_{x\to a}\frac{f(x)}{x-a}$$

and

$$g'(a) = \lim_{x\to a}\frac{g(x)-g(a)}{x-a} = \lim_{x\to a}\frac{g(x)}{x-a} \quad (2)$$

Consequently,

$$\lim_{x\to a}\frac{f(x)}{g(x)} = \lim_{x\to a}\left[\frac{\frac{f(x)}{x-a}}{\frac{g(x)}{x-a}}\right] = \frac{\lim_{x\to a}\frac{f(x)}{x-a}}{\lim_{x\to a}\frac{g(x)}{x-a}}$$

$$= \frac{f'(a)}{g'(a)}, \text{ using (2).}$$

Theorem 2.7. (L'Hospital Rule for $\frac{0}{0}$ form). Let f and g be functions such that

(i) $\lim_{x\to a} f(x) = \lim_{x\to a} g(x) = 0$,

(ii) $f'(x)$ and $g'(x)$ exist and $g'(x) \neq 0$ for $x \in (a-\delta, a+\delta)$, $\delta > 0$ except possibly at $x = a$, and

(iii) $\lim_{x\to a} \frac{f'(x)}{g'(x)}$ exists.

Then,

$$\lim_{x\to a} \frac{f(x)}{g(x)} = \lim_{X\to a} \frac{f'(x)}{g'(x)}, \quad g'(a) \neq 0.$$

Proof: Under the given hypothesis, f and g are continuous at $x = a$ and so,

$$f(a) = \lim_{x\to a} f(x) = 0 \text{ and } g(a) = \lim_{X\to a} g(x) = 0.$$

By Taylor's Theorem truncated at the first-derivative term, we have

$$f(a+h) = f(a) + hf'(a+\theta_1 h) = hf'(a+\theta_1 h),$$
$$0 < \theta_1 < 1$$

and

$$g(a+h) = g(a) + hg'(a+\theta_2 h) = hg'(a+\theta_2 h),$$
$$0 < \theta_2 < 1.$$

Therefore,

$$\lim_{x\to a} \frac{f(x)}{g(x)} = \lim_{h\to 0} \frac{f(a+h)}{g(a+h)} = \lim_{h\to 0} \frac{f'(a+\theta_1 h)}{g'(a+\theta_2 h)}$$

$$= \frac{f'(a)}{g'(a)}, \; g'(a) \neq 0 = \lim_{x\to 0} \frac{f'(x)}{g'(x)}.$$

Hence,

$$\lim_{x\to a} \frac{f(x)}{g(x)} = \lim_{x\to a} \frac{f'(x)}{g'(x)}, \quad g'(a) \neq 0.$$

Obviously, this relation fails if $g'(a) = 0$. If $g'(a) = 0$ and if $f'(a) \neq 0$, then

$$\lim_{x\to a} \frac{f'(x)}{g'(x)} = +\infty \text{ or } -\infty.$$

If $f'(a) = g'(a) = 0$, then by Taylor's Theorem (assuming second-order derivatives of f and g exist), we have

$$f(a+h) = f(a) + hf'(a) + \frac{h^2}{2!} f''(a+\theta_3 h),$$
$$0 < \theta_3 < 1$$

$$= \frac{h^2}{2!} f''(a+\theta_3 h)$$

and

$$g(a+h) = g(a) + hg'(a) + \frac{h^2}{2!} g''(a+\theta_4 h),$$

Therefore, $\quad 0 < \theta_4 < 1$

$$\lim_{x\to a} \frac{f(x)}{g(x)} = \lim_{h\to 0} \frac{f(a+h)}{g(a+h)} = \lim_{h\to 0} \frac{f'(a+\theta_3 h)}{g'(a+\theta_4 h)}$$

$$= \frac{f''(a)}{g''(a)} = \lim_{x\to a} \frac{f''(x)}{g''(x)}, g''(a) \neq 0.$$

Hence, in general, if

$$f(a) = f'(a) = f'(a) = \ldots = f^{(n-1)}(a) = 0,$$

$$g(a) = g'(a) = g''(a) = \ldots = g^{(n-1)}(a) = 0,$$

and $g^{(n)}(a) \neq 0$, then using Taylor's Theorem with a remainder after n terms, we have

$$\lim_{x\to a} \frac{f(x)}{g(x)} = \lim_{x\to a} \frac{f^{(n)}(x)}{g^{(n)}(x)}, \; g^{(n)}(a) \neq 0,$$

which is called the *generalized L'Hospital Rule.*

Remark 2.2. L'Hospital Rule is also applicable if $\lim_{x\to\infty} f(x) = \lim_{x\to\infty} g(x) = 0$. The proof follows the same lines as the proof given in Remark after Theorem 2.8.

EXAMPLE 2.44

Evaluate $\lim_{x\to 0} \frac{e^x + \log\left(\frac{1-x}{e}\right)}{\tan x - x}$.

Solution. Since $\frac{1}{e} = \log 1 - \log e = 0 - 1 = -1$,

$\lim_{x\to a} \frac{e^x + \log\left(\frac{1-x}{e}\right)}{\tan x - x}$ is of $\frac{0}{0}$ form:

Therefore, using L'Hospital Rule, we have

$$\lim_{x\to a} \frac{e^x + \log\left(\frac{1-x}{e}\right)}{\tan x - x} = \lim_{x\to 0} \frac{e^x + \frac{\frac{1}{1-x}}{\frac{e}{e}}\left(-\frac{1}{e}\right)}{\sec^2 x - 1}$$

$$= \lim_{x\to 0}\frac{e^x - \frac{1}{1-x}}{\tan^2 x}\left(\frac{0}{0}\text{ form}\right)$$

$$= \lim_{x\to 0}\frac{e^x - \frac{1}{1-x}}{x^2\left(\frac{\tan x}{x}\right)^2}$$

$$= \lim_{x\to 0}\frac{e^x - \left(\frac{1}{1-x}\right)}{x^2} \text{ since } \lim_{x\to 0}\frac{\tan x}{x} = 1.$$

$$= \lim_{x\to 0}\frac{e^x - (1-x)^{-2}}{2x}\left(\frac{0}{0}\text{ form}\right) = \lim_{x\to 0} e^x - \frac{2}{(1-x)^2}$$

$$= \frac{e^0 - 2}{2} = \frac{1-2}{2} = -\frac{1}{2}$$

EXAMPLE 2.45

Evaluate $\lim_{x\to 0}\frac{a^x - x^a}{a^a - x^x}$.

Solution. We have

$$\lim_{x\to a}\frac{a^x - x^a}{a^a - x^X}\left(\frac{0}{0}\text{ form}\right)$$

$$= \lim_{x\to a}\frac{a^x \log a - ax^{a-1}}{-x^x(1+\log x)} = \frac{a^a \log a - a^a}{-a^a(1+\log a)} = \frac{1-\log a}{1+\log a}$$

$$= \frac{\log e - \log a}{\log e + \log a} = \frac{\log \frac{e}{a}}{\log ea}.$$

EXAMPLE 2.46

Evaluate $\lim_{x\to a}\frac{(1+x)^{\frac{1}{x}} - e}{x}$.

Solution. Since $\lim_{x\to 0}(1+x)^{\frac{1}{x}} = e$, the given limit is of $\frac{0}{0}$ form. Therefore, using L'Hospital Rule, we have

$$\lim_{x\to 0}\frac{(1+x)^{\frac{1}{x}} - e}{x} = \lim_{x\to 0}\frac{\frac{d}{dx}(1+x)^{\frac{1}{x}}}{1}. \qquad (1)$$

To find $\frac{d}{dx}(1+x)^{\frac{1}{x}}$, let $y = (1+x)^{\frac{1}{x}}$. Then,

$$\log y = \frac{1}{x}\log(1+x)\cdot$$

Differentiating with respect to x yields

$$\frac{1}{y}\frac{dy}{dx} = \frac{1}{x(1+x)} - \frac{1}{x^2}\log(1+x)$$

or

$$\frac{dy}{dx} = y\left[\frac{1}{x(1+x)} - \frac{1}{x^2}\log(1+x)\right]$$

$$= (1+x)^{\frac{1}{x}}\left[\frac{x-(1+x)\log(1+x)}{x^2(1+x)}\right].$$

Thus, (1) reduces to

$$\lim_{x\to 0}\frac{(1+x)^{\frac{1}{x}} - e}{x} = \lim_{x\to 0}\frac{(1+x)^{\frac{1}{x}}[x-(1+x)\log(1+x)]}{x^2(1+x)}$$

$$= \lim_{x\to 0}(1+x)^{\frac{1}{x}}\lim_{x\to 0}\frac{x-(1+x)\log(1+x)}{x^2+x^3}$$

$$\left(\frac{0}{0}\text{ form}\right)$$

$$= e\lim_{x\to 0}\frac{1-\left[\frac{1+x}{1+x} + \log(1+x)\right]}{2x+3x^2}$$

$$= -e\lim_{x\to 0}\frac{\log(1+x)}{2x+3x^2}\left(\frac{0}{0}\text{ form}\right)$$

$$= -e\lim_{x\to 0}\frac{1}{(1+x)(2+6x)} = -\frac{e}{2}$$

EXAMPLE 2.47

Evaluate $\lim_{x\to 0}\frac{e^x - e^{-x} - 2x}{x - \sin x}$.

Solution. We have, by L'Hospital Rule,

$$\lim_{x\to 0}\frac{e^x - e^{-x} - 2x}{x - \sin x}\left(\frac{0}{0}\text{ form}\right)$$

$$= \lim_{x\to 0}\frac{e^x + e^{-x} - 2}{1 - \cos x}\left(\frac{0}{0}\text{ form}\right)$$

$$= \lim_{x\to 0}\frac{e^x - e^{-x}}{\sin x}\left(\frac{0}{0}\text{ form}\right)$$

$$= \lim_{x\to 0}\frac{e^x + e^{-x}}{\cos x} = \frac{1+1}{1} = 2$$

EXAMPLE 2.48

Evaluate $\lim_{x\to a}\frac{x^a - a^x}{x^x - a^a}$.

Solution. Using L'Hospital Rule, we have

$$\lim_{x\to a}\frac{x^a - a^x}{x^x - a^a}\left(\frac{0}{0}\text{ form}\right)$$

$$= \lim_{x\to a}\frac{ax^{a-1} - a^x \log a}{x^x(1+\log x) - 0} = \frac{a^a - a^a \log a}{a^a(1+\log a)}$$

$$= \frac{a^a(1-\log a)}{a^a(1+\log a)} = \frac{1-\log a}{1+\log a}.$$

EXAMPLE 2.49

Evaluate $\lim_{x\to 1} \frac{x^x - x}{-1+x-\log x}$.

Solution. We have

$$\lim_{x\to 1} \frac{x^x - x}{-1+x-\log x}\left(\frac{0}{0} \text{ form}\right)$$

$$= \lim_{x\to 1} \frac{x^x(1+\log x)-1}{1-\frac{1}{x}}\left(\frac{0}{0} \text{ form}\right)$$

$$= \lim_{x\to 1} \frac{x^x(1+\log x)(1+\log x)+x^x.\frac{1}{x}}{\frac{1}{x^2}} = \frac{1+1}{1} = 2.$$

EXAMPLE 2.50

Evaluate $\lim_{x\to a} \frac{e^x - e^{-x} - 2\log(1+x)}{x \sin x}$.

Solution. Use of L'Hospital Rule yields

$$\lim_{x\to a} \frac{e^x - e^{-x} - 2\log(1+x)}{x \sin x}\left(\frac{0}{0} \text{ form}\right)$$

$$= \lim_{x\to a} \frac{e^x + e^{-x} - \frac{2}{1+x}}{x\cos x + \sin x}\left(\frac{0}{0} \text{ form}\right)$$

$$= \lim_{x\to a} \frac{e^x + e^{-x} - \frac{2}{(1+x)^2}}{-x\sin x + \cos x + \cos x} = \frac{2}{2} = 1.$$

Theorem 2.8. (L'Hospital Rule for the indeterminate form $\frac{\infty}{\infty}$). Let $\lim_{x\to a} f(x) = \infty$, $\lim_{x\to a} g(x) = \infty$, and $\lim_{X\to a} \frac{f'(x)}{g'(x)}$ exist (finite or infinite). Then,

$$\lim_{x\to a} \frac{f(x)}{g(x)} = \lim_{x\to a} \frac{f'(x)}{g'(x)}.$$

Proof: Since $\lim_{x\to a} f(x) = \lim_{x\to a} g(x) = \infty$, we have

$$\lim_{x\to a} \frac{1}{f(x)} = \lim_{x\to a} \frac{1}{g(x)} = 0.$$

Therefore, using L'Hospital Rule for $\frac{0}{0}$ form, we obtain

$$\lim_{X\to a} \frac{f(x)}{g(x)} = \lim_{x\to a} \frac{\frac{1}{g(x)}}{\frac{1}{f(x)}}\left(\frac{0}{0} \text{ form}\right)$$

$$= \lim_{x\to a} \frac{\frac{-g'(x)}{[g(x)]^2}}{\frac{f'(x)}{[f(x)]^2}} = \lim_{x\to a} \frac{g'(x)}{f'(x)}\left[\frac{f(x)}{g(x)}\right]^2. \quad (1)$$

Since $\lim_{x\to a} \frac{f'(x)}{g'(x)}$ exists by the given hypothesis, it follows from (1) that $\lim_{x\to a} \frac{f(x)}{g(x)}$ exists.

Let $\lim_{x\to a} \frac{f(x)}{g(x)} = l$. Then three cases arise.

Case 1. If l is neither zero nor infinite, then (1) reduces to

$$l = \lim_{X\to a} \frac{g'(x)}{f'(x)} l^2 \text{ or } \frac{1}{l} = \lim_{X\to a} \frac{g'(x)}{f'(x)} \text{ or } l = \lim_{x\to a} \frac{f'(x)}{g'(x)},$$

that is,

$$\lim_{x\to a} \frac{f(x)}{g(x)} = \lim_{x\to a} \frac{f'(x)}{g'(x)}$$

Case 2. If $l = 0$, then,

$$l+1 = \lim_{x\to a} \frac{f(x)}{g(x)} + 1 = \lim_{x\to a} \frac{f(x)+g(x)}{g(x)}.$$

Thus, $\lim_{x\to a} \frac{f(x)+g(x)}{g(x)}$ is neither zero nor infinity. Therefore, by case 1,

$$l+1 = \lim_{x\to a} \frac{f'(x)+g'(x)}{g'(x)} = \lim_{x\to a} \frac{f'(x)}{g'(x)} + 1$$

or

$$l = \lim_{X\to a} \frac{f'(x)}{g'(x)}, \text{ that is, } \lim_{x\to a} \frac{f(x)}{g(x)} = \lim_{X\to a} \frac{f'(x)}{g'(x)}.$$

Case 3. If $l = \infty$, then $\lim_{x\to a} \frac{g(x)}{f(x)} = 0$. Therefore, by case 2, we have

$$0 = \lim_{x\to a} \frac{g(x)}{f(x)} = \lim_{x\to a} \frac{g'(x)}{f'(x)}$$

or

$$\infty = \lim_{x\to a} \frac{f'(x)}{g'(x)} \quad \text{or } l = \lim_{x\to a} \frac{f'(x)}{g'(x)},$$

that is,

$$\lim_{x\to a} \frac{f(x)}{g(x)} = \lim_{x\to a} \frac{f'(x)}{g'(x)}.$$

Remarks 2.3. The L'Hospital Rule, proved above, can be extended to the case where $x \to \infty$. Thus, if $\lim_{x\to\infty} f(x) = \infty, \lim_{x\to\infty} g(x) = \infty$, and $\lim_{x\to\infty} \frac{f'(x)}{g'(x)}$ exists, then

$$\lim_{x\to\infty} \frac{f(x)}{g(x)} = \lim_{x\to\infty} \frac{f'(x)}{g'(x)}.$$

In fact, substituting $x = \frac{1}{z}$, we have $z \to 0$ as $x \to \infty$ and therefore,

$$\lim_{z\to 0} f\left(\frac{1}{z}\right) = \infty \text{ and } \lim_{z\to 0} g\left(\frac{1}{z}\right) = \infty.$$

Applying L'Hospital Rule to $\frac{f\left(\frac{1}{z}\right)}{g\left(\frac{1}{z}\right)}$, we have

$$\lim_{x\to\infty} \frac{f(x)}{g(x)} = \lim_{z\to 0} \frac{f\left(\frac{1}{z}\right)}{g\left(\frac{1}{z}\right)}$$

$$= \lim_{z\to 0} \frac{f'\left(\frac{1}{z}\right)\left(-\frac{1}{z^2}\right)}{g'\left(\frac{1}{z}\right)\left(-\frac{1}{z^2}\right)}$$

$$= \lim_{z\to 0} \frac{f'\left(\frac{1}{z}\right)}{g'\left(\frac{1}{z}\right)} = \lim_{x\to\infty} \frac{f'(x)}{g'(x)}.$$

2. If we have $\lim_{x\to a} \frac{f(x)}{g(x)}$ in $\frac{\infty}{\infty}$ form, we should change it to $\frac{0}{0}$ form to avoid differentiation of numerators and denominators again and again.

EXAMPLE 2.51

Evaluate $\lim_{x\to 1-} \frac{\log(1-x)}{\cot(\pi x)}$.

Solution. We have

$$\lim_{x\to 1} \frac{\log(1-x)}{\cot(\pi x)} \left(\frac{\infty}{\infty} \text{ form}\right), \text{ since } \log 0 = -\infty$$

$$= \lim_{x\to 1-} \frac{\frac{-1}{1-x}}{-\pi \cos ec^2(\pi x)}$$

$$= \lim_{x\to 1} \frac{\sin^2(\pi x)}{-\pi(1-x)} \left(\frac{0}{0} \text{ form}\right)$$

$$= \lim_{x\to 1-} \frac{2\pi \sin \pi x \cos \pi x}{-\pi} = 0.$$

EXAMPLE 2.52

Evaluate $\lim_{x\to\frac{\pi}{2}} \frac{\log\left(x-\frac{\pi}{2}\right)}{\tan x}$.

Solution. We have

$$\lim_{x\to\frac{\pi}{2}} \frac{\log\left(x-\frac{\pi}{2}\right)}{\tan x} \left(\frac{\infty}{\infty} \text{ } form\right)$$

$$= \lim_{x\to\frac{\pi}{2}} \frac{\frac{1}{x-\frac{\pi}{2}}}{\sec^2 x} \left(\frac{\infty}{\infty} \text{ form}\right) = \lim_{x\to\frac{\pi}{2}} \frac{\frac{1}{x-\frac{\pi}{2}}}{\frac{1}{\cos^2 x}}$$

$$= \lim_{x\to\frac{\pi}{2}} \frac{\cos^2 x}{x-\frac{\pi}{2}} \left(\frac{0}{0} \text{ form}\right)$$

$$= \lim_{x\to\frac{\pi}{2}} \frac{-2\cos x \sin x}{1} = 0.$$

EXAMPLE 2.53

Evaluate $\lim_{x\to 0} \frac{\log x}{\cot x}$.

Solution. We note that

$$\lim_{x\to 0} \frac{\log x}{\cot x} \left(\frac{\infty}{\infty} \text{ form}\right)$$

$$= \lim_{x\to 0} \frac{\frac{1}{x}}{-\text{cosec}^2 x} \left(\frac{\infty}{\infty} \text{ form}\right)$$

$$= \lim_{x\to 0} -\frac{\sin^2 x}{x} \left(\frac{0}{0} \text{ form}\right)$$

$$= \lim_{x\to 0} \frac{2\sin x \cos x}{1} = 0.$$

EXAMPLE 2.54

Show that $f(x) = x \log \sin x$, $x \neq 0$, $f(0) = 0$ is continuous at $x = 0$.

Solution. To show that f is continuous at $x = 0$, we have to prove that $\lim_{x\to 0} f(x) = f(0) = 0$. We note that

$$\lim_{x\to 0} f(x) = \lim_{x\to 0} x \log \sin x$$

$$= \lim_{x\to 0} \frac{\log \sin x}{\frac{1}{x}} \left(\frac{\infty}{\infty} \text{ form}\right)$$

$$= \lim_{x\to 0} \frac{\cot x}{-\frac{1}{x^2}} = \lim_{x\to 0} - x.\frac{x}{\tan x}$$

$$= -\lim_{x\to 0} x.\lim_{x\to 0} \frac{x}{\tan x} = 0 = f(0).$$

Hence f is continuous at $x = 0$.

EXAMPLE 2.55

Evaluate $\lim_{x\to 0} \dfrac{\log \log(1-x^2)}{\log \log \cos x}$.

Solution. We have

$$\lim_{x\to 0} \frac{\log \log(1-x^2)}{\log \log \cos x} \left(\frac{\infty}{\infty} \text{form}\right)$$

$$= \lim_{x\to 0} \frac{\frac{1}{\log(1-x^2)} \cdot \frac{(-2x)}{1-x^2}}{\frac{1}{\log \cos x} \cdot \frac{-\sin x}{\cos x}}$$

$$= \lim_{x\to 0} \frac{2x \cos x \log \cos x}{(1-x^2) \sin x \log(1-x^2)}$$

$$= \lim_{x\to 0} \frac{x}{\sin x} \lim_{x\to 0} \frac{2x \cos x \log \cos x}{(1-x^2) \log(1-x^2)}$$

$$= \lim_{x\to 0} \frac{2x \cos x \log \cos x}{(1-x^2) \log(1-x^2)} \left(\frac{\infty}{\infty} \text{ form}\right)$$

$$= \lim_{x\to 0} \frac{-2 \sin x \log \cos x + 2 \cos x \frac{1}{\cos x}(- \sin x)}{(1-x^2).\frac{(-2x)}{1-x^2} + \log(1-x^2)(-2x)}$$

$$= \lim_{x\to 0} \frac{-2 \sin x \log \cos x - 2 \sin x}{(-2x) - 2x \log(1-x^2)} \left(\frac{0}{0} \text{ form}\right)$$

$$= \lim_{x\to 0} \frac{2 \frac{\sin^2 x}{\cos x} - 2 \cos x \log \cos x - 2 \cos x}{-2 \log(1-x^2) - 2x.\frac{(-2x)}{1-x^2}}$$

$$= \frac{-2}{-2} = 1.$$

Indeterminate form $\infty - \infty$.

The indeterminate form $\infty - \infty$ arises when we want to evaluate $\lim_{x\to a} [f(x) - g(x)]$ and when $\lim_{x\to a} f(x) = \lim_{x\to a} g(x) = \infty$. In such a case, we may write

$$\lim_{x\to a} [f(x) - g(x)] = \lim_{x\to a} \frac{\frac{1}{g(x)} - \frac{1}{f(x)}}{\frac{1}{f(x)g(x)}} \left(\frac{0}{0} \text{ form}\right),$$

which can be evaluated using L'Hospital Rule for $\frac{0}{0}$ form.

EXAMPLE 2.56

Evaluate $\lim_{x\to 2} \left[\frac{1}{x-2} - \frac{1}{\log (x-1)}\right]$.

Solution. We have

$$\lim_{x\to 2} \left[\frac{1}{x-2} - \frac{1}{\log(x-1)}\right] \quad (\infty - \infty \text{ form})$$

$$= \lim_{x\to 2} \left[\frac{\log(x-1) - (x-2)}{(x-2) \log(x-1)}\right] \left(\frac{0}{0} \text{ form}\right)$$

$$= \lim_{x\to 2} \left[\frac{\frac{1}{x-1} - 1}{\frac{x-2}{x-1} + \log (x-1)}\right]$$

$$= \lim_{x\to 2} \left[\frac{2-x}{x-2+(x-1) \log(x-1)}\right] \left(\frac{0}{0} \text{ form}\right)$$

$$= \lim_{x\to 2} \left[\frac{-1}{1+1+ \log(x-1}\right] = -\frac{1}{2}.$$

EXAMPLE 2.57

Evaluate $\lim_{x\to 0} (\frac{1}{x} - \cot x)$.

Solution. We have

$$\lim_{x\to 0} \left(\frac{1}{x} - \cot x\right) \quad (\infty - \infty \text{ form})$$

$$= \lim_{x\to 0} \left(\frac{1}{x} - \frac{\cos x}{\sin x}\right)$$

$$= \lim_{x\to 0} \frac{\sin x - x \cos x}{x \sin x} \left(\frac{0}{0} \text{ form}\right)$$

$$= \lim_{x\to 0} \frac{x}{\sin x}.\lim_{x\to 0} \frac{\sin x - x \cos x}{x^2}$$

$$= \lim_{x\to 0} \frac{\sin x - x \cos x}{x^2} \left(\frac{0}{0} \text{ form}\right)$$

$$= \lim_{x\to 0}\frac{\cos x + x\sin x - \cos x}{2x}$$

$$= \lim_{x\to 0}\frac{\sin x}{2} = 0\,.$$

EXAMPLE 2.58

Evaluate $\lim_{x\to 0}\left(\frac{1}{x^2}-\frac{1}{\sin^2 x}\right)$.

Solution. To determine the required limit, we have

$$\lim_{x\to 0}\left(\frac{1}{x^2}-\frac{1}{\sin^2 x}\right)\quad (\infty-\infty \text{ form})$$

$$= \lim_{x\to 0}\frac{\sin^2 x - x^2}{x^2\sin^2 x}\left(\frac{0}{0}\text{ form}\right)$$

$$= \lim_{x\to 0}\frac{x^2}{\sin^2 x}\lim_{x\to 0}\frac{\sin^2 x - x^2}{x^4}$$

$$= \lim_{x\to 0}\left(\frac{x}{\sin x}\right)^2\lim_{x\to 0}\frac{\sin^2 x - x^2}{x^4}$$

$$= \lim_{x\to 0}\frac{\sin^2 x - x^2}{x^4}\left(\frac{0}{0}\text{ form}\right)$$

$$= \lim_{x\to 0}\frac{2\sin x\cos x - 2x}{4x^3}\lim_{x\to 0}\frac{\sin 2x - 2x}{4x^3}\left(\frac{0}{0}\text{ form}\right)$$

$$= \lim_{x\to 0}\frac{2\cos 2x - 2}{12x^2}\left(\frac{0}{0}\text{ form}\right)$$

$$= \lim_{x\to 0}\frac{-2\sin 2x}{12x} = \lim_{x\to 0}\frac{-\sin 2x}{6x}$$

$$= -\frac{1}{6}\lim_{x\to 0}\frac{2\sin x\cos x}{x}$$

$$= \frac{-1}{3}\lim_{x\to 0}\frac{\sin x}{x}\lim_{x\to 0}\cos x = -\frac{1}{3}.$$

EXAMPLE 2.59

Evaluate $\lim_{x\to 0}\left(\frac{1}{x^2}-\cot^2 x\right)$.

Solution. We have

$$\lim_{x\to 0}\left(\frac{1}{x^2}-\cot^2 x\right)\ (\infty-\infty \text{ form})$$

$$= \lim_{x\to 0}\left(\frac{1}{x^2}-\frac{1}{\tan^2 x}\right) = \lim_{x\to 0}\frac{\tan^2 x - x^2}{x^2\tan^2 x}\ \left(\frac{0}{0}\text{ form}\right)$$

$$= \lim_{x\to 0}\frac{\tan^2 x^2}{x^4\left(\frac{\tan x}{x}\right)^2} = \lim_{x\to 0}\frac{1}{\left(\frac{\tan x}{x}\right)^2}\lim_{x\to 0}\frac{\tan^2 x - x^2}{x^4}$$

$$= \lim_{x\to 0}\frac{\tan^2 x - x^2}{x^4} = \lim_{x\to 0}\frac{\left(x+\frac{x^3}{3}+\frac{2}{15}x^5+\ldots\right)^2 - x^2}{x^4}$$

$$= \lim_{x\to 0}\frac{\left[x^2+\frac{2x^4}{3}+\text{terms having higher power of } x\right]-x^2}{x^4}$$

$$= \lim_{x\to 0}\left[\frac{2}{3}+\text{ terms having power of } x\right] = \frac{2}{3}.$$

EXAMPLE 2.60

Evaluate $\lim_{x\to 0}\left(\operatorname{cos} ecx - \frac{1}{x}\right)$.

Solution. We have

$$\lim_{x\to 0}\left(\operatorname{cos ec}\, x - \tfrac{1}{x}\right)\ (\infty-\infty \text{ form})$$

$$= \lim_{x\to 0}\frac{x-\sin x}{x\sin x}\left(\frac{0}{0}\text{ form}\right)$$

$$= \lim_{x\to 0}\frac{x-\sin x}{x^2}\left(\frac{x}{\sin x}\right) = \lim_{x\to 0}\left(\frac{x}{\sin x}\right)\lim_{X\to 0}\frac{x-\sin x}{x^2}$$

$$= \lim_{x\to 0}\frac{x-\sin x}{x^2}\left(\frac{0}{0}\text{ form}\right)$$

$$= \lim_{x\to 0}\frac{1-\cos x}{2x}\left(\frac{0}{0}\text{ form}\right) = \lim_{x\to 0}\frac{\sin x}{2} = 0.$$

The Indeterminate Form $0\times\infty$.

If $\lim_{x\to 0} f(x) = 0$ and $\lim_{x\to 0} g(x) = \infty$, then $\lim_{x\to 0} f(x)g(x)$ is indeterminate form of $0\times\infty$. To find this limit, the product $f(x)\,g(x)$ is written in one of the following form:

(i) $f(x)g(x) = \frac{f(x)}{\frac{1}{g(x)}}\left(\frac{0}{0}\text{ form}\right)$

(ii) $f(x)g(x) = \frac{g(x)}{\frac{1}{f(x)}}\left(\frac{\infty}{\infty}\text{ form}\right)$.

EXAMPLE 2.61

Evaluate $\lim_{X\to\infty} x\tan\frac{1}{x}$.

Solution. For the given limit, we have

$$\lim_{x\to\infty} x \tan\frac{1}{x} \quad (0\times\infty \text{ form})$$

$$= \lim_{x\to\infty}\frac{\tan\frac{1}{x}}{\frac{1}{x}}\left(\frac{0}{0}\text{ form}\right) = \lim_{h\to 0}\frac{\tan h}{h} = 1,\ h = \frac{1}{x}.$$

EXAMPLE 2.62

Evaluate $\lim_{x\to 0+} x \log x$.

Solution. We have

$$\lim_{x\to 0+} x \log x \ (0\times\infty \text{ form})$$

$$= \lim_{x\to 0+}\frac{\log x}{\frac{1}{x}}\left(\frac{\infty}{\infty}\text{ form}\right) = \lim_{x\to 0+}\frac{\frac{1}{x}}{\frac{-1}{x^2}}$$

$$= \lim_{x\to 0+}(-x) = 0.$$

The Indeterminate Forms $0^0, 1^\infty$, **and** ∞^0.

These indeterminate forms occur when evaluation of limits of function of the type $[f(x)]^{g(x)}$ is required when $x \to a$. Thus,

(i) if $\lim_{x\to a} f(x) = \lim_{x\to a} g(x) = 0$, then $\lim_{x\to a}[f(x)]^{g(x)}$ is of 0^0 form.

(ii) if $\lim_{x\to a} f(x) = 1$ and $\lim_{x\to a} g(x) = \infty$, then $\lim_{x\to a}[f(x)]^{g(x)}$ is of 1^∞ form.

(iii) if $\lim_{x\to a} f(x) = \infty$ and $\lim_{x\to a} g(x) = 0$, then $\lim_{x\to a}[f(x)]^{g(x)}$ is of ∞^0 form.

For $f(x) > 0$, let $y = [f(x)]^{g(x)}$ Then,

$$\log y = g(x) \log f(x)$$

and so,

$$\lim_{x\to a} \log y = \lim_{x\to a} g(x) \log f(x) \quad (0\times\infty \text{ form})$$

Since lim (log y) = log (lim y), lim y can be found.

EXAMPLE 2.63

Evaluate $\lim_{x\to 0}(x-a)^{x-a}$.

Solution. We have

$$\lim_{x\to a}(x-a)^{x-a}\ \left(0^0 \text{ form}\right).$$

Put $y = (x-a)^{x-a}$. Then

$$\log y = (x-a)\log(x-a) = \frac{\log(x-a)}{\frac{1}{x-a}}.$$

Therefore,

$$\lim_{x\to a} \log y = \lim_{x\to a}\frac{\log(x-a)}{\frac{1}{x-a}}$$

$$= \lim_{x\to a}\frac{\frac{1}{x-a}}{-\frac{1}{(x-a)^2}} = \lim_{x\to a} -(x-a) = 0.$$

Thus,

$$\lim_{x\to a} \log y = \log\left(\lim_{x\to a} y\right) = 0 \text{ or } \lim_{x\to a} y = e^0 = 1,$$

that is,

$$\lim_{x\to 0}(x-a)^{(x-a)} = 1.$$

EXAMPLE 2.64

Evaluate $\lim_{x\to 0}\left(\frac{\tan x}{x}\right)^{\frac{1}{x^2}}$.

Solution. The given limit is of 1^∞ form. So, let

$$y = \left(\frac{\tan x}{x}\right)^{\frac{1}{x^2}},$$

and we have

$$\log y = \frac{1}{x^2}\log\frac{\tan x}{x}.$$

Thus,

$$\lim_{x\to 0}\log y = \lim_{x\to 0}\frac{\log\frac{\tan x}{x}}{x^2}$$

$$\left(\frac{0}{0}\text{ form since } \lim_{x\to 0}\frac{\tan x}{x} = 1\right)$$

$$= \lim_{x\to 0}\frac{\frac{\sec^2 x}{\tan x} - \frac{1}{x}}{2x} = \lim_{x\to 0}\frac{x\sec^2 x - \tan x}{2x^2 \tan x}\left(\frac{0}{0}\text{ form}\right)$$

$$= \lim_{x\to 0}\frac{\sec^2 \tan x}{2\tan x + x\sec^2 x} = \lim_{x\to 0}\frac{\tan x}{\sin 2x + x}\left(\frac{0}{0}\text{ form}\right)$$

$$= \lim_{x\to 0}\frac{\sec^2 x}{2\cos 2x + 1} = \frac{1}{3}.$$

Hence,

$$\lim_{x\to 0} \log y = \log\left(\lim_{x\to 0} y\right) = \frac{1}{3},$$

which implies

$$\lim_{x\to 0} y = e^{\frac{1}{3}} \text{ or } \lim_{x\to 0}\left(\frac{\tan x}{x}\right)^{\frac{1}{x^2}} = e^{\frac{1}{3}}.$$

EXAMPLE 2.65

Evaluate $\lim_{x\to 0}(\cot x)^{\frac{1}{\log x}}$.

Solution. The given limit is of the form ∞^0. So, let $y = (\cot x)^{\frac{1}{\log x}}$ and we have

$$\log y = \frac{1}{\log x} \log \cot x.$$

Thus,

$$\lim_{x\to 0} \log y = \lim_{x\to 0} \frac{\log \cot x}{\log x} \left(\frac{\infty}{\infty} \text{ form}\right)$$

$$= \lim_{x\to 0} \frac{\frac{-\operatorname{cosec}^2 x}{\cot x}}{\frac{1}{x}} = \lim_{x\to 0} \frac{-x}{\sin x \cos x}$$

$$= \lim_{x\to 0} -\left(\frac{x}{\sin x}\right) \lim_{x\to 0} \frac{1}{\cos x} = -1.$$

Therefore,

$$\lim_{x\to 0} \log y = \log\left(\lim_{x\to 0} y\right) = -1 \text{ or } \lim_{x\to 0} y = e^{-1} = \frac{1}{e}$$

or

$$\lim_{x\to 0}(\cot x)^{\frac{1}{\log x}} = \frac{1}{e}.$$

EXAMPLE 2.66

Evaluate $\lim_{x\to 0}\left(\frac{\sin x}{x}\right)^{\frac{1}{x^2}}$

Solution. Since $\lim_{x\to 0} \frac{\sin x}{x} = 1$, the given limit is of the form 1^∞. So, let

$$y = \left(\frac{\sin x}{x}\right)^{\frac{1}{x^2}},$$

which yields

$$\log y = \frac{1}{x^2} \log \frac{\sin x}{x}$$

and so,

$$\lim_{x\to 0} \log y = \lim_{x\to 0} \frac{\log \frac{\sin x}{x^2}}{x} \left(\frac{0}{0} \text{ form}\right)$$

$$= \lim_{x\to 0} \frac{\frac{x}{\sin x}\left[\frac{x\cos x - \sin x}{x^2}\right]}{2x}$$

$$= \lim_{x\to 0} \frac{x\cos x - \sin x}{2x^3} \left(\frac{0}{0} \text{ form}\right)$$

$$= \lim_{x\to 0} \frac{x(-\sin x) + \cos x - \cos x}{6x^2}$$

$$= \lim_{x\to 0} \frac{-\sin x}{6x} = \lim_{x\to 0} \frac{-\cos x}{6} = -\frac{1}{6}.$$

Thus,

$$\lim_{x\to 0} \log y = \log\left(\lim_{x\to 0} y\right) = -\frac{1}{6}$$

or

$$\lim_{x\to 0} y = e^{\frac{1}{6}} \text{ or } \lim_{x\to 0}\left(\frac{\sin x}{x}\right)^{\frac{1}{x^2}} = e^{-\frac{1}{6}}.$$

EXAMPLE 2.67

Evaluate $\lim_{x\to 0}(\cos x)^{\frac{1}{x^2}}$.

Solution. The given limit is of 1^∞ form. Let $y = (\cos x)^{\frac{1}{x^2}}$. Therefore,

$$\log y = \frac{\log \cos x}{x^2}$$

and so,

$$\lim_{x\to 0} \log y = \lim_{x\to 0} \frac{\log \cos x}{x^2} \left(\frac{0}{0} \text{ form}\right)$$

$$= \lim_{x\to 0} -\frac{\tan x}{2x} = -\frac{1}{2}.$$

Thus,

$$\lim_{x\to 0} \log y = \log\left(\lim_{x\to 0} y\right) = -\frac{1}{2},$$

which implies

$$\lim_{x\to 0} y = e^{-\frac{1}{2}} \text{ or } \lim_{x\to 0}(\cos x)^{\frac{1}{x^2}} = e^{-\frac{1}{2}}.$$

EXERCISES

Successive Derivatives

1. If $x^3 + y^3 = 3axy$, show that
$$\frac{d^2y}{dx^2} = -\frac{2a^2xy}{(y^2-ax)^3}.$$

2. If $y = e^{ax} \sin bx$, show that
$$\frac{d^2y}{dx^2} - 2a\frac{dy}{dx} + (a^2+b^2)y = 0.$$

3. If $y = \tan^{-1}(\sinh x)$, show that
$y_2 - y_1^2 \tan y = 0$.

4. Find the *n*th differential coefficient of $\frac{1}{x^2+a^2}$

Ans: $y_n = (-1)^n n! a^{-(n+2)} \sin(n+1)\phi \sin^{n+1}\phi$, where $\phi = \tan^{-1}\left(\frac{a}{x}\right)$.

5. Find the *n*th derivative of $e^{2x}\cos^2 x \sin x$.

Ans: $y_n = \frac{1}{4}\left[[(13)]^{\frac{n}{2}} \sin\left(3x + n\tan^{-1}\left(\frac{3}{2}\right)\right) + (5)^{\frac{n}{2}} \sin\left(x + n\tan^{-1}\left(\frac{1}{2}\right)\right)\right]$.

6. Find the *n*th derivative of $\frac{x^2}{(x+2)(2-3x)}$

Ans: $\frac{(-1)^n}{2}\left[\frac{1}{(x+2)^{n+1}} - \frac{(-3)^{n-1}}{(2-3x)^{n+1}}\right]$.

7. Find the *n*th derivative of $\frac{x}{2x^2+3x+1}$

Ans: $(-1)^n n!\left[\frac{1}{(x+1)^{n+1}} - \frac{2^n}{(2x+1)^{n+1}}\right]$.

8. Find the *n*th derivative of $\sin^2 x\cos^3 x$

Ans: $\frac{1}{16}\left[2\cos\left(x+\frac{n\pi}{2}\right) - 3^n\cos\left(3x+\frac{n\pi}{2}\right) - 5^n\cos\left(5x+\frac{n\pi}{2}\right)\right]$.

9. Find the *n*th derivative of $\tan^{-1}\frac{1+x}{1-x}$

Ans: $(-1)^{n-1}(n-1)!\sin n\phi \sin^n\phi$, $\phi = \cot^{-1}x$.

10. If $y = A\sin mx + B\cos mx$, show that
$y_2 + m^2 y = 0$.

11. If $p^2 = a^2\cos^2\theta + b^2\sin^2\theta$, show that
$$\frac{d^2p}{d\theta^2} + p = \frac{a^2b^2}{p^3}.$$

Leibnitz's Theorem

12. Find the *n*th differential coefficient of $e^x \log x$.

Ans: $e^x\left[\log x + n_{c_1}x^{-1} - n_{c_2}x^{-2} + n_{c_3}.2!x^{-3} - \ldots + (-1)^{n-1}.(n-1)!x^{-n}\right]$.

13. If $y = (x^2-1)^n$, show that
$(x^2-1)y_{n+2} + 2xy_{n+1} - n(n+1)y_n = 0$.

14. If $y = x^2e^x$, show that
$$y_n = \frac{1}{2}n(n-1)y_2 - n(n-2)y_1 + \frac{1}{2}(n-1)(n-2)y.$$

Hint: By Leibnitz's Theorem $y_n = x^2e^x + 2nxe^x + n(n-1)e^x$. Find y_1 and y_2. Substitute the values of x^2e^x, $2xe^x$ and e^x from y, y_1 and y_2 respectively in y_n and get the required result.

15. If $y = (\sin^{-1}x)^2$, show that

(i) $(1-x^2)\frac{d^2y}{dx^2} - x\frac{dy}{dx} - 2 = 0$,

(ii) $(1-x^2)y_{n+2} - (2n+1)xy_{n+1} - n^2y_n = 0$.

16. If $y = \sin^{-1}x$, show that
$(y_n)_0 = (n-2)^2(n-4)^2(y_{n-4})_0$.

Deduce that

(i) $(y_n)_0 = (n-2)^2(n-4)^2 \ldots 5^2.3^2.1$ for odd n,

(ii) $(y_n)_0 = 0$ for even n.

17. If $y = e^{a\sin^{-1}x}$ find $(y_n)_0$.

Hint: From Example 2.11, we have $(y_{n+2})_0 = (n^2 + a^2)(y_n)_0$.

Ans: $(y_n)_0 = \left[(n-2)^2 + a^2\right]...(3^2 + a^2)$
$\times (1^2 + a^2).a$ if n is odd

$(y_n)_0 = \left[(n-2)^2 + a^2\right]...(4^2 + a^2)$
$\times (2^2 + a^2) - a^2$ if n in even

18. If $y^{\frac{1}{m}} + y^{\frac{1}{m}} = 2x$, show that
$(x^2-1)\,y_{n+2} + (2n+1)\,xy_{n+1}$
$+ (n^2 - m^2)\,y_n = 0.$

19. If $y = \log\left[x + \sqrt{1+x^2}\right]$, find $(y_n)_0$.

Ans: $(y_n)_0 = (-1)^{\frac{n-1}{2}}(n-2)^2(n-4)^2...$
$\times 3^2.1^2$ if n is odd
$(y_n)_0 = 0$ if n is even

20. Verify Rolle's Theorem for $f(x) = x(x+3)e^{\frac{-x}{2}}$ in [–3, 0]

21. In the Mean Value Theorem: $f(x+h) = f(x) + hf'(x+\theta h)$, show that $\theta = \frac{1}{2}$ for $f(x) = ax^2 + bx + c$ in [0,1].

Hint: Since $f'(x) = 2ax + b$, the given theorem
$f(x+h) - f(x) = hf'(x+\theta h)$ implies
$a(x+h)^2 + b(x+h) + c - (ax^2 + bx + c)$

$= h[2a(x+\theta h) + b]$

$\Rightarrow ah^2 + 2axh + bh = 2axh + 2a\theta h^2 + bh$

$\Rightarrow ah^2 = 2a\theta h^2 \Rightarrow \theta = \dfrac{1}{2}.$

22. If x is positive, show that $x > \log(1+x) > x - \frac{x^2}{2}$.

23. Using Lagrange's Mean Value Theorem, show that

$$\frac{\pi}{3} - \frac{1}{5\sqrt{3}} > \cos^{-1}\frac{3}{5} > \frac{\pi}{3} - \frac{1}{8}.$$

24. If $f(x) = \sin^{-1} x$, $0 < a < b < 1$, use Mean Value Theorem to prove that

$$\frac{b-a}{\sqrt{1-a^2}} < \sin^{-1} b - \sin^{-1} a < \frac{b-a}{\sqrt{1-b^2}}.$$

Hint: Consider $f(x) = \sin–1\ x$. Then, the Mean Value Theorem implies

$$\frac{\sin^{-1}b - \sin^{-1}a}{b-a} = \frac{1}{\sqrt{1-\xi^2}},\quad a < \xi < b,$$

since $\xi > a$, $\frac{1}{\sqrt{1-\xi^2}} > \frac{1}{\sqrt{1-a^2}}$ and as $\xi > b$, $\frac{1}{\sqrt{1-\xi^2}} > \frac{1}{\sqrt{1-b^2}}$. Hence,

$$\frac{1}{1-a^2} < \frac{\sin^{-1}b - \sin^{-1}a}{b-a} < \frac{1}{1-b^2},$$

which gives the required result.

25. Assuming f'' to be continuous in [a, b], show that

$$f(c) - \frac{b-c}{b-a}f(a) - \frac{c-a}{b-a}f(b)$$
$$= \frac{1}{2}(c-a)(c-b)f''(\xi),$$

Where c and ξ both lie in [a, b].

26. Using Lagrange's Mean Value Theorem, show that

$$\frac{x}{1+x} < \log(1+x) < x, x > 0.$$

Hint: Consider $f(x) = \log(1+x)$ in [0, x]. Then, by Lagrange's Mean Value Theorem Theorem,

$$\frac{f(x) - f(0)}{x - 0} = f'(\theta x) \text{ or}$$

$$\frac{1}{x}\log(1+x) = \frac{1}{1+\theta x}.$$

But $0<\theta<1$ and $x>0 \Rightarrow \theta x<x \Rightarrow 1+\theta x<1+x$ implies $\frac{1}{1+\theta x}>\frac{1}{1+x} \Rightarrow \frac{x}{1+\theta x}>\frac{x}{1+x}$. Thus, $\log(1+x)>\frac{x}{1+x}$. Similarly, prove the second half of the inequality.

27. If in the Cauchy's Mean Value Theorem, $f(x)=e^x$ and $F(x)=e^{-x}$, show that c is the arithmetic mean between a and b.

Hint:

$$\frac{e^b-e^a}{e^{-b}-e^{-a}}=\frac{f'(c)}{F'(c)}=\frac{e^c}{e^{-c}}=-e^{2c} \Rightarrow -e^{a+b}$$

$$=-e^{2c} \Rightarrow c=\frac{a+b}{2}.$$

28. Expand log sin $(x+h)$ in power of h by Taylor's Theorem.

29. Show that $\sin(x+h)=\sin x+h+\cos x - \frac{h^2}{2!}\sin x-\ldots$

30. Show that

$$\cos\left(\frac{\pi}{4}+h\right)=\frac{1}{\sqrt{2}}\left(1-h-\frac{h^2}{2!}+\frac{h^3}{3!}+\ldots\right).$$

31. Calculate the approximate value of $\sqrt{10}$ to four decimal places.

32. Using Maclaurin's Theorem, find the expansion of log (sec x + tan Ix).

Ans. $x+\frac{x^3}{6}+\frac{x^5}{24}+\ldots$

33. Expand $y=\sin(e^x-1)$ by Maclaurin's Theorem.

Ans. $x-\frac{x^2}{2}-\frac{5}{24}x^4-\ldots$

34. Expand $\frac{\sin^{-1}x}{\sqrt{1-x^2}}$ in power of x up to three terms.

Ans. $x+\frac{2}{3}x^3+\frac{2.4}{3.5}x^5+\ldots$

35. Apply Maclaurin's Theorem to prove that

$$e^{ax}\cos bx=1+ax+\frac{a^2-b^2}{2}x^2 +\frac{a(a^2-3b^2)}{3!}x^3+\ldots+\frac{(a^2+b^2)^{\frac{n}{2}}}{n!} x^n\cos\left[n\tan^{-1}\left(\frac{b}{a}\right)\right]+\ldots$$

36. Apply Maclaurin's Theorem to obtain terms up to x^4 in the expansion of log $(1+\sin^2 x)$.

Ans. $x-\frac{1}{2}x^2+\frac{2}{3}x^3-\frac{7}{12}x^4+\ldots$

37. Show that

$$e^{ax}\sin x=x+x^2-\frac{2}{3!}x^3-\frac{2^2}{5!}x^5-\ldots +\sin\left(\frac{n\pi}{4}\right)\frac{2^{\frac{n}{2}}}{n!}x^n+\ldots$$

38. Expand $\left[x+\sqrt{1+x^2}\right]^m$ in ascending power of x.

Ans. $1+mx+\frac{m^2}{2!}x^2+\frac{m(m^2-1^2)}{3!}x^3 +\frac{m^2(m^2-2^2)}{4!}x^4+\frac{m(m^2-1^2)(m^2-3^2)}{5!}x^5 +\ldots$

Indeterminate form $\frac{0}{0}$.

39. Evaluate the following limits:

(i) $\lim\limits_{x\to 0}\frac{x-\sin x}{\tan^2 x}$.

Ans. $\frac{1}{6}$.

(ii) $\lim\limits_{x\to 0}\log(1-x)\cot\frac{\pi}{2}x$.

Ans. $-\frac{2}{\pi}$.

(iii) $\lim\limits_{x\to 0}\frac{1+\sin x-\cos x+\log(1-x)}{x\tan^2 x}$.

Ans. 1.

(iv) $\lim\limits_{x\to 0}\frac{x\cos x-\log(1+x)}{x^2}$.

Ans. $\frac{1}{2}$.

(v) $\lim\limits_{x\to 0}\frac{e^x\sin x-x-x^2}{x^2+x\log(1-x)}$.

Ans. $-\frac{2}{3}$.

(vi) $\lim\limits_{x\to 0}\frac{e^x\sin x-x-x^2}{x^3}$.

Ans. $\frac{1}{3}$.

(vii) $\lim\limits_{x\to 0}\frac{\cosh x-\cos x}{x\sin x}$.

Ans. 1.

(viii) $\lim\limits_{x\to\infty}\frac{2^{\frac{1}{x}}-3^{\frac{1}{x}}}{\log\frac{x}{x-1}}$

Ans. $\log\frac{2}{3}$.

40. If $\lim\limits_{x\to 0}\frac{x(1+a\cos x)-b\sin x}{x^3}=1$, find the value of a and b.

Hint: It is of $\frac{0}{0}$ form. After differentiation, the denominator becomes zero for $x = 0$. So, to find the limit the numerator should also be zero. This will yield $1 + a - b = 0$. Again differentiating twice we get the denominator 6 at $x = 0$ with the numerator $b - 3a$. But the limit is 1. Thus, $\frac{b-3a}{6} = 1$ or $b - 3a = 6$. Solving for a and b, we get $a = -\frac{5}{2}$ and $b = -\frac{3}{2}$.

41. Point out the fallacy in the following use of L'Hospital Rule:

$$\lim_{x\to 2}\frac{x^3-7x+6}{x^2-5x+6} = \lim_{x\to 2}\frac{3x^2-7}{2x-5} = \lim_{x\to 2}\frac{6x}{2} = 6.$$

Ans. $\lim_{x\to 2}\frac{3x^2-7}{2x-5}$ is not of $\frac{0}{0}$ form. So,we cannot use L' Hospital Rule further.The answer is −5.

Indeterminate form $\frac{\infty}{\infty}$.

42. Evaluate the following:

(i) $\lim_{x\to 0}\frac{\log\sin x}{\cot x}$. **Ans.** 0.

(ii) $\lim_{x\to\frac{\pi}{2}}\frac{\tan x}{\tan 3x}$. **Ans.** 3.

(iii) $\lim_{x\to 0+}\frac{\operatorname{cosec} x}{\log x}$. **Ans.** $-\infty$.

(iv) $\lim_{x\to\infty}\frac{x^n}{e^x}$, where n is a positive integer. **Ans.** 0.

(v) $\lim_{x\to 0+}\frac{\cot x}{\log\sin x}$. **Ans.** $-\infty$.

(vi) $\lim_{x\to 0+}\log_x \sin x$. **Ans.** 1.

(vii) $\lim_{x\to 0}\frac{\log x^2}{\log\cot^2 x}$. **Ans.** −1.

(viii) $\lim_{x\to a}\frac{\log(x-a)}{\log\left(e^x-e^a\right)}$. **Ans.** 1.

Indeterminate form $\infty - \infty$.

43. Evaluate

(i) $\lim_{x\to 1}\left(\frac{1}{x^2-1}-\frac{2}{x^4-1}\right)$. **Ans.** $\frac{1}{2}$.

(ii) $\lim_{x\to\frac{\pi}{2}}(\sec x - \tan x)$. **Ans.** 0.

(iii) $\lim_{x\to 0}(\operatorname{cosec} x - \cot x)$. **Ans.** 0.

(iv) $\lim_{x\to 0}\left(\frac{1}{x}-\frac{1}{x^2}\log(1+x)\right)$. **Ans.** $\frac{1}{2}$.

(v) $\lim_{x\to 0}\left(\frac{1}{x^2}-\frac{1}{x\tan x}\right)$. **Ans.** $\frac{1}{3}$.

(vi) $\lim_{x\to 0}\left(\frac{1}{x^2}-\frac{\operatorname{cosec} x}{x}\right)$. **Ans.** $-\frac{1}{6}$.

(vii) $\lim_{x\to 0}\left(\frac{1}{e^x-1}-\frac{1}{x}\right)$. **Ans.** $-\frac{1}{2}$.

(viii) $\lim_{x\to 1}\left(\frac{x}{x-1}-\frac{1}{\log x}\right)$. **Ans.** $\frac{1}{2}$.

Indeterminate form $0\times\infty$.

44. Evaluate

(i) $\lim_{x\to 0+} x^m \log x,\ m > 0$. **Ans.** 0.

(ii) $\lim_{x\to\frac{\pi}{2}}(1-\sin x)\tan x$. **Ans.** 0.

(iii) $\lim_{x\to\infty} 2^x \sin\frac{a}{2^x}, a \neq 0$. **Ans.** a.

(iv) $\lim_{x\to\infty} x\sin\frac{1}{x}$. **Ans.** 1.

(v) [illegible] $_{x\to 0}$ $\left(\frac{\pi}{2}-x\right)$ **Ans.** 1.

(vi) $\lim\limits_{x\to 0} x \cot x$.

Ans. 1.

Indeterminate form $0^0, 1^\infty$, and ∞^0.

45. Evaluate

(i) $\lim\limits_{x\to 0}\left(\frac{\sin x}{x}\right)^{\frac{1}{x}}$

Hint: It is of 1^∞ form. Let $y=\left(\frac{\sin x}{x}\right)^{\frac{1}{x}}$, then $\log y = \frac{1}{x}\log\frac{\sin x}{x}$. Therefore, $\lim\limits_{x\to 0}\log y = \lim\limits_{x\to 0}\frac{\log\frac{\sin x}{x}}{x}$ and so on.

(ii) $\lim\limits_{x\to\infty}\left(\frac{\pi}{2}-\tan^{-1}x\right)^{\frac{1}{x}}$.

Ans. 1.

(iii) $\lim\limits_{x\to a}\left(2-\frac{x}{a}\right)^{\tan\frac{\pi x}{2a}}$.

Ans. $e^{\frac{2}{\pi}}$.

(iv) $\lim\limits_{x\to 0+}(\operatorname{cosec} x)^{\frac{1}{\log x}}$.

Ans. $\frac{1}{e}$.

(v) $\lim\limits_{x\to 0}\left(\frac{\sinh x}{x}\right)^{\frac{1}{x^2}}$.

Ans. $e^{\frac{1}{6}}$.

(vi) $\lim\limits_{x\to 0}(\cos x)^{\frac{1}{x^2}}$.

Ans. $e^{-\frac{1}{2}}$.

(vii) $\lim\limits_{x\to\infty}\left(1+\frac{3}{x}\right)^{x}$.

Ans. e^3.

(viii) $\lim\limits_{x\to\infty}(1+x)^{\frac{1}{x}}$.

Ans. 1.

(ix) $\lim\limits_{x\to 0}\left[\frac{1}{2}\left(a^x+b^x\right)\right]^{\frac{1}{x}}$.

Ans. $\sqrt{ab}$.

(x) $\lim\limits_{x\to 0}\frac{(1+x)^{\frac{1}{x}}-e+\frac{1}{2}ex-\frac{11}{24}ex^2}{x^2}$.

Ans. $\frac{-7e}{16}$.

3 Curvature

Let P be any point on a given curve and Q a neighboring point of P such that the arc PQ is concave towards its chord. Let the normals at P and Q intersect at N.

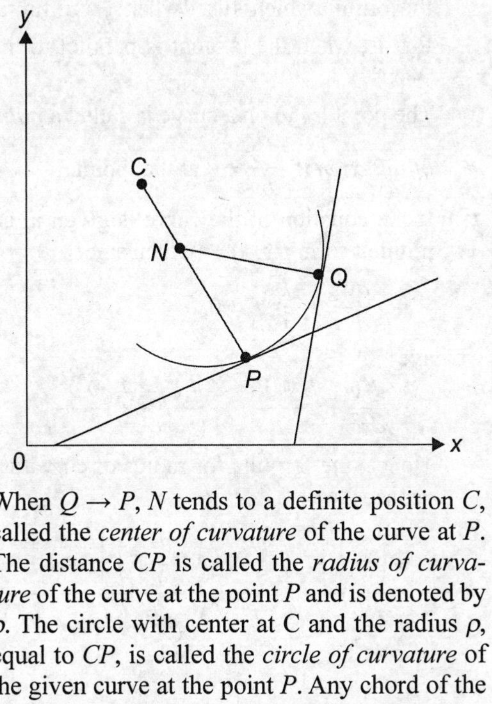

When $Q \rightarrow P$, N tends to a definite position C, called the *center of curvature* of the curve at P. The distance CP is called the *radius of curvature* of the curve at the point P and is denoted by ρ. The circle with center at C and the radius ρ, equal to CP, is called the *circle of curvature* of the given curve at the point P. Any chord of the circle of curvature drawn through the point P is called the *chord of curvature*. The reciprocal of the radius of curvature is called the *curvature of the curve* at the point P and is denoted by K.

3.1 RADIUS OF CURVATURE OF INTRINSIC CURVES

Let A be a fixed point on a given curve and P and Q be two neighboring points on the curve. Let arcs AP and PQ be s and δs, respectively. Let the tangents at P and Q make angles ψ and $\psi + \delta\psi$ with the x-axis so that the angle between the tangents at P and Q is $\delta\psi$. Let R be the point of intersection of the tangents at P and Q and let N be the intersection of normals at P and Q. We have

$$\angle PNQ = \angle SRT = \delta\psi.$$

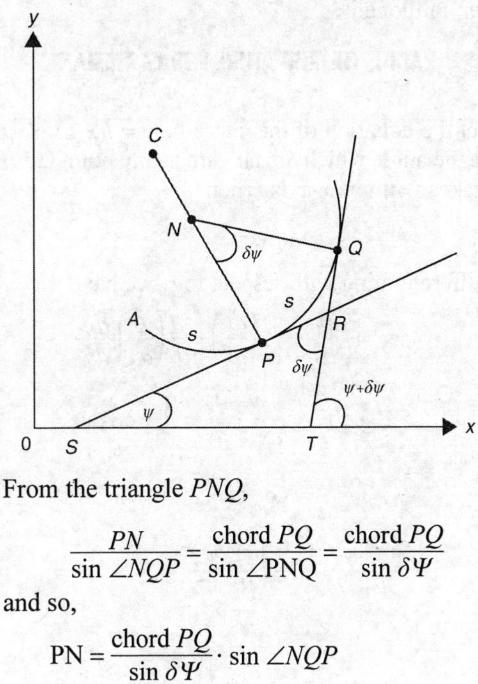

From the triangle PNQ,

$$\frac{PN}{\sin \angle NQP} = \frac{\text{chord } PQ}{\sin \angle \text{PNQ}} = \frac{\text{chord } PQ}{\sin \delta\Psi}$$

and so,

$$\text{PN} = \frac{\text{chord } PQ}{\sin \delta\Psi} \cdot \sin \angle NQP$$

$$= \frac{\text{chord } PQ}{\delta s} \frac{\delta s}{\delta\Psi} \frac{\delta\Psi}{\sin \delta\Psi} \sin \angle NQP.$$

If $Q \rightarrow P$, $\delta\psi \rightarrow 0$, and $\delta s \rightarrow 0$, and the chord PQ tends to the tangent at P and QN tends to the normal at P. Consequently, $\angle NQP \rightarrow \frac{\pi}{2}$.

Suppose N approaches C as Q approaches P. Then the radius of curvature at P is given by

$$\rho = \lim_{Q\to P} PN$$

$$= \lim_{Q\to P}\left(\frac{\text{chord } PQ}{\text{arc } PQ}\right)\left(\lim_{\delta\psi\to 0}\frac{\delta s}{\delta\psi}\right)\left(\lim_{\delta\psi\to 0}\frac{\delta\psi}{\sin\delta\psi}\right) \times\left(\lim_{Q\to P}\sin\angle NQP\right)$$

$$= 1\cdot\frac{ds}{d\psi}\cdot 1\cdot\sin\frac{\pi}{2} = \frac{ds}{d\psi}.$$

Consequently,

$$K = \frac{1}{\rho} = \frac{d\psi}{ds}.$$

Since the relation between arc length s and ψ is called the *intrinsic equation of a curve*, the formula $\rho = \frac{ds}{d\psi}$ is known as the *intrinsic formula for the radius of curvature* of the curve. Since the unit of $\delta\psi$ is radian, the unit of curvature is radian per unit length.

3.2 RADIUS OF CURVATURE FOR CARTESIAN CURVES

Let the equation of the curve be $y = f(x)$. Let ψ be the angle which the tangent at any point (x, y) makes with the x-axis. Then

$$\tan\psi = \frac{dy}{dx}.$$

Differentiating with respect to s, we have

$$\sec^2\psi\frac{d\psi}{ds} = \frac{d}{ds}\left(\frac{dy}{dx}\right) = \frac{d}{dx}\left(\frac{dy}{dx}\right)\cdot\frac{dx}{ds}$$

or

$$\frac{1}{\rho}\sec^2\psi = \frac{d^2y}{dx^2}\cdot\frac{dx}{ds} = \frac{d^2y}{dx^2}\cos\psi,$$

since

$$\frac{dx}{ds} = \cos\psi$$

or

$$\rho = \frac{\sec^3\psi}{\frac{d^2y}{dx^2}} = \frac{(1+\tan^2\psi)^{\frac{3}{2}}}{\frac{d^2y}{dx^2}}$$

$$= \frac{\left[1+\left(\frac{dy}{dx}\right)^2\right]^{\frac{3}{2}}}{\frac{d^2y}{dx^2}} = \frac{(1+y_1^2)^{\frac{3}{2}}}{y_2}.$$

Remark 3.1

(i) If the tangent at the point is parallel to the y-axis, then $\frac{dy}{dx}$ is infinite and so the above formula for ρ does not hold good in that case.

(ii) The value ρ is positive or negative in accordance with y_2 being positive or negative. However, we will ignore the negative sign as we shall be interested only in finding the length of the radius of curvature.

(iii) The value of ρ depends on the curve and not on the coordinate axes chosen. Therefore, interchanging x and y in the above formula, we get

$$\rho = \frac{\left[1+\left(\frac{dx}{dy}\right)^2\right]^{\frac{3}{2}}}{\frac{d^2x}{dy^2}},$$

the formula which is used when $\frac{dy}{dx}$ is infinite, that is, when the tangent is parallel to the y-axis.

(iv) The point (x, y) on a curve is called a *point of inflexion* if $\frac{d^2y}{dx^2} = 0$ at that point.

(v) If the equation of the curve is given in an implicit form $f(x, y) = 0$, then since

$$\frac{dy}{dx} = -\frac{f_x}{f_y},\ f_y \neq 0,$$

we have

$$\frac{d^2y}{dx^2} = \frac{f_{xx}(f_y)^2 - 2f_x f_y f_{xy} + f_{yy}(f_x)^2}{(f_y)^3}.$$

Hence, the formula for radius of curvature reduces to

$$\rho = \frac{\left[1+\left(\frac{dy}{dx}\right)^2\right]^{\frac{3}{2}}}{\frac{d^2y}{dx^2}} = -\frac{\left[1+\left(\frac{-f_x}{f_y}\right)^2\right]^{\frac{3}{2}}(f_y)^3}{f_{xx}(f_y)^2 - 2f_x f_y f_{xy} + f_{yy}(f_x)^2}$$

$$= \frac{[(f_x)^2 + (f_y)^2]^{\frac{3}{2}}}{f_{xx}(f_y)^2 - 2f_x f_y f_{xy} + f_{yy}(f_x)^2} \text{ (in magnitude).}$$

EXAMPLE 3.1

Find the radius of curvature at the point (s, ψ) on the following curves:

(i) $s = c\log\sec\psi$.

(ii) $s = a\log\tan\left(\frac{\pi}{4}+\frac{\psi}{2}\right)$.

(iii) $s = a \log \cot\left(\frac{\pi}{4} - \frac{\psi}{2}\right) + a\frac{\sin\psi}{\cos^2\psi}$.
(iv) $s^2 = 8ay$.

Solution.

(i) We have $s = c \log \sec \psi$. Therefore,

$$\rho = \frac{ds}{d\psi} = c \cdot \frac{1}{\sec\psi} \sec\psi \tan\psi = c\tan\psi.$$

(ii) We are given that

$$s = a \log \tan\left(\frac{\pi}{4} + \frac{\psi}{2}\right).$$

Therefore,

$$\begin{aligned}\rho &= \frac{ds}{d\psi}\\ &= a\frac{1}{\tan\left(\frac{\pi}{4}+\frac{\psi}{2}\right)} \cdot \sec^2\left(\frac{\pi}{4}+\frac{\psi}{2}\right)\cdot\frac{1}{2}\\ &= a\frac{\cos\left(\frac{\pi}{4}+\frac{\psi}{2}\right)}{\sin\left(\frac{\pi}{4}+\frac{\psi}{2}\right)}\cdot\frac{1}{\cos^2\left(\frac{\pi}{4}+\frac{\psi}{2}\right)}\cdot\frac{1}{2}\\ &= \frac{a}{2\sin\left(\frac{\pi}{4}+\frac{\psi}{2}\right)\cos\left(\frac{\pi}{4}+\frac{\psi}{2}\right)}\\ &= \frac{a}{\sin 2\left(\frac{\pi}{4}+\frac{\psi}{2}\right)}\\ &= \frac{a}{\sin\left(\frac{\pi}{2}+\psi\right)}\\ &= \frac{a}{\cos\psi}\\ &= a\sec\psi.\end{aligned}$$

(iii) Since $s = a\log\cot\left(\frac{\pi}{4}-\frac{\psi}{2}\right) + a\frac{\sin\psi}{\cos^2\psi}$, we have

$$\rho = \frac{ds}{d\psi} = \frac{d}{d\psi}\left[a\log\cot\left(\frac{\pi}{4}-\frac{\psi}{2}\right) + a\sin\psi\sec^2\psi\right]$$

$$= a\cdot\frac{1}{\cot\left(\frac{\pi}{4}-\frac{\psi}{2}\right)}\left[-\text{cosec}^2\left(\frac{\pi}{4}-\frac{\psi}{2}\right)\right]\left(-\frac{1}{2}\right)$$
$$+ a\sin\psi\,(2\sec\psi\sec\psi\tan\psi)$$
$$+ a\cos\psi\sec^2\psi$$

$$= \frac{a}{2\sin\left(\frac{\pi}{4}-\frac{\psi}{2}\right)\cos\left(\frac{\pi}{4}-\frac{\psi}{2}\right)} + \frac{2a\sin^2\psi}{\cos^3\psi} + \frac{a}{\cos\psi}$$

$$= \frac{a}{\sin 2\left(\frac{\pi}{4}-\frac{\psi}{2}\right)} + \frac{2a\sin^2\psi}{\cos^3\psi} + \frac{a}{\cos\psi}$$

$$= \frac{a}{\sin\left(\frac{\pi}{2}-\psi\right)} + \frac{2a\sin^2\psi}{\cos^3\psi} + \frac{a}{\cos\psi}$$

$$= \frac{2a}{\cos\psi} + \frac{2a\sin^2\psi}{\cos^3\psi}$$

$$= \frac{2a}{\cos^3\psi}(\cos^2\psi + \sin^2\psi) + = 2a\sec^3\psi.$$

(iv) We are given that

$$s^2 = 8\,a\,y.$$

Differentiating with respect to s, we get

$$2s = 8a\frac{dy}{ds} = 8a\sin\psi \text{ since } \sin\psi = \frac{dy}{ds}$$

or

$$s = 4a\sin\psi.$$

Therefore,

$$\rho = \frac{ds}{d\psi} = 4a\cos\psi = 4a\sqrt{1-\sin^2\psi}$$

$$= 4a\sqrt{1-\frac{s^2}{16a^2}} = 4a\sqrt{1-\frac{8ay}{16a^2}} = 4a\sqrt{1-\frac{y}{2a}}.$$

EXAMPLE 3.2

Show that for any curve the following relation holds:

$$\frac{1}{\rho} = \frac{d}{dx}\left(\frac{dy}{ds}\right).$$

Solution. We know that

$$\frac{dy}{ds} = \sin\psi.$$

Therefore,

$$\frac{d}{dx}\left(\frac{dy}{ds}\right) = \frac{d}{dx}[\sin\psi] = \cos\psi\frac{d\psi}{dx} = \cos\psi\frac{d\psi}{ds}\cdot\frac{ds}{dx}$$

$$= \cos\psi\cdot\frac{1}{\rho}\sec\psi,\ \sec\psi = \frac{ds}{dx}$$

$$= \frac{1}{\rho}, \text{ which proves the result.}$$

EXAMPLE 3.3

Find the radius of curvature of the curve, $s = ae^{\frac{x}{a}}$.

Solution. The equation of the curve is $s = ae^{\frac{x}{a}}$.

Differentiating with respect to x, we get

$$\frac{ds}{dx} = a\,e^{\frac{x}{a}}\cdot\frac{1}{a} = e^{\frac{s}{a}} = \frac{s}{a}$$

or

$$s = a\frac{ds}{dx} = a\sec\psi, \text{ since } \sec\psi = \frac{ds}{dx}.$$

Therefore,

$$\rho = \frac{ds}{d\psi}$$
$$= a \sec \psi \tan \psi$$
$$= s \tan \psi = s\sqrt{\sec^2 \psi - 1}$$
$$= s\sqrt{\frac{s^2}{a^2} - 1} = \frac{s}{a}\sqrt{s^2 - a^2}.$$

EXAMPLE 3.4

Find the radius of curvature of the following curves:

(i) $x^{\frac{2}{3}} + y^{\frac{2}{3}} = a^{\frac{2}{3}}$ at (x, y).

(ii) $xy = c^2$ at (x, y).

(iii) $y = 4 \sin x - \sin 2x$ at $x = \frac{\pi}{2}$.

(iv) $x^3 + y^3 = 3axy$ at the point $\left(\frac{3a}{2}, \frac{3a}{2}\right)$.

Solution.

(i) The equation of the given curve is

$$x^{\frac{2}{3}} + y^{\frac{2}{3}} = a^{\frac{2}{3}}.$$

Any point on this curve may be taken as $(a \cos^3 t, a \sin^3 t)$ where t is a parameter. Then,

$$\frac{dx}{dt} = -3a \cos^2 t \sin t \text{ and } \frac{dy}{dt} = 3a \sin^2 t \cos t.$$

Therefore,

$$\frac{dy}{dx} = \frac{dy/dt}{dx/dt} = -\frac{3a \sin^2 t \cos t}{3a \cos^2 t \sin t} = -\tan t$$

and

$$\frac{d^2y}{dx^2} = \frac{d}{dx}(-\tan t) = \frac{d}{dt}(-\tan t)\frac{dt}{dx} = -\sec^2 t \frac{dt}{dx}$$
$$= \sec^2 t \frac{1}{-3a \cos^2 t \sin t} = \frac{1}{3a}\sec^4 t \operatorname{cosec} t.$$

Hence, ρ at the point $(a \cos^3 t, a \sin^3 t)$ is given by

$$\rho = \frac{\left[1 + \left(\frac{dy}{dx}\right)^2\right]^{\frac{3}{2}}}{\frac{d^2y}{dx^2}} = \frac{3a(1 + \tan^2 t)^{\frac{3}{2}}}{\sec^4 t \operatorname{cosec} t}$$
$$= \frac{3a \sec^3 t}{\sec^4 t \operatorname{cosec} t} = 3a \sin t \cos t.$$

But $x = a \cos^3 t$ and $y = a \sin^3 t$ imply $\cos t = \left(\frac{x}{a}\right)^{\frac{1}{3}}$ and $\sin t = \left(\frac{y}{a}\right)^{\frac{1}{3}}$.

Hence, $\rho = 3a\left(\frac{x}{a}\right)^{\frac{1}{3}}\left(\frac{y}{a}\right)^{\frac{1}{3}} = 3a^{\frac{1}{3}}x^{\frac{1}{3}}y^{\frac{1}{3}}.$

(ii) The given curve is

$$xy = c^2 \quad \text{or} \quad y = \frac{c^2}{x}.$$

Therefore,

$$\frac{dy}{dx} = -\frac{c^2}{x^2} \quad \text{and} \quad \frac{d^2y}{dx^2} = \frac{2c^2}{x^3}.$$

Hence,

$$\rho = \frac{\left[1 + \left(\frac{dy}{dx}\right)^2\right]^{\frac{3}{2}}}{\frac{d^2y}{dx^2}} = \frac{\left(1 + \frac{c^4}{x^4}\right)^{\frac{3}{2}}}{\frac{2c^2}{x^3}}$$
$$= \frac{(x^4 + c^4)^{\frac{3}{2}}}{2c^2x^3} = \frac{(x^4 + x^2y^2)^{\frac{3}{2}}}{2c^2x^3}, \text{ since } xy = c^2$$
$$= (x^2 + y^2)^{\frac{3}{2}}/2c^2 = \frac{r^3}{2c^2}, \text{ where } r^2 = x^2 + y^2.$$

(iii) The equation of the given curve is

$$y = 4 \sin x \sin 2x$$

Therefore,

$$\frac{dy}{dx} = 4 \cos x - 2 \cos 2x \text{ and}$$
$$\frac{d^2y}{dx^2} = -4 \sin x + 4 \sin 2x$$

and so,

$$\left(\frac{dy}{dx}\right)_{x=\frac{\pi}{2}} = 4 \cos\frac{\rho}{2} - 2 \cos \rho = 2 \text{ and}$$
$$\left(\frac{d^2y}{dx^2}\right)_{x=\frac{\pi}{2}} = -4 \sin\frac{\rho}{2} + 4 \sin \rho = -4.$$

Hence,

$$\rho = \frac{\left[1 + \left(\frac{dy}{dx}\right)^2\right]^{\frac{3}{2}}}{\frac{d^2y}{dx^2}} = \frac{(1 + 4)^{\frac{3}{2}}}{-4}$$
$$= \frac{5\sqrt{5}}{4} \text{ (ignoring sign).}$$

(iv) The given curve is

$$x^3 + y^3 = 3axy.$$

Differentiating both sides with respect to x, we get

$$3x^2 + 3y^2\frac{dy}{dx} = 3a\left(x\frac{dy}{dx} + y\right)$$

or

$$(y^2 - ax)\frac{dy}{dx} = ay - x^2 \qquad (1)$$

or

$$\frac{dy}{dx} = \frac{ay - x^2}{y^2 - ax}$$

so that,

$$\left(\frac{dy}{dx}\right)_{\left(\frac{3a}{2},\frac{3a}{2}\right)} = -1. \qquad (2)$$

Further, differentiating (1) again, we get

$$(y^2 - ax)\frac{d^2y}{dx^2} + \left(2y\frac{dy}{dx} - a\right)\frac{dy}{dx} = a\frac{dy}{dx} - 2x$$

or

$$(ax - y^2)\frac{d^2y}{dx^2} = 2x + 2y\left(\frac{dy}{dx}\right)^2 - 2a\frac{dy}{dx}.$$

Therefore at the point $\left(\frac{3a}{2}, \frac{3a}{2}\right)$, we have

$$\left(\frac{d^2y}{dx^2}\right)_{\left(\frac{3a}{2},\frac{3a}{2}\right)} = -\frac{32}{3a}, \text{ [using (2)]}.$$

Hence, the radius of curvature ρ at $\left(\frac{3a}{2}, \frac{3a}{2}\right)$ is

$$\rho = \frac{\left[1 + \left(\frac{dy}{dx}\right)^2\right]^{\frac{3}{2}}}{\frac{d^2y}{dx^2}} = \frac{(1+1)^{\frac{3}{2}}}{-\frac{32}{3a}}$$

$$= \frac{3a}{8\sqrt{2}} \text{(in magnitude)}.$$

EXAMPLE 3.5

Find the radius of curvature of the following curves:

(i) $y = c\cosh\left(\frac{x}{c}\right)$ at $(0, c)$.

(ii) $\sqrt{x} + \sqrt{y} = \sqrt{a}$ at (x, y).

Solution.

(i) The equation of the curve is

$$y = c\cosh\left(\frac{x}{c}\right).$$

Therefore,

$$\frac{dy}{dx} = c\sinh\left(\frac{x}{c}\right)\frac{1}{c} = \sinh\left(\frac{x}{c}\right)$$

and

$$\frac{d^2y}{dx^2} = \frac{1}{c}\cosh\left(\frac{x}{c}\right).$$

Hence, the radius of curvature is

$$\rho = \frac{\left[1 + \left(\frac{dy}{dx}\right)^2\right]^{\frac{3}{2}}}{\frac{d^2y}{dx^2}} = \frac{\left[1 + \sinh^2\left(\frac{x}{c}\right)\right]^{\frac{3}{2}}}{\frac{1}{c}\cosh\left(\frac{x}{c}\right)}$$

$$= \frac{c\left[\cosh^2\left(\frac{x}{c}\right)\right]^{\frac{3}{2}}}{\cosh\left(\frac{x}{c}\right)}$$

$$= c\cosh^2\left(\frac{x}{c}\right)$$

$$= c\left(\frac{y}{c}\right)^2 = \frac{y^2}{c}.$$

Thus, ρ at $(0, c)$ is

$$\rho = \frac{c^2}{c} = c.$$

(ii) The equation of the curve is

$$\sqrt{x} + \sqrt{y} = \sqrt{a}.$$

Differentiating with respect to x, we get

$$\frac{1}{2}x^{-\frac{1}{2}} + \frac{1}{2}y^{-\frac{1}{2}}\frac{dy}{dx} = 0,$$

which yields

$$\frac{dy}{dx} = \frac{x^{-\frac{1}{2}}}{y^{-\frac{1}{2}}} = -y^{\frac{1}{2}}x^{-\frac{1}{2}}.$$

Differentiating again with respect to x, we have

$$\frac{d^2y}{dx^2} = -\left(\frac{1}{2}y^{-\frac{1}{2}}\frac{dy}{dx}\right)x^{-\frac{1}{2}} - y^{\frac{1}{2}} - \left(\frac{1}{2}x^{-\frac{3}{2}}\right)$$

$$= -\frac{1}{2}y^{-\frac{1}{2}}x^{-\frac{1}{2}}\left(-y^{\frac{1}{2}}x^{-\frac{1}{2}}\right) + \frac{1}{2}y^{\frac{1}{2}}x^{-\frac{3}{2}}$$

$$= \frac{1}{2x} + \frac{y^{\frac{1}{2}}}{2x^{\frac{3}{2}}} = \frac{1}{2x}\left(1 + \frac{y^{\frac{1}{2}}}{x^{\frac{1}{2}}}\right)$$

$$= \frac{1}{2x}\left(\frac{\sqrt{x}+\sqrt{y}}{\sqrt{x}}\right) = \frac{\sqrt{a}}{2x^{\frac{3}{2}}}.$$

Hence,

$$\rho = \frac{\left[1 + \left(\frac{dy}{dx}\right)^2\right]^{\frac{3}{2}}}{\frac{d^2y}{dx^2}} = \frac{\left[1 + \frac{y}{x}\right]^{\frac{3}{2}}}{\frac{\sqrt{a}}{2x^{\frac{3}{2}}}} = \frac{2(x+y)^{\frac{3}{2}}}{\sqrt{a}}.$$

EXAMPLE 3.6

Prove that for the ellipse $\frac{x^2}{a^2} + \frac{y^2}{b^2} = 1$, the radius of curvature $\rho = \frac{a^2b^2}{p^3}$, where p is the perpendicular from the center to the tangent at (x, y).

Solution. The equation of the ellipse is $\frac{x^2}{a^2} + \frac{y^2}{b^2} = 1$, which implies $b^2x^2 + a^2y^2 = a^2b^2$.

Differentiating with respect to x, we get

$$\frac{2x}{a^2} + \frac{2y}{b^2}\frac{dy}{dx} = 0$$

or

$$\frac{dy}{dx} = -\frac{b^2}{a^2}\left(\frac{x}{y}\right).$$

Differentiating again with respect to x, we have

$$\frac{d^2y}{dx^2} = -\frac{b^2}{a^2}\left(\frac{y - x\frac{dy}{dx}}{y^2}\right) = -\frac{b^2}{a^2y^2}\left[y - x\left(-\frac{b^2x}{a^2y}\right)\right]$$

$$= -\frac{b^2}{a^2y^2}\left[\frac{a^2y^2 + b^2x^2}{a^2y}\right] = -\frac{b^2}{a^2y^2}\left[\frac{a^2b^2}{a^2y}\right] = -\frac{b^4}{a^2y^3}.$$

Therefore ρ at the point (x, y) is

$$\rho = \frac{\left[1 + \left(-\frac{b^2x}{a^2y}\right)^2\right]^{\frac{3}{2}}}{-\frac{b^4}{a^2y^3}}$$

$$= \frac{[a^4y^2 + b^4x^2]^{\frac{3}{2}}}{a^4b^4} \text{ (in magnitude).}$$

The equation of the tangent to the given ellipse at (x, y) is $\frac{Xx}{a^2} + \frac{Yy}{b^2} = 1$. Therefore, p = length of the perpendicular from the center (0, 0) to the tangent

$$= \frac{1}{\sqrt{\left(\frac{x}{a^2}\right)^2 + \left(\frac{y}{b^2}\right)^2}} = \frac{a^2b^2}{\sqrt{b^4x^2 + a^4y^2}}.$$

Thus,
$$p^3 = \frac{a^6b^6}{(b^4x^2 + a^4y^2)^{\frac{3}{2}}}$$

so that,

$$(b^4x^2 + a^4y^2)^{\frac{3}{2}} = \frac{a^6b^6}{p^3}.$$

Substituting this value in the expression for ρ, we get

$$\rho = \frac{\frac{a^6b^6}{p^3}}{a^4b^4} = \frac{a^2b^2}{p^3}.$$

3.3 RADIUS OF CURVATURE FOR PARAMETRIC CURVES

Let the parametric equation of the curve be

$$x = f(t) \text{ and } y = \phi(t).$$

Then if $x' = \frac{dx}{dt}$ and $y' = \frac{dy}{dt}$, we have

$$\frac{dy}{dx} = \frac{dy/dt}{dx/dt} = \frac{y'}{x'}$$

and

$$\frac{d^2y}{dx^2} = \frac{d}{dx}\left(\frac{dy}{dx}\right) = \frac{d}{dx}\left(\frac{y'}{x'}\right) = \left[\frac{d}{dt}\left(\frac{y'}{x'}\right)\right]\frac{dt}{dx}$$

$$= \frac{x'y'' - y'x''}{(x')^2}\cdot\frac{1}{x'} = \frac{x'y'' - y'x''}{(x')^3}.$$

Therefore,

$$\rho = \frac{\left[1 + \left(\frac{dy}{dx}\right)^2\right]^{\frac{3}{2}}}{\frac{d^2y}{dx^2}}$$

$$= \frac{\left[1 + \left(\frac{y'}{x'}\right)^2\right]^{\frac{3}{2}}}{\frac{x'y'' - y'x''}{-x'^3}}$$

$$= \frac{(x'^2 + y'^2)^{\frac{3}{2}}}{x'y'' - y'x''}, \quad x'y'' - y'x'' \neq 0.$$

EXAMPLE 3.7

Solve Example 3.6 using parametric equation of the ellipse.

Solution. Any point on the ellipse $\frac{x^2}{a^2} + \frac{y^2}{b^2} = 1$ is $(a \cos \theta, b \sin \theta)$ and the parametric equation of the ellipse is

$$x = a \cos \theta \text{ and } y = b \sin \theta.$$

Therefore,

$$x' = -a \sin \theta \text{ and } y' = b \cos \theta,$$
$$x'' = -a \cos \theta \text{ and } y'' = -b \sin \theta.$$

Hence, ρ at $(a \cos \theta, b \sin \theta)$ is

$$\rho = \frac{(x'^2 + y'^2)^{\frac{3}{2}}}{x'y'' - y'x''} = \frac{(a^2 \sin^2 \theta + b^2 \cos^2 \theta)^{\frac{3}{2}}}{ab \sin^2 \theta + ab \cos^2 \theta}$$

$$= \frac{(a^2 \sin^2 \theta + b^2 \cos^2 \theta)^{\frac{3}{2}}}{ab}. \quad (1)$$

Now the equation of tangent to the ellipse at $(a \cos \theta, b \sin \theta)$ is

$$\frac{xa \cos \theta}{a^2} + \frac{yb \sin \theta}{b^2} = 1 \text{ or } \frac{x}{a}\cos \theta + \frac{y}{b}\sin \theta = 1.$$

Therefore,

p = length of the perpendicular from the center (0, 0) on the tangent

$$= \frac{1}{\sqrt{\frac{\cos^2 \theta}{a^2} + \frac{\sin^2 \theta}{b^2}}} = \frac{ab}{\sqrt{a^2 \sin^2 \theta + b^2 \cos^2 \theta}}$$

Thus,

$$p^3 = \frac{a^3b^3}{(a^2\sin^2\theta + b^2\cos^2\theta)^{\frac{3}{2}}}.$$

Putting this value in (1), we get

$$\rho = \frac{a^2b^2}{p^3}.$$

EXAMPLE 3.8

Find the radius of curvature at any point θ of the curve

$$x = a(\theta - \sin\theta) \text{ and } y = a(1 - \cos\theta).$$

Solution. The equation of the curve is

$$x = a(\theta - \sin\theta) \text{ and } y = a(1 - \cos\theta).$$

Therefore,

$$x' = a(1 - \cos\theta) = 2a\sin^2\frac{\theta}{2} \text{ and}$$

$$y' = a\sin\theta = 2a\sin\frac{\theta}{2}\cos\frac{\theta}{2}.$$

Thus,

$$\frac{dy}{dx} = \frac{y'}{x'} = \frac{2a\sin\frac{\theta}{2}\cos\frac{\theta}{2}}{2a\sin^2\frac{\theta}{2}} = \cot\frac{\theta}{2}.$$

Further,

$$\frac{d^2y}{dx^2} = \frac{d}{dx}\left(\frac{dy}{dx}\right) = \frac{d}{dx}\left(\cot\frac{\theta}{2}\right)$$

$$= \frac{d}{d\theta}\left(\cot\frac{\theta}{2}\right)\frac{d\theta}{dx}$$

$$= -\frac{1}{2}\operatorname{cosec}^2\frac{\theta}{2}\cdot\frac{1}{2a\sin^2\frac{\theta}{2}}$$

$$= -\frac{1}{4a}\operatorname{cosec}^4\frac{\theta}{2}.$$

Hence,

$$\rho = \frac{\left[1 + \left(\frac{dy}{dx}\right)^2\right]^{\frac{3}{2}}}{\frac{d^2y}{dx^2}} = \frac{\left(1 + \cot^2\frac{\theta}{2}\right)^{\frac{3}{2}}}{-\frac{1}{4}\operatorname{cosec}^4\frac{\theta}{2}}$$

$$= -\frac{4a}{\operatorname{cosec}\frac{\theta}{2}} = 4a\sin\frac{\theta}{2} \text{ (in magnitude).}$$

EXAMPLE 3.9

Show that the radius of curvature at any point of the cycloid

$x = a(t + \sin t)$ and $y = a(1 - \cos t)$ is $4a\cos\frac{t}{2}$.

Solution. The equation of the curve is

$$x = a(t + \sin t) \text{ and } y = a(1 - \cos t).$$

Therefore,

$$x' = a(1 + \cos t) \text{ and } y' = a\sin t,$$

$$x'' = -a\sin t \text{ and } y'' = a\cos t.$$

Hence,

$$\rho = \frac{(x'^2 + y'^2)^{\frac{3}{2}}}{x'y'' - y'x''}$$

$$= \frac{[a^2(1 + \cos t)^2 + a^2\sin^2 t]^{\frac{3}{2}}}{a(1 + \cos t)\, a\cos t - a\sin t\,(-a\sin t)}$$

$$= \frac{a\,[2(1 + \cos t)]^{\frac{3}{2}}}{1 + \cos t} = (2\sqrt{2})a(1 + \cos t)^{\frac{1}{2}}$$

$$= (2\sqrt{2})a\left(2\cos^2\frac{t}{2}\right)^{\frac{1}{2}} = 4a\cos\frac{t}{2}.$$

EXAMPLE 3.10

If ρ_1 and ρ_2 are the radii of curvatures at the extremities of a focal chord of the parabola $y^2 = 4ax$, prove that

$$\rho_1^{-\frac{2}{3}} + \rho_2^{-\frac{2}{3}} = (2a)^{-\frac{2}{3}}.$$

Solution. The parametric equation of the parabola $y^2 = 4ax$ is

$$x = at^2 \text{ and } y = 2at.$$

Therefore,

$$x' = 2at \text{ and } y' = 2a,$$

$$x'' = 2a \text{ and } y'' = 0.$$

Therefore, ρ at the point $(at^2, 2at)$ is

$$\rho = \frac{[x'^2 + y'^2]^{\frac{3}{2}}}{x'y'' - y'x''} = \frac{[4a^2t^2 + 4a^2]^{\frac{3}{2}}}{-4a^2}$$

$$= \frac{8a^3(t^2 + 1)^{\frac{3}{2}}}{-4a^2} = 2a(1 + t^2)^{\frac{3}{2}} \text{ (in magnitude).}$$

Let the extremities of the focal chord be

$$P(at_1^2, 2at_1) \text{ and } Q(at_2^2, 2at_2).$$

Then, by the property of focal chord, $t_1t_2 = -1$.

Further, by the above-obtained expression for ρ,

we have

$$\rho_1 = \text{radius of curvature at } P(at_1^2, 2at_1)$$

$$= 2a(1 + t_1^2)^{\frac{3}{2}} \text{ and}$$

$$\rho_2 = \text{radius of curvature at } Q(at_2^2, 2at_2)$$

$$= 2a(1 + t_2^2)^{\frac{3}{2}}.$$

Therefore, since $t_1^2 t_2^2 = 1$, we have

$$\rho_1^{-\frac{2}{3}} + \rho_2^{-\frac{2}{3}} = (2a)^{-\frac{2}{3}}\left[\frac{1}{1-t_1^2} + \frac{1}{1-t_2^2}\right]$$

$$= (2a)^{-\frac{2}{3}}\left[\frac{t_1^2 + t_1^2 + 2}{t_1^2 + t_1^2 + 2}\right] = (2a)^{-\frac{2}{3}}.$$

EXAMPLE 3.11

If ρ_1 and ρ_2 are the radii of curvatures at the extremities of two conjugate semi-diameters of an ellipse $\frac{x^2}{a^2} + \frac{y^2}{b^2} = 1$, show that

$$\left(\rho_1^{\frac{2}{3}} + \rho_2^{\frac{2}{3}}\right)(ab)^{\frac{2}{3}} = a^2 + b^2.$$

Solution. The parametric equations of the ellipse are $x = a\cos\theta$ and $y = b\sin\theta$. As in Example 3.7, we have

$$\rho = \frac{(a^2\sin^2\theta + b^2\cos^2\theta)^{\frac{3}{2}}}{ab}$$

and so,

$$\rho ab = (a^2\sin^2\theta + b^2\cos^2\theta)^{\frac{3}{2}}$$

or

$$(ab)^{\frac{2}{3}}\rho^{\frac{2}{3}} = a^2\sin^2\theta + b^2\cos^2\theta.$$

If C is the center of the ellipse and CP and CQ be a pair of conjugate semi-diameters of the ellipse, then the eccentric angles of P and Q are θ and $\theta + \frac{\pi}{2}$, respectively. Thus, if ρ is the radius of curvature at P $(a\cos\theta, b\sin\theta)$, then the curvature at Q is obtained by replacing θ by $\theta + \frac{\pi}{2}$. Thus, if ρ_1 and ρ_2 are the radii of curvatures at P and Q, then from the above expressions we get

$$(ab)^{\frac{2}{3}}\rho_1^{\frac{2}{3}} = a^2\sin^2\theta + b^2\cos^2\theta$$

$$(ab)^{\frac{2}{3}}\rho_2^{\frac{2}{3}} = a^2\sin^2\left(\theta + \frac{\pi}{2}\right) + b^2\cos^2\left(\theta + \frac{\pi}{2}\right)$$

$$= a^2\cos^2\theta + b^2\sin^2\theta.$$

Adding these two relations, we get

$$\left(\rho_1^{\frac{2}{3}} + \rho_2^{\frac{2}{3}}(ab)^{\frac{2}{3}}\right) = a^2(\sin^2\theta + \cos^2\theta) + b^2(\sin^2\theta + \cos^2\theta) = a^2 + b^2.$$

EXAMPLE 3.12

Show that the radius of curvature at a point P on an ellipse $\frac{x^2}{a^2} + \frac{y^2}{b^2} = 1$ is given by $\rho = \frac{(CQ)^3}{ab}$, where CQ is the semi-diameter conjugate to CP.

Solution. If P $(a\cos\theta, b\sin\theta)$ is the point on the ellipse, then the point Q shall be $Q\left(a\cos\left(\theta + \frac{\pi}{2}\right), b\sin\left(\theta + \frac{\pi}{2}\right)\right) = Q(-a\sin\theta\, b\cos\theta)$. Therefore,

$$CQ = \sqrt{a^2\sin^2\theta + b^2\cos^2\theta} \quad \text{or}$$

$$(CQ)^3 = (a^2\sin^2\theta + b^2\cos^2\theta)^{\frac{3}{2}}.$$

Hence, from Example 3.7, we have

$$\rho = \frac{(a^2\sin^2\theta + b^2\cos^2\theta)^{\frac{3}{2}}}{ab} = \frac{(CQ)^3}{ab}.$$

EXAMPLE 3.13

Show that the radius of curvature at the end of the major axis of the ellipse $\frac{x^2}{a^2} + \frac{y^2}{b^2} = 1$ is equal to the semi-latus rectum of the ellipse.

Solution. As in the above examples, we have

$$\rho = \frac{(a^2\sin^2\theta + b^2\cos^2\theta)^{\frac{3}{2}}}{ab}.$$

Now at the ends of the major axis, $\theta = 0$ and π. Therefore, putting $\theta = 0$, we get

$$\rho = \frac{(b^2)^{\frac{3}{2}}}{ab} = \frac{b^2}{a} = \text{Semi-latus rectum of the ellipse.}$$

3.4 RADIUS OF CURVATURE FOR PEDAL CURVES

Let $P(r, \theta)$ be a given point on the curve such that the tangent to the curve at P makes an angle ψ with the x-axis. Let ϕ be the angle between the tangent and the radius vector. Then $\psi = \theta + \phi$.

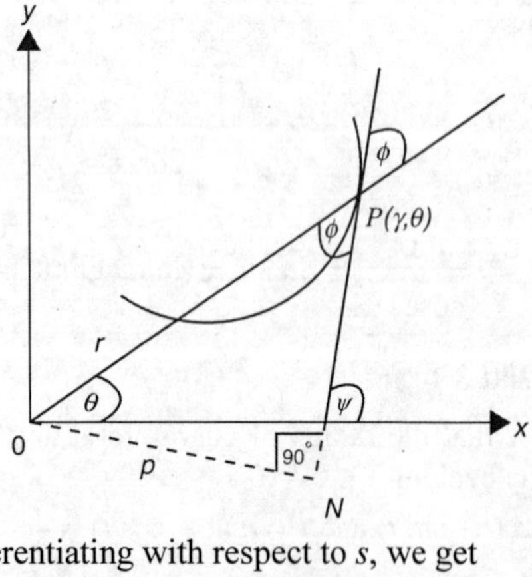

Differentiating with respect to s, we get

$$\frac{d\psi}{ds} = \frac{d\theta}{ds} + \frac{d\phi}{ds}.$$

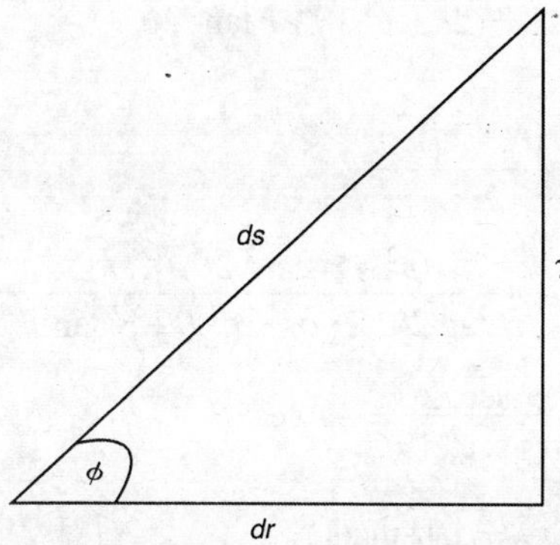

Also $p = r\sin\phi$. Differentiating with respect to r, we get

$$\frac{dp}{dr} = \sin\phi + r\cos\phi\,\frac{d\phi}{dr} = r\frac{d\theta}{ds} + r\,\frac{dr}{ds}\cdot\frac{d\phi}{dr}$$

$$= r\left[\frac{d\theta}{ds} + \frac{d\phi}{ds}\right] = r\,\frac{d\psi}{ds} = \frac{r}{\rho},\ \rho = \frac{ds}{d\psi}.$$

Hence,

$$\rho = r\,\frac{dr}{dp}.$$

3.5 RADIUS OF CURVATURE FOR POLAR CURVES

From the figure of article 3.4, we have $\psi = \theta + \phi$ and so differentiation yields

$$\frac{d\psi}{ds} = \frac{d\theta}{ds} + \frac{d\phi}{ds}$$

or

$$\frac{1}{\rho} = \frac{d\theta}{ds} + \frac{d\phi}{d\theta}\cdot\frac{d\theta}{ds}. \qquad (1)$$

But,

$$\tan\phi = r\,\frac{d\theta}{dr} = \frac{r}{\frac{dr}{d\theta}} = \frac{r}{r_1},\ \text{where } r_1 = \frac{dr}{d\theta}.$$

Differentiating with respect to θ, we get

$$\sec^2\phi\,\frac{d\phi}{d\theta} = \frac{r_1 r_1 - rr_2}{r_1^2},\ \text{where } r_2 = \frac{d^2 r}{d\theta^2}.$$

Therefore,

$$\frac{d\phi}{d\theta} = \frac{r_1^2 - rr_2}{r_1^2\sec^2\phi} = \frac{r_1^2 - rr_2}{(1+\tan^2\phi)}$$

$$= \frac{r_1^2 - rr_2}{r_1^2\left(1+\frac{r^2}{r_1^2}\right)},\ \text{since } \tan\phi = \frac{rd\theta}{dr} = \frac{r}{r_1}.$$

$$= \frac{r_1^2 - rr_2}{r_1^2 + r^2}.$$

Moreover, since $\sin\phi = r\dfrac{d\theta}{ds}$, we have

$$\frac{d\theta}{ds} = \frac{\sin\phi}{r} = \frac{1}{r}\cdot\frac{r}{\sqrt{r^2+r_1^2}} = \frac{1}{\sqrt{r^2+r_1^2}}.$$

Hence from (1), we have

$$\frac{1}{\rho} = \frac{1}{\sqrt{r^2+r_1^2}}\left[1+\frac{r_1^2 - rr_2}{r_1^2+r^2}\right]$$

$$= \frac{r^2 - rr_2 + 2r_1^2}{(r^2+r_1^2)^{\frac{3}{2}}}$$

which yields

$$\rho = \frac{(r^2+r_1^2)^{\frac{3}{2}}}{r^2 - rr_2 + 2r_1^2}.$$

3.5.1 Second Method

The relation between p, r, and θ is

$$\frac{1}{p^2} = \frac{1}{r^2} + \frac{1}{r^4}r_1^2,\ r_1 = \frac{dr}{d\theta}.$$

Differentiating with respect to r, we get

$$-\frac{2}{p^3}\frac{dp}{dr} = -\frac{2}{r^3} - \frac{4}{r^5}r_1^2 + \frac{2}{r^4}r_1\,\frac{d}{dr}r_1$$

$$= -\frac{2}{r^3} - \frac{4}{r^5}r_1^2 + \frac{2}{r^4}r_1\left[\frac{d}{d\theta}\left(\frac{dr}{d\theta}\right)\right]\frac{d\theta}{dr}$$

$$= -\frac{2}{r^3} - \frac{4}{r^5}r_1^2 + \frac{2}{r^4}r_1\cdot r_2\cdot\frac{1}{r_1}$$

$$= -\frac{2}{r^3} - \frac{4}{r^5}r_1^2 + \frac{2}{r^4}r_2.$$

Therefore,

$$\frac{1}{p^3}\frac{dp}{dr} = \frac{1}{r^3} + \frac{2}{r^5}r_1^2 - \frac{1}{r^4}r^2$$

$$= \frac{r^2 + 2r_1^{\,2} - rr_2}{r^5}. \qquad (1)$$

Since,

$$\frac{1}{p^3} = \left[\left(\frac{1}{r^2} + \frac{1}{r^4}r_1^2\right)\right]^{\frac{3}{2}} \quad\text{and}\quad \frac{dp}{dr} = \frac{r}{\rho},$$

we have,

$$\rho = \frac{r^6\left[\left(\frac{1}{r^2} + \frac{1}{r^4}r_1^2\right)^2\right]^{\frac{3}{2}}}{r^2 + 2r_1^2 - rr_2} = \frac{(r^2 + r_1^2)^{\frac{3}{2}}}{r^2 + 2r_1^2 - rr_2}.$$

Corollary. Let the equation of the curve be $\frac{1}{r} = f(\theta)$. Put $u = \frac{1}{r}$ or $r = \frac{1}{u}$. Then,

$$\frac{dr}{d\theta} = -\frac{1}{u^2}\cdot\frac{du}{d\theta} \quad \text{and}$$

$$\frac{d^2r}{d\theta^2} = -\frac{1}{u^2}\cdot\frac{d^2u}{d\theta^2} + \frac{2}{u^3}\left(\frac{du}{d\theta}\right)^2.$$

Therefore,

$$\rho = \frac{\left[r^2 + \left(\frac{dr}{d\theta}\right)^2\right]^{\frac{3}{2}}}{r^2 + 2\left(\frac{dr}{d\theta}\right)^2 - r\frac{d^2r}{d\theta^2}}$$

$$= \frac{\left[\frac{1}{u^2} + \frac{1}{u^4}\left(\frac{du}{d\theta}\right)^2\right]^{\frac{3}{2}}}{\frac{1}{u^2} + 2\cdot\frac{1}{u^4}\left(\frac{du}{d\theta}\right)^2 - \frac{1}{u}\left[-\frac{1}{u^2}\frac{d^2u}{d\theta^2} + \frac{2}{u^3}\left(\frac{du}{d\theta}\right)^2\right]}$$

$$= \frac{\left[\frac{1}{u^6}\left[u^2 + \left(\frac{du}{d\theta}\right)^2\right]\right]^{\frac{3}{2}}}{\frac{1}{u^2} + \frac{1}{u^3}\frac{d^2u}{d\theta^2}} = \frac{\left[u^2 + \left(\frac{du}{d\theta}\right)^2\right]^{\frac{3}{2}}}{u^2\left(u + \frac{d^2u}{d\theta^2}\right)}$$

$$= \frac{\left[u^2 + u_1^2\right]^{\frac{3}{2}}}{u^3(u + u_2)},$$

where

$$u_1 = \frac{du}{d\theta} \text{ and } u_2 = \frac{d^2u}{d\theta^2}.$$

EXAMPLE 3.14

Find the radius of curvature at the point (r, θ) of the following curves:

(i) $r^2 \cos 2\theta = a^2$. (ii) $r^n = a^n \sin n\theta$.
(iii) $r = a(1 - \cos\theta)$.

Solution.

(i) The equation of the curve is

$$r^2 \cos 2\theta = a^2. \tag{1}$$

Taking logarithm, we get

$$2\log r + \log\cos 2\theta = \log a^2. \tag{2}$$

Differentiating (2) with respect to θ, we get

$$\frac{2}{r}\frac{dr}{d\theta} + \frac{-2\sin 2\theta}{\cos 2\theta} = 0 \quad \text{or} \quad r_1 = \frac{dr}{d\theta} = r\tan 2\theta.$$

Differentiating again with respect to θ, we get

$$r_2 = \frac{d^2r}{d\theta^2} = 2r\sec^2 2\theta + \tan 2\theta\frac{dr}{d\theta}$$

$$= 2r\sec^2 2\theta + r\tan^2 2\theta.$$

Hence,

$$\rho = \frac{(r^2 + r_1^2)^{\frac{3}{2}}}{r^2 + 2r_1^2 - rr_2}$$

$$= \frac{(r^2 + r^2\tan^2 2\theta)^{\frac{3}{2}}}{r^2 + 2r^2\tan^2 2\theta - (2r^2\sec^2\theta + r^2\tan^2 2\theta)}$$

$$= -\frac{r^3\sec^3 2\theta}{r^2\sec^2 2\theta} = -r\sec 2\theta$$

$$= -r\left(\frac{r^2}{a^2}\right), \text{ using(1)}$$

$$= \frac{r^3}{a^2} \text{ (in magnitude).}$$

(ii) The given curve is

$$r^n = a^n \sin n\theta \tag{1}$$

Taking log, we get

$$n\log r = n\log a + \log\sin n\theta.$$

Differentiating with respect to θ, we get

$$\frac{n}{r}\frac{dr}{d\theta} = n\cot n\theta.$$

Therefore,

$$r_1 = \frac{dr}{d\theta} = r\cot n\theta.$$

Differentiating once more, we get

$$r_2 = \frac{d^2r}{d\theta^2} = nr\,\text{cosec}^2\,n\theta + \cot n\theta\frac{dr}{d\theta}$$

$$= -nr\,\text{cosec}^2 n\theta + r\cot^2 n\theta.$$

Hence,

$$\rho = \frac{(r^2 + r_1^2)^{\frac{3}{2}}}{r^2 + 2r_1^2 - rr_2}$$

$$= \frac{(r^2 + r^2\cot^2 n\theta)^{\frac{3}{2}}}{r^2 + 2r^2\cot^2 n\theta - r(-nr\,\text{cosec}^2 n\theta + r\cot^2 n\theta)}$$

$$= \frac{r^3\,\text{cosec}^3\,n\theta}{r^2 + r^2\cot^2 n\theta + nr^2\,\text{cosec}^2\,n\theta}$$

$$= \frac{r^3\text{cosec}^3 n\theta}{r^2(1 + \cot^2 n\theta) + nr^2\,\text{cosec}^2\,n\theta}$$

$$= \frac{r^3\,\text{cosec}^3\,n\theta}{r^2\text{cosec}^2 n\theta\,(1 + n)}$$

$$= \frac{r\,\text{cosec}\,n\theta}{n + 1}$$

$$= \frac{a^n}{(n + 1)r^{n-1}}, \text{ using (1).}$$

(iii) The equation of the given curve is

$$r = a(1 - \cos\theta). \quad (1)$$

Differentiating with respect to θ we get

$$r_1 = \frac{dr}{d\theta} = a\sin\theta \text{ and } r_2 = \frac{d^2r}{d\theta^2} = a\cos\theta.$$

Therefore,

$$\rho = \frac{(r^2 + r_1^2)^{\frac{3}{2}}}{r^2 + 2r_1^2 - rr_2}$$

$$= \frac{[a^2(1-\cos\theta)^2 + a^2\sin^2\theta]^{\frac{3}{2}}}{a^2(1-\cos\theta)^2 + 2a^2\sin^2\theta - a^2\cos\theta(1-\cos\theta)} \quad \text{using (1)}$$

$$= \frac{a^3(1+\cos^2\theta + \sin^2\theta - 2\cos\theta)^{\frac{3}{2}}}{a^2[1+\cos^2\theta - 2\cos\theta + 2\sin^2\theta - \cos\theta + \cos^2\theta]}$$

$$= \frac{2^{\frac{3}{2}}a(1-\cos\theta)^{\frac{3}{2}}}{3(1-\cos\theta)} = \frac{2^{\frac{3}{2}}a\sqrt{r}}{3\sqrt{a}} = \frac{2}{3}\sqrt{2ar}.$$

It also follows that the radius of curvature at any point of the cardioid $r = a(1 - \cos\theta)$ varies as $\sqrt{r}$.

EXAMPLE 3.15

Show that the radius of curvature at any point of the cardioid $r = a(1 + \cos\theta)$ is $\frac{2}{3}\sqrt{2ar}$ and that $\frac{\rho^2}{r}$ is constant.

Solution. The equation of the curve is

$$r = a(1 + \cos\theta).$$

Therefore,

$$r_1 = \frac{dr}{d\theta} = -a\sin\theta \text{ and } r_2 = \frac{d^2r}{d\theta^2} = -a\cos\theta.$$

Therefore,

$$\rho = \frac{(r^2 + r_1^2)^{\frac{3}{2}}}{r^2 + 2r_1^2 - rr_2}$$

$$= \frac{[a^2(1+\cos\theta)^2 + a^2\sin^2\theta]^{\frac{3}{2}}}{a^2(1+\cos\theta)^2 + 2a^2\sin^2\theta + a^2\cos\theta(1+\cos\theta)}$$

$$= \frac{a^3(1+\cos^2\theta + \sin^2\theta + 2\cos\theta)^{\frac{3}{2}}}{a^2[1+\cos^2\theta + 2\cos\theta + 2\sin^2\theta + \cos\theta + \cos^2\theta]}$$

$$= \frac{2^{\frac{3}{2}}a(1+\cos\theta)^{\frac{3}{2}}}{3(1+\cos\theta)} = \frac{2^{\frac{3}{2}}a\sqrt{r}}{3\sqrt{a}}$$

$$= \frac{2}{3}\sqrt{2ar} = \frac{4a}{3}\cos\frac{\theta}{2} \text{ (in terms of } \theta).$$

It follows that ρ varies as $\sqrt{r}$. Further,

$$\rho^2 = \frac{8ar}{9} \quad \text{or} \quad \frac{\rho^2}{r} = \frac{8a}{9} \text{ (constant)}.$$

EXAMPLE 3.16

Find the radius of curvature of the curve

$$\frac{l}{r} = 1 + e\cos\theta.$$

Solution. The equation of the given curve is

$$\frac{l}{r} = 1 + e\cos\theta.$$

Let $\frac{1}{r} = u$ Then the equation of the curve becomes

$$u = \frac{1}{l}(1 + e\cos\theta).$$

Therefore,

$$u_1 = \frac{du}{d\theta} = \frac{1}{l}(-e\sin\theta),$$

$$u_2 = \frac{d^2u}{d\theta^2} = \frac{1}{l}(-e\cos\theta).$$

Hence, the radius of curvature is

$$\rho = \frac{(u^2 + u_1^2)}{u^3(u + u_2)}$$

$$= \frac{\left(\frac{1}{l^3}\right)[(1+e\cos\theta)^2 + e^2\sin^2\theta]^{\frac{3}{2}}}{\left(\frac{1}{l^4}\right)(1+e\cos\theta)^3[(1+e\cos\theta) - e\cos\theta]}$$

$$= \frac{l[1 + e^2 + 2e\cos\theta]^{\frac{3}{2}}}{(1+e\cos\theta)^3}.$$

EXAMPLE 3.17

If ρ is the radius of curvature and $\tan\phi = r\frac{d\theta}{dr}$, show that

$$\frac{r}{\rho} = \sin\phi\left(1 + \frac{d\phi}{d\theta}\right) \text{ for any curve.}$$

Solution. We know that

$$\psi = \theta + \phi.$$

Differentiating with respect to s, we get

$$\frac{d\psi}{ds} = \frac{d\theta}{ds} + \frac{d\phi}{ds} = \frac{d\theta}{ds} + \frac{d\phi}{d\theta}\cdot\frac{d\theta}{ds}$$

$$= \frac{d\theta}{ds}\left(1 + \frac{d\phi}{d\theta}\right),$$

that is, $\frac{1}{\rho} = \frac{\sin\phi}{r}\left(1 + \frac{d\phi}{d\theta}\right),$

$$\text{since } \rho = \frac{ds}{d\psi} \text{ and } \sin\phi = r\frac{d\theta}{ds}.$$

Hence,

$$\frac{r}{\rho}=\sin\phi\left(1+\frac{d\phi}{d\theta}\right) \text{ so that } \rho=\frac{r\operatorname{cosec}\phi}{1+\frac{d\phi}{d\theta}}.$$

EXAMPLE 3.18

If ρ_1 and ρ_2 are the radii of curvatures at the extremities of any chord of the cardioid $r = a(1 + \cos\theta)$ which passes through the pole, show that

$$\rho_1^2+\rho_2^2=\frac{16a^2}{9}.$$

Solution. As in Example 3.15, the radius of curvature at any point of the cardioid $r = a(1 + \cos\theta)$ is

$$\rho=\frac{2}{3}\sqrt{2ar} \text{ so that } \rho^2=\frac{8}{9}ar.$$

If PQ is any chord of the cardioid passing through the pole and if P and Q are the points (r_1, θ_1) and (r_2, θ_2) respectively, then $\theta_2 = \pi + \theta_1$,Hence, if ρ_1 and ρ_2 are the radii of curvatures at P and Q, respectively, then

$$\rho_1^2=\frac{8}{9}ar_1 \text{ and } \rho_2^2=\frac{8}{9}ar_2.$$

Therefore,

$$\begin{aligned}\rho_1^2+\rho_2^2&=\frac{8a}{9}[r_1+r_2]\\&=\frac{8a}{9}[a(1+\cos\theta_1)+a(1+\cos\theta_2)]\\&=\frac{8a^2}{9}[2+\cos\theta_1+\cos\theta_2]\\&=\frac{8a^2}{9}[2+\cos\theta_1+\cos(\pi+\theta_1)]\\&=\frac{8a^2}{9}[2+\cos\theta_1-\cos\theta_1]=\frac{16a^2}{9}.\end{aligned}$$

EXAMPLE 3.19

Find the radius of curvature at any point (p, r) of the following curves.

(i) ellipse $\frac{1}{p^2}=\frac{1}{a^2}+\frac{1}{b^2}-\frac{r^2}{a^2b^2}$.

(ii) $pa^2=r^3$.

(iii) $pa^n=r^{m+1}$

Solution.

(i) The equation of the ellipse is

$$\frac{1}{p^2}=\frac{1}{a^2}+\frac{1}{b^2}-\frac{r^2}{a^2b^2}.$$

Differentiating with respect to p, we get

$$-\frac{2}{p^3}=-\frac{2r}{a^2b^2}\cdot\frac{dr}{dp} \quad\text{or}\quad r\frac{dr}{dp}=\frac{a^2b^2}{p^3}.$$

Hence,

$$\rho=r\frac{dr}{dp}=\frac{a^2b^2}{p^3}.$$

(ii) The equation of the curve is

$$pa^2=r^3.$$

Differentiating with respect to p we have

$$a^2=3r^2\frac{dr}{dp} \quad\text{or}\quad r\frac{dr}{dp}=\frac{a^2}{3r}.$$

Hence,

$$\rho=r\frac{dr}{dp}=\frac{a^2}{3r}$$

(iii) The given equation is

$$pa^n=r^{m+1}.$$

Therefore, differentiation with respect to p yields

$$a^n=(m+1)r^m\frac{dr}{dp} \quad\text{or}\quad r\frac{dr}{dp}=\frac{a^n}{(m+1)r^{m-1}}.$$

Hence,

$$\rho=r\frac{dr}{dp}=\frac{a^n}{(m+1)r^{m-1}}.$$

Thus, radius of curvature varies inversely as the $(m - 1)$th power of the radius vector.

EXAMPLE 3.20

Show that the radius of curvature of the equiangular spiral $r = ae^{\theta\cot\alpha}$ at any point (r, θ) is $\rho = r\operatorname{cosec}\alpha$

Solution. The given curve is

$$r=ae^{\theta\cot\alpha}.$$

Differentiating with respect to θ, we have

$$r_1=\frac{dr}{d\theta}=ae^{\theta\cot\alpha}\cdot\cot\alpha=r\cot\alpha \text{ and}$$

$$r_2=\frac{d^2r}{d\theta^2}=r\cot^2\alpha.$$

Therefore,

$$\rho=\frac{(r+r_1^2)^{\frac{3}{2}}}{r^2+2r_1^2-rr_2}=\frac{(r^2+r^2\cot^2\alpha)^{\frac{3}{2}}}{r^2+2r^2\cot^2\alpha-r^2\cot^2\alpha}$$

$$=\frac{r^3(1+\cot^2\alpha)^{\frac{3}{2}}}{r^2(1+\cot^2\alpha)}=r(\operatorname{cosec}^2\alpha)^{\frac{1}{2}}=r\operatorname{cosec}\alpha.$$

3.6 RADIUS OF CURVATURE AT THE ORIGIN

The radius of curvature at the origin is determined by the following methods:

1. Newton's Method: Suppose that a curve passes through the origin and the axis of x is tangent at the origin. Then, slope of the tangent at the origin $=\left(\frac{dy}{dx}\right)_{(0,\,0)}$

$$= y_1\,(0, 0).$$

Since the tangent at the origin is x-axis, the slope of the tangent should be zero. Therefore $y_1\,(0, 0) = 0$. The radius of curvature at the origin is

$$\rho \text{ at } (0, 0) = \frac{[1 + y_1^2\,(0, 0)]^{\frac{3}{2}}}{y_2\,(0, 0)} = \frac{1}{y_2\,(0, 0)}. \quad (1)$$

On the other hand,

$$\lim_{\substack{x\to 0\\ y\to 0}} \frac{x^2}{2y} = \lim_{\substack{x\to 0\\ y\to 0}} \frac{2x}{2y_1} \quad \left(\frac{0}{0}\text{ form}\right)$$

$$= \lim_{\substack{x\to 0\\ y\to 0}} \frac{1}{y_2} = \frac{1}{y_2(0, 0)}. \quad (2)$$

From (1) and (2), it follows that

$$\rho\, at\, (0, 0) = \lim_{\substack{x\to 0\\ y\to 0}} \frac{x^2}{2y}.$$

Similarly, if the curve passes through the origin and y-axis is tangent to it at the origin, then

$$\rho \text{ at } (0, 0) = \lim_{\substack{x\to 0\\ y\to 0}} \frac{y^2}{2x}.$$

If the initial line is tangent at the pole (origin), then

$$\rho = \lim_{\substack{x\to 0\\ y\to 0}} \frac{x^2}{2y} = \lim_{\substack{r\to 0\\ \theta\to 0}} \frac{r^2\cos^2\theta}{2r\sin\theta}$$

$$= \lim_{\substack{r\to 0\\ \theta\to 0}} \frac{r}{2\theta}\cdot\frac{\theta}{\sin\theta}\cdot\cos^2\theta = \lim_{\substack{r\to 0\\ \theta\to 0}} \frac{r}{2\theta}\left(\frac{0}{0}\text{ form}\right)$$

$$= \lim_{\substack{r\to 0\\ \theta\to 0}} \frac{\frac{dr}{d\theta}}{2} = \frac{1}{2}\cdot\frac{dr}{d\theta}.$$

2. Method of Expansion: This method is used when neither x-nor y-axis is tangent to the curve at the origin. Suppose that the equation of the curve is $y = f(x)$. Since the curve passes through the origin, $f(0) = 0$. By Maclaurin's expansion, we have

$$y = f(0) + x f'(0) + \frac{x^2}{2!} f''(0) + \frac{x^3}{3!} f'''(0) + \ldots$$

$$= xf'(0) + \frac{x^2}{2!} f''(0) + \frac{x^3}{3!} f'''(0) + \ldots$$

$$= px + \frac{1}{2!} qx^2 + \frac{1}{3!} rx^3 + \ldots \quad (1)$$

where,

$$p = f'(0) = y_1\,(0),$$
$$q = f''(0) = y_2\,(0),$$
$$r = f'''(0) = y_3\,(0), \text{ and so on.}$$

Differentiating (1) with respect to x, we get

$$y_1 = p + qx + \frac{1}{2} rx^2 + \ldots \text{ and}$$
$$y_2 = q + rx + \ldots.$$

Thus,

$$y_1\,(0, 0) = p,\ y_2\,(0, 0) = q.$$

Hence,

$$\rho \text{ at the origin} = \frac{[1 + y_1^2\,(0, 0)]^{\frac{3}{2}}}{y_2(0)} = \frac{(1 + p^2)^{\frac{3}{2}}}{q}.$$

EXAMPLE 3.21

Find the radius of curvature at the origin for the curve

$5x^3 + 7y^2x^2 + 4x^2y + xy^2 + 2x^2 + 3xy + y^2 + 4x = 0.$

Solution. The given curve passes through the origin. The lowest-degree term in the given equation is $4x$. Therefore, the tangent at the origin is $4x = 0$ or $x = 0$, that is, the y-axis. Therefore, by Newton's Method,

$$\rho \text{ at the origin} = \lim_{\substack{x\to 0\\ y\to 0}} \frac{y^2}{2x}.$$

Dividing the equation of the curve throughout by $2x$, we get

$$\frac{5}{2}x^2 + \frac{7y^2x^2}{2x} + 2xy + \frac{y^2}{2} + x + \frac{3}{2}y + \frac{y^2}{2x} + 2 = 0.$$

Taking limit as $x \to 0$ and $y \to 0$, we get

$$\lim_{\substack{x\to 0\\ y\to 0}} \frac{y^2}{2x} + 2 = 0$$

or

$$\rho = -2 = 2 \text{ (numerically)}.$$

EXAMPLE 3.22

Show that the radii of curvatures at the origin on the curve $x^3 + y^3 = 3axy$ is each equal to $\frac{3a}{2}$.

Solution. The equation of the given curve is

$$x^3 + y^3 = 3axy.$$

The curve passes through the origin. Further, the tangent at the origin is given by $3axy = 0$. Thus, $x = 0$ and $y = 0$ are the tangents at the origin.

Therefore, the radii of curvatures are given by

$$\rho_1 \text{ at the origin} = \lim_{\substack{x\to 0\\ y\to 0}} \frac{x^2}{2y}.$$

and

$$\rho_2 \text{ at the origin} = \lim_{\substack{x\to 0\\ y\to 0}} \frac{y^2}{2x}.$$

Dividing the given equation throughout by $2xy$, we get

$$\frac{x^2}{2y} + \frac{y^2}{2x} = \frac{3a}{2}.$$

Taking limit as $x \to 0$ and $y \to 0$, we have

$$\lim_{\substack{x\to 0\\ y\to 0}} \frac{x^2}{2y} + \lim_{\substack{x\to 0\\ y\to 0}} \frac{y^2}{2x} = \frac{3a}{2}$$

or

$$\lim_{\substack{x\to 0\\ y\to 0}} \frac{x^2}{2y} + \lim_{\substack{x\to 0\\ y\to 0}} \frac{1}{4} xy \frac{2y}{x^2} = \frac{3a}{2}$$

or

$$\rho_1 + 0 \frac{1}{\rho_1} = \frac{3a}{2} \quad \text{or} \quad \rho_1 = \frac{3a}{2}.$$

Similarly, we can show that $\rho_2 = \frac{3a}{2}$.

EXAMPLE 3.23

Show that the radius of curvature of the curve

$$y^2 = x^2 \frac{a+x}{a-x}$$

is $\pm a\sqrt{2}$ (or $a\sqrt{2}$ in magnitude).

Solution. The curve is

$$y^2(a-x) = x^2(a+x).$$

Obviously, the curve passes through the origin. But equating to zero the lowest-degree terms, we get

$$a(y^2 - x^2) = 0.$$

Therefore $y = \pm x$ is the tangent at the origin. Hence, Newton's method cannot be applied to this curve. So let

$$y = px + \frac{1}{2!}qx^2 + \frac{1}{3!}rx^3 + \ldots$$

Substituting this value of y in the given equation of the curve, we get

$$\left(px + \frac{1}{2}qx^2 + \frac{1}{3}!\, rx^3 + \ldots\right)^2 (a-x)$$
$$= x^2(a+x).$$

Equating coefficients of x^2 and x^3 on both sides, we get

$$ap^2 = a \text{ and } -p^2 + apq = 1,$$

which yield

$$p = 1, q = \frac{2}{a} \quad \text{and} \quad p = -1, q = \frac{-2}{a}.$$

Therefore,

$$\rho_1 = \frac{(1+p^2)^{\frac{3}{2}}}{q} = \frac{2^{\frac{3}{2}}}{\frac{2}{a}} = \sqrt{2}a$$

and

$$\rho_2 = \frac{(1+1)^{\frac{3}{2}}}{-\frac{2}{a}} = -\sqrt{2}a = \sqrt{2}a \text{ (in magnitude).}$$

EXAMPLE 3.24

Find the radius of curvature at the origin of the cycloid

$$x = a(\theta + \sin\theta) \text{ and } y = a(1 - \cos\theta).$$

Solution. The given curve is

$$x = a(\theta + \sin\theta) \text{ and } y = a(1 - \cos\theta).$$

Therefore, differentiation with respect to θ yields

$$\frac{dx}{d\theta} = a(1 + \cos\theta) \quad \text{and} \quad \frac{dy}{d\theta} = a\sin\theta.$$

Then,

$$\frac{dy}{dx} = \frac{dy/d\theta}{dx/d\theta} = \frac{a\sin\theta}{a(1+\cos\theta)} = \frac{2\sin\frac{\theta}{2}\cos\frac{\theta}{2}}{2\cos^2\frac{\theta}{2}}$$
$$= \tan\frac{\theta}{2}.$$

Since $\theta = 0$ at $(0, 0)$, we have $\left(\frac{dy}{dx}\right)_{(0,0)} = 0$.

Therefore, x-axis is tangent to the curve at the origin and so, the Newton's method is applicable. Therefore, ρ at the origin is given by

$$\rho(0,0) = \lim_{\substack{x\to 0\\ y\to 0}} \frac{x^2}{2y}$$

$$= \lim_{\theta\to 0}\left[\frac{a^2(\theta + \sin\theta)^2}{2a(1-\cos\theta)}\right]\left(\frac{0}{0}\text{ form}\right)$$

$$= \lim_{\theta\to 0}\left[\frac{a}{2}\frac{\{2(\theta + \sin\theta)(1+\cos\theta)\}}{\sin\theta}\right]\left(\frac{0}{0}\text{ form}\right)$$

$$= \lim_{\theta\to 0}\left[\frac{a(\theta+\sin\theta)(-\sin\theta) + (1+\cos\theta)^2}{\cos\theta}\right]$$

$$= 4a.$$

3.7 CENTER OF CURVATURE

Let $P(x, y)$ be any point on the curve and let PT be the tangent at P making an angle ψ with the positive direction of x-axis. Let $C(X, Y)$ be the center of curvature corresponding to $P(x, y)$.

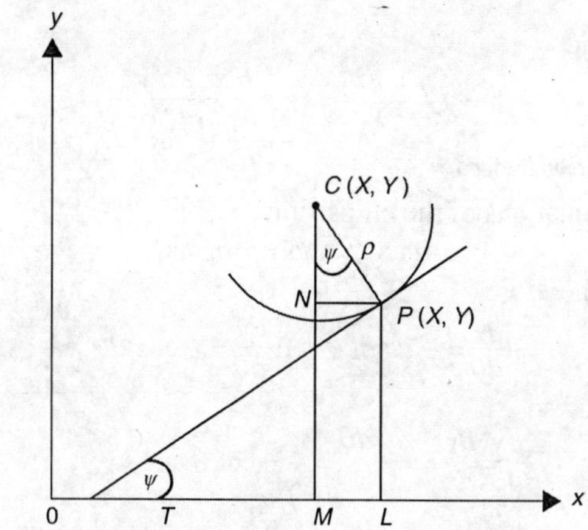

Draw PL and CM perpendicular to the x-axis from P and C, respectively. Draw $PN \perp CM$. Then

$$X = OM = OL - ML = OL - NP$$
$$= x - \rho \sin \psi \qquad (1)$$

and

$$Y = CM = NM + CN = NM + \rho \cos \psi$$
$$= y + \rho \cos\psi \qquad (2)$$

But,

$$\tan \psi = \frac{dy}{dx} = y_1.$$

Therefore,

$$\sec \psi = \sqrt{1 + \tan^2 \psi}$$
$$= \sqrt{1 + y_1^2}$$

or

$$\cos \psi = \frac{1}{\sqrt{1 + y_1^2}}$$

and so,

$$\sin \psi = \sqrt{1 - \cos^2 \psi} = \sqrt{1 - \frac{1}{1 + y_1^2}}$$
$$= \frac{y_1}{\sqrt{1 + y_1^2}}.$$

Also,

$$\rho = \frac{\left(1 + y_1^2\right)^{\frac{3}{2}}}{y_2}.$$

Hence (1) yields

$$X = x - \frac{\left(1 + y_1^2\right)^{\frac{3}{2}}}{y_2} \cdot \frac{y_1}{\sqrt{1 + y_1^2}} = x - \frac{y_1\left(1 + y_1^2\right)}{y_2}$$

and

$$Y = y + \frac{\left(1 + y_1^2\right)^{\frac{3}{2}}}{y_2} \cdot \frac{1}{\sqrt{1 + y_1^2}} = y + \frac{1 + y_1^2}{y_2}.$$

Also (1) and (2) can be expressed as

$$X = x - \rho \sin\psi = x - \frac{ds}{d\psi} \cdot \frac{dy}{ds} = x - \frac{dy}{d\psi}$$

and

$$Y = y + \rho \cos\psi = y + \frac{ds}{d\psi} \cdot \frac{dx}{ds} = y + \frac{dx}{d\psi}.$$

3.8 EVOLUTES AND INVOLUTES

The locus of the center of curvature for a curve is called the *evolute* of the curve and the curve itself is called the *involute* of its evolute.

3.9 EQUATION OF THE CIRCLE OF CURVATURE

Let (X, Y) be the coordinates of the center of curvature and ρ the radius of curvature at any point (x, y) of the curve. Then the equation of the circle of curvature is

$$(x - X)^2 + (y - Y)^2 = \rho^2.$$

EXAMPLE 3.25

Find the center of curvature of the following curves.

(i) $y = x^3 - 6x^2 + 3x + 1$ at $(1, -1)$.

(ii) the parabola $y^2 = 4ax$ at (x, y). Also, find the equation of the evolute of the given parabola.

(iii) the ellipse $\frac{x^2}{a^2} + \frac{y^2}{b^2} = 1$ at (x, y). Also, find the evolute.

Solution.

(i) The given curve is

$$y = x^3 - 6x^2 + 3x + 1.$$

Therefore,

$$y_1 = \frac{dy}{dx} = 3x^2 - 12x + 3 \text{ and so } y_1(1, -1) = -6,$$

$$y_2 = \frac{d^2y}{dx^2} = 6x - 12 \text{ and so } y_2(1, -1) = -6,$$

$$X = x - \frac{y_1(1 + y_1^2)}{y_2}$$
$$= 1 - \frac{(-6)[1 + (-6)^2]}{-6} = -36, \text{ and}$$

$$Y = -1 + \frac{1 + (-6)^2}{-6} = -\frac{43}{6}.$$

Hence, the center of curvature is

$$(X, Y) = \left(-36, -\frac{43}{6}\right).$$

(ii) The equation of the given parabola is

$$y^2 = 4ax.$$

Therefore,

$$y_1 = \frac{2a}{y} = \frac{2a}{\sqrt{4ax}} = \sqrt{\frac{a}{x}} \text{ and } y_2 = -\frac{1}{2}\sqrt{a}\,x^{-\frac{3}{2}}.$$

Therefore,

$$X = \frac{x - \sqrt{\frac{a}{x}}\left(1 + \frac{a}{x}\right)}{-\frac{1}{2}\sqrt{a}\,x^{-\frac{3}{2}}}$$

$$= x + 2\,(x + a)$$

$$= 3x + 2a \qquad (1)$$

and

$$Y = \frac{y + 1 + \frac{a}{x}}{-\frac{1}{2}\sqrt{a}\,x^{-\frac{3}{2}}}$$

$$= 2\sqrt{a}\sqrt{x} - \frac{2(x + a)}{\sqrt{a}\,x^{-\frac{1}{2}}}$$

$$= 2\sqrt{a}\sqrt{x}\left(1 - \frac{x + a}{a}\right)$$

$$= -2\,\frac{x^{\frac{3}{2}}}{\sqrt{a}}. \qquad (2)$$

Hence, the center of curvature of the given parabola is

$$(X, Y) = \left(3x + 2a, -\frac{2x^{\frac{3}{2}}}{\sqrt{a}}\right).$$

From (1), we have $x = \frac{X - 2a}{3}$. Putting this value in (2), we get

$$Y = -\frac{2(X - 2a)^{\frac{3}{2}}}{\sqrt{a}} \quad \text{or} \quad aY^2 = 4\left(\frac{X - 2a}{3}\right)^3$$

or

$$27aY^2 = 4\,(X - 2a)^3.$$

Therefore, the locus of the center of curvature (X, Y) is

$$27ay^2 = 4\,(x - 2a)^3,$$

which is the required evolute.

(iii) The equation of the ellipse is

$$\frac{x^2}{a^2} + \frac{y^2}{b^2} = 1.$$

Therefore,

$$y_1 = -\frac{b^2x}{a^2y} \quad \text{and} \quad y_2 = -\frac{b^4}{a^2y^3}.$$

Therefore,

$$X = x - \frac{y_1(1 + y_1^2)}{y_2} = x - \frac{b^4x^2 + a^4y^2}{a^4b^2}$$

and

$$Y = y + \frac{1 + y_1^2}{y_2} = y + \frac{b^4x^2 + a^4y^2}{a^2b^4}.$$

Second Method

Equations of the ellipse are

$$x = a\cos\theta \text{ and } y = b\sin\theta.$$

Therefore,

$$\frac{dx}{d\theta} = -a\sin\theta \text{ and } \frac{dy}{d\theta} = b\cos\theta$$

and so,

$$y_1 = \frac{dy}{dx} = \frac{dy/d\theta}{dx/d\theta} = -\frac{b}{a}\cot\theta \text{ and}$$

$$y_2 = \frac{d^2y}{dx^2} = \frac{d}{dx}\left(\frac{dy}{dx}\right) = \left[\frac{d}{d\theta}\left(-\frac{b}{a}\cot\theta\right)\right]\frac{d\theta}{dx}$$

$$= \frac{b}{a}\operatorname{cosec}^2\theta\left(\frac{1}{-a\sin\theta}\right) = -\frac{b}{a^2}\operatorname{cosec}^3\theta.$$

Therefore,

$$X = x - \frac{y_1(1 + y_1^2)}{y_2}$$

$$= x - \frac{-\frac{b}{a}\cot\theta\left(1 + \frac{b^2}{a^2}\cot^2\theta\right)}{-\frac{b}{a^2}\operatorname{cosec}^3\theta}$$

$$= a\cos\theta - \frac{(a^2\sin^2\theta + b^2\cos^2\theta)}{a}\cos\theta$$

$$= \frac{a^2 - b^2}{a}\cos^3\theta \qquad (1)$$

and

$$Y = y + \frac{1 + y_1^2}{y_2} = y + \frac{1 + \frac{b^2}{a^2}\cot^2\theta}{-\frac{b}{a^2}\operatorname{cosec}^3\theta}$$

$$= b\sin\theta - \frac{\sin\theta}{b}(a^2\sin^2\theta + b^2\cos^2\theta)$$

$$= \frac{1}{b}\sin\theta\,[b^2 - a^2\sin^2\theta - b^2\cos^2\theta]$$

$$= \frac{b^2 - a^2}{b}\sin^3\theta = -\frac{a^2 - b^2}{b}\sin^3\theta \quad (2)$$

Hence the center of curvature is

$$(X, Y) = \left(\frac{a^2 - b^2}{a}\cos^3\theta, -\frac{a^2 - b^2}{b}\sin^3 t\right).$$

From (1) and (2), we have

$$aX = (a^2 - b^2)\cos^3\theta \text{ or } (aX)^{\frac{2}{3}} = (a^2 - b^2)^{\frac{2}{3}}\cos^2\theta$$

and

$$bY = -(a^2 - b^2)\sin^3\theta \quad \text{or}$$
$$(bY)^{\frac{2}{3}} = -(a^2 - b^2)^{\frac{2}{3}}\sin^3\theta.$$

Adding both we get

$$(aX)^{\frac{2}{3}} + (bY)^{\frac{2}{3}} = (a^2 - b^2)^{\frac{2}{3}}.$$

Therefore, the evolute of the ellipse is

$$(ax)^{\frac{2}{3}} + (by)^{\frac{2}{3}} = (a^2 - b^2)^{\frac{2}{3}}.$$

EXAMPLE 3.26

Show that the evolute of the cycloid $x = a(\theta - \sin\theta)$ and $y = a(1 - \cos\theta)$ lies on an equal cycloid.

Solution. For the given cycloid, we have

$$\frac{dx}{d\theta} = a(1 - \cos\theta) \text{ and } \frac{dy}{d\theta} = a\sin\theta.$$

Therefore,

$$y_1 = \frac{dy}{dx} = \frac{dy/d\theta}{dx/d\theta} = \frac{a\sin\theta}{a(1-\cos\theta)} = \cot\frac{\theta}{2}$$

and

$$y_2 = \frac{d}{dx}(y_1) = \frac{d}{d\theta}(y_1)\frac{d\theta}{dx}$$
$$= -\frac{1}{2}\operatorname{cosec}^2\frac{\theta}{2}\cdot\frac{1}{a(1-\cos\theta)}$$
$$= \frac{-1}{4a\sin^4\frac{\theta}{2}}.$$

Therefore,

$$X = x - \frac{y_1(1+y_1^2)}{y_2}$$
$$= a(\theta - \sin\theta) - \frac{\cot\frac{\theta}{2}\left[1 + \cot^2\frac{\theta}{2}\right]}{-\frac{1}{4}\operatorname{cosec}^4\frac{\theta}{2}}$$
$$= a(\theta - \sin\theta) + 4a\cos\frac{\theta}{2}\sin\frac{\theta}{2}$$
$$= a\theta - a\sin\theta + 2a\sin\theta = a(\theta + \sin\theta) \quad (1)$$

and

$$Y = y + \frac{1+y_1^2}{y_2}$$
$$= a(1-\cos\theta) - 4a\sin^2\frac{\theta}{2}$$
$$= a - a\cos\theta - 2a + 2a\cos\theta$$
$$= -a(1-\cos\theta). \quad (2)$$

Hence, the center of curvature of the given cycloid is

$$(X, Y) = (a(\theta + \sin\theta), -a(1-\cos\theta)).$$

The locus of (X, Y) is

$$x = a(\theta + \sin\theta) \text{ and } y = -a(1 - \cos\theta),$$

which is another equal cycloid.

EXAMPLE 3.27

Show that the evolute of the rectangular hyperbola $xy = c^2$ is the curve

$$(x+y)^{\frac{2}{3}} - (x-y)^{\frac{2}{3}} = (4c)^{\frac{2}{3}}.$$

Solution. We have

$$xy = c^2 \quad \text{or} \quad y = \frac{c^2}{x}.$$

Therefore,

$$y_1 = \frac{dy}{dx} = -\frac{c^2}{x^2} \quad \text{and} \quad y_2 = \frac{d^2y}{dx^2} = \frac{2c^2}{x^3}.$$

Let (X, Y) be the coordinate of the center of curvature.

Then,

$$X = x - \frac{y_1(1+y_1^2)}{y_2} = x - \frac{\frac{-c^2}{x^2}\left(1+\frac{c^4}{x^4}\right)}{\frac{2c^2}{x^3}}$$
$$= x + \frac{x^4+c^4}{2x^3} = \frac{2x^4+x^4+c^4}{2x^3}$$
$$= \frac{3x^4+x^2y^2}{2x^3}, \text{ since } xy = c^2$$
$$= \frac{3x}{2} + \frac{y^2}{2x}$$

and

$$Y = y + \frac{1+y_1^2}{y_2} = y + \frac{1+\frac{c^4}{x^4}}{\frac{2c^2}{x^3}} = y + \frac{x^4+c^4}{2c^2x}$$
$$= y + \frac{x^4+x^2y^2}{2x^2y}, \text{ since } xy = c^2$$
$$= y + \frac{x^2+y^2}{2y} = \frac{3y^2+x^2}{2y} = \frac{3}{2}y + \frac{x^2}{2y}.$$

Hence, the center of curvature is

$$(X, Y) = \left(\frac{3x}{2} + \frac{y^2}{2x}, \frac{3y}{2} + \frac{x^2}{2y}\right).$$

Further, we note that

$$X + Y = \frac{1}{2xy}[x^3 + y^3 + 3x^2y + 3xy^2]$$
$$= \frac{1}{2c^2}(x+y)^3$$

and so,

$$(X+Y)^{\frac{2}{3}} = \frac{1}{(2c^2)^{\frac{2}{3}}}(x+y)^2.$$

Similarly,

$$(X-Y)^{\frac{2}{3}} = -\frac{1}{(2c^2))^{\frac{2}{3}}}(x-y)^2.$$

Therefore,

$$(X+Y)^{\frac{2}{3}} - (X-Y)^{\frac{2}{3}}$$

$$= \frac{1}{(2c^2)^{\frac{2}{3}}}[(x+y)^2 + (x-y)^2]$$

$$= \frac{1}{(2c^2)^{\frac{2}{3}}}(4xy) = \frac{1}{(2c^2)^{\frac{2}{3}}}(4c^2)$$

$$= (4c)^{\frac{2}{3}}.$$

The locus of (X, Y) is therefore,

$$(x+y)^{\frac{2}{3}} - (x-y)^{\frac{2}{3}} = (4c)^{\frac{2}{3}}\text{s}.$$

Remark 3.2

The question can also be solved by taking a parametric equation of the curve as $x = ct$ and $y = \frac{c}{t}$, and then eliminating t from the expression for X and Y.

EXAMPLE 3.28

Show that the circle of curvature at the origin of the parabola $y = mx + \frac{x^2}{a}$ is

$$x^2 + y^2 = a(1+m^2)(y - mx).$$

Solution. The equation of the given parabola is

$$y = mx + \frac{x^2}{a}.$$

Therefore,

$$y_1 = m + \frac{2x}{a} = \frac{am + 2x}{a}$$

so that $y_1(0, 0) = m$ and

$$y_2 = \frac{2}{a} \text{ so that } y_2(0, 0) = \frac{2}{a}.$$

Let (X, Y) be the center of curvature at the origin. Then,

$$X = x - \frac{y_1(1+y_1^2)}{y_2} = 0 - \frac{m(1+m^2)}{\frac{2}{a}} = \frac{-am(1+m^2)}{2}$$

and

$$Y = y + \frac{1+y_1^2}{y_2} = 0 + \frac{1+m^2}{\frac{2}{a}} = \frac{a(1+m^2)}{2}.$$

Also,

$$\rho \text{ at } (0, 0) = \frac{(1+y_1^2))^{\frac{3}{2}}}{y_2} = \frac{(1+m^2)^{\frac{3}{2}}}{\frac{2}{a}}$$

$$= \frac{a(1+m^2)^{\frac{3}{2}}}{2}$$

Therefore, the equation of the circle of curvature is

$$(x-X)^2 + (y-Y)^2 = \rho^2$$

or

$$\left[x + \frac{am(1+m^2)}{2}\right]^2 + \left[y - \frac{a(1+m^2)}{2}\right]^2$$

$$= \frac{a^2(1+m^2)^3}{4}$$

or

$$4x^2 + 4y^2 = a(1+m^2) + [a(1+m^2)^2 - am^2(1+m^2)$$

$$- a(1+m^2) + 4(y - mx)]$$

$$= 4a(1+m^2)(y - mx),$$

which yields

$$x^2 + y^2 = a(1+m^2)(y - mx).$$

3.10 CHORDS OF CURVATURE PARALLEL TO THE COORDINATE AXES

Let $P(X, Y)$ be a point on the curve and let the tangent at P make an angle ψ with the positive direction of x-axis. Then $CP = \rho$ and $PD = 2\rho$ Let PA and PB be the chords of curvature parallel to the x- and y-axis, respectively.

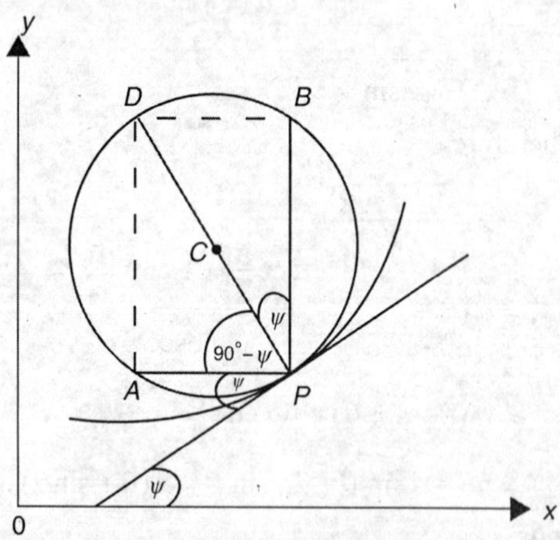

The chord PA makes an angle $\frac{\pi}{2} - \psi$ with CP and the chord PB makes an angle ψ with CP. Let C_x denote the length of PA and C_y denote the length of the chord PB. Then

$$C_x = PD\cos\left(\frac{\pi}{2} - \psi\right) = 2\rho\sin\psi$$

$$= \frac{2(1+y_1^2)^{\frac{3}{2}}}{y_2} \cdot \frac{y_1}{\sqrt{1+y_1^2}} = \frac{2y_1(1+y_1^2)}{y_2}$$

and

$$C_y = PD\cos\psi = 2\rho\cos\psi$$

$$= \frac{2(1+y_1^2)^{\frac{3}{2}}}{y_2}\cdot\frac{1}{\sqrt{1+y_1^2}} = \frac{2(1+y_1^2)}{y_2}.$$

3.11 CHORD OF CURVATURE IN POLAR COORDINATES

1. Chord of Curvature Through the Pole (Origin)

Let P (X,Y) be a point on the curve and let the tangent at P make an angle ϕ with the radius vector OP. Then PA, the chord of curvature through the pole O makes an angle $\frac{\pi}{2} - \phi$ with CP, the radius of curvature.

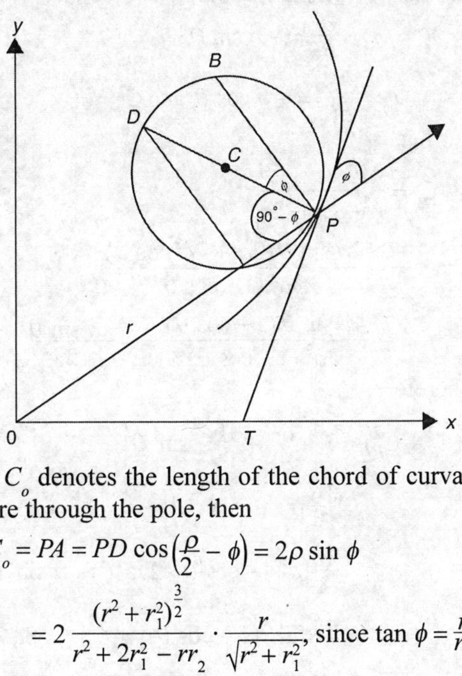

If C_o denotes the length of the chord of curvature through the pole, then

$$C_o = PA = PD\cos\left(\frac{\rho}{2} - \phi\right) = 2\rho\sin\phi$$

$$= 2\frac{(r^2+r_1^2)^{\frac{3}{2}}}{r^2+2r_1^2-rr_2}\cdot\frac{r}{\sqrt{r^2+r_1^2}}, \text{ since } \tan\phi = \frac{r}{r_1}$$

$$= \frac{2r\,(r^2+r_1^2)}{r^2+2r_1^2-rr^2}.$$

2. Chord of Curvature Perpendicular to the Radius Vector

The chord PB of curvature perpendicular to the radius vector makes an angle ϕ with CP. Therefore, the length C_p of this chord is given by

$$C_p = PB = PD\cos\phi = 2\rho\sin\phi$$

$$= \frac{2(r^2+r_1^2)^{\frac{3}{2}}}{r^2+2r_1^2-rr^2}\cdot\frac{r_1}{\sqrt{r^2+r_1^2}}$$

$$= \frac{2r_1(r^2+r_1^2)}{r^2+r_1^2-rr_2}.$$

EXAMPLE 3.29

Show that in the curve $y = a\log\sec\frac{x}{a}$, the chord of curvature parallel to the axis of y is of constant length.

Solution. The equation of the curve is

$$y = a\log\sec\frac{x}{a}.$$

Therefore,

$$y_1 = \frac{dy}{dx} = a\frac{1}{\sec\frac{x}{a}}\sec\frac{x}{a}\tan\frac{x}{a}\cdot\frac{1}{a} = \tan\frac{x}{a},$$

$$y_2 = \frac{d^2y}{dx^2} = \frac{1}{a}\sec^2\frac{x}{a}.$$

Therefore, the length of chord of curvature parallel to the y-axis is

$$C_y = \frac{2(1+y_1^2)}{y_2} = \frac{2(1+\tan^2\frac{x}{a})}{\frac{1}{a}\sec^2\frac{x}{a}}$$

$$= 2a \text{ (constant)}.$$

EXAMPLE 3.30

Find the length of the chord of curvature through the pole of the cardioid $r = a(1+\cos\theta)$.

Solution. The cardioid is $r = a(1+\cos\theta)$.

Therefore,

$$r_1 = \frac{dr}{d\theta} = -a\sin\theta \text{ and } r_2 = \frac{d^2r}{d\theta^2} = -a\cos\theta.$$

Hence,

$$C_o = \frac{2r(r^2+r_1^2)}{r^2+2r_1^2-rr_2}$$

$$= \frac{2r[a^2(1+\cos\theta)^2 + a^2\sin^2\theta]}{a^2(1+\cos\theta)^2 + 2a^2\sin^2\theta + a^2(1+\cos\theta)\cos\theta}$$

$$= \frac{2r\,(a^2 + a^2\cos^2\theta + 2a^2\cos\theta + a^2\sin^2\theta)}{3a^2 + 3a^2\cos\theta}$$

$$= \frac{2r[2a^2 + 2a^2\cos\theta]}{3a^2(1+\cos\theta)} = \frac{4r^2a}{3ar} = \frac{4r}{3}.$$

EXAMPLE 3.31

If C_x and C_y are the chords of curvature parallel to the axes at any point of the curve $y = ae^{\frac{x}{a}}$, show that

$$\frac{1}{C_x^2} + \frac{1}{C_y^2} = \frac{1}{2aC_x}.$$

Solution. We know that

$$C_x = \frac{2y_1(1+y_1^2)}{y_2} \text{ and } C_y = \frac{2(1+y_1^2)}{y_2}.$$

The given curve is $y = ae^{\frac{x}{a}}$. Therefore,

$$y_1 = \frac{dy}{dx} = ae^{\frac{x}{a}} \cdot \frac{1}{a} = e^{\frac{x}{a}} \text{ and } y_2 = \frac{d^2y}{dx^2} = e^{\frac{x}{a}} \cdot \frac{1}{a}.$$

Hence,

$$C_x = \frac{2e^{\frac{x}{a}}\left(1+e^{\frac{2x}{a}}\right)}{\frac{1}{a}e^{\frac{x}{a}}} = 2a\left(1+e^{\frac{2x}{a}}\right)$$

and

$$C_y = \frac{2\left(1+e^{\frac{2x}{a}}\right)}{\frac{1}{a}e^{\frac{x}{a}}} = \frac{2a}{e^{\frac{x}{a}}}\left(1+e^{\frac{2x}{a}}\right).$$

Then,

$$\frac{1}{C_x^2} + \frac{1}{C_y^2} = \frac{1}{4a^2\left(1+e^{\frac{2x}{a}}\right)^2} + \frac{1}{\frac{4a^2}{e^{\frac{2x}{a}}}\left(1+e^{\frac{2x}{a}}\right)^2}$$

$$= \frac{1}{4a^2\left(1+e^{\frac{2a}{x}}\right)^2}\left(1+e^{\frac{2x}{a}}\right).$$

$$= \frac{1}{4a^2\left(1+e^{\frac{2x}{a}}\right)} = \frac{1}{2aC_x}.$$

EXAMPLE 3.32

Find the length of the chord of curvature through the pole of the curve $r^n = a^n \cos n\theta$.

Solution. The equation of the given curve is $r^n = a^n \cos n\theta$. Differentiating with respect to θ, we have

$$nr^{n-1}\frac{dr}{d\theta} = -na^n \sin n\theta$$

or

$$r_1 = \frac{dr}{d\theta} = -\frac{a^n \sin n\theta}{r^{n-1}} = -r\frac{a^n \sin n\theta}{r^n}$$

$$= -r\frac{a^n \sin n\theta}{a^n \cos n\theta} = -r\tan n\theta \text{ and}$$

$$r_2 = -r_1 \tan n\theta - rn\sec^2 n\theta$$

$$= r\tan^2 n\theta - rn\sec^2 n\theta.$$

Therefore,

$$C_o = \frac{2r(r^2+r_1^2)}{r^2 - rr_2 + 2r_1^2}$$

$$= \frac{2r(r^2 + r^2\tan n\theta)}{r^2 - r(r\tan^2 n\theta - rn\sec^2 n\theta) + 2r^2\tan^2\theta}$$

$$= \frac{2r^3\sec^2 n\theta}{r^2[1-\tan^2 n\theta + n\sec^2 n\theta + 2\tan^2 n\theta]}$$

$$= \frac{2r\sec^2 n\theta}{1+\tan^2 n\theta + n\sec^2 n\theta}$$

$$= \frac{2r\sec^2 n\theta}{\sec^2 n\theta + n\sec^2 n\theta} = \frac{2r}{n+1}.$$

EXAMPLE 3.33

If C_o and C_p denote the length of the chords of curvature of the cardioid $r = a(1+\cos\theta)$, along and perpendicular to the radius vector through any point, show that

$$3(C_o^2 + C_p^2) = 8aC_o.$$

Solution. From Example 3.30, we note that

$$r_1 = \frac{dr}{d\theta} = -a\sin\theta,$$

$$r_2 = \frac{d^2r}{d\theta^2} = -a\cos\theta, \text{ and } C_o = \frac{4r}{3}.$$

Further,

$$C_p = \frac{2r_1(r^2+r_1^2)}{r^2 - rr_2 + 2r_1^2}$$

$$= -\frac{2a\sin\theta(2a^2 + 2a^2\cos\theta)}{3a^2(1+\cos\theta)}$$

$$= -\frac{4a^2\sin\theta(1+\cos\theta)}{3a^2(1+\cos\theta)} = -\frac{4a\sin\theta}{3}.$$

Therefore,

$$3(C_o^2 + C_p^2) = 3\left[\frac{16r^2}{9} + \frac{16a^2}{9}\sin^2\theta\right]$$

$$= \frac{16}{3}[r^2 + a^2\sin^2\theta]$$

$$= \frac{16}{3}[a^2(1+\cos^2\theta + 2\cos\theta) + a^2\sin^2\theta]$$

$$= \frac{16}{3}[2a^2(1+\cos\theta)] = \frac{32ra}{3} = 8aC_o.$$

EXAMPLE 3.34

Prove that the points on the curve $r = f(\theta)$, is the circle of curvature at which it passes through the origin (pole) are given by the equation

$$f(\theta) + f''(\theta) = 0.$$

Solution. Since the circle of curvature passes through the pole, the chord of curvature through the pole is r. But the chord of curvature C_o is given by

$$C_o = \frac{2r(r^2+r_1^2)}{r^2 - rr_2 + 2r_1^2}.$$

Therefore,

$$r = \frac{2r(r^2 + r_1^2)}{r^2 - rr_2 + 2r_1^2}$$

or

$$r^2 - rr_2 + 2r_1^2 = 2\,(r^2 + r_1^2)$$

or

$$r^2 + rr_2 = 0 \text{ or } r\,(r + r_2) = 0$$

and so, $r + r_2 = 0$. Hence,

$$f(\theta) + f''(\theta) = 0.$$

EXAMPLE 3.35

Show that in any curve the chord of curvature perpendicular to the radius vector is

$$\frac{2\rho}{r}\sqrt{r^2 - p^2}.$$

Solution. The chord of curvature perpendicular to the radius vector is

$$C_p = 2\rho\cos\phi = 2\rho\sqrt{1 - \sin^2\phi}$$

$$= 2\rho\sqrt{1 - \left(\frac{p}{r}\right)^2}, \text{ since } p = r\sin\phi$$

$$= \frac{2\rho}{r}\sqrt{r^2 - p^2}.$$

3.12 MISCELLANEOUS EXAMPLES

EXAMPLE 3.36

Find the curvature of the curve $2x^2 + 2y^2 + 5x - 2y + 1 = 0$.

Solution. Let

$$f(x, y) = 2x^2 + 2y^2 + 5x - 2y + 1.$$

Then,

$$f_x = 4x + 5, \quad f_y = 4y - 2, \quad f_{xx} = 4,$$

$$f_{yy} = 4, \qquad f_{xy} = 0.$$

Therefore,

$$\rho = \frac{[(f_x)^2 + (f_y)^2]^{\frac{3}{2}}}{f_{xx}(f_y)^2 - 2f_x f_y f_{xy} + f_{yy}(f_x)^2}$$

$$= \frac{[(4x+5)^2 + (4y-2)^2]^{\frac{3}{2}}}{4\,(4y-2)^2 + 0 + 4\,(4x+5)^2}$$

$$= \frac{1}{4}\,[\,(4x+5)^2 + (4y-2)^2]^{\frac{1}{2}}.$$

The curvature is given by

$$K = \frac{1}{\rho} = \frac{4}{[(4x+5)^2 + (4y-2)^2]^{\frac{1}{2}}}.$$

EXAMPLE 3.37

Find the points on the parabola $y^2 = 4x$ at which the radius of curvature is $4\sqrt{2}$.

Solution. The parametric equations of the parabola $y^2 = 4x$ are $x = t^2$ and $y = 2t$.
Then, (see Example 3.10), ρ at $(t^2, 2t)$ is given by

$$\rho = 2\,(1 + t^2)^{\frac{1}{2}}.$$

But $\rho = 4\sqrt{2}$ (given).
Therefore,

$$2\,(1 + t^2)^{\frac{1}{2}} = 4\sqrt{2}$$

or

$$4\,(1 + t^2) = 32$$

or

$$t = \sqrt{7}.$$

Hence, the required point is $(t^2, 2t) = (7, 2\sqrt{7})$.

EXAMPLE 3.38

For the curve $y = \frac{ax}{a+x}$, show that

$$\left(\frac{2\rho}{a}\right)^{\frac{2}{3}} = \left(\frac{x}{y}\right)^2 + \left(\frac{y}{x}\right)^2.$$

Solution. We have $y = \frac{ax}{a+x}$ or $\frac{y}{x} = \frac{a}{a+x}$.
Then,

$$\frac{dy}{dx} = \frac{(a+x)a - ax}{(a+x)^2} = \frac{a^2}{(a+x)^2} = \left(\frac{y}{x}\right)^2$$

and

$$\frac{d^2y}{dx^2} = -\frac{2a^2}{(a+x)^3} = -\frac{2}{a}\left(\frac{y}{x}\right)^3.$$

Therefore, the radius of curvature ρ is given by

$$\rho = \frac{\left[1 + \left(\frac{dy}{dx}\right)^2\right]^{\frac{3}{2}}}{\frac{d^2y}{dx^2}} = \frac{\left[1 + \left(\frac{y}{x}\right)^4\right]^{\frac{3}{2}}}{\frac{2}{a}\left(\frac{y}{x}\right)^3} \text{ (in magnitude)}$$

or

$$\frac{2\rho}{a} = \frac{\left[1 + \left(\frac{y}{x}\right)^4\right]^{\frac{3}{2}}}{\left(\frac{y}{x}\right)^3}$$

or

$$\left(\frac{2\rho}{a}\right)^{\frac{2}{3}} = \frac{1 + \left(\frac{y}{x}\right)^4}{\left(\frac{y}{x}\right)^2} = \left(\frac{x}{y}\right)^2 + \left(\frac{y}{x}\right)^2.$$

EXAMPLE 3.39

With the usual notation, prove that

$$\frac{1}{p^2} = \frac{1}{r^2} + \frac{1}{r^4}\left(\frac{dr}{d\theta}\right)^2.$$

Solution. From figure of article 3.4, we have

$$p = r \sin \phi.$$

Therefore,

$$\frac{1}{p^2} = \frac{1}{r^2} \cdot \operatorname{cosec}^2 \phi = \frac{1}{r^2}(1 + \cot^2 \phi)$$

$$= \frac{1}{r^2}\left[1 + \frac{1}{r^2}\left(\frac{dr}{d\theta}\right)^2\right], \text{ since } \tan \phi = \frac{rd\theta}{dr}$$

$$= \frac{1}{r^2} + \frac{1}{r^4}\left(\frac{dr}{d\theta}\right)^2.$$

EXAMPLE 3.40

For the cycloid $x = a(\theta - \sin \theta)$, $y = a(1 - \cos \theta)$, find $\frac{ds}{dx}$ and $\frac{ds}{dy}$.

Solution. For the points $P(x, y)$ and $Q(x + \delta x, y + \delta y)$ in the figure of article 3.1, we have

$$(\text{chord } PQ)^2 = (\delta x)^2 + (\delta y)^2 \qquad (1)$$

or

$$\left(\frac{\text{chord } PQ}{\delta x}\right)^2 = 1 + \left(\frac{\delta y}{\delta x}\right)^2$$

or

$$\left(\frac{\text{chord } PQ}{\text{arc } PQ} \cdot \frac{\text{arc } PQ}{\delta x^2}\right)^2 = \left[1 + \left(\frac{\delta y}{\delta x}\right)^2\right]$$

or

$$\left(\frac{\text{chord } PQ}{\text{arc } PQ}\right)^2 \left(\frac{\delta s}{\delta x}\right)^2 = \left[1 + \left(\frac{\delta y}{\delta x}\right)^2\right].$$

Letting $Q \to P$, that is, $\delta x \to 0$, we get

$$\lim_{Q \to P}\left(\frac{\text{chord } PQ}{\text{arc } PQ}\right)^2 \cdot \lim_{\delta x \to 0}\left(\frac{\delta s}{\delta x}\right)^2 = \lim_{\delta x \to 0}\left[1 + \left(\frac{\delta y}{\delta x}\right)^2\right]$$

or

$$1\left(\frac{ds}{dx}\right)^2 = \left[1 + \left(\frac{dy}{dx}\right)^2\right]$$

or

$$\frac{ds}{dx} = \left[1 + \left(\frac{dy}{dx}\right)^2\right]^{\frac{1}{2}}. \qquad (2)$$

Similarly, if we divide (1) throughout by $(\delta y)^2$, we get

$$\frac{ds}{dx} = \left[1 + \left(\frac{dx}{dy}\right)^2\right]^{\frac{1}{2}}. \qquad (3)$$

We shall use (3) and (4) to solve our problem. We have

$$x = a(\theta - \sin \theta) \quad \text{and} \quad y = a(1 - \cos \theta).$$

Therefore, as in Example 3.8, we have

$$\frac{dx}{d\theta} = 2a\sin^2 \frac{\theta}{2} \quad \text{and} \quad \frac{dy}{d\theta} = 2a \sin \frac{\theta}{2} \cos \frac{\theta}{2},$$

which yield

$$\frac{dy}{dx} = \cot \frac{\theta}{2}.$$

Hence, from (2), we get

$$\frac{ds}{dx} = \sqrt{1 + \cot^2 \frac{\theta}{2}} = \operatorname{cosec} \frac{\theta}{2}.$$

Similarly, from (3), we have

$$\frac{ds}{dy} = \sqrt{1 + \tan^2 \frac{\theta}{2}} = \sec \frac{\theta}{2}.$$

EXAMPLE 3.41

Find the radius of curvature of $\sqrt{a} = \sqrt{r}\cos\frac{\theta}{2}$ at (r, θ).

Solution. We have

$$r^{-\frac{1}{2}} = a^{-\frac{1}{2}} \cos \frac{\theta}{2}.$$

Taking log, we get

$$-\frac{1}{2} \log r = -\frac{1}{2} \log a + \log \cos \frac{\theta}{2}.$$

Differentiating, we get

$$-\frac{1}{2r}\frac{dr}{d\theta} = -\frac{1}{2} \tan \frac{\theta}{2} \quad \text{or} \quad \frac{dr}{d\theta} = r \tan \frac{\theta}{2}.$$

Differentiating once more, we get

$$\frac{d^2r}{d\theta^2} = \frac{1}{2} r\sec^2 \frac{\theta}{2} + \tan \frac{\theta}{2}\frac{dr}{d\theta}$$

$$= \frac{1}{2} r\sec^2 \frac{\theta}{2} + r \tan^2 \frac{\theta}{2}.$$

Therefore,

$$\rho = \frac{(r^2 + r_1^2)^{\frac{3}{2}}}{r^2 + 2r_1^2 - rr_2}$$

$$= \frac{\left(r^2 + r^2\tan^2 \frac{\theta}{2}\right)^{\frac{3}{2}}}{r^2 + 2r^2 \tan^2 \frac{\theta}{2} - r\left(\frac{1}{2} r\sec^2 \frac{\theta}{2} + r\tan^2 \frac{\theta}{2}\right)}$$

$$= \frac{r^3 \sec^3 \frac{\theta}{2}}{r^2 + r^2\tan^2 \frac{\theta}{2} - \frac{1}{2} r^2 \sec^2 \frac{\theta}{2}}$$

$$= \frac{r^3 \sec^3 \frac{\theta}{2}}{r^2 \sec^2 \frac{\theta}{2} - \frac{1}{2} r^2 \sec^2 \frac{\theta}{2}}$$

$$= 2r \sec\frac{\theta}{2} = 2r \frac{\sqrt{r}}{\sqrt{a}}$$

$$= 2r^{\frac{3}{2}}a^{-\frac{1}{2}}.$$

EXAMPLE 3.42

Find the envelope of the straight lines $\frac{x}{a}+\frac{y}{b}=1$, where $a^2+b^2=4$.

Solution. We have (using $b=\sqrt{4-a^2}$)

$$f(x, y, a)=\frac{x}{a}+\frac{y}{\sqrt{4-a^2}}-1=0.$$

Therefore,

$$\frac{df}{da}=-\frac{x}{a^2}+y\left[\frac{a}{(4-a^2)^{\frac{3}{2}}}\right]=0$$

or

$$\frac{a^3}{x}=\frac{(4-a^2)^{\frac{3}{2}}}{y}$$

or

$$\frac{a}{x^{\frac{1}{3}}}=\frac{(4-a^2)^{\frac{1}{2}}}{y^{\frac{1}{3}}}=\frac{\sqrt{a^2+(4-a^2)}}{\sqrt{x^{\frac{2}{3}}+y^{\frac{2}{3}}}}=\frac{2}{\sqrt{x^{\frac{2}{3}}+y^{\frac{2}{3}}}}.$$

Therefore

$$a=\frac{2x^{\frac{1}{3}}}{\sqrt{x^{\frac{2}{3}}+y^{\frac{2}{3}}}},$$

Putting these values of a and b in $\frac{x}{a}+\frac{y}{b}=1$, we get

$$\left(x^{\frac{2}{3}}+y^{\frac{2}{3}}\right)^{\frac{3}{2}}=2$$

and so,

$$x^{\frac{2}{3}}+y^{\frac{2}{3}}=2^{\frac{2}{3}}.$$

EXAMPLE 3.43

Find the curvature of $x^2+y^2+4x-6y-1=0$.

Solution. We have $x^2+y^2+4x-6y-1=0$ and so $y^2=1+6y-4x-x^2$. Differentiating this equation, we get

$$\frac{dy}{dx}=\frac{x+2}{3-y}. \qquad (1)$$

Further, Differentiating (1), we get

$$\frac{d^2y}{dx^2}=\frac{5-y+x}{(3-y)^2}.$$

Then,

$$\rho=\frac{(1+y_1^2)}{y_2}=\frac{\left[1+\left(\frac{x+2}{3-y}\right)^2\right]^{\frac{3}{2}}}{\frac{(5-y+x)}{(3-y)^2}}$$

$$=\frac{[(3-y)^2+(x+2)^{\frac{3}{2}}]}{(3-y)(5-y+x)}.$$

Further, curvature $K=\frac{1}{\rho}$, that is,

$$K=\frac{(3-y)(5-y+x)}{(3-y)^2+(x+2)^2}.$$

EXAMPLE 3.44

Find the envelope of $y\cot^2\alpha+x-\alpha\operatorname{cosec}^2\alpha=0$ where α is the parameter.

Solution. We have $y\cot^2\alpha+x-a\operatorname{cosec}^2\alpha=0$, where α is a parameter.
Differentiating w.r.t. α, we get

$$2y\cot\alpha(-\operatorname{cosec}^2\alpha)$$
$$-2a\operatorname{cosec}\alpha(-\operatorname{cosec}\alpha\cdot\cot\alpha)=0$$
$$\Rightarrow(2a-2y)(\operatorname{cosec}^2\alpha\cdot\cot\alpha)=0$$
$$\Rightarrow\operatorname{cosec}^2\alpha\cdot\cot\alpha=0$$
$$\Rightarrow(1+\cot^2\alpha).\cot\alpha=0$$
$$\Rightarrow\cot\alpha+\cot^3\alpha=0$$
$$\Rightarrow\cot^2\alpha=-1.$$

Substituting in given equations, we get

$$y(-1)+x-a(1+\cot^2\alpha)=0$$
$$\Rightarrow y(-1)+x-a(1-1)=0$$
$$\Rightarrow x-y=0, \text{ which is the required envelope.}$$

EXAMPLE 3.45

Find the radius of curvature of the curve $xy^2=a^3x^3$ at $(a, 0)$.

Solution. Let $f(x, y)=xy^2+x^3-a^3$.
Then,

$$f_x=y^2+3x^2,\quad f_y=2xy,$$
$$f_{xx}=6x,\quad f_{yy}=2x,\quad f_{xy}=2y.$$

Therefore,

$$f_x(a,0)=3a^2,\quad f_y(a,0)=0,\quad f_{xx}(a,0)=6a,$$
$$f_{yy}(a,0)=2a,\quad f_{xy}(a,0)=0.$$

Hence,

$$\rho(a,0)=\frac{[(f_x)^2+(f_y)^2]^{\frac{3}{2}}}{f_{xx}(f_y)^2-2f_xf_yf_{xy}+f_{yy}(f_x)^2}$$

at $(a, 0)$ in magnitude

$$=\frac{[(3a^2)^2]^2}{0-0+2a(3a^2)^2}=\frac{27a^6}{18a^5}=\frac{3a}{2}.$$

EXAMPLE 3.46

Obtain the equation of the evolute of the curve $x = a(\cos\theta + \theta\sin\theta), y = a(\sin\theta - \theta\cos\theta)$.

Solution. Evolute is the locus of the center of curvature. Therefore, we find the center of curvature of the given curve. We have

$x = a(\cos\theta + \theta\sin\theta), y = a(\sin\theta - \theta\cos\theta)$.

Therefore,

$$\frac{dx}{d\theta} = a(-\sin\theta + \sin\theta + \theta\cos\theta) = a\theta\cos\theta,$$

$$\frac{dy}{d\theta} = a(\cos\theta + \sin\theta - \cos\theta) = a\theta\sin\theta,$$

and

$$y_1 = \frac{dy}{dx} = \frac{dy/d\theta}{dx/d\theta} = \frac{a\theta\sin\theta}{a\theta\cos\theta} = \tan\theta,$$

$$y_2 = \frac{d^2y}{dx^2} = \frac{d}{dx}\left(\frac{dy}{dx}\right) = \left[\frac{d}{d\theta}(\tan\theta)\right]\frac{d\theta}{dx}$$

$$= \sec^2\theta\left(\frac{1}{a\theta\cos\theta}\right) = \frac{1}{a\theta}\sec^3\theta.$$

Therefore, if (X, Y) is the center of curvature, we have

$$X = x - \frac{y_1(1+y_1^2)}{y_2} = a(\cos\theta + \theta\sin\theta)$$

$$- \frac{[\tan\theta(1+\tan^2\theta)]\,a\theta}{\sec^3\theta}$$

$$= a(\cos\theta + \theta\sin\theta) - \frac{(\sin\theta)a\theta}{\cos^3\theta\sec^3\theta}$$

$$= a\cos\theta \qquad (1)$$

and

$$Y = y + \frac{(1+y_1^2)}{y_2} = a(\sin\theta - \theta\cos\theta)$$

$$+ \frac{(1+\tan^2\theta)\,a\theta}{\sec^3\theta}$$

$$= a(\sin\theta - \theta\cos\theta) + a\theta\cos\theta = a\sin\theta. \qquad (2)$$

From (1) and (2), it follows that

$$X^2 + Y^2 = a^2.$$

Hence, the evolute of the curve, being the locus of (X, Y), is given by

$$x^2 + y^2 = a^2.$$

EXAMPLE 3.47

Find the circle of curvature at (0, 0) for $x + y = x^2 + y^2 + x^3$.

Solution. Differentiating the given equation with respect to x, we have

$$y_1 = \frac{dy}{dx} = \frac{3x^2 + 2x - 1}{1 - 2y} = -1 \text{ at } (0, 0)$$

$$y_2 = \frac{d^2y}{dx^2}$$

$$= \frac{(1-2y)(6x+2) - (3x^2+3x-1)(-2y_1)}{(1-2y)^2}$$

$$= 4 \text{ at } (0, 0).$$

Therefore, the radius of curvature is

$$\rho = \frac{(1+y_1^2)^{\frac{3}{2}}}{y_2} = \frac{(1+1)^{\frac{3}{2}}}{4} = \frac{2\sqrt{2}}{4} = \frac{1}{\sqrt{2}}.$$

The center of curvature at (0, 0) is given by

$$X = x - \frac{y_1(1+y_1^2)}{y_2} = 0 - \frac{(-1)[1+1]}{4} = \frac{1}{2}$$

$$Y = y + \frac{1+y_1^2}{y_2} = 0 + \frac{2}{4} = \frac{1}{2}.$$

Therefore, the required circle of curvature at (0, 0) of the given curve is

$$(x - X)^2 + (y - Y)^2 = \rho^2$$

or

$$\left(x - \frac{1}{2}\right)^2 + \left(y - \frac{1}{2}\right)^2 = \left(\frac{1}{\sqrt{2}}\right)^2$$

or

$$(x^2 + y^2 - x - y) + \frac{1}{2} = \frac{1}{2}$$

or

$$x^2 + y^2 - x - y = 0.$$

EXAMPLE 3.48

Find the equation of the circle of curvature of the curve $x = a(\cos\theta + \theta\sin\theta), y = a(\sin\theta - \theta\cos\theta)$.

Solution. Differentiating the given equation with respect to θ, we have

$$\frac{dx}{d\theta} = a(-\sin\theta + \theta\cos\theta + \sin\theta) = a\theta\cos\theta$$

and

$$\frac{dy}{d\theta} = a(\cos\theta + \theta\sin\theta - \cos\theta) = a\theta\sin\theta.$$

Therefore,

$$y_1 = \frac{dy}{dx} = \frac{dy/d\theta}{dx/d\theta} = \frac{a\theta\sin\theta}{a\theta\cos\theta} = \tan\theta$$

and
$$y_2 = \frac{d^2y}{dx^2} = \frac{dy_1}{d\theta}\cdot\frac{d\theta}{dx} = \sec^2\theta\left(\frac{1}{a\theta\cos\theta}\right)$$
$$= \frac{1}{a\theta}\sec^3\theta.$$

The center of curvature (X, Y) is given by
$$X = x - \frac{y_1(1+y_1^2)}{y_2}$$
$$= a(\cos\theta + \theta\sin\theta) - \frac{\tan\theta\sec^2\theta}{\sec^3\theta}(a\theta)$$
$$= a(\cos\theta + \theta\sin\theta) - a\theta\sin\theta = a\cos\theta,$$
$$Y = y + \frac{y_1(1+y_1^2)}{y_2}$$
$$= a(\sin\theta - \theta\cos\theta) + a\theta\cos\theta$$
$$= a\sin\theta.$$

Hence, the equation of the circle of curvature is
$$(x - X)^2 + (y - Y)^2 = \rho^2$$
or
$$(x - a\cos\theta)^2 + (y - a\sin\theta)^2 = a^2\theta^2.$$
or
$$x^2 + y^2 + a^2 - 2a(x\cos\theta + y\sin\theta) = a^2\theta^2.$$

EXAMPLE 3.49

For the curve $\sqrt{x} + \sqrt{y} = 1$, find the equation of the circle of curvature at $\left(\frac{1}{4}, \frac{1}{4}\right)$.

Solution. The given curve is $\sqrt{x} + \sqrt{y} = 1$. From Example 3.5 (ii), we have
$$y_1 = \frac{dy}{dx} = -y^{\frac{1}{2}}x^{-\frac{1}{2}},$$
$$y_2 = \frac{1}{2x^{\frac{3}{2}}}, \text{ and}$$
$$\rho(x, y) = 2(x+y)^{\frac{3}{2}}.$$

Therefore,
$$y_1\left(\frac{1}{4}, \frac{1}{4}\right) = -\frac{\frac{1}{2}}{\frac{1}{2}} = -1,$$
$$y_2\left(\frac{1}{4}, \frac{1}{4}\right) = \frac{1}{2\left(\frac{1}{4}\right)^{\frac{3}{2}}} = 4, \text{ and}$$
$$\rho\left(\frac{1}{4}, \frac{1}{4}\right) = 2\left(\frac{1}{4} + \frac{1}{4}\right)^{\frac{3}{2}} = \frac{1}{\sqrt{2}}.$$

If (X, Y) denotes the center of curvature, then
$$X = x - \frac{y_1(1+y_1^2)}{y_2} = \frac{1}{4} - \frac{-1(1+1)}{4} = \frac{3}{4},$$
$$Y = y + \frac{1+y_1^2}{y_2} = \frac{1}{4} + \frac{2}{4} = \frac{3}{4}.$$

Therefore, the equation of the circle of curvature is
$$(x - X)^2 + (y - Y)^2 = \rho^2$$
or
$$\left(x - \frac{3}{4}\right)^2 + \left(y - \frac{3}{4}\right)^2 = \frac{1}{2}.$$

EXAMPLE 3.50

Find the equation of the envelope of $\frac{x}{a} + \frac{y}{b} = 1$, where the parameters a and b are connected by the relation $a^2 + b^2 = c^2$ and c is a constant.

Solution. Assume that a and b are functions of parameter t. Then differentiating $\frac{x}{a} + \frac{y}{b} = 1$ and $a^2 + b^2 = c^2$ with respect to t, taking x and y as constants, we have, respectively,
$$\frac{x}{a^2}\frac{da}{dt} + \frac{y}{b^2}\frac{db}{dt} = 0 \text{ and so,}$$
$$\frac{\frac{da}{dt}}{\frac{db}{dt}} = -\frac{\frac{y}{b^2}}{\frac{x}{a^2}} \quad (1)$$
$$a\frac{da}{dt} + b\frac{db}{dt} = 0 \text{ and so,}$$
$$\frac{\frac{da}{dt}}{\frac{db}{dt}} = -\frac{b}{a}. \quad (2)$$

From (1) and (2), we get
$$\frac{\frac{y}{b^2}}{\frac{x}{a^2}} = \frac{b}{a} \quad \text{or} \quad \frac{\frac{x}{a^2}}{\frac{y}{b^2}} = \frac{a}{b}.$$

Therefore,
$$\frac{\frac{x}{a}}{a^2} = \frac{\frac{y}{b}}{b^2} = \frac{\frac{x}{a} + \frac{y}{b}}{a^2 + b^2} = \frac{1}{c^2},$$

which yields
$$a^3 = c^2x \quad \text{and} \quad b^3 = c^2y.$$

Putting these values in $a^2 + b^2 = c^2$, we get the envelope $x^{\frac{2}{3}} + y^{\frac{2}{3}} = c^{\frac{2}{3}}$.

EXAMPLE 3.51

Find the envelope of $\frac{x}{l}+\frac{y}{m}=1$ where l and m are connected by $\frac{l}{a}+\frac{m}{b}=1$ and a, b are constants.

Solution. We assume that l and m are functions of a parameter t. Differentiating $\frac{x}{l}+\frac{y}{m}=1$ and $\frac{l}{a}+\frac{m}{b}=1$ with respect to t taking x and y as constants, we have

$$\frac{x}{l^2}\frac{dl}{dt}+\frac{y}{m^2}\frac{dm}{dt}=0$$

and

$$\frac{1}{a}\frac{dl}{dt}+\frac{1}{b}\frac{dm}{dt}=0$$

or

$$\frac{\frac{dl}{dt}}{\frac{dm}{dt}}=-\frac{\frac{y}{m^2}}{\frac{x}{l^2}} \quad (1)$$

and

$$\frac{\frac{dl}{dt}}{\frac{dm}{dt}}=-\frac{a}{b} \quad (2)$$

From (1) and (2), we have

$$\frac{\frac{y}{m^2}}{\frac{x}{l^2}}=\frac{a}{b} \quad \text{or} \quad \frac{\frac{x}{l^2}}{\frac{y}{m^2}}=\frac{b}{a}.$$

Therefore,

$$\frac{\frac{x}{l}}{\frac{l}{a}}=\frac{\frac{y}{m}}{\frac{m}{b}}=\frac{\frac{x}{l}+\frac{y}{m}}{\frac{l}{a}+\frac{m}{b}}=\frac{1}{1}=1.$$

Hence,

$$\frac{ax}{l^2}=1\text{, which yields } l=\sqrt{ax}$$

and

$$\frac{by}{m^2}=1\text{, which yields } m=\sqrt{by}.$$

Putting these values of l and m in $\frac{l}{a}+\frac{m}{b}=1$, we get

$$\sqrt{\frac{x}{a}}+\sqrt{\frac{y}{b}}=1.$$

EXAMPLE 3.52

Find the envelope of $\frac{x}{a}+\frac{y}{b}=1$, where $a^m b^n = c^{m+n}$.

Solution. Assume that a and b are functions of a parameter t. Then differentiating $\frac{x}{a}+\frac{y}{b}=1$, and $a^m b^n = c^{m+n}$ with respect to t, treating x and y as constants, we have respectively

$$\frac{x}{a^2}\frac{da}{dt}+\frac{y}{b^2}\frac{db}{dt}=0 \quad \text{or} \quad \frac{\frac{da}{dt}}{\frac{db}{dt}}=-\frac{\frac{y}{b^2}}{\frac{x}{a^2}}. \quad (1)$$

and

$$mb^n a^{n-1}\frac{da}{dt}+na^m b^{n-1}\frac{db}{dt}=0 \quad \text{or}$$

or

$$\frac{\frac{da}{dt}}{\frac{db}{dt}}=-\frac{na^m b^{n-1}}{mb^n a^{m-1}}. \quad (2)$$

From (1) and (2), we get

$$\frac{\frac{y}{b^2}}{\frac{x}{a^2}}=\frac{na^m b^{n-1}}{mb^n a^{m-1}} \quad \text{or} \quad \frac{\frac{x}{a^2}}{\frac{x}{b^2}}=\frac{mb^n a^{m-1}}{na^m b^{n-1}}. \quad (3)$$

Relation (3) yields

$$\frac{\frac{x}{a}}{m}=\frac{\frac{y}{b}}{n}=\frac{\frac{x}{a}+\frac{y}{b}}{m+n}=\frac{1}{m+n}.$$

These relations yield

$$a=\frac{m+n}{m}x \quad \text{and} \quad b=\frac{m+n}{n}y.$$

Putting the values of a and b in $a^m b^n = c^{m+n}$, we get

$$x^m y^n\left(\frac{m+n}{m}\right)^m\left(\frac{m+n}{n}\right)^n=c^{m+n},$$

or

$$x^m y^n = m^m n^n\left(\frac{c}{m+n}\right)^{m+n}.$$

EXERCISES

1. Find the radius of curvature at any point (x, y) of the following curves:

   (i) $ay^2=x^3$ **Ans.** (i) $\frac{\sqrt{x}(4a+9x)^{\frac{3}{2}}}{6a}$.

   (ii) $xy=c^2$ **Ans.** (ii) $\frac{x^2+y^2}{2c^2}$.

2. Find the radius of curvature at a given point of the following curves:

   (i) $y=e^x$ at the point where it cross the y-axis.

   Hint: The point is $x=0$ and $y=e^0=1$.

   (ii) $y=\frac{c}{2}\left(e^{\frac{x}{c}}+e^{-\frac{x}{c}}\right)$ at (x, y).

   Hint: The curve is $y=c\cosh\frac{x}{c}$.

(iii) $\sqrt{x} + \sqrt{y} = \sqrt{a}$ at $\left(\frac{a}{4}, \frac{a}{4}\right)$.

(iv) $y^2 = \frac{a^2(a-x)}{x}$ at $(a, 0)$.

Ans. (i) $2\sqrt{2}$ (ii) $\frac{y^2}{c}$ (iii) $\frac{a}{\sqrt{2}}$ (iv) $\frac{a}{2}$.

3. Show that the radius of curvature at any point of the astroid $x^{\frac{2}{3}} + y^{\frac{2}{3}} = a^{\frac{2}{3}}$ is three times the length of the perpendicular from the origin to the tangent at that point.

Hint: If parametric equations are $x = a\cos^2 t$ and $y = a\sin^3 t$, then $\rho = 3a\sin t\cos t$, tangent is $y - a\sin^3 t = \frac{dy}{dx}(x - a\cos^3 t)$, and length of perpendicular from (0, 0) on tangent is sin t cos t.

4. If ρ is the radius of curvature of the parabola $y^2 = 4ax$ at a point P, show that ρ^2 varies as $(SP)^3$, where S is the focus.

Hint: ρ at $(at^2, 2at)$ is $2a(1 + t^2)^{\frac{3}{2}}$, $S(a, 0)$ is the focus and so,

$SP = \sqrt{(at^2 - a)^2 + (2at - 0)^2} = a(1 + t^2)$.

Then $\rho^2 = \frac{4}{a}(SP)^3$.

5. Show that for the curve $x = a\cos\theta(1 + \sin\theta)$ and $y = a\sin\theta(1 + \cos\theta)$, the radius of curvature at $\theta = -\frac{\rho}{4}$ is a.

6. Prove that for the curve $r = a(\sec 3\psi - 1)$ the radius of curvature is $\rho = 3a\tan\psi\sec 3\psi$ and hence, show that $3a\frac{dy}{dx}\cdot\frac{d^2y}{dx^2} = 1$.

7. Prove that for the curve $4x = \theta + \sin\theta$ and $4y = 1 - \cos\theta$, the radius of curvature is $\rho = \cos\frac{\theta}{2}$.

8. Find the radius of curvature at the point θ of the curve $x = a(\cos\theta + \log\tan\frac{\theta}{2})$ and $y = a\sin\theta$.

Hint: $\frac{dx}{d\theta} = \frac{a\cos^2\theta}{\sin\theta}, \frac{dy}{d\theta} = a\cos\theta, \frac{dy}{dx} = \tan\theta$, and $\frac{d^2y}{dx^2} = \frac{\sin\theta}{a\cos^4\theta}$.

Ans. $a\cos\theta$.

9. Find the radius of curvature of the point (p, r) of the following curves:

(i) $r^3 = 2ap^2$.

(ii) $p^2 = \frac{r^4}{r^2 + a^2}$.

Ans. (i) $\frac{2}{3}\sqrt{2ar}$ (ii) $\frac{(r^2 + a^2)^{\frac{3}{2}}}{r^2 + 2a^2}$.

10. Find the radius of curvature at any point of the curve $r^n = a^n\cos n\theta$.

Hint: $r_1 = -r\tan n\theta$ and $r_2 = r\tan^2\theta - nr\sec^2 n\theta$. Putting in $\rho = \frac{(r^2 + r_1^2)}{r^2 + 2r_1^2 - rr_2}$, we get $\rho = \frac{a^n}{(n+1)r^{n-1}}$.

11. Find the radius of curvature at (r, θ) of the curve $r = 6\left(1 - \sin^2\frac{\theta}{2}\right)$.

Ans. $\rho = 4\cos\frac{\theta}{2}$.

12. Show that the radius of curvature at the point (r, θ) on the curve

$\theta = \frac{1}{a}(r^2 - a^2)^{\frac{1}{2}} - \cos^{-1}\left(\frac{a}{r}\right)$ is $\rho = (r^2 - a^2)^{\frac{1}{2}}$.

Hint: $\frac{d\theta}{dr} = \frac{(r^2 - a^2)^{\frac{1}{2}}}{ar}$ and so, $\frac{dr}{d\theta} = \frac{ar}{\sqrt{r^2 - a^2}}$.

Also, $\frac{1}{p^2} = \frac{1}{r^2} + \frac{1}{r^4}\left(\frac{dr}{d\theta}\right)^2 = \frac{1}{r^2 - a^2}$ and so pedal equation is $p^2 = r^2 - a^2$. Differentiating, we get $2p = 2r\frac{dr}{dp}$. Then $\rho = r\frac{dr}{dp} = p = \sqrt{r^2 - a^2}$.

13. Show that the curvatures of the curves $r = a\theta$ and $r\theta = a$ at their common point are in the ratio 3:1.

Hint: Eliminating r between the two equations, we get $\theta^2 = 1$. Thus, at the common point $\theta^2 = 1$, for the first curve $r_1 = a$ and $r_2 = 0$ and so, ρ_1 (at $\theta^2 = 1$) $= \frac{2a\sqrt{2}}{3}$. For the second curve $r_1 = -\frac{a}{\theta^2}$ and $r_2 = \frac{2a}{\theta^3}$ and so, ρ_2 (at $\theta^2 = 1$) $= 2a\sqrt{2}$.

14. Prove that for any curve $\frac{d^2r}{ds^2} = \frac{\sin^2\phi}{r} - \frac{\sin\phi}{\rho}$

Hint: $\frac{dr}{ds} = \cos\phi$, $\rho = \frac{ds}{d\psi}$, $\phi = \psi - \theta$, and $\sin\phi = r\frac{d\theta}{ds}$.

15. Find the radius of curvature for the curve

$$x = c\log\{s + \sqrt{s^2 + c^2}\} \text{ and}$$
$$y = \sqrt{s^2 + c^2}.$$

Hint: $\frac{dx}{ds} = -\frac{c}{\sqrt{s^2 + c^2}}$ and $\frac{dy}{ds} = \frac{s}{s^2 + c^2}\frac{d^2y}{ds^2} =$

$\frac{d}{ds}\left(\frac{dy}{ds}\right) = \frac{c^2}{(s^2 + c^2)^{\frac{3}{2}}}$ and so, $\rho = \frac{\frac{dx}{ds}}{\frac{d^2y}{ds^2}} = \frac{y^2}{c}$.

16. Find the radius of curvature at the origin of the following curves:

(i) $x^3 - y^3 - 2x^2 + 6y = 0.$

(ii) $x^3 - 2x^2y + 3xy^2 - 4y^3 + 5x^2 - 6xy + 7y^2 - 8y = 0.$

Ans. (i) $\rho(0, 0) = \frac{3}{2}$ (ii) $\rho(0, 0) = \frac{4}{5}$.

17. Find the radius of curvature at the pole for the curve $r = a\sin n\theta$.

Ans. $\rho = \frac{1}{2}\frac{dr}{d\theta} = \frac{na}{2}$.

18. Find the radius of curvature at the origin for the curve $a(y^2 - x^2) = x^3$.

Ans. $2\sqrt{2a}$.

19. Find the coordinates of the center of curvature for the curve $a^2y = x^3$.

Ans. $\left(\frac{x}{2}\left(1 - \frac{9x^4}{a^4}\right), \left(\frac{5x^3}{2a^3} + \frac{a^2}{6x}\right)\right)$.

20. Show that the evolute of the tractrix $x = c\cos t + c\log\tan\frac{t}{2}$ and $y = c\sin t$ is the catenary $y = c\cosh\frac{x}{c}$.

21. Find the circle of curvature for the curve $x^3 + y^3 = 3xy$ at the point $\left(\frac{3}{2}, \frac{3}{2}\right)$.

Ans. $8(x^2 + y^2) - 21(x + y) + 27 = 0$.

22. Find the length of the chord of curvature through the pole for the curves $r = ae^{m\theta}$.

Ans. $2r$.

23. Find the coordinates of the center of curvature at $(at^2, 2at)$ on the parabola $y^2 = 4ax$.

Ans. $(2a + 3at^2, -2at^3)$.

24. Show that in the curve $y = a\cosh\left(\frac{x}{a}\right)$, the chord of curvature parallel to the axis of x is of length $a\sinh\left(\frac{2x}{a}\right)$.

Hint: $y_1 = \sinh\frac{x}{a}$ and $y_2 = \frac{1}{a}\cosh\frac{x}{a}$.

Therefore, $C_x = \frac{2y_1(1 + y_1^2)}{y_2} = a\sinh\frac{2x}{a}$.

25. If C_x and C_y are the chords of curvature parallel to the axis of x and y, respectively, at any point of the curve $y = a\cosh\frac{x}{a}$, show that

$$4a^2(C_x^2 + C_y^2) = C_y^4.$$

4 Asymptotes and Curve Tracing

The aim of this chapter is to study the shape of a *plane curve* $y=f(x)$. For this purpose, we must investigate the variation of the function f, in the case of unlimited increase and absolute value and of x or y, or both, of a variable point (x, y) on the curve. The study of such variation of the function requires the concept of an asymptote. Before defining an asymptote to a curve, let us define finite and infinite branches of a plane curve as follows:

Consider the equation of the ellipse $\frac{x^2}{a^2}+\frac{y^2}{b^2}=1$.

Solving this equation, we get

$$y=b\sqrt{1-\frac{x^2}{a^2}} \text{ or } y=-b\sqrt{1-\frac{x^2}{a^2}}.$$

The first equation represents the upper half of the ellipse while the second equation represents the lower half of the ellipse. Thus, the earlier equation represents two branches of the ellipse. Further, both these branches lie within the finite part of the xy-plane bounded by $x = \pm a$ and $y = \pm b$. Hence, both these branches of the ellipse are *finite*.

Consider now the equation of the hyperbola $\frac{x^2}{a^2}-\frac{y^2}{b^2}=1$. Its solution is

$$y=\frac{b}{a}\sqrt{x^2-a^2} \text{ or } y=-\frac{b}{a}\sqrt{x^2-a^2}.$$

Therefore, y tends to $\pm\infty$ as $x \to \pm\infty$. Hence, both branches this hyperbola extend to infinity and are therefore called the *infinite branches* of the rectangular hyperbola.

A variable point $P(x, y)$ moves along a curve to infinity if the distance of the point from the origin increases without bound. In other words, a point $P(x, y)$ on an infinite branch of a curve is said to tend to infinity along the curve if either x or y, or both, tend to infinity as $P(x, y)$ moves along the branch of the curve.

Now we are in a position to define an asymptote to a curve.

A straight line, at a finite distance from the origin, is said to be a *rectilinear asymptote* (or simply *asymptote*) of an infinite branch of a curve if the perpendicular distance of a point P on that branch from this straight line tends to zero as P tends to infinity along the branch of the curve.

For example, the line AB will be asymptote of the curve in the following figure if the perpendicular distance PM from the point P to the line AB tends to zero as P tends to infinity along the curve.

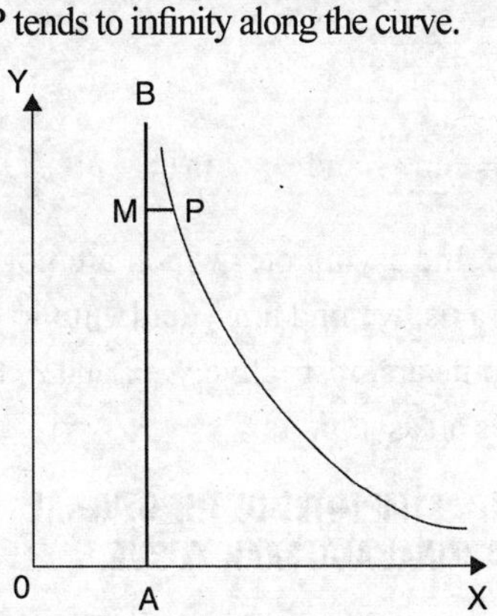

4.1 DETERMINATION OF ASYMPTOTES WHEN THE EQUATION OF THE CURVE IN CARTESIAN FORM IS GIVEN

Let

$$y = mx + c \tag{1}$$

be the equation of a straight line. Let $P(x, y)$ be an arbitrary point on the infinite branch of the curve $f(x, y) = 0$. We wish to find the values of m and c so that (1) is an asymptote to the curve. Let $PM = p$ be the perpendicular distance of the point $P(x, y)$ from (1). Then

$$p = \frac{y - mx - c}{\sqrt{1+m^2}}.$$

The abscissa x must tend to infinity as the point $P(x, y)$ recedes to infinity along this line. Thus, $p \to 0$ as $x \to \infty$. Therefore,

$$\lim_{X\to\infty}(y - mx - c) = 0$$

or

$$\lim_{X\to\infty}(y - mx) = c.$$

On the other hand,

$$\frac{y}{x} - m = (y - mx)\frac{1}{x}.$$

Therefore,

$$\lim_{X\to\infty}\left(\frac{y}{x} - m\right) = \lim_{X\to\infty}(y - mx)\lim_{X\to\infty}\frac{1}{x} = c(0) = 0$$

or

$$\lim_{X\to\infty}\frac{y}{x} = m.$$

Hence,

$$m = \lim_{X\to\infty}\frac{y}{x} \text{ and } c = \lim_{X\to\infty}(y - mx).$$

Thus, to find asymptotes which are not parallel to the y-axis, we find $\lim\limits_{x\to\infty}\frac{y}{x}$ and $\lim\limits_{x\to\infty}(y - mx)$. If these limits are, respectively, m and c, then $y = mx + c$ is an asymptote.

4.2 THE ASYMPTOTES OF THE GENERAL RATIONAL ALGEBRAIC CURVE

Let $f(x, y) = 0$ be the equation of any rational algebraic curve of the nth degree. Arranging this equation in groups of homogeneous terms in x and y, we get

$$\begin{aligned}&(a_0x^n + a_1x^{n-1}y + a_2x^{n-2}y^2 + \ldots + a_ny^n)\\ &\quad + \ldots + (b_1x^{n-1} + b_2x^{n-2}y + \ldots + b_nx^{n-1})\\ &\quad + (c_2x^{n-2} + c_3x^{n-3}y + \ldots + c_ny^{n-2})\\ &\quad + \ldots + (k_0x + k_1y) + K = 0.\end{aligned}$$

This equation can be written as

$$x^n\phi_n\left(\frac{y}{x}\right) + x^{n-1}\phi_{n-1}\left(\frac{y}{x}\right) + x^{n-2}\phi_{n-2}\left(\frac{y}{x}\right) + \ldots + x\phi_1\left(\frac{y}{x}\right) + \phi_0\left(\frac{y}{x}\right) = 0, \quad (1)$$

where $\phi_r\left(\frac{y}{x}\right)$ is a polynomial in $\frac{y}{x}$ of degree r. Suppose $y = mx + c$ as an asymptote of the curve, where m and c are finite. We have to find m and c. Dividing both sides of equation (1) by x^n, we get

$$\phi_n\left(\frac{y}{x}\right) + \frac{1}{x}\phi_{n-1}\left(\frac{y}{x}\right) + \frac{1}{x^2}\phi_{n-2}\left(\frac{y}{x}\right) + \ldots = 0.$$

Proceeding to limits as $x \to \infty$ so that $\lim\limits_{x\to\infty}\frac{y}{x} = m$, we have

$$\lim_{x\to\infty}\left[\phi_n\left(\frac{y}{x}\right) + \frac{1}{x}\phi_{n-1}\left(\frac{y}{x}\right) + \frac{1}{x^2}\phi_{n-2}\left(\frac{y}{x}\right) + \ldots\right] = 0,$$

which yields

$$\phi_n(m) = 0. \quad (2)$$

Solving the equation (2), we get the slope m of the asymptote $y = mx + c$. But $\lim\limits_{x\to\infty}(y - mx) = c$. Let $y - mx = p$ so that $x \to \infty, p \to c$. But $y–mx = p$ implies $\frac{y}{x} = m + \frac{p}{x}$. Substituting this value of $\frac{y}{x}$ in equation (1), we have

$$x^n\phi_n\left(m + \frac{p}{x}\right) + x^{n-1}\phi_{n-1}\left(m + \frac{p}{x}\right) + x^{n-2}\phi_{n-2}\left(m + \frac{p}{x}\right) + \ldots = 0. \quad (3)$$

Taylor's Theorem expansion of equation (3) yields

$$\begin{aligned}&x^n\left[\phi_n(m) + \frac{p}{x}\phi_n'(m) + \frac{p^2}{2x^2}\phi_n''(m) + \ldots\right]\\ &\quad + x^{n-1}\left[\phi_{n-1}(m) + \frac{p}{x}\phi_{n-1}'(m) + \ldots\right]\\ &\quad + x^{n-2}\left[\phi_{n-2}(m) + \frac{p}{x}\phi_{n-2}'(m) + \ldots\right] + \ldots = 0.\end{aligned}$$

Using equation (2), the said equation reduces to

$$x^{n-1}\left[p\phi_n'(m) + \phi_{n-1}(m)\right] + x^{n-2}\left[\frac{p^2}{2!}\phi_n''(m) + p\phi_{n-1}''(m) + \phi_{n-2}(m)\right] + \ldots = 0.$$

or

$$p\phi_n'(m)+\phi_{n-1}(m)+\frac{1}{x}\left[\frac{p^2}{2!}\phi_n''(m)+p\phi_{n-1}'(m)\right.$$

$$\left.+\phi_{n-2}(m)\right]+\ldots=0. \quad (4)$$

Since $x \to \infty, p \to c,$ we have

$$c\phi_n'(m)+\phi_{n-1}(m)=0. \quad (5)$$

Case (i): If $\phi_n(m)$ has no repeated root, then $\phi'_n(m) \neq 0$. Hence, in that case, equation (5) implies

$$c=-\frac{\phi_{n-1}(m)}{\phi_n'(m)}. \quad (6)$$

If $m_1, m_2, m_3,\ldots$ are the distinct roots of $\phi_n(m)=0$ and $c_1, c_2, c_3,\ldots$ are the corresponding values of c determined by equation (6), then the asymptotes are $y=m_1x+c_1, y=m_2x+c_2, y=m_3x+c_3,\ldots$

Case (ii): If $\phi'_n(m) = 0$, that is, $\phi'_n(m)$ has a repeated root and if $\phi'_{n-1}(m) \neq 0$, then equation (6) implies that c is undefined. Hence, there exists no asymptote to the curve in this case.

Case (iii): If $\phi'_n(m)=\phi_{n-1}(m)=0$. Then equation (5) reduces to an identity and equation (4) reduces to

$$\frac{p^2}{2!}\phi_n''(m)+p\phi_{n-1}'(m)+\phi_{n-2}(m)+\frac{1}{x}[\ldots]+\ldots=0.$$

As $x \to \infty, p \to c$ we have

$$\frac{c^2}{2}\phi_n''(m)+c\phi_{n-1}'(m)+\phi_{n-2}(m)=0.$$

If $\phi''_n(m) \neq 0$, then this last quadratic in c gives two values of c. Therefore, there are two asymptotes $y=mx+c_1$ and $y=mx+c_2$, corresponding to the slope m. Thus, in this case, we have two parallel asymptotes.

Remark 4.1

(i) Since the degree of $\phi_n(m) = 0$ is n at the most, the number of asymptotes, real or imaginary, which are not parallel to y-axis, cannot exceed n. In case the curve has asymptotes parallel to y-axis, then the degree of $\phi_n(m)$ is smaller than n by at least the number of asymptotes parallel to y-axis. Thus, the total number of asymptotes cannot exceed the degree n of the curve.

(ii) Asymptotes parallel to y-axis cannot be found by the said method as the equation of a straight line parallel to y-axis cannot be put in the form $y = mx + c$.

4.3 ASYMPTOTES PARALLEL TO COORDINATE AXES

(i) Asymptotes parallel to y-axis of a rational algebraic curve: Let $f(x, y) = 0$ be the equation of any algebraic curve of the mth degree. Arranging the equation in descending powers of y, we get

$$y^m\phi_0(x)+y^{m-1}\phi_1(x)+y^{m-2}\phi_2(x)$$
$$+\ldots+\phi_m(x)=0, \quad (1)$$

where $\phi_0(x), \phi_1(x), \phi_2(x),\ldots$ are polynomials in x. Dividing the equation (1) by y^m, we get

$$\phi_0(x)+\frac{1}{y}\phi_1(x)+\frac{1}{y^2}\phi_2(x)$$
$$+\ldots+\frac{1}{y^m}\phi_m(x)=0. \quad (2)$$

If $x = c$ be an asymptote of the curve parallel to y-axis then $\lim\limits_{y\to\infty} x = c$, where (x, y) lies on the curve (1). Therefore,

$$\lim_{y\to\infty}\left[\phi_0(x)+\frac{1}{y}\phi_1(x)+\frac{1}{y^2}\phi_2(x)+\ldots\right]=0$$

or $\phi_0(c)=0$

so that c is a root of the equation $\phi_0(x) = 0$. If $c_1,c_2,\ldots$ are the roots of $\phi_0(x)=0$, then $(x-c_1)$, $(x-c_2),\ldots$ are the factors of $\phi_0(x)$. Also $\phi_0(x)$ is the coefficient of the highest power of y, that is, of y^m in equation (1). Thus, we have the following simple rule to determine the asymptotes parallel to y-axis.

The asymptotes parallel to the y-axis are obtained by equating to zero the coefficient of the highest power of y in the given equation of the curve. In case the coefficient of the highest power of y is a constant or if its linear factors are imaginary, then there will be no asymptotes parallel to the y-axis.

(ii) Asymploles parallel to the *x*-axis of a rational algebraic curve: Proceeding exactly as in case (i) mentioned earlier, we arrive at the following rule to determine the asymptotes parallel to the x-axis:

The asymptotes parallel to the x-axis are obtained by equating to zero the coefficient of the highest power of x in the given equation of the curve. In case the coefficient of the highest power of x is a constant or if its linear factors are imaginary, then there will be no asymptotes parallel to the x-axis.

4.4 WORKING RULE FOR FINDING ASYMPTOTES OF RATIONAL ALGEBRAIC CURVE

In view of the mentioned discussion, we arrive at the following working rule for finding the asymptotes of rational algebraic curves:

1. A curve of degree n may have atmost n asymptotes.
2. The asymptotes parallel to the y-axis are obtained by equating to zero the coefficient of the highest power of y in the given equation of the curve. In case the coefficient of the highest power of y is a constant or if its linear factors are imaginary, then there will be no asymptotes parallel to the y-axis.

   The asymptotes parallel to the x-axis are obtained by equating to zero the coefficient of the highest power of x in the given equation of the curve. In case the coefficient of the highest power of x is a constant or if its linear factors are imaginary, then there will be no asymptotes parallel to the x-axis.

If $y = mx + c$ is an asymptote not parallel to the y-axis, then the values of m and c are found as follows:

(i) Find $\phi_n(m)$ by putting $x = 1$, $y = m$ in the highest-degree terms of the given equation of the curve. Solve the equation $\phi_0(m) = 0$ for slope (m). If some values are imaginary, reject them.

(ii) Find $\phi_{n-1}(m)$ by putting $x = 1$, $y = m$ in the next lower-degree terms of the equation of the curve. Similarly $\phi_{n-2}(m)$ may be found taking $x = 1, y = m$ in the next lower-degree terms in the curve and so on.

(iii) If $m_1, m_2,\ldots$ are the real roots of $\phi_n(m)$, then the corresponding values of c, that is, $c_1, c_2,\ldots$ are given by

$$c = -\frac{\phi_{n-1}(m)}{\phi'_n(m)}, m = m_1, m_2, \ldots$$

Then the required asymptotes are

$$y = m_1x + c_1, y = m_2x + c_2, \ldots$$

(iv) If $\phi'_n(m) = 0$ for some m but $\phi_{n-1}(m) \neq 0$, then there will be no asymptote corresponding to that value of m.

(v) If $\phi'_n(m) = 0$ and $\phi_{n-1}(m) = 0$ for some value of m, then the value of c is determined from

$$\frac{c^2}{2!}\phi'_n(m) + \frac{c}{1!}\phi'_{n-1}(m) + \phi_{n-2}(m) = 0.$$

This equation will yield two values of c and thus, we will get atmost two parallel asymptotes corresponding to this value of m, provided $\phi''_n(m) \neq 0$.

(vi) Similarly, if $\phi''_n(m) = \phi'_{n-1}(m) = \phi_{n-2}(m) = 0$, then the value of c is determined from

$$\frac{c^3}{3!}\phi''_n(m) + \frac{c^2}{2!}\phi''_{n-1}(m) + \frac{c}{1!}\phi'_{n-2}(m) + \phi_{n-3}(m) = 0.$$

In this case, we get atmost three parallel asymptotes corresponding to this value of m.

EXAMPLE 4.1

Find the asymptotes of the curve

$$y^2(x^2 - a^2) = x^2(x^2 - 4a^2).$$

Solution. The equation of the curve is

$$y^2(x^2 - a^2) = x^2(x^2 - 4a^2)$$

or

$$y^2x^2 - x^4 - a^2y^2 + 4a^2x^2 = 0.$$

Since the degree of the curve is 4, it cannot have more than four asymptotes. Equating to zero, the coefficient of the highest power of y, the asymptote parallel to the y-axis is given by $x^2 - a^2 = 0$. Thus, the asymptotes parallel to the y-axis are $x = \pm a$.

Since the coefficient of the highest power of x in the given equation is constant, there is no asymptote parallel to the x-axis.

To find the oblique asymptotes, we put $x = 1$ and $y = m$ in the highest-degree term, that is fourth- degree term $y^2x^2 - x^4$ in the given equation and get $\phi_4(m) = m^2 - 1$. Therefore, slopes of the asymptotes are given by

$$\phi_4(m) = m^2 - 1 = 0.$$

Hence, $m = \pm 1$. Again putting $y = m$ and $x = 1$ in the next highest-degree term, that is, third-degree term, we have $\phi_3(m) = 0$ (since there is no term of degree 3).

Now c is given by

$$c = -\frac{\phi_3(m)}{\phi_4'(m)} = \frac{0}{2m} = 0.$$

Therefore, the oblique asymptotes are $y = x + 0$ and $y = -x + 0$.

Hence, all the four asymptotes of the given curve are $x = \pm a$ and $y = \pm x$.

EXAMPLE 4.2

Find all the asymptotes of the curve

$$f(x, y) = y^3 - xy^2 - x^2y + x^3 + x^2 - y^2 - 1 = 0.$$

Solution. The given curve is of degree 3 and so, it may have atmost three asymptotes. Since the coefficients of the highest power of x and y are constants, the curve has no asymptote parallel to the coordinate axes.

To find the oblique asymptotes, we put $x = 1$ and $y = m$ in the expression containing third-degree terms of $f(x, y)$. Thereby we get

$$\phi_3(m) = m^3 - m^2 - m + 1 = 0.$$

This equation yields $m = 1, 1, -1$. Further, putting $x = 1, y = m$ in the next highest-degree term, we get

$$\phi_2(m) = 1 - m^2.$$

Therefore for $m = -1$, the expression

$$c = -\frac{1 - m^2}{3m^2 - 2m - 1}$$

yields $c = 0$ and the corresponding asymptote is

$$y = -x + 0 \text{ or } y + x = 0.$$

For $m = 1$, the denominator is zero and so, c cannot be determined by the preceding formula. Putting $x = 1$, $y = m$ in the first-degree terms, we have $\phi_1(m) = 0$ (since there is no first-degree term). Now for $m = 1$, the constant c is given by

$$\frac{c^2}{2}\phi_3''(m) + c\phi_2'(m) + \phi_1(m) = 0$$

$$\text{or} \quad (3m-1)c^2 - 2mc = 0$$

$$\text{or} \quad 2c^2 - 2c = 0 \text{ for } m = 1$$

$$\text{or} \quad c(c - 1) = 0.$$

Hence, $c = 0$ and $c = 1$. So the two parallel asymptotes corresponding to $m = 1$ are $y = x$ and $y = x + 1$. Therefore, the asymptotes to the curve are

$$y + x = 0, y = x \text{ and } y = x + 1.$$

EXAMPLE 4.3

Find the asymptotes of the curve

$$y^2 (x - 2a) = x^3 - a^3.$$

Solution. The degree of the curve is 3. So, there cannot be more than three asymptotes. There is no asymptote parallel to the x-axis. The asymptote parallel to the y-axis is given by $x-2a = 0$, that is, $x = 2a$.

To find the oblique asymptotes, we put $x = 1, y = m$ in the third-degree term and get $\phi_3(m) = m^2-1$ and so, the slope m is given by

$$\phi_3(m) = m^2-1 = 0.$$

Thus, $m = \pm 1$. Further, putting $x = 1$, $y = m$ in the second-degree terms, we get $\phi_2(m) = -2am^2$. Therefore for $m = 1$ and $m = -1$, the expression

$$c = -\frac{\phi_2(m)}{\phi_3'(m)} = \frac{2am^2}{2m} = am$$

yields $c = a$, and $-a$ respectively. Hence, the oblique asymptotes are

$$y = x + a \text{ and } y = -x-a.$$

Hence, the three asymptotes of the curve are

$$x = 2a, x-y + a = 0, \text{ and } x + y + a = 0.$$

EXAMPLE 4.4

Find the asymptotes of the curve

$$x^3 + 3x^2y - 4y^3 - x + y + 3 = 0.$$

Solution. There is no asymptote parallel to the coordinate axes. To find the oblique asymptotes, we have

$$\phi_3(m) = 1 + 3m - 4m^3$$

and so, the slope m is given by
$\phi_3(m) = 1 + 3m - 4m^3 = 0.$

Therefore, $m = 1, -\frac{1}{2}, \text{and} -\frac{1}{2}$. For $m = 1$, the value of c is given by

$$c = -\frac{\phi_2(m)}{\phi_3''(m)} = -\frac{0}{-12m^2 + 3} = 0.$$

Thus, the asymptote corresponding to $m = 1$ is $y = x$ or $x - y = 0$.

For $m = -\frac{1}{2}, \phi_3'(m) = 0.$ So we find $\phi_1(m)$, which is equal to $\phi_1(m) = -1 + m$.

Hence, c is given by,

$$\frac{c^2\phi_3''(m)}{2} + c\phi_2'(m) + \phi_1(m) = 0$$

or

$$6c^2 - \frac{3}{2} = 0 \text{ or } c^2 = \frac{1}{4} \text{ or } c = \pm\frac{1}{2}.$$

Thus, the asymptotes corresponding to $m = -\frac{1}{2}$ are

$$y = -\frac{1}{2}x + \frac{1}{2} \text{ and } y = -\frac{1}{2}x - \frac{1}{2}$$

or

$$x + 2y - 1 = 0 \text{ and } x + 2y + 1 = 0.$$

Hence, the three asymptotes of the curve are $x - y = 0$, $x + 2y - 1 = 0$, and $x + 2y + 1 = 0$.

EXAMPLE 4.5

Find the asymptotes of the curve

$(x - y)^2 (x^2 + y^2) - 10 (x - y) x^2 + 12y^2 + 2x + y = 0.$

Solution. The equation of the given curve is

$(x - y)^2 (x^2 + y^2) - 10 (x - y) x^2 + 12y^2 + 2x + y = 0.$

The coefficients of x^4 and y^4 are constant. Therefore, the curve has no asymptotes parallel to the axes. Putting $x = 1$ and $y = m$ in the fourth-, third- and second-degree terms, we have

$\phi_4(m) = (1 - m)^2(1 + m^2) = m^4 - 2m^3 + 2m^2 - 2m + 1$

$\phi_3(m) = 10 (m-1)$, and $\phi_2(m) = 12m^2$.

The slopes of the asymptotes are given by

$$\phi_4(m) = (1 - m)^2(1 + m^2) = 0.$$

Therefore, $m = 1, 1$ are the real roots. Further we have

$$\phi'_n(m) = 4m^3 - 6m^2 + 4m - 2,$$

so that $\phi'_4(m) = 0$ for $m = 1$. Therefore, values of c are given by

$$\frac{c^2}{2}\phi_4''(m) + c\phi_3'(m) + \phi_2(m) = 0,$$

that is,

$$\frac{c^2}{2}(12m^2 - 12m + 4) + 10c + 12m^2 = 0.$$

For $m = 1$, this equation yields

$$2c^2 + 10c + 12 = 0 \text{ or } c^2 + 5c + 6 = 0$$

This equation gives $c = -2, -3$. Putting the values of m and c in $y = mx + c$, the asymptotes are given by

$$y = x - 2 \text{ and } y = x - 3.$$

EXAMPLE 4.6

Find the asymptotes of the curve

$(x + y)^2 (x + 2y) + 2(x + y)^2 - (x + 9y) - 2 = 0.$

Solution. Since the coefficients of the highest-degree term of x and y are constant, the given curve does not have asymptotes parallel to the axes.

To find the oblique asymptotes, we put $x = 1$ and $y = m$ in third-, second- and first-degree terms and get

$$\phi_3(m) = (1+m)^2(1+2m) = 2m^3 + 5m^2 + 4m + 1$$

$$\phi_2(m) = 2(1+m)^2 = 2m^2 + 4m + 1$$

$$\phi_1(m) = -(1+9m) = -9m - 1.$$

Thus,

$$\phi_3'(m) = 6m^2 + 10m + 4,$$

$$\phi_3''(3) = 12m + 10, \text{ and}$$

$$\phi_2'(m) = 4m + 4.$$

The slopes of the asymptotes are given by

$$\phi_3(m) = (1+m)^2(1+2m) = 0,$$

which yields $m = -1, -1$, and $-\frac{1}{2}$. The value of c is given by

$$c = -\frac{\phi_2(m)}{\phi_3'(m)} = -\frac{2m^2 + 4m + 1}{6m^2 + 10m + 4}.$$

For $m = -1$, $\phi'_3(m) = 0$ and so, c cannot be found from this equation. For $m = -\frac{1}{2}$ we have $c = -1$. Thus, the asymptotes corresponding to $m = -\frac{1}{2}$ is

$$y = -\frac{1}{2}x - 1 \text{ or } x + 2y + 2 = 0.$$

For $m = -1$, the value of c is calculated from the relation

$$\frac{c^2}{2}\phi_3''(m) + c\phi_2'(m) + \phi_1(m) = 0$$

or

$$\frac{c^2}{2}(12m + 10) + c(4m + 4) - 9m - 1 = 0$$

or

$$c^2 (6m + 5) + c (4m + 4) - 9m - 1 = 0$$

or

$$c^2 (-1) + 9 - 1 = 0$$

or

$$c^2 = 8, \text{ which yields } c = \pm 2\sqrt{2}.$$

Thus, the two parallel asymptotes corresponding to the slope $m = -1$ are

$$y = -x + 2\sqrt{2} \text{ and } y = -x - 2\sqrt{2}.$$

Hence, the asymptotes of the curve are $x + 2y + 2 = 0$,

$$y + x = 2\sqrt{2}, \text{ and } y + x = -2\sqrt{2}.$$

EXAMPLE 4.7

Find the asymptotes of the curve

$$6x^2 + xy - 2y^2 + x + 2y + 1 = 0$$

Solution. Since the coefficients of the highest powers of x and y are constants, there is no asymptotes parallel to the axes. To find the oblique asymptotes, we put $x = 1$ and $y = m$ in second- and first-degree terms and get

$$\phi_2(m) = 6 - 2m^2 + m,\ \phi_1(m) = 2m + 1$$

and

$$\phi'_2(m) = -4m + 1.$$

The slopes of the asymptotes are given by

$$\phi_2(m) = 6 - 2m^2 + m = 0$$

and so, $m = 2, -\frac{3}{2}$. The value of c is given by

$$c = -\frac{\phi_1(m)}{\phi_2'(m)} = -\frac{2m+1}{-4m+1}.$$

For $m = 2$ and $m = -\frac{3}{2}$, the value of c are $\frac{5}{7}$ and $\frac{2}{7}$ respectively.

Therefore, the asymptotes are

$$y = 2x + \frac{5}{7} \text{ and } y = -\frac{3}{2}x + \frac{2}{7}$$

or

$$14x - 7y + 5 = 0 \text{ and } 21x + 14y - 4 = 0$$

EXAMPLE 4.8

Find the asymptotes of the curve $\frac{a^2}{x^2} - \frac{b^2}{y^2} = 1$.

Solution. The equation of the given curve is

$$\frac{a^2}{x^2} - \frac{b^2}{y^2} = 1$$

or

$$x^2y^2 - a^2y^2 + b^2x^2 = 0.$$

Since the curve is of degree 4, it cannot have more than four asymptotes.

Equating the coefficient of the highest power of x to zero, we get $y^2 + b^2 = 0$, which yields imaginary asymptotes.

Equating the coefficient of the highest power of y to zero, we get

$$x^2 - a^2 = 0 \text{ or } (x - a)(x + b) = 0.$$

Hence, the asymptotes parallel to the y-axis are $x = a$ and $x = -a$. Thus, the only real asymptotes are $x - a = 0$ and $x + a = 0$.

4.5 INTERSECTION OF A CURVE AND ITS ASYMPTOTES

We have seen that the equation of a curve of degree n can be expressed in the form

$$x^n\phi_n\left(\frac{y}{x}\right) + x^{n-1}\phi_{n-1}\left(\frac{y}{x}\right) + x^{n-2}\phi_{n-2}\left(\frac{y}{x}\right) + \ldots = 0. \quad (1)$$

Let

$$y = mx + c \quad (2)$$

be an asymptote to the curve (1). Eliminating y from (1) and (2), we get

$$x^n\phi_n\left(m + \frac{c}{x}\right) + x^{n-1}\phi_{n-1}\left(m + \frac{c}{x}\right) + x^{n-2}\phi_{n-2}\left(m + \frac{c}{x}\right) + \ldots = 0.$$

Expanding by Taylor's Theorem, we get

$$x^n\left[\phi_n(m) + \frac{c}{x}\phi_n'(m) + \frac{c^2}{2!x^2}\phi_n''(m) + \ldots\right] + x^{n-1}\left[\phi_{n-1}(m) + \frac{c}{x}\phi_{n-1}'(m) + \frac{c^2}{2!x^2}\phi_{n-2}''(m) + \ldots\right] + x^{n-2}\left[\phi_{n-2}(m) + \frac{c}{x}\phi_{n-2}'(m) + \frac{c^2}{2!x^2}\phi_{n-2}''(m) + \ldots\right] = 0, \quad (3)$$

that is,

$$x^n\phi_n(m) + x^{n-1}\left[\phi_n'(m) + \phi_{n-1}(m)\right] + x^{n-2}\left[\frac{c^2}{2!}\phi_n''(m) + c\phi_{n-1}'(m) + \phi_{n-2}(m)\right] + \ldots = 0.$$

But equation (2) being an asymptote of equation (1), the values of m and c are given by

$$\phi_n(m) = 0 \text{ and } c\phi_n'(m) + \phi_{n-1}(m) = 0.$$

Hence, equation (3) reduces to

$$x^{n-2}\left[\frac{c^2}{2!}\phi_n''(m) + c\phi_{n-1}'(m) + \phi_{n-2}(m)\right] + \ldots = 0,$$

which is of degree $n - 2$ and so, yields $(n - 2)$ values of x. Hence, the asymptote (2) cuts the curve (1) in $(n - 2)$ points.

If the curve has n asymptotes, then they all will intersect the curve in n(n − 2) points.

Further, if the equation of the curve of the nth degree can be put in the form $F_n + F_{n-2} = 0$, where F_{n-2} is of degree $n-2$ at the most and F_n consists of n distinct linear factors, then the *$n(n-2)$ points of intersection of the curve $F_n + F_{n-2} = 0$ and its n asymptotes (given by $F_n = 0$) lie on the curve $F_{n-2} = 0$.*

EXAMPLE 4.9

Find the asymptotes of the curve

$$x^2y - xy^2 + xy + y^2 + x - y = 0$$

and show that they cut the curve in three points that lie on the straight line $x + y = 0$.

Solution. Equating to zero the coefficient of highest power of x, we get $y = 0$. Thus, x-axis is an asymptote to the given curve. Similarly, equating to zero the coefficient of the highest power of y, we get $-x + 1 = 0$ or $x = 1$. Thus, $x = 1$ is the asymptote parallel to y-axis. To find the oblique asymptotes, we put $x = 1$ and $y = m$ in the third- and second- degree terms and get

$$\phi_3(m) = m - m^2,\ \phi_2(m) = m + m^2, \text{ and}$$
$$\phi'_3(m) = 1 - 2m.$$

Then the slopes of the asymptotes are given by

$$\phi_3(m) = m - m^2 = 0,$$

which implies $m = 0$ and $m = 1$. The values of c are given by

$$c = -\frac{\phi_2(m)}{\phi_3'(m)} = -\frac{m + m^2}{1 - 2m}.$$

Thus, the values of c corresponding to $m = 0$ and $m = 1$ are $c = 0$ and $c = 2$, respectively. Therefore, the oblique asymptotes are $y = 0$ and $y = x + 2$. Hence, the asymptotes of the curve are

$$y = 0,\ x = 1, \text{ and } x - y + 2 = 0.$$

The joint equation of the asymptotes is

$$(x - 1)\, y\, (x - y + 2) = 0$$

or

$$x^2y - xy^2 + xy + y^2 - 2y = 0.$$

On the other hand, the equation of the curve can be written as

$$(x^2y - xy^2 + xy + y^2 - 2y) + y + x = 0,$$

which is of the form $F_n + F_{n-2} = 0$. Hence, the points of intersection which are $n(n-2) = 3(1) = 3$ in number lie on the curve $F_{n-2} = x + y = 0$, which is a straight line.

EXAMPLE 4.10

Show that the four asymptotes of the curve

$$(x^2 - y^2)(y^2 - 4x^2) + 6x^3 - 5x^2y$$
$$- 3xy^2 + 2y^3 - x^2 + 3xy - 1 = 0$$

cut the curve in eight points which lie on the circle $x^2 + y^2 = 1$.

Solution. Substituting $x = 1$ and $y = m$ in the fourth- and third-degree terms, respectively, we get

$$\phi_4(m) = (1 - m^2)(m^2 - 4) \text{ and}$$
$$\phi_3(m) = 6 - 5m - 3m^2 + 2m^3.$$

Thus,

$$\phi'_4(m) = 10m - 4m^3.$$

The slopes of the asymptotes are given by

$$\phi_4(m) = (1 - m^2)(m^2 - 4) = 0$$

and so, m = ± 1, and ± 2. The value of c is given by the expression

$$c = -\frac{\phi_3(m)}{\phi_4'(m)} = \frac{6 - 5m - 3m^2 + 2m^3}{4m^3 - 10m}.$$

The value of c corresponding to $m = 1, -1, 2,$ and -2 are respectively 0,1,0, and 1. Hence, the asymptotes are

$$y = x,\ y = -x + 1,\ y = 2x, \text{ and } y = -2x + 1.$$

Since the degree of the given curve is 4, the number point of intersection is equal to $(n - 2) = 4(4 - 2) = 8$.

The joint equation of the asymptotes is

$$(y - x)(y + x - 1)(y - 2x)(y + 2x - 1) = 0$$

or

$$(y^2 - x^2)(y^2 - 4x^2) - 6x^3 + 5x^2y$$
$$+ 3xy^2 - 2y^3 + y^2 - 3xy + 2x^2 = 0$$

or

$$(x^2 - y^2)(y^2 - 4x^2) - 6x^3 + 5x^2y$$
$$- 3xy^2 + 2y^3 - y^2 + 3xy - 2x^2 = 0$$

The given equation of the curve can be written as

$$(x^2 - y^2)(y^2 - 4x^2) + 6x^3 - 5x^2y - 3xy^2 + 2y^3$$
$$- y^2 + 3xy - 2x^2 + (x^2 + y^2 - 1) = 0,$$

which is of the form $F_n + F_{n-2} = 0$. Hence, the points of intersection lie on $F_{n-2} = 0$, that is, on the circle $x^2 + y^2 - 1 = 0$.

EXAMPLE 4.11

Find the equation of the cubic which has the same asymptotes as the curve $x^3 - 6x^2y + 11xy^2 - 6y^2 + x + y + 1 = 0$, and which touches the axis of y at the origin and passes through the point (3, 2).

Solution. The equation of the curve is

$$x^3 - 6x^2y + 11xy^2 - 6y^2 + x + y + 1 = 0.$$

The curve has no asymptote parallel to the axes. To find the oblique asymptotes, we have

$$\phi_3(m) = 1 - 6m + 11m^2 - 6m^2$$
$$= (1 - m)(1 - 2m)(1 - 3m),$$
$$\phi_2(m) = 0,\ \phi'_3(m) = 10m - 6.$$

The slopes of the asymptotes are given by

$$\phi_3(m) = (1-m)(1 - 2m)(1-3m) = 0$$

and so, $m = 1, \frac{1}{2}$, and $\frac{1}{3}$. Further,

$$c = -\frac{\phi_2(m)}{\phi_3'(m)} = 0.$$

Therefore, the asymptotes are

$$y = x, y = \frac{x}{2}, \text{ and } y = \frac{x}{3}.$$

The joint equation of the asymptotes is

$$(x - y)(x - 2y)(x - 3y) = 0.$$

The most general equation of any curve having these asymptotes is

$$F_n + F_{n-2} = 0, \text{ that is, } F_3 + F_1 = 0$$

or

$$(x - y)(x - 2y)(x - 3y) + ax + by + k = 0,$$

since F_1 is of degree 1.

Since the curve passes through the origin, putting $x = 0$, $y = 0$, in the preceding equation, we get $k = 0$. Thus, the equation of the curve becomes

$$(x - y)(x - 2y)(x - 3y) + ax + by = 0. \qquad (1)$$

Equating to zero, the lowest-degree term in (1), we get $ax + by = 0$ as the equation of the tangent at the origin. But y-axis, that is, $x = 0$ is tangent at the origin. Therefore, $b = 0$ and the equation of the curve reduces to

$$(x - y)(x - 2y)(x - 3y) + ax = 0.$$

Since the curve passes through (3, 2), we have

$$(3 - 2)(3 - 4)(3 - 6) + 3a = 0$$

and so, $a = -1$. Hence, the required curve is

$$(x - y)(x - 2y)(x - 3y) - x = 0.$$

or

$$x^3 - 6x^2y + 11xy^2 - 6y^3 - x = 0.$$

EXAMPLE 4.12

Show that the eight points of the curve

$$x^4 + 5x^2y^2 + 4y^4 + x^2 - y^2 + x + y + 1 = 0$$

and its asymptotes lie on a rectangular hyperbola.

Solution. The equation of the curve is of degree 4. Therefore, the number of points of intersection with the asymptotes is $n(n - 2) = 4(4 - 2) = 8$. Further, the equation of the given curve can be written as

$$(x^2-y^2)(x^2-4y^2) + x^2-y^2 + x + y + 1 = 0$$

or

$$F_n + F_{n-2} = 0,$$

where

$F_n = (x^2-y^2)(x^2-4y^2)$ is of degree 4 and
$F_{n-2} = x^2 - y^2 + x + y + 1 = 0$ is of degree 2.

The asymptotes are given by $F_n = 0$, that is, by $(x^2-y^2)(x^2-4y^2) = 0$. Thus, the asymptotes are $x = \pm y$ and $x = \pm 2y$. The equation $F_{n-2} = 0$, that is, $x^2 - y^2 + x + y + 1 = 0$ is the equation of the curve on which the points of intersection of the asymptotes and the given curve lie. The conic $x^2 - y^2 + x + y + 1 = 0$ is a hyperbola since the sum of the coefficients of x^2 and y^2 is zero. Hence, the eight points of intersection of the given curve with its asymptotes lie on a rectangular hyperbola.

EXAMPLE 4.13

Find the asymptotes of the curve

$$x = \frac{a(1-t^2)}{1+t^2},\quad y = \frac{at(1-t^2)}{1+t^2}.$$

Solution. The equation of the curve is given in parametric form. We eliminate t by dividing and get

$$\frac{x}{y} = \frac{1}{t} \text{ so that } t = \frac{y}{x}.$$

Substituting this value of t in $x = \frac{a(1-t^2)}{1+t^2}$, we obtain

$$x = \frac{a(x^2 - y^2)}{x^2 + y^2}$$

or

$$y^2(a + x) = x^2(a - x). \qquad (1)$$

Equating to zero the highest power of y in the equation (1) of the curve, we have $x + a = 0$. Hence, $x + a = 0$ is the asymptote parallel to the y-axis. To find the oblique asymptotes, we put $x = 1$ and $y = m$ in the highest-degree term of $f(x, y)$ to get

$$\phi_3(m) = m^2 + 1 = 0.$$

But the roots of the equation $m^2 + 1 = 0$ are imaginary. Therefore, there is no oblique asymptote. Hence, the only asymptote is $x + a = 0$.

4.6 ASYMPTOTES BY EXPANSION

Let the equation of the given curve be of the form

$$y = mx + c + \frac{A}{x} + \frac{B}{x^2} + \frac{C}{x^3} + \ldots \qquad (1)$$

Dividing both sides by x, we get

$$\frac{y}{x}=m+\frac{c}{x}+\frac{A}{x^2}+\frac{B}{x^3}+\frac{C}{x^4}+\ldots$$

Taking limit as $x\to\infty$, we have

$$\lim_{x\to\infty}\left(\frac{y}{x}\right)=m. \qquad (2)$$

The equation (1) can also be written as

$$y-mx=c+\frac{A}{x}+\frac{B}{x^2}+\frac{C}{x^3}+\ldots$$

Taking limit as $x\to\infty$, we get

$$\lim_{x\to\infty}(y-mx)=c. \qquad (3)$$

It follows (see Article 4.1) from (2) and (3) that $y=mx+c$ is an asymptote of the curve (1). Hence, $y=mx+c$ is an asymptote of a curve, whose equation can be expressed in the form (1) given earlier.

For example, consider the curve

$$f(x, y) = 2x^3 + x^2(2-y) + x + 1 = 0.$$

The given equation can be written as

$$x^2y = 2x^3 + 2x^2 + x + 1$$

or

$$y=2x+2+\frac{1}{x}+\frac{1}{x^2}.$$

Hence, $y=2x+2$ is an asymptote of the given curve.

4.7 ASYMPTOTES OF THE POLAR CURVES

If α is a root of the equation $f(\theta)=0$, then $r\sin(\theta-\alpha)=\frac{1}{f'(\alpha)}$ is an asymptote of the polar curve $\frac{1}{r}=f(\theta)$.

Thus, to find the asymptotes of a polar curve, first write down the equation of the curve in the form $\frac{1}{r}=f(\theta)$. Then find the roots of the equation $f(\theta)=0$. If the roots are θ_1, θ_2, θ_3,…, find $f'(\theta)$ at $\theta=\theta_1, \theta_2, \theta_3,\ldots$. Then the asymptotes of the curve shall be

$$r\sin(\theta-\theta_1)=\frac{1}{f'(\theta_1)},$$

$$r\sin(\theta-\theta_2)=\frac{1}{f'(\theta_2)}, \text{and so on.}$$

EXAMPLE 4.14

Find the asymptotes of the curve

$$r\sin\theta=2\cos 2\theta.$$

Solution. The equation of the given curve can be written as

$$\frac{1}{r}=\frac{\sin\theta}{2\cos\theta}=f(\theta).$$

Therefore, $f(\theta)=0$ yields $\sin\theta=0$ and so, $\theta=n\pi$, where n is an integer. Since

$$f'(\theta)=\frac{1}{2}\left[\frac{\cos 2\theta\cos\theta-\sin\theta(-2\sin 2\theta)}{\cos^2 2\theta}\right],$$

we have

$$\frac{1}{f'(n\pi)}=\frac{2\cos^2(2n\pi)}{\cos(2n\pi)\cos 2n\pi+2\sin n\pi\sin 2n\pi}$$

$$=\frac{2}{\cos n\pi}=\frac{2}{(-1)^n}.$$

Hence, the required asymptotes are

$$r\sin(\theta-n\pi)=\frac{2}{(-1)^n}$$

or

$$-r\sin(n\pi-\theta)=\frac{2}{(-1)^n}$$

or

$$-r\left[(-1)^{n-1}\sin\theta\right]=\frac{2}{(-1)^n}$$

or

$$r\sin\theta=2.$$

EXAMPLE 4.15

Show that the curve $r=\frac{a}{1-\cos\theta}$ has no asymptotes.

Solution. The equation of the given curve can be written in the form

$$\frac{1}{r}=\frac{1-\cos\theta}{a}=f(\theta).$$

Then $f(\theta)=0$ implies $\cos\theta=1$ and so, $\theta=2n\pi$, where n is an integer. Further,

$$f'(\theta)=\frac{1}{a}\sin\theta$$

and so,

$$f'(2n\pi)=\frac{1}{a}\sin(2n\pi)=0.$$

We know that if α is a root of the equation $f(\theta)=0$, then asymptote corresponding to this asymptotic direction α is given by

$$f(\alpha).r\sin(\theta-\alpha)=1.$$

So for $\alpha = 2n\pi$, the equation of the asymptote is

$$f'(2n\pi).r \sin(\theta - 2n\pi) = 1.$$

But, we have shown that $f'(2n\pi) = 0$. Thus, 0 = 1, which is impossible. Hence, there is no asymptote to the given curve.

EXAMPLE 4.16

Find the asymptotes of the curve

$$r = a(\sec\theta + \tan\theta).$$

Solution. We are given that

$$r = a\left(\frac{1}{\cos\theta} + \frac{\sin\theta}{\cos\theta}\right) = \frac{a(1+\sin\theta)}{\cos\theta}.$$

Thus,

$$\frac{1}{r} = \frac{\cos\theta}{a(1+\sin\theta)} = f(\theta).$$

But $f(\theta) = 0$ yields $\frac{\cos\theta}{a(1+\sin\theta)} = 0$ or $\cos\theta = 0$ or $\theta = (2n+1)\frac{\pi}{2}$.

Also,

$$f'(\theta) = \frac{1}{a}\left[\frac{(1+\sin\theta)(-\sin\theta) - \cos\theta\cos\theta}{(1+\sin\theta)^2}\right]$$

$$= -\frac{(\sin\theta + 1)}{a(1+\sin\theta)^2}.$$

Therefore,

$$f'\left[(2n+1)\frac{\pi}{2}\right] = -\frac{1}{a}\frac{\sin(2n+1)\frac{\pi}{2}+1}{\left[1+\sin(2n+1)\frac{\pi}{2}\right]^2}$$

$$= -\frac{1}{a}\frac{(-1)^n+1}{\left[1+(-1)^n\right]^2}$$

and so, the asymptotes are

$$r\sin\left[\theta - (2n+1)\frac{\pi}{2}\right] = \frac{1}{f'\left[(2n+1)\frac{\pi}{2}\right]}$$

or

$$-r\sin\left((n\pi + \frac{\pi}{2} - \theta)\right) = -\frac{a[1+(-1)^n]^2}{(-1)^n+1}$$

or

$$(-1)^n r\sin\left(\frac{\pi}{2} - \theta\right) = \frac{a[1+(-1)^n]^2}{(-1)^n+1}$$

or

$$r\cos\theta = \frac{a[1+(-1)^n]^2}{1+(-1)^n} = a[1+(-1)^n]$$

Putting $n = 0, 1, 2, \ldots$, the asymptotes of the curve are given by

$$r\cos\theta = 2a \text{ and } r\cos\theta = 0.$$

Thus, we note that there are only two asymptotes of the given curve.

EXAMPLE 4.17

Find the asymptotes of the curve $r = a\tan\theta$.

Solution. The equation of the given curve may be written as

$$\frac{1}{r} = \frac{1}{a}\frac{\cos\theta}{\sin\theta} = f(\theta).$$

Therefore, $f(\theta) = 0$ implies $\cos\theta = 0$ and so, $\theta = (2n+1)\frac{\pi}{2}$. Also

$$f'(\theta) = -\frac{1}{a}\operatorname{cosec}^2\theta.$$

Therefore,

$$f'\left[(2n+1)\frac{\pi}{2}\right] = -\frac{1}{a\left[\sin(2n+1)\frac{\pi}{2}\right]^2} = \frac{-1}{a(-1)^{2n}}$$

Thus,

$$\frac{1}{f'\left[(2n+1)\frac{\pi}{2}\right]} = a(-1)^{2n-1} = \pm a.$$

The asymptotes are now given by

$$r\sin\left(\theta - (2n+1)\frac{\pi}{2}\right) = \pm a.$$

Proceeding as in the earlier example, we get the asymptotes as

$$r\cos\theta = a \text{ and } r\cos\theta = -a.$$

EXAMPLE 4.18

Find the asymptotes of the following curves:

(i) $r\theta = a$

(ii) $r = \frac{2a}{1-2\cos\theta}$

(iii) $r = \sin n\theta = a$.

Solution. (i) From the given equation, we get

$$\frac{1}{r} = \frac{\theta}{a} = f(\theta).$$

Therefore, $f(\theta) = 0$ yields $\frac{\theta}{a} = 0$ or $\theta = 0$. Also

$$f'(\theta) = \frac{1}{a} \text{ and so, } \frac{1}{f'(\theta)} = a.$$

Thus, the asymptotes are given by

$$r \sin(\theta - 0) = \frac{1}{f'(0)} = a \text{ or } r \sin\theta = a.$$

(ii) From the given equation, we get

$$\frac{1}{r} = \frac{1 - 2\cos\theta}{2a} = f(\theta).$$

Therefore, $f(\theta) = 0$ gives $1 - 2\cos\theta = 0$ or $\cos\theta = \frac{1}{2}$ and so, $\theta = 2n\pi \pm \frac{\pi}{3}$, where n is an integer. Further,

$$f'(\theta) = \frac{1}{2a}(2\sin\theta) = \frac{\sin\theta}{a}.$$

This gives

$$f'\left(2n\pi \pm \frac{\pi}{3}\right) = \frac{1}{a}\sin\left(2n\pi \pm \frac{\pi}{3}\right) = \pm\frac{1}{a}\sin\frac{\pi}{3}$$

$$= \pm\frac{\sqrt{3}}{2a}.$$

Hence, the asymptotes are given by

$$r \sin\left[\theta - \left(2n\pi \pm \frac{\pi}{3}\right)\right] = \frac{1}{f'\left(2n\pi \pm \frac{\pi}{3}\right)} = \pm\frac{2a}{\sqrt{3}}$$

or on simplification,

$$r \sin\left(\theta - \frac{\pi}{3}\right) = \frac{2a}{\sqrt{3}} \text{ and } r \sin\left(\theta + \frac{\pi}{3}\right) = -\frac{2a}{\sqrt{3}}.$$

(iii) The equation of the curve may be written as

$$\frac{1}{r} = \frac{\sin n\theta}{a} = f(\theta).$$

Therefore, $f(\theta) = 0$ implies that $\sin n\theta = 0$ and so, $n\theta = m\pi$, where m is an integer. Thus, $\theta = \frac{m\pi}{n}$. Also,

$$f'(\theta) = \frac{n\cos n\theta}{a}$$

and so,

$$f'\left(\frac{m\pi}{n}\right) = \frac{n\cos m\pi}{a}.$$

Hence, the asymptotes are given by

$$r \sin\left(\theta - \frac{m\pi}{n}\right) = \frac{1}{f'\left(\frac{m\pi}{n}\right)} = \frac{a}{n\cos m\pi},$$

where m is an integer.

4.8 CIRCULAR ASYMPTOTES

Let the equation of a curve be $r = f(\theta)$. If $\lim_{\theta\to\infty} f(\theta) = a$, then the circle $r = a$ is called the *circular asymptote* of the curve $r = f(\theta)$.

EXAMPLE 4.19

Find the circular asymptotes of the curves

(i) $r(e^\theta - 1) = a(e^\theta + 1)$.

(ii) $r(\theta + \sin\theta) = 2\theta + \cos\theta$.

(iii) $r = \frac{a\theta}{\theta - 1}$.

Solution. (i) The given equation is

$$r(e^\theta - 1) = a(e^\theta + 1)$$

or

$$r = \frac{a(e^\theta - 1)}{e^\theta - 1} = f(\theta).$$

Now

$$\lim_{\theta\to\infty} \frac{a(e^\theta - 1)}{e^\theta - 1} = a \lim_{\theta\to\infty} \frac{1 + e^{-\theta}}{1 - e^{-\theta}} = a.$$

Hence, $r = a$ is the circular asymptote.

(ii) The equation of the given curve is

$$r = \frac{2\theta + \cos\theta}{\theta + \sin\theta} = f(\theta).$$

Further,

$$\lim_{\theta\to\infty} f(\theta) = \lim_{\theta\to\infty} \frac{2\theta + \cos\theta}{\theta + \sin\theta} = \lim_{\theta\to\infty} \frac{2 + \frac{1}{\theta}\theta}{1 + \frac{\sin\theta}{\theta}}$$

$$= \frac{2}{1 + 0} = 2.$$

Hence, $r = 2$ is the required circular asymptote.

(iii) The given equation is

$$r = \frac{a\theta}{\theta - 1}$$

and

$$\lim_{\theta\to\infty} \frac{a\theta}{\theta - 1} = a \lim_{\theta\to\infty} \frac{1}{1 - \frac{1}{\theta}} = a.$$

Hence, $r = a$ is the circular asymptote of the given curve.

4.9 CONCAVITY, CONVEXITY AND SINGULAR POINTS

Consider the curve $y = f(x)$, which is the graph of a single-valued differentiable function in a plane. The curve is said to be *convex upward* or *concave downward* on the interval (a, b) if all points of the curve lie below any tangent to it on this interval. We say that the curve is *convex downward* or *concave upward* on the interval (c, d) if all points of the curve lie above any tangent to it on this interval. Generally, a convex upward curve is called a *convex curve* and a curve convex down is called a *concave curve*. For example, the curves in figures (a) and (b) are respectively convex and concave curves.

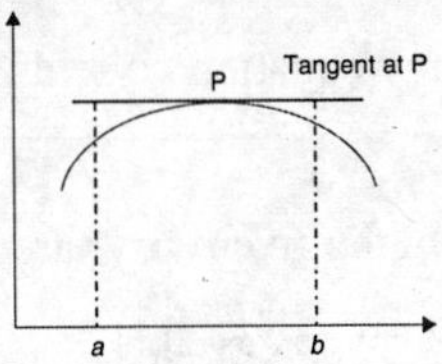

(a) Convex Curve

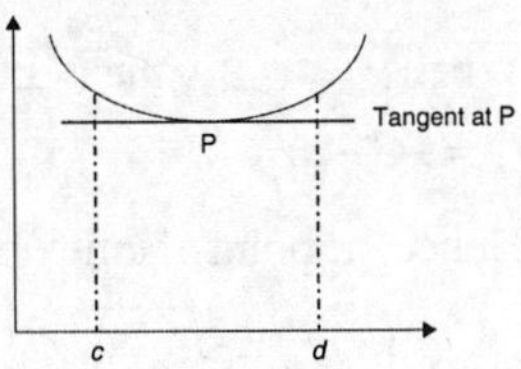

(b) Concave Curve

The following theorems tell us whether the given curve is convex or concave in some given interval.

Theorem 4.1. If at all points of an interval (a, b) the second derivative of the function $f(x)$ is negative, that is, $f''(x) < 0$, then the curve $y = f(x)$ is convex on that interval.

Theorem 4.2. If at all points of an interval (c, d) the second derivative of the function $f(x)$ is positive, that is, $f''(x) > 0$, then the curve $y = f(x)$ is concave on that interval.

A point P on a continuous curve $y = f(x)$ is said to be a *point of inflexion* if the curve is convex on one side and concave on the other side of P with respect to any line, not passing through the point P.

In other words, the point that separates the convex part of a continuous curve from the concave part is called the point of inflexion.

The following theorem gives the sufficient conditions for a given point of a curve to be a point of inflexion.

Theorem 4.3. Let $y = f(x)$ be a continuous curve. If $f''(p) = 0$ or $f''(p)$ does not exist and if the derivative $f''(x)$ changes sign when passing through $x = p$, then the point of the curve with abscissa $x = p$ is the point of inflexion.

Thus at a point of inflexion P, $f'(x)$ is positive on one side of P and negative on the other side. The above theorem implies that at a point of inflexion $f''(x) = 0$ and $f'''(x) \neq 0$.

For example, the point P, in the figure shown below is a point of inflexion.

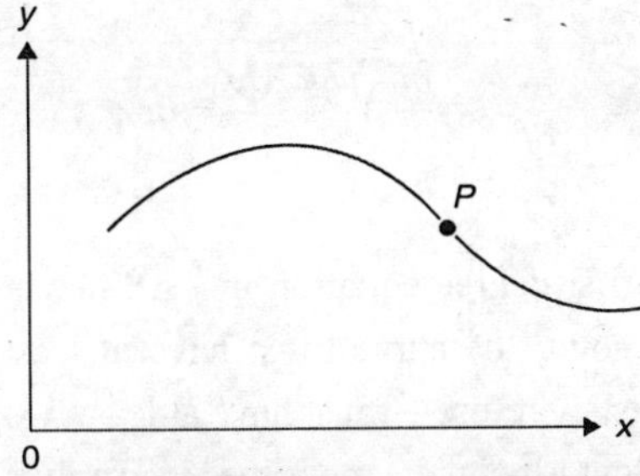

A point through which more than one branches of a curve pass is called a *multiple point* on the curve.

If two branches of curve pass through a point, then that point is called a *double point*. If r branches of a curve pass through a point, then that point is called a multiple *point of order r*.

If two branches of a curve through a double point are real and have different tangents, then the double point is called a *node*.

For example, the curve in the figure below has a node at the origin.

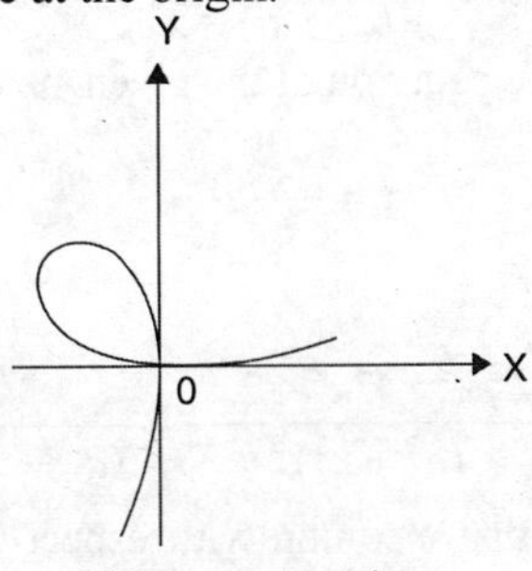

(Origin as a Node)

If two branches through a double point P are real and have coincident tangents, then P is called a *cusp*.

For example, the curve in the figure below has a cusp at the origin.

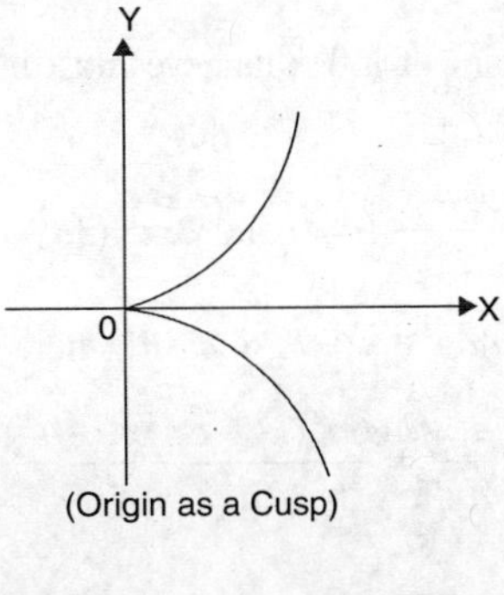

(Origin as a Cusp)

Let $P(x,y)$ be any point on the curve $f(x, y) = 0$. The slope $\frac{dy}{dx}$ of the tangent at P is given by

$$\frac{dy}{dx} = -\frac{\frac{\partial f}{\partial x}}{\frac{\partial f}{\partial y}} \text{ or } \frac{\partial f}{\partial x} + \frac{\partial f}{\partial y}\cdot\frac{dy}{dx} = 0,$$

which is a first degree equation in $\frac{dy}{dx}$. Since at a multiple point, the curve must have at least two tangents, therefore $\frac{dy}{dx}$ must have at least two values at a double point. It is possible if and only if

$$\frac{\partial f}{\partial x} = 0 \text{ and } \frac{\partial f}{\partial y} = 0.$$

Hence the necessary and sufficient conditions for the existence of multiple points are

$$\frac{\partial f}{\partial x} = 0 \text{ and } \frac{\partial f}{\partial y} = 0.$$

EXAMPLE 4.20

Find the points of inflexion of the curve

$$y\,(a^2 + x^2) = x^3.$$

Solution. The equation of the given curve is

$$y = \frac{x^3}{a^2 + x^2}.$$

Therefore

$$\frac{dy}{dx} = \frac{(a^2+x^2)3x^2 - 2x^4}{(a^2+x^2)^2} = \frac{x^4 + 3a^2x^2}{(a^2+x^2)^2}.$$

Differentiating once more with respect to x, we get

$$\frac{d^2y}{dx^2} = \frac{x(6a^4 + 10a^2x^2 + 4x^4 - 12a^2x^2 - 4x^4)}{(a^2+x^2)^3}$$

$$= \frac{2xa^2(3a^2 - x^2)}{(a^2+x^2)^3}.$$

At the point of inflextion, we must have $\frac{d^2y}{dx^2} = 0$ and so

$$\frac{2xa^2(3a^2 - x^2)}{(a^2+x^2)^3} = 0 \text{ and } 2xa^2(3a^2 - x^2) = 0,$$

which yields $x = 0$, $\pm\sqrt{3}a$. Further,

$$\frac{d^3y}{dx^3} = \frac{6a^2(x^4 - 6a^2x^2 + a^4)}{(a^2+x^2)^4}.$$

If $x = 0$, then $\frac{d^3y}{dx^3} = \frac{6a^6}{a^8} = \frac{6}{a^2} \neq 0.$

If $x = \sqrt{3a}$, then $\frac{d^3y}{dx^3} = -\frac{3}{4}a^2 \neq 0.$

If $x = -\sqrt{3a}$, then $\frac{d^3y}{dx^3} = -\frac{3}{4}a^2 \neq 0.$

Thus all the three values of x corresponds to the points of inflexion.

When $x = 0$, the given equation yields $y = 0$.

When $x = \sqrt{3}a$, the given equation yields $y = \frac{3\sqrt{3}a}{4}$

When $x = -\sqrt{3}a$, the given equation yields $y = -\frac{3\sqrt{3}}{4}a$.

Hence the point of inflexion of the given curve are

$$(0,0), \left(\sqrt{3}a, \frac{3\sqrt{3}}{4}a\right) \text{and} \left(-\sqrt{3}a, \frac{3\sqrt{3}}{4}a\right).$$

EXAMPLE 4.21

Does the curve $y = x^4$ have points of inflexion?

Solution. The equation of the given curve is $y = x^4$. Differentiating with respect to x, we have

$$\frac{dy}{dx} = 4x^3, \frac{d^2y}{dx^2} = 12x^2, \frac{d^3y}{dx^3} = 24x.$$

Then for the points of inflexion, we must have

$$\frac{d^2y}{dx^2} = 0, \text{that is, } 12x^2 = 0.$$

which yields $x = 0$. But

for $x < 0, \frac{d^2y}{dx^2} > 0$ and therefore the curve is concave,

for $x > 0, \frac{d^2y}{dx^2} > 0$ and therefore the curve is concave. Since the second derivative does not change sign passing through $x = 0$, the curve has no points of inflexion.

EXAMPLE 4.22

Find the points of inflexion on the curve $y^2 = x\,(x+1)^2$.

Solution. The given curve is symmetrical about x axis and gives

$$y = \pm x^{\frac{1}{2}}(x+1).$$

So, we can proceed with

$$y = x^{\frac{1}{2}}(x+1).$$

Then

$$\frac{dy}{dx} = x^{\frac{1}{2}} + \frac{1}{2}x^{-\frac{1}{2}}(x+1) = \frac{3x+1}{2x^{\frac{1}{2}}},$$

$$\frac{d^2y}{dx^2} = \frac{1}{2}\left[\frac{3x^{\frac{1}{2}} - (3x+1)\frac{1}{2}x^{\frac{1}{2}}}{x}\right] = \frac{3x-1}{4x^{\frac{3}{2}}}.$$

To determine the point of inflexion, we put $\frac{d^2y}{dx^2}$ equal to 0. Therefore $3x-1=0$ or $x=\frac{1}{3}$. Further

$$\frac{d^2y}{dx^3} = \frac{3}{8x^{\frac{5}{2}}}(1-x) \neq 0 \text{ at } x = \frac{1}{3}.$$

Therefore the curve has point of inflexion corresponding to $x=\frac{1}{3}$. Putting $x=\frac{1}{3}$ in the equation of the curve, we have $y=\pm\frac{4}{3\sqrt{3}}$. Hence the points of inflexion on the curve are

$$\left(\frac{1}{3}, \frac{4}{3\sqrt{3}}\right) \text{ and } \left(\frac{1}{3}, -\frac{4}{3\sqrt{3}}\right).$$

EXAMPLE 4.23

Find the points of inflexion and the intervals of convexity and concavity of the Gaussian curve $y=e^{-x^2}$.

Solution. The equation of the Gaussian curve is

$y=e^{-x^2}$. Therefore

$$\frac{dy}{dx} = -2xe^{-x^2}, \quad \frac{d^2y}{dx^2} = 2e^{-x^2}[2x^2-1].$$

For the existence of points of inflexion, we must have $\frac{d^2y}{dx^2}=0$, which yields $x=\pm\frac{1}{\sqrt{2}}$.

Now, since

$$\text{for } x < \frac{1}{\sqrt{2}}, \text{ we have } \frac{d^2y}{dx^2} < 0$$

$$\text{for } x > \frac{1}{\sqrt{2}}, \text{ we have } \frac{d^2y}{dx^2} > 0,$$

therefore the point of inflexion exists for $x=\frac{1}{\sqrt{2}}$. Putting $x=\frac{1}{\sqrt{2}}$ in the given equation, $y=e^{-\frac{1}{2}}$. Therefore $\left(\frac{1}{\sqrt{2}}, e^{-\frac{1}{2}}\right)$ is a point of inflexion on the curve.

Also

$$\text{for } x < -\frac{1}{\sqrt{2}}, \text{ we have } \frac{d^2y}{dx^2} > 0$$

$$\text{for } x > -\frac{1}{\sqrt{2}}, \text{ we have } \frac{d^2y}{dx^2} < 0.$$

Thus another point of inflexion exists for the value $x=-\frac{1}{\sqrt{2}}$. Putting $x=-\frac{1}{\sqrt{2}}$ in the equation of the Gaussian curve, we get $y=e^{-\frac{1}{2}}$. Hence the second point of inflexion is $\left(-\frac{1}{\sqrt{2}}, e^{-\frac{1}{2}}\right)$.

EXAMPLE 4.24

Determine whether the curve $y=e^x$ is concave or convex.

Solution. The given exponential curve is $y=e^x$. Then $\frac{dy}{dx}=e^x$, $\frac{dy^2}{dx^2}=e^x>0$, for the all values of x. Hence the curve is everywhere concave.

EXAMPLE 4.25

Determine the existence and nature of the double points on the curve

$$f(x,y) = y^2 - (x-2)^2(x-1) = 0.$$

Solution. We have

$$f(x,y) = y^2 - (x-2)^2(x-1) = 0,$$

$$\frac{\partial f}{\partial x} = -(x-2)(3x-4),$$

$$\frac{\partial f}{\partial y} = 2y.$$

Now for the existence of double points, we must have

$$\frac{\partial f}{\partial x} = \frac{\partial f}{\partial y} = 0.$$

Hence

$$(x-2)(3x-4) = 0 \text{ and } 2y = 0,$$

which yield

$$x = 2, \frac{4}{3} \text{ and } y = 0.$$

Thus the possible double points are (2, 0) and $\left(\frac{4}{3}, 0\right)$. But, only (2, 0) satisfies the equation of the curve. To find the nature of the double point (2,0), we shift the origin to (2, 0). The equation reduces to

$$y^2 = (x+2-2)^2(x+2-1) = x^2(x+1)$$
$$= x^3 + x^2.$$

Equating to zero the lowest degree term, we get $y^2 - x^2 = 0$, which gives $y = \pm x$ as the tangent at (2,0). Therefore, at the double point (2,0), there are two real and district tangents. Hence the double point (2, 0) is a node on the given curve.

EXAMPLE 4.26

Does the curve $x^4 - ax^2y + axy^2 + a^2y^2 = 0$ have a node on the origin?

Solution. Equating to zero the lowest degree term in the equation of the given curve, we have

$$a^2y^2 = 0, \text{ which yields } y = 0, 0.$$

Therefore there are two real and coincident tangents at the origin. Hence the given curve has a cusp or conjugate point at the origin and not a node.

4.10 CURVE TRACING (CARTESIAN EQUATIONS)

The aim of this section is to find the appropriate shape of a curve whose equation is given. We shall examine the following properties of the curves to trace it.

1. Symmetry: (i) If the equation of a curve remains unaltered when y is changed to $-y$, then the curve is *symmetrical about the x-axis*. In other words, if the equation of a curve consists of even powers of y, then the curve is symmetrical about the x-axis. For example, the parabola $y^2 = 4ax$ is symmetrical about the x-axis.

(ii) If the equation of a curve remains unaltered when x is changed to $-x$, then the curve is *symmetrical about the y-axis*. Thus, a curve is symmetrical about the y-axis, if its equation consists of even powers of x. For example, the curve $x^2 + y^2 = a^2$ is symmetrical about the y-axis.

(iii) If the equation of a curve remains unchanged when x is replaced by $-x$ and y is replaced by $-y$, then the curve is *symmetrical in the opposite quadrants*. For example, the curve $xy = c^2$ is symmetrical in the opposite quadrants.

(iv) If the equation of a curve remains unaltered when x and y are interchanged, then the curve is *symmetrical about the line* $y = x$. For example, the folium of Descarte's $x^3 + y^3 = 3axy$ is symmetrical about the line $y = x$.

2. Origin: (i) If the equation of a curve does not contain a constant term, then the curve passes through the origin. In other words, a *curve passes through the origin* if (0, 0) *satisfies the equation of the curve.*

(ii) If the curve passes through the origin, find the *equation of the tangents at the origin by equating to zero the lowest-degree terms* in the equation of the curve. In case there is only one tangent, determine whether the curve lies below or above the tangent in the neighbourhood of the origin. If there are two tangents at the origin, then the *origin is a double point*; if the two tangents are real and distinct, then the origin is a *node*; if the two tangents are real and coincident, then the origin is *cusp*; if the two tangents are imaginary, then the origin is a *conjugate point* or an *isolated point*.

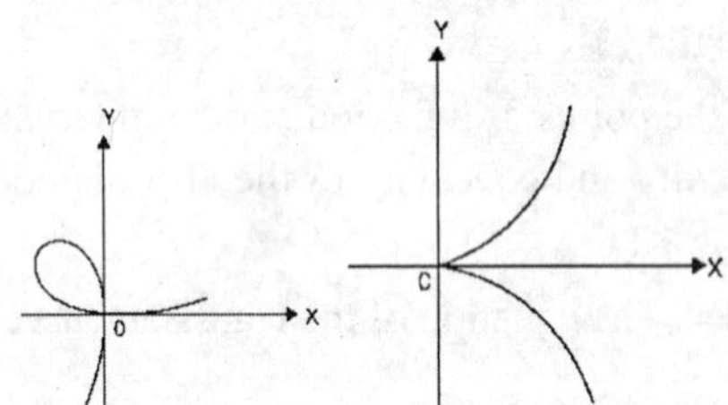

(Origin as a Node) (Origin as a Cusp)

3. Intersection with the Coordinate Axes: To find the points where the curve cuts the coordinate axes, we put $y = 0$ in the equation of the curve to find where the curve cuts the x-axis. Similarly, we put $x = 0$ in the equation to find where the curve cuts the y-axis.

4. Asymptotes: Determine the asymptotes of the curve parallel to the axes and the oblique asymptotes.

5. Sign of the Derivative: Determine the points where the derivative $\frac{dy}{dx}$ vanishes or becomes infinite. This step will yield the points where the tangent is parallel or perpendicular to the x-axis.

6. Points of Inflexion: A point P on a curve is said to be a *point of inflexion* if the curve is concave on one side and convex on the other side of P with respect to any line AB, not passing through the point P.

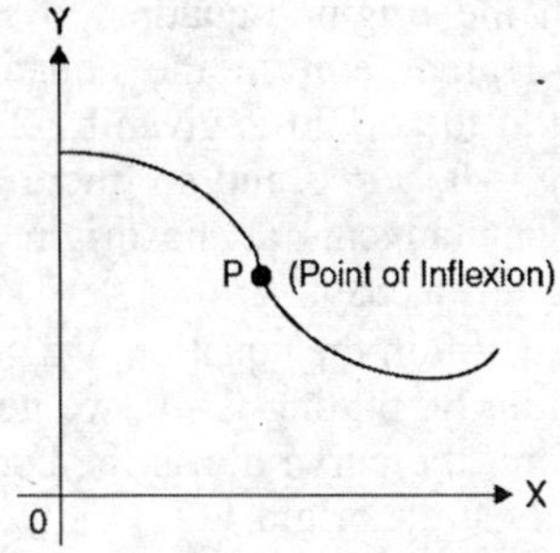

There will be a point of inflexion at a point P on the curve if $\frac{d^2y}{dx^2}=0$ but $\frac{d^3y}{dx^3}\neq 0$.

7. Region, Where the Curve Does Not Exist: Find out if there is any region of the plane such that no part of the curve lies in it. This is done by solving the given equation for one variable in terms of the other. The curve will not exist for those values of one variable which make the other variable imaginary.

EXAMPLE 4.27

Trace the curve

$$a^2y^2 = x^2(a^2 - x^2).$$

Solution. The equation of the curve is

$$a^2y^2 = x^2(a^2 - x^2).$$

We observe the following:

(i) Since powers of both x and y are even, it follows that the curve is symmetrical about both the axes.

(ii) Since the equation does not contain constant terms, the curve passes through the origin. To find the tangent at the origin, we equate to zero the lowest-degree terms in the given equation. Thus, the tangents at the origin are given by

$a^2y^2 - a^2x^2 = 0 \text{ or } y = \pm x.$

Since tangents are distinct, the origin is a node.

(iii) Putting $y = 0$ in the given equation, we get $x = 0$ and $x = \pm a$. Therefore, the curve crosses the x-axis at (0,0), (a,0), and ($-a$,0).

(iv) Shifting the origin to (a, 0), the given equation reduces to

$$a^2y^2 = (x+a)^2\left[a^2 - (x+a)^2\right]$$

or

$$a^2y^2 = (x+a)^2(-2ax - x^2).$$

Equating to zero the lowest-degree term, the tangent at the new origin is given by $4a^2x^2 = 0$ or $x = 0$. Thus, the tangent at (a, 0) is parallel to the y-axis.

(v) The given equation can be written as

$$y^2 = \frac{x^2(a^2 - x^2)}{a^2}.$$

When $x = 0$, $y = 0$ and when $x = a$, $y = a$. When $0 < x < a$, y is real and so, the curve exists in this region. When $x > a$, y^2 is negative and so, y is imaginary. Hence, the curve does not exist in the region $x > a$.

(vi) The given curve has no asymptote. Hence, the shape of the curve is as shown in the following figure:

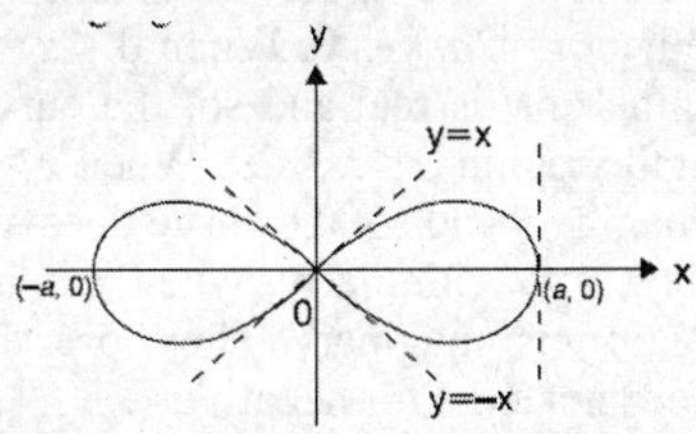

EXAMPLE 4.28

Trace the curve

$$xy^2 = 4a^2(2a - x) \text{ (Witch of Agnesi).}$$

Solution. We note that

(i) The curve is symmetrical about the x-axis because the equation contains even powers of y.

(ii) Since the equation consists of a constant term, $8a^3$, the curve does not pass through the origin.

(iii) Putting $y = 0$ in the equation, we get $x = 2a$. Therefore, the curve crosses the x-axis at ($2a$, 0). When $x = 0$, we do not get any value of y. Therefore, the curve does not meet the y-axis.

Shifting *the* origin to ($2a$, 0), the equation of the curve reduces to

$$(x+2a)y^2 = 4a^2(2a - x - 2a)$$

or

$$y^2x + 2ay^2 + 4a^2x = 0\cdot$$

Equating to zero, the lowest-degree terms of this equation, the equation of the tangent at this new origin is given by

$$4a^2x = 0 \ \ or \ \ x = 0.$$

Hence, the tangent at the point $(2a, 0)$ to the curve is parallel to y-axis.

(iv) Equating to zero the coefficient of highest power of y, the asymptote parallel to the y-axis is $x = 0$, that is, the y-axis. Further, the curve has no other real asymptote.

(v) The equation of the given curve can be written as

$$y^2 = \frac{4a^2(2a-x)}{x}.$$

Therefore, when $x \to 0$, y approaches ∞ and so, the line $x = 0$ is an asymptote. When $x = 2a$, $y = 0$. When $0 < x < 2a$, the value of y is real and so, the curve exists in the region $0 < x < 2a$. When $x > 2a$, y is imaginary and so, the curve does not exists for $x > 2a$. Similarly, when x is negative, again y is imaginary. Therefore, the curve does not exist for negative x.

In view of the mentioned points, the shape of the curve is as shown in the following figure.

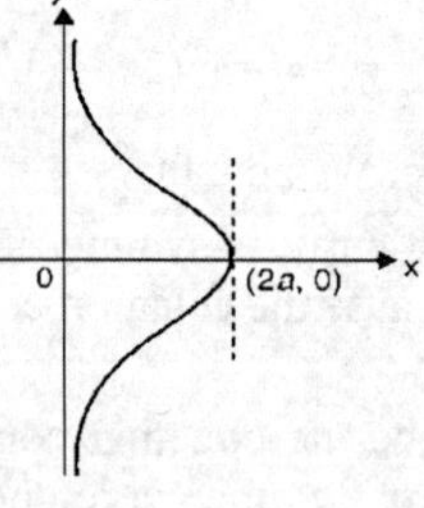

EXAMPLE 4.29

Trace the curve

$$y^2(2a-x) = x^3 \text{ (Cissoid).}$$

Solution. We note that

(i) Since the powers of y in the given equation of the curve are even, the curve is symmetrical about the x-axis.

(ii) Since the equation of the curve does not contain a constant term, the curve passes through the origin. Equating to zero the lowest-degree term in the equation, the tangent at the origin is given by $2ay^2 = 0$. Thus, $y = 0$, $y = 0$ and so, there are two coincident tangents at the origin. Hence, the origin is a *cusp*.

(iii) Putting $x = 0$ in the equation, we get $y = 0$ and similarly, putting $y = 0$, we get $x = 0$. Therefore, the curve meets the coordinate axes only at the origin.

(iv) Equating to zero the highest power of y in the equation of the curve, the asymptote parallel to the y-axis is $x = 2a$. The curve does not have an asymptote parallel to the x-axis or any other oblique asymptote.

(v) The given equation can be written as

$$y^2 = \frac{x^3}{2a-x}.$$

When $x \to 2a$, $y^2 \to \infty$ and so, $x = 2a$ is an asymptote. If $x > 2a$, y is imaginary. Therefore, the curve does not exist beyond $x = 2a$. When $0 < x < 2a$, y^2 is positive and so, y is real. Therefore, the curve exists in the region $0 < x < 2a$. When $x < 0$, again y is imaginary. Therefore, the curve does not exist for a negative x.

In view of the said observations, the shape of the curve is as shown in the following figure:

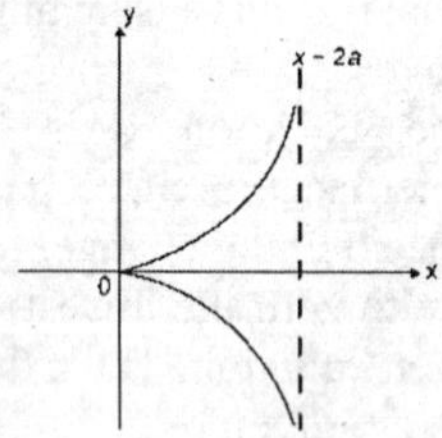

EXAMPLE 4.30

Trace the curve

$$x^3 + y^3 = 3axy \text{ (Folium of Descartes).}$$

Solution: We observe that

(i) The curve is not symmetrical about the axes. However, the equation of the curve remains unaltered if x and y are interchanged. Hence, the curve is symmetrical about the line $y = x$. It meets this line at $\left(\frac{3a}{2}, \frac{3a}{2}\right)$.

(ii) Since the equation does not contain a constant term, the curve passes through the origin. Equating to zero the lowest-degree term, we get $3axy = 0$. Hence, $x = 0, y = 0$ are the tangents at the origin. Thus, both y- and x-axis are tangents to the curve at the origin. Since there are two real and distinct tangents at the origin, the origin is a *node* of the curve.

(iii) The curve intersects the coordinate axes only at the origin.

(iv) If, in the equation of the curve, we take both x and y as negative, then the right-hand side becomes positive while the left-hand side is negative. Therefore, we cannot take both x and y as negative. Thus, the curve does not lie in the third quadrant.

(v) There is no asymptote parallel to the axes. Further, putting $x = 1, y = m$ in the highest-degree term, we have

$$\phi_3(m) = m^3 + 1$$

The slope of the asymptotes are given by $m^3 + 1 = 0$. The real root of this equation is $m = -1$. Also, putting $x = 1, y = m$ in the second-degree term, we have

$$\phi_2(m) = -3am$$

and further,

$$\phi_3'(m) = 3m^2.$$

Therefore,

$$c = -\frac{\phi_2(m)}{\phi_3'(m)} = \frac{3am}{3m^2} = \frac{a}{m}.$$

For $m = -1$, we have $c = -a$. Hence, the oblique asymptote is

$$y = -x - a \ \ or \ \ x + y + a = 0.$$

In view of the earlier facts, the shape of the curve is as shown in the following figure:

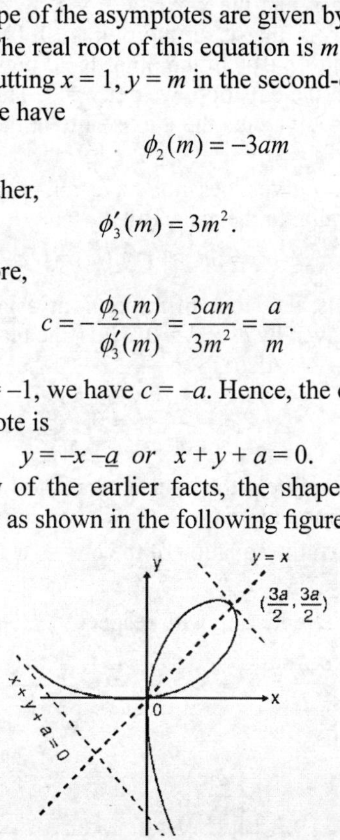

EXAMPLE 4.31

Trace the curve

$$y^2 (a + x) = x^2 (a - x).$$

Solution. We note that

(i) The equation of the curve does not alter if y is changed to $-y$. Therefore, the curve is symmetrical about the x-axis.

(ii) Since the equation does not contain a constant term, the curve passes through the origin. The tangents at the origin are given by

$$ay^2 - ax^2 = 0 \ \ or \ \ y = \pm x.$$

Thus, there are two real and distinct tangents at the origin. Therefore, the origin is a *node*.

(iii) Putting $y = 0$, we have $x^2 (a - x) = 0$ and so, the curve intersects the x-axis at $x = 0$ and $x = a$, that is, at the points (0, 0) and (a, 0). Putting $x = 0$, we get $y = 0$. Thus, the curve intersects the y-axis only at (0, 0). Shifting the origin to (a, 0), the equation of the curve reduces to

$$y^2(2a + x) = -x\left(x^2 + 2ax + a^2\right).$$

Equating to zero the lowest-degree term, we get $a^2 x = 0$. Hence, at the new origin, $x = 0$ is the tangent. Thus, the tangent at (a, 0) is parallel to the y-axis.

(iv) The equation of the curve can be written as

$$y^2 = \frac{x^2(a - x)}{a + x}.$$

When x lies in $0 < x < a$, y^2 is positive and so, the curve exists in this region. But when $x > a$, y^2 is negative and so, y is imaginary. Thus, the curve does not exist in the region $x > a$. Further, if $x \to -a$, then $y^2 \to \infty$ and so, $x = -a$ is an asymptote of the curve. If $-a < x < 0$, y^2 is positive and therefore, the curve exists in $-a < x < 0$. When $x < -a$, y^2 is negative and so, the curve does not lie in the region $x < -a$.

(v) Equating to zero the coefficient of the highest power of y in the equation of the curve, we have $x + a = 0$. Thus, $x + a = 0$ is the asymptote parallel to the y-axis. To see whether oblique asymptotes are there or not, we have $\phi_3(m) = m^2 + 1$. But the roots of $m^2 + 1 = 0$ are imaginary. Hence, there is no oblique asymptote.

Thus, the shape of the curve is as shown in the following figure:

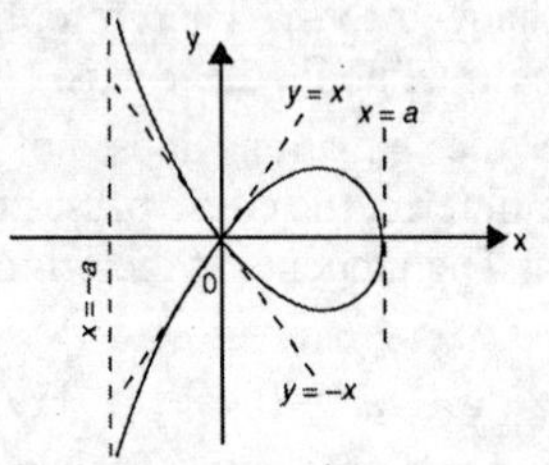

EXAMPLE 4.32

Trace the curve

$$x = (y-1)(y-2)(y-3).$$

Solution. We note that

(i) The equation of the curve has odd powers of x and y. Therefore, the curve is not symmetrical about the axes. It is also not symmetrical about $y = x$ or in the opposite quadrants.

(ii) The curve does not pass through the origin.

(iii) Putting $x = 0$ in the given equation, we get $y = 1, 2$ and 3. Thus, the curve cuts the y-axis at $y = 1, 2$ and 3. Similarly, putting $y = 0$, we see that the curve cuts the x-axis at $x = -6$.

(iv) The curve has no linear asymptotes since $y \to \pm\infty$, $x \to \pm\infty$.

(v) When $0 < y < 1$, then all the factors are negative and so, x is negative. When $1 < y < 2$, x is positive. Similarly, when $2 < y < 3$, then x is negative. At $y = 3$, $x = 0$. When $y > 3$, x is positive. When $y < 0$, x is negative. Hence, the shape of the curve is as shown in the following figure:

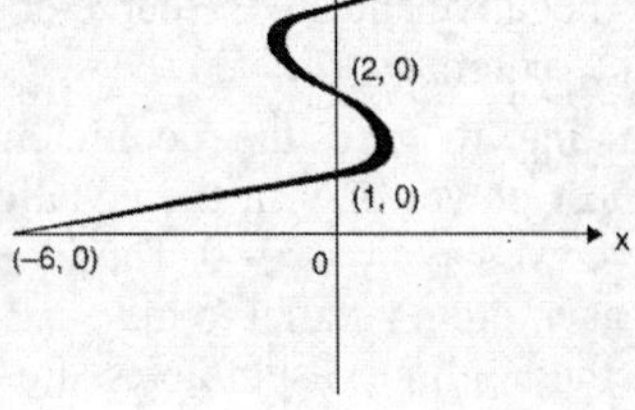

EXAMPLE 4.33

Trace the curve

$$x^3 + y^3 = a^2x.$$

Solution. We note the following characteristics of the given curve:

(i) Since the equation of the curve contains odd powers of x and y, the curve is not symmetrical about the axes. But if we change the sign of both x and y, then the equation remains unaltered. Therefore, the curve is symmetrical in the opposite quadrants.

(ii) Since the equation of the curve does not have a constant term, the curve passes through the origin. The tangent at the origin is given by $a^2 x = 0$. Thus, $x = 0$, that is, y-axis is tangent to the curve at the origin.

(iii) Putting $y = 0$ in the equation, we get $x(x^2 - a^2) = 0$ or $x(x-a)(x+a) = 0$. Hence, the curve cuts the x-axis at $x = 0$, $x = a$, and $x = -a$, that is, at the points $(0, 0)$, $(a, 0)$, and $(-a, 0)$. On the other hand putting $x = 0$ in the equation, we get $y = 0$. Therefore, the curve cuts the y-axis only at the origin $(0, 0)$.

(iv) The curve does not have any asymptote parallel to the axes. But

$$\phi_3(m) = m^3 + 1, \ \phi_2(m) = 0.$$

Thus, the slope of the oblique asymptotes is given by $m^3 + 1 = 0$. Thus, the real root is $m = -1$. Also

$$c = -\frac{\phi_2(m)}{\phi_3(m)} = 0.$$

Therefore, the curve has an oblique asymptote $y = -x$.

(v) From the equation of the curve, we have $y^3 = a^2 x - x^3$.

Differentiating with respect to x, we get

$$3y^2 \frac{dy}{dx} = a^2 - 3x^2 \text{ or } \frac{dy}{dx} = \frac{a^2 - 3x^2}{3y^2}.$$

Thus,

$$\left(\frac{dy}{dx}\right)_{(a,0)} = -\infty$$

and so, the tangent at $(a, 0)$ is perpendicular to the x-axis. Similarly, $\left(\frac{dy}{dx}\right)_{(-a,0)} = -\infty$ and so, the tangent at $(-a, 0)$ is also perpendicular to the x-axis.

Also we note that $\frac{dy}{dx} = 0$ implies $x = \pm\frac{a}{\sqrt{3}}$. Therefore, the tangents at these points are parallel to the x-axis.

(vi) Also $y^3 = a^2 x - x^3 = x(a^2 - x^2)$ implies that y^3 is positive in the region $0 < x < a$. But y^3 is negative in the region $x > a$. The earlier facts imply that the shape of the given curve is as shown in the following figure:

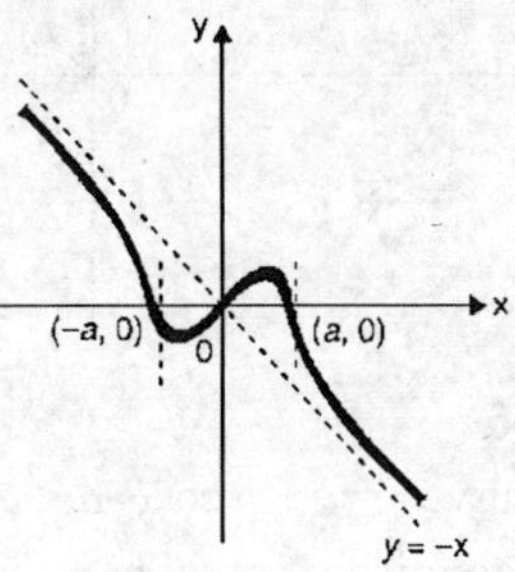

4.11 CURVE TRACING (POLAR EQUATIONS)

To trace a curve with a polar form of equation, we adopt the following procedure:

1. Symmetry: If the equation of the curve does not change when θ is changed into $-\theta$ the curve is *symmetrical about the initial line.*

If the equation of the curve remains unchanged by changing r into $-r$, then *the curve is symmetrical about the pole and the pole is the center of the curve.*

If the equation of the curve remains unchanged when θ is changed to $-\theta$ and r is changed in to $-r$, then the curve is symmetrical about the line $\theta = \frac{\pi}{2}$.

2. Pole: By putting $r = 0$, if we find some real value of θ, then the curve passes through the pole which otherwise not. Further, putting $r = 0$, the real value of θ, if exists, gives the tangent to the curve at the pole.

3. Asymptotes: Find the asymptotes using the method to determine asymptotes of a polar curve.

4. Special Points on the Curve: Solve the equation of the curve for r and find how r varies as θ increases from 0 to ∞ and also as θ decreases from 0 to $-\infty$. Form a table with the corresponding values of r and θ. The points so obtained will help in tracing the curve.

5. Region: Find the region, where the curve does not exist. If r is imaginary in $\alpha < \theta < \beta$, then the curve does not exist in the region bounded by the lines $\theta = \alpha$ and $\theta = \beta$.

6. Value of tanϕ: Find $\tan\phi$, that is, $r\frac{d\theta}{dr}$, which will indicate the direction of the tangent at any point. If for $\theta = \alpha$, $\phi = 0$ then $\theta = \alpha$ will be tangent to the curve at the point $\theta = \alpha$. On the other hand if for $\theta = \alpha$, $\phi = \frac{\pi}{2}$, then at the point $\theta = \alpha$, the tangent will be perpendicular to the radius vector $\theta = \alpha$.

7. Cartesian Form of the Equation of the Curve: It is useful sometimes to convert the given equation from polar form to cartesian form using the relations $x = r\cos\theta$ and $y = r\sin\theta$.

EXAMPLE 4.34

Trace the curve $r = a\sin 3\theta$.

Solution. We note that

(i) The curve is not symmetrical about the initial line. But if we change θ to $-\theta$ and r to $-r$, then the equation of the curve remains unchanged. Therefore, the curve is symmetrical about the line $\theta = \frac{\pi}{2}$.

(ii) Putting $r = 0$, we get $\sin 3\theta = 0$. Thus, $3\theta = 0, \pi$ or $\theta = 0, \frac{\pi}{3}$. Thus, the curve passes through the pole, and the lines $\theta = 0$ and $\theta = \frac{\pi}{3}$ are tangents to the curve at the pole.

(iii) r is maximum when $\sin 3\theta = 1$ or $3\theta = \frac{\pi}{2}$ or $\theta = \frac{\pi}{6}$. The maximum value of r is a.

(iv) We have $\frac{dr}{d\theta} = 3a\cos 3\theta$ and so, $\tan\phi = r\frac{d\theta}{dr} = \frac{1}{3}\tan 3\theta$. Thus, $\phi = \frac{\pi}{2}$ when $3\theta = \frac{\pi}{2}$ or $\theta = \frac{\pi}{6}$, and the tangent is perpendicular to the radius vector $\theta = \frac{\pi}{6}$.

(v) Some points on the curve are given below:

θ :	0	$\frac{\pi}{6}$	$\frac{\pi}{3}$	$\frac{\pi}{2}$	$\frac{2\pi}{3}$	$\frac{5\pi}{6}$	π
r :	0	a	0	$-a$	0	a	0

One loop of the curve lies in the region $0 < \theta < \frac{\pi}{3}$. The second loop lies in the region $\frac{\pi}{3} < \theta < \frac{2\pi}{3}$ in the opposite direction because r is negative there. The third loop lies in the region $\frac{2\pi}{3} < \theta < \pi$ as r is positive (equal to a) there.

When θ increases from π to 2π, we get again the same branches of the curve. Hence, the shape of the curve is shown in the following figure:

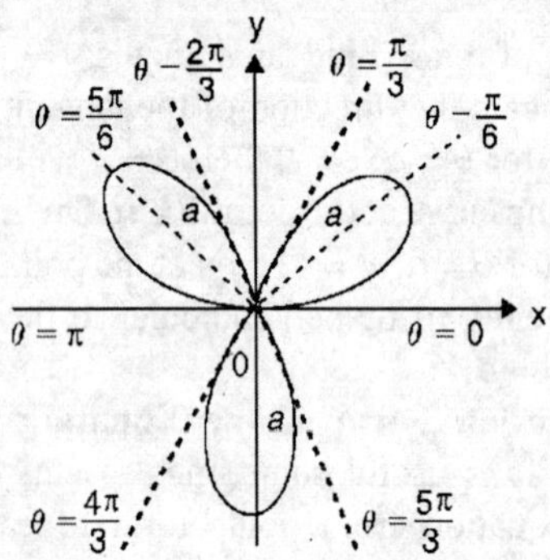

EXAMPLE 4.35

Trace the curve $r = a(1-\cos\theta)$ (Cardioid).

Solution. The equation of the given curve is $r = a(1-\cos\theta)$. We note the following characteristics of the curve:

(i) The equation of the curve remains unchanged when θ is changed to $-\theta$. Therefore, the curve is symmetrical about the initial line.

(ii) When $r = 0$, we have $1-\cos\theta = 0$ or $\theta = 0$. Hence, the curve passes through the pole and the line $\theta = 0$ is tangent to the curve at the pole.

(iii) The curve cuts the line $\theta = \pi$ at $(2a, \pi)$.

(iv) $\frac{dr}{d\theta} = a\sin\theta$ and so, $\tan\phi = r\frac{d\theta}{dr} = \frac{r}{a\sin\theta} = \frac{a(1-\cos\theta)}{a\sin\theta} = \tan\frac{\theta}{2}$. If $\frac{\theta}{2} = \frac{\pi}{2}$, then $\phi = 90^{\circ}$.

Thus, at the point $\theta = \pi$, the tangent to the curve is perpendicular to the radius vector.

(v) The values of θ and r are:

$$\theta : 0 \quad \frac{\pi}{3} \quad \frac{\pi}{2} \quad \frac{2\pi}{3} \quad \pi$$
$$r : 0 \quad \frac{a}{3} \quad a \quad \frac{3a}{3} \quad 2a$$

We observe that as θ increases from 0 to π, r increases from 0 to $2a$. Further, r is never greater than $2a$. Hence, no portion of the curve lies to the left of the tangent at $(2a, 0)$. Since $|r| \le 2a$, the curve lies entirely within the circle $r = 2a$.

(vi) There is no asymptote to the curve because for any finite value of θ, r does not tend to infinity.

Hence, the shape of the curve is an shown in the following figure:

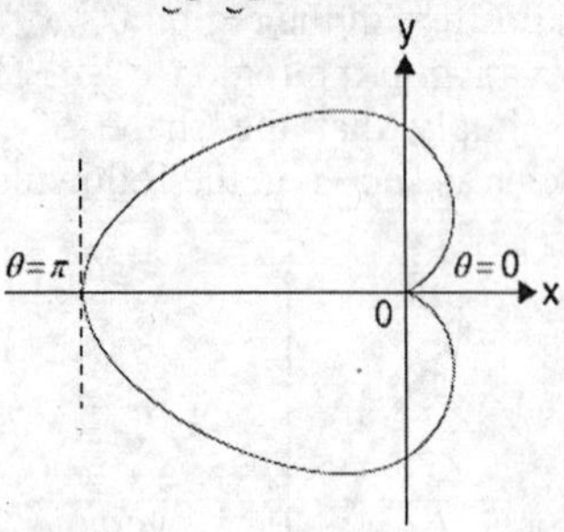

EXAMPLE 4.36

Trace the curve $r = a + b\cos\theta$, $a < b$ (Limacon).

Solution. The given curve has the following characteristics:

(i) Since the equation of the curve remains unaltered when θ is changed to $-\theta$, it follows that the curve is symmetrical about the initial line.

(ii) $r = 0$ when $a + b\cos\theta = 0$ or $\theta = \cos^{-1}\left(-\frac{a}{b}\right)$. Since $\frac{a}{b} < 1, \cos^{-1}\left(-\frac{a}{b}\right)$ is real. Therefore, the curve passes through the pole and the radius vector $\theta = \cos^{-1}\left(-\frac{a}{b}\right)$ is tangent to the curve at the pole.

(iii) We note that r is maximum when $\cos\theta = 1$, that is when $\theta = 0$. Thus, the maximum value of r is $a + b$. Thus, the entire curve lies within the circle $r = a + b$. Similarly, r is minimum when $\cos\theta = -1$, that is when $\theta = \pi$. Thus, the mimimum value of r is $a - b$, which is negative.

(iv) $\frac{dr}{d\theta} = -b\sin\theta$ and so, $\tan\phi = r\frac{d\theta}{dr} = -\frac{r}{b\sin\theta} = -\frac{a+b\cos\theta}{b\sin\theta}$. Thus, $\phi = 90^{\circ}$ when $\theta = 0, \pi$. Hence, at the points $\theta = 0$ and $\theta = \pi$, the tangent is perpendicular to the radius vector.

(v) The following table gives the value of r corresponding to the value of θ:

θ: 0 $\frac{\pi}{2}$ $\cos^{-1}\left(-\frac{a}{b}\right)$ $\cos^{-1}\left(-\frac{a}{b}\right) < \theta < \pi$ θ

r: $a + b$ a 0 negative $a - b$

(vi) Since r is not infinite for any value of θ, the given curve has no asymptote:

Hence, the shape of the curve is as shown in the following figure:

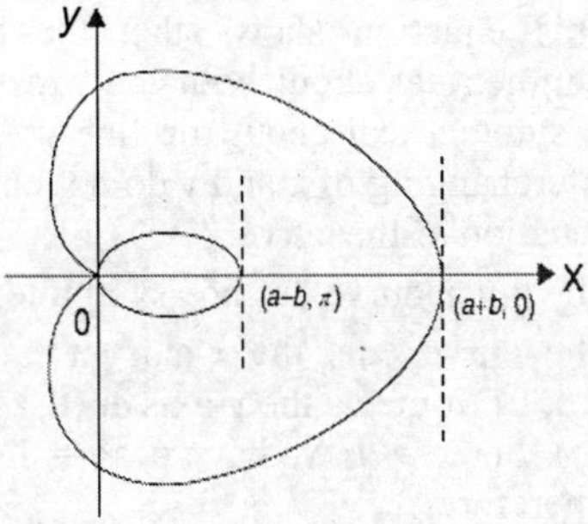

EXAMPLE 4.37

Trace the curve $r^2 \cos 2\theta = a^2$.

Solution. The equation of the given curve can be written as

$$r^2 (\cos^2\theta - \sin^2\theta) = a^2$$

or

$$x^2 - y^2 = a^2 \text{ since } x = r\cos\theta, y = r\sin\theta.$$

Therefore, the given curve is a rectangular hyperbola. We note that

(i) The curve is symmetrical about both the axes.

(ii) It does not pass through the origin.

(iii) It cuts the x-axis at $(a, 0)$ and $(-a, 0)$. But it does not meet y-axis.

(iv) Shifting the origin to $(a, 0)$, we get

$(x + a)^2 - y^2 = a^2$ or $x^2 - y^2 + 2ax = 0$.

Therefore, the tangent at $(a, 0)$ is given by $2ax = 0$ and so, the tangent at $(a, 0)$ is $x = 0$, the line parallel to the y-axis.

(v) The curve has no asymptote parallel to coordinate axes. The oblique asymptote (verify) are $y = x$ and $y = -x$.

(vi) The equation of the curve can be written as $y^2 = x^2 - a^2$. When $0 < x < a$, the y^2 is negative and so, y is imaginary. Therefore, the curve does not lie in the region $0 < x < a$. But when $x > a$, y^2 is positive and so, y is real. Thus, the curve exists in the region $x > a$. Further, when $x \to \infty$, $y^2 \to \infty$.

In view of the mentioned facts, the shape of the curve is as shown in the following figure:

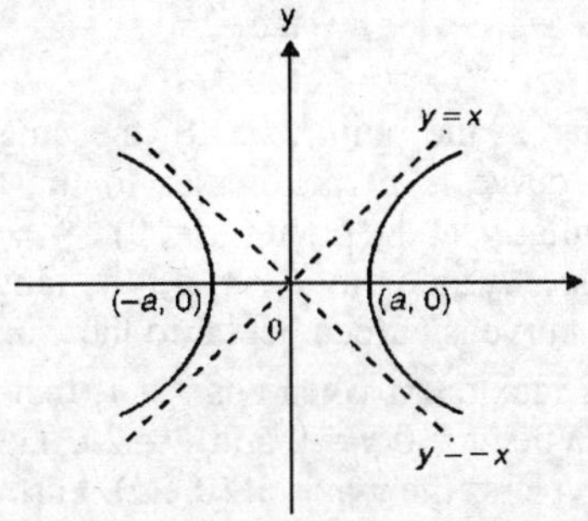

4.12 CURVE TRACING (PARAMETRIC EQUATIONS)

If the equation of the curve is given in a parametric form, $x = f(t)$ and $y = \phi(t)$, then eliminate the parameter and obtain a cartesian equation of the curve. Then, trace the curve as dealt with in case of cartesian equations.

In case the parameter is not eliminated easily, a series of values are given to t and the corresponding values of x, y, and $\frac{dy}{dx}$ are found. Then we plot the different points and find the slope of the tangents at these points by the values of $\frac{dy}{dx}$ at the points.

EXAMPLE 4.38

Trace the curve $x = a(t + \sin t)$ $y = a(1 + \cos t)$.

Solution. We note that

(i) Since $y = a(1 + \cos t)$ is an even function of t, the curve is symmetrical about the y-axis.

(ii) We have $y = 0$ when $\cos t = -1$, that is when $t = -\pi, \pi$. When $t = \pi$, we have $x = a\pi$. When $t = -\pi$, $x = -a\pi$. Thus, the curve meets the x-axis at $(a\pi, 0)$ and $(-a\pi, 0)$.

(iii) Differentiating the given equation, we get

$$\frac{dx}{dt} = a(1 + \cos t), \frac{dy}{dt} = -a\sin t.$$

Therefore,

$$\frac{dy}{dx} = \frac{\frac{dy}{dt}}{\frac{dx}{dt}} = -\frac{a\sin t}{a(1 + \cos t)}$$

$$= -\frac{2a \sin\frac{t}{2} \cos\frac{t}{2}}{2a\cos^2\frac{t}{2}} = -\tan\frac{t}{2}.$$

Now

$$\left(\frac{dy}{dx}\right)_{t=x} = -\tan\frac{\pi}{2} = -\infty$$

Thus, at the point $(a\pi, 0)$, the tangent to the curve is perpendicular to the x-axis. Similarly, at the point $(-a\pi, 0)$, $\frac{dy}{dt} = \infty$ and hence, at the point $(-a\pi, 0)$, the tangent to the curve is perpendicular to the x-axis.

(iv) y is maximum when $\cos t = 1$, that is, $t = 0$. When $t = 0$ $x = 0$ and $y = 2a$. Thus, the curve cuts the y-axis at $(0, 2a)$. Further,

$$\left(\frac{dy}{dx}\right)_{t=0} = 0$$

and so, at the point $(0, 2a)$, the tangent to the curve is parallel to the x-axis.

(v) It is clear from the equation that y cannot be negative. Further, no portion of the curve lies in the region $y > 2a$.

(vi) There is no asymptote parallel to the axes.

(vii) The values of x, y corresponding to the values of t are as follows:

t	$-\pi$	$-\frac{\pi}{2}$	0	$\frac{\pi}{2}$	π
x	$-a\pi$	$-a\left(\frac{\pi}{2}+1\right)$	0	$a\left(\frac{\pi}{2}+1\right)$	$a\pi$
y	0	a	$2a$	a	0

Hence, the shape of the curve is as shown in the following figure:

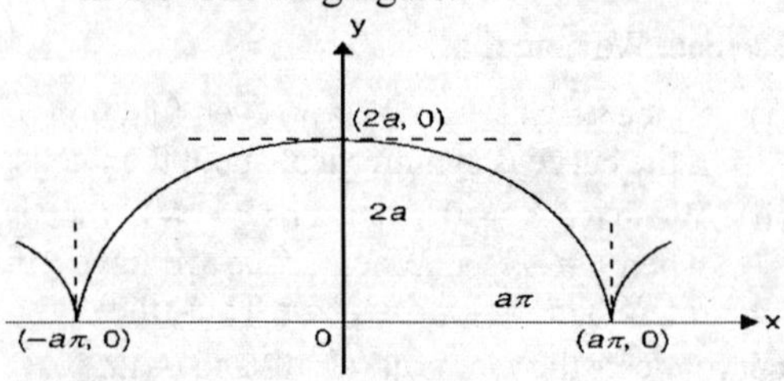

EXAMPLE 4.39

Trace the curve $x^{\frac{2}{3}} + y^{\frac{2}{3}} = a^{\frac{2}{3}}$.

Solution.

(i) The parametric equation of the curve are $x = a\cos^3 t$, $y = a\sin^3 t$.

Therefore,

$$|x| \leq a \text{ and } y \leq \text{a}.$$

This implies that the curve lies within the square bounded by the lines $x = \pm a$, $y = \pm a$.

(ii) The equation of the curve can be written as

$$\left(\frac{x^2}{a^2}\right)^{\frac{1}{3}} + \left(\frac{y^2}{a^2}\right)^{\frac{1}{3}} = 1.$$

This equation shows that the curve is symmetrical about both the axes. Also it is symmetrical about the line $y = x$ since interchanging of x and y do not change the equation of the curve.

(iii) The given curve has no asymptote.

(iv) The curve cuts the x-axis at $(a, 0)$ and $(-a, 0)$. It meets the y-axis at $(0, a)$ and $(0, -a)$. For $x = a$, we have $\cos^3 t = 1$ or $t = 0$. Therefore,

$$\left(\frac{dy}{dx}\right)_{t=0} = \left(\frac{\frac{dy}{dt}}{\frac{dx}{dt}}\right)_{t=0} = (-\tan t)_{t=0} = 0.$$

Hence, at the point $(a, 0)$, the x-axis is the tangent to the curve.

Similarly, at the point $(0,a)$, the y-axis is the tangent to the curve.

Hence, the shape of the curve is as shown in the following figure.

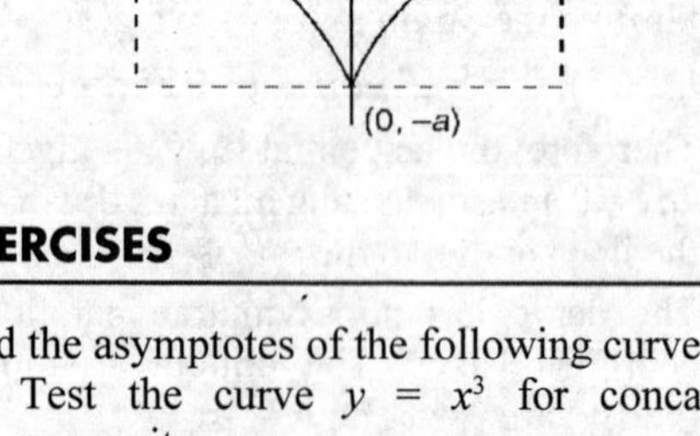

EXERCISES

Find the asymptotes of the following curves:

1. Test the curve $y = x^3$ for concavity/convexity.

   Ans. Concave for $x > 0$ convex for $x < 0$.

2. Find the points of inflexion on the curve $y(a^2 + x^2) = a^2x$.

   Ans. $(0,0)$, $(0,0)$, $\left(\sqrt{3}a, \frac{\sqrt{3}a}{4}\right)$, $\left(-\sqrt{3}a, \frac{\sqrt{3}a}{4}\right)$.

3. Show that the points of inflexion on the curve $y^2 = (x - a)^2 (x - b)$ lie on the line $3x + a = 4b$.

Hint: $\frac{d^2y}{dx^2} = 0$ yields $3x+a = 4b$

4. Find the points of inflexion on the curve x = a(2θ – sinθ), y = a(2–cosθ).

Ans. $\left[a\left(4n\pi \pm \frac{2\pi}{3} \mp \frac{\sqrt{3}}{2}\right), \frac{3a}{2}\right]$

5. Show that origin is a node on the curve $\frac{a^2}{x^2} - \frac{b^2}{y^2} = 1$.

Find the asymptotes of the following curves:

6. $y^3 + x^2y + 2xy^2 - y+1=0$

Ans. $y = 0, y + x-1 = 0, y + x + 1 = 0$.

7. $x^3 + y^3 - 3axy = 0$

Ans. $x + y + a = 0$

8. $x^3 - 2y^3 + 2x^2y - xy^2 + xy - y^2 + 1 = 0$

Ans. $x-y = 0, x + y + 1 = 0, x + 2y -1 = 0$

9. $3x^3 + y -7xy^2 + 2y^3 - 14xy+7y^2+ 4x + 5y = 0$

Ans. $6x-6y-7 = 0$, $y = 3x -1$, $3x + 6y + 5 = 0$

10. $y^3 = x^3 + ax^2$

Ans. $y = x + \frac{a}{3}$

11. $y^3-3x^2y + xy^2 - 3x^3 + 2y^2 + 2xy + 4x + 5y + 6 = 0$

Ans. $y = x - 2, y = x\sqrt{3} - 1, y = -x\sqrt{3} - 1$

12. $x^2y^2 = a^2(x^2 + y^2)$

Ans. $x = \pm a, y = \pm a$

13. $x^2y^3 + x^3y^2 = x^3 + y^3$

Ans. $x = \pm 1, y = \pm 1, y = -x$

14. $x^3 + 2x^2y + xy^2 - x^2 - xy + 2 = 0$

Ans. $x = 0$, $x+ y = 0$, $x+ y-1= 0$

15. $x^3 + x^2y -xy^2- y^3- 3x - y - 1 = 0$

Ans. $y = x, y = -x+1, x + y + 1 = 0$

16. $(x^2 - y^2)(x +2y + 1) + x + y + 1 = 0$

Ans. $x + y = 0, x - y = 0, x + 2y + 1 = 0$

17. (i) $y^2 (x - 2) = x^2(y - 1)$

Ans. $x = 2, y = 1, y = x + 1$

(ii) $x = \frac{a(t+t^3)}{1+t^4}, y = \frac{a(t-t^3)}{1+t^4}$.

Hint: Eliminating t, we get $(x^2 + y^2)^2 = a^2(x^2-y^2)$

Ans. No asymptote.

18. Show that the asymptotes of the curve $x^2y^2 - a^2(x^2+y^2)-a^3(x + y) + a^4 = 0$ form a square and that the curve passes through two angular points of that square.

Hint: The four asymptotes are $x = \pm a$, $y = \pm a$. They form a square of length $2a$. The angular points are (a, a), $(a, -a)$, $(-a, a)$, and $(-a, -a)$. The curve passes through two angular points $(-a, a)$, and $(-a, -a)$.

19. Show that the points of intersection of the curve

$2y^3 - 2xy - 4xy^2 + 4x^3 - 14xy + 6y^2 + 4x^2 + 6y + 1 = 0$

and its asymptotes lie on the straight line $8x + 2y + 1 = 0$.

20. Show that the asymptotes of the cubic $x^3 - 2y^3 + xy(2x - y) + y(x - y) + 1 = 0$ cut the curve again in three points which lie on the straight line $x - y + 1 = 0$.

Hint: The asymptotes are $y = x$, $y = -x-1$, and $y = -\frac{1}{2}x + \frac{1}{2}$. Their joint equation is $x^2 - 2y^3 + 2x^2y -x y^2 + xy - y^2- x + y = 0$. Subtracting this equation from the equation of the curve, we get $x - y + 1 = 0$.

21. Show that the point of intersection of the curve $4(x^4 + y^4) - 17x^2y^2-4x(4y^2 - x^2) + 2(x^2 - 2) = 0$ and its asymptotes lie on the ellipse $x^2 + 4y^2 = 4$.

22. Find the equation of the hyperbola passing through the origin and having asymptotes $x + y -1 = 0$ and $x - y + 2 = 0$.

Hint: joint equation of asymptotes is $F_2 = (x - y -1)(x - y -2) = 0$. Equation of the curve is $F_n + F_{n-2} = 0$, that is, $F_2 + F_0 = 0$. Thus, F_0 is of a zero degree and so, is a constant. Thus, the equation of the curve is $(x + y - 1) (x - y + 2) + k = 0$. It passes through the origin. So $k = 2$. Hence, the curve is $(x + y -1) (x - y + 2) + 2 = 0$ or $x^2 - y^2 + x + 3y = 0$.

23. Find the asymptotes of the curve $xy(x^2 - y^2) + x^2 + y^2 + a^2$ and show that the eight points of intersection of the curve and its asymptotes lie on a circle with the origin at the center.

24. Find the equation of the cubic which has the same asymptotes as the curve $x^3 - 6x^2y + 11xy^3 + 4x + 5y + 7 = 0$ and which passes through the points (0,0,)(0,2), and (2,0).

Ans. The joint equation of the asymptote is $(x - y)(x - 2y)(x - 3y) = 0$.

The cubic is $x^3 - 6x^2y + 11xy^2 + -6y^3 - 4x + 24y = 0$.

25. Find the equation of the straight line on which lie the three points of intersection of the curve

$x^3 + 2x^2y - xy^2 - 4y^2 + 2xy + y - 1 = 0$ and its asymptotes.

Ans. $x + 3y = 1$.

26. Find the asymptotes of the following polar curves:

(i) $r\theta\cos\theta = a\cos2\theta$

Ans. $r\cos\theta = \frac{2a}{(2k+1)\pi}$

(ii) $r = a\,\mathrm{cosec}\theta + b$

Ans. $r\sin\theta = a$

(iii) $r = \frac{a}{\log\theta}$

Ans. $r\sin(\theta - 1) = a$

(iv) $r = a\sec\theta + b\tan\theta$

Ans. $r = \cos\theta = a + b,\ r\cos\theta = a - b$

(v) $r(1 - e^{\theta}) = a$

Ans. $r\sin\theta = -a$

(vi) $r = \frac{2a\theta}{\theta^2 - H}$

Ans. $r\sin\theta = -a$

27. Find the circular asymptote of the curves:

(i) $r = \frac{3\theta^2 + 2\theta + 1}{2\theta^2 + \theta + 1}$

Ans. $r = \frac{3}{2}$

(ii) $r = \frac{6\theta^2 + 5\theta - 1}{2\theta^2 - 3\theta + 7}$

Ans. $r = 3$

28. Trace the curve $x^2y^2 = a^2(x^2 + y^2)$

Ans.

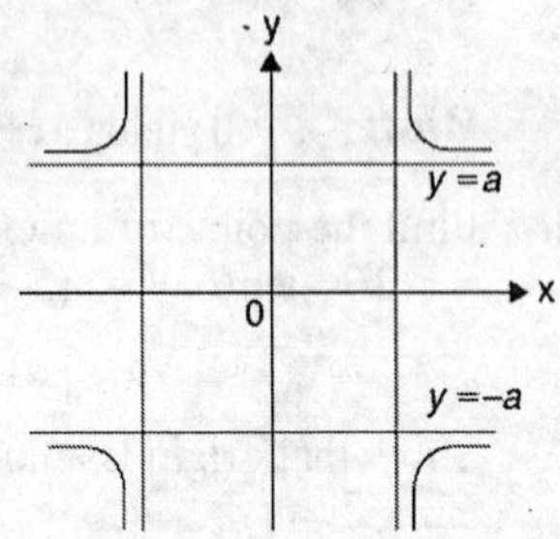

29. Trace the curve $y^2(x^2 + y^2) + a^2(x^2 - y^2) = 0$

Ans.

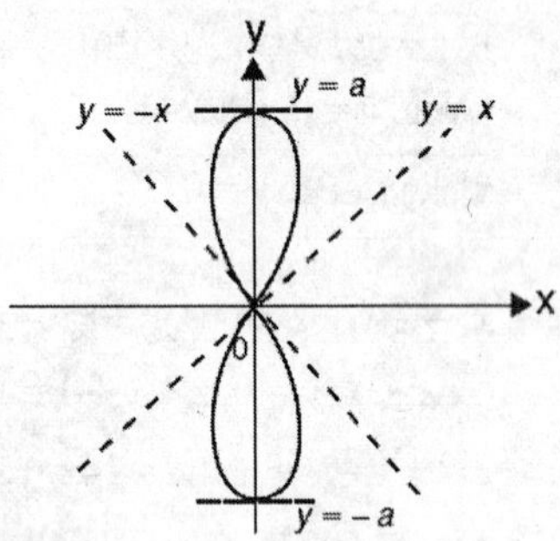

30. Trace the curve $ay^2 = x(x - a)^2$

Ans.

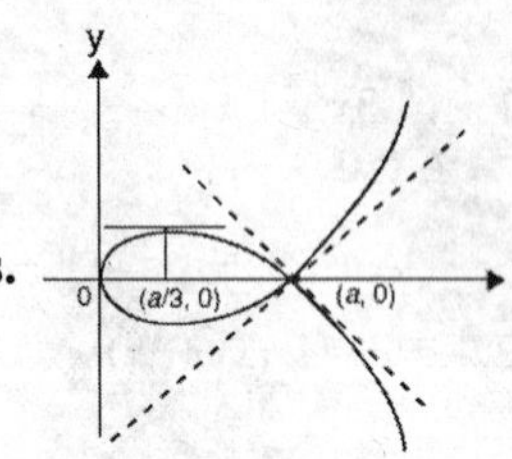

31. Trace the curve $xy = a^2(a - x)$

Ans:

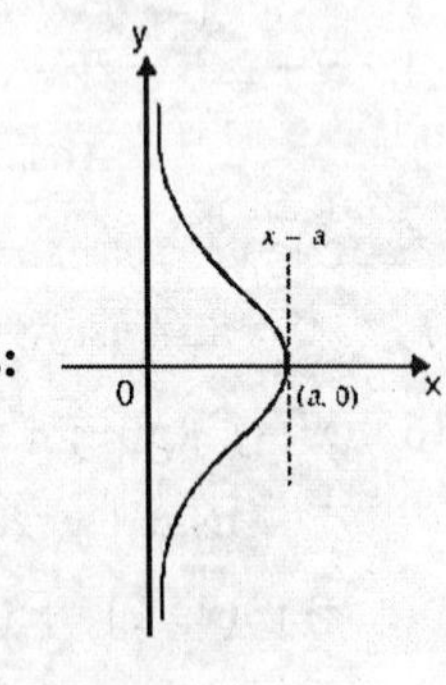

32. Trace the curve $y^2(a - x) = x^2(a + x)$

Ans:

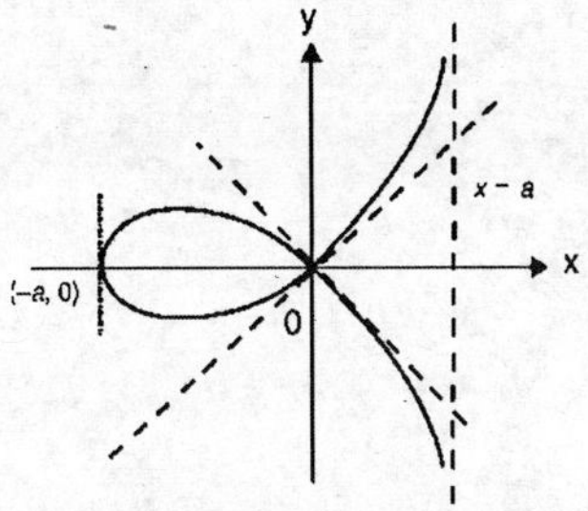

33. Trace the curve $r = a\,(\cos\theta + \sec\theta)$

Hint: $r^2 = ar\,(\cos\theta + \sec\theta)$. Therefore, cartesian form is $y^2(x - a) = x^2(2a - x)$

Ans:

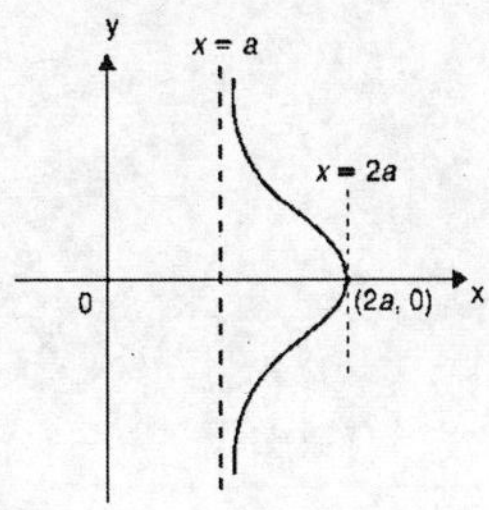

34. Trace the curve $r = \frac{a\sin^2\theta}{\cos\theta}$.

Hint: $r^2 = \frac{ar^2\sin^2\theta}{r_3\cos\theta} = \frac{ay^2}{x}$ or $x^2 + y^2 = \frac{ay^2}{x}$ or $y^2\,(a - x) = x$

Ans:

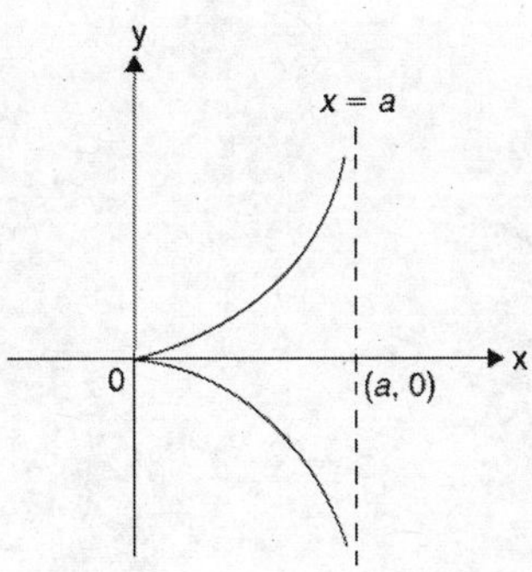

35. Trace the curve $r = a\cos 2\theta$.

Ans:

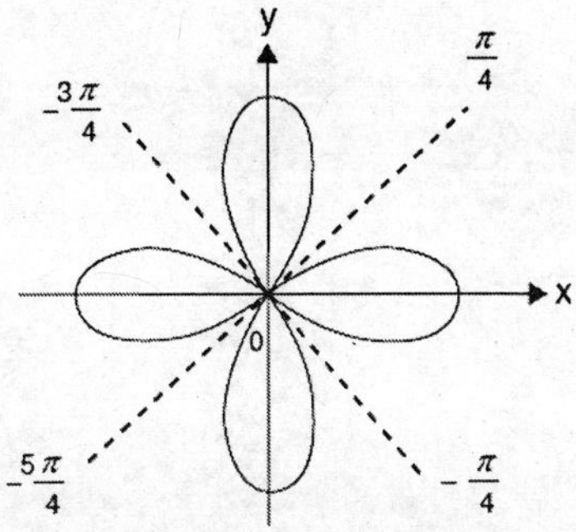

36. Trace the curve x = a(θ –sinθ), $y = a\,(1 + \cos\theta)$.

Ans:

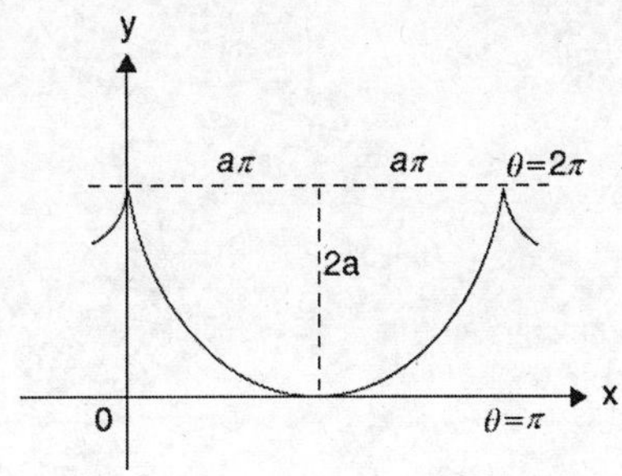

5 Functions of Several Variables

Let n be a positive integer and $\Re$ be the set of real numbers. Then, $\Re^n$ is the set of all n-tuples $(x_1, x_2,\ldots, x_n)$, $x_n \in \Re$. Thus,

$\Re$ is the set of all real numbers called, the *real line*,

$\Re^2 = \Re \times \Re = \{(x, y): x, y\ \Re\}$ is a *two-dimensional Cartesian plane*,

$\Re^3 = \Re \times \Re \times \Re = \{(x, y, z): x, y, z\ \Re\}$ is a *three-dimensional Euclidean space*.

Let A be a nonempty subset of $\Re^n$. Then, a function $f: A \to \Re$ is called a *real-valued function of* n *variables* defined on the set A. Thus, f maps $(x_1, x_2, \ldots, x_n)$, $x_i \in \Re$ into a unique real number $f(x_1, x_2, \ldots, x_n)$.

A function f of n variables $x_1, x_2,\ldots, x_n$ is said to tend to a limit λ as $(x_1, x_2,\ldots, x_n) \to (a_n, a_2,\ldots, a_n)$ if given $\varepsilon > 0$, however small, there exists a real number $\delta > 0$ such that

$$|f(x_1, x_2, \ldots, x_n) - \lambda| < \varepsilon \text{ whenever}$$

$$|(x_1, x_2, \ldots, x_n) - (a_1, a_2, \ldots, a_n)| < \delta$$

or

$$|f(x_1, x_2, \ldots, x_n) - \lambda| < \varepsilon,$$

whenever

$$0 < \sqrt{(x_1 - a_1)^2 + (x_2 - a_2)^2 + \cdots + (x_n - a_n)^2} < \delta.$$

In this case, we write

$$\lim_{(x_1, x_2,\ldots, x_n) \to (a_1, a_2,\ldots, a_n)} f(x_1, x_2, \ldots, x_n) = \lambda.$$

EXAMPLE 5.1

Show that $\lim\limits_{\substack{x \to 0 \\ y \to 0}} \dfrac{xy}{x^2 + y^2}$ does not exist.

Solution. We want to evaluate

$$\lim_{(x, y)\to(0, 0)} \frac{xy}{x^2 + y^2}.$$

If we put $y = mx$, then

$$\lim_{(x, y) \to (0, 0)} \frac{xy}{x^2 + y^2} = \lim_{x \to 0} \frac{mx^2}{x^2 + m^2x^2} = \lim_{x \to 0} \frac{m}{1 + m^2}.$$

Thus the limit depends upon the value of m and so is not unique for various path followed from (x, y) to $(0, 0)$. Hence the limit does not exist.

EXAMPLE 5.2

Show that

$$\lim_{(x,y) \to (0, 0)} xy\, \frac{x^2 - y^2}{x^2 + y^2} = 0.$$

Solution. Putting $x = r \cos \theta$, $y = r \sin \theta$, we have

$$\left| xy\, \frac{x^2 - y^2}{x^2 + y^2} - 0 \right| = |r^2 \sin \theta \cos \theta \cos 2\theta| = \left| \frac{r^2}{4} \sin 4\theta \right| \le \frac{r^2}{4} = \frac{x^2 + y^2}{4}.$$

If we choose $\dfrac{x^2}{4} < \dfrac{\varepsilon}{2}$ and $\dfrac{y^2}{4} < \dfrac{\varepsilon}{2}$ or $x < \sqrt{2\varepsilon}$, $y < \sqrt{2\varepsilon}$ and take $\delta = \sqrt{2\varepsilon}$ then

$$\left| xy \frac{x^2 y^2}{x^2 + y^2} - 0 \right| < \varepsilon \text{ whenever } |x - 0| < \delta,\ |y - 0| < \delta.$$

Hence for $\varepsilon > 0$, there exists $\delta > 0$ such that

$$\left| xy\, \frac{x^2 - y^2}{x^2 + y^2} - 0 \right| < \varepsilon \text{ whenever}$$

$$|x - 0| < \delta,\ |y - 0| < \delta.$$

Hence

$$\lim_{(x, y) \to (0, 0)} xy\, \frac{x^2 - y^2}{x^2 + y^2} = 0.$$

In what follows, we shall generally deal with functions of two variables.

5.1 CONTINUITY OF A FUNCTION OF TWO VARIABLES

A function $f(x, y)$ is said to be *continuous* at the point (a, b) of its domain if for every $\varepsilon > 0$ there exists a positive number δ such that

$|f(x, y) - f(a, b)| < \varepsilon$ whenever
$$|x-a| < \delta, \quad |y-b| < \delta,$$

Thus, $f(x, y)$ is continuous at a point(a, b) if
$$\lim_{(x,y)\to(a,b)} f(x, y) = f(a, b).$$

EXAMPLE 5.3

Investigate the continuity of the following functions at the origin.

$$\text{(i) } f(x, y) = \begin{cases} \dfrac{xy}{\sqrt{x^2+y^2}} & \text{if } (x, y) \neq (0, 0) \\ 0 & \text{if } (x, y) = (0, 0) \end{cases}$$

$$\text{(ii) } f(x, y) = \begin{cases} \dfrac{x^2-y^2}{x^2+y^2} & \text{if } (x, y) \neq (0, 0) \\ 0 & \text{if } (x, y) = (0, 0). \end{cases}$$

Solution. Putting $x = r\cos\theta$, $y = r\sin\theta$, we have

$$|f(x, y) - f(0, 0)| = \left|\frac{xy}{\sqrt{x^2+y^2}} - 0\right|$$
$$= r|\cos\theta\sin\theta| \leq r = x^2 + y^2.$$

If we choose $x^2 < \dfrac{\varepsilon^2}{2}, y^2 < \dfrac{\varepsilon^2}{2}$ or $x < \dfrac{\varepsilon}{\sqrt{2}}, y < \dfrac{\varepsilon}{\sqrt{2}}$ and take $\delta = \dfrac{\varepsilon}{\sqrt{2}}$, then

$$|f(x, y) - (0, 0)| < \varepsilon^2 \text{ whenever}$$
$$|x - 0| < \frac{\varepsilon}{\sqrt{2}} \text{ and } |y - 0| < \frac{\varepsilon}{\sqrt{2}}.$$

Hence the given function is continuous at the origin

(ii) Putting $y = mx$, we get

$$f(x, y) = \frac{x^2-y^2}{x^2+y^2} = \frac{x^2-m^2x^2}{x^2+m^2x^2} = \frac{1-m^2}{1+m^2}.$$

If we take $m \neq 1$, then

$$\lim_{(x,y)\to(0,0)} f(x, y) \neq 0.$$

But $f(0, 0) = 0$ (given). Hence $\lim_{(x,y)\to(0,0)} f(x, y) \neq f(0, 0)$, which implies that f is not continuous at $(0, 0)$.

5.2 DIFFERENTIABILITY OF A FUNCTION OF TWO VARIABLES

Let $u(x, y)$ be a function of two variables and let Δx and Δy be the increments given to x and y, respectively. A function $u(x, y)$ is said to be *differentiable* at the point (x, y) if it possesses a determinate value in the neighborhood of this point and if
$$\Delta u = A\Delta x + B\Delta y + \varepsilon\rho,$$
where $\rho = |\Delta x| + |\Delta y|$, $\varepsilon \to 0$ as $\rho \to 0$ and A, B are independent of Δx and Δy.

In the preceding definition, ρ may be replaced by η, where $\eta = \sqrt{(\Delta x)^2 + (\Delta y)^2}$.

If the increment ratio
$$\frac{u(x+\Delta x, y) - u(x, y)}{\Delta x}$$
tends to a unique limit as $\Delta x \to 0$, then that limit is called the *partial derivative of u with respect to* x and is written as $\dfrac{\partial u}{\partial x}$ or u_x.

Similarly $\dfrac{\partial u}{\partial y}$ can be defined. Thus, if a derivative of a function of several independent variables is found with respect to any one of the independent variables, treating the other as constant, it is said to be a partial derivative of the function with respect to that variable.

5.3 THE DIFFERENTIAL COEFFICIENTS

If in the equation
$$\Delta u = A\,\Delta x + B\,\Delta y + \varepsilon\rho,$$
we suppose that $\Delta y = 0$, then on the assumption that u is differentiable at the point (x, y), we have
$$\Delta u = u(x + \Delta x, y) - u(x, y) = A\,\Delta x + \varepsilon|\Delta x|.$$
Division by Δx yields
$$\frac{u(x+\Delta x, y) - u(x, y)}{\Delta x} = A \pm \varepsilon.$$
Taking the limit as $\Delta x \to 0$, we get
$$\frac{\partial u}{\partial x} = A, \text{ since } \varepsilon \to 0 \text{ as } \Delta x \to 0.$$
Similarly, we can show that $\dfrac{\partial u}{\partial y} = B$. Thus, if $u = u(x, y)$ is differentiable, the partial derivatives $\dfrac{\partial u}{\partial x}$ and $\dfrac{\partial u}{\partial y}$ are, respectively the differential coefficients A and B. Hence,
$$\Delta u = \frac{\partial u}{\partial x}\Delta x + \frac{\partial u}{\partial y}\Delta y + \varepsilon\rho.$$

The differential of the dependent variable du is defined to be the principal part of Δu. Hence,

$$\Delta u = du + \varepsilon\rho.$$

Now as in the case of functions of one variable, the differentials of the independent variables are identical with the arbitrary increments of these variables, that is, $dx = \Delta x$, $dy = \Delta y$. Therefore,

$$du = \frac{\partial u}{\partial x}dx + \frac{\partial u}{\partial y}dy.$$

5.4 DISTINCTION BETWEEN DERIVATIVES AND DIFFERENTIAL COEFFICIENTS

We know that the necessary and sufficient condition for a function $y = f(x)$ to be differentiable at a point x is that it possesses a finite derivative at that point. Thus, for functions of one variable, the existence of derivative $f'(x)$ implies the differentiability of f at that point. But for a function of more than one variable this is not true. We have seen earlier that if $f(x, y)$ is differentiable at (x, y), then the partial derivative of f with respect to x and y exist and are equal to the differential coefficients A and B, respectively. However, the partial derivative may exist at a point when the function is not differentiable at that point. In other words, the *partial derivatives need not always be the differential coefficients.*

EXAMPLE 5.4

Show that the function $f(x, y) = \dfrac{x^3 - y^3}{x^2 + y^2}$, where x and y are not simultaneously zero and $f(0, 0) = 0$, is not differentiable at (0, 0) but the partial derivatives at (0, 0) exist.

Solution. Suppose that the given function is differentiable at the origin. Then, by definition,

$$f(h, k) - f(0, 0) = Ah + Bk + \varepsilon\eta, \quad (1)$$

where $\eta = \sqrt{h^2 + k^2}$ and $\varepsilon \to 0$ as $\eta \to 0$. Putting $h = \eta\cos\theta$, $k = \eta\sin\theta$ in (1) and dividing throughout by η, we get

$$\cos^3\theta - \sin^3\theta = A\cos\theta + B\sin\theta + \varepsilon.$$

Since $\varepsilon \to 0$ as $\eta \to 0$, we take the limit as $\eta \to 0$ and get

$$\cos^2\theta - \sin^3\theta = A\cos\theta + b\sin\theta,$$

which is impossible since θ is arbitrary. Hence, the function is *not differentiable* at (0, 0). On the other hand,

$$f_x(0, 0) = \lim_{h\to 0}\frac{f(h, 0) - f(0, 0)}{h} = \lim_{h\to 0}\frac{h - 0}{h} = 1 \text{ and}$$

$$f_y(0, 0) = \lim_{k\to 0}\frac{f(0, k) - f(0, 0)}{k} = \lim_{k\to 0}\frac{-k - 0}{k} = -1.$$

Hence, the partial derivatives at (0, 0) exist.

5.5 HIGHER-ORDER PARTIAL DERIVATIVES

Partial derivatives are also, in general, functions of x and y which may possess partial derivatives with respect to both independent variables. Thus,

(i) $\dfrac{\partial}{\partial x}\left(\dfrac{\partial u}{\partial x}\right) = \lim\limits_{\Delta x\to 0}\dfrac{u_x(x + \Delta x, y) - u_x(x, y)}{\Delta x}$ and

(ii) $\dfrac{\partial}{\partial y}\left(\dfrac{\partial u}{\partial x}\right) = \lim\limits_{\Delta y\to 0}\dfrac{u_x(x, y + \Delta y) - u_x(x, y)}{\Delta y}$,

provided that each of these limits exist. The second-order partial derivatives are denoted by $\dfrac{\partial^2 u}{\partial x^2}$ or u_{xx} and $\dfrac{\partial^2 u}{\partial y\partial x}$ or u_{yx}. Similarly, we may define higher order partial derivatives of $\dfrac{\partial u}{\partial y}$.

EXAMPLE 5.5

If

$$f(x, y) = \frac{xy(x^2 - y^2)}{x^2 + y^2}, \; f(0, 0) = 0,$$

show that $f_{xy}(0, 0) \neq f_{yx}(0, 0)$.

Solution. When (x, y) is not the origin, then

$$\frac{\partial f}{\partial x} = y\left[\frac{x^2 - y^2}{x^2 + y^2} + \frac{4x^2y^2}{(x^2 + y^2)^2}\right], \quad (1)$$

$$\frac{\partial f}{\partial y} = x\left[\frac{x^2 - y^2}{x^2 + y^2} + \frac{4x^2y^2}{(x^2 + y^2)^2}\right]. \quad (2)$$

On the other hand,

$$f_x(0, 0) = \lim_{h\to 0}\frac{f(h, 0) - f(0, 0)}{h} = 0,$$

$$f_y(0, 0) = \lim_{k\to 0}\frac{f(0, k) - f(0, 0)}{k} = 0.$$

From (1) and (2), we have

$$f_x(0, y) = -y\ (y \neq 0) \text{ and } f_y(x, 0) = x\ (x \neq 0).$$

Therefore,

$$f_{xy}(0, 0) = \lim_{h\to 0} \frac{f_y(h, 0) - f_y(0, 0)}{h} = \lim_{h\to 0} \frac{h}{h} = 1.$$

and

$$f_{yx}(0, 0) = \lim_{k\to 0} \frac{f_x(0, k) - f_x(0, 0)}{k} = \lim_{k\to 0} \frac{-k}{k} = -1.$$

Hence,

$$f_{xy}(0, 0) \neq f_{yx}(0, 0).$$

The question arises: Under what conditions, $f_{xy}(a, b) = f_{yx}(a, b)$? The following theorems answer this question.

Theorem 5.1. (Young). If (i) f_x and f_y exist in the neighborhood of the point (a, b) and (ii) f_x and f_y are differentiable at (a, b), then $f_{xy}(a, b) = f_{yx}(a, b)$.

Theorem 5.2. (Schwarz). If (i) f_x, f_y, and f_{yx} all exist in the neighborhood of the point (a, b) and (ii) f_{yx} is continuous at (a, b), then f_{xy} also exists at (a, b) and $f_{xy}(a, b) = f_{yx}(a, b)$.

In this chapter, we shall assume in the examples involving f_{xy} and f_{yx} that the partial derivatives of the first two orders of the given function are continuous so that $f_{xy} = f_{yx}$.

EXAMPLE 5.6

If $z = x \log y$, show that

$$\frac{\partial^2 z}{\partial x \partial y} = \frac{\partial^2 z}{\partial y \partial x}.$$

Solution. The given function is $z = x \log y$. Differentiating z partially with respect to x, taking y as constant, we get $\frac{\partial z}{\partial x} = \log y$ and then,

$$\frac{\partial^2 z}{\partial y \partial x} = \frac{\partial}{\partial y}\left(\frac{\partial z}{\partial x}\right) = \frac{\partial}{\partial y}(\log y) = \frac{1}{y}.$$

Differentiating z with respect to y, taking x as constant, we get $\frac{\partial z}{\partial y} = \frac{x}{y}$ and so,

$$\frac{\partial^2 z}{\partial x \partial y} = \frac{\partial}{\partial x}\left(\frac{\partial z}{\partial y}\right) = \frac{\partial}{\partial x}\left(\frac{x}{y}\right) = \frac{1}{y}.$$

Hence,

$$\frac{\partial^2 z}{\partial y \partial x} = \frac{\partial^2 z}{\partial x \partial y} = \frac{1}{y}.$$

EXAMPLE 5.7

If $\frac{1}{u} = \sqrt{x^2 + y^2 + z^2}$, show that

$$\frac{\partial^2 u}{\partial x^2} + \frac{\partial^2 u}{\partial y^2} + \frac{\partial^2 u}{\partial z^2} = 0.$$

Solution. We are given that

$$u = (x^2 + y^2 + z^2)^{-\frac{1}{2}}.$$

Therefore,

$$\frac{\partial u}{\partial x} = -\frac{1}{2}(x^2 + y^2 + z^2)^{-\frac{3}{2}}(2x)$$
$$= -x(x^2 + y^2 + z^2)^{-\frac{3}{2}}.$$

Differentiating partially once again with respect to x, we get

$$\frac{\partial^2 u}{\partial x^2} = \frac{3x}{2}(x^2 + y^2 + z^2)^{-\frac{5}{2}}(2x) - (x^2 + y^2 + z^2)^{-\frac{3}{2}}$$
$$= (x^2 + y^2 + z^2)^{-\frac{5}{2}}[3x^2 - (x^2 + y^2 + z^2)]$$
$$= u^5(2x^2 - y^2 - z^2).$$

Similarly, by symmetry in the given function, we have

$$\frac{\partial^2 u}{\partial y^2} = u^5(2y^2 - z^2 - x^2),$$
$$\frac{\partial^2 u}{\partial z^2} = u^5(2z^2 - x^2 - y^2).$$

Adding, we get

$$\frac{\partial^2 u}{\partial x^2} + \frac{\partial^2 u}{\partial y^2} + \frac{\partial^2 u}{\partial z^2} = u^5(2x^2 - y^2 - z^2 + 2y^2 - z^2 - x^2 + 2z^2 - x^2 - y^2)$$
$$= u^5(0) = 0$$

EXAMPLE 5.8

If $x = r\cos\theta$, $y = r\sin\theta$, show that

$$\frac{\partial^2 \theta}{\partial x^2} + \frac{\partial^2 \theta}{\partial y^2} = 0, \quad x, y \neq 0.$$

Solution. We have

$$x = r\cos\theta, \ y = r\sin\theta.$$

Therefore,

$$r^2 = x^2 + y^2 \text{ and } \theta = \tan^{-1}\frac{y}{x}.$$

Differentiating partially with respect to x, we get

$$\frac{\partial \theta}{\partial x} = \frac{\partial}{\partial x}\left(\tan^{-1}\frac{y}{x}\right) = \frac{1}{1 + \left(\frac{y}{x}\right)^2}\left(-\frac{y}{x^2}\right) = -\frac{y}{x^2 + y^2}$$

and

$$\frac{\partial^2 \theta}{\partial x^2} = \frac{\partial}{\partial x}\left(\frac{-y}{x^2+y^2}\right) = \frac{2xy}{(x^2+y^2)^2}.$$

Similarly, differentiating partially with respect to y, we get

$$\frac{\partial \theta}{\partial y} = \frac{\partial}{\partial y}\left(\tan^{-1}\frac{y}{x}\right) = \frac{1}{1+\left(\frac{y}{x}\right)^2}\left(\frac{1}{x}\right)$$
$$= \frac{x}{x^2+y^2}$$

and

$$\frac{\partial^2 \theta}{\partial y^2} = \frac{\partial}{\partial y}\left(\frac{\partial \theta}{\partial y}\right) = \frac{-2xy}{(x^2+y^2)^2}.$$

Hence,

$$\frac{\partial^2 \theta}{\partial x^2} + \frac{\partial^2 y}{\partial y} = 0.$$

The result holds for $x \neq 0$, $y \neq 0$, otherwise $\frac{\partial^2 \theta}{\partial x^2}$ and $\frac{\partial^2 \theta}{\partial y^2}$ are of the form $\left(\frac{0}{0}\right)$.

EXAMPLE 5.9

If $u = \sin^{-1}\frac{x}{y} + \tan^{-1}\frac{y}{x}$, show that

$$x\frac{\partial u}{\partial x} + y\frac{\partial u}{\partial y} = 0.$$

Solution. We have

$$u = \sin^{-1}\frac{x}{y} + \tan^{-1}\frac{y}{x}.$$

Therefore, differentiating partially with respect to x, we get

$$\frac{\partial u}{\partial x} = \frac{1}{\sqrt{1-\left(\frac{x}{y}\right)^2}}\cdot\frac{1}{y} + \frac{1}{1+\left(\frac{y}{x}\right)^2}\left(-\frac{y}{x^2}\right)$$
$$= \frac{1}{\sqrt{y^2-x^2}} - \frac{y}{x^2+y^2}$$

and so,

$$x\frac{\partial u}{\partial x} = \frac{x}{\sqrt{y^2-x^2}} - \frac{xy}{x^2+y^2}. \qquad (1)$$

Now, differentiating partially with respect to y, we get

$$\frac{\partial u}{\partial y} = \frac{1}{\sqrt{1-\left(\frac{x}{y}\right)^2}}\left(\frac{-x}{y^2}\right) + \frac{1}{1+\left(\frac{y}{x}\right)^2}\left(\frac{1}{x}\right)$$
$$= \frac{-x}{y\sqrt{y^2-x^2}} + \frac{x}{x^2+y^2}$$

and so,

$$y\cdot\frac{\partial u}{\partial y} = \frac{-x}{\sqrt{y^2-x^2}} + \frac{xy}{x^2+y^2}. \qquad (2)$$

Adding (1) and (2), we have

$$x\frac{\partial u}{\partial x} + y\frac{\partial u}{\partial y} = 0.$$

EXAMPLE 5.10

If $u = x^2\tan^{-1}\frac{y}{x} - y^2\tan^{-1}\frac{x}{y}$, show that

$$\frac{\partial^2 u}{\partial x\partial y} = \frac{x^2-y^2}{x^2+y^2}.$$

Solution. Differentiating partially the given function with respect to y, we get

$$\frac{\partial u}{\partial y} = x^2\left[\frac{1}{1+\left(\frac{y}{x}\right)^2}\left(\frac{1}{x}\right)\right] - 2y\tan^{-1}\frac{x}{y}$$
$$-y^2\left[\frac{1}{1+\left(\frac{x}{y}\right)^2}\left(\frac{-x}{y^2}\right)\right]$$
$$= \frac{x^3}{x^2+y^2} - 2y\tan^{-1}\frac{x}{y} + \frac{xy^2}{x^2+y^2}$$
$$= x - 2y\tan^{-1}\frac{x}{y}.$$

Now, differentiating partially with respect to x, we get

$$\frac{\partial^2 u}{\partial x\partial y} = 1 - 2y\left[\frac{1}{1+\left(\frac{x}{y}\right)^2}\left(\frac{1}{y}\right)\right]$$
$$= 1 - \frac{2y^2}{x^2+y^2} = \frac{x^2-y^2}{x^2+y^2}.$$

EXAMPLE 5.11

If $u = \log(x^3+y^3+z^3-3xyz)$, show that

$$\left(\frac{\partial}{\partial x}+\frac{\partial}{\partial y}+\frac{\partial}{\partial z}\right)^2 u = -\frac{9}{(x+y+z)^2}.$$

Solution. We are given that

$$u = \log(x^3+y^3+z^3-3xyz).$$

Differentiating partially with respect to x, y, and z, we get

$$\frac{\partial u}{\partial x} = \frac{3x^2-3yz}{x^3+y^3+z^3-3xyz},$$
$$\frac{\partial u}{\partial y} = \frac{3y^2-3xz}{x^3+y^3+z^3-3xyz}, \text{ and}$$
$$\frac{\partial u}{\partial z} = \frac{3z^2-3xy}{x^3+y^3+z^3-3xyz}.$$

Therefore,
$$\left(\frac{\partial}{\partial x}+\frac{\partial}{\partial y}+\frac{\partial}{\partial z}\right)^2 u$$
$$=\left(\frac{\partial}{\partial x}+\frac{\partial}{\partial y}+\frac{\partial}{\partial z}\right)\left(\frac{\partial}{\partial x}+\frac{\partial}{\partial y}+\frac{\partial}{\partial z}\right)u$$
$$=\left(\frac{\partial}{\partial x}+\frac{\partial}{\partial y}+\frac{\partial}{\partial z}\right)\left(\frac{\partial u}{\partial x}+\frac{\partial u}{\partial y}+\frac{\partial u}{\partial z}\right)$$
$$=\left(\frac{\partial}{\partial x}+\frac{\partial}{\partial y}+\frac{\partial}{\partial z}\right)\left[\frac{3(x^2+y^2+z^2-yz-zx-xy)}{x^3+y^3+z^3-3xyz}\right]$$
$$=\left(\frac{\partial}{\partial x}+\frac{\partial}{\partial y}+\frac{\partial}{\partial z}\right)\left(\frac{3}{x+y+z}\right)$$
$$=3\left[\frac{\partial}{\partial x}\left(\frac{1}{x+y+z}\right)+\frac{\partial}{\partial y}\left(\frac{1}{x+y+z}\right)+\frac{\partial}{\partial z}\left(\frac{1}{x+y+z}\right)\right]$$
$$=3\left[\frac{-1}{(x+y+z)^2}-\frac{1}{(x+y+z)^2}-\frac{1}{(x+y+z)^2}\right]$$
$$=\frac{-9}{(x+y+z)^2}.$$

EXAMPLE 5.12

If $u=f(r)$, where $r^2=x^2+y^2$, show that
$$\frac{\partial^2 u}{\partial x^2}+\frac{\partial^2 u}{\partial y^2}=f''(r)+\frac{1}{r}f'(r).$$

Solution. We have $r^2=x^2+y^2$. Therefore, partial differentiation with respect to x and y yields
$$2r\frac{\partial r}{\partial x}=2x \quad \text{and} \quad 2r\frac{\partial r}{\partial y}=2y$$
or
$$\frac{\partial r}{\partial x}=\frac{x}{r} \quad \text{and} \quad \frac{\partial r}{\partial y}=\frac{y}{r}.$$
But $u=f(r)$. Therefore,
$$\frac{\partial u}{\partial x}=\frac{\partial u}{\partial r}\cdot\frac{\partial r}{\partial x}=f'(r)\frac{\partial r}{\partial x}=\frac{x}{r}f'(r),$$
$$\frac{\partial^2 u}{\partial x^2}=\frac{\partial}{\partial x}\left(\frac{x}{r}f'(r)\right)=\frac{1}{r}f'(r)$$
$$+xf'(r)\left(-\frac{1}{r^2}\frac{\partial r}{\partial x}\right)+\frac{x}{r}f''(r)\frac{\partial r}{\partial x}$$
$$=\frac{1}{r}f'(r)-\frac{x^2}{r^3}f'(r)+\frac{x^2}{r^2}f''(r). \quad (1)$$
Similarly, due to symmetry, we have
$$\frac{\partial^2 u}{\partial y^2}=\frac{1}{r}f'(r)-\frac{y^2}{r^3}f'(r)+\frac{y^2}{r^2}f''(r). \quad (2)$$
Adding (1) and (2), we get
$$\frac{\partial^2 u}{\partial x^2}+\frac{\partial^2 u}{\partial y^2}=\frac{2}{r}f'(r)-\frac{x^2+y^2}{r^3}f'(r)+\frac{x^2+y^2}{r^2}f''(r)$$
$$=\frac{2}{r}f'(r)-\frac{1}{r}f'(r)+f''(r)$$
$$=\frac{1}{r}f'(r)+f''(r).$$

EXAMPLE 5.13

If $z=f(x+ct)+\phi\,(x-ct)$, show that
$$\frac{\partial^2 z}{\partial t^2}=c^2\frac{\partial^2 z}{\partial x^2}.$$

Solution. We are given that
$$z=f(x+ct)+\phi\,(x-ct).$$
Therefore,
$$\frac{\partial z}{\partial x}=f'\,(x+ct)\cdot\frac{\partial}{\partial x}(x+xt)+\phi'\,(x-ct)\frac{\partial}{\partial x}(x-ct)$$
$$=f'\,(x+ct)+\phi'\,(x-ct)$$
and
$$\frac{\partial^2 z}{\partial x^2}=f''\,(x+ct)+\phi''\,(x-ct).$$
On the other hand,
$$\frac{\partial z}{\partial t}=f'\,(x+ct)\cdot\frac{\partial}{\partial t}(x+ct)+\phi'\,(x-ct)\frac{\partial}{\partial t}(x-ct)$$
$$=cf'\,(x+ct)-c\phi'\,(x-ct)$$
and so,
$$\frac{\partial^2 z}{\partial t^2}=c^2 f''\,(x+ct)+c^2\phi''\,(x-ct)$$
$$=c^2[f''\,(x+ct)+\phi''\,(x-ct)]=c^2\frac{\partial^2 z}{\partial x^2}.$$

EXAMPLE 5.14

If $x=r\cos\theta$, $y=\sin\theta$, show that
$$\frac{\partial^2 r}{\partial x^2}+\frac{\partial^2 r}{\partial y^2}=\frac{1}{r}\left[\left(\frac{\partial r}{\partial x}\right)^2+\left(\frac{\partial r}{\partial y}\right)^2\right].$$

Solution. If $x=r\cos\theta, y=\sin\theta$, then $r^2=x^2+y^2$. Therefore, partial differentiation with respect to x yields
$$\frac{\partial r}{\partial x}=\frac{x}{r},$$

$$\frac{\partial^2 r}{\partial x^2} = \frac{r - x\dfrac{\partial r}{\partial x}}{r^2} = \frac{r^2 - x^2}{r^3}$$
$$= \frac{x^2 + y^2 - x^2}{r^3} = \frac{y^2}{r^3}.$$

Similarly, partial differentiation with respect to y gives

$$\frac{\partial r}{\partial y} = \frac{y}{r},$$
$$\frac{\partial^2 r}{\partial y^2} = \frac{r - y\dfrac{\partial r}{\partial y}}{r^2} = \frac{r^2 - y^2}{r^3} = \frac{x^2 + y^2 - y^2}{r^3} = \frac{x^2}{r^3}.$$

Therefore,

$$\frac{\partial^2 r}{\partial x^2} + \frac{\partial^2 r}{\partial y^2} = \frac{x^2 + y^2}{r^3} = \frac{r^2}{r^3} = \frac{1}{r}.$$

On the other hand,

$$\frac{1}{r}\left[\left(\frac{\partial r}{\partial x}\right)^2 + \left(\frac{\partial r}{\partial y}\right)^2\right] = \frac{1}{r}\left[\frac{x^2}{r^2} + \frac{y^2}{r^2}\right]$$
$$= \frac{1}{r}\left(\frac{r^2}{r^2}\right) = \frac{1}{r}.$$

Hence,

$$\frac{\partial^2 r}{\partial x^2} + \frac{\partial^2 r}{\partial y^2} = \frac{1}{r}\left(\frac{\partial r}{\partial x}\right)^2 + \left(\frac{\partial r}{\partial y}\right)^2.$$

EXAMPLE 5.15

If $x^x y^y z^z = c$, show that at $x = y = z$,

$$\frac{\partial^2 z}{\partial x \partial y} = -[x \log ex]^{-1}.$$

Solution. We have $x^x y^y z^z = c$. Here z can be regarded as a function of x and y. Taking logarithm of both sides, we get

$$x \log x + y \log y + z \log z = \log c. \quad (1)$$

Since z is a function of x and y, differentiating (1) partially with respect to x (taking y as constant), we get

$$x \cdot \frac{1}{x} + \log x + \left[z\,\frac{1}{z} + \log z\right]\frac{\partial z}{\partial x} = 0$$

or

$$1 + \log x + (1 + \log z)\frac{\partial z}{\partial x} = 0.$$

Therefore,

$$\frac{\partial z}{\partial x} = -\frac{1 + \log x}{1 + \log z} \quad (2)$$

Similarly, differentiating (1) partially with respect to y, we get

$$1 + \log y + (1 + \log z)\frac{\partial z}{\partial y} = 0$$

or

$$\frac{\partial z}{\partial y} = -\frac{1 + \log y}{1 + \log z}. \quad (3)$$

Differentiating partially (3) with respect to x, we get

$$\frac{\partial^2 z}{\partial x \partial y} = \frac{\partial}{\partial x}\left(-\frac{1 + \log y}{1 + \log z}\right)$$
$$= -(1 + \log y)\,\frac{\partial}{\partial x}\left(\frac{1}{1 + \log z}\right)$$
$$= -(1 + \log y)\left[-(1 + \log z)^{-2} \cdot \frac{1}{z} \cdot \frac{\partial z}{\partial x}\right]$$
$$= \frac{1 + \log y}{z\,(1 + \log z)^2}\left[-\frac{1 + \log x}{1 + \log z}\right], \text{ using (2).}$$

Taking $x = y = z$, we get

$$\frac{\partial^2 z}{\partial x \partial y} = -\frac{(1 + \log x)^2}{x\,(1 + \log x)^3} = -\frac{1}{x\,(1 + \log x)}$$
$$= -\frac{1}{x\,(\log e + \log x)} = -\frac{1}{x \log ex} = -[x \log ex]^{-1}.$$

EXAMPLE 5.16

If $\dfrac{x^2}{a^2 + u} + \dfrac{y^2}{b^2 + u} + \dfrac{z^2}{c^2 + u} = 1$, show that

$$\left(\frac{\partial u}{\partial x}\right)^2 + \left(\frac{\partial u}{\partial y}\right)^2 + \left(\frac{\partial u}{\partial z}\right)^2 = 2\left(x\frac{\partial u}{\partial x} + y\frac{\partial u}{\partial y} + z\frac{\partial u}{\partial z}\right).$$

Solution. Here u is a function of three variables x, y, and z. Differentiating the given equation partially with respect to x, we get

$$(a^2 + u)2x - \frac{x^2}{(a^2 + u)^2} \cdot \frac{\partial u}{\partial x} - \frac{y^2}{(b^2 + u)^2} \cdot \frac{\partial u}{\partial x}$$
$$- \frac{z^2}{(c^2 + u)^2} \cdot \frac{\partial u}{\partial x} = 0$$

or

$$\frac{2x}{a^2 + u} - \left[\frac{x^2}{(a^2 + u)^2} + \frac{y^2}{(b^2 + u)^2} + \frac{z^2}{(c^2 + u)^2}\right]\frac{\partial u}{\partial x} = 0$$

or

$$\frac{\partial u}{\partial x} = \frac{2x}{(a^2 + u)\left[\dfrac{x^2}{(a^2 + u)^2} + \dfrac{y^2}{(b^2 + u)^2} + \dfrac{z^2}{(c^2 + u)^2}\right]}.$$

Thus,

$$\left(\frac{\partial u}{\partial x}\right)^2$$

$$= \frac{4x^2}{(a^2+u)^2\left[\frac{x^2}{(a^2+u)^2}+\frac{y^2}{(b^2+u)^2}+\frac{z^2}{(c^2+u)^2}\right]^2}$$

$$= \frac{4x^2}{(a^2+u)^2\left[\sum \frac{x^2}{(a^2+u)^2}\right]^2}. \quad (1)$$

Similarly, due to symmetry,

$$\left(\frac{\partial u}{\partial y}\right)^2 = \frac{4y^2}{(b^2+u)^2\left[\sum \frac{y^2}{(b^2+u)^2}\right]^2}, \quad (2)$$

$$\left(\frac{\partial u}{\partial z}\right)^2 = \frac{4z^2}{(c^2+u)^2\left[\sum \frac{z^2}{(c^2+u)^2}\right]^2}. \quad (3)$$

Adding (1), (2), and (3), we get

$$\left(\frac{\partial u}{\partial x}\right)^2+\left(\frac{\partial u}{\partial y}\right)^2+\left(\frac{\partial u}{\partial z}\right)^2 = \frac{4}{\sum \frac{x^2}{(a^2+u)^2}}. \quad (4)$$

On the other hand,

$$2x\frac{\partial u}{\partial x} = \frac{4x^2}{(a^2+u)\sum \frac{x^2}{(a^2+u)^2}}.$$

Since $\sum \frac{x^2}{a^2+u} = 1$ (given), we have

$$\sum 2x\frac{\partial u}{\partial x} = \frac{4\sum \frac{x^2}{a^2+u}}{\sum \frac{x^2}{(a^2+u)^2}} = \frac{4}{\sum \frac{x^2}{(a^2+u)^2}}. \quad (5)$$

From (4) and (5), we get

$$\left(\frac{\partial u}{\partial x}\right)^2+\left(\frac{\partial u}{\partial y}\right)^2+\left(\frac{\partial u}{\partial z}\right)^2 = 2\left(x\frac{\partial u}{\partial x}+y\frac{\partial u}{\partial y}+z\frac{\partial u}{\partial z}\right).$$

EXAMPLE 5.17

If $r^2 = x^2+y^2+z^2$ and $V = r^m$, show that

$$V_{xx}+V_{yy}+V_{zz} = m\,(m+1)\,r^{m-2}.$$

Solution. We have $r^2 = x^2+y^2+z^2$. Differentiating partially with respect to x, we get $2r\frac{\partial r}{\partial x} = 2x$ or $\frac{\partial r}{\partial x} = \frac{x}{r}$.

It is given that $V = r^m$. Therefore,

$$\frac{\partial V}{\partial x} = \frac{\partial V}{\partial r}\cdot\frac{\partial r}{\partial x} = mr^{m-1}\left(\frac{x}{r}\right) = mxr^{m-2}$$

and

$$\frac{\partial^2 V}{\partial x^2} = m\left[r^{m-2}+x\,(m-2)\,r^{m-3}\frac{\partial r}{\partial x}\right]$$
$$= m\,[r^{m-2}+(m-2)\,x^2r^{m-4}].$$

Similarly, due to symmetry, we have

$$\frac{\partial^2 V}{\partial y^2} = m\,[r^{m-2}+(m-2)\,y^2\,r^{m-4}],$$

$$\frac{\partial^2 V}{\partial z^2} = m\,[r^{m-2}+(m-2)\,z^2r^{m-4}].$$

Hence,

$$\frac{\partial^2 V}{\partial x^2}+\frac{\partial^2 V}{\partial y^2}+\frac{\partial^2 V}{\partial z^2} = 3mr^{m-2}+m\,(m-2) \times (x^2+y^2+z^2)\,r^{m-4}$$
$$= 3mr^{m-2}+m\,(m-2)\,r^{m-2}$$
$$= r^{m-2}\,[3m+m^2-2m]$$
$$= m\,(m+1)\,r^{m-2}.$$

5.6 ENVELOPES AND EVOLUTES

Let α be a parameter which can take all real values and let $f(x,y,\alpha)=0$ be a family of curves. Suppose that P is a point of intersection of two members $f(x, y, \alpha) = 0$ and $f(x, y, \alpha+\delta\alpha) = 0$ of this family. As $\delta\alpha \to 0$, let P tends to a definite point Q on the member α. The locus of Q (for varying value of α) is called the *envelope of the family*. Thus,

"The envelope of a one parameter family of curves is the locus of the limiting positions of the points of intersection of any two members of the family when one of them tends to coincide with the other which is kept fixed."

The coordinates of the points of intersection of the curves $f(x, y, \alpha) = 0$ and $f(x, y, \alpha+\delta\alpha) = 0$ satisfy the equations

$$f(x,y,\alpha)=0 \text{ and } f(x,y,\alpha+\delta\alpha)-f(x,y,\alpha)=0$$

and therefore, they also satisfy

$$f(x,y,\alpha)=0 \text{ and } \frac{f(x,y,\alpha+\delta\alpha)-f(x,y,\alpha)}{\delta\alpha}=0.$$

Taking limit as $\delta\alpha \to 0$, it follows that the coordinates of the limiting positions of the point of intersection of the curves $f(x,y,\alpha)=0$ and $f(x, y, \alpha+\delta\alpha)$ satisfy the equations

$$f(x, y, \alpha) = 0 \text{ and } \frac{\partial f}{\partial \alpha} = 0.$$

Hence, the equation of the envelope of the family of curves $f(x, y, \alpha) = 0$, where α is a parameter, is determined by eliminating α between the equations $f(x, y, \alpha) = 0$ and $\frac{\partial}{\partial \alpha} f(x, y, \alpha) = 0$.

The *evolute* of a curve is the envelope of the normals to that curve.

EXAMPLE 5.18

Find the envelope of the family of straight lines $y = mx + \frac{a}{m}$, the parameter being m.

Solution. We have

$$y = mx + \frac{a}{m}. \qquad (1)$$

Differentiating with respect to m, we obtain

$$0 = x - \frac{a}{m^2} \quad \text{or} \quad m = \left(\frac{a}{x}\right)^{\frac{1}{2}}.$$

Putting this value of m in (1), we get

$$y = \left(\frac{a}{x}\right)^{\frac{1}{2}} x + a\left(\frac{a}{x}\right)^{-\frac{1}{2}} = \frac{xa^{\frac{1}{2}}}{x^{\frac{1}{2}}} + \frac{ax^{\frac{1}{2}}}{a^{\frac{1}{2}}} = 2a^{\frac{1}{2}} \cdot x^{\frac{1}{2}},$$

and so, the parabola $y^2 = 4ax$ is the envelope of the family.

EXAMPLE 5.19

Find the envelope of the straight lines $x \cos \alpha + y \sin \alpha = l \sin \alpha \cos \alpha$, where the parameter is the angle α.

Solution. Dividing throughout by $\sin \alpha \cos \alpha$ we have

$$x \operatorname{cosec} \alpha + y \sec \alpha = l \qquad (1)$$

Differentiating partially with respect to α, we get

$$x(-\cos \alpha \cot \alpha) + y(\sec \alpha \tan \alpha) = 0$$

or

$$\tan \alpha = \frac{x^{\frac{1}{3}}}{y^{\frac{1}{3}}}.$$

This yields

$$\operatorname{cosec} \alpha = \sqrt{1 + \cot^2 \alpha} = \sqrt{1 + \frac{y^{\frac{2}{3}}}{x^{\frac{2}{3}}}} = \sqrt{\frac{x^{\frac{2}{3}} + y^{\frac{2}{3}}}{x^{\frac{2}{3}}}}$$

and

$$\sec \alpha = \sqrt{1 + \tan^2 \alpha} = \sqrt{1 + \frac{x^{\frac{2}{3}}}{y^{\frac{2}{3}}}} = \sqrt{\frac{x^{\frac{2}{3}} + y^{\frac{2}{3}}}{x^{\frac{2}{3}}}}.$$

Putting these values in (1), we get

$$x \frac{\sqrt{x^{\frac{2}{3}} + y^{\frac{2}{3}}}}{x^{\frac{1}{3}}} + y \frac{\sqrt{x^{\frac{2}{3}} + y^{\frac{2}{3}}}}{x^{\frac{1}{3}}} = l$$

or

$$x^{\frac{2}{3}} + y^{\frac{2}{3}} = l^{\frac{2}{3}} \text{ (astroid).}$$

EXAMPLE 5.20

Find the envelope of the family of straight lines $y = mx + \sqrt{a^2 m^2 + b^2}$, the parameter being m.

Solution. The given family of straight lines is

$$y = mx + \sqrt{a^2 m^2 + b^2}$$

or

$$(y - mx)^2 = a^2 m^2 + b^2. \qquad (1)$$

Differentiating partially with respect to m, we get

$$2(y - mx)(-x) = 2a^2 m$$

or

$$-2xy + 2mx^2 = 2a^2 m$$

or

$$m(a^2 - x^2) = -xy$$

or

$$m = -\frac{xy}{a^2 - x^2} = \frac{xy}{x^2 - a^2}.$$

Putting this value of m in (1), we get

$$\left(y - \frac{x^2 y}{x^2 - a^2}\right)^2 = a^2 \frac{x^2 y^2}{(x^2 - a^2)^2} + b^2$$

or

$$a^4 y^2 = a^2 x^2 y^2 + b^2 (x^2 - a^2)^2$$

or

$$0 = a^2 y^2 (x^2 - a^2) + b^2 (x^2 - a^2)^2$$

or

$$0 = a^2 y^2 + b^2 (x^2 - a^2) = a^2 y^2 + b^2 x^2 - a^2 b^2$$

or

$$b^2 x^2 + a^2 y^2 = a^2 b^2$$

or

$$\frac{x^2}{a^2} + \frac{y^2}{b^2} = 1 \text{ (ellipse)}$$

EXAMPLE 5.21

Find the evolute of the parabola $y = 4ax$.

Solution. The equation of parabola is $y^2 = 4ax$. Any normal to the parabola is

$$y = mx - 2am - am^3. \qquad (1)$$

Differentiating partially with respect to m, we get

$$0 = x - 2a - 3am^2,$$

which yields

$$m = \sqrt{\frac{x - 2a}{3a}}.$$

Putting this value of m in (1), we have

$$y = = \sqrt{\frac{x-2a}{3a}}\left[x-2a-\frac{a\,(x-2a)}{3a}\right]$$

or

$$27ay^2 = 4\,(x-2a)^3.$$

EXAMPLE 5.22

Find the evolute of the ellipse $\frac{x^2}{a^2}+\frac{y^2}{b^2}=1.$

Solution. The parametric equation of the ellipse is

$x = a\cos\theta,\ y = b\sin\theta$, where θ is a parameter.

Therefore,

$$\frac{dx}{d\theta} = -a\sin\theta \text{ and } \frac{dy}{d\theta} = b\cos\theta.$$

Thus,

$$\frac{dy}{dx} = \frac{\frac{dy}{d\theta}}{\frac{dx}{d\theta}} = -\frac{b\cos\theta}{a\sin\theta}.$$

Therefore, slope of the normal to the ellipse at $(a\cos\theta, b\sin\theta)$ is

$$-\frac{dx}{dy} = -\frac{a\sin\theta}{b\cos\theta}.$$

Hence, the equation of the normal to the ellipse at $(a\cos\theta, b\sin\theta)$ is

$$y - b\sin\theta = \frac{a\sin\theta}{b\cos\theta}(x - a\cos\theta)$$

or

$$\frac{ax}{\cos\theta} - \frac{by}{\sin\theta} = a^2 - b^2. \tag{1}$$

Differentiating (1) partially with respect to θ, we get

$$\frac{ax\sin\theta}{\cos^2\theta} + \frac{by\cos\theta}{\sin^2\theta} = 0. \tag{2}$$

The equation (2) gives

$$\tan^3\theta = -\frac{by}{ax} \quad\text{or}\quad \tan\theta = -\frac{(by)^{\frac{1}{3}}}{(ax)^{\frac{1}{3}}}.$$

Therefore,

$$\sin\theta = \frac{(by)^{\frac{1}{3}}}{\sqrt{(ax)^{\frac{2}{3}}+(by)^{\frac{2}{3}}}} \text{ and}$$

$$\cos\theta = -\frac{(ax)^{\frac{1}{3}}}{\sqrt{(ax)^{\frac{2}{3}}+(by)^{\frac{2}{3}}}}$$

or

$$\sin\theta = -\frac{(by)^{\frac{1}{3}}}{\sqrt{(ax)^{\frac{2}{3}}+(by)^{\frac{2}{3}}}} \text{ and}$$

$$\cos\theta = \frac{(ax)^{\frac{1}{3}}}{\sqrt{(ax)^{\frac{2}{3}}+(by)^{\frac{2}{3}}}}.$$

Substituting these values of $\sin\theta$ and $\cos\theta$ in (1) we get

$$\pm\left[(ax)^{\frac{2}{3}}+(by)^{\frac{2}{3}}\right]^{\frac{3}{2}} = a^2 - b^2$$

or

$$(ax)^{\frac{2}{3}}+(by)^{\frac{2}{3}} = (a^2-b^2)^{\frac{2}{3}}.$$

5.7 HOMOGENEOUS FUNCTIONS AND EULER'S THEOREM

An expression of the form

$$a_0x^n + a_1x^{n-1}y + a_2x^{n-2}y^2 + \ldots + a_ny^n$$

in which every term is of nth degree is called a *homogeneous function of degree n.* This can also be rewritten as

$$x^n\left[a_0 + a_1\left(\frac{y}{x}\right) + a_2\left(\frac{y}{x}\right)^2 + \ldots + a_{n-1}\left(\frac{y}{x}\right)^{n-1} + a_n\left(\frac{y}{x}\right)^n\right]$$

or $x^n\phi\left(\frac{y}{x}\right)$. Thus, we can define a homogeneous function as follows:

A function $f(x, y)$, which can be expressed as $x^n\phi\left(\frac{y}{x}\right)$, is called a *homogeneous function of degree n in x and y.*

To check whether a function $f(x, y)$ is homogeneous or not, we put tx for x and ty for y in it. If $f(tx, ty) = t^n f(x, y)$, then the function $f(x, y)$ is a homogeneous function of degree n which is otherwise a nonhomogeneous function.

Let $u = x^n f\left(\frac{y}{x}\right)$ be a homogeneous function of x and y of degree n. Then,

$$\frac{\partial u}{\partial x} = nx^{n-1}f\left(\frac{y}{x}\right) + x^n f'\left(\frac{y}{x}\right)\cdot\left(-\frac{y}{x^2}\right)$$

$$= x^{n-1}\left[nf\left(\frac{y}{x}\right) - \left(\frac{y}{x}\right)f'\left(\frac{y}{x}\right)\right] = x^{n-1}\phi\left(\frac{y}{x}\right),$$

which is a homogeneous function of degree $n-1$. Similarly,

$$\frac{\partial u}{\partial y} = x^n f'\left(\frac{y}{x}\right)\cdot\frac{1}{x} = x^{n-1}f'\left(\frac{y}{x}\right) = x^{n-1}\,\psi\left(\frac{y}{x}\right),$$

which is a homogeneous function of degree $n-1$. It follows, therefore, that "If u is a homogeneous function of x and y of degree n, then u_x and u_y are also homogeneous functions of x and y of degree $n-1$."

Theorem 5.3 (Euler's Theorem). If u is a homogeneous function of x and y of degree n, then

$$x\frac{\partial u}{\partial x}+y\frac{\partial u}{\partial y}=nu.$$

Proof: Since u is a homogeneous function of x and y of degree n, it can be expressed as

$$u=x^n f\left(\frac{y}{x}\right).$$

Differentiating partially with respect to x, we have

$$\frac{\partial u}{\partial x}=nx^{n-1}f\left(\frac{y}{x}\right)+x^n f'\left(\frac{y}{x}\right)\left(-\frac{y}{x^2}\right)$$

$$=nx^{n-1}f\left(\frac{y}{x}\right)-yx^{n-2}f'\left(\frac{y}{x}\right).$$

Similarly, the differentiation with respect to y yields

$$\frac{\partial u}{\partial y}=x^n f'\left(\frac{y}{x}\right)\left(\frac{1}{x}\right)=x^{n-1}f'\left(\frac{y}{x}\right).$$

Therefore,

$$x\frac{\partial u}{\partial x}+y\frac{\partial u}{\partial y}=nx^n f\left(\frac{y}{x}\right)-yx^{n-1}f'\left(\frac{y}{x}\right)$$

$$+x^{n-1}yf'\left(\frac{y}{x}\right)$$

Remark 5.1. Euler's theorem on homogeneous functions also holds good for functions of n variables. Thus, "If $u\ (x_1, x_2, \ldots x_n)$ is a homogeneous function of $x_1, x_2, \ldots x_n$ of degree n, then

$$x_1\frac{\partial u}{\partial x_1}+x_2\frac{\partial u}{\partial x_2}+\ldots+x_n\frac{\partial u}{\partial x_n}=nu."$$

EXAMPLE 5.23

If $u=\tan^{-1}\left(\dfrac{x^3+y^3}{x+y}\right)$, show that

$$x\frac{\partial u}{\partial x}+y\frac{\partial u}{\partial y}=\sin 2u.$$

Solution. The given function is

$$u=\tan^{-1}\left(\frac{x^3+y^3}{x+y}\right).$$

Therefore,

$$\tan u=\left(\frac{x^3+y^3}{x+y}\right)=x^2\,\frac{1+\left(\frac{y}{x}\right)^3}{1+\frac{y}{x}}=z.$$

Therefore, z is a homogeneous function of degree 2 in x and y. Hence, by Euler's theorem,

$$x\frac{\partial z}{\partial x}+y\frac{\partial z}{\partial y}=2z. \qquad (1)$$

But,

$$\frac{\partial z}{\partial x}=\sec^2 u\,\frac{\partial u}{\partial x}\quad\text{and}\quad\frac{\partial z}{\partial y}=\sec^2 u\,\frac{\partial u}{\partial y}.$$

Therefore, (1) reduces to

$$x\sec^2 u\,\frac{\partial u}{\partial x}+y\sec^2 u\,\frac{\partial u}{\partial y}=2\tan u$$

or

$$x\frac{\partial u}{\partial x}+y\frac{\partial u}{\partial y}=\frac{2\tan u}{\sec^2 u}=2\sin u\cos u=\sin 2u.$$

EXAMPLE 5.24

If $u=\sin^{-1}\dfrac{x^2+y^2}{x+y}$, show that

$$x\frac{\partial u}{\partial x}+y\frac{\partial u}{\partial y}=\tan u.$$

Solution. We have

$$u=\sin^{-1}\frac{x^2+y^2}{x+y}.$$

Therefore,

$$\sin u=\frac{x^2+y^2}{x+y}=z.$$

Thus,

$$z=\sin u=x\,\frac{1+\left(\frac{y}{x}\right)^2}{1+\frac{y}{x}},$$

and so, z is a homogeneous function of degree 1 in x and y. Hence by Euler's theorem, we have

$$x\frac{\partial z}{\partial x}+y\frac{\partial z}{\partial y}=z. \qquad (1)$$

But

$$\frac{\partial z}{\partial x}=\cos u\,\frac{\partial u}{\partial x}\quad\text{and}\quad\frac{\partial z}{\partial y}=\cos u\,\frac{\partial u}{\partial y}.$$

Therefore, (1) reduces to

$$x\cos u\,\frac{\partial u}{\partial x}+y\cos u\,\frac{\partial u}{\partial y}=\sin u$$

or

$$x\frac{\partial u}{\partial x}+y\frac{\partial u}{\partial y}=\tan u.$$

EXAMPLE 5.25

If $u=xf\left(\dfrac{y}{x}\right)+g\left(\dfrac{y}{x}\right)$, show that

$$x^2\frac{\partial^2 u}{\partial x^2}+2xy\frac{\partial^2 u}{\partial x\partial y}+y^2\frac{\partial^2 u}{\partial y^2}=0.$$

Solution. Let $u_1=xf\left(\frac{y}{x}\right)$ and $u_2=g\left(\frac{y}{x}\right)$ so that $u=u_1+u_2$. Then u_1 is a homogeneous function of degree 1 and u_2 is a homogeneous function of degree 0. Therefore, by Euler's theorem, we have

$$x\frac{\partial u_1}{\partial x}+y\frac{\partial u_1}{\partial y}=u_1 \quad \text{and} \quad x\frac{\partial u_2}{\partial x}+y\frac{\partial u_2}{\partial y}=0. \quad (1)$$

But

$$\begin{aligned} x\frac{\partial u}{\partial x}+y\frac{\partial u}{\partial y} &= x\frac{\partial}{\partial x}(u_1+u_2)+y\frac{\partial}{\partial y}(u_1+u_2) \\ &= x\frac{\partial u_1}{\partial x}+x\frac{\partial u_2}{\partial x}+y\frac{\partial u_1}{\partial y}+y\frac{\partial u_2}{\partial y} \\ &= \left(x\frac{\partial u_1}{\partial x}+y\frac{\partial u_1}{\partial y}\right)+\left(x\frac{\partial u_2}{\partial x}+y\frac{\partial u_2}{\partial y}\right) \\ &= u_1+0=u_1 \text{ [using (1)].} \quad (2) \end{aligned}$$

Differentiating (2) partially with respect to x and y, we get

$$x\frac{\partial^2 u}{\partial x^2}+\frac{\partial u}{\partial x}+y\frac{\partial^2 u}{\partial x\partial y}=\frac{\partial u_1}{\partial x} \quad (3)$$

and

$$x\frac{\partial^2 u}{\partial y\partial x}+\frac{\partial u}{\partial x}+y\frac{\partial^2 u}{\partial y^2}=\frac{\partial u_1}{\partial y}. \quad (4)$$

Multiplying (3) by x and (4) by y and then adding both, we get

$$x^2\frac{\partial^2 u}{\partial x^2}+2xy\frac{\partial^2 u}{\partial x\partial y}+y^2\frac{\partial^2 u}{\partial y^2}+x\frac{\partial u}{\partial x}+y\frac{\partial u}{\partial y} = x\frac{\partial u_1}{\partial x}+y\frac{\partial u_1}{\partial y}.$$

Using (1) and (2), the last expression reduces to

$$x^2\frac{\partial^2 u}{\partial x^2}+2xy\frac{\partial^2 u}{\partial x\partial y}+y^2\frac{\partial^2 u}{\partial y^2}+u_1=u_1$$

or

$$x^2\frac{\partial^2 u}{\partial x^2}+2xy\frac{\partial^2 u}{\partial x\partial y}+y^2\frac{\partial^2 u}{\partial y^2}=0.$$

EXAMPLE 5.26

If u is a homogeneous function of x and y of degree n, show that

$$x^2\frac{\partial^2 u}{\partial x^2}+2xy\frac{\partial^2 u}{\partial x\partial y}+y^2\frac{\partial^2 u}{\partial y^2}=n(n-1)\,u.$$

Solution. By Euler's theorem, we have

$$x\frac{\partial u}{\partial x}+y\frac{\partial u}{\partial y}=n\,u. \quad (1)$$

Differentiating (1) partially with respect to x, we get

$$x\frac{\partial^2 u}{\partial x^2}+\frac{\partial u}{\partial x}+y\frac{\partial^2 u}{\partial x\partial y}=n\frac{\partial u}{\partial x}$$

or

$$x\frac{\partial^2 u}{\partial x^2}+y\frac{\partial u}{\partial x\partial y}=n\frac{\partial u}{\partial x}-\frac{\partial u}{\partial x}=(n-1)\frac{\partial u}{\partial x}. \quad (2)$$

Again differentiating (1) partially with respect to y, we have

$$x\frac{\partial^2 u}{\partial y\partial x}+y\frac{\partial^2 u}{\partial y^2}+\frac{\partial u}{\partial y}=n\frac{\partial u}{\partial y}$$

or

$$x\frac{\partial^2 u}{\partial y\partial x}+y\frac{\partial^2 u}{\partial y^2}=(n-1)\frac{\partial u}{\partial y}. \quad (3)$$

Multiplying (2) by x and (3) by y and then adding both, we have

$$\begin{aligned} x^2\frac{\partial^2 u}{\partial x^2}+2xy\frac{\partial^2 u}{\partial x\partial y}+y^2\frac{\partial^2 u}{\partial y^2} &= (n-1)\left(x\frac{\partial u}{\partial x}+y\frac{\partial u}{\partial y}\right) \\ &= (n-1)\,nu \text{ [using (1)].} \end{aligned}$$

EXAMPLE 5.27

If $\sin u=\frac{x^2y^2}{x+y}$, show that

$$x\frac{\partial u}{\partial x}+y\frac{\partial u}{\partial y}=3\tan u.$$

Solution. We have

$$\sin u=\frac{x^2y^2}{x+y}=\frac{x^3\left(\frac{y}{x}\right)^2}{1+\frac{y}{x}}=v.$$

We observe that v is a homogeneous function of degree 3. Therefore, by Euler's theorem,

$$x\frac{\partial v}{\partial x}+y\frac{\partial v}{\partial y}=3\,v. \quad (1)$$

Since $v=\sin u$, we have

$$\frac{\partial v}{\partial x}=\cos u\frac{\partial u}{\partial x} \quad \text{and} \quad \frac{\partial v}{\partial y}=\cos u\frac{\partial u}{\partial y}.$$

Hence, (1) reduces to

$$x\cos u\frac{\partial u}{\partial x}+y\cos u\frac{\partial u}{\partial y}=3\sin u$$

or

$$x\frac{\partial u}{\partial x}+y\frac{\partial u}{\partial y}=3\tan u.$$

EXAMPLE 5.28

If $u=\sin^{-1}\frac{x}{y}+\tan^{-1}\frac{y}{x}$, show that

$$x\frac{\partial u}{\partial x}+y\frac{\partial u}{\partial y}=0.$$

Solution. We have

$$u=\sin^{-1}\frac{x}{y}+\tan^{-1}\frac{y}{x}=\sin^{-1}\frac{1}{\frac{y}{x}}+\tan^{-1}\frac{y}{x}=x^0 f\left(\frac{y}{x}\right).$$

Therefore, u is a homogeneous function of degree 0 in x and y. Hence by Euler's theorem, we have

$$x\frac{\partial u}{\partial x}+y\frac{\partial u}{\partial y}=0\cdot u=0.$$

For verification of the result, see Example 5.9.

EXAMPLE 5.29

If $u=\tan^{-1}\dfrac{x^3+y^3}{x-y}$, show that

$$x^2\frac{\partial^2 u}{\partial x^2}+2xy\frac{\partial^2 u}{\partial x\partial y}+y^2\frac{\partial^2 u}{\partial y^2}=\sin 4u-\sin 2u.$$

Solution. We have

$$\tan u=\frac{x^3+y^3}{x-y}=x^2\,\frac{1+\left(\frac{y}{x}\right)^3}{1-\frac{y}{x}}=v.$$

Then v is a homogeneous function of degree 2 in x and y. Hence, by Euler's theorem,

$$x\frac{\partial v}{\partial x}+y\frac{\partial v}{\partial y}=2v.$$

But

$$\frac{\partial v}{\partial x}=\sec^2 u\,\frac{\partial u}{\partial x},\ \frac{\partial v}{\partial y}=\sec^2 u\,\frac{\partial u}{\partial y}.$$

Therefore,

$$x\sec^2 u\,\frac{\partial u}{\partial x}+y\sec^2 u\,\frac{\partial u}{\partial y}=2\tan u$$

or

$$x\frac{\partial u}{\partial x}+y\frac{\partial u}{\partial y}=2\sin u\cos u=\sin 2u. \quad (1)$$

Differentiating (1) partially with respect to x and y, respectively, we get

$$x\frac{\partial^2 u}{\partial x^2}+\frac{\partial u}{\partial x}+y\frac{\partial^2 u}{\partial x\partial y}=2\cos 2u\,\frac{\partial u}{\partial x} \quad (2)$$

and

$$x\frac{\partial^2 u}{\partial y\partial x}+\frac{\partial u}{\partial y}+y\frac{\partial^2 u}{\partial y^2}=2\cos 2u\,\frac{\partial u}{\partial y}. \quad (3)$$

Multiplying (2) by x and (3) by y and then adding both, we get

$$x^2\frac{\partial^2 u}{\partial x^2}+2xy\frac{\partial^2 u}{\partial x\partial y}+y^2\frac{\partial^2 u}{\partial y^2}+x\frac{\partial u}{\partial x}+y\frac{\partial u}{\partial y}$$
$$=2\cos 2u\left(x\frac{\partial u}{\partial x}+y\frac{\partial u}{\partial y}\right)$$

or using (1), we have

$$x^2\frac{\partial^2 u}{\partial x^2}+2xy\frac{\partial^2 u}{\partial x\partial y}+y^2\frac{\partial^2 u}{\partial y^2}+\sin 2u=2\cos 2u\sin 2u$$

or

$$x^2\frac{\partial^2 u}{\partial x^2}+2xy\frac{\partial^2 u}{\partial x\partial y}+y^2\frac{\partial^2 u}{\partial y^2}$$
$$=2\cos 2u\sin 2u-\sin 2u$$
$$=\sin 4u-\sin 2u.$$

EXAMPLE 5.30

If $u=\dfrac{x}{y+z}+\dfrac{y}{z+x}+\dfrac{z}{x+y}$, show that

$$x\frac{\partial u}{\partial x}+y\frac{\partial u}{\partial y}+z\frac{\partial u}{\partial z}=0.$$

Solution. Replacing x by tx, y by ty, and z by tz, we get

$$u(tx,ty,tz)=\frac{tx}{ty+tz}+\frac{ty}{tz+tx}+\frac{tz}{tx+ty}$$
$$=1.u(x,y,z)=t^0u(x,y,z).$$

Hence, $u(x, y, z)$ is a homogeneous function of degree zero in x, y, and z. Hence, by Euler's theorem,

$$x\frac{\partial u}{\partial x}+y\frac{\partial u}{\partial y}+z\frac{\partial u}{\partial z}=nu=0\cdot u=0.$$

EXAMPLE 5.31

If $u=e^{x^2+y^2}$, show that

$$x\frac{\partial u}{\partial x}+y\frac{\partial u}{\partial y}=2u\log u.$$

Solution. We have $u=e^{x^2+y^2}$ and so,

$$\log u=x^2+y^2=x^2\left[1+\left(\frac{y}{x}\right)^2\right]=v.$$

Thus, v is a homogeneous function of degree 2 in x and y. Hence, by Euler's theorem,

$$x\frac{\partial v}{\partial x}+y\frac{\partial v}{\partial y}=2v.$$

But

$$\frac{\partial v}{\partial x}=\frac{1}{u}\frac{\partial u}{\partial x},\ \frac{\partial v}{\partial y}=\frac{1}{u}\frac{\partial u}{\partial y}.$$

Therefore,

$$\frac{x}{u}\frac{\partial u}{\partial x}+\frac{y}{u}\frac{\partial u}{\partial y}=2\log u$$

or

$$x\frac{\partial u}{\partial x}+y\frac{\partial u}{\partial y}=2u\log u.$$

EXAMPLE 5.32

If $u=\sin^{-1}\dfrac{x+y}{\sqrt{x}+\sqrt{y}}$, show that

$$x\frac{\partial u}{\partial x}+y\frac{\partial u}{\partial y}=\frac{1}{2}\tan u.$$

Solution. We have

$$\sin u = \frac{x+y}{\sqrt{x}+\sqrt{y}} = x^{\frac{1}{2}} \frac{1+\dfrac{y}{x}}{1+\sqrt{\dfrac{y}{x}}} = v.$$

Therefore, v is a homogeneous function of degree $\frac{1}{2}$ in x and y. Hence, by Euler's theorem,

$$x\frac{\partial v}{\partial x} + y\frac{\partial v}{\partial y} = \frac{1}{2}v.$$

But

$$\frac{\partial v}{\partial x} = \cos u \frac{\partial u}{\partial x} \quad \text{and} \quad \frac{\partial v}{\partial y} = \cos u \frac{\partial u}{\partial y}.$$

Therefore,

$$x \cos u \frac{\partial u}{\partial x} + y \cos u \frac{\partial u}{\partial y} = \frac{1}{2}\sin u$$

or

$$x\frac{\partial u}{\partial x} + y\frac{\partial u}{\partial y} = \frac{1}{2}\tan u.$$

EXAMPLE 5.33

If $u = \operatorname{cosec}^{-1}\left[\dfrac{x^{\frac{1}{2}}+y^{\frac{1}{2}}}{x^{\frac{1}{3}}+y^{\frac{1}{3}}}\right]^{\frac{1}{2}}$, show that

$$x^2\frac{\partial^2 u}{\partial x^2} + 2xy\frac{\partial^2 u}{\partial x \partial y} + y^2\frac{\partial^2 u}{\partial y^2} = \frac{\tan u}{144}(13+\tan^2 u).$$

Solution. We have

$$\operatorname{cosec} u = \left[\frac{x^{\frac{1}{2}}+y^{\frac{1}{2}}}{x^{\frac{1}{3}}+y^{\frac{1}{3}}}\right]^{\frac{1}{2}} = \left[\frac{x^{\frac{1}{2}}\left(1+\left(\frac{y}{x}\right)^{\frac{1}{2}}\right)}{x^{\frac{1}{3}}\left(1+\left(\frac{y}{x}\right)^{\frac{1}{3}}\right)}\right]^{\frac{1}{2}}$$

$$= x^{\frac{1}{12}}\left[\frac{1+\left(\frac{y}{x}\right)^{\frac{1}{2}}}{1+\left(\frac{y}{x}\right)^{\frac{1}{3}}}\right]^{\frac{1}{2}} = v.$$

Therefore, v is a homogeneous function of degree $\frac{1}{12}$ in x and y. Hence, by Euler's theorem,

$$x\frac{\partial v}{\partial x} + y\frac{\partial v}{\partial y} = \frac{1}{12}v.$$

But

$$\frac{\partial v}{\partial x} = -\operatorname{cosec} u \cot u \frac{\partial u}{\partial x},$$

$$\frac{\partial v}{\partial y} = -\operatorname{cosec} u \cot u \frac{\partial u}{\partial y}.$$

Therefore,

$$-x\operatorname{cosec} u \cot u\frac{\partial u}{\partial x} - y\operatorname{cosec} u \cot u\frac{\partial u}{\partial y} = \frac{1}{12}\operatorname{cosec} u$$

or

$$x \cot u \frac{\partial u}{\partial x} + y \cot u \frac{\partial u}{\partial y} = -\frac{1}{12}$$

or

$$x\frac{\partial u}{\partial x} + y\frac{\partial u}{\partial y} = -\frac{1}{12}\tan u. \qquad (1)$$

Differentiating (1) partially with respect to x, we get

$$x\frac{\partial^2 u}{\partial x^2} + \frac{\partial u}{\partial x} + y\frac{\partial^2 u}{\partial x \partial y} = -\frac{1}{12}\sec^2 u \frac{\partial u}{\partial x}$$

or

$$x\frac{\partial^2 u}{\partial x^2} + y\frac{\partial^2 u}{\partial x \partial y} = \left(-\frac{1}{12}\sec^2 u - 1\right)\frac{\partial u}{\partial x}. \qquad (2)$$

Similarly, differentiating (1) partially with respect to y, we get

$$y\frac{\partial^2 u}{\partial y^2} + x\frac{\partial^2 u}{\partial y \partial x} = \left(\frac{1}{12}\sec^2 u - 1\right)\frac{\partial u}{\partial y}. \qquad (3)$$

Multiplying (2) by x and (3) by y and then adding both, we get

$$x^2\frac{\partial^2 u}{\partial x^2} + 2xy\frac{\partial^2 u}{\partial x \partial y} + y^2\frac{\partial^2 u}{\partial y^2}$$

$$= \left(-\frac{1}{12}\sec^2 u - 1\right)\left(x\frac{\partial u}{\partial x} + y\frac{\partial u}{\partial y}\right)$$

$$= \left(-\frac{1}{12}\sec^2 u - 1\right)\left(-\frac{1}{12}\tan u\right)$$

$$= \left(\frac{1}{12}\sec^2 u - 1\right)\left(\frac{1}{12}\tan u\right)$$

$$= \left[\frac{1}{12}(1+\tan^2 u) - 1\right]\left(\frac{1}{12}\tan u\right)$$

$$= \frac{1}{144}\tan u\,(13+\tan^2 u).$$

EXAMPLE 5.34

If $z = x^n f_1\left(\dfrac{y}{x}\right) + y^{-n} f_2\left(\dfrac{x}{y}\right)$, show that

$$x^2\frac{\partial^2 z}{\partial x^2} + 2xy\frac{\partial^2 z}{\partial x \partial y} + y^2\frac{\partial^2 z}{\partial y^2} + x\frac{\partial z}{\partial x} + y\frac{\partial z}{\partial y} = n^2 z.$$

Solution. Let $u_1 = x^n f_1\left(\dfrac{y}{x}\right)$ and $u_2 = y^{-n} f_2\left(\dfrac{x}{y}\right)$ so that $z = u_1 + u_2$. Then, u_1 is a homogeneous function of degree n and u_2 is a homogeneous function of degree n. Therefore, byEuler's theorem, we have

$$x\frac{\partial u_1}{\partial x} + y\frac{\partial u_1}{\partial y} = nu_1 \text{ and } x\frac{\partial u_2}{\partial x} + y\frac{\partial u_2}{\partial y} = -nu_2. (1)$$

But

$$x\frac{\partial z}{\partial x} + y\frac{\partial z}{\partial y} = x\frac{\partial}{\partial x}(u_1+u_2) + y\frac{\partial}{\partial y}(u_1+u_2)$$

$$= \left(x\frac{\partial u_1}{\partial x} + y\frac{\partial u_1}{\partial y}\right) + x\left(\frac{\partial u_2}{\partial x} + y\frac{\partial u_2}{\partial y}\right)$$

$$= n(u_1 - u_2), \text{ using (1).} \qquad (2)$$

Differentiating (2) partially with respect to x and y, we get

$$x\frac{\partial^2 z}{\partial x^2} + \frac{\partial z}{\partial x} + y\frac{\partial^2 z}{\partial x \partial y} = n\left(\frac{\partial u_1}{\partial x} - \frac{\partial u_2}{\partial x}\right) \qquad (3)$$

and

$$x\frac{\partial^2 z}{\partial x \partial y} + \frac{\partial z}{\partial y} + y\frac{\partial^2 z}{\partial y^2} = n\left(\frac{\partial u_1}{\partial y} - \frac{\partial u_2}{\partial y}\right). \qquad (4)$$

Multiplying (3) by x and (4) by y and then adding both, we get

$$x^2\frac{\partial^2 z}{\partial x^2} + 2xy\frac{\partial^2 z}{\partial x \partial y} + y^2\frac{\partial^2 z}{\partial y^2} + x\frac{\partial z}{\partial x} + y\frac{\partial z}{\partial y}$$

$$= nx\left(\frac{\partial u_1}{\partial x} - \frac{\partial u_2}{\partial x}\right) + ny\left(\frac{\partial u_1}{\partial y} - \frac{\partial u_2}{\partial y}\right)$$

$$= n\left[\left(x\frac{\partial u_1}{\partial x} + y\frac{\partial u_1}{\partial y}\right) - \left(x\frac{\partial u_2}{\partial x} + y\frac{\partial u_2}{\partial y}\right)\right]$$

$$= n\,[nu_1 + nu_2] = n^2(u_1 + u_2) = n^2 z.$$

5.8 DIFFERENTIATION OF COMPOSITE FUNCTIONS

Let $x = \phi(t)$ and $y = \psi(t)$. If $u = f(x, y)$, then u is called a *composite function of the single variable t*. Similarly, if $x = \phi(u, v)$, $y = \psi(u, v)$, and $F = f(x, y)$, then F is called a *composite function of two variables u and v.*

Theorem 5.4. If $u = f(x, y)$, where $x = \phi(t)$ and $y = \psi(t)$, then

$$\frac{du}{dt} = \frac{\partial u}{\partial x}\cdot\frac{dx}{dt} + \frac{\partial u}{\partial y}\cdot\frac{dy}{dt}.$$

($\frac{du}{dt}$ is called the *total differential coefficient* or *total derivative of u with respect to t*).

Proof: Let δt be the small increment given to t and δx, δy, and δu be the corresponding increments in x, y, and u, respectively. We have

$$u = f(x, y), \qquad (1)$$

$$u + \delta u = f(x + \delta x, y + \delta y). \qquad (2)$$

Subtracting (1) from (2), we get

$$\delta u = f(x + \delta x, y + \delta y) - f(x, y)$$

$$= [f(x + \delta x, y + \delta y) - f(x, y + \delta y)] + [f(x, y + \delta y) - f(x, y)].$$

Thus,

$$\frac{\delta u}{\delta t} = \frac{f(x + \delta x, y + \delta y) - f(x, y + \delta y)}{\delta t} + \frac{f(x, y + \delta y) - f(x, y)}{\delta t}$$

$$= \frac{f(x + \delta x, y + \delta y) - f(x, y + \delta y)}{\delta x}\cdot\frac{\delta x}{\delta t} + \frac{f(x, y + \delta y) - f(x, y)}{\delta y}\cdot\frac{\delta y}{\delta t}.$$

If $\delta t \to 0$, δx and δy both tend to zero and so,

$$\lim_{\delta t \to 0}\frac{\delta u}{dt}$$

$$= \lim_{\delta x \to 0}\frac{f(x + \delta x, y + \delta y - f(x, y + \delta y)}{\delta x}\lim_{\delta t \to 0}\frac{\delta x}{\delta t} + \lim_{\delta y \to 0}\frac{f(x, y + \delta y) - f(x, y)}{\delta y}\lim_{\delta t \to 0}\frac{\delta y}{\delta t}$$

$$= \frac{\partial f}{\partial x}\cdot\lim_{\delta t \to 0}\frac{\delta x}{\delta t} + \frac{\partial f}{\partial y}\cdot\lim_{\delta t \to 0}\frac{\delta y}{\delta t}.$$

or

$$\frac{du}{dt} = \frac{\partial f}{\partial x}\cdot\frac{dx}{dt} + \frac{\partial f}{\partial y}\frac{dy}{dt} = \frac{\partial u}{\partial x}\frac{dx}{dt} + \frac{\partial u}{\partial y}\frac{dy}{dt}. \qquad (3)$$

Corollary 1. If u is a function of x and y and y is a function of x, then

$$\frac{du}{dx} = \frac{\partial u}{\partial x} + \frac{\partial u}{\partial y}\cdot\frac{dy}{dx}.$$

The result follows taking $t = x$ in (3).

Corollary 2. If $z = f(x, y)$, $x = \phi(u, v)$, and $y = \psi(u, v)$, then

$$\frac{\partial z}{\partial u} = \frac{\partial z}{\partial x}\cdot\frac{\partial x}{\partial u} + \frac{\partial z}{\partial y}\cdot\frac{\partial y}{\partial u}$$

and

$$\frac{\partial z}{\partial v} = \frac{\partial z}{\partial x}\cdot\frac{\partial x}{\partial v} + \frac{\partial z}{\partial y}\cdot\frac{\partial y}{\partial v}.$$

Corollary 3. If $u = f(x, y) = c$, then by Corollary 1, we have

$$0 = \frac{\partial f}{\partial x} + \frac{\partial f}{\partial y}\cdot\frac{dy}{dx}$$

or

$$\frac{dy}{dx} = -\frac{\frac{\partial f}{\partial x}}{\frac{\partial f}{\partial y}}.$$

Differentiating once more with respect to x, we get

$$\frac{d^2y}{dx^2} = -\frac{f_{xx}f_y^2 - 2f_x f_y f_{xy} + f_{yy}f_x^2}{f_y^3}.$$

EXAMPLE 5.35

If $x^y + y^x = a^b$, find $\frac{dy}{dx}$.

Solution. Let

$$f(x, y) = x^y + y^x - a^b.$$

Since from the given relation $x^y + y^x - a^b = 0$, it follows that $f(x, y) = 0$. Therefore,

$$\frac{dy}{dx} = -\frac{\frac{\partial f}{\partial x}}{\frac{\partial f}{\partial y}} = \frac{yx^{y-1} + y^x \log y}{x^y \log x + xy^{x-1}}.$$

EXAMPLE 5.36

If $f(x, y) = 0$ and $\phi(y, z) = 0$, show that

$$\frac{\partial f}{\partial y} \cdot \frac{\partial \phi}{\partial z} \cdot \frac{dz}{dx} = \frac{\partial f}{\partial x} \cdot \frac{\partial \phi}{\partial y}.$$

Solution. Since $f(x, y) = 0$, we have

$$\frac{dy}{dx} = -\frac{\frac{\partial f}{\partial x}}{\frac{\partial f}{\partial y}}. \quad (1)$$

Again, since $\phi(y, z) = 0$, we have

$$\frac{dz}{dy} = -\frac{\frac{\partial \phi}{\partial y}}{\frac{\partial \phi}{\partial z}}. \quad (2)$$

Multiplication of (1) and (2) yields

$$\frac{dy}{dx} \cdot \frac{dz}{dy} = \frac{\frac{\partial f}{\partial x}}{\frac{\partial f}{\partial y}} \cdot \frac{\frac{\partial \phi}{\partial y}}{\frac{\partial \phi}{\partial z}}$$

or

$$\frac{dz}{dx} = \frac{\frac{\partial f}{\partial x} \cdot \frac{\partial \phi}{\partial y}}{\frac{\partial f}{\partial y} \cdot \frac{\partial \phi}{\partial z}}$$

or

$$\frac{dz}{dx} \cdot \frac{\partial f}{\partial y} \cdot \frac{\partial \phi}{\partial z} = \frac{\partial f}{\partial x} \cdot \frac{\partial \phi}{\partial y}.$$

EXAMPLE 5.37

If z is a function of x and y and $x = e^u + e^{-v}$ and $y = e^{-u} - e^v$, show that

$$\frac{\partial z}{\partial u} - \frac{\partial z}{\partial v} = x\frac{\partial z}{\partial x} - y\frac{\partial z}{\partial y}.$$

Solution. The function z is a composite function of u and v. Therefore,

$$\frac{\partial z}{\partial u} = \frac{\partial z}{\partial x} \cdot \frac{\partial x}{\partial u} + \frac{\partial z}{\partial y} \cdot \frac{\partial y}{\partial u} = \frac{\partial z}{\partial x} \cdot e^u - \frac{\partial z}{\partial y} \cdot e^{-u} \quad (1)$$

and

$$\frac{\partial z}{\partial v} = \frac{\partial z}{\partial x} \cdot \frac{\partial x}{\partial v} + \frac{\partial z}{\partial y} \cdot \frac{\partial y}{\partial v} = -\frac{\partial z}{\partial x} \cdot e^{-v} - \frac{\partial z}{\partial y} \cdot e^{v}. \quad (2)$$

Subtracting (2) from (1), we have

$$\frac{\partial z}{\partial u} - \frac{\partial z}{\partial v} = (e^u - e^{-v})\frac{\partial z}{\partial x} - (e^{-u} + e^v)\frac{\partial z}{\partial y}$$

$$= x\frac{\partial z}{\partial x} - y\frac{\partial z}{\partial y}.$$

EXAMPLE 5.38

Prove that $\frac{\partial^2 u}{\partial x^2} + \frac{\partial^2 u}{\partial y^2} = \frac{\partial^2 u}{\partial \zeta^2} + \frac{\partial^2 u}{\partial \eta^2}$, where

$$x = \zeta \cos\alpha - \eta \sin\alpha, \; y = \zeta \sin\alpha + \eta \cos\alpha.$$

Solution. We have

$$x = \zeta \cos\alpha - \eta \sin\alpha, \; y = \zeta \sin\alpha + \eta \cos\alpha.$$

Now

$$\frac{\partial u}{\partial \zeta} = \frac{\partial u}{\partial x} \cdot \frac{\partial x}{\partial \zeta} + \frac{\partial u}{\partial y} \cdot \frac{\partial y}{\partial \zeta} = \cos\alpha \frac{\partial u}{\partial x} + \sin\alpha \frac{\partial u}{\partial y},$$

which yields

$$\frac{\partial}{\partial \zeta}(u) = \left(\cos\alpha \frac{\partial}{\partial x} + \sin\alpha \frac{\partial}{\partial y}\right)(u)$$

and so,

$$\frac{\partial}{\partial \zeta} = \cos\alpha \frac{\partial}{\partial x} + \sin\alpha \frac{\partial}{\partial y}.$$

Thus,

$$\frac{\partial^2 u}{\partial \zeta^2} = \frac{\partial}{\partial \zeta}\left(\frac{\partial u}{\partial \zeta}\right) = \left(\cos\alpha \frac{\partial}{\partial x} + \sin\alpha \frac{\partial}{\partial y}\right) \times \left(\cos\alpha \frac{\partial u}{\partial x} + \sin\alpha \frac{\partial u}{\partial y}\right)$$

$$= \cos\alpha \frac{\partial}{\partial x}\left(\cos\alpha \frac{\partial u}{\partial x} + \sin\alpha \frac{\partial u}{\partial y}\right) + \sin\alpha \frac{\partial}{\partial y}\left(\cos\alpha \frac{\partial u}{\partial x} + \sin\alpha \frac{\partial u}{\partial y}\right)$$

$$= \cos^2\alpha \frac{\partial^2 u}{\partial x^2} + 2\sin\alpha\cos\alpha \frac{\partial^2 u}{\partial x \partial y} + \sin^2\alpha \frac{\partial^2 u}{\partial y^2}. \quad (1)$$

On the other hand,

$$\frac{\partial u}{\partial \eta} = \frac{\partial u}{\partial x} \cdot \frac{\partial x}{\partial \eta} + \frac{\partial u}{\partial y} \cdot \frac{\partial y}{\partial \eta} = -\sin\alpha \frac{\partial u}{\partial x} + \cos\alpha \frac{\partial u}{\partial y},$$

which gives

$$\frac{\partial}{\partial \eta} = \left(-\sin\alpha \frac{\partial}{\partial x} + \cos\alpha \frac{\partial}{\partial y}\right).$$

Therefore,

$$\frac{\partial^2 u}{\partial \eta^2} = \frac{\partial}{\partial \eta}\left(\frac{\partial u}{\partial \eta}\right)$$

$$= \left(-\sin\alpha \frac{\partial}{\partial x} + \cos\alpha \frac{\partial}{\partial y}\right)$$

$$\times \left(-\sin\alpha \frac{\partial u}{\partial x} + \cos\alpha \frac{\partial u}{\partial y}\right)$$

$$= -\sin\alpha \frac{\partial}{\partial x}\left(-\sin\alpha \frac{\partial u}{\partial x} + \cos\alpha \frac{\partial u}{\partial y}\right)$$

$$+ \cos\alpha \frac{\partial}{\partial y}\left(-\sin\alpha \frac{\partial u}{\partial x} + \cos\alpha \frac{\partial u}{\partial y}\right)$$

$$= \sin^2\alpha \frac{\partial^2 u}{\partial x^2} - 2\sin\alpha\cos\alpha \frac{\partial^2 u}{\partial x \partial y} + \cos^2\alpha \frac{\partial^2 u}{\partial y^2}. \quad (2)$$

Adding (1) and (2), we get

$$\frac{\partial^2 u}{\partial \zeta^2} + \frac{\partial^2 u}{\partial \eta^2} = \frac{\partial^2 u}{\partial x^2} + \frac{\partial^2 u}{\partial y^2}.$$

EXAMPLE 5.39

If $y^3 - 3ax^2 + x^3 = 0$, show that

$$\frac{d^2y}{dx^2} + \frac{2a^2x^2}{y^5} = 0.$$

Solution. Let

$$f(x, y) = y^3 - 3ax^2 + x^3 = 0.$$

Therefore, by Corollary 3 of Theorem 5.4

$$\frac{d^2y}{dx^2} = -\frac{f_{xx}f_y^2 - 2f_xf_yf_{xy} + f_{yy}f_x^2}{f_y^3}. \quad (1)$$

But in the present case,

$f_x(x, y) = -6ax + 3x^2$, $f_y(x, y) = 3y^2$, $f_{xy}(x, y) = 0$, $f_{xx}(x, y) = -6a + 6x$, and $f_{yy}(x, y) = 6y$.

Therefore, (1) reduces to

$$\frac{d^2y}{dx^2} = -\frac{6(x-a)\,9y^4 + (3x^2 - 6ax)^2\,6y}{27y^6} = -2\frac{a^2x^2}{y^5}.$$

Hence,

$$\frac{d^2y}{dx^2} + 2\frac{a^2x^2}{y^5} = 0.$$

EXAMPLE 5.40

If $u = f(y - z, z - x, z - y)$, show that

$$\frac{\partial u}{\partial x} + \frac{\partial u}{\partial y} + \frac{\partial u}{\partial z} = 0.$$

Solution. Substituting $y - z, = A$, $z - x = B$, and $x - y = C$, we have

$$u = f(A, B, C).$$

Thus, u is a composite function of A, B, and C. Therefore,

$$\frac{\partial u}{\partial x} = \frac{\partial u}{\partial A}\cdot\frac{\partial A}{\partial x} + \frac{\partial u}{\partial B}\cdot\frac{\partial B}{\partial x} + \frac{\partial u}{\partial C}\cdot\frac{\partial C}{\partial x}$$

$$= \frac{\partial u}{\partial A}(0) + \frac{\partial u}{\partial B}(-1) + \frac{\partial u}{\partial C}(1) = -\frac{\partial u}{\partial B} + \frac{\partial u}{\partial C}, (1)$$

$$\frac{\partial u}{\partial y} = \frac{\partial u}{\partial A}\cdot\frac{\partial A}{\partial y} + \frac{\partial u}{\partial B}\cdot\frac{\partial B}{\partial y} + \frac{\partial u}{\partial C}\cdot\frac{\partial C}{\partial y} = \frac{\partial u}{\partial A} - \frac{\partial u}{\partial C}, (2)$$

$$\frac{\partial u}{\partial z} = \frac{\partial u}{\partial A}\cdot\frac{\partial A}{\partial z} + \frac{\partial u}{\partial B}\cdot\frac{\partial B}{\partial z} + \frac{\partial u}{\partial C}\cdot\frac{\partial C}{\partial z} = -\frac{\partial u}{\partial A} + \frac{\partial u}{\partial B} \ (3)$$

Adding (1), (2), and (3), we have

$$\frac{\partial u}{\partial x} + \frac{\partial u}{\partial y} + \frac{\partial u}{\partial z} = 0.$$

EXAMPLE 5.41

If $u = f(r, s, t)$ and $r = \frac{x}{y}$, $s = \frac{y}{z}$, $t = \frac{z}{x}$, show that

$$x\frac{\partial u}{\partial x} + y\frac{\partial u}{\partial y} + z\frac{\partial u}{\partial z} = 0$$

Solution. The function $u = f(r, s, t)$ is a composite function of r, s, and t. We have,

$$\frac{\partial u}{\partial x} = \frac{\partial u}{\partial r}\cdot\frac{\partial r}{\partial x} + \frac{\partial u}{\partial s}\cdot\frac{\partial s}{\partial x} + \frac{\partial u}{\partial t}\cdot\frac{\partial t}{\partial x}$$

$$= \frac{\partial u}{\partial r}\left(\frac{1}{y}\right) + \frac{\partial u}{\partial s}(0) + \frac{\partial u}{\partial t}\left(-\frac{z}{x^2}\right)$$

$$= \frac{\partial u}{y\partial r} - \frac{z}{x^2}\frac{\partial u}{\partial t},$$

$$\frac{\partial u}{\partial y} = \frac{\partial u}{\partial r}\cdot\frac{\partial r}{\partial y} + \frac{\partial u}{\partial s}\cdot\frac{\partial s}{\partial y} + \frac{\partial u}{\partial t}\cdot\frac{\partial t}{\partial y}$$

$$= -\frac{x}{y^2}\frac{\partial u}{\partial r} + \frac{1}{z}\frac{\partial u}{\partial s}, \text{ and}$$

$$\frac{\partial u}{\partial z} = \frac{\partial u}{\partial r}\cdot\frac{\partial r}{\partial z} + \frac{\partial u}{\partial s}\cdot\frac{\partial s}{\partial z} + \frac{\partial u}{\partial t}\cdot\frac{\partial t}{\partial z}$$

$$= -\frac{y}{z^2}\frac{\partial u}{\partial s} + \frac{1}{x}\frac{\partial u}{\partial t}.$$

Therefore,

$$x\frac{\partial u}{\partial x} + y\frac{\partial u}{\partial y} + z\frac{\partial u}{\partial z} = \frac{x}{y}\frac{\partial u}{\partial r} - \frac{z}{x}\frac{\partial u}{\partial t} - \frac{x}{y}\frac{\partial u}{\partial r}$$

$$+\frac{y}{z}\frac{\partial u}{\partial s} - \frac{y}{z}\frac{\partial u}{\partial s} + \frac{z}{x}\frac{\partial u}{\partial t} = 0.$$

EXAMPLE 5.42

If $f(x, y) = \theta(u, v)$, where $u = x^2 - y^2$ and $v = 2xy$, show that

$$\frac{\partial^2 f}{\partial x^2} + \frac{\partial^2 f}{\partial y^2} = 4(x^2 + y^2)\left(\frac{\partial^2\theta}{\partial u^2} + \frac{\partial^2\theta}{\partial v^2}\right).$$

Solution. Here $f(x, y)$ is composite function of u and v. Therefore,

$$\frac{\partial f}{\partial x} = \frac{\partial f}{\partial u}\cdot\frac{\partial u}{\partial x} + \frac{\partial f}{\partial v}\cdot\frac{\partial v}{\partial x} = \frac{\partial\theta}{\partial u}(2x) + \frac{\partial\theta}{\partial v}(2y)$$

$$= 2\left(x\frac{\partial}{\partial u} + y\frac{\partial}{\partial v}\right)\theta,$$

which yields

$$\frac{\partial}{\partial x} = 2\left(x\frac{\partial}{\partial u} + y\frac{\partial}{\partial v}\right).$$

Therefore,

$$\frac{\partial^2 f}{\partial x^2} = \frac{\partial}{\partial x}\left(\frac{\partial f}{\partial x}\right) = 2\frac{\partial}{\partial x}\left[x\frac{\partial\theta}{\partial u} + y\frac{\partial\theta}{\partial v}\right]$$

$$= 4\left(x\frac{\partial}{\partial u} + y\frac{\partial}{\partial v}\right)\left(x\frac{\partial\theta}{\partial u} + y\frac{\partial\theta}{\partial v}\right)$$

$$= 4\left(x^2\frac{\partial^2\theta}{\partial u^2} + xy\frac{\partial^2\theta}{\partial u\partial v} + xy\frac{\partial^2\theta}{\partial v\partial u} + y^2\frac{\partial^2\theta}{\partial u^2}\right)$$

$$= 4\left(x^2\frac{\partial^2\theta}{\partial u^2} + 2xy\frac{\partial^2\theta}{\partial u\partial v} + y^2\frac{\partial^2\theta}{\partial v^2}\right). \quad (1)$$

Further,

$$\frac{\partial f}{\partial y} = \frac{\partial f}{\partial u}\cdot\frac{\partial u}{\partial y} + \frac{\partial f}{\partial v}\cdot\frac{\partial v}{\partial y} = \frac{\partial\theta}{\partial u}(-2y) + \frac{\partial\theta}{\partial v}(2x)$$

$$= -2\left(y\frac{\partial\theta}{\partial u} - x\frac{\partial\theta}{\partial v}\right)$$

and so,

$$\frac{\partial}{\partial y} = -2\left(y\frac{\partial}{\partial u} - x\frac{\partial}{\partial v}\right).$$

Therefore,

$$\frac{\partial^2 f}{\partial y^2} = \frac{\partial}{\partial y}\left(\frac{\partial f}{\partial y}\right) = 4\left(y\frac{\partial}{\partial u} - x\frac{\partial}{\partial v}\right)\left(y\frac{\partial\theta}{\partial u} - x\frac{\partial\theta}{\partial v}\right)$$

$$= 4\left(y^2\frac{\partial^2\theta}{\partial u^2} - xy\frac{\partial^2\theta}{\partial u\partial v} - xy\frac{\partial^2\theta}{\partial v\partial u} + x^2\frac{\partial^2\theta}{\partial v^2}\right)$$

$$= 4\left(y^2\frac{\partial^2\theta}{\partial u} - 2xy\frac{\partial^2\theta}{\partial x\partial y} + x^2\frac{\partial^2\theta}{\partial v^2}\right), \quad (2)$$

Adding (1) and (2), we get

$$\frac{\partial^2 f}{\partial x^2} + \frac{\partial^2 f}{\partial y^2} = 4(x^2 + y^2)\left[\frac{\partial^2\theta}{\partial u^2} + \frac{\partial^2\theta}{\partial v^2}\right].$$

EXAMPLE 5.43

Let $u(x, y)$ be a function possessing continuous partial derivatives of the first two orders and let $x + y = 2e^\theta\cos\phi$ and $x - y = 2ie^\theta\sin\phi$. Show that

$$\frac{\partial^2 u}{\partial\theta^2} + \frac{\partial^2 u}{\partial\phi^2} = 4xy\frac{\partial^2 u}{\partial x\partial y}.$$

Solution. The function u is a composite function of θ and ϕ. We are given that $x + y = 2e^\theta\cos\phi$ and $x - y = 2ie^\theta\sin\phi$. Solving these equations, we get

$$x = e^\theta(\cos\phi + i\sin\phi) = e^\theta e^{i\phi} \text{ and}$$

$$y = e^\phi(\cos\phi - i\sin\phi) = e^\theta e^{-\phi}.$$

Therefore,

$$\frac{\partial u}{\partial\theta} = \frac{\partial u}{\partial x}\frac{\partial x}{\partial\theta} + \frac{\partial u}{\partial y}\frac{\partial y}{\partial\theta} = e^\theta\left(e^{i\phi}\frac{\partial u}{\partial x} + e^{-i\phi}\frac{\partial u}{\partial y}\right)$$

$$= e^\theta\left(e^{i\phi}\frac{\partial}{\partial x} + e^{-i\phi}\frac{\partial}{\partial y}\right)(u),$$

and so,

$$\frac{\partial}{\partial\theta} = e^\theta\left(e^{i\phi}\frac{\partial}{\partial x} + e^{-i\phi}\frac{\partial}{\partial y}\right).$$

Therefore,

$$\frac{\partial^2 u}{\partial\theta^2} = \frac{\partial}{\partial\theta}\left(\frac{\partial u}{\partial\theta}\right) = e^{2\phi}\left(e^{i\theta}\frac{\partial}{\partial x} + e^{-i\phi}\frac{\partial}{\partial y}\right)$$

$$\times\left(e^{i\phi}\frac{\partial u}{\partial x} + e^{-i\phi}\frac{\partial u}{\partial y}\right)$$

$$= e^{2\theta}\left[e^{2i\phi}\frac{\partial^2 u}{\partial x^2} + 2\frac{\partial^2 u}{\partial x\partial y} + e^{-2i\phi}\frac{\partial^2 u}{\partial y^2}\right].$$

Similarly,

$$\frac{\partial^2 u}{\partial \phi} = -e^{2\theta}\left[e^{2i\phi}\frac{\partial^2 u}{\partial x^2} - 2\frac{\partial^2 u}{\partial x \partial y} + e^{-2i\phi}\frac{\partial^2 u}{\partial y^2}\right].$$

Adding the earlier two expressions, we get

$$\frac{\partial^2 u}{\partial \theta^2} + \frac{\partial^2 u}{\partial \phi^2} = 4e^{2\theta}\left(\frac{\partial^2 u}{\partial x \partial y}\right) = 4xy\frac{\partial^2 u}{\partial x \partial y}.$$

5.9 TRANSFORMATION FROM CARTESIAN TO POLAR COORDINATES AND VICE VERSA

In this section, we deal with some important transformations of relations from one coordinate system to another coordinate system.

EXAMPLE 5.44

Transform the Laplace equation $\frac{\partial^2 V}{\partial x^2} + \frac{\partial^2 V}{\partial y^2} = 0$ into polar coordinates.

Solution. The function V is a function of x and y, where x and y are expressed in polar coordinates as $x = r\cos\theta$ and $y = r\sin\theta$. Thus, V is a function of r and θ also. We have $r = \sqrt{x^2+y^2}$ and $\theta = \tan^{-1}\left(\frac{y}{x}\right)$.

Therefore,

$$\frac{\partial r}{\partial x} = \frac{x}{\sqrt{x^2+y^2}} = \frac{r\cos\theta}{r} = \cos\theta,$$

$$\frac{\partial r}{\partial y} = \frac{y}{\sqrt{x^2+y^2}} = \frac{r\sin\theta}{r} = \sin\theta,$$

$$\frac{\partial \theta}{\partial x} = \frac{1}{1+\frac{y^2}{x^2}}\left(-\frac{y}{x^2}\right) = -\frac{y}{x^2+y^2}$$

$$= -\frac{r\sin\theta}{r^2} = -\frac{\sin\theta}{r}, \text{ and}$$

$$\frac{\partial \theta}{\partial y} = \frac{1}{1+\frac{y^2}{x^2}}\left(\frac{1}{x}\right) = \frac{x}{x^2+y^2} = \frac{r\cos\theta}{r^2} = \frac{\cos\theta}{r}.$$

Now,

$$\frac{\partial V}{\partial x} = \frac{\partial V}{\partial r}\cdot\frac{\partial r}{\partial x} + \frac{\partial V}{\partial \theta}\cdot\frac{\partial \theta}{\partial x} = \cos\theta\frac{\partial V}{\partial r} - \frac{\sin\theta}{r}\frac{\partial V}{\partial \theta} \text{ and}$$

$$\frac{\partial V}{\partial y} = \frac{\partial V}{\partial r}\cdot\frac{\partial r}{\partial y} + \frac{\partial V}{\partial \theta}\cdot\frac{\partial \theta}{\partial y} = \sin\theta\frac{\partial V}{\partial r} + \frac{\cos\theta}{r}\frac{\partial V}{\partial \theta}.$$

Therefore,

$$\frac{\partial}{\partial x} = \left(\cos\theta\frac{\partial}{\partial r} - \frac{\sin\theta}{r}\frac{\partial}{\partial \theta}\right) \text{ and}$$

$$\frac{\partial}{\partial y} = \left(\sin\theta\frac{\partial}{\partial r} + \frac{\cos\theta}{r}\frac{\partial}{\partial \theta}\right).$$

Hence,

$$\frac{\partial^2 V}{\partial x^2} = \left(\cos\theta\frac{\partial}{\partial r} - \frac{\sin\theta}{r}\frac{\partial}{\partial \theta}\right) \times \left(\cos\theta\frac{\partial V}{\partial r} - \frac{\sin\theta}{r}\frac{\partial V}{\partial \theta}\right)$$

$$= \cos\theta\frac{\partial}{\partial r}\left(\cos\theta\frac{\partial V}{\partial r} - \frac{\sin\theta}{r}\frac{\partial V}{\partial \theta}\right) - \frac{\sin\theta}{r}\frac{\partial}{\partial \theta}\left(\cos\theta\frac{\partial V}{\partial r} - \frac{\sin\theta}{r}\frac{\partial V}{\partial \theta}\right)$$

$$= \cos\theta\left(\cos\theta\frac{\partial^2 V}{\partial r^2} + \frac{\sin\theta}{r^2}\frac{\partial V}{\partial \theta} - \frac{\sin\theta}{r}\frac{\partial^2 V}{\partial r\partial \theta}\right) - \frac{\sin\theta}{r}\left(\cos\theta\frac{\partial^2 V}{\partial \theta\partial r} - \sin\theta\frac{\partial V}{\partial r} - \frac{\cos\theta}{r}\frac{\partial V}{\partial \theta} - \frac{\sin\theta}{r}\frac{\partial^2 V}{\partial \theta^2}\right)$$

$$= \cos^2\theta\frac{\partial^2 V}{\partial r^2} - 2\frac{\sin\theta\cos\theta}{r^2}\frac{\partial^2 V}{\partial r\partial \theta} + \frac{\sin\theta}{r^2}\frac{\partial^2 V}{\partial \theta^2} + \frac{\sin^2\theta}{r}\frac{\partial V}{\partial r} + \frac{2\sin\theta}{r^2}\frac{\partial V}{\partial \theta} \quad (1)$$

and

$$\frac{\partial^2 V}{\partial y^2} = \left(\sin\theta\frac{\partial}{\partial r} + \frac{\cos\theta}{r}\frac{\partial}{\partial \theta}\right)\left(\sin\theta\frac{\partial V}{\partial r} + \frac{\cos\theta}{r}\frac{\partial V}{\partial \theta}\right)$$

$$= \sin\theta\frac{\partial}{\partial r}\left(\sin\theta\frac{\partial V}{\partial r} + \frac{\cos\theta}{r}\frac{\partial V}{\partial \theta}\right) + \frac{\cos\theta}{r}\frac{\partial}{\partial \theta}\left(\sin\theta\frac{\partial V}{\partial r} + \frac{\cos\theta}{r}\frac{\partial V}{\partial \theta}\right)$$

$$= \sin\theta\left(\sin\theta\frac{\partial^2 V}{\partial r^2} - \frac{\cos\theta}{r^2}\frac{\partial V}{\partial \theta} + \frac{\cos\theta}{r}\frac{\partial^2 V}{\partial r\partial \theta}\right) + \frac{\cos\theta}{r}\left(\sin\theta\frac{\partial^2 V}{\partial \theta\partial r} + \cos\theta\frac{\partial V}{\partial r} - \frac{\sin\theta}{r}\frac{\partial V}{\partial \theta} + \frac{\cos\theta}{r}\frac{\partial^2 V}{\partial \theta}\right)$$

$$= \sin^2\theta\frac{\partial^2 V}{\partial r^2} + \frac{\sin\theta\cos\theta}{r}\frac{\partial^2 V}{\partial r\partial \theta} + \frac{\cos\theta\sin\theta}{r^2}\frac{\partial V}{\partial \theta} + \frac{\cos\theta\sin\theta}{r}\frac{\partial^2 V}{\partial \theta\partial r} + \frac{\cos^2\theta}{r^2}\cdot\frac{\partial^2 V}{\partial \theta^2} + \frac{\cos^2\theta}{r}\frac{\partial V}{\partial r} - \frac{\sin\theta\cos\theta}{r^2}\frac{\partial V}{\partial \theta}. \quad (2)$$

Adding (1) and (2), we obtain

$$\frac{\partial^2 V}{\partial x^2} + \frac{\partial^2 V}{\partial y^2} = \frac{\partial^2 V}{\partial r^2} + \frac{1}{r}\cdot\frac{\partial V}{\partial r} + \frac{1}{r^2}\frac{\partial^2 V}{\partial \theta^2},$$

which is the required result.

EXAMPLE 5.45

Transform the expression

$$\left(x\frac{\partial V}{\partial x}+y\frac{\partial V}{\partial y}\right)^2+(a^2-x^2-y^2)\left\{\left(\frac{\partial V}{\partial x}\right)^2+\left(\frac{\partial V}{\partial y}\right)^2\right\}$$

by substituting $x = r\cos\theta$ and $y = r\sin\theta$.

Solution. If V is a function of x, y, then

$$\frac{\partial V}{\partial r}=\frac{\partial V}{\partial x}\cdot\frac{\partial x}{\partial r}+\frac{\partial V}{\partial y}\cdot\frac{\partial y}{\partial r}=\frac{x}{r}\frac{\partial V}{\partial x}+\frac{y}{r}\frac{\partial V}{\partial y}$$

or

$$r\frac{\partial V}{\partial r}=x\frac{\partial V}{\partial x}+y\frac{\partial V}{\partial y}=\left(x\frac{\partial}{\partial x}+y\frac{\partial}{\partial y}\right)V.$$

Thus,

$$r\frac{\partial}{\partial r}=x\frac{\partial}{\partial x}+y\frac{\partial}{\partial y}\cdot$$

Similarly,

$$\frac{\partial}{\partial\theta}=x\frac{\partial}{\partial y}-y\frac{\partial}{\partial x}\cdot$$

Now,

$$\frac{\partial V}{\partial x}=\frac{\partial V}{\partial r}\cdot\frac{\partial r}{\partial x}+\frac{\partial V}{\partial\theta}\cdot\frac{\partial\theta}{\partial x}=\cos\theta\frac{\partial V}{\partial r}-\frac{\sin\theta}{r}\frac{\partial V}{\partial\theta}\text{ and}$$

$$\frac{\partial V}{\partial y}=\sin\theta\frac{\partial V}{\partial r}+\frac{\cos\theta}{r}\frac{\partial V}{\partial\theta}\cdot$$

Therefore,

$$\left(\frac{\partial V}{\partial x}\right)^2+\left(\frac{\partial V}{\partial y}\right)^2=\left(\frac{\partial V}{\partial r}\right)^2+\frac{1}{r^2}\left(\frac{\partial V}{\partial\theta}\right)^2,$$

and the given expression is equal to

$$\left(r\frac{\partial V}{\partial r}\right)^2+(a^2-r^2)\left[\left(\frac{\partial V}{\partial r}\right)^2+\frac{1}{r^2}\left(\frac{\partial V}{\partial\theta}\right)^2\right]$$
$$=a^2\left(\frac{\partial V}{\partial r}\right)^2+\left(\frac{a^2}{r^2}-1\right)\left(\frac{\partial V}{\partial\theta}\right)^2.$$

EXAMPLE 5.46

If $x = r\cos\theta$ and $y = r\sin\theta$, prove that

$$(x^2-y^2)\left(\frac{\partial^2u}{\partial x^2}-\frac{\partial^2u}{\partial y^2}\right)+4xy\frac{\partial^2u}{\partial x\partial y}$$
$$=r^2\frac{\partial^2u}{\partial r^2}-r\frac{\partial u}{\partial r}-\frac{\partial^2u}{\partial\theta^2},$$

where u is any twice-differentiable function of x and y.

Solution. We have

$$\frac{\partial u}{\partial r}=\frac{\partial u}{\partial x}\frac{\partial x}{\partial r}+\frac{\partial u}{\partial y}\cdot\frac{\partial y}{\partial r}$$
$$=\cos\theta\frac{\partial u}{\partial x}+\sin\theta\frac{\partial u}{\partial y}=\frac{x}{r}\frac{\partial u}{\partial x}+\frac{y}{r}\frac{\partial u}{\partial y}$$

and so,

$$r\frac{\partial u}{\partial r}=x\frac{\partial u}{\partial x}+y\frac{\partial u}{\partial y}\cdot\qquad(1)$$

Therefore,

$$r\frac{\partial}{\partial r}\left(r\frac{\partial u}{\partial r}\right)$$
$$=\left(x\frac{\partial}{\partial x}+y\frac{\partial}{\partial y}\right)\left(x\frac{\partial u}{\partial x}+y\frac{\partial u}{\partial y}\right)$$
$$=x\frac{\partial}{\partial x}\left(x\frac{\partial u}{\partial x}+y\frac{\partial u}{\partial y}\right)+y\frac{\partial}{\partial y}\left(x\frac{\partial u}{\partial x}+y\frac{\partial u}{\partial y}\right)$$
$$=x^2\frac{\partial^2u}{\partial x^2}+xy\frac{\partial^2u}{\partial x\partial y}+xy\frac{\partial^2u}{\partial y\partial x}+y^2\frac{\partial^2u}{\partial y^2}+x\frac{\partial u}{\partial x}+y\frac{\partial u}{\partial y}$$
$$=x^2\frac{\partial^2u}{\partial x^2}+2xy\frac{\partial^2u}{\partial x\partial y}+y^2\frac{\partial^2u}{\partial y^2}+x\frac{\partial y}{\partial x}+y\frac{\partial u}{\partial y}\cdot$$

Therefore,

$$r^2\frac{\partial^2u}{\partial r^2}-r\frac{\partial u}{\partial r}=x^2\frac{\partial^2u}{\partial x^2}+2xy\frac{\partial^2u}{\partial x\partial y}$$
$$+y^2\frac{\partial^2u}{\partial y^2},\text{ using (1).}\qquad(2)$$

Again,

$$\frac{\partial u}{\partial\theta}=\frac{\partial u}{\partial x}\frac{\partial x}{\partial\theta}+\frac{\partial u}{\partial y}\frac{\partial y}{\partial\theta}=x\frac{\partial u}{\partial y}-y\frac{\partial u}{\partial x}\cdot$$

Therefore,

$$\frac{\partial^2u}{\partial\theta^2}=\left(x\frac{\partial}{\partial y}-y\frac{\partial}{\partial x}\right)\left(x\frac{\partial u}{\partial y}-y\frac{\partial u}{\partial x}\right)$$
$$=x\frac{\partial}{\partial y}\left(x\frac{\partial u}{\partial y}-y\frac{\partial u}{\partial x}\right)-y\frac{\partial}{\partial x}\left(x\frac{\partial u}{\partial y}-y\frac{\partial u}{\partial x}\right)$$
$$=x^2\frac{\partial^2u}{\partial y^2}-2xy\frac{\partial^2u}{\partial y\partial x}+y^2\frac{\partial^2u}{\partial x^2}-x\frac{\partial u}{\partial x}-y\frac{\partial u}{\partial y}\cdot\qquad(3)$$

Subtracting (3) from (2), we get the required result.

EXAMPLE 5.47

If $x = r\cos\theta$, $y = r\sin\theta$, and $z = f(x, y)$, prove that

$$\frac{\partial z}{\partial x}=\frac{\partial z}{\partial r}\cos\theta-\frac{1}{r}\frac{\partial z}{\partial\theta}\sin\theta.$$

Solution. Since $x = r\cos\theta$ and $y = r\sin\theta$, we have

$r=\sqrt{x^2+y^2}$ and $\theta=\tan^{-1}\left(\frac{y}{x}\right)$. Therefore,

$$\frac{\partial r}{\partial x}=\frac{x}{r}=\frac{r\cos\theta}{r}=\cos\theta,$$
$$\frac{\partial r}{\partial y}=\frac{y}{r}=\frac{r\sin\theta}{r}=\sin\theta,$$
$$\frac{\partial\theta}{\partial x}=-\frac{y}{x^2+y^2}=\frac{-r\sin\theta}{r^2}=\frac{-\sin\theta}{r},\text{ and}$$
$$\frac{\partial\theta}{\partial y}=\frac{y}{x^2+y^2}=\frac{r\cos\theta}{r^2}=\frac{\cos\theta}{r}.$$

Now,

$$\frac{\partial z}{\partial x}=\frac{\partial z}{\partial r}\cdot\frac{\partial r}{\partial x}+\frac{\partial z}{\partial \theta}\cdot\frac{\partial \theta}{\partial x}=\cos\theta\frac{\partial z}{\partial r}-\frac{\sin\theta}{r}\frac{\partial z}{\partial \theta}.$$

5.10 TAYLOR'S THEOREM FOR FUNCTIONS OF SEVERAL VARIABLES

In view of Taylor's theorem for functions of one variable, it is not unnatural to expect the possibility of expanding a function of more than one variable $f(x+h, y+k, z+l,\ldots)$ in a series of ascending powers of $h, k, l, \ldots$. To fix the ideas, we consider here a function of two variables only, the reasoning in the general case being preciselythe same.

Theorem 5.5. (Taylor). If $f(x, y)$ and all its partial derivatives of order n are finite and continuous for all points (x, y) in the domain $a \le x \le a+h$ and $b \le y \le b+k$, then $f(a+h, b+k)=f(a,b)+df(a,b)+\frac{1}{2!}d^2f(a, b)+\ldots+\frac{1}{(n-1)!}d^{n-1}f(a, b)+R_n$,

where

$$R_n=\frac{1}{n!}d^nf(a+\theta h, b+\theta k),\ 0<\theta<1$$

and

$$df=h\frac{\partial f}{\partial x}+k\frac{\partial f}{\partial y}.$$

Proof: Consider a circular domain of center (a, b) and radius large enough for the point $(a+h, b+k)$ to be also with in the domain. The partial derivatives of the order n of $f(x, y)$ are continuous in the domain. Write

$$x=a+ht \quad \text{and} \quad y=b+kt$$

so that as t ranges from 0 to 1, the point (x, y) moves along the line joining the point (a, b) to the point $(a+h, b+k)$. Then

$$f(x, y)=f(a+ht, b+kt)=\phi(t). \qquad (1)$$

Now,

$$\phi'(t)=\frac{\partial f}{\partial x}\cdot\frac{dx}{dt}+\frac{\partial f}{\partial y}\cdot\frac{dy}{dt}=h\frac{\partial f}{\partial x}+k\frac{\partial f}{\partial y}=df.$$

Similarly,

$$\phi''(t)=d^2f,\ldots,\phi^{(n)}(t)=d^nf.$$

Thus, $\phi(t)$ and its n derivatives are continuous functions of t in the interval $0\le t\le 1$ and so, by Maclaurin's theorem for function of one variable, we have

$$\phi(t)=\phi(0)+t\phi'(0)+\frac{t^2}{2!}\phi''(0)$$

$$+\ldots+\frac{t^n}{n!}\phi^{(n)}(\theta t), \qquad (2)$$

where $0<\theta<1$. Further, the relation (1) yields

$$\phi(1)=f(a+h, b+k),$$
$$\phi(0)=f(a, b),$$
$$\phi'(0)=df(a, b),$$
$$\phi''(0)=d^2f(a, b), \text{ and}$$
$$\phi^{(n)}(\theta t)=d^nf(a+\theta h, b+\theta k).$$

Hence, (2) yields

$$f(a+h, b+k)=f(a, b)+df(a, b)+\frac{1}{2!}d^2f(a, b)+\ldots+\frac{1}{(n-1)!}d^{n-1}f(a, b)+R_n,$$

where

$$R_n=\frac{1}{n!}d^nf(a+\theta h, b+\theta k),\ 0<\theta<1.$$

Remark 5.2. Taylor's expansion does not necessarily hold if the partial derivatives of the order n of the function are not continuous in the domain. The theorem can be extended easily to any number of variables.

Theorem 5.6. (Maclaurin). Under the conditions of the Taylor's theorem

$$f(x, y)=f(0, 0)+df(0, 0)+\frac{1}{2!}d^2f(0, 0)+\ldots+\frac{1}{(n-1)!}d^{n-1}f(0, 0)+R_n,$$

where

$$R_n=\frac{1}{n!}d^nf(\theta x, \theta y),\ 0<\theta<1.$$

Proof By Taylor's theorem, we have

$$f(a+h, b+k)=f(a, b)+df(a, b)+\frac{1}{2!}d^2f(a, b)+\ldots+\frac{1}{(n-1)!}d^{n-1}f(a, b)+R_n,$$

where

$$R_n=\frac{1}{n!}d^nf(a+\theta h, b+\theta k),\ 0<\theta<1.$$

Putting

$$a=b=0,\ h=x, \text{ and } k=y, \text{ we get}$$

$$f(x, y)=f(0, 0)+df(0, 0)+\frac{1}{2!}d^2f(0, 0)+\ldots+\frac{1}{(n-1)!}d^{n-1}f(0, 0)+R_n,$$

where

$$R_n=\frac{1}{n!}d^nf(\theta x, \theta y),\ 0<\theta<1.$$

EXAMPLE 5.48

If $f(x, y)=(|xy|)^{\frac{1}{2}}$, prove that the Taylor's expansion about the point (x, x) is not valid in any domain which includes the origin. Give reasons.

Solution. We are given that

$$f(x, y) = (|xy|)^{\frac{1}{2}}.$$

Therefore,

$$f_x(0, 0) = \lim_{h\to 0} \frac{f(h, 0) - f(0, 0)}{h} = \lim_{h\to 0} \frac{0}{h} = 0,$$

$$f_y(0, 0) = \lim_{k\to 0} \frac{f(0, k) - f(0, 0)}{k} = \lim_{k\to 0} \frac{0}{k} = 0.$$

Also,

$$f_x(x, y) = \begin{cases} \frac{1}{2}\left(\left|\frac{y}{x}\right|\right)^{\frac{1}{2}}, & x > 0 \\ -\frac{1}{2}\left(\left|\frac{y}{x}\right|\right)^{\frac{1}{2}}, & x < 0 \end{cases}$$

and

$$f_y(x, y) = \begin{cases} \frac{1}{2}\left(\left|\frac{x}{y}\right|\right)^{\frac{1}{2}}, & y > 0 \\ -\frac{1}{2}\left(\left|\frac{x}{y}\right|\right)^{\frac{1}{2}}, & y < 0 \end{cases}$$

Thus,

$$f_x(x, x) = f_y(x, x) = \begin{cases} \frac{1}{2}, & x > 0 \\ -\frac{1}{2}, & x < 0. \end{cases}$$

If Taylor's expansion about (x, x) for $n = 1$ were possible, then we should have

$$f(x + h, x + h) = f(x, x) + h\,[f_x(x + \theta h, x + \theta h) + f_y(x + \theta h, x + \theta h)]$$

or

$$|x + h| = \begin{cases} |x| + h & \text{if } x + \theta h > 0 \\ |x| - h & \text{if } x - \theta h < 0 \\ |x| & \text{if } x - \theta h = 0 \end{cases} \quad (1)$$

Now if the domain $(x, x; x + h, x + h)$ includes the origin, then x and $x + h$ are of opposite signs. Thus, either $|x+h| = x+h, |x| = -x$ or $|x+h| = -(x+h), |x| = x$. But under these conditions, none of the inequalities (1) holds. Hence, the expansion is not valid.

EXAMPLE 5.49

Expand $x^2y + 3y - 2$ in powers of $(x - 1)$ and $(y + 2)$ using Taylor's theorem for several variables.

Solution. For all points in the domain $a \le x < a + h$ and $b \le y \le b + k$, the Taylor's theorem asserts that

$$f(x, y) = f(a + h, b + k) = f(a, b) + df(a, b) + \frac{1}{2!}d^2f(a, b) + \frac{1}{3!}d^3f(a, b) + \frac{1}{4!}d^4f(a, b) + \ldots$$

In the present example, $a = 1$ and $b = -2$. Thus,

$f(x,y) = x^2y + 3y - 2$ which yields $f(1, -2) = -10$,
$f_x(x, y) = 2xy$ which yields $f_x(1, -2) = -4$,
$f_y(x, y) = x^2 + 3$ which yields $f_y(1, -2) = 4$,
$f_{xx}(x, y) = 2y$ which yields $f_{xx}(1, -2) = -4$,
$f_{xy}(x, y) = 2x$ which yields $f_{xy}(1, -2) = 2$,
$f_{yy}(x, y) = 0$ which yields $f_{yy}(1, -2) = 0$,
$f_{xxx}(x, y) = 0$ which yields $f_{xxx}(1, -2) = 0$,
$f_{yyy}(x, y) = 0$ which yields $f_{yyy}(1, -2) = 0$,

and

$$f_{yxx}(1, -2) = f_{xxy}(1, -2) = 2.$$

All other higher derivatives are zero. Hence,

$$\begin{aligned} f(x, y) &= f(a, b) + \left[(x - 1)\frac{\partial f}{\partial x}(1, -2) + (y + 2)\frac{\partial f}{\partial y}(1, -2)\right] \\ &\quad + \frac{1}{2!}\left[(x - 1)\frac{\partial}{\partial x} + (y + 2)\frac{\partial}{\partial y}\right]^2 f(1, -2) \\ &\quad + \frac{1}{3!}\left[(x - 1)\frac{\partial}{\partial x} + (y + 2)\frac{\partial}{\partial y}\right]^3 f(1, -2) \\ &= -10 - 4(x - 1) + 4(y + 2) \\ &\quad + \frac{1}{2!}[-4(x - 1)^2 + 4(x - 1)(y + 2)] \\ &\quad + \frac{1}{3!}[(x - 1)^3(0) + 3(x - 1)^2(y + 2)(2) \\ &\quad + 3(x - 1)(y + 2)^2(0) + (y + 2)^2(0)] \\ &= -10 - 4(x - 1) + 4(y + 2) - 2(x - 1)^2 \\ &\quad + 2(x - 1)(y + 2) + (x - 1)^2(y + 2). \end{aligned}$$

EXAMPLE 5.50

Expand $\sin(xy)$ in power of $(x - 1)$ and $\left(y - \frac{\pi}{2}\right)$ up to and including second-degree terms.

Solution. We want to expand $\sin(xy)$ about the point $\left(1, \frac{\pi}{2}\right)$. By Taylor's theorem, we have

$$\begin{aligned} f(x, y) &= f(a + h, b + k) \\ &= f(a, b) + \left[h\frac{\partial}{\partial x} + k\frac{\partial}{\partial y}\right] f(a, b) \end{aligned}$$

$$+\frac{1}{2}\left[h\frac{\partial}{\partial x}+k\frac{\partial}{\partial y}\right]^2 f(a, b)+...$$

But

$f(x, y) = \sin xy$ implies $f\left(1, \frac{\pi}{2}\right) = 1,$

$f_x(x, y) = y \cos xy$ implies $f_x\left(1, \frac{\pi}{2}\right) = 0,$

$f_y(x, y) = x \cos xy$ implies $f_y\left(1, \frac{\pi}{2}\right) = 0,$

$f_{xx}(x, y) = -y^2 \sin xy$ implies $f_{xx}\left(1, \frac{\pi}{2}\right) = -\frac{\pi^2}{4},$

$f_{xy}(x, y) = -xy \sin xy$ implies $f_{xy}\left(1, \frac{\pi}{2}\right) = \left(y - \frac{\pi}{2}\right),$

and

$f_{yy}(x, y) = -x^2 \sin xy$ implies $f_{yy}\left(1, \frac{\pi}{2}\right) = -1.$

Hence,

$$f(x, y) = 1 - \frac{\pi^2}{8}(x-1)^2 - \frac{\pi}{2}(x-1)\left(y - \frac{\pi}{2}\right) - \frac{1}{2}\left(y - \frac{\pi}{2}\right)^2.$$

EXAMPLE 5.51

Expand e^{xy} at $(1, 1)$ in powers of $(x-1)$ and $(y-1)$.

Solution. We have $f(x, y) = e^{xy}$. By Taylor's theorem, we have

$$f(x, y) = f(a+h, b+k)$$
$$= f(a, b) + \left(h\frac{\partial}{\partial x} + k\frac{\partial}{\partial y}\right) f(a, b) + \frac{1}{2!}\left(h\frac{\partial}{\partial x} + k\frac{\partial}{\partial y}\right)^2 f(a, b) + \ldots.$$

But,

$f(x, y) = e^{xy}$ implies $f(1, 1) = e,$

$f_x(x, y) = y e^{xy}$ implies $f_x(1, 1) = e,$

$f_y(x, y) = x e^{xy}$ implies $f_y(1, 1) = e,$

$f_{xx}(x, y) = y^2 e^{xy}$ implies $f_{xx}(1, 1) = e,$

$f_{xy}(x, y) = xye^{xy} + e^{xy}$ implies $f_{xy}(1, 1) = e + e = 2e,$

and

$f_{yy}(x, y) = x^2 e^{xy}$ implies $f_{yy}(1, 1) = e.$

We have $h = x - a = x - 1$ and $k = y - b = y - 1$. Hence,

$$f(x, y) = f(1, 1) + (x-a) f_x(1, 1) + (y-k) \times f_y(1, 1) + \frac{1}{2!}(x-1)^2 f_{xx}(1, 1) + (y-1)^2 f_{yy}(1, 1) + 2(x-1)(y-1) f_{xy}(1, 1)$$

$$= e + (x-1)e + (y-1)e + \frac{1}{2!}[(x-1)^2 e + 4(x-1)(y-1)e + (y-1)^2 e] + \ldots$$

$$= e\left\{1 + (x-1) + (y-1) + \frac{1}{2!}(x-1)^2 + 4(x-1)(y-1) + (y-1)^2\right\} + \ldots$$

EXAMPLE 5.52

Expand $e^{ax} \sin by$ in power of x and y as far as terms of third degree.

Solution. We have $f(x, y) = e^{ax} \sin by$. By Taylor's theorem for function of two variables,

$$f(x, y) = f(a+h, b+k)$$
$$= f(a, b) + \left(h\frac{\partial}{\partial x} + k\frac{\partial}{\partial y}\right) f(a, b) + \frac{1}{2!}\left(h\frac{\partial}{\partial x} + k\frac{\partial}{\partial y}\right)^2 f(a, b) + \frac{1}{3!}\left(h\frac{\partial}{\partial x} + k\frac{\partial}{\partial y}\right)^3 f(a, b) + \ldots.$$

We wish to expand the function about $(0, 0)$. So $h = x - 0 = x$ and $k = y - 0 = y$.

$$f(x, y) = f(0+h, 0+k)$$
$$= f(0, 0) + \left(x\frac{\partial}{\partial x} + y\frac{\partial}{\partial y}\right) f(0, 0) + \frac{1}{2!}\left(x\frac{\partial}{\partial x} + y\frac{\partial}{\partial y}\right)^2 f(0, 0) + \frac{1}{3!}\left(x\frac{\partial}{\partial x} + y\frac{\partial}{\partial y}\right)^3 f(0, 0) + \ldots$$

But,

$f(x, y) = e^{ax} \sin by$ implies $f(0, 0) = 0,$

$f_x(x, y) = ae^{ax} \sin by$ implies $f_x(0, 0) = 0,$

$f_y(x, y) = be^{ax} \cos by$ implies $f_y(0, 0) = b,$

$f_{xx}(x, y) = a^2 e^{ax} \sin by$ implies $f_{xx}(0, 0) = 0,$

$f_{xy}(x, y) = abe^{ax} \cos by$ implies $f_{xy}(0, 0) = ab,$

$f_{yy}(x, y) = -b^2 e^{ax} \sin by$ implies $f_{yy}(x, y) = 0,$

$f_{xxx}(x, y) = a^3 e^{ax} \sin by$ implies $f_{xxx}(0, 0) = 0,$

$f_{xxy}(x, y) = a^2 be^{ax} \cos by$ implies $f_{xxy}(0, 0) = a^2 b,$

$f_{xyy}(x, y) = -b^2 ae^{ax} \sin by$ implies $f_{xyy}(0, 0) = 0,$

$f_{yyy}(x, y) = -b^3 e^{ax} \cos by$ implies $f_{yyy}(0, 0) = -b^3,$

and so on. Hence,

$$f(x, y) = by + abxy + \frac{1}{3!}(3a^2bx^2y - b^3y^3) + \ldots$$

EXAMPLE 5.53

Expand $f(x, y) = \tan^{-1}\frac{y}{x}$ in the neighborhood of (1, 1) up to third-degree terms. Hence, compute $f(1.1, 0.9)$ approximately.

Solution. We note that

$$f(x, y) = \tan^{-1}\frac{y}{x} \text{ implies } f(1, 1) = \frac{\pi}{4},$$

$$f_x(x, y) = \frac{-y}{x^2 + y^2} \text{ implies } f_x(1, 1) = -\frac{1}{2},$$

$$f_y(x, y) = \frac{x}{x^2 + y^2} \text{ implies } f_y(1, 1) = \frac{1}{2},$$

$$f_{xx}(x, y) = \frac{2xy}{(x^2 + y^2)^2} \text{ implies } f_x(1, 1) = \frac{1}{2},$$

$$f_{xy}(x, y) = \frac{y^2 - x^2}{(x^2 + y^2)^2} \text{ implies } f_{xy}(1, 1) = 0,$$

$$f_{yy}(x, y) = -\frac{2xy}{(x^2 + y^2)^2} \text{ implies } f_{yy}(1, 1) = -\frac{1}{2},$$

$$f_{xxx}(x, y) = \frac{2y(y^2 - 3x^2)}{(x^2 + y^2)^3} \text{ implies } f_{xxx}(1, 1) = -\frac{1}{2},$$

$$f_{xxy}(x, y) = \frac{2x(x^2 - 3y^2)}{(x^2 + y^2)^3} \text{ implies } f_{xxy}(1, 1) = -\frac{1}{2},$$

$$f_{xyy}(x, y) = \frac{2y(3x^2 - y^2)}{(x^2 + y^2)^3} \text{ implies } f_{xyy}(1, 1) = \frac{1}{2}, \text{ and}$$

$$f_{yyy}(x, y) = \frac{2x(3y^2 - x^2)}{(x^2 + y^2)^3} \text{ implies } f_{yyy}(1, 1) = \frac{1}{2}.$$

Therefore, by Taylor's theorem, we have

$$f(x, y) = \tan^{-1}\frac{y}{x}$$

$$= f(1, 1) + \left[(x-1)\frac{\partial}{\partial x} + (y-1)\frac{\partial}{\partial y}\right] f(1, 1)$$

$$+ \frac{1}{2!}\left[(x-1)\frac{\partial}{\partial x} + (y-1)\frac{\partial}{\partial y}\right]^2 f(1, 1)$$

$$+ \frac{1}{3!}\left[(x-1)\frac{\partial}{\partial x} + (y-1)\frac{\partial}{\partial y}\right]^3 f(1, 1) + \ldots$$

$$= \frac{\pi}{4} + \left[(x-1)\left(-\frac{1}{2}\right) + (y-1)\left(\frac{1}{2}\right)\right]$$

$$+ \frac{1}{2}\left[(x-1)^2\left(\frac{1}{2}\right) + 2(x-1)(y-1)(0) + (y-1)^2\left(-\frac{1}{2}\right)\right]$$

$$+ \frac{1}{3!}\left[(x-1)^3\left(-\frac{1}{2}\right) + 3(x-1)^2(y-1)\left(-\frac{1}{2}\right) + 3(x-1)(y-1)^2\left(\frac{1}{2}\right) + (y-1)^3\left(\frac{1}{2}\right)\right] + \ldots$$

$$= \frac{\pi}{4} - \frac{1}{2}\left[(x-1) - (y-1)\right] + \frac{1}{4}\left[(x-1)^2 + (y-1)^2\right]$$

$$- \frac{1}{12}\left[(x-1)^3 + 3(x-1)^2(y-1) - 3(x-1)(y-1)^3 - (y-1)^3\right] + \ldots.$$

Putting $x = 1.1$ and $y = 0.9$, we get

$$f(1.1, 0.9) = 0.6857.$$

5.11 APPROXIMATION OF ERRORS

In numerical computation, the quantity [True value – Approximate Value] is called the *error*.

We come across the following types of errors in numerical computation.

1. *Inherent Error* (*initial error*). Inherent error is the quantity which is already present in the statement (data) of the problem before its solution This type of error arises due to the use of approximate value in the given data because there are limitations of the mathematical tables and calculators. This type of error can also be there due to mistakes by human. For example, some one can write, by mistake, 67 instead of 76. The error in this case is called *transposing* error.
2. *Round–off Error.* This error arises due to rounding off the numbers during computation and occur due to the limitation of computing aids. However, this type of error can be minimized by
   (i) Avoiding the subtraction of nearly equal numbers or division by a small number.
   (ii) Retaining at least one more significant figure at each step of calculation.
3. *Truncation Error.* If is the error caused by using approximate formulas during computation such as the one that arise when a function $f(x)$ is evaluated from an infinite series for x after truncating it at certain stage.

For example, we will see that in Newton–Raphson Method for finding the roots of an equation, if x is the true value of the root of $f(x) = 0$ and x_0 and h are approximate value and correction respectively, then by Taylor's Theorem,

$$f(x_0 + h) = f(x_0) + hf'(x_0) + \frac{h^2}{2!}f''(x_0) + \ldots + = 0.$$

To find the correction h, we truncate the series just after first derivative. Therefore some error occurs due to this truncation.

4. *Absolute Error*. If x is the true value of a quantity and x_0 is the approximate value, then $|x - x_0|$ is called the *absolute error*.
5. *Relative Error*. If x is the true value of a quantity and x_0 is the approximate value, then $\left(\frac{x - x_0}{x}\right)$ is called the *relative error*.
6. *Percentage Error*. If x is the true value of quantity and x_0 is the approximate value, then $\left(\frac{x-x_0}{x}\right) \times 100$ is called the *percentage error*. Thus, percentage error is 100 times the relative error.

5.12 GENERAL FORMULA FOR ERRORS

Let

$$u = f(u_1, u_2, \ldots, u_n) \tag{1}$$

be a function of $u_1, u_2, \ldots, u_n$ which are subject to the errors $\Delta u_1, \Delta u_2, \ldots, \Delta u_n$, respectively.

Let Δu be the error in u caused by the errors Δu_1, $\Delta u_2, \ldots, \Delta u_n$ in $u_1, u_2, \ldots, u_n$, respectively. Then

$$u + \Delta u = f(u_1 + \Delta u_1, u_2 + \Delta u_2, \ldots, u_n + \Delta u_n) \tag{2}$$

Expanding the right hand side of (2) by Taylor's Theorem for a function of several variables, we have

$$\begin{aligned} u + \Delta u &= f(u_1, u_2, \ldots, u_n) \\ &\quad + \left(\Delta u_1 \frac{\partial}{\partial u_1} + \ldots + \Delta u_n \frac{\partial}{\partial u_n}\right) f \\ &\quad + \frac{1}{2}\left(\Delta u_1 \frac{\partial}{\partial u_1} + \ldots + \Delta u_n \frac{\partial}{\partial u_n}\right)^2 f + \ldots \end{aligned}$$

Since the errors are relatively small, we neglect the squares, product and higher powers and have

$$\begin{aligned} u + \Delta u &= f(u_1, u_2, \ldots, u_n) \\ &\quad + \left(\Delta u_1 \frac{\partial}{\partial u_1} + \ldots + \Delta u_n \frac{\partial}{\partial u_n}\right) f \end{aligned} \tag{3}$$

Subtracting (1) from (3), we have

$$\Delta u = \frac{\partial f}{\partial u_1} \Delta u_1 + \frac{\partial f}{\partial u_2} \Delta u_2 + \ldots + \frac{\partial f}{\partial u_n} \Delta u_n$$

or

$$\Delta u = \frac{\partial u}{\partial u_1} \Delta u_1 + \frac{\partial u}{\partial u_2} \Delta u_2 + \ldots + \frac{\partial u}{\partial u_n} \Delta u_n,$$

which is known as ***general formula for error***. We note that the right hand side is simply the total derivative of the function u.

For a relative error E_r of the function u, we have

$$E_r = \frac{\Delta u}{u} = \frac{\partial u}{\partial u_1} \frac{\Delta u_1}{u} + \frac{\partial u}{\partial u_2} \frac{\Delta u_2}{u} + \ldots + \frac{\partial u}{\partial u_n} \frac{\Delta u_n}{u}.$$

EXAMPLE 5.54

If $u = \frac{5xy^2}{z^3}$ and error in x, y, z are 0.001, compute the relative maximum error $(E_r)_{\max}$ in u when $x = y = z = 1$.

Solution. We have $u = \frac{5xy^2}{z^3}$. Therefore

$$\frac{\partial u}{\partial x} = \frac{5y^2}{z^3}, \frac{\partial u}{\partial y} = \frac{10xy}{z^3}, \frac{\partial u}{\partial z} = -\frac{15xy^2}{z^4}$$

and so

$$\Delta u = \frac{5y^2}{z^3} \Delta x + \frac{10xy}{z^3} \Delta y - \frac{15xy^2}{z^4} \Delta z$$

But it is given that $\Delta x = \Delta y = \Delta z = 0.001$ and $x = y = z = 1$. Therefore

$$\begin{aligned} (\Delta u)_{\max} &\approx \left|\frac{5y^2}{z^3} \Delta x\right| + \left|\frac{10xy}{z^3} \Delta y\right| + \left|\frac{15xy^2}{z^4} \Delta z\right| \\ &= 5\,(0.001) + 10\,(0.001) + 15\,(0.001) \\ &= 0.03. \end{aligned}$$

Thus the relative maximum error $(E_r)_{\max}$ is given by

$$(E_r)_{\max} = \frac{(\Delta u)_{\max}}{u} = \frac{0.03}{u} = \frac{0.03}{5} = 0.006.$$

EXAMPLE 5.55

The diameter and altitude of a can in the shape of a right circular cylinder are measured as 4 cm and 6 cm respectively. The possible error in each measurement is 0.1 cm. Find approximately the maximum possible error in the values computed for the volume and the lateral surface.

Solution. If x and y denote the diameter and the height of the can, then volume of the can is given by $V = \frac{\pi}{4} x^2 y$ and so

$$\frac{\partial V}{\partial x} = \frac{\pi}{2} xy \text{ and } \frac{\partial V}{\partial y} = \frac{\pi}{4} x^2.$$

Therefore the error formula yields

$$\Delta V = \frac{\partial V}{\partial x} \Delta x + \frac{\partial V}{\partial y} \Delta y = \frac{\pi}{2} (xy\,\Delta x) + \frac{\pi}{4} (x^2 \Delta y).$$

Putting $x = 4$, $y = 6$, $\Delta x = \Delta y = 0.1$, we get

$$\Delta V = (1.2)\,\pi + (0.4)\,\pi = 1.6\,\pi \text{ cm}^3.$$

Further, the lateral surface is given by $S = \pi xy$ and so

$$\frac{\partial S}{\partial x} = \pi y \quad \text{and} \quad \frac{\partial S}{\partial y} = \pi x.$$

Therefore

$$\Delta S = \pi (y\Delta x + x\Delta y).$$

Putting the values of x, y, Δx and Δy, we get

$$\Delta S = \pi (0.6 + 0.4) = \pi \text{ cm}^2.$$

EXAMPLE 5.56

The height h and the semi-vertical angle α of a cone are measured and from them the total area A of the cone (including the base) is calculated. If h and α are in error by small quantities δh *and* $\delta \alpha$ respectively, find corresponding error in the area. Show further that $\alpha = \frac{\pi}{6}$, an error of 1 percent in h will be approximately compensated by an error of -0.33 degree in α.

Solution. Radius of the base $= r = h \tan \alpha$. Further, slant height $= l = h \sec \alpha$. Therefore,

$$\begin{aligned}\text{Total area} &= \pi r^2 + \pi r l = \pi r(r + l)\\ &= \pi h \tan \alpha (h \tan \alpha + h \sec \alpha)\\ &= \pi h^2 (\tan^2 \alpha + \sec \alpha \tan \alpha).\end{aligned}$$

Then the error in A is given by

$$\begin{aligned}\delta A &= \frac{\partial A}{\partial h}\delta h + \frac{\partial A}{\partial \alpha}\delta \alpha\\ &= 2\pi h (\tan^2 \alpha + \sec \alpha \tan \alpha)\, \delta h\\ &\quad + \pi h^2 (2 \tan \alpha \sec^2 \alpha + \sec^3 \alpha\\ &\quad + \sec \alpha \tan^2 \alpha)\, \delta \alpha\end{aligned}$$

For the second part of the question,

$$\alpha = \frac{\pi}{6},\ \delta h = \frac{h}{100}.$$

Therefore

$$\begin{aligned}\delta A &= 2\pi h \left[\frac{1}{3} + \frac{2}{3}\right] \frac{h}{100}\\ &\quad + \pi h^2 \left(\frac{2}{\sqrt{3}}\left(\frac{4}{3}\right) + \frac{8}{3\sqrt{3}} + \frac{2}{3\sqrt{3}}\right) \delta \alpha\\ &= \frac{\pi h^2}{50} + 2\sqrt{3}\, \pi h^2\, \delta \alpha \qquad (1)\end{aligned}$$

But after compensation $\delta A = 0$. Therefore (1) implies

$$\delta \alpha = -\frac{1}{100\sqrt{3}} \text{ radians} = -\frac{57.3°}{173.2} = -0.33°.$$

EXAMPLE 5.57

The time T of a complete oscillation of a simple pendulum of length L is governed by the equation $T = 2\pi \sqrt{L/g}$, g is constant. Find the approximate error in the calculated value of T corresponding to the error of 2% in the value of L.

Solution. We have

$$T = 2\pi \sqrt{\frac{L}{g}}.$$

Taking logarithm, we get

$$\log T = \log 2\pi + \frac{1}{2} \log l - \frac{1}{2} \log g \qquad (1)$$

Differentiating (1), we get

$$\frac{1}{T}\delta T = \frac{1}{2}\frac{\delta L}{L} - \frac{1}{2}\frac{\delta g}{g}$$

or

$$\begin{aligned}\frac{\delta T}{T} \times 100 &= \frac{1}{2}\left[\frac{\delta L}{L} \times 100 - \frac{1}{2}\frac{\delta g}{g} \times 100\right]\\ &= \frac{1}{2}[2 - 0] = 1.\end{aligned}$$

Hence the approximate error is 1%.

EXAMPLE 5.58

The radius of circle is found to be 100 cm. Find the approximate error in the area of the circle due to an error of 1 mm in the measured value of the radius. Also, find relative error.

Solution.

$$\text{Area of circle is } A = \pi r^2 \qquad (1)$$

On differentiation

$$\delta A = 2\pi r\, \delta r$$

where $r = 100$ cm, and $\delta r = 1$ mm $= 0.1$ cm

$$\begin{aligned}\therefore\ \delta A &= 2\pi \times 100 \times 0.1\\ &= 20\pi \text{ sq. cm.}\end{aligned}$$

Now taking log both sides of Eq. (1), we have

$$\log A = \log \pi + 2 \log r$$

On differentitation, we get

$$\frac{1}{A}\delta A = 0 + \frac{2}{r}\delta r$$

Putting $r = 100$ cm and $\delta r = 0.1$ cm, we have

$$\text{Relative Error} = \frac{\delta A}{A} = \frac{2}{100} \times 0.1 = \frac{2}{1000} = 0.002 \text{ cm}$$

EXAMPLE 5.59

The radius of a sphere is found to be 10 cm with a possible error of 0.02 cm. What is relative error in computing the volume?

Solution.
We know that the volume of sphere is

$$V = \frac{4}{3}\pi r^3$$

i.e., $\log V = \log\left(\frac{4}{3}\pi\right) + 3\log r$ (1)

Given $r = 10$ cm. and $\delta r = 0.02$ cm
On differentitation of Eq. (1), we have

$$\frac{1}{v}\delta V = 0 + \frac{3}{r}\delta r$$

$$\text{Relative Error } \frac{\delta r}{V} = 3 \times \frac{0.02}{10}$$

$$= 0.006 \text{ cm.}$$

EXAMPLE 5.60

Find the percentage error in the area of an ellipse when an error of 1% is made in measuring the major and minor axis.

Solution. The area of an ellipse is given by

$$A = \pi xy$$

where x and y are the semi-major and semi-minor axis.

Taking both side log

$$\log A = \log\pi + \log x + \log y$$

and so

$$\frac{1}{A}\delta A = 0 + \frac{1}{x}\delta x + \frac{1}{y}\delta y$$

Now, multiplying throughout by 100, we have

$$\frac{1}{A}\delta A \times 100 = \frac{\delta x}{x} \times 100 + \frac{\delta y}{y} \times 100$$

Hence
Percentage Error = 1% + 1% = 2%.

EXAMPLE 5.61

Find the percentage error in the area of rectangle when an error of 1% is made in measuring its length and breadth.

Solution. Let x and y be the length of sides the area of rectangle $A = xy$.

Taking log of both sides, we get

$$\log A = \log x + \log y$$

Therefore

$$\frac{1}{A}\delta A = \frac{1}{x}\delta x + \frac{1}{y}\delta y$$

On multiplying both side by 100, we get

$$\frac{\delta A}{A} \times 100 = \frac{\delta x}{x} 100 + \frac{\delta y}{y} \times 100$$

Hence
Percentage Error in A = Percentage Error in x + Percentage Error in y

$$= 1\% + 1\%$$
$$= 2\%$$

Percentage Error in $A = 2\%$

EXAMPLE 5.62

Deflectionofabeamvariesasthecubeofthedepth. What is the percentage error in the calculated deflection due to on error of 1.75% in the measurement of depth?

Solution. Let y be the deflection and D be the depth of beam then $y = KD^3$, where K is a constant.

Now taking log of both sides, we have

$$\log y = \log k + 3\log D$$

On differention, we have

$$\frac{1}{y}\delta y = 0 + \frac{3}{D}\delta D.$$

Therefore

$$\frac{\delta y}{y} \times 100 = 3 \times \frac{\delta D}{D} 100$$
$$= 3 \times (\% \text{ Error in depth})$$
$$= 3 \times 1.75$$
$$= 5.25\%,$$

which is required percentage error.

EXAMPLE 5.63

The indicated horse-power I of an engine is calculated from the formula $I = \frac{PLAN}{33000}$. Where $A = \frac{\pi}{4}d^2$. Assuming that errors of r percent may have been made in measuring P, L, N and d. Find the greatest possible error in I.

Solution. Here,

$$I = \frac{PLAN}{33000}$$

Putting

$$A = \frac{\pi}{4}d^2 \text{ we get}$$

$$I = \frac{\pi}{4}\frac{PLN}{33000}$$

or

$$d^2 = \frac{\pi PLd^2N}{132000}$$

Taking log on both sides, we have

$$\log I = \log\pi + \log P + \log L + 2\log d + \log N - \log 132000$$

On differentitation, we have

$$\frac{1}{I}\delta I = 0 + \frac{1}{P}\delta P + \frac{1}{L}\delta L + 2\frac{\delta d}{d} + \frac{\delta N}{N} - 0$$

Multiplying both sides by 100, we get

$$\frac{\delta I}{I}\times 100=\frac{\delta P}{P}\times 100+\frac{\delta L}{L}\times 100+2\,\frac{\delta d}{d}\times 100+\frac{\delta N}{N}\times 100$$

Now putting,

$$100\times\frac{\delta P}{P}=100\times\frac{\delta L}{L}=100\times\frac{\delta d}{d}=100\times\frac{\delta N}{N}=r.$$

Therefore, percentage error is

$$\therefore \quad \frac{\delta I}{I}\times 100=r+r+2r+r$$
$$=5r\%$$

Thus the greatest possible error in $I = 5r\%$.

EXAMPLE 5.64

The side a and the opposite angle A of a triangle ABC remains constant, show that when the others sides and angles are slightly varied,

$$\frac{\delta b}{\cos b}+\frac{\delta c}{\cos c}=0$$

Solution. We know that by the sine formula,

$$\frac{a}{\sin A}=\frac{b}{\sin B}=\frac{c}{\sin C}$$

$$\therefore \quad a\sin B=b\sin A \qquad (1)$$

$$\text{and} \quad a\sin C=c\sin A \qquad (2)$$

Since a and A are constants, we have from Eq. (1) and (2) on taking differentials

$$a\cos B\,\delta B=\sin A\,\delta b$$

$$\text{and} \quad a\cos C\,\delta C=\sin A\,\delta c$$

$$\therefore \quad \frac{\delta b}{\cos B}=\frac{a\delta B}{\sin A} \quad \text{and} \quad \frac{\delta c}{\cos C}=\frac{a\delta c}{\sin A}$$

$$\therefore \quad \frac{\delta b}{\cos B}+\frac{\delta c}{\cos C}=\frac{a}{\sin A}(\delta B+\delta C) \qquad (3)$$

Since $A+B+C=180°$ and A is a constant, we note that

$$\delta A+\delta B+\delta C=0\Rightarrow \delta B=-\delta C$$

Hence (3) yields

$$\frac{\delta b}{\cos B}+\frac{\delta c}{\cos C}=0$$

EXAMPLE 5.65

Find the approximate value of cube root of 127.

Solution. Let $f(x)=x^{\frac{1}{3}}$

Now

$$f(x+\delta x)-f(x)=f'(x)\,\delta x$$

or

$$f(x+\delta x)-f(x)=\frac{1}{3x^{\frac{2}{3}}}\cdot\delta x$$

We may write $127 = 125 + 2$

Taking $x = 125$ and $\delta x = 2$, then we have

$$f(127)-f(125)=\frac{1}{3(125)^{\frac{2}{3}}}\times 2$$

Therefore

$$f(127)=f(125)+\frac{2}{3\times 25}$$
$$=(125)^{\frac{1}{3}}+\frac{2}{75}$$
$$=5+\frac{2}{75}=\frac{377}{75}$$
$$=\frac{377}{75}$$

5.13 TANGENT PLANE AND NORMAL TO A SURFACE

Let $P(x,y,z)$ and $Q(x+\delta x,\ y+\delta y,\ z+\delta z)$ be two neighboring points on given surface $F(x,y,z) = 0$. Let δs and δc be the length of the are PQ and that of the chord PQ. Then $\lim_{Q\to P}\frac{\delta s}{\delta c}=1$.

The direction cosines of the chord PQ are

$$\frac{\delta x}{\delta c},\frac{\delta y}{\delta c},\frac{\delta z}{\delta c} \ or\ \frac{\delta x}{\delta s}.\frac{\delta s}{\delta c},\frac{\delta y}{\delta s}.\frac{\delta s}{\delta c},\frac{\delta z}{\delta s}.\frac{\delta s}{\delta c}$$

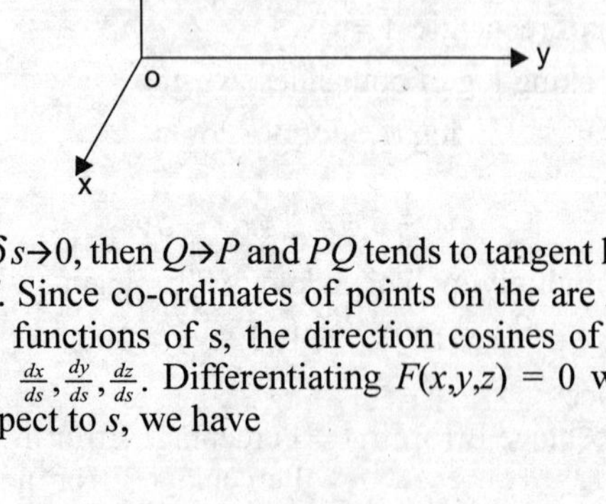

If $\delta s\to 0$, then $Q\to P$ and PQ tends to tangent line PT. Since co-ordinates of points on the are PQ are functions of s, the direction cosines of PT are $\frac{dx}{ds},\frac{dy}{ds},\frac{dz}{ds}$. Differentiating $F(x,y,z) = 0$ with respect to s, we have

$$\frac{\partial F}{\partial x}\cdot\frac{dx}{ds}+\frac{\partial F}{\partial y}\cdot\frac{dy}{ds}+\frac{\partial F}{\partial z}\cdot\frac{dz}{ds}=0. \qquad (1)$$

Expression (1) shows that the tangent line PT with direction cosines $\frac{dx}{ds},\frac{dy}{ds},\frac{dz}{ds}$ is perpendicular to the line having direction ratios $\frac{\partial F}{\partial x},\frac{\partial F}{\partial y},\frac{\partial F}{\partial z}$. Corresponding to different curves joining P and Q, there are distinct tangent lines at P to which the line with direction ratios $\frac{\partial F}{\partial x},\frac{\partial F}{\partial y},\frac{\partial F}{\partial z}$ is perpendicular. Thus all tangent lines at P lie in a plane through P perpendicular to the line with direction ratios $\frac{\partial F}{\partial x},\frac{\partial F}{\partial y},\frac{\partial F}{\partial z}$. Hence the equation of the tangent plane to the surface $F(x,y,z) = 0$ is given by

$$\frac{\partial F}{\partial x}(X-x)+\frac{\partial F}{\partial x}(Y-y)+\frac{\partial F}{\partial z}(Z-z)=0.$$

Then the equation of the normal to the surface at P is given by

$$\frac{X-x}{\frac{\partial F}{\partial x}}=\frac{Y-y}{\frac{\partial F}{\partial y}}=\frac{Z-z}{\frac{\partial F}{\partial z}}.$$

EXAMPLE 5.66

Find the equations of the tangent plane and the normal to the surface $2xz^2 - 3xy - 4x = 7$ at the point $(1, -1, 2)$.

Solution. The given surface is

$$F(x, y, z) = 2xz^2 - 3xy - 4x - 7.$$

Therefore,

$$\frac{\partial F}{\partial x}=2z^2-3y-4=7 \text{ at } (1,-1,2)$$

$$\frac{\partial F}{\partial y}=-3x=-3 \text{ at } (1,-1,2)$$

$$\frac{\partial F}{\partial z}=4xz=8 \text{ at } (1,-1,2).$$

Therefore equation of the tangent plane at $(1, -1, 2)$ is

$$7(x-1)-3(y+1)+8(z-2)=0$$

or

$$7x-3y+8z-26=0.$$

The equation of the normal to the surface at $(1, -1, 2)$ is $\frac{x-1}{7}=\frac{y+1}{-3}=\frac{z-2}{8}$.

EXAMPLE 5.67

Find the equations of the tangent plane and the normal to the surface $z^2 = 4(x^2 + y^2 + 1)$ at the point $(2, 2, 6)$.

Solution. The given surface is

$$F(x, y, z) = 4x^2 + 4y^2 - z^2 + 4.$$

Therefore,

$$\frac{\partial F}{\partial x}=8x=16 \text{ at } (2,2,6)$$

$$\frac{\partial F}{\partial y}=8y-16 \text{ at } (2,2,6)$$

$$\frac{\partial F}{\partial z}=-2z=-12 \text{ at } (2,2,6).$$

Therefore the equation of the tangent plane is

$$16(x-2)+16(y-2)-12(z-6)=0$$

or

$$4x+4y-3z+2=0.$$

The equation of the normal to the surface is

$$\frac{x-2}{18}=\frac{y-2}{16}=\frac{z-6}{-12}$$

or

$$\frac{x-2}{4}=\frac{y-2}{4}=\frac{z-6}{-3}.$$

EXAMPLE 5.68

Find the equation of the tangent plane to the surface $2x^2 + y^2 - 3 + 2z = 0$ at $(2, 1, -3)$.

Solution. We have

$$F(x, y, z) = 2x^2 + y^2 + 2z - 3.$$

Therefore,

$$\frac{\partial F}{\partial x}=4x,\ \frac{\partial F}{\partial y}=2y,\ \frac{\partial F}{\partial z}=2.$$

At $(2, 1, -3)$, we have

$$\frac{\partial F}{\partial x}=8,\ \frac{\partial F}{\partial y}=2,\ \frac{\partial F}{\partial z}=2.$$

Therefore, the equation of the tangent plane to the given surface at the point $(2, 1, -3)$ is

$$8(x-2)+2(y-1)+2(z+3)=0$$

or

$$4x + 2y + 3z - 6 = 0$$

EXAMPLE 5.69

Show that the plane $3x + 12y - 6x - 17 = 0$ touches the conicoid $3x^2 - 6y^2 + 9z^2 + 17 = 0$. Find also the point of contact.

Solution. Comparing the equation of the plane with $lx + my + nz = p$, we get

$$l = 3, m = 12, n = -6, p = 17.$$

The equation of the given conicoid in standard form is

$$-\frac{3}{17}x^2 + \frac{6}{17}y^2 - \frac{9}{17}z^2 = 1$$

$$(ax^2 + by^2 + cz^2 = 1 \text{ form}).$$

The plane will touch the conicoid if

$$\frac{l^2}{a} + \frac{m^2}{b} + \frac{n^2}{c} = p^2 \text{ (\textit{condition} of tangency)}$$

or if

$$\frac{9}{\frac{-3}{17}} + \frac{144}{\frac{6}{17}} + \frac{36}{\frac{-9}{17}} = (17)^2$$

or if

$$-51 + 408 - 68 = 289, \text{ which is true.}$$

Now let $P(x_1, y_1, z_1)$ be the point of contact. The equation of the tangent plane to the conicoid at (x_1, y_1, z_1) is

$$3xx_1 - 6yy_1 + 9zz_1 = -17.$$

Comparing it with $3x + 12y - 62 = 17$, we get

$$\frac{x_1}{1} = -\frac{6y_1}{12} = \frac{9z_1}{-6} = \frac{-17}{17}.$$

Therefore $x_1 = -1, y_1 = 2, z_1 = \frac{2}{3}$. Hence, the point of contact is $(-1, 2, \frac{2}{3})$.

5.14 JACOBIANS

If $u_1, u_2, \ldots, u_n$ are n functions of n variables $x_1, x_2, \ldots, x_n$, then the determinant

$$\begin{vmatrix} \frac{\partial u_1}{\partial x_1} & \frac{\partial u_1}{\partial x_2} & \cdots & \cdots & \frac{\partial u_1}{\partial x_n} \\ \frac{\partial u_2}{\partial x_1} & \frac{\partial u_2}{\partial x_2} & \cdots & \cdots & \frac{\partial u_2}{\partial x_n} \\ \cdots & \cdots & \cdots & \cdots & \cdots \\ \cdots & \cdots & \cdots & \cdots & \cdots \\ \frac{\partial u_n}{\partial x_1} & \frac{\partial u_n}{\partial x_2} & \cdots & \cdots & \frac{\partial u_n}{\partial x_n} \end{vmatrix}$$

is called the *Jacobian* of $u_1, u_2, \ldots, u_n$ with regard to $x_1, x_2, \ldots, x_n$. This determinant is often denoted by

$$\frac{\partial(u_1, u_2, \ldots, u_n)}{\partial(x_1, x_2, \ldots, x_n)} \text{ or } J(u_1, u_2, \ldots, u_n).$$

5.15 PROPERTIES OF JACOBIAN

Theorem 5.7. If U, V are functions of u and v, where u and v are themselves functions of x and y, then

$$\frac{\partial(U, V)}{\partial(x, y)} = \frac{\partial(U, V)}{\partial(u, v)} \cdot \frac{\partial(u, v)}{\partial(x, y)}.$$

Proof: Let

$U = f(u, v)$, $V = F(u, v)$, $u = \phi(x, y)$, and $v = \psi(x, y)$.

Then,

$$\frac{\partial U}{\partial x} = \frac{\partial U}{\partial u} \cdot \frac{\partial u}{\partial x} + \frac{\partial U}{\partial v} \cdot \frac{\partial v}{\partial x},$$

$$\frac{\partial U}{\partial y} = \frac{\partial U}{\partial u} \cdot \frac{\partial u}{\partial y} + \frac{\partial U}{\partial v} \cdot \frac{\partial v}{\partial y},$$

$$\frac{\partial V}{\partial x} = \frac{\partial V}{\partial u} \cdot \frac{\partial u}{\partial x} + \frac{\partial V}{\partial v} \cdot \frac{\partial v}{\partial x},$$

$$\frac{\partial V}{\partial y} = \frac{\partial V}{\partial u} \cdot \frac{\partial u}{\partial y} + \frac{\partial V}{\partial v} \cdot \frac{\partial v}{\partial y}.$$

and so,

$$\frac{\partial(U, V)}{\partial(u, v)} \cdot \frac{\partial(u, v)}{\partial(x, y)}$$

$$= \begin{vmatrix} \frac{\partial U}{\partial u} & \frac{\partial U}{\partial v} \\ \frac{\partial V}{\partial u} & \frac{\partial V}{\partial v} \end{vmatrix} \begin{vmatrix} \frac{\partial u}{\partial x} & \frac{\partial u}{\partial y} \\ \frac{\partial v}{\partial x} & \frac{\partial v}{\partial y} \end{vmatrix}$$

$$= \begin{vmatrix} \frac{\partial U}{\partial u} \cdot \frac{\partial u}{\partial x} + \frac{\partial U}{\partial v} \cdot \frac{\partial v}{\partial x} & \frac{\partial U}{\partial u} \cdot \frac{\partial u}{\partial y} + \frac{\partial u}{\partial v} \cdot \frac{\partial v}{\partial y} \\ \frac{\partial V}{\partial u} \cdot \frac{\partial u}{\partial x} + \frac{\partial V}{\partial v} \cdot \frac{\partial v}{\partial x} & \frac{\partial V}{\partial u} \cdot \frac{\partial u}{\partial y} + \frac{\partial V}{\partial v} \cdot \frac{\partial v}{\partial y} \end{vmatrix}$$

$$= \begin{vmatrix} \frac{\partial U}{\partial x} & \frac{\partial U}{\partial y} \\ \frac{\partial V}{\partial x} & \frac{\partial V}{\partial y} \end{vmatrix}$$

$$= \frac{\partial (U, V)}{\partial (x, y)}.$$

Theorem 5.8. If J is the Jacobian of the system u, v with regard to x, y, and J' is the Jacobian of x, y with regard to u, v, then $JJ' = 1$.

Proof: Let $u = f(x, y)$ and $v = F(x, y)$. Suppose that these are solved for x and y giving $x = \phi(u,v)$ and $y = \psi(u, v)$. Differentiating $u = f(x, y)$ with respect to u and v, we have

$$1 = \frac{\partial u}{\partial x}\cdot\frac{\partial x}{\partial u} + \frac{\partial u}{\partial y}\cdot\frac{\partial y}{\partial u} \quad\text{and}\quad 0 = \frac{\partial u}{\partial x}\cdot\frac{\partial x}{\partial v} + \frac{\partial u}{\partial y}\cdot\frac{\partial y}{\partial v}.$$

Similarly, differentiating $v = F(x, y)$ with respect to u and v, we get

$$0 = \frac{\partial v}{\partial x}\cdot\frac{\partial x}{\partial u} + \frac{\partial v}{\partial y}\cdot\frac{\partial y}{\partial u} \quad\text{and}\quad 1 = \frac{\partial v}{\partial x}\cdot\frac{\partial x}{\partial v} + \frac{\partial v}{\partial y}\cdot\frac{\partial y}{\partial v}.$$

Therefore,

$$JJ' = \begin{vmatrix} \frac{\partial u}{\partial x} & \frac{\partial u}{\partial y} \\ \frac{\partial v}{\partial x} & \frac{\partial v}{\partial y} \end{vmatrix} \begin{vmatrix} \frac{\partial x}{\partial u} & \frac{\partial x}{\partial v} \\ \frac{\partial y}{\partial u} & \frac{\partial v}{\partial v} \end{vmatrix}$$

$$= \begin{vmatrix} \frac{\partial u}{\partial x}\cdot\frac{\partial u}{\partial x} + \frac{\partial u}{\partial v}\cdot\frac{\partial y}{\partial u} & \frac{\partial u}{\partial x}\cdot\frac{\partial x}{\partial v} + \frac{\partial u}{\partial y}\cdot\frac{\partial v}{\partial v} \\ \frac{\partial v}{\partial x}\cdot\frac{\partial x}{\partial u} + \frac{\partial v}{\partial y}\cdot\frac{\partial y}{\partial u} & \frac{\partial y}{\partial x}\cdot\frac{\partial x}{\partial v} + \frac{\partial v}{\partial y}\cdot\frac{\partial y}{\partial v} \end{vmatrix}$$

$$= \begin{vmatrix} 1 & 0 \\ 0 & 1 \end{vmatrix} = 1.$$

EXAMPLE 5.70

If $x = r\cos\theta$ and $y = r\sin\theta$, show that

$$\text{(i)}\ \frac{\partial(x, y)}{\partial(r, \theta)} = r \quad\text{and}\quad \text{(ii)}\ \frac{\partial(r, \theta)}{\partial(x, y)} = \frac{1}{r}.$$

Solution. (i) We have

$$\frac{\partial(x, y)}{\partial(r, \theta)} = \begin{vmatrix} \frac{\partial x}{\partial r} & \frac{\partial x}{\partial \theta} \\ \frac{\partial y}{\partial r} & \frac{\partial y}{\partial \theta} \end{vmatrix} = \begin{vmatrix} \cos\theta & -r\sin\theta \\ \sin\theta & r\cos\theta \end{vmatrix}$$

$$= r\cos^2\theta + r\sin^2\theta = r.$$

(ii) We have

$$r^2 = x^2 + y^2 \text{ and } \tan\theta = \frac{y}{x}.$$

Differentiating partially with respect to x and y, we get

$$2r\frac{\partial r}{\partial x} = 2x \text{ and so, } \frac{\partial r}{\partial x} = \frac{x}{r},$$

$$2r\frac{\partial r}{\partial y} = 2y \text{ and so, } \frac{\partial r}{\partial y} = \frac{y}{r},$$

$$\sec^2\theta\frac{\partial\theta}{\partial x} = -\frac{y}{x^2} \text{ and so, } \frac{\partial\theta}{\partial x} = -\frac{y}{x^2\sec^2\theta}$$

$$= -\frac{y}{r^2\cos^2\theta\sec^2\theta} = -\frac{y}{r^2},$$

$$\sec^2\theta\frac{\partial\theta}{\partial y} = \frac{1}{x}, \text{ and so, } \frac{\partial\theta}{\partial y} = \frac{1}{x\sec^2\theta} = \frac{\cos^2\theta}{x}$$

$$= \frac{x^2}{r^2}\frac{1}{x} = \frac{x}{r^2},$$

Therefore,

$$\frac{\partial(r, \theta)}{\partial(x, y)} = \begin{vmatrix} \frac{\partial r}{\partial x} & \frac{\partial r}{\partial y} \\ \frac{\partial \theta}{\partial r} & \frac{\partial \theta}{\partial y} \end{vmatrix} = \begin{vmatrix} \frac{x}{r} & \frac{y}{r} \\ -\frac{y}{r^2} & \frac{x}{r^2} \end{vmatrix}$$

$$= \frac{x^2}{r^3} + \frac{y^2}{r^3} = \frac{x^2 + y^2}{r^3} = \frac{r^2}{r^3} = \frac{1}{r}.$$

EXAMPLE 5.71

If $x = r\sin\theta\cos\phi$, $y = r\sin\theta\sin\phi$, and $z = r\cos\theta$, show that

$$\frac{\partial(x, y, z)}{\partial(r, \theta, \phi)} = r^2\sin\theta.$$

Solution. We have

$$\frac{\partial(x, y, z)}{\partial(r, \theta, \phi)} = \begin{vmatrix} \frac{\partial x}{\partial r} & \frac{\partial x}{\partial \theta} & \frac{\partial x}{\partial \phi} \\ \frac{\partial y}{\partial r} & \frac{\partial y}{\partial \theta} & \frac{\partial y}{\partial \phi} \\ \frac{\partial z}{\partial r} & \frac{\partial z}{\partial \theta} & \frac{\partial z}{\partial \phi} \end{vmatrix}$$

$$= \begin{vmatrix} \sin\theta\cos\phi & r\cos\theta\cos\phi & -r\sin\theta\sin\phi \\ \sin\theta\sin\phi & r\cos\theta\sin\phi & r\sin\theta\cos\phi \\ \cos\theta & -r\sin\theta & 0 \end{vmatrix}$$

$$= \cos\theta\,(r^2\sin\theta\cos\theta\cos^2\phi + r^2\sin\theta\cos\theta\sin^2\phi) + r\sin\theta\,(r\sin^2\theta\cos^2\phi + r\sin^2\theta\sin^2\phi)$$

$$= r^2\sin\theta\cos^2\theta + r^2\sin^3\theta$$

$$= r^2\sin\theta\,(\cos^2\theta + \sin^2\theta) = r^2\sin\theta.$$

EXAMPLE 5.72

If $u = x + y + z$, $uv = y + z$, and $uvw = z$, show that

$$\frac{\partial(x, y, z)}{\partial(u, v, w)} = u^2 v.$$

Solution. We have

$z = uvw,$
$y = uv - z = uv - uvw$ and
$x = u - y - z = u - uv + uvw - uvw = u - uv.$
Therefore,

$$\frac{\partial (x, y, z)}{\partial (u, v, w)} = \begin{vmatrix} \frac{\partial x}{\partial u} & \frac{\partial x}{\partial v} & \frac{\partial x}{\partial w} \\ \frac{\partial y}{\partial u} & \frac{\partial y}{\partial v} & \frac{\partial y}{\partial w} \\ \frac{\partial z}{\partial u} & \frac{\partial z}{\partial v} & \frac{\partial z}{\partial w} \end{vmatrix} = \begin{vmatrix} 1-v & -u & 0 \\ v - vw & u - uw & -uv \\ vw & uw & uv \end{vmatrix}$$

$$= \begin{vmatrix} 1-v & -u & 0 \\ v & u & 0 \\ vw & uw & uv \end{vmatrix} = uv \begin{vmatrix} 1-v & -u \\ v & u \end{vmatrix}$$

$$= uv\,(u - uv + uv) = u^2 v.$$

EXAMPLE 5.73

If $u_1 = \frac{x_2 x_3}{x_1}$, $u_2 = \frac{x_3 x_1}{x_2}$, and $u_3 = \frac{x_1 x_2}{x_3}$, show that

$$\frac{\partial (u_1, u_2, u_3)}{\partial (x_1, x_2, x_3)} = 4.$$

Solution. We have

$$\frac{\partial (u_1, u_2, u_3)}{\partial (x_1, x_2, x_3)} = \begin{vmatrix} \frac{\partial u_1}{\partial x_1} & \frac{\partial u_1}{\partial x_2} & \frac{\partial u_1}{\partial x_3} \\ \frac{\partial u_2}{\partial x_1} & \frac{\partial u_2}{\partial x_2} & \frac{\partial u_2}{\partial x_3} \\ \frac{\partial u_3}{\partial x_1} & \frac{\partial u_3}{\partial x_2} & \frac{\partial u_3}{\partial x_3} \end{vmatrix}$$

$$= \begin{vmatrix} -\frac{x_2 x_3}{x_1^2} & \frac{x_3}{x_1} & \frac{x_2}{x_1} \\ \frac{x_3}{x_2} & -\frac{x_3 x_1}{x_2^2} & \frac{x_1}{x_2} \\ \frac{x_2}{x_3} & \frac{x_1}{x_3} & -\frac{x_1 x_2}{x_3^2} \end{vmatrix}$$

$$= \frac{1}{x_1^2 x_2^2 x_3^2} \begin{vmatrix} -x_2 x_3 & x_3 x_1 & x_1 x_2 \\ x_2 x_3 & -x_3 x_1 & x_1 x_2 \\ x_2 x_3 & x_3 x_1 & -x_1 x_2 \end{vmatrix}$$

$$= \frac{1}{x_1^2 x_2^2 x_3^2} \begin{vmatrix} 0 & 0 & 2x_1 x_2 \\ x_2 x_3 & -x_3 x_1 & x_1 x_2 \\ x_2 x_3 & x_3 x_1 & -x_1 x_2 \end{vmatrix}$$

using $R_1 \to R_1 + R_2$

$$= \frac{2x_1 x_2}{x_1^2 x_2^2 x_3^2} \begin{vmatrix} x_2 x_3 & -x_3 x_1 \\ x_2 x_3 & x_3 x_1 \end{vmatrix}$$

$$= \frac{2x_1 x_2}{x_1^2 x_2^2 x_3^2}(2x_1 x_2 x_3^2) = 4.$$

EXAMPLE 5.74

If $u = \frac{y^2}{2x}$ and $v = \frac{x^2 + y^2}{2x}$, find $\frac{\partial (u, v)}{\partial (x, y)}$.

Solution. We have

$$\frac{\partial (u, v)}{\partial (x, y)} = \begin{vmatrix} \frac{\partial x}{\partial u} & \frac{\partial u}{\partial y} \\ \frac{\partial v}{\partial x} & \frac{\partial v}{\partial y} \end{vmatrix} = \begin{vmatrix} -\frac{y^2}{2x^2} & \frac{y}{x} \\ \frac{1}{2} - \frac{y^2}{2x^2} & \frac{y}{x} \end{vmatrix}$$

$$= -\frac{y^3}{2x^3} - \frac{y}{2x} + \frac{y^3}{2x^3} = -\frac{y}{2x}.$$

EXAMPLE 5.75

If $u = \frac{x + y}{1 - xy}$ and $v = \tan^{-1}x + \tan^{-1}y$, find $\frac{\partial (u, v)}{\partial (x, y)}$.

Solution. We have

$$\frac{\partial u}{\partial x} = \frac{(1 - xy) - (x + y)(-y)}{(1 - xy)^2} = \frac{1 - xy + xy + y^2}{(1 - xy)^2}$$

$$= \frac{1 + y^2}{(1 - xy)^2},$$

$$\frac{\partial u}{\partial y} = \frac{(1 - xy) - (x + y)(-x)}{(1 - xy)^2} = \frac{1 - xy + x^2 + xy}{(1 - xy)^2}$$

$$= \frac{1 + x^2}{(1 - xy)^2},$$

$$\frac{\partial v}{\partial x} = \frac{1}{1 + x^2}, \text{ and } \frac{\partial v}{\partial y} = \frac{1}{1 + y^2}.$$

Therefore,

$$\frac{\partial (u, v)}{\partial (x, y)} = \begin{vmatrix} \frac{1 + y^2}{(1 - xy)^2} & \frac{1 + x^2}{(1 - xy)^2} \\ \frac{1}{1 + x^2} & \frac{1}{1 + y^2} \end{vmatrix}$$

$$= \frac{1}{(1 - xy)^2} - \frac{1}{(1 - xy)^2} = 0.$$

EXAMPLE 5.76

If $u = 2xy$, $v = x^2 - y^2$, $x = r\cos\theta$, and $y = r\sin\theta$, find $\frac{\partial (u, v)}{\partial (r, \theta)}$.

Solution. We have

$$\frac{\partial(u,v)}{\partial(r,\theta)} = \frac{\partial(u,v)}{\partial(x,y)} \cdot \frac{\partial(x,y)}{\partial(r,\theta)}$$

$$= \begin{vmatrix} \frac{\partial u}{\partial x} & \frac{\partial u}{\partial y} \\ \frac{\partial v}{\partial x} & \frac{\partial v}{\partial y} \end{vmatrix} \cdot \begin{vmatrix} \frac{\partial x}{\partial r} & \frac{\partial x}{\partial \theta} \\ \frac{\partial v}{\partial r} & \frac{\partial v}{\partial \theta} \end{vmatrix}$$

$$= \begin{vmatrix} 2y & 2x \\ 2x & -2y \end{vmatrix} \cdot \begin{vmatrix} \cos\theta & -r\sin\theta \\ \sin\theta & r\cos\theta \end{vmatrix}$$

$$= -4(y^2 + x^2)\cdot r = -4r^3.$$

5.16 NECESSARY AND SUFFICIENT CONDITIONS FOR JACOBIAN TO VANISH

The following two theorems, stated without proof, provide necessary and sufficient condition for the Jacobian to vanish.

Theorem 5.9. If $u_1, u_2, \ldots, u_n$ are n differentiable functions of the n independent variables $x_1, x_2, \ldots, x_n$ and there exists an identical, differentiable functional relation $\phi(u_1, u_2, \ldots, u_n) = 0$, which does not involve x_i explicitly, then the Jacobian $\frac{\partial(u_1, u_2, \ldots, u_n)}{\partial(x_1, x_2, \ldots, x_n)}$ vanishes identically provided that ϕ as a function of the u_i has no stationary values in the domain considered.

Theorem 5.10 If $u_1, u_2, \ldots, u_n$ are n functions of the n variables $x_1, x_2, \ldots, x_n$, say, $u_m = f_m(x_1, x_2, \ldots, x_n)$, $m = 1, 2, \ldots, n$, and if $\frac{\partial(u_1, u_2, \ldots, u_n)}{\partial(x_1, x_2, \ldots, x_n)} = 0$, then if all the differential coefficients are continuous, there exists a functional relation connecting some or all of the variables and which is independent of $x_1, x_2, \ldots, x_n$.

EXAMPLE 5.77

If $u = x + 2y + z$, $v = x - 2y + 3z$, and $w = 2xy - xz + 4yz - 2z^2$, show that $\frac{\partial(u, v, w)}{\partial(x, y, z)} = 0$ and find a relation between u, v, and w.

Solution. We have

$$\frac{\partial(u,v,w)}{\partial(x,y,z)} = \begin{vmatrix} \frac{\partial u}{\partial x} & \frac{\partial u}{\partial y} & \frac{\partial u}{\partial z} \\ \frac{\partial v}{\partial x} & \frac{\partial v}{\partial y} & \frac{\partial v}{\partial z} \\ \frac{\partial w}{\partial x} & \frac{\partial w}{\partial y} & \frac{\partial w}{\partial z} \end{vmatrix}$$

$$= \begin{vmatrix} 1 & 2 & 1 \\ 1 & -2 & 3 \\ 2y - z & 2x + 4z & -x + 4y - 4z \end{vmatrix}$$

$$= \begin{vmatrix} 1 & 0 & 0 \\ 1 & -4 & 2 \\ 2y - z & 2x + 6z - 4y & -x - 2y - 3z \end{vmatrix} = 0.$$

Hence, a relation between u, v, and w exists. Now,

$$u + v = 2x + 4z = 2(x + 2z)$$
$$u - v = 4y - 2z = 2(2y - z)$$
$$w = x(2y - z) + 2z(2y - z)$$
$$= (x + 2z)(2y - z) = \frac{1}{4}(u + v)(u - v).$$

Therefore,

$$4w = (u + v)(u - v)$$

is the required relation connecting u, v, and w.

EXAMPLE 5.78

If $f(0) = 0$ and $f'(x) = \frac{1}{1 + x^2}$, show that

$$f(x) + f(y) = f\left(\frac{x + y}{1 - xy}\right).$$

Solution. Suppose that

$$u = f(x) + f(y) \text{ and } v = \frac{x + y}{1 - xy}.$$

Then,

$$\frac{\partial(u,v)}{\partial(x,y)} = \begin{vmatrix} \frac{\partial u}{\partial x} & \frac{\partial u}{\partial y} \\ \frac{\partial v}{\partial x} & \frac{\partial v}{\partial y} \end{vmatrix} = \begin{vmatrix} \frac{1}{1+x^2} & \frac{1}{1+y^2} \\ \frac{1+y^2}{(1-xy)^2} & \frac{1+x^2}{(1-xy)^2} \end{vmatrix} = 0.$$

Therefore, u and v are connected by a functional relation. Let $u = \phi(v)$, that is,

$$f(x) + f(y) = \phi\left(\frac{x + y}{1 - xy}\right).$$

Putting $y = 0$, we get

$f(x) + f(0) = \phi(x)$ or $f(x) = \phi(x)$, since $f(0) = 0$.

Hence,

$$f(x) + f(y) = f\left(\frac{x+y}{1-xy}\right).$$

5.17 DIFFERENTIATION UNDER THE INTEGRAL SIGN

In the following theorem of Leibnitz, we shall show that under suitable conditions, the derivative of the integral and the integral of the derivative are equal. The result is useful to determine the value of a definite integral by differentiating the integrand with respect to a quantity of which the limits of integration are independent.

Theorem 5.11. (Leibnitz's Rule): Let $f(x, \alpha)$ and $f_x(x, \alpha)$ be continuous functions of x and α. Then,

$$\frac{d}{d\alpha}\left[\int_a^b f(x, \alpha)\, dx\right] = \int_a^b \frac{\partial}{\partial \alpha}[f(x, \alpha)]\, dx,$$

where the limits a and b are independent of α.

Proof:

Let $F(\alpha) = \int_a^b f(x, \alpha)\, dx$. Then

$$F(\alpha + \delta\alpha) = \int_a^b f(x, \alpha + \delta\,\alpha)\, dx$$

and so,

$$F(\alpha + \delta\alpha) - F(\alpha) = \int_a^b [f(x, \alpha + \delta\alpha) - f(x, \alpha)]\, dx. \quad (1)$$

But by Lagrange's mean value theorem, we have

$f(\alpha + \delta\alpha) - f(\alpha) = \delta\alpha \frac{\partial}{\partial \alpha} f(x, \alpha + \delta\alpha), 0 < \theta < 1.$

Hence, (1) reduces to

$$\frac{F(\alpha + \delta\alpha) - F(\alpha)}{\delta\alpha} = \int_a^b \frac{\partial}{\partial \alpha} f(x, \alpha + \theta\delta\alpha)\, dx.$$

Therefore,

$$\lim_{\delta\alpha \to 0} \frac{F(\alpha + \delta\,\alpha) - F(\alpha)}{d\alpha} = \int_a^b \frac{\partial}{\partial \alpha} f(x, \alpha)\, dx$$

or

$$\frac{d}{d\alpha}[F(\alpha)] = \frac{d}{d\alpha}\left[\int_a^b f(x, \alpha)\, dx\right] = \int_a^b \frac{\partial}{\partial \alpha} f(x, \alpha)\, dx.$$

Remark 5.3 If the limits of integration a and b are not independent of α, then

$$\frac{d}{d\alpha}\left[\int_a^b f(x, \alpha)\, dx\right] = \int \frac{\partial}{\partial \alpha}[f(x, \alpha)]dx + \frac{db}{d\alpha} f(b, \alpha) - \frac{da}{d\alpha} f(a, \alpha).$$

EXAMPLE 5.79

Show that

$$\int_0^\infty \frac{\tan^{-1} ax}{x(1+x^2)}\, dx = \frac{\pi}{2} \log(1+a), a \geq 0.$$

Solution. Let

$$F(a) = \int_0^\infty \frac{\tan^{-1} ax}{x(1+x^2)}\, dx.$$

Then by Leibnitz's Rule,

$$\frac{d}{da}\left[\int_0^\infty \frac{\tan^{-1} ax}{x(1+x^2)}\, dx\right]$$

$$= \int_0^\infty \frac{\partial}{\partial a}\left[\frac{\tan^{-1} ax}{x(1+x^2)}\right] dx$$

$$= \int_0^\infty \frac{1}{x(1+x^2)} \cdot \frac{1}{1+a^2x^2} \cdot x dx$$

$$= \int_0^\infty \frac{dx}{(1+x^2)(1+a^2x^2)}$$

$$= \frac{1}{1-a^2} \int_0^\infty \left[\frac{1}{1+x^2} - \frac{a^2}{1+a^2x^2}\right] dx$$

(by partial fractions)

$$= \frac{1}{1-a^2}[\tan^{-1} x]_0^\infty - \frac{a^2}{1-a^2}\int_0^\infty \frac{dx}{1+a^2x^2}$$

$$= \frac{\pi}{2(1-a^2)} - \frac{1}{1-a^2}\int_0^\infty \frac{dx}{x^2 + \frac{1}{a^2}}$$

$$= \frac{\pi}{2(1-a^2)} - \frac{1}{1-a^2} \cdot \frac{1}{\frac{1}{a}}\left[\tan^{-1}\frac{x}{\frac{1}{a}}\right]_0^\infty$$

$$= \frac{\pi}{2(1-a^2)} - \frac{a}{1-a^2}\frac{\pi}{2}$$

$$= \frac{\pi}{2(1-a^2)}[1-a] = \frac{\pi}{2(1+a)}.$$

Integrating both sides with respect to a, we get

$$F(a)=\int_0^\infty \frac{\tan^{-1} ax}{x(1+x^2)}dx=\frac{\pi}{2}\log(1+a)+c. \quad (1)$$

Also $F(0) = 0$. Therefore, (1) yields $0=\frac{\pi}{2}\log 1 + c$ and so, $c = 0$. Hence,

$$\int_0^\infty \frac{\tan^{-1}ax}{x(1+x^2)}dx=\frac{\pi}{2}\log(1+a).$$

EXAMPLE 5.80

Evaluate $\int_0^1 \frac{x^\alpha-1}{\log x}dx$, $\alpha \geq 0$ using differentiation under the integral sign.

Solution. Let $F(\alpha) = \int_0^1 \frac{x^\alpha - 1}{\log x}dx$. Then by Leibnitz's Rule,

$$\frac{d}{d\alpha}\left[\int_0^1 \frac{x^\alpha-1}{\log x}dx\right]=\int_0^1 \frac{\partial}{\partial a}\left[\frac{x^\alpha-1}{\log x}\right]dx$$

$$=\int_0^1 \frac{1}{\log x}\cdot x^\alpha \log x\,dx$$

$$=\int_0^1 x^\alpha\,dx=\left[\frac{x^{\alpha+1}}{\alpha+1}\right]_0^1=\frac{1}{1+\alpha}.$$

Integrating with respect to a, we get

$$F(\alpha)=\int_0^1 \frac{x^\alpha-1}{\log x}dx$$

$= \log(1+\alpha) + c$ (a constant of integration)

But when $\alpha = 0$, $F(0) = \int_0 0\,dx = 0$. Therefore, $0 = \log 1 + c = c$. Hence,

$$F(\alpha)=\int_0^1 \frac{x^\alpha-1}{\log x}dx=\log(1+\alpha).$$

EXAMPLE 5.81

Evaluate the integral

$$\int_0^\infty e^{-x}\frac{\sin \alpha x}{x}dx.$$

Solution. We cannot compute this integral directly because the anti-derivative of the function $e^{-x}\frac{\sin \alpha x}{x}$ is not expressible in terms of elementary functions. So we use Leibnitz's Rule to evaluate it. Let

$$F(\alpha)=\int_0^\infty e^{-x}\cdot\frac{\sin \alpha x}{x}dx.$$

Then by Leibnitz's Rule, we have

$$\frac{d}{d\alpha}F(\alpha)=\frac{d}{d\alpha}\left[\int_0^\infty e^{-x}\frac{\sin \alpha}{x}dx\right]$$

$$=\int_0^\infty \frac{\partial}{\partial \alpha}\left[e^{-x}\frac{\sin \alpha x}{x}\right]dx$$

$$=\int_0^\infty e^{-x}\frac{1}{x}\cos \alpha x\cdot x\,dx$$

$$=\int_0^\infty e^{-x}\cos \alpha x\,dx=\frac{1}{1+\alpha^2}.$$

Integrating, we get

$$F(\alpha)=\tan^{-1}\alpha+c. \quad (1)$$

But,

$$F(0)=\int_0^\infty e^{-x}\frac{\sin 0x}{x}dx=\int_0^\infty 0\,dx=0.$$

Therefore, (1) yields

$$0=\tan^{-1}0+c \text{ and so, } c=0.$$

Hence,

$$F(\alpha)=\int_0^\infty e^{-x}\frac{\sin \alpha x}{x}dx=\tan^{-1}\alpha.$$

EXAMPLE 5.82

Evaluate $\int_0^a \frac{\log(1+ax)}{1+x^2}dx$ and hence, show that

$$\int_0^1 \frac{\log(1+x)}{1+x^2}dx=\frac{\pi}{8}\log 2.$$

Solution. We note that the limits of integration are not independent of the parameter a. Therefore, the formula mentioned in Remark after Theorem 5.11 is applicable. Let

$$F(a)=\int_0^a \frac{\log(1+ax)}{1+x^2}dx.$$

Then,

$$\frac{d}{da}\int_0^a\left[\frac{\log(1+ax)}{1+x^2}dx\right]$$

$$=\int_0^a \frac{\partial}{\partial a}\left[\frac{\log(1+ax)}{1+x^2}\right]dx+\frac{\log(1+a^2)}{1+a^2}\frac{d}{da}(a)$$

$$-\frac{\log(1+a.0)}{1+0}\cdot\frac{d}{da}(0)$$

$$=\int_0^a \frac{x}{(1+x^2)(1+ax)}dx+\frac{\log(1+a^2)}{1+a^2}$$

$$=\frac{1}{1+a^2}\int_0^a\left[\frac{-a}{1+ax}+\frac{x+a}{1+x^2}\right]dx$$

$$+\frac{\log(1+a^2)}{1+a^2}\text{ (by partial fractions)}$$

$$=\frac{1}{1+a^2}\left[-\log(1+ax)+\frac{1}{2}\log(1+x^2)\right.$$

$$\left.+a\tan^{-1}x\right]_0^a+\frac{\log(1+a^2)}{1+a^2}$$

$$=\frac{1}{1+a^2}\left[-\frac{1}{2}\log(1+a^2)+a\tan^{-1}a\right]$$

$$+\frac{\log(1+a^2)}{1+a^2}$$

$$=\frac{1}{1+a^2}\left[\frac{1}{2}\log(1+a^2)+a\tan^{-1}a\right].$$

Integrating with respect to a, we get

$$F(a)=\int_0^a\frac{\log(1+ax)}{1+x^2}dx$$

$$=\frac{1}{2}\int\log(1+a^2)\frac{1}{1+a^2}da$$

$$+\int\frac{a\tan^{-1}a}{1+a^2}da+c$$

$$=\frac{1}{2}\left[\log(1+a^2)\tan^{-1}a\right.$$

$$\left.-\int\frac{2a}{1+a^2}\tan^{-1}a\,da\right]$$

$$+\int\frac{a\tan^{-1}a}{1+a^2}da+c$$

$$=\frac{1}{2}\log(1+a^2)\tan^{-1}a+c.$$

Substituting $a=0$, we have $F(0)=0$. Therefore,

$$0=\frac{1}{2}\log 1\tan^{-1}0+c \text{ and so, } c=0.$$

Hence,

$$F(a)=\int_0^a\frac{\log(1+ax)}{1+x^2}dx=\frac{1}{2}\log(1+a^2)\tan^{-1}a.$$

Substituting $a=1$, we get

$$\int_0^1\frac{\log(1+x)}{1+x^2}dx=\frac{\pi}{8}\log 2.$$

EXAMPLE 5.83

Prove that

$$\int_{\frac{\pi}{2}-\alpha}^{\frac{\pi}{2}}\sin\theta\cos^{-1}(\cos\alpha\,\text{cosec}\,\theta)d\theta=\frac{\pi}{2}(1-\cos\alpha).$$

Solution. Here the limits involve the parameter α. Let

$$F(\alpha)=\int_{\frac{\pi}{2}-\alpha}^{\frac{\pi}{2}}\sin\theta\cos^{-1}(\cos\alpha\,\text{cosec}\,\theta)\,d\theta.$$

Then,

$$\frac{d}{d\alpha}[F(\alpha)]$$

$$=\int_{\frac{\pi}{2}-\alpha}^{\frac{\pi}{2}}\frac{\partial}{\partial\alpha}[\sin\theta\cos^{-1}(\cos\alpha\,\text{cosec}\,\theta)]d\theta$$

$$+\frac{d}{d\alpha}\left(\frac{\pi}{2}\right)\left[\sin\frac{\pi}{2}\cos^{-1}\left(\cos\alpha\,\text{cosec}\,\frac{\pi}{2}\right)\right]$$

$$-\frac{d}{d\alpha}\left(\frac{\pi}{2}-\alpha\right)\left[\sin\left(\frac{\pi}{2}-\alpha\right)\right.$$

$$\left.\times\cos^{-1}\left(\cos\alpha\,\text{cosec}\left(\frac{\pi}{2}-\alpha\right)\right)\right]$$

$$=\int_{\frac{\pi}{2}-\alpha}^{\frac{\pi}{2}}\frac{\sin\alpha\,d\theta}{\sqrt{1-\cos^2\alpha\,\text{cosec}^2\theta}}+\sin\left(\frac{\pi}{2}-\alpha\right)$$

$$\times\cos^{-1}\left[\cos\alpha\,\text{cosec}\left(\frac{\pi}{2}-\alpha\right)\right]$$

$$=\int_{\frac{\pi}{2}-\alpha}^{\frac{\pi}{2}}\frac{\sin\alpha\sin\theta}{\sqrt{\sin^2\theta-\cos^2\alpha}}d\theta+\cos\alpha\cos^{-1}(1)$$

$$=\int_0^{\sin\alpha}\frac{\sin\alpha}{\sqrt{\sin^2\alpha-t^2}}dt \text{ taking } \cos\theta=t$$

$$=\left[\sin\alpha\sin^{-1}\left(\frac{t}{\sin\alpha}\right)\right]_0^{\sin\alpha}=\frac{\pi}{2}\sin\alpha.$$

Integrating with respect to α, we get

$$F(\alpha)=-\frac{\pi}{2}\cos\alpha+c.$$

But $F(0)=0$, therefore,

$$0 = -\frac{\pi}{2} + c \quad \text{or} \quad c = \frac{\pi}{2}.$$

Hence,

$$F(\alpha) = -\frac{\pi}{2}\cos\alpha + \frac{\pi}{2} = \frac{\pi}{2}[1 - \cos\alpha].$$

EXAMPLE 5.84

If $y = \int_0^x f(t)\sin[k(x-t)]\,dt$, show that it satisfies the differential equation $\frac{d^2y}{dx^2} + k^2y = kf(x)$.

Solution. We have

$$y = \int_0^x f(t)\sin[k(x-t)]\,dt.$$

The upper limit in this integral involves the parameter x. So, using Leibnitz's Rule, we have

$$\begin{aligned}\frac{dy}{dx} &= \int_0^x \frac{\partial}{\partial x}[f(t)\sin[k(x-t)]\,dt \\ &\quad + f(x)\sin[k(x-x)]\frac{d}{dx}(x) \\ &\quad - f(0)\sin[k(x-0)]\frac{d}{dx}(0) \\ &= \int_0^x kf(t)\cos[k(x-t)]\,dt.\end{aligned}$$

Using once more the Leibnitz's Rule, we get

$$\begin{aligned}\frac{d^2y}{dx^2} &= \int_0^x \frac{\partial}{\partial x}[kf(t)\cos[k(x-t)]\,dt \\ &\quad + kf(x)\cos[k(x-x)]\frac{d}{dx}(x) \\ &\quad - kf(0)\cos[k(x-0)]\frac{d}{dx}(0). \\ &= -k^2\int_0^x f(t)\sin[k(x-t)]\,dt + kf(x) \\ &= -k^2y + kf(x).\end{aligned}$$

Hence,

$$\frac{d^2y}{dx^2} + k^2y = kf(x).$$

EXAMPLE 5.85

By successive use of Leibnitz's Rule to $\int_0^1 x^m\,dx$, evaluate $\int_0^1 x^m(\log x)^n\,dx$.

Solution. We have

$$I = \int_0^1 x^m dx = \left|\frac{x^{m-1}}{m+1}\right|_0^1 = \frac{1}{m+1}.$$

Therefore, using Leibnitz's Rule, we get

$$\frac{d}{dm}\left(\frac{1}{m+1}\right) = \int_0^1 \frac{\partial}{\partial m}(x^m)\,dx$$

or

$$-\left(\frac{1}{m+1}\right)^2 = \int_0^1 x^m \log x dx.$$

Applying again the Leibnitz's Rule, we get

$$\frac{d}{dm}\left(\frac{-1}{(m+1)^2}\right) = \int_0^1 \frac{\partial}{\partial m}(x^m\log x)\,dx$$

or

$$\frac{(-1)(-2)}{(m+1)^3} = \int_0^1 x^m(\log x)^2\,dx.$$

Repeated use of Leibnitz's Rule yields

$$\frac{(-1)(-2)(-3)}{(m+1)^4} = \int_0^1 x^m(\log x)^3\,dx$$

....

....

$$\frac{(-1)(-2)(-3)\ldots(-n)}{(m+1)^{n+1}} = \int_0^1 x^m(\log x)^n\,dx$$

or

$$\frac{(-1)^n\,n!}{(m+1)^{n+1}} = \int_0^1 x^m(\log x)^n\,dx.$$

5.18 MISCELLANEOUS EXAMPLES

EXAMPLE 5.86

If $z = f(x, y)$ and u, v are two variables such that $u = lx + my$, $v = ly - mx$. Prove that

$$\frac{\partial^2 z}{\partial x^2} + \frac{\partial^2 z}{\partial y^2} = (l^2 + m^2)\left[\frac{\partial^2 z}{\partial u^2} + \frac{\partial^2 z}{\partial v^2}\right]$$

Solution. We have

$$u = lx + my, \quad v = ly - mx,$$

$$\frac{\partial u}{\partial x} = l, \frac{\partial v}{\partial x} = -m$$

$$\frac{\partial u}{\partial y} = m, \frac{\partial v}{\partial y} = l.$$

Therefore

$$\frac{\partial z}{\partial x} = \frac{\partial z}{\partial u}\cdot\frac{\partial u}{\partial x} + \frac{\partial z}{\partial v}\cdot\frac{\partial v}{\partial x} = l\frac{\partial z}{\partial u} - m\frac{\partial z}{\partial v} \qquad (1)$$

$$\frac{\partial z}{\partial y} = \frac{\partial z}{\partial u}\cdot\frac{\partial u}{\partial y} + \frac{\partial z}{\partial v}\cdot\frac{\partial v}{\partial y} = m\frac{\partial z}{\partial u} + l\frac{\partial z}{\partial v} \quad (2)$$

From (1) and (2), we have

$$\frac{\partial}{\partial x} = l\frac{\partial}{\partial u} - m\frac{\partial}{\partial v}$$

and

$$\frac{\partial}{\partial y} = m\frac{\partial}{\partial u} + l\frac{\partial}{\partial v}.$$

Therefore

$$\frac{\partial^2 z}{\partial x^2} = \frac{\partial}{\partial x}\left(\frac{\partial z}{\partial x}\right) = \left(l\frac{\partial}{\partial u} - m\frac{\partial}{\partial v}\right)\left(l\frac{\partial z}{\partial u} - m\frac{\partial z}{\partial v}\right)$$

$$= l\frac{\partial}{\partial u}\left(l\frac{\partial z}{\partial u} - m\frac{\partial z}{\partial v}\right) - m\frac{\partial}{\partial v}\left(l\frac{\partial z}{\partial u} - m\frac{\partial z}{\partial v}\right)$$

$$= l^2\frac{\partial^2 z}{\partial u^2} - lm\frac{\partial^2 z}{\partial u\partial v} - lm\frac{\partial^2 z}{\partial v\partial u} + m^2\frac{\partial^2 z}{\partial v^2}$$

$$= l^2\frac{\partial^2 z}{\partial u^2} + m^2\frac{\partial^2 z}{\partial v^2} - 2lm\frac{\partial^2 z}{\partial u\partial v} \quad (3)$$

and

$$\frac{\partial^2 z}{\partial y^2} = \frac{\partial}{\partial y}\left(\frac{\partial z}{\partial y}\right) = \left(m\frac{\partial}{\partial u} + l\frac{\partial}{\partial v}\right)\left(m\frac{\partial z}{\partial u} + l\frac{\partial z}{\partial v}\right)$$

$$= m\frac{\partial}{\partial u}\left(m\frac{\partial z}{\partial u} + l\frac{\partial z}{\partial v}\right) + l\frac{\partial}{\partial v}\left(m\frac{\partial z}{\partial u} + l\frac{\partial z}{\partial v}\right)$$

$$= m^2\frac{\partial^2 z}{\partial u^2} + lm\frac{\partial^2 z}{\partial u\partial v} + lm\frac{\partial^2 z}{\partial v\partial u} + l^2\frac{\partial^2 z}{\partial v^2}$$

$$= l^2\frac{\partial^2 z}{\partial v^2} + m^2\frac{\partial^2 z}{\partial u^2} + 2lm\frac{\partial^2 z}{\partial u\partial v}. \quad (4)$$

Adding (3) and (4), we get

$$\frac{\partial^2 z}{\partial x^2} + \frac{\partial^2 z}{\partial y^2} = (l^2 + m^2)\left(\frac{\partial^2 z}{\partial u^2} + \frac{\partial^2 z}{\partial v^2}\right).$$

EXAMPLE 5.87

(a) If $f(x, y) = (1 - 2xy + y^2)^{-\frac{1}{2}}$, show that

$$\frac{\partial}{\partial x}\left[(1-x^2)\frac{\partial f}{\partial x}\right] + \frac{\partial}{\partial y}\left[y^2\frac{\partial f}{\partial y}\right] = 0.$$

(b) If $V = f(2x - 3y, 3y - 4z, 4z - 2x)$, compute the value of $6V_x + 4V_y + 3V_z$.

Solution. (a) We have, $f = (1 - 2xy + y^2)^{-\frac{1}{2}}$. Then

$$\frac{\partial f}{\partial x} = -\frac{1}{2}(1 - 2xy + y^2)^{-\frac{3}{2}}(-2y)$$

$$= y(1 - 2xy + y^2)^{-\frac{3}{2}},$$

$$\frac{\partial^2 f}{\partial x^2} = -\frac{3}{2}y(1 - 2xy + y^2)^{-\frac{5}{2}}(-2y)$$

$$= 3y^2(1 - 2xy + y^2)^{-\frac{5}{2}}.$$

Therefore

$$\frac{\partial}{\partial x}\left[(1-x^2)\frac{\partial f}{\partial x}\right]$$

$$= \frac{\partial}{\partial x}(1-x^2)\frac{\partial f}{\partial x} + (1-x^2)\frac{\partial^2 f}{\partial x^2}$$

$$= (-2x)\frac{\partial f}{\partial x} + (1-x^2)\frac{\partial^2 f}{\partial x^2}$$

$$= -2xy(1 - 2xy + y^2)^{-\frac{3}{2}}$$

$$+ 3(1-x^2)y^2(1 - 2xy + y^2)^{-\frac{5}{2}}. \quad (1)$$

Similarly, differentiating partially with respect to y, we get

$$\frac{\partial f}{\partial y} = (x - y)(1 - 2xy + y^2)^{-\frac{3}{2}},$$

$$\frac{\partial^2 f}{\partial y^2} = -(1 - 2xy + y^2)^{-\frac{3}{2}}$$

$$+ 3(x-y)^2(1 - 2xy + y^2)^{-\frac{5}{2}}.$$

Therefore,

$$\frac{\partial}{\partial y}\left[y^2\frac{\partial f}{\partial y}\right] = \frac{\partial}{\partial y}(y^2)\frac{\partial f}{\partial y} + y^2\frac{\partial^2 f}{\partial y^2}$$

$$= 2y(x-y)(1 - 2xy + y^2)^{-\frac{3}{2}}$$

$$+ y^2\Big[-(1 - 2xy + y^2)^{-\frac{3}{2}}$$

$$+ 3(x-y)^2(1 - 3xy + y^2)^{-\frac{5}{2}}\Big]$$

$$= y(1 - 2xy + y^2)^{-\frac{3}{2}}[3y(x-y)^2$$

$$\times (1 - 2xy + y^2)^{-1} + (2x - 3y)] \quad (2)$$

Adding (1) and (2), we get the required result.

(b) We have

$$V = f(2x - 3y, 3y - 4z, 4z - 2x).$$

Let

$$r = 2x - 3y,\ s = 3y - 4z \text{ and } t = 4z - 2x.$$

Then

$$V = f(r, s, t).$$

Further,

$$\frac{\partial V}{\partial x} = \frac{\partial V}{\partial r}\cdot\frac{\partial r}{\partial x} + \frac{\partial V}{\partial s}\frac{\partial s}{\partial x} + \frac{\partial V}{\partial t}\cdot\frac{\partial t}{\partial x}$$

$$= 2\frac{\partial V}{\partial r} + 0 - 2\frac{\partial V}{\partial t} = 2\frac{\partial V}{\partial r} - 2\frac{\partial V}{\partial t} \qquad (1)$$

$$\frac{\partial V}{\partial y} = \frac{\partial V}{\partial r}\cdot\frac{\partial r}{\partial y} + \frac{\partial V}{\partial s}\frac{\partial s}{\partial y} + \frac{\partial V}{\partial t}\cdot\frac{\partial t}{\partial y}$$

$$= -3\frac{\partial V}{\partial r} + 3\frac{\partial V}{\partial s} + 0 = 3\frac{\partial V}{\partial r} + 3\frac{\partial V}{\partial s} \qquad (2)$$

and

$$\frac{\partial V}{\partial z} = \frac{\partial V}{\partial r}\cdot\frac{\partial r}{\partial z} + \frac{\partial V}{\partial s}\frac{\partial s}{\partial z} + \frac{\partial V}{\partial t}\cdot\frac{\partial t}{\partial z}$$

$$= 0 - 4\frac{\partial V}{\partial s} + 4\frac{\partial V}{\partial t} = -4\frac{\partial V}{\partial s} + 4\frac{\partial V}{\partial t}.$$

The relations (1), (2) and (3) yields

$$6V_x + 4V_y + 3V_z = 6\left(2\frac{\partial V}{\partial r} - 2\frac{\partial V}{\partial t}\right) + 4\left(-3\frac{\partial V}{\partial r} + 3\frac{\partial V}{\partial s}\right) + 3\left(-4\frac{\partial V}{\partial s} + 4\frac{\partial V}{\partial t}\right) = 0.$$

EXAMPLE 5.88

(a) If $u = \sin^{-1}\left(\dfrac{3x^2 + 4y^2}{3x + 4y}\right)$, prove that $x\dfrac{\partial u}{\partial x} + y\dfrac{\partial u}{\partial y} = \tan u.$

(b) If $u = x^3 + y^3 + z^3 + 3xyz$, show that

$$x\frac{\partial u}{\partial x} + y\frac{\partial u}{\partial y} + z\frac{\partial u}{\partial z} = 3u.$$

(c) If $u = \log\left(\dfrac{x^2 + y^2}{x + y}\right)$, prove that $x\dfrac{\partial u}{\partial x} + y\dfrac{\partial u}{\partial y} = 1.$

(d) If $u = x^2\tan^{-1}\left(\dfrac{y}{x}\right) - y^2\tan^{-1}\left(\dfrac{x}{y}\right)$, show that $x^2\dfrac{\partial^2 u}{\partial x^2} + 2xy\dfrac{\partial^2 u}{\partial x\,\partial y} + y^2\dfrac{\partial^2 u}{\partial y^2} = 2u.$

Solution. (a) We have

$$\sin u = \frac{3x^2 + 4y^2}{3x + 4y} = z, \text{ say.}$$

Thus

$$z = \frac{3x^2}{3x}\left(\frac{1 + \frac{4}{3}\left(\frac{y}{x}\right)^2}{1 + \frac{4}{3}\left(\frac{y}{x}\right)}\right) = x\left(\frac{1 + \frac{4}{3}\left(\frac{y}{x}\right)^2}{1 + \frac{4}{3}\left(\frac{y}{x}\right)}\right),$$

and so z is a homogeneous function of degree 1 in x and y. Hence, by Euler's Theorem, we have

$$x\frac{\partial z}{\partial x} + y\frac{\partial z}{\partial y} = z. \qquad (1)$$

But

$$\frac{\partial z}{\partial x} = \cos u\frac{\partial u}{\partial x} \quad\text{and}\quad \frac{\partial z}{\partial y} = \cos u\frac{\partial u}{\partial y}.$$

Hence (1) reduces to

$$x\cos u\frac{\partial u}{\partial x} + y\cos u\frac{\partial u}{\partial y} = \sin u$$

or

$$x\frac{\partial u}{\partial x} + y\frac{\partial u}{\partial y} = \tan u.$$

(b) We have

$$u = x^3 + y^3 + z^3 + 3xyz.$$

Replacing x by tx, y by ty and z by tz, we get

$$u(tx, ty, tz) = t^3x^3 + t^3y^3 + t^3z^3 + 3tx\,ty\,tz = t^3(x^3 + y^3 + z^3 + 3xyz) = t^3u(x, y, z).$$

Hence, $u(x, y, z)$ is a homogeneous function of degree 3 in $u(x, y, z)$. Therefore, by Euler's Theorem,

$$x\frac{\partial u}{\partial z} + y\frac{\partial u}{\partial y} + z\frac{\partial u}{\partial z} = 3u.$$

(c) We have

$$u = \log\left(\frac{x^2 + y^2}{x + y}\right).$$

Therefore

$$e^u = \frac{x^2 + y^2}{x + y},$$

which is homogeneous function of degree 1 in x and y. Therefore, by Euler's Theorem, we have

$$x\frac{\partial}{\partial x}(e^u) + y\frac{\partial}{\partial y}(e^u) = e^u$$

or

$$xe^u\frac{\partial u}{\partial x} + ye^u\frac{\partial u}{\partial y} = e^u$$

or

$$x\frac{\partial u}{\partial x} + y\frac{\partial u}{\partial y} = 1.$$

(d) We have $u = u_1 + u_2$, where

$$u_1 = x^2\tan^{-1}\left(\frac{y}{x}\right),\ u_2 = y^2\tan^{-1}\left(\frac{x}{y}\right).$$

we note that u_1 and u_2 are both homogeneous functions of degree 2. Therefore (see Example 5.26), we have

$$x^2\frac{\partial^2 u_1}{\partial x^2}+2xy\frac{\partial^2 u_1}{\partial x\,\partial y}+y^2\frac{\partial^2 u_1}{\partial y^2}=2(2-1)u_1$$

$$=2u_1 \qquad (1)$$

and

$$x^2\frac{\partial^2 u_2}{\partial x_2}+2xy\frac{\partial^2 u_2}{\partial x\,\partial y}+y^2\frac{\partial^2 u_2}{\partial y^2}=2(2-1)u_2$$

$$=2u_2 \qquad (2)$$

Adding (1) and (2) we get

$$x^2\left(\frac{\partial^2 u_1}{\partial x_2}+\frac{\partial^2 u_2}{\partial x_2}\right)+2xy\left(\frac{\partial^2 u_1}{\partial x\,\partial y}+\frac{\partial^2 u_2}{\partial x\,\partial y}\right)$$

$$+\ y^2\left(\frac{\partial^2 u_1}{\partial y^2}+\frac{\partial^2 u_2}{\partial y^2}\right)=2(u_1+u_2)$$

or

$$x^2\frac{\partial^2 u}{\partial x_2}+2xy\,\frac{\partial^2 u}{\partial x\,\partial y}+y^2\frac{\partial^2 u}{\partial y^2}=2u.$$

EXAMPLE 5.89

(a) If $u = x + y$ and $y = uv$, find the Jacobian $\dfrac{\partial\,(x, y)}{\partial\,(u, v)}$.

(b) Show that the functions $u = x^2 + y^2 + z^2$, $v = x + y + z$, $w = yz + zx + xy$ are not independent of one another.

Solution. (a) We are given that $u = x + y$, $y = uv$. Therefore

$$x = u - y = u - uv \quad \text{and} \quad y = uv.$$

Therefore

$$\frac{\partial\,(x, y)}{\partial\,(u, v)}=\begin{vmatrix}\dfrac{\partial x}{\partial u} & \dfrac{\partial x}{\partial v}\\[2mm] \dfrac{\partial y}{\partial u} & \dfrac{\partial y}{\partial v}\end{vmatrix}=\begin{vmatrix}1-v & -u\\ v & u\end{vmatrix}$$

$$= u\,(1 - v) + uv = u.$$

(b) We have

$$u = x^2 + y^2 + z^2,\quad v = x + y + z,\quad w = yz + zx + xy.$$

Then

$$\frac{\partial\,(u, v, w)}{\partial\,(x, y, z)}=\begin{vmatrix}\dfrac{\partial u}{\partial x} & \dfrac{\partial u}{\partial y} & \dfrac{\partial u}{\partial z}\\[2mm] \dfrac{\partial v}{\partial x} & \dfrac{\partial v}{\partial y} & \dfrac{\partial v}{\partial z}\\[2mm] \dfrac{\partial w}{\partial x} & \dfrac{\partial w}{\partial y} & \dfrac{\partial w}{\partial z}\end{vmatrix}=\begin{vmatrix}2x & 2y & 2z\\ 1 & 1 & 1\\ z+y & z+x & x+y\end{vmatrix}$$

$$= 2x\,[(x + y) - (z + x)] - 2y\,[(x + y) - (z + y)] + 2z\,[(z + x) - (z + y)] = 0.$$

Since, Jacobian $J\,(u, v, w) = 0$, there exists a functional relation connecting some or all of the variables x, y and z. Hence u, v, w are not independent.

EXAMPLE 5.90

(a) If $x = e^u \cos v$ and $y = e^u \sin v$, show that $J{\cdot}J' = 1$.

(b) Verify the chain rule for Jacobians if $x = u$, $y = u \tan v$, $z = w$.

Solution. (a) We have

$$x = e^u \cos v,\quad y = e^u \sin v.$$

Then

$$J=\frac{\partial\,(x, y)}{\partial\,(u, v)}=\begin{vmatrix}\dfrac{\partial x}{\partial u} & \dfrac{\partial x}{\partial v}\\[2mm] \dfrac{\partial y}{\partial u} & \dfrac{\partial y}{\partial v}\end{vmatrix}=\begin{vmatrix}e^u \cos v & -e^u \sin v\\ e^u \sin v & e^u \cos u\end{vmatrix}$$

$$= e^{2u}\cos^2 u + e^{2u}\sin^2 v = e^{2u}.$$

One the other hand the given equations yield

$$x^2 + y^2 = e^{2u} \quad \text{and} \quad v = \tan^{-1}\frac{y}{x}.$$

Therefore

$$2e^{2u}\frac{\partial u}{\partial x} = 2x \text{ which yields } \frac{\partial u}{\partial x} = \frac{x}{e^{2u}},$$

$$2e^{2u}\frac{\partial u}{\partial y} = 2y \text{ which yields } \frac{\partial u}{\partial y} = \frac{y}{e^{2u}},$$

$$\frac{\partial v}{\partial x} = \frac{-y}{x^2 + y^2} = -\frac{y}{e^{2u}},\quad \frac{\partial v}{\partial y} = \frac{x}{x^2 + y^2} = \frac{x}{e^{2u}}.$$

Therefore

$$J'=\frac{\partial\,(u, v)}{\partial\,(x, y)}=\begin{vmatrix}\dfrac{x}{e^{2u}} & \dfrac{y}{e^{2u}}\\[2mm] \dfrac{-y}{e^{2u}} & \dfrac{x}{e^{2u}}\end{vmatrix}=\frac{2x^2}{e^{4u}}+\frac{2y^2}{e^{4u}}=\frac{e^{2u}}{e^{4u}}=\frac{1}{e^{2u}}.$$

Hence $JJ' = 1$.

(b) We are given thet

$$x = u,\ y = u \tan v,\ z = w \qquad (1)$$

Then

$$J = \frac{\partial (x, y, z)}{\partial (u, v, w)} = \begin{vmatrix} \frac{\partial x}{\partial u} & \frac{\partial x}{\partial v} & \frac{\partial x}{\partial w} \\ \frac{\partial y}{\partial u} & \frac{\partial y}{\partial v} & \frac{\partial y}{\partial w} \\ \frac{\partial z}{\partial u} & \frac{\partial z}{\partial v} & \frac{\partial z}{\partial w} \end{vmatrix}$$

$$= \begin{vmatrix} 1 & 0 & 0 \\ \tan v & u \sec^2 v & 0 \\ 0 & 0 & 1 \end{vmatrix} = u \sec^2 v.$$

Also, from (1), we have

$$u = x,\ v = \tan^{-1}\frac{y}{x} \text{ and } w = z.$$

Therefore

$$J' = \frac{\partial (u, v, w)}{\partial (x, y, z)}$$

$$= \begin{vmatrix} \frac{\partial u}{\partial y} & \frac{\partial u}{\partial y} & \frac{\partial u}{\partial z} \\ \frac{\partial v}{\partial x} & \frac{\partial v}{\partial y} & \frac{\partial v}{\partial z} \\ \frac{\partial w}{\partial x} & \frac{\partial w}{\partial y} & \frac{\partial w}{\partial z} \end{vmatrix} = \begin{vmatrix} 1 & 0 & 0 \\ \frac{-y}{x^2+y^2} & \frac{x}{x^2+y^2} & 0 \\ 0 & 0 & 1 \end{vmatrix}$$

$$= \frac{x}{x^2+y^2} = \frac{1}{x\left[1 + \left(\frac{y}{x}\right)^2\right]}$$

$$= \frac{1}{u(1+\tan^2 v)}, \text{ since } \frac{y}{x} = \tan v$$

$$= \frac{1}{u \sec^2 v}.$$

Hence

$JJ' = 1$, which proves the chain rule.

EXAMPLE 5.91

Assuming $\int_0^\infty e^{-x^2}\, dx = \frac{\sqrt{\pi}}{2}$, prove that

$$\int_0^\infty e^{-x^2} \cos 2ax\, dx = \frac{\sqrt{\pi}}{2} e^{-a^2}.$$

Solution. Let

$$F(a) = \int_0^\infty e^{-x^2} \cos 2ax\, dx.$$

Then

$$F'(a) = \int_0^\infty e^{-x^2} \frac{\partial}{\partial a} (\cos 2ax)\, dx$$

$$= \int_0^\infty -2xe^{-x^2} \sin 2ax\, dx$$

$$= [e^{-x^2} \sin 2ax]_0^\infty - 2a \int e^{-x^2} \cos 2ax\, dx$$

$$= -2aF(a).$$

Therefore

$$\frac{F'(a)}{F(a)} = -2a.$$

Integrating, we get,

$$\log F(a) = -\frac{2a^2}{2} = -a^2 + \log c$$

or

$$\log \frac{F(a)}{c} = -a^2$$

or

$$F(a) = c\, e^{-a^2}.$$

Putting $a = 0$, we get $c = F(0) = \int_0^\infty e^{-x^2}\, dx = \frac{\sqrt{\pi}}{2}$.

Hence

$$F(a) = \int_0^\infty e^{-x^2} \cos 2ax\, dx = \frac{\sqrt{\pi}}{2} e^{-a^2}.$$

5.19 EXTREME VALUES

A function $f(x, y)$ of two independent variables x and y is said to have an *extreme value* at the point (a, b) if the increment

$$\Delta f = f(a + h, b + k) - f(a, b)$$

preserves the same sign for all values of h and k whose moduli do not exceed a sufficiently small positive number η.

If Δf is negative, then the extreme value is a *maximum* and if Δf is positive, then the extreme value is a *minimum.*

Necessary and Sufficient Conditions for Extreme Values

By Taylor's theorem, we have

$$\Delta f = f(a + b, b + k) - f(a, b)$$

$$= h\frac{\partial f}{\partial x}(a, b) + k\frac{\partial f}{\partial y}(a, b)$$

+ terms of second and higher order.

Now by taking h and k sufficiently small, the first-order terms can be made to govern the sign of the right-hand side and therefore, of the left-hand side of the previous expansion. Hence, the change in the sign of h and k would change the sign of the left-hand side, that is, of Δf. But if the sign of Δf changes, $f(x, y)$ cannot have an extreme point at (a, b). Hence, as a first condition for the extreme value, we must have

$$h\frac{\partial f}{\partial x}(a, b) + k\frac{\partial f}{\partial y}(a, b) = 0.$$

Since the arbitrary increments h and k are independent of each other, we must have

$$\frac{\partial f}{\partial x}(a, b) = 0 \text{ and } \frac{\partial f}{\partial y}(a, b) = 0,$$

which are **necessary conditions for the existence of extreme points**. However, these are not sufficient conditions for the existence of extreme points. Further, a point (a, b) is called a *stationary point* if $f_x(a, b) = f_y(a, b) = 0$. The value $f(a, b)$ is called a *stationary value*.

To find sufficient conditions, let (a, b) be an interior point of the domain of f such that f admits the second-order continuous partial derivatives in the neighborhood of (a, b). Suppose that $f_x(a, b)$ $f_y(a, b) = 0$. We further, suppose that

$$r = \frac{\partial^2 f}{\partial x^2},\ s = \frac{\partial^2 f}{\partial x \partial y}, \text{ and}$$

$$t = \frac{\partial^2 f}{\partial y^2}, \text{ when } x = a \text{ and } y = b.$$

Thus,

$$f_{xx}(a, b) = r,\ f_{xy}(a, b) = s, \text{ and } f_{yy}(a, b) = t.$$

If $(a + b, b + k)$ is any point in the neighborhood of (a, b), then by Taylor's theorem, we have

$$\begin{aligned}\Delta f &= f(a + b, b + k) - f(a, b)\\ &= hf_x(a, b) + kf_y(a, b) + \frac{1}{2}[h^2 f_{xx}(a, b)\\ &\quad + 2hkf_{xy}(a, b) + k^2 f_{yy}(a, b)] + R_3\\ &= \frac{1}{2}[rh^2 + 2hks + tk^2] + R_3,\end{aligned}$$

where R_3 consists of terms of third and higher orders of small quantities. Thus, by taking h and k sufficiently small, now the second order terms can be made to govern the sign of the right-hand side and therefore, of the left-hand side of the previous expansion. But

$$\begin{aligned}\frac{1}{2}[rh^2 + 2hks + tk^2] &= \frac{1}{2r}[r^2h^2 + 2hkrs + rtk^2]\\ &= \frac{1}{2r}[r^2h^2 + 2hkrs\\ &\quad + rtk^2 + k^2s^2 - k^2s^2]\\ &= \frac{1}{2r}[(rh + sk)^2 + k^2(rt - s^2)].\end{aligned}$$

Since $(rh + sk)^2$ is always positive, it follows that Δf is positive if $rt - s^2$ is positive. Now $rt - s^2 >$ 0 if both r and t have the same sign. Thus, the sign of Δf shall be that of r. Therefore, if $rt - s^2$ is positive, we have a maximum or a minimum accordingly, as both r and t are either negative or positive. This condition was first pointed out by Lagrange and is known as *Lagrange's condition.* However, if $rt = s^2$, then $rh^2 + 2hks + tk^2$ becomes $\frac{1}{r}(hr + ks)^2$ and is, therefore, of the same sign as r or t unless

$$\frac{h}{k} = -\frac{s}{r}, \text{ say, for which } (hr + ks)^2 \text{ vanishes.}$$

In such a case, we must consider terms of higher order in the expansion of $f(a + h, b + k)$. Thus, we may state that

1. The value $f(a, b)$ is an extreme value of $f(x, y)$ if $f_x(a, b) = f_y(a, b) = 0$ and if $rt - s^2 > 0$. The value is *maximum* or *minimum* accordingly as $f_{xx}(a, b)$ or $f_{yy}(a, b)$ is negative or positive.
2. If $rt - s^2 < 0$, then $f(x, y)$ has no extreme value at (a, b). The point (a, b) is a *saddle point* in this case.
3. If $rt - s^2 = 0$, the case is doubtful and requires terms of higher order in the expansion of the function.

EXAMPLE 5.92

Show that the function $f(x, y) = y^2 + x^2y + x^4$ has a minimum value at the origin.

Solution. We have

$$f(x, y) = y^2 + x^2y + x^4.$$

Therefore,

$$\begin{aligned}f_x &= 2xy + 4x^3 \text{ which yields } f_x(0, 0) = 0,\\ f_y &= 2y + x^2 \text{ which yields } f_y(0, 0) = 0,\\ f_{xx} &= 2y + 12x^2 \text{ which yields } f_{xx}(0, 0) = 0,\\ f_{yy} &= 2 \text{ which yields } f_{yy}(0, 0) = 2, \text{ and}\\ f_{xy} &= 2x \text{ which yields } f_{xy}(0, 0) = 0.\end{aligned}$$

Hence, at the origin, we have $rt - s^2 = 0$. Thus, further investigation is needed in the case. We write

$$f(x, y) = y^2 + x^2y + x^4 = \left(y + \frac{1}{2}x^2\right)^2 + \frac{3x^4}{4}.$$

Then,

$$\Delta f = f(h, k) - f(0, 0) = \left(k + \frac{h^2}{2}\right)^2 + \frac{3h^2}{4},$$

which is always greater than zero for all values of h and k. Hence, $f(x, y)$ has a *minimum value at the origin.*

EXAMPLE 5.93

Show that the function

$$u = xy + \frac{a^3}{x} + \frac{a^3}{y}$$

has a minimum value at (a, a).

Solution. We have

$$u = xy + \frac{a^3}{x} + \frac{a^3}{y}.$$

Therefore,

$$\frac{\partial u}{\partial x} = y - \frac{a^3}{x^2} \text{ yields } f_x(a, a) = 0,$$

$$\frac{\partial u}{\partial y} = x - \frac{a^3}{y^2} \text{ yields } f_y(a, a) = 0,$$

$$\frac{\partial^2 u}{\partial x^2} = \frac{2a^3}{x^3} \text{ yields } f_{xx}(a, a) = 2,$$

$$\frac{\partial^2 u}{\partial x \partial y} = 1 \text{ yields } f_{xy}(a, a) = 1, \text{ and}$$

$$\frac{\partial^2 u}{\partial y^2} = \frac{2a^3}{y^3} \text{ and so, } f_{yy}(a, a) = 2.$$

We observe that $rt - s^2 = 4 - 1 = 3$ (positive) and r and t too positive. Therefore, u has the minimum at (a, a). Thus, the minimum value of u is

$$u(a, a) = a^2 + a^2 + a^2 = 3a^2.$$

EXAMPLE 5.94

Show that the function $f(x, y) = 2x^4 - 3x^2y + y^2$ does not have a maximum or a minimum at $(0, 0)$.

Solution. The given function is

$$f(x, y) = 2x^4 - 3x^2y + y^2.$$

Therefore,

$$\frac{\partial f}{\partial x} = 8x^3 - 6xy, \text{ which implies } f_x(0, 0) = 0,$$

$$\frac{\partial f}{\partial y} = -3x^2 + 2y, \text{ which implies } f_y(0, 0) = 0,$$

$$\frac{\partial^2 f}{\partial x^2} = 24x^2 - 6y, \text{ which implies } f_{xx}(0, 0) = 0,$$

$$\frac{\partial^2 f}{\partial x \partial y} = -6x, \text{ which implies } f_{xy}(0, 0) = 0, \text{ and}$$

$$\frac{\partial^2 f}{\partial y^2} = 2, \text{ which implies } f_{yy}(0, 0) = 2.$$

Thus, $rt - s^2 = 0$ and so, further investigation is required. We have

$$f(x, y) = (x^2 - y)(2x^2 - y), \; f(0, 0) = 0.$$

Therefore,

$$\Delta f = f(x, y) - f(0, 0) = (x^2 - y)(2x^2 - y).$$

Thus, Δf is positive, for $y < 0$ or $x^2 > y > 0$ and Δf is negative, for $y > x^2 > \frac{y}{2} > 0$. Thus, Δf does not keep the same sign in the neighborhood of (0, 0). Hence, the function does not have a maximum or a minimum at (0, 0).

EXAMPLE 5.95

Examine the function $\sin x + \sin y + \sin(x + y)$ for extreme points.

Solution. The given function is $f(x, y) = \sin x + \sin y + \sin(x + y)$. Therefore,

$$f_x = \cos x + \cos(x + y),$$

$$f_y = \cos y + \cos(x + y),$$

$$f_{xx} = -\sin x - \sin(x + y),$$

$$f_{xy} = -\sin(x + y), \text{ and}$$

$$f_{yy} = -\sin y - \sin(x + y).$$

For extreme points, we must have $f_x = f_y = 0$ and so,

$$\cos x + \cos(x + y) = 0 \qquad (1)$$

and

$$\cos y + \cos(x + y) = 0 \qquad (2)$$

Subtracting (2) from (1), we get $\cos x = \cos y$ and so, $x = y$. Also then, $\cos x + \cos 2x = 0$ which yields $\cos 2x = -\cos x = \cos(\pi - x)$ and so, $2x = \pi - x$ or $x = \frac{\pi}{3}$. Thus, $\left(\frac{\pi}{3}, \frac{\pi}{3}\right)$ is a stationary point. Now

$$r = f_{xx}\left(\frac{\pi}{3}, \frac{\pi}{3}\right) = -\frac{\sqrt{3}}{2} - \frac{\sqrt{3}}{2} = -\sqrt{3} \text{ (negative)},$$

$$s = f_{xy}\left(\frac{\pi}{3}, \frac{\pi}{3}\right) = \frac{\sqrt{3}}{2}$$

$$t = f_{yy}\left(\frac{\pi}{3}, \frac{\pi}{3}\right) = \frac{\sqrt{3}}{2} - \frac{\sqrt{3}}{2} = -\sqrt{3} \text{ (negative)}.$$

Thus,

$$rt - s^2 = 3 - \frac{3}{4} = \frac{9}{4} \text{ (positive)}$$

and r is negative. Hence, the given function has a maximum value at $\left(\frac{\pi}{3}, \frac{\pi}{3}\right)$ given by

$$f\left(\frac{\pi}{3}, \frac{\pi}{3}\right) = \sin\frac{\pi}{3} + \sin\frac{\pi}{3} + \sin\frac{2\pi}{3}$$
$$= \frac{\sqrt{3}}{2} + \frac{\sqrt{3}}{2} + \frac{\sqrt{3}}{2} = \frac{3\sqrt{3}}{2}.$$

EXAMPLE 5.96

Examine the following surface for high- and low points:

$$z = x^2 + xy + 3x + 2y + 5.$$

Solution. We have

$$\frac{\partial z}{\partial x} = 2x + y + 3, \frac{\partial z}{\partial y} = x + 2,$$
$$\frac{\partial^2 z}{\partial x^2} = 2, \frac{\partial^2 z}{\partial x \partial y} = 1, \text{ and } \frac{\partial^2 z}{\partial y^2} = 0.$$

For an extreme point, we must have $\frac{\partial z}{\partial x} = \frac{\partial z}{\partial y} = 0$ and so,

$$2x + y + 3 = 0 \text{ and } x + 2 = 0.$$

Solving these equations, we get $x = -2$, $y = 1$. Thus, z can have a maximum or a minimum only at $(-2, 1)$. Further,

$$r = \frac{\partial^2 z}{\partial x^2}(-2, 1) = 2,$$
$$s = \frac{\partial^2 z}{\partial x \partial y}(-2, 1) = 1, \text{ and}$$
$$t = \frac{\partial^2 z}{\partial y^2}(-2, 1) = 0.$$

Therefore, $rt - s^2 = -1$ (negative) and so, the stationary value of z at $(-2, 1)$ is neither a maximum nor a minimum. Hence, the surface has no high- or low point.

EXAMPLE 5.97

Locate the stationary points of $x^4 + y^4 - 2x^2 + 4xy - 2y^2$ and determine their nature.

Solution. We have

$$f(x, y) = x^4 + y^4 - 2x^2 + 4xy - 2y^2.$$

Therefore,

$$f_x = 4x^3 - 4x + 4y \text{ and } f_y = 4y^3 + 4x - 4y.$$

The stationary points are given by

$$f_x = 4x^3 - 4x + 4y = 0 \quad (1)$$
$$f_y = 4y^3 + 4x - 4y = 0 \quad (2)$$

Adding (1) and (2), we get

$$x^3 + y^3 = 0 \text{ or } (x + y)(x^2 - xy + y^2) = 0.$$

Therefore, either $y = -x$ or $x^2 - xy + y^2 = 0$. Putting $y = -x$ in (1), we get $x(x^2 - 2) = 0$, which yields $x = 0, \sqrt{2}$, or $-\sqrt{2}$. The value of y corresponding to these values are $0, -\sqrt{2}$, and $\sqrt{2}$. Thus, the points $(0, 0)$, $(\sqrt{2}, -\sqrt{2})$, and $(-\sqrt{2}, \sqrt{2})$ satisfy (1) and (2). On the other hand, from equation (1) and $x^2 - xy + y^2 = 0$ we get $(0, 0)$ as the only real root. Thus, the stationary points are

$$(0, 0), (\sqrt{2}, -\sqrt{2}), \text{ and } (-\sqrt{2}, \sqrt{2}).$$

Also,

$$f_{xx} = 12x^2 - 4, f_{xy} = 4, \text{ and } f_{yy} = 12y^2 - 4.$$

At (0,0), we have

$$r = f_{xx}(0, 0) = -4,$$
$$s = f_{xy}(0, 0) = 4, \text{ and}$$
$$t = f_{yy}(0, 0) = -4,$$

and so, $rt - s^2 = 0$. Thus, at $(0, 0)$, the case is doubtful. The given equation can be written as

$$f(x, y) = x^4 + y^4 - 2(x - y)^2.$$

So,

$$f(0, 0) = 0 \text{ and } f(h, k) = h^4 + k^4 - 2(h - k)^2.$$

We observe that for small quantities of h and k,

$$\Delta f = f(h, k) - f(0, 0) = h^4 + k^4 - 2(h - k)^2$$

is greater than 0, if $h = k$ and less than 0, if $h \neq k$. Since Δf does not preserve the sign, the function has no extreme value at the origin.

At $(\sqrt{2}, -\sqrt{2})$, we have $r = 20$, $s = 4$, and $t = 0$ so that $rt - s^2 = 384$ (positive). Since r is positive, $f(x, y)$ has a minimum at this point.

At $(-\sqrt{2}, \sqrt{2})$, we have $r = 20$, $s = 4$, and $t = 20$. Thus, $rt - s^2$ is positive. Since r is positive, $f(x, y)$ has a minimum at $(-\sqrt{2}, \sqrt{2})$ also.

EXAMPLE 5.98

Find the minimum value of $x^2 + y^2 + z^2$ when $ax + by + cz = p$.

Solution. Let $f(x, y, z) = x^2 + y^2 + z^2$. From the relation $ax + by + cz = p$, we get $z = \frac{p - ax - by}{c}$.

Putting this value of z in $f(x, y, z)$, we get

$$f(x, y, z) = x^2 + y^2 + \left(\frac{p - ax - by}{c}\right)^2$$

as a function of two variables x and y. Then,

$$f_x = 2x - \frac{2a}{c^2}(p - ax - by)$$

and

$$f_y = 2y - \frac{2b}{c^2}(p - ax - by).$$

For extreme points, we must have $f_x = f_y = 0$. Thus,

$$2x - \frac{2a}{c^2}(p - ax - by) = 0$$

and

$$2y - \frac{2b}{c^2}(p - ax - by) = 0.$$

Solving these equations, we get

$$x = \frac{ap}{a^2 + b^2 + c^2} \text{ and } y = \frac{bp}{a^2 + b^2 + c^2}.$$

Now,

$$f_{xx} = 2 + \frac{2a^2}{c^2}, f_{xy} = \frac{2ab}{c^2}, \text{ and } f_{yy} = 2 + \frac{2b^2}{c^2},$$

so that

$$rt - s^2 = 4\left(1 + \frac{a^2}{c^2}\right)\left(1 + \frac{b^2}{c^2}\right) - \frac{4a^2b^2}{c^4}$$

$$= 4\left(1 + \frac{a^2}{c^2} + \frac{b^2}{c^2}\right) \text{ (positive).}$$

Also, $r = f_{xx}$ is positive. Therefore, $f(x, y)$ has a minimum at $\left(\frac{ap}{a^2 + b^2 + c^2}, \frac{bp}{a^2 + b^2 + c^2}\right)$ and the minimum value is

$$\text{Min. } f(x, y, z) = \frac{p^2}{a^2 + b^2 + c^2}.$$

EXAMPLE 5.99

Show that the function

$$f(x, y) = x^2 - 2xy + y^2 + x^3 - y^3 + x^5$$

has neither a maximum nor a minimum at (0, 0).

Solution. For the given function,

$$f_x = 2x - 2y + 3x^2 + 5x^4 \quad \text{and } f_y = -2x + 2y - 3y^2.$$

$$f_{xx} = 2 + 6x + 20x^3 \quad \text{and } f_{xy} = -2, f_{yy} = 2 - 6y.$$

For a stationary value of $f(x, y)$, we must have $f_x = f_y = 0$. Thus,

$$2x - 2y + 3y^2 + 5x^4 = 0 \text{ and } -2x + 2y - 3y^2 = 0.$$

The origin (0, 0) satisfies these equations. Further,

$$r = f_{xx}(0,0) = 2,\ s = f_{xy}(0,0) = -2,\ t = f_{yy}(0,0) = 2,$$

and so, $rt - s^2 = 0$. Hence, further investigations are required. We rewrite the equation as

$$f(x, y) = (x - y)^2 + (x - y)(x^2 + xy + y^2) + x^5.$$

We note that $f(0, 0) = 0$. But,

$$\Delta f = f(h, k) - f(0, 0) = f(h, k)$$

$$= (h - k)^2 + (h - k)(h^2 + hk + k^2) + k^5.$$

In the neighborhood of (0, 0), if $h = k$, then $\Delta f = k^5$; which is positive, when $k > 0$ and negative, when $k < 0$. Thus, Δf does not keep the same sign in the neighborhood of (0, 0). Hence, $f(x, y)$ cannot have a maximum or a minimum at the point (0, 0).

EXAMPLE 5.100

Find the dimensions of the rectangular box, open at the top, of maximum capacity whose surface is 432 sq. cm.

Solution. Let x, y, and z cm be the dimensions of the box and S be its surface. Then

$$S = xy + 2yz + 2zx = 432 \text{ (given)} \quad (1)$$

and

$$V = xyz. \quad (2)$$

We have to maximize V. From (1), we have

$$z = \frac{432 - xy}{2y + 2x}. \quad (3)$$

Therefore, (2) reduces to

$$V = xy\left(\frac{432 - xy}{2y + 2x}\right) = \frac{432xy - x^2y^2}{2y + 2x}.$$

Now,

$$\frac{\partial V}{\partial x} = \frac{(2y + 2x)(432y - 2xy^2) - 2(432xy - x^2y^2)}{(2y + 2x)^2}$$

$$= \frac{864y^2 - 4xy^3 - 2x^2y^2}{(2x + 2y)^2},$$

$$\frac{\partial V}{\partial y}=\frac{(2x+2y)(432x-2x^2y)-2(432xy-x^2y^2)}{(2y+2x)^2}$$

$$=\frac{864x^2-4x^3y-2x^2y^2}{(2x+2y)^2}.$$

For stationary points, we must have $\frac{\partial V}{\partial x}=\frac{\partial V}{\partial y}=0$. So,

$$864-4xy-2x^2=0 \quad (4)$$

$$864-4xy-2y^2=0. \quad (5)$$

Subtracting (5) from (4), we get $y=\pm x$. Substituting $x=y$ in (5), we get

$$864-4y^2-2y^2=0 \text{ or } y^2=\frac{864}{6}=124.$$

Thus, $x=y=12$ and (3) implies $z=6$. It can be verified that $rt-s^2>0$ and that r is positive for these values. Hence, the dimensions of the box are $x=y=12$cm and $z=6$cm.

EXAMPLE 5.101

Examine $x^3y^2(1-x-y)$ for extreme points.

Solution. We have

$$f(x,y)=x^3y^2(1-x-y).$$

Therefore,

$$\frac{\partial f}{\partial x}=3x^2y^2(1-x-y)+x^2y^2(-1)$$

$$=3x^2y^2-4x^3y^2-3x^2y^3 \text{ and}$$

$$\frac{\partial f}{\partial y}=2x^3y(1-x-y)+x^2y^2(-1)$$

$$=2x^3y-2x^4y-3x^3y^2.$$

For a maximum or a minimum of f, we must have $\frac{\partial f}{\partial x}=\frac{\partial f}{\partial y}=0$. Therefore,

$$x^2y^2(3-4x-3y)=0 \text{ and } x^3y(2-2x-3y)=0.$$

Solving these equations, we get the stationary points $(0,0)$ and $\left(\frac{1}{2},\frac{1}{3}\right)$. Further,

$$r=\frac{\partial^2 f}{\partial x^2}=6xy^2-12x^2y^2-6xy^3=6xy^2(1-2x-y),$$

$$s=\frac{\partial^2 f}{\partial x\partial y}=x^2y(6-8x-9y), \text{ and}$$

$$t=\frac{\partial^2 f}{\partial y^2}=2x^3(1-x-3y).$$

Therefore,

(i) at (0,0), $r=0$, $t=0$, and $s=0$, and so, $rt-s^2=0$.

But,

$$\Delta f=f(h,k)-f(0,0)=h^3k^2(1-h-k).$$

Sign is governed by h^3k^2 which is positive, if $h>0$ and negative, if $h<0$. Since Δf does not keep the same sign in the neighborhood of (0, 0), the given function does not have a maximum or a minimum value at (0, 0).

(ii) at $\left(\frac{1}{2},\frac{1}{3}\right)$, we have

$$r=\frac{\partial^2 f}{\partial x^2}\left(\frac{1}{2},\frac{1}{3}\right)=-\frac{1}{9},$$

$$s=\frac{\partial^2 f}{\partial x\partial y}\left(\frac{1}{2},\frac{1}{3}\right)=-\frac{1}{12}, \text{ and}$$

$$t=\frac{\partial^2 f}{\partial y^2}\left(\frac{1}{2},\frac{1}{3}\right)=-\frac{1}{8}.$$

Therefore,

$$rt-s^2=\left(-\frac{1}{9}\right)\left(-\frac{1}{8}\right)-\left(-\frac{1}{12}\right)^2=\frac{1}{72}-\frac{1}{144}$$

$$=\frac{1}{144} \text{ (positive)}.$$

But r is negative. Hence, $f(x,y)$ has a maximum at $\left(\frac{1}{2},\frac{1}{3}\right)$. The maximum value is

$$f\left(\frac{1}{2},\frac{1}{3}\right)=\frac{1}{8}\cdot\frac{1}{9}\left(1-\frac{1}{2}-\frac{1}{3}\right)=\frac{1}{432}.$$

EXAMPLE 5.102

Find the points where the function x^3+y^3-3axy has a maximum or a minimum.

Solution. We have

$$f_x=3x^2-3ay, f_y=3y^2-3ax,$$

$$f_{xx}=6x, f_{yy}=6y, \text{ and } f_{xy}=-3a.$$

For extreme points, we have $f_x=f_y=0$ and so,

$$3x^2-3ay=0 \text{ and } 3y^2-3ax=0.$$

Solving the earlier equations, we get two stationary points (0, 0) and (a, a). Further,

$$rt-s^2=36xy-9a^2.$$

At (0, 0), $rt-s^2=-9a^2$ (negative). Therefore, there is no extreme point at the origin.

At (a, a), we have $rt-s^2=36a^2-9a^2=27a^2>0$. Also r at (a, a) is equal to $6a$. If a is positive, then r is positive and $f(x, y)$ will have a minimum at (a, a). If a is negative, then r is negative and so, $f(x, y)$ will have a maximum at (a, a) for $a < 0$.

EXAMPLE 5.103

Prove that the rectangular solid of maximum volume which can be inscribed in a sphere is a cube.

Solution. Let x, y, and z be the length, breadth, and height of a rectangular solid. Then, the volume of the solid is

$$V = xyz. \qquad (1)$$

Now each diagonal of the rectangular solid passes through the center of the sphere. Therefore, each diagonal is the diameter of the sphere, that is,

$$\sqrt{x^2+y^2+z^2} = d$$

or

$$x^2+y^2+z^2 = d^2$$

or

$$z = \sqrt{d^2-x^2-y^2}. \qquad (2)$$

Therefore, (1) reduces to

$$V = xy\sqrt{d^2-x^2-y^2}$$

or

$$V^2 = x^2y^2(d^2-x^2-y^2)$$
$$= x^2y^2d^2 - x^4y^2 - x^2y^4 = f(x, y).$$

Then,

$$\frac{\partial f}{\partial x} = 2xy^2d^2 - 4x^3y^2 - 2xy^4 = 2xy^2(d^2-2x^2-y^2),$$

$$\frac{\partial f}{\partial y} = 2x^2yd^2 - 2x^4y - 4x^2y^3 = 2x^2y(d^2-x^2-2y^2),$$

$$\frac{\partial^2 f}{\partial x^2} = 2d^2y^2 - 12x^2y^2 - 2y^4,$$

$$\frac{\partial^2 f}{\partial y^2} = 2d^2x^2 - 12x^2y^2 - 2x^4, \text{ and}$$

$$\frac{\partial^2 f}{\partial x \partial y} = 4xyd^2 - 8x^3y - 8xy^3.$$

For stationary points, we have $\frac{\partial f}{\partial x} = \frac{\partial f}{\partial y} = 0$. Therefore,

$$d^2 - 2x^2 - y^2 = 0 \text{ and}$$
$$d^2 - x^2 - 2y^2 = 0. \qquad (3)$$

Solving the preceding equations, we get $y = x$. Substituting $y = x$ in (3), we get $x = \frac{d}{\sqrt{3}}$. Thus, $x = y = \frac{d}{\sqrt{3}}\cdot$ Hence, from (2), we have $z = \frac{d}{\sqrt{3}}\cdot$ Thus, the stationary point is $\left(\frac{d}{\sqrt{3}}, \frac{d}{\sqrt{3}}\right)$. At $\left(\frac{d}{\sqrt{3}}, \frac{d}{\sqrt{3}}\right)$, $r = -\frac{8d^4}{9}$ (negative), $s = -\frac{4d^4}{9}$, and $t = -\frac{8d^4}{9}\cdot$

Therefore, $rt - s^2 = \frac{64d^8}{81} - \frac{16d^8}{81} = \frac{48d^8}{81} = \frac{16d^8}{27} > 0.$

Since r is negative, it follows that $f(x, y)$ or V^2 has a maximum value at $\left(\frac{d}{\sqrt{3}}, \frac{d}{\sqrt{3}}\right)$. Hence, V is maximum when $x = y = z$. Consequently, the solid is a cube.

EXAMPLE 5.104

A rectangular box, open at the top, is to have a volume of 32 cubic feet. Determine the dimensions of the box requiring least material for its construction.

Solution. Let S be the surface, and x, y, and z in feet be the edges of the box. Then,

$$S = xy + 2yz + 2zx \qquad (1)$$

and

$$V = xyz = 32 \text{ cubic feet (given)}. \qquad (2)$$

From (2), we have $z = \frac{32}{xy}$ and so,

$$S = xy + 2(y+x)\frac{32}{xy} = xy + 64\left(\frac{1}{x} + \frac{1}{y}\right).$$

Then,

$$\frac{\partial S}{\partial x} = y - \frac{64}{x^2}, \frac{\partial S}{\partial y} = x - \frac{64}{y^2},$$

$$\frac{\partial^2 S}{\partial x^2} = \frac{128}{x^3}, \frac{\partial^2 S}{\partial x \partial y} = 1, \text{ and } \frac{\partial^2 S}{\partial y^2} = \frac{128}{y^3}.$$

The stationary values are given by

$$\frac{\partial S}{\partial x} = y - \frac{64}{x^2} = 0 \text{ and } \frac{\partial S}{\partial y} = x - \frac{64}{y^2} = 0.$$

Solving these equations, we get $x = y = 4$. Putting these values in (1), we get $z = 2$. Further, at (4, 4), we have $rt - s^2 = 3$ (positive) and r at (4, 4) is 2 (positive). Therefore, S is minimum for (4, 4). The dimensions of the box are $x = 4$, $y = 4$, $z = 2$.

EXAMPLE 5.105

Find the points on the surface $z^2 = xy + 1$ nearest to the origin.

Solution. If r is the distance from (0, 0, 0) of any point (x, y, z) on the given surface, then

$$r^2 = (x-0)^2 + (y-0)^2 + (z-0)^2 = x^2 + y^2 + z^2$$

$$= x^2 + y^2 + xy + 1, \text{ using the equation of the given surface.}$$

Thus, we have a function of two variables given by

$$r^2 = x^2 + y^2 + xy + 1 = f(x, y), \text{ say.}$$

Then,

$$\frac{\partial f}{\partial x} = 2x + y, \frac{\partial f}{\partial y} = 2y + x,$$

$$\frac{\partial^2 f}{\partial x^2} = 2, \frac{\partial^2 f}{\partial y^2} = 2, \text{ and } \frac{\partial^2 f}{\partial x \partial y} = 1.$$

The stationary points are given by $\frac{\partial f}{\partial x} = 0$ and $\frac{\partial f}{\partial y} = 0$ and therefore,

$$2x + y = 0 \text{ and } 2y + x = 0.$$

Solving the preceding equations, we get $x = y = 0$ and then, $z^2 = xy + 1$ yields $z = \pm 1$. Thus, the stationary points are $(0, 0, \pm 1)$. Further, at these points, $r = 2$, $s = 1$, and $t = 2$ and so, $rt - s^2 = 3$ (positive). Since r is positive, the value is minimum at $(0, 0, \pm 1)$.

5.20 LAGRANGE'S METHOD OF UNDETERMINED MULTIPLIERS

Let $u = \phi(x_1, x_2, \ldots, x_n)$ be a function of n variables $x_1, x_2, \ldots, x_n$, which are connected by m equations

$$f_1(x_1, x_2, \ldots, x_n) = 0, f_2(x_1, x_2, \ldots, x_n) = 0, \ldots, f_m(x_1, x_2, \ldots, x_n) = 0,$$

so that only $n - m$ of the variables are independent. For a maximum or a minimum value of u, we must have

$$du = \frac{\partial u}{\partial x_1} dx_1 + \frac{\partial u}{\partial x_2} dx_2 + \frac{\partial u}{\partial x_3} dx_3 + \ldots + \frac{\partial u}{\partial x_n} dx_n = 0.$$

Also, differentiating the given m equations connecting the variables, we have

$$df_1 = \frac{\partial f_1}{\partial x_1} dx_1 + \frac{\partial f_1}{\partial x^2} dx_2 + \frac{\partial f_1}{\partial x_3} dx_3 + \ldots + \frac{\partial f_1}{\partial x_n} dx_n = 0$$

$$df_2 = \frac{\partial f_2}{\partial x_1} dx_1 + \frac{\partial f_2}{\partial x_2} dx_2 + \frac{\partial f_2}{\partial x_3} dx_3 + \ldots + \frac{\partial f_2}{\partial x_n} dx_n = 0$$

....

....

$$df_m = \frac{\partial f_m}{\partial x_1} dx_1 + \frac{\partial f_m}{\partial x_2} dx_2 + \frac{\partial f_m}{\partial x_3} dx_3 + \ldots + \frac{\partial f_m}{\partial x_n} dx_n = 0.$$

Multiplying the earlier $(m + 1)$ equations, obtained on differentiation, by $1, \lambda_1, \lambda_2, \ldots, \lambda_m$, respectively, and then adding all, we get an equation which may be written as

$$P_1 dx_1 + P_2 dx_2 + P_3 dx_3 + \ldots + P_n dx_n = 0, \quad (1)$$

where,

$$P_r = \frac{\partial u}{\partial x_r} + \lambda_1 \frac{\partial f_1}{\partial x_r} + \lambda_2 \frac{\partial f_2}{\partial x_r} + \lambda_3 \frac{\partial f_3}{\partial x_r} + \ldots + \lambda_m \frac{\partial f_m}{\partial x_r}.$$

The m quantities $\lambda_1, \lambda_2, \ldots, \lambda_m$ are at our choice. Let us choose them so as to satisfy the m linear equations.

$$P_1 = P_2 = P_3 = \ldots = P_m = 0.$$

Then, the equation (1) reduces to

$$P_{m+1} dx_{m+1} + P_{m+2} dx_{m+2} + \ldots + P_n dx_n = 0.$$

It is indifferent which of the n – m of the n variables are regarded as independent. So, suppose that the variables $x_{m+1}, x_{m+2}, \ldots, x_n$ are independent, then as the $n - m$ quantities $dx_{m+1}, d_{m+2}, \ldots, dx_n$ are all independent, their coefficients must be separately zero. Thus, we obtain the additional $n - m$ equations as follows:

$$P_{m+1} = 0, P_{m+2} = 0, \ldots, P_n = 0.$$

In this way, we get $(m + n)$ equations

$$f_1 = 0, f_2 = 0, \ldots, f_m = 0 \text{ and}$$

$$P_1 = 0, P_2 = 0, \ldots, P_n = 0,$$

which together with relation $u = \phi(x_1, x_2, \ldots, x_n)$ determine the m multipliers $\lambda_1, \lambda_2, \ldots,$

λ_m and the values of n variables $x_1, x_2, \dots, x_n$ for which the maximum and minimum values of u are possible.

The drawback of the Lagrange's method of undetermined multipliers is that it does not determine the nature of the stationary point.

EXAMPLE 5.105

Find the point of the circle $x^2 + y^2 + z^2 = k^2$ and $lx + my + nz = 0$ at which the function $u = ax^2 + by^2 + cz^2 + 2fyz + 2gzx + 2hxy$ attains its greatest and the least value.

Solution. We have

$$u = ax^2 + by^2 + cz^2 + 2fyz + 2gzx + 2hxy, \quad (1)$$

$$f_1 = lx + my + nz, \text{ and} \quad (2)$$

$$f_2 = x^2 + y^2 + z^2 - k^2. \quad (3)$$

For extreme points, we must have $du = 0$. So,

$$(ax + gz + hy)\,dx + (hx + by + fz)\,dy + (gx + fy + cz)\,dz = 0. \quad (4)$$

Also differentiating (2) and (3), we get

$$ldx + mdy + ndz = 0, \text{ and} \quad (5)$$

$$xdx + ydy + zdz = 0. \quad (6)$$

Multiplying (4), (5), and (6) by 1, λ_1, and λ_2, respectively, and then by adding all and equating to zero the coefficients of dx, dy, and dz, we get

$$ax + hy + gz + \lambda_1 l + \lambda_2 x = 0, \quad (7)$$

$$hx + by + fz + \lambda_1 m + \lambda_2 y = 0, \text{ and} \quad (8)$$

$$gx + fy + cz + \lambda_1 n + \lambda_2 z = 0. \quad (9)$$

Multiplying (7), (8), and (9) by x, y, and z, respectively, and then adding all, we get

$$u + \lambda_2 = 0 \quad \text{or} \quad \lambda_2 = -u.$$

Putting $\lambda_2 = -u$ in (7), (8), and (9), we obtain

$$(a-u)\,x + hy + gz + \lambda_1 l = 0, \quad (10)$$

$$hx + (b-u)\,y + fz + \lambda_1 m = 0, \text{ and} \quad (11)$$

$$gx + f_y + (c-u)\,z + \lambda_1 n = 0. \quad (12)$$

Also,

$$lx + my + nz + \lambda_1 \cdot 0 = 0. \quad (13)$$

Eliminating x, y, z, and λ_1 from (10), (11), (12), and (13), we get

$$\begin{vmatrix} a-u & h & g & l \\ h & b-u & f & m \\ g & f & c-u & n \\ l & m & n & 0 \end{vmatrix} = 0,$$

which gives the maximum or minimum value of u.

EXAMPLE 5.106

Prove that the volume of the greatest rectangular parallelopiped that can be inscribed in the ellipsoid

$$\frac{x^2}{a^2} + \frac{y^2}{b^2} + \frac{z^2}{c^2} = 1 \text{ is } \frac{8abc}{3\sqrt{3}}.$$

Solution. Let (x, y, z) denote the coordinates of the vertex of the rectangular parallelopiped which lies in the positive octant and let V denote its volume. Volume V is given by $V = 8xyz$. Its maximum value is to be determined under the condition that it is inscribed in the ellipsoid $\frac{x^2}{a^2} + \frac{y^2}{b^2} + \frac{z^2}{c^2} = 1$. Thus, we have

$$V = 8xyz, \text{ and} \quad (1)$$

$$f_1 = \frac{x^2}{a^2} + \frac{y^2}{b^2} + \frac{z^2}{c^2} - 1. \quad (2)$$

For an extreme value, we must have

$$dV = yzdx + zxdy + xydz. \quad (3)$$

Also differentiating (2), we get

$$df_1 = \frac{x}{a^2}\,dx + \frac{y}{b^2}\,dy + \frac{z}{c^2}\,dz. \quad (4)$$

Multiplying (3) and (4) by 1 and λ, respectively, and then adding both and equating the coefficients of dx, dy, and dz to zero, we get

$$yz + \frac{\lambda x}{a^2} = 0, \quad (5)$$

$$zx + \frac{\lambda y}{b^2} = 0, \quad (6)$$

and

$$xy + \frac{\lambda z}{c^2} = 0. \quad (7)$$

From (5), (6), and (7), we get

$$\lambda = -\frac{a^2yz}{x} = -\frac{b^2zx}{y} = -\frac{c^2xy}{z}$$

and so,

$$\frac{a^2yz}{x} = \frac{b^2zx}{y} = \frac{c^2xy}{z}.$$

Dividing throughout by xyz, we get

$$\frac{a^2}{x^2} = \frac{b^2}{y^2} = \frac{c^2}{z^2}.$$

Then, equation (2) yields

$$3\frac{x^2}{a^2} = 1 \quad \text{or} \quad x = \frac{a}{\sqrt{3}},$$

$$3\frac{y^2}{b^2} = 1 \quad \text{or} \quad y = \frac{b}{\sqrt{3}}, \text{ and}$$

$$3\frac{z^2}{c^2} = 1 \quad \text{or} \quad z = \frac{c}{\sqrt{3}}.$$

Thus, the stationary value is at the point

$$\left(\frac{a}{\sqrt{3}}, \frac{b}{\sqrt{3}}, \frac{c}{\sqrt{3}}\right).$$

Differentiating partially the equation (2) with respect to x, taking y as constant, we get

$$\frac{2x}{a^2} + \frac{2z}{c^2}\frac{\partial z}{\partial x} = 0 \text{ and so, } \frac{\partial z}{\partial x} = -\frac{c^2x}{a^2z}.$$

Now,

$$\frac{\partial V}{\partial x} = 8yz + 8xy\frac{\partial z}{\partial x} = 8yz + 8xy\left(-\frac{c^2x}{a^2z}\right)$$

$$= 8yz - \frac{8c^2x^2y}{a^2z}$$

and so,

$$\frac{\partial^2 V}{\partial x^2} = 8y\left(-\frac{c^2x}{a^2z}\right) - \frac{16c^2xy}{a^2z} - \frac{8c^2x^2y}{a^2z}\cdot\frac{c^2x}{a^2z},$$

which is negative. Hence, V is maximum at

$$\left(\frac{a}{\sqrt{3}}, \frac{b}{\sqrt{3}}, \frac{c}{\sqrt{3}}\right) \text{ and}$$

$$\text{Max } V = \frac{8abc}{3\sqrt{3}}.$$

EXAMPLE 5.107

Solve Example 5.100 using Lagrange's method of undetermined multipliers.

Solution. We have

$$S = xy + 2yz + 2zx \quad \text{and} \tag{1}$$

$$V = xyz = 32. \tag{2}$$

For S to be minimum, we must have

$$dS = (y + 2z)\, dx + (x + 2z)\, dy + 2\,(x + y)\, dz = 0. \tag{3}$$

Also, from (2), since V is constant, we have

$$yz\,dx + zx\,dy + xy\,dz = 0. \tag{4}$$

Multiplying (3) by 1 and (4) by λ and then adding both and equating to zero the coefficients of dx, dy, and dz, we get

$$(y + 2z) + \lambda yz = 0 \tag{5}$$

$$(x + 2z) + \lambda xz = 0 \tag{6}$$

$$2x + 2y + \lambda xy = 0. \tag{7}$$

Multiplying (5) by x and (6) by y and subtracting, we get

$$2zx - 2zy = 0 \text{ or } x = y,$$

since $z = 0$ is not admissible due to the fact that depth cannot be zero.

Similarly, from the equations (6) and (7), we get $y = 2z$. Thus, for a stationary value, the dimensions of the box are

$$x = y = 2z = 4, \text{ [using (2)]}.$$

Proceeding, as in Example 5.102, we note that $\frac{\partial^2 f}{\partial x^2} = 2$ (positive) and $rt - s^2 > 0$. Thus, at (4, 4, 2), S has a minimum. Hence, the required dimensions are $x = 4$, $y = 4$, and $z = 2$.

EXAMPLE 5.108

Investigate the maximum- and minimum radii vector of the sector of "surface of elasticity" $(x^2 + y^2 + z^2) = a^2x^2 + b^2y^2 + z^2c^2$, made by the plane $lx + my + nz = 0$.

Solution. On differentiating, we get

$$xdx + ydy + ndz = 0 \tag{1}$$

$$a^2\,xdx + b^2\,ydy + c^2\,zdz = 0 \tag{2}$$

$$ldx + mdy + ndz = 0. \tag{3}$$

Multiplying (1), (2), and (3) by 1, λ_1, and λ_2, respectively, and adding and equating to zero the coefficients of dx, dy, and dz, we get

$$x + a^2x\lambda_1 + l\lambda_2 = 0 \tag{4}$$

$$y + b^2y\lambda_1 + m\lambda_2 = 0 \qquad (5)$$

$$z + c^2z\lambda_1 + n\lambda_2 = 0. \qquad (6)$$

Multiplying (4), (5), and (6) by x, y, and z, respectively, and adding we get

$$(x^2+y^2+z^2) + (a^2x^2+b^2y^2+c^2z^2)\,\lambda_1 + (lx+my+nz)\,\lambda_2 = 0$$

or

$$r^2 + \lambda_1 r^4 = 0 \text{ or } \lambda_1 = -\frac{1}{r^2}.$$

Putting this value of λ_1 in (4), (5), and (6), we get

$$x = \frac{\lambda_2 lr^2}{a^2 - r^2},\ y = \frac{\lambda_2 mr^2}{b^2 - r^2}, \text{ and } z = \frac{\lambda_2 nr^2}{c^2 - r^2}.$$

Substituting these values of x, y, and z in $lx + my$ $nz = 0$, we get

$$\frac{\lambda_2 l^2r^2}{a^2 - r^2} + \frac{\lambda_2 m^2r^2}{b^2 - r^2} + \frac{\lambda_2 n^2r^2}{c^2 - r^2} = 0,$$

or

$$\frac{l^2}{a^2 - r^2} + \frac{m^2}{b^2 - r^2} + \frac{n^2}{c^2 - r^2} = 0,$$

which is an equation in r giving the required values.

EXAMPLE 5.109

Find the length of the axes of the section of the ellipsoid

$$\frac{x^2}{a^2} + \frac{y^2}{b^2} + \frac{z^2}{c^2} = 1 \text{ by the plane } lx + my + nz = 0.$$

Solution. We have to find the extreme values of the function $r^2 = x^2 + y^2 + z^2$ subject to the conditions

$$\frac{x^2}{a^2} + \frac{y^2}{b^2} + \frac{z^2}{c^2} = 1 \text{ and } lx + my + nz = 0.$$

Differentiation yields

$$x\,dx + y\,dy + z\,dz = 0 \qquad (1)$$

$$\frac{x}{a^2}\,dx + \frac{y}{b^2}\,dy + \frac{z}{c^2}\,dz = 0 \qquad (2)$$

$$l dx + m dy + n dz = 0. \qquad (3)$$

Multiplying (1), (2), and (3) by 1, λ_1, and λ_2, respectively, adding and then equating to zero the coefficients of dx, dy, and dz, we get

$$x + \lambda_1 \frac{x}{a^2} + \lambda_2 l = 0, \qquad (4)$$

$$y + \lambda_1 \frac{y}{a^2} + \lambda_2 m = 0, \text{ and} \qquad (5)$$

$$z + \lambda_1 \frac{z}{a^2} + \lambda_2 n = 0. \qquad (6)$$

Multiplying (4), (5), and (6) by x, y, and z and adding, we obtain

$$(x^2+y^2+z^2) + \lambda_1\left(\frac{x_2}{a^2} + \frac{y^2}{b^2} + \frac{z^2}{c^2}\right) + \lambda_2\,(lx+my+nz) = 0$$

or

$$r^2 + \lambda_1 = 0, \text{ which gives } \lambda_1 = -r^2.$$

Hence, from (4), (5), and (6), we have

$$x = \frac{\lambda_2 l}{\frac{r^2}{a^2} - 1},\ y = \frac{\lambda_2 m}{\frac{r^2}{b^2} - 1}, \text{ and } z = \frac{\lambda_2 n}{\frac{r^2}{c^2} - 1}.$$

Putting these values of x, y, and z in $lx + my + nz = 0$, we get

$$\lambda^2\left(\frac{l^2a^2}{r^2 - a^2} + \frac{m^2b^2}{r^2 - b^2} + \frac{n^2c^2}{r^2 - c^2}\right) = 0.$$

Since $\lambda_2 \neq 0$, the equation giving the values of r^2, the squares of the length of the semi-axes, is

$$\frac{l^2a^2}{r^2 - a^2} + \frac{m^2b^2}{r^2 - b^2} + \frac{n^2c^2}{r^2 - c^2} = 0.$$

EXAMPLE 5.110

If a, b, and, c are positive and

$$u = \frac{a^2x^2 + b^2y^2 + c^2z^2}{x^2y^2z^2},$$

$$ax^2 + by^2 + cz^2 = 1,$$

show that a stationary value of u is given by

$$x^2 = \frac{\mu}{2a\,(\mu + a)},\ y^2 = \frac{\mu}{2b\,(\mu + b)}, \text{ and } z^2 = \frac{\mu}{2c\,(\mu + c)},$$

where μ is the positive root of the cubic

$$\mu^3 - (bc + ca + ab)\,\mu - 2abc = 0.$$

Solution. We have

$$u = \frac{a^2x^2 + b^2y^2 + c^2z^2}{x^2y^2z^2}, \qquad (1)$$

$$ax^2 + by^2 + cz^2 = 1 \qquad (2)$$

Differentiating (1), we get

$$\sum \frac{1}{x^3}\left(\frac{b^2}{z^2}+\frac{c^2}{y^2}\right)dx = 0,$$

which on multiplication by $x^2y^2z^2$ yields

$$\sum \frac{1}{x}\left(b^2y^2 + c^2z^2\right)\,dx = 0. \quad (3)$$

Differentiating (2), we get

$$\sum axdx = 0. \quad (4)$$

Using Lagrange's multipliers 1 and μ, we get

$$\frac{1}{x}(b^2y^2 + c^2z^2) = \mu\, ax \text{ or } b^2y^2 + c^2z^2 = \mu\, ax^2, \quad (5)$$

$$\frac{1}{y}(c^2z^2 + a^2x^2) = \mu\, by \text{ or } c^2z^2 + a^2x^2 = \mu\, by^2, \quad (6)$$

and

$$\frac{1}{z}(a^2x^2 + b^2y^2) = \mu cz \text{ or } a^2x^2 + b^2y^2 = \mu cz^2. \quad (7)$$

Then, (6) + (7) − (5) yields

$$2a^2x^2 = \mu\,(by^2 + cz^2 - ax^2)$$

$$= \mu\,(1 - 2ax^2), \text{ using (2).}$$

Thus,

$$2a\,(a+\mu)\,x^2 = \mu \quad \text{or} \quad x^2 = \frac{\mu}{2a\,(\mu + a)}.$$

Similarly, we obtain

$$y^2 = \frac{\mu}{2b\,(\mu + b)} \quad \text{and} \quad z^2 = \frac{\mu}{2c\,(\mu + c)}.$$

Substituting these values of x^2, y^2, and z^2 in (2), we have

$$\frac{\mu}{2\,(a+\mu)} + \frac{\mu}{2\,(b+\mu)} + \frac{\mu}{2\,(c+\mu)} = 1$$

or

$$\mu^3 - (bc + ca + ab)\,\mu - 2abc = 0. \quad (8)$$

Since a, b, and c are positive, any one of (5), (6), or (7) shows that μ must be positive. Hence, μ is a positive root of (8).

EXAMPLE 5.111

Find the stationary points of $x^2 - xy + y^2 - 2x + y$.

Solution. We have $f(x, y) = x^2 - xy + y^2 - 2x + y$,

$$f_x = 2x - y - 2, f_y = -x + 2y + 1.$$

Therefore the stationary points are given by

$$f_x = f_y = 0$$

and so

$$2x - y - 2 = 0 \quad \text{and} \quad -x + y + 1 = 0.$$

Solving these equations, we get the stationary point as $\left(\frac{3}{5}, \frac{-4}{5}\right)$.

EXAMPLE 5.112

The temperature $u(x, y, z)$ at any point in space is $u = 400xyz^2$. Find the highest temperature on surface of the sphere $x^2 + y^2 + z^2 = 1$.

Solution. We have

$$u(x, y, z) = 400\,xyz^2 = 400\,xy\,(1 - y^2 - x^2)$$

$$= 400\,xy - 400\,xy^3 - 400\,x^3y,$$

which is a function of two variables x and y. Then

$$u_x = 400y - 400y^3 - 1200x^2y,$$

$$u_y = 400x - 1200xy^2 - 400x^3.$$

For extreme points, we must have $u_x = u_y = 0$. Thus, $1 - y^2 - 3x^2 = 0$ and $1 - 3y^2 - x^2 = 0$. Solving these equations, we get $x = \pm\frac{1}{2}$, $y = \pm\frac{1}{2}$. Thus, we have four stationary points $\left(\frac{1}{2}, \frac{1}{2}\right)$, $\left(-\frac{1}{2}, -\frac{1}{2}\right)$, $\left(\frac{1}{2}, -\frac{1}{2}\right)$ and $\left(-\frac{1}{2}, \frac{1}{2}\right)$. Also,

$$u_{xx} = -2400xy,\ u_{xy} = 400 - 1200y^2 - 1200x^2$$

$$u_{yy} = -2400xy.$$

At $\left(\frac{1}{2}, \frac{1}{2}\right)$ and $\left(-\frac{1}{2}, -\frac{1}{2}\right)$, $rt - s^2$ is positive and r is negative. Therefore, maximum exist at these points. Further, $x = \frac{1}{2}$, $y = \frac{1}{2}$, give $z = \sqrt{1 - x^2 - y^2} = \frac{1}{\sqrt{2}}$. Therefore,

$$\max u(x, y, z) = 400\left(\frac{1}{2}\right)\left(\frac{1}{2}\right)\left(\frac{1}{2}\right) = 50.$$

EXAMPLE 5.113

A flat circular plate is heated so that the temperature at any point (x, y) is $u(x, y) = x^2 + 2y^2 - x$. Find the coldest point on the plate.

Solution. We have $u(x, y) = x^2 + 2y^2 - x$, so that

$$u_x = 2x - 1,\ u_y = 4y.$$

Then $u_x = u_y = 0$ imply $x = \frac{1}{2}$, $y = 0$. Also $u_{xx} = 2$, $u_{yy} = 4$ and $u_{xy} = 0$. Then $rt - s^2 = 8$ (+ ve) and $r = 2$. Therefore u is minimum at $\left(\frac{1}{2}, 0\right)$. Therefore the coldest point is $\left(\frac{1}{2}, 0\right)$.

EXAMPLE 5.114

Find the minimum value of $x^2 + y^2 + z^2$ when $x + y + z = 3a$.

Solution. Special case of Example 5.94 (putting $a = b = c = 1$ and $p = 3a$ in that example).

EXERCISES

1. Let $f: \Re^2 \to \Re$ be a continuous function. Define $\phi: \Re^2 \to \Re$ by

$$\phi(x, y) = \begin{cases} f(x, y) \text{ if } (x, y) \neq (0, 0) \\ f(x, y) + 1 \text{ if } (x, y) = (0, 0). \end{cases}$$

Show that ϕ is not continuous at (0, 0).

Hint: $\phi(x, y) = f(x, y) + 1$ for (0, 0) and so, $\phi(0, 0) = f(0, 0) + 1$. Since f is continuous, it is continuous at (0, 0) also. So $\lim_{(x,y)\to(0,0)} f(x, y) = f(0,0)$. Then, $\lim_{(x,y)\to(0,0)} \phi(x,y) = \lim_{(x,y)\to(0,0)} f(x,y) = f(0, 0)$. Thus, $\lim_{(x,y)\to(0,0)} \phi(x,y) \neq \phi(0,0)$. Hence, ϕ is not continuous at (0, 0).

2. Show that $f(x, y) = x^2 + y - 1$ is continuous at $(1, -2)$.

3. Prove that the function $f(x, y) = (|xy|)^{\frac{1}{2}}$ is not differentiable at the point (0, 0) but that $\frac{\partial f}{\partial x}$ and $\frac{\partial f}{\partial y}$ exist at the origin and have value zero. Deduce that these two partial derivatives are continuous except at the origin.

Hint: If $f(x, y)$ is differentiable at (0, 0), then

$$\Delta f = f(h, k) = h\frac{\partial f}{\partial x} + k\frac{\partial f}{\partial y} + \varepsilon\sqrt{h^2 + k^2},$$

where $\varepsilon \to 0$ as $\sqrt{h^2 + k^2} \to 0$. Since $f_x = f_y = 0$ at (0, 0), $f_1(h, k) = 0 + 0 + \varepsilon\sqrt{h^2 + k^2}$ and so, $\varepsilon = \frac{[|hk|]^{\frac{1}{2}}}{\sqrt{h^2 + k^2}}$. Put $h = \rho\cos\theta$ and $k = \rho\sin\theta$, so that $\sqrt{h^2 + k^2} = \rho$. Then, $\varepsilon = \sqrt{|\sin\theta\cos\theta|}$ and so, $\lim_{\rho\to 0} \varepsilon = \sqrt{|\sin\theta\cos\theta|} = 0$, which is absurd.

Hence, f is not differentiable at (0, 0).

4. If $u = x\phi\left(\frac{y}{x}\right) + \psi\left(\frac{y}{x}\right)$, prove that

$$x^2\frac{\partial^2 u}{\partial x^2} + 2xy\frac{\partial^2 u}{\partial x\partial y} + y^2\frac{\partial^2 u}{\partial y^2} = 0.$$

5. If $u = x^y$, show that $\frac{\partial^3 u}{\partial x^2\partial y} = \frac{\partial^3 u}{\partial x\partial y\partial x}$.

6. If $z = f\left(\frac{y}{x}\right)$, show that $x\frac{\partial z}{\partial x} + y\frac{\partial z}{\partial y} = 0$.

7. If $z = \frac{x^2 + y^2}{x + y}$, show that

$$\left(\frac{\partial z}{\partial x} - \frac{\partial z}{\partial y}\right)^2 = 4\left(1 - \frac{\partial z}{\partial x} - \frac{\partial z}{\partial y}\right).$$

8. If $u = (1 - 2xy + y^2)^{-\frac{1}{2}}$, show that

$$\frac{\partial}{\partial x}\left\{(1 - x^2)\frac{\partial u}{\partial x}\right\} + \frac{\partial}{\partial y}\left\{y^2\frac{\partial u}{\partial y}\right\} = 0.$$

9. Find the value of $\frac{1}{a^2}\frac{\partial^2 z}{\partial x^2} + \frac{1}{b^2}\frac{\partial^2 z}{\partial y^2}$, where $a^2x^2 + b^2y^2 - c^2z^2 = 0$.

Ans. $\frac{1}{c^2 z}$.

10. If $u = \frac{y}{z} + \frac{z}{x} + \frac{x}{y}$, show that $x\frac{\partial u}{\partial x} + y\frac{\partial u}{\partial y} + z\frac{\partial u}{\partial z} = 0$.

11. If $u = e^x(x\cos y - y\sin y)$, show that $u_{xx} + u_{yy} = 0$.

12. Find the envelope of a system of concentric and coaxial ellipses of constant area.

Hint: Area of the ellipse $\frac{x^2}{a^2}+\frac{y^2}{b^2}=1$ is πab, given to be constant say equal to πc^2 and so, $ab=c^2$ or $b=\frac{c^2}{a}$. Putting $b=\frac{c^2}{a}$ in $\frac{x^2}{a^2}+\frac{y^2}{b^2}=1$, we get $\frac{x^2}{a^2}+\frac{y^2}{c^4a^2}=0$. Differentiating partially with respect to a gives $a=\frac{c^2x}{y}$. Putting this value of a in $\frac{x^2}{a^2}+\frac{y^2}{c^4a^2}=0$, we get $2xy=c^2$, which is the required equation of the envelope.

13. Find the envelope of the straight line $\frac{x}{a}+\frac{y}{b}=1$, where $a^2+b^2=c^2$.

Ans. $\sqrt{x}+\sqrt{y}=\sqrt{c}$.

14. Show that the evolute of the rectangular hyperbola $xy=c^2$ is

$$(x+y)^{\frac{2}{3}}-(x-y)^{\frac{2}{3}}=(4c)^{\frac{2}{3}}.$$

15. If $u=\log\frac{x^4+y^4}{x+y}$, show that $x\frac{\partial u}{\partial x}+y\frac{\partial u}{\partial y}=3$.

Hint: $e^u=\frac{x^4+y^4}{x+y}$ (A homogeneous function of degree 3 in x and y).

16. If $u=\sin^{-1}\left(\frac{x+y}{\sqrt{x}+\sqrt{y}}\right)$, show that

$$x\frac{\partial u}{\partial x}+y\frac{\partial u}{\partial y}=\frac{1}{2}\tan u.$$

Hint: $\sin u=\frac{x+y}{\sqrt{x}+\sqrt{y}}$ is a homogeneous function of degree $\frac{1}{2}$.

17. If $u=\sin^{-1}\left(\frac{x+y}{\sqrt{x}+\sqrt{y}}\right)$, show that

$$x^2\frac{\partial^2 u}{\partial x^2}+2xy\frac{\partial^2 u}{\partial x\partial y}+y^2\frac{\partial^2 u}{\partial y^2}=\frac{-\sin u\cos 2u}{4\cos^3 u}.$$

18. If $u=\tan^{-1}\left(\frac{y^2}{x}\right)$, show that

$$x^2\frac{\partial^2 u}{\partial x^2}+2xy\frac{\partial^2 u}{\partial x\partial y}+y^2\frac{\partial^2 u}{\partial y^2}=-\sin 2u\sin^2 u.$$

19. If $u=\sin^{-1}\left(\frac{x^3+y^3+z^3}{ax+by+cz}\right)$, show that

$$x\frac{\partial u}{\partial x}+y\frac{\partial u}{\partial y}+z\frac{\partial u}{\partial z}=2\tan u.$$

20. If $u=\cos^{-1}\frac{x+y}{\sqrt{x}+\sqrt{y}}$, show that

$$x\frac{\partial u}{\partial x}+y\frac{\partial u}{\partial y}=-\frac{1}{2}\cot u.$$

21. If $u=\tan^{-1}\left(\frac{y}{x}\right)$, where $x=e^t-e^{-t}$ and $y=e^t+e^{-t}$, find $\frac{du}{dt}$.

Ans. $\frac{-2}{e^{2x}+e^{-2x}}$.

22. If $z=e^{ax+by}f(ax-by)$, show that

$$b\frac{\partial z}{\partial x}+a\frac{\partial z}{\partial y}=2ab\,z.$$

23. If $\sqrt{1-x^2}+\sqrt{1-y^2}=a(x-y)$, find $\frac{dy}{dx}$.

Ans. $\sqrt{\frac{1-y^2}{1-x^2}}$.

24. If $y^3-3ax^2+x^3=0$, show that

$$\frac{d^2y}{dx^2}+\frac{2a^2x^2}{y^4}=0.$$

25. If $x=r\cos\theta$ and $y=r\sin\theta$, show that

$$\frac{\partial^2 r}{\partial x^2}+\frac{\partial^2 r}{\partial y^2}=\frac{1}{r}\left[\left(\frac{\partial r}{\partial x}\right)^2+\left(\frac{\partial r}{\partial y}\right)^2\right].$$

26. If $x+y=2e^\theta\cos\phi$ and $x-y=2\sqrt{-1}\,e^\theta\sin\phi$, show that

$$\frac{\partial^2 V}{\partial\theta^2}+\frac{\partial^2 V}{\partial\phi^2}=4xy\frac{\partial^2 V}{\partial x\partial y}.$$

27. Expand $e^x\log(1+y)$ in a Taylor's series in the neighborhood of the point (0, 0).

Ans. $y+xy-\frac{y^2}{2}+\frac{x^2y}{2}-\frac{xy^2}{2}+\frac{1}{3}y^2-\ldots$.

28. Expand $e^x\cos y$ in powers of x and y up to third-degree terms.

Ans. $1+x+\frac{1}{2}(x^2-y^2)+\frac{1}{6}(x^3-3xy^2)+\ldots$.

29. show that for $0<\theta<1$, $\sin x\sin y=xy-\frac{1}{6}[(x^3+3xy^2)\cos\theta x\sin\theta y+(y^3+3x^2y)\sin\theta x\cos\theta y]$.

Hint: Use Maclaurin's theorem.

30. If $x=c\cos u\cosh v$ and $y=c\sin u\sinh v$, show that

$$\frac{\partial(x,y)}{\partial(u,v)}=\frac{1}{2}c^2(\cos 2u-\cosh 2v).$$

31. If $u=\frac{x}{y-z}$, $v=\frac{y}{z-x}$, and $w=\frac{z}{z-y}$, show that

$$\frac{\partial(u,v,w)}{\partial(x,y,z)}=0.$$

32. If $x = r\cos\theta$, $y = r\sin\theta$, and $z = z$, evaluate $\dfrac{\partial(x, y, z)}{\partial(r, \theta, z)}$.

Ans. r.

33. If $x = uv$ and $y = \dfrac{u+v}{u-v}$, determine $\dfrac{\partial(u, v)}{\partial(x, y)}$.

Ans. $\dfrac{(u-v)^2}{4uv}$.

34. The roots of the equation in λ

$$(\lambda - x)^3 + (\lambda - y)^3 + (\lambda - z)^3 = 0$$

are u, v, and w. Show that

$$\frac{\partial(u, v, w)}{\partial(x, y, z)} = -2\frac{(y-z)(z-x)(x-y)}{(v-w)(w-u)(u-v)}.$$

Hint: The equation simplifies to

$$\lambda^3 - (x+y+z)\lambda^2 + (x^2+y^2+z^2)\lambda - \frac{1}{3}(x^3+y^3+z^3) = 0.$$

Let $x+y+z=\xi, x^2+y^2+z^2=\eta$, and $x^3+y^3+z^3=\xi$. Then $u+v+w=\xi, uv+vw+uw=\eta$, and $uvw = \xi$. Find $\dfrac{\partial(\xi, \eta, \zeta)}{\partial(x, y, z)}$ and $\dfrac{\partial(\xi, \eta, \zeta)}{\partial(u, v, w)}$. Then

$$\frac{\partial(u, v, w)}{\partial(x, y, z)} = \frac{\partial(u, v, w)}{\partial(\xi, \eta, \zeta)} \cdot \frac{\partial(\xi, \eta, \zeta)}{\partial(x, y, z)}.$$

35. If $U = x + y - z$, $V = x - y + z$, and $W = x^2 + y^2 + z^2 - 2yz$, show that U, V and W are connected by a functional relation, and find that functional relation.

Hint: Show that $\dfrac{\partial(U, V, W)}{\partial(x, y, z)} = 0$. Further, $U + V = 2x$, $U - V = 2(y - z)$. Then $(U + V)^2 + (U - V)^2 = 4(x^2 + y^2 + z^2 - 2yz) = 4W$.

36. If $u = x^2 - 2y$, $v = x + y + z$, and $w = x - 2y = x - 2y + 3z$ find $\dfrac{\partial(u, v, w)}{\partial(x, y, z)}$.

Ans. $10x - 2$.

37. Evaluate $\int_0^{\pi} \log(1 + a\cos x)\,dx$ using Leibnitz 's Role

Ans. $\pi\log\left(\dfrac{1}{2} + \dfrac{1}{2}\sqrt{1-a^2}\right)$.

38. Show that $\displaystyle\int_0^{\infty} \frac{\log(1 + \sin a\cos x)}{\cos x}\,dx = \pi a$.

39. Show that $\displaystyle\int_0^{\infty} \frac{e^{-x}}{x}(1 - e^{-ax})\,dx = \log(1 + a)$, $(a > -1)$.

40. Differentiating $\displaystyle\int_0^{x} \frac{dx}{x^2 + a^2} = \frac{1}{a}\tan^{-1}\frac{x}{a}$ under the integral sign, find the value of $\displaystyle\int_0^{x} \frac{dx}{(x^2 + a^2)^2}$.

41. Evaluate $\displaystyle\int_0^{\frac{\pi}{2}} \frac{dx}{(a^2\cos^2 x + b^2\sin^2 x)^2}$.

Hint: Putting $\tan x = t$, we get

$$\int_0^{\frac{\pi}{2}} \frac{dx}{a^2\cos^2 x + b^2\sin^2 x} = \frac{\pi}{2ab}.$$

Use Leibnitz's Rule, first differentiating with respect to a and then with respect to b. We shall get

$$\int_0^{\frac{\pi}{2}} \frac{\cos^2 x\,dx}{(a^2\cos^2 x + b^2\sin^2 x)^2} = \frac{\pi}{4a^3 b} \text{ and}$$

$$\int_0^{\frac{\pi}{2}} \frac{\sin^2 x\,dx}{(a^2\cos^2 x + b^2\sin^2 x)^2} = \frac{\pi}{4ab^3}.$$

Adding these two results, we get the value of the given integral as $\dfrac{\pi(a^2 + b^2)}{4a^3 b^3}$.

42. If the perimeter of a triangle is constant, show that its area is maximum when the triangle is equilateral.

Hint: $2s = a + b + c \quad c = 2s - a - b$ and

$$\Delta = \sqrt{s(s-a)(s-b)(s-c)}$$
$$= \sqrt{s(s-a)(s-b)(a+b-s)}$$

Take $f(a, b) = \Delta^2 = s(s-a)(s-b)(a+b-s)$ and find f_a and f_b etc, and proceed.

43. Find the points (x, y), where the function $xy(1 - x - y)$ is either maximum or minimum.

Ans. $\left(\frac{1}{3}, \frac{1}{3}\right)$.

44. Find a point within a triangle such that the sum of the squares of its distances from the three vertices is the minimum.

Ans. $\left(\frac{x_1 + x_2 + x_3}{3}, \frac{y_1 + y_2 + y_3}{3}\right)$ (centroid of the triangle).

45. Find the point on the plane $2x + 3y - z = 12$ that is nearest to the origin.

Hint: Distance form the origin is

$$l = \sqrt{x^2 + y^2 + (2x + 3y - 12)^2}.$$

Put $f(x, y) = x_2 + y_2 + (2x + 3y - 12)^2$ and proceed.

Ans. $\left(\frac{12}{7}, \frac{18}{7}, \frac{-6}{7}\right)$.

46. Discuss the maxima and minima of

$$u = 2 \sin \frac{1}{2}(x + y) \cos \frac{1}{2}(x - y) + \cos (x + y).$$

47. Find the extreme values of $2(x^2 - y^2) - x^4 + y^4$.

Ans. Max at $(\pm 1, 0)$ and Min at $(0, \pm 1)$.

48. Find the maximum value of $x^m y^n z^p$ subject to the condition $x + y + z = a$.
Hint: Use Lagrange's method of undetermined multipliers.

Ans. $x = \frac{ma}{m+n+p}, y = \frac{na}{m+n+p}$, and $z = \frac{pa}{m+n+p}$.

Max. value $= \frac{m^m n^n p^{m+n+p}}{(m+n+p)^{m+n+p}}$.

49. Divide 24 into three parts such that the continued product of the first, square of the second, and the cube of the third part may be maximum.

Hint: Find the Max of xy^2z^3 subject to the condition $x + y + z = 24$. Also, can be obtained from Exercise 7 by putting $a = 24$, $m = 1$, $n = 2$, and $p = 3$.

50. The temperature T at any point (x, y, z) in space is $T = 400xyz^2$. Find the highest temperature on the surface of the unit sphere $x^2 + y^2 + z^2 = 1$.

Ans. 50.

6 Tangents and Normals

6.1 INTRODUCTION

We know that a *tangent line* to a curve C at a point P is the line, in the limiting position, if it exists, of the secant through the point P and a variable point Q on C as Q moves along C closer to P. Also we know that the slope of tangent to the curve $y = f(x)$ at (x, y) is $f'(x)$.

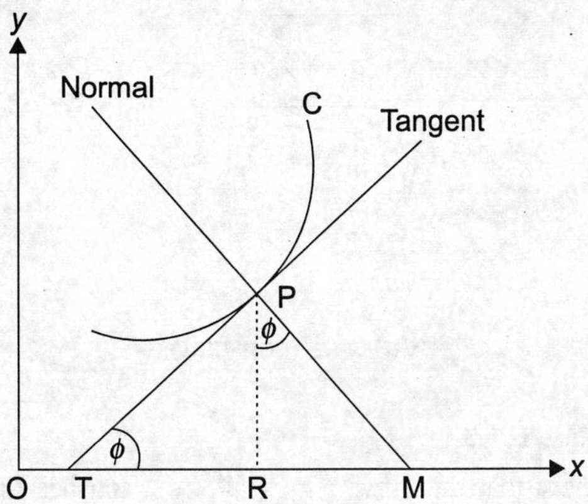

In the figure above, the line PT is tangent to the curve at P, TR is called *sub-tangent* and RM is called *sub – normal*.

6.2 EQUATION OF THE TANGENT AT A POINT OF A CURVE

Let $P(x_1, y_1)$ be a point in a curve C. Then equation of a line passing through $P(x_1, y_1)$ and with slope m is given by

$$y - y_1 = m(x - x_1),$$

where $m = \tan\phi$. If the line is tangent to the curve at P, then $m = \tan\phi = \frac{dy}{dx}$.

Hence, the equation of the tangent at P is

$$y - y_1 = \frac{dy}{dx}(x - x_1). \qquad (1)$$

On the x – axis, we have $y = 0$ and so from (1)

$$x = x_1 - \frac{y}{\frac{dy}{dx}} = x_1 - \frac{y\,dx}{dy},$$

which is *intercept cut off by the tangent on the x-axis*.

Similarly, on the y-axis, we have $x = 0$ and so

$$y = y_1 = x\frac{dy}{dx}$$

is the *intercept cut off by the tangent on the y-axis*.

6.3 EQUATION OF THE NORMAL AT A POINT $P(x_1, y_1)$ OF A CURVE

We know that the normal to the curve C at a point P is the line that passes through P and is perpendicular to the tangent line to the curve at the point P. Thus if m is the slope of the tangent line, then slope m' of the normal satisfies

$$m' = -\frac{1}{m} = -\frac{1}{\frac{dy}{dx}}.$$

Hence the equation of the normal at $P(x_1, y_1)$ is

$$y - y_1 = -\frac{dx}{dy}(x - x_1).$$

EXAMPLE 6.1

Prove that $\frac{x}{a} + \frac{y}{b} = 1$ touches the curve $y = be^{-\frac{x}{a}}$ at the point where the curve crosses the axis of y.

Solution. The given curve is

$$y = be^{-\frac{x}{a}}.$$

Differentiating with respect to x, we have

$$\frac{dy}{dx} = -\frac{b}{a}e^{-\frac{x}{a}}. \qquad (1)$$

The given curve cross the axis of y at the point $(0, b)$. The slope at this point is

$$\frac{dy}{dx} = -\frac{b}{a}e^{0} = -\frac{b}{a}.$$

Thus the equation of the tangent at $(0, b)$ is

or $$y - b = -\frac{b}{a}(x - 0)$$

or $$\frac{y}{b} - 1 = -\frac{x}{a}$$

or $$\frac{x}{a} + \frac{y}{b} = 1.$$

Hence the line $\frac{x}{a} + \frac{y}{b} = 1$ touches the curve at the point $(0, b)$, which is the point at which the curve crosses the y – axis.

EXAMPLE 6.2

Find the equation of the tangent at any point (x, y) to the curve $x^{\frac{2}{3}} + y^{\frac{2}{3}} = a^{\frac{2}{3}}$. Show that the portion of the tangent intercepted between the axes is of constant length.

Solution. The give curve is

$$x^{\frac{2}{3}} + y^{\frac{2}{3}} = a^{\frac{2}{3}}.$$

Differentiating with respect to x, we get

$$\frac{2}{3}x^{-\frac{1}{3}} + \frac{2}{3}y^{-\frac{1}{3}}\frac{dy}{dx} = 0$$

or

$$\frac{dy}{dx} = -\left(\frac{y}{x}\right)^{\frac{1}{3}} \text{ (slope of the tangent).}$$

Hence the equation of the tangent at (x, y) is

$$Y - y = -\left(\frac{y}{x}\right)^{\frac{1}{3}}(X - x). \qquad (1)$$

For intercept on the x – axis, $y = 0$ and so (1) yields

$$X = x + x^{\frac{1}{3}}y^{\frac{2}{3}}$$

$$= x^{\frac{1}{3}}\left(x^{\frac{2}{3}} + y^{\frac{2}{3}}\right) \qquad (2)$$

For intercept of the y – axis, $x = 0$ and so (1) yields

$$Y = y + x^{\frac{2}{3}}y^{\frac{1}{3}}$$

$$= y^{\frac{1}{3}}\left(x^{\frac{2}{3}} + y^{\frac{2}{3}}\right). \qquad (3)$$

Hence the portion of the tangent intercepted between the axes is equal to

$$\sqrt{X^2 + Y^2} = \sqrt{x^{\frac{2}{3}}\left(x^{\frac{2}{3}} + y^{\frac{2}{3}}\right)^2 + y^{\frac{2}{3}}\left(x^{\frac{2}{3}} + y^{\frac{2}{3}}\right)}$$

$$= \sqrt{x^{\frac{2}{3}}a^{\frac{4}{3}} + y^{\frac{2}{3}}a^{\frac{4}{3}}}$$

$$= \sqrt{a^{\frac{4}{3}}\left(x^{\frac{2}{3}} + y^{\frac{2}{3}}\right)}$$

$$= \sqrt{a^{\frac{4}{3}}.a^{\frac{2}{3}}}$$

$$= \sqrt{a^2} = a \text{ (constant).}$$

EXAMPLE 6.3

Show that the curve $\left(\frac{x}{a}\right)^n + \left(\frac{y}{b}\right)^n = 2$ touches the straight line $\frac{x}{a} + \frac{y}{b} = 2$ at the point (a, b), whatever be the value of n.

Solution. The equation of the curve is

$$\left(\frac{x}{a}\right)^n + \left(\frac{y}{b}\right)^n = 2.$$

Differentiating we get

$$\frac{n}{a^n}x^{n-1} + \frac{n}{b^n}y^{n-1}\frac{dy}{dx} = 0$$

or

$$\frac{dy}{dx} = -\frac{b^n}{a^n}.\frac{x^{n-1}}{y^{n-1}}.$$

Therefore $\left(\dfrac{dy}{dx}\right)_{(a,b)} = -\dfrac{b}{a}$.

Hence the equation of the tangent to the curve at the point (a, b) is

$$y - b = -\frac{b}{a}(x - a)$$

or

$$\frac{x}{a} + \frac{y}{b} = 2.$$

Hence the line $\frac{x}{a} + \frac{y}{b} = 2$ touches the given curve at (a, b) for any arbitrary value of n.

EXAMPLE 6.4

The straight line $\frac{x}{a} + \frac{y}{b} = 2$ touches the curve $\left(\frac{x}{a}\right)^n + \left(\frac{y}{b}\right)^n = 2$ for all values of n. Find the point of contact.

Solution. Let (x_1, y_1) be the point of contact. Since the point of contact lies on both the curve and the tangent, we have

$$\left(\frac{x_1}{a}\right)^n + \left(\frac{y_1}{b}\right)^n = 2 \text{ for all } n. \quad (1)$$

and

$$\frac{x_1}{a} + \frac{y_1}{b} = 2. \quad (2)$$

The only solution of (1) and (2) is $x_1 = a, y_1 = b$. Hence (a, b) is the required point of contact.

EXAMPLE 6.5

Tangents are drawn from the origin to the curve $y = \sin x$. Show that the point of contact lies on $x^2y^2 = x^2 - y^2$.

Solution. The equation of the curve is

$$y = \sin x \quad (1)$$

and so

$$\frac{dy}{dx} = \cos x.$$

The equation of the tangent at (x, y) is

$$Y - y = \cos x(X - x).$$

Since tangent passes through the origin, we have

$$0 - y = \cos x(0 - x)$$

or

$$\frac{y}{x} = \cos x. \quad (2)$$

The point of contact lie on the locus given by (1) and (2). Squaring and adding (1) and (2), we get

$$y^2 + \frac{y^2}{x^2} = 1 \quad \text{or} \quad x^2y^2 = x^2 - y^2.$$

EXAMPLE 6.6

Find the equation of the tangent and normal to the curve $y(x-2)(x-3) - x + 7 = 0$ at the point where it cuts the x axis.

Solution. The given curve cuts the x – axis at $x = 7$, $y = 0$. Thus we wish to find the equation of the tangent and normal to the given curve at the point (7, 0).

Differentiating the equation of the given curve, we have

$$2xy + x^2\frac{dy}{dx} - 5\left[y + x\frac{dy}{dx}\right] + 6\frac{dy}{dx} - 1 = 0$$

or

$$\frac{dy}{dx}[x^2 - 5x + 6] = 1 - 2xy + 5y$$

or

$$\frac{dy}{dx} = \frac{1 - 2xy + 5y}{x^2 - 5x + 6}.$$

Therefore

$$\left(\frac{dy}{dx}\right)_{(7,0)} = \frac{1}{49 - 35 + 6} = \frac{1}{20}.$$

Hence the equation of the tangent at (7, 0) is

$$y - 0 = \frac{1}{20}(x - 7)$$

or

$$x - 20y = 7.$$

For normal, the slope is – 20. Therefore the equation of the normal is

$$y - 0 = -20(x - 7)$$

or

$$20x + y = 140.$$

EXAMPLE 6.7

At what point is the tangent to the curve $y = \log x$ parallel to the chord joining the points (0, 0) and (0, 1).

Solution. The chord joining (0, 0) and (0, 1) is $y-$ axis. The given curve is $y = \log x$. Differentiating, we get

$$\frac{dy}{dx} = \frac{1}{x}.$$

Therefore slope of the tangent is $\frac{1}{x}$. But it is parallel to the $y-$ axis. But slope of $y-$ axis is $\tan\frac{\pi}{2} = \infty$. Therefore $\frac{1}{x} = \infty$ or $x = 0$. Putting $x = 0$ in the equation $y = \log x$, we get $y = -\infty$. Hence the point of contact is $(0, -\infty)$.

EXAMPLE 6.8

Find the equation of the tangent at the point (x, y) to the curve

$$\left(\frac{x}{a}\right)^m + \left(\frac{y}{b}\right)^m = 1.$$

Solution. The given curve is

$$\left(\frac{x}{a}\right)^m + \left(\frac{y}{b}\right)^m = 1.$$

Differentiating both sides with respect to x, we get

$$\frac{m}{a^m}x^{m-1} + \frac{m}{b^m}y^{m-1}\frac{dy}{dx} = 0$$

or

$$\frac{dy}{dx} = -\frac{b^m}{a^m}.\frac{x^{m-1}}{y^{m-1}}.$$

Therefore, the equation of the tangent is

$$Y - y = -\frac{b^m}{a^m}.\frac{x^{m-1}}{y^{m-1}}(X - x)$$

or

$$\frac{X x^{m-1}}{a^m} + \frac{Y y^{m-1}}{b^m} = \frac{x^m}{a^m} + \frac{y^m}{b^m} = 1,$$

since (x, y) lies on the curve. Hence the equation of the tangent is

$$\frac{Xx^{m-1}}{a^m} + \frac{Y y^{m-1}}{b^m} = 1.$$

EXAMPLE 6.9

If the normal to the curve $x^{\frac{2}{3}} + y^{\frac{2}{3}} = a^{\frac{2}{3}}$ makes an angle ϕ with the axis of x, show that its equation is

$$y\cos\phi - x\sin\phi = a\cos 2\phi.$$

Solution. The equation of the curve is

$$x^{\frac{2}{3}} + y^{\frac{2}{3}} = a^{\frac{2}{3}}. \qquad (1)$$

Differentiating both sides with respects to x, we have

$$\frac{2}{3}x^{-\frac{1}{3}} + \frac{2}{3}y^{-\frac{1}{3}}\frac{dy}{dx} = 0$$

or

$$\frac{dy}{dx} = -\frac{y^{\frac{1}{3}}}{x^{\frac{1}{3}}}.$$

Thus the slope of the normal is $\frac{x^{\frac{1}{3}}}{y^{\frac{1}{3}}}$. But it is given that slope of the normal is $\tan\phi$. Therefore

$$\frac{x^{\frac{1}{3}}}{y^{\frac{1}{3}}} = \tan\phi. \qquad (2)$$

Squaring (2), we get

$$y^{\frac{2}{3}}\tan^2\phi = x^{\frac{2}{3}}.$$

Putting this value of $x^{\frac{2}{3}}$ in (1), we get

$$y^{\frac{2}{3}}\tan^2\phi + y^{\frac{2}{3}} = a^{\frac{2}{3}}$$

or

$$y^{\frac{2}{3}} = a^{\frac{2}{3}}\cos^2\phi$$

or

$$y = a\cos^3\phi.$$

Putting this value of y in (2), we get

$$x^{\frac{1}{3}} = a^{\frac{1}{3}}\sin\phi$$

or

$$x = a\sin^3\phi.$$

Thus $\tan\phi$ is the slope of the normal at $(a\sin^3\phi, a\cos^3\phi)$. Hence the equation of the normal at this point is

$$y - a\cos^3\phi = \tan\phi(x - a\sin^3\phi)$$

$$= \frac{\sin\phi}{\cos\phi}(x - a\sin^3\phi)$$

or

$$y\cos\phi - a\cos^4\phi = x\sin\phi - a\sin^4\phi$$

or

$$x\sin\phi - y\cos\phi + a(\cos^2\phi - \sin^2\phi)(\cos^2\phi + \sin^2\phi) = 0$$

or

$$x\sin\phi - y\cos\phi + a\cos 2\phi = 0\cdot$$

6.4 LENGTHS OF TANGENT, NORMAL, SUB-TANGENT AND SUBNORMAL AT ANY POINT OF A CURVE

(i) **Length of the tangent.** The length of the normal is given by

$$PT = TR\operatorname{cosec}\phi$$
$$= y\sqrt{1+\cot^2\phi}$$
$$= y\sqrt{1+\left(\frac{dx}{dy}\right)^2},\ \text{since}\tan\phi = \frac{dy}{dx}$$

(ii) **Length of the normal**
We have

$$PM = PR\sec\phi$$
$$= y\sqrt{1+\tan^2\phi}$$
$$= y\sqrt{1+\left(\frac{dy}{dx}\right)^2}\cdot$$

(iii) **Length of the sub-tangent**
We have

$$TR = PR\cot\phi$$
$$= y\frac{dx}{dy}\cdot$$

(iv) **Length of the sub-normal**
We have

$$RM = PR\tan\phi$$
$$= y\frac{dy}{dx}\cdot$$

EXAMPLE 6.10

Show that the sub-tangent at any point of the curve $x^m y^n = a^{m+n}$ varies as the abscissa of the point of contact.

Solution. The equation of the curve is

$$x^m y^m = a^{m+n}\cdot$$

Taking log of both sides, we have

$$m\log x + n\log y = (m+n)\log a\cdot$$

Differentiating both sides with respect to x, we get

$$\frac{m}{x}+\frac{n}{y}\frac{dy}{dx} = 0$$

or

$$\frac{dy}{dx} = -\frac{my}{nx}\cdot$$

Therefore the length of the sub-tangent is equal is

$$y\frac{dx}{dy} = -\frac{n}{m}x\cdot$$

Thus, we note that

$$\frac{\text{length of the sub}-\text{tangent}}{x} = -\frac{n}{m}\ \text{(constant)}.$$

Hence the result.

EXAMPLE 6.11

Find the lengths of the tangent, sub-tangent and sub-normal at the point $t = \frac{\pi}{2}$ on the cycloid $x = a(t+\sin t)$, $y = a(1-\cos t)\cdot$

Solution. The equation of the curve is

$$x = a(t+\sin t)\ ,\ \ y = a(1-\cos t)\cdot$$

Therefore

$$\frac{dx}{dt} = a(1+\cos t),\ \text{and}\ \frac{dy}{dt} = a(\sin t)\text{:}$$

Thus

$$\frac{dy}{dx} = \frac{\frac{dy}{dt}}{\frac{dx}{dt}} = \frac{a\sin t}{a(1+\cos t)}$$
$$= \frac{2a\sin\frac{t}{2}\cos\frac{t}{2}}{2a\cos^2\frac{t}{2}} = \tan\frac{t}{2}$$

and so

$$\left(\frac{dy}{dx}\right)_{t=\frac{\pi}{2}} = 1\cdot$$

Also taking $t=\frac{\pi}{2}$, we get $y_2 = a$ from the equation of the curve. Hence

(i) Length of tangent $= y\sqrt{1+\left(\frac{dx}{dy}\right)^2}$

$$= a\sqrt{1+1} = a\sqrt{2}.$$

(ii) Length of sub – tangent $= y\frac{dx}{dy}$

$$= a.1 = a.$$

(iii) Length of the normal $= y\sqrt{1+\left(\frac{dy}{dx}\right)^2}$

$$= a\sqrt{1+1} = a\sqrt{2}.$$

(iv) Length of the sub – normal $= y\frac{dy}{dx}$

$$= a.1 = a.$$

EXAMPLE 6.12

In the catenary $y = c\cosh\frac{x}{c}$, show that the length of the sub-tangent is $c\ \coth\frac{x}{c}$ and that of the sub-normal is $\frac{c}{2}\sinh 2x$.

Solution. The equation of the curve is

$$y = c\ \cosh\frac{x}{c}\cdot$$

Therefore

$$\frac{dy}{dx} = \sinh\frac{x}{c}\cdot$$

Hence

$$\text{Length of the sub – tangent} = y\frac{dx}{dy} = \frac{c\cosh\frac{x}{c}}{\sinh\frac{x}{c}}$$

$$= c\coth\frac{x}{c}$$

$$\text{Length of the sub-normal} = y\frac{dy}{dx}$$

$$= c\cosh\frac{x}{c}\sinh\frac{x}{c}$$

$$= \frac{c}{2}\sinh 2x\cdot$$

EXERCISES

1. Find the equation of the tangent at $t=\frac{\pi}{2}$ to the cycloid $x = a(t-\sin t)$, $y = a(1-\cos t)$.

 Ans. $y - x = 2a - \frac{\pi a}{2}$

2. Find the condition for the line $x\cos\theta +$ $y\sin\theta = p$ to touches the curve $\frac{x^m}{a^m}+\frac{y^m}{b^m}=1$.

 Hint. Find equation of the tangent and equate it to $x\cos\theta + y\sin\theta = p$ .Find x and y from that equation.Then (x, y) lies on the curve.

 Ans. $(a\cos\theta)^{\frac{m}{m-1}} + (b-\sin\theta)^{\frac{m}{m-1}} = p^{\frac{m}{m-1}}$.

3. Show that the sum of intercepts on the axes of any tangent to the curve $\sqrt{x}+\sqrt{y} = a$ is constant.

4. Find the equation of the normal to the parabola $y^2 = 4ax$ in the form

 Ans. $y = mx - 2am - am^2$,

 where m denotes the slope of the normal.

5. If α and β be the intercepts on the axes of x and y respectively cut off by the tangent to the curve $\left(\frac{x}{a}\right)^n + \left(\frac{y}{b}\right)^n = 1$, show that

$$\left(\frac{a}{\alpha}\right)^{\frac{n}{n-1}} + \left(\frac{b}{\beta}\right)^{\frac{n}{n-1}} = 1$$

6. For the catenary $y = c\cosh\frac{x}{c}$, show that the length of the normal is $\frac{y^2}{c}$.

7. Find the lengths of tangent, normal, sub-tangent and sub-normal at the point θ on the cycloid $x = a(\theta - \sin\theta)$, $y = a(1-\cos\theta)$.

 Ans. $2a\sin\frac{\theta}{2}\tan\frac{\theta}{2}$

 normal : $2a\sin\frac{\theta}{2}$

 sub – tangent : $2a\sin^3\frac{\theta}{2}\sec\frac{\theta}{2}$

 sub – normal : $a\sin\theta$.

8. Show that the sub-normal to the curve $xy = c^2$ varies as the cube of the ordinate.

7 Beta and Gamma Functions

The beta and gamma functions, also called Euler's Integrals, are the improper integrals, which are extremely useful in the evaluation of integrals.

7.1 BETA FUNCTION

The integral $\int_0^1 x^{m-1}(1-x)^{n-1}\,dx$, which converges for $m > 0$ and $n > 0$ is called the *beta function* and is denoted by $\beta(m, n)$. Thus,

$$\beta(m, n) = \int_0^1 x^{m-1}(1-x)^{n-1}\,dx,\ m > 0,\ n > 0.$$

Beta function is also known as *Eulerian Integral of First Kind.*

As an illustration, consider the integral $\int_0^1 x^{\frac{1}{2}}(1-x)^4 dx$. We can write this integral as

$$\int_0^1 x^{\frac{3}{2}-1}(1-x)^{5-1}dx,$$

which is a beta function, denoted by $\beta\left(\frac{3}{2}, 5\right)$. But, on the other hand, the integral $\int_0^1 x^{\frac{1}{2}}(1-x)^{-3}dx$ is *not* a beta function as $n-1=-3$ implies $n=-2$ (negative).

7.2 PROPERTIES OF BETA FUNCTION

1. We have

$$\beta(m, n) = \int_0^1 x^{m-1}(1-x)^{n-1}\,dx$$

$$= \int_0^1 (1-x)^{m-1}(1-(1-x))^{n-1}dx, \text{ since } \int_0^a f(x)\,dx = \int_0^a f(a-x)\,dx,$$

$$= \int_0^1 x^{n-1}(1-x)^{m-1}\,dx$$

$$= \beta(n, m),\ m > 0,\ n > 0.$$

Thus,

$$\beta(m, n) = \beta(n, m),\ m > 0,\ n > 0.$$

2. We have

$$\beta(m, n) = \int_0^1 x^{m-1}(1-x)^{n-1}\,dx. \tag{1}$$

Putting $x = \sin^2\theta$, we get $dx = 2\sin\theta\cos\theta\,d\theta$ and therefore, (1) becomes

$$\beta(m, n) = \int_0^{\frac{\pi}{2}} (\sin^2\theta)^{m-1}(1-\sin^2\theta)^{n-1} \cdot 2\sin\theta\cos\theta\,d\theta$$

$$= 2\int_0^{\frac{\pi}{2}} \sin^{2m-2}\theta\cos^{2n-2}\theta\sin\theta\cos\theta\,d\theta$$

$$= 2\int_0^{\frac{\pi}{2}} \sin^{2m-1}\theta\cos^{2n-1}\theta\,d\theta.$$

Thus,

$$\beta(m, n) = \int_0^1 x^{m-1}(1-x)^{n-1}\,dx = 2\int_0^{\frac{\pi}{2}} \sin^{2m-1}\theta\cos^{2n-1}\theta\,d\theta.$$

3. Let m and n be positive integers. By definition,

$$\beta(m, n) = \int_0^1 x^{m-1}(1-x)^{n-1}\,dx,\ m > 0,\ n > 0.$$

Integration by parts yields

$$\beta(m, n) = \left[x^{m-1}\frac{(1-x)^n}{n(-1)}\right]_0^1$$

$$-\int_0^1 (m-n)\, x^{m-2}\left[\frac{(1-x)^n}{n(-1)}\right]dx$$

$$= \frac{m-1}{n}\int_0^1 x^{m-2}(1-x)^n\, dx$$

$$= \frac{m-1}{n}\beta(m-1, n+1).$$

Similarly,

$$\beta(m-1, n+1) = \frac{m-2}{n+1}\beta(m-2, n+2)$$

$$\beta(2, m+n-2) = \frac{1}{m+n-2}\beta(1, m+n-1).$$

Multiplying the preceding equations, we get

$$\beta(m, n) = \frac{(m-1)(m-2)\,.....\,(2)(1)}{n(n+1)(n+2)\,.....\,(m+n-2)} \times \beta(1, m+n-1)$$

$$= \frac{(m-1)!\,[1.2.3....(n-1)]}{1.2....(n-1)\,n\,(n+1)(n+2)...(m+n-2)} \times \int_0^1 x^{1-1}(1-x)^{m+n-2}\, dx$$

$$= \frac{(m-1)!\,(n-1)!}{(m+n-2)!}\int_0^1 (1-x)^{m+n-2}\, dx$$

$$= \frac{(m-1)!\,(n-1)!}{(m+n-2)!}\left[\frac{(1-x)^{m+n-1}}{(m+n-1)(-1)}\right]_0^1$$

$$= \frac{(m-1)!\,(n-1)!}{(m+n-2)!}\cdot\frac{1}{m+n-1}$$

$$= \frac{(m-1)!\,(n-1)!}{(m+n-1)!}.$$

Hence, if m and n are positive integers, then

$$\beta(m, n) = \frac{(m-1)!\,(n-1)!}{(m+n-1)!}.$$

4. Put $x = \frac{t}{1+t}$ so that $dx = \frac{1}{(1+t)^2}dt$. Then the expression for beta function reduces to

$$\beta(m, n) = \int_0^\infty \left(\frac{t}{1+t}\right)^{m-1}\left(1-\frac{t}{1+t}\right)^{n-1}\cdot\frac{1}{(1+t)^2}dt$$

$$= \int_0^\infty \frac{t^{m-1}}{(1+t)^{m-1}}\cdot\frac{1}{(1+t)^{n-1}}\cdot\frac{1}{(1+t)^2}dt$$

$$= \int_0^\infty \frac{t^{m-1}}{(1+t)^{m+n}}dt = \int_0^\infty \frac{x^{m-1}}{(1+x)^{m+n}}dx.$$

Hence,

$$\beta(m, n) = \int_0^\infty \frac{x^{m-1}}{(1+x)^{m+n}}dx.$$

Since $\beta(m, n) = \beta(n, m)$, we have

$$\beta(m, n) = \int_0^\infty \frac{x^{m-1}}{(1+x)^{m+n}}dx = \int_0^\infty \frac{x^{n-1}}{(1+x)^{m+n}}dx.$$

2. From the property (4), we have

$$\beta(m, n) = \int_0^\infty \frac{x^{m-1}}{(1+x)^{m+n}}dx$$

$$= \int_0^1 \frac{x^{m-1}}{(1+x)^{m+n}}dx + \int_1^\infty \frac{x^{m-1}}{(1+x)^{m+n}}dx$$

$$= I_1 + I_2, \text{ say.} \tag{2}$$

In I_2, put $x = \frac{1}{t}$ so that $dx = -\frac{1}{t^2}dt$ and so,

$$I_2 = \int_1^0 \frac{\left(\frac{1}{t}\right)^{m-1}}{\left(1+\frac{1}{t}\right)^{m+n}}\left(-\frac{1}{t^2}\right)dt$$

$$= \int_0^1 \frac{t^{m+n}}{t^{m-1}(t+1)^{m+n}}\cdot\frac{1}{t^2}dt$$

$$= \int_0^1 \frac{t^{n-1}}{(1+t)^{m+n}}dt = \int_0^1 \frac{x^{n-1}}{(1+x)^{m+n}}dx.$$

Hence, (2) reduces to

$$\beta(m, n) = \int_0^1 \frac{x^{m-1}}{(1+x)^{m+n}}dx + \int_0^1 \frac{x^{n-1}}{(1+x)^{m+n}}dx$$

$$= \int_0^1 \frac{x^{m-1}+x^{n-1}}{(1+x)^{m+n}}dx,\ m>0,\ n>0.$$

EXAMPLE 7.1

Show that

$$\int_0^\infty \frac{x^{m-1}+x^{n-1}}{(1+x)^{m+n}}dx = 2\beta(m, n).$$

Solution. We know that

$$\beta(m, n) = \int_0^\infty \frac{x^{m-1}}{(1+x)^{m+n}}dx. \tag{3}$$

Since $\beta(m, n) = \beta(n, m)$, we have

$$\beta(m, n) = \int_0^\infty \frac{x^{n-1}}{(1+x)^{m+n}} dx. \qquad (4)$$

Adding (3) and (4), we get

$$2\beta(m, n) = \int_0^\infty \frac{x^{m-1} + x^{n-1}}{(1+x)^{m+n}} dx.$$

EXAMPLE 7.2

Show that

$$\int_0^a x^{m-1}(a-x)^{n-1}\, dx = a^{m+n-1}\beta(m, n).$$

Solution. Putting $x = ay$, we get

$$\int_0^a x^{m-1}(a-x)^{n-1}\, dx$$

$$= \int_0^1 (ay)^{m-1}(a-ay)^{n-1}.\, a\, dy$$

$$= \int_0^1 (ay)^{m-1}a^{n-1}(1-y)^{n-1}.\, a\, dy$$

$$= \int_0^1 a^{m-1+n-1+1} y^{m-1}(1-y)^{n-1} dy$$

$$= a^{m+n-1}\int_0^1 y^{m-1}(1-y)^{n-1} dy$$

$$= a^{m+n-1}\beta(m, n).$$

EXAMPLE 7.3

Show that

$$\frac{\beta(m+1, n)}{\beta(m, n)} = \frac{m}{m+n}.$$

Solution. We have

$$\beta(m+1, n) = \int_0^1 x^m(1-x)^{n-1} dx$$

$$= \int_0^1 (1-x)^m x^{n-1}\, dx, \text{ since } \beta(m+1, n) = \beta(n, m+1)$$

$$= \left[(1-x)^m \frac{x^n}{n}\right]_0^1 - \int_0^1 m(1-x)^{m-1}(-1)\cdot\frac{x^n}{n}\, dx$$

$$= \frac{m}{n}\int_0^1 x^{n-1}\cdot x(1-x)^{m-1}\, dx$$

$$= \frac{m}{n}\int_0^1 x^{n-1}[1-(1-x)](1-x)^{m-1}\, dx$$

$$= \frac{m}{n}\left[\int_0^1 x^{n-1}(1-x)^{m-1}\, dx - \int_0^1 x^{n-1}(1-x)^m\, dx\right]$$

$$= \frac{m}{n}[\beta(n, m) - \beta(n, m+1)]$$

$$= \frac{m}{n}\beta(m, n) - \frac{m}{n}\beta(m+1, n).$$

Thus,

$$\left(1 + \frac{m}{n}\right)\beta(m+1, n) = \frac{m}{n}\beta(m, n)$$

or

$$(n+m)\,\beta(m+1, n) = m\,\beta(m, n)$$

or

$$\frac{\beta(m+1, n)}{\beta(m, n)} = \frac{m}{m+n}.$$

EXAMPLE 7.4

Prove that

$$\int_0^{\frac{\pi}{2}} \sin^m\theta\cos^n\theta d\theta = \frac{1}{2}\beta\left(\frac{m+1}{2}, \frac{n+1}{2}\right),$$

$$m > -1 \text{ and } n > -1.$$

Solution. We have

$$\int_0^{\frac{\pi}{2}} \sin^m\theta\cos^n\theta\, d\theta$$

$$= \int_0^{\frac{\pi}{2}} \sin^{m-1}\theta\cos^{n-1}\theta\cdot\sin\theta\cos\theta\, d\theta$$

$$= \int_0^{\frac{\pi}{2}} \sin^{m-1}\theta\,(1-\sin^2\theta)^{\frac{n-1}{2}}\sin\theta\cos\theta\, d\theta.$$

Putting $\sin^2\theta = x$ so that $2\sin\theta\cos\theta\, d\theta = dx$, we get

$$\int_0^{\frac{\pi}{2}} \sin^m\theta\cos^n\theta\, d\theta$$

$$= \frac{1}{2}\int_0^1 x^{\frac{m-1}{2}}(1-x)^{\frac{n-1}{2}} dx = \frac{1}{2}\int_0^1 x^{\frac{m+1}{2}-1}(1-x)^{\frac{n+1}{2}-1} dx$$

$$= \frac{1}{2}\beta\left(\frac{m+1}{2}, \frac{n+1}{2}\right), \; m > -1 \text{ and } n > -1.$$

EXAMPLE 7.5

Show that

$$\beta(m, n) = \beta(m+1, n) + \beta(m, n+1).$$

Solution. By definition,

$$\begin{aligned}\beta(m+1, n) + \beta(m, n+1) &= \int_0^1 x^m (1-x)^{n-1}\, dx + \int_0^1 x^{m-1} (1-x)^n\, dx \\ &= \int_0^1 x^{m-1} (1-x)^{n-1}[x + (1-x)]\, dx \\ &= \int_0^1 x^{m-1}(1-x)^{n-1}\, dx = \beta(m, n).\end{aligned}$$

EXAMPLE 7.6

Express $\int_0^\infty \frac{x^{m-1}}{(a+bx)^{m+n}} dx$, $m, n, a, b > 0$ in terms of beta function.

Solution. Put $bx = ay$ so that $dx = \frac{a}{b}\, dy$ in the given integral. This gives

$$\int_0^\infty \frac{x^{m-1}}{(a+bx)^{m+n}} dx = \int_0^\infty \frac{\left(\frac{ay}{b}\right)^{m-1}}{(a+ay)^{m+n}} \cdot \frac{a}{b}\, dy$$

$$= \frac{1}{a^n b^m} \int_0^\infty \frac{y^{m-1}}{(1+y)^{m+n}} dy = \frac{1}{a^n b^m} \beta(m, n),$$

using property (4) of beta function.

EXAMPLE 7.7

Show that $\beta\left(m, \frac{1}{2}\right) = 2^{2m-1}\beta(m, m)$.

Solution. We know [see property (2)] that

$$\beta(m, n) = 2\int_0^{\frac{\pi}{2}} \sin^{2m-1}\theta \cos^{2n-1}\theta\, d\theta. \qquad (5)$$

Putting $n = \frac{1}{2}$, we get

$$\beta\left(m, \frac{1}{2}\right) = 2\int_0^{\frac{\pi}{2}} \sin^{2m-1}\theta\, d\theta. \qquad (6)$$

Now, putting $n = m$ in (5), we have

$$\begin{aligned}\beta(m, m) &= 2\int_0^{\frac{\pi}{2}} \sin^{2m-1}\theta \cos^{2m-1}\theta\, d\theta \\ &= 2\int_0^{\frac{\pi}{2}} (\sin\theta \cos\theta)^{2m-1}\, d\theta \\ &= 2\int_0^{\frac{\pi}{2}} \left(\frac{1}{2}\sin 2\theta\right)^{2m-1} d\theta \\ &= \frac{1}{2^{2m-1}} \cdot 2\int_0^{\frac{\pi}{2}} \sin^{2m-1} 2\theta\, d\theta \\ &= \frac{1}{2^{2m-1}} \int_0^{\pi} \sin^{2m-1}\phi\, d\phi,\ \phi = 2\theta \\ &= \frac{1}{2^{2m-1}} \cdot 2\int_0^{\frac{\pi}{2}} \sin^{2m-1}\phi\, d\phi \\ &= \frac{1}{2^{2m-1}} \beta\left(m, \frac{1}{2}\right), \text{ using (6)},\end{aligned}$$

and so,

$$\beta\left(m, \frac{1}{2}\right) = 2^{2m-1}\beta(m, m).$$

7.3 GAMMA FUNCTION

The *gamma function* is defined as the definite integral

$$\Gamma(n) = \int_0^\infty e^{-x} x^{n-1}\, dx,\ n > 0.$$

The gamma function is also known as *Euler's Integral of Second Kind.*

7.4 PROPERTIES OF GAMMA FUNCTION

1. We have

$$\Gamma(n+1) = \int_0^\infty e^{-x} x^n\, dx = \left[-x^n e^{-x}\right]_0^\infty + n\int_0^\infty x^{n-1} e^{-x}\, dx$$

$$= n\int_0^\infty e^{-x} x^{n-1}\, dx = n\Gamma(n).$$

Hence,

$$\Gamma(n+1) = n\Gamma(n),$$

which is called the *recurrence formula* for $\Gamma(n)$.

2. Let n be a positive integer. By property (1), we have

$$\begin{aligned}\Gamma(n+1) &= n\Gamma(n) = n(n-1)\Gamma(n-1) \\ &= n(n-1)(n-2)\Gamma(n-2) \\ &= n(n-1)(n-2)\ldots 3.2.1\,\Gamma(1) \\ &= n!\Gamma(1).\end{aligned}$$

But, by definition,

$$\Gamma(1) = \int_0^\infty e^{-x}\, dx = \left[-e^{-x}\right]_0^\infty = 1.$$

Hence,
$\Gamma(n+1) = n!$, when n is a positive integer.
If we take $n = 0$, then

$$0! = \Gamma(1) = 1,$$

and so, gamma function defines 0!
Further, from the relation $\Gamma(n+1) = n\Gamma(n)$, we deduce that

$$\Gamma(2) = 1\cdot\Gamma(1) = 1!,$$
$$\Gamma(3) = 2\cdot\Gamma(2) = 2.1 = 2!,$$
$$\Gamma(4) = 3\cdot\Gamma(3) = 3.2.1 = 3!, \text{ and so on.}$$

Moreover, $\Gamma(0) = \infty$ and $\Gamma(-n) = -\infty$ if $n > 0$.
Also,

$$\Gamma(n) = \frac{\Gamma(n+1)}{n},\ n \neq 0$$
$$= \frac{(n+1)\,\Gamma(n+1)}{n(n+1)}$$
$$= \frac{\Gamma(n+2)}{n(n+1)},\ n \neq 0 \text{ and } n \neq -1$$
$$= \frac{(n+2)\,\Gamma(n+2)}{n\,(n+1)\,(n+2)}$$
$$= \frac{\Gamma(n+3)}{n(n+1)\,(n+2)},\ n \neq 0,\ n \neq -1, \text{ and } n \neq -2$$

$$\ldots\ldots$$

$$= \frac{\Gamma(n+k+1)}{n(n+1)\,(n+2)\,\ldots\,(n+k)},\ n \neq 0,\ n \neq -1,\ n \neq -2, \text{ and } n \neq -k.$$

Thus, $\Gamma(n)$ for $n < 0$ is defined, where k is a least-positive integer such that $n + k + 1 > 0$.

7.5 RELATION BETWEEN BETA AND GAMMA FUNCTIONS

We know that

$$\Gamma(m) = \int_0^\infty e^{-t} t^{m-1} dt.$$

Putting $t = x^2$ so that $dt = 2x\,dx$, we get

$$\Gamma(m) = 2\int_0^\infty e^{-x^2} x^{2m-1}\,dx. \tag{7}$$

Similarly, we can have

$$\Gamma(n) = 2\int_0^\infty e^{-y^2} y^{2n-1} dy.$$

Therefore,

$$\Gamma(m)\,\Gamma(n) = 4\int_0^\infty e^{-x^2} x^{2m-1}\,dx \cdot \int_0^\infty e^{-y^2} y^{2n-1} dy$$
$$= 4\int_0^\infty\int_0^\infty e^{-(x^2+y^2)} x^{2m-1} y^{2n-1}\,dx\,dy.$$

Taking $x = r\cos\theta$ and $y = r\sin\theta$, we have

$$dx\,dy = r d\theta\,dr.$$

Therefore,

$$\Gamma(m)\Gamma(n) = 4\int_0^{\frac{\pi}{2}}\int_0^\infty e^{-r^2} r^{2(m+n)-1} \cos^{2m-1}\theta \times \sin^{2n-1}\theta\,dr d\theta$$
$$= 2\int_0^{\frac{\pi}{2}}\left[2\int_0^\infty e^{-r^2} r^{2(m+n)-1} dr\right] \times \cos^{2m-1}\theta\,\sin^{2n-1}\theta\,d\theta$$
$$= \Gamma(m+n)\left[2\int_0^{\frac{\pi}{2}} \cos^{2m-1}\theta\,\sin^{2n-1}\theta\,d\theta\right], \text{ using (7)}$$

$= \Gamma(m+n)\,\beta(m, n)$ using property (2) of beta function.

Hence,

$$\beta(m, n) = \frac{\Gamma(m)\Gamma(n)}{\Gamma(m+n)}.$$

EXAMPLE 7.8

Show that $\Gamma\left(\frac{1}{2}\right) = \sqrt{\pi}$.

Solution. We know that

$$\beta(m, n) = \frac{\Gamma(m)\Gamma(n)}{\Gamma(m+n)}.$$

Putting $m = n = \frac{1}{2}$, we get

$$\beta\left(\frac{1}{2}, \frac{1}{2}\right) = \frac{\Gamma\left(\frac{1}{2}\right)\Gamma\left(\frac{1}{2}\right)}{\Gamma(1)}$$
$$= \left[\Gamma\left(\frac{1}{2}\right)\right]^2.$$

Thus,

$$\left[\Gamma\left(\frac{1}{2}\right)\right]^2 = \beta\left(\frac{1}{2}, \frac{1}{2}\right) = \int_0^1 x^{\frac{1}{2}-1}(1-x)^{\frac{1}{2}-1} dx$$
$$= \int_0^1 \frac{dx}{\sqrt{x}\sqrt{1-x}} = \int_0^1 \frac{dx}{\sqrt{x - x^2}}$$

$$= \int_0^1 \frac{dx}{\sqrt{\frac{1}{4} - \left(x - \frac{1}{2}\right)^2}} = \left[\sin^{-1}\left(\frac{x - \frac{1}{2}}{\frac{1}{2}}\right)\right]_0^1$$

$$= \sin^{-1}(1) - \sin^{-1}(-1) = \frac{\pi}{2} - \left(-\frac{\pi}{2}\right) = \pi.$$

Hence,

$$\Gamma\left(\frac{1}{2}\right) = \sqrt{\pi}.$$

Second Method

We know that (see Example 7.4)

$$\int_0^{\frac{\pi}{2}} \sin^m \theta \cos^n \theta \, d\theta = \frac{1}{2} \beta\left(\frac{m+1}{2}, \frac{n+1}{2}\right)$$

$$= \frac{1}{2} \frac{\Gamma\left(\frac{m+1}{2}\right)\Gamma\left(\frac{n+1}{2}\right)}{\Gamma\left(\frac{m+n+2}{2}\right)}.$$

Putting $m = n = 0$, we get

$$\int_0^{\frac{\pi}{2}} d\theta = \frac{\left[\Gamma\left(\frac{1}{2}\right)\right]^2}{2\Gamma(1)} = \frac{\left[\Gamma\left(\frac{1}{2}\right)\right]^2}{2}.$$

Thus,

$$\frac{\pi}{2} = \frac{\left[\Gamma\left(\frac{1}{2}\right)\right]^2}{2}.$$

Hence,

$$\Gamma\left(\frac{1}{2}\right) = \sqrt{\pi}.$$

EXAMPLE 7.9

Express the integrals $\int_0^{\frac{\pi}{2}} \sqrt{\tan \theta} \, d\theta$ and $\int_0^{\frac{\pi}{2}} \sqrt{\cot \theta} \, d\theta$ in terms of gamma function.

Solution. We have

$$\int_0^{\frac{\pi}{2}} \sqrt{\tan \theta} \, d\theta = \int_0^{\frac{\pi}{2}} \frac{\sin^{\frac{1}{2}} \theta}{\cos^{\frac{1}{2}} \theta} d\theta = \int_0^{\frac{\pi}{2}} \sin^{\frac{1}{2}} \theta \cos^{-\frac{1}{2}} \theta \, d\theta$$

$$= \frac{\Gamma\left(\frac{\frac{1}{2}+1}{2}\right)\Gamma\left(\frac{-\frac{1}{2}+1}{2}\right)}{2\Gamma\left(\frac{\frac{1}{2}-\frac{1}{2}+2}{2}\right)} = \frac{\Gamma\left(\frac{3}{4}\right)\Gamma\left(\frac{1}{4}\right)}{2\Gamma(1)}$$

$$= \left(\frac{1}{2}\right)\Gamma\left(\frac{3}{4}\right)\Gamma\left(\frac{1}{4}\right).$$

Similarly, we can show that

$$\int_0^{\frac{\pi}{2}} \sqrt{\cot \theta} \, d\theta = \frac{1}{2} \Gamma\left(\frac{3}{4}\right)\Gamma\left(\frac{1}{4}\right).$$

Example 7.10

Show that $\Gamma(n)\,\Gamma(1-n) = \frac{\pi}{\sin n\pi}$, $0 < n < 1$.

(Euler's Reflection Formula)

Solution. We know that

$$\beta(m, n) = \int_0^{\infty} \frac{x^{n-1}}{(1+x)^{m+n}} dx.$$

Also,

$$\beta(m, n) = \frac{\Gamma(m)\Gamma(n)}{\Gamma(m+n)}, \; m > 0 \text{ and } n > 0.$$

Therefore,

$$\frac{\Gamma(m)\,\Gamma(n)}{\Gamma(m+n)} = \int_0^{\infty} \frac{x^{n-1}}{(1+x)^{m+n}} dx.$$

Putting $m = 1 - n$ so that $m > 0$ implies $n < 1$, we get

$$\frac{\Gamma(n)\,\Gamma(1-n)}{\Gamma(1)} = \int_0^{\infty} \frac{x^{n-1}}{(1+x)} dx$$

or

$$\Gamma(n)\,\Gamma(1-n) = \int_0^{\infty} \frac{x^{n-1}}{1+x} dx = \frac{\pi}{\sin n\pi}, \; 0 < n < 1.$$

EXAMPLE 7.11

Show that

$$\int_0^{\frac{\pi}{2}} \sin^{2m-1} \theta \cos^{2n-1} \theta \, d\theta = \frac{\Gamma(m)\,\Gamma(n)}{2\Gamma(m+n)}.$$

Hence, evaluate $\int_0^{\frac{\pi}{2}} \sin^p \theta \, d\theta$ and $\int_0^{\frac{\pi}{2}} \cos^p \theta \, d\theta$.

Solution. We know that

$$2 \int_0^{\frac{\pi}{2}} \sin^{2m-1} \theta \cos^{2n-1} \theta \, d\theta = \beta(m, n)$$

or

$$\int_0^{\frac{\pi}{2}} \sin^{2m-1} \theta \cos^{2n-1} \theta \, d\theta = \frac{1}{2} \cdot \beta(m, n).$$

But, $\beta(m, n) = \frac{\Gamma(m)\Gamma(n)}{\Gamma(m+n)}$. Therefore,

$$\int_0^{\frac{\pi}{2}} \sin^{2m-1}\theta \cos^{2n-1}\theta\, d\theta = \frac{\Gamma(m)\Gamma(n)}{2\Gamma(m+n)}.$$

If we put $2m - 1 = p$ and $2n - 1 = q$, then this result reduces to

$$\int_0^{\frac{\pi}{2}} \sin^p\theta \cos^q\theta\, d\theta = \frac{\Gamma\left(\frac{p+1}{2}\right)\Gamma\left(\frac{q+1}{2}\right)}{2\Gamma\left(\frac{p+q+2}{2}\right)}. \quad (8)$$

Putting $q = 0$ in (8), we get

$$\int_0^{\frac{\pi}{2}} \sin^p\theta\, d\theta = \frac{\Gamma\left(\frac{p+1}{2}\right)\Gamma\left(\frac{1}{2}\right)}{2\Gamma\left(\frac{p+2}{2}\right)} = \frac{\Gamma\left(\frac{p+1}{2}\right)}{\Gamma\left(\frac{p+2}{2}\right)}\frac{\sqrt{\pi}}{2}.$$

Similarly, taking $p = 0$, we get

$$\int_0^{\frac{\pi}{2}} \cos^q\theta\, d\theta = \frac{\Gamma\left(\frac{(q+1)}{2}\right)}{\Gamma\left(\frac{q+2}{2}\right)}\frac{\sqrt{\pi}}{2}.$$

EXAMPLE 7.12

Show that

$$\Gamma(m)\,\Gamma\left(m+\frac{1}{2}\right) = \frac{\sqrt{\pi}}{2^{2m-1}}\Gamma(2m)$$

(Duplication Formula).

Solution. In Example 7.7, we have shown that

$$\beta\left(m, \frac{1}{2}\right) = 2^{2m-1}\beta(m, m).$$

Converting into gamma function, we get

$$\frac{\Gamma(m)\Gamma\left(\frac{1}{2}\right)}{\Gamma\left(m+\frac{1}{2}\right)} = 2^{2m-1}\frac{\Gamma(m)\,\Gamma(m)}{\Gamma(2m)}.$$

Since $\Gamma\left(\frac{1}{2}\right) = \sqrt{\pi}$, we get

$$\frac{\sqrt{\pi}}{\Gamma\left(m+\frac{1}{2}\right)} = 2^{2m-1}\frac{\Gamma(m)}{\Gamma(2m)}$$

or

$$\Gamma(m)\,\Gamma\left(m+\frac{1}{2}\right) = \frac{\sqrt{\pi}}{2^{2m-1}}\Gamma(2m).$$

EXAMPLE 7.13

Show that

$$\int_0^\infty e^{-ax}x^{n-1}\,dx = \frac{\Gamma(n)}{a^n},$$

where a and n are positive. Deduce that

(i) $\int_0^\infty e^{-ax}x^{n-1}\cos bx\,dx = \frac{\Gamma(n)}{r^n}\cos n\theta$

(ii) $\int_0^\infty e^{-ax}x^{n-1}\sin bx\,dx = \frac{\Gamma(n)}{r^n}\sin n\theta,$

where $r^2 = a^2 + b^2$ and $\theta = \tan^{-1}\frac{b}{a}$. Also evaluate

$$\int_0^\infty e^{-ax}\cos bx\,dx \text{ and } \int_0^\infty e^{-ax}\sin bx\,dx.$$

Solution. Put $ax = z$, so that $adx = dz$, to get

$$\int_0^\infty e^{-ax}x^{n-1}\,dx = \int_0^\infty e^{-z}\left(\frac{z}{a}\right)^{n-1}\cdot\frac{dz}{a}$$

$$= \frac{1}{a^n}\int_0^\infty e^{-z}z^{n-1}dz = \frac{\Gamma(n)}{a^n}. \quad (9)$$

Replacing a by $a + ib$ in (9), we get

$$\int_0^\infty e^{-(a+ib)x}x^{n-1}\,dx = \frac{\Gamma(n)}{(a+ib)^n}. \quad (10)$$

But as,

$$e^{-(a+ib)x} = e^{-ax}\cdot e^{-ibx} = e^{-ax}(\cos bx - i\sin bx)$$

and taking $a = r\cos\theta$ and $b = r\sin\theta$, De-Moivre's Theorem implies

$$(a+ib)^n = (r\cos\theta + ir\sin\theta)^n$$
$$= r^n(\cos\theta + i\sin\theta)^n$$
$$= r^n(\cos n\theta + i\sin\theta).$$

Therefore, (10) reduces to

$$\int_0^\infty [e^{-ax}(\cos bx - i\sin bx)]x^{n-1}\,dx$$

$$= \frac{\Gamma(n)}{r^n(\cos n\theta + i\sin n\theta)} = \frac{\Gamma(n)}{r^n}(\cos n\theta + i\sin n\theta)^{-1}$$

$$= \frac{\Gamma(n)}{r^n}(\cos n\theta - i\sin n\theta).$$

Equating real-and imaginary parts on both sides, we get

$$\int_0^\infty e^{-ax}x^{n-1}\cos bx\,dx = \frac{\Gamma(n)}{r^n}\cos n\theta$$

and

$$\int_0^\infty e^{-ax}x^{n-1}\sin bx\,dx = \frac{\Gamma(n)}{r^n}\sin n\theta.$$

If we put $n = 1$, then

$$\int_0^{\frac{\pi}{2}} e^{-ax}\cos bx\,dx = \frac{\Gamma(1)}{r}\cos\theta = \frac{r\cos\theta}{r^2} = \frac{a}{a^2+b^2}$$

and

$$\int_0^{\pi/2} e^{-ax}\sin bx\,dx = \frac{\Gamma(1)}{r}\sin\theta = \frac{r\sin\theta}{r^2} = \frac{b}{a^2+b^2}.$$

EXAMPLE 7.14

Show that

$$\Gamma\left(n+\frac{1}{2}\right) = \frac{\sqrt{\pi}\,\Gamma(2n+1)}{2^{2n}\Gamma(n+1)}.$$

Hence, deduce that $\Gamma\left(\frac{1}{4}\right)\Gamma\left(\frac{3}{4}\right) = \pi\sqrt{2}$.

Solution. We know that

$$\beta(m,n) = 2\int_0^{\frac{\pi}{2}} \sin^{2m-1}\theta\cos^{2n-1}\theta\,d\theta.$$

Therefore, $\beta\left(n+\frac{1}{2}, n+\frac{1}{2}\right)$

$$= 2\int_0^{\frac{\pi}{2}} \sin^{2n}\theta\cos^{2n}\theta\,d\theta = 2\int_0^{\frac{\pi}{2}}\left(\frac{\sin 2\theta}{2}\right)^{2n} d\theta$$

$$= \frac{1}{2^{2n-1}}\int_0^{\frac{\pi}{2}} \sin^{2n} 2\theta\,d\theta = \frac{1}{2^{2n}}\cdot\int_0^{\pi} \sin^{2n}\phi\,d\phi,\ \phi = 2\theta$$

$$= \frac{1}{2^{2n}}\cdot 2\int_0^{\frac{\pi}{2}} \sin^{2n}\phi\,d\phi = \frac{1}{2^{2n-1}}\left[\frac{\Gamma\left(\frac{2n+1}{2}\right)}{\Gamma\left(\frac{2n+2}{2}\right)}\frac{\sqrt{\pi}}{2}\right]$$

(see Example 7.11)

$$= \frac{\sqrt{\pi}\,\Gamma\left(n+\frac{1}{2}\right)}{2^{2n}\Gamma(n+1)}. \tag{11}$$

Also,

$$\beta\left(n+\frac{1}{2}, n+\frac{1}{2}\right) = \frac{\Gamma\left(n+\frac{1}{2}\right)\Gamma\left(n+\frac{1}{2}\right)}{\Gamma(2n+1)}$$

$$= \frac{\left[\Gamma\left(n+\frac{1}{2}\right)\right]^2}{\Gamma(2n+1)}. \tag{12}$$

From (11) and (12), we have

$$\Gamma\left(n+\frac{1}{2}\right) = \frac{\sqrt{\pi}}{2^{2n}}\cdot\frac{\Gamma(2n+1)}{\Gamma(n+1)}.$$

Further, putting $n = \frac{1}{4}$, we have

$$\Gamma\left(\frac{3}{4}\right) = \frac{\sqrt{\pi}}{\sqrt{2}}\cdot\frac{\Gamma\left(\frac{3}{2}\right)}{\Gamma\left(\frac{5}{4}\right)} = \sqrt{\frac{\pi}{2}}\,\frac{\frac{1}{2}\Gamma\left(\frac{1}{2}\right)}{\frac{1}{4}\Gamma\left(\frac{1}{4}\right)}.$$

Hence,

$$\Gamma\left(\frac{3}{4}\right)\Gamma\left(\frac{1}{4}\right) = \pi\sqrt{2}.$$

Example 7.15

Show that

$$\int_0^{\frac{\pi}{2}}\sqrt{\tan\theta}\,d\theta = \int_0^{\frac{\pi}{2}}\sqrt{\cot\theta}\,d\theta = \frac{1}{2}\Gamma\left(\frac{1}{4}\right)\Gamma\left(\frac{3}{4}\right) = \frac{\pi}{\sqrt{2}}.$$

Solution. In Example 7.9, we have proved that

$$\int_0^{\frac{\pi}{2}}\sqrt{\tan\theta}\,d\theta = \int_0^{\frac{\pi}{2}}\sqrt{\cot\theta}\,d\theta = \frac{1}{2}\Gamma\left(\frac{1}{4}\right)\Gamma\left(\frac{3}{4}\right).$$

But,

$$\Gamma\left(\frac{1}{4}\right)\Gamma\left(\frac{3}{4}\right) = \pi\sqrt{2} \text{ (Example 7.14.)}$$

Hence,

$$\int_0^{\frac{\pi}{2}}\sqrt{\tan\theta}\,d\theta = \int_0^{\frac{\pi}{2}}\sqrt{\cot\theta}\,d\theta = \frac{\pi}{\sqrt{2}}.$$

EXAMPLE 7.16

Show that

$$\int_0^1 y^{q-1}\left(\log\frac{1}{y}\right)^{p-1} dy = \frac{\Gamma(p)}{q^p},\ p>0,\ q>0.$$

Solution. Putting $\log\frac{1}{y} = x$, we have $\frac{1}{y} = e^x$ or $y = e^{-x}$ and so, $dy = -e^{-x}\,dx$. Therefore,

$$\int_0^1 y^{q-1}\left(\log\frac{1}{y}\right)^{p-1} dy$$

$$= \int_0^{\infty} e^{-(q-1)x}x^{p-1}(-e^{-x})\,dx$$

$$= \int_0^{\infty} e^{-qx}x^{p-1}\,dx$$

$$= \int_0^{\infty} e^{-t}\left(\frac{t}{q}\right)^{p-1}\cdot\frac{dt}{q}, \text{ putting } qx = t$$

$$= \frac{1}{q^p}\int_0^{\infty} e^{-t}t^{p-1}dt = \frac{\Gamma(p)}{q^p}.$$

EXAMPLE 7.17

Show that

$$\int_0^{\frac{\pi}{2}}\sqrt{\sin\theta}\,d\theta\cdot\int_0^{\frac{\pi}{2}}\frac{d\theta}{\sqrt{\sin\theta}} = \pi.$$

Solution. We have

$$\int_0^{\frac{\pi}{2}} \sqrt{\sin\theta}\, d\theta \cdot \int_0^{\frac{\pi}{2}} \frac{d\theta}{\sqrt{\sin\theta}}$$

$$= \int_0^{\frac{\pi}{2}} \sin^{\frac{1}{2}} \theta\, d\theta \cdot \int_0^{\frac{\pi}{2}} \sin^{-\frac{1}{2}} \theta\, d\theta.$$

$$= \frac{\Gamma\left(\frac{\frac{1}{2}+1}{2}\right)}{\Gamma\left(\frac{\frac{1}{2}+2}{2}\right)} \cdot \frac{\sqrt{\pi}}{2} \cdot \frac{\Gamma\left(\frac{-\frac{1}{2}+1}{2}\right)}{\Gamma\left(\frac{-\frac{1}{2}+2}{2}\right)} \cdot \frac{\sqrt{\pi}}{2}$$

$$= \frac{\Gamma\left(\frac{3}{4}\right)\Gamma\left(\frac{1}{4}\right)}{\Gamma\left(\frac{5}{4}\right)\Gamma\left(\frac{3}{4}\right)} \cdot \frac{\pi}{4} \qquad \text{(see Example 7.11)}$$

$$= \frac{\Gamma\left(\frac{1}{4}\right)}{\Gamma\left(\frac{5}{4}\right)} \cdot \frac{\pi}{4} = \frac{\Gamma\left(\frac{1}{4}\right)}{\frac{1}{4}\Gamma\left(\frac{1}{4}\right)} \cdot \frac{\pi}{4} = \pi.$$

EXAMPLE 7.18

Prove that

$$\int_0^1 \left(\log \frac{1}{y}\right)^{n-1} dy = \Gamma(n),\ n > 0.$$

Solution. Putting $\log \frac{1}{y} = x$, that is, $\frac{1}{y} = e^x$ or $y = e^{-x}$, we have $dy = -e^{-x}\, dx$. Hence,

$$\int_0^1 \left(\log \frac{1}{y}\right)^{n-1} dy = -\int_\infty^0 x^{n-1} e^{-x}\, dx$$

$$= \int_0^\infty e^{-x} x^{n-1}\, dx$$

$$= \Gamma(n),\ n > 0.$$

EXAMPLE 7.19

Evaluate $\int_0^1 x^4 \left(\log \frac{1}{x}\right)^3 dx.$

Solution. Putting $\log \frac{1}{x} = y$, that is, $\frac{1}{x} = e^y$ or $x = e^{-y}$, we have $dx = -e^y\, dy$. Therefore,

$$\int_0^1 x^4 \left(\log \frac{1}{x}\right)^3 dx = -\int_\infty^0 e^{-4y} \cdot y^3 \cdot e^{-y} dy = \int_0^\infty e^{-5y} \cdot y^3 dy$$

$$= \int_0^\infty e^{-t} \cdot \frac{t^3}{125} \cdot \frac{dt}{5}, \text{ putting } 5y = t$$

$$= \frac{1}{625} \int_0^\infty e^{-t} \cdot t^3 dt = \frac{1}{625} \Gamma(3) = \frac{6}{625}.$$

EXAMPLE 7.20

Show that

$$\int_0^1 x^m (\log x)^n\, dx = \frac{(-1)^n n!}{(m+n)^{n+1}}.$$

Solution. Putting $\log x = -y$, we have $x = e^{-y}$ and so, $dx = -e^{-y}$. When $x = 1$, $y = 0$ and when $x \to 0$, $y \to \infty$. Therefore,

$$\int_0^1 x^m (\log x)^n\, dx$$

$$= \int_0^\infty e^{-my} (-y)^n (-e^{-y}) dy = (-1)^n \int_0^\infty e^{-(m+1)y} \cdot y^n dy$$

$$= (-1)^n \int_0^\infty e^{-t} \cdot \left(\frac{t}{m+1}\right)^n \frac{dt}{m+1},$$

putting $(m + 1)y = t$

$$= \frac{(-1)^n}{(m+1)^{n+1}} \int_0^\infty e^{-t} t^n\, dt = \frac{(-1)^n}{(m+1)^{n+1}} \Gamma(n+1)$$

$$= \frac{(-1)^n n!}{(m+1)^{n+1}}.$$

EXAMPLE 7.21

Prove that

$$\int_0^1 \frac{dx}{\sqrt{1+x^4}} = \frac{1}{4\sqrt{2}} \beta\left(\frac{1}{4}, \frac{1}{2}\right).$$

Solution. Putting $x^2 = \tan\theta$, we have $2x dx = \sec^2 \theta\, d\theta$.

Therefore,

$$\int_0^1 \frac{dx}{\sqrt{1+x^4}} = \int_0^{\frac{\pi}{4}} \frac{\sec^2\theta}{\sqrt{2\tan\theta}} \cdot \frac{1}{\sqrt{1+\tan^2\theta}}\, d\theta$$

$$= \frac{1}{2} \int_0^{\frac{\pi}{4}} \frac{\sec^2\theta}{\sqrt{\tan\theta}} \cdot \frac{1}{\sec\theta}\, d\theta = \frac{1}{2} \int_0^{\frac{\pi}{4}} \frac{d\theta}{\sin^{\frac{1}{2}}\theta \cos^{\frac{1}{2}}\theta}$$

$$= \frac{1}{2} \int_0^{\frac{\pi}{4}} \frac{d\theta}{(\sin\theta \quad \cos\theta)^{\frac{1}{2}}} = \frac{1}{2} \int_0^{\frac{\pi}{4}} \frac{d\theta}{\left(\frac{\sin 2\theta}{2}\right)^{\frac{1}{2}}}$$

$$= \frac{1}{\sqrt{2}} \int_0^{\frac{\pi}{4}} \frac{d\theta}{\sqrt{\sin 2\theta}} = \frac{1}{\sqrt{2}} \int_0^{\frac{\pi}{2}} \frac{d\phi}{2\sqrt{\sin\phi}},\ \phi = 2\theta$$

$$= \frac{1}{2\sqrt{2}} \int_0^{\frac{\pi}{2}} \sin^{-\frac{1}{2}} \phi \cos^0 \phi \, d\phi$$

$$= \frac{2}{4\sqrt{2}} \int_0^{\frac{\pi}{2}} \sin^{-\frac{1}{2}} \phi \cos^0 \phi \, d\phi$$

$$= \frac{1}{4\sqrt{2}} \beta\left(\frac{1}{4}, \frac{1}{2}\right),$$

since $\beta(m, n) = 2 \int_0^{\frac{\pi}{2}} \sin^{2m-1} \phi \cos^{2n-1} \phi \, d\phi$.

EXAMPLE 7.22

Show that

$$\int_0^{\frac{\pi}{2}} (\sqrt{\tan \theta} + \sqrt{\sec\theta})d\theta = \frac{1}{2}\Gamma\left(\frac{1}{4}\right)\left[\Gamma\left(\frac{3}{4}\right) + \frac{\sqrt{\pi}}{\Gamma\left(\frac{3}{4}\right)}\right].$$

Solution. We have

$$\int_0^{\frac{\pi}{2}} (\sqrt{\tan\theta} + \sqrt{\sec \theta})d\theta$$

$$= \int_0^{\frac{\pi}{2}} \sqrt{\tan\theta} \, d\theta + \int_0^{\frac{\pi}{2}} \sqrt{\sec\theta} \, d\theta$$

$$= \int_0^{\frac{\pi}{2}} \sin^{\frac{1}{2}}\theta \cos^{-\frac{1}{2}}\theta \, d\theta + \int_0^{\frac{\pi}{2}} \sin^0 \theta \cos^{-\frac{1}{2}}\theta \, d\theta$$

$$= \frac{2}{2} \int_0^{\frac{\pi}{2}} \sin^{\frac{1}{2}} \theta \cos^{-\frac{1}{2}}\theta \, d\theta + \frac{2}{2} \int_0^{\frac{\pi}{2}} \sin^0 \theta \cos^{-\frac{1}{2}}\theta \, d\theta$$

$$= \frac{1}{2} \beta\left(\frac{3}{4}, \frac{1}{4}\right) + \frac{1}{2}\beta\left(\frac{1}{2}, \frac{1}{4}\right)$$

$$= \frac{1}{2}\left[\frac{\Gamma\left(\frac{3}{4}\right)\Gamma\left(\frac{1}{4}\right)}{\Gamma(1)} + \frac{\Gamma\left(\frac{1}{2}\right)\Gamma\left(\frac{1}{4}\right)}{\Gamma\left(\frac{3}{4}\right)}\right]$$

$$= \frac{1}{2}\Gamma\left(\frac{1}{4}\right)\left[\Gamma\left(\frac{3}{4}\right) + \frac{\Gamma\left(\frac{1}{2}\right)}{\Gamma\left(\frac{3}{4}\right)}\right]$$

$$= \frac{1}{2}\Gamma\left(\frac{1}{4}\right)\left[\Gamma\left(\frac{3}{4}\right) + \frac{\sqrt{\pi}}{\Gamma\left(\frac{3}{4}\right)}\right].$$

7.6 DIRICHLET'S AND LIOUVILLE'S THEOREMS

The following theorems of Dirichlet and Liouville are useful in evaluating multiple integrals.

Theorem 7.1 (Dirichlet). If V is the region, where $x \geq 0, y \geq 0, z \geq 0$, and $x + y + z \leq 1$, then

$$\iiint_V x^{p-1}y^{q-1}z^{r-1} \, dx \, dy \, dz = \frac{\Gamma(p)\Gamma(q)\Gamma(r)}{\Gamma(p+q+r+1)}.$$

(The Dirichlet's Theorem can be extended to a finite number of variables).

Proof: Since $x + y + z \leq 1$, we have $y + z \leq 1 - x = a$.

Therefore,

$$\iiint_V x^{p-1}y^{q-1}z^{r-1} \, dx \, dy \, dz$$

$$= \int_0^1 \int_x^{1-x} \int_0^{1-x-y} x^{p-1}y^{q-1}z^{r-1} \, dx \, dy \, dz$$

$$= \int_0^1 x^{p-1}\left[\int_0^a \int_0^{a-y} y^{q-1}z^{r-1} dz \, dy\right] dx. \quad (13)$$

Let

$$I = \int_0^a \int_0^{a-y} y^{q-1}z^{r-1} dz \, dy.$$

Putting $y = aY$ and $z = aZ$, this integral reduces to

$$I = \int_D (aY)^{q-1}(aZ)^{r-1}.a^2 dZ \, dY,$$

where D is the domain where $X \geq 0$, $Y \geq 0$, and $Y + Z \geq 1$. Thus,

$$I = a^{q+r} \int_0^1 \int_0^{1-Y} Y^{q-1} Z^{r-1} dZ \, dY$$

$$= a^{q+r} \int_0^1 Y^{q-1}\left[\frac{Z^r}{r}\right]_0^{1-Y} dY$$

$$= \frac{a^{q+r}}{r} \int_0^1 Y^{q-1} (1 - Y)^r dY$$

$$= \frac{a^{q+r}}{r} \beta(q, r + 1) = \frac{a^{q+r}}{r} \frac{\Gamma(q)\Gamma(r+1)}{\Gamma(q+r+1)}$$

$$= a^{q+r} \frac{\Gamma(q)\,\Gamma(r)}{\Gamma(q+r+1)}.$$

Hence, (13) yields

$$\iiint_V x^{p-1}y^{q-1}z^{r-1}\,dx\,dy\,dz$$

$$=\int_0^1 \frac{\Gamma(q)\,\Gamma(r)}{\Gamma(q+r+1)}x^{p-1}a^{q+r}\,dx$$

$$=\frac{\Gamma(q)\Gamma(r)}{\Gamma(q+r+1)}\int x^{p-1}(1-x)^{q+r}\,dx, \text{ since } a=1-x$$

$$=\frac{\Gamma(q)\Gamma(r)}{\Gamma(q+r+1)}\beta(p,q+r+1)$$

$$=\frac{\Gamma(q)\Gamma(r)}{\Gamma(q+r+1)}\cdot\frac{\Gamma(p)\Gamma(q+r+1)}{\Gamma(p+q+r+1)}=\frac{\Gamma(p)\Gamma(q)\Gamma(r)}{\Gamma(p+q+r+1)}.$$

Remark 7.1.

If $x+y+z\le h$, then by putting $\frac{x}{h}=X$, $\frac{y}{h}=Y$, and $\frac{z}{h}=Z$, we have $X+Y+Z\le\frac{h}{h}=1$ and so, the Dirichlet's Theorem takes the form

$$\iiint_V x^{p-1}y^{q-1}z^{r-1}\,dx\,dy\,dz$$

$$=\frac{\Gamma(p)\Gamma(q)\Gamma(r)}{\Gamma(q+r+1)}\cdot h^{p+q+r}.$$

Theorem 7.2 (Liouville). If x, y, and z are all positive such that $h_1<x+y+z<h_2$, then

$$\iiint f(x+y+z)\,x^{p-1}y^{q-1}z^{r-1}\,dx\,dy\,dz$$

$$=\frac{\Gamma(p)\Gamma(q)\,\Gamma(r)}{\Gamma(p+q+r)}\int_{h_1}^{h_2} f(h)h^{p+q+r-1}dh.$$

(Proof, not provided here, is a slight modification of the proof of Dirichlet's Theorem).

EXAMPLE 7.23

Evaluate $\iiint x\,y\,z\,dx\,dy\,dz$ taken throughout the ellipsoid

$$\frac{x^2}{a^2}+\frac{y^2}{b^2}+\frac{z^2}{c^2}\le 1.$$

Solution. Put $\frac{x^2}{a^2}=X$, $\frac{y^2}{b^2}=Y$, and $\frac{z^2}{c^2}=Z$ to get

$$x=aX^{\frac{1}{2}},\ y=bY^{\frac{1}{2}},\text{ and } z=cZ^{\frac{1}{2}}$$

and

$$x\,dx=\frac{a^2}{2}\,dX,\ ydy=\frac{b^2}{2}\,dY,\text{ and } zdz=\frac{c^2}{2}\,dZ.$$

The condition, under this substitution, becomes

$$X+Y+Z\le 1.$$

Therefore, for the first quadrant,

$$\iiint xyz\,dx\,dy\,dz$$

$$=\iiint(x\,dx)\,(ydy)\,(zdz)$$

$$=\iiint\left(\frac{a^2}{2}\,dX\right)\left(\frac{b^2}{2}\,dY\right)\left(\frac{c^2}{2}\,dZ\right)$$

$$=\frac{a^2b^2c^2}{8}\iiint X^{1-1}Y^{1-1}Z^{1-1}dX\,dY\,dZ$$

$$=\frac{a^2b^2c^2}{8}\,\frac{\Gamma(1)\,\Gamma(1)\,\Gamma(1)}{\Gamma(1+1+1+1)},$$

by Dirichlet's Theorem

$$=\frac{a^2b^2c^2}{8}\cdot\frac{1}{\Gamma(4)}=\frac{a^2b^2c^2}{8.3!}=\frac{a^2b^2c^2}{48}.$$

Therefore, value of the integral for the whole of the ellipsoid is $8\left(\frac{a^2b^2c^2}{48}\right)=\frac{a^2b^2c^2}{6}$.

EXAMPLE 7.24

The plane $\frac{x}{a}+\frac{y}{b}+\frac{z}{c}=1$ meets the axes in A, B, and C. Find the volume of the tetrahedron $OABC$.

Solution. We wish to evaluate

$$\iiint dx\,dy\,dz$$

under the condition $\frac{x}{a}+\frac{y}{b}+\frac{z}{c}=1$. Putting $\frac{x}{a}=X$, $\frac{y}{b}=Y$, and $\frac{z}{c}=Z$, we get $X+Y+Z=1$. Also $dx=adX$, $dy=bdY$, and $dz=cdZ$. Therefore, using Dirichlet's Theorem, the required volume of the tetrahedron is

$$V=\iiint dx\,dy\,dz$$

$$=\iiint abc\,dX\,dY\,dZ$$

$$=abc\iiint X^{1-1}Y^{1-1}Z^{1-1}dX\,dY\,dZ$$

$$=abc\,\frac{\Gamma(1)\Gamma(1)\Gamma(1)}{\Gamma(1+1+1+1)}$$

$$=\frac{abc}{\Gamma(4)}=\frac{abc}{3!}=\frac{abc}{6}.$$

EXAMPLE 7.25

Evaluate $\iiint x^{l-1} y^{m-1} z^{n-1}\, dx\, dy\, dz$, where $x > 0$, $y > 0$, and $z > 0$ under the condition $\left(\frac{x}{a}\right)^p + \left(\frac{y}{b}\right)^q + \left(\frac{z}{c}\right)^r \le 1$.

Solution. Put $\left(\frac{x}{a}\right)^p = X$, $\left(\frac{y}{b}\right)^q = Y$, and $\left(\frac{z}{c}\right)^r = Z$ so that

$$dx = \frac{a}{p} X^{\frac{1}{p}-1} dX,$$

$$dy = \frac{b}{q} Y^{\frac{1}{q}-1} dY, \text{ and}$$

$$dz = \frac{c}{r} Z^{\frac{1}{r}-1} dz.$$

Therefore,

$$\iiint x^{l-1} y^{m-1} z^{n-1}\, dx\, dy\, dz$$

$$= \iiint \left(aX^{\frac{1}{p}}\right)^{l-1} \left(bY^{\frac{1}{q}}\right)^{m-1} \left(cZ^{\frac{1}{r}}\right)^{n-1}$$

$$\times \frac{abc}{pqr} X^{\frac{1}{p}-1} Y^{\frac{1}{q}-1} Z^{\frac{1}{r}-1} dX\, dY\, dZ$$

$$= \frac{a^l b^m c^n}{pqr} \iiint X^{\frac{l}{p}-1} Y^{\frac{m}{q}-1} Z^{\frac{n}{r}-1} dX\, dY\, dZ$$

$$= \frac{a^l b^m c^n}{pqr} \cdot \frac{\Gamma\left(\frac{l}{p}\right)\Gamma\left(\frac{m}{q}\right)\Gamma\left(\frac{n}{r}\right)}{\Gamma\left(1 + \frac{l}{p} + \frac{m}{q} + \frac{n}{r}\right)}.$$

EXAMPLE 7.26

Show that $\iint x^{m-1} y^{n-1}\, dx\, dy$ over the positive octant of the ellipse $\frac{x^2}{a^2} + \frac{y^2}{b^2} = 1$ is $\frac{a^m b^n}{2n}\, \beta \left(\frac{m}{2}, \frac{n}{2} + 1\right)$.

Solution. Putting $\frac{x^2}{a^2} = X$ and $\frac{y^2}{b^2} = Y$, we get $x = a\sqrt{X}$ and $y = b\sqrt{Y}$ and $dx = \frac{a}{2} X^{-\frac{1}{2}} dX$ and $dy = \frac{b}{2} Y^{-\frac{1}{2}} dY$. Therefore,

$$\iint x^{m-1} y^{n-1}\, dxdy$$

$$= \iint a^{m-1} X^{\frac{m-1}{2}} b^{n-1} Y^{\frac{n-1}{2}} \frac{a}{2} X^{-\frac{1}{2}} \frac{b}{2} Y^{-\frac{1}{2}} dXdY$$

$$= \frac{a^m b^n}{4} \iint X^{\frac{m}{2}-1} Y^{\frac{n}{2}-1} dX\, dY$$

$$= \frac{a^m b^n}{4} \cdot \frac{\Gamma\left(\frac{m}{2}\right)\Gamma\left(\frac{n}{2}\right)}{\Gamma\left(1 + \frac{m}{2} + \frac{n}{2}\right)} = \frac{a^m b^n}{\frac{4n}{2}} \frac{\Gamma\left(\frac{m}{2}\right)\Gamma\left(\frac{n}{2}\right)\frac{n}{2}}{\Gamma\left(\frac{m+n}{2} + 1\right)}$$

$$= \frac{a^m b^n}{2n} \cdot \frac{\Gamma\left(\frac{m}{2}\right)\Gamma\left(\frac{n}{2} + 1\right)}{\Gamma\left(\frac{m+n}{2} + 1\right)} = \frac{a^m b^n}{2n}\, \beta\left(\frac{m}{2}, \frac{n}{2} + 1\right).$$

7.7 MISCELLANEOUS EXAMPLES

EXAMPLE 7.27

Evaluate $\int_0^\infty \frac{x\, dx}{1 + x^6}$ using Beta and Gamma functions.

Solution. Putting $x^6 = t$, we have $x = t^{\frac{1}{6}}$ and so $dx = \frac{1}{6} t^{-\frac{5}{6}}$. When $x = 0$, $t = 0$ and when $x = \infty$, $t = \infty$. Therefore,

$$\int_0^\infty \frac{x\, dx}{1 + x^6} = \frac{1}{6} \int_0^\infty \frac{t^{\frac{1}{6}}}{1 + t} t^{-\frac{5}{6}} dt$$

$$= \frac{1}{6} \int_0^\infty \frac{t^{-\frac{2}{3}}}{1 + t} dt = \frac{1}{6} \int_0^\infty \frac{t^{\frac{1}{3}-1}}{(1 + t)^{\frac{1}{3}+\frac{2}{3}}} dt$$

$$= \frac{1}{6}\, \beta\left(\frac{1}{3}, \frac{2}{3}\right),$$

$$\text{since } \beta(m, n) = \int_0^\infty \frac{x^{m-1}}{(1 + x)^{m+n}} dx$$

$$= \frac{1}{6} \frac{\Gamma\left(\frac{1}{3}\right)\Gamma\left(\frac{2}{3}\right)}{\Gamma\left(\frac{1}{3} + \frac{2}{3}\right)} = \frac{1}{6} \frac{\Gamma\left(\frac{1}{3}\right)\Gamma\left(\frac{2}{3}\right)}{\Gamma(1)}$$

$$= \frac{1}{6} \Gamma\left(\frac{1}{3}\right)\Gamma\left(\frac{2}{3}\right)$$

$$= \frac{1}{6}\left(\frac{\pi}{\sin \frac{\pi}{3}}\right),$$

using $\Gamma(n)\, \Gamma(1 - n) = \frac{\pi}{\sin n\pi}$ $(0 < n)$.

EXAMPLE 7.28

Evaluate $\int_0^a x^4 \sqrt{a^2 - x^2}\, dx$ using Gamma function.

Solution. Putting $x = a \sin\theta$, we get $dx = a\cos\theta\, d\theta$ and $\sqrt{a^2 - x^2} = \sqrt{a^2 - a^2 \sin^2\theta} = a\cos\theta$. When $x = 0$, $\theta = 0$ and when $x = a$, $\theta = \frac{\pi}{2}$. Therefore

$$I = \int_0^a x^4 \sqrt{a^2 - x^2}\, dx$$

$$= \int_0^{\frac{\pi}{2}} a^4 \sin^4\theta\,(a\cos\theta)\,(a\cos\theta)\,d\theta$$

$$= a^6 \int_0^{\frac{\pi}{2}} \sin^4\theta\cos^2\theta\,d\theta.$$

Since $\int_0^{\frac{\pi}{2}} \sin^m\theta\cos^n\theta\,d\theta = \frac{1}{2}\,\frac{\Gamma\left(\frac{m+1}{2}\right)\Gamma\left(\frac{n+1}{2}\right)}{\Gamma\left(\frac{m+n+2}{2}\right)}$, we have

$$I = a^6\left[\frac{\Gamma\left(\frac{5}{2}\right)\Gamma\left(\frac{3}{2}\right)}{2\Gamma(4)}\right]$$

$$= \frac{a^6}{2}\left[\frac{\frac{3}{2}\cdot\frac{1}{2}\sqrt{\pi}\cdot\frac{1}{2}\sqrt{\pi}}{3!}\right] = \frac{\pi a^6}{32}.$$

EXAMPLE 7.29

Show that $\int_0^1 \frac{x^n}{\sqrt{1-x^2}}\,dx = \frac{2.4.\ \ldots\ (n-1)}{1.3.\ 5.\ \ldots\ n}$, where n is odd integer.

Solution. Putting $x = \sin\theta$, we have $dx = \cos\theta\,d\theta$. Therefore

$$\int_0^1 \frac{x^n}{\sqrt{1-x^2}}\,dx = \int_0^{\frac{\pi}{2}} \frac{\sin^n\theta}{\cos\theta}\cos\theta\,d\theta$$

$$= \int_0^{\frac{\pi}{2}} \sin^n\theta\,d\theta$$

$$= \frac{\Gamma\left(\frac{n+1}{2}\right)}{2\Gamma\left(\frac{n+2}{2}\right)}\sqrt{\pi} \text{ (see Example 7.11).}$$

Since n is odd, we take $n = 2m+1$. Therefore

$$\int_0^1 \frac{x^n}{\sqrt{1-x^2}}\,dx$$

$$= \frac{\Gamma(m+1)}{2\,\Gamma\left(m+\frac{3}{2}\right)}\,\frac{\sqrt{\pi}}{2}$$

$$= \frac{m(m-1)(m-2)\ldots 3.2.1}{\left(m+\frac{1}{2}\right)\left(m-\frac{1}{2}\right)\left(m-\frac{3}{2}\right)\ldots\frac{3}{2}\cdot\frac{1}{2}\Gamma\left(\frac{1}{2}\right)}\,\frac{\sqrt{\pi}}{2}$$

$$= \frac{2m\,(2m-2)\,(2m-4)\ldots 6.4.2}{(2m+1)\,(2m-1)\,(2m-3)\ldots 3.1}$$

$$= \frac{(n-1)\,(n-3)\,(n-5)\ldots 6.4.2}{n\,(n-2)\ldots 5.3.1},\ (n \text{ odd})$$

which was to be established.

EXAMPLE 7.30

Show that $\int_0^\infty x^m e^{-a^2x^2}\,dx = \frac{1}{2a^{m+1}}\,\Gamma\left(\frac{m+1}{2}\right)$.

Solution. Putting $ax = \sqrt{z}$, we have $a\,dx = \frac{1}{2}z^{-\frac{1}{2}}\,dz$ Therefore

$$\int_0^\infty x^m e^{-a^2x^2}\,dx = \frac{1}{2}a\int_0^\infty e^{-z}\left(\frac{\sqrt{z}}{a}\right)^m z^{-\frac{1}{2}}\,dz$$

$$= \frac{1}{2a^{m+1}}\int_0^\infty e^{-z}z^{\frac{m-1}{2}}\,dz$$

$$= \frac{1}{2a^{m+1}}\int_0^\infty e^{-z}z^{\frac{m+1}{2}-1}\,dz$$

$$= \frac{1}{2a^{m+1}}\Gamma\left(\frac{m+1}{2}\right).$$

EXERCISES

1. Show that
   (i) $\beta(2.5, 1.5) = \frac{\pi}{16}$.
   (ii) $\beta\left(\frac{9}{2}, \frac{7}{2}\right) = \frac{5\pi}{2048}$.
   (iii) $\Gamma\left(\frac{3}{4}\right)\Gamma\left(\frac{1}{4}\right) = \pi\sqrt{2}$.
2. Show that
   $$\int_0^\infty \sqrt{x}e^{-x^3}\,dx = \frac{\sqrt{\pi}}{3}.$$
3. Show that
   $$\int_0^\infty x^{2n-1}e^{-ax^2}\,dx = \frac{\Gamma(n)}{2a^n}.$$
4. Show that
   $$\int_0^\infty \frac{x^a}{a^x}\,dx = \frac{\Gamma(a+1)}{(\log a)^{a+1}},\ \text{if } a > 1.$$
5. Show that
   $$\int_0^{\frac{\pi}{2}} \sin^3 x\cos^{5/2} x\,dx = \frac{8}{77}.$$

6. Show that

$$\int_a^b (x-a)^m (b-x)^n\, dx = (b-a)^{m+n+1}\beta(m+1, n+1).$$

7. Prove that

$$\int_0^\infty x^n e^{-a^2x^2}\, dx = \frac{1}{2a^{n+1}}\Gamma\left(\frac{n+1}{2}\right), n > -1.$$

8. Show that

$$\int_0^1 \frac{x^2 dx}{\sqrt{1-x^4}} \cdot \int_0^1 \frac{dx}{\sqrt{1+x^4}} = \frac{\pi}{4\sqrt{2}}.$$

9. Prove that

$$\int_0^1 \frac{x dx}{\sqrt{1-x^5}} = \frac{1}{5}\beta\left(\frac{2}{5}, \frac{1}{2}\right).$$

10. Show that

$$\int_0^1 \frac{dx}{\sqrt{1-x^n}} = \frac{\sqrt{\pi}}{n} \cdot \frac{\Gamma\left(\frac{1}{n}\right)}{\Gamma\left(\frac{1}{n}+\frac{1}{2}\right)}.$$

11. Express the integrals in terms of gamma function:

(i) $\int_0^\infty x^{p-1} e^{-kx}\, dx, k > 0$ **Ans.** (i) $\frac{\Gamma(n)}{k^n}$.

(ii) $\int_0^1 \left(\log \frac{1}{x}\right)^{n-1} dx$ (iii) $\int_0^\infty e^{-y} y^{n-1} dy = \Gamma(n)$.

(iv) $\int_0^\infty e^{-x^2}\, dx$ (v) $\frac{1}{2}\int_0^\infty e^{-t} t^{-\frac{1}{2}}\, dt = \frac{1}{2}\Gamma\left(\frac{1}{2}\right)$.

12. Show that $\int_0^\infty x^3 e^{-x^2}\, dx = \frac{1}{9}\Gamma\left(\frac{1}{3}\right)$.

13. Show that $\int_0^\infty x^{2n-1} e^{-ax^2}\, dx = \frac{\Gamma(n)}{2a^n}$.

14. Show that $y\beta(x+1, y) = x\beta(x, y+1)$.

15. Show that $\int_0^\infty \sqrt{x}\, e^{-x^2}\, dx \cdot \int_0^\infty \frac{e^{-x^2}}{\sqrt{x}} dx = \frac{\pi}{2\sqrt{2}}$.

16. The plane $\frac{x}{a} + \frac{y}{b} + \frac{z}{c} = 1$ meets the axes in A, B, and C, respectively. Find the mass of the tetrahedron $OABC$ if the density at any point is $\rho = \mu\, xyz$.

Hint: Mass $= \iiint \rho\, dx\, dy\, dz,\ 0 \le \frac{x}{a} + \frac{y}{b} + \frac{z}{c} \le 1$

$= \iiint \mu xyz\, dx\, dy\, dz$

Put $\frac{x}{a} = X, \frac{y}{b} = Y$, and $\frac{z}{c} = Z$ and proceed.

Ans. $\frac{\mu a^2 b^2 c^2}{720}$.

17. Show that the volume of the solid bounded by the coordinate planes and the surface $\sqrt{\frac{x}{a}} + \sqrt{\frac{y}{b}} + \sqrt{\frac{z}{c}} = 1$ is $\frac{abc}{90}$.

18. Find the volume of the ellipsoid $\frac{x^2}{a^2} + \frac{y^2}{b^2} + \frac{z^2}{c^2} = 1$.

Hint: $V = 8 \iiint dx\, dy\, dz$.

Put $\frac{x^2}{a^2} = X, \frac{y^2}{b^2} = Y$, and $\frac{z^2}{c^2} = Z$,

and use Dirichlet's Theorem to get

$$V = \frac{4\pi}{3} abc.$$

19. Show that the entire volume of the solid

$$\left(\frac{x}{a}\right)^{\frac{2}{3}} + \left(\frac{y}{b}\right)^{\frac{2}{3}} + \left(\frac{z}{c}\right)^{\frac{2}{3}} = 1 \text{ is } \frac{4}{35}\pi\, abc.$$

8 Reduction Formulas

A *reduction formula* for a given integral is an integral which is of the same type as the given integral but of a lower degree (or order). The reduction formula is used when the given integral cannot be evaluated otherwise. The repeated application of the reduction formula helps us to evaluate the given integral. In what follows, we shall observe that the reduction formulas are obtained by repeated application of integration by parts.

8.1 REDUCTION FORMULAS FOR $\int \sin^n x dx$ AND $\int \cos^n x\, dx$

(i) Integration by parts yields

$$\int \sin^n x dx = \int \sin^{n-1} x . \sin x dx$$

$$= \sin^{n-1} x(-\cos x)$$

$$-\int (n-1)\sin^{n-2} x \cos x(-\cos x)dx$$

$$= -\sin^{n-1} x \cos x$$

$$+(n-1)\int \sin^{n-2} x \cos^2 xdx$$

$$= -\sin^{n-1} x \cos x$$

$$+(n-1)\int \sin^{n-2} x(1-\sin^2 x)dx$$

$$= -\sin^{n-1} x \cos x + (n-1)$$

$$\times \int \sin^{n-2} xdx - (n-1)\int \sin^n xdx.$$

Therefore, transposing the last term to the left-hand side, we get

$$n\int \sin^n xdx = -\sin^{n-1} x \cos x + (n-1)\int \sin^{n-2} xdx.$$

Hence,

$$\int \sin^n xdx = \frac{-\sin^{n-1} x \cos x}{n} + \frac{n-1}{n}\int \sin^{n-2} xdx$$

(ii) Following the steps of part (i), we get

$$\int \cos^n xdx = \frac{\sin x \cos^{n-1} x}{n} + \frac{n-1}{n}\int \cos^{n-2} xdx$$

Deduction: We have

$$\int_0^{\frac{\pi}{2}} \sin^n xdx$$

$$= \int_0^{\frac{\pi}{2}} \cos^n xdx$$

$$= \begin{cases} \dfrac{(n-1)(n-3)(n-5)\ldots 2}{n(n-2)(n-4)3}, & \text{if } n \text{ is odd.} \\ \dfrac{(n-1)(n-3)(n-5)\ldots 2}{n(n-2)(n-4)3}\dfrac{\pi}{2}, & \text{if } n \text{ is even.} \end{cases}$$

Proof: Using reduction formula for $\int \sin^n xdx$ we have

$$I_n = \int_0^{\frac{\pi}{2}} \sin^n xdx.$$

$$= -\left|\frac{\sin^{n-1} x \cos x}{n}\right|_0^{\frac{\pi}{2}} + \frac{n-1}{n}\int_0^{\frac{\pi}{2}} \sin^{n-2} xdx^3$$

$$= \frac{n-1}{n} I_{n-2}.$$

When n is odd, we have

$$I_{n-2} = \frac{n-3}{n-2} I_{n-4},\ I_{n-4} = \frac{n-5}{n-4} I_{n-6},$$

$$I_5=\frac{4}{5}I_3,\text{ and }I_3=\frac{2}{3}I_1=\frac{2}{3}\int_0^{\frac{\pi}{2}}\sin x dx$$

$$=\frac{2}{3}[-\cos x]_0^{\frac{\pi}{2}}=\frac{2}{3}.$$

Hence,

$$I_n=\frac{(n-1)(n-3)(n-5)\ldots 2}{n(n-2)(n-4)\ldots 3}.$$

When n is even, we note that

$$I_{n-2}=\frac{n-3}{n-2}I_{n-4},\ I_{n-4}=\frac{n-5}{n-4}I_{n-6},$$

$$I_4=\frac{3}{4}I_2,\text{ and }I_2=\frac{1}{4}I_0$$

$$=\frac{1}{2}\int_0^{\frac{\pi}{2}}\sin^0 x dx=\frac{1}{2}.\frac{\pi}{2}.$$

Hence,

$$I_n=\frac{(n-1)(n-3)(n-5)\ldots 3.1}{n(n-2)(n-4)\ldots 4.2}\frac{\pi}{2}.$$

EXAMPLE 8.1

Evaluate $\int \cos^5 x dx$

Solution. Putting $\sin x=t$, we get $\cos x dx=$ dt. So,

$$\int \cos^5 x dx=\int \cos^4 x\cos x\, dx$$

$$=\int (1-\sin^2 x)^2\cos x dx$$

$$=\int (1-t^2)^2 dt$$

$$=\int (1+t^4-2t^2)dt$$

$$=t+\frac{t^5}{5}-2\frac{t^3}{3}$$

$$=\sin x-\frac{2}{3}\sin^3 x+\frac{1}{5}\sin^5 x.$$

EXAMPLE 8.2

Evaluate $\int_0^{\frac{\pi}{2}}\sin^6 x\, dx.$

Solution. For even n, we have

$$\int_0^{\frac{\pi}{2}}\sin^n x dx=\frac{(n-1)(n-3)(n-5)\ldots 3.1}{n(n-2)(n-4)\ldots 2}.\frac{\pi}{2}.$$

Therefore,

$$\int_0^{\frac{\pi}{2}}\sin^6 x\, dx=\frac{(6-1)(6-3)(6-5)}{6(6-2)(6-4)}.\frac{\pi}{2}$$

$$=\frac{5.3.1}{6.4.2}\frac{\pi}{2}=\frac{5\pi}{32}.$$

EXAMPLE 8.3

Evaluate $\int_0^{\infty}\frac{dx}{(a^2+x^2)^n}.$

Solution. Putting $x=a\tan\theta$, we have $dx=a\sec^2\theta d\theta$. When $x=0$, $\theta=0$ and when $x=\infty$, $\theta=\frac{\pi}{2}$. Therefore,

$$\int_0^{\infty}\frac{dx}{(a^2+x^2)^2}=\int_0^{\frac{\pi}{2}}\frac{a\sec^2\theta d\theta}{[a^2(1+\tan^2\theta)]^n}=\int_0^{\frac{\pi}{2}}\frac{a\sec^2\theta d\theta}{a^{2n}\sec^{2n}\theta}$$

$$=\frac{1}{a^{2n-1}}\int_0^{\frac{\pi}{2}}\cos^2\theta d\theta(\text{even power of cosine})$$

$$=\frac{1}{a^{2n-1}}.\frac{(2n-3)(2n-5)\ldots 3.1}{(2n-2)(2n-4)\ldots 4.2}.\frac{\pi}{2}.$$

EXAMPLE 8.4

Evaluate $\int_0^{\frac{\pi}{2}}\cos^7 x\, dx$

Solution. We have, for odd power of cosine,

$$\int_0^{\frac{\pi}{2}}\cos^n x\, dx=\frac{(n-1)(n-3)(n-5)\ldots 2}{n(n-2)(n-1)\ldots 3}.$$

Therefore, $\int_0^{\frac{\pi}{2}}\cos^7 x\, dx=\frac{6.4.2}{7.5.3}=\frac{16}{35}.$

EXAMPLE 8.5

Evaluate $\int_0^{2a} x^{\frac{7}{2}}(2a-x)^{-\frac{1}{2}}dx.$

Solution. Putting $x=2a\sin^2\theta$, we get $dx=4a\sin\theta\cos\theta d\theta$. Therefore, the given integral reduces to

$$\int_0^{\frac{\pi}{2}}(2a\sin^2\theta)^{\frac{7}{2}}.\frac{4a\sin\theta\cos\theta}{\sqrt{2a}\sqrt{1-\sin^2\theta}}d\theta$$

$$= \int_0^{\frac{\pi}{2}} 2^5 a^4 \sin^8 \theta d\theta = 32a^4 \int_0^{\frac{\pi}{2}} \sin^8 \theta d\theta$$

$$= 32a^4 \frac{7.5.3.1}{8.6.4.2}.\frac{\pi}{2} = \frac{35\pi a^4}{8}.$$

8.2 REDUCTION FORMULA FOR $\int \sin^m x \cos^n x\, dx$

Integration by parts yields

$$\int \sin^m x \cos^n x\, dx$$

$$= \int \sin^{m-1} x.\cos^n x \sin x dx$$

$$= \sin^{m-1} x\left[-\frac{\cos^{n+1} x}{n+1}\right]$$

$$-\int (m-1)\sin^{m-2} x \cos x\left[-\frac{\cos^{n-1} x}{n+1}\right] dx$$

$$= -\frac{\sin^{m-1} x \cos^{n+1} x}{n+1} + \frac{m-1}{n+1}\int \sin^{m-2} x \cos^{n+2} x dx$$

$$= -\frac{\sin^{m-1} x \cos^{n+1} x}{n+1} + \frac{m-1}{n+1}\int \sin^{m-2} x \cos^{n} x dx$$

$$-\frac{m-1}{n+1}\int \sin^m x \cos^n x\, dx.$$

Hence,

$$\int \sin^m x \cos^n x dx = -\frac{\sin^{m-1} x \cos^{n+1} x}{m+n}$$

$$+\frac{m-1}{m+n}\int \sin^{m-2} x \cos^n x dx,$$

which is the required reduction formula.

Deduction: We have

$$\int_0^{\frac{\pi}{2}} \sin^m x \cos^n x\, dx.$$

$$= \begin{cases} \dfrac{(m-1)(m-3)\ldots(n-1)(n-3)\ldots}{(m+n)(m+n-2)(m+n-4)\ldots} \\ \dfrac{(m-1)(m-3)\ldots(n-1)(n-3)\ldots}{(m+n)(m+n-2)(m+n-4)\ldots}.\dfrac{\pi}{2}, \end{cases}$$

if both m and n are even.

Proof: Using reduction formula for $\int \sin^m x \cos^n x\, dx$, we have

$$I_{m,n} = \int_0^{\frac{\pi}{2}} \sin^m x \cos^n x\, dx = \left[-\frac{\sin^{m-1} x \cos^{n+1} x}{m+n}\right]_0^{\frac{\pi}{2}}$$

$$+\frac{m-1}{m+n}\int_0^{\frac{\pi}{2}} \sin^{m-2} x \cos^n x\, dx = \frac{m-1}{m+n} I_{m-2,n}.$$

When m is odd, we get

$$I_{m-2,n} = \frac{m-3}{m+n-2} I_{m-4,n},$$

$$I_{m-4,n} = \frac{m-5}{m+n-4} I_{m-6,n},$$

$$I_{5,n} = \frac{4}{n+5} I_{3,n}, \text{ and}$$

$$I_{3,n} = \frac{2}{n+3} I_{1,n} = \frac{2}{n+3}\int_0^{\frac{\pi}{2}} \sin x \cos^n x\, dx$$

$$= \frac{2}{n+3}\left[-\frac{\cos^{n+1} x}{n+1}\right]_0^{\frac{\pi}{2}} = \frac{2}{(n+3)(n+1)}.$$

Hence, $I_{m,n}$

$$= \frac{(m-1)(m-3)(m-5)\ldots 4.2}{(m+n)(m+n-2)(m+n-4)\ldots(n+3)(n+1)}.$$

When m is even, we have

$$I_{m-2,n} = \frac{m-3}{m+n-2} I_{m-4,n},$$

$$I_{m-4,n} = \frac{m-5}{m+n-4} I_{m-6,n},$$

$$I_{4,n} = \frac{3}{n+4} I_{2,n}, \text{ and}$$

$$I_{2,n} = \frac{1}{n+2} I_{0,n} = \frac{1}{n+2}\int_0^{\frac{\pi}{2}} \cos^n x\, dx.$$

Hence,

$$I_{m,n} = \frac{(m-1)(m-3)(m-5)\ldots 1}{(m+m)(m+n-2)(m+n-4)\ldots(n+2)}$$

$$\times\int_0^{\pi} \cos^n x\, dx$$

$$= \begin{cases} \dfrac{(m-1)(m-3)(m-5)\ldots1}{(m+n)(m+n-2)(m+n-4)\ldots(n+2)} \\ \quad .\dfrac{(n-1)(n-3)\ldots}{n(n-2)\ldots} \text{ if } n \text{ is odd.} \\ \dfrac{(m-1)(m-3)(m-5)\ldots1}{(m+n)(m+n-2)(m+n-4)\ldots(n+2)} \\ \quad .\dfrac{(n-1)(n-3)\ldots}{n(n-2)\ldots}\dfrac{\pi}{2} \text{ if } n \text{ is even.} \end{cases}$$

Hence,

$$I_{m,n} = \begin{cases} \dfrac{(m-1)(m-3)(m-5)\ldots1}{(m+n)(m+n-2)(m+n-4)\ldots(n+2)} \\ \quad .\dfrac{(n-1)(n-3)\ldots}{n(n-2)\ldots} \text{ if at least one} \\ \quad \text{of } m \text{ and } n \text{ is odd.} \\ \dfrac{(m-1)(m-3)(m-5)\ldots1}{(m+n)(m+n-2)(m+n-4)\ldots(n+2)} \\ \quad .\dfrac{(n-1)(n-3)\ldots}{n(n-2)\ldots}\dfrac{\pi}{2} \text{ if both } m \\ \quad \text{and } n \text{ are even.} \end{cases}$$

EXAMPLE 8.6

Evaluate (i) $\int_0^{\frac{\pi}{2}} \sin^3 x \cos^4 x\, dx$ and

(ii) $\int_0^{\frac{\pi}{6}} \cos^4 3\theta \sin^3 6\theta\, d\theta.$

Solution. (i) Since $m = 3$ is odd, we have

$$\int_0^{\frac{\pi}{2}} \sin^3 x \cos^4 x\, dx = \frac{(3-1)(4-1)(4-3)}{7.5.3} = \frac{2}{35}.$$

(ii) We have

$$\int_0^{\frac{\pi}{6}} \cos^4 3\theta \sin^3 6\theta\, d\theta$$

$$= \int_0^{\frac{\pi}{6}} \cos^4 3\theta (2\sin 3\theta \cos 3\theta)^3\, d\theta$$

$$= 8\int_0^{\frac{\pi}{6}} \sin^3 3\theta \cos^7 3\theta\, d\theta.$$

Putting $3\theta = x$, we get $3\, d\theta = dx$, and so,

$$\int_0^{\frac{\pi}{6}} \cos^4 3\theta \sin^3 6\theta\, d\theta = \frac{8}{3}\int_0^{\frac{\pi}{2}} \sin^3 x \cos^7 x\, dx$$

$$= \frac{8}{3}\left[\frac{2.6.4.2}{10.8.6.4.2}\right]$$

$$= \frac{1}{15}.$$

EXAMPLE 8.7

Evaluate $\int_0^{2a} x^{\frac{5}{2}} \sqrt{2a - x}\, dx.$

Solution. Substituting $x = 2a\sin^2\theta$, we have $dx = 4a\sin\theta\cos\theta\, d\theta.$

Therefore,

$$\int_0^{2a} x^{\frac{5}{2}} \sqrt{2a - x}\, dx$$

$$= \int_0^{\frac{\pi}{2}} \sqrt{2a}\,(2a\sin^2\theta)^{\frac{5}{2}} (4a\sin\theta\cos\theta)\cos\theta\, d\theta$$

$$= 32a^4 \int_0^{\frac{\pi}{2}} \sin^6\theta\cos^2\theta\, d\theta$$

$$= 32a^4\, \frac{5.3.1.1}{8.6.4.2}\frac{\pi}{2} = \frac{5\pi a^4}{8}.$$

EXAMPLE 8.8

Evaluate $\int_0^1 x^6 \sqrt{1 - x^2}\, dx.$

Solution. Putting $x = \sin\theta$, we get $dx = \cos\theta\, d\theta$. Therefore, the given integral reduces to

$$\int_0^{\frac{\pi}{2}} \sin^6\theta.\cos\theta.\cos\theta\, d\theta = \int_0^{\frac{\pi}{2}} \sin^6\theta.\cos^2\theta\, d\theta$$

$$= \frac{5.3.1.1}{8.6.4.2}.\frac{\pi}{2} = \frac{5\pi}{256}.$$

EXAMPLE 8. 9

Evaluate

(i) $\int_0^{\pi} \sin^4\frac{\theta}{2}\cos^3\frac{\theta}{2}\, d\theta$ and (ii) $\int_0^{\frac{\pi}{2}} \sin^{15} x\cos^3 x\, dx.$

Solution.

(i) Putting $\frac{\theta}{2}=\phi$, we get $d\theta = 2d\phi$. Hence, the integral becomes

$$2\int_0^{\frac{\pi}{2}} \sin^4\phi\cos^3\phi\, d\phi = 2\frac{3.1.2}{7.5.3.1} = \frac{4}{35}.$$

(ii) We have

$$\int_0^{\frac{\pi}{2}} \sin^{15}x\cos^3 x\, dx = \frac{14.12.10.8.6.4.2.2}{18.16.14.12.10.8.6.4.2} = \frac{1}{144}.$$

8.3 REDUCTION FORMULAS FOR $\int \tan^n x\, dx$ AND $\int \sec^n x\, dx$

(i) Let

$$\begin{aligned} I_n &= \int \tan^n x\, dx = \int \tan^{n-2}x\tan^2 x\, dx \\ &= \int \tan^{n-2}x(\sec^2 x - 1)dx \\ &= \int \tan^{n-2}x\sec^2 x dx - \int \tan^{n-2}x\, dx \\ &= \frac{\tan^{n-1}x}{n-1} - I_{n-2}. \end{aligned}$$

(ii) Let

$$\begin{aligned} I_n &= \int \sec^n xdx \\ &= \int \sec^{n-2}x\sec^2 x\, dx \\ &= \int \sec^{n-2}x\tan x - \int (n-2)\sec^{n-3}x \\ &\qquad .\sec x\tan x.\tan x dx \\ &= \sec^{n-2}x\tan x - (n-2)\int \sec^{n-2}x\tan^2 x\, dx. \\ &= \sec^{n-2}x\tan x - (n-2)\int \sec^{n-2}x(\sec^2 x - 1)\, dx \\ &= \sec^{n-2}x\tan x - (n-2)\int (\sec^n x - \sec^{n-2}x)\, dx \\ &= \sec^{n-2}x\tan x - (n-2)I_n + (n-2)I_{n-2}. \end{aligned}$$

Hence,

$$(n-1)I_n = \sec^{n-2}x\tan x + (n-2)I_{n-2}$$

or $$I_n = \frac{\sec^{n-2}x\tan x}{n-1} + \frac{n-2}{n-1}I_{n-2}.$$

EXAMPLE 8.10

Obtain a reduction formula for $\text{cosec}^n x dx$.

Solution. We have

$$\begin{aligned} I_n &= \int \text{cosec}^n x dx = \int \text{cosec}^{n-2}x\,\text{cosec}^2 x dx \\ &= -\text{cosec}^{n-2}x\cot x \\ &\quad -\int (n-2)\,\text{cosec}^{n-3}x\,\text{cosec}\, x\cot x\cot x\, dx \\ &= -\text{cosec}^{n-2}x\cot x - (n-2) \\ &\quad \times\int \text{cosec}^{n-2}x\cot^2 x\, dx. \\ &= -\text{cosec}^{n-2}x\cot x - (n-2) \\ &\quad \times\int \text{cosec}^{n-2}x(\text{cosec}^2 x - 1)\, dx \\ &= -\text{cosec}^{n-2}x\cot x - (n-2) \\ &\quad \times\int \text{cosec}^n x\, dx + (n-2)\int \text{cosec}^{n-2}x\, dx \\ &= -\text{cosec}^{n-2}x\cot x - (n-2)I_n + (n-2)I_{n-2}. \end{aligned}$$

Hence,

$$(n-1)I_n = -\text{cosec}^{n-2}\cot x + (n-2)I_{n-2}$$

or $$I_n = -\frac{\text{cosec}^{n-2}\cot x}{n-1} + \frac{n-2}{n-1}I_{n-2}.$$

EXAMPLE 8.11

Evaluate $\int \sec^4 x\, dx$.

Solution. The reduction formula for $\int \sec^n x\, dx$ is

$$I_n = \int \sec^n x\, dx = \frac{\sec^{n-2}x\tan x}{n-1} + \frac{n-2}{n-1}I_{n-2}.$$

Therefore,

$$\begin{aligned} I_4 &= \frac{\sec^2 x\tan x}{3} + \frac{2}{3}\int \sec^2 x\, dx. \\ &= \frac{\sec^2 x\tan x}{3} + \frac{2}{3}\tan x \\ &= \frac{1}{3}[\tan x\sec^2 x + 2\tan x]. \end{aligned}$$

EXAMPLE 8.12

If $I_n = \int_0^{\frac{\pi}{4}} \tan^n x\, dx$ show that $(n-1)(I_n + I_{n-2}) = 1$. Hence, evaluate I_5.

Solution. We have

$$I_n = \int_0^{\frac{\pi}{4}} \tan^n x\,dx = \int_0^{\frac{\pi}{4}} \tan^{n-2} x \tan^2 x\,dx$$

$$= \int_0^{\frac{\pi}{4}} \tan^{n-2} x(\sec^2 x - 1)\,dx$$

$$= \int_0^{\frac{\pi}{4}} \tan^{n-2} x \sec^2 x\,dx - \int_0^{\frac{\pi}{4}} \tan^{n-2} x\,dx$$

$$= \left[\frac{\tan^{n-1} x}{n-1}\right]_0^{\frac{\pi}{4}} - I_{n-2}.$$

Hence,

$$(n-1)I_n + (n-1)I_{n-2} = [\tan^{n-1} x]_0^{\frac{\pi}{4}} = 1$$

or

$$(n-1)(I_n + I_{n-2}) = 1. \qquad (1)$$

Putting $n = 5$ in (1) we get

$$4(I_5 + I_3) = 1 \text{ or } I_5 = \tfrac{1}{4} - I_3.$$

But, putting $n = 3$ in (1), we have

$$2(I_3 + I_1) = 1 \text{ or } I_3 = \tfrac{1}{2} - I_1. \qquad (2)$$

Now,

$$I_1 = \int_0^{\frac{\pi}{4}} \tan x\,dx = [-\log\cos x]_0^{\frac{\pi}{4}} = -\log\frac{1}{\sqrt{2}}$$

$$= -[\log 1 - \log\sqrt{2}] = \log\sqrt{2} = \tfrac{1}{2}\log 2.$$

Therefore, (2) yields

$$I_3 = \frac{1}{2} - \frac{1}{2}\log 2.$$

Hence,

$$I_5 = \frac{1}{4} - \frac{1}{2} + \frac{1}{2}\log 2 = \frac{1}{2}\left[\log 2 - \frac{1}{2}\right].$$

8.4 REDUCTION FORMULAS FOR $\int x^n \sin mx\,dx$ AND $\int x^n \cos mx\,dx$

(i) Put $I_n = \int x^n \sin mx\,dx$. Then integration by parts yields the following reduction formulas:

$$I_n = x^n\left(\frac{-\cos mx}{m}\right) - \int nx^{n-1}\left(\frac{-\cos mx}{m}\right)dx$$

$$= -\frac{x^n \cos mx}{m} + \frac{n}{m}\int x^{n-1}\cos mx\,dx$$

$$= -\frac{x^n \cos mx}{m} + \frac{n}{m}\left[\left(x^{n-1}\frac{\sin mx}{m}\right) - \int (n-1)x^{n-2}\frac{\sin mx}{x}dx\right]$$

$$= -\frac{x^n \cos mx}{m} + \frac{n}{m^2}x^{n-1}\sin mx - \frac{n(n-1)}{m^2}\int x^{n-2}\sin mx\,dx$$

$$= -\frac{x^n \cos mx}{m} + \frac{n}{m^2}x^{n-1}\sin mx - \frac{n(n-1)}{m^2}I_{n-2}.$$

(ii) Let $I_n = \int x^n \cos mx dx$. Then proceeding exactly as in part (i), we get the following reduction formulas:

$$I_n = \frac{x^n \sin mx}{m} + \frac{n}{m^2}x^{n-1}\cos mx - \frac{n(n-1)}{m^2}I_{n-2}.$$

EXAMPLE 8.13

If $I_n = \int_0^{\frac{\pi}{2}} x^n \sin x dx, n > 1$, show that

$$I_n + n(n-1)I_{n-2} = n\left(\frac{\pi}{2}\right)^{n-1}.$$

Hence, evaluate I_5.

Solution. The reduction formula for $\int x^n \sin^{mx} dx$ is

$$I_n = -\frac{x^n \cos mx}{m} + \frac{n}{m^2}x^{n-1}\sin mx - \frac{n(n-1)}{m^2}I_{n-2}.$$

Here $m = 1$ and the limits are from 0 to $\frac{\pi}{2}$. Therefore,

$$I_n = [-x^n \cos x + nx^{n-1}\sin x]_0^{\frac{\pi}{2}} - n(n-1)I_{n-2}$$

or

$$I_n + n(n-1)I_{n-2} = n\left(\frac{\pi}{2}\right)^{n-1}. \quad (1)$$

Putting $n = 5$ in (1), we get

$$I_5 + 20I_3 = 5\left(\frac{\pi}{2}\right)^4. \quad (2)$$

Putting $n = 3$ in (1), we have

$$I_3 + 6I_1 = 3\left(\frac{\pi}{2}\right)^2. \quad (3)$$

But,

$$I_1 = \int_0^{\frac{\pi}{2}} x\sin x\,dx = [-x\cos x]_0^{\frac{\pi}{2}} - \int_0^{\frac{\pi}{2}} \cos x\,dx$$

$$= [\sin x]_0^{\frac{\pi}{2}} = 1.$$

Therefore, (3) yields

$$I_3 = 3\frac{\pi^2}{4} - 6.$$

Now putting this value of I_3 in (2), we get

$$I_5 = \frac{5\pi^4}{16} - 20I_3 = \frac{5\pi^4}{16} - 20\left(\frac{3\pi^2}{4} - 6\right)$$

$$= \frac{5\pi^4}{16} - 15\pi^2 + 120.$$

8.5 REDUCTION FORMULAS FOR $\int x^n e^{ax}\,dx$ AND $\int x^m (\log x)^n\,dx$

(i) Let

$$I_n = \int x^n e^{ax}\,dx.$$

Then integration by parts yields the following reduction formulas:

$$I_n = x^n\frac{e^{ax}}{a} - \int nx^{n-1}\frac{e^{ax}}{a}\,dx = \frac{x^n e^{ax}}{a} - \frac{n}{a}I_{n-1}.$$

(ii) Let

$$I_{m,n} = \int x^m (\log x)^n\,dx.$$

Then integrating by parts, we get

$$I_{m,n} = (\log x)^n\frac{x^{m+1}}{m+1} - \int n(\log x)^{n-1}.\frac{1}{x}\frac{x^{m+1}}{m+1}\,dx$$

$$= \frac{x^{m+1}}{m+1}(\log x)^n - \frac{n}{m+1}\int x^m(\log x)^{n-1}\,dx$$

$$= \frac{x^{m+1}}{m+1}(\log x)^n - \frac{n}{m+1}I_{m,n-1},$$

which is the required reduction formula.

EXAMPLE 8.14

Evaluate $\int_0^\infty e^{-x}x^n\,dx$, where n is a positive integer.

Solution. We have

$$\int_0^\infty e^{-x}x^n\,dx = \Gamma(n+1)$$

$$= n! \text{ since } n \text{ is a positive integer.}$$

EXAMPLE 8.15

If $I_n = \int_0^a e^{-x}x^n$, show that

$$I_n - (n+a)I_{n-1} + a(n-1)I_{n-2} = 0.$$

Solution. We have

$$I_n = \left[x^n\frac{e^{-x}}{-1}\right]_0^a - \int_0^a nx^{n-1}\frac{e^{-x}}{-1}\,dx$$

$$= -a^n e^{-a} + n\int_0^a x^{n-1}e^{-x}\,dx$$

$$= -a^n e^{-a} + (n+a)\int_0^a x^{n-1}e^{-x}\,dx - a\int_0^a x^{n-1}e^{-x}\,dx$$

$$= -a^n e^{-a} + (n+a)I_{n-1}$$

$$-a\left\{\left[\frac{x^{n-1}e^{-x}}{-1}\right]_0^a + \int_0^a (n-1)x^{n-2}e^{-x}\,dx\right\}$$

$$= -a^n e^{-a} + (n+a)I_{n-1} - a(n-1)I_{n-2}$$

$$+a(a^{n-1}e^{-a})$$

$$= (n+a)I_{n-1} - a(n-1)I_{n-2}.$$

Hence,

$$I_n-(n+a)I_{n-1}+a(n-1)I_{n-2}=0.$$

8.6 REDUCTION FORMULA FOR $\int \cos^m x \sin nx\,dx$

Let

$$I_{mn}=\int \cos^m x\sin nx\,dx.$$

Integrating by parts, we get

$$I_{mn}=-\cos^m x\frac{\cos nx}{n}$$

$$-\int m\cos^{n-1}x(-\sin x)\left(-\frac{\cos nx}{n}\right)^{dx}$$

$$=-\frac{1}{n}\cos^m x\cos nx$$

$$-\frac{m}{n}\int \cos^{m-1}x\cos nx\sin x\,dx$$

$$=\frac{1}{n}\cos^m x\cos nx$$

$$-\frac{m}{n}\int \cos^{m-1}x[\sin nx\cos x-\sin(n-1)x]\,dx$$

$$=-\frac{1}{n}\cos^m x\cos nx-\frac{m}{n}[I_{m,n}-I_{m-1,n-1}].$$

Thus,

$$\left(1+\frac{m}{n}\right)I_{m,n}=-\frac{1}{n}\cos^m x\cos nx+\frac{m}{n}I_{m-1,n-1},$$

which in tum implies

$$I_{m,n}=-\frac{\cos^m x\cos nx}{m+n}+\frac{m}{m+n}I_{m-1,n-1}.$$

8.7 REDUCTION FORMULA FOR $\int \frac{dx}{(x^2+a^2)^n}$

We write

$$\int\frac{dx}{(x^2+a^2)^n}=\int 1.\frac{1}{(x^2+a^2)^n}dx \qquad (1)$$

Integrating (1) by parts, we get

$$\int\frac{dx}{(x^2+a^2)^n}=\frac{x}{(x^2+a^2)^n}$$

$$-\int x(-n)(x^2+a^2)^{-n-1}2x\,dx$$

$$=\frac{x}{(x^2+a^2)^n}+2n\int\frac{x^2}{(x^2+a^2)^{n+1}}dx$$

$$=\frac{x}{(x^2+a^2)^n}+2n\int\frac{x^2+a^2-a^2}{(x^2+a^2)^{n+1}}dx$$

$$=\frac{x}{(x^2+a^2)^n}+2n\int\frac{1}{(x^2+a^2)^n}dx$$

$$-2na^2\int\frac{dx}{(x^2+a^2)^{n+1}}.$$

Thus

$$2na^2\int\frac{dx}{(x^2+a^2)^{n+1}}=\frac{x}{(x^2+a^2)^n}+(2n-1)$$

$$\times\int\frac{dx}{(x^2+a^2)^n}$$

or

$$\int\frac{dx}{(x^2+a^2)^{n+1}}=\frac{x}{2na^2(x^2+a^2)^n}$$

$$+\frac{2n-1}{2na^2}\int\frac{dx}{(x^2+a^2)^n} \qquad (2)$$

Changing n to $n–1$, the expression (2) reduces to the required Reduction Formula:

$$\int\frac{dx}{(x^2+a^2)^n}=\frac{x}{2(n-1)a^2(x^2+a^2)^{n-1}}$$

$$+\frac{2n-3}{2(n-1)a^2}\int\frac{dx}{(x^2+a^2)^{n-1}}. \qquad (3)$$

EXAMPLE 8.16

Evaluate the integral $\int\frac{dx}{(x^2+1)^3}$.

Solution. By the reduction formula, we have

$$\int\frac{dx}{(x^2+1)^3}=\frac{x}{2(3-1)1(x^2+1)^2}$$

$$+\frac{3}{2.2.1}\int\frac{dx}{(x^2-1)^2}$$

$$=\frac{x}{4(x^2+1)^2}+\frac{3}{4}\int\frac{dx}{(x^2+1)^2}. \qquad (1)$$

Further, the reduction formula applied to the integral on the right hand side of (1) yields

$$\int \frac{dx}{(x^2+1)^2}dx = \frac{x}{2(x^2+1)} + \frac{1}{2}\int \frac{dx}{x^2+1}$$

$$= \frac{x}{2(x^2+1)} + \frac{1}{2}\tan^{-1}x.$$

Hence (1) reduces to

$$\int \frac{dx}{(x^2+1)^3} = \frac{x}{4(x^2+1)} + \frac{3}{4}\left[\frac{x}{2(x^2+1)} + \frac{1}{2}\tan^{-1}x\right]$$

$$= \frac{x}{4(x^2+1)^2} + \frac{3x}{8(x^2+1)} + \frac{3}{8}\tan^{-1}x.$$

EXERCISES

1. Evaluate (i) $\int_0^{\frac{\pi}{6}} \sin^5 3\theta\, d\theta$ and (ii) $\int_0^1 x^4(1-x^2)^{\frac{3}{2}}dx$.

   Ans. (i) $\frac{8}{45}$ (ii) $\frac{3\pi}{256}$.

2. Evaluate $\int_0^1 x^5 \sin^{-1}x\,dx$.

   Ans. $\frac{11\pi}{192}$.

3. Evaluate $\int \cos^{\frac{3}{4}} x \sin^5 x\,dx$.

   Ans. $\frac{1}{2}\log\frac{x^2-x+1}{x^2+x+1}$.

4. Evaluate $\int_0^1 x^{\frac{3}{2}}(1-x)^{\frac{3}{2}}dx$.

   Ans. $\frac{3\pi}{128}$.

5. Evaluate $\int \sec^5 x\,dx$.

   Ans. $\frac{1}{4}\sec^3 x \tan x + \frac{3}{8}\sec x \tan x + \frac{3}{8}\log(\sec x + \tan x)$.

6. Evaluate $\int_0^a (a^2+x^2)^{\frac{5}{2}}dx$.

   Ans. $\left[\frac{67\sqrt{2}}{48} + \frac{5}{16}\log(1+\sqrt{2})\right]a^6$.

7. If $I_n = \int_0^{\frac{\pi}{4}} \tan^n \theta\, d\theta$, show that $n[I_{n-1} + I_{n+1}] = 1$.

8. Show that $\int_0^{\frac{\pi}{2}} \cos^m x \cos nx\, dx = \frac{m}{m+n}\int_0^{\frac{\pi}{2}} \cos^{m-1} x \cos(n-1)x\,dx$.

9. If $I_n = \int_0^{\frac{\pi}{2}} x \sin^n x\,dx \ (n>1)$, show that $n^2 I_n = n(n-1)I_{n-2} + 1$. Hence, evaluate I_5.

   Ans. $\frac{149}{225}$.

10. Prove that $\int_0^{\frac{\pi}{2}} \cos^{n-2}x \sin nx\,dx = \frac{1}{n+1}, n>1$.

11. If $I_{m,n} = \int_0^{\frac{\pi}{2}} \cos^m x \cos nx\, dx$, show that $I_{m,n} = \frac{m(m-1)}{m^2-n^2}I_{m-2,n}$.

12. Evaluate $\int \frac{dx}{(x^2+1)^4}$

    Ans. $\frac{x}{6(x^2+1)^3} + \frac{5x}{24(x^2+1)^2} + \frac{5x}{16(x^2+1)} + \frac{5}{16}\tan^{-1}x$

13. Show that

$$\int_0^\infty \frac{dx}{(x^2+1)^n} = \frac{2n-3}{2n-2}\int_0^\infty \frac{dx}{(x^2+a^2)^{n-1}}$$

and deduce that

$$\int_0^\infty \frac{dx}{(x^2+1)^3} = \frac{\pi}{2}.\frac{1.3.5.7}{2.4.6.8}.$$

9 Quadrature and Rectification

In this chapter, we discuss the processes to find area of a given bounded region of a curve and to find length of an arc of a curve between two given points. Of course, these concepts are nothing but applications of definite integral.

9.1 QUADRATURE

The process of finding the area of a bounded region of a curve is called *quadrature*.

9.1.1 Area of a Curve Given by the Cartesian Equation

Let AB be the arc of the curve $y = f(x)$, between two ordinates $x = a$ and $x = b$ represented by AC and BD respectively.

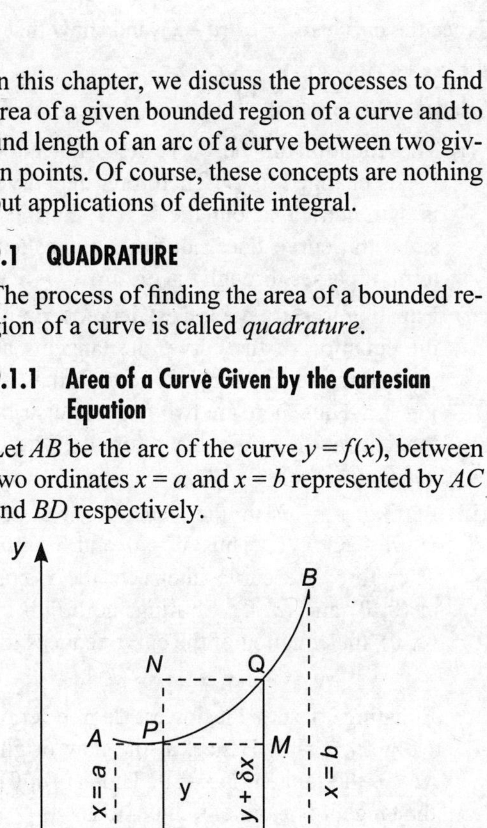

Let $P(x, y)$ and $Q(x+\delta x, y+\delta y)$ be two neighboring points on the given curve. Draw PR and QS perpendiculars to the x-axis. Then,

$$PR = x,\ QS = y+\delta y \text{ and } RS = \delta x.$$

Let A be the area $ACRP$ and $A+\delta A$ be the area $ACSQ$. Then the area $PRSQ$ is given by $A+\delta A - A = \delta A$. We observe that δA lies between areas of the rectangles $PRSM$ and $NRSQ$, that is, δA lies between $y\delta x$ and $(y+\delta y)\delta x$. Thus,

$$(y+\delta y)\delta x > \delta A > y\delta x$$

or

$$y+\delta y > \frac{\delta A}{\delta x} > y$$

Taking limit as $Q \to P$, that is, as $\delta x \to 0$ and $\delta y \to 0$, we get

$$\frac{dA}{dx} = y = f(x)$$

or

$$dA = ydx. \tag{1}$$

Integrating both sides of (1) between the limits $x = a$ and $x = b$, we get

$$\int_a^b dA = \int_a^b ydx = \int_a^b f(x)dx$$

or

$$|A|_a^b = \int_a^b ydx = \int_a^b f(x)dx$$

or

$$\text{Area } ACDB = \int_a^b ydx = \int_a^b f(x)dx \tag{2}$$

Remark 9.1.

1. The area bounded by a curve, the x-axis, and the two ordinates and given by (2) is called the area under the curve.

2. It can be shown similarly that the area bounded by the curve $x = f(y)$, the axis of y and the abscissae $y = c$ and $y = d$ is given by

$$|A|_c^d = \int_c^d x\,dy = \int_c^d f(y)\,dy. \quad (3)$$

3. The area is considered to be positive if its boundary is described in the anticlockwise direction and it will be considered negative if its boundary is described in the clockwise direction.

EXAMPLE 9.1

Find the area of

(i) quadrant of the ellipse $\frac{x^2}{a^2}+\frac{y^2}{b^2}=1$

(ii) whole area of the ellipse $\frac{x^2}{a^2}+\frac{y^2}{b^2}=1$

Solution. (i) The equation of the ellipse is

$$\frac{x^2}{a^2}+\frac{y^2}{b^2}=1.$$

which yields

$$y=\frac{b}{a}\sqrt{a^2-x^2} \text{ (for the first quadrant).}$$

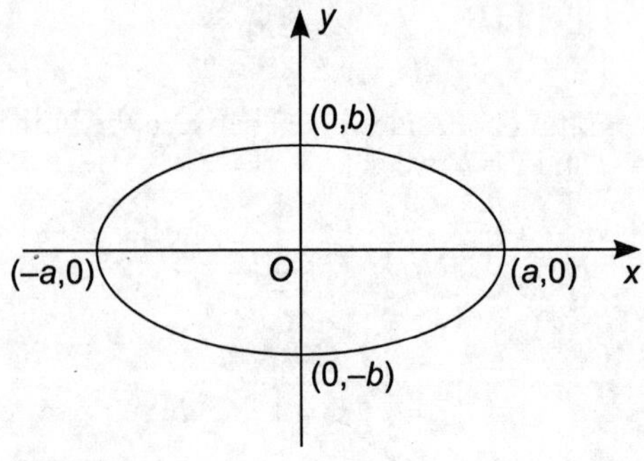

Therefore,

Area of quadrant

$$=\int_0^a y\,dx$$

$$=\frac{b}{a}\int_0^a \sqrt{a^2-x^2}\,dx$$

$$=\frac{b}{a}\left|\frac{x}{2}\sqrt{a^2-x^2}+\frac{a^2}{2}\sin^{-1}\frac{x}{a}\right|_0^a$$

$$=\frac{b}{a}\left[\frac{a^2}{2}\sin^{-1}1\right]$$

$$=\frac{b}{a}\left(\frac{a^2}{2}\right)\left(\frac{\pi}{2}\right)=\frac{\pi ab}{4}.$$

(ii) The whole area of the ellipse is 4 times the area of one quadrant. Hence,

$$\text{Area of the ellipse} = 4\left(\frac{\pi ab}{4}\right)=\pi ab.$$

EXAMPLE 9.2

Trace the curve $ay^2 = x^2(a-x)$ and show that the area of its loop is $\frac{8}{15}a^2$.

Solution. We note that

(i) The equation of the curve does not alter if y is changed to $-y$. Therefore, the curve is symmetrical about the x-axis. Further, since the curve does not have a constant term, it passes through the origin.

(ii) Equating to zero, the lowest degree term in the equation of the curve, the tangents at the origin are given by $y^2 - x^2 = 0$, that is, $y = \pm x$. Thus there are two real and district tangents at the origin. Therefore the origin is a node.

(iii) Putting $y = 0$ in the equation, we get $x^2(a-x) = 0$. Thus $x = 0$ and $x = a$. Therefore, the curve intersects the x-axis at (0, 0) and (a, 0). Shifting the origin to (a, 0), the equation of the curve reduces to

$$ay^2 = -x(x^2+2ax+a^2).$$

Equating to zero the lowest degree term, we get $a^2x = 0$. Hence, at the new origin (a, 0), the tangent is $x = 0$. Thus at (a, 0), the tangent is parallel to the y-axis.

(iv) The equation of the curve can be written as $y^2 = \frac{x^2(a-x)}{a}$. When $0 < x < a$, y^2 is positive and so the curve exists in the region $0 < x < a$. But when $x > a$, y^2 is negative and so y becomes imaginary. Thus, the curve does not exist in the region $x > a$.

(v) Equating to zero, the coefficient of highest power of y in the given equation, we see that above that there is no asymptote to the curve. Thus, the shape of the curve is as shown below:

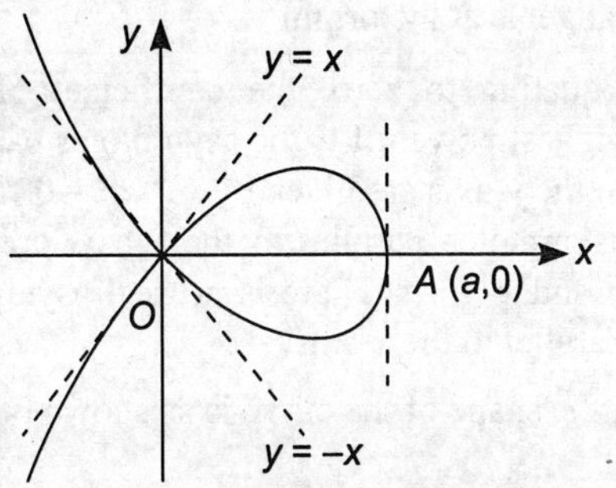

From the given equation, we have

$$y = \frac{x\sqrt{a-x}}{\sqrt{a}}.$$

Therefore,

$$\text{Area of the loop} = 2\int_0^a y\,dx$$

$$= 2\int_0^a \frac{x\sqrt{a-x}}{\sqrt{a}}\,dx$$

$$= 4a^2\int_0^{\frac{\pi}{2}} \sin^3\theta\cos^2\theta\,d\theta,$$

$$x = a\sin^2\theta$$

$$= 4a^2\frac{2.1}{5.3.1} = \frac{8}{15}a^2.$$

EXAMPLE 9.3

Trace the curve $xy^2 = 4a^2(2a-x)$ and find the area bounded by this curve and its asymptote.

Solution. The shape of the curve is

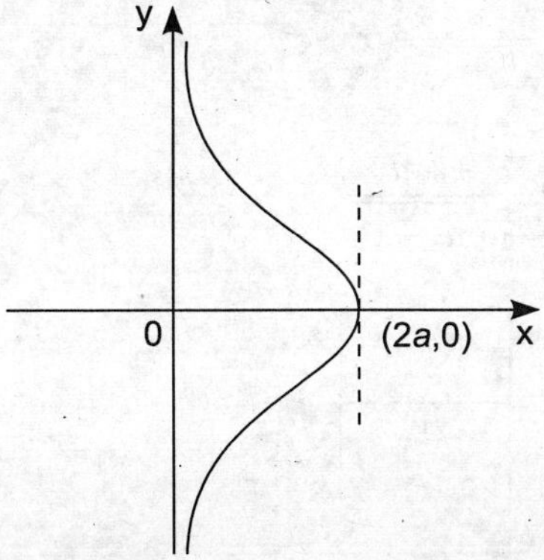

Further, from the given equation, we have

$$y = \frac{2a\sqrt{2a-x}}{\sqrt{x}}.$$

Also, $x = 0$, that is, y-axis is the asymptote of the curve. Hence, the required area is

$$A = 2\int_0^{2a} y\,dx$$

$$= 2\int_0^{2a} \frac{2a\sqrt{2a-x}}{\sqrt{x}}dx$$

$$= 4a\int_0^{\frac{\pi}{2}} 4a\cos^2\theta\,d\theta, \text{putting } x = 2a\sin^2\theta,$$

$$= 16a^2\int_0^{\frac{\pi}{2}} \cos^2\theta\,d\theta$$

$$= 16a^2\left[\frac{1}{2}.\frac{\pi}{2}\right] = 4\pi a^2.$$

EXAMPLE 9.4

Trace the curve $a^2y^2 = x^2(a^2-x^2)$ and find its whole area.

Solution. From the given equation, we have

$$y = \frac{x\sqrt{a^2-x^2}}{a}.$$

Also,

Whole area = 4 times the area of half loop.

$$= 4\int_0^a \frac{x\sqrt{a^2-x^2}}{a}dx$$

$$= 4a\int_0^{\frac{\pi}{2}} \cos^2\theta\sin\theta\,d\theta,$$

Putting $x = a\sin\theta$

$$= 4a^2\left[\frac{1}{3.1}\right] = \frac{4}{3}a^2.$$

EXAMPLE 9.5

Find the area enclosed by the curve
$a^2x^2 = y^3(2a-y)$.

Solution. We note that

(i) The curve is symmetrical about y-axis and passes through the origin.

(ii) Tangents at the origin are given by $a^2x^2 = 0$, that is, $x = 0$. Thus, the origin is a cusp.

(iii) The curve cuts the y-axis at (0, 0) and (0, 2a).

(iv) The curve exists in the region $0 < y < 2a$ but not for $y > 2a$ and $y < 0$.

(v) Shifting the origin to (0, 2a), we see that $y = 2a$ is tangent at (0, 2a).

Hence, the shape of the curve is as shown below:

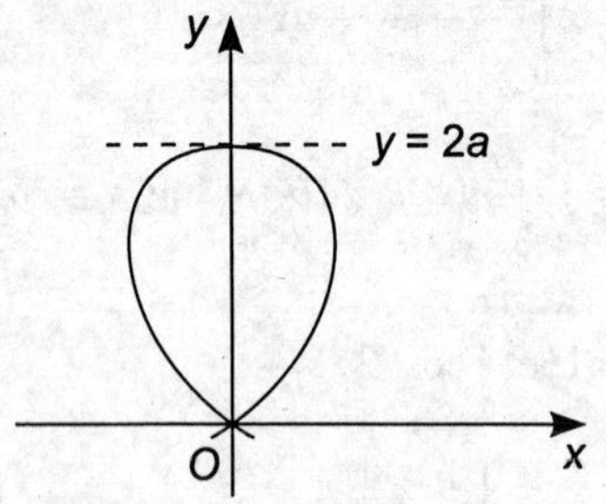

From the given equation, we have

$$x = \frac{y^{\frac{3}{2}}\sqrt{2a-y}}{a}.$$

Therefore,

Required Area = twice the area of half loop

$$= 2\int_0^{2a} x\,dy$$

$$= 2\int_0^{2a} \frac{y^{\frac{3}{2}}\sqrt{2a-y}}{a}dy$$

$$= 32a^2\int_0^{\frac{\pi}{2}} \sin^4\theta\, d\theta,$$

putting $x = 2a\sin^2\theta$

$$= 32a^2\left[\frac{3.1.1}{6.4.2}.\frac{\pi}{2}\right],$$

using reduction formula

$$= \pi a^2.$$

EXAMPLE 9.6

Find the area enclosed between the curve $x^2y^2 = a^2(y^2 - x^2)$ and its asymptotes.

Solution. To trace the curve, we note that

(i) The given curve is symmetrical about both axes and passes through the origin.

(ii) The tangent at the origin are given by $a(y^2 - a^2) = 0$. Thus $y = x$ and $y = -x$ are tangents at the origin

(iii) Equating to zero, the coefficient of the highest power of y, the asymptotes parallel to the y-axis are given by $x^2 - a^2 = 0$. Thus, asymptotes parallel to the y-axis are $x = a$ and $x = -a$. There is no real asymptote parallel to the x-axis.

Thus, the shape of the curve is as shown below:

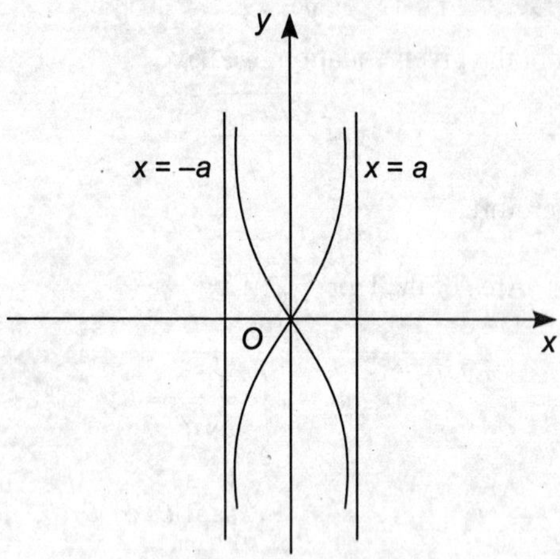

From the given equation, we have

$$y = \sqrt{\frac{a^2x^2}{a^2-x^2}}.$$

Then the required area is given by

A = 4 times the area enclosed in first quadrant

$$= 4\int_0^a \frac{ax}{\sqrt{a^2-x^2}}dx$$

$$= -2a\int_0^a \frac{-2x\,dx}{\sqrt{a^2-x^2}}$$

$$= -2a\left|\frac{(a^2-x^2)^{\frac{1}{2}}}{\frac{1}{2}}\right|_0^a = -4a(0-a) = 4a^2.$$

EXAMPLE 9.7

Find the common area between the curves $y^2 = 4ax$ and $x^2 = 4ay$.

Solution. Solving the given equations of the parabolas, we note that $(4a, 4a)$ and $(0, 0)$ are the points of their intersections.

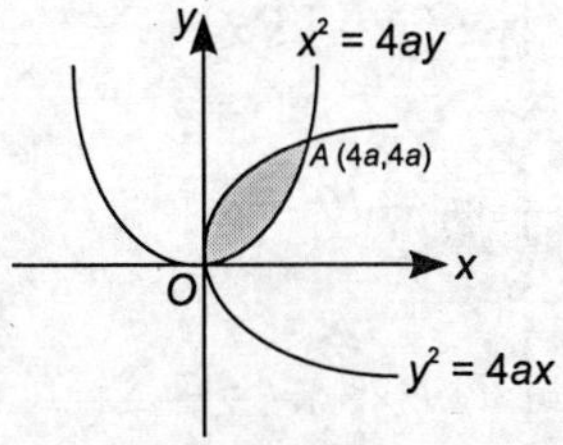

Then the required area is given by

$$A = \int_0^{4a} y_1 dx - \int_0^{4a} y_2 dx$$

$$= \int_0^{4a} \sqrt{4ax}\,dx - \int_0^{4a} \frac{x^2}{4a} dx$$

$$= 2\sqrt{a}\left|\frac{x^{\frac{3}{2}}}{\frac{3}{2}}\right|_0^{4a} - \frac{1}{4a}\left|\frac{x^3}{3}\right|_0^{4a}$$

$$= \frac{32}{3}a^2 - \frac{16}{3}a^2 = \frac{16}{3}a^2.$$

EXAMPLE 9.8

Find the area common to the curves $y^2 = ax$ and $x^2 + y^2 = 4ax$.

Solution. The equation $y^2 = ax$ represents a parabola with vertex at the origin, axis along x-axis and latus rectum a. On the other hand, $x^2 + y^2 = 4ax$ represents a circle with center $(2a, 0)$ and radius $2a$. Both curves are symmetrical about x-axis. Their points of intersection are $x = 0$ and $x = 3a$ as shown in the figure below:

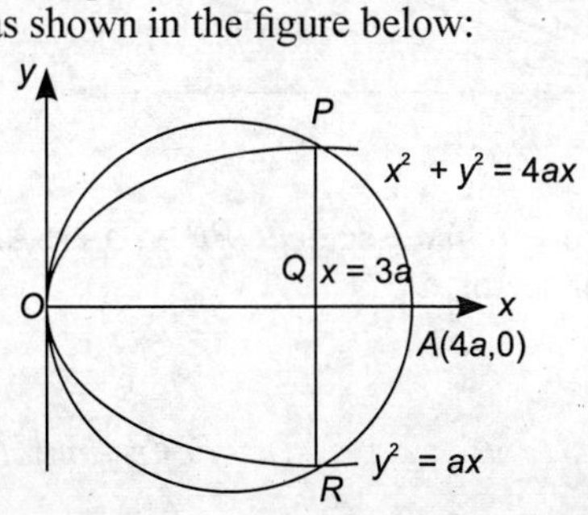

Therefore,

Common area

$$= 2[Area\,OPQ + Area\,PQA]$$

$$= 2\left[\int_0^{3a} \sqrt{ax}\,dx + \int_{3a}^{4a} \sqrt{4ax - x^2}\,dx\right]$$

$$= 2\sqrt{a}\int_0^{3a} \sqrt{x}\,dx + 2\int_{3a}^{4a} [4a^2 - (x-2a)^2]^{\frac{1}{2}}\,dx$$

$$= 2\sqrt{a}\left[\frac{2}{3}x^{\frac{3}{2}}\right]_0^{3a}$$

$$+2\left[\frac{1}{2}(x-2a)\sqrt{4a^2 - (x-2a)^2}\right.$$

$$\left.+\frac{4a^2}{2}\sin\frac{x-2a}{2a}\right]_{3a}^{4a}$$

$$= 4a\sqrt{3a^2} - \sqrt{3a^2} + \frac{4}{3}\pi a^2$$

$$= a^2\left(3\sqrt{3} + \frac{4}{3}\pi\right).$$

EXAMPLE 9.9

Find the area common to the circle $x^2 + y^2 = 4$ and the ellipse $x^2 + 4y^2 = 9$.

Solution. The point of intersection of the two curves is $x = \sqrt{\frac{7}{3}}$ (for the first quadrant). Further, $y = \frac{1}{2}\sqrt{9 - x^2}$ from the equation of ellipse and $y = \sqrt{4 - x^2}$ from the equation of the circle.

Therefore,

Area common to the curves

$$= 4 \times [Area\ O\ APB]$$

$$= 4\ [Area\ O\ M\ PB + Area M\ AP]$$

$$= 4\left[\int_0^{\sqrt{\frac{7}{3}}} \frac{1}{2}\sqrt{9-x^2}\,dx + \int_{\sqrt{\frac{7}{3}}}^{2} \sqrt{4-x^2}\,dx\right]$$

$$= 2\left[\frac{x}{2}\sqrt{9-x^2} + \frac{9}{2}\sin^{-1}\frac{x}{3}\right]_0^{\sqrt{\frac{7}{3}}}$$

$$+4\left[\frac{x}{2}\sqrt{4-x^2} + \frac{4}{2}\sin^{-1}\frac{x}{3}\right]_{\sqrt{\frac{7}{3}}}^{2}$$

$$= \sqrt{\frac{7}{3}}\cdot\sqrt{\frac{20}{3}} - 2\sqrt{\frac{5}{3}} + 9\sin^{-1}\sqrt{\frac{7}{57}} + \frac{8\pi}{2}$$

$$-8\sin^{-1}\sqrt{\frac{7}{12}}$$

$$= 4\pi + 9\sin^{-1}\sqrt{\frac{7}{27}} - 8\sin^{-1}\sqrt{\frac{7}{12}}.$$

EXAMPLE 9.10

Find the area included between the cycloid $x = a(\theta - \sin\theta)$, $y = a(1 - \cos\theta)$ and its base.

Solution. We note that $y = 0$ if $\cos\theta = 1$, that is, $\theta = 0$. When $\theta = 0$. we have $x = 0$. Thus, the curve passes through the origin. Further, y-axis is tangent at (0, 0). Also y is maximum when $\cos\theta = -1$, that is $\theta = \pi$. When $\theta = \pi$, we get $x = a\pi$ and $y = 2a$. The shape of the curve is as shown below:

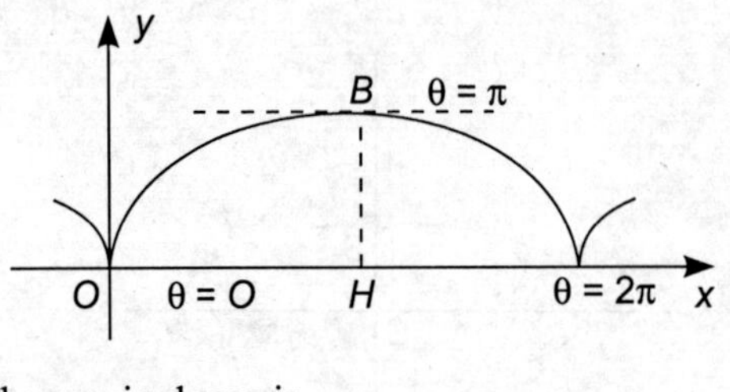

The required area is

$$= 2\int_0^{\pi} y\,dx = 2\int_0^{\pi} y\cdot\frac{dx}{d\theta}\,d\theta$$

$$= 2\int_0^{\pi} a(1-\cos\theta).a(1-\cos\theta)\,d\theta$$

$$= 2a^2\int_0^{\pi}(1-\cos\theta)^2\,d\theta$$

$$= 2a^2\int_0^{\pi}\left(2\sin^2\frac{\theta}{2}\right)^2 d\theta$$

$$= 8a^2\int_0^{\pi}\sin^4\frac{\theta}{2}\,d\theta$$

$$= 8a^2\int_0^{\frac{\pi}{2}}\sin^4 t\,(2dt),\ t = \frac{\theta}{2}$$

$$= 16a^2\int_0^{\frac{\pi}{2}}\sin^4 t\,dt$$

$$= 16a^2\left[\frac{3.1}{4.2}\cdot\frac{\pi}{2}\right]$$

$$= 3\pi a^2.$$

9.1.2 Area of a Curve Given by Polar Equation

Let AB be the curve $r = f(\theta)$ and let OA and OB be the radii vectors $\theta = \alpha$ and $\theta = \beta$ Let $P(r, \theta)$ be any point on the curve and $Q(r + \delta r, \theta + \delta\theta)$ be its neighboring point. Let the areas of the sectors AOP and AOQ be A and $A + \delta A$ respectively. Then,

$$\text{Curvilinear area } OPQO = A + \delta A - A = \delta A.$$

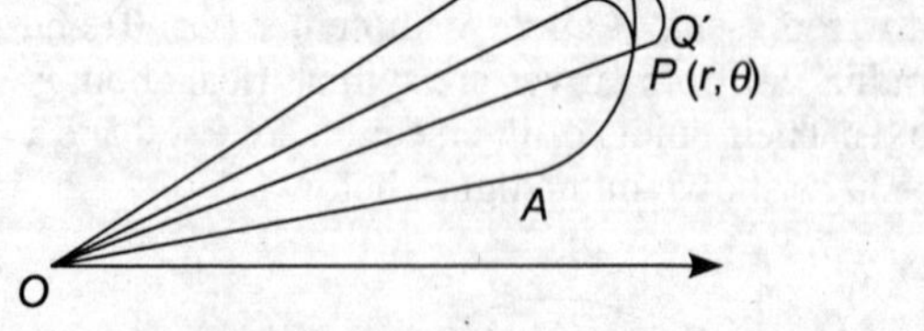

Thus,

Area of circular sector $OPP' < \delta A <$ Area of circular sector OQQ'

or

$$\frac{1}{2}r(r\delta\theta) < \delta A < \frac{1}{2}(r+\delta r)(r+\delta r)\delta\theta$$

or

$$\frac{1}{2}r^2\delta\theta < \delta A < \frac{1}{2}(r+\delta r)^2\,\delta\theta$$

or

$$\frac{1}{2}r^2 < \frac{\delta A}{\delta\theta} < \frac{1}{2}(r+\delta r)^2.$$

Taking limit as $\delta\theta \to 0$. we have $\delta r \to 0$ and so

$$\frac{dA}{d\theta} = \frac{1}{2}r^2.$$

Integrating both sides, between the limits α and β, we have

$$\int_\alpha^\beta \frac{dA}{d\theta}d\theta = \int_\alpha^\beta \frac{1}{2}r^2 d\theta$$

or

$$[A]_\alpha^\beta = \frac{1}{2}\int_\alpha^\beta r^2 d\theta.$$

EXAMPLE 9.11

Find the area of the cardioid $r = a(1+\cos\theta)$.

Solution. The given curve is symmetrical about the initial line. We have $r=0$ when $\theta=\pi$. Thus, $\theta=\pi$ is tangent at the pole. Further, r is maximum when $\cos\theta=1$, that is, $\theta=0$ and then $r=2a$. When θ increases from 0 to π, r decreases from $2a$ from 0. The shape of the curve is shown below:

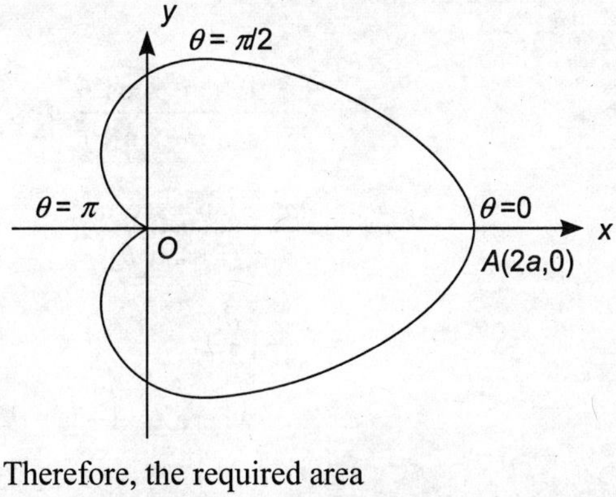

Therefore, the required area

$= 2 \times$ area of the upper half of the cardioid

$$= 2\int_0^\pi \frac{1}{2}r^2 d\theta$$

$$= \int_0^\pi a^2(1+\cos\theta)^2 d\theta$$

$$= a^2\int_0^\pi \left(2\cos^2\frac{\theta}{2}\right)^2 d\theta$$

$$= 4a^2\int_0^\pi \cos^4\frac{\theta}{2}d\theta$$

$$= 8a^2\int_0^{\frac{\pi}{2}} \cos^4\phi\, d\phi,\ \phi=\frac{\theta}{2}$$

$$= 8a^2\left[\frac{3.1}{4.2}.\frac{\pi}{2}\right] = \frac{3}{2}\pi a^2.$$

EXAMPLE 9.12

Trace the curve $r = a(1-\cos\theta)$ and find its area.

Solution. The shape of the curve is

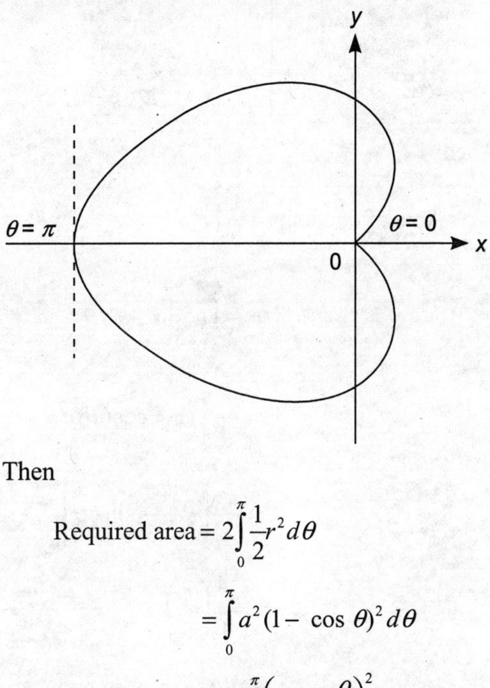

Then

$$\text{Required area} = 2\int_0^\pi \frac{1}{2}r^2 d\theta$$

$$= \int_0^\pi a^2(1-\cos\theta)^2 d\theta$$

$$= a^2\int_0^\pi \left(2\sin^2\frac{\theta}{2}\right)^2 d\theta$$

$$= 4a^2\int_0^{\frac{\pi}{2}} \sin^4\phi\, d\phi,\ \phi=\frac{\theta}{2}$$

$$= 8a^2\left[\frac{3.1}{4.2}\cdot\frac{\pi}{2}\right]$$

$$= \frac{3}{2}\pi a^2.$$

EXAMPLE 9.13

Find the area of one loop of the curve $r = a\sin 3\theta$.

Solution. The shape of the curve is

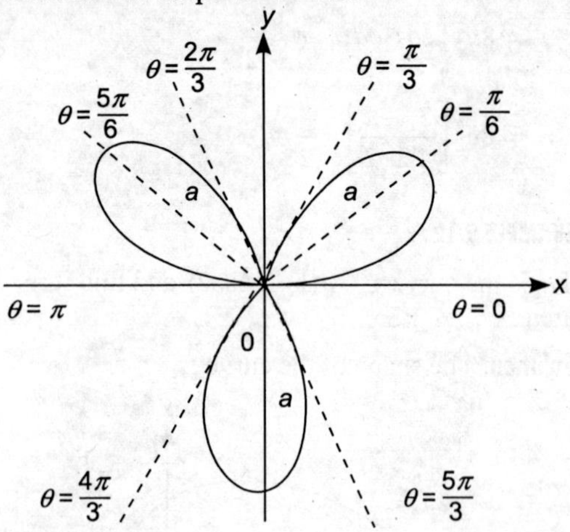

Then

$$\text{Area of one loop} = \frac{1}{2}\int_0^{\frac{\pi}{3}} r^2 d\theta$$

$$= \frac{1}{2}\int_0^{\frac{\pi}{3}} a^2\sin^2 3\theta\, d\theta$$

$$= \frac{a^2}{4}\int_0^{\frac{\pi}{3}} (1 - \cos 6\theta)\, d\theta$$

$$= \frac{a^2}{4}\left[\theta - \frac{1}{6}\sin 6\theta\right]_0^{\frac{\pi}{3}}$$

$$= \frac{a^2}{4}\left(\frac{\pi}{3}\right)$$

$$= \frac{\pi a^2}{12}.$$

Also,

Total area enclosed by the curve

$$= 3 \times (\text{Area of one loop})$$

$$= 3\left(\frac{\pi a^2}{12}\right)$$

$$= \frac{\pi a^2}{4}.$$

EXAMPLE 9.14

Find the area enclosed by the loop of the curve $x^3 + y^3 = 3axy$ (Folium of Descartes).

Solution. The shape of the curve is

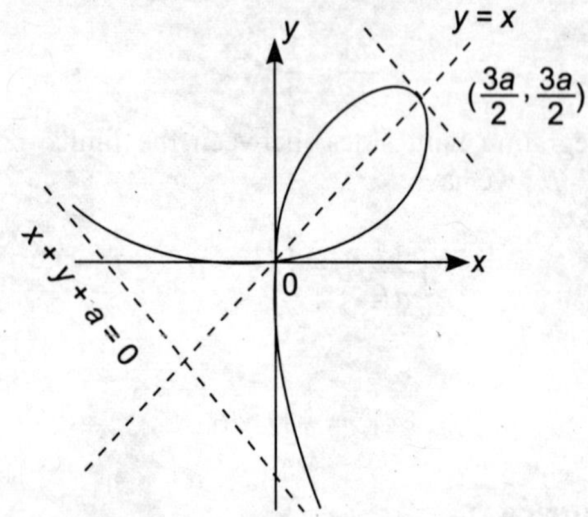

Putting $x = \cos\theta$, $y = r\sin\theta$. we see that

$$r = \frac{3a\sin\theta\cos\theta}{\cos^3\theta + \sin^3\theta}.$$

We note that $r = 0$, when $3a\sin\theta\cos\theta = 0$, that is, $\theta = 0, \frac{\pi}{2}$. Therefore,

$$\text{Area of the loop} = \frac{1}{2}\int_0^{\frac{\pi}{2}} r^2 d\theta$$

$$= \frac{1}{2}\int_0^{\frac{\pi}{2}} \frac{9a^2\sin^2\theta\cos^2\theta}{(\cos^3\theta + \sin^3\theta)^2} d\theta$$

$$= \frac{9a^2}{2}\int_0^{\frac{\pi}{2}} \frac{\tan^2\theta\sec^2\theta}{(1 + \tan^3\theta)^2} d\theta$$

$$= \frac{3a^2}{2}\int_1^{\infty} \frac{dt}{t^2},$$

where $t = 1 + \tan^3\theta$

$$= \frac{3a^2}{2}\left[\frac{-1}{t}\right]_1^{\infty}$$

$$= \frac{3}{2}a^2.$$

EXAMPLE 9.15

Find the area outside the circle $r = 2a\cos\theta$ and inside the cardioid $r = a(1+\cos\theta)$.

Solution. From the given equations, we have

$$2a\cos\theta = a(1+\cos\theta).$$

which yields $\cos\theta = 1$, that is, $\theta = 0$. Since $2a\cos\theta \le a(1+\cos\theta)$, therefore the circle lies entirely within the cardioid. Thus,

Required area = Area of the cardioide – Area of the circle

$$= 2\int_0^{\pi} \frac{1}{2} r^2 d\theta - \pi a^2.$$

$$= \frac{3}{2}\pi a^2 - \pi a^2$$

$$= \frac{1}{2}\pi a^2.$$

9.2 RECTIFICATION

The process of finding the length of an arc of a curve between two given points is called *rectification.*

9.2.1 Length of a Curve

(i) Cartesian form: Let s denote the length of the arc of a curve $y = f(x)$ between any two points. Then s is a function of x and from a result from Differential calculus, we have

$$\frac{ds}{dx} = \sqrt{1+\left(\frac{dy}{dx}\right)^2}\, dx.$$

Integrating between the limits a and b, we have

$$\int_a^b \frac{ds}{dx} dx = \int_a^b \sqrt{1+\left(\frac{dy}{dx}\right)^2}\, dx$$

or

$$[s]_a^b = \int_a^b \sqrt{1+\left(\frac{dy}{dx}\right)^2}\, dx \qquad (1)$$

Hence, the length s of the arc between $x = a$ and $x = b$ is given by the value of the integral on the right hand side of the formula (1).

Similarly, the length of the arc of the curve $x = f(y)$ between $y = a$ and $y = b$ is equal to

$$[s]_a^b = \int_a^b \sqrt{1+\left(\frac{dx}{dy}\right)^2}\, dy. \qquad (2)$$

(ii) Parametric form: If the equation of the curve is given in the parametric form $x = f(t)$, $y = \phi(t)$. Then the arc length s is a function of t and

$$\frac{ds}{st} = \sqrt{\left(\frac{dx}{dt}\right)^2 + \left(\frac{dy}{dt}\right)^2}$$

or

$$ds = \sqrt{\left(\frac{dx}{dt}\right)^2 + \left(\frac{dy}{dt}\right)^2}\, dt.$$

Integrating between the limits t_1 and t_2 m we have

$$s = \int_{t_1}^{t_2} \sqrt{\left(\frac{dx}{dt}\right)^2 + \left(\frac{dy}{dt}\right)^2}\, dt. \qquad (3)$$

(iii) Polar form: If the equation of the curve is given in the polar form $r = f(\theta)$ and we measure the arc length s in the direction of θ increasing, then

$$\frac{ds}{d\theta} = \sqrt{r^2 + \left(\frac{ds}{d\theta}\right)^2}$$

or

$$ds = \sqrt{r^2 + \left(\frac{dr}{d\theta}\right)^2}\, d\theta.$$

Integrating between the limits θ_1 and θ_2, we have

$$s = \int_{\theta_1}^{\theta_2} \sqrt{r^2 + \left(\frac{dr}{d\theta}\right)^2}\, d\theta. \qquad (4)$$

In case the curve is given in the form $\theta = f(r)$. then

$$s = \int_{r_1}^{r_2} \sqrt{1 + \left(r\frac{d\theta}{dr}\right)^2}\, dr. \qquad (5)$$

(iv) Pedal form: If $p = f(r)$ is equation of the curve and r_1 and r_2 be the values of r at two given points of the curve, then

$$\frac{ds}{dr} = \frac{r}{\sqrt{r^2 - p^2}}$$

or

$$ds = \frac{r}{\sqrt{r^2 - p^2}} sr\cdot$$

Integrating between the limits r_1 and r_2, we get

$$s = \int_{r_1}^{r_2} \frac{r}{\sqrt{r^2 - p^2}} dr\cdot$$

EXAMPLE 9.16

Find the length of arc of the parabola $y^2 = 4ax$.

(i) from the vertex to an extremity of the latus rectum

(ii) cut off by the latus rectum.

Solution. Let O be the vertex and L be an extremity of the latus rectum (line through the focus and perpendicular to the x-axis). The co-ordinates of the extremities of latus rectum of the parabola $y^2 = 4ax$ are $(a,2a)$ and $(a,-2a)$.

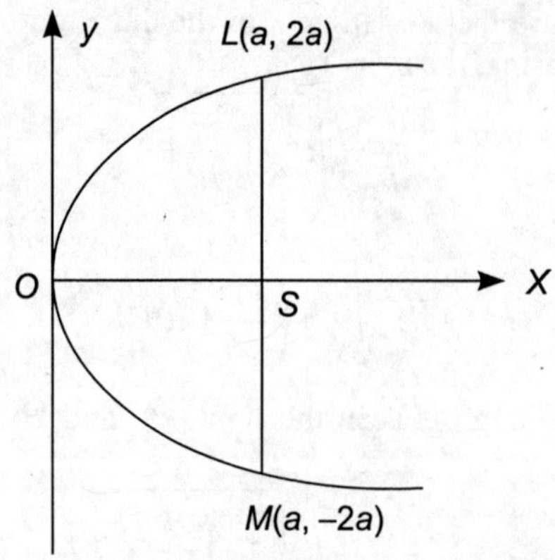

Then, differentiating the given equation, we have $2y\frac{dy}{dx} = 4a$, that is, $\frac{dx}{dy} = \frac{y}{2a}$. Therefore,

(i) $$\text{Length } OL = \int_0^{2a} \sqrt{1+\left(\frac{dx}{dy}\right)^2}\, dy$$

$$= \int_0^{2a} \sqrt{1+\frac{y^2}{4a^2}}\, dy$$

$$= \frac{1}{2a}\int_0^{2a} \sqrt{y^2+4a^2}\, dy$$

$$= \frac{1}{2a}\left[\frac{1}{2}y\sqrt{y^2+4a^2} + 2a^2\sin^{-1}\frac{y}{2a}\right]_0^{2a}$$

$$= \frac{1}{2a}\left[a\sqrt{8a^2} - 2a^2 \sinh^{-1} 1 - 0\right]$$

$$= a\left[\sqrt{2} + \sin h^{-1} 1\right]$$

$$= a\left[\sqrt{2} + \log(1+\sqrt{1+1})\right]$$

$$= a\left[\sqrt{2} + \log(1+\sqrt{2})\right].$$

(ii) Length of the arc cut off by the latus rectum is equal to

$$2 \times arc\, OL = 2a[\sqrt{2} + \log(1+\sqrt{2}).$$

EXAMPLE 9.17

Find the perimeter of the circle $x^1 + y^1 = a^1$.

Solution. The center of the given circle is (0,0) and its radius is a. It is symmetrical about both axes. Therefore,

Perimeter = 4 × Arc in the first quadrant from the point (0, 0) to the point $(a, 0)$.

Differentiating the given equation, we get

$$2x + 2y\frac{dy}{dx} = 0$$

or

$$\frac{dy}{dx} = -\frac{x}{y} = -\frac{x}{\sqrt{a^2 - x^2}}.$$

Hence,

$$\text{Perimeter} = 4\int_0^a \sqrt{1+\left(\frac{dy}{dx}\right)^2}\, dx$$

$$= 4\int_0^a \sqrt{1+\frac{x^2}{a^2-x^2}}\, dx$$

$$= 4a\int_0^a \frac{dx}{\sqrt{a^2-x^2}}$$

$$= 4a\left[\sin^{-1}\left(\frac{x}{a}\right)\right]_0^a$$

$$= 4a[\sin^{-1}1 - \sin^{-1}0]$$

$$= 4a\left[\frac{\pi}{2} - 0\right] = 2\pi a.$$

EXAMPLE 9.18

Find the length of the loop of the curve $3ay^2 = x(x-a)^2$.

Solution. The given curve is symmetrical about the x-axis. It cuts the x-axis at (0, 0) and $(a, 0)$. Thus, the loop varies from $x = 0$ to $x = a$. The given equation is $3ay^2 = x(x-a)^2$. Taking logarithm, we get

$$\log 3a + 2\log y = \log x + 2\log(x-a).$$

Differentiating with respect to x, we get

$$\frac{2}{y}\frac{dy}{dx} = \frac{1}{x} + \frac{2}{x-a} = \frac{3x-a}{x(x-a)}$$

or

$$\frac{dy}{dx} = \frac{3x-a}{x(x-a)} \cdot \frac{y}{2}.$$

Thus,

$$\left(\frac{dy}{dx}\right)^2 = \frac{(3x-a)^2}{x^2(x-a)^2} \cdot \frac{y^2}{4}$$

$$= \frac{(3x-a)^2}{x^2(x-a)^2} \cdot \frac{x(x-a)^2}{12a}$$

$$= \frac{(3x-a)^2}{12ax}.$$

Therefore,

Total length of the loop

$= 2 \times$ length of the loop in first quadrant

$$= 2\int_0^a \sqrt{1 + \left(\frac{dy}{dx}\right)^2}\, dx$$

$$= 2\int_0^a \frac{3x+a}{\sqrt{12ax}}\, dx$$

$$= \frac{1}{\sqrt{3a}} \int_0^a (3\sqrt{x} + ax^{\frac{1}{2}})\, dx$$

$$= \frac{1}{\sqrt{3a}}\left[3\frac{x^{\frac{3}{2}}}{\frac{3}{2}} + ax^{\frac{1}{2}}\right]_0^a$$

$$= \frac{1}{\sqrt{3a}}\left[4a^{\frac{3}{2}}\right] = \frac{4a}{\sqrt{3}}.$$

EXAMPLE 9.19

Show that the length of the asteroid $x^{\frac{2}{3}} + y^{\frac{2}{3}} = a^{\frac{2}{3}}$ is $6a$.

Solution. The shape of the curve is

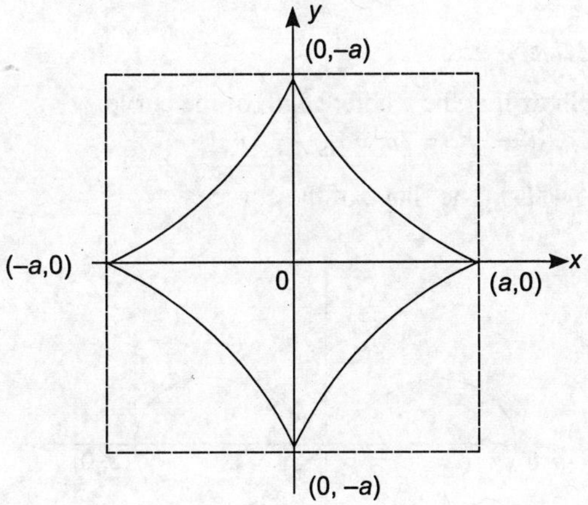

Differentiating the given equation with respect to x, we get

$$\frac{2}{3}x^{-\frac{1}{3}} + \frac{2}{3}y^{\frac{1}{3}}\frac{dy}{dx} = 0$$

or $$\frac{dy}{dx} = -\left(\frac{y}{x}\right)^{\frac{1}{3}}.$$

Then,

Required length

$= 4 \times$ length of curve in first quadrant.

$$= 4\int_0^a \sqrt{1 + \left(\frac{dy}{dx}\right)^2}\, dx$$

$$= 4\int_0^a \left(1 + \frac{y^{\frac{2}{3}}}{x^{\frac{2}{3}}}\right)^{\frac{1}{2}} dx$$

$$= 4\int_0^a \frac{\left(x^{\frac{2}{3}} + y^{\frac{2}{3}}\right)}{x^{\frac{1}{3}}} dx$$

$$= 4\int_0^a \frac{\left[(a)^{\frac{2}{3}}\right]^{\frac{1}{2}}}{x^{\frac{1}{3}}} dx, \text{ using the given equation}$$

$$= 4a^{\frac{1}{3}}\int_0^a x^{-\frac{1}{3}}\,dx$$

$$= 4a^{\frac{1}{3}}\left[\frac{3}{2}x^{\frac{2}{3}}\right]_0^a$$

$$= 4a^{\frac{1}{3}}\left[\frac{3}{2}a^{\frac{2}{3}}\right] = 6a\cdot$$

EXAMPLE 9.20

Show that the whole length of the curve $x^2(a^2 - x^2) = 8a^2y^2$ is $\pi a\sqrt{2}$.

Solution. The shape of the curve is

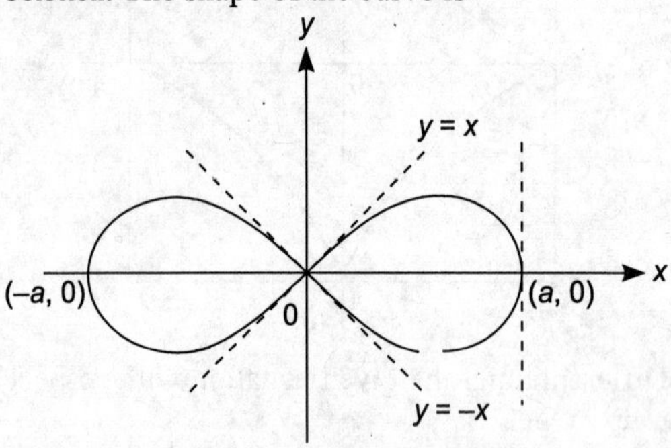

The whole curve lies between the lines $x = -a$ and $x = a$. The given equation yields

$$y = \frac{1}{\sqrt{8a}}x\sqrt{a^2 - x^2}\cdot$$

Therefore,

$$\frac{dy}{dx} = \frac{1}{\sqrt{8a}}\left[\frac{a^2 - 2x^2}{\sqrt{a^2 - x^2}}\right]$$

Then,

Whole length of the curve

= 4 × length in the first quadrant

$$= 4\int_0^a \sqrt{1 + \left(\frac{dy}{dx}\right)^2}\,dx$$

$$= 4\int_0^a \sqrt{1 + \frac{(a^2 - 2x^2)^2}{8a^2(a^2 - x^2)}}\,dx$$

$$= \frac{42}{\sqrt{2}a}\int_0^a \frac{3a^2 - 2x^2}{\sqrt{a^2 - x^2}}\,dx$$

$$= \sqrt{2}a\int_0^a \frac{2(a^2 - x^2) + a^2}{\sqrt{a^2 - x^2}}\,dx$$

$$= \frac{\sqrt{2}}{a}\int_0^a \left(2\sqrt{a^2 - x^2} + \frac{a^2}{\sqrt{a^2 - x^2}}\right)dx$$

$$= \frac{\sqrt{2}}{a}\left[2\left(xa^2 - x^2 2 + \frac{a^2}{2}\sin^{-1}\frac{x}{a}\right) + a^2\sin^{-1}\frac{x}{a}\right]_0^a$$

$$= \frac{\sqrt{2}}{a}\left[x(a^2 - x^2) + 2a^2\sin^{-1}\frac{x}{a}\right]_0^a$$

$$= \frac{\sqrt{2}}{a}[2a^2\sin^{-1}1]$$

$$= \frac{\sqrt{2}}{a}\left[2a^2.\frac{\pi}{2}\right]$$

$$= \pi a\sqrt{2}.$$

EXAMPLE 9.21

Rectify the cycloid $x = a(t + \sin t)$, $y = a(1 + \cos t)\cdot$

Solution. The shape of the curve is

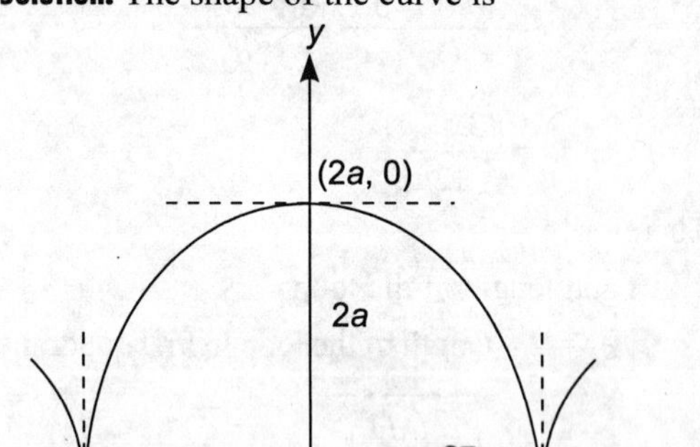

Further,

$$\frac{dx}{dt} = a(1 + \cos t)\,and\,\frac{dy}{dt} = -a\sin t\cdot$$

Therefore,

$$\left(\frac{dx}{dt}\right)^2 + \left(\frac{dy}{dt}\right)^2 = a^2(1 + \cos t)^2 + a^2\sin^2 t$$

$$= 2a^2(1+\cos t)$$

$$= 4a^2\cos^2\frac{t}{2}.$$

Therefore,

$$\text{Length of cycloid} = 2\int_0^{\pi}\sqrt{\left(\frac{dx}{dt}\right)^2+\left(\frac{dy}{dt}\right)^2}\,dt$$

$$= 2\int_0^{\pi}\sqrt{4a^2\cos^2\frac{t}{2}}\,dt$$

$$= 4a\int_0^{\pi}\cos\frac{t}{2}\,dt$$

$$= 4a\left[\frac{\sin\frac{t}{2}}{\frac{1}{2}}\right]_0^{\pi}$$

$$= 8a\left[\sin\frac{\pi}{2}\right] = 8a.$$

EXAMPLE 9.22

Find the whole length of the cardioid

$$r = a(1+\cos\theta)$$

Solution. For the shape of the curve, please see Example 9.11. Also,

$$\frac{dr}{d\theta} = -a\sin\theta.$$

Therefore,

Length of whole cardioid

$$= 2\times\text{length of the upper half}$$

$$= 2\int_0^{\pi}\sqrt{r^2+\left(\frac{dr}{d\theta}\right)^2}\,d\theta$$

$$= 2\int_0^{\pi}\sqrt{a^2(1+\cos\theta)^2+a^2\sin^2\theta}\,d\theta$$

$$= 2a\int_0^{\pi}\sqrt{2(1+\cos\theta)}\,d\theta$$

$$= 2a\int_0^{\pi}2\cos\frac{\theta}{2}\,d\theta$$

$$= 4a\left[\frac{\sin\frac{\theta}{2}}{\frac{1}{2}}\right]_0^{\pi} = 8a\sin\frac{\pi}{2} = 8a.$$

Note: If we find the length of the upper half from 0 to $\frac{\pi}{3}$. then

$$\text{Length} = \int_0^{\frac{\pi}{3}}2a\cos\frac{\theta}{2}\,d\theta = 2a\left[\frac{\sin\frac{\theta}{2}}{\frac{1}{2}}\right]_0^{\frac{\pi}{3}}$$

$$= 4a\left[\sin\frac{\pi}{6}\right] = \frac{4a}{2}$$

$$= 2a = \text{half the length of upper point.}$$

Thus, $\theta = \frac{\pi}{3}$ bisects the arc of upper half of the cardioid.

EXAMPLE 9.23

Find the perimeter of the cardioid $r = a(1-\cos\theta)$ and show that the arc of the upper half is bisected by $\theta = \frac{2\pi}{3}$.

Solution. For the shape of the curve, please see Example 9.12. Due to symmetry, the perimeter of the given cardioid is twice the length of the upper half of the cardioid. We have

$$\frac{dr}{d\theta} = a\sin\theta$$

and so

$$r^2+\left(\frac{dr}{d\theta}\right)^2 = a^2(1-\cos\theta)^2+a^2\sin^2\theta$$

$$= 4a^2\sin^2\frac{\theta}{2}.$$

Therefore,

$$\text{Required perimeter} = 2\int_0^{\pi}\sqrt{4a^2\sin^2\frac{\theta}{2}}\,d\theta$$

$$= 4a\int_0^{\pi}\sin\frac{\theta}{2}\,d\theta$$

$$= 4a\left[\frac{-\cos\frac{\theta}{2}}{\frac{1}{2}}\right]_0^{\pi}$$

$$= 8a.$$

Thus,

$$\text{Length of the upper half} = 4a.$$

On the other hand, the arc length of the cardioid from the point $\theta = 0$ to the point $\theta = \frac{2\pi}{3}$ is equal to

$$2a\int_0^{\frac{2\pi}{3}} \sin\frac{\theta}{2}\,d\theta = 2a\left[\frac{-\cos\frac{\theta}{2}}{\frac{1}{2}}\right]_0^{\frac{2\pi}{3}}$$

$$= -4a\left[\cos\frac{\pi}{3} - \cos\theta\right]$$

$$= -4a\left(\frac{1}{2}-1\right)$$

$$= 2a.$$

Hence, $\theta = \frac{2\pi}{3}$ bisects the arc length of the upper half of the cardioid.

EXAMPLE 9.24

Find the length of the arc of the equiangular spiral $r = ae^{\theta\cot\alpha}$ between the points for which radii vectors are r_1 and r_2.

Solution. Differentiating the given equation, we have

$$\frac{dr}{d\theta} = ae^{\theta\cot\alpha}.\cot\alpha = r\cot\alpha$$

or

$$\frac{d\theta}{dr} = \frac{1}{r\cot\alpha} \text{ or } r\frac{d\theta}{dr} = \tan\alpha.$$

Therefore,

$$\text{Required length} = \int_{r_1}^{r_2}\sqrt{1+\left(r\frac{d\theta}{dr}\right)^2}\,dr$$

$$= \int_{r_1}^{r_2}\sqrt{1+\tan^2\alpha}\,dr$$

$$= \sec\alpha\int_{r_1}^{r_2} dr = \sec\alpha[r]_{r_1}^{r_2}$$

$$= \sec\alpha[r_2 - r_1].$$

EXERCISES

Quadrature

1. Find the area of a loop of the curve $x^4 + y^4 = 4a^2xy$.

   Ans. $\frac{\pi a^2}{2}$.

2. Show that the area of a loop of the curve $a^4y^2 = x^4(a^2 - x^2)$ is $\frac{\pi a^2}{8}$.

3. Find the area bounded by the curve $xy^2 = 4a^2(2a - x)$ and its asymptote.

   Ans. $4\pi a^2$.

4. Find the area included between $y^2 = 4ax$ and $y = mx$.

   Ans. $\frac{8a^2}{3m^3}$.

5. Show that the larger of the two areas into which the circle $x^2 + y^2 = 64a^2$ is divided by the parabola $y^2 = 12ax$ is $\frac{16}{3}a^2(8\pi - \sqrt{3})$.

6. Show that the area bounded by the Cissoid $x = a\sin^2 t$, $y = \frac{a\sin^3 t}{\cos t}$ and its asymptote is $\frac{3\pi a^2}{4}$.

7. Find the area enclosed by the spiral $r\theta^{\frac{1}{2}} = a$ between $\theta = \alpha$ to $\theta = \beta$.

   Ans. $\frac{1}{2}a^2\log\left(\frac{\beta}{\alpha}\right)$.

8. Find the area of the loop of the curve $r = a\theta\cos\theta$ between $\theta = 0$ and $\theta = \frac{\pi}{2}$.

   Ans. $\frac{\pi a^2}{96}(\pi^2 - 6)$.

9. Find the whole area of the curve $r = \sin 2\theta$.

   Ans. $\frac{1}{2}\pi a^2$.

10. Find the area of the curve $r^2 = a^2\cos 2\theta$.

    Ans. a^2.

11. Find the area common to the circles $r = a\sqrt{2}$ and $r = 2a\cos\theta$.

    Ans. $a^2(\pi - 1)$.

12. Find the area of the portion included between the cardioids $r = a(1+\cos\theta)$ and $r = a(1-\cos\theta)$.

Ans. $2a^2\left(\frac{3\pi}{4}-2\right)$.

Rectification

13. Find the perimeter of the loop of the curve $3ay^2 = x^2(a-x)$.

Ans. $\frac{4a}{\sqrt{3}}$.

14. Show that the length of the arc of the parabola $y^2 = 4ax$ intercepted between the points of intersection of the parabola and the straight line $3y = 8x$ is $a\left(\log 2+\frac{15}{16}\right)$:

15. Find the whole length of the curve $x = a\cos^3 t,\ y = b\sin^3 t$.

Ans. $4\frac{b^2+ab+a^2}{b+a}$.

16. Find the length of the arc of the curve $x = e^t \sin t$, $y = e^t \cos t$ between $t = 0$ and $t = \frac{\pi}{2}$.

Ans. $\sqrt{2}\left[e^{\frac{\pi}{2}}-1\right]$:

17. Find the length of the cardioid $r = a(1-\cos\theta)$ lying outside the circle $r = a\cos\theta$.

Ans. $4a\sqrt{3}$.

18. Show that the perimeter of the circle $r = a\cos\theta$ is approximately $2\pi a\left(1+\frac{b^2}{4a^2}\right)$.

19. Find the length of a loop of the curve $r^3 = \sin 3\theta$.

Ans. $2\int_0^1 \frac{dr}{\sqrt{1-r^6}}$.

10 Centre of Gravity and Moment of Inertia

10.1 CENTRE OF GRAVITY

A point fixed in the body through which the weight of the body passes in whatever position the body may be placed is called the *centre of gravity* of that body.

(1) Centre of Gravity of a System of Particles: Consider a system of particles of masses m_1, m_2, ..., m_n situated at the points (x_1, y_1), (x_2, y_2), ..., $(x_n y_n)$ respectively. Then the centre of gravity (C.G) of this system is the point $(\bar{x}, \bar{y})$, where

$$\bar{x} = \frac{\sum_{i=1}^{n} m_i x_i}{\sum_{i=1}^{n} m_i}, \quad \bar{y} = \frac{\sum_{i=1}^{n} m_i y_i}{\sum_{i=1}^{n} m_i} \qquad (1)$$

Due to relation (1), the centre of gravity is also known as *centre of mass* or *Centre of Inertia*

(2) Centre of Gravity of a Curve: Let δs be a small portion of the curve between the points (x, y) and $(x+\delta x, y+\delta y)$. Let ρ be the mass per unit length of the matter of the curve $y = f(x)$. Then mass of the small portion is $\delta m = \rho \delta s$. Therefore

$$\bar{x} = \frac{\int x.\rho ds}{\int \rho ds}, \quad \bar{y} = \frac{\int y.\rho ds}{\int \rho ds} \qquad (2)$$

where

$$ds = \sqrt{1+\left(\frac{dy}{dx}\right)^2}$$ in case of Cartesian co-ordinates.

$$ds = \sqrt{\left(\frac{dx}{dt}\right)^2 + \left(\frac{dy}{dt}\right)^2}$$ in case of parametric co-ordinates.

and

$$ds = \sqrt{r^2 + \left(\frac{dr}{d\theta}\right)^2}$$ in case of polar co-ordinates.

When ρ is constant, then (2) reduces to

$$\bar{x} = \frac{\int xdA}{\int dA}, \quad \bar{y} = \frac{\int ydA}{\int dA}.$$

(3) Centre of Gravity of an Area: If δA is an elementary area an ρ is the mass per unit area, then mass of the elementary area is $\delta m = \rho \delta A$. Therefore, the centre of gravity is given by

$$\bar{x} = \frac{\int x\rho dA}{\int \rho dA} \text{ and } \bar{y} = \frac{\int y\rho dA}{\int \rho dA}.$$

For uniform area, ρ is constant and so in that case,

$$\bar{x} = \frac{\int xdA}{\int dA}, \quad \bar{y} = \frac{\int ydA}{\int dA}.$$

Special Cases

(i) When the area is bounded by the curve $y = f(x)$, x-axis and the two ordinates $x = a$ and $x = b$, then the area of the elementary strip of width δx is $\delta A = y\delta x$. The centre of gravity of the elementary strip is $\left(x, \frac{y}{2}\right)$. Therefore centre of gravity of the area is given by

$$\bar{x} = \frac{\int_a^b xydx}{\int_a^b y\,dx} \text{ and } \bar{y} = \frac{\int_a^b \frac{y}{2}}{\int_a^b y\,dx}.$$

(ii) When the area is bounded by the curve $x = f(y)$, y-axis, and the ordinates $y = a$ and $y = b$, then the area of the elementary strip of width δy is $\delta A = x\delta y$. Therefore centre of gravity of the area is given by

$$\overline{x} = \frac{\int_a^b \frac{x}{2} xdy}{\int_a^b xdy} = \frac{\frac{1}{2}\int_a^b x^2 dy}{\int_a^b xdy}$$

$$\overline{y} = \frac{\int yxdy}{\int xdy}.$$

(iii) When the area is sectorial area bounded by a curve $r = f(\theta)$ and two radii vectors $\theta = \alpha$ and $\theta = \beta$. Then the area of the elementary sartorial area is $\delta A = \frac{1}{2}r^2\delta\theta$. The centre of gravity of this elementary sectorial area is $(\frac{2}{3}r\cos\theta,\ \frac{2}{3}r\sin\theta)$. Therefore centre of gravity of the sartorial area is given by

$$\overline{x} = \frac{2}{3}\frac{\int_\alpha^\beta r^3\cos\theta\, d\theta}{\int_\alpha^\beta r^2 d\theta},$$

$$\overline{y} = \frac{2}{3}\frac{\int_\alpha^\beta r^3\sin\theta\, d\theta}{\int_\alpha^\beta r^2 d\theta}.$$

(4) Centre of Gravity of Volume of Revolution: Let δV be the element of volume of body with uniform density ρ. Then mass of this element is $\delta m = \rho dV$. Therefore the centre of gravity $(\overline{x}, \overline{y})$ is given by

$$\overline{x} = \frac{\int x\rho dV}{\int \rho dV} = \frac{\int x dV}{\int dV}$$

$$\overline{y} = \frac{\int y\rho dV}{\int \rho dV} = \frac{\int ydV}{\int dV}.$$

In case of the volume of revolution about x-axis, the volume of the circular disc of thickness δx and radii y is $\delta V = \pi y^2\delta x$. The centre of gravity lies on the x-axis. Therefore the centre of gravity $(\overline{x}, \overline{y})$

$$\overline{x} = \frac{\int x\cdot\pi y^2 dx}{\int \pi y^2 dx}, \overline{y} = 0.$$

In case of the volume of revolution about y-axis, the centre of gravity is

$$\overline{x} = 0, \overline{y} = \frac{\int y\pi x^2 dy}{\int \pi x^2 dy}$$

(5) Centre of Gravity of a Surface of Revolution: For surface of revolution about the x-axis, the centre of gravity is given by

$$\overline{x} = \frac{\int x\cdot 2\pi yds}{\int 2\pi yds} = \frac{\int xyds}{\int yds},$$

$$\overline{y} = 0.$$

For surface of revolution about y-axis, we have

$$\overline{x} = 0 \text{ and } \overline{y} = \frac{\int xyds}{\int xds}.$$

EXAMPLE 10.1

Find the centre of gravity of the arc of the astroid $x^{\frac{2}{3}} + y^{\frac{2}{3}} = a^{\frac{2}{3}}$ in the first quadrant.

Solution. The parametric equations of the curve are $x = a\cos^3\theta, y = a\sin^3\theta$.

Therefore

$$\frac{dx}{d\theta} = -3a\cos^2\theta\sin\theta,$$

$$\frac{dy}{d\theta} = 3a\sin^2\theta\cos\theta,$$

$$\frac{ds}{s\theta} = \sqrt{\left(\frac{dx}{d\theta}\right)^2 + \left(\frac{dy}{d\theta}\right)^2}.$$

Therefore the centre of gravity is

$$\overline{x} = \frac{\int_0^{\frac{\pi}{2}} xds}{\int_0^{\frac{\pi}{2}} ds} = \frac{\int_0^{\frac{\pi}{2}} x\frac{ds}{d\theta}d\theta}{\int_0^{\frac{\pi}{2}} \frac{ds}{d\theta}\cdot d\theta}$$

$$= a\frac{\int_0^{\frac{\pi}{2}} \cos^4\theta \sin\theta\, d\theta}{\int_0^{\frac{\pi}{2}} \cos\theta \sin\theta\, d\theta} = \frac{2a}{5},$$

and similarly,

$$\overline{y} = \frac{2a}{5},$$

EXAMPLE 10.2

Find the centre of gravity of a uniform lamina bounded by the co-ordinate axes and the arc of the ellipse $\frac{x^2}{a^2}+\frac{y^2}{b^2}=1$ in the first quadrant.

Solution. From the equation of the ellipse, we have

$$y = \frac{b}{a}\sqrt{a^2-x^2}.$$

Therefore the centre of gravity of the ellipse in the first quadrant is given by

$$\overline{x} = \frac{\int_0^a xyds}{\int_0^a ydx} = \frac{\int_0^a x\frac{b}{a}\sqrt{a^2-x^2}dx}{\int_0^{\frac{\pi}{2}} \frac{b}{a}\cdot\sqrt{a^2-x^2}dx} = \frac{4a}{3\pi}.$$

$$\overline{y} = \frac{\frac{1}{2}\int_0^a y.ydx}{\int_0^a ydx} = \frac{\frac{b^2}{2a^2}\int_0^a \frac{b}{a}\sqrt{a^2-x^2}dx}{\frac{b}{a}\int_0^a \frac{b}{a}\cdot\sqrt{a^2-x^2}dx} = \frac{4b}{3\pi}.$$

Hence

$$(\overline{x},\overline{y}) = \left(\frac{4a}{3\pi},\frac{4b}{3\pi}\right).$$

EXAMPLE 10.3

Find the centre of gravity of the arc of the cardioid $r = a(1+\cos\theta)$.

Solution. We have

$$r = a + a\cos\theta,$$

$$\frac{dr}{d\theta} = -a\sin\theta,$$

$$\frac{ds}{d\theta} = \sqrt{r^2+\left(\frac{dr}{d\theta}\right)^2}$$

$$= \sqrt{a^2(1+\cos\theta)^2 + a^2\sin^2\theta}$$

$$= 2a\cos\frac{\theta}{2}.$$

Therefore

$$\overline{x} = \frac{\int xds}{\int ds} = \frac{\int_0^{\pi} r\cos\theta(2a\cos\frac{\theta}{2})d\theta}{\int_0^{\pi} 2a\cos\frac{\theta}{2}d\theta}$$

$$= \frac{a\int_0^{\pi}(1+\cos\theta)\cos\theta\cos\frac{\theta}{2}d\theta}{2}$$

$$= \frac{a}{2}\left[\int_0^{\pi} 2\cos^2\frac{\theta}{2}\left(\cos^2\frac{\theta}{2}-1\right)\cos\frac{\theta}{2}d\theta\right]$$

$$= \frac{4a}{5}\text{(using the substitution } \tfrac{\theta}{2}=\phi).$$

EXAMPLE 10.4

Find the centre of gravity of a uniform hemispherical shell.

Solution. Hemi-sphere is obtained by revolving the circle $x^2+y^2=a^2$ about the x-axis. Thus

$$\frac{dy}{dx} = -\frac{x}{y}$$

$$\frac{ds}{dx} = \sqrt{1+\left(\frac{dy}{dx}\right)^2}, = \sqrt{\frac{x^2+y^2}{y^2}}$$

$$= \frac{a}{y}.$$

Hence the centre of gravity $(\overline{x},\overline{y})$ is given by

$$\overline{x} = \frac{\int_0^a x\cdot 2\pi yds}{\int 2\pi yds}$$

$$= \frac{\int_0^a xy\frac{ds}{dx}dx}{\int_0^a y\frac{ds}{dx}dx}$$

$$= \frac{\int_0^a ax\,dx}{\int_0^a a\,dx} = \frac{\int_0^a x\,dx}{\int dx}$$

$$= \frac{\left[\frac{x^2}{2}\right]_0^a}{[x]_0^a} = \frac{a}{2}.$$

10.2 MOMENT OF INERTIA

Let a particle of mass m be at a distance r from a given line. Then mr^2 is called the *moment of inertia* (M.I) of that particle about the given line.

The moment of inertia of a system of particles of masses $m_1, m_2, \ldots m_n$ placed at distances $r_1, r_2, \ldots, r_n$ is given by $\sum_{i=1}^{n} m_i r_1^2$.

Further, if m is the mass of a body and I_G is the moment of inertia of the body about an axis through the centre of gravity of the body, then moment of inertia I of that body about an axis parallel to the axis through centre of gravity and at a distance d from centre of gravity is given by

$$I = I_G + md^2.$$

Similarly, if OX and OY are perpendicular axes in a plane and I_x and I_y are moment of inertia of a plane area about OX and OY, then moment of inertia of that plane about an axis perpendicular to both OX and OY is given by

$$I = I_x + I_y.$$

EXAMPLE 10.5

Find the moment of inertia of a thin uniform rod of length $2a$ about a perpendicular axis through its centre of gravity and about an perpendicular axis on the extremities of the rod.

Solution. Let M be the mass of the rod AB of length $2a$ and let O be the centre of AB.

We first find moment of inertia about the line OY perpendicular to AB (see Fig. 10.1). If ρ is the mass per unit length of the rod then the mass of an element length δx is $\rho\delta x$. Therefore the $M.I$ of the element about OY is $\rho\delta x.x^2$.

Hence,

Moment of inertia about OY $= \lim \sum \rho x^2 \delta x$

$$= \int_{-a}^{a} \rho x^2 dx$$

$$= \rho \int_{-a}^{a} x^2 dx$$

$$= \frac{2}{3}\rho a^3$$

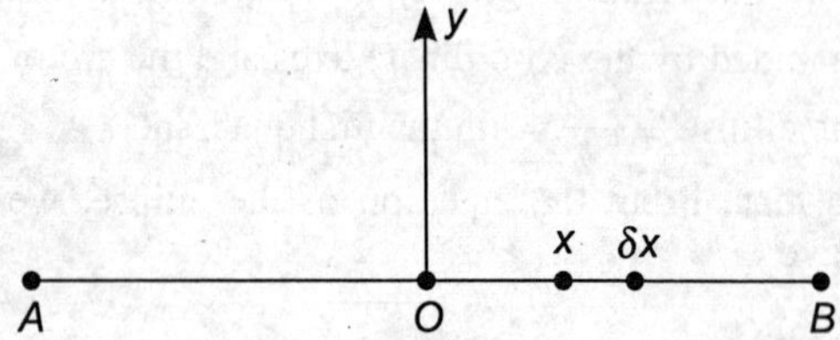

Figure 10.1

Further,

Moment of inertia about OY' through the extremity A is given by

$$I = I_G + Md^2 = \frac{1}{3}Ma^2 + Ma^2 = \frac{4}{3}Ma^2.$$

EXAMPLE 10.6

Find the moment of inertia of a circular disc about an axis perpendicular to the disc at its centre and about a tangent.

Solution. Let ρ be the mass per unit area of the circular disc of radius a. Thus mass of the disc is $M = \pi a^2 \rho$. Let OY be the axis perpendicular to the plane of the disc through the centre O (see Fig. 10.2). Then mass of the elementary ring of radius x and thickness δx is $2\rho\pi x\delta x$. Therefore, moment of inertia of this elementary ring about OY is $2\pi x \delta x \rho \cdot x^2$.

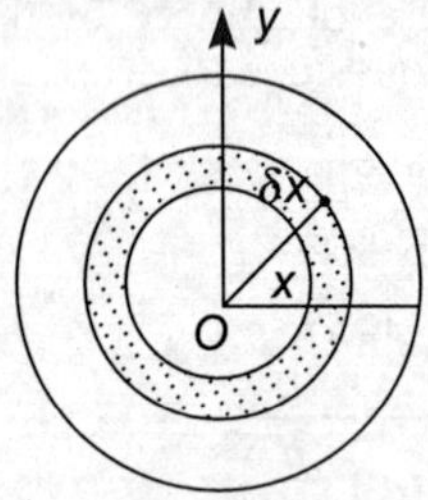

Figure 10.2

Hence moment of inertia of the disc about OY is

$$I = \int_a^a 2\pi x^3 \rho dx$$

$$= \frac{1}{2}(\pi \rho a^4)$$

$$= (\pi a^2 \rho).\frac{1}{2}a^2$$

$$= \frac{1}{2}Ma^2.$$

Then moment of inertia about a diameter is $\frac{1}{2}$ (moment of inertia about the perpendicular to the disc) $= \frac{Ma^2}{4}$. Then using the concept of parallel axis, the moment of inertia of the circular disc about a tangent to the disc is $\frac{Ma^2}{4} + Ma^2 = \frac{5Ma^2}{4}$.

EXAMPLE 10.7

Find moment of inertia of a uniform rectangular lamina about axis through the centre.

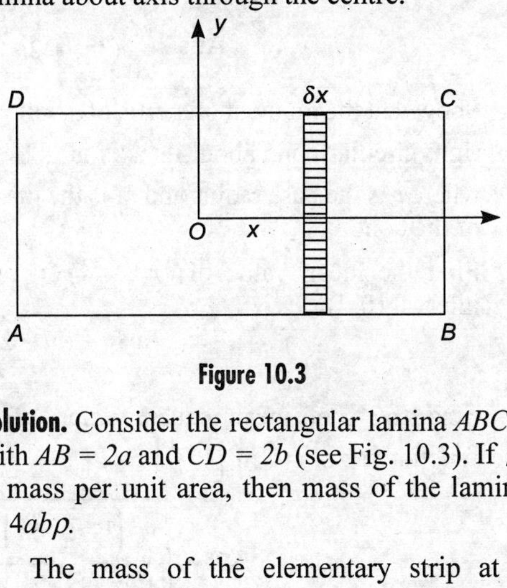

Figure 10.3

Solution. Consider the rectangular lamina $ABCD$ with $AB = 2a$ and $CD = 2b$ (see Fig. 10.3). If ρ is mass per unit area, then mass of the lamina is $4ab\rho$.

The mass of the elementary strip at a distance x from the centre of the lamina and perpendicular to OX is $2b\rho\delta x$. Therefore its moment of inertia is $2b\rho\delta x.\frac{b^2}{3}$. Hence the moment of inertia of the lamina about OX is

$$\frac{2}{3}\int_{-a}^{a} b^3 \rho dx = \frac{4}{3}ab^3\rho = \frac{Mb^2}{3}.$$

Similarly, taking strip perpendicular to OY, the moment of inertia of the lamina about OY is also $\frac{Mb^2}{3}$.

EXAMPLE 10.8

Find the moment of inertia of a solid right circular cone about it axis.

Solution. Consider the right circular cone with vertex O, height h, radius r as shown in Figure 10.4. Its mass is $\frac{1}{3}\pi r^2 h\rho$.

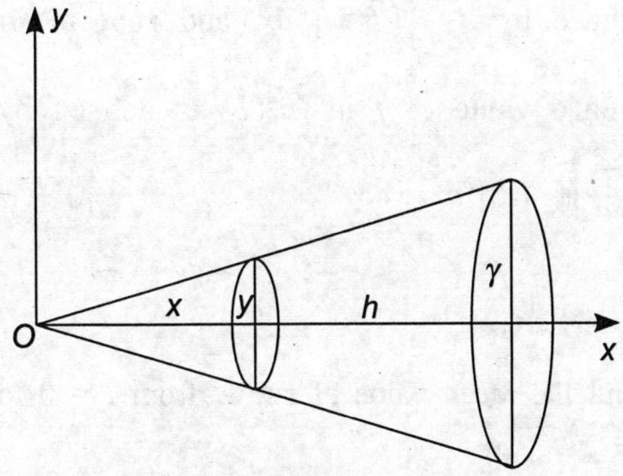

Figure 10.4

Let y be the radius of an elementary disc perpendicular to the x-axis at a distance x from O and of width δx. Then

$$\frac{y}{r} = \frac{x}{h} \text{ or } y = \frac{rx}{h}.$$

Mass of this strip is $\pi y^2 6x\rho_2$. Its moment of inertia about the x-axis is $\pi y^2 \delta x \rho \cdot \frac{y^2}{2}$. Therefore moment of inertia of the solid right circular cone is

$$\int_0^h \pi y^2 6x\rho \cdot \frac{y^2}{2} dx = \frac{\rho\pi}{2}\int_0^b \left(\frac{rx}{h}\right)^4 dx$$

$$= \frac{\pi\rho hr^4}{2\pi}\int_0^b x^4 dx$$

$$= \frac{\pi\rho hr^4}{10} = \frac{3Mr^2}{10}.$$

10.3 MEAN VALUES OF A FUNCTION

Let $f(x)$ be a continuous function defined on (a, b). Then *the mean value off* in (a, b) is defined by $\frac{\int_a^b f(x)dx}{b-a}$.

Thus mean value of f is equal to

$$\frac{\text{area enclosed by f from } a \text{ to } b}{\text{length of the interval } (a,b)}.$$

Further, **Mean Square Value** of f is (a, b) is defined by $\frac{1}{b-a}\int_a^b [f(x)]^2\,dx$ and **root mean square value** of f in (a, b) is defined by $\sqrt{\frac{1}{b-a}\int_a^b [f(x)]^2 dx}$.

EXAMPLE 10.9

Find the mean value of $\sin^2 wt$ from $t = 0$ to $t = \frac{2\pi}{w}$.

Solution. The required mean value is

$$\frac{1}{\frac{2\pi}{w}-0}\int_0^{\frac{2\pi}{w}} \sin^2 wt\,dt = \frac{w}{2\pi}\int_0^{\frac{2\pi}{w}} \frac{1}{2}(1-\cos 2wt)dt$$

$$= \frac{w}{4\pi}\left[\int_0^{\frac{2\pi}{w}} dt - \int_0^{\frac{2\pi}{w}} \cos 2wt\right] dt$$

$$= \frac{w}{4\pi}\left[\frac{2\pi}{w} - \frac{1}{2w}(\sin 2wt)\right]_0^{\frac{2\pi}{w}}$$

$$= \frac{w}{4\pi}\left[\frac{2\pi}{w}\right] = \frac{1}{2}.$$

EXERCISES

1. Find the centre of gravity of a uniform solid cone of height h.

   Hint: $\bar{x} = \frac{\int_0^h \pi x y^2}{\int_0^h \pi y^2} dx$ **Ans.** $\frac{3}{4}h$

2. Find the C.G. of the area of a loop of the curve $r = \sin 2\theta$ in the first quadrant.

   Ans. $(\bar{x}, \bar{y}) = \left(\frac{128}{105\pi}, \frac{128}{105\pi}\right)$

3. Find the C.G. of a uniform solid hemisphere

   Ans. $\bar{x} = \frac{3a}{8}, \bar{y} = 0$

4. Find the moment of inertia of a solid right circular cone about the diameter of its base

   Ans. $\frac{M}{20}(3r^2 + 2h^2)$

5. Show that the moment of inertia of a hollow right circular cone about its axis is $\frac{1}{2}Mr^2$, where r is the base radius and M is the mass of the cone.

6. Find the mean value of $kx(1 - x)$ in the interval $(0, 1)$.

   Ans. $\frac{k}{6}(3l - 2)$

7. Find the root mean square value of a function $e^{-kt} \sin wt$ in $\left[0, \frac{\pi}{w}\right]$.

   Ans. $\sqrt{\frac{w^2\left(1-e^{-2k\frac{\pi}{w}}\right)}{4k\pi(w^2+k^2)}}$.

11 Volumes and Surfaces of Solids of Revolution

The body generated by the revolution of a plane area, about a fixed line lying in its own plane, is called a *solid of revolution*. On the other hand, the surface generated by the boundary of the plane area is called the *surface of revolution*. The fixed line, lying in the plane of plane area, about which the plane area revolves is called the *axis of revolution*. The section of a solid of revolution by a plane, perpendicular to the axis of revolution, is a circle having its center on the axis of revolution.

11.1 VOLUME OF THE SOLID OF REVOLUTION (CARTESIAN EQUATIONS)

(A) Revolution About x-axis: Let AB be the arc of the curve $y = f(x)$ lying between the ordinates $x = a$ and $x = b$. We assume that f is continuous in (a, b) and that it does not meet the x-axis. Let $P(xy)$ and $Q(x + \delta x, y + \delta y)$ be any two neighboring points on the curve $y = f(x)$. Draw PM ⊥ OX and QN ⊥ OX Draw PR ⊥ QN and SQ ⊥ PM.

Let V be the volume of the solid generated by the revolution about the x-axis of the area ACMP. As x increases, MP moves toward the right and the volume increases. Let the volume of revolution obtained by revolving the area ACNQ about x-axis be $V+\delta V$ so that the volume of the solid generated by the revolution of the strip PMNQ about the x-axis is δV. We now have $PM = y$ and $QN = y+\delta y$ and $MN = x+\delta x - x = \delta x$. Then, the volume of the solid generated by revolving the area PMNR is $\pi y^2 \delta x$ and the volume of the solid generated by revolving the area PMNQ is $\pi(y+\delta y)^2 \delta x$.

We further observe that the volume of the solid generated by the revolution of the area PMNQP lies between the volumes of the right circular cylinder generated by the revolution of the areas PMNR and PMNQ. Thus, δV lies between $\pi y^2 dx$ and $\pi(y+\delta y)^2 \delta x$.

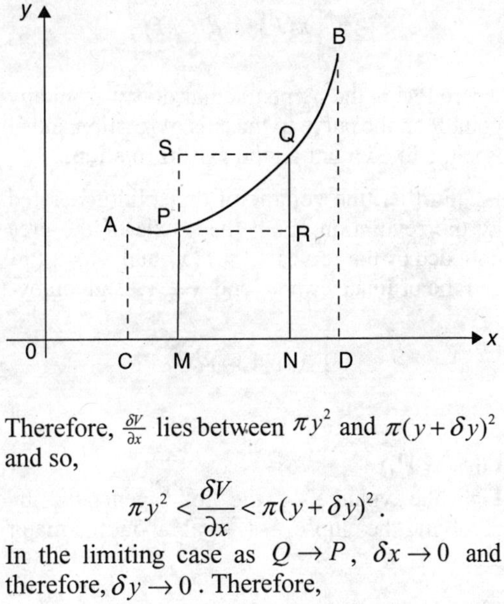

Therefore, $\frac{\delta V}{\partial x}$ lies between πy^2 and $\pi(y+\delta y)^2$ and so,

$$\pi y^2 < \frac{\delta V}{\partial x} < \pi(y+\delta y)^2.$$

In the limiting case as $Q \to P$, $\delta x \to 0$ and therefore, $\delta y \to 0$. Therefore,

$$\frac{dV}{dx} = \pi y^2 \text{ or } dV = \pi y^2 dx.$$

Hence,

$$\int_a^b \pi y^2 dx = \int_a^b dV = V(b) - V(a)$$

= Volume generated by the area ACDB–0. It follows, therefore, that the volume of the solid generated by the revolution about the x-axis of the area, bounded by the curve $y = f(x)$, the x-axis, and the ordinates $x = a$ and $x = b$, is given by

$$V = \int_a^b \pi y^2 dx.$$

Similarly, it can be shown that the volume of the solid generated by the revolution about y-axis of the area bounded by the curve $f(x)$, the y-axis, and the abscissae $y = a$ and $y = b$ is $\int_a^b \pi x^2 dy$.

If the axis rotation is neither x-axis nor y-axis but a line CD, then the volume of the solid generated by the revolution about CD of the area bounded by the curve AB, the axis CD, and the perpendiculars AC and BD on the axis is

$$V = \int_{OC}^{OD} \pi (PM)^2 d(OM),$$

where PM is the perpendicular drawn from any point P on the curve to the axis of rotation and O is some fixed point on the axis of rotation.

Further, the volume of the solid generated by the revolution about the x-axis of the area bounded by the curves $y_1 = f(x)$ and $y_2 = g(x)$ and the ordinates $x = a$ and $x = b$, is given by

$$\int_a^b \pi (y_1^2 - y_2^2) dx.$$

EXAMPLE 11.1

Find the volume of the solid generated by revolving the ellipse $\frac{x^2}{a^2} + \frac{y^2}{b^2} = 1$ about the major axis.

Solution. The ellipse $\frac{x^2}{a^2} + \frac{y^2}{b^2} = 1$ is symmetrical about y-axis. Therefore, the volume generated by the ellipse by the revolution about the x-axis is twice the volume generated by the area in the first quadrant about the x-axis.

Therefore,

$$\text{Required volume} = 2\int_0^a \pi y^2 dx$$

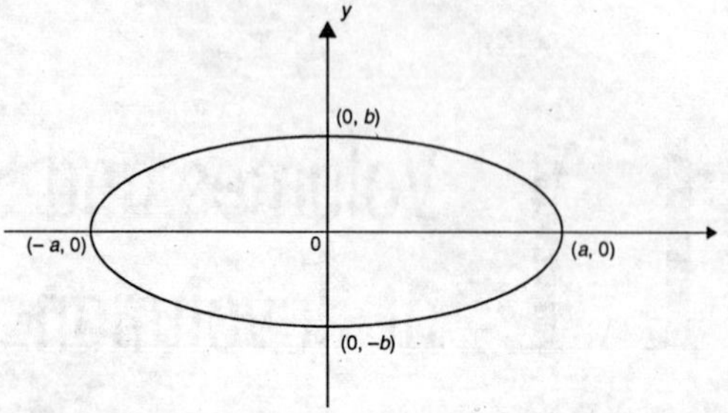

$$= 2\int_0^a \pi \left[\frac{b^2}{a^2}(a^2 - x^2) \right] dx$$

$$= \frac{2\pi b^2}{a^2} \int_0^a (a^2 - x^2) dx$$

$$= \frac{2\pi b^2}{a^2} \left[a^3 - \frac{a^3}{3} \right]$$

$$= \frac{4}{3} \pi a b^2.$$

EXAMPLE 11.2

Find the volume of the solid generated by the revolution of the curve $y(a^2 + x^2) = a^3$ about its asymptote.

Solution. The given curve is $y(a^2 + x^2) = a^3$ which is symmetrical about y-axis, cuts the y-axis at $y = a$, and does not cut the x-axis. The asymptote parallel to the x-axis is $y = 0$, that is, x-axis. Due to symmetry, the required volume is twice that of the volume generated by half of the curve in the first quadrant. Thus,

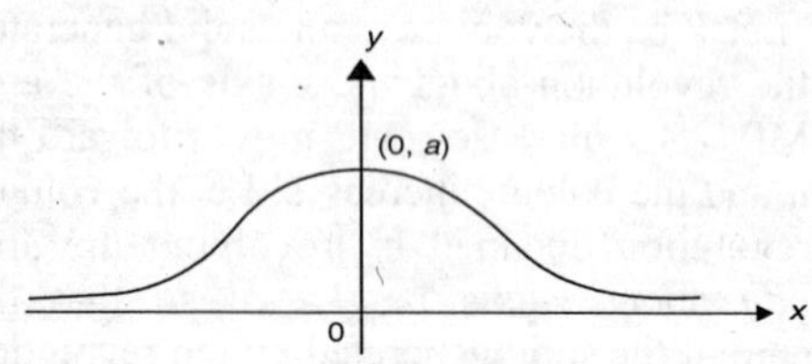

$$\text{Required volume} = 2\int_0^\infty \pi y^2 dx$$

$$= 2\pi \int_0^\infty \left(\frac{a^3}{a^2 + x^2} \right)^2 dx.$$

Put $x = a\tan\theta$ so that $dx = a\sec^2\theta d\theta$. Further, when $x = 0$, $\theta = 0$ and when $x = \infty$, $\theta = \frac{\pi}{2}$.

Therefore,

$$\text{Required volume} = 2\pi a^6 \int_0^{\frac{\pi}{2}} \frac{a\sec^2\theta}{a^4\sec^4\theta} d\theta$$

$$= 2\pi a^3 \int_0^{\frac{\pi}{2}} \cos^2\theta d\theta$$

$$= 2\pi a^3 \cdot \frac{1}{2} \cdot \frac{\pi}{2} = \frac{1}{2}\pi^2 a^3.$$

EXAMPLE 11.3

The curve $y^2(a+x) = x^2(3a-x)$ revolves about the axis of x. Find the volume generated by the loop.

Solution. The given curve is symmetrical about the x-axis, cuts the x-axis at $(3a, 0)$, and passes through the origin. The required volume, generated by the loop, is given by

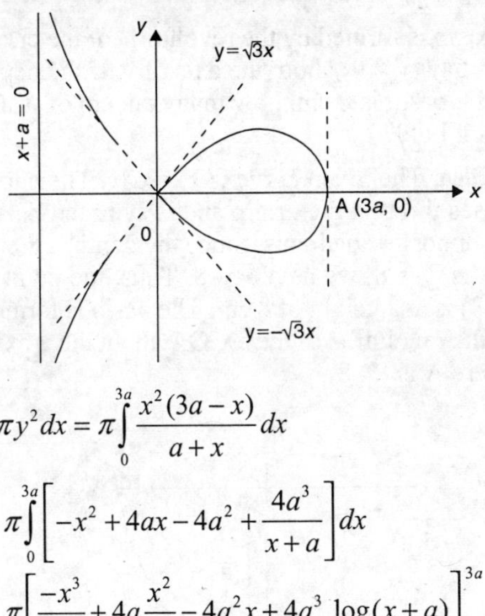

$$\int_0^{3a} \pi y^2 dx = \pi \int_0^{3a} \frac{x^2(3a-x)}{a+x} dx$$

$$= \pi \int_0^{3a} \left[-x^2 + 4ax - 4a^2 + \frac{4a^3}{x+a} \right] dx$$

$$= \pi \left[\frac{-x^3}{3} + 4a\frac{x^2}{2} - 4a^2 x + 4a^3 \log(x+a) \right]_0^{3a}$$

$$= \pi\left[-9a^3 + 18a^3 - 12a^3 + 4a^3(\log 4a - \log a)\right]$$

$$= \pi\left[-3a^3 + 4a^3 \log 4\right] = \pi a^3 (8\log 2 - 3).$$

EXAMPLE 11.4

Find the volume of the solid generated by the revolution of the cissoid $y^2(2a-x) = x^3$ about its asymptote.

Solution. The given cissoid $y^2(2a-x) = x^3$ is symmetrical about the x-axis, passing through the origin, and the asymptote parallel to the y-axis is given by $2a - x = 0$, that is, $x = 2a$. Further, if $P(x, y)$ is any point on the curve and PM is perpendicular to the asymptote, then PM = NA, where PN is perpendicular from P to the x-axis. Thus, PM = NA = OA − ON = $2a - x$ Also AM = NP = y.

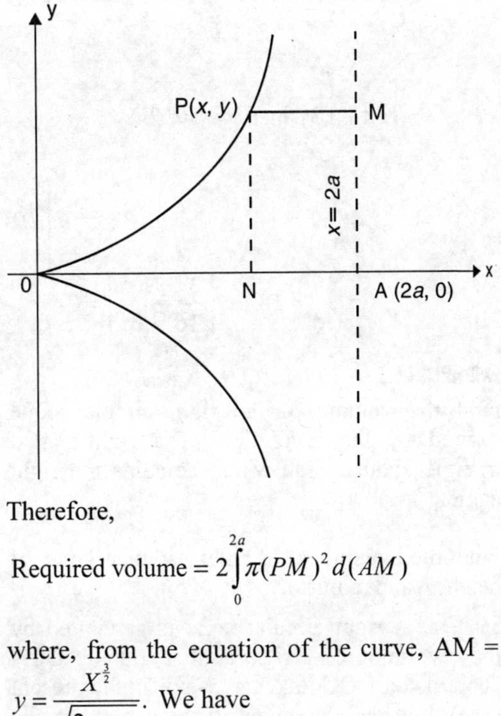

Therefore,

$$\text{Required volume} = 2\int_0^{2a} \pi (PM)^2 d(AM)$$

where, from the equation of the curve, AM = $y = \frac{X^{\frac{3}{2}}}{\sqrt{2a-x}}$. We have

$d(AM) = dy$

$$= \frac{\sqrt{2a-x}.\frac{3}{2}x^{\frac{1}{2}} - x^{\frac{3}{2}}\frac{1}{2}(2a-x)^{\frac{1}{2}}(-1)}{2a-x} dx\cdot$$

$$= \frac{x^{\frac{1}{2}}(6a-3x+x)}{2(2a-x)^{\frac{3}{2}}} dx = \frac{\sqrt{x}(3a-x)}{(2a-x)^{\frac{3}{2}}} dx\cdot$$

Hence,
Required volume

$$= 2\pi \int (2a-x)^2 \cdot \frac{\sqrt{x}(3a-x)}{(2a-x)^{\frac{3}{2}}} dx\,.$$

Putting $x = 2a\sin^2\theta$, we have $dx = 4a\sin\theta\cos\theta\,dx$. When $x = 0$, $\theta = 0$ and when $x = 2a$, $\theta = \frac{\pi}{2}$. Hence,

$$\text{Required volume} = 2\pi\int_0^{\frac{\pi}{2}}\sqrt{2a}\sin\theta\sqrt{2a}\cos\theta \cdot(3a-2a\sin^2\theta)\cdot(4a\sin\theta\cos\theta)d\theta$$

$$= 16\pi a^3\int_0^{\frac{\pi}{2}}(3-2\sin^2\theta)\times\sin^2\theta\cos^2\theta d\theta$$

$$= 16\pi a^3\left[3\int_0^{\frac{\pi}{2}}\sin^2\theta\cos^2\theta d\theta - 2\int_0^{\frac{\pi}{2}}\sin^4\theta\cos^2\theta\, d\theta\right]$$

$$= 16\pi a^3\left[\frac{3\pi}{16}-\frac{\pi}{16}\right] = 2\pi^2 a^3.$$

EXAMPLE 11.5

Find the volume of the right circular cone formed by the revolution of a right-angled triangle about a side which contains the right angle.

or

Find the volume of a right circular cone of height h and radius a.

Solution. A right circular cone is generated by the revolution of a right-angled triangle OCA about its base OC (taken as x-axis in the present case). Let h be the height of the cone and a be the radius of its circular base.

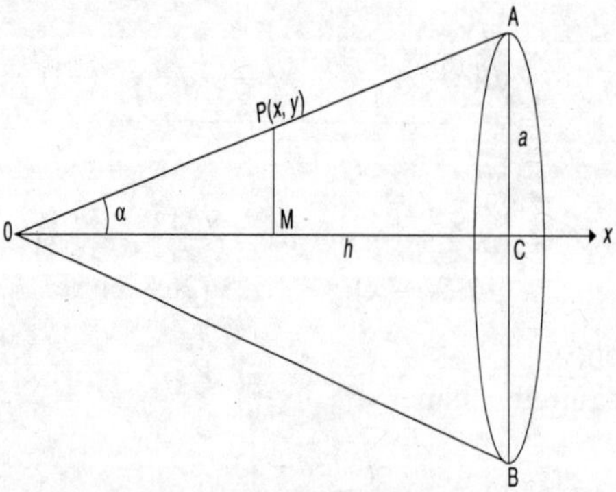

In ΔOMP, we have $\tan\alpha = \frac{PM}{OM} = \frac{y}{x}$ and so,

$$y = x\tan\alpha. \tag{1}$$

Therefore, the required volume of the cone is given by

$$\int_0^h \pi y^2 dx = \int_0^h \pi x^2\tan^2\alpha\, dx, \text{ using (1)}$$

$$= \pi\tan^2\alpha\int_0^h x^2 dx$$

$$= \pi\tan^2\alpha\frac{h^3}{3}$$

$$= \frac{\pi h^3}{3}\left(\frac{AC}{OC}\right)^2$$

$$= \frac{\pi h^3}{3}\left(\frac{a}{h}\right)^2$$

$$= \frac{1}{3}\pi a^2 h.$$

EXAMPLE 11.6

A basin is formed by the revolution of the curve $x^3 = 64y$ $(y > 0)$ about the axis of y. If the depth of the basin is 8 cm, how many cu. cm of water will it hold?

Solution. The given curve is $x^3 = 64y$. The curve passes through the origin and is symmetrical in the opposite quadrants. Since the depth is 8 cm, that is, $y = 8$, we have $x = 8$. Thus, the point A (8, 8) is at a height of 8 cm. The basin is formed by the revolution of the arc OA about the y-axis, where A is (8, 8).

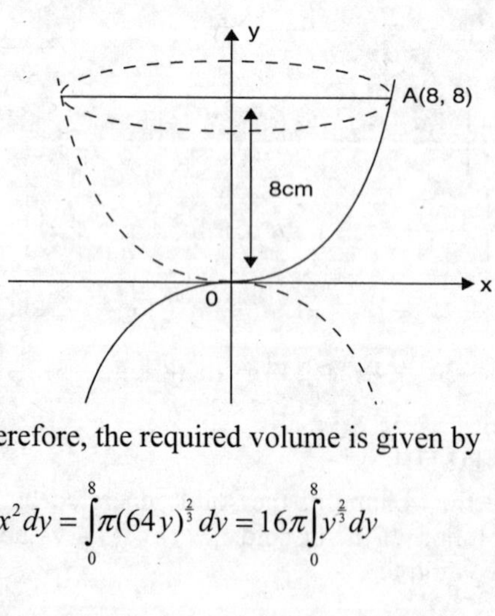

Therefore, the required volume is given by

$$\int_0^8 \pi x^2 dy = \int_0^8 \pi(64y)^{\frac{2}{3}}dy = 16\pi\int_0^8 y^{\frac{2}{3}}dy$$

$$= 16\pi\left[\frac{y^{\frac{5}{3}}}{\frac{5}{3}}\right]_0^8 = \frac{48\pi}{5}(8)^{\frac{5}{3}}$$

$$= \frac{48\pi}{5}(32) = \frac{1536\pi}{5} \text{ cu cm.}$$

EXAMPLE 11.7

Show that the volume of the solid generated by the revolution of the curve $(a-x)y^2 = a^2x$ about its asymptote is $\frac{1}{2}\pi^2a^3$.

Solution. The given curve $(a-x)y^2 = a^2x$ is symmetrical about the x-axis, passes through the origin, and its asymptotes parallel to the y-axis is $a - x = 0$ or $x = a$.

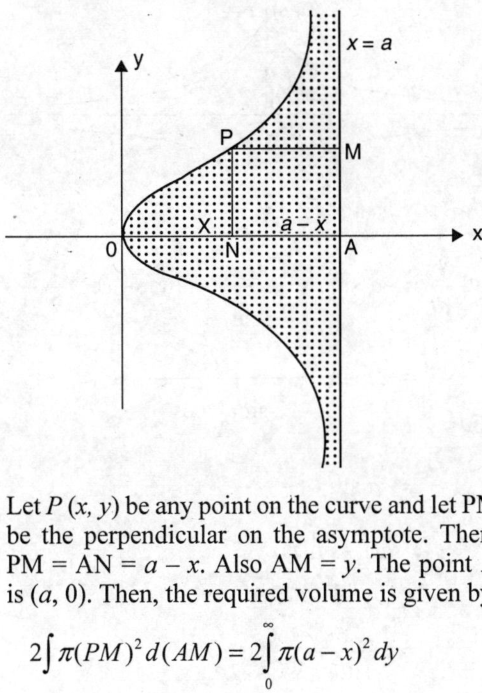

Let $P(x, y)$ be any point on the curve and let PM be the perpendicular on the asymptote. Then, PM = AN = $a - x$. Also AM = y. The point A is $(a, 0)$. Then, the required volume is given by

$$2\int \pi(PM)^2 d(AM) = 2\int_0^{\infty} \pi(a-x)^2 dy$$

$$= 2\pi\int_0^{\infty}\left(a - \frac{ay^2}{a^2+y^2}\right)^2 dy,$$

using equation of the curve

$$= 2\pi\int_0^{\infty}\left(\frac{a^3}{a^2+b^2}\right)^2 dy = 2\pi\int_0^{\infty}\frac{a^6}{(a^2+y^2)^2}dy.$$

Put $y = a\tan\theta$ so that $dy = a\sec^2\theta\, d\theta$.

Therefore,

$$\text{Required volume} = 2\pi\int_0^{\frac{\pi}{2}}\frac{a^6}{a^4\sec^4\theta}\cdot a\sec^2\theta\, d\theta$$

$$= 2\pi a^3\int_0^{\frac{\pi}{2}}\cos^2\theta\, d\theta = 2\pi a^3\cdot\frac{1}{2}\cdot\frac{\pi}{2} = \frac{1}{2}\pi^2a^3.$$

EXAMPLE 11.8

Find the volume of the reel-shaped solid formed by the revolution about the y-axis, of the part of the parabola $y^2 = 4ax$, cut off by the latus rectum.

Solution. Let O be the vertex and L one extremity of the latus rectum of the given parabola $y^2 = 4ax$. For the arc OL, y varies from 0 to $2a$.

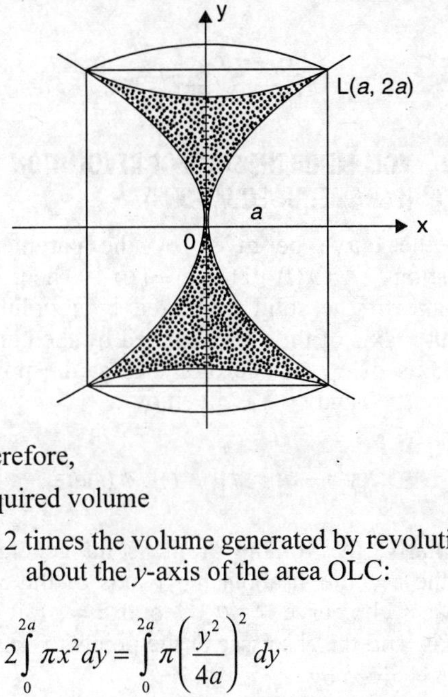

Therefore,

Required volume

= 2 times the volume generated by revolution about the y-axis of the area OLC:

$$= 2\int_0^{2a}\pi x^2 dy = \int_0^{2a}\pi\left(\frac{y^2}{4a}\right)^2 dy$$

$$= \frac{\pi}{8a^2}\int_0^{2a}y^4 dy = \frac{\pi}{8a^2}\cdot\frac{32a^5}{5} = \frac{4\pi a^3}{5}.$$

EXAMPLE 11.9

Find the volume of the solid generated by the revolution of the plane area bounded by $y^2 = 9x$ and $y = 3x$ about the x-axis.

Solution. The parabola $y^2 = 9x$ is symmetrical about the x-axis. The line $y = 3x$ cuts the parabola at $P(1, 3)$. Therefore, the required volume is given by

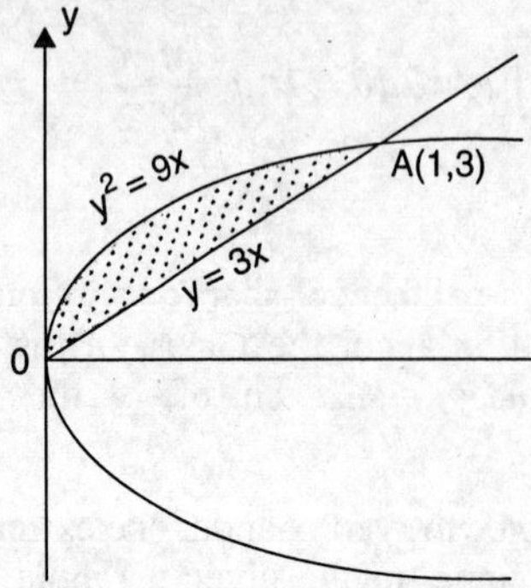

$$\int_0^1 \pi(y_1^2 - y_2^2)dx = \pi\int_0^1 (9x - 9x^2)dx$$

$$= \pi\left[\frac{9x^2}{2} - \frac{9x^3}{3}\right]_0^1 = \pi\left(\frac{9}{2} - 3\right) = \frac{3\pi}{2}.$$

11.2 VOLUME OF THE SOLID OF REVOLUTION (PARAMETRIC EQUATIONS)

Let the curve be given by the parametric equations $x = \phi(t)$ and $y = \psi(t)$. Then, the volume of the solid generated by revolution about x-axis of the area bounded by the curve, the axis of x, and the ordinates at the points where $t = a$ and $t = b$ is given by

$$\int_a^b \pi y^2 \frac{dx}{dt} dt = \pi\int_a^b [\psi(t)]^2 . \phi'(t) dt.$$

Similarly, the volume of the solid generated by the revolution about the y-axis of the area between the curve $x = \phi(t)$ and $y = \psi(t)$, the y-axis, and the abscissae at the points $t = a$ and $t = b$ is given by

$$\int_a^b \pi x^2 \frac{dy}{dt} dt = \pi\int_a^b [\phi(t)]^2 . \psi'(t) dt.$$

EXAMPLE 11.10

Find the volume of the solid generated by the revolution of the tractrix $x = a \cos t + \frac{1}{2} a \log \tan^2 \frac{t}{2}$ and $y = a \sin t$ about its asymptote.

Solution. The equations of the tractrix are

$$x = a \cos t + \frac{1}{2} a \log \tan^2 \tfrac{t}{2} \text{ and } y = a \sin t.$$

The given curve is symmetrical about both the axes and the asymptote to the curve is $y = 0$, that is, x-axis. For the portion of the curve lying in the second quadrant, y varies from a to 0, t varies from $\frac{\pi}{2}$ to 0, and x varies from 0 to $-\infty$. Thus,

$$\text{Required volume} = 2\int_{-\infty}^{0} \pi y^2 dx$$

$$= 2\int_{-\infty}^{\frac{\pi}{2}} \pi y^2 \frac{dx}{dt} \cdot dt.$$

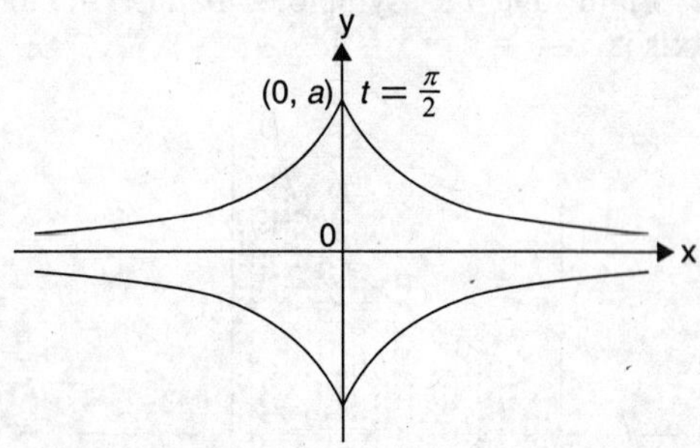

But, $$\frac{dx}{dt} = -a \sin t + \frac{1}{2} . \frac{a}{\tan^2 \frac{t}{2}} .2 \tan \frac{t}{2} \sec^2 \frac{t}{2} . \frac{1}{2}$$

$$= -a \sin t + \frac{a}{2 \sin \frac{t}{2} \cos \frac{t}{2}}$$

$$= -a \sin t + \frac{a}{\sin t}$$

$$= \frac{a(1 - \sin^2 t)}{\sin t}$$

$$= a \frac{\cos^2 t}{\sin t}.$$

Therefore,

Required volume

$$= 2\pi \int_0^{\frac{\pi}{2}} a^2 \sin^2 t . \frac{a \cos^2 t}{\sin t} dt$$

$$= 2\pi a^3 \int_0^{\frac{\pi}{2}} \cos^2 t \sin t dt = 2\pi a^3 . \frac{1}{3.1} = \frac{2}{3}\pi a^3.$$

EXAMPLE 11.11

Find the volume of the solid formed by revolving the cycloid $x = a(\theta - \sin\theta)$ and $y = a(1 - \cos\theta)$ about its base.

Solution. The given equations of the cycloid are

$x = a(\theta - \sin\theta)$ and $y = a(1 - \cos\theta)$.

The cycloid is symmetrical about the line through the point, where $\theta = \pi$, perpendicular to x-axis. For the first half, OBH of the cycloid, θ varies from 0 to π.

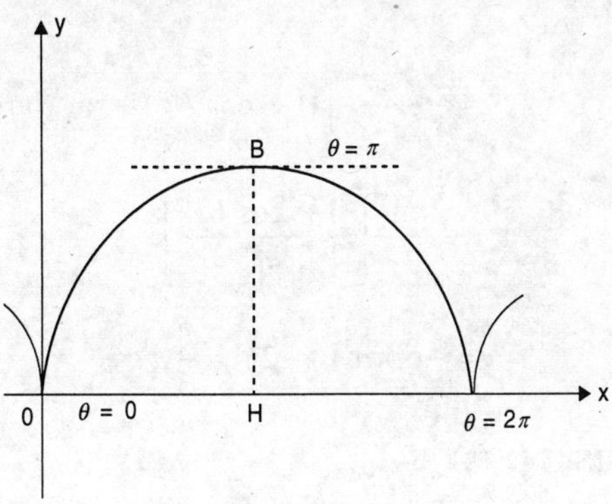

The required volume is given by

$$2\int_0^{\pi} \pi y^2 \frac{dx}{d\theta}.d\theta$$

$$= 2\pi\int_0^{\pi} a^2(1 - \cos\theta)^2.a(1 - \cos\theta)d\theta$$

$$= 2\pi a^2\int_0^{\pi}\left(2\sin^2\frac{\theta}{2}\right)^2.a\left(2\sin^2\frac{\theta}{2}\right)d\theta$$

$$= 2\pi a^3\int_0^{\frac{\pi}{2}}(2\sin^2 t)^3 2dt, \frac{\theta}{2} = t$$

$$= 32\pi a^3\int_0^{\frac{\pi}{2}}\sin^6 t dt = 32\pi a^3.\frac{5.3.1}{6.4.2}.\frac{\pi}{2} = 5\pi^2 a^3.$$

EXAMPLE 11.12

Find the volume of the spindle-shaped solid generated by revolving the asteroid $x^{\frac{2}{3}} + y^{\frac{2}{3}} = a^{\frac{2}{3}}$ about the x-axis.

Solution. The parametric equations of the asteroid are

$$x = a\cos^3 t \text{ and } y = a\sin^3 t.$$

The curve is symmetrical about both the axes. It cuts the axes at $t = 0$ and $t = \frac{\pi}{2}$. For the portion of the curve in the first quadrant, t varies from 0 to $\frac{\pi}{2}$. Therefore,

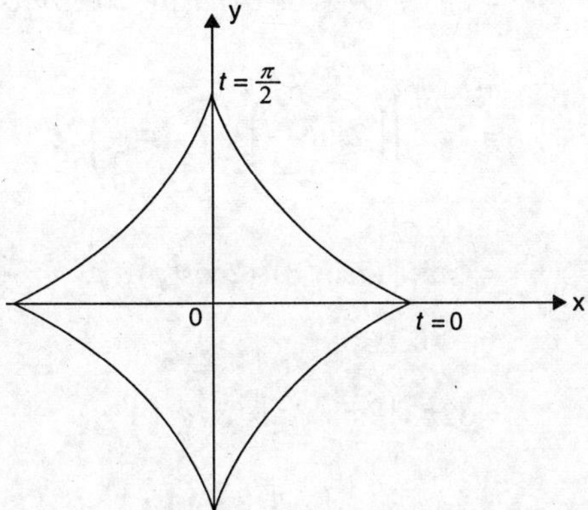

$$\text{Required volume} = 2\int_0^{\frac{\pi}{2}}\pi y^2\frac{dx}{dt}dt = 2\pi\int_0^{\frac{\pi}{2}} y^2\frac{dx}{dt}dt$$

$$= 2\pi\int_0^{\frac{\pi}{2}} a^2\sin^6 t.(-3a\cos^2 t\sin t)dt$$

$$= -6\pi a^3\int_0^{\frac{\pi}{2}}\sin^7 t.\cos^2 t dt = -6\pi a^3.\frac{6.4.2.1}{9.7.5.3.1}$$

$$= \frac{-32}{105}\pi a^3 = \frac{32}{105}\pi a^3 \text{ (in magnitude).}$$

EXAMPLE 11.13

Find the volume of the solid formed by the revolution of an arch of the cycloid $x = a(\theta + \sin\theta)$ and $y = a(1 - \cos\theta)$ about the tangent at the vertex.

Solution. The cycloid $x = a(\theta + \sin\theta)$ and $y = a(1 - \cos\theta)$ is symmetrical about the y-axis. Further, tangent at the vertex is x-axis. For the portion of the curve in the first quadrant, θ varies from 0 to π. Therefore,

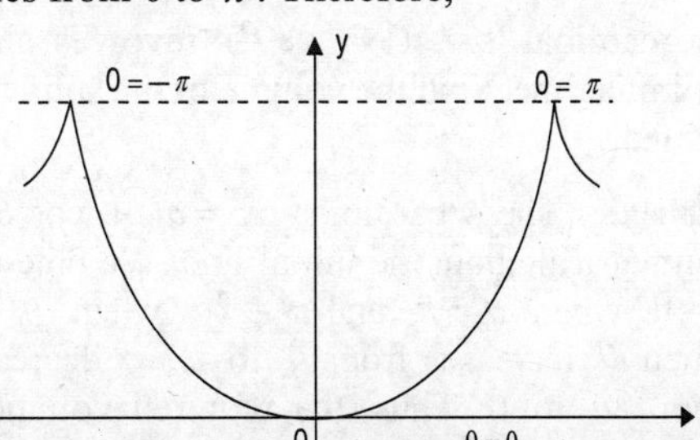

$$\text{Required volume} = 2\int_0^{\pi} \pi y^2 \frac{dx}{d\theta} d\theta$$

$$= 2\pi \int_0^{\pi} a^2(1-\cos\theta)^2 . a(1+\cos\theta) d\theta$$

$$= 2\pi a^3 \int_0^{\pi} \left(2\sin^2\frac{\theta}{2}\right)^2 \left(2\cos^2\frac{\theta}{2}\right) d\theta$$

$$= 4\pi a^3 \int_0^{\pi} (2\sin^2\phi)^2 (2\cos^2\phi) . d\phi, \ \phi = \frac{\theta}{2}$$

$$= 32\pi a \int_0^{\frac{\pi}{2}} {}^3 \sin^4\phi \cos^2\phi \, d\phi$$

$$= 32\pi a^3 . \frac{3.1}{6.4.2} . \frac{\pi}{2} = \pi^2 a^3 .$$

11.3 VOLUME OF THE SOLID OF REVOLUTION (POLAR CURVES)

The volume of the solid generated by the revolution of the area bounded by the curve $r = f(\theta)$ and the radii vectors $\theta = \alpha$ and $\theta = \beta$ about the initial line $(\theta = 0)$ is given by

$$\int_\alpha^\beta \frac{2}{3}\pi r^3 \sin\theta \, d\theta .$$

Similarly, the volume of the solid generated by the revolution of the area bounded by the curve $r = f(\theta)$ and the radii vectors $\theta = \alpha$ and $\theta = \beta$ about the line $\theta = \frac{\pi}{2}$ is given by

$$\int_\alpha^\beta \frac{2}{3}\pi r^3 \cos\theta \, d\theta .$$

EXAMPLE 11.14

The cardioid $r = a(1+\cos\theta)$ revolves about the initial line. Find the volume of the solid thus formed.

Solution. The cardioid $r = a(1+\cos\theta)$ is symmetrical about the initial line. We note that $r = 0$ when $\theta = \pi$ and $r = 2a$ when $\theta = 0$. When θ increases from 0 to π, r decreases from $2a$ to 0. Thus, the required volume is given by

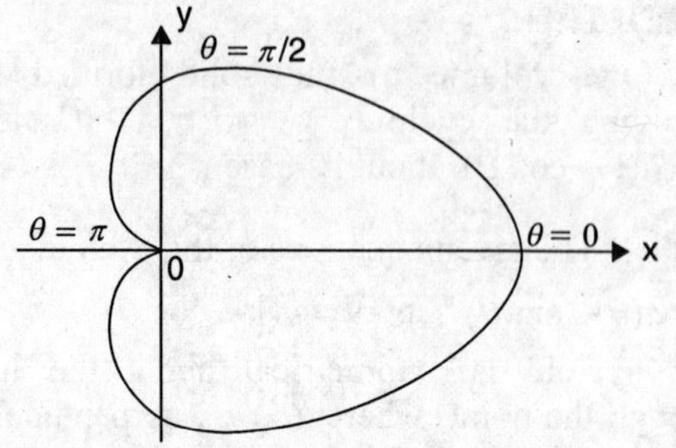

$$\int_0^{\pi} \frac{2}{3}\pi r^3 \sin\theta \, d\theta = \frac{2\pi}{3} \int_0^{\pi} a^3 (1+\cos\theta)^3 \sin\theta \, d\theta$$

$$= -\frac{2\pi a^3}{3} \int_0^{\pi} (1+\cos\theta)^3 (-\sin\theta) d\theta$$

$$-\frac{2\pi a^3}{3} \left[\frac{(1+\cos\theta)^4}{4} \right]_0^{\pi}$$

$$= -\frac{1}{6}\pi a^3 (0 - 2^4) = \frac{8}{3}\pi a^3 .$$

EXAMPLE 11.15

Find the volume of the solid generated by revolving one loop of the lemniscate $r^2 = a^2 \cos 2\theta$ about the line $\theta = \frac{\pi}{2}$.

Solution. The given curve $r^2 = a^2 \cos 2\theta$ is symmetrical about the initial line. Putting $r = 0$ in the equation of the curve, we get $\cos 2\theta = 0$ or $\theta = \pm\frac{\pi}{4}$. The revolution is about $\theta = \frac{\pi}{2}$.

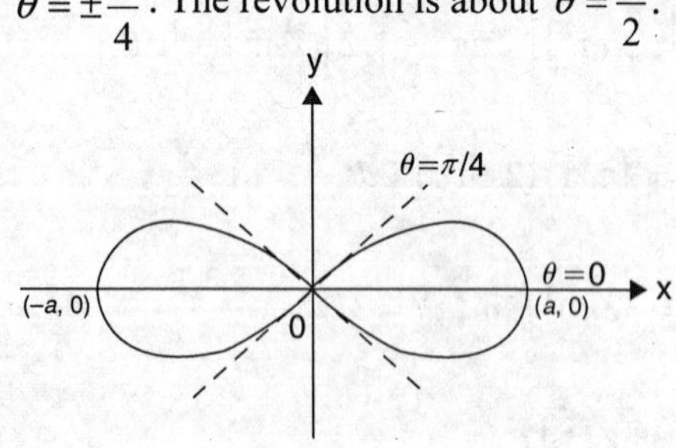

Therefore, the required volume is given by

$$2\int_0^{\frac{\pi}{4}} \frac{2}{3}\pi r^3 \cos\theta \, d\theta$$

$$= \frac{4\pi}{3} \int_0^{\frac{\pi}{4}} a^3 (\cos 2\theta)^{\frac{3}{2}} \cos\theta \, d\theta$$

$$= \frac{4\pi a^3}{3}\int_0^{\frac{\pi}{4}} (1-2\sin^2\theta)^{\frac{3}{2}} \cos\theta d\theta$$

$$= \frac{4\pi a^3}{3}\int_0^{\frac{\pi}{4}} (1-\sin^2\phi)^{\frac{3}{2}} \frac{1}{\sqrt{2}} \cos\phi d\phi,$$

$\sqrt{2}\sin\theta = \sin\phi$

$$= \frac{4\pi a^3}{3\sqrt{2}}\int_0^{\frac{\pi}{2}} \cos^4\phi d\phi = \frac{4\pi a^3}{3\sqrt{2}}\cdot\frac{3.1}{4.2}\cdot\frac{\pi}{2} = \frac{\pi^2 a^3}{4\sqrt{2}}.$$

EXAMPLE 11.16

Show that if the area, lying within the cardioid $r = 2a(1+\cos\theta)$ and not included in the parabola $r(1+\cos\theta) = 2a$, revolves about the initial line, the volume generated is $18\pi a^3$.

Solution. Both the curves are symmetrical about the initial line. Eliminating r from the given equations, we get

$$2a(1+\cos\theta) = \frac{2a}{1+\cos\theta} \text{ or}$$

$$(1+\cos\theta)^2 = 1 \text{ or } \cos\theta(2+\cos\theta) = 0.$$

Since $\cos\theta$ cannot be -2, we have $\cos\theta = 0$ and so, $\theta = \pm\frac{\pi}{2}$. So, for the upper half of the area, θ varies from 0 to $\frac{\pi}{2}$.

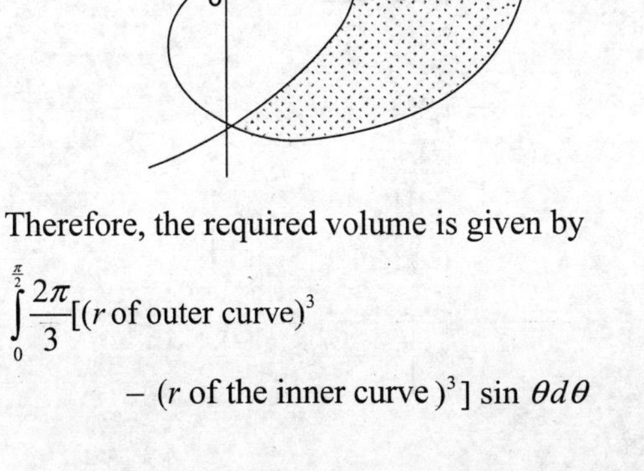

Therefore, the required volume is given by

$$\int_0^{\frac{\pi}{2}} \frac{2\pi}{3}[(r \text{ of outer curve})^3$$

$$- (r \text{ of the inner curve })^3] \sin\theta d\theta$$

$$= \frac{2\pi}{3}\int_0^{\frac{\pi}{2}}\left[\{2a(1+\cos\theta)\}^3 - \left(\frac{2a}{1+\cos\theta}\right)^3\right]\sin\theta d\theta$$

$$= \frac{16\pi a^3}{3}\left[-\frac{(1+\cos\theta)^4}{4} + \frac{(1+\cos\theta)^{-2}}{-2}\right]_0^{\frac{\pi}{2}}$$

$$= \frac{16\pi a^3}{3}\left[\frac{15}{4} - \frac{3}{8}\right] = 18\pi a^3.$$

11.4 SURFACE OF THE SOLID OF REVOLUTION (CARTESIAN EQUATIONS)

Let AB be the arc of the curve $y = f(x)$ included between the ordinates $x = a$ and $x = b$. Assume that the curve does not cut the x-axis. Let $P(x, y)$ and $Q(x+\delta x, y+\delta y)$ be any two neighboring points on the curve. Let length of the arc AP be s and that of the arc AQ be $s+\delta s$ so that the arc PQ is of length δs. Draw the ordinates PM and QN. Let S denote the curved surface of the solid generated by the revolution of the area CMPA about the x-axis. The curved surface of the solid generated by the revolution of the area MNQP is δS. Without any loss of generality, we assume that the curved surface of the solid generated by the revolution of the area MNQP about the x-axis lies between the curved surfaces of the right circular cylinders whose radii are PM and NQ and which are of the same thickness (height) δs.

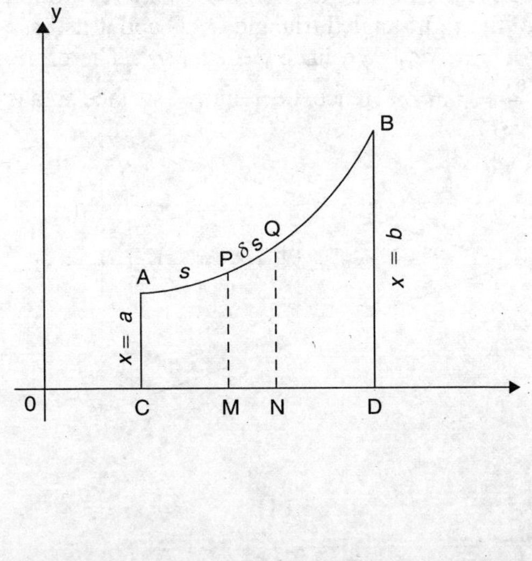

Therefore, δS lies between $2\pi y\delta s$ and $2\pi(y+\delta y)\delta s$, that is,

$$2\pi y\delta s < \delta S < 2\pi(y+\delta y)\delta s$$

or $$2\pi y < \frac{\delta S}{\delta s} < 2\pi(y+\delta y).$$

As $Q \to P$, that is, as $\delta s \to 0$, δy will tend to 0. Therefore,

$$\frac{dS}{ds} = 2\pi y \text{ or } dS = 2\pi y ds.$$

Hence,

$$\int_{x=a}^{x=b} 2\pi y ds = \int_{x=a}^{x=b} dS = [S]_a^b.$$

Hence, the required curved surface is given by

$$\int_a^b 2\pi y ds = \int_a^b 2\pi y \frac{ds}{dx} dx,$$

where

$$\frac{ds}{dx} = \sqrt{1+\left(\frac{dy}{dx}\right)^2}.$$

Similarly, the curved surface of the solid generated by the revolution about the y-axis, of the area bounded by the curve $x = f(y)$, the lines $y = a$ and $y = b$, and the y-axis, is

$$2\pi\int_0^b xds.$$

EXAMPLE 11.17

Find the area of the surface of a cone with a semi-vertical angle α and a base as a circle of radius a.

Solution. The cone is obtained by the revolution of the right-angled triangle OAC about its base OC (x-axis). We have $y = x \tan\alpha$. Therefore, $\frac{dy}{dx} = \tan\alpha$. Hence, the required surface area is

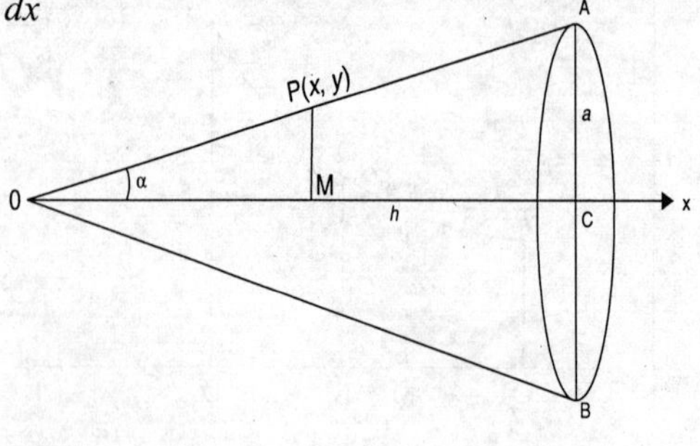

$$\int_0^h 2\pi y\left[1+h\left(\frac{dy}{dx}\right)^2\right]^{\frac{1}{2}} dx$$

$$= 2\pi\int_0^h x\tan\alpha[1+\tan^2\alpha]^{\frac{1}{2}} dx$$

$$= 2\pi\int_0^h x \tan\alpha \sec\alpha dx = 2\pi \tan\alpha \sec\alpha\int_0^h xdx$$

$$= 2\pi \tan\alpha \sec\alpha.\frac{h^2}{2}$$

$$= \pi \tan\alpha \sec\alpha\left(\frac{a^2}{\tan^2\alpha}\right), \tan\alpha = \frac{a}{h}$$

$$= \pi a^2 \operatorname{cosec}\alpha.$$

EXAMPLE 11.18

Find the surface generated by the revolution of an arc of the catenary $y = c\cosh\frac{x}{c}$ from $x = 0$. $x = 0$ to any point (x, y) about the axis of x.

Solution. The equation of the catenary is

$$y = c\cosh\frac{x}{c}.$$

Therefore,

$$\frac{dy}{dx} = c\sinh\left(\frac{x}{c}\right)\cdot\frac{1}{c} = \sinh\left(\frac{x}{c}\right)$$

and so,

$$\frac{ds}{dx} = \sqrt{1+\left(\frac{dy}{dx}\right)^2} = \sqrt{1+\sinh^2\frac{x}{c}} = \cosh\frac{x}{c}.$$

Hence, the required surface area is

$$\int_0^x 2\pi y\frac{ds}{dx}dx = \int_0^x 2\pi c\cosh\frac{x}{c}.\cosh\frac{x}{c}dx$$

$$= \int_0^x 2\pi c\cosh^2\frac{x}{c}dx$$

$$= \pi c\int_0^x\left[1+\cosh\frac{2x}{c}\right]dx$$

$$= \pi c\left[x+\frac{c}{2}\sinh\frac{2x}{c}\right].$$

EXAMPLE 11.19

Find the area of the surface formed by the revolution of $y^2 = 4ax$ about the x-axis, by an arc from the vertex to one end of the latus rectum.

Solution. The equation of the parabola is $x^2 = 4ax$. Differentiating with respect to x, we get

$$2y\frac{dy}{dx} = 4a \text{ or } \frac{dy}{dx} = \frac{2a}{y}.$$

Therefore,

$$\frac{ds}{dx} = \sqrt{1+\left(\frac{dy}{dx}\right)^2} = \sqrt{1+\frac{4a^2}{y^2}} = \sqrt{1+\frac{4a^2}{4ax}}$$

$$= \sqrt{\frac{x+a}{x}}.$$

For the arc from the vertex O to L, the end of the latus rectum, x varies from 0 to a. Therefore,

$$\text{Required surface} = \int_0^a 2\pi y \frac{ds}{dx}dx$$

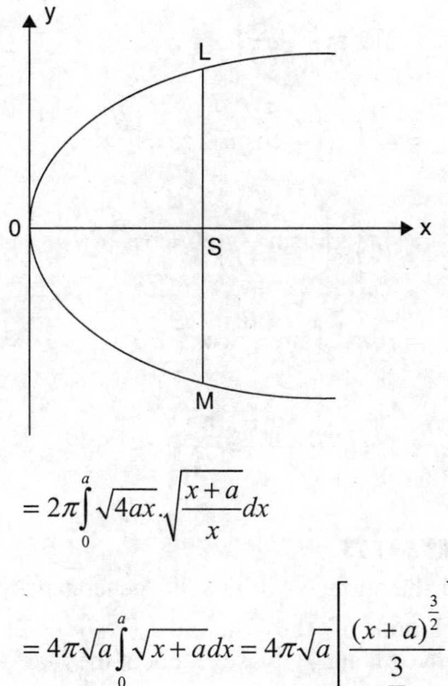

$$= 2\pi\int_0^a \sqrt{4ax}.\sqrt{\frac{x+a}{x}}dx$$

$$= 4\pi\sqrt{a}\int_0^a \sqrt{x+a}\,dx = 4\pi\sqrt{a}\left[\frac{(x+a)^{\frac{3}{2}}}{\frac{3}{2}}\right]_0^a$$

$$= \frac{8\pi\sqrt{a}}{3}\left[(2a)^{\frac{3}{2}} - a^{\frac{3}{2}}\right] = \frac{8\pi a^2}{3}(2\sqrt{2}-1).$$

11.5 SURFACE OF THE SOLID OF REVOLUTION (PARAMETRIC EQUATIONS)

The area of the surface of the solid generated by the revolution about the x-axis, of the area bounded by the curve $x = f(t), y = \phi(t)$, the x-axis, and the ordinates at the points, where $t = a$ and $t = b$ is

$$\int_a^b 2\pi y \frac{ds}{dt}dt,$$

where

$$\frac{ds}{dt} = \sqrt{\left(\frac{dx}{dt}\right)^2 + \left(\frac{dy}{dt}\right)^2}$$

EXAMPLE 11.20

Show that the surface generated by revolution of the tractrix $x = a\cos t + \frac{a}{2}\log\tan^2\frac{t}{2}$ and $y = a\sin t$ about its asymptote is equal to the surface of a sphere of radius a.

Solution. The equations of the tractrix are

$$x = a\cos t + \frac{a}{2}\log\tan^2\frac{t}{2} \text{ and } y = a\sin t.$$

This curve is symmetrical about both the axes and its asymptote is $y = 0$, that is, x-axis. We have (see Example 11.11)

$$\frac{ds}{dt} = \sqrt{\left(\frac{dx}{dt}\right)^2 + \left(\frac{dy}{dt}\right)^2} = \sqrt{\frac{a^2\cos^4 t}{\sin^2 t} + a^2\cos^2 t}$$

$$= \sqrt{\frac{a^2\cos^2 t(\cos^2 t + \sin^2 t)}{\sin^2 t}} = \frac{a\cos t}{\sin t}.$$

For the curve in the second quadrant, t varies from 0 to $\frac{\pi}{2}$. Hence, the required surface is given by

$$2\int_0^{\frac{\pi}{2}} 2\pi y\frac{ds}{dt}dt$$

$$= 4\pi\int_0^{\frac{\pi}{2}} a\sin t.\frac{a\cos t}{\sin t}dt$$

$$= 4\pi a^2\int_0^{\frac{\pi}{2}} \cos t\,dt = 4\pi a^2[\sin t]_0^{\frac{\pi}{2}} = 4\pi a^2$$

= the surface of a sphere of radius a.

EXAMPLE 11.21

Find the surface of the solid generated by revolving the loop of the curve $9y^2 = x(x-3)^2$ about the x-axis.

Solution. The parametric equations of the curve are

$$x = t^2 \text{ and } y = t - \frac{1}{3}t^3.$$

We have

$$\frac{dx}{dt} = 2t \text{ and } \frac{dy}{dt} = 1 - \frac{1}{3}3t^2 = 1 - t^2.$$

Therefore,

$$\frac{ds}{dt} = \sqrt{\left(\frac{dx}{dt}\right)^2 + \left(\frac{dy}{dt}\right)^2} = \sqrt{4t^2 + (1-t^2)^2} = 1 + t^2.$$

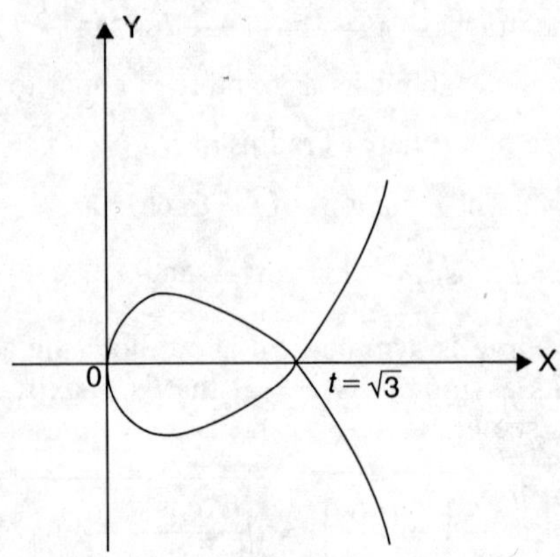

The curve cuts the x-axis at 0 and 3. So, as x varies from 0 to 3, t varies from 0 to $\sqrt{3}$. Hence, the required surface is

$$\int_0^{\sqrt{3}} 2\pi y \frac{ds}{dt} dt = 2\pi \int_0^{\sqrt{3}} \left(t - \frac{t^3}{3}\right)(1+t^2) dt$$

$$= 2\pi \int_0^{\sqrt{3}} \left[t + \frac{2t^3}{3} - \frac{t^5}{3}\right] dt = 2\pi \left[\frac{t^2}{2} + \frac{2t^4}{12} - \frac{t^6}{18}\right]_0^{\sqrt{3}}$$

$$= 2\pi \left[\frac{3}{2} + \frac{18}{12} - \frac{27}{18}\right] = 3\pi.$$

EXAMPLE 11.22

Find the surface area of the solid generated by revolving the cycloid $x = a(\theta + \sin\theta)$ and $y = a(1 - \cos\theta)$ about the tangent at the vertex.

Solution. The curve (see Example 11.14) is symmetrical about the y-axis and tangent at the vertex is x-axis. We have

$$\frac{dx}{d\theta} = a(1 + \cos\theta) \text{ and } \frac{dy}{d\theta} = a\sin\theta.$$

Therefore,

$$\frac{ds}{d\theta} = \sqrt{\left(\frac{dx}{d\theta}\right)^2 + \left(\frac{dy}{d\theta}\right)^2}$$

$$= \sqrt{a^2(1+\cos\theta)^2 + a^2\sin^2\theta}$$

$$= a\sqrt{1 + \cos^2\theta + 2\cos\theta + \sin^2\theta}$$

$$= a\sqrt{2(1+\cos\theta)}$$

$$= a\sqrt{2.2\cos^2\frac{\theta}{2}}$$

$$= 2a\cos\frac{\theta}{2}.$$

Therefore,

Required surface

$$= 2\int_0^{\pi} 2\pi y \frac{ds}{d\theta} d\theta$$

$$= 4\pi \int_0^{\pi} a(1 - \cos\theta) 2a\cos\frac{\theta}{2} d\theta$$

$$= 4\pi \int_0^{a} a\left(2\sin^2\frac{\theta}{2}\right).2a\cos\frac{\theta}{2} d\theta$$

$$= 16\pi a^2 \int_0^{\pi} \sin^2\frac{\theta}{2} \cos\frac{\theta}{2} d\theta$$

$$= 16a^2 \left|\frac{\sin^3\frac{\theta}{2}}{3.\frac{1}{2}}\right|_0^{\pi} = \frac{32}{3}\pi a^2.$$

EXAMPLE 11.23

Find the surface of the solid generated by the revolution of the asteroid $x^{\frac{2}{3}} + y^{\frac{2}{3}} = a^{\frac{2}{3}}$ or $x = a\cos^3 t$ and $y = a\sin^3 t$ about the x-axis.

Solution. As in the Example 11.13, we have

$$\frac{dx}{dt} = -3a\cos^2 t \sin t \text{ and } \frac{dy}{dt} = 3a\sin^2 t \cos t$$

Therefore,

$$\frac{ds}{dt} = \sqrt{9a^2\cos^4 t\sin^2 t + 9a^2\sin^4 t\cos^2 t}$$

$$= \sqrt{9a^2\cos^2 t\cos^2 t(\cos^2 t + \sin^2 t)}$$

$$= 3a \sin t \cos t.$$

Hence, the required surface is given by

$$2\int_0^{\frac{\pi}{2}} 2\pi y \frac{ds}{dt}.dt = 4\pi\int_0^{\frac{\pi}{2}} a\sin^3 t.3a \sin t \cos t dt$$

$$= 12\pi a^2 \int_0^{\frac{\pi}{2}} \sin^4 t \cos t dt$$

$$= 12\pi a^2 \left[\frac{\sin^5 t}{5}\right]_0^{\frac{\pi}{2}}$$

$$= \frac{12}{5}\pi a^2(1-0) = \frac{12}{5}\pi a^2.$$

11.6 SURFACE OF THE SOLID OF REVOLUTION (POLAR CURVES)

The curved surface of the solid generated by the revolution, about the initial line, of the area bounded by the curve $r = f(\theta)$ and the radii vector $\theta = \alpha$ and $\theta = \beta$, is

$$\int_\alpha^\beta 2\pi y \frac{ds}{d\theta}.d\theta,$$

where $\dfrac{ds}{d\theta} = \sqrt{r^2 + \left(\dfrac{dr}{d\theta}\right)^2}$, $y = r \sin \theta$.

EXAMPLE 11.24

Find the area of the surface of revolution formed by revolving the curve $r = 2a \cos \theta$ about the initial line.

Solution. The equation of the curve is $r = 2a \cos \theta$. Therefore, $\dfrac{dr}{d\theta} = -2a \sin \theta$. The curve is symmetrical about the initial line.

For the upper half of the circle, θ varies from 0 to $\dfrac{\pi}{2}$.

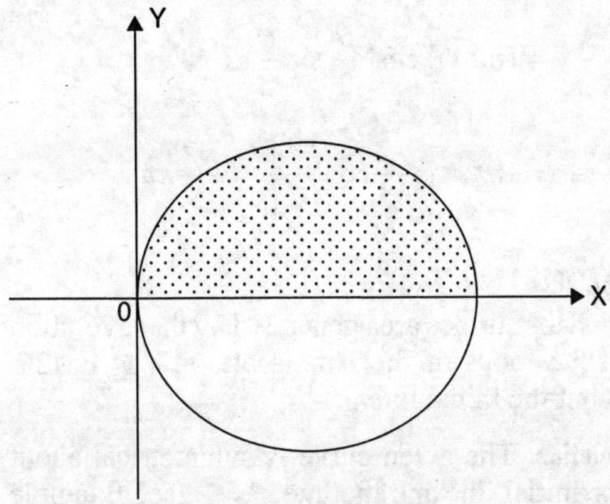

Therefore, the required surface area is equal to

$$\int_0^{\frac{\pi}{2}} 2\pi y \frac{ds}{d\theta}.d\theta = \int_0^{\frac{\pi}{2}} 2\pi r \sin \theta \left[r^2 + \left(\frac{dr}{d\theta}\right)^2\right]^{\frac{1}{2}} d\theta$$

$$= \int_0^{\frac{\pi}{2}} 4a\pi \cos \theta \sin \theta\sqrt{4a^2\cos^2\theta + 4a^2\sin^2\theta}\, d\theta$$

$$= 8\pi a^2 \int_0^{\frac{\pi}{2}} \cos\theta \sin \theta d\theta = 8\pi a^2 \left[\frac{\sin^2\theta}{2}\right]_0^{\frac{\pi}{2}}$$

$$= 4\pi a^2.$$

EXAMPLE 11.25

Find the surface of the solid formed by the revolution of the cardioid $r = a(1 + \cos \theta)$ about the initial line.

Solution. We have $r = a(1 + \cos \theta)$. Therefore, $\dfrac{dr}{d\theta} = -a \sin \theta$. For the upper half of the curve (see Example 11.15), θ varies from 0 to π.

Hence, the required surface area is

$$\int_0^\pi 2\pi y \frac{ds}{d\theta}.d\theta = 2\pi\int_0^\pi r \sin \theta \left[r^2 + \left(\frac{dr}{d\theta}\right)^2\right]^{\frac{1}{2}} d\theta$$

$$= 2\pi\int_0^\pi a(1 + \cos \theta) \sin \theta[a^2(1 + \cos \theta)^2$$

$$+ a^2\sin^2\theta]^{\frac{1}{2}} d\theta$$

$$= 16\pi a^2 \int_0^\pi \cos^3\frac{\theta}{2} \sin\frac{\theta}{2} \cos\frac{\theta}{2} d\theta$$

$$= 16\pi a^2 \int_0^{\pi} \cos^4 \frac{\theta}{2} \sin\frac{\theta}{2} d\theta$$

$$= 16\pi a^2 \left[-\frac{2}{5}\cos^5 \frac{\theta}{2} \right]_0^{\pi} = \frac{32}{5}\pi a^2.$$

EXAMPLE 11.26

Find the surface area generated by the revolution of the loops of the lemniscate $r^2 = a^2 \cos 2\theta$ about the initial line.

Solution. The given curve is symmetrical about the initial line and the line $\theta = \frac{\pi}{2}$ (see Example 11.16). It consists oftwo loops and the tangents at the pole are $\theta = -\frac{\pi}{4}$ and $\theta = \frac{\pi}{4}$. For half of the loop in the first quadrant, θ varies from 0 to $\theta = \frac{\pi}{4}$. Further,

$$\frac{dr}{d\theta} = \frac{a \sin 2\theta}{\sqrt{\cos 2\theta}} \text{ and } \frac{ds}{d\theta} = \sqrt{r^2 + \left(\frac{dr}{d\theta}\right)^2}$$

$$= \sqrt{a^2 \frac{\cos^2 2\theta + \sin^2 2\theta}{\cos 2\theta}} = \frac{a}{\sqrt{\cos 2\theta}}.$$

Therefore, the required surface is given by

$$2\int_0^{\frac{\pi}{4}} 2\pi y \frac{ds}{d\theta} d\theta = 4\pi \int_0^{\frac{\pi}{4}} r \sin\theta \frac{a}{\sqrt{\cos 2\theta}} d\theta$$

$$= 4\pi \int_0^{\frac{\pi}{4}} a\sqrt{\cos 2\theta} \sin\theta . \frac{a}{\sqrt{\cos 2\theta}} d\theta$$

$$= 4\pi a^2 \int_0^{\frac{\pi}{4}} \sin\theta \, d\theta$$

$$= 4\pi a^2 [-\cos\theta]_0^{\frac{\pi}{4}}$$

$$= 2\pi a^2 (2 - \sqrt{2}).$$

EXERCISES

1. Show that the volume of the solid formed by the revolution of the loop of the curve $x^2(a-x)$ about the x-axis is $2\pi(\log 2 - \frac{2}{3})a^3$
2. Find the volume of a sphere of radius a.

   Hint: The sphere is the solid of revolution generated by the revolution of a semicircular area about its bounding diameter. Thus, $V = \int_{-a}^{a} \pi y^2 dx$, $x^2 + y^2 = a^2$.

   Ans. $\frac{4}{3}\pi a^3$.
3. Find the volume of the solid generated by revolving the curve $xy^2 = 4(2-x)$ about the y-axis.

   Ans. $4\pi^2$.
4. Find the volume when the loop of the curve $y^2 = x(2x-1)^2$ revolves about the x-axis.

   Ans. $\frac{\pi}{48}$.
5. Find the volume generated by revolving the portion of the parabola $y^2 = 4ax$ cut off by its latus rectum about the axis of the parabola

   Hint: $V = \int_0^a \pi y^2 dx$.

   Ans. $2\pi a^3$.
6. Find the volume of the paraboloid generated by the revolution of the parabola $y^2 = 4ax$ about the x-axis from $x = 0$ to $x = b$.

   Ans. $2ab^2\pi$.
7. The area bounded by the parabola $y^2 = 4x$ and the straight line $4x - 3y + 2 = 0$ is rotated about the y-axis. Find the volume of the solid so generated.

   Ans. $\frac{\pi}{20}$.
8. Find the volume of the reel formed by the revolution of the cycloid $x = a(t + \sin t)$ and $y = a(1 - \cos t)$ about the tangent at the vertex.

   Ans. $\pi^2 a^3$.
9. Prove that the volume of the solid generated by the revolution about the x-axis of the loop of the curve $9y^2 = x(x-3)^2$ or $x = t^2$ and $y = t - \frac{1}{3}t^3$.

   Ans. $\frac{3\pi}{4}$.
10. Find the volume of the solid generated by the revolution of the cycloid $x = a(\theta - \sin\theta)$ and $y = a(1 - \cos\theta)$ about y-axis.

    Ans. $6\pi^3 a^3$.

11. Find the volume of the solid generated by revolving the lemniscate $r^2 = a^2 \cos 2\theta$ about the line $\theta = \frac{\pi}{2}$.

Ans. $\frac{\sqrt{2}\pi^2 a^3}{8}$

12. Show that the volume of the solid formed by the revolution of the curve $r = a + b\cos\theta$ $(a > b)$ about the initial line is $\frac{4}{3}\pi a(a^2 + b^2)$.

13. Find the surface of the solid generated by the revolution of the ellipse $x^2 + 4y^2 = 16$ about its major axis

Ans. $8\pi[1 + \frac{4\pi}{3\sqrt{3}}]$.

14. Find the surface area of the solid generated by revolving the cycloid $x = a(\theta - \sin\theta)$ and $y = a(1 - \cos\theta)$ about the x-axis.

Ans. $\frac{64}{3}\pi a^2$.

15. Show that the surface area of the solid generated by revolving one complete arc of the cycloid $x = a(\theta - \sin\theta)$ and $y = a(1 - \cos\theta)$ about the line $y = 2a$ is $\frac{32}{3}\pi a^2$.

16. Find the area of the surface of the solid formed by the revolution of the cardioid $r = a\ (1 - \cos\theta)$ about the initial line.

Ans. $\frac{32\pi a^2}{5}$.

17. The lemniscate $r^2 = a^2 \cos 2\theta$ revolves about a tangent at the pole. Show that the surface of the solid so generated is $4\pi a^2$.

12 Multiple Integrals

The aim of this chapter is to study double- and triple integrals along with their applications. Thus, we shall consider here the integrals of the functions of two- and three variables.

12.1 DOUBLE INTEGRALS

The notion of a double integral is an extension of the concept of a definite integral on the real line to the case of two-dimensional space. Let $f(x, y)$ be a continuous function of two independent variables x and y inside and on the boundary of a region R. Divide the region R into subdomains $R_1, R_2, \ldots, R_n$ of areas $\delta R_1, \delta R_2, \ldots, \delta R_n$, respectively. Let (x_i, y_i) be an arbitrary point inside the ith elementary area, δR_i. Consider the sum

$$\begin{aligned} S_n &= f(x_i, y_i)\delta R_1 + f(x_2, y_2)\,\delta R_2 + \ldots \\ &\quad + f(x_i, y_i)\,\delta R_i + \ldots + f(x_n, y_n)\delta R_n. \\ &= \sum_{i=1}^{n} f(x_i, y_i)\delta R_i. \end{aligned}$$

When $n \to \infty$, the number of subregions increases indefinitely such that the largest of the areas δR_i approaches zero. The $\lim\limits_{\substack{n\to\infty \\ \delta R_i \to 0}} S_n$, if exists, is called the *double integral* of the function $f(x,y)$ over the region (domain) R and is denoted by

$$\iint_R f(x, y)\, dR.$$

If the region R is divided into rectangular meshes by a network of lines parallel to the coordinate axes and if dx and dy be the length and breadth of a rectangular mess, then $dxdy$ is an element of area in Cartesian coordinates. In such a case, we have

$$\iint_R f(x, y)\, dR = \iint_R f(x, y)\, dx\, dy.$$

We now state, without proof, two theorems that provide sufficient conditions for the existence of a double integral over a closed region R.

Theorem 12.1. Let ϕ and ψ be two continuous functions defined on a closed interval $[a, b]$ such that $\phi(x) \le \psi(x)$ for all $x \in [a, b]$. Let f be a continuous function defined over $R = \{(x, y): a \le x \le b,\ \phi(x) \le y \le \psi(x)\}$. Then, $\iint_R f(x, y)dxdy$ and $\int_b^a \left[\int_{\phi(x)}^{\psi(x)} f(x, y)\, dy\right] dx$ exist and are equal.

Theorem 12.2. Let ϕ and ψ be two continuous functions defined on a closed interval $[c, d]$ such that $\phi(y) \le \psi(y)$ for $y \in [c, d]$. Let f be a continuous function defined over $R = \{(x, y) : c \le y \le d,\ \phi(y) \le x \le \psi(y)\}$. Then, $\iint_R f(x, y)\, dxdy$ and $\int_c^d \left[\int_{\phi(y)}^{\psi(y)} f(x, y)\, dx\right] dy$ exist and are equal.

EXAMPLE 12.1

Show that

$$\int_0^1 \left[\int_0^1 \frac{x-y}{(x+y)^3}\, dx\right] dy \ne \int_0^1 \left[\int_0^1 \frac{x-y}{(x+y)^3}\, dy\right] dx.$$

Solution. We have

$$\int_0^1 \left[\int_0^1 \frac{x-y}{(x+y)^3}\, dx\right] dy$$

$$= \int_0^1 \left[\int_0^1 \frac{x+y-2y}{(x+y)^3} \, dx \right] dy$$

$$= \int_0^1 \left[\int_0^1 \left\{ \frac{1}{(x+y)^2} - \frac{2y}{(x+y)^3} \right\} dx \right] dy$$

$$= \int_0^1 \left[\frac{-1}{x+y} + \frac{2y}{2(x+y)^2} \right]_0^1 dy$$

$$= \int_0^1 \left[\frac{-1}{(1+y)^2} \right] dy = \left[\frac{1}{1+y} \right]_0^1 = -\frac{1}{2}.$$

Similarly, we can show that

$$\int_0^1 \left[\int_0^1 \frac{x-y}{(x+y)^3} \, dy \right] dx = \frac{1}{2}.$$

Hence,

$$\int_0^1 \left[\int_0^1 \frac{x-y}{(x+y)^3} \, dx \right] dy \neq \int_0^1 \left[\int_0^1 \frac{x-y}{(x+y)^3} \, dy \right] dx.$$

The reason is that the function $f(x, y) = \dfrac{x-y}{(x+y)^3}$ is not continuous in $R = \{(x, y): 0 \le x \le 1, 0 \le y \le 1\}$.

12.2 PROPERTIES OF A DOUBLE INTEGRAL

1. Let $K \neq 0$ be any real number. Then,

$$\iint_R K f(x, y) \, dx \, dy = K \iint_R f(x, y) \, dx \, dy.$$

2. The double integral of the algebraic sum of a finite number of functions f_i is equal to the sum of the double integrals taken for each function. Thus,

$$\iint_R [f_1(x, y) + f_2(x, y) + \ldots + f_n(x, y)] \, dx \, dy$$
$$= \iint_R f_1(x, y) \, dx \, dy + \iint_R f_2(x, y) \, dx \, dy$$
$$+ \ldots + \iint_R f_n(x, y) \, dx \, dy.$$

3. If the region R is partitioned into two regions R_1 and R_2, then

$$\iint_R f(x, y) \, dx \, dy = \iint_{R_1} f(x, y) \, dx \, dy$$
$$+ \iint_{R_2} f(x, y) \, dx \, dy.$$

12.3 EVALUATION OF DOUBLE INTEGRALS (CARTESIAN COORDINATES)

The double integrals can be evaluated using Theorems 12.1 and 12.2. In fact,

(i) If the limits in the inner integral are functions of x, then we evaluate $\int_{\phi(x)}^{\psi(x)} f(x, y) \, dy$, first taking x as a constant and then evaluate the integrand (function of x), obtained in the first step, integrating it with respect to x between the limits a and b. Thus,

$$\int_a^b \int_{\phi(x)}^{\psi(x)} f(x, y) \, dx \, dy = \int_a^b \left[\int_{\phi(x)}^{\psi(x)} f(x, y) \, dy \right] dx.$$

(ii) If the limits in the inner integral are functions of y, then we evaluate $\int_{\phi(y)}^{\psi(y)} f(x, y) \, dx$, first taking y as a constant and then evaluate the integrand (function of y), obtained as a result of the first step, integrating it with respect to y between the limits c and d. Thus,

$$\int_c^d \int_{\phi(y)}^{\psi(y)} f(x, y) \, dx \, dy = \int_c^d \left[\int_{\phi(y)}^{\psi(y)} f(x, y) \, dx \right] dy.$$

EXAMPLE 12.2

Show that

$$\int_1^2 \left[\int_3^4 (xy + e^y) \, dx \right] dy = \int_3^4 \left[\int_1^2 (xy + e^y) \, dy \right] dx.$$

Solution. The function $f(x, y) = xy + e^y$ is a continuous function over the rectangle $R = \{(x, y): 1 \le x \le 2, 3 \le y \le 4\}$. Therefore, the values of these integrals are equal. In fact, we note that

$$\int_3^4 (xy + e^y) \, dx = \frac{7}{2} y + e^y$$

and so

$$\int_1^2 \int_3^4 (xy + e^y) \, dx \, dy = \int_1^2 \left[\frac{7}{2} y + e^y \right] dy$$

$$= \frac{21}{4} + e^2 - e.$$

One the other hand,

$$\int_1^2 (xy + e^y)\, dy = \frac{3}{2}x + e^2 - e$$

and so,

$$\int_3^4 \left[\int_1^2 (xy + e^y) dy\right] dx = \int_3^4 \left[\frac{3}{2}x + e^2 - e\right] dx$$

$$= \frac{21}{4} + e^2 - e$$

Hence, the result.

EXAMPLE 12.3

Evaluate $\iint x^2y^2\, dx\, dy$ over the circle $x^2 + y^2 \le 1$.

Solution. Since $x^2 + y^2 \le 1$, it follows that

$$x^2 \le 1 \quad \text{and} \quad y^2 \le 1 - x^2$$

or

$$|x| \le 1 \quad \text{and} \quad |y| \le \sqrt{1 - x^2}$$

or

$$-1 \le x \le 1 \quad \text{and} \quad -\sqrt{1 - x^2} \le y \le \sqrt{1 - x^2}.$$

The integrand $f(x, y) = x^2y^2$ is continuous over the region

$$R = \{(x, y): -1 \le x \le 1, -\sqrt{1 - x^2} \le y \le \sqrt{1 - x^2}\}.$$

Therefore,

$$\iint_R x^2y^2\, dx\, dy = \int_{-1}^{1} \left[\int_{-\sqrt{1-x^2}}^{\sqrt{1-x^2}} x^2y^2\, dy\right] dx$$

$$= \int_{-1}^{1} x^2 \left\{\frac{y^3}{3}\right\}_{-\sqrt{1-x^2}}^{\sqrt{1-x^2}} dx = \int_{-1}^{1} \frac{2}{3} x^2 (1 - x^2)^{\frac{3}{2}}\, dx$$

$$= \frac{4}{3}\int_0^1 x^2(1 - x^2)^{\frac{3}{2}}\, dx, \text{ since integrand is even}$$

$$= \frac{4}{3}\int_0^{\frac{\pi}{2}} \sin^2\theta \cos^4\theta\, d\theta, \text{ substituting } x = \sin\theta$$

$$= \frac{4}{3} \cdot \frac{1\ 3.1}{6.4.2} \cdot \frac{\pi}{2} = \frac{\pi}{24}.$$

EXAMPLE 12.4

Evaluate the double integral of the function $f(x, y) = 1 + x + y$ over a region bounded by $y = -x$, $x = \sqrt{y}$, and $y = 2$.

Solution. The region R is bounded by $y = -x$, the parabola $y^2 = x$, and the line $y = 2$. Thus, the limits of integration for x are $x = -y$, $x = \sqrt{y}$, $y = 0$, and $y = 2$. Therefore,

$$\iint_R (1 + x + y)\, dx\, dy = \int_0^2 \left[\int_{-y}^{\sqrt{y}} (1 + x + y)\, dx\right] dy$$

$$= \int_0^2 \left[x + \frac{x^2}{2} + xy\right]_{-y}^{\sqrt{y}} dy$$

$$= \int_0^2 \left[\left(\sqrt{y} + \frac{y}{2} + y\sqrt{y}\right) - \left(-y + \frac{y^2}{2} - y^2\right)\right] dy$$

$$= \int_0^1 \left(\frac{y^2}{2} + \frac{3y}{4} + \sqrt{y} + y\sqrt{y}\right) dy$$

$$= \left[\frac{y^3}{2} + \frac{3y^2}{4} + \frac{2}{3}y^{\frac{3}{2}} + \frac{2}{5}y^{\frac{5}{2}}\right]_0^2$$

$$= \frac{13}{3} + \frac{44}{15}\sqrt{2}.$$

EXAMPLE 12.5

Evaluate $\iint_R y\, dx\, dy$, where R is the region bounded by the parabolas $y^2 = 4x$ and $x^2 = 4y$.

Solution. The given parabolas are $y^2 = 4x$ and $x^2 = 4y$. Solving these equations, we get $x = 0$ and $y = 4$. The corresponding values of y are $y = 0$ and $y = 4$. Both the curves pass through the origin and the points of intersection are (0, 0) and (4, 4).

Thus, the limits of integration are $x = 0$ to $x = 4$ and $y = \frac{x^2}{4}$ to $y = 2\sqrt{x}$. Thus,

$$R = \left\{(x, y) : 0 \le x \le 4; \frac{x^2}{4} \le y \le 2\sqrt{x}\right\}.$$

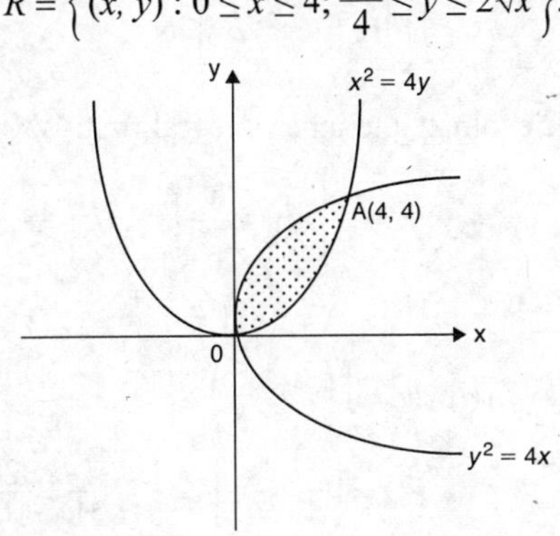

Therefore,

$$\iint_R y\, dx\, dy = \int_0^4 \left[\int_{\frac{x^2}{4}}^{2\sqrt{x}} y\, dy\right] dx$$

$$= \int_0^4 \left[\frac{y^2}{2}\right]_{\frac{x^2}{4}}^{2\sqrt{x}} dx = \int_0^4 \left(\frac{4x}{2} - \frac{x^4}{32}\right) dx$$

$$= \int_0^4 \left(2x - \frac{x^4}{32}\right) dx = \left[x^2 - \frac{x^5}{160}\right]_0^4 = \frac{48}{5}.$$

EXAMPLE 12.6

Calculate the double integral

$$\int_0^1 \int_0^{x^2} (x^2 + y^2)\, dx\, dy$$

and determine the region of integration.

Solution. The region of integration is bounded by the lines $x = 0$, $x = 1$, $y = 0$, and the parabola $x^2 = y$. Thus, the region is

$$R = \{(x, y) : 0 \le x \le 1; 0 \le y \le x^2\},$$

and is shown in the following figure:

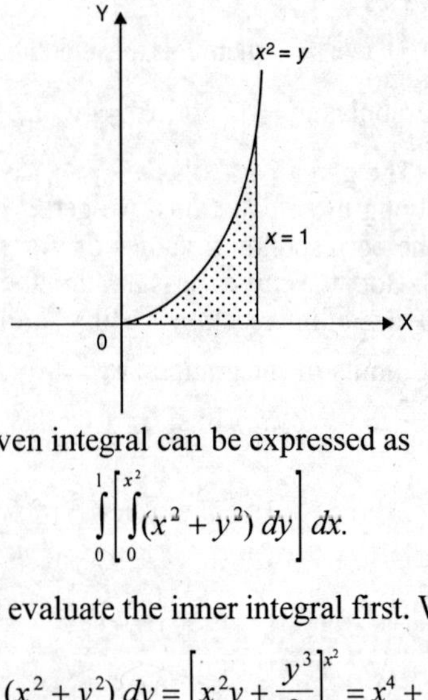

The given integral can be expressed as

$$\int_0^1 \left[\int_0^{x^2} (x^2 + y^2)\, dy\right] dx.$$

So, we evaluate the inner integral first. We have

$$\int_0^{x^2} (x^2 + y^2)\, dy = \left[x^2 y + \frac{y^3}{3}\right]_0^{x^2} = x^4 + \frac{x^6}{3}.$$

Therefore,

$$\iint_R (x^2 + y^2)\, dx\, dy = \int_0^1 \left(x^4 + \frac{x^6}{3}\right) dx$$

$$= \left[\frac{x^5}{5} + \frac{x^7}{21}\right]_1^0 = \frac{1}{5} + \frac{1}{21} = \frac{26}{105}.$$

EXAMPLE 12.7

Evaluate the double integral $\iint_R e^{x+y}\, dR$ over the region R, which lies between two squares with their center at the origin and with sides parallel to the axes of coordinates, if each side of the inner square is equal to 2 and that of the outer square is 4.

Solution. The region R is irregular. However, the straight lines $x = -1$ and $x = 1$ divide this region into four regular subregions R_1, R_2, R_3, and R_4.

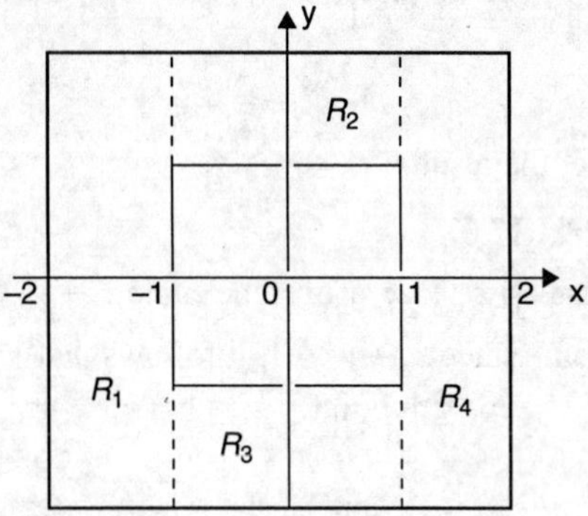

Therefore,

$$\iint_R e^{x+y}\, dR = \iint_{R_1} e^{x+y}\, dx\, dy + \iint_{R_2} e^{x+y}\, dx\, dy$$

$$+ \iint_{R_3} e^{x+y}\, dx\, dy + \iint_{R_4} e^{x+y}\, dx\, dy$$

$$= \int_{-2}^{-1} \left[\int_{-2}^{2} e^{x+y}\, dy\right] dx + \int_{-1}^{1} \left[\int_{1}^{2} e^{x+y}\, dy\right] dx$$

$$+ \int_{-1}^{1} \left[\int_{-2}^{-1} e^{x+y}\, dy\right] dx + \int_{1}^{2} \left[\int_{-1}^{2} e^{x+y}\, dy\right] dx$$

$$= (e^2 - e^{-2})(e^{-1} - e^{-2}) + (e^2 - e)(e - e^{-1})$$

$$+ (e^{-1} - e^{-2})(e - e^{-1}) + (e^2 - e^{-2})(e^2 - e)$$

$$= (e^3 - e^{-3})(e - e^{-1}) = 4 \sinh 3 \sinh 1.$$

EXAMPLE 12.8

Evaluate $\iint xy\, dx\, dy$ over the positive quadrant of the circle $x^2 + y^2 = a^2$.

Solution. The region of integration is

$$R = \left\{(x, y): 0 \le x \le a; 0 \le y \le \sqrt{a^2 - x^2}\right\}.$$

The integrand $f(x, y) = xy$ is continuous over R. Therefore,

$$\iint_R xy\, dx\, dy = \int_0^a \left[\int_0^{\sqrt{a^2 - x^2}} xy\, dy\right] dx$$

$$= \int_0^a \left[x\frac{y^2}{2}\right]_0^{\sqrt{a^2-x^2}} = \frac{1}{2}\int_0^a x[a^2 - x^2]\, dx$$

$$= \frac{1}{2}\left[\int_0^a a^2x\, dx - \int_0^a x^3 dx\right]$$

$$= \frac{1}{2}\left[a^2\frac{x^2}{2}\right]_0^a - \frac{1}{2}\left[\frac{x^4}{4}\right]_0^a = \frac{a^4}{4} - \frac{a^4}{8} = \frac{a^4}{8}.$$

EXAMPLE 12.9

Evaluate $\iint_R xy\; dx\; dy$, where R is the domain bounded by the x-axis, ordinate $x = 2a$, and the curve $x^2 = 4ay$.

Solution. The region of integration is

$$R = \left\{(x, y) : 0 \le x \le 2a;\ 0 \le y \le \frac{x^2}{4a}\right\}.$$

The region is bounded by $y = 0$, $x = 2a$, and the parabola $x^2 = 4ay$.

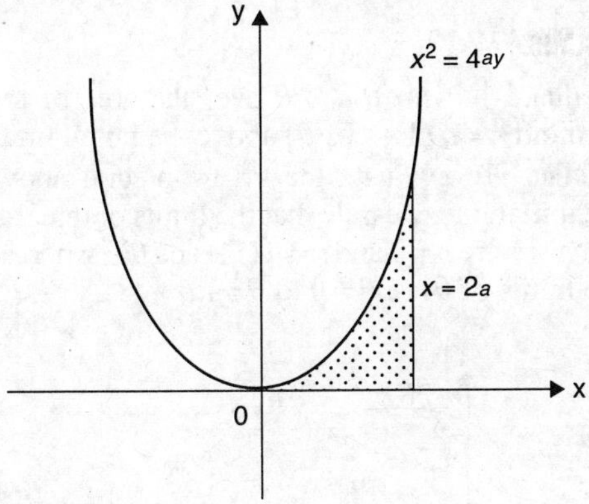

Therefore,

$$\iint_R xy\; dx\, dy = \int_0^{2a}\int_0^{\frac{x^2}{4a}} xy\; dy\, dx$$

$$= \int_0^{2a}\left\{x\left[\frac{y^2}{2}\right]_0^{\frac{x^2}{4a}}\right\}dx = \int_0^{2a}\frac{x^5}{32a^2}\, dx$$

$$= \frac{1}{32a^2}\left[\frac{x^6}{6}\right]_0^{2a} = \frac{1}{32a^2}\left[\frac{64a^6}{6}\right] = \frac{a^4}{3}.$$

EXAMPLE 12.10

Evaluate $\int_0^1\int_x^{\sqrt{x}} (x^2 + y^2)\; dy\; dx.$

Solution. The region of integration is bounded by $x = 0$, $x = 1$, $y = x$, and the parabola $y^2 = x$. Thus,

$$R = \{(x, y) : 0 \le x \le 1;\ x \le y \le \sqrt{x}\}.$$

Therefore,

$$\int_0^1\int_x^{\sqrt{x}}(x^2 + y^2)\; dy\; dx = \int_0^1\left[\int_x^{\sqrt{x}}(x^2 + y^2)\; dy\right] dx$$

$$= \int_0^1\left\{\left[x^2y + \frac{y^3}{3}\right]_x^{\sqrt{x}}\right\} dx$$

$$= \int_0^1\left[\left(x^{\frac{5}{2}} + \frac{x^{\frac{3}{2}}}{3}\right) - \left(x^3 + \frac{x^3}{3}\right)\right] dx$$

$$= \int_0^1 x^{\frac{5}{2}} + \frac{x^{\frac{3}{2}}}{3} - \frac{4x^3}{3}\, dx.$$

$$= \left[\frac{x^{\frac{7}{2}}}{\frac{7}{2}} + \frac{x^{\frac{5}{2}}}{\frac{15}{2}} - \frac{4x^4}{12}\right]_0^1 = \frac{2}{7} + \frac{2}{15} - \frac{1}{3} = \frac{9}{105} = \frac{3}{35}.$$

EXAMPLE 12.11

Evaluate $\iint (x + y)^2 dx\; dy$ over the area bounded by the ellipse $\frac{x^2}{a^2} + \frac{y^2}{b^2} = 1.$

Solution. The equation of the ellipse is $\frac{x^2}{a^2} + \frac{y^2}{b^2} = 1.$

Therefore,

$$\frac{y^2}{b^2} = 1 - \frac{x^2}{a} \quad \text{or} \quad \frac{y}{b} = \pm\sqrt{1 - \frac{x^2}{a^2}} \quad \text{or}$$

$$y = \pm\frac{b}{a}\sqrt{a^2 - x^2}.$$

The ellipse cuts the x-axis at $x = \pm\, a$. Therefore, the region of integration is

$$R = \left\{(x, y):\ -a \le x \le a;\ -\frac{b}{a}\sqrt{a^2 - x^2} \le y \le \frac{b}{a}\sqrt{a^2 - x^2}\right\}.$$

Since $x^2 + y^2$ is an even function of y and xy is an odd function of y, we have

$$\iint_R (x + y)^2\, dx\; dy$$

$$= \int_{-a}^{a}\left[\int_{-\frac{b}{a}\sqrt{a^2-x^2}}^{\frac{b}{a}\sqrt{a^2-x^2}} (x^2 + y^2 + 2xy)\; dy\right] dx$$

$$= \int_{-a}^{a} \left[\int_{-\frac{b}{a}\sqrt{a^2-x^2}}^{\frac{b}{a}\sqrt{a^2-x^2}} (x^2+y^2)\, dy \right] dx$$

$$+ \int_{-a}^{a} \left[\int_{-\frac{b}{a}\sqrt{a^2-x^2}}^{\frac{b}{a}\sqrt{a^2-x^2}} 2xy\, dy \right] dx$$

$$= \int_{-a}^{a} \left[2 \int_{0}^{\frac{b}{a}\sqrt{a^2-x^2}} (x^2+y^2)\, dy \right] dx + 0$$

$$= 2 \int_{-a}^{a} \left[x^2 y + \frac{y^3}{3} \right]_0^{\frac{b}{a}\sqrt{a^2-x^2}} dx$$

$$= 2 \int_{-a}^{a} \left[\frac{x^2 b}{a} \sqrt{a^2-x^2} + \frac{b^3}{3a^3} (a^2-x^2)^{\frac{3}{2}} \right] dx$$

$$= 4 \int_{-a}^{a} \left[\frac{b}{a} x^2 \sqrt{a^2-x^2} + \frac{b^3}{3a^3} (a^2-x^2)^{\frac{3}{2}} \right] dx.$$

Substituting $x = a \sin \theta$, we have $dx = a \cos \theta\, d\theta$ and so,

$$\iint_R (x+y)^2\, dx\, dy$$

$$= 4 \int_0^{\frac{\pi}{2}} \left[a^3 b \sin^2 \theta \cos^2 \theta + \frac{ab^3}{3} \cos^4 \theta \right] d\theta$$

$$= 4 \left[a^3 b \cdot \frac{1}{4.2} \cdot \frac{\pi}{2} + \frac{ab^3}{3} \cdot \frac{3.1}{4.2} \cdot \frac{\pi}{2} \right]$$

$$= \frac{\pi}{4} (a^3 b + ab^3) = \frac{\pi\, ab}{4} (a^2 + b^2).$$

EXAMPLE 12.12

Evaluate $\displaystyle\int_0^1 \int_0^{\sqrt{1+x^2}} \frac{1}{1+x^2+y^2}\, dy\, dx.$

Solution. We have

$$\int_0^1 \int_0^{\sqrt{1+x^2}} \frac{1}{1+x^2+y^2}\, dy\, dx.$$

$$= \int_0^1 \left[\int_0^{\sqrt{1+x^2}} \frac{1}{1+x^2+y^2}\, dy \right] dx.$$

$$= \int_0^1 \left[\int_0^{\sqrt{1+x^2}} \frac{1}{(1+x^2)+y^2}\, dy \right] dx.$$

$$= \int_0^1 \frac{1}{\sqrt{1+x^2}} \left[\tan^{-1} \frac{y}{\sqrt{1+x^2}} \right]_0^{\sqrt{1+x^2}} dx,$$

$$\text{since } \int \frac{dx}{a^2+x^2} = \frac{1}{a} \tan^{-1} \frac{x}{a}$$

$$= \int_0^1 \frac{1}{\sqrt{1+x^2}} [\tan^{-1} 1 - \tan^{-1} 0] dx$$

$$= \frac{\pi}{4} \int_0^1 \frac{dx}{\sqrt{1+x^2}} = \frac{\pi}{4} \left[\log \left(\frac{x+\sqrt{1+x^2}}{1} \right) \right]_0^1$$

$$= \frac{\pi}{4} [\log (1+\sqrt{2}) - \log 1] = \frac{\pi}{4} \log (1+\sqrt{2}).$$

12.4 EVALUATION OF DOUBLE INTEGRALS (POLAR COORDINATES)

We wish to evaluate $\displaystyle\int_{\theta_1}^{\theta_2} \int_{r_1}^{r_2} f(r, \theta)\, dr\, d\theta$ over the region bounded by the straight lines $\theta = \theta_1$ and $\theta = \theta_2$ and the curves $r = r_1$ and $r = r_2$. To do so, we first integrate $f(r, \theta)$ with respect to r between the limits $r = r_1$ and $r = r_2$. The resulting integrand is then integrated with respect to θ between the limits $\theta = \theta_1$ and $\theta = \theta_2$.

EXAMPLE 12.13

Evaluate $\iint r \sin \theta\, dr\, d\theta$ over the area of the cardioid $r = a(1 + \cos \theta)$ above the initial line.

Solution. To evaluate the given integral above the initial line, we note that the limits of integration for r are $r = 0$ and $r = a(1+\cos\theta)$, whereas the limits of θ are $\theta = 0$ to $\theta = \pi$.

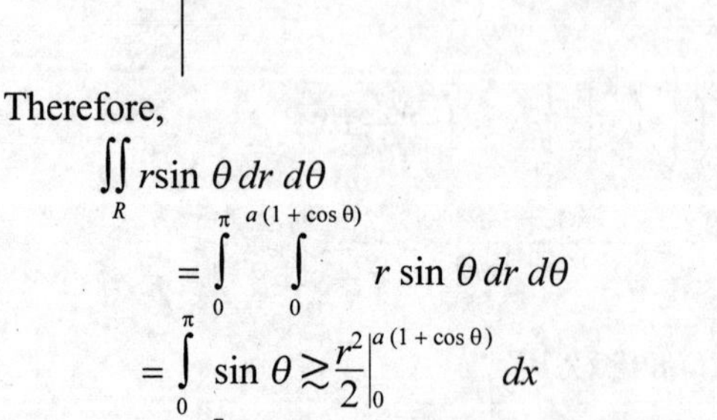

Therefore,

$$\iint_R r \sin \theta\, dr\, d\theta$$

$$= \int_0^{\pi} \int_0^{a(1+\cos\theta)} r \sin \theta\, dr\, d\theta$$

$$= \int_0^{\pi} \sin \theta \gtrsim \left. \frac{r^2}{2} \right|_0^{a(1+\cos\theta)} dx$$

$$= \frac{1}{2} \int_0^{\pi} a^2 (1+\cos\theta)^2 \sin \theta\, d\theta$$

$$= \frac{a^2}{2}\int_0^{\pi}\left(2\cos^2\frac{\theta}{2}\right)^2 2\sin\frac{\theta}{2}\cos\frac{\theta}{2}\,d\theta$$

$$= 4a^2\int_0^{\pi}\cos^5\frac{\theta}{2}\sin\frac{\theta}{2}\,d\theta$$

$$= 4a^2\int_0^{\frac{\pi}{2}}\cos^5\phi\sin\phi . 2d\phi,\ \theta = 2\phi$$

$$= 8a^2\int_0^{\frac{\pi}{2}}\cos^5\phi\sin\phi\,d\,\phi$$

$$= -\,8a^2\int_0^{\frac{\pi}{2}}\cos^5\phi\,(-\sin\phi)\,d\phi$$

$$= -\,8a^2\left[\frac{\cos^6\phi}{6}\right]_0^{\frac{\pi}{2}} = \frac{4a^2}{3}.$$

EXAMPLE 12.14

Calculate $\iint r^3 dr\, d\theta$ over the area included between the circles $r = 2\sin\theta$ and $r = 4\sin\theta$.

Solution. The region of integration is between the circles as shown in the following figure:

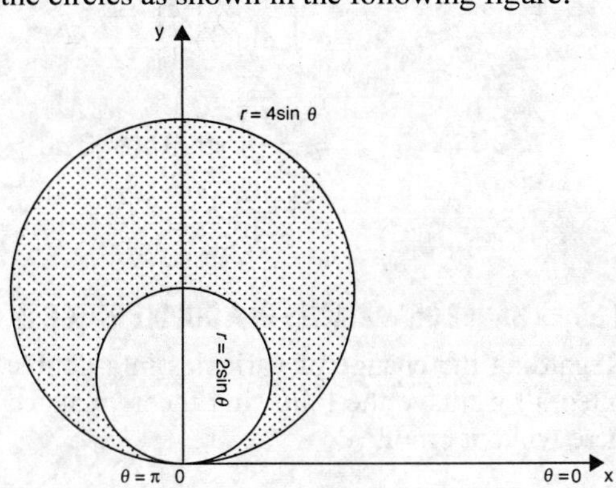

Therefore,

$$\iint r^3 dr\, d\theta = \int_0^{\pi}\int_{2\sin\theta}^{4\sin\theta} r^3 dr\, d\theta = \int_0^{\pi}\left[\frac{r^4}{4}\right]_{2\sin\theta}^{4\sin\theta} d\theta$$

$$= \frac{1}{4}\int_0^{\pi}(256\sin^4\theta - 16\sin^4\theta)\,d\theta$$

$$= 60\int_0^{\pi}\sin^4\theta\,d\theta = 120\int_0^{\frac{\pi}{2}}\sin^4\phi\,d\,\phi,\ \theta = 2\phi$$

$$= 120\cdot\frac{3.1}{4.2}\cdot\frac{\pi}{2} = 45\,\frac{\pi}{2}.$$

EXAMPLE 12.15

Evaluate $\iint \frac{r\,dr\,d\theta}{\sqrt{a^2 + r^2}}$ over the loop of the lemnis cates $r^2 = a^2\cos 2\theta$.

Solution. From the figure of the curve, we note that in the region of integration, r varies from 0 to $a\sqrt{\cos 2\theta}$ and θ varies from $-\frac{\pi}{4}$ to $\frac{\pi}{4}$

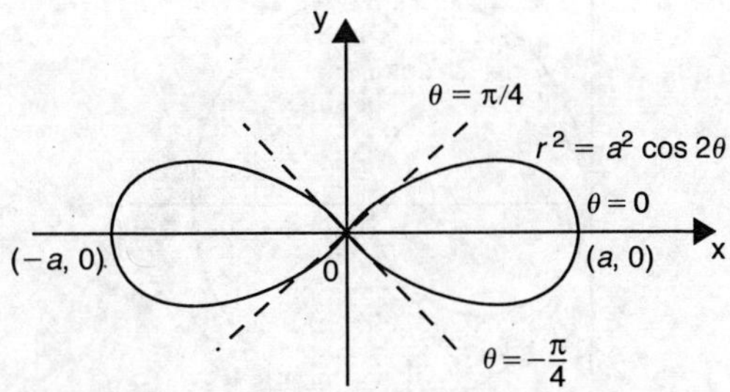

Therefore,

$$\iint \frac{rdr\,d\theta}{\sqrt{a^2 + r^2}} = \int_{-\frac{\pi}{4}}^{\frac{\pi}{4}}\left[\int_0^{a\sqrt{\cos 2\theta}} \frac{r}{\sqrt{a^2 + r^2}}\,dr\right]d\theta$$

$$= \int_{-\frac{\pi}{4}}^{\frac{\pi}{4}}\left[\frac{1}{2}\int_0^{a\sqrt{\cos 2\theta}} 2r\,(a^2 + r^2)^{-\frac{1}{2}}\,dr\right]d\theta$$

$$= \frac{1}{2}\int_{-\frac{\pi}{4}}^{\frac{\pi}{4}}\left|\frac{(a^2 + r^2)^{\frac{1}{2}}}{\frac{1}{2}}\right|_0^{a\sqrt{\cos 2\theta}} d\theta$$

$$= \int_{-\frac{\pi}{4}}^{\frac{\pi}{4}} a\,[(1 + \cos 2\theta)^{\frac{1}{2}} - 1]\,d\theta$$

$$= a\int_{-\frac{\pi}{4}}^{\frac{\pi}{4}} a\,[(2\cos^2\theta)^{\frac{1}{2}} - 1]\,d\theta$$

$$= a\int_{-\frac{\pi}{4}}^{\frac{\pi}{4}}(\sqrt{2}\cos\theta - 1)\,d\theta$$

$$= 2a\int_{-\frac{\pi}{4}}^{\frac{\pi}{4}}(\sqrt{2}\cos\theta - 1)\,d\theta$$

$$= 2a\,|\sqrt{2}\sin\theta - \theta|_0^{\frac{\pi}{4}} = 2a\left(1 - \frac{\pi}{4}\right).$$

EXAMPLE 12.16

Evaluate $\iint r^3 dr\, d\theta$ over the area included between the circles $r = 2a\cos\theta$, $r = 2b\cos\theta$, where $b < a$.

Solution. The region of integration between the given circles is shown in the following figure:

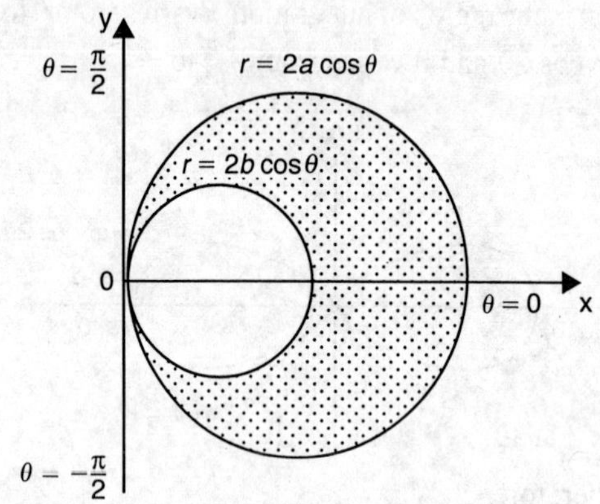

In the region of integration, r varies from $2b\cos\theta$ to $2a\cos\theta$, whereas θ varies from $-\frac{\pi}{2}$ to $\frac{\pi}{2}$. Therefore,

$$\iint r^3 dr\, d\theta = \int_{-\frac{\pi}{2}}^{\frac{\pi}{2}} \left[\int_{2b\cos\theta}^{2a\cos\theta} r^3 dr \right] d\theta$$

$$= \int_{-\frac{\pi}{2}}^{\frac{\pi}{2}} \left[\frac{r^4}{4} \right]_{2b\cos\theta}^{2a\cos\theta} d\theta$$

$$= \frac{1}{4} \int_{-\frac{\pi}{2}}^{\frac{\pi}{2}} [16a^4\cos^4\theta - 16b^4\cos^4\theta]\, d\theta$$

$$= 4 \int_{-\frac{\pi}{2}}^{\frac{\pi}{2}} (a^4 - b^4)\cos^4\theta\, d\theta$$

$$= 8 \int_{0}^{\frac{\pi}{2}} (a^4 - b^4)\cos^4\theta\, d\theta$$

$$= 8(a^4 - b^4)\cdot \frac{3.1}{4.2}\cdot\frac{\pi}{2} = \frac{3\pi}{2}(a^4 - b^4).$$

EXAMPLE 12.17

Show that $\iint r^2 \sin\theta\, dr\, d\theta = \frac{2a^3}{3}$, where R is the region bounded by the semi-circle $r = 2a\cos\theta$, above the initial line.

Solution. The region of integration is shown in the following figure. In this region r varies from 0 to $2a\cos\theta$, whereas θ varies from 0 to $\frac{\pi}{2}$.

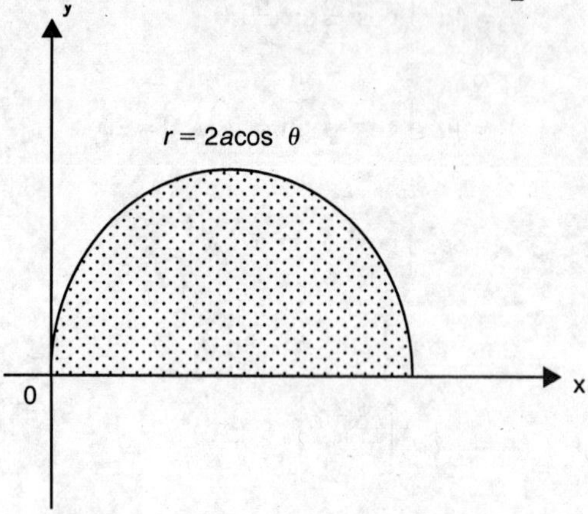

Therefore,

$$\iint_R r^2 \sin\theta\, dr\, d\theta = \int_0^{\frac{\pi}{2}} \left[\int_0^{2a\cos\theta} r^2 \sin\theta\, dr \right] d\theta$$

$$= \int_0^{\frac{\pi}{2}} \sin\theta \left| \frac{r^3}{3} \right|_0^{2a\cos\theta} d\theta$$

$$= \frac{8a^3}{3} \int_0^{\frac{\pi}{2}} \sin\theta\cos^3\theta\, d\theta$$

$$= \frac{8a^3}{3}\cdot\frac{2}{4.2} = \frac{2a^3}{3}.$$

12.5 CHANGE OF VARIABLES IN A DOUBLE INTEGRAL

Regarding the change of variables in a double integral, we have the following theorem stated here without proof.

Theorem 12.3 Let D be a domain in $\Re^2$ and $f: D \to \Re$ be integrable over D. Suppose that D be mapped on to a set A of the uv-plane by the transformation $x = \phi(u, v)$ and $y = \psi(u, v)$, where ϕ and ψ have continuous partial derivatives in A and the Jacobian $\frac{\partial(x, y)}{\partial(u, v)} \neq 0$ for all $(u, v) \in A$. Then,

$$\iint_D f(x, y)\, dx\, dy = \iint_A f(\phi(u, v), \psi(u, v))\, |J|\, du dv.$$

Deduction. If there is a change in variable from Cartesian- to polar coordinates, then $x = r\cos\theta$, $y = r\sin\theta$, and

$$J = \begin{vmatrix} \frac{\partial x}{\partial r} & \frac{\partial x}{\partial \theta} \\ \frac{\partial y}{\partial r} & \frac{\partial y}{\partial \theta} \end{vmatrix} = \begin{vmatrix} \cos\theta & r\sin\theta \\ \sin\theta & r\cos\theta \end{vmatrix}$$

$$= r(\cos^2\theta + \sin^2\theta) = r.$$

Therefore,

$$\iint_D f(x, y)\, dx\, dy = \iint_A f(r\cos\theta, r\sin\theta) r\, d\theta\, dr.$$

EXAMPLE 12.18

Calculate the integral

$$\iint \frac{(x-y)^2}{x^2+y^2} dx\, dy$$

over the circle $x^2 + y^2 \le 1$.

Solution. The value of the given integral is four times the value of the integral taken over the positive quadrant of the circle $x^2 + y^2 = 1$. Substituting $x = r\cos\theta$ and $y = r\sin\theta$, the given integral is equal to

$$4\int_0^1\int_0^{\frac{\pi}{2}} \frac{1}{r^2}(r\cos\theta - r\sin\theta)^2 \frac{\partial(x, y)}{\partial(r, \theta)} dr\, d\theta$$

$$= 4\int_0^1 r \int_0^{\frac{\pi}{2}} (\cos^2\theta + \sin^2\theta - 2\sin\theta\cos\theta)\, dr\, d\theta$$

$$= 4\int_0^{\frac{\pi}{2}} (1 - \sin 2\theta)\left[\int_0^1 r dr\right] d\theta$$

$$= 2\int_0^{\frac{\pi}{2}} (1 - \sin 2\theta)\, d\theta$$

$$= 2\left[\theta + \frac{\cos 2\theta}{2}\right]_0^{\frac{\pi}{2}} = \pi - 2.$$

EXAMPLE 12.19

Evaluate $\int_0^\infty\int_0^\infty e^{-(x^2+y^2)} dy\, dx$ by changing topolar coordinates. Hence, deduce that $\int_0^\infty e^{-x^2} dx = \frac{\sqrt{\pi}}{2}$.

Solution. In the given integral, both x and y vary from 0 to ∞. Hence, the region of integration is xy-plane. Changing to polar coordinates by substituting $x = r\cos\theta$ and $y = r\sin\theta$, we get $x^2 + y^2 = r^2$; and in the region of integration, r varies from 0 to ∞ and θ varies from 0 to $\frac{\pi}{2}$. Thus,

$$\int_0^\infty\int_0^\infty e^{-(x^2+y^2)} dy\, dx = \int_0^{\frac{\pi}{2}}\int_0^\infty e^{-r^2} \frac{\partial(x, y)}{\partial(r, \theta)} dr\, d\theta$$

$$= \int_0^{\frac{\pi}{2}}\int_0^\infty re^{-r^2} dr\, d\theta$$

$$= \int_0^{\frac{\pi}{2}}\int_0^\infty e^{-t}. \frac{dt}{2} d\theta,\ r^2 = t$$

$$= \frac{1}{2}\int_0^{\frac{\pi}{2}} \left|\frac{e^{-t}}{-1}\right|_0^\infty d\theta$$

$$= \frac{-1}{2}\int_0^{\frac{\pi}{2}} (0 - 1)\, d\theta = \frac{\pi}{4}. \qquad (1)$$

On the other hand,

$$\int_0^\infty\int_0^\infty e^{-(x^2+y^2)} dy\, dx = \int_0^\infty e^{-x^2} dx \int_0^\infty e^{-y^2} dy$$

$$= \left[\int_0^\infty e^{-x^2} dx\right]^2.$$

Hence, (1) implies

$$\int_0^\infty e^{-x^2} dx = \frac{\sqrt{\pi}}{2}.$$

EXAMPLE 12.20

Evaluate $\iint \sqrt{\frac{a^2b^2 - b^2x^2 - a^2y^2}{a^2b^2 + b^2x^2 + a^2y^2}}\, dx\, dy$ over the positive quadrant of the ellipse $\frac{x^2}{a^2} + \frac{y^2}{b^2} = 1$.

Solution. Substituting $\frac{x}{a} = X$ and $\frac{y}{b} = Y$, the problem reduces to the evaluation of $\iint ab \sqrt{\frac{1 - X^2 - Y^2}{1 + X^2 + Y^2}}\, dx\, dy$ over the positive quadrant of the circle $X^2 + Y^2 = 1$. Substituting $X = r\cos\theta$ and $Y = r\sin\theta$, we have $dXdY = rdr\, d\theta$. In the region of integration r varies from 0 to1 and θ varies from 0 to $\frac{\pi}{2}$ Hence, the given integral reduces to

$$ab\int_0^{\frac{\pi}{2}}\int_0^1 \sqrt{\frac{1 - r^2}{1 + r^2}}\, rdr\, d\theta$$

$$= ab\int_0^1 \sqrt{\frac{1 - r^2}{1 + r^2}}\, rdr \int_0^{\frac{\pi}{2}} d\theta$$

$$= \frac{ab\pi}{2}\int_0^1 \sqrt{\frac{1-r^2}{1+r^2}}\, r dr$$

$$= \frac{ab\pi}{2}\int_0^{\frac{\pi}{2}} \sqrt{\frac{1-\sin t}{1+\sin t}}\,\frac{1}{2}\cos t\, dt,\ r^2 = \sin t$$

$$= \frac{\pi\, ab}{4}\int_0^{\frac{\pi}{2}} \sqrt{\frac{1-\sin t}{1+\sin t}}\,\frac{\sqrt{1-\sin t}}{\sqrt{1-\sin t}}\cos t\, dt$$

$$= \frac{\pi\, ab}{4}\int_0^{\frac{\pi}{2}} \frac{1-\sin t}{\cos t}\cos t\, dt$$

$$= \frac{\pi\, ab}{4}\int_0^{\frac{\pi}{2}} (1-\sin t)\, dt$$

$$= \frac{\pi\, ab}{4}[t+\cos t]_0^{\frac{\pi}{2}} = \frac{\pi\, ab}{4}\left[\frac{\pi}{2}-1\right]$$

$$= \frac{\pi\, ab}{8}(\pi-2).$$

Remark 12.1 Substituting $a = b = 1$, the problem reduces to the evaluation of $\iint \sqrt{\frac{1-x^2-y^2}{1+x^2+y^2}}\,dx\,dy$ over the positive quadrant of the circle $x^2+y^2 = 1$. The value of the integral in that case is $\frac{\pi^2}{8}-\frac{\pi}{4}$.

EXAMPLE 12.21

Evaluate $\iint_{x^2+y^2\le 1} \sin \pi (x^2+y^2)\, dx\, dy.$

Solution. Changing to polar form yields

$$\iint_{x^2+y^2\le 1} \sin \pi (x^2+y^2)\, dx\, dy = \int_0^{2\pi}\int_0^1 \sin(\pi r^2)\, r dr\, d\theta$$

$$= \int_0^{2\pi}\int_0^1 \sin(\pi t)\frac{dt}{2}\, d\theta$$

$$= \frac{1}{2}\int_0^{2\pi} \left| -\frac{\cos \pi t}{\pi}\right|_0^1 d\theta$$

$$= -\frac{1}{2\pi}\int_0^{2\pi} [\cos \pi - \cos 0]\, d\theta$$

$$= \frac{2}{2\pi}\int_0^{2\pi} d\theta = 2.$$

EXAMPLE 12.22

Evaluate $\int_0^2 \int_0^{\sqrt{2x-x^2}} \frac{x}{\sqrt{x^2+y^2}}\, dy\, dx$ by changing to polar coordinates.

Solution. The limits of y are from 0 to $\sqrt{2x-x^2}$. But $y = \sqrt{2x-x^2}$ implies $y^2 = 2x - x^2$ or $x^2+y^2 = 2x$. Thus, the region of integration is bounded by $x = 0$, $x = 2$, $y = 0$, and $x^2+y^2 = 2x$. Changing to polar coordinates $x^2+y^2 = 2x$ transforms to $r^2 = 2r\cos\theta$ or $r = 2\cos\theta$ or $r = 2\cos\theta$.

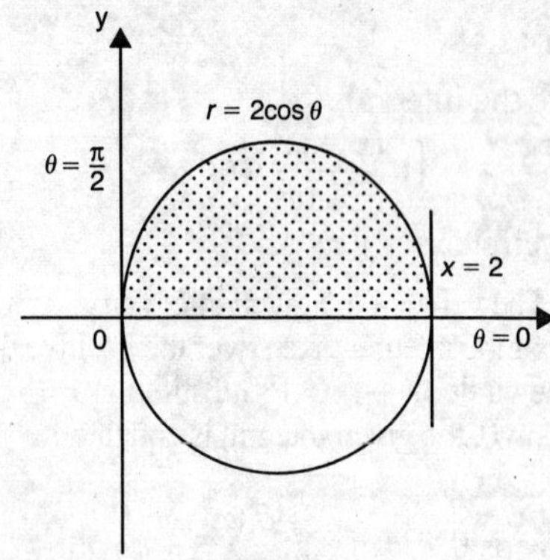

For the region of integration, r varies from 0 to $2\cos\theta$, whereas θ varies from 0 to $\frac{\pi}{2}$. Therefore,

$$\int_0^2 \int_0^{\sqrt{2x-x^2}} \frac{x}{\sqrt{x^2+y^2}}\, dy\, dx$$

$$= \int_0^{\frac{\pi}{2}}\int_0^{2\cos\theta} \frac{r\cos\theta}{r}\, r\, dr\, d\theta$$

$$= \int_0^{\frac{\pi}{2}} \cos\theta \int_0^{2\cos\theta} r\, dr\, d\theta$$

$$= \int_0^{\frac{\pi}{2}} \cos\theta \left|\frac{r^2}{2}\right|_0^{2\cos\theta} d\theta$$

$$= 2\int_0^{\frac{\pi}{2}} \cos^3\theta\, d\theta = 2\,\frac{2}{3} = \frac{4}{3}.$$

EXAMPLE 12.23

Transform to the Cartesian form and hence, evaluate the integral $\int_0^{\pi}\int_0^{a} r^3 \sin\theta\cos\theta\, dr\, d\theta.$

Solution. We are given that

$$I = \int_0^{\pi}\int_0^{a} r^3 \sin\theta \cos\theta \, dr\, d\theta.$$

Put $x = r\cos\theta$ and $y = r\sin\theta$ so that $dx\,dy = r\,dr\,d\theta$. The region of integration is shown in the following figure:

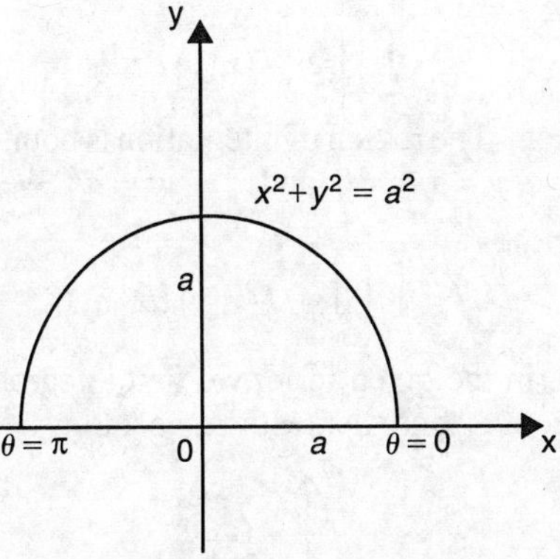

Therefore, the Cartesian form of the given integral is

$$I = \int_{-a}^{a}\int_0^{a} xy\, dx\, dy.$$

Further,

$$I = \int_0^{\pi}\int_0^{a} r^3 \sin\theta\cos\theta\, dr\, d\theta = \int_0^{\pi} \sin\theta\cos\theta \left|\frac{r^4}{4}\right|_0^a d\theta$$

$$= \frac{a^4}{4}\int_0^{\pi} \sin\theta\cos\theta\, d\theta = \frac{a^4}{4}\left|\frac{\sin^2\theta}{2}\right|_0^{\pi} = 0.$$

EXAMPLE 12.24

Use the transformation $x + y = u$ and $y = uv$ to show that $\int_0^1\int_0^{1-x} e^{\frac{y}{x+y}}\, dy\, dx = \frac{e-1}{2}.$

Solution. We have

$x = u - y = u - uv = u(1 - v)$ and $y = uv$.

Therefore,

$$\frac{\partial(x, y)}{\partial(u, v)} = \begin{vmatrix} \frac{\partial x}{\partial r} & \frac{\partial x}{\partial v} \\ \frac{\partial y}{\partial u} & \frac{\partial y}{\partial v} \end{vmatrix} = \begin{vmatrix} 1\ v & u \\ v & u \end{vmatrix}$$

$$= u(1 - v) - (-uv) = u.$$

The Jacobian vanishes when $u = 0$, that is, when $x = y = 0$, but not otherwise. Also the origin (0, 0) corresponds to the whole line $u = 0$ of the uv-plane so that the correspondence ceases to be one-to-one. In order to exclude (0, 0), we note that the given integral exists as the limit, when $h \to 0$ of the integral over the region is bounded by

$$x + y = 1,\ x = h, \text{ and } y = 0 \text{ where } h > 0.$$

The transformed region is then bounded by the lines

$$u = 1,\ v = 0, \text{ and } u(1 - v) = h.$$

When $h \to 0$, the new region of the uv-plane tends, as its limit, to the square bounded by the lines $u = 1$, $v = 1$, $u = 0$, and $v = 0$. Thus, the region of integration in the xy- and uv-planes are as shown in the following figures:

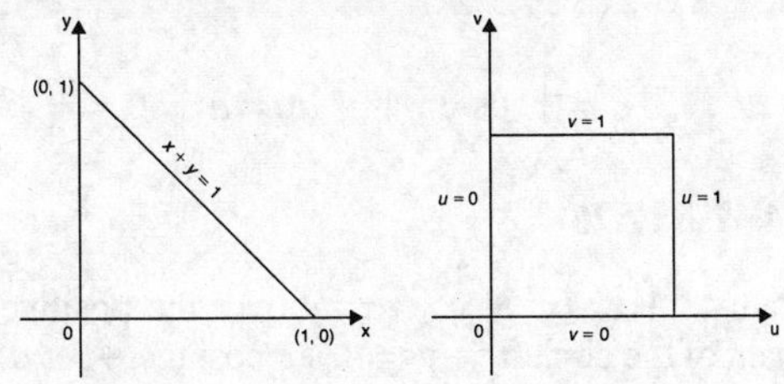

Therefore,

$$\int_0^1\int_0^{1-x} e^{\frac{y}{x+y}}\, dy\, dx = \int_0^1\int_0^1 e^{\frac{uv}{u}}.u\, du\, dv$$

$$= \int_0^1 e^v \int_0^1 u\, du\, dv = \int_0^1 e^v\left[\frac{u^2}{2}\right]_0^1 dv$$

$$= \frac{1}{2}\int_0^1 e^v\, dv = \frac{1}{2}[e^v]_0^1 = \frac{1}{2}(e - 1).$$

EXAMPLE 12.25

Evaluate $\int_0^a\int_y^a \frac{x\, dx\, dy}{x^2 + y^2}$ by changing into polar coordinates.

Solution. The region of integration is shown in the following figure:

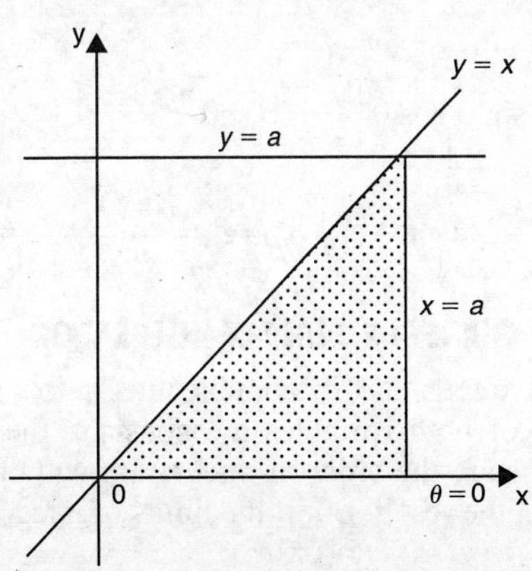

The region is bounded by $x = y$, $x = a$, $y = 0$, and $y = a$. Changing topolar coordinates, we have $x = r\cos\theta$, $y = r\sin\theta$, and $dx\,dy = r\,dr\,d\theta$. Further, in the region of integration θ varies from 0 to $\frac{\pi}{4}$. Also, $x = a$ implies $r\cos\theta = a$ or $r = \frac{a}{\cos\theta}$. Therefore, r varies from 0 to $\frac{a}{\cos\theta}$. Hence,

$$\int_0^a\int_y^a \frac{x\,dx\,dy}{x^2+y^2} = \int_0^{\frac{\pi}{4}}\int_0^{\frac{a}{\cos\theta}} \frac{r\cos\theta}{r^2}\,r\,dr\,d\theta$$

$$= \int_0^{\frac{\pi}{4}} \cos\theta\,[r]_0^{\frac{a}{\cos\theta}}\,d\theta = a\int_0^{\frac{\pi}{4}} d\theta = \frac{a\pi}{4}.$$

EXAMPLE 12.26

Evaluate $\iint xy\,(x^2+y^2)^{\frac{n}{2}}\,dx\,dy$ over the positive octant of the circle $x^2+y^2=4$, supposing $n+3>0$.

Solution. The region of integration is bounded by $x = 0$, $x = 2$, $y = 0$, and $y = 2$. Changing to polar coordinates, we have $x = r\cos\theta$ and $y = r\sin\theta$ and so, $rd\theta\,dr = dxdy$. The limits of integration in the first quadrant of the given circle are now $r = 0$ to $r = 2$ and $\theta = 0$ to $\theta = \frac{\pi}{2}$. Hence,

$$\iint xy\,(x^2+y^2)^{\frac{n}{2}}\,dx\,dy$$

$$= \int_0^{\frac{\pi}{2}}\int_0^2 r\cos\theta . r\sin\theta\,(r^2)^{\frac{n}{2}} . r\,dr\,d\theta$$

$$= \int_0^{\frac{\pi}{2}} \sin\theta\cos\theta\left[\int_0^2 r^{n+3}\,dr\right]d\theta$$

$$= \int_0^{\frac{\pi}{2}} \sin\theta\cos\theta\left[\frac{r^{n+4}}{n+4}\right]_0^2 d\theta$$

$$= \frac{2^{n+4}}{n+4} = \int_0^{\frac{\pi}{2}} \sin\theta\cos\theta\,d\theta = \frac{2^{n+4}}{n+4}\cdot\frac{1.1}{2} = \frac{2^{n+3}}{n+4}.$$

12.6 CHANGE OF ORDER OF INTEGRATION

We have seen that, in a double integration, if the limits of both variables are constant, then we can change the order of integration without affecting the result. But if the limits of integration are variable, a change in the order of integration requires a change in the limits of integration. Some integrals are easily evaluated by changing the order of integration in them.

EXAMPLE 12.27

Change the order of integration in the integral

$$I = \int_{-a}^{a}\int_0^{\sqrt{a^2-y^2}} f(x, y)\,dx\,dy.$$

Solution. The region of integration is bounded by $y = -a$, $y = a$, $x = 0$, and $x^2+y^2=a^2$. We have

$$I = \int_{-a}^{a}\left[\int_0^{\sqrt{a^2-y^2}} f(x, y)\,dx\right]dy.$$

Thus, in the given form, we first integrate with respect to x and then with respect to y.

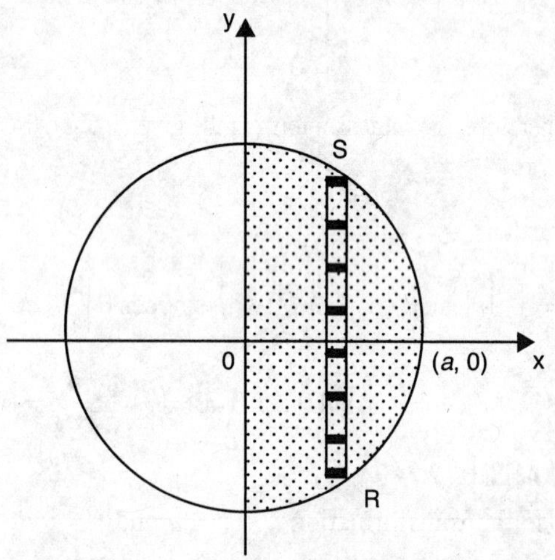

On changing the order of integration, we first integrate with respect to y, along a vertical ship RS, which extends from $y = -\sqrt{a^2-x^2}$ to $y = \sqrt{a^2-x^2}$. To cover the whole region of integration, we then integrate with respect to x from $x = 0$ to $x = a$. Thus,

$$I = \int_{-a}^{a} dx \int_{-\sqrt{a^2-x^2}}^{\sqrt{a^2-x^2}} f(x, y)\,dy$$

$$= \int_{-a}^{a}\left[\int_{-\sqrt{a^2-x^2}}^{\sqrt{a^2-x^2}} f(x, y)\,dy\right]dx.$$

EXAMPLE 12.28

Change the order of integration in $I = \int_0^1 \int_{x^2}^{2-x} xy dx\, dy$ and hence, evaluate the same.

Solution. For a given integral, the region of integration is bounded by $x = 0$, $x = 1$, $y = x^2$, (parabola), and the line $y = 2 - x$. Thus, the region of integration OABO is as shown in the following figure:

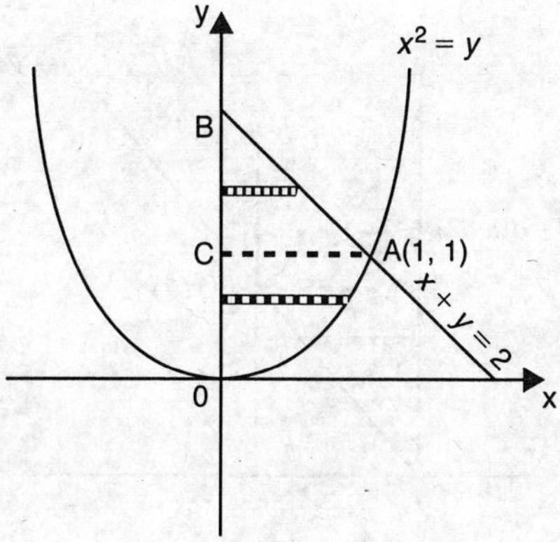

In the given form of the integral, we have to integrate first with respect to y and then with respect to x. Therefore, on changing the order of integration, we first integrate the integrand, with respect to x and then, with respect to y. The integration with respect to x requires the splitting-up of the region $OABO$ into two parts $OACO$ and the triangle ABC. For the subregion $OACO$, the limit of integration are from $x = 0$ to $x = \sqrt{y}$ and $y = 0$ to $y = 1$. Thus, the contribution to the integral I from the region $OACO$ is

$$I_1 = \int_0^1 \left[\int_0^{\sqrt{y}} xy\, dx \right] dy.$$

For the subregion ABC, the limits of integration are from $x = 0$ to $x = 2 - y$ and $y = 1$ to $y = 2$. Thus, the contribution to I from the subregion ABC is

$$I_2 = \int_1^2 \left[\int_0^{2-y} xy\, dx \right] dy.$$

Hence, on changing the order of integration, we get

$$I = \int_0^1 \left[\int_0^{\sqrt{y}} xy\, dx \right] dy + \int_1^2 \left[\int_0^{2-y} xy\, dx \right] dy$$

$$= \int_0^1 \left[y \frac{x^2}{2} \right]_0^{\sqrt{y}} dy + \int_1^2 \left[\frac{yx^2}{2} \right]_0^{2-y} dy$$

$$= \frac{1}{2} \int_0^1 y^2 dy + \frac{1}{2} \int_1^2 y(2-y)^2 dy$$

$$= \frac{1}{2} \left[\frac{y^3}{3} \right]_0^1 + \frac{1}{2} \left[\frac{y^4}{4} + 4\frac{y^2}{2} - 4\frac{y^3}{3} \right]_1^2$$

$$= \frac{1}{6} + \frac{5}{24} = \frac{3}{8}.$$

EXAMPLE 12.29

Changing the order of integration, find the value of the integral $\int_0^\infty \int_x^\infty \frac{e^{-y}}{y} dy\, dx$.

Solution. The region of integration is bounded by $x = 0$ and $y = x$. The limits of x are from 0 to ∞ and those of y are from x to ∞. The region of integration is shown in the following figure:

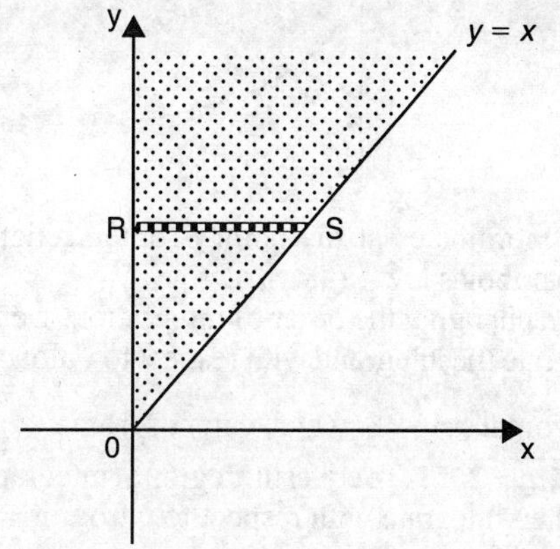

On changing the order of integration, we first integrate the integrand, with respect to x, along a horizontal strip RS, which extends from $x = 0$ to $x = y$. To cover the region of integration, we then integrate, with respect to y, from $y = 0$ to $y = \infty$. Thus,

$$I = \int_0^\infty \left[\int_0^y \frac{e^{-y}}{y} dx \right] dy = \int_0^\infty \frac{e^{-y}}{y} [x]_0^y\, dy$$

$$= \int_0^\infty e^{-y} dy = [-e^{-y}]_0^\infty = -\left[\frac{1}{e^y} \right]_0^\infty$$

$$= -(0 - 1) = 1.$$

EXAMPLE 12.30

Change the order of integration in the integral

$\int_0^{4a}\int_{\frac{x^2}{4a}}^{2\sqrt{ax}} dy\,dx$ and evaluate.

Solution. The given integral is $\int_0^{4a}\int_{\frac{x^2}{4a}}^{2\sqrt{ax}} dy\,dx$. The integration is first carried out with respect to y and then with respect to x. The region of integration is bounded by $x = 0$, $x = 4a$, and the parabolas $x^2 = 4ay$ and $y^2 = 4ax$. Thus, the region of integration is as shown in the following figure:

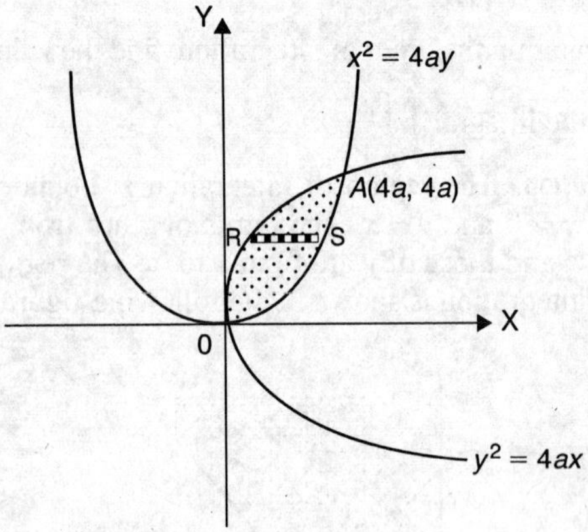

The coordinates at the point of intersection of the parabolas are A $(4a, 4a)$.

On changing the order of integration, we first integrate the integrand with respect to x along the horizontal strip RS, which extends from $x = \frac{y^2}{4a}$ to $x = \sqrt{4ay} = 2\sqrt{ay}$. To cover the region of integration, we then integrate with respect to y from $y = 0$ to $y = 4a$. Thus,

$$\int_0^{4a}\int_{\frac{x^2}{4a}}^{2\sqrt{ax}} dy\,dx = \int_0^{4a}\left[\int_{\frac{y^2}{4a}}^{2\sqrt{ay}} dx\right] dy = \int_0^{4a} [x]_{\frac{y^2}{4a}}^{2\sqrt{ay}}\, dy$$

$$= \int_0^{4a}\left[2\sqrt{ay} - \frac{y^2}{4a}\right] dy = \left[\frac{2\sqrt{a}y^{\frac{3}{2}}}{\frac{3}{2}} - \frac{y^3}{12a}\right]_0^{4a}$$

$$= \frac{32a^2}{3} - \frac{16a^2}{3} = \frac{16a^2}{3}.$$

EXAMPLE 12.31

Evaluate the integral

$$\int_0^1\int_x^{\sqrt{2-x^2}} \frac{x\,dy\,dx}{\sqrt{x^2+y^2}}$$

by changing the order of integration.

Solution. For the given integral, the region of integration is bounded by $x = 0$, $x = 1$, $y = x$, and the circle $x^2 + y^2 = 2$. Thus, the region of integration is as shown in the following figure:

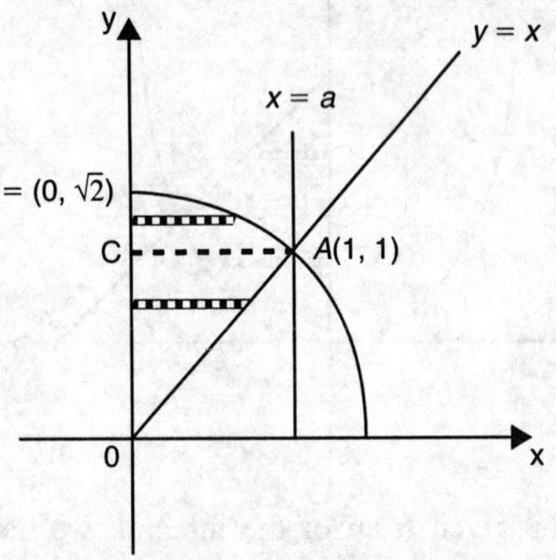

The point of intersection of the circle $x^2 + y^2 = 2$ and $x = y$ is A (1, 1). Draw $AC \,||\, OX$. Thus, the region of integration is divided intotwo-subregions $ABCA$ and ACO.

On changing the order of integration, we first integrate with respect to x, along the strips parallel to the x-axis.

In the subregion $ABCA$, the strip extends from $x = 0$ to $x = \sqrt{2 - y^2}$ To cover the subregion, we then integrate with respect to y from $y = 1$ to $y = \sqrt{2}$. Thus, the contribution tothe integral due to this subregion is

$$I_1 = \int_1^{\sqrt{2}}\left[\int_0^{\sqrt{2-y^2}} \frac{x}{\sqrt{x^2+y^2}}\, dx\right] dy.$$

On the other hand, in the subregion ACO, the strip extends from $x = 0$ to $x = y$. To cover this subregion, we then integrate with respect to y from $y = 0$ to $y = 1$. Thus, the contribution to the integral by this subregion is

$$I_2 = \int_0^1\left[\int_0^y \frac{y}{\sqrt{x^2+y^2}}\, dx\right] dy.$$

Hence, the given integral is equal to

$$I = I_1 + I_2 = \int_1^{\sqrt{2}}\left[\int_0^{\sqrt{2-y^2}} \frac{x}{\sqrt{x^2+y^2}}dx\right]dy$$

$$+ \int_0^1\left[\int_0^y \frac{x}{\sqrt{x^2+y^2}}dx\right]dy$$

$$= \int_1^{\sqrt{2}}\left[(x^2+y^2)^{\frac{1}{2}}\right]_0^{\sqrt{2-y^2}} dy + \int_0^1\left[(x^2+y^2)^{\frac{1}{2}}\right]_0^y dy$$

$$= \int_1^{\sqrt{2}}(\sqrt{2} - y)\,dy + \int_0^1(\sqrt{2y} - y)\,dy$$

$$= \left[\sqrt{2y} - \frac{y^2}{2}\right]_1^{\sqrt{2}} + (\sqrt{2} - 1)\left[\frac{y^2}{2}\right]_0^1$$

$$= \frac{2-\sqrt{2}}{2} = 1 - \frac{1}{\sqrt{2}}.$$

EXAMPLE 12.32

Evaluate the integral $\int_0^3 \int_1^{\sqrt{4-y}} (x + y)\, dx\, dy$ by changing the order of integration.

Solution. The region of integration is bounded by $x = 1$, $x^2 = 4 - y$, $y = 0$, and $y = 3$, as shown in the following figure:

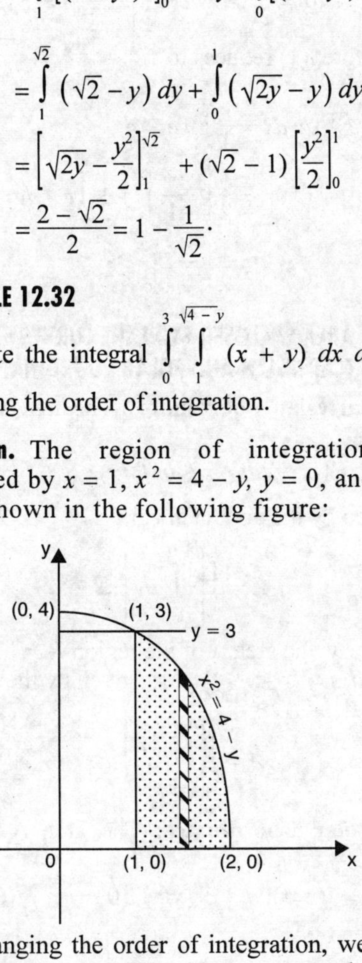

On changing the order of integration, we first integrate the integrand, with respect to y, by taking the strip parallel to the axis of y. In the region of integration, y varies from 0 to $4 - x^2$ and x varies from 1 to 2. Therefore,

$$\int_0^3 \int_0^{\sqrt{4-y}} (x + y)\, dx\, dy$$

$$= \int_1^2\left[\int_0^{4-x^2}(x+y)\,dy\right]dx = \int_1^2\left[xy + \frac{y^2}{2}\right]_0^{4-x^2} dx$$

$$= \int_1^2\left[4x - x^3 + 8 + \frac{x^4}{2} - 4x^2\right]dx$$

$$= \left[4\frac{x^2}{2} - \frac{x^4}{4} + 8x + \frac{x^5}{10} - \frac{4x^3}{3}\right]_1^2$$

$$= \left(8 - 4 + 16 + \frac{32}{10} - \frac{32}{3}\right)$$

$$- \left(2 - \frac{1}{4} + 8 + \frac{1}{10} - \frac{4}{3}\right) = \frac{241}{60}.$$

EXAMPLE 12.33

Change the order of integration in

$$I = \int_0^{2a} \int_{\sqrt{2ax-x^2}}^{\sqrt{2ax}} \phi(x, y)\,dy\,dx.$$

Solution. The region of integration is bounded by $x = 0$ and $x = 2a$, the circle $x^2 + y^2 = 2ax$, and the parabola $y^2 = 2ax$. The equation of the circle can be written as $(x - a)^2 + y^2 = a^2$ and so, has the center at $(a, 0)$. The region of integration is as shown in the following figure:

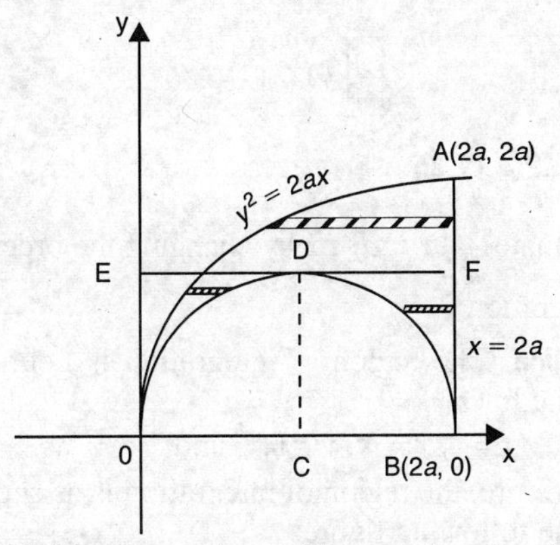

We divide the region of integration intothree parts by drawing the line EDF through D parallel to the x-axis. On changing the order of integration, we first integrate the integrand, with respect to x and then integrate the resultant integrand, with respect to y.

So, we draw horizontal strips parallel to the x-axis.

In the subregion OEDO, x varies from $\frac{y^2}{2a}$ to $a - \sqrt{a^2 - y^2}$ and y varies from 0 to a. Thus, the contribution to the integral due to this subregion is

$$I_1 = \int_0^a \left[\int_{\frac{y^2}{2a}}^{a-\sqrt{a^2-y^2}} \phi(x, y)\, dx \right] dy.$$

Similarly, the contribution to the integral due to the subregion DBFD is

$$I_2 = \int_0^a \left[\int_{a+\sqrt{a^2-y^2}}^{2a} \phi(x, y)\, dx \right] dy,$$

and the contribution to the integral due to the subregion AEFA is

$$I_3 = \int_0^a \left[\int_{\frac{y^2}{2a}}^{2a} \phi(x, y)\, dx \right] dy.$$

Hence,

$$I = \int_0^a \int_{\frac{y^2}{2a}}^{a-\sqrt{a^2-y^2}} \phi(x, y)\, dx\, dy$$

$$+ \int_0^a \int_{a+\sqrt{a^2-y^2}}^{2a} \phi(x, y)\, dx\, dy$$

$$+ \int_0^{2a} \int_{\frac{y^2}{2a}}^{2a} \phi(x, y)\, dx\, dy.$$

EXAMPLE 12.34

Evaluate $\int_0^\infty \int_0^x e^{-\frac{x^2}{y}}\, dy\, dx$ by changing the order of integration.

Solution. The region of integration is bounded by the lines

$x = 0,\ x = \infty,\ y = 0,$ and $y = x.$

Therefore, the region of integration is as shown in the following figure:

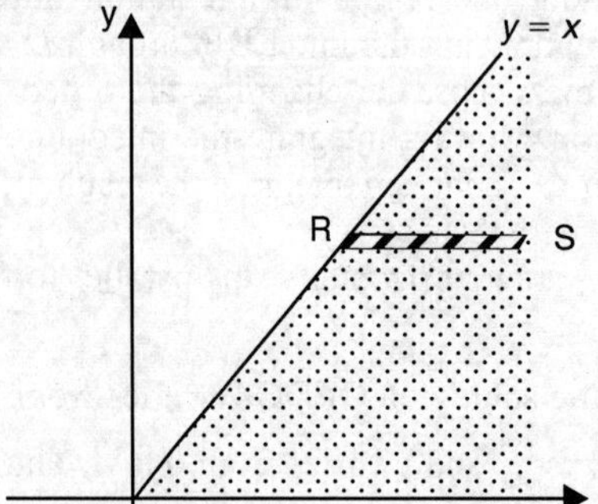

On changing the order of integration, we first integrate, with respect to x and then, with respect to y. Thus,

$$\int_0^\infty \int_0^x x e^{-\frac{x^2}{y}}\, dy\, dx = \int_0^\infty \left[\int_y^\infty x\, e^{-\frac{x^2}{y}}\, dx \right] dy. \qquad (1)$$

We first evaluate the inner integral. Substituting $x^2 = t$, we have $2x\,dx = dt$. When $x = y$, $t = y^2$ and when $x = \infty$, $t = \infty$, Therefore,

$$\int_y^\infty x e^{-\frac{x^2}{y}}\, dx = \frac{1}{2} \int_{y^2}^\infty e^{-\frac{t}{y}} dt = \frac{1}{2} \left[\frac{e^{-\frac{t}{y}}}{-\frac{1}{y}} \right]_{y^2}^\infty = \frac{1}{2} y\, e^{-y}.$$

Therefore, (1) reduces to

$$\int_0^\infty \int_0^x x\, e^{-\frac{x^2}{y}}\, dy\, dx = \frac{1}{2} \int_0^\infty y\, e^{-y}\, dy$$

$$= \frac{1}{2} \left[\frac{y\, e^{-y}}{-1} \right]_0^\infty + \frac{1}{2} \int_0^\infty e^{-y}\, dy$$

$$= \frac{1}{2} \left[\frac{e^{-y}}{-1} \right]_0^\infty = \frac{1}{2}.$$

12.7 AREA ENCLOSED BY PLANE CURVES (CARTESIAN AND POLAR COORDINATES)

(A) Cartesian Coordinates: The area A of the region

$$R = \{(x, y): a \le x \le b; f_1(x) \le y \le f_2(x)\}$$

is given by the double integral

$$A = \int_a^b \left[\int_{f_1(x)}^{f_2(x)} dy \right] dx.$$

Similarly, the area A of the region $R = \{(x, y): c \le y \le d; f_1(y) \le x \le f_2(y)\}$ is given by the double integral

$$A = \int_c^d \left[\int_{f_1(y)}^{f_2(y)} dx \right] dy.$$

(B) Polar Coordinates: The area A of the region

$$R = \{(r, \theta); \alpha \le \theta \le \beta; f_1(\theta) \le r \le f_2(\theta)\}$$

is given by

$$A = \int_\alpha^\beta \left[\int_{f_1(\theta)}^{f_2(\theta)} r\, dr \right] d\theta.$$

Similarly, the area A of the region $R = \{(r, \theta); r_1 \le r \le r^2; f_1(r) \le \theta \le f_2(r)\}$ is given by

$$A = \int_{r_1}^{r_2}\left[\int_{f_1(r)}^{f_2(r)} d\theta\right] r\,dr.$$

EXAMPLE 12.35

Find the area of a plate in the form of a quadrant of the ellipse $\frac{x^2}{a^2}+\frac{y^2}{b^2}=1$. Hence, find the area enclosed by the given ellipse.

Solution. From the figure, we note that the required area is bounded by $x = 0$, $x = a$, $y = 0$, and $y=\frac{b}{a}\sqrt{a^2-x^2}$.

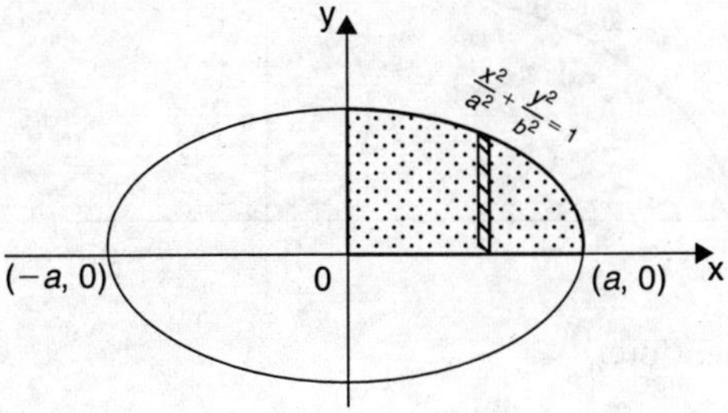

Thus,

$$A=\int_0^a\int_0^{\frac{b}{a}\sqrt{a^2-x^2}} dy\,dx=\int_0^a [y]_0^{\frac{b}{a}\sqrt{a^2-x^2}}\,dx$$

$$=\frac{b}{a}\int_0^a \sqrt{a^2-x^2}\,dx$$

$$=\frac{b}{a}\left[\frac{x\sqrt{a^2-x^2}}{2}+\frac{a^2}{2}\sin^{-1}\frac{x}{a}\right]_0^a$$

$$=\frac{b}{a}\left[\left(0+\frac{a^2}{2}\sin^{-1}1\right)-0\right]$$

$$=\frac{ba}{2}\sin^{-1}1=\frac{ba}{2}\cdot\frac{\pi}{2}$$

$$=\frac{\pi ab}{4}\text{ sq units.}$$

Hence, the total area enclosed by the given ellipse is four times the area enclosed by the plate in the form of one quadrant $= \pi ab$ sq units.

EXAMPLE 12.36

Find the area lying between the parabola $y = 4x - x^2$ and the line $y = x$.

Solution. The parabola passes through the origin. Solving $y = 4x - x^2$ and $y = x$ for x, we get $x = 0$ and $x = 3$. Thus, the curves $y = 4x - x^2$ and $y = x$ intersect at $x = 0$ and $x = 3$. When $0 < x < 3$, $4x < x^2$ is greater than x. Therefore, the region of integration is as shown in the following figure:

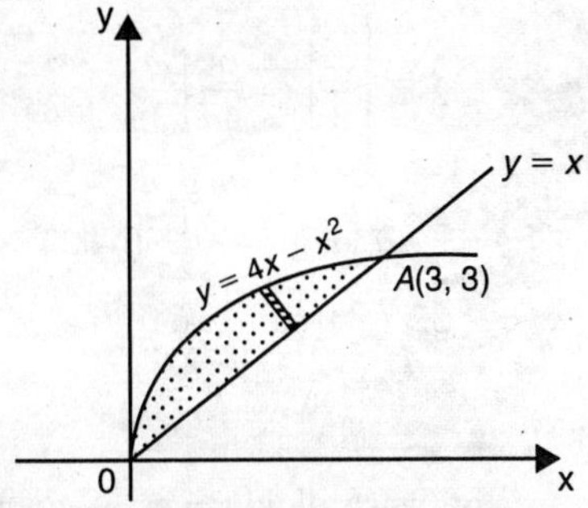

Thus, the required area lies between $y = x$, $y = 4x - x^2$, $x = 0$, and $x = 3$. Therefore,

$$A=\int_0^3\left[\int_x^{4x-x^2} dy\right]dx=\int_0^3 [y]_x^{4x-x^2}\,dx=\int_0^3 (3x-x^2)\,dx$$

$$=\left[\frac{3x^2}{2}-\frac{x^3}{3}\right]_0^3=\frac{27}{2}-\frac{27}{3}=\frac{9}{2}.$$

EXAMPLE 12.37

Find the area lying between the parabola $y^2 = 4ax$ and $x^2 = 4ay$.

Solution. Solving the equation of the given parabola, we have O(0, 0) and $A(4a, 4a)$ as the points of intersection. The region of integration is shown in the following figure:

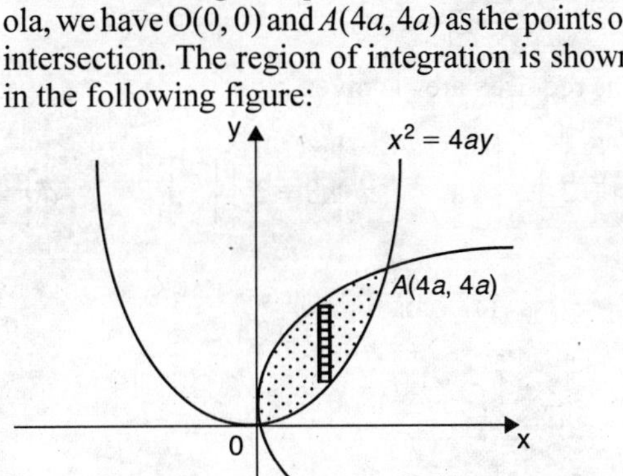

Therefore, the required area is

$$A=\int_0^{4a}\left[\int_{\frac{x^2}{4a}}^{2\sqrt{ax}} dy\right]dx=\int_0^{4a}[y]_{\frac{x^2}{4a}}^{2\sqrt{ax}}$$

$$=\int_0^{4a}\left[2\sqrt{ax}-\frac{x^2}{4a}\right]dx$$

$$= 2\sqrt{a}\int_0^{4a}\sqrt{x}\,dx - \frac{1}{4a}\int_0^{4a}x^2dx$$

$$= 2\sqrt{a}\left[\frac{x^{\frac{3}{2}}}{\frac{3}{2}}\right]_0^{4a} - \frac{1}{4a}\left[\frac{x^3}{3}\right]_0^{4a}$$

$$= \frac{4}{3}\sqrt{a}\left(8a^{\frac{3}{2}}\right) - \frac{1}{12a}(64a^3)$$

$$= \frac{32a^2}{3} - \frac{16a^2}{3} = \frac{16}{3}a^2.$$

EXAMPLE 12.38

Find the area of the cardioid $r = a(1 + \cos\theta)$.

Solution. The curve passes through the origin and cuts the x-axis at $x = 2a$. Clearly, θ varies from 0 to π and r varies from 0 to $a(1 + \cos\theta)$ in the upper-half part of the integration region.

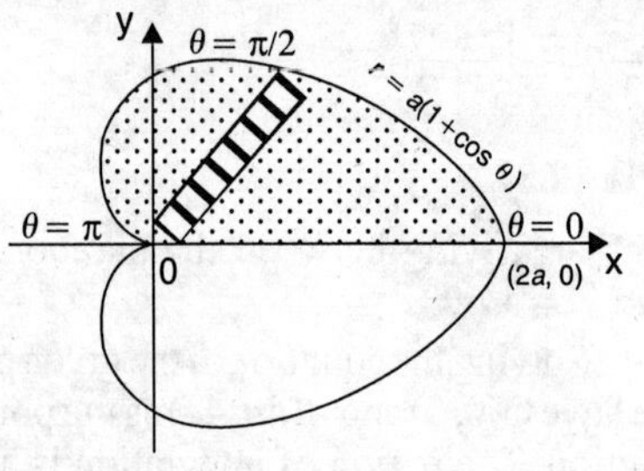

The required area is given by

$$A = 2\int_0^{\pi}\left[\int_0^{a(1+\cos\theta)} r\,dr\right]d\theta = 2\int_0^{\pi}\left[\frac{r^2}{2}\right]_0^{a(1+\cos\theta)}d\theta$$

$$= \int_0^{\pi} a^2(1+\cos\theta)^2\,d\theta$$

$$= 4a^2\int_0^{\pi}\left(\cos^2\frac{\theta}{2}\right)^2 d\theta$$

$$= 4a^2\int_0^{\pi}\cos^4\frac{\theta}{2}\,d\theta$$

$$= 8a^2\int_0^{\frac{\pi}{2}}\cos^4\phi\,d\theta,\ \theta = 2\phi$$

$$= 8a^2\cdot\frac{3}{4.2}\cdot\frac{\pi}{2} = \frac{3\pi a^2}{2}.$$

EXAMPLE 12.39

Find the area lying inside the circle $r = a\sin\theta$ and outside the cardioid $r = a(1 - \cos\theta)$.

Solution. We have $r = a\sin\theta$ and $r = a(1 - \cos\theta)$. Therefore, $a\sin\theta = a(1 - \cos\theta)$, which yields $\sin\theta + \cos\theta = 1$ or $\sin^2\theta + \cos^2\theta + 2\sin\theta\cos\theta = 1$ or $\sin 2\theta = 0$. Hence, $2\theta = 0$ and π and so, $\theta = 0$ or $\frac{\pi}{2}$. Further, from the region of integration, it is clear that r varies from $a(1 - \cos\theta)$ to $a\sin\theta$.

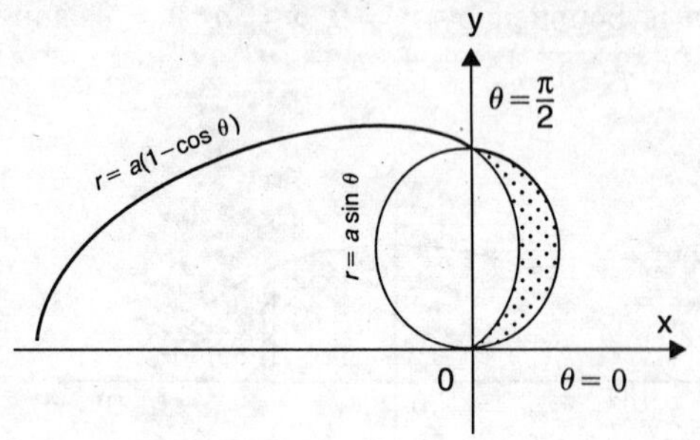

Therefore,

$$A = \int_0^{\frac{\pi}{2}}\left[\int_{a(1-\cos\theta)}^{a\sin\theta} dr\right] r\,d\theta = \int_0^{\frac{\pi}{2}}\left[\frac{r^2}{2}\right]_{a(1-\cos\theta)}^{a\sin\theta} d\theta$$

$$= \frac{a^2}{2}\int_0^{\frac{\pi}{2}}[\sin^2\theta - (1-\cos\theta)^2]d\theta$$

$$= \frac{a^2}{2}\int_0^{\frac{\pi}{2}}[\sin^2\theta - \cos^2\theta + 2\cos\theta - 1]\,d\theta$$

$$= \frac{a^2}{2}\int_0^{\frac{\pi}{2}}(-2\cos^2\theta + 2\cos\theta)\,d\theta$$

$$= a^2\left[-\frac{1}{2}\cdot\frac{\pi}{2} + 1\right] = a^2\left(1 - \frac{\pi}{4}\right)$$

EXAMPLE 12.40

Find the area of one loop of the lemniscates $r^2 = a^2\cos 2\theta$.

Solution. The region of integration is shown in the following figure:

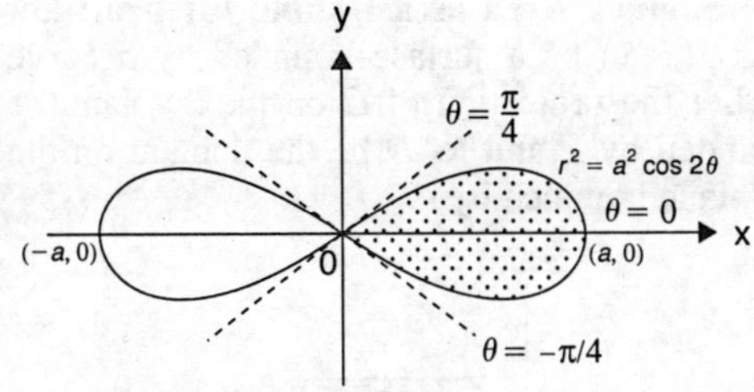

The required area is given by

$$A = 2\int_0^{\frac{\pi}{4}}\left[\int_0^{a\sqrt{\cos 2\theta}} r\,dr\right]d\theta = 2\int_0^{\frac{\pi}{4}}\left[\frac{r^2}{2}\right]_0^{a\sqrt{\cos 2\theta}} d\theta$$

$$= a^2\int_0^{\frac{\pi}{4}} \cos 2\theta\, d\theta = \frac{a^2}{2}\,[\sin 2\theta]_0^{\frac{\pi}{4}} = \frac{a^2}{2}.$$

EXAMPLE 12.41

Find the area included between the curve $r = a$ $(\sec\theta + \cos\theta)$ and its asymptote.

Solution. The curve $r = a(\sec\theta + \cos\theta)$ is symmetrical about the initial line. The equation of the asymptote is $r = a\sec\theta$.

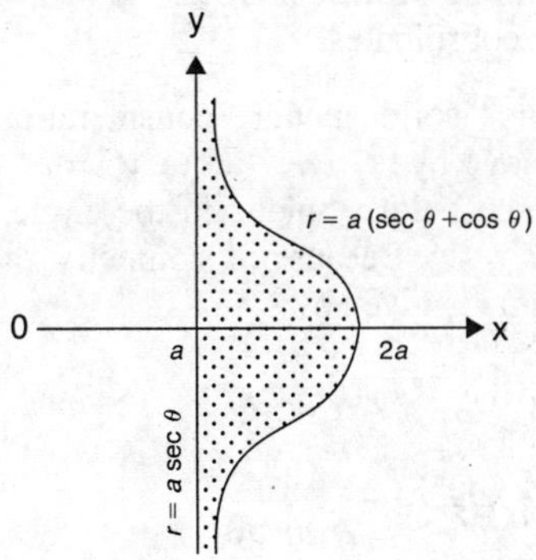

The required area is

$$A = 2\int_0^{\frac{\pi}{2}}\left[\int_{a\sec\theta}^{a(\sec\theta+\cos\theta)} r\,dr\right]d\theta$$

$$= 2\int_0^{\frac{\pi}{2}}\left[\frac{r^2}{2}\right]_{a\sec\theta}^{a(\sec\theta+\cos\theta)} d\theta$$

$$= \int_0^{\frac{\pi}{2}} [a^2(\sec\theta + \cos\theta)^2 - a^2\sec^2\theta]\, d\theta$$

$$= a^2\int_0^{\frac{\pi}{2}} (\sec^2\theta + \cos^2\theta + 2 - \sec^2\theta)\, d\theta$$

$$= a^2\int_0^{\frac{\pi}{2}} (\cos^2\theta + 2)\, d\theta = a^2\left[\frac{1}{2}\cdot\frac{\pi}{2} + \frac{2\pi}{2}\right] = \frac{5\pi a^2}{4}$$

EXAMPLE 12.42

Find the area bounded by the parabolas $y^2 = 4 - x$ and $y^2 = 4 - 4x$.

Solution. The required area is given by

$$A = 2\int_0^2\left(\int_{\frac{4-y^2}{4}}^{4-y^2} dx\right)dy = 2\int_0^2\left[4 - \frac{y^2}{4} - \frac{4-y^2}{4}\right]dy$$

$$= 2\int_0^2\left(3 - \frac{3}{4}y^2\right)dy = 2\left[3y - \frac{y^3}{4}\right]_0^2$$

$$= 2\,[6 - 2] = 8.$$

12.8 VOLUME AND SURFACE AREA AS DOUBLE INTEGRALS

(A) Volume as a Double Integral: Consider a surface $z = f(x, y)$. Let the region S be the orthogonal projection of the portion S' of $z = f(x, y)$ on the xy-plane. Divide S into elementary rectangles of area $\delta x\,\delta y$ by drawing lines parallel tothe x- and y-axis. On each of these rectangles, erect a prism which has a length parallel to Oz. Then, the volume of the prism between S' and S is $z\,\delta x\,\delta y$.

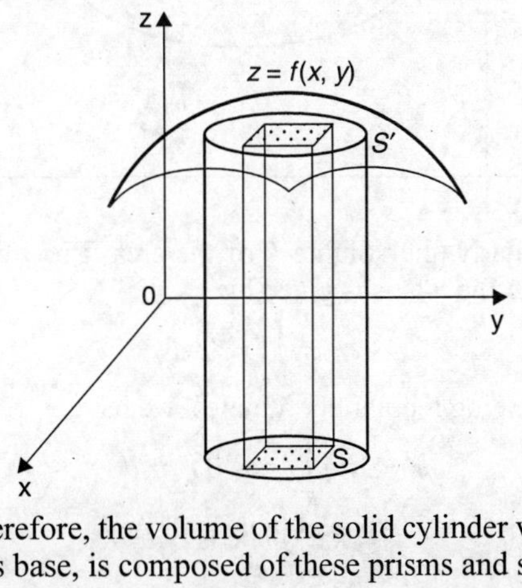

Therefore, the volume of the solid cylinder with S as base, is composed of these prisms and so,

$$V = \lim_{\substack{\delta x \to 0 \\ \delta y \to 0}} \sum z\, \delta x\, \delta y = \iint z\, dx\, dy$$

$$= \iint_S f(x, y)\, dx\, dy.$$

In the polar coordinates, the region S is divided intoelements of area $r\,\delta r\,\delta\theta$ and so, the volume in that case is given by

$$V = \iint_S f(r\cos\theta, r\sin\theta)\, r dr\, d\theta.$$

(B) Volumes of Solids of Revolution: Let $P(x, y)$ be a point in a plane area R. Suppose that the elementary area $\delta x\, \delta y$ at $P(x, y)$ revolves about the x-axis. This will generate a ring of radius y. The elementary volume of this ring is $\delta V = 2\pi y\, \delta y\, \delta x$. Hence, the total volume of the solid formed by the revolution of the area R about the x-axis is given by

$$V = 2\pi \iint_R y\, dy\, dx.$$

Changing to polar coordinates, we get

$$V = 2\pi \iint_R r\sin\theta\, r\, dr\, d\theta$$

$$= 2\pi \iint_R r^2 \sin\theta\, dr\, d\theta.$$

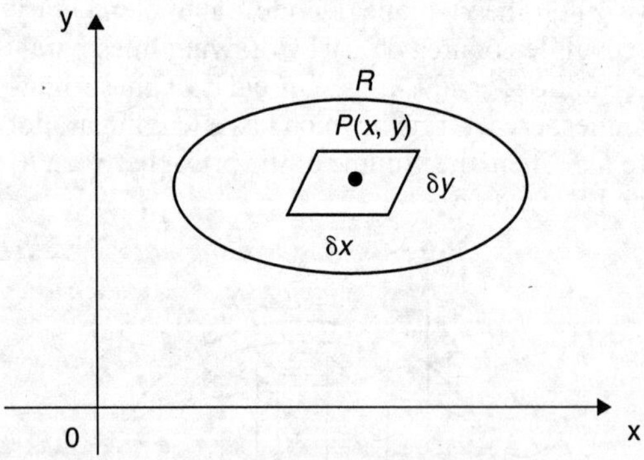

Similarly, the volume V of the area R revolved about the y-axis is given by

$$V = 2\pi \iint_R x\, dx\, dy.$$

Changing topolar coordinates, we have

$$V = 2\pi \iint_R r\cos\theta\, r\, dr\, d\theta$$

$$= 2\pi \iint_R r^2 \cos\theta\, dr\, d\theta.$$

(C) Surface Area as a Double Integral: Let $z = \psi(x, y)$ be a surface bounded by *a* curve C. Let the projection of C on the xy-plane be bounded by Γ and let D be the domain on the xy-plane bounded by Γ.

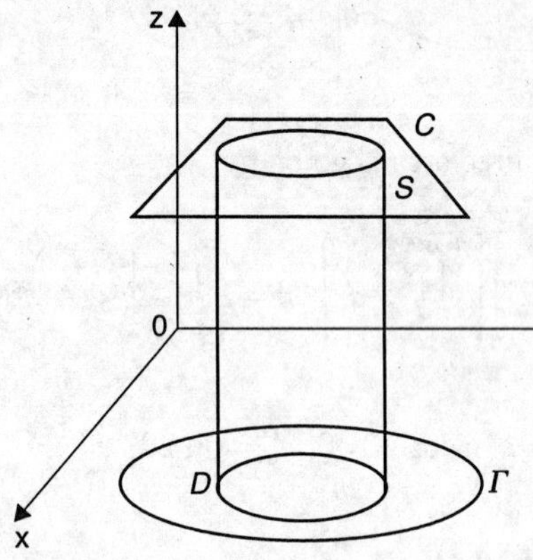

Then, the area of the surface S is given by

$$S = \iint_D \sqrt{1 + \left(\frac{\partial z}{\partial x}\right)^2 + \left(\frac{\partial z}{\partial y}\right)^2}\, dx\, dy.$$

EXAMPLE 12.43

Find the volume of the sphere $x^2 + y^2 + z^2 = a^2$ using polar coordinates.

Solution. The solid under consideration is bounded above by $z^2 = a^2 - (x^2 + y^2) = a^2 - r^2$. The sphere cuts the xy-plane in the circle $x^2 + y^2 = a^2$ or $r^2 = a^2$. Because of symmetry, the required volume is given by

$$V = 2\int_0^{2\pi}\int_0^{a} \sqrt{a^2 - r^2}\, r\, dr\, d\theta$$

$$= \int_0^{2\pi}\int_0^{a} \sqrt{a^2 - r^2} \cdot 2r\, dr\, d\theta$$

$$= \int_0^{2\pi} \left[\frac{(a^2 - r^2)^{\frac{3}{2}}}{\frac{3}{2}}\right]_0^a d\theta = \frac{2a^3}{3}\int_0^{2\pi} d\theta = \frac{4\pi a^3}{3}.$$

EXAMPLE 12.44

Find the volume of the ellipsoid $\frac{x^2}{a^2} + \frac{y^2}{b^2} + \frac{z^2}{c^2} = 1.$

Solution. Duetosymmetry,thevolumeofthegiven ellipsoid is eight times the volume of the portion of the ellipsoid in the first octant. For the positive octant, the given equation yields

$$z = c\sqrt{1 - \frac{x^2}{a^2} - \frac{y^2}{b^2}}.$$

The region in this octant is bounded by

$$x = 0, x = a, y = 0, \quad \text{and} \quad y = b\sqrt{1 - \frac{x^2}{a^2}}.$$

Hence, the required volume is given by

$$V = 8\int_0^a \int_0^{b\sqrt{1-\frac{x}{a^2}}} z\,dy\,dx$$

$$= 8\int_0^a \int_0^{b\sqrt{1-\frac{x}{a^2}}} c\sqrt{1 - \frac{x^2}{a^2} - \frac{y^2}{b^2}}\,dy\,dx$$

$$= \int_0^a \int_0^{b\sqrt{1-\frac{x}{a^2}}} c\left[\left(1 - \frac{x^2}{a^2}\right) - \frac{y^2}{b^2}\right]^{\frac{1}{2}} dy\,dx.$$

Substituting $\frac{y}{b} = \sqrt{1 - \frac{x^2}{a^2}}$ sin θ, we get $dy = b\sqrt{1 - \frac{x^2}{a^2}}\cos\theta\,d\theta$ (as x is a constant). Therefore,

$$V = 8\int_0^a \int_0^{\frac{\pi}{2}} c\left[\left(1 - \frac{x^2}{a^2}\right) - \left(1 - \frac{x^2}{a^2}\right)\sin^2\theta\right]^{\frac{1}{2}}$$

$$\times\, b\sqrt{1 - \frac{x^2}{a^2}}\cos\theta\,d\theta$$

$$= 8bc\int_0^a \left(1 - \frac{x^2}{a^2}\right)\int_0^{\frac{\pi}{2}} [1 - \sin^2\theta]^{\frac{1}{2}}\cos\theta\,d\theta$$

$$= 8bc\int_0^a \left(1 - \frac{x^2}{a^2}\right)\int_0^{\frac{\pi}{2}} \cos^2\theta\,d\theta$$

$$= 8bc\int_0^a \left(1 - \frac{x^2}{a^2}\right)\cdot\frac{1}{2}\cdot\frac{\pi}{2}$$

$$= 2\pi\,bc\left[x - \frac{x^3}{3a^2}\right]_0^a = \frac{4}{3}\pi\,abc.$$

EXAMPLE 12.45

Find the volume contained between the ellipsoid $\frac{x^2}{a^2} + \frac{y^2}{b^2} + \frac{z^2}{c^2} = 1$ and the cylinder $\frac{x^2}{a^2} + \frac{y^2}{b^2} = \frac{x}{a}$.

Solution. The equation of the given elliptical cylinder is

$$\frac{x^2}{a^2} + \frac{y^2}{b^2} = \frac{x}{a},$$

Substituting $\frac{x}{a} = r\cos\theta$ and $\frac{y}{b} = r\sin\theta$, this equation yields

$$r^2 = r\cos\theta \text{ or } r = \cos\theta.$$

The required volume is given by

$$V = 4\iint c\sqrt{1 - \frac{x^2}{a^2} - \frac{y^2}{b^2}}\,dy\,dx$$

$$= 4abc\int_0^{\frac{\pi}{2}}\int_0^{\cos\theta} \sqrt{1 - r^2}\,r\,dr\,d\theta$$

$$= -\frac{4\,abc}{2}\int_0^{\frac{\pi}{2}}\left[\frac{(1 - r^2)^{\frac{3}{2}}}{\frac{3}{2}}\right]_0^{\cos\theta} d\theta$$

$$= -\frac{4abc}{2}\int_0^{\frac{\pi}{2}}(\sin^3\theta - 1)\,d\theta$$

$$= -\frac{4abc}{3}\left[\frac{2}{3} - \frac{\pi}{2}\right] = \frac{2}{9}abc\,[3\pi - 4].$$

EXAMPLE 12.46

Find the volume common toa sphere $x^2 + y^2 + z^2 = a^2$ and a circular cylinder $x^2 + y^2 = ax$. (Particular case of Example 12.45, taking $a = b = c$).

Solution. The required volume is the part of the sphere lying within the cylinder and is given by

$$V = 4\iint_R z\,dy\,dx = 4\iint_R \sqrt{a^2 - x^2 - y^2}\,dy\,dx,$$

where R is the half of the circle lying in the first quadrant. Substituting $x = r\cos\theta$ and $y = r\sin\theta$, the equation $x^2 + y^2 = ax$ yields

$$r^2 = a\,r\cos\theta \quad \text{or} \quad r = a\cos\theta.$$

Thus, the region of integration is bounded by

$$r = 0,\ r = a\cos\theta,\ \theta = 0, \quad \text{and} \quad \theta = \frac{\pi}{2}$$

Hence,

$$V = 4\int_0^{\frac{\pi}{2}}\int_0^{a\cos\theta} \sqrt{a^2 - r^2}\,r\,dr\,d\theta$$

$$= \frac{4}{-2}\int_0^{\frac{\pi}{2}}\int_0^{a\cos\theta} \sqrt{a^2 - r^2}\,(-2r)\,dr\,d\theta$$

$$= -2\int_0^{\frac{\pi}{2}} \left[\frac{(a^2-r^2)^{\frac{3}{2}}}{\frac{3}{2}}\right]_0^{a\cos\theta} d\theta$$

$$= -\frac{4}{3}\int_0^{\frac{\pi}{2}} (a^3 \sin^3\theta - a^3)\, d\theta$$

$$= -\frac{4a^3}{3}\left[\frac{2}{3} - \frac{\pi}{2}\right] = \frac{2a^3}{9}(3\pi - 4).$$

EXAMPLE 12.47

Find the volume bounded by the cylinder $x^2 + y^2 = 4$ and the planes $y + z = 4$ and $z = 0$.

Solution. Tofind the required volume, $z = 4 - y$ is to be integrated over the circle $x^2 + y^2 = 4$ in the xy-plane.

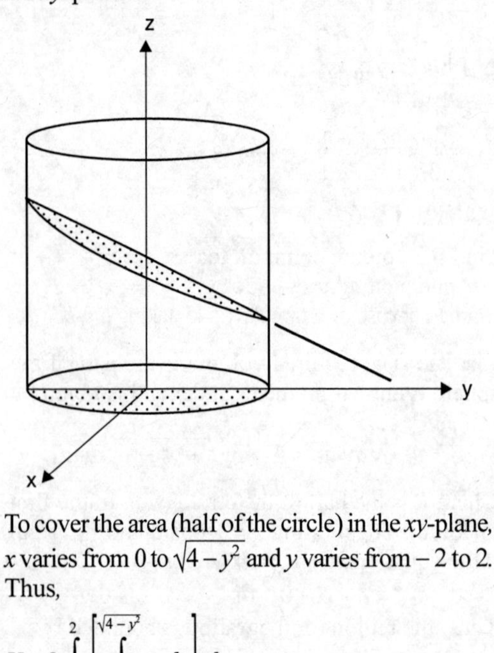

To cover the area (half of the circle) in the xy-plane, x varies from 0 to $\sqrt{4-y^2}$ and y varies from -2 to 2. Thus,

$$V = 2\int_{-2}^{2}\left[\int_0^{\sqrt{4-y^2}} z\, dx\right] dy$$

$$= 2\int_{-2}^{2}\left[\int_0^{\sqrt{4-y^2}} (4-y)\, dx\right] dy$$

$$= 2\int_{-2}^{2} (4-y)\left[x\right]_0^{\sqrt{4-y^2}} dy$$

$$= 2\int_{-2}^{2} (4-y)\sqrt{4-y^2}\, dy$$

$$= 2\left[4\int_{-2}^{2}\sqrt{4-y^2}\, dy - \int_{-2}^{2} y\sqrt{4-y^2}\, dy\right]$$

$$= 8\int_{-2}^{2}\sqrt{4-y^2}\, dy, \text{ second integrand being odd}$$

$$= 16\int_0^2\sqrt{4-y^2}\, dy, \text{ because of even integrand}$$

$$= 16\left[\frac{y\sqrt{4-y^2}}{2} + \frac{4}{2}\sin^{-1}\frac{y}{2}\right]_0^2$$

$$= 16\,[2\sin^{-1} 1] = \frac{32\pi}{2} = 16\pi.$$

EXAMPLE 12.48

Find the volume of the solid bounded above by the parabolic cylinder $z = 4 - y^2$ and bounded below by the elliptic paraboloid $z = x^2 + 3y^2$.

Solution. The two surfaces intersect in a space curve, whose projection on the xy-plane is the ellipse $x^2 + 4y^2 = 4$ or $\frac{x^2}{4} + y^2 = 1$. Substituting $x = 2r\cos\theta$ and $y = r\sin\theta$, the ellipse becomes $r^2 = 1$.

Further,

$$z = \phi_1(x) = 4 - y^2 \text{ and}$$
$$z = \phi_2(x) = x^2 + 3y^2.$$

Therefore,

$$\phi_1(x) - \phi_2(x) = 4 - y^2 - x^2 - 3y^2$$
$$= 4 - 4y^2 - x^2 = 4(1 - r^2).$$

Also,

$$J = \frac{\partial(x, y)}{\partial(r, \theta)} = 2.$$

Since the solid is symmetrical about x- and y-axis, we have

$$V = 4\int_0^{\frac{\pi}{2}}\int_0^1 4(1-r^2)\, 2r\, dr\, d\theta$$

$$= 32\int_0^{\frac{\pi}{2}}\int_0^1 (r - r^3)\, dr\, d\theta$$

$$= 32\int_0^{\frac{\pi}{2}}\left[\frac{r^2}{2} - \frac{r^4}{4}\right]_0^1 d\theta = \frac{32\pi}{8} = 4\pi.$$

EXAMPLE 12.49

Find the volume bounded by xy-plane, the cylinder $x^2 + y^2 = 1$, and the plane $x + y + z = 3$.

Solution. We have to integrate $z = 3 - x - y$ over the circle $x^2 + y^2 = 1$. Substituting $x = r \cos \theta$ and $y = r \sin \theta$, so that $x^2 + y^2 = r^2$, the integrand reduces to $3 - r\cos\theta - r\sin\theta = 3 - r(\cos\theta + \sin\theta)$ and the circle $x^2 + y^2 = 1$ reduces to $r^2 = 1$. Thus, to cover half of the region, r varies from 0 to 1 and θ varies from 0 to $\frac{\pi}{2}$. Hence,

$$V = 4\int_0^{\frac{\pi}{2}}\left[\int_0^1 \{3 - r(\cos\theta + \sin\theta)\}r\,dr\right]d\theta$$

$$= 4\int_0^{\frac{\pi}{2}}\left[3\frac{r^2}{2} - \frac{r^3}{3}(\cos\theta + \sin\theta)\right]_0^1 d\theta$$

$$= 4\int_0^{\frac{\pi}{2}}\left[\frac{3}{2} - \frac{1}{3}(\cos\theta + \sin\theta)\right]d\theta$$

$$= \int_0^{\frac{\pi}{2}} 6d\theta - \frac{4}{3}\int_0^{\frac{\pi}{2}}(\cos\theta + \sin\theta)d\theta$$

$$= \frac{6\pi}{2} - \frac{4}{3}\left[\sin\theta - \cos\theta\right]_0^{\frac{\pi}{2}}$$

$$= \frac{6\pi}{2} - \frac{4}{3}[1 + 1] = 3\pi - \frac{8}{3}.$$

EXAMPLE 12.50

Find the volume common to the cylinders $x^2 + y^2 = a^2$ and $x^2 + z^2 = a^2$.

Solution. The required volume is given by

$$V = 8\int_0^a\left[\int_0^{\sqrt{a^2 - r^2}} \sqrt{a^2 - x^2}\,dy\right]dx$$

$$= 8\int_0^a \sqrt{a^2 - x^2}\,[y]_0^{\sqrt{a^2 - x^2}}\,dx$$

$$= 8\int_0^a (a^2 - x^2)\,dx = \left[8a^2x - \frac{x^3}{3}\right]_0^a = \frac{16a^3}{3}.$$

EXAMPLE 12.51

Prove that the volume, enclosed between $x^2 + y^2 = 2ax$ and $z^2 = 2ax$ is $\frac{128a^3}{15}$.

Solution. To find the required volume, $z = \sqrt{2ax}$ is to be integrated over the curve $x^2 + y^2 = 2ax$ in the xy plane. Changing to polar coordinates by substituting $x = r\cos\theta$ and $y = r\sin\theta$, the required volume is given by

$$V = 4\int_0^{\frac{\pi}{2}}\int_0^{2a\cos\theta} \sqrt{2ar\cos\theta}\; r\,dr\,d\theta$$

$$= 4\sqrt{2a}\int_0^{\frac{\pi}{2}}\int_0^{2a\cos\theta} \sqrt{r}\sqrt{\cos\theta}\; r\,dr\,d\theta$$

$$= 4\sqrt{2a}\int_0^{\frac{\pi}{2}}\sqrt{\cos\theta}\left[\int_0^{2a\cos\theta} r^{\frac{3}{2}}\,dr\right]d\theta$$

$$= 4\sqrt{2a}\int_0^{\frac{\pi}{2}}\sqrt{\cos\theta}\left[\frac{r^{\frac{5}{2}}}{\frac{5}{2}}\right]_0^{2a\cos\theta}$$

$$= \frac{8\sqrt{2a}}{5}\int_0^{\frac{\pi}{2}}(2a)^{\frac{5}{2}}\sqrt{\cos\theta}\,(\cos\theta)^{\frac{5}{2}}\,d\theta$$

$$= \frac{8}{5}(2a)^3\int_0^{\frac{\pi}{2}}\cos^3\theta\,d\theta = \frac{64a^3}{5}\cdot\frac{2}{3} = \frac{128a^3}{15}.$$

EXAMPLE 12.52

Find, by double integration, the volume generated by revolving the cardioid $r = a(1 + \cos\theta)$ about the initial line.

Solution. We observe that the upper and lower halves of the cardioid $r = a(1 + \cos\theta)$ generate the same volume. Therefore, it is sufficient to consider the revolution of the upper-half cardioid only, for which r varies from 0 to $a(1 + \cos\theta)$ and θ varies from 0 to π.

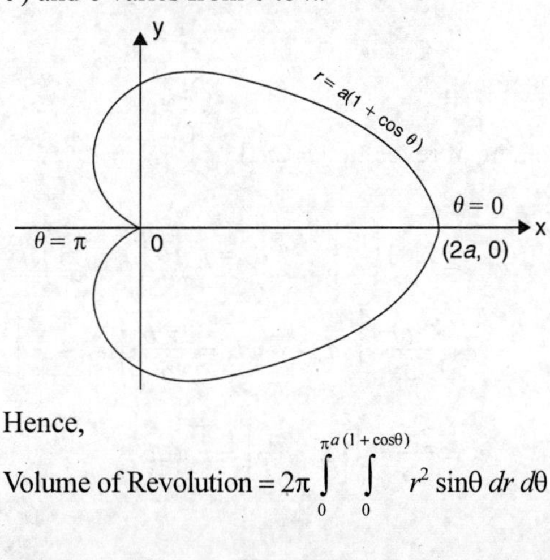

Hence,

$$\text{Volume of Revolution} = 2\pi\int_0^{\pi}\int_0^{a(1+\cos\theta)} r^2\sin\theta\,dr\,d\theta$$

$$= 2\pi \int_0^{\pi} \sin\theta \left[\frac{r^3}{3}\right]_0^{a(1+\cos\theta)}$$

$$= \frac{2\pi a^3}{3} \int_0^{\pi} \sin\theta (1+\cos\theta)^3 \, d\theta$$

$$= -\frac{2\pi a^3}{3} \left[\frac{(1+\cos\theta)^4}{4}\right]_0^{\pi}$$

$$= -\frac{2\pi a^3}{3}(-2^4) = \frac{8}{3}\pi a^3.$$

EXAMPLE 12.53

Find the volume of the solid generated by revolving the ellipse $\frac{x^2}{a^2} + \frac{y^2}{b^2} = 1$ about the x-axis.

Solution. Due tosymmetry, it is sufficient to calculate the volume obtained on revolving the upper half of the ellipse. For this, x varies from $-a$ to a and y varies from 0 to $b\sqrt{1 - \frac{x^2}{a^2}}$.

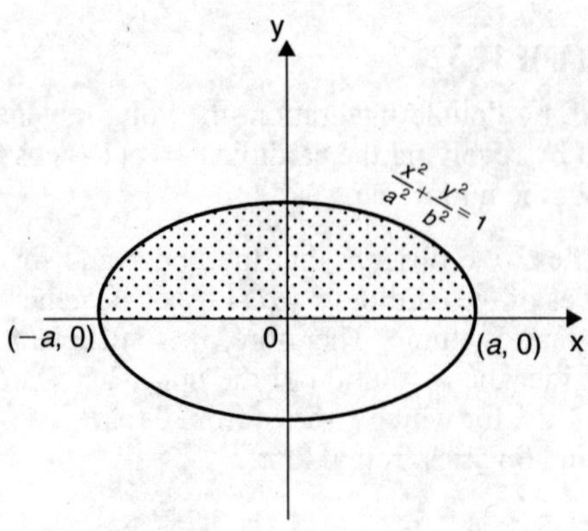

Therefore,

$$\text{volume of revolution} = 2\pi \int_{-a}^{a} \int_0^{b\sqrt{1-\frac{x^2}{a^2}}} y \, dy \, dx$$

$$= 2\pi \int_{-a}^{a} \left[\frac{y^2}{2}\right]_0^{b\sqrt{1-\frac{x^2}{a^2}}} dy = \pi \int_{-a}^{a} b^2\left(1 - \frac{x^2}{a^2}\right) dx$$

$$= \frac{2\pi b^2}{a^2} \int_0^{a} (a^2 - x^2) \, dx = \frac{2\pi b^2}{a^2} \left[a^2 x - \frac{x^3}{3}\right]_0^{a}$$

$$= \frac{2\pi b^2}{a^2} \int_0^{a} \left[a^3 - \frac{a^3}{3}\right] = \frac{4}{3}\pi a b^2.$$

EXAMPLE 12.54

Find the area of the surface of the paraboloid $x^2 + y^2 = z$, which lies between the planes $z = 0$ and $z = 1$.

Solution. The required surface area is given by

$$S = \iint \sqrt{1 + \left(\frac{\partial z}{\partial x}\right)^2 + \left(\frac{\partial z}{\partial y}\right)^2} \, dx \, dy.$$

But,

$$\frac{\partial z}{\partial x} = 2x \quad \text{and} \quad \frac{\partial z}{\partial y} = 2y.$$

Therefore,

$$S = \iint \sqrt{1 + 4(x^2 + y^2)} \, dx \, dy$$

$$= \iint \sqrt{1 + 4r^2} \, r \, dr \, d\theta$$

(changing to polar coordinates).

To find the limits, we see that the projection on the *plane* $z = 1$ is the circle $x^2 + y^2 = 1$ or $r^2 = 1$ and this circle lies between $\theta = 0$ and $\theta = 2\pi$. Hence,

$$S = \int_0^{\frac{\pi}{2}} \int_0^{1} \sqrt{1 + 4r^2} \, r \, d\theta \, dr$$

$$= \frac{1}{8} \int_0^{\frac{\pi}{2}} \int_0^{1} \sqrt{1 + 4r^2} \, 8r \, d\theta \, dr$$

$$= \frac{1}{8} \int_0^{\frac{\pi}{2}} \left[\frac{(1 + 4r^2)^{\frac{3}{2}}}{\frac{3}{2}}\right]_0^1 d\theta = \frac{1}{12} \int_0^{2\pi} (5\sqrt{5} - 1) \, d\theta$$

$$= \frac{5\sqrt{5} - 1}{12} [\theta]_0^{2\pi} = \frac{\pi}{6} [5\sqrt{5} - 1].$$

EXAMPLE 12.55

Compute the surface area of the sphere $x^2 + y^2 + z^2 = a^2$.

Solution. The surface area of the sphere is twice the surface area of the upper-half sphere $z = \sqrt{a^2 - x^2 - y^2}$. We have

$$\frac{\partial z}{\partial x} = -\frac{x}{\sqrt{a^2 - x^2 - y^2}} \quad \text{and}$$

$$\frac{\partial z}{\partial y} = -\frac{y}{\sqrt{a^2 - x^2 - y^2}}.$$

Therefore,

$$S = \iint \sqrt{1 + \left(\frac{\partial z}{\partial x}\right)^2 + \left(\frac{\partial z}{\partial y}\right)^2}\, dx\, dy.$$

$$= \iint \frac{a}{\sqrt{a^2 - x^2 - y^2}} dx\, dy.$$

The domain of integration is the circle $x^2 + y^2 = a^2$ on the xy-plane. Therefore,

$$S = 2\int_{-a}^{a} \left[\int_{-\sqrt{a^2-x^2}}^{\sqrt{a^2-x^2}} \frac{a}{\sqrt{a^2 - x^2 - y^2}}\, dy \right] dx.$$

Changing topolar coordinates, we have

$$S = 2\int_{0}^{\frac{\pi}{2}} \left[\int_{0}^{a} \frac{a}{\sqrt{a^2 - r^2}}\, r\, dr \right] d\theta$$

$$= \int_{0}^{2\pi} \left[\int_{0}^{a} \frac{a}{\sqrt{a^2 - r^2}}\, 2r\, dr \right] d\theta = 4\pi a^2.$$

EXAMPLE 12.56

Find the area of the spherical surface $x^2 + y^2 + z^2 = a^2$ inside the cylinder $x^2 + y^2 = ax$.

Solution. We have

$$S = 4\iint \sqrt{1 + \left(\frac{\partial z}{\partial x}\right)^2 + \left(\frac{\partial z}{\partial y}\right)^2}\, dx\, dy$$

$$= 4\iint \frac{adx\, dy}{\sqrt{a^2 - x^2 - y^2}} \text{ over } x^2 + y^2 = ax$$

$$= 4a \int_{0}^{\frac{\pi}{2}} \int_{0}^{a\cos\theta} \frac{r\, dr\, d\theta}{\sqrt{a^2 - r^2}},\ x = r\cos\theta,\ y = r\sin\theta$$

$$= 2a^2(\pi - 2).$$

EXAMPLE 12.57

Find the area of that part of the cylinder $x^2 + y^2 = a^2$, which is cut off by the cylinder $x^2 + z^2 = a^2$.

Solution. The equation of the surface has the form $y = \sqrt{a^2 - x^2}$ so that

$$\frac{\partial y}{\partial x} = -\frac{x}{\sqrt{a^2 - x^2}},\ \frac{\partial y}{\partial z} = 0$$

and

$$\sqrt{1 + \left(\frac{\partial y}{\partial x}\right)^2 + \left(\frac{\partial y}{\partial z}\right)^2} = \frac{a}{\sqrt{a^2 - x^2}}.$$

The domain of integration is a quarter circle $x^2 + z^2 = a^2$ where $x \geq 0$ and $z \geq 0$ on the xz plane. Therefore,

$$S = 8\int_{0}^{a} \left[\int_{0}^{\sqrt{a^2-x^2}} \frac{a}{\sqrt{a^2 - x^2}}\, dz \right] dx = 8a^2.$$

EXAMPLE 12.58

Find the area of the paraboloid $2z = \frac{x^2}{a} + \frac{y^2}{b}$ inside the cylinder $\frac{x^2}{a^2} + \frac{y^2}{b^2} = 1$

Solution. The required area is

$$S = 4\iint \sqrt{1 + \left(\frac{\partial z}{\partial x}\right)^2 + \left(\frac{\partial z}{\partial y}\right)^2}\, dx\, dy,$$

where the integration extends over the positive octant of the ellipse $\frac{x^2}{a^2} + \frac{y^2}{b^2} = 1$.

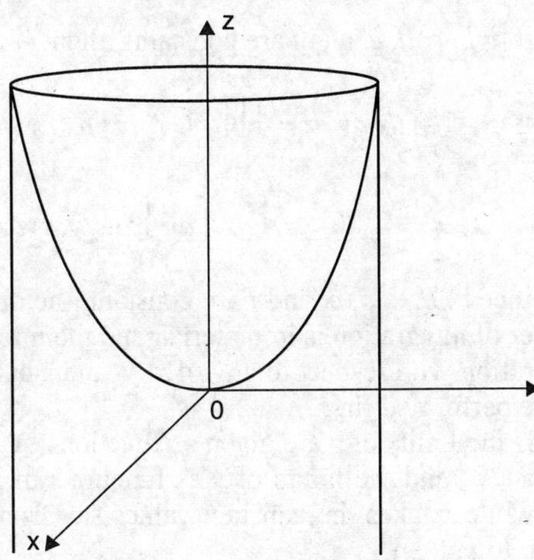

We have $\frac{\partial z}{\partial x} = \frac{x}{a}$ and $\frac{\partial z}{\partial y} = \frac{y}{b}$. Therefore,

$$S = 4\iint \left(1 + \frac{x^2}{a^2} + \frac{y^2}{b^2}\right)^{\frac{1}{2}} dx\, dy$$

$$= 4ab \iint (1 + \xi^2 + \eta^2)\, d\xi\, d\eta, \text{ where}$$

$$x = a\xi,\ y = b\xi, \text{ so that } \xi^2 + \eta^2 = 1$$

$$= 4ab \int_{0}^{\frac{\pi}{2}} \int_{0}^{1} (1 + r^2)\, r\, dr\, d\theta,$$

$\xi = r\cos\theta,\ \eta = r\sin\theta$

$$= \frac{2}{3}\pi ab\left(2^{\frac{3}{2}} - 1\right).$$

12.9 TRIPLE INTEGRALS AND THEIR EVALUATION

Let $f(x, y, z)$ be a continuous function in a finite region V of $\Re^3$. Divide the region V into n subregions of respective volumes $\delta V_1, \delta V_2, \ldots, \delta V_n$. If (x_i, y_i, z_i) be an arbitrary point of the ith subregion, then

$$\lim_{\substack{n\to\infty \\ \delta V_i \to 0}} \sum_{i=1}^{n} f(x_i, y_i, z_i)\,\delta V_i,$$

if exists, is called the triple integral of $f(x,y,z)$ over the region V, and is denoted by $\iiint_V f(x, y, z)dV$ or $\iiint_V f(x, y, z)\,dx\,dy\,dz$.

Evaluation of Triple Integrals

(a) If the region V is specified by the inequalities,

$$a \le x \le b,\ c \le y \le d, \quad \text{and} \quad e \le z \le f,$$

and if $a, b, c, d, e,$ and f are constants, then

$$\iiint_V f(x, y, z)\,dx\,dy\,dz = \int_a^b\int_c^d\int_e^f f(x, y, z)\,dx\,dy\,dz$$

$$= \int_a^b dx \int_c^d dy \int_e^f f(x, y, z)\,dz.$$

Since $a, b, c, d, e,$ and f are constant, the order of integration is immaterial, and the integration with respect to any of x, y, and z can be performed first.

(b) If the limits of z are given as functions of x and y, and the limits of y as functions of x while x takes the constant values say from a to b, then

$$\iiint_V f(x, y, z)\,dx\,dy\,dz$$

$$= \int_a^b \left[\int_{y_1(x)}^{y_2(x)} \left\{\int_{z_1(x,y)}^{z_2(x,y)} f(x, y, z)\,dz\right\} dy\right] dx.$$

Thus, the integration with respect to z is performed first regarding x and y as constants, then the integration with respect to y is performed regarding x as constant and in the last, the integration with respect to x is performed.

EXAMPLE 12.59

Evaluate

$$I = \int_0^{\log 2}\int_0^{x}\int_0^{x+\log y} e^{x+y+z}\,dz\,dy\,dx.$$

Solution. We have

$$I = \int_0^{\log 2}\int_0^{x} [e^{x+y+z}]_0^{x+\log y}\,dy\,dx$$

$$= \int_0^{\log 2}\int_0^{x} [e^{x+y+x+\log y} - e^{x+y}]\,dy\,dx$$

$$= \int_0^{\log 2}\int_0^{x} [e^{2x}\cdot e^{y}\cdot e^{\log y} - e^{x}\cdot e^{y}]\,dy\,dx$$

$$= \int_0^{\log 2}\int_0^{x} [e^{2x}ye^{y} - e^{x}\cdot e^{y}]\,dy\,dx$$

$$= \int_0^{\log 2}\left[\int_0^{x} e^{2x}ye^{y}dy - \int_0^{x} e^{x}e^{y}dy\right]dx$$

$$= \int_0^{\log 2}\left[e^{2x}\{ye^{y}\}_0^{x} - e^{2x}\int_0^{x} e^{y}dy - e^{x}\int_0^{x} e^{y}dy\right]dx$$

$$= \int_0^{\log 2} [e^{2x}\cdot xe^{x} - e^{2x}(e^{x} - 1) - e^{x}(e^{x} - 1)]\,dx$$

$$= \int_0^{\log 2} [xe^{3x} - e^{3x} + e^{x}]\,dx$$

$$= \int_0^{\log 2} xe^{3x}dx - \int_0^{\log 2} e^{3x}dx + \int_0^{\log 2} e^{x}dx$$

$$= \frac{1}{3}[xe^{3x}]_0^{\log 2} - \frac{1}{3}\int_0^{\log 2} e^{3x}dx - \int_0^{\log 2} e^{3x}dx + \int_0^{\log 2} e^{x}dx$$

$$= \frac{1}{3}\log 2\, e^{3\log 2} - \frac{4}{3}\left[\frac{e^{3x}}{3}\right]_0^{\log 2} + \left[e^{x}\right]_0^{\log 2}$$

$$= \frac{1}{3}\log 2\, e^{\log 8} - \frac{4}{9}(e^{\log 8} - 1) + (e^{\log 2} - 1)$$

$$= \frac{8}{3}\log 2 - \frac{4}{9}(8 - 1) + (2 - 1)$$

$$= \frac{8}{3}\log 2 - \frac{28}{9} + 1 = \frac{8}{3}\log 2 - \frac{19}{9}.$$

EXAMPLE 12.60

Evaluate

$$I = \int_1^e \int_1^{\log y} \int_1^{e^x} \log z \, dz \, dx \, dy.$$

Solution. We have

$$I = \int_1^e \int_1^{\log y} \left[\int_1^{e^x} \log z \cdot 1 \, dz \right] dx \, dy$$

$$= \int_1^e \int_1^{\log y} \left\{ [z \log z]_0^{e^x} - \int_1^{e^x} z \cdot \frac{1}{z} \, dz \right\} dx \, dy$$

$$= \int_1^e \int_1^{\log y} [e^x \log e^x - 0 - e^x + 1] \, dx \, dy$$

$$= \int_1^e \int_1^{\log y} [(x-1)e^x + 1] \, dx \, dy$$

$$= \int_1^e \left[\int_1^{\log y} (x-1)e^x \, dx + \int_1^{\log y} dx \right] dy$$

$$= \int_1^e \left[\int_1^{\log y} (x-1)e^x \, dx + \log y - 1 \right] dy$$

$$= \int_1^e \left\{ [(x-1)e^x]_1^{\log y} - \int_1^{\log y} e^x \, dx + \log y - 1 \right\} dy$$

$$= \int_1^e [(\log y - 1)e^{\log y} - (e^{\log y} - e) + \log y - 1] \, dy$$

$$= \int_1^e [y(\log y - 1) - (y - e) + \log y - 1] dy$$

$$= \int_1^e [(y+1)\log y - 2y + e - 1] \, dy$$

$$= \left[\log y \left(\frac{y^2}{2} + y \right) \right]_1^e - \int_1^e \frac{1}{y} \left(\frac{y^2}{2} + y \right) dy$$

$$- \left[2\frac{y^2}{2} \right]_1^e + (e-1)\,[y]_1^e$$

$$= \frac{e^2}{2} + e - \int_1^e \left(\frac{y}{2} + 1 \right) dy - (e^2 - 1) + (e-1)^2$$

$$= \frac{e^2}{2} + e - \left(\frac{y^2}{4} + y \right)_1^e - 2e + 2$$

$$= \frac{e^2}{2} + e - \left(\frac{e^2}{4} + e \right) - \left(\frac{1}{4} + 1 \right) - 2e + 2$$

$$= \frac{e^2}{4} - 2e + \frac{13}{4} = \frac{1}{4}(e^2 - 8e + 13).$$

EXAMPLE 12.61

Evaluate the integral

$$\int_0^1 \int_0^{\sqrt{1-x^2}} \int_0^{\sqrt{1-x^2-y^2}} xyz \, dz \, dy \, dx.$$

Solution. The given triple integral is

$$I = \int_0^1 xy \int_0^{\sqrt{1-x^2}} \left[\frac{z^2}{2} \right]_0^{\sqrt{1-x^2-y^2}} dy \, dx$$

$$= \int_0^1 \int_0^{\sqrt{1-x^2}} \frac{1}{2} xy(1 - x^2 - y^2) \, dy \, dx$$

$$= \int_0^1 \frac{1}{2} x \left[\int_0^{\sqrt{1-x^2}} (y - x^2 y - y^3) \, dy \right] dx$$

$$= \frac{1}{2} \int_0^1 x \left[\frac{y^2}{2} - x^2 \frac{y^2}{2} - \frac{y^4}{4} \right]_0^{\sqrt{1-x^2}} dx$$

$$= \frac{1}{8} \int_0^1 x(1 - x^2)^2 \, dx$$

$$= \frac{1}{8} \int_0^{\frac{\pi}{2}} \sin\theta \, (1 - \sin^2\theta)^2 \cos\theta \, d\theta, \; x = \sin\theta$$

$$= \frac{1}{8} \int_0^{\frac{\pi}{2}} \sin\theta \cos^5\theta \, d\theta = \frac{1}{8} \cdot \frac{1.4.2}{6.4.2} = \frac{1}{48}.$$

EXAMPLE 12.62

Evaluate

$$I = \int_0^a \int_0^{a-x} \int_0^{a-x-y} x^2 \, dx \, dy \, dz.$$

Solution. We have

$$I = \int_0^a \int_0^{a-x} x^2 \left[\int_0^{a-x-y} dz \, dx \right] dy.$$

$$= \int_0^a \int_0^{a-x} x^2 [z]_0^{a-x-y}\, dx\, dy$$

$$= \int_0^a x^2 \left[\int_0^{a-x} (a-x-y)\, dy \right] dx$$

$$= \int_0^a x^2 \left[ay - xy - \frac{y^2}{2} \right]_0^{a-x} dx$$

$$= \int_0^a x^2 \left[a^2 - ax - ax + x^2 - \frac{(a-x)^2}{2} \right] dx$$

$$= \frac{1}{2} \int_0^a (x^2a^2 - 2ax^3 + x^4)\, dx$$

$$= \frac{1}{2} \left[a^2 \frac{x^3}{3} - 2a\frac{x^4}{4} + \frac{x^5}{5} \right]_0^a = \frac{a^5}{60}.$$

EXAMPLE 12.63

Evaluate $\iiint \frac{dx\, dy\, dz}{(x+y+z+1)^3}$ over a tetrahedron bounded by coordinate planes and the plane $x + y + z = 1$.

Solution. The region of integration is bounded by the coordinate planes $x = 0$, $y = 0$, and $z = 0$ and the plane $x + y + z = 1$. Thus,

$$R = \{(x, y, z); x > 0, y > 0, z > 0, x + y + z < 1\}$$
$$= \{(x, y, z); 0 < x < 1, 0 < y < 1 - x, 0 < z < 1 - x - y\}.$$

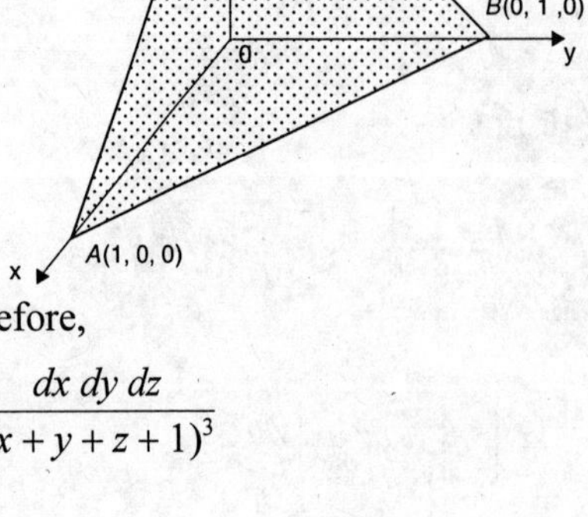

Therefore,

$$\iiint_R \frac{dx\, dy\, dz}{(x+y+z+1)^3}$$

$$= \int_0^1 \int_0^{1-x} \int_0^{1-x-y} \frac{1}{(x+y+z+1)^3}\, dz\, dy\, dx$$

$$= \int_0^1 \int_0^{1-x} \left[\int_0^{1-x-y} (x+y+z+1)^{-3}\, dz \right] dy\, dx$$

$$= \int_0^1 \int_0^{1-x} \left[\frac{(x+y+z+1)^{-2}}{-2} \right]_0^{1-x-y} dy\, dx$$

$$= \frac{1}{2} \int_0^1 \int_0^{1-x} \left[\frac{1}{(x+y+1)^2} - \frac{1}{4} \right] dy\, dx$$

$$= \frac{1}{2} \int_0^1 \left[\frac{-1}{(x+y+1)} - \frac{y}{4} \right]_0^{1-x} dx$$

$$= \frac{1}{2} \int_0^1 \left(-\frac{1}{2} - \frac{1-x}{4} + \frac{1}{x+1} \right) dx$$

$$= \frac{1}{2} \int_0^1 \left[-\frac{3}{4} + \frac{x}{4} + \frac{1}{x+1} \right] dx$$

$$= \frac{1}{2} \left[-\frac{3x}{4} + \frac{x^2}{8} + \log(x+1) \right]_0^1$$

$$= \frac{1}{2} \left[-\frac{3}{4} + \frac{1}{8} + \log 2 \right] = \frac{1}{2} \log 2 - \frac{5}{16}.$$

EXAMPLE 12.64

Evaluate $\int_0^a \int_0^x \int_0^{x+y} e^{x+y+z}\, dz\, dy\, dx$.

Solution. We have

$$\int_0^a \int_0^x \int_0^{x+y} e^{x+y+z}\, dz\, dy\, dx.$$

$$= \int_0^a \int_0^x [e^{x+y+z}]_0^{x+y}\, dy\, dx$$

$$= \int_0^a \int_0^x [e^{2(x+y)} - e^{x+y}]\, dy\, dx$$

$$= \int_0^a \left[\int_0^x e^{2(x+y)}\, dy - \int_0^x e^{x+y}\, dy \right] dx$$

$$= \int_0^a \left\{ \left[\frac{e^{2(x+y)}}{2} \right]_0^x - [e^{x+y}]_0^x \right\} dx$$

$$= \int_0^a \left[\frac{e^{4x}}{2} - \frac{e^{2x}}{2} - e^{2x} + e^x \right] dx$$

$$= \left[\frac{e^{4x}}{8} - \frac{e^{2x}}{4} - \frac{e^{2x}}{2} + e^x \right]_0^a$$

$$= \frac{e^{4a}}{8} - \frac{3e^{2a}}{4} + e^a.$$

EXAMPLE 12.65

Evaluate the triple integral

$$I = \int_0^a \int_0^{\sqrt{a^2-x^2}} \int_0^{\sqrt{a^2-x^2-y^2}} \frac{1}{\sqrt{a^2-x^2-y^2-z^2}}\, dz\, dy\, dx.$$

Solution. We have

$$I = \int_0^a \int_0^{\sqrt{a^2-x^2}} \left[\int_0^{\sqrt{a^2-x^2-y^2}} \frac{1}{\sqrt{(a^2-x^2-y^2)-z^2}}\, dz \right] dy\, dx$$

$$= \int_0^a \int_0^{\sqrt{a^2-x^2}} \left[\sin^{-1} \frac{z}{\sqrt{a^2-x^2-y^2}} \right]_0^{\sqrt{a^2-x^2-y^2}} dy\, dx$$

$$= \int_0^a \left[\int_0^{\sqrt{a^2-x^2}} [\sin^{-1} 1]\, dy \right] dx$$

$$= \frac{\pi}{2} \int_0^a [y]_0^{\sqrt{a^2-x^2}}\, dx = \frac{\pi}{2} \int_0^a \sqrt{a^2-x^2}\, dx$$

$$= \frac{\pi}{2} \left[\frac{x\sqrt{a^2-x^2}}{2} + \frac{a^2}{2} \sin^{-1} \frac{x}{a} \right]_0^a$$

$$= \frac{\pi}{2} \left[0 + \frac{a^2}{2} \sin^{-1} 1 - 0 \right] = \frac{\pi a^2}{4} \sin^{-1} 1$$

$$= \frac{\pi a^2}{4} \cdot \frac{\pi}{2} = \frac{\pi^2 a^2}{8}.$$

Note: The earlier example may be restated as "Evaluate $\iiint \frac{dx\, dy\, dz}{\sqrt{a^2-x^2-y^2-z^2}}$ over the positive octant of the sphere $x^2+y^2+z^2=a^2$".

EXAMPLE 12.66

Evaluate

$$I = \int_{-c}^{c} \int_{-b}^{b} \int_{-a}^{a} (x^2+y^2+z^2)\, dz\, dy\, dx$$

Solution. We observe that the integrand $x^2 + y^2 + z^2$ is symmetrical in x, y, and z. Therefore, the limits of integration can be assigned as per our preference. We have

$$I = \int_{-c}^{c} \int_{-b}^{b} \int_{-a}^{a} (x^2+y^2+z^2)\, dx\, dy\, dz$$

$$= 2 \int_{-c}^{c} \int_{-b}^{b} \left[\int_0^a (x^2+y^2+z^2)\, dx \right] dy\, dz,$$

since $x^2+y^2+z^2$ is even in x

$$= 2 \int_{-c}^{c} \int_{-b}^{b} \left[\frac{x^3}{3} + y^2x + z^2x \right]_0^a dy\, dz$$

$$= 2 \int_{-c}^{c} \int_{-b}^{b} \left[\frac{a^3}{3} + ay^2 + az^2 \right] dy\, dz$$

$$= 4 \int_{-c}^{c} \left[\int_0^b \left(\frac{a^3}{3} + ay^2 + az^2 \right) dy \right] dz,$$

since integrand is even in y

$$= 4 \int_{-c}^{c} \left[\frac{a^3y}{3} + \frac{ay^3}{3} + az^2y \right]_0^b dz$$

$$= 4 \int_{-c}^{c} \left[\frac{ba^3}{3} + \frac{ab^3}{3} + abz^2 \right] dz$$

$$= 8 \int_0^c \left[\frac{ba^3}{3} + \frac{ab^3}{3} + abz^2 \right] dz,$$

since integrand is even in z

$$= 8 \left[\frac{ba^3}{3} z + \frac{ab^3z}{3} + \frac{abz^3}{3} \right]_0^c$$

$$= 8 \left[\frac{ba^3c}{3} + \frac{ab^3c}{3} + \frac{abc^3}{3} \right]$$

$$= \frac{8abc}{3} [a^2+b^2+c^2].$$

EXAMPLE 12.67

Evaluate $\iiint xyz\ dx\ dy\ dz$ over the ellipsoid $\frac{x^2}{a^2}+\frac{y^2}{b^2}+\frac{z^2}{c^2}=1$.

Solution. The region of integration is bounded by

$$z=-c\sqrt{1-\frac{x^2}{a^2}-\frac{y^2}{b^2}} \quad \text{and} \quad z=c\sqrt{1-\frac{x^2}{a^2}-\frac{y^2}{b^2}}.$$

The projection on the xy-plane is the ellipse $\frac{x^2}{a^2}+\frac{y^2}{b^2}=1$. Hence, the limits of integration for y and x are from $y=-b\sqrt{1-\frac{x^2}{a^2}}$ to $y=b\sqrt{1-\frac{x^2}{a^2}}$ and $x=-a$ to $x=a$. Thus,

$$R=\left\{(x, y, z); -a\le x\le a, -b\sqrt{1-\frac{x^2}{a^2}}\le y\le b\sqrt{1-\frac{x^2}{a^2}}, -c\sqrt{1-\frac{x^2}{a^2}-\frac{y^2}{b^2}}\le z\le c\sqrt{1-\frac{x^2}{a^2}-\frac{y^2}{b^2}}\right\}$$

Hence,

$$\iiint_R xyz\ dx\ dy\ dz = \int_{-a}^{a}\int_{-b\sqrt{1-\frac{x^2}{a^2}}}^{b\sqrt{1-\frac{x^2}{a^2}}} xy\left[\int_{-c\sqrt{1-\frac{x^2}{a^2}-\frac{y^2}{b^2}}}^{c\sqrt{1-\frac{x^2}{a^2}-\frac{y^2}{b^2}}} z\ dz\right] dy\ dx$$

$= 0$, since the integrand z is an odd function.

12.10 CHANGE TO SPHERICAL POLAR COORDINATES FROM CARTESIAN COORDINATES IN A TRIPLE INTEGRAL

Let $P(x, y, z)$ be any point in $\mathfrak{R}^3$. Then, the position of this point is determined by the following three numbers:

(i) The distance $r=\sqrt{x^2+y^2+z^2}$ of $P(x, y, z)$ from the origin $(0, 0, 0)$.

(ii) The polar distance θ, where θ is the angle between the radius vector OP and the positive direction of z-axis.

(iii) The angle ϕ, which the projection of the radius vector OP on the xy-plane makes with the x-axis.

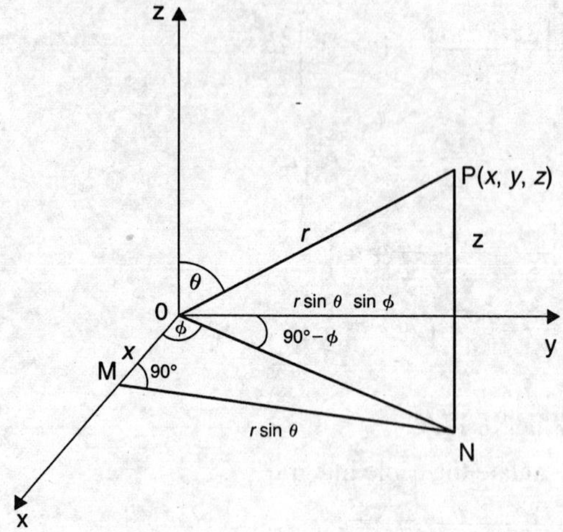

Then,

$$x=r\sin\theta\cos\phi,\ y=r\sin\theta\sin\phi,\ z=r\cos\theta$$

and so, $x^2+y^2+z^2=r^2$. Under these transformations, the region

$$R=(x, y, z); x^2+y^2+z^2\le a^2\}$$

is mapped intothe region

$$R'=\{(r, \theta, \phi); 0\le r\le a,\ 0\le\theta\le\pi,\ 0\le\phi\le 2\pi\}.$$

Also,

$$\frac{\partial(x, y, z)}{\partial(r, \theta, \phi)}=\begin{vmatrix}\frac{\partial x}{\partial r} & \frac{\partial x}{\partial \theta} & \frac{\partial x}{\partial \phi}\\ \frac{\partial y}{\partial r} & \frac{\partial y}{\partial \theta} & \frac{\partial y}{\partial \phi}\\ \frac{\partial z}{\partial r} & \frac{\partial z}{\partial \theta} & \frac{\partial z}{\partial \phi}\end{vmatrix}$$

$$=\begin{vmatrix}\sin\theta\cos\phi & r\cos\theta\cos\phi & -r\sin\theta\sin\phi\\ \sin\theta\sin\phi & r\cos\theta\sin\phi & r\sin\theta\cos\phi\\ \cos\theta & -r\sin\theta & 0\end{vmatrix}$$

$$=r^2\sin\theta.$$

Hence,

$$I=\iiint_R f(x, y, z)\ dx\ dy\ dz$$

$$=\iiint_{R'} f(r\sin\theta\cos\phi, r\sin\theta\sin\phi, r\cos\theta) \times r^2\sin\theta\ dr\ d\theta\ d\phi.$$

The polar spherical coordinates are useful when the region of integration is a sphere or a part of it. If the region of integration is a whole sphere, then $0 \le r \le a$, $0 \le \theta \le \pi$, and $0 \le \phi \le 2\pi$. But if the region of integration is the positive octant of the sphere, then $0 \le r \le a$, $0 \le \theta \le \pi$, and $0 \le \phi \le \frac{\pi}{2}$.

Remark 12.2. If the region of integration is a right circular cylinder, then the Cartesian coordinates are changed to *cylindrical polar coordinates* (r, θ, z) because the position of $P(x, y, z)$ is determined by r, θ, and z as shown in the following figure:

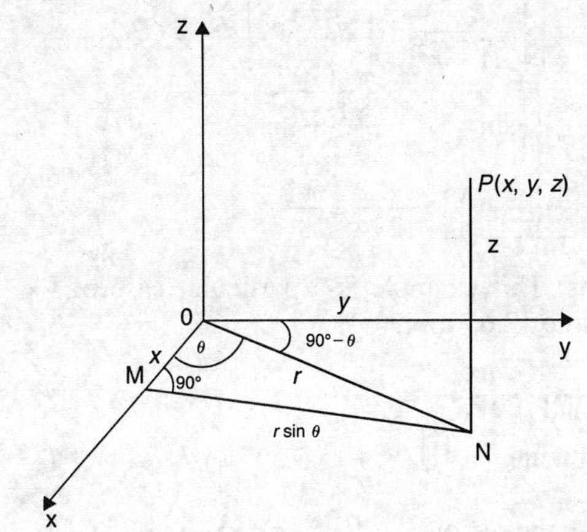

Then,

$$x = r\cos\theta,\ y = r\sin\theta, \text{ and } z = z,$$

and

$$\frac{\partial(x,y,z)}{\partial(r,\theta,z)} = \begin{vmatrix} \frac{\partial x}{\partial r} & \frac{\partial x}{\partial \theta} & \frac{\partial x}{\partial \phi} \\ \frac{\partial y}{\partial r} & \frac{\partial y}{\partial \theta} & \frac{\partial y}{\partial \phi} \\ \frac{\partial z}{\partial r} & \frac{\partial z}{\partial \theta} & \frac{\partial z}{\partial \phi} \end{vmatrix} = \begin{vmatrix} \cos\phi & -r\sin\theta & 0 \\ \sin\theta & r\cos\theta & 0 \\ 0 & 0 & 1 \end{vmatrix}$$

$$= r\cos^2\theta + r\sin^2\theta = r.$$

Hence,

$$I = \iiint_R f(x, y, z)\, dx\, dy\, dz$$

$$= \iiint_R f(r\cos\theta, r\sin\theta, z)\, r\, dr\, d\theta\, dz.$$

EXAMPLE 12.68

Evaluate $I = \iiint z(x^2 + y^2)\, dx\, dy\, dz$ over $x^2 + y^2 \le 1$ and $2 \le z \le 3$.

Solution. The region of integration is

$$V = \{(x, y, z);\ x^2 + y^2 < 1,\ 2 < z < 3\}.$$

Using the transformation

$x = r\cos\theta$, $y = r\sin\theta$, *and* $z = z$ (cylindrical polar coordinates), we have

$$I = \int_0^1 \int_0^{2\pi} \left[\int_2^3 zr^2 \cdot r\, dz\right] d\theta\, dr = \int_0^1 r^3 \int_0^{2\pi} \left[\frac{z^2}{2}\right]_2^3 d\theta\, dr$$

$$= \int_0^1 r^3 \int_0^{2\pi} \left[\frac{9}{2} - \frac{4}{2}\right] d\theta\, dr = \frac{5}{2}\int_0^1 r^3 [\theta]_0^{2\pi}\, dr$$

$$= 5\pi \int_0^1 r^3 dr = 5\pi \left[\frac{r^4}{4}\right]_0^1 = \frac{5\pi}{4}.$$

EXAMPLE 12.69

Evaluate $I = I = \iiint \sqrt{1 - \frac{x^2}{a^2} - \frac{y^2}{b^2} - \frac{z^2}{c^2}}\, dx\, dy\, dz$ over the region

$$V = \{(x, y, z);\ x > 0, y > 0, z > 0, \frac{x^2}{a^2} + \frac{y^2}{b^2} + \frac{z^2}{c^2} < 1\}$$

Solution. Substituting $\frac{x}{a} = X$, $\frac{y}{b} = Y$, and $\frac{z}{c} = Z$ so that $dx = a dX$, $dy = b dY$, $dz = c dZ$ and hence, $dx\, dy\, dz = abc\, dX\, dY\, dZ$. Therefore,

$$I = abc \iiint (1 - X^2 - Y^2 - Z^2)^{\frac{1}{2}}\, dx\, dy\, dz,$$

over the region

$$V' = \{(X, Y, Z);\ X \ge 0,\ Y \ge 0,\ Z \ge 0,\ X^2 + Y^2 + Z^2 \le 1\}.$$

Using spherical polar coordinates,

$X = r\sin\theta\cos\phi$, $Y = r\sin\theta\sin\phi$, and $Z = r\cos\theta$,

the region of integration becomes

$$V'' = \left\{(r, \theta, \phi);\ 0 \le r \le 1, 0 < 0, \le \frac{\pi}{2}, 0 < \phi \le \frac{\pi}{2}\right\}$$

Hence,

$$I = abc \int_0^1 \int_0^{\frac{\pi}{2}} \int_0^{\frac{\pi}{2}} (1 - r^2)^{\frac{1}{2}} r^2 \sin\theta\, dr\, d\theta\, d\phi$$

$$= abc \int_0^1 r^2 (1 - r^2)^{\frac{1}{2}} \int_0^{\frac{\pi}{2}} \sin\theta \left[\int_0^{\frac{\pi}{2}} d\phi\right] d\theta\, dr$$

$$= abc \int_0^1 r^2(1-r^2)^{\frac{1}{2}} \int_0^{\frac{\pi}{2}} \sin\theta\, [\phi]_0^{\frac{\pi}{2}}\, d\theta\, dr$$

$$= \frac{abc\pi}{2} \int_0^1 r^2(1-r^2)^{\frac{1}{2}} \left[\int_0^{\frac{\pi}{2}} \sin\theta\, d\theta \right] dr$$

$$= \frac{abc\pi}{2} \int_0^1 r^2(1-r^2)^{\frac{1}{2}} [-\cos\theta]_0^{\frac{\pi}{2}}\, dr$$

$$= \frac{abc\pi}{2} \int_0^1 r^2(1-r^2)^{\frac{1}{2}}\, dr. \qquad (1)$$

But, substituting $r = \sin t$ so that $dr = \cos t\, dt$, we have

$$\int_0^1 r^2(1-r^2)^{\frac{1}{2}}\, dr = \int_0^{\frac{\pi}{2}} \sin^2 t \sqrt{1-\sin^2 t} + \cos t\, dt$$

$$= \int_0^{\frac{\pi}{2}} \sin^2 t \cos^2 t\, dt = \frac{1}{4.2}\cdot\frac{\pi}{2} = \frac{\pi}{16}.$$

Hence (1) reduces to

$$\mathrm{I} = \frac{\pi\, abc}{2}\left(\frac{\pi}{16}\right) = \frac{\pi^2 abc}{32}$$

EXAMPLE 12.70

Evaluate

$$I = \int_0^1 \int_0^{\sqrt{1-x^2}} \int_0^{\sqrt{1-x^2-y^2}} \frac{dz\, dy\, dx}{\sqrt{1-x^2-y^2-z^2}}$$

by changing tospherical polar coordinates.

Solution. The region of integration is

$$V = \{(x, y, z);\ 0 \le x \le 1,\ 0 \le y \le \sqrt{1-x^2},\ 0 \le z \le \sqrt{1-x^2-y^2}\}.$$

Now, we transform the region by using spherical polar coordinates, by substituting $x = r \sin\theta \cos\phi$, $y = r\sin\theta\sin\phi$, and $z = r\cos\theta$. The transformed region is

$$V' = \left\{ (r, \theta, \phi);\ 0 \le r \le 1,\ 0 < \theta \le \frac{\pi}{2},\ 0 \le \phi \le \frac{\pi}{2} \right\}.$$

Therefore,

$$I = \int_0^1 \int_0^{\frac{\pi}{2}} \int_0^{\frac{\pi}{2}} \frac{r^2 \sin\theta}{\sqrt{1-r^2}}\, dr\, d\theta\, d\phi$$

$$= \int_0^1 \frac{r^2}{\sqrt{1-r^2}} \int_0^{\frac{\pi}{2}} \sin\theta \left[\int_0^{\frac{\pi}{2}} d\phi \right] d\theta\, dr$$

$$= \int_0^1 \frac{r^2}{\sqrt{1-r^2}} \int_0^{\frac{\pi}{2}} \sin\theta\, [\phi]_0^{\frac{\pi}{2}}\, d\theta\, dr$$

$$= \frac{\pi}{2} \int_0^1 \frac{r^2}{\sqrt{1-r^2}} \left[\int_0^{\frac{\pi}{2}} \sin\theta\, d\theta \right] dr$$

$$= \frac{\pi}{2} \int_0^1 \frac{r^2}{\sqrt{1-r^2}} [-\cos\theta]_0^{\frac{\pi}{2}}\, dr$$

$$= \frac{\pi}{2} \int_0^1 \frac{r^2}{\sqrt{1-r^2}}\, dr = \frac{\pi}{2} \int_0^1 \frac{1-(1-r^2)}{\sqrt{1-r^2}}\, dr$$

$$= \frac{\pi}{2} \left[\int_0^1 \frac{dr}{\sqrt{1-r^2}} - \int_0^1 \sqrt{1-r^2}\, dr \right]$$

$$= \frac{\pi}{2} \left\{ \left[\sin^{-1} r - \left(\frac{r\sqrt{1-r}}{2} + \frac{1}{2}\sin^{-1} r \right) \right]_0^1 \right\}$$

$$= \frac{\pi}{2}\left\{ \frac{\pi}{2} - \frac{\pi}{4} \right\} = \frac{\pi}{2}\cdot\frac{\pi}{4} = \frac{\pi^2}{8}.$$

Note: This example is a particular case of Example 12.65 for $a = 1$.

EXAMPLE 12.71

Evaluate $I = \iiint_V (x^2 + y^2 + z^2)^m\, dx\, dy\, dz,\ m > 0$ over the region $V = \{(x, y, z);\ x^2 + y^2 + z^2 \le 1\}$.

Solution. The given region of integration is

$$V = \{(x, y, z);\ x^2 + y^2 + z^2 \le 1\}.$$

Changing tospherical polar coordinates by substituting

$$x = r\sin\theta\cos\phi,\ y = r\sin\theta\sin\phi, \text{ and}$$
$$z = r\cos\theta,$$

we get $x^2 + y^2 + z^2 = r^2$ and $\dfrac{\partial(x, y, z)}{\partial(r, \theta, \phi)} = r^2 \sin\theta$.

Therefore, the region of integration reduces to $V' = \{(r, \theta, \phi);\ 0 \le r \le 1,\ 0 \le \theta \le \pi,\ 0 \le \phi \le 2\pi\}$.

Hence,

$$I = \int_0^1 \int_0^{\pi} \int_0^{2\pi} r^{2m+2} \sin\theta\, dr\, d\theta\, d\phi$$

$$= \int_0^1 r^{2m+2} \int_0^{\pi} \sin\theta \left[\int_0^{2\pi} d\phi \right] d\theta\, dr$$

$$= \int_0^1 r^{2m+2} \int_0^{\pi} \sin\theta\, [\phi]_0^{2\pi}\, d\theta\, dr$$

$$= 2\pi \int_0^1 r^{2m+2} \left[\int_0^{\pi} \sin\theta \, d\theta \right] dr$$

$$= 2\pi \int_0^1 r^{2m+2} [-\cos\theta]_0^{\pi} \, dr$$

$$= 4\pi \int_0^1 r^{2m+2} \, dr = 4\pi \left[\frac{r^{2m+3}}{2m+3} \right]_0^1$$

$$= \frac{4\pi}{2m+3}.$$

12.11 VOLUME AS A TRIPLE INTEGRAL

In Cartesian coordinates, the volume of a region V is given by the triple integral

$$\iiint dx\, dy\, dz,$$

where the limits of integration are chosen to cover the entire region V.

In spherical polar coordinates, the volume of a region V is given by the triple integral

$$\iiint r^2 \sin\theta \, dr\, d\theta\, d\phi,$$

where the limits of integration are chosen to cover the entire region V.

In cylindrical coordinates, the volume of a region V is given by the triple integral

$$\iiint r\, dr\, d\theta\, dz,$$

where the limits of integration are chosen to cover the entire region V.

EXAMPLE 12.72

Find the volume of the sphere $x^2 + y^2 + z^2 = r^2$.

Solution. The required volume is given by

$$V = 8 \iiint dx\, dy\, dz,$$

taken over the positive octant of the given sphere. Changing to spherical polar coordinates, we put $x = r\sin\theta\cos\phi$, $y = r\sin\theta\sin\phi$, and $z = r\cos\theta$. So, $x^2 + y^2 + z^2 = r^2$. In the positive octant, we have

$$0 \le r \le a,\ 0 \le \theta \le \frac{\pi}{2}, \text{ and } 0 \le \phi \le \frac{\pi}{2}.$$

Therefore,

$$V = 8 \int_0^a \int_0^{\frac{\pi}{2}} \int_0^{\frac{\pi}{2}} r^2 \sin\theta \, dr\, d\theta\, d\phi$$

$$= 8 \int_0^a r^2 \int_0^{\frac{\pi}{2}} \sin\theta \left[\int_0^{\frac{\pi}{2}} d\phi \right] d\theta\, dr$$

$$= 8 \int_0^a r^2 \int_0^{\frac{\pi}{2}} \sin\theta \, [\phi]_0^{\frac{\pi}{2}} \, d\theta\, dr$$

$$= 4\pi \int_0^a r^2 \left[\int_0^{\frac{\pi}{2}} \sin\theta \, d\theta \right] dr$$

$$= 4\pi \int_0^a r^2 [-\cos\phi]_0^{\frac{\pi}{2}} \, dr = 4\pi \int_0^a r^2 \, dr$$

$$= 4\pi \left[\frac{r^3}{3} \right]_0^a = \frac{4\pi a^3}{3}.$$

EXAMPLE 12.73

Find the volume of the ellipsoid $\frac{x^2}{a^2} + \frac{y^2}{b^2} + \frac{z^2}{c^2} = 1.$

Solution. Substituting $\frac{x}{a} = X$, $\frac{y}{b} = Y$, and $\frac{z}{c} = Z$, we have

$dx = adX$, $dy = adY$, and $z = adZ$. Therefore, the volume is given by

$$V = abc \iiint dX\, dY\, dZ,$$

taken over $X^2 + Y^2 + Z^2 = 1$. Changing tospherical polar coordinates by substituting $X = r\sin\theta \cos\phi$, $y = r\sin\theta\sin\phi$, and $z = r\cos\theta$, we have

$$V = 8abc \int_0^1 \int_0^{\frac{\pi}{2}} \int_0^{\frac{\pi}{2}} r^2 \sin\theta \, dr\, d\theta\, d\phi$$

$$= 8abc \int_0^1 r^2 \int_0^{\frac{\pi}{2}} \sin\theta \left[\int_0^{\frac{\pi}{2}} d\phi \right] d\theta\, dr$$

$$= 8abc \int_0^1 r^2 \int_0^{\frac{\pi}{2}} \sin\theta \, [\phi]_0^{\frac{\pi}{2}} \, d\theta\, dr$$

$$= 4\pi\, abc \int_0^1 r^2 \left[\int_0^{\frac{\pi}{2}} \sin\theta \, d\theta \right] dr$$

$$= 4\pi\, abc \int_0^1 r^2 [-\cos\theta]_0^{\frac{\pi}{2}} \, dr = 4\pi\, abc \int_0^1 r^2 \, dr$$

$$= 4\pi\, abc \left[\frac{r^3}{3} \right]_0^1 = \frac{4}{3}\pi\, abc.$$

EXAMPLE 12.74

Find the volume of the solid bounded by the surface

$$\left(\frac{x}{a}\right)^{\frac{2}{3}} + \left(\frac{y}{b}\right)^{\frac{2}{3}} + \left(\frac{z}{c}\right)^{\frac{2}{3}} = 1.$$

Solution. Substituting $\left(\frac{x}{a}\right)^{\frac{2}{3}} = X$, $\left(\frac{y}{b}\right)^{\frac{2}{3}} = Y$, and

$\left(\frac{z}{c}\right)^{\frac{2}{3}} = Z$, that is, $x = aX^3$, $y = bY^3$, and $z = cZ^3$, we get $dx = 3aX^2\, dX$, $dy = 3bY^2 dY$, and $dz = 3cZ^2\, dZ$. Then, the required volume is given by

$$V = \iiint_V dx\, dy\, dz$$
$$= \iiint 27abc\, X^2Y^2Z^2\, dX\, dY\, dZ,$$

taken throughout the sphere $X^2 + Y^2 + Z^2 = 1$. Changing to spherical polar coordinates (r, θ, ϕ), we have

$$V = 8 \iiint_{V'} 27abc\, r^2 \sin^2\theta \cos^2\phi$$
$$\times r^2 \sin^2\theta \sin^2\phi\, r^2 \cos^2\theta \cdot r^2 \sin\theta\, dr\, d\theta\, d\phi,$$

where

$$V' = \left\{(r, \theta, \phi);\, 0 \le r \le 1,\, 0 \le \theta \le \frac{\pi}{2},\, 0 \le \phi \le \frac{\pi}{2}\right\}.$$

Thus,

$$V = 216abc \int_0^1 r^8 \int_0^{\frac{\pi}{2}} \sin^5\theta \cos^2\theta$$
$$\times \cdot \left[\int_0^{\frac{\pi}{2}} \sin^2\phi \cos^2\phi\, d\phi\right] d\theta\, dr$$
$$= 216abc \int_0^1 r^8 \left[\int_0^{\frac{\pi}{2}} \sin^5\theta \cos^2\theta \cdot \frac{1}{4.2} \cdot \frac{\pi}{2}\, d\theta\right] dr$$
$$= \frac{216}{16}\pi\, abc \int_0^1 r^8 \left[\int_0^{\frac{\pi}{2}} \sin^5\theta \cos^2\theta\, d\theta\right] dr$$
$$= \frac{216}{16}\pi abc \int_0^1 r^8 \left[\frac{4.2.1}{7.5.3.1}\right] dr$$
$$= \frac{216}{210}\pi abc \left[\frac{r^9}{9}\right]_0^1$$
$$= \frac{216\pi abc}{1890}$$
$$= \frac{4}{35}\pi abc.$$

Second Method: For the positive octant, we have

$$V = \iiint dx\, dy\, dz.$$

Substituting $\left(\frac{x}{a}\right)^{\frac{2}{3}} = X$, $\left(\frac{y}{b}\right)^{\frac{2}{3}} = Y$, and $\left(\frac{z}{c}\right)^{\frac{2}{3}} = Z$, we have $dx = \frac{3}{2}aX^{\frac{1}{2}}\, dX$, $dy = \frac{3}{2}bY^{\frac{1}{2}}\, dY$, $dz = \frac{3}{2}cZ^{\frac{1}{2}} dZ$, and $X + Y + Z \le 1$. Therefore, by Dirichlet's Theorem,

$$V = \iiint \frac{27}{8} abc\, X^{\frac{1}{2}}\, Y^{\frac{1}{2}}\, Z^{\frac{1}{2}}\, dX\, dY\, dZ$$
$$= \frac{27}{8} abc \iiint X^{\frac{3}{2}-1}\, Y^{\frac{3}{2}-1}\, Z^{\frac{3}{2}-1}\, dX\, dY\, dZ$$
$$= \frac{27}{8} abc\, \frac{\Gamma\left(\frac{3}{2}\right)\Gamma\left(\frac{3}{2}\right)\Gamma\left(\frac{3}{2}\right)}{\Gamma\left(1 + \frac{3}{2} + \frac{3}{2} + \frac{3}{2}\right)}$$
$$= \frac{27abc}{8}\, \frac{\left(\Gamma\left(\frac{3}{2}\right)\right)^3}{\Gamma\left(\frac{9}{2} + 1\right)}$$
$$= \frac{27}{8} abc\, \frac{\left(\frac{1}{2} \cdot \sqrt{\pi}\right)^2}{\frac{9}{2} \cdot \frac{7}{2} \cdot \frac{5}{2} \cdot \frac{3}{2} \cdot \frac{1}{2} \cdot \sqrt{\pi}}$$
$$= \frac{\pi}{8} abc \cdot \frac{4}{35}.$$

Hence, the total volume is $= 8\, \frac{\pi abc}{8} \cdot \frac{4}{35} = \frac{4\pi\, abc}{35}$.

EXAMPLE 12.75

Find the volume of the portion cut off from the sphere $x^2 + y^2 + z^2 = a^2$ by the cylinder $x^2 + y^2 = ax$.

Solution. The required volume is

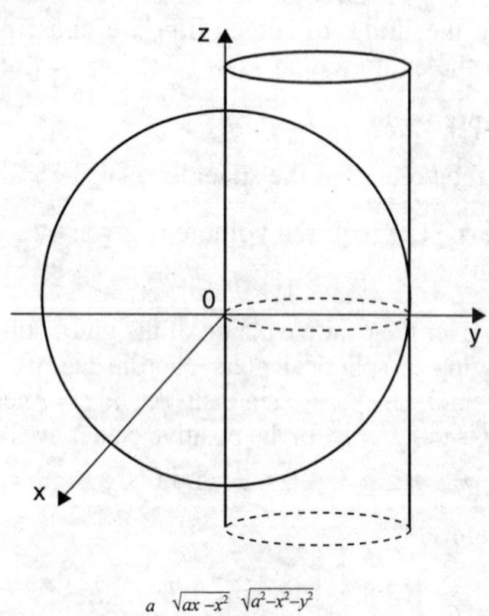

$$V = 4 \int_0^a \int_0^{\sqrt{ax - x^2}} \int_0^{\sqrt{a^2 - x^2 - y^2}} dx\, dy\, dz$$

$$= 4\int_0^a \int_0^{\sqrt{ax-x^2}} (a^2 - x^2 - y^2)^{\frac{1}{2}}\, dy\, dx$$

$$= 4\int_0^{\frac{\pi}{2}} \int_0^{a\cos\theta} \sqrt{a^2 - r^2}\, r\, dr\, d\theta,$$

changing to polar coordinates

$$= \frac{4}{2}\int_0^{\frac{\pi}{2}} \int_0^{a\cos\theta} 2r\sqrt{a^2 - r^2}\, dr\, d\theta$$

$$= 2\int_0^{\frac{\pi}{2}} \left[\frac{-(a^2 - r^2)^{\frac{3}{2}}}{\frac{3}{2}}\right]_0^{a\cos\theta} d\theta$$

$$= \frac{4}{3}\int_0^{\frac{\pi}{2}} [-a^3 \sin^3\theta + a^3]\, d\theta$$

$$= \frac{4}{3}a^3\left[-\frac{2}{3} + \frac{\pi}{2}\right]$$

$$= \frac{2}{3}a^3\left(\pi - \frac{4}{3}\right).$$

EXAMPLE 12.76

Prove that the volume of the wedge intercepted between the cylinder $x^2 + y^2 = 2ax$ and the planes $z = mx$ and $z = nx$ is $\pi(m-n)a^3$.

Solution. The required volume is given by

$$V = 2\int_0^{2a} \int_0^{\sqrt{2ax-x^2}} \int_{nx}^{mx} dx\, dy\, dz$$

$$= 2\int_0^{2a} \int_0^{\sqrt{2ax-x^2}} [z]_{nx}^{mx}\, dy\, dx$$

$$= 2\int_0^{2a} \int_0^{\sqrt{2ax-x^2}} (m-n)\, x\, dy\, dx$$

$$= 2\int_0^{2a} (m-n)\, x[y]_0^{\sqrt{2ax-x^2}}\, dx$$

$$= 2(m-n)\int_0^{2a} x\sqrt{2ax - x^2}\, dx$$

$$= 2(m-n)\int_0^{2a} x^{\frac{3}{2}}\sqrt{2a - x}\, dx.$$

Substituting $x = 2a\sin^2\theta$, we get $dx = 4a\sin\theta\cos\theta\, d\theta$. The limits of integration are $\theta = 0$ to $\theta = \frac{\pi}{2}$. Therefore,

$$V = 2(m-n)\int_0^{\frac{\pi}{2}} 16a^3 \sin^4\theta \cos^2\theta\, d\theta$$

$$= 32(m-n)\cdot\frac{3.1}{6.4.2}\cdot\frac{\pi}{2} = \pi(m-n)a^3.$$

EXAMPLE 12.77

The axes of two right circular cylinders of the same radius a, intersect at right angles. Prove that the volume inside both the cylinders is $\frac{16a^3}{3}$.

Solution. Let the equations of the cylinders be

$$x^2 + y^2 = a^2 \text{ and } x^2 + z^2 = a^2.$$

Therefore, the region of integration is defined by

$$-\sqrt{a^2 - x^2} \le z \le \sqrt{a^2 - x^2},$$
$$-\sqrt{a^2 - x^2} \le y \le \sqrt{a^2 - x^2}, \text{ and}$$
$$-a \le x \le a.$$

Hence, the required volume is

$$V = \int_{-a}^{a} \int_{-\sqrt{a^2-x^2}}^{\sqrt{a^2-x^2}} \int_{-\sqrt{a^2-x^2}}^{\sqrt{a^2-x^2}} dx\, dy\, dz$$

$$= 8\int_0^a \int_0^{\sqrt{a^2-x^2}} \int_0^{\sqrt{a^2-x^2}} dx\, dy\, dz$$

$$= 8\int_0^a \int_0^{\sqrt{a^2-x^2}} \sqrt{a^2 - x^2}\, dy\, dx$$

$$= 8\int_0^a \sqrt{a^2 - x^2}\, [y]_0^{\sqrt{a^2-x^2}}\, dx$$

$$= 8\int_0^a (a^2 - x^2)\, dx$$

$$= 8\left[a^2x - \frac{x^3}{3}\right]_0^a = 8\left[a^3 - \frac{a^3}{3}\right] = \frac{16a^3}{3}.$$

EXAMPLE 12.78

Find the volume in the positive octant bounded by the coordinate planes and the plane. $x + 2y + 3z = 4$.

Solution. The region of integration is

$$V = \left\{ (x, y, z); 0 \le x \le 4, 0 \le y \le \frac{4-x}{2}, 0 \le z \le \frac{4-x-2y}{3} \right\}.$$

Therefore, the required volume is

$$V = \int_0^4 \int_0^{\frac{4-x}{2}} \left[\int_0^{\frac{4-x-2y}{3}} dz \right] dy\, dx$$

$$= \frac{1}{3} \int_0^4 \int_0^{\frac{4-x}{2}} (4 - x - 2y)\, dy\, dx$$

$$= \frac{1}{3} \int_0^4 \left(4y - xy - 2\frac{y^2}{2} \right)_0^{\frac{4-x}{2}} dx$$

$$= \frac{1}{3} \int_0^4 \left(8 - 2x - 2x + \frac{x^2}{2} - 4 - \frac{x^2}{4} + 2x \right) dx$$

$$= \frac{1}{3} \int_0^4 \left(4 - 2x + \frac{x^2}{4} \right) dx = \frac{1}{3} \left(4x - x^2 + \frac{x^3}{12} \right)_0^4$$

$$= \frac{1}{3} \left[16 - 16 + \frac{64}{12} \right] = \frac{16}{9}.$$

EXAMPLE 12.79

Find the volume of the tetrahedron bounded by the planes $x = 0, y = 0, z = 0$, and $\frac{x}{a} + \frac{y}{b} + \frac{z}{c} = 1$; $a, b, c \ge 0$.

Solution. The region of integration is bounded by four planes $x = 0$, $y = 0$, $z = 0$, and $\frac{x}{a} + \frac{y}{b} + \frac{z}{c} = 1$. It is bounded below and above by $z = 0$ and $z = c\left(1 - \frac{x}{a} - \frac{y}{b}\right)$. Its projection on the xy-plane is a triangle bounded by $x = 0$, $y = 0$, and $\frac{x}{a} + \frac{y}{b} = 1$. Therefore, the region is

$$V = \left\{ (x, y, z);\ 0 \le x \le a,\ 0 \le y \le b\left(1 - \frac{x}{a}\right), 0 \le Z \le c\left(1 - \frac{x}{a} - \frac{x}{b}\right). \right\}$$

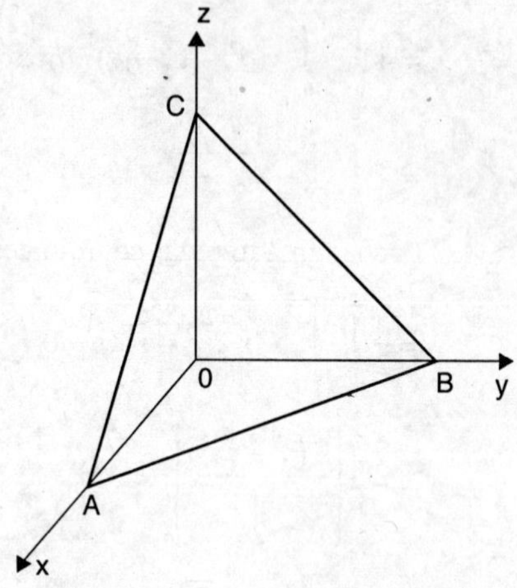

Hence, the volume of the tetrahedron OABC is

$$V = \iiint_V dx\, dy\, dz = \int_0^a \int_0^{b\left(1-\frac{x}{a}\right)} \int_0^{c\left(1-\frac{x}{a}-\frac{y}{b}\right)} dz\, dy\, dx$$

$$= \int_0^a \int_0^{b\left(1-\frac{x}{a}\right)} [z]_0^{c\left(1-\frac{x}{a}-\frac{y}{b}\right)} dy\, dx$$

$$= c \int_0^a \left[\int_0^{b\left(1-\frac{x}{a}\right)} \left(1 - \frac{x}{a} - \frac{y}{b}\right) dy \right] dx$$

$$= c \int_0^a \left[\frac{\left(1 - \frac{x}{a} - \frac{y}{b}\right)^2}{2\left(-\frac{1}{b}\right)} \right]_0^{b\left(1-\frac{x}{a}\right)} dx = \frac{bc}{2} \int_0^a \left(1 - \frac{x}{a}\right)^2 dx$$

$$= \frac{bc}{2} \left[\frac{\left(1 - \frac{x}{a}\right)^3}{3\left(-\frac{1}{a}\right)} \right]_0^a = \frac{abc}{6} [0-1]$$

$$= \frac{abc}{6} \text{ cu units.}$$

Second Method: Substituting $\frac{x}{a} = u$, $\frac{y}{b} = v$, and $\frac{z}{c} = w$, we have $u \ge o, v \ge 0, w \ge 0$, and $u + v + w \le 1$. Therefore

$$V = \iiint_V dx\, dy\, dz$$

$$= \iiint_{V'} (a\, du).(b\, du).(c\, du),\ u + v + w \le 1$$

$$= abc \iiint_{V'} 0\, u^{1-1}\, v^{1-1}\, w^{1-1}\, du\, dv\, dw$$

$$= abc\frac{\Gamma(1)\Gamma(1)\Gamma(1)}{\Gamma(1+1+1+1)}, \text{ by Dirichlet's Theorem}$$

$$= \frac{abc}{3!} = \frac{abc}{6}.$$

EXAMPLE 12.80

Find the volume of the portion cut off from a sphere $x^2 + y^2 + z^2 = a^2$ by a cone $x^2 + y^2 = z^2$.

Solution. The origin is the center of the sphere and the vertex of the cone $x^2 + y^2 = z^2$. Therefore, the volume is symmetrical about the plane $z = 0$. Hence,

$$V = 2\iiint dx\, dy\, dz.$$

Changing the coordinates to spherical polar, $x^2 + y^2 + z^2 = a^2$ reduces to $r^2 = a^2$ or $r = a$. Further, $x^2 + y^2 = z^2$ reduces to

$$r^2 \sin^2\theta \cos^2\phi + r^2 \sin^2\theta \sin^2\phi = r^2 \cos^2\theta$$

or

$$r^2 \sin^2\theta (\cos^2\phi + \sin^2\phi) = r^2 \cos^2\theta$$

or

$$\sin^2\theta = \cos^2\theta, \text{ which yields } \theta = \frac{\pi}{4}.$$

Thus, θ varies from 0 to $\frac{\pi}{4}$ and ϕ varies from 0 toπ. Therefore,

$$V = 2\int_0^a \int_0^{\frac{\pi}{4}} \int_0^{2\pi} r^2 \sin\theta\, dr\, d\theta\, d\phi$$

$$= 2\int_0^a r^2 \int_0^{\frac{\pi}{4}} \sin\theta\, [\phi]_0^{2\pi}\, d\theta\, dr$$

$$= 4\pi \int_0^a r^2 \int_0^{\frac{\pi}{4}} \sin\theta\, d\theta$$

$$= 4\pi \int_0^a r^2[-\cos\theta]_0^{\frac{\pi}{4}}\, dr - 4\pi \int_0^a r^2\left(1 - \frac{1}{\sqrt{2}}\right) dr$$

$$= 4\pi\left(1 + \frac{1}{\sqrt{2}}\right)\left[\frac{r^3}{3}\right]_0^a = \frac{4\pi a^3}{3}\left(1 - \frac{1}{\sqrt{2}}\right)$$

$$= \frac{2\pi a^3}{3}(2 - \sqrt{2}).$$

12.12 MISCELLANEOUS EXAMPLES

EXAMPLE 12.81

Evaluate $\int_4^3 \int_1^2 (x + y)^{-2}\, dx\, dy$.

Solution.

$$\int_4^3 \left[\int_1^2 (x + y)^{-2}\, dx\right] dy$$

$$= \int_4^3 \left[\frac{(x + y)^{-1}}{-1}\right]_0^2 dy$$

$$= -\int_4^3 \left[\frac{1}{2 + y} - \frac{1}{1 + y}\right] dy$$

$$= -[\log(2 + y) - \log(1 + y)]_4^3$$

$$= \log 4 + \log 6 - 2\log 5$$

$$= \log\frac{24}{25}.$$

EXAMPLE 12.82

Evaluate the integral $\int_0^a \int_a^y \frac{x}{x^2 + y^2}\, dxdy$ by changing the order of integration.

Solution. The given integral is

$$I = \int_0^a \int_a^y \frac{x}{x^2 + y^2}\, dxdy.$$

The region of integration is bounded by the lines $x = y$, $x = a$, $y = 0$ and $y = a$. Thus the region of integration is shown in the figure below:

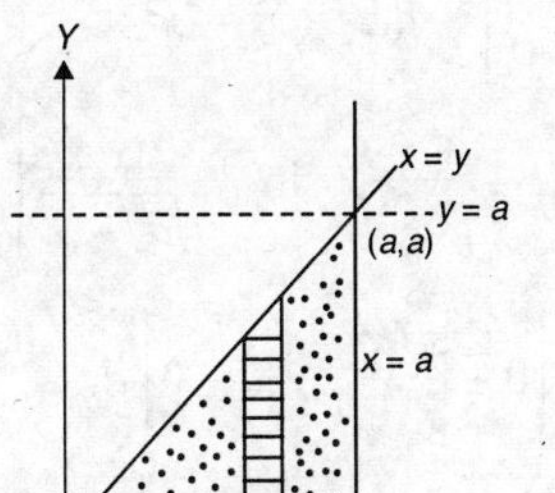

On changing the order of integration, we first integrate with respect to y along the strip parallel to y axis. The strip extends from $y = 0$ to $y = x$. To cover the whole region, we then integrate with respect to x from $x = 0$ to $x = a$. Hence

$$I = \int_0^a \int_0^x \frac{x}{x^2 + y^2}\, dx\, dy = \int_0^a x\left[\frac{1}{x}\tan^{-1}\frac{y}{x}\right]_0^x dx$$

$$= \int_0^a \frac{\pi}{4}\, dx = \frac{\pi}{4}a.$$

EXAMPLE 12.83

By changing the order of integrations, evaluate

$$\int_0^a \int_0^{\sqrt{a^2-x^2}} (a^2 - x^2 - y^2)\, dy\, dx.$$

Solution. The region of integration is bounded by $x = 0$, $x = a$, $y = 0$ and the circle $x^2 + y^2 = a^2$.

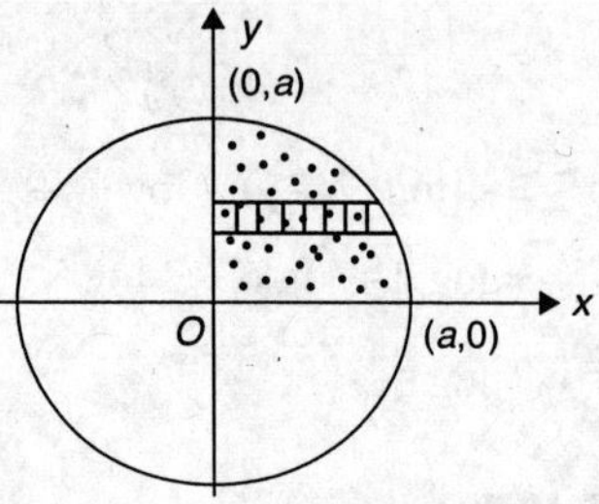

After changing the order of integration, we have to integrate the integrand first with respect to x and then with respect to y. We take a strip parallel to x-axis. The limit of x varies from 0 to $\sqrt{a^2 - y^2}$. To cover the whole region, the limits of y will vary from 0 to a. Hence the given integral is

$$I = \int_0^a \int_0^{\sqrt{a^2-y^2}} [(a^2 - y^2) - x^2]\, dxdy$$

$$= \int_0^a \left[\frac{x}{2}\sqrt{(a^2 - y^2) - x^2} + \frac{a^2 - y^2}{2} \sin^{-1} \frac{x}{\sqrt{a^2 - y^2}}\right]_0^{\sqrt{a^2-y^2}} dy$$

$$= \int_0^a \left[0 + \frac{\pi (a^2 - y^2)}{4} - 0\right] dy$$

$$= \frac{\pi}{4}\int_0^a (a^2 - y^2)\, dy = \frac{\pi}{4}\left[a^2 y - \frac{y^3}{3}\right]_0^a$$

$$= \frac{\pi}{4}\left[a^3 - \frac{a^3}{3}\right] = \frac{\pi}{6} a^3.$$

EXAMPLE 12.84

Changing the order of integration of $\int_0^\infty \int_0^\infty e^{-xy} \sin nx\, dx\, dy$, show that $\int_0^\infty \left(\frac{\sin nx}{x}\right) dx = \frac{\pi}{2}.$

Solution.
We have

$$I = \int_0^\infty \int_0^\infty e^{-xy} \sin nx\, dx\, dy$$

$$= \int_0^\infty \sin nx \left[\int_0^\infty e^{-xy}\, dy\right] dx$$

$$= \int_0^\infty \sin nx \left[\frac{e^{-xy}}{-x}\right]_0^\infty dx$$

$$= \int_0^\infty \frac{\sin nx}{x}\, dx.$$

On the other hand,

$$I = \int_0^\infty \int_0^\infty e^{-xy} \sin nx\, dx\, dy$$

$$= \int_0^\infty \left[\int_0^\infty e^{-xy} \sin nx\, dx\right] dy$$

$$= \int_0^\infty \left[\frac{e^{-xy}}{n^2 + y^2}(n \cos nx + y \sin nx)\right]_0^\infty dy$$

$$= \int_0^\infty \frac{n}{n^2 + y^2}\, dy = \left[\tan^{-1} \frac{y}{n}\right]_0^\infty = \frac{\pi}{2} \qquad (2)$$

From (1) and (2), it follows that

$$= \int_0^\infty \frac{\sin nx}{x}\, dx = \frac{\pi}{2}.$$

EXAMPLE 12.85

By transforming into polar co-ordinates, evaluate $\iint \frac{x^2 y^2}{x^2 + y^2}\, dxdy$ over the annular region between the circles $x^2 + y^2 = a^2$ and $x^2 + y^2 = b^2$, where $b > a$.

Solution. Putting $x = r\cos\theta$, $y = r \sin\theta$, we have $dx\, dy = rdrd\theta$. Therefore

$$\iint \frac{x^2 y^2}{x^2 + y^2}\, dxdy = \int_a^b \int_0^{2\pi} \frac{r^4 \sin^2\theta \cos^2\theta}{r^2}\, rdr\, d\theta$$

$$= \frac{1}{16}(b^4 - a^4) \int_0^{2\pi} \left(\frac{1 - \cos 4\theta}{2}\right) d\theta$$

$$= \frac{b^4 - a^4}{16}\left[\frac{1}{2}\theta - \frac{1}{8}\sin 4\theta\right]_0^{2\pi}$$

$$= \frac{\pi}{16}(b^4 - a^4).$$

EXAMPLE 12.86

Evaluate the integral $\int_0^1 \int_0^{\sqrt{1-x^2}} \int_{\sqrt{x^2+y^2}}^0 \frac{dz\, dy\, dx}{\sqrt{x^2+y^2+z^2}}$.

Solution. We have

$$I = \int_0^1 \int_0^{\sqrt{1-x^2}} \int_{\sqrt{x^2+y^2}}^0 \frac{1}{\sqrt{x^2+y^2+z^2}}\, dz\, dy\, dx$$

$$= \int_0^1 \int_0^{\sqrt{1-x^2}} \int_{\sqrt{x^2+y^2}}^0 \frac{1}{\sqrt{(\sqrt{x^2+y^2})^2+z^2}}\, dz\, dy\, dx$$

$$= \int_0^1 \int_0^{\sqrt{1-x^2}} \left[\log\left|z + \sqrt{x^2+y^2+z^2}\right|\right]_{\sqrt{x^2+y^2}}^0 dxdy$$

$$= \int_0^1 \int_0^{\sqrt{1-x^2}} \left[\log\left|\sqrt{x^2+y^2}\right| - \log\left|\sqrt{x^2+y^2} + \sqrt{2(x^2+y^2)}\right|\right] dxdy$$

$$= -(\log(\sqrt{2}+1)) \int_0^1 \int_0^{\sqrt{1-x^2}} dy\, dx$$

$$= -(\log(\sqrt{2}+1)) \int_0^1 \sqrt{1-x^2}\, dx$$

$$= -(\log(\sqrt{2}+1)) \left[\frac{1}{2}x\sqrt{1-x^2} + \frac{1}{2}\sin^{-1}\frac{x}{1}\right]_0^1$$

$$= -(\log(\sqrt{2}+1)) \left[\frac{1}{2}\sin^{-1} 1\right] = -\frac{\pi}{4}\log(\sqrt{2}+1).$$

EXAMPLE 12.87

Find the area in the first quadrant enclosed by the curve $\left(\frac{x}{a}\right)^{\alpha} + \left(\frac{y}{b}\right)^{\beta} = 1$, where $\alpha > 0, \beta > 0$.

Solution. The equation of the curve is

$$\left(\frac{x}{a}\right)^{\alpha} + \left(\frac{y}{b}\right)^{\beta} = 1,\ \alpha, \beta > 0.$$

The parametric form of the curve is

$x = a\cos^{\frac{2}{\alpha}} t,\ y = b\sin^{\frac{2}{\beta}} t.$

Therefore, the required area is

$$A = \int_{\frac{\pi}{2}}^0 y\, dx = \int_{\frac{\pi}{2}}^0 y\frac{dx}{dt}\, dt$$

$$= \int_{\frac{\pi}{2}}^0 \left(b\sin^{\frac{2}{\beta}} t\right)\left(-\frac{2a}{\alpha}\cos^{\left(\frac{2}{\alpha-1}\right)} t\right)\sin t\, dt$$

$$= \frac{2ab}{\alpha}\int_0^{\frac{\pi}{2}} \sin^{\left(\frac{2}{\beta+1}\right)} t\cos^{\left(\frac{2}{\alpha-1}\right)} t\, dt$$

$$= \frac{2ab}{\alpha}\left[\frac{\Gamma\left(\frac{\left(\frac{2}{\beta}+2\right)}{2}\right)\Gamma\left(\frac{\left(\frac{2}{\alpha}-1+1\right)}{2}\right)}{2\Gamma\left(\frac{\frac{2}{\beta}+1+\frac{2}{\alpha}-1+2}{2}\right)}\right]$$

$$= \frac{2ab}{2\alpha\beta}\left[\frac{\Gamma\left(\frac{1}{\alpha}\right)\Gamma\left(\frac{1}{\beta}\right)}{\Gamma\left(\frac{1}{\alpha}+\frac{1}{\beta}+1\right)}\right]$$

$$= \frac{ab}{\alpha+\beta}\,\frac{\Gamma\left(\frac{1}{\alpha}\right)\Gamma\left(\frac{1}{\beta}\right)}{F\left(\frac{1}{\alpha}+\frac{1}{\beta}\right)}$$

EXERCISES

Evaluation of a Double Integral

1. Evaluate $\iint_D (4 - x^2 - y^2)\, dx\, dy$ if the region D is bounded by the lines $x = 0$, $x = 1$, $y = 0$, and $y = \frac{3}{2}$.

   Ans. $\frac{35}{8}$.

2. Evaluate $\iint e^{2x+3y}\, dx$ over the triangle bounded by $x = 0$, $y = 0$, and $x + y = 1$.

   Ans. $\frac{1}{6}(2e+1)(e-1)^2$.

3. Evaluate $\int_0^a \int_0^{\sqrt{a^2-y^2}} (a^2 - x^2 - y^2)^{\frac{1}{2}}\, dx\, dy$.

   Ans. $\frac{\pi a^3}{6}$.

4. Evaluate $\int_1^a \int_1^b \frac{dx\, dy}{xy}$.

   Ans. $\log b \log a$.

5. Evaluate $\int_0^1 \int_0^{\sqrt{1+x^2}} \frac{dx\, dy}{1 + x^2 + y^2}$.

Hint: $I = \int_0^1 \frac{1}{\sqrt{1+x^2}} \left[\tan^{-1} \frac{y}{\sqrt{1+x^2}}\right]_0^{\sqrt{1+x^2}} dx$

$= \int_0^1 \frac{1}{\sqrt{1+x^2}} [\tan^{-1} 1 - \tan^{-1} 0]dx$

$= \frac{\pi}{4} \int_0^1 \frac{1}{\sqrt{1+x^2}} dx$

$= \frac{\pi}{4} [\log \{x + \sqrt{1+x^2}\}]_0^1$

$= \frac{\pi}{4} \log[1 + \sqrt{2}].$

6. Evaluate $\iint x^2y^2 dx\, dy$ over the region bounded by $x = 0$, $y = 0$, and $x^2 + y^2 = 1$.

Ans. $\int_0^1 \int_0^{\sqrt{1-y^2}} x^2 y^2 dx\, dy = \frac{\pi}{96}.$

7. Evaluate $\iint_R y\, dx\, dy$, where R is the region in the first quadrant bounded by the ellipse $\frac{x^2}{a^2} + \frac{y^2}{b^2} = 1.$

Hint: $R = \left\{ (x,y); 0 \le x \le; a, 0 \le y \le b\sqrt{1 - \frac{x^2}{a^2}} \right\}$

Ans. $\frac{ab^2}{3}.$

8. Evaluate $\iint xy\,(x + y)\, dx\, dy$ over the area between $y = x^2$ and $y = x$.

Ans. $\frac{3}{56}.$

9. Evaluate $\iint_D xy\, dx\, dy$, where A is the region common to the circle $x^2 + y^2 = x$ and $x^2 + y^2 = y$.

Hint: The points of intersection of two circles are (0, 0) and $\left(\frac{1}{2}, \frac{1}{2}\right)$. The limits are $x = 0$ to $x = \frac{1}{2}$ and $y = \frac{1 - \sqrt{1-4x^2}}{2}$ to $y = \sqrt{x - x^2}$.

Ans. $\frac{1}{96}.$

10. Evaluate $\int_0^{\pi} \int_a^{(1+\cos\theta)} r^2 \cos\theta\, d\theta\, dr.$

Ans. $\frac{5\pi a^3}{8}.$

11. Evaluate $\iint r \sin\theta\, dr\, d\theta$ over the cardioid $r = a\,(1 - \cos\theta)$ above the initial line.

Hint: $\iint_R r \sin\theta\, dr d\theta = \int_0^{\pi} \int_0^{a(1-\cos\theta)} r \sin\theta\, dr\, d\theta$

$= \int_0^{\pi} \sin\theta \left[\frac{r^2}{2}\right]_0^{a(1-\cos\theta)} d\theta$

$= \frac{a^2}{2} \int_0^{\pi} \sin\theta\, (1 - \cos\theta)^2\, d\theta$

$= \frac{a^2}{2} \left[\frac{(1-\cos\theta)^3}{3}\right]_0^{\pi} = \frac{4a^2}{3}.$

12. Evaluate $\iint r^2 d\theta\, dr$ over the area of the circle $r = a\cos\theta$.

Ans. $\frac{4a^3}{9}.$

13. Evaluate $\int_0^{\frac{\pi}{2}} \int_0^{a\cos\theta} r\sqrt{a^2 - r^2}\, dr\, d\theta.$

Ans. $\frac{a^3}{18}(3\pi - 4).$

14. Evaluate $\iint \frac{r\, dr\, d\theta}{\sqrt{a^2 + r^2}}$ over one loop of $r^2 = a^2 \cos^2 2\theta$.

Ans. $\frac{a}{2}(4 - \pi).$

Change of Variable in a Double Integral

15. Transform the following double integral to polar coordinates and hence, evaluate the same.

$$I = \int_0^a \int_0^{\sqrt{a^2 - y^2}} (a^2 - x^2 - y^2)\, dx\, dy.$$

Hint: $I = \int_0^{\frac{\pi}{2}} \int_0^1 (a^2 - r^2)\, r\, dr\, d\theta.$

Ans. $\frac{\pi a^4}{8}.$

16. Changing to polar coordinates, evaluate

$\iint\sqrt{\frac{1-x^2-y^2}{1+x^2+y^2}}\,dx\,dy$ over the positive quadrant of the circle $x^2+y^2=1$.

Hint: $I=\int_0^{\frac{\pi}{2}}\int_0^1\sqrt{\frac{1-r^2}{1+r^2}}\,r\,dr\,d\theta=\int_0^{\frac{\pi}{2}}\int_0^1\frac{1-r^2}{\sqrt{1-r^4}}\times r\,dr\,d\theta$.

Ans. $\frac{\pi}{4}\left(\frac{\pi}{2}-1\right)$.

17. Using the transformation $x+y=u$ and $y=uv$, show that $\iint[xy(1-x-y)]^{\frac{1}{2}}\,dx\,dy$, taken over the area of triangle bounded by the lines $x=0$, $y=0$, and $x+y=1$, is $\frac{2\pi}{105}$.

Hint: $x=u-y=u-uv$ and $y=uv$. Therefore, Jacobian $J=u$. So, $dx\,dy=u\,du\,dv$. Further, $0\le u\le 1$ and $0\le v\le 1$. Thus,

$$I=\int_0^1\int_0^1 u(1-u)^{\frac{1}{2}}v^{\frac{1}{2}}(1-v)^{\frac{1}{2}}u\,du\,dv$$

$$=\int_0^1 u^2(1-u)^{\frac{1}{2}}du.\int_0^1 v^{\frac{1}{2}}(1-v)^{\frac{1}{2}}dv$$

$$=\int_0^1 u^{3-1}(1-u)^{\frac{3}{2}-1}du.\int_0^1 v^{\frac{3}{2}-1}(1-v)^{\frac{3}{2}-1}dv$$

$=\beta\left(3,\frac{3}{2}\right)\beta\left(\frac{3}{2},\frac{3}{2}\right)$, convert to gamma function.

18. Evaluate $\iint\sqrt{a^2-x^2-y^2}\,dx\,dy$ over the semi-circle $x^2+y^2=ax$ in the positive quadrant.

Hint: $I=\int_0^{\frac{\pi}{2}}\int_0^{a\cos\theta}\sqrt{a^2-r^2}\,r\,dr\,d\theta$.

Ans. $\frac{a^3}{3}\left(\frac{\pi}{2}-\frac{2}{3}\right)$.

19. Evaluate $\iint\sqrt{\frac{1-\frac{x^2}{a^2}-\frac{y^2}{b^2}}{1+\frac{x^2}{a^2}+\frac{y^2}{b^2}}}\,dx\,dy$ over the positive quadrant of the ellipse $\frac{x^2}{a^2}+\frac{y^2}{b^2}=1$.

Hint: Another from of Example 12.20.

20. Change $\int_0^{4a}\int_{\frac{y^2}{4a}}^{y}\frac{x^2-y^2}{x^2+y^2}\,dx\,dy$ in polar coordinates and hence, evaluate the same.

Ans. $\int_{\frac{\pi}{4}}^{\frac{\pi}{2}}\int_0^{\frac{4a\cos\theta}{\sin^2\theta}} r\cos 2\theta\,d\theta\,dr=8a^2\left(\frac{\pi}{2}-\frac{5}{3}\right)$.

21. Changing to polar coordinates, evaluate $\iint xy\,(x^2+y^2)^{\frac{3}{2}}\,dx\,dy$ over the positive quadrant of the circle $x^2+y^2=1$.

Ans. $\frac{1}{14}$.

Change of Order of Integration

22. Change the order of integration in

$$\int_0^1\int_x^{x(2-x)} f(x,y)\,dx\,dy.$$

Ans. $\int_0^1\int_{1-\sqrt{1-y}}^{y} f(x,y)\,dy\,dx$.

23. Changing the order of integration, evaluate the integral $\int_0^a\int_{\frac{x^2}{a}}^{2a-x} xy\,dy\,dx$.

Ans. $\frac{5a^4}{6}$.

24. Changing the order of integration, evaluate

$$\int_0^a\int_y^a\frac{x\,dx\,dy}{x^2+y^2}.$$

Ans. $\frac{\pi a}{4}$.

25. Change the order of integration

$\int_0^a\int_x^{\frac{a^2}{x}}(x+y)\,dx\,dy$. and hence evaluate it.

Ans. ∞.

26. Evaluate the integral $\int_0^b\int_0^{\frac{a}{b}\sqrt{b^2-y^2}} xy\,dy\,dx$.

Ans. $\frac{a^2b^2}{8}$.

27. Changing the order of integration, evaluate the integral $\int_0^{\frac{a}{\sqrt{2}}}\int_y^{\sqrt{a^2-y^2}}\log(x^2+y^2)\,dx\,dy,\ a>0$.

Ans. $\frac{\pi a^4}{4}\left(\log a-\frac{1}{2}\right)$.

Area Enclosed by Curves

28. Find the area bounded by the parabola $y = x^2$ and the line $y = 2x + 3$.

Ans. $\frac{32}{3}$.

29. Find the area of the region bounded by the lines, $x = -2$ and $x = 2$ and the circle, $x^2 + y^2 = 9$.

Ans. $4\sqrt{5} + 18 \sin^{-1}\frac{2}{3}$ sq. units.

30. Find the area of the cardioid $r = a(1 - \cos\theta)$.

Ans. Area $= 2\int_0^{\pi}\int_0^{a(1-\cos\theta)} r\,dr\,d\theta = \frac{3\pi a^2}{2}$ sq. units.

31. Find the area outside the circle $r = a$ and inside the cardioid. $r = a(1 + \cos\theta)$.

Ans. $\frac{\pi a^2}{2}$.

32. Find by double integration, the area lying inside the cardioid $r = a(1 + \cos\theta)$ and outside the parabola $r(1 + \cos\theta). = 1$.
Hint: Eliminating r between the two equations, we get $\cos^2\theta + 2\cos\theta = 0$, which implies $\theta = \pm\frac{\pi}{2}$.
Then Area $= \int_{-\frac{\pi}{2}}^{\frac{\pi}{2}}\int_{\frac{1}{1+\cos\theta}}^{1+\cos\theta} r\,dr\,d\theta = \frac{9\pi + 16}{12}$.

33. Find, using double integration, the smaller of the areas bounded by the circle $x^2 + y^2 = 9$ and the line $x + y = 3$.

Ans. $\frac{9}{4}(\pi - 2)$.

Volume and Surface Areas as Double Integrals

34. Find the volume of the solid region under the surface $z = 3 - x^2 - 2y^2$ for $x^2 + y^2 \le 1$.

Ans. $\frac{9\pi}{4}$.

35. Using double integration, find the volume of the tetrahedron bounded by the coordinate planes and the plane $\frac{x}{a} + \frac{y}{b} + \frac{z}{c} = 1$.

Ans. $\int_0^a\left[\int_0^{b\left(1-\frac{x}{a}\right)} c\left(1 - \frac{x}{a} - \frac{y}{b}\right) dy\right] dx = \frac{abc}{6}$.

36. Find the volume of the region bounded by the surfaces $y = x^2$ and $x = y^2$ and the planes $z = 0$ and $z = 3$.

Hint: $V = \int_0^1\int_{x^2}^{\sqrt{x}} 3\,dy\,dx = 1$.

37. Calculate the volume of the solid bounded by the surfaces $x = 0, y = 0, z = 0$, and $x + y + z = 1$.

Ans. $\frac{1}{6}$.

38. Find the volume of the cylinder $x^2 + y^2 - 2ax = 0$ intercepted between the paraboloid $x^2 + y^2 = 2az$ and the xy-plane.

Ans. $3\pi a^3$.

39. Find the volume bounded by the xy-plane, the paraboloid $2z = x^2 + y^2 = 2ax$, and the cylinder $x^2 + y^2 = 4$.

Ans. 4π.

40. Find the volume common to the surface $y^2 + z^2 = 4ax$ and $x^2 + y^2 = 2ax$, the axis being rectangular.
Hint:

$$V = \int_0^{2a}\int_{-\sqrt{2ax - x^2}}^{\sqrt{2ax - x^2}} \sqrt{4ax - y^2}\,dy\,dx$$

$$= 2\int_0^{2a}\int_0^{\sqrt{2ax - x^2}} \sqrt{4ax - y^2}\,dy\,dx.$$

Ans. $\frac{2}{3}(3\pi + 8)a^3$.

41. Find the volume of the sphere $x^2 + y^2 + z^2 = 9$.

Ans. 36π.

42. 42. Find the area of the surface $z^2 = 2xy$ included between $x = 0, x = a, y = 0$, and $y = b$.

Ans. $\frac{2\sqrt{2}}{3}\sqrt{ab}\,(a + b)$.

43. Find the area of the portion of the sphere $x^2 + y^2 + z^2 = 9$ lying inside the cylinder $x^2 + y^2 = 3y$.

Hint: $z^2 = 9 - x^2 - y^2$. Then

$1+\left(\frac{\partial z}{\partial x}\right)^2+\left(\frac{\partial z}{\partial y}\right)^2=\frac{9}{9-x^2-y^2}$. Change to polar coordinates. Surface area

$$=4\int_0^{\frac{\pi}{2}}\int_0^{3\sin\theta}\frac{3}{\sqrt{9-r^2}}\,dr\,d\theta=18\pi-36.$$

44. Find the area of the portion of the cylinder $x^2+y^2=4y$ lying inside the sphere $x^2+y^2+z^2=16$.

Ans. 64 sq. units.

45. Using double integration, find the volume generated by the revolution of the cardioid $r=a(1-\cos\theta)$ about its axis.
Hint: Volume of revolution

$$=2\pi\int_0^{\pi}\int_0^{a(1-\cos\theta)}r^2\sin\theta\,dr\,d\theta=\frac{8}{3}\pi a^3.$$

46. Find the volume generated by revolving the ellipse $\frac{x^2}{a^2}+\frac{y^2}{b^2}=1$ about the y-axis.

Ans. $\frac{4}{3}\pi a^2 b$.

Evaluation of Triple Integral

47. Evaluate $\int_0^1\int_0^{1-x}\int_0^{1-x-y} xyz\,dx\,dy\,dz$.

Ans. $\frac{1}{720}$.

48. Evaluate $\iiint z^2 dx\,dy\,dz$ over the sphere. $x^2+y^2+z^2=1$.

Ans. $\frac{4\pi}{15}$.

49. Evaluate $\int_1^3\int_{\frac{1}{x}}^{1}\int_0^{\sqrt{xy}} xyz\,dz\,dy\,dx$.

Ans. $\frac{1}{3}\left(\frac{13}{3}-\frac{1}{2}\log 3\right)$.

50. Evaluate $\int_0^{\frac{\pi}{2}}\int_0^{a\cos\theta}\int_0^{\sqrt{a^2-r^2}} r\,dz\,dr\,d\theta$.

Ans. $a^3\left(\frac{\pi}{6}-\frac{\pi}{9}\right)$.

51. Evaluate $\iiint_V dx\,dy\,dz$, where $V=\{(x,y,z);\ x^2+y^2+z^2\le a^2,\ 0\le z\le h\}$, using cylindrical polar coordinates.

Ans. $\int_0^{\frac{\pi}{2}}\int_0^{2\pi}\int_0^{h} r\,dz\,d\theta\,dr=\pi a^2 h$.

Volume as a Triple Integral

52. Find the volume bounded by the surface $x^2+y^2=a^2$ and $\frac{x^2}{p}+\frac{y^2}{q}=2z,\ p>0,\ q>0$.

Ans. $\frac{\pi a^4}{8}\left(\frac{1}{p}+\frac{1}{q}\right)$.

53. Find the volume of the paraboloid of revolution $x^2+y^2=4z$ cut off by the plane $z=4$.

Hint: $V=4\int_0^4\int_0^{\sqrt{16-x^2}}\int_{\frac{x^2+y^2}{4}}^{4} dx\,dy\,dz=32\pi$.

54. Find the volume bounded above by the sphere $x^2+y^2+z^2=2a^2$ and below by the paraboloid $az=x^2+y^2$.

Ans. $\pi a^3\left(\frac{4\sqrt{2}}{3}-\frac{7}{6}\right)$.

55. Show that the volume enclosed by the cylinder $x^2+y^2=2ax$ and $z^2=2ax$ is $\frac{128a^3}{15}$.

56. Show that the volume of the wedge intercepted between the cylinder $x^2+y^2=2ax$ and the planes $z=x$ and $z=2x$ is πa^3.
Hint: See Example 12.76.

13 Vector Calculus

We know that *scalar* is a quantity that is characterized solely by magnitude whereas *vector* is a quantity which is characterized by both magnitude and direction. For example, time, mass, and temperature are scalar quantities whereas displacement, velocity, and force are vector quantities. We represent a vector by an arrow over it. Geometrically, we represent a vector $\vec{a}$ by a directed line segment $\vec{PQ}$, where $\vec{a}$ has direction from P to Q. The point P is called the *initial point* and the point Q is called the *terminal point* of $\vec{a}$. The length $|\vec{PQ}|$ of this line segment is the *magnitude* of $\vec{a}$. Two vectors $\vec{a}$ and $\vec{b}$ are said to be *equal* if they have the same magnitude and direction. The product of a vector $\vec{a}$ and a scalar m is a vector $m\vec{a}$ with magnitude $|m|$ times the magnitude of $\vec{a}$ with direction, the same or opposite to that of $\vec{a}$, according as $m > 0$ or $m < 0$. In particular, if $m = 0$, then $m\vec{a}$ is a null vector $\vec{0}$. A vector with unit magnitude is called a *unit vector*. If $\vec{a}$ is non-zero vector, then $\frac{\vec{a}}{|\vec{a}|} = \frac{\vec{a}}{a}$ is a unit vector having the same direction as that of $\vec{a}$ and is denoted by $\hat{a}$.

If $\vec{a}, \vec{b}$ and $\vec{c}$ are vectors and m and n are scalars (real or complex), then addition and scalar multiplication of vectors satisfy the following properties:

(i) $\vec{a}+\vec{b}=\vec{b}+\vec{a}$ (Commutative law for addition).

(ii) $\vec{a}+\left(\vec{b}+\vec{c}\right)=\left(\vec{a}+\vec{b}\right)+\vec{c}$ (Associative law for addition).

(iii) $m\left(\vec{a}+\vec{b}\right)=m\vec{a}+n\vec{a}$ (Distributive law for addition).

(iv) $(m+n)\vec{a}=m\vec{a}+n\vec{a}$ (Distributive law for scalars).

(v) $\vec{a}+\vec{0}=\vec{a}=\vec{0}+\vec{a}$ (Existence of identity for addition).

(vi) $\vec{a}+(-\vec{a})=\vec{0}=(-\vec{a})+\vec{a}$

(Existence of inverse for addition).

(vii) $|m\vec{a}|=|m|\ |\vec{a}|$.

(viii) $m(n\vec{a})=(mn)\vec{a}$.

(ix) $n(m\vec{a})=m(n\vec{a})$.

The unit vectors in the directions of positive *x*-, *y*-, and *z*- axes of a three-dimensional, rectangular coordinate system are called the *rectangular unit vectors* and are denoted, respectively, by $\hat{i}, \hat{j},$ and $\hat{k}$.

Let a_1, a_2, and a_3 be the rectangular coordinates of the terminal point of vector $\vec{a}$ with the initial point at the origin O of a rectangular coordinate system in three dimensions. Then, the vectors $a_1\hat{i}, a_2\hat{j},$ and $a_3\hat{k}$ are called *rectangular component vectors* or simply *component vectors* of $\vec{a}$ in the *x*, *y*, and *z* directions, respectively.

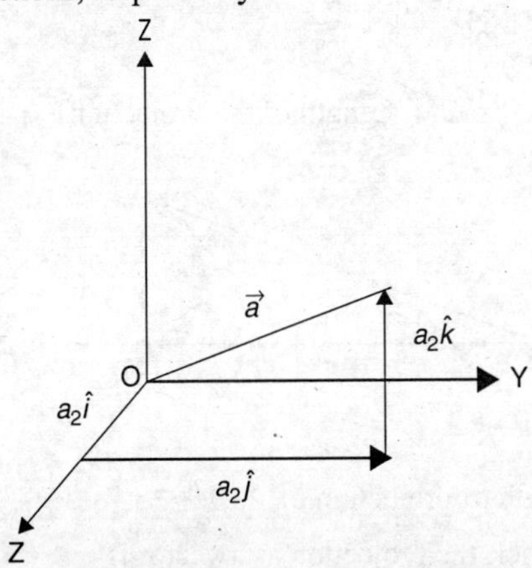

The resultant (sum) of $a_1\hat{i}, a_2\hat{j}$, and $a_3\hat{k}$ is the vector $\vec{a}$ and so,

$$\vec{a} = a_1\hat{i} + a_2\hat{j} + a_3\hat{k}.$$

Further, the magnitude of $\vec{a}$ is

$$|\vec{a}| = \sqrt{a_1^2 + a_2^2 + a_3^2}.$$

In particular, the *radius vector* or *position vector* $\vec{r}$ from O to the point (x, y, z) in a three-dimension space is expressed as

$$\vec{r} = x\hat{i} + y\hat{j} + z\hat{k}$$

and

$$r = |\vec{r}| = \sqrt{x^2 + y^2 + z^2}.$$

Scalar Product or Dot Product or Inner Product of Vectors

The *scalar product* or *dot product* or *inner product* of two vector $\vec{a}$ and $\vec{b}$ is a scalar defined by

$$\vec{a}.\vec{b} = |\vec{a}||\vec{b}|\cos\theta,$$

where θ is the angle between the vectors $\vec{a}$ and $\vec{b}$ and $0 \leq \theta \leq \pi$.

Projection of a Vector on another Vector.

Let $\vec{OM}$ be the projection of the vector $\vec{a}$ on the vector $\vec{b}$. Then

$$\vec{a}.\vec{b} = |\vec{a}||\vec{b}|\cos\theta$$

or

$$\frac{\vec{a}.\vec{b}}{|\vec{b}|} = |\vec{a}|\cos\theta$$

$= OM$, Length of the projection.

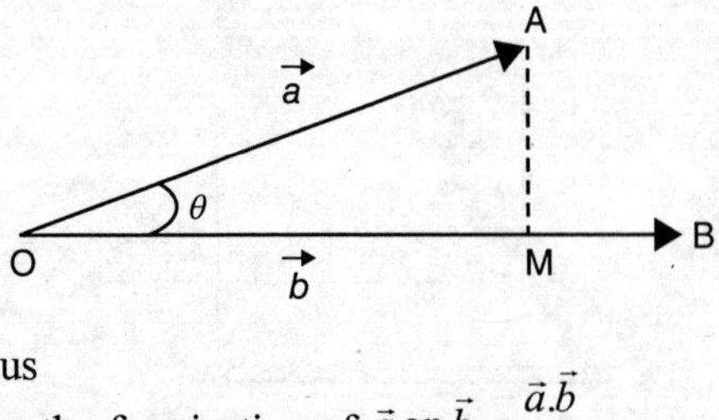

Thus

Length of projection of $\vec{a}$ on $\vec{b} = \dfrac{\vec{a}.\vec{b}}{|b|}$.

Further, the projection as a vector is

$$\vec{OM} = OM\,\vec{e_b} = \frac{\vec{a}.\vec{b}}{|b|}\left(\frac{\vec{b}}{|\vec{b}|}\right) = \frac{\vec{a}.\vec{b}}{|\vec{b}|^2}\vec{b}.$$

Geometrical Interpretation of Scalar Product

Let $\vec{a} = \vec{OL}, \vec{b} = \vec{OM}$ and let θ be the angle between $\vec{a}$ and $\vec{b}$. Then

$$\vec{a}.\vec{b} = |\vec{a}||\vec{b}|\cos\theta$$

$$= |\vec{a}|(\text{Projection of } \vec{b} \text{ in the direction of } \vec{a})$$

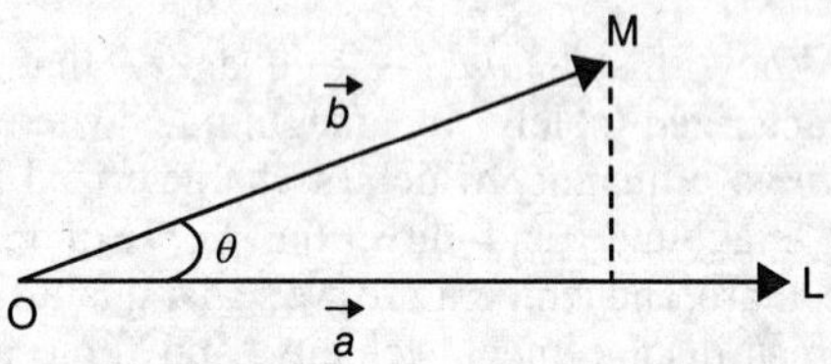

Thus, "*The scalar (dot) product of two vectors* $\vec{a}$ *and* $\vec{b}$ *is the product of the length of the vector* $\vec{a}$ *and the length of the projection of the vector* $\vec{b}$ *in the direction of the vector* $\vec{a}$"

Properties of Scalar Product. The scalar product of two vectors $\vec{a}$ and $\vec{b}$ satisfies the following properties.

(i) $\vec{a}.\vec{b} = \vec{b}.\vec{a}$.

In fact,

$$\vec{a}.\vec{b} = |\vec{a}||\vec{b}|\cos\theta = |\vec{b}||\vec{a}|\cos\theta = \vec{b}.\vec{a}.$$

(ii) *Two vectors are perpendicular if and only if their scalar product is zero.*

In fact, assume first that $\vec{a}$ and $\vec{b}$ are perpendicular. Then angle between them is $\theta = 90°$ and so

$$\vec{a}.\vec{b} = |\vec{a}||\vec{b}|\cos\theta = |\vec{a}||\vec{b}|\cos 90° = 0$$

Conversely, if $\vec{a}.\vec{b} = \vec{b} = 0$, then

$$|\vec{a}|.|\vec{b}|\cos\theta = 0 \Rightarrow \cos\theta = 0 \Rightarrow \theta = 90°$$

and so $\vec{a}$ and $\vec{b}$ are perpendicular to each other. In particular, if $\hat{i}, \hat{j}$ and $\hat{k}$ *are mutually perpendicular unit vectors, then*

$$\hat{i}.\hat{j} = \hat{j}.\hat{k} = \hat{k}.\hat{i} = (1)(1)\ \cos\ 90° = 0$$

And

$$\hat{i}.\hat{i} = \hat{j}.\hat{j} = \hat{k}.\hat{k} = (1)(1)\ \cos\ 0° = 1.$$

(iii) $\vec{a}.(\vec{b}+\vec{c}) = \vec{a}.\vec{b}+\vec{a}.\vec{c}.$

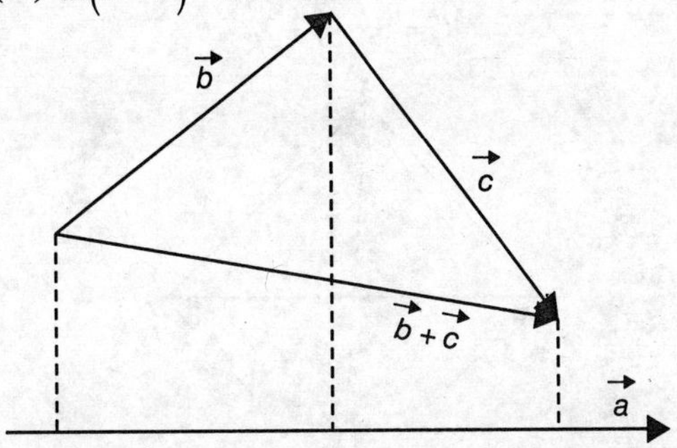

In fact, let $\overrightarrow{e_a}$ be a unit vector in the direction of $\vec{a}$. Then Projection of $\vec{b}+\vec{c}$ on $\vec{a}$ = projection of $\vec{b}$ on $\vec{a}$ $\vec{a}$ + Projection of $\vec{c}$ on $\vec{a}$.

or

$$(\vec{b}+\vec{c}).\overrightarrow{e_a} = \vec{b}.\overrightarrow{e_a} + \vec{a}.\overrightarrow{e_a}$$

or

$$(\vec{b}+\vec{c}).|\vec{a}|\overrightarrow{e_a} = \vec{b}.|\vec{a}|\overrightarrow{e_a} + \vec{c}.|\vec{a}|.\overrightarrow{e_a}$$

or

$$(\vec{b}+\vec{c}).\vec{a} = \vec{b}.\vec{a}+\vec{c}.\vec{a}$$

or by property (i),

$$\vec{a}.(\vec{b}+\vec{c}) = \vec{a}.\vec{b}+\vec{a}.\vec{c}.$$

(iv) If $\vec{a} = a_1\hat{i}+a_2\hat{j}+a_3\hat{k}$ and $\hat{b} = b_1\hat{i}+b_2\hat{j}+b_3\hat{k}$, then

$$\vec{a}.\vec{b} = a_1b_1+a_2b_2+a_3b_3.$$

In fact, using (iii), we have

$$\begin{aligned}\vec{a}.\vec{b} &= (a_1\hat{i}+a_2\hat{j}+a_3\hat{k})\cdot(b_1\hat{i}+b_2\hat{j}+b_3\hat{k})\\ &= a_1\hat{i}\cdot(b_1\hat{i}+b_2\hat{j}+b_3\hat{k})\\ &\quad +a_2\hat{j}.(b_1\hat{i}+b_2\hat{j}+b_3\hat{k})\\ &\quad +a_3\hat{k}.(b_1\hat{i}+b_2\hat{j}+b_3\hat{k})\\ &= a_1b_1\hat{i}\cdot\hat{i}+a_1b_2\hat{i}.\hat{j}+a_1b_3\hat{i}.\hat{k}\\ &\quad +a_2b_1\hat{j}.\hat{i}+a_2b_2\hat{j}.\hat{j}+a_2b_3\hat{j}.\hat{k}\\ &\quad +a_3b_1\hat{k}.\hat{i}+a_3b_2\hat{k}.\hat{j}+a_2b_3\hat{k}.\hat{k}\\ &= a_1b_1+a_2b_2+a_3b_3,\end{aligned}$$

since all other terms are zero by property (ii). As a deduction, we have

$$\begin{aligned}\vec{a}.\vec{a} &= |\vec{a}||\vec{a}|\cos 0^\circ = |\vec{a}|^2\\ &= a_1^2+a_2^2+a_3^2.\end{aligned}$$

(v) For any scalar t

$$t(\vec{a}.\vec{b}) = (t\vec{a}.\vec{b})\vec{a}.(t\vec{b}) = (\vec{a}.\vec{b})t.$$

Proof: follows directly from the definition of dot product.

EXAMPLE 13.1

Find the angle between the vectors

$$\vec{a} = 3\hat{i}+4\hat{j} \ and \ \vec{b} = 6\hat{i}-3\hat{j}+2\hat{k}.$$

Solution. By dot product, we have

$$\vec{a}.\vec{b} = |\vec{a}||\vec{b}|\cos\theta^\circ$$

or

$$\cos\theta = \frac{\vec{a}.\vec{b}}{|\vec{a}||\vec{b}|} = \frac{3(6)-4(3)+4(2)}{\sqrt{3^2+4^2}\sqrt{36+9+4}} = \frac{2}{5(7)} = \frac{2}{35}$$

and so $\theta = \cos^{-1}\left(\frac{2}{35}\right).$

EXAMPLE 13.2

Find the value of a for which the vectors $3\hat{i}+2\hat{j}+9\hat{k}$ and $\hat{i}+a\hat{j}+3\hat{k}$ are perpendicular.

Solution. The given vectors will be perpendicular if their dot product is zero, that is, if

$$(3\hat{i}+2\hat{j}+9\hat{k}).(\hat{i}+a\hat{j}+3\hat{k}) = 0$$

or

$$3(1)+2(a)+9(3) = 0$$

which yields $a = -15.$

EXAMPLE 13.3

(Cauchy-Schwarz Inequality). Show that $(\vec{a}.\vec{b})^2 \le |\vec{a}|^2|\vec{b}|^2$, that is

$$(a_1b_1+a_2b_2+a_3b_3)^2 \le (a_1^2+a_2^2+a_3^2) \times (b_1^2+b_2^2+b_3^2).$$

Solution. Since $\cos^2\theta \le 1$, we have

$$(\vec{a}.\vec{b})^2 \le \left[|\vec{a}||\vec{b}|\cos\theta\right]^2 \le |\vec{a}|^2|\vec{b}|^2$$

or

$$(a_1b_1 + a_2b_2 + a_3b_3)^2$$
$$= (a_1^2 + a_2^2 + a_3^2)(b_1^2 + b_2^2 + b_3^2).$$

EXAMPLE 13.4

Determine the projection of the vector $\hat{i} - 2\hat{j} + \hat{k}$ on the vector $6\hat{i} - 3\hat{j} + 2\hat{k}$.

Solution. Let $\hat{a} = \hat{i} - 2\hat{j} + \hat{k}$ and $b = 6\hat{i} - 3\hat{j} + 2\hat{k}$. Then, the unit vector in the direction of $\vec{b}$ is

$$\vec{e_b} = \frac{\vec{b}}{|\vec{b}|} = \frac{6\hat{i} - 3\hat{j} + 2\hat{k}}{\sqrt{36+9+4}} = \frac{6}{7}\hat{i} - \frac{3}{7}\hat{j} + \frac{2}{7}\hat{k}.$$

Therefore, the length of projection of $\vec{a}$ on the vector $\vec{b}$ is

$$\vec{a}.\vec{e_b} = \left(\hat{i} - 2\hat{j} + \hat{k}\right).\left(\frac{6}{7}\hat{i} - \frac{3}{7}\hat{j} + \frac{2}{7}\hat{k}\right)$$

$$= \frac{6}{7} + \frac{6}{7} + \frac{2}{7} = 2.$$

The projection (as a vector) is given

$$\frac{\vec{a}.\vec{b}}{|\vec{b}|^2}\vec{b} = \frac{6+6+2}{7^2}\left(6\hat{i} - 3\hat{j} + 2\hat{k}\right).$$

$$= \frac{2}{7}\left(6\hat{i} - 3\hat{j} + 2\hat{k}\right) = \frac{12}{7}\hat{i} - \frac{6}{7}\hat{j} + \frac{4}{7}\hat{k}.$$

EXAMPLE 13.5

If $\vec{a}$ and $\vec{b}$ are two non-zero vectors, show that

$$|\vec{a}+\vec{b}|^2 + |\vec{a}-\vec{b}|^2 = 2|\vec{a}|^2 + 2|\vec{b}|^2.$$

Solution. We have

$$|\vec{a}+\vec{b}|^2 + |\vec{a}-\vec{b}|^2$$

$$(\vec{a}+\vec{b}).(\vec{a}+\vec{b}) + (\vec{a}-\vec{b}).(\vec{a}-\vec{b})$$

$$= \vec{a}.\vec{a} + \vec{a}.\vec{b} + \vec{b}.\vec{a} + \vec{b}.\vec{b} + \vec{a}.\vec{a} - \vec{a}.\vec{b} - \vec{b}.\vec{a} + \vec{b}.\vec{b}$$

$$= 2(\vec{a}.\vec{a}) + 2(\vec{b}.\vec{b}) = 2|\vec{a}|^2 + 2|\vec{b}|^2.$$

EXAMPLE 13.6

Find the sides and angles of the triangle whose vertices are $\hat{i} - 2\hat{j} + 2\hat{k}, 2\hat{i} + \hat{j} - \hat{k}\ 3\ \hat{i} - \hat{j} + 2\hat{k}$.

Solution. The position vector of the vertices A, B and C are

$$\vec{OA} = \hat{i} - 2\hat{j} + 2\hat{k},\ \vec{OB} = 2\hat{i} + \hat{j} - \hat{k} \text{ and}$$
$$\vec{OC} = 3\hat{i} - \hat{j} + 2\hat{k}.$$

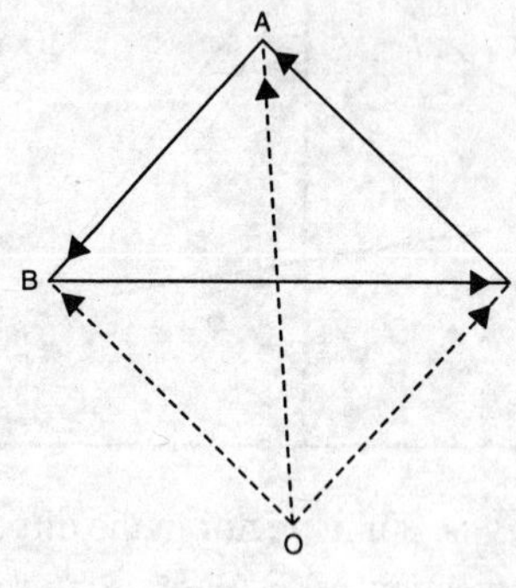

Then

(i) $\overrightarrow{OA} + \overrightarrow{AB} \rightarrow= \overrightarrow{OB}$

or

$$\overrightarrow{AB} = \overrightarrow{OB} - \overrightarrow{OA} = \hat{i} + 3\hat{j} - 3\hat{k}.$$

Therefore $AB = \sqrt{1+9+9} = \sqrt{19}$.

(ii) $\overrightarrow{OC} + \overrightarrow{CA} = \overrightarrow{OA}$

or

$$\overrightarrow{CA} = \overrightarrow{OA} - \overrightarrow{OC} = -2\hat{i} - \hat{j}.$$

Therefore $CA = \sqrt{4+1} = \sqrt{5}$.

(iii) $\vec{OB} + \vec{BC} = \vec{OC}$

or

$$\overrightarrow{BC} = \overrightarrow{OC} - \overrightarrow{OB} = \hat{i} - 2\hat{j} + 3\hat{k}.$$

Therefore $BC = \sqrt{1+4+9} = \sqrt{14}$.

Further, angle between AB and BC is given by

$$\cos\theta = \frac{\overrightarrow{AB}.\overrightarrow{BC}}{|\overrightarrow{AB}||\overrightarrow{BC}|}$$

$$= \frac{\left(\hat{i} + 3\hat{j} - 3\hat{k}\right)\cdot\left(\hat{i} - 2\hat{j} + 3\hat{k}\right)}{|\hat{i} + 3\hat{j} - 3\hat{k}||\hat{i} - 2\hat{j} + 3\hat{k}|}$$

$$= \frac{1-6-9}{\sqrt{19}\sqrt{14}} = \frac{-14}{\sqrt{19}\sqrt{14}} = -\sqrt{\frac{14}{19}}.$$

Hence $\theta = \cos^{-1}\sqrt{\frac{14}{19}}$.

Similarly, the angles between the vectors $\overrightarrow{BC}$ and $\overrightarrow{CA}$ and between $\overrightarrow{CA}$ and $\overrightarrow{AB}$ are given respectively by $\cos^{-1}\sqrt{\frac{5}{19}}$ and $\cos^{-1}(0)$.

EXAMPLE 13.7

Show that the triangle formed by the vectors $\vec{a} = 3\hat{i} - 2\hat{j} + \hat{k}$, $\vec{b} = \hat{i} - 3\hat{j} + 5\hat{k}$ and $\vec{a} = 2\hat{i} + \hat{j} - 4\hat{k}$ is a right triangle.

Solution. We note that

$$\vec{a}.\vec{b} = 3 + 6 + 5 = 14$$
$$\vec{b}.\vec{c} = 2 - 3 - 20 = -21$$
$$\vec{c}.\vec{a} = 6 - 2 - 4 = 0.$$

Since dot product of $\vec{a}$ and $\vec{c}$ is zero, it follows that they are perpendiculars to each other. Thus the angle between two sides formed by the vector $\vec{a}$ and $\vec{c}$ is 90°. Hence the triangle formed by the given vectors is a right triangle.

EXAMPLE 13.8

If $\vec{a}$ and $\vec{b}$ are two vectors, show that

(i) $|\vec{a}+\vec{b}| \le |\vec{a}| + |\vec{b}|$

(ii) $\big||\vec{a}| - |\vec{b}|\big| \le |\vec{a} - \vec{b}|$.

Solution. (i) Let θ be the angle between $\vec{a}$ and $\vec{b}$. Then

$$|\vec{a}+\vec{b}|^2 = (\vec{a}+\vec{b}).(\vec{a}+\vec{b}) = \vec{a}.\vec{a} + \vec{a}.\vec{b} + \vec{b}.\vec{a} + \vec{b}.\vec{b}$$
$$= |\vec{a}|^2 + |\vec{b}|^2 + 2\vec{a}.\vec{b}, \text{since } \vec{a}.\vec{b} = \vec{b}.\vec{a}$$
$$= |\vec{a}|^2 + |\vec{b}|^2 + 2|\vec{a}|.|\vec{b}|\cos\theta$$
$$\le |\vec{a}|^2 + |\vec{b}|^2 + 2|\vec{a}||\vec{b}|, \text{since } \cos\theta \le 1$$
$$= (|\vec{a}| + |\vec{b}|)^2.$$

Hence

$$|\vec{a}+\vec{b}| \le |\vec{a}| + |\vec{b}|.$$

The equality in this relation holds if and only if cos $\theta = 1$, that is, if and only if $\theta = 0$, that is, if and only if the vectors are parallel and in the same direction. In other words, equality holds if and only if the vectors $\vec{a}$ and $\vec{b}$ are *like vectors*.

(ii) To prove (ii) We note that

$$|\vec{a}| = |\vec{a} - \vec{b} + \vec{b}| \le |\vec{a} - \vec{b}| + |\vec{b}|$$

using (i) proved above

or $\quad |\vec{a}| - |\vec{b}| \le |\vec{a} - \vec{b}|. \qquad (1)$

Interchanging $\vec{a}$ and $\vec{b}$, we get

$$|\vec{b}| - |\vec{a}| \le |\vec{b} - \vec{a}| = |\vec{a} - \vec{b}|$$

or

$$|\vec{a} - \vec{b}| \le |\vec{a}| - |\vec{b}| \qquad (2)$$

Combining (1) and (2), we get

$$|\vec{a} - \vec{b}| \le |\vec{a}| - |\vec{b}| \le |\vec{a} - \vec{b}|$$

or

$$\big||\vec{a}| - |\vec{b}|\big| \le |\vec{a} - \vec{b}|.$$

Vector Product or Cross Product of Two Vectors. The *vector product* or *cross product* of two vectors $\vec{a}$ and $\vec{b}$ is a vector defined by

$$\vec{a} \times \vec{b} = |\vec{a}||\vec{b}|\sin\theta\,\hat{e},$$

Where θ in the angle between $\vec{a}$ and $\vec{b}$ such that $0 \le \theta \le \pi$ and $\hat{e}$ in the unit vector perpendicular to both $\vec{a}$ and $\vec{b}$. The direction of $\vec{a} \times \vec{b}$ is perpendicular to the plane of $\vec{a}$ and $\vec{b}$ such that $\vec{a}$, $\vec{b}$ and $\vec{a} \times \vec{b}$ form a right-handed triad of vectors.

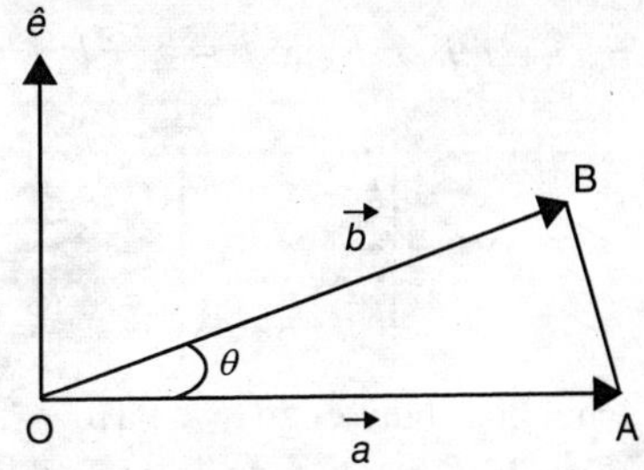

In particular, if $\vec{a} = \vec{b}$ or if $\vec{a}$ is parallel to $\vec{b}$, then $\vec{a} \times \vec{b} = 0$.

Geometrical Interpretation of $\vec{a} \times \vec{b}$. Let $\hat{e}$ be a unit vector normal to the plane of the triangle OAB having $\vec{a}$ and $\vec{b}$ as its adjacent sides. Then

$$\vec{a} \times \vec{b} = |\vec{a}|\,|\vec{b}|\sin\theta\,\hat{e} = 2\left[\frac{1}{2}|\vec{a}|\,|\vec{b}|\sin\theta\right]\hat{e}$$
$$= 2\Delta\overrightarrow{OAB}.$$

Thus $\vec{a} \times \vec{b}$ represents twice the vector area of the triangle having $\vec{a}$ and $\vec{b}$ as adjacent sides.

Properties of Vector Product. The vector product satisfies the following properties: (i) *The vector product is anti-commutative*. In fact, we have

$$\vec{b} \times \vec{a} = |\vec{b}||\vec{a}|\sin\theta(-\hat{e})$$

$$= -|\vec{a}||\vec{b}|\sin\theta\,\hat{e} = -(\vec{a}\times\vec{b})$$

Hence

$$\vec{a}\times\vec{b} = -(\vec{b}\times\vec{a})\text{(anti-commutative law)}$$

(ii) *The vector product is distributive over vector addition.*

In fact, if $\vec{a},\vec{b}$ and $\vec{c}$ are three vectors, then

$$\vec{a}\times(\vec{b}\times\vec{c}) = \vec{a}\times\vec{b}+\vec{a}\times\vec{c} \text{ (distributive law)}$$

(iii) For a scalar t, we have

$$t(\bar{a}\times\bar{b}) = (t\,\bar{a})\times\bar{b} = \bar{a}\times(t\,\bar{b}) = (\bar{a}\times\bar{b})t$$

(iv) If $\hat{i}$, $\hat{j}$ and $\hat{k}$ are three mutually perpendicular unit vectors, then

$$\hat{i}\times\hat{i} = \hat{j}\times\hat{j} = \hat{k}\times\hat{k} = \vec{0}$$

$$\hat{i}\times\hat{j} = \hat{k},\ \hat{j}\times\hat{k} = \hat{i},\ \hat{k}\times\hat{i} = \hat{j}$$

$$\hat{j}\times\hat{i} = -\hat{k},\ \hat{k}\times\hat{j} = -\hat{i},\ \hat{i}\times\hat{k} = -\hat{j}.$$

(v) If $\vec{a} = a_1\hat{i}+a_2\hat{j}+a_3\hat{k}$ and $\vec{b} = b_1\hat{i}+b_2\hat{j}+b_3\hat{k}$, then

$$\vec{a}\times\vec{b} = \begin{vmatrix}\hat{i} & \hat{j} & \hat{k}\\ a_1 & a_2 & a_3\\ b_1 & b_2 & b_3\end{vmatrix}.$$

In fact, using distributive law, we have

$$\vec{a}\times\vec{b} = \left(a_1\hat{i}\times a_2\hat{j}+a_3\hat{k}\right)\times\left(b_1\hat{i}\times b_2\hat{j}+b_3\hat{k}\right)$$

$$= a_1\hat{i}\times\left(b_1\hat{i}\times b_2\hat{j}+b_3\hat{k}\right)+a_2\hat{j}$$

$$\times\left(b_1\hat{i}\times b_2\hat{j}+b_3\hat{k}\right)$$

$$+a_3\hat{k}\times\left(b_1\hat{i}\times b_2\hat{j}+b_3\hat{k}\right)$$

$$= a_1b_1\left(\hat{i}\times\hat{i}\right)+a_1b_2\left(\hat{i}\times\hat{j}\right)+a_1b_3\left(\hat{i}\times\hat{k}\right)$$

$$+a_2b_1\left(\hat{j}\times\hat{i}\right)+a_2b_2\left(\hat{j}\times\hat{j}\right)+a_2b_3\left(\hat{j}\times\hat{k}\right)$$

$$+a_3b_1\left(\hat{k}\times\hat{i}\right)+a_3b_2\left(\hat{k}\times\hat{j}\right)+a_3b_3\left(\hat{k}\times\hat{k}\right)$$

$$= a_1b_2\hat{k}-a_1b_3\hat{j}-a_2b_1\hat{k}+a_2b_3\hat{i}$$

$$+a_3b_1\hat{j}-a_3b_2\hat{i}$$

$$= (a_2b_3-a_3b_2)\hat{i}+(a_3b_1-a_1b_3)\hat{j}$$

$$+(a_1b_2-a_2b_1)\hat{k}$$

$$= \begin{vmatrix}\hat{i} & \hat{j} & \hat{k}\\ a_1 & a_2 & a_3\\ b_1 & b_2 & b_3\end{vmatrix}.$$

EXAMPLE 13.9

Show that for vector $\vec{a}$ and $\vec{b}$,

$$|\vec{a}\times\vec{b}|^2+|\vec{a}.\vec{b}|^2 = |\vec{a}|^2|\vec{b}|^2$$

Solution. We have

$$|\vec{a}\times\vec{b}|^2+|\vec{a}.\vec{b}|^2$$

$$= \left||\vec{a}||\vec{b}|\sin\theta\,\hat{e}\right|^2+\left||\vec{a}||\vec{b}|\cos\theta\right|^2$$

$$= |\vec{a}|^2|\vec{b}|^2\sin^2\theta+|\vec{a}|^2|\vec{b}|^2\cos^2\theta$$

$$|\vec{a}|^2|\vec{b}|^2(\sin^2\theta+\cos^2\theta) = |\vec{a}|^2|\vec{b}|^2.$$

EXAMPLE 13.10

Find a unit vector perpendicular to both $\vec{a} = 2\hat{i}+\hat{j}-\hat{k}$ and $\vec{b} = \hat{i}+\hat{j}+2\hat{k}$.

Solution. The vector $\vec{a}\times\vec{b}$ is perpendicular to both $\vec{a}$ and $\vec{b}$. But

$$\vec{a}\times\vec{b} = \left(2\hat{i}+\hat{j}-\hat{k}\right)\times\left(\hat{i}+\hat{j}+2\hat{k}\right)$$

$$= \begin{vmatrix}\hat{i} & \hat{j} & \hat{k}\\ 2 & 1 & -1\\ 1 & -1 & 2\end{vmatrix}$$

$$= \hat{i}(2-1)-\hat{j}(4+1)+\hat{k}(-2-1)$$

$$= \hat{i}-5\hat{j}-3\hat{k}.$$

Therefore unit vector in the direction of $\vec{a}\times\vec{b}$ is

$$\hat{e} = \frac{\hat{i}-5\hat{j}-3\hat{k}}{|\hat{i}-5\hat{j}-3\hat{k}|} = \frac{\hat{i}-5\hat{j}-3\hat{k}}{\sqrt{1+25+9}} = \frac{\hat{i}-5\hat{j}-3\hat{k}}{\sqrt{35}}.$$

EXAMPLE 13.11

Show that the area of a parallelogram with sides $\vec{a}$ and $\vec{b}$ is $|\vec{a}\times\vec{b}|$.

Solution. Let θ be the angle between $\vec{a}$ and $\vec{b}$. Then Area of the parallelogram = *(height h)* $|\vec{b}|$

$$=|\vec{a}|\sin\theta|\vec{b}|=|\vec{a}\times\vec{b}|.$$

EXAMPLE 13.12

In ΔABC with sides $\vec{a}$, $\vec{b}$ and $\vec{c}$, prove the law of sines:

$$\frac{a}{\sin A}=\frac{b}{\sin B}=\frac{c}{\sin C}.$$

Solution. Let $\vec{a}$, $\vec{b}$ and $\vec{c}$ be the sides of the triangle as shown below:

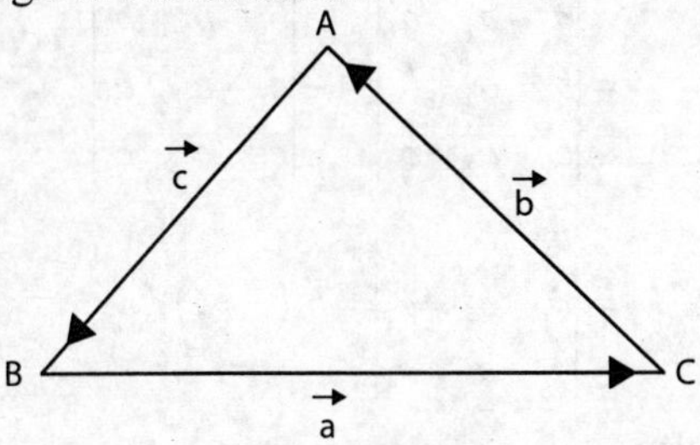

Then

$$\vec{a}+\vec{b}+\vec{c}=\vec{0}.$$

Therefore

$$\vec{a}\times(\vec{a}+\vec{b}+\vec{c})=\vec{a}\times\vec{0}$$

or

$$\vec{a}\times\vec{a}+\vec{a}\times\vec{b}+\vec{a}\times\vec{c}=\vec{0}$$

or

$$\vec{a}\times\vec{b}+\vec{a}\times\vec{c}=\vec{0} \qquad (1)$$

Similarly,

$$\vec{b}\times(\vec{a}+\vec{b}+\vec{c})=\vec{b}\times\vec{0}$$

or

$$\vec{b}\times\vec{a}+\vec{b}\times\vec{b}+\vec{b}\times\vec{c}=\vec{0}$$

or

$$-\vec{a}\times\vec{b}+\vec{b}\times\vec{c}=\vec{0}. \qquad (2)$$

and

$$\vec{c}\times(\vec{a}+\vec{b}+\vec{c})=\vec{c}\times\vec{0}$$

or

$$\vec{c}\times\vec{a}+\vec{c}\times\vec{b}+\vec{c}\times\vec{c}=\vec{0}$$

or

$$\vec{c}\times\vec{a}-\vec{b}\times\vec{c}=\vec{0}. \qquad (3)$$

From (1), (2) and (3), we have

$$\vec{a}\times\vec{b}=\vec{b}\times\vec{c}=\vec{c}\times\vec{a}$$

or

$$|\vec{a}|\,|\vec{b}|\sin C=|\vec{b}||\vec{c}|\sin A=|\vec{c}||\vec{a}|\sin B$$

or

Dividing throughout by $|\vec{a}|\,|\vec{b}|\,|\vec{c}|$,, we get

$$\frac{\sin C}{|\vec{c}|}=\frac{\sin A}{|\vec{a}|}=\frac{\sin B}{|\vec{b}|}$$

or

$$\frac{a}{\sin A}=\frac{b}{\sin B}=\frac{c}{\sin C}.$$

EXAMPLE 13.13

If $\vec{a}=3\hat{i}-\hat{j}+2\hat{k}$, $\vec{b}=2\hat{i}+\hat{j}-\hat{k}$ and $\vec{c}=\hat{i}-2\hat{j}+2\hat{k}$, determine $(\vec{a}\times\vec{b})\times\vec{c}, \vec{a}\times(\vec{b}\times\vec{c})$ and $a.(\vec{b}\times\vec{c})$.

Solution. We have

$$\vec{a}\times\vec{b}=\begin{vmatrix}\hat{i} & \hat{j} & \hat{k}\\ 3 & -1 & 2\\ 2 & 1 & -1\end{vmatrix}=-\hat{i}+7\hat{j}+5\hat{k},$$

$$\vec{b}\times\vec{c}=\begin{vmatrix}\hat{i} & \hat{j} & \hat{k}\\ 2 & 1 & -1\\ 1 & -2 & 2\end{vmatrix}=-5\hat{j}-5\hat{k}.$$

Therefore

$$(\vec{a}\times\vec{b})\times\vec{c}=\begin{vmatrix}\hat{i} & \hat{j} & \hat{k}\\ -1 & 7 & 5\\ 1 & -3 & 2\end{vmatrix}=-\hat{i}+7\hat{j}+5\hat{k}$$

$$=\hat{i}(14+10)-\hat{j}(-2-5)+\hat{k}(2-7)$$

$$=24\hat{i}+7\hat{j}-5\hat{k},$$

$$\vec{a}\times(\vec{b}\times\vec{c})=\begin{vmatrix}\hat{i} & \hat{j} & \hat{k}\\ 3 & -1 & 2\\ 0 & -5 & -5\end{vmatrix}$$

$$=\hat{i}(5+10)-\hat{j}(-15)+\hat{k}(-15)$$

$$=15\hat{i}+15\hat{j}-15\hat{k}.$$

Thus, we observe that

$$\vec{a}\times(\vec{b}\times\vec{c})\neq(\vec{a}\times\vec{b})\times\vec{c}.$$

Furthermore,

$$\vec{a}.(\vec{b}\times\vec{c}) = (3\hat{i}-\hat{j}+2\hat{k})\cdot(-5\hat{j}-5\hat{k}).$$
$$= 5 - 10 = -5.$$

Triple Product of Vectors. Let $\vec{a}, \vec{b}$ and $\vec{c}$ be three vectors. Then the products of the form $(\vec{a}.\vec{b})\vec{c}, \vec{a}.(\vec{b}\times\vec{c})$ and $\vec{a}\times(\vec{b}\times\vec{c})$ are called *triple products* of the vectors $\vec{a}, \vec{b}$ and $\vec{c}$.

Note that $(\vec{a}.\vec{b}).\vec{c}$ is not defined.

These products follow the following laws:

(i) $(\vec{a}.\vec{b})\vec{c} \neq \vec{a}(\vec{b}.\vec{c})$.

In fact, if $\vec{a} = a_1\hat{i}+a_2\hat{j}+a_3\hat{k}$

$\vec{b} = b_1\hat{i}+b_2\hat{j}+b_3\hat{k}$ and

$\vec{c} = c_1\hat{i}+c_2\hat{j}+c_3\hat{k}$,

then

$$(\vec{a}.\vec{b})\vec{c}$$
$$= (a_1b_1+a_2b_2+a_3b_3)(c_1\hat{i}+c_2\hat{j}+c_3\hat{k})$$
$$= (a_1b_1c_1+a_2b_2c_1+a_3b_3c_1)\hat{i}$$
$$+(a_1b_1c_2+a_2b_2c_2+a_3b_3c_3)\hat{j}$$
$$+(a_1b_1c_3+a_2b_2c_2+a_3b_3c_3)\hat{k},$$

and

$$\vec{a}(\vec{b}.\vec{c}) = (a_1\hat{i}+a_2\hat{j}+a_3\hat{k})(b_1c_1+b_2c_2+b_3c_3)$$
$$= (a_1b_1c_1+a_1b_2c_2+a_1b_3c_3)\hat{i}$$
$$+(a_2b_1c_1+a_2b_2c_2+a_2b_3c_3)\hat{j}$$
$$+(a_3b_1c_1+a_3b_2c_2+a_3b_3c_3)\hat{k}.$$

We note that

$$(\vec{a}.\vec{b})\vec{c} \neq \vec{a}(\vec{b}.\vec{c}).$$

(ii) $\vec{a}.(\vec{b}\times\vec{c}) = \vec{b}.(\vec{c}\times\vec{a}) = \vec{c}.(\vec{a}\times\vec{b})$.

In fact,

$$\vec{a}.(\vec{b}\times\vec{c}) = (a_1\hat{i}+a_2\hat{j}+a_3\hat{k})\cdot\begin{vmatrix}\hat{i} & \hat{j} & \hat{k}\\ b_1 & b_2 & b_3\\ c_1 & c_2 & c_3\end{vmatrix}$$
$$= (a_1\hat{i}+a_2\hat{j}+a_3\hat{k})$$
$$\times\left[(b_2c_3-b_3c_2)\hat{i}+(b_3c_1-b_1c_3)\hat{j}\right.$$
$$\left.+(b_1c_2-b_2c_1)\hat{k}\right]$$
$$= a_1(b_2c_3-b_3c_2)+a_2(b_3c_1-b_1c_3)$$
$$+a_3(b_1c_2-b_2c_1)$$
$$= \begin{vmatrix}a_1 & a_2 & a_3\\ b_1 & b_2 & b_3\\ c_1 & c_2 & c_3\end{vmatrix} = \Delta, \text{ say.}$$

But, by property of determinant,

$$\Delta = \begin{vmatrix}a_1 & a_2 & a_3\\ b_1 & b_2 & b_3\\ c_1 & c_2 & c_3\end{vmatrix} = -\begin{vmatrix}b_1 & b_2 & b_3\\ a_1 & a_2 & a_3\\ c_1 & c_2 & c_3\end{vmatrix}$$
$$= \begin{vmatrix}b_1 & b_2 & b_3\\ c_1 & c_2 & c_3\\ a_1 & a_2 & a_3\end{vmatrix} = \vec{b}.(\vec{c}\times\vec{a}),$$

and also,

$$\Delta = \begin{vmatrix}a_1 & a_2 & a_3\\ b_1 & b_2 & b_3\\ c_1 & c_2 & c_3\end{vmatrix} = -\begin{vmatrix}c_1 & c_2 & c_3\\ b_1 & b_2 & b_3\\ a_1 & a_2 & a_3\end{vmatrix}$$
$$= \begin{vmatrix}c_1 & c_2 & c_3\\ a_1 & a_2 & a_3\\ b_1 & b_2 & b_3\end{vmatrix} = \vec{c}.(\vec{a}\times\vec{b}).$$

Hence

$$a\cdot(\vec{b}\times\vec{c}) = b\cdot(\vec{c}\times\vec{a}) = c\cdot(\vec{a}\times\vec{b}).$$

(iii) $\vec{a}.(\vec{b}\times\vec{c}) = (\vec{a}\times\vec{b}).\vec{c}$

In fact

$\vec{a}.(\vec{b}\times\vec{c}) = \vec{c}.(\vec{a}\times\vec{b})$ (proved above)

$= (\vec{a}\times\vec{b}).\vec{c}$ (commutation of dot product).

(iv) $\vec{a}\times(\vec{b}\times\vec{c}) \neq (\vec{b}\times\vec{c})\times\vec{c}$.

In fact,

$$\vec{a}\times(\vec{b}\times\vec{c})$$
$$= \vec{a}\times\begin{vmatrix}\hat{i} & \hat{j} & \hat{k}\\ b_1 & b_2 & b_3\\ c_1 & c_2 & c_3\end{vmatrix}$$

$$= \vec{a}\times\left[(b_2c_3-c_2b_3)\hat{i}-(b_1c_3-c_1b_3)\hat{j} +(b_1c_2-b_2c_1)\hat{k}\right]$$

$$= \begin{vmatrix} \hat{i} & \hat{j} & \hat{k} \\ a_1 & a_2 & a_3 \\ b_2c_3-b_3c_2 & -b_1c_3+b_3c_1 & b_1c_2-b_2c_1 \end{vmatrix},$$

where as

$$(\vec{a}\times\vec{b})\times\vec{c}$$

$$= \begin{vmatrix} \hat{i} & \hat{j} & \hat{k} \\ a_1 & a_2 & a_3 \\ b_1 & b_2 & b_3 \end{vmatrix} \times\left(c_1\hat{i}+c_2\hat{j}+c_3\hat{k}\right)$$

$$= \left[(a_2b_3-a_3b_2)\hat{i}-(a_1b_3-a_3b_1)\hat{j} +(a_1b_2-a_2b_1)\hat{k}\right]\times\left(c_1\hat{i}+c_2\hat{j}+c_3\hat{k}\right)$$

$$= \begin{vmatrix} \hat{i} & \hat{j} & \hat{k} \\ a_2b_3-a_3b_2 & -a_1b_3+a_3b_1 & a_1b_2-a_2b_1 \\ c_1 & c_2 & c_3 \end{vmatrix}$$

Expanding these determinants, we find that

$$\vec{a}\times(\vec{b}\times\vec{c}) \neq (\vec{a}\times\vec{b})\times\vec{c}.$$

(v) $\vec{a}\times(\vec{b}\times\vec{c}) = (\vec{a}.\vec{c})\vec{b}-(\vec{a}.\vec{b})\vec{c}$

and

$$(\vec{a}\times\vec{b})\times\vec{c} = (\vec{a}.\vec{c})\vec{b}-(\vec{b}.\vec{c})\vec{a}.$$

In fact, we note that, as in part iv, the left hand side of first relation is

$$\vec{a}\times(\vec{b}\times\vec{c})$$

$$= (a_1i+a_2j+a_3k)\times\begin{vmatrix} \hat{i} & \hat{j} & \hat{k} \\ b_1 & b_2 & b_3 \\ c_1 & c_2 & c_3 \end{vmatrix}$$

$$= \begin{vmatrix} \hat{i} & \hat{j} & \hat{k} \\ a_1 & a_2 & a_3 \\ (b_2c_3-b_3c_2) & (b_3c_1-b_1c_3) & (b_1c_2-b_2c_1) \end{vmatrix}$$

$$= \left[(a_2b_1c_2-a_2b_2c_1-a_3b_3c_1+a_3b_1c_3)\hat{i}\right] +(a_3b_2c_3-a_3b_3c_2-a_1b_1c_2-a_1b_2c_1)\hat{j} +(a_1b_3c_1-a_1b_1c_3-a_2b_2c_3+a_2b_3c_2)\hat{k}.$$

Also $\vec{b}(\vec{a}.\vec{c})-(\vec{a}.\vec{b})\vec{c}$

$$= \left(b_1\hat{i}+b_2\hat{j}+b_3\hat{k}\right)(a_1c_1+a_2c_2+a_3c_3) -(a_1b_1+c_2b_2+a_3b_3)\left(c_1\hat{i}+c_2\hat{j}+c_3\hat{k}\right)$$

$$= \left[(a_2b_1c_2-a_2b_2c_1-a_3b_3c_1+a_3b_1c_3)\hat{i}\right] +(a_3b_2c_3-a_3b_3c_2-a_1b_1c_2-a_1b_2c_1)\hat{j} +(a_1b_3c_1-a_1b_1c_3-a_2b_2c_3+a_2b_3c_2)\hat{k}.$$

Hence

$$\vec{a}\times(\vec{b}\times\vec{c}) = (\vec{a}.\vec{c})\vec{b}-(\vec{a}.\vec{b})\vec{c}.$$

The second relation can be established similarly.

The product $\vec{a}.(\vec{b}\times\vec{c})$ is called the *scalar triple product* or *box product* and is denoted by [*abc*].

The product $\vec{a}\times(\vec{b}\times\vec{c})$ is called the *vector triple product.*

EXAMPLE 13.14

Show that

$$\vec{a}\times(\vec{b}\times\vec{c})+\vec{b}\times(\vec{c}\times\vec{a})+\vec{c}\times(\vec{a}\times\vec{b}) = \vec{0}.$$

Solution. By property (v), we have

$$\vec{a}\times(\vec{b}\times\vec{c}) = \vec{b}(\vec{a}.\vec{c})-\vec{c}(\vec{a}.\vec{b}), \quad (1)$$

$$\vec{b}\times(\vec{c}\times\vec{a}) = \vec{c}(\vec{b}.\vec{a})-\vec{a}(\vec{b}.\vec{c}), \quad (2)$$

$$\vec{c}\times(\vec{a}\times\vec{b}) = \vec{a}(\vec{c}.\vec{b})-\vec{b}(\vec{c}.\vec{a}). \quad (3)$$

Adding(l), (2), (2) and (3), we have

$$\vec{a}\times(\vec{b}\times\vec{c})+\vec{b}\times(\vec{c}\times\vec{a})+\vec{c}\times(\vec{a}\vec{b}) = \vec{0}.$$

Geometrical Interpretation of Scalar Triple Product

The scalar triple product $\vec{a}.(\vec{b}\times\vec{a})$ *equals the volume of the parallelepiped having the vectors* $\vec{a}$, $\vec{b}$ *and* $\vec{c}$ *as its three coterminus edges.*

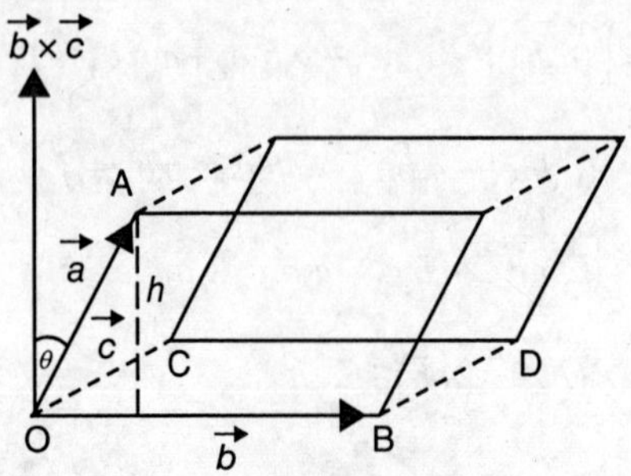

The vector $\vec{b}\times\vec{c}$ is perpendicular to the vectors $\vec{b}$ *and* $\vec{c}$. Further, $|\vec{b}\times\vec{c}|$ gives the area of the parallelogram OBCD having adjacent sides $\vec{b}$ and $\vec{c}$. If h is the height of the parallelepiped, then

$$h = |\vec{a}|\,|\cos\theta|$$

Therefore

Volume of the parallelepiped

$$= h\times \text{ area of parallelogram OBCD}$$

$$= |\vec{a}|\,|\cos\theta|\,|\vec{b}\times\vec{c}|$$

$$= \left|\vec{a}.\left(\vec{b}\times\vec{c}\right)\right|.$$

EXAMPLE 13.15

Show that three vectors $\vec{a}, \vec{b}$ and $\vec{c}$ are coplanar if and only if $[abc] = 0$.

Solution. Suppose that the vectors $\vec{a}, \vec{b}$ and $\vec{c}$ are coplanar. Since $\vec{a}\times\vec{b}$ is perpendicular to the plane of $\vec{a}$ and $\vec{b}$ and since $\vec{a},\vec{b},\vec{c}$ are coplanar, it follows the $\vec{a}\times\vec{b}$ is perpendicular to $\vec{c}$. Hence $\left(\vec{a}\times\vec{b}\right).\vec{c} = 0$

or $\qquad [abc] = 0.$

Conversely, suppose that $[abc] = 0$, that is $\left(\vec{a}\times\vec{b}\right).\vec{c} = 0$. Therefore $\vec{a}\times\vec{b}$ is perpendicular to $\vec{c}$. But $\vec{a}\times\vec{b}$ is perpendicular to the plane of $\vec{a}$ and $\vec{b}$. Hence $\vec{c}$ lies in the plane of $\vec{a}$ and $\vec{b}$. Hence $\vec{a}, \vec{b}$ and $\vec{c}$ are coplanar.

EXAMPLE 13.16

Find the value of λ which makes the vectors $\hat{i}-\hat{j}+\hat{k},\ 2\hat{i}+\hat{j}-\hat{k}$ and $\lambda\hat{i}-\hat{j}+\lambda\hat{k}$ coplanar

Solution. The necessary and sufficient condition for three vectors $\vec{a}, \vec{b}$ and $\vec{c}$ to be coplanar is that the box product $[abc] = 0$. But

$$[abc] = \vec{a}.\left(\vec{b}\times\vec{c}\right) = \begin{vmatrix} a_1 & a_2 & a_3 \\ b_1 & b_2 & b_3 \\ c_1 & c_2 & c_3 \end{vmatrix} = 0$$

Thus for coplanarity, we must have

$$\begin{vmatrix} 1 & -1 & 1 \\ 2 & 1 & -1 \\ \lambda & -1 & \lambda \end{vmatrix} = 0$$

or

$$1(\lambda-1)-1(2\lambda+\lambda)+1(-2-\lambda) = 0$$

or

$$3\lambda-3 = 0 \text{ or } \lambda=1.$$

EXAMPLE 13.17

Find the volume of a parallelepiped whose three coterminus edges are represented by

$$\vec{a} = 2\hat{i}-3\hat{j}+4\hat{k},\ \vec{b} = \hat{i}-2\hat{j}-\hat{k} \text{ and}$$

$$\vec{c} = 3\hat{i}-\hat{j}+2\hat{k}.$$

Solution. The volume of the parallelepiped with the given edges is

$$V = \left[\vec{a}\ \vec{b}\ \vec{c}\right] = \vec{a}.\left(\vec{b}\times\vec{c}\right)$$

$$= \begin{vmatrix} a_1 & a_2 & a_3 \\ b_1 & b_2 & b_3 \\ c_1 & c_2 & c_3 \end{vmatrix} = \begin{vmatrix} 2 & -3 & 4 \\ 1 & 2 & -1 \\ 3 & -1 & 2 \end{vmatrix}$$

$$= 2(4-1)+3(2+3)+4(-1-6)$$

$$= 6+15-28 = -7 \text{ cubic units.}$$

The negative sign shows that $\vec{a}, \vec{b}, \vec{c}$, constitute a left hand triad of vectors. Thus $V = 7$ cubic units

EXAMPLE 13.18

Find the volume of a tetrahedron whose coterminus edges OA, OB and OC are represented by vectors $\vec{a}, \vec{b}$ and $\vec{c}$

Solution. Let OABC be the tetrahedron in which the edges OA, OB and OC are represented by $\vec{a}, \vec{b}$ and $\vec{c}$

respectively.

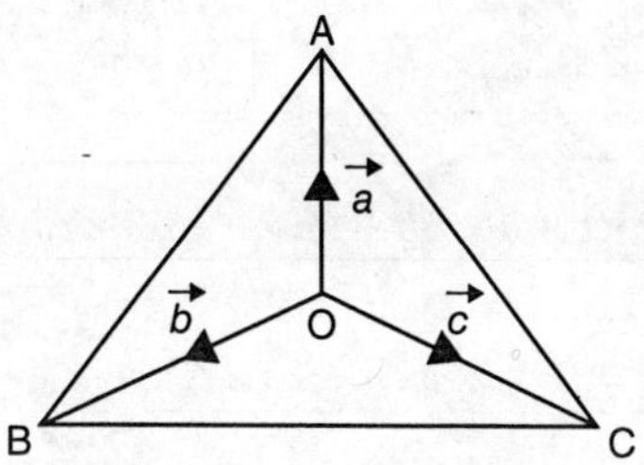

Then

$$\text{Area of } \Delta OBC = \frac{1}{2}\left(\vec{b}\times\vec{c}\right) \text{ and}$$

Volume of the tetrahedron $OABC$

$$= \frac{1}{3}\text{height} \times (\text{area of the base})$$

$$= \frac{1}{6}\vec{a}.\left(\vec{b}\times\vec{c}\right) = \frac{1}{6}\left[\vec{a}\,\vec{b}\,\vec{c}\right].$$

EXAMPLE 13.19

Find volume of the tetrahedron formed by the points (1,1,1), (2,1,3),(3,2,2) and (3,3,4).

Solution. Let A(1, 1, 1), B(2, 1, 3), C(3, 2, 2) and D(3, 3, 4) be the given points. Then

$$\overrightarrow{AB} = \left(2\hat{i}+\hat{j}+3\hat{k}\right)-\left(\hat{i}+\hat{j}+\hat{k}\right) = \hat{i}+2\hat{k}$$

$$\overrightarrow{AC} = \left(3\hat{i}+2\hat{j}+2\hat{k}\right)-\left(\hat{i}+\hat{j}+\hat{k}\right) = 2\hat{i}+\hat{j}+\hat{k}$$

$$\overrightarrow{AD} = \left(3\hat{i}+3\hat{j}+4\hat{k}\right)-\left(\hat{i}+\hat{j}+\hat{k}\right) = 2\hat{i}+2\hat{j}+3\hat{k}.$$

Hence

Volume of tetrahedron

$$= \frac{1}{6}\left[\overrightarrow{AB}\ \overrightarrow{AC}\ \overrightarrow{AD}\right]$$

$$= \frac{1}{6}\begin{vmatrix} 1 & 0 & 2 \\ 2 & 1 & 1 \\ 2 & 2 & 3 \end{vmatrix} = \frac{1}{6}[1+0+4]$$

$$= \frac{5}{6} \text{ cubic units.}$$

13.1 DIFFERENTIATION OF A VECTOR

A vector $\vec{r}$ is said to be a vector function of a scalar variable t if to each value of t there corresponds a value of $\vec{r}$.

A vector function is denoted by $\vec{r} = \vec{r}(t)$ or $\vec{r} = \vec{f}(t)$. For example, the position vector $\vec{r}$ of a particle moving along a curved path is a vector function of time t. In rectangular coordinate system, the vector function $\vec{f}$ can be expressed in a component form as

$$\vec{r} = \vec{f}(t) = f_1\hat{i}+f_2\hat{j}+f_3\hat{k},$$

where f_1, f_2, and f_3 are scalar functions of t and are called components of $\vec{f}$.

Let $\vec{r} = \vec{f}(t)$ be a vector function of the scalar variable t. If Δt denotes a small increment in t and $\Delta\vec{r}$ the corresponding increment in $\vec{r}$, then

$$\frac{d\vec{r}}{dt} = \lim_{\Delta t\to 0}\frac{\Delta\vec{r}}{\Delta t} = \lim_{\Delta t\to 0}\frac{\vec{f}(t+\Delta t)-\vec{f}(t)}{\Delta t},$$

if exists, is called the *ordinary derivative* of $\vec{r}$ with respect to the scalar t.

Since $\frac{d\vec{r}}{dt}$ is itself a vector depending on t, we can further consider its derivative with respect to t. If this derivative exists, it is denoted by $\frac{d^2\vec{r}}{dt^2}$.

Similarly, higher derivatives of $\vec{r}$ can be defined.

Geometric Significance of $\frac{d\vec{r}}{dt}$: Let $\vec{r} = \vec{f}(t)$ be the vector equation of a curve C in space. Let P and Q be two neighboring points on C with position vectors $\vec{r}$ and $\vec{r}+\delta\vec{r}$. Then, $\overrightarrow{OP} = \vec{r}, \overrightarrow{OQ} = \vec{r}+\delta\vec{r}$ and so,

$$\vec{PQ} = \vec{OQ}-\vec{OP} = \vec{r}+\delta\vec{r}-\vec{r} = \delta\vec{r}.$$

Therefore, $\frac{\delta\vec{r}}{\delta t}$ is directed along the chord PQ. As $\delta t \to 0$, that is, as $Q \to P$, the chord PQ tends to the tangent to the curve C at P. Hence, $\frac{d\vec{r}}{dt} = \lim_{\delta t\to 0}\frac{d\vec{r}}{\delta t}$ is a vector along the tangent to the curve at P.

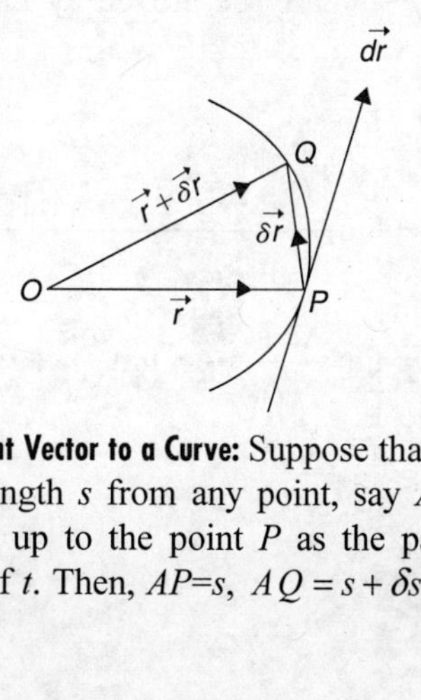

Unit Tangent Vector to a Curve: Suppose that we take an arc length s from any point, say A, on the curve C, up to the point P as the parameter, instead of t. Then, $AP=s$, $AQ = s+\delta s$, and so,

$PQ = \delta s$. In this case, $\frac{d\vec{r}}{ds}$ will be vector along the tangent at P. Further,

$$\left|\frac{d\vec{r}}{ds}\right| = \lim_{\delta s \to 0}\left|\frac{\delta \vec{r}}{\delta s}\right| = \lim_{Q \to P}\frac{\text{chord } PQ}{\text{Arc } PQ} = 1.$$

Hence, $\frac{d\vec{r}}{ds}$ is the unit vector $\hat{t}$ along the tangent at P.

Theorem 13.1.

If $\vec{a}, \vec{b}$, and $\vec{c}$ are differentiable vector functions of a scalar t and ϕ is a differentiable scalar function of t, then

(i) $\frac{d}{dt}\left(\vec{a} \pm \vec{b}\right) = \frac{d\vec{a}}{dt} + \frac{d\vec{b}}{dt}$.

(ii) $\frac{d}{dt}\left(\vec{a} \cdot \vec{b}\right) = \vec{a} \cdot \frac{d\vec{b}}{dt} + \frac{d\vec{a}}{dt} \cdot \vec{b}$.

(iii) $\frac{d}{dt}\left(\vec{a} \times \vec{b}\right) = \vec{a} \times \frac{d\vec{b}}{dt} + \frac{d\vec{a}}{dt} \times \vec{b}$.

(iv) $\frac{d}{dt}(\phi\vec{a}) = \phi\frac{d\vec{a}}{dt} + \frac{d\phi}{dt}\vec{a}$.

(v) $\frac{d}{dt}\left[\vec{a} \cdot \left(\vec{b} \times \vec{c}\right)\right] = \vec{a} \cdot \left(\vec{b} \times \frac{d\vec{c}}{dt}\right) + \vec{a} \cdot \left(\frac{d\vec{b}}{dt} \times \vec{c}\right) + \frac{d\vec{a}}{dt} \cdot (\vec{b} \times \vec{c})$.

(vi) $\frac{d}{dt}\left[\vec{a} \times \left(\vec{b} \times \vec{c}\right)\right] = \vec{a} \times \left(\vec{b} \times \frac{d\vec{c}}{dt}\right) + \vec{a} \times \left(\frac{d\vec{b}}{dt} \times \vec{c}\right) + \frac{d\vec{a}}{dt} \times \left(\vec{b} \times \vec{c}\right)$.

Proof: We prove (i), (iii), and (v). The other parts may similarly be proved by the readers themselves.

(i) $\frac{d}{dt}\left(\vec{a} + \vec{b}\right)$

$$= \lim_{\Delta t \to 0}\frac{\left[\left(\vec{a} + \Delta\vec{a}\right) + \left(\vec{b} + \Delta\vec{b}\right)\right] - \left(\vec{a} + \vec{b}\right)}{\Delta t}$$

$$= \lim_{\Delta t \to 0}\frac{\Delta\vec{a} + \Delta\vec{b}}{\Delta t} = \lim_{\Delta t \to 0}\frac{\Delta\vec{a}}{\Delta t} + \lim_{\Delta t \to 0}\frac{\Delta\vec{b}}{\Delta t}$$

$$= \frac{d\vec{a}}{dt} + \frac{d\vec{b}}{dt}.$$

(iii) $\frac{d}{dt}\left(\vec{a} \times \vec{b}\right)$

$$= \lim_{\Delta t \to 0}\frac{\left[\left(\vec{a} + \Delta\vec{a}\right) \times \left(\vec{b} + \Delta\vec{b}\right)\right] - \vec{a} \times \vec{b}}{\Delta t}$$

$$= \lim_{\Delta t \to 0}\frac{\vec{a} \times \vec{b} + \vec{a} \times \Delta\vec{b} + \Delta\vec{a} \times \vec{b} + \Delta\vec{a} \times \Delta\vec{b} - \vec{a} \times \vec{b}}{\Delta t}$$

$$= \lim_{\Delta t \to 0}\frac{\vec{a} \times \Delta\vec{b} + \Delta\vec{a} \times \vec{b} + \Delta\vec{a} \times \Delta\vec{b}}{\Delta t}$$

$$= \lim_{\Delta t \to 0}\left[\vec{a} \times \frac{\Delta\vec{b}}{\Delta t} + \frac{\Delta\vec{a}}{\Delta t} \times \vec{b} + \frac{\Delta\vec{a}}{\Delta t} \times \Delta\vec{b}\right]$$

$$= \vec{a} \times \frac{d\vec{b}}{dt} + \frac{d\vec{a}}{dt} \times \vec{b} + \frac{d\vec{a}}{dt} \times \vec{0}$$

$$= \vec{a} \times \frac{d\vec{b}}{dt} + \frac{d\vec{a}}{dt} \times \vec{b} + \vec{0} = \vec{a} \times \frac{d\vec{b}}{dt} + \frac{d\vec{a}}{dt} \times \vec{b}.$$

(v) Using (ii) and (iii), we have

$$\frac{d}{dt}\left[\vec{a} \cdot \left(\vec{b} \times \vec{c}\right)\right]$$

$$= \vec{a} \cdot \frac{d}{dt}\left(\vec{b} \times \vec{c}\right) + \frac{d\vec{a}}{dt} \cdot \left(\vec{b} \times \vec{c}\right)$$

$$= \vec{a} \cdot \left[\vec{b} \times \frac{d\vec{c}}{dt} + \frac{d\vec{b}}{dt} \times \vec{c}\right] + \frac{d\vec{a}}{dt} \cdot \left(\vec{b} \times \vec{c}\right)$$

$$= \vec{a} \cdot \left[\vec{b} \times \frac{d\vec{c}}{dt}\right] + \vec{a} \cdot \left[\frac{d\vec{b}}{dt} \times \vec{c}\right] + \frac{d\vec{a}}{dt} \cdot \left(\vec{b} \times \vec{c}\right).$$

Theorem 13.2.

The derivative of a constant vector is the zero vector.

Proof: We know that a vector is called constant if its magnitude and direction do not change. Let $\vec{c}$ be a constant vector and let $\vec{r} = \vec{c}$. Then, $\vec{r} + \delta\vec{r} = \vec{c}$ and so, $\delta\vec{r} = 0$. Thus, $\frac{\delta\vec{r}}{\delta t} = \frac{\vec{0}}{\delta t} = \vec{0}$ and hence, $\frac{d\vec{r}}{dt} = \lim_{\delta t \to 0}\frac{\delta\vec{r}}{\delta t} = \lim_{\delta t \to 0}\vec{0} = \vec{0}$.

Theorem 13.3.

A vector function $\vec{f}$ of a scalar variable t is constant if and only if $\frac{d\vec{f}}{dt} = \vec{0}$.

Proof: If $\vec{f}$ is a constant vector, then, by Theorem 13.2, $\frac{d\vec{f}}{dt} = \vec{0}$.

Conversely, suppose that $\frac{d\vec{f}}{dt} = \vec{0}$. If f_1, f_2 and f_3 are the components of $\vec{f}$ along x-, y-, and z-axes, then $\vec{f} = f_1\hat{i} + f_2\hat{j} + f_3\hat{k}$. Hence,

$$\vec{0} = \frac{d\vec{f}}{dt} = \frac{df_1}{dt}\hat{i} + \frac{df_2}{dt}\hat{j} + \frac{df_3}{dt}\hat{k}.$$

Therefore, equality of two vectors implies

$$\frac{df_1}{dt} = \frac{df_2}{dt} = \frac{df_3}{dt} = 0$$

Therefore, f_1, f_2 and f_3 are constant scalars, independent of t. Hence, f is a constant vector function.

Theorem 13.4.

A vector function $\vec{f}$ of a scalar variable t has a constant magnitude if and only if $\vec{f}.\frac{d\vec{f}}{dt} = 0$.

Proof: First, suppose that the vector function $\vec{f}$ has a constant magnitude c. Then,

$$\vec{f}\cdot\vec{f} = |\vec{f}|^2 = c^2$$

and so,

$$\frac{d}{dt}(\vec{f}\cdot\vec{f}) = \frac{d}{dt}(c^2) = 0$$

But,

$$\frac{d}{dt}(\vec{f}\cdot\vec{f}) = \vec{f}\cdot\frac{d\vec{f}}{dt} + \frac{d\vec{f}}{dt}\cdot\vec{f} = 2\vec{f}\cdot\frac{d\vec{f}}{dt}.$$

Hence,

$$2\vec{f}\cdot\frac{d\vec{f}}{dt} = 0 \text{ or } \vec{f}\cdot\frac{d\vec{f}}{dt} = 0.$$

Conversely, suppose that $\vec{f}.\frac{d\vec{f}}{dt} = 0$. Therefore, $2\vec{f}\cdot\frac{d\vec{f}}{dt} = 0$ or $\vec{f}\cdot\frac{d\vec{f}}{dt} + \frac{d\vec{f}}{dt}\cdot\vec{f} = 0$ or $\frac{d}{dt}\left(\vec{f},\vec{f}\right) = 0$, which implies $\vec{f}\cdot\vec{f}$ is constant = c^2, say.

Hence, $|\vec{f}|^2 = c^2$ or $|\vec{f}| = c$, , that is, $\vec{f}$ has a constant magnitude.

Theorem 13.5.

The necessary and sufficient condition for a vector function $\vec{f}$ of a scalar variable t to have a constant direction is that $\vec{f}\times\frac{d\vec{f}}{dt} = \vec{0}$.

Proof: Let $\vec{F}$ be a vector function of modulus unity for all t. Let $\left|\vec{f}\right| = f$. Then, $\vec{f} = f\,\vec{\text{F}}$.

The Condition is Necessary: Suppose that $\vec{f}$ has a constant direction. Since $\vec{f} = f\vec{F}$, it follows that $\vec{f}$ and $\vec{F}$ have the same direction. Thus, $\vec{F}$ has a constant magnitude, equal to unity and a constant direction too and so, is a constant vector. Therefore, $\frac{d\vec{F}}{dt} = 0$. Differentiating $\vec{f} = f\vec{F}$ with respect to t, we have

Now,
$$\frac{d\vec{f}}{dt} = \frac{df}{dt}\vec{F} + f\frac{d\vec{F}}{dt}.$$

$$\vec{f}\times\frac{d\vec{f}}{dt} = (f\vec{F})\times\left[\frac{df}{dt}\vec{F} + f\frac{d\vec{F}}{dt}\right]$$

$$= f\frac{df}{dt}\vec{F}\times\vec{F} + f^2\vec{F}\times\frac{d\vec{F}}{dt}$$

$$= \vec{0} + f^2\vec{F}\times\frac{d\vec{F}}{dt},$$

since $\vec{F}\times\vec{F} = \vec{0}$

$$= f^2\vec{F}\times\frac{d\vec{F}}{dt}$$

$$= f^2\vec{F}\times\vec{0} = \vec{0},$$

since $\frac{d\vec{F}}{df} = 0$ (as shown earlier).

The Condition is Sufficient: Suppose that $\vec{f}\times\frac{d\vec{f}}{dt} = \vec{0}$. Therefore, as shown previously, $f^2\vec{F}\times\frac{d\vec{F}}{dt} = \vec{0}$ and so $\vec{F}\times\frac{d\vec{F}}{dt} = \vec{0}$. Also, since $\vec{F}$ is of constant magnitude, $\vec{F}.\frac{d\vec{F}}{dt} = \vec{0}$. These two facts imply that $\frac{d\vec{F}}{dt} = \vec{0}$. Therefore, $\vec{F}$ is a constant vector. But magnitude of F is constant (unity). Therefore, $\vec{F}$ has a constant direction. But $\vec{f} = f\vec{F}$. Therefore, direction of $\vec{f}$ is also constant.

Corollary 13.1: The derivative of a vector function of a scalar variable t having a constant direction is collinear with it.

Proof: Since $\vec{f}$ has a constant direction, $\vec{f}\times\frac{d\vec{f}}{dt} = 0$ and so, $\vec{f}$ and $\frac{d\vec{f}}{dt}$ are collinear. This completes the proof of the corollary.

From Theorems 13.3−13.5, we conclude that

(i) $\frac{d\vec{f}}{dt} = \vec{0}$ if and only if $\vec{f}$ is a constant vector function in both magnitude and direction

(ii) $\vec{f}.\frac{d\vec{f}}{dt}=0$ if and only if $\vec{f}$ has a constant magnitude.

(iii) $\vec{f}\times\frac{d\vec{f}}{dt}=0$ if and only if $\vec{f}$ has a constant direction.

Theorem 13.6.

If $\vec{f}=f_1\hat{i}+f_2\hat{j}+f_3\hat{k}$ is a vector function of the scalar variable t, then

$$\frac{d\vec{f}}{dt}=f_1'(t)\hat{i}+f_2'(t)\hat{j}+f_3'(t)\hat{k}.$$

Proof: We have

$$\vec{f}=f_1\hat{i}+f_2\hat{j}+f_3\hat{k},$$

where f_1, f_2, and f_3 are scalar functions of t. Therefore,

$$\frac{d\vec{f}}{dt}=\frac{d}{dt}(f_1\hat{i})+\frac{d}{dt}(f_2\hat{j})+\frac{d}{dt}(f_3\hat{k})$$

$$=f_1\frac{d\hat{i}}{dt}+\frac{df_1}{dt}\hat{i}+f_2\frac{d\hat{j}}{dt}+\frac{df_2}{dt}\hat{j}+f_3\frac{d\hat{k}}{dt}+\frac{df_3}{dt}\hat{k}$$

$$=\vec{0}+\frac{df_1}{dt}\hat{i}+\vec{0}+\frac{df_2}{dt}\hat{j}+\vec{0}+\frac{df_3}{dt}\hat{k}$$

$$=\frac{df_1}{dt}\hat{i}+\frac{df_2}{dt}\hat{j}+\frac{df_3}{dt}\hat{k}.$$

Thus, to differentiate a vector, it is sufficient to differentiate its components.

Velocity and Acceleration. Let $\vec{r}$ be the position vector of a moving particle P, and let $\delta\vec{r}$ be the displacement of the particle in time δt, where t denotes time. Then, the vector $\frac{\delta\vec{r}}{\delta t}$ denotes the average velocity of the particle during the interval δt of time. Therefore, the velocity vector $\vec{v}$ of the particle at P is given by

$$\vec{v}=\lim_{\delta t\to 0}\frac{\delta\vec{r}}{\delta t}=\frac{d\vec{r}}{dt},$$

and its direction is along the tangent at P. Further, if $\delta\vec{v}$ is the change in velocity $\vec{v}$ during the time interval δt, then the rate of change of velocity, that is, $\frac{\delta\vec{v}}{\delta t}$ is the average acceleration of the particle during the interval δt. Thus, the acceleration of the particle at P is

$$\vec{a}=\lim_{\delta t\to 0}\frac{\delta\vec{v}}{\delta t}=\frac{d\vec{v}}{dt}=\frac{d}{dt}\left(\frac{d\vec{r}}{dt}\right)=\frac{d^2\vec{r}}{dt^2}.$$

Tangential and Normal Acceleration. Let $\vec{r}$ be the position vector of a point P moving in a plane curve at any time t. Then the velocity of the moving point is given by

$$\vec{v}=\frac{d\vec{r}}{dt}=\frac{d\vec{r}}{ds}\cdot\frac{ds}{dt}.$$

But $\frac{d\vec{r}}{ds}=\hat{T}$ is a unit vector along the tangent at P. Therefore, $\vec{v}=\frac{ds}{dt}\hat{T}$. Thus, the acceleration is

$$\frac{d\vec{v}}{dt}=\frac{d}{dt}\left(\frac{ds}{dt}\hat{T}\right)=\frac{d^2s}{dt^2}\hat{T}+\frac{ds}{dt}\cdot\frac{d\hat{T}}{dt}. \quad (1)$$

But,

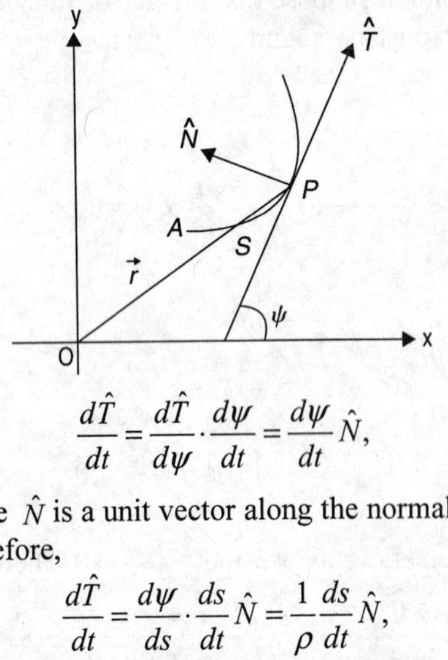

$$\frac{d\hat{T}}{dt}=\frac{d\hat{T}}{d\psi}\cdot\frac{d\psi}{dt}=\frac{d\psi}{dt}\hat{N},$$

where $\hat{N}$ is a unit vector along the normal at P. Therefore,

$$\frac{d\hat{T}}{dt}=\frac{d\psi}{ds}\cdot\frac{ds}{dt}\hat{N}=\frac{1}{\rho}\frac{ds}{dt}\hat{N},$$

where ρ is the radius of curvature at P. Hence, (1) reduces to

$$\frac{d\vec{v}}{dt}=\frac{d^2s}{dt^2}\hat{T}+\frac{1}{\rho}\left(\frac{ds}{dt}\right)^2\hat{N}=\frac{dv}{dt}\hat{T}+\frac{v^2}{\rho}\hat{N}.$$

Therefore,

Tangential acceleration $=\frac{dv}{dt}=\frac{d^2s}{dt^2}$ and

Normal acceleration $=\frac{v^2}{\rho}$.

Radial and Transverse Acceleration of a Moving Particle.

Let $\vec{r}$ be the position vector of a moving particle $P(r, \theta)$. Suppose that $\hat{R}$ and $\hat{T}$ are the unit vector in radial- and transverse directions, respectively. Then, $\hat{r} = r\hat{R}$ and

$$\text{velocity } \vec{v} = \frac{d\vec{r}}{dt} = \frac{d}{dt}(r\hat{R}) = \frac{dr}{dt}\hat{R} + r\frac{d\hat{R}}{dt}$$

$$= \frac{dr}{dt}\hat{R} + r\frac{d\hat{R}}{d\theta}.\frac{d\theta}{dt} = \frac{dr}{dt}\hat{R} + r\frac{d\theta}{dt}\hat{T}.$$

Therefore, the components of the velocity in the radial- and transverse directions are

$$v_R = \frac{dr}{dt} \text{ and } v_T = r\frac{d\theta}{dt}.$$

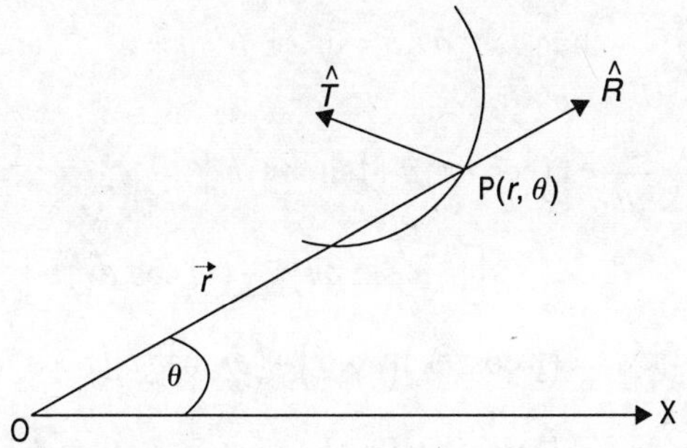

Further, since $\frac{dR}{d\theta} = \hat{T}$ and $\frac{d\hat{T}}{d\theta} = -\hat{R}$, we have acceleration $\vec{a}$

$$= \frac{d\vec{v}}{dt} = \frac{d}{dt}\left(\frac{dr}{dt}\hat{R} + r\frac{d\theta}{dt}\hat{T}\right)$$

$$= \frac{d^2r}{dt^2}\hat{R} + \frac{dr}{dt}.\frac{d\hat{R}}{dt} + \frac{dr}{dt}.\frac{d\theta}{dt}\hat{T} + r\frac{d^2\theta}{dt^2}\hat{T} + r\frac{d\theta}{dt}.\frac{d\hat{T}}{dt}$$

$$= \frac{d^2r}{dt^2}\hat{R} + \frac{dr}{dt}\frac{d\hat{R}}{d\theta}.\frac{d\theta}{dt} + \left(\frac{dr}{dt}\frac{d\theta}{dt} + r\frac{d^2\theta}{dt^2}\right)\hat{T}$$

$$+ r\frac{d\theta}{dt}\frac{d\hat{T}}{d\theta}.\frac{d\theta}{dt}$$

$$= \frac{d^2r}{dt^2}\hat{R} + \frac{dr}{dt}.\frac{d\theta}{dt}\hat{T} + \left(\frac{dr}{dt}.\frac{d\theta}{dt} + r\frac{d^2\theta}{dt^2}\right)\hat{T}$$

$$- r\left(\frac{d\theta}{dt}\right)^2\hat{R}$$

$$= \left[\frac{d^2r}{dt^2} - r\left(\frac{d\theta}{dt}\right)^2\right]\hat{R} + \left[2\frac{dr}{dt}\frac{d\theta}{dt} + r\frac{d^2\theta}{dt^2}\right]\hat{T}.$$

Hence,

$$\text{Radial acceleration} = a_R = \frac{d^2r}{dt^2} - r\left(\frac{d\theta}{dt}\right)^2$$

and

Transverse acceleration

$$= a_T = 2\frac{dr}{dt}.\frac{d\theta}{dt} + r\frac{d^2\theta}{dt^2}.$$

EXAMPLE 13.20

If $a = \sin\theta\hat{i} + \cos\theta\hat{j} + \theta\hat{k}$, $\vec{b} = \cos\theta\hat{i} - \sin\theta\hat{j} - 3\hat{k}$, and $\vec{c} = 2\vec{i} + 3\vec{j} - \vec{k}$, find $\frac{d}{d\theta}\left(\vec{a}\times\left(\vec{b}\times\vec{c}\right)\right)$ at $\theta = 0$.

Solution. We are given that

$a = \sin\theta\hat{i} + \cos\theta\hat{j} + \theta\hat{k}$,

$\vec{b} = \cos\theta\hat{i} - \sin\theta\hat{j} - 3\hat{k}$, and $\vec{c} = 2\vec{i} + 3\vec{j} - \vec{k}$.

Therefore,

$$\vec{b}\times\vec{c} = \begin{vmatrix} \hat{i} & \hat{j} & \hat{k} \\ \cos\theta & -\sin\theta & -3 \\ 2 & 3 & -1 \end{vmatrix}$$

$$= (\sin\theta + 9)\hat{i} - (-\cos\theta + 6)\hat{j} + (3\cos\theta + 2\sin\theta)\hat{k}.$$

Then, $\vec{a}\times(\vec{b}\times\vec{c})$

$$= \begin{vmatrix} \hat{i} & \hat{j} & \hat{k} \\ \sin\theta & \cos\theta & \theta \\ 9 + \sin\theta & -\cos\theta + 6 & 3\cos\theta + 2\sin\theta \end{vmatrix}$$

$$= (3\cos^2\theta + \sin 2\theta - \theta\cos\theta + 6\theta)\hat{i}$$

$$-\left(\frac{3}{2}\sin 2\theta + 2\sin^2\theta - 9\theta - \theta\sin\theta\right)\hat{j}$$

$$+(-6\sin\theta - 9\cos\theta)\hat{k}.$$

Therefore,

$$\frac{d}{d\theta}\left[\vec{a}\times\left(\vec{b}\times\vec{c}\right)\right]$$

$$= (-6\cos\theta\sin\theta + 2\cos 2\theta - \cos\theta + \theta\sin\theta + 6)\hat{i}$$

$$-(3\cos 2\theta + 4\sin\theta\cos\theta - 9 - \theta\cos\theta - \sin\theta)\hat{j}$$

$$+(-6\cos\theta + 9\sin\theta)\hat{k}.$$

Putting $\theta = 0$, we get

$$\frac{d}{d\theta}\left[\vec{a}\times\left(\vec{b}\times\vec{c}\right)\right] = (2-1+6)\hat{i} - (3-9)\hat{j} - 6\hat{k}$$

$$= 7\hat{i} + 6\hat{j} - 6\hat{k}.$$

EXAMPLE 13.21

If $\vec{r} = (\cos nt)\hat{i} + (\sin nt)\hat{j}$, where n is a constant and t varies, show that $\vec{r}\times\frac{d\vec{r}}{dt} = n\hat{k}.$

Solution. We have

$$\vec{r} = (\cos nt)\hat{i} + (\sin nt)\hat{j}.$$

Therefore,

$$\frac{d\vec{r}}{dr} = (-n\sin nt)\hat{i} + (n\cos nt)\hat{j}.$$

Therefore,

$$\vec{r}\times\frac{d\vec{r}}{dt} = \begin{vmatrix} \hat{i} & \hat{j} & \hat{k} \\ \cos nt & \sin nt & 0 \\ -n\sin nt & n\cos nt & 0 \end{vmatrix}$$

$$= \hat{k}(n\cos^2 nt + n\sin^2 nt)$$

$$= n\hat{k}.$$

EXAMPLE 13.22

If $\vec{a}$ and $\vec{b}$ are constant vectors, ω is a constant scalar, and $\vec{r} = \vec{a}\sin\omega t + \vec{b}\cos\omega t$, show that

$$\frac{d^2\vec{r}}{dt^2} = +\omega^2\vec{r} = \vec{0} \text{ and}$$

$$\vec{r}\times\frac{d\vec{r}}{dt} = -\omega\vec{a}\times\vec{b}.$$

Solution. (i) Since $\vec{a}$ and $\vec{b}$ are constant vectors, we have

$$\frac{d\vec{a}}{dt} = \vec{0} \text{ and } \frac{d\vec{b}}{dt} = \vec{0}. \qquad (1)$$

Now it is given that $\vec{r} = \vec{a}\sin\omega t + \vec{b}\cos\omega t.$ Therefore,

$$\frac{d\vec{r}}{dt} = \sin\omega t\frac{d\vec{a}}{dt} + \frac{d}{dt}(\sin\omega t)\vec{a}$$

$$+ \cos\omega t\frac{d\vec{b}}{dt} + \frac{d}{dt}(\cos\omega t)\vec{b}$$

$$= 0 + \frac{d}{dt}(\sin\omega t)\vec{a} + 0 + \frac{d}{dt}(\cos\omega t)\vec{b}, \text{ using}(1)$$

$$= (\omega\cos\omega t)\vec{a} - (\omega\sin\omega t)\vec{b}$$

and

$$\frac{d^2\vec{r}}{dt^2} = (-\omega^2\sin\omega t)\vec{a} - (\omega^2\cos\omega t)\vec{b}$$

$$= -\omega^2\left(\vec{a}\sin\omega t + \vec{b}\cos\omega t\right) = -\omega^2\vec{r}.$$

Hence,

$$\frac{d^2\vec{r}}{dt^2} + \omega^2\vec{r} = \vec{0}.$$

(ii) Since $\vec{a}\times\vec{a} = \vec{0}$, $\vec{b}\times\vec{b} = 0$, and $\vec{a}\times\vec{b} = -\vec{b}\times\vec{a}$, we have

$$\vec{r}\times\frac{d\vec{r}}{dt} = \left[(\cos\omega t)\vec{a} + (\sin\omega t)\vec{b}\right]$$

$$\times\left[(-\omega\sin\omega t)\vec{a} + (\omega\cos\omega t)\vec{b}\right]$$

$$= \left(\omega\cos^2\omega t\right)\left(\vec{a}\times\vec{b}\right) - \left(\omega\sin^2\omega t\right)\left(\vec{b}\times\vec{a}\right)$$

$$= \left(\omega\cos^2\omega t\right)\left(\vec{a}\times\vec{b}\right) + \left(\omega\sin^2\omega t\right)\left(\vec{a}\times\vec{b}\right)$$

$$= \left[\omega\left(\cos^2\omega t + \sin^2\omega t\right)\right]\left(\vec{a}\times\vec{b}\right)$$

$$= \omega\left(\vec{a}\times\vec{b}\right).$$

EXAMPLE 13.23

Show that

$$\frac{d}{dt}\left(\vec{a}\times\frac{d\vec{b}}{dt} - \frac{d\vec{a}}{dt}\times\vec{b}\right) = \vec{a}\times\frac{d^2\vec{b}}{dt^2} - \frac{d^2\vec{a}}{dt^2}\times\vec{b}.$$

Solution. We have

$$\frac{d}{dt}\left(\vec{a}\times\frac{d\vec{b}}{dt} - \frac{d\vec{a}}{dt}\times\vec{b}\right)$$

$$= \frac{d\vec{a}}{dt}\times\frac{d\vec{b}}{dt} + \vec{a}\times\frac{d^2\vec{b}}{dt^2} - \frac{d^2\vec{a}}{dt^2}\times\vec{b} - \frac{d\vec{a}}{dt}\times\frac{d\vec{b}}{dt}$$

$$= \vec{a}\times\frac{d^2\vec{b}}{dt^2} - \frac{d^2\vec{a}}{dt^2}\times\vec{b},$$

which proves our assertion.

EXAMPLE 13.24

Let $\vec{r} = t^2\hat{i} - 3t\hat{j} + (2t+1)\hat{k}$, find at $t = 0$, the value of $\left|\frac{d^2\vec{r}}{dt^2}\right|$.

Solution. Let $\vec{r} = t^2\hat{i} - 3t\hat{j} + (2t+1)\hat{k}$. Then,

$$\frac{d\vec{r}}{dt} = \frac{d}{dt}(t^2)\hat{i} - \frac{d}{dt}(3t)\hat{j} + \frac{d}{dt}(2t+1)\hat{k}$$

$$= 2t\hat{i} - 3\hat{j} + 2\hat{k} \text{ and}$$

$$\frac{d^2\vec{r}}{dt^2} = \frac{d}{dt}(2t)\hat{i} - \frac{d}{dt}(3)\hat{j} + \frac{d}{dt}(2)\hat{k} = 2\hat{i}.$$

When $t = 0$, we have $\frac{d^2\vec{r}}{dt^2} = 2\hat{i}$.. Further,

$$\left|\frac{d^2\vec{r}}{dt^2}\right| = \sqrt{2^2 + 0^2 + 0^2} = 2.$$

EXAMPLE 13.25

If $\vec{a} = 5t^2\hat{i} + t\hat{j} - t^2\hat{k}$. and $\vec{b} =$ sin $t\hat{i}$ – cos $t\hat{j}$, find $\frac{d}{dt}(\vec{a}.\vec{b})$ and $\frac{d}{dt}(\vec{a}\times\vec{b})$.

Solution. Let $\vec{a} = 5t^2\hat{i} + t\hat{j} - t^2\hat{k}$ and $\vec{b} =$ sin $t\hat{i}$ – cos $t\hat{j}$.

Then,

$$\vec{a}.\vec{b} = 5t^2 \sin t - t\cos t.$$

Therefore,

$$\frac{d}{dt}(\vec{a}.\vec{b}) = \frac{d}{dt} = (5t^2 \sin t - t\cos t)$$

$$= 5t^2 \cos t + 10t \sin t + t\sin t - \cos t$$

$$= (5t^2 - 1)\cos t + 11t \sin t$$

Also,

$$\vec{a}\times\vec{b} = \begin{vmatrix} \hat{i} & \hat{j} & \hat{k} \\ 5t^2 & t & -t \\ sint & -cost & 0 \end{vmatrix}$$

$$= (-t^2 \cos t)\hat{i} - (t^3 \sin t)\hat{j}$$

$$+(-5t^2 \cos t - t\sin t)\hat{k}.$$

Therefore,

$$\frac{d}{dt}(\vec{a}\times\vec{b})$$

$$= \frac{d}{dt}(-t^3 \cos t)\hat{i} - \frac{d}{dt}(t^3 \sin t)\hat{j}$$

$$+\frac{d}{dt}(-5t^2 \cos t - t\sin t)\hat{k}$$

$$= (t^3\sin t - 3t^2\cos t)\hat{i} - (t^3\cos t + 3t^2\sin t)\hat{j}$$

$$+ (5t^2\sin t - 11t\cos t - \sin t)\hat{k}.$$

EXAMPLE 13.26

Find a unit tangent vector to any point on the curve $x = a\cos\omega t$, $y = a\sin\omega t$, and $z = bt$, where a, b, and ω are constants.

Solution. Let $\vec{r}$ be the position vector of any point (x, y, z) on the given curve. Then,

$$\vec{r} = x\hat{i} + y\hat{j} + z\hat{k}$$

$$= (a\cos\omega t)\hat{i} + (a\sin\omega t)\hat{j} + (bt)\hat{k}.$$

Therefore,

$$\frac{d\vec{r}}{dt} = (-a\omega\sin\omega t)\hat{i} + (a\omega\cos\omega t)\hat{j} + b\hat{k}.$$

The vector $\frac{d\vec{r}}{dt}$ is along the tangent at the point (x, y, z) to the given curve. Hence, unit tangent vector is given by

$$\vec{T} = \frac{\frac{d\vec{r}}{dt}}{\left|\frac{d\vec{r}}{dt}\right|} = \frac{(-a\omega\sin\omega t)\hat{i} + (a\omega\cos\omega t)\hat{j} + b\hat{k}}{\sqrt{a^2\omega^2 + b^2}}.$$

EXAMPLE 13.27

A particle moves along the curve $x = 3t^2, y = t^2 - 2t$ and $z = t^3$. Find its velocity and acceleration at $t = 1$ in the direction of $\hat{i} + \hat{j} - \hat{k}$.

Solution. Let $\vec{r}$ be the position vector of any point (x, y, z) on the given curve. Then,

$$\vec{r} = x\hat{i} + y\hat{j} + z\hat{k} = 3t^2\hat{i} + (t^2 - 2t)\hat{j} + t^3\hat{k}$$

and so, the velocity and acceleration of the particle are, respectively,

$$\vec{v} = \frac{d\vec{r}}{dt} = 6t\hat{i} + (2t-2)\hat{j} + 3t^2\hat{k} = 6\hat{i} + 3\hat{k} \text{ at } t = 1$$

and

$$\vec{a} = \frac{d^2\vec{r}}{dt^2} = 6\hat{i} + 2\hat{j} + 6t\hat{k} = 6\hat{i} + 2\hat{j} + 6\hat{k} \text{ at } t = 1.$$

The unit vector in the direction of $\hat{i} + \hat{j} - \hat{k}$ is

$$\hat{n} = \frac{\hat{i} + \hat{j} - \hat{k}}{|\hat{i} + \hat{j} - \hat{k}|} = \frac{\hat{i} + \hat{j} - \hat{k}}{\sqrt{3}}.$$

Therefore, the components of velocity and acceleration in the direction of $\hat{i} + \hat{j} - \hat{k}$ are

$$\vec{v}.\hat{n} = (6\hat{i} + 3\hat{k}) \cdot \frac{(\hat{i} + \hat{j} - \hat{k})}{\sqrt{3}} = \frac{3}{\sqrt{3}} = \sqrt{3} \text{ and}$$

$$\vec{a}.\hat{n} = (6\hat{i} + 2\hat{j} + 3\hat{k}) \cdot \frac{(\hat{i} + \hat{j} - \hat{k})}{\sqrt{3}} = \frac{2}{\sqrt{3}} = \frac{2\sqrt{3}}{3}.$$

EXAMPLE 13.28

Find the angle between the tangents to the curve $\vec{r} = t^2\hat{i} + 2t\hat{j} - t^3\hat{k}$ at the points $t = \pm 1$.

Solution. We have

$$\vec{r} = t^2\hat{i} + 2t\hat{j} - t^3\hat{k}.$$

Therefore, the vector along the tangent at any point is

$$\frac{d\vec{r}}{dt} = 2t\hat{i} + 2\hat{j} - 3t^2\hat{k}.$$

Thus, the vectors along the tangents at $t = \pm 1$ are

$$\vec{T}_1 = 2\hat{i} + 2\hat{j} - 3\hat{k} \text{ and } \vec{T}_2 = -2\hat{i} + 2\hat{j} - 3\hat{k}.$$

The angle θ between the tangents is given by

$$\cos\theta = \frac{\vec{T}_1.\vec{T}_2}{|\vec{T}_1||\vec{T}|} = \frac{2(-2) + 2(2) - 3(-3)}{\sqrt{4+4-9}.\sqrt{4+4+9}} = \frac{9}{17}.$$

Hence,

$$\theta = \cos^{-1}\left(\frac{9}{17}\right).$$

EXAMPLE 13.29

A particle moves along the curve

$$\vec{r} = (t^3 - 4t)\hat{i} + (t^2 + 4t)\hat{j} + (8t^2 - 3t^3)\hat{k},$$

where t denotes time. Find the magnitude of acceleration along the tangent and normal at time t =2.

Solution. The curve is

$$\vec{r} = (t^3 - 4t)\hat{i} + (t^2 + 4t)\hat{j} + (8t^2 - 3t^3)\hat{k}.$$

Therefore,

$$\vec{v} = \frac{d\vec{r}}{dt} = (3t^2 - 4)\hat{i} + (2t + 4)\hat{j} + (16t - 9t^2)\hat{k}$$

$$= 8\hat{i} + 8\hat{j} - 4\hat{k} \text{ at } t = 2$$

and

$$\text{acceleration } \vec{a} = \frac{d^2\vec{r}}{dt^2} = 6t\hat{i} + 2\hat{j} + (16 - 18t)\hat{k}$$

$$= 12\hat{i} + 2\hat{j} - 20\hat{k} \text{ at } t = 2.$$

The velocity is along the tangent to the curve. Therefore,

Component of $\vec{a}$ along the tangent

$$= \vec{a}.\frac{\vec{v}}{|\vec{v}|} = (12\hat{i} + 2\hat{j} - 20\hat{k})\frac{8\hat{i} + 8\hat{j} - 4\hat{k}}{|8\hat{i} + 8\hat{j} - 4\hat{k}|}$$

$$= (12\hat{i} + 2\hat{j} - 20\hat{k})\frac{8\hat{i} + 8\hat{j} - 4\hat{k}}{\sqrt{64 + 64 + 16}}$$

$$= \frac{96 + 16 + 80}{12} = 16$$

and

Component of $\vec{a}$ along the normal

$$= |\,\vec{a} - \text{resolved part of } \vec{a} \text{ along the tangent}\,|$$

$$= \left|12\hat{i} + 2\hat{j} - 20\hat{k} - 16\frac{8\hat{i} + 8\hat{j} - 4\hat{k}}{\sqrt{64 + 64 + 16}}\right|$$

$$= \frac{1}{3}|\,4\hat{i} - 26\hat{j} - 44\hat{k}\,|$$

$$= \frac{1}{3}[\sqrt{16 + 676 + 1936}] = 2\sqrt{73}.$$

13.2 PARTIAL DERIVATIVES OF A VECTOR FUNCTION

Let $\vec{f}$ be a vector function of $x, y,$ and z. Let $\delta\vec{f}$ be the change in $\vec{f}$ corresponding to a small change δx in x. Then, $\delta\vec{f} = \vec{f}(x + \delta x, y, z) - \vec{f}(x, y, z)$ The limit

$$\lim_{\delta x \to 0} \frac{\delta\vec{f}}{\delta x} = \lim_{\delta x} \frac{\vec{f}(x + \delta x, y, z) - \vec{f}(x, y, z)}{\delta x},$$

if it exists, is called the *partial derivative of the vector function* $\vec{f}$ *with respect to* x and is denoted by $\frac{\partial\vec{f}}{\partial x}$ or $\vec{f}_x$.

Similarly, the partial derivatives of $\vec{f}$ with respect to y and z are defined by

$$\vec{f}_y = \lim_{\delta y \to 0} \frac{\vec{f}(x, y+\delta y, z) - \vec{f}(x, y, z)}{\delta y}$$

and

$$\vec{f}_z = \lim_{\delta z \to 0} \frac{\vec{f}(x, y, z+\delta z) - \vec{f}(x, y, z)}{\delta z},$$

provided these limits exist.

If $\vec{f}$ and $\vec{g}$ are differentiable vector functions of the independent variables x, y, and z and ϕ is a differentiable scalar function of x, y, and z, then

(i) $\frac{\partial}{\partial x}\left(\vec{f} + \vec{g}\right) = \frac{\partial \vec{f}}{\partial x} + \frac{\partial \vec{g}}{\partial x}$.

(ii) $\frac{\partial}{\partial x}\left(\phi \vec{f}\right) = \phi \frac{\partial \vec{f}}{\partial x} + \frac{\partial \phi}{\partial x}.\vec{f}$

(iii) $\frac{\partial}{\partial x}(\vec{f}.\vec{g}) = \vec{f}.\frac{\partial \vec{g}}{\partial x} + \frac{\partial \vec{f}}{\partial x}.\vec{g}$.

(iv) $\frac{\partial}{\partial x}\left(\vec{f} \times \vec{g}\right) = \vec{f} \times \frac{\partial \vec{g}}{\partial x} + \frac{\partial \vec{f}}{\partial x} \times \vec{g}$.

Similar expressions for partial derivatives with respect to y and z are valid. Higher partial derivatives of $\vec{f}$ may also be defined in the same way. For example,

$$\vec{f}_{xx} = \frac{\partial^2 \vec{f}}{\partial x^2} = \frac{\partial}{\partial x}\left(\frac{\partial \vec{f}}{\partial x}\right).$$

EXAMPLE 13.30

If $\vec{f} = xyz\hat{i} + xz^2\hat{j} - y^3\hat{k}$, find $\frac{\partial f}{\partial x \partial y}$ at the origin.

Solution. We have

$$\vec{f} = xyz\hat{i} + xz^2\hat{j} - y^3\hat{k}.$$

Therefore,

$$\frac{\partial \vec{f}}{\partial y} = xz\hat{i} + \vec{0} - 3y^2\hat{k} \text{ and}$$

$$\frac{\partial^2 \vec{f}}{\partial x \partial y} = z\hat{i} = \vec{0} \text{ at } (0, 0, 0).$$

EXAMPLE 13.31

If $\vec{a} = x^2yz\hat{i} - 2xz^3\hat{j} + xz^2\hat{k}$ and $\vec{b} = 2z\hat{i} + y\hat{j} - x^2\hat{k}$, find the value of $\frac{\partial^2}{\partial x^2}\left(\vec{a} \times \vec{b}\right)$ at the point (1, 0, 1).

Solution. We have

$$\vec{a} = x^2yz\hat{i} - 2xz^3\hat{j} + xz^2\hat{k} \text{ and } \vec{b} = 2z\hat{i} + y\hat{j} - x^2\hat{k}.$$

Therefore,

$$\vec{a} \times \vec{b} = \begin{vmatrix} \hat{i} & \hat{j} & \hat{k} \\ x^2yz & -2xz^3 & xz^2 \\ 2z & y & -x^2 \end{vmatrix}$$

$$= (2x^3z^3 - xyz^2)\hat{i} - (-x^4yz - 2xz^3)\hat{j} + (x^2y^2z + 4xz^4)\hat{k}.$$

Hence,

$$\frac{\partial}{\partial x}(\vec{a} \times \vec{b}) = (6x^2z^3 - yz^2)\hat{i} - (-4x^3yz - 2z^3)\hat{j} + (2xy^2z + 4z^4)\hat{k}$$

and

$$\frac{\partial^2}{\partial x^2}\left(\vec{a} \times \vec{b}\right) = \left(12xz^3\right)\hat{i} - \left(-12x^2yz\right)\hat{j} + \left(2y^2z\right)\hat{k}$$

$$= 12\hat{i} \text{ at } (1, 0, 1).$$

13.3 GRADIENT OF A SCALAR FIELD

A variable quantity whose value at any point in a region of space depends upon the position of the point is called a *point function*. If for each point P(*x, y, z*) of a region R, there corresponds a scalar $\phi(x, y, z)$, then ϕ is called a *scalar-point function* for the region R. The region R is then called a *scalar field*. For example, the temperature at any point within or on the surface of the earth is a scalar-point function. Similarly, atmospheric pressure in the space is a scalar-point function. On the other hand, if for each point P(*x, y, z*), of a region R, there exists a vector $\vec{f}(x, y, z)$, then $\vec{f}$ is called a *vector-point function* and the region R is then called a *vector field*. For example, the gravitational force is a vector-point function.

Let $f(x, y, z)$, be a scalar-point function. Then, the points satisfying an equation of the type $f(x, y, z) = c$ (constant) constitute a family of surface in a three-dimensional space. The surfaces of this family are called *level surfaces*. Since the value of the function *f* at any point of the surface is the same, these surfaces are also called *iso-f-surfaces*.

The operator ∇, defined by

$$\nabla = \hat{i}\frac{\partial}{\partial x} + \hat{j}\frac{\partial}{\partial y} + \hat{k}\frac{\partial}{\partial z},$$

is called the *vector differential operator* and is read as *del* or *nabla*.

Let ϕ be a scalar function defined and differentiable at each point (x, y, z) in a certain region of space. Then, the vector defined by

$$\nabla\phi = \left(\hat{i}\frac{\partial}{\partial x} + \hat{j}\frac{\partial}{\partial y} + \hat{k}\frac{\partial}{\partial z}\right)\phi,$$

$$= \hat{i}\frac{\partial\phi}{\partial x} + \hat{j}\frac{\partial\phi}{\partial y} + \hat{k}\frac{\partial\phi}{\partial z}$$

is called the *gradient of the scalar function* ϕ and is denoted by *grad* ϕ or $\nabla\phi$.

Thus, *grad* ϕ is a vector with components $\frac{\partial\phi}{\partial x}$, $\frac{\partial\phi}{\partial y}$, and $\frac{\partial\phi}{\partial z}$. We note that ϕ is a scalar-point function, whereas $\nabla\phi$ is a vector-point function.

13.4 GEOMETRICAL INTERPRETATION OF A GRADIENT

Let $\vec{r} = x\hat{i} + y\hat{j} + z\hat{k}$ be the position vector of a point P through which a level surface $\phi(x, y, z) = c$ (constant) passes. Then, differentiating $\phi(x, y, z) = c$ with respect to t, we get

$$\frac{d\phi}{dt} = 0 \text{ or } \frac{\partial\phi}{\partial x}.\frac{dx}{dt} + \frac{\partial\phi}{\partial y}.\frac{dy}{dt} + \frac{\partial\phi}{\partial z}.\frac{dz}{dt} = 0$$

or

$$\left(\frac{\partial\phi}{\partial x}\hat{i} + \frac{\partial\phi}{\partial y}\hat{j} + \frac{\partial\phi}{\partial z}\hat{k}\right).\left(\frac{dx}{dt}\hat{i} + \frac{dy}{dt}\hat{j} + \frac{dz}{dt}\hat{k}\right) = 0$$

or

$$\nabla\phi.\frac{d\vec{r}}{dt} = 0.$$

Since $\frac{d\vec{r}}{dt}$ is the vector tangent to the curve at P and since P is an arbitrary point on $\phi(x, y, z) = c$, it follows that $\nabla\phi$ is perpendicular to $\phi(x, y, z) = c$ at every point. Hence, $\nabla\phi$ *is normal to the surface* $\phi(x, y, z) = c$.

13.5 PROPERTIES OF A GRADIENT

The following theorem illustrates the properties satisfied by a gradient.

Theorem 13.7.

If ϕ and ψ are two scalar-point functions, and c is a constant, then,

(i) $\nabla(\phi \pm \psi) = \nabla\phi \pm \nabla\psi$.

(ii) $\nabla(\phi\psi) = \phi\nabla\psi + \psi\nabla\phi$.

(iii) $\nabla\left(\frac{\theta}{\psi}\right) = \frac{\psi\nabla\phi - \phi\nabla\psi}{\psi^2}$, provided that $\psi \neq 0$.

(iv) $\nabla(c\phi) = c\nabla\phi$.

(v) $\nabla\phi$ is a constant if and only if ϕ is a constant.

Proof: (i). By the definition of a gradient, we have

$$\nabla(\phi \pm \psi) = \left(\hat{i}\frac{\partial}{\partial x} + \hat{j}\frac{\partial}{\partial y} + \hat{k}\frac{\partial}{\partial z}\right)(\phi \pm \psi)$$

$$= \hat{i}\frac{\partial}{\partial x}(\phi \pm \psi) + \hat{j}\frac{\partial}{\partial y}(\phi \pm \psi) + \hat{k}\frac{\partial}{\partial z}(\phi \pm \psi)$$

$$= \left(\hat{i}\frac{\partial}{\partial x} + \hat{j}\frac{\partial}{\partial y} + \hat{k}\frac{\partial}{\partial z}\right)\phi \pm \left(\hat{i}\frac{\partial}{\partial x} + \hat{j}\frac{\partial}{\partial y} + \hat{k}\frac{\partial}{\partial z}\right)\psi$$

$$= \nabla\phi \pm \nabla\psi$$

(ii) $$\nabla(\phi\psi) = \left(\hat{i}\frac{\partial}{\partial x} + \hat{j}\frac{\partial}{\partial y} + \hat{k}\frac{\partial}{\partial z}\right)(\phi\psi)$$

$$= \hat{i}\frac{\partial}{\partial x}(\phi\psi) + \hat{j}\frac{\partial}{\partial y}(\phi\psi) + \hat{k}\frac{\partial}{\partial z}(\phi\psi)$$

$$= \hat{i}\left[\phi\frac{\partial\psi}{\partial x} + \psi\frac{\partial\phi}{\partial x}\right] + \hat{j}\left[\phi\frac{\partial\psi}{\partial y} + \psi\frac{\partial\phi}{\partial y}\right] + \hat{k}\left[\phi\frac{\partial\psi}{\partial z} + \psi\frac{\partial\phi}{\partial z}\right]$$

$$= \phi\left[\hat{i}\frac{\partial\psi}{\partial x} + \hat{j}\frac{\partial\psi}{\partial y} + \hat{k}\frac{\partial\psi}{\partial z}\right] + \psi\left[\hat{i}\frac{\partial\phi}{\partial x} + \hat{j}\frac{\partial\phi}{\partial y} + \hat{k}\frac{\partial\phi}{\partial z}\right]$$

$$= \phi\nabla\psi + \psi\nabla\phi$$

(iii) $$\nabla\left(\frac{\phi}{\psi}\right) = \left(\hat{i}\frac{\partial}{\partial x} + \hat{j}\frac{\partial}{\partial y} + \hat{k}\frac{\partial}{\partial z}\right)\left(\frac{\phi}{\psi}\right)$$

$$= \hat{i}\frac{\partial}{\partial x}\left(\frac{\phi}{\psi}\right) + \hat{j}\frac{\partial}{\partial y}\left(\frac{\phi}{\psi}\right) + \hat{k}\frac{\partial}{\partial z}\left(\frac{\phi}{\psi}\right)$$

$$= \hat{i}\left[\frac{\psi\frac{\partial\phi}{\partial x}-\phi\frac{\partial\psi}{\partial x}}{\psi^2}\right]+\hat{j}\left[\frac{\psi\frac{\partial\phi}{\partial y}-\phi\frac{\partial\psi}{\partial y}}{\psi^2}\right]$$

$$+\hat{k}\left[\frac{\psi\frac{\partial\phi}{\partial z}-\phi\frac{\partial\psi}{\partial z}}{\psi^2}\right]$$

$$=\psi\frac{\left[\hat{i}\frac{\partial\phi}{\partial x}+\hat{j}\frac{\partial\phi}{\partial y}+\hat{k}\frac{\partial\phi}{\partial z}\right]}{\psi^2}$$

$$-\phi\frac{\left[\hat{i}\frac{\partial\psi}{\partial x}+\hat{j}\frac{\partial\psi}{\partial y}+\hat{k}\frac{\partial\psi}{\partial z}\right]}{\psi^2}$$

$$=\frac{\psi\nabla\phi-\phi\nabla\psi}{\psi^2},\ \psi\neq 0.$$

(iv) We have

$$\nabla(c\phi)=\left(\hat{i}\frac{\partial}{\partial x}+\hat{j}\frac{\partial}{\partial y}+\hat{k}\frac{\partial}{\partial z}\right)(c\phi)$$

$$=\hat{i}\frac{\partial}{\partial x}(c\phi)+\hat{j}\frac{\partial}{\partial y}(c\phi)+\hat{k}\frac{\partial}{\partial z}(c\phi)$$

$$=c\hat{i}\frac{\partial}{\partial x}\phi+c\hat{j}\frac{\partial}{\partial y}\phi+c\hat{k}\frac{\partial}{\partial z}\phi$$

$$=c\left(\hat{i}\frac{\partial\phi}{\partial x}+\hat{j}\frac{\partial\phi}{\partial y}+\hat{k}\frac{\partial\phi}{\partial z}\right)=c\nabla\phi.$$

(v) We note that

$$\nabla\phi=\vec{0}\Leftrightarrow\hat{i}\frac{\partial\phi}{\partial x}+\hat{j}\frac{\partial\phi}{\partial y}+\hat{k}\frac{\partial\phi}{\partial z}=\vec{0}$$

$$\Leftrightarrow\frac{\partial\phi}{\partial x}=0,\ \frac{\partial\phi}{\partial y}=0,\ \frac{\partial\phi}{\partial z}=0.$$

$\Leftrightarrow \phi$ is a constant.

13.6 DIRECTIONAL DERIVATIVES

Let A be any given point in the region of definition of a scalar-point function ϕ. Let P be a point on any line drawn on one side of A. Then $\lim_{P\to A}\frac{\phi(P)-\phi(A)}{AP}$, if exists, is called the *directional derivative of the scalar-point function* ϕ at A in the direction of AP. The length of AP is regarded as positive. The direction derivative in the direction of AP′, where P′ is a point on the other side of A, is negative of that in the direction of AP.

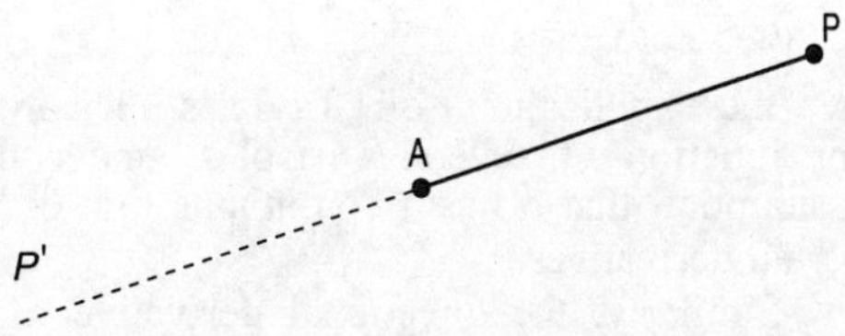

The *directional derivative of the vector function* $\vec{f}$ at A in the direction of AP is defined as $\lim_{P\to A}\frac{\vec{f}(P)-\vec{f}(A)}{AP}$, provided the limit exists.

13.6.1 Directional Derivatives along Coordinate Axes

Let $A(x, y, z)$ be a point and let $P(x+\ \delta x, y, z)$ be a point on a line drawn through A and parallel to the positive direction of x-axis. Then, $AP=\delta x>0$. Therefore, directional derivative of a scalar-point function at A along AP is defined as

$$\lim_{P\to A}\frac{\phi(P)-\phi(A)}{AP}=\lim_{\delta x\to 0}\frac{\phi(x+\delta x,y,z)-\phi(x,y,z)}{\delta x}$$

$$=\frac{\partial\phi}{\partial x}.$$

Thus, the directional derivative of a scalar-point function ϕ along the x-axis is the *partial derivative of ϕ with respect to x*.

Similarly, directional derivatives of ϕ along y- and z-axis are, respectively, $\frac{\partial\phi}{\partial y}$ and $\frac{\partial\phi}{\partial z}$.

The directional derivatives of a vector-point function $\vec{f}$ along the coordinate axes are similarly $\frac{\partial\vec{f}}{\partial x},\frac{\partial\vec{f}}{\partial y}$, and $\frac{\partial\vec{f}}{\partial z}$, respectively.

Further, if l, m and n are direction cosines of $AP = r$, then the coordinates of P are $x + lr$, $y + mr$, and $z + nr$ and so, the directional derivative of the scalar-point function ϕ along AP becomes

$$\lim_{P\to A}\frac{\phi(P)-\phi(A)}{AP}$$

$$=\lim_{r\to 0}\frac{\phi(x+lr,y+mr,z+nr)-\phi(x,y,z)}{r}$$

$$=\lim_{r\to 0}\frac{\phi(x,y,z)+\left(lr\frac{\partial\phi}{\partial x}+mr\frac{\partial\phi}{\partial y}+nr\frac{\partial\phi}{\partial z}\right)+\ldots-\phi(x,y,z)}{r}$$

$$=l\frac{\partial\phi}{\partial x}+m\frac{\partial\phi}{\partial y}+n\frac{\partial\phi}{\partial z},$$

by the application of Taylor's Theorem for function of several variables, under the assumption that ϕ has a continuous first-order partial derivatives.

Similarly, the directional derivative of a vector-point function $\vec{f}$ along any line with direction cosines l, m and n is $l\frac{\partial\vec{f}}{\partial x}+m\frac{\partial\vec{f}}{\partial y}+n\frac{\partial\vec{f}}{\partial z}$.

Theorem 13.8

The directional derivative of a scalar-point function ϕ along the direction of unit vector $\vec{a}$ is $\nabla\phi.\hat{a}$.

Proof: The unit vector $\hat{a}$ along a line whose direction cosines are l, m and n is

$$\hat{a}=l\hat{i}+m\hat{j}+n\hat{k}.$$

Therefore,

$$\nabla\phi.\hat{a}=\left(\hat{i}\frac{\partial\phi}{\partial x}+\hat{j}\frac{\partial\phi}{\partial y}+\hat{k}\frac{\partial\phi}{\partial z}\right)\left(l\hat{i}+m\hat{j}+n\hat{k}\right)$$

$$=l\frac{\partial\phi}{\partial x}+m\frac{\partial\phi}{\partial y}+n\frac{\partial\phi}{\partial z},$$

which is nothing but directional derivative of ϕ in the direction of the unit vector $\hat{a}$.

Theorem 13.9

Grad ϕ is a vector in the direction of which the maximum value of the directional derivative of ϕ occurs. Hence, the directional derivative is maximum along the normal to the surface and the maximum value is

$$|grad\phi|=|\nabla\phi|.$$

Proof: Recall that $\vec{a}.\vec{b}=|a||b|\cos\theta$, where θ is the angle between the vectors $\vec{a}$ and $\vec{b}$. Since $(grad\phi).\hat{a}$ gives the directional derivative in the direction of unit vector $\hat{a}$, that is, the rate of change of ϕ (x, y, z) in the direction of the unit vector $\hat{a}$, it follows that the rate of change of ϕ (x, y, z) is zero along directions perpendicular to grad ϕ (since $\cos\frac{\pi}{2}=0$) and is maximum along the direction parallel to grad ϕ. Since grad ϕ acts along the normal direction to the level surface of ϕ (x, y, z), the directional derivative is maximum along the normal to the surface. The maximum value is $|grad\phi|=|\nabla\phi|$.

EXAMPLE 13.32

If $\vec{r}=x\hat{i}+y\hat{j}+z\hat{k}$ and $|\vec{r}|=r$,, show that (i) $\nabla f(r)=f'(r)\nabla r$ and (ii) $\nabla f(r)\times\vec{r}=\vec{0}$.

Solution. (i) By the definition of gradient,

$$\nabla f(r)=\hat{i}\frac{\partial}{\partial x}f(r)+\hat{j}\frac{\partial}{\partial y}f(r)+\hat{k}\frac{\partial}{\partial z}f(r)$$

$$=\hat{i}f'(r)\frac{\partial r}{\partial x}+\hat{j}f'(r)\frac{\partial r}{\partial y}+\hat{k}f'(r)\frac{\partial r}{\partial z}$$

$$=f'(r)\left(\hat{i}\frac{\partial r}{\partial x}+\hat{j}\frac{\partial r}{\partial y}+\hat{k}\frac{\partial r}{\partial z}\right)=f'(r)\nabla r.$$

(ii) As in part (i), we have

$$\nabla f(r)=f'(r)\left(\hat{i}\frac{\partial r}{\partial x}+\hat{j}\frac{\partial r}{\partial y}+\hat{k}\frac{\partial r}{\partial z}\right).$$

Since $r=|\vec{r}|=\sqrt{x^2+y^2+z^2}$, we have

$$\frac{\partial r}{\partial x}=\frac{1}{2(x^2+y^2+z^2)^{\frac{1}{2}}}(2x)=\frac{x}{\sqrt{x^2+y^2+z^2}}=\frac{x}{r},$$

and similarly, $\frac{\partial r}{\partial x}=\frac{y}{r}$ and $\frac{\partial r}{\partial x}=\frac{z}{r}$. Therefore,

$$\nabla f(r)=f'(r)\left(\hat{i}\frac{x}{r}+\hat{j}\frac{y}{r}+\hat{k}\frac{z}{r}\right)=f'(r)\frac{\vec{r}}{r}.$$

Hence, $\nabla f(r)\times\vec{r}=\frac{f'(r)}{r}(\vec{r}\times\vec{r})=\vec{0}.$

EXAMPLE 13.33

If $f(\vec{r})=x^2yz^2$, find ∇f (at the point (1,2,3). Hence calculate

(i) the directional derivative of $f(\vec{r})$ at (1, 2, 3) in the direction of the vector (–2, 3, 6).

(ii) the maximum rate of change of the function at (1,2,3) and its direction.

Solution. Since $\frac{\partial f}{\partial x}=2xyz^2$, $\frac{\partial f}{\partial y}=x^2z^2$, and $\frac{\partial f}{\partial z}=2x^2yz$, we have

$$\nabla f = 2xyz^2\hat{i} + x^2z^2\hat{j} + 2x^2yz\hat{k}.$$

Therefore, at the point (1, 2, 3),

$$\text{grad } f = \nabla f = 36\hat{i} + 9\hat{j} + 12\hat{k}.$$

(i) The unit vector $\hat{a}$ in the direction of the vector (–2, 3, –6) is

$$\frac{-2\hat{i} + 3\hat{j} - 6\hat{k}}{\sqrt{4+9+36}} = \frac{-2}{7}\hat{i} + \frac{3}{7}\hat{j} - \frac{6}{7}\hat{k}.$$

Therefore, the directional derivative at (1, 2, 3) in the direction of the vector (–2, 3, –6) is

$$\nabla f.\hat{a} = (36\hat{i} + 9\hat{j} + 12\hat{k}).(-\frac{2}{7}\hat{i} + \frac{3}{7}\hat{j} - \frac{6}{7}\hat{k})$$

$$= -\frac{72}{7} + \frac{27}{7} - \frac{72}{7} = -\frac{117}{7}.$$

(ii) The maximum rate of change of the function at (1, 2, 3) occurs along the direction parallel to ∇f at (1, 2, 3), that is, parallel to $36\hat{i} + 9\hat{j} + 12\hat{k}$. The unit vector in that direction is

$$\frac{36\hat{i} + 9\hat{j} + 12\hat{k}}{\sqrt{1296+144+81}} = \frac{36\hat{i} + 9\hat{j} + 12\hat{k}}{39} = \frac{12\hat{i} + 3\hat{j} + 4\hat{k}}{13}.$$

The maximum rate of change of $f(\vec{r})$ is $|\text{grad } f| = \sqrt{1296+144+81} = 39.$

EXAMPLE 13.34

If $\vec{r}$ is the usual position vector $\vec{r} = x\hat{i} + y\hat{j} + z\hat{k}$ with $|\vec{r}| = r$, evaluate

(*i*) ∇r, (*ii*) $\nabla\left(\frac{1}{r}\right)$, (*iii*) ∇r^n, and (iv) $\nabla\left(\frac{1}{r^2}\right)$.

Solution. Since $|\vec{r}| = r = \sqrt{x^2 + y^2 + z^2}$, we have $\frac{\partial r}{\partial x} = \frac{x}{r}$, $\frac{\partial r}{\partial y} = \frac{y}{r}$, and $\frac{\partial r}{\partial z} = \frac{z}{r}$.

Therefore,

(i) $$\nabla(r) = \left(\hat{i}\frac{\partial}{\partial x} + \hat{j}\frac{\partial}{\partial y} + \hat{k}\frac{\partial}{\partial z}\right)(r) = \hat{i}\frac{\partial r}{\partial x} + \hat{j}\frac{\partial r}{\partial y} + \hat{k}\frac{\partial r}{\partial z}$$

$$= \hat{i}\frac{x}{r} + \hat{j}\frac{y}{r} + \hat{k}\frac{z}{r} = \frac{x\hat{i} + y\hat{j} + z\hat{k}}{r} = \frac{\hat{r}}{r}.$$

(ii) $$\nabla(\frac{1}{r}) = (\hat{i}\frac{\partial}{\partial x} + \hat{j}\frac{\partial}{\partial y} + \hat{k}\frac{\partial}{\partial z})(\frac{1}{r})$$

$$= \hat{i}\frac{\partial}{\partial x}\left(\frac{1}{r}\right) + \hat{j}\frac{\partial}{\partial y}\left(\frac{1}{r}\right) + \hat{k}\frac{\partial}{\partial z}\left(\frac{1}{r}\right)$$

$$= \hat{i}\left(-\frac{1}{r^2}\frac{\partial r}{\partial x}\right) + \hat{j}\left(-\frac{1}{r^2}\frac{\partial r}{\partial y}\right) + \hat{k}\left(-\frac{1}{r^2}\frac{\partial r}{\partial z}\right)$$

$$= \hat{i}\left(-\frac{1}{r^2}.\frac{x}{r}\right) + \hat{j}\left(-\frac{1}{r^2}.\frac{y}{r}\right) + \hat{k}\left(-\frac{1}{r^2}.\frac{z}{r}\right)$$

$$= -\frac{1}{r^3}\left(x\hat{i} + y\hat{j} + z\hat{k}\right) = -\frac{\vec{r}}{r^3}.$$

(iii) $$\nabla r^n = \left(\hat{i}\frac{\partial}{\partial x} + \hat{j}\frac{\partial}{\partial y} + \hat{k}\frac{\partial}{\partial z}\right)(r^n)$$

$$= \hat{i}\frac{\partial}{\partial x}(r^n) + \hat{j}\frac{\partial}{\partial y}(r^n) + \hat{k}\frac{\partial}{\partial z}(r^n)$$

$$= \hat{i}\left(nr^{n-1}\frac{\partial r}{\partial x}\right) + \hat{j}\left(nr^{n-1}\frac{\partial r}{\partial y}\right) + \hat{k}\left(nr^{n-1}\frac{\partial r}{\partial z}\right)$$

$$= \hat{i}\left(nr^{n-1}.\frac{x}{r}\right) + \hat{j}\left(nr^{n-1}.\frac{y}{r}\right) + \hat{k}\left(nr^{n-1}.\frac{z}{r}\right)$$

$$= nr^{n-2}(x\hat{i} + y\hat{j} + z\hat{k}) = nr^{n-2}\vec{r}.$$

(iv) $$\nabla\left(\frac{1}{r^2}\right) = \left(\hat{i}\frac{\partial}{\partial x} + \hat{j}\frac{\partial}{\partial y} + \hat{k}\frac{\partial}{\partial z}\right)\left(\frac{1}{r^2}\right)$$

$$= \hat{i}\frac{\partial}{\partial x}\left(\frac{1}{r^2}\right) + \hat{j}\frac{\partial}{\partial y}\left(\frac{1}{r^2}\right) + \hat{k}\frac{\partial}{\partial z}\left(\frac{1}{r^2}\right)$$

$$= \hat{i}\left(-\frac{2}{r^3}\frac{\partial r}{\partial x}\right) + \hat{j}\left(-\frac{2}{r^3}\frac{\partial r}{\partial y}\right) + \hat{k}\left(-\frac{2}{r^3}\frac{\partial r}{\partial z}\right)$$

$$= \hat{i}\left(-\frac{2}{r^3}.\frac{x}{r}\right) + \hat{j}\left(-\frac{2}{r^3}.\frac{y}{r}\right) + \hat{k}\left(-\frac{2}{r^3}.\frac{z}{r}\right)$$

$$= -\frac{2}{r^4}(x\hat{i} + y\hat{j} + z\hat{k}) = -\frac{2\vec{r}}{r^4}.$$

EXAMPLE 13.35

Find the directional derivative of $f(x,y,z) = xy^2 + yz^3$ at the point (2, –1, 1) in the direction of the vector $\hat{i}+2\hat{j}+2\hat{k}$.

Solution. We have

$$\nabla f = \left(\hat{i}\frac{\partial}{\partial x}+\hat{j}\frac{\partial}{\partial y}+\hat{k}\frac{\partial}{\partial z}\right)(xy^2+yz^3)$$

$$= y^2\hat{i}+(2xy+z^3)\hat{j}+(3yz^2)\hat{k}$$

$$= \hat{i}-3\hat{j}-3\hat{k} \text{ at the point } (2, -1, 1).$$

The unit vector in the direction of the vector $\hat{i}+2\hat{j}+2\hat{k}$ is

$$\hat{a} = \frac{\hat{i}+2\hat{j}+2\hat{k}}{\sqrt{1+4+4}} = \frac{1}{3}(\hat{i}+2\hat{j}+2\hat{k}).$$

Therefore, the directional derivative of f at (2, –1, 1) in the direction of $\hat{i}+2\hat{j}+2\hat{k}$ is

$$\nabla\phi.\hat{a} = \left(\hat{i}-3\hat{j}-3\hat{k}\right).\frac{1}{3}\left(\hat{i}+2\hat{j}+2\hat{k}\right)$$

$$= \frac{1}{3}(1-6-6) = -\frac{11}{3}.$$

EXAMPLE 13.36

Find the directional derivative of ϕ (x, y, z) $= xy^2 + yz^3$ at the point (2, –1, 1) in the direction of the normal to the surface $x \log z - y^2 + 4 = 0$ at (2, –1, 1).

Solution. We have $\phi = xy^2+yz^3$, $\frac{\partial\phi}{\partial x} = y^2$, $\frac{\partial\phi}{\partial y} = 2xy+z^3$, and $\frac{\partial\phi}{\partial z} = 3yz^2$. Therefore, as in Example 13.16,

$$\nabla\phi = \hat{i}-3\hat{j}-3\hat{k} \text{ at } (2, -1, 1).$$

On the other hand, $\nabla(x \log z - y^2 + 4)$

$$= \left(\hat{i}\frac{\partial}{\partial x}+\hat{j}\frac{\partial}{\partial y}+\hat{k}\frac{\partial}{\partial z}\right)(x \log z - y^2+4)$$

$$= \hat{i}\frac{\partial}{\partial x}(x \log z - y^2+4)+\hat{j}\frac{\partial}{\partial y}(x \log z - y^2+4)$$

$$+\hat{k}\frac{\partial}{\partial z}(x \log z - y^2+4)$$

$$= \log z\ \hat{i} - 2y\hat{j} + \frac{x}{Z}\hat{k} = -4\hat{j}-\hat{k} \text{ at } (-1, 2, 1).$$

But $\nabla\left(x\log z = y^2+4\right)$ is normal to the surface $x\log z = y^2+4$. Unit vector along $\nabla\left(x\log z = y^2+4\right)$ is

$$\hat{a} = \frac{-4\hat{j}-\hat{k}}{\sqrt{16+1}} = \frac{-4\hat{j}-\hat{k}}{\sqrt{17}}.$$

Therefore, the directional derivative of ϕ at (2, –1, 1) in the direction of the normal to the surface $x\log z = y^2+4$ at (–1, 2, 1) is

$$\nabla\phi.\hat{a} = (\hat{i}-3\hat{j}-3\hat{k})\left(\frac{-4\hat{j}-\hat{k}}{\sqrt{17}}\right) = \frac{12+3}{\sqrt{17}} = \frac{15}{\sqrt{17}}.$$

EXAMPLE 13.37

Find the angle between the surfaces $x^2+y^2+z^2=9$ and $z = x^2+y^2-3$ at the point (2, –1, 2).

Solution. Let $\phi(x,y,z) = x^2+y^2+z^2-9$ and $\psi(x,y,z) = x^2+y^2-3-z$. Then, the angle between the surfaces at the given point (2,–1,2) is the angle between the normal to the surfaces at that point.

Also $\nabla\phi$ and $\nabla\psi$ are along the normal to ϕ and ψ, respectively. But,

$$\nabla\phi = \left(\hat{i}\frac{\partial}{\partial x}+\hat{j}\frac{\partial}{\partial y}+\hat{k}\frac{\partial}{\partial z}\right)(x^2+y^2+z^2-9).$$

$$= 2x\hat{i}+2y\hat{j}+2z\hat{k}$$

$$= 4\hat{i}-2\hat{j}+4\hat{k} \text{ at the point } (2,-1,2)$$

and

$$\nabla\psi = \left(\hat{i}\frac{\partial}{\partial x}+\hat{j}\frac{\partial}{\partial y}+\hat{k}\frac{\partial}{\partial z}\right)(x^2+y^2-3-z)$$

$$= 2x\hat{i}+2y\hat{j}-\hat{k}$$

$$= 4\hat{i}-2\hat{j}-\hat{k} \text{ at the point } (2,-1,2).$$

If θ is the angle between $\nabla\phi$ and $\nabla\psi$, then

$$\cos\theta = \frac{\nabla\phi.\nabla\psi}{|\nabla\phi||\nabla\psi|} = \frac{(4\hat{i}-2\hat{j}+4\hat{k}).(4\hat{i}-2\hat{j}-\hat{k})}{\sqrt{16+4+16}\sqrt{16+4+1}}$$

$$= \frac{16+4-4}{6\sqrt{21}} = \frac{8}{3\sqrt{21}}.$$

Hence, the required angle is $\theta = \cos^{-1}\left(\frac{8}{3\sqrt{21}}\right)$.

EXAMPLE 13.38

In what direction from (3, 1,–2) is the directional derivative of $\phi = x^2y^2z^4$ maximum and what is its magnitude?

Solution. The directional derivative at a given point of a given surface ϕ is maximum along the normal to the surface and grad ϕ acts along the normal. Therefore, the directional derivative is maximum along $\nabla\phi$. We have

$$\nabla\phi = \left(\hat{i}\frac{\partial}{\partial x} + \hat{j}\frac{\partial}{\partial y} + \hat{k}\frac{\partial}{\partial z}\right)(x^2y^2z^4)$$

$$= (2xy^2z^4)\hat{i} + (2x^2yz^4)\hat{j} + (4x^2y^2z^3)\hat{k}$$

$$= 96\hat{i} + 288\hat{j} - 288\hat{k} \text{ at the point } (3, 1, -1).$$

Thus, the directional derivative is maximum in the direction of $96\hat{i} + 288\hat{j} - 288\hat{k}$. The magnitude of the maximum directional derivative is $|\nabla\phi| = 96\sqrt{1+9+9} = 96\sqrt{19}$.

EXAMPLE 13.39

Find the angles between the normal to the surface $xy = z^2$ at the points (4, 1, 2) and (3, 3, –3).

Solution. Let $\phi(x, y, z) = xy - z^2$. Since $\nabla\phi$ is along the normal, it is sufficient to find angle between $\nabla\phi$ at (4, 1, 2) and $\nabla\phi$ at (3, 3, –3). Now,

$$\nabla\phi = \left(\hat{i}\frac{\partial}{\partial x} + \hat{j}\frac{\partial}{\partial y} + \hat{k}\frac{\partial}{\partial z}\right)(xy - z^2)$$

$$= y\hat{i} + x\hat{j} - 2z\hat{k}.$$

Therefore,

$\nabla\phi$ at (4, 1, 2) is $\hat{i} + 4\hat{j} - 4\hat{k}$ and
$\nabla\phi$ at (3, 3, –3) is $3\hat{i} + 3\hat{j} + 6\hat{k}$.

Hence, the required angle θ is the angle between $\hat{i} + 4\hat{j} - 4\hat{k}$ and $3\hat{i} + 3\hat{j} + 6\hat{k}$. Therefore,

$$\cos\theta = \frac{(\hat{i} + 4\hat{j} - 4\hat{k}).(3\hat{i} + 3\hat{j} + 6\hat{k})}{|\hat{i} + 4\hat{j} - 4\hat{k}||3\hat{i} + 3\hat{j} + 6\hat{k}|} = \frac{-9}{\sqrt{33}\sqrt{54}}$$

$$= \frac{-1}{\sqrt{22}}.$$

Hence,

$$\theta = \cos^{-1}\left(-\frac{1}{\sqrt{22}}\right).$$

EXAMPLE 13.40

Find the constants a and b so that the surface $ax^2 - byz = (a+2)x$ is orthogonal to the surface $4x^2y + z^3 = 4$ at the point (1, −1, 2).

Solution. The two given surfaces will be orthogonal if the angle between the normal to the surfaces at the point (1, −1, 2) is $\frac{\pi}{2}$. Since $\nabla\phi$ acts along the normal, it is sufficient to find $\nabla\phi$ and $\nabla\psi$ at (4, 1,2) where $\phi = ax^2 - byz - (a+2)x$ and $\psi = 4x^2y + z^3 - 4$. We have

$$\nabla\phi = \left(\hat{i}\frac{\partial}{\partial x} + \hat{j}\frac{\partial}{\partial y} + \hat{k}\frac{\partial}{\partial z}\right)(ax^2 - byz - (a+2)x)$$

$$= \hat{i}(2ax - a - 2) + \hat{j}(-bz) + \hat{k}(-by)$$

$$= (a-2)\hat{i} - 2b\hat{j} + b\hat{k} \text{ at } (1, -1, 2),$$

and

$$\nabla\psi = \left(\hat{i}\frac{\partial}{\partial x} + \hat{j}\frac{\partial}{\partial y} + \hat{k}\frac{\partial}{\partial z}\right)(4x^2y + z^3 - 4)$$

$$= \hat{i}(8xy) + \hat{j}(4x^2) + \hat{k}(3z^2)$$

$$= -8\hat{i} + 4\hat{j} + 12\hat{k} \text{ at } (1, -1, 2).$$

Since $\theta = \dfrac{\pi}{2}$, we have

$$\cos\frac{\pi}{2} = \frac{((a-2)\hat{i} - 2b\hat{j} + b\hat{k}).(-8\hat{i} + 4\hat{j} + 12\hat{k})}{|(a-2)\hat{i} - 2b\hat{j} + b\hat{k}||-8\hat{i} + 4\hat{j} + 12\hat{k}|}.$$

Hence,

$$((a-2)\hat{i} - 2b\hat{j} + b\hat{k}).(-8\hat{i} + 4\hat{j} + 12\hat{k}) = 0$$

or

$$(a-2)(-8) - 8b + 12b = 0$$

or

$$-8a + 4b = -16. \qquad (1)$$

Since the points (1, −1, 2) lie on both surfaces ϕ and ψ, we have from the surface ϕ,

$$a + 2b = a + 2 \text{ or } b = 1. \qquad (2)$$

Putting the value of b from (2) in (1), we get

$$-8a = -16 - 4 \text{ or } a = \frac{20}{8} = \frac{5}{2}.$$

Hence, $a=\frac{5}{2}$ and $b=1$.

EXAMPLE 13.41

A paraboloid of revolution has the equation $2z=x^2+y^2$. Find the equation of the normal and the tangent plane to the surface at the point (1, 3, 5).

Solution. Let $\phi=x^2+y^2-2z$. Then, $\nabla\phi$ gives the vector normal to the surface. Thus, the normal vector to the surface is

$$\nabla\phi=\left(\hat{i}\frac{\partial}{\partial x}+\hat{j}\frac{\partial}{\partial y}+\hat{k}\frac{\partial}{\partial z}\right)(x^2+y^2-2z)$$

$$=2x\hat{i}+2y\hat{j}-2\hat{k}$$

$$=2\hat{i}+6\hat{j}-2\hat{k}\text{, at the point (1, 3, 5).}$$

Therefore, the unit normal vector at the point (1,3,5) is

$$\hat{a}=\frac{2\hat{i}+6\hat{j}-2\hat{k}}{\sqrt{4+36+4}}=\frac{\hat{i}+3\hat{j}-\hat{k}}{\sqrt{11}}.$$

The equation of the line through the point (1, 3, 5) in the direction of this normal vector is

$$\frac{x-1}{1}=\frac{y-3}{3}=\frac{z-5}{-1}.$$

Therefore, the equation of the tangent plane to the surface at the point (1,3,5) is

$$1\,(x-1)+3(y-3)+(-1)(z-5)=0 \text{ or}$$

$$x+3y-z=5.$$

EXAMPLE 13.42

Find the angle between the tangent planes to the surfaces $x\log z=y^2-1$ and $x^2y=2-z$ at the point (1, 1, 1).

Solution. The required angle will be the angle between the vectors normal to the given surfaces at the given point. The normal vectors to the surfaces $\phi=x\log z-y^2+1$ and $\psi=x^2y-2+z$ are given by

$$\nabla\phi=\left(\hat{i}\frac{\partial}{\partial x}+\hat{j}\frac{\partial}{\partial y}+\hat{k}\frac{\partial}{\partial z}\right)(x\log z-y^2+1)$$

$$=\log z\,\hat{i}-2y\hat{j}+\frac{x}{z}\hat{k}$$

$$=-2\hat{j}+\hat{k}\text{ at the point (1, 1, 1)}$$

and

$$\nabla\psi=\left(\hat{i}\frac{\partial}{\partial x}+\hat{j}\frac{\partial}{\partial y}+\hat{k}\frac{\partial}{\partial z}\right)(x^2y-2+z)$$

$$=2xy\hat{i}+x^2\hat{j}+\hat{k}$$

$$=2\hat{i}+\hat{j}+\hat{k}\text{ at the point (1, 1, 1).}$$

Therefore, the required angle is given by

$$\cos\theta=\frac{(-2\hat{j}+\hat{k}).(2\hat{i}+\hat{j}+\hat{k})}{\sqrt{5}.\sqrt{6}}=\frac{-2+1}{\sqrt{30}}=\frac{-1}{\sqrt{30}}.$$

Hence,

$$\theta=\cos^{-1}\left(\frac{-1}{\sqrt{30}}\right).$$

13.7 DIVERGENCE OF A VECTOR-POINT FUNCTION

If we want to consider the rate of change of a vector-point function $\vec{f}$, there are two ways of operating the vector operator ∇ to the vector $\vec{f}$. Thus, we have two cases to consider, namely, $\nabla\cdot\vec{f}$ and $\nabla\times\vec{f}$. These two cases lead us to the two concepts called *Divergence of a Vector Function* and *curl of a Vector Function*. If we consider a vector field as a fluid flow, then at every point in the flow, we need to measure the *rate of flow* of the fluid from that point and the *amount of spin* possessed by the particles of the fluid at that point. The above two concepts provide respectively, the two measures called divergence of $\vec{f}$ and curl of $\vec{f}$.

Let $\vec{f}=f_1\hat{i}+f_2\hat{j}+f_3\hat{k}$ be a vector function, where f_1, f_2, and f_3 are scalar-point functions, which is defined and differentiable at each point of the region of space. Then, the *divergence* of *f*, denoted by $\nabla.\vec{f}$ or div $\vec{f}$, is a scalar given by

$$\nabla\cdot\vec{f}=\left(\hat{i}\frac{\partial}{\partial x}+\hat{j}\frac{\partial}{\partial y}+\hat{k}\frac{\partial}{\partial z}\right).(f_1\hat{i}+f_2\hat{j}+f_3\hat{k})$$

$$=\frac{\partial f_1}{\partial x}+\frac{\partial f_2}{\partial y}+\frac{\partial f_3}{\partial z}.$$

The vector $\vec{f}$ is called *Solenoidal* if $\nabla\cdot\vec{f}=0$.

13.8 PHYSICAL INTERPRETATION OF DIVERGENCE

Consider the steady motion of the fluid having velocity $\vec{v}=v_x\hat{i}+v_y\hat{j}+v_z\hat{k}$ at a point $P(x,y,z)$ Consider a small parallelopiped with edges δx,

δy, and δz parallel to the axes, with one of its comer at $P(x,y,z)$. The mass of the fluid entering through the face PQRS per unit time is $v_y \delta x \delta z$ and the mass of the fluid that flows out through the opposite face ABCD is $v_{y+\delta y}$ $\delta x \delta z$. Therefore, the change in the mass of fluid flowing across these two faces is equal to

$$v_{y+\delta y}\,\delta x\,\delta z - v_y\,\delta x\,\delta z = \left(v_y + \frac{\partial v_y}{\partial y}.\delta y\right)\delta x\,\delta z$$

$$-v_y \delta x \delta z \quad = \frac{\partial v_y}{\partial y}\delta y \delta x \delta z .$$

Similarly, the changes in the mass of the fluid for the other two pairs of faces are

$$\frac{\partial v_X}{\partial x}\delta x \delta y \delta z \text{ and } \frac{\partial v_Z}{\partial z}\delta x \delta y \delta z .$$

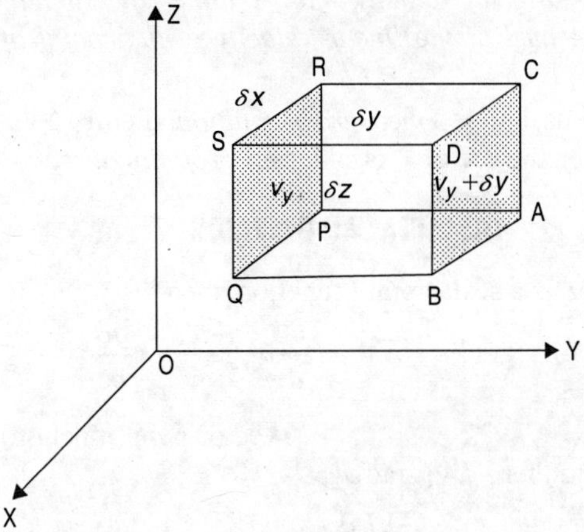

Therefore, the total change in the mass of the fluid inside the parallelopiped per unit time is equal to

$$\left(\frac{\partial v_x}{\partial x}+\frac{\partial v_y}{\partial y}+\frac{\partial v_Z}{\partial z}\right)\delta x \delta y \delta z.$$

Hence, the rate of change of the mass of the fluid per unit time per unit volume is

$$\frac{\partial v_x}{\partial x}+\frac{\partial v_y}{\partial y}+\frac{\partial v_Z}{\partial z} = \nabla \cdot \vec{v},$$

by the definition of divergence. Hence, div $\vec{v}$ gives the rate at which the fluid (the vector field) is flowing away at a point of the fluid.

EXAMPLE 13.43

Find div $\vec{v}$, where $\vec{v} = 3x^2 y\hat{i} + z\hat{j} + x^2\hat{k}$.

Solution. We know that

$$div\vec{v} = \frac{\partial v_1}{\partial x}+\frac{\partial v_2}{\partial y}+\frac{\partial v_3}{\partial z}.$$

Here, $v_1 = 3x^2 y$, $v_2 = z$, and $v_3 = x^2$.
Therefore,

$$\text{div } \vec{v} = 6xy.$$

EXAMPLE 13.44

Find the value of the constant λ such that the vector field defined by

$$\vec{f} = (2x^2y^2 + z^2)\hat{i} + (3xy^3 - x^2 z)\hat{j} + (\lambda xy^2 z + xy)\hat{k} \text{ is solenoidal.}$$

Solution. We have
$f_1 = 2x^2y^2 + z^2$, $f_2 = 3xy^3 - x^2 z$, and
$f_3 = \lambda xy^2 z + xy$.
Therefore,

$$div\, \vec{f} = \frac{\partial f_1}{\partial x}+\frac{\partial f_2}{\partial y}+\frac{\partial f_3}{\partial z} = 4xy^2 + 9xy^2 + \lambda xy^2 .$$

The vector field shall be a solenoidal if $div\ \vec{f} = 0$. So, we must have

$4xy^2 + 9xy^2 + \lambda xy^2 = 0$,

which yields $\lambda = -13$.

EXAMPLE 13.45

Find $div\, \vec{f}$, where $\vec{f} = grad(x^3 + y^3 + z^3 - 3xyz)$.

Solution. We have

$$\vec{f} = \nabla(x^3 + y^3 + z^3 - 3xyz)$$

$$= \left(\hat{i}\frac{\partial}{\partial x} + \hat{j}\frac{\partial}{\partial y} + \hat{k}\frac{\partial}{\partial z}\right)(x^3 + y^3 + z^3 - 3xyz)$$

$$= (3x^2 - 3yz)\hat{i} + (3y^2 - 3xz)\hat{j} + (3z^2 - 3xy)\hat{k}$$

$$= f_1\hat{i} + f_2\hat{j} + f_3\hat{k}, \text{ say.}$$

Then,

$$div\vec{f} = \frac{\partial f_1}{\partial x}+\frac{\partial f_2}{\partial y}+\frac{\partial f_3}{\partial z}$$

$$= 6x + 6y + 6z = 6(x + y + z).$$

EXAMPLE 13.46

Find div $(3x^2\hat{i}+5xy^2\hat{j}+xyz^3\hat{k})$ at the point. (1, 2, 3).

Solution. Let

$$\vec{f}=3x^2\hat{i}+5xy^2\hat{j}+xyz^3\hat{k}$$
$$=f_1\hat{i}+f_2\hat{j}+f_3\hat{k}, \text{ say.}$$

Then,

$$div\ \vec{f}=\frac{\partial f_1}{\partial x}+\frac{\partial f_2}{\partial y}+\frac{\partial f_3}{\partial z}=6x+10xy+3xyz^2$$
$$=6+20+54=80 \text{ at } (1, 2, 3).$$

13.9 CURL OF A VECTOR-POINT FUNCTION

Let $\vec{f}=f_1\hat{i}+f_2\hat{j}+f_3\hat{k}$ be a vector-point function, where f_1, f_2, and f_3 are scalar-point functions. If f is defined and differentiable at each point (x, y, z) of the region of space, then the *curl* (or *rotation*) of $\vec{f}$, denoted, by curl $\vec{f}$ $\nabla\times\vec{f}$, or rot $\vec{f}$ is defined by Curl $\vec{f}=\nabla\times\vec{f}$

$$=\left(\hat{i}\frac{\partial}{\partial x}+\hat{j}\frac{\partial}{\partial y}+\hat{k}\frac{\partial}{\partial z}\right)\times(f_1\hat{i}+f_2\hat{j}+f_3\hat{k})$$

$$=\begin{vmatrix}\hat{i} & \hat{j} & \hat{k}\\ \frac{\partial}{\partial x} & \frac{\partial}{\partial y} & \frac{\partial}{\partial z}\\ f_1 & f_2 & f_3\end{vmatrix}$$

$$=\left(\frac{\partial f_3}{\partial y}-\frac{\partial f_2}{\partial z}\right)\hat{i}+\left(\frac{\partial f_1}{\partial z}-\frac{\partial f_3}{\partial x}\right)\hat{j}$$
$$+\left(\frac{\partial f_2}{\partial x}-\frac{\partial f_1}{\partial y}\right)\hat{k}.$$

Obviously, *curl $\vec{f}$ is a vector-point function.*

13.10 PHYSICAL INTERPRETATION OF CURL

Consider a rigid body rotating about a fixed axis through the origin with angular velocity $\vec{\omega}=\omega_1\hat{i}+\omega_2\hat{j}+\omega_3\hat{k}$. Let $\vec{r}=x\hat{i}+y\hat{j}+z\hat{k}$ be the position vector of any point $P(x, y, z)$ on the body. Then, the velocity $\vec{v}$ of P is given by

$$\vec{v}=\vec{\omega}\times\vec{r}=\begin{vmatrix}\hat{i} & \hat{j} & \hat{k}\\ \omega_1 & \omega_2 & \omega_3\\ x & y & z\end{vmatrix}$$

$$=(\omega_2 z-\omega_3 y)\hat{i}+(\omega_3 x-\omega_1 z)\hat{j}+(\omega_1 y-\omega_2 x)\hat{k}.$$

Therefore,

Curl $\vec{v}=\nabla\times\vec{v}$

$$=\begin{vmatrix}\hat{i} & \hat{j} & \hat{k}\\ \frac{\partial}{\partial x} & \frac{\partial}{\partial y} & \frac{\partial}{\partial z}\\ \omega_2 z-\omega_3 y & \omega_3 x-\omega_1 z & \omega_1 y-\omega_2 x\end{vmatrix}$$

$=2(\omega_1\hat{i}+\omega_2\hat{j}+\omega_3\hat{k})$, since ω_1, ω_2, and ω_3 are constants

$=2\vec{\omega}$.

Hence, $\vec{\omega}=\frac{1}{2}$ curl $\vec{v}$. It follows, therefore, that the *angular velocity at any point is equal to half the curl of the linear velocity at that point of the body.*

Thus, curl is a *measure of rotation.* If curl $\vec{v}=0$, then the vector $\vec{v}$ is called an *irrotational vector*.

13.11 THE LAPLACIAN OPERATOR ∇^2

If ϕ is a scalar-point function, then

$$\text{grad }\phi=\nabla\phi=\hat{i}\frac{\partial\phi}{\partial x}+\hat{j}\frac{\partial\phi}{\partial y}+\hat{k}\frac{\partial\phi}{\partial z}$$

(vector-point function)

and then, $\text{div}[\text{grad }\phi]=\nabla\cdot\nabla\phi$

$$=\frac{\partial}{\partial x}\left(\frac{\partial\phi}{\partial x}\right)+\frac{\partial}{\partial y}\left(\frac{\partial\phi}{\partial y}\right)+\frac{\partial}{\partial z}\left(\frac{\partial\phi}{\partial z}\right)$$
$$=\frac{\partial^2\phi}{\partial x^2}+\frac{\partial^2\phi}{\partial y^2}+\frac{\partial^2\phi}{\partial z^2}$$
$$=\left(\frac{\partial^2}{\partial x^2}+\frac{\partial^2}{\partial y^2}+\frac{\partial^2}{\partial z^2}\right)\phi=\nabla^2\phi,$$

where $\nabla^2=\frac{\partial^2}{\partial x^2}+\frac{\partial^2}{\partial y^2}+\frac{\partial^2}{\partial z^2}$ is called *Laplacian operator*.

A scalar-point function possessing second-order continuous partial derivatives and satisfying the *Laplacian equation* $\nabla^2\phi=0$ is called a *harmonic function*.

EXAMPLE 13.47

Find curl $\vec{F}$, where $\vec{F} = grad(x^3 + y^3 + z^3 - 3xyz)$.

Solution. We have

$$\vec{F} = \text{grad}(x^3 + y^3 + z^3 - 3xyz)$$

$$= \left(\hat{i}\frac{\partial}{\partial x} + \hat{j}\frac{\partial}{\partial y} + \hat{k}\frac{\partial}{\partial z}\right)(x^3 + y^3 + z^3 - 3xyz)$$

$$= (3x^2 - 3yz)\hat{i} + (3y^2 - 3xz)\hat{j} + (3z^2 - 3xy)\hat{k}.$$

Therefore,

$$\text{Curl } \vec{F} = \nabla \times \vec{F} = \left(\hat{i}\frac{\partial}{\partial x} + \hat{j}\frac{\partial}{\partial y} + \hat{j}\frac{\partial}{\partial z}\right) \times \vec{F}$$

$$= \begin{vmatrix} \hat{i} & \hat{j} & \hat{k} \\ \frac{\partial}{\partial x} & \frac{\partial}{\partial y} & \frac{\partial}{\partial z} \\ 3x^2 - 3yz & 3y^2 - 3xz & 3z^2 - 3xy \end{vmatrix}$$

$$= \hat{i}(-3x + 3x) + \hat{j}(-3y + 3y) + \hat{k}(-3z + 3z)$$

$$= \vec{0}.$$

EXAMPLE 13.48

Show that the vector $\vec{v} = (yz)\hat{i} + (zx)\hat{j} + (xy)\hat{k}$ is irrotational.

Solution. It is sufficient to show that the curl $\vec{v} = \vec{0}$. We have

$$\text{Curl } \vec{v} = \nabla \times \vec{v} = \left(\hat{i}\frac{\partial}{\partial x} + \hat{j}\frac{\partial}{\partial y} + \hat{k}\frac{\partial}{\partial z}\right) \times \left[(yz)\hat{i} + (zx)\hat{j} + (xy)\hat{k}\right]$$

$$= \begin{vmatrix} \hat{i} & \hat{j} & \hat{k} \\ \frac{\partial}{\partial x} & \frac{\partial}{\partial y} & \frac{\partial}{\partial z} \\ yz & zx & xy \end{vmatrix}$$

$$= \hat{i}(x - x) + \hat{j}(y - y) + \hat{k}(z - z) = \vec{0}.$$

EXAMPLE 13.49

If $\vec{r} = x\hat{i} + y\hat{j} + z\hat{k}$, show that curl $\vec{r} = \vec{0}$.

Solution. We have

$$\text{curl } \vec{r} = \nabla \times \vec{r} = \begin{vmatrix} \hat{i} & \hat{j} & \hat{k} \\ \frac{\partial}{\partial x} & \frac{\partial}{\partial y} & \frac{\partial}{\partial z} \\ x & y & z \end{vmatrix}$$

$$= \hat{i}\left[\frac{\partial}{\partial y}(z) - \frac{\partial}{\partial z}(y)\right] + \hat{j}\left[\frac{\partial}{\partial z}(x) - \frac{\partial}{\partial x}(z)\right]$$

$$+ \hat{k}\left[\frac{\partial}{\partial x}(y) - \frac{\partial}{\partial y}(x)\right]$$

$$= \hat{i}(0) + \hat{j}(0) + \hat{k}(0) = \vec{0}.$$

EXAMPLE 13.50

Show that curl curl $\vec{f} = \vec{0}$, where $\vec{f} = z\hat{i} + x\hat{j} + y\hat{k}$.

Solution. Let $\vec{f} = z\hat{i} + x\hat{j} + y\hat{k}$. Then

$$\text{curl } \vec{f} = \begin{vmatrix} \hat{i} & \hat{j} & \hat{k} \\ \frac{\partial}{\partial x} & \frac{\partial}{\partial y} & \frac{\partial}{\partial z} \\ z & x & y \end{vmatrix}$$

$$= \left[\frac{\partial y}{\partial y} - \frac{\partial x}{\partial z}\right]\hat{i} + \left[\frac{\partial z}{\partial z} - \frac{\partial y}{\partial x}\right]\hat{j} + \left[\frac{\partial x}{\partial x} - \frac{\partial z}{\partial y}\right]\hat{k}$$

$$= (1 - 0)\hat{i} + (1 - 0)\hat{j} + (1 - 0)\hat{k}$$

$$= \hat{i} + \hat{j} + \hat{k}.$$

Hence,

$$curl\ curl\ \vec{f} = \begin{vmatrix} \hat{i} & \hat{j} & \hat{k} \\ \frac{\partial}{\partial x} & \frac{\partial}{\partial y} & \frac{\partial}{\partial z} \\ 1 & 1 & 1 \end{vmatrix} = \left[\frac{\partial}{\partial y}(1) - \frac{\partial}{\partial z}(1)\right]\hat{i}$$

$$+ \left[\frac{\partial}{\partial z}(1) - \frac{\partial}{\partial x}(1)\right]$$

$$\hat{j} + \left[\frac{\partial}{\partial x}(1) - \frac{\partial}{\partial y}(1)\right]\hat{k}$$

$$= 0\hat{i} + 0\hat{j} + 0\hat{k} = \vec{0}.$$

EXAMPLE 13.51

If all second-order derivatives of ϕ and $\vec{v}$ are continuous, show that

(i) curl $(\text{grad } \phi) = \vec{0}$,

(ii) curl (curl $\vec{v}$) = grad div $\vec{v} - \nabla^2\vec{v}$,

(iii) div (curl $\vec{v}$)=0, and

(iv) grad (div $\vec{v}$)

$$(\text{div } \vec{v}) = \left(\hat{i}\frac{\partial}{\partial x} + \hat{j}\frac{\partial}{\partial y} + \hat{k}\frac{\partial}{\partial z}\right)\left(\frac{\partial v_1}{\partial x} + \frac{\partial v_2}{\partial y} + \frac{\partial v_3}{\partial z}\right).$$

Solution. (i) We have

$$\text{Curl (grad } \phi) = \nabla \times \nabla \phi$$

$$= \left(\hat{i}\frac{\partial}{\partial x} + \hat{j}\frac{\partial}{\partial y} + \hat{k}\frac{\partial}{\partial z}\right) \times \left(\hat{i}\frac{\partial \phi}{\partial x} + \hat{j}\frac{\partial \phi}{\partial y} + \hat{k}\frac{\partial \phi}{\partial z}\right)$$

$$= \begin{vmatrix} \hat{i} & \hat{j} & \hat{k} \\ \frac{\partial}{\partial x} & \frac{\partial}{\partial y} & \frac{\partial}{\partial z} \\ \frac{\partial \phi}{\partial x} & \frac{\partial \phi}{\partial y} & \frac{\partial \phi}{\partial z} \end{vmatrix}$$

$$= \hat{i}\left(\frac{\partial^2 \phi}{\partial y \partial z} - \frac{\partial^2 \phi}{\partial z \partial y}\right) + \hat{j}\left(\frac{\partial^2 \phi}{\partial z \partial x} - \frac{\partial^2 \phi}{\partial x \partial z}\right) + \hat{k}\left(\frac{\partial^2 \phi}{\partial x \partial y} - \frac{\partial^2 \phi}{\partial y \partial x}\right)$$

$$= \hat{i}(0) + \hat{j}(0) + \hat{k}(0) = \vec{0}.$$

(ii) If $\vec{v} = v_1\hat{i} + v_2\hat{j} + v_3\hat{k}$, then

$$\text{curl } \vec{v} = \nabla \times \vec{v} = \begin{vmatrix} \hat{i} & \hat{j} & \hat{k} \\ \frac{\partial}{\partial x} & \frac{\partial}{\partial y} & \frac{\partial}{\partial z} \\ v_1 & v_2 & v_3 \end{vmatrix}$$

$$= \sum i\left(\frac{\partial v_3}{\partial y} - \frac{\partial v_2}{\partial z}\right)$$

and so,

$$\text{curl (curl } \vec{v}) = \nabla \times (\nabla \times \vec{v})$$

$$= \begin{vmatrix} \hat{i} & \hat{j} & \hat{k} \\ \frac{\partial}{\partial x} & \frac{\partial}{\partial y} & \frac{\partial}{\partial z} \\ \frac{\partial v_3}{\partial y} - \frac{\partial v_2}{\partial z} & \frac{\partial v_1}{\partial z} - \frac{\partial v_3}{\partial x} & \frac{\partial v_2}{\partial x} - \frac{\partial v_1}{\partial y} \end{vmatrix}$$

$$= \sum \hat{i}\left[\frac{\partial}{\partial y}\left(\frac{\partial v_2}{\partial x} - \frac{\partial v_1}{\partial y}\right) - \frac{\partial}{\partial z}\left(\frac{\partial v_1}{\partial z} - \frac{\partial v_3}{\partial x}\right)\right]$$

$$= \sum \hat{i}\left[\frac{\partial}{\partial x}\left(\frac{\partial v_2}{\partial y} + \frac{\partial v_3}{\partial z}\right) - \left(\frac{\partial^2 v_1}{\partial y^2} + \frac{\partial^2 v_1}{\partial z^2}\right)\right]$$

$$= \sum \hat{i}\left[\frac{\partial}{\partial x}\left(\frac{\partial v_1}{\partial x} + \frac{\partial v_2}{\partial y} + \frac{\partial v_3}{\partial z}\right) - \left(\frac{\partial^2 v_1}{\partial x^2} + \frac{\partial^2 v_2}{\partial y^2} + \frac{\partial^2 v_3}{\partial z^2}\right)\right]$$

$$= \sum \hat{i}\frac{\partial}{\partial x}(div\vec{v}) - \sum \nabla^2 \hat{i} v_1$$

$$= \left(\hat{i}\frac{\partial}{\partial x} + \hat{j}\frac{\partial}{\partial y} + \hat{k}\frac{\partial}{\partial z}\right)(\text{div } \vec{v})$$

$$-\nabla^2(\hat{i}v_1 + \hat{j}v_2 + \hat{k}v_3) = \text{grad div } \vec{v} - \nabla^2 \vec{v}.$$

(iii) As in (ii),

$$\text{curl } \vec{v} = \hat{i}\left(\frac{\partial v_3}{\partial y} - \frac{\partial v_2}{\partial z}\right) + \hat{j}\left(\frac{\partial v_1}{\partial z} - \frac{\partial v_3}{\partial x}\right) + \hat{k}\left(\frac{\partial v_2}{\partial x} - \frac{\partial v_1}{\partial y}\right).$$

Therefore,

$$\text{div (curl } \vec{v}) = \nabla \cdot (\nabla \times \vec{v})$$

$$= \frac{\partial}{\partial x}\left[\left(\frac{\partial v_3}{\partial y} - \frac{\partial v_2}{\partial z}\right) + \frac{\partial}{\partial y}\left(\frac{\partial v_1}{\partial z} - \frac{\partial v_3}{\partial x}\right)\right] + \frac{\partial}{\partial z}\left(\frac{\partial v_2}{\partial x} - \frac{\partial v_1}{\partial y}\right)$$

$$= \frac{\partial^2 v_3}{\partial x \partial y} - \frac{\partial^2 v_2}{\partial x \partial z} + \frac{\partial^2 v_1}{\partial y \partial z} - \frac{\partial^2 v_3}{\partial y \partial x} + \frac{\partial^2 v_2}{\partial z \partial x} - \frac{\partial^2 v_1}{\partial z \partial y}$$

$$= \frac{\partial^2 v_3}{\partial x \partial y} - \frac{\partial^2 v_2}{\partial x \partial z} + \frac{\partial^2 v_1}{\partial y \partial z} - \frac{\partial^2 v_3}{\partial x \partial y} + \frac{\partial^2 v_2}{\partial x \partial z} - \frac{\partial^2 v_1}{\partial y \partial z} \text{ since } \vec{v} \text{ is continuous}$$

$$= 0.$$

(iv) If $\vec{v} = v_1\hat{i} + v_2\hat{j} + v_3\hat{k}$, then

$$\text{div } \vec{v} = \frac{\partial v_1}{\partial x} + \frac{\partial v_2}{\partial y} + \frac{\partial v_3}{\partial z}.$$

Therefore,

$$\text{grad}(\text{div } \vec{v}) = \left(\hat{i}\frac{\partial}{\partial x} + \hat{j}\frac{\partial}{\partial y} + \hat{k}\frac{\partial}{\partial z}\right)\left(\frac{\partial v_1}{\partial x} + \frac{\partial v_2}{\partial y} + \frac{\partial v_3}{\partial z}\right).$$

EXAMPLE 13.52

If $\vec{a}$ is a constant vector and $\vec{r} = x\hat{i} + y\hat{j} + z\hat{k}$, show that

(i) $\text{div}(\vec{a}\times\vec{r}) = 0$ and (ii) curl $(\vec{a}\times\vec{r}) = 2\vec{a}$.

Solution. (i) We have $\vec{a} = a_1\hat{i} + a_2\hat{j} + a_3\hat{k}$. Therefore,

$$\vec{a}\times\vec{r} = \begin{vmatrix} \hat{i} & \hat{j} & \hat{k} \\ a_1 & a_2 & a_3 \\ x & y & z \end{vmatrix}$$

$$= \hat{i}(a_2 z - a_3 y) - \hat{j}(a_1 z - a_3 x) + \hat{k}(a_1 y - a_2 x).$$

Hence,

$$div(\vec{v}\times\vec{r}) = \frac{\partial}{\partial x}(a_2 z - a_3 y) - \frac{\partial}{\partial y}(a_1 z - a_3 x)$$

$$+\frac{\partial}{\partial z}(a_1 y - a_2 x) = 0 - 0 + 0 = 0$$

(ii) curl $(\vec{a}\times\vec{r}) = \nabla\times(\vec{a}\times\vec{r})$

$$= \begin{vmatrix} \hat{i} & \hat{j} & \hat{k} \\ \frac{\partial}{\partial x} & \frac{\partial}{\partial y} & \frac{\partial}{\partial z} \\ a_2 z - a_3 y & a_3 x - a_1 z & a_1 y - a_2 x \end{vmatrix}$$

$$= \hat{i}(a_1 + a_1) + \hat{j}(a_2 + a_2) + \hat{k}(a_3 + a_3)$$

$$= 2(a_1\hat{i} + a_2\hat{j} + a_3\hat{k}) = 2\vec{a}.$$

EXAMPLE 13.53

Determine curl curl $\vec{v}$ if $\vec{v} = x^2y\hat{i} + y^2z\hat{j} + z^2y\hat{k}$.

Solution. Let $\vec{v} = x^2y\hat{i} + y^2z\hat{j} + z^2y\hat{k}$. Then

$$\text{curl } \vec{v} = \nabla\times\vec{v} = \begin{vmatrix} \hat{i} & \hat{j} & \hat{k} \\ \frac{\partial}{\partial x} & \frac{\partial}{\partial y} & \frac{\partial}{\partial z} \\ x^2y & y^2z & z^2y \end{vmatrix}$$

$$= \hat{i}(z^2 - y^2) + \hat{j}(0) + \hat{k}(-x^2)$$

$$= \hat{i}(z^2 - y^2) + 0\hat{j} - x^2\hat{k}$$

Hence,

$$\text{curl curl } \vec{v} = \begin{vmatrix} \hat{i} & \hat{j} & \hat{k} \\ \frac{\partial}{\partial x} & \frac{\partial}{\partial y} & \frac{\partial}{\partial z} \\ z^2 - y^2 & 0 & -x^2 \end{vmatrix}$$

$$= (2z + 2x)\hat{j} + 2y\hat{k}$$

EXAMPLE 13.54

Show that $r^n\vec{r}$ is irrotational.

Solution. It is sufficient to show that curl $r^n\vec{r} = \vec{0}$. We have

$$\text{curl } r^n\vec{r} = \nabla\times r^n\vec{r}$$

$$= \nabla\times\left[r^n\left(x\hat{i} + y\hat{j} + z\hat{k}\right)\right]$$

$$= \begin{vmatrix} \hat{i} & \hat{j} & \hat{k} \\ \frac{\partial}{\partial x} & \frac{\partial}{\partial y} & \frac{\partial}{\partial z} \\ r^n x & r^n y & r^n z \end{vmatrix}$$

$$= \left[\frac{\partial}{\partial x}(r^n z) - \frac{\partial}{\partial z}(r^n y)\right]\hat{i}$$

$$+\left[\frac{\partial}{\partial z}(r^n x) - \frac{\partial}{\partial x}(r^n z)\right]\hat{j}$$

$$+\left[\frac{\partial}{\partial x}(r^n y) - \frac{\partial}{\partial y}(r^n x)\right]\hat{k}$$

$$= \left[znr^{n-1}\frac{\partial r}{\partial y} - ynr^{n-1}\frac{\partial r}{\partial z}\right]\hat{i}$$

$$+\left[xnr^{n-1}\frac{\partial r}{\partial z} - znr^{n-1}\frac{\partial r}{\partial x}\right]\hat{j}$$

$$+\left[ynr^{n-1}\frac{\partial r}{\partial x} - xnr^{n-1}\frac{\partial r}{\partial y}\right]\hat{k}$$

$$= \left[znr^{n-1}\frac{y}{r} - ynr^{n-1}\frac{z}{r}\right]\hat{i}$$

$$+\left[xnr^{n-1}\frac{z}{r} - znr^{n-1}\frac{x}{r}\right]\hat{j}$$

$$+\left[ynr^{n-1}\frac{x}{r} - xnr^{n-1}\frac{y}{r}\right]\hat{k}$$

$$= 0\hat{i} + 0\hat{j} + 0\hat{k} = \vec{0}.$$

Hence, $r^n\vec{r}$ is irrotational.

13.12 PROPERTIES OF DIVERGENCE AND CURL

(A) **Properties of Divergence.** Let $\vec{f}$ and $\vec{g}$ be two Vector-point functions and ϕ a scalar-point function. Then, the divergence has the following properties.

(i) $\text{div}(\vec{f}+\vec{g}) = \nabla\cdot(\vec{f}+\vec{g}) = \nabla\cdot\vec{f}+\nabla\cdot\vec{g}$.

(ii) $\text{div}(\phi\vec{f}) = \nabla\cdot(\phi\vec{f}) = (\nabla\phi).\vec{f}+\phi(\nabla\cdot\vec{f})$

$$= (grad\ \phi)\cdot\vec{f}+\phi\,\text{div}\vec{f}.$$

(iii) $\text{div}\vec{f} = 0$ if $\vec{f}$ is a constant vector.

(iv) $\text{div}(\vec{f}\times\vec{g}) = \vec{g}.$ curl $\vec{f} - \vec{f}.curl\ \vec{g}$.

Proof: Let $\vec{f} = f_1\hat{i}+f_2\hat{j}+f_3\hat{k}$ and $\vec{g} = g_1\hat{i}+g_2\hat{j}+g_3\hat{k}$.

(i) We have

$$\vec{f}+\vec{g} = (f_1+g_1)\hat{i}+(f_2+g_2)\hat{j}+(f_3+g_3)\hat{k}.$$

Therefore,

$$\nabla\cdot(\vec{f}+\vec{g}) = \left(\hat{i}\frac{\partial}{\partial x}+\hat{j}\frac{\partial}{\partial y}+\hat{k}\frac{\partial}{\partial z}\right).$$

$$\left[(f_1+g_1)\hat{i}+(f_2+g_2)\hat{j}+(f_3+g_3)\hat{k}\right]$$

$$= \frac{\partial}{\partial x}(f_1+g_1)+\frac{\partial}{\partial y}(f_2+g_2)+\frac{\partial}{\partial z}(f_3+g_3)$$

$$= \left(\frac{\partial f_1}{\partial x}+\frac{\partial f_2}{\partial y}+\frac{\partial f_3}{\partial z}\right)$$

$$+\left(\frac{\partial g_1}{\partial x}+\frac{\partial g_2}{\partial y}+\frac{\partial g_3}{\partial z}\right)$$

$$= \nabla\cdot\vec{f}+\nabla\cdot\vec{g}$$

(ii) $\nabla\cdot(\phi\vec{f})$

$$= \left(\hat{i}\frac{\partial}{\partial x}+\hat{j}\frac{\partial}{\partial y}+\hat{k}\frac{\partial}{\partial z}\right).(\phi f_1\hat{i}+\phi f_2\hat{j}+\phi f_3\hat{k})$$

$$= \frac{\partial}{\partial x}(\phi f_1)+\frac{\partial}{\partial y}(\phi f_2)+\frac{\partial}{\partial z}(\phi f_3)$$

$$= \left(\frac{\partial\phi}{\partial x}f_1+\phi\frac{\partial f_1}{\partial x}\right)+\left(\frac{\partial\phi}{\partial y}f_2+\phi\frac{\partial f_2}{\partial y}\right)$$

$$+\left(\frac{\partial\phi}{\partial z}f_3+\phi\frac{\partial f_3}{\partial z}\right)$$

$$= \left(\frac{\partial\phi}{\partial x}f_1+\frac{\partial\phi}{\partial y}f_2+\frac{\partial\phi}{\partial z}f_3\right)$$

$$+\left(\phi\frac{\partial f_1}{\partial x}+\phi\frac{\partial f_2}{\partial y}+\phi\frac{\partial f_3}{\partial z}\right)$$

$$= \left(\frac{\partial\phi}{\partial x}\hat{i}+\frac{\partial\phi}{\partial y}\hat{j}+\frac{\partial\phi}{\partial z}\hat{k}\right)\cdot(f_1\hat{i}+f_2\hat{j}+f_3\hat{k})$$

$$+\phi\left(\frac{\partial f_1}{\partial x}+\frac{\partial f_2}{\partial y}+\frac{\partial f_3}{\partial z}\right)$$

$$= (\nabla\phi)\cdot\vec{f}+\phi(\nabla\cdot\vec{f})$$

$$= (\text{grad }\phi).\vec{f}+\phi\,\text{div}\vec{f}.$$

(iii) Let $\vec{f} = f_1\hat{i}+f_2\hat{j}+f_3\hat{k}$.. Then

$$\nabla\cdot\vec{f} = \frac{\partial f_1}{\partial x}+\frac{\partial f_2}{\partial y}+\frac{\partial f_3}{\partial z} = 0,\ \text{since } \vec{f} \text{ is constant.}$$

(iv) $\text{div}(\vec{f}\times\vec{g}) = \nabla\cdot(\vec{f}\times\vec{g})$

$$= \left(\hat{i}\frac{\partial}{\partial x}+\hat{j}\frac{\partial}{\partial y}+\hat{k}\frac{\partial}{\partial z}\right).(\vec{f}\times\vec{g})$$

$$= \sum\hat{i}.\frac{\partial}{\partial x}(\vec{f}\times\vec{g})$$

$$= \sum\hat{i}.\left(\frac{\partial\vec{f}}{\partial x}\times\vec{g}+\vec{f}\times\frac{\partial\vec{g}}{\partial x}\right)$$

$$= \sum\hat{i}.\left(\frac{\partial\vec{f}}{\partial x}\times\vec{g}\right)$$

$$+\sum\hat{i}.\left(\vec{f}\times\frac{\partial\vec{g}}{\partial x}\right)$$

$$= \sum\left(\hat{i}\times\frac{\partial\vec{f}}{\partial x}\right).\vec{g}$$

$$-\sum\hat{i}.\left(\frac{\partial\vec{g}}{\partial x}\times\vec{f}\right)$$

$$= \sum\left(\hat{i}\times\frac{\partial\vec{f}}{\partial x}\right).\vec{g}$$

$$-\sum\left(\hat{i}\times\frac{\partial \vec{g}}{\partial x}\right).\vec{f}$$

$$=\left(\nabla\times\vec{f}\right).\vec{g}-\left(\nabla\times\vec{g}\right).\vec{f}$$

$$=\vec{g}.\left(\nabla\times\vec{f}\right)-\vec{f}\left(\nabla\times\vec{g}\right)$$

(commutativity of dot product)

$$=\vec{g}\ curl\ \vec{f}-\vec{f}.curl\ \vec{g}.$$

(B) **Properties of Curl.** Let $\vec{f}$ and $\vec{g}$ be two vector-point functions and ϕ a scalar-point function, all having continuous second-order partial derivatives. Then,

(i) curl $\left(\vec{f}+\vec{g}\right)=$ curl $\vec{f}+$ curl $\vec{g}$.

(ii) curl $\left(\phi\vec{f}\right)=(\text{grad }\phi)\times\vec{f}+\phi$ curl $\vec{f}$.

(iii) curl $\vec{f}=\vec{0}$, if $\vec{f}$ is a constant vector.

(iv) $\nabla\times\left(\vec{f}\times\vec{g}\right)=\left(\vec{g}.\nabla\right)\vec{f}-\left(\vec{f}.\vec{\nabla}\right)\vec{g}$ $+\vec{f}\left(\nabla\cdot\vec{g}\right)-\left(\nabla\cdot\vec{f}\right)\vec{g}.$

Proof: If $\vec{f}=f_1\hat{i}+f_2\hat{j}+f_3\hat{k}$ and $\vec{g}=g_1\hat{i}+g_2\hat{j}+g_3\hat{k}$, then

(i) $\vec{f}+\vec{g}=\left(f_1+g_1\right)\hat{i}+\left(f_2+g_2\right)\hat{j}+\left(f_3+g_3\right)\hat{k}$

and so,

$$curl\ \left(\vec{f}+\vec{g}\right)=\nabla\times\left(\vec{f}+\vec{g}\right)$$

$$=\begin{vmatrix}\hat{i} & \hat{j} & \hat{k}\\ \frac{\partial}{\partial x} & \frac{\partial}{\partial y} & \frac{\partial}{\partial z}\\ f_1+g_1 & f_2+g_2 & f_3+g_3\end{vmatrix}$$

$$=\left[\frac{\partial}{\partial y}\left(f_3+g_3\right)-\frac{\partial}{\partial z}\left(f_2+g_2\right)\right]\hat{i}$$

$$+\left[\frac{\partial}{\partial z}\left(f_1+g_1\right)-\frac{\partial}{\partial x}\left(f_3+g_3\right)\right]\hat{j}$$

$$+\left[\frac{\partial}{\partial x}\left(f_2+g_2\right)-\frac{\partial}{\partial y}\left(f_1+g_1\right)\right]\hat{k}$$

$$=\left[\left(\frac{\partial f_3}{\partial y}-\frac{\partial f_2}{\partial z}\right)\hat{i}+\left(\frac{\partial f_1}{\partial z}-\frac{\partial f_3}{\partial x}\right)\hat{j}\right.$$

$$\left.+\left(\frac{\partial f_2}{\partial x}-\frac{\partial f_1}{\partial y}\right)\hat{k}\right]$$

$$+\left[\left(\frac{\partial g_3}{\partial y}-\frac{\partial g_2}{\partial z}\right)\hat{i}+\left(\frac{\partial g_1}{\partial z}-\frac{\partial g_3}{\partial x}\right)\hat{j}\right.$$

$$\left.+\left(\frac{\partial g_2}{\partial x}-\frac{\partial g_1}{\partial y}\right)\hat{k}\right]$$

$$=\nabla\times\vec{f}+\nabla\times\vec{g}=\text{curl}\,\vec{f}+\text{curl}\,\vec{g}.$$

(ii) $curl\ \left(\phi\vec{f}\right)=\nabla\times\left(\phi\vec{f}\right)$

$$=\nabla\times\left(\phi f_1\hat{i}+\phi f_2\hat{j}+\phi f_3\hat{k}\right)$$

$$=\begin{vmatrix}\hat{i} & \hat{j} & \hat{k}\\ \frac{\partial}{\partial x} & \frac{\partial}{\partial y} & \frac{\partial}{\partial z}\\ \phi f_1 & \phi f_2 & \phi f_3\end{vmatrix}$$

$$=\left[\frac{\partial}{\partial y}(\phi f_3)-\frac{\partial}{\partial z}(\phi f_2)\right]\hat{i}$$

$$+\left[\frac{\partial}{\partial z}(\phi f_1)-\frac{\partial}{\partial x}(\phi f_3)\right]\hat{j}$$

$$+\left[\frac{\partial}{\partial x}(\phi f_2)-\frac{\partial}{\partial y}(\phi f_1)\right]\hat{k}$$

$$=\left[\phi\frac{\partial f_3}{\partial y}+\frac{\partial\phi}{\partial y}f_3-\phi\frac{\partial f_2}{\partial z}-\frac{\partial\phi}{\partial z}f_2\right]\hat{i}$$

$$+\left[\phi\frac{\partial f_1}{\partial y}+\frac{\partial\phi}{\partial z}f_1-\phi\frac{\partial f_3}{\partial x}-\frac{\partial\phi}{\partial x}f_3\right]\hat{j}$$

$$+\left[\phi\frac{\partial f_2}{\partial z}+\frac{\partial\phi}{\partial x}f_2-\phi\frac{\partial f_1}{\partial y}-\frac{\partial\phi}{\partial y}f_1\right]\hat{k}$$

$$=\phi\left[\left(\frac{\partial f_3}{\partial y}-\frac{\partial f_2}{\partial z}\right)\hat{i}+\left(\frac{\partial f_1}{\partial z}-\frac{\partial f_3}{\partial x}\right)\hat{j}+\left(\frac{\partial f_2}{\partial x}-\frac{\partial f_1}{\partial y}\right)\hat{k}\right]$$

$$+\left[\left(\frac{\partial\phi}{\partial y}f_3-\frac{\partial\phi}{\partial z}f_2\right)\hat{i}+\left(\frac{\partial\phi}{\partial z}f_1-\frac{\partial\phi}{\partial x}f_3\right)\hat{j}+\left(\frac{\partial\phi}{\partial x}f_2-\frac{\partial\phi}{\partial y}f_1\right)\hat{k}\right]$$

$$=\phi\left(\nabla\times\vec{f}\right)+\begin{vmatrix}\hat{i} & \hat{j} & \hat{k}\\ \frac{\partial\phi}{\partial x} & \frac{\partial\phi}{\partial y} & \frac{\partial\phi}{\partial z}\\ f_1 & f_2 & f_3\end{vmatrix}$$

$$=\phi\left(\nabla\times\vec{f}\right)+(\nabla\phi)\times\vec{f}$$

$$=\phi\ \text{curl}=\phi\ \text{curl}\,\vec{f}+\left(\text{grad}\ \phi\right)\times\vec{f}.$$

(iii) Let $\vec{f} = f_1\hat{i} + f_2\hat{j} + f_3\hat{k}$ be a constant vector. Then,

$$curl\ \vec{f} = \nabla \times \vec{f}$$

$$= \left(\hat{i}\frac{\partial}{\partial x} + \hat{j}\frac{\partial}{\partial y} + \hat{k}\frac{\partial}{\partial z}\right) \times \left(f_1\hat{i} + f_2\hat{j} + f_3\hat{k}\right)$$

$$= \begin{vmatrix} \hat{i} & \hat{j} & \hat{k} \\ \frac{\partial}{\partial x} & \frac{\partial}{\partial y} & \frac{\partial}{\partial z} \\ f_1 & f_2 & f_3 \end{vmatrix}$$

$$= \hat{i}\left(\frac{\partial f_3}{\partial y} - \frac{\partial f_2}{\partial z}\right) + \hat{j}\left(\frac{\partial f_1}{\partial z} - \frac{\partial f_3}{\partial x}\right) + \hat{k}\left(\frac{\partial f_2}{\partial x} - \frac{\partial f_1}{\partial y}\right)$$

$$= \hat{i}(0) + \hat{j}(0) + \hat{k}(0) = \vec{0}.$$

(iv) If $\vec{f} = f_1\hat{i} + f_2\hat{j} + f_3\hat{k}$ and $g_1\hat{i} + g_2\hat{j} + g_3\hat{k}$, then

$$\nabla \times \left(\vec{f} \times \vec{g}\right) = \left(\hat{i}\frac{\partial}{\partial x} + \hat{j}\frac{\partial}{\partial y} + \hat{k}\right) \times \left(\vec{f} \times \vec{g}\right)$$

$$= \hat{i} \times \frac{\partial}{\partial x}\left(\vec{f} \times \vec{g}\right) + \hat{j} \times \frac{\partial}{\partial y}\left(\vec{f} \times \vec{g}\right) + \hat{k} \times \frac{\partial}{\partial z}\left(\vec{f} \times \vec{g}\right)$$

$$= \hat{i} \times \left[\frac{\partial \vec{f}}{\partial x} \times \vec{g} + \vec{f} \times \frac{\partial \vec{g}}{\partial x}\right] + \hat{j} \times \left[\frac{\partial \vec{f}}{\partial y} \times \vec{g} + \vec{f} \times \frac{\partial \vec{g}}{\partial y}\right] + \hat{k} \times \left[\frac{\partial \vec{f}}{\partial z} \times \vec{g} + \vec{f} \times \frac{\partial \vec{g}}{\partial z}\right]$$

$$= \sum \hat{i} \times \left(\frac{\partial \vec{f}}{\partial x} \times \vec{g}\right) + \sum \hat{i} \times \left(\vec{f} \times \frac{\partial \vec{g}}{\partial x}\right)$$

$$= \sum \left[\left(\hat{i}.\vec{g}\right)\frac{\partial \vec{f}}{\partial x} - \left(\hat{i}.\frac{\partial \vec{f}}{\partial x}\right)\vec{g}\right] + \sum \left[\left(\hat{i}.\frac{\partial \vec{g}}{\partial x}\right)\vec{f} - \left(\hat{i}.\vec{f}\right)\frac{\partial \vec{g}}{\partial x}\right]$$

(by property of vector triple product)

$$= \sum g_1 \frac{\partial \vec{f}}{\partial x} - \left(\sum \frac{\partial f_1}{\partial x}\right)\vec{g} + \left(\sum \frac{\partial g_1}{\partial x}\right)\vec{f} - \sum f_1 \frac{\partial \vec{g}}{\partial x}$$

$$= \left(g_1\frac{\partial}{\partial x} + g_2\frac{\partial}{\partial y} + g_3\frac{\partial}{\partial z}\right)\vec{f} - \left(\nabla \cdot \vec{f}\right)\vec{g} + \left(\nabla \cdot \vec{g}\right)\vec{f} - \left(f_1\frac{\partial}{\partial x} + f_2\frac{\partial}{\partial y} + f_3\frac{\partial}{\partial z}\right)\vec{g}$$

$$= \left(\vec{g}.\nabla\right)\vec{f} - \left(\nabla \cdot \vec{f}\right)\vec{g} + \left(\nabla \cdot \vec{g}\right)\vec{f} - \left(\vec{f}.\nabla\right)\vec{g}.$$

EXAMPLE 13.55

Show that

$$\text{div (grad } r^n) = \nabla \cdot (\nabla r^n) = \nabla^2 (r^n) = n(n+1)r^{n-2}.$$

Deduce that $\nabla^2\left(\frac{1}{r}\right) = 0$.

Solution. From the definition of the Laplacian operator ∇^2, we have

$$\text{div (grad } r^n) = \nabla \cdot \left(\nabla r^n\right) = \nabla^2\left(r^n\right). = \left(\frac{\partial^2}{\partial x^2} + \frac{\partial^2}{\partial y^2} + \frac{\partial^2}{\partial z^2}\right)\left(r^n\right). \quad (1)$$

But,

$$\frac{\partial^2}{\partial x^2}\left(r^n\right) = \frac{\partial}{\partial x}\left(\frac{\partial}{\partial x} r^n\right) = \frac{\partial}{\partial x}\left(nr^{n-1}\frac{\partial r}{\partial x}\right)$$

$$= \frac{\partial}{\partial x}\left(nr^{n-1}.\frac{x}{r}\right), \text{ since } \frac{\partial r}{\partial x} = \frac{x}{r}$$

$$= \frac{\partial}{\partial x}\left(nr^{n-2}x\right) = n\left[(n-2)r^{n-3}\frac{\partial r}{\partial x}.x + r^{n-2}\right]$$

$$= n(n-2)r^{n-3}.\frac{x}{r}.x + nr^{n-2}$$

$$= n(n-2)r^{n-4}x^2 + nr^{n-2}.$$

Similarly,

$$\frac{\partial^2}{\partial y^2}(r^n) = n(n-2)r^{n-4}y^2 + nr^{n-2} \text{ and}$$

$$\frac{\partial^2}{\partial z^2}(r^n) = n(n-2)r^{n-4}z^2 + n\, r^{n-2}.$$

Hence, (1) reduces to

$$\nabla^2(r^n) = n\left[(n-2)r^{n-4}(x^2+y^2+z^2) + 3r^{n-2}\right]$$

$$= n\left[(n-2)r^{n-2} + 3r^{n-2}\theta\right] = n(n+1)r^{n-2}.$$

Putting $n=-1$, we get $\nabla^2\left(\frac{1}{r}\right) = 0$.

Second Method: From Example 13.34, we have $\nabla r^n = nr^{n-2}\vec{r}$. Therefore,

$$\nabla^2(r^n) = \nabla\cdot(\nabla r^n)$$

$$= \nabla\cdot(nr^{n-2}\vec{r})$$

$$\vec{r} = 0\cdot \qquad (2)$$

But, $div(\phi \bar{r}) = \nabla\cdot(\phi\bar{r}) = (\nabla\phi).\bar{r} + \phi(\nabla\cdot\bar{r})$.

Therefore, (2) becomes

$$\nabla^2(r^n) = n[(\nabla r^{n-2}).\bar{r} + r^{n-2}(\nabla\cdot\bar{r})]$$

$$= n[(n-2)r^{n-4}\,\bar{r}.\bar{r} + 3r^{n-2}] \text{ since } \nabla\cdot\bar{r} = 3$$

$$= n[(n-2)r^{n-4}r^2 + 3r^{n-2}], \text{ since } \bar{r}.\bar{r} = r^2$$

$$= n(n+1)r^{n-2}.$$

EXAMPLE 13.56

Show that curl $(grad\,\vec{r}) = \nabla\times\nabla r^n = \vec{0}$.

Solution. We have seen that

$$\nabla r^n = nr^{n-2}\vec{r} = nr^{n-2}\left(x\hat{i} + y\hat{j} + z\hat{k}\right).$$

Therefore,

$$\nabla\times\nabla r^n = \begin{vmatrix} \hat{i} & \hat{j} & \hat{k} \\ \frac{\partial}{\partial x} & \frac{\partial}{\partial y} & \frac{\partial}{\partial z} \\ nr^{n-2}x & nr^{n-2}y & nr^{n-2}z \end{vmatrix}$$

$$= \sum\left[\frac{\partial}{\partial y}(nr^{n-2}z) - \frac{\partial}{\partial z}(nr^{n-2}y)\right]\hat{i}$$

$$= \sum\left[nz(n-2)r^{n-3}\frac{\partial r}{\partial y} - ny(n-2)r^{n-3}\frac{\partial r}{\partial z}\right]\hat{i}$$

$$= \sum\left[nz(n-2)r^{n-3}\frac{y}{r} - ny(n-2)r^{n-3}\frac{z}{r}\right]\hat{i}$$

$$= \sum\left[n(n-2)r^{n-4}yz - n(n-2)r^{n-4}yz\right]\hat{i}$$

$$= 0\hat{i} + 0\hat{j} + 0\hat{k} = \vec{0}.$$

EXAMPLE 13.57

If $\vec{f}$ and $\vec{g}$ are irrotational, show that $\vec{f}\times\vec{g}$ is solenoidal.

Solution. Since $\vec{f}$ and $\vec{g}$ are irrotational, we have

$$\nabla\times\vec{f} = \vec{0} \text{ and } \nabla\times\vec{g} = \vec{0}.$$

Now,

$$\text{div}(\vec{f}\times\vec{g}) = \vec{g}.(\nabla\times\vec{f}) - \vec{f}.(\nabla\times\vec{g})$$

$$= \vec{g}.\vec{0} - \vec{f}.\vec{0} = \vec{0}.$$

Hence, $\vec{f}\times\vec{g}$ is solenoidal.

EXAMPLE 13.58

Show that

$$\nabla^2 f(r) = f''(r) + \frac{2}{r}f'(r),$$

where $\vec{r} = x\hat{i} + y\hat{j} + z\hat{k}$.

Solution. We have $r^2 = x^2 + y^2 + z^2$ and so, $\frac{\partial r}{\partial x} = \frac{x}{r}$, $\frac{\partial r}{\partial y} = \frac{y}{r}$, and $\frac{\partial r}{\partial z} = \frac{z}{r}$. Then,

$$\text{grad } r = \left(i\frac{\partial}{\partial x} + j\frac{\partial}{\partial y} + k\frac{\partial}{\partial z}\right)r = \hat{i}\frac{\partial r}{\partial x} + \hat{j}\frac{\partial r}{\partial y} + \hat{k}\frac{\partial r}{\partial z}$$

$$= \hat{i}\frac{x}{r} + \hat{j}\frac{y}{r} + \hat{k}\frac{z}{r} = \frac{1}{r}\left(x\hat{i} + y\hat{j} + z\hat{k}\right) = \frac{\vec{r}}{r}. \quad (1)$$

Also, by Example 13.32,

$$\text{grad} f(r) = f'(r)\nabla r = f'(r)\text{grad } r. \quad (2)$$

Therefore,

$$\nabla^2 f(r) = \nabla \cdot (\nabla f(r)) = \text{div}(\text{grad} f(r)) = \text{div}(f'(r)\nabla r)$$

$$= div\left[\frac{1}{r}f'(r)\vec{r}\right], \text{ using (1)}$$

$$= \frac{1}{r}f'(r)\text{div } \vec{r} + \vec{r}.\text{grad}\left(\frac{1}{r}f'(r)\right),$$

by divergent property

$$= \frac{3}{r}f'(r) + \vec{r}.\left[\frac{d}{dr}\left(\frac{1}{r}f'(r)\right)\text{grad } r\right],$$

using (2) and $\text{div } \vec{r} = 3$.

$$= \frac{3}{r}f'(r) + \vec{r}\left[-\frac{1}{r^2}f'(r) + \frac{1}{r}f''(r)\right]\frac{\vec{r}}{r},$$

using (l)

$$= \frac{3}{r}f'(r) + \frac{\vec{r}.\vec{r}}{r}\left[-\frac{1}{r^2}f'(r) + \frac{1}{r}f''(r)\right]$$

$$= \frac{3}{r}f'(r) + \frac{r^2}{r}\left[-\frac{1}{r^2}f'(r) + \frac{1}{r}f''(r)\right]$$

$$= \frac{3}{r}f'(r) + f''(r) - \frac{1}{r}f'(r) = \frac{2}{r}f'(r) + f''(r).$$

EXAMPLE 13.59

Show that $\text{div}\left(r^n\vec{r}\right) = (n+3)r^n$.

Solution. We know that

$$\text{div}\left(\phi\vec{f}\right) = (\text{grad } \phi).\vec{f} + \phi\,\text{div}\,\vec{f}.$$

Therefore,

$$\text{div}\left(r^n\vec{r}\right) = \left(\text{grad } r^n\right).\vec{r} + r^n \text{ div } \vec{r}.$$

But, $\text{grad } r^n = nr^{n-2}\vec{r}$ and $\text{div}\,\vec{r} = 3$. Therefore,

$$\text{div}\left(r^n\vec{r}\right) = nr^{n-2}\vec{r}.\vec{r} + 3r^n = nr^{n-2}r^2 + 3r^n$$

$= (n+3)\, r^n$.

EXAMPLE 13.60

Show that $\text{curl}\,(\phi\,\text{grad}\,\phi) = \vec{0}$.

Solution. We have

$$\text{grad } \phi = \nabla\phi = \frac{\partial\phi}{\partial x}\hat{i} + \frac{\partial\phi}{\partial y}\hat{j} + \frac{\partial\phi}{\partial z}\hat{k}$$

and so,

$$\phi\,\text{grad } \phi = \phi\frac{\partial\phi}{\partial x}\hat{i} + \phi\frac{\partial\phi}{\partial y}\hat{j} + \phi\frac{\partial\phi}{\partial z}\hat{k}.$$

Hence,

$$\text{curl } (\phi\text{grad } \phi) = \begin{vmatrix} \hat{i} & \hat{j} & \hat{k} \\ \frac{\partial}{\partial x} & \frac{\partial}{\partial y} & \frac{\partial}{\partial z} \\ \phi\frac{\partial\phi}{\partial x} & \phi\frac{\partial\phi}{\partial y} & \phi\frac{\partial\phi}{\partial z} \end{vmatrix}$$

$$= \sum\left[\frac{\partial}{\partial y}\left(\phi\frac{\partial\phi}{\partial z}\right) - \frac{\partial}{\partial z}\left(\phi\frac{\partial\phi}{\partial y}\right)\right]\hat{i}$$

$$= \sum\left[\frac{\partial\phi}{\partial y}\cdot\frac{\partial\phi}{\partial z} + \phi\frac{\partial^2\phi}{\partial y\partial z} - \frac{\partial\phi}{\partial z}\cdot\frac{\partial\phi}{\partial y} - \phi\frac{\partial^2\phi}{\partial z\partial y}\right]\hat{i}$$

$$= \sum 0\hat{i} = 0\hat{i} + 0\hat{j} + 0\hat{k} = \vec{0}.$$

Also, it follows that ϕ grad ϕ is irrotational.

EXAMPLE 13.61

Show that the vector $f(r)\vec{r}$ is irrotational.

Solution. A vector $\vec{f}$ is irrotational if $curl\vec{f} = \vec{0}$. Also, we know that

$$\vec{r}\times\vec{r} = 0, \text{ curl } \vec{r} = 0 \text{ and}$$

$$grad f(r) = f'(r)\, grad\, r = f'(r)\frac{\vec{r}}{r}.$$

Therefore,

$$\text{curl } [f(r)\vec{r}] = [\text{grad} f(r)]\times\vec{r} + f(r) \text{ curl } \vec{r}$$

$$= [f'(r)\text{grad } r]]\times\vec{r} + \vec{0}$$

$$= f'(r)\frac{\vec{r}}{r}\times\vec{r} + \vec{0}$$

$$= \frac{1}{r}f'(r)(\vec{r}\times\vec{r}) = \vec{0}.$$

Hence, $f(r)\vec{r}$ is irrotational.

13.13 INTEGRATION OF VECTOR FUNCTIONS

If $\vec{f}$ and $\vec{F}$ are two vector functions of the scalar variable t such that $\frac{d\vec{F}}{dt}=\vec{f}$, then $\vec{F}$ is called the *indefinite integral of* $\vec{f}$ with respect to t. Thus,

$$\int \vec{f}\, dt = \vec{F}.$$

If $\vec{f} = f_1\hat{i} + f_2\hat{j} + f_3\hat{k}$, where $f_1 f_2$, and f_3 are scalar functions of the scalar t, then

$$\int \vec{f}\, dt = \hat{i}\int f_1 dt + \hat{j}\int f_2 dt + \hat{k}\int f_3 dt.$$

Hence, to integrate a vector function, we integrate its components.

EXAMPLE 13.62

The acceleration of a particle at any time $t \geq 0$ is given by

$$\vec{a} = \frac{d\vec{v}}{dt} = 12\cos 2t\,\hat{i} - 8\sin 2t\,\hat{j} + 16t\,\hat{k}.$$

If the velocity $\vec{v}$ and displacement $\vec{r}$ are zero at $t = 0$, find $\vec{v}$ and $\vec{r}$ at any time t.

Solution. We have

$$\vec{a} = \frac{d\vec{v}}{dt} = 12\cos 2t\,\hat{i} - 8\sin 2t\,\hat{j} + 16t\,\hat{k}.$$

Integration with respect to t yields

$$\vec{v} = 6\sin 2t\,\hat{i} + 4\cos 2t\,\hat{j} + 8t^2\,\hat{k} + \vec{c},$$

w here $\vec{c}$ is a constant of integration. But $\vec{v} = \vec{0}$ when $t = 0$. Therefore,

$$\vec{0} = 0\hat{i} + 4\hat{j} + 0\hat{k} + \hat{c}$$

and so, $\vec{c} = -4\hat{j}$. Therefore,

$$\vec{v} = 6\sin 2t\,\hat{i} + (4\cos 2t - 4)\,\hat{j} + 8t^2\hat{k}$$

or

$$\frac{d\vec{r}}{dt} = 6\sin 2t\,\hat{i} + (4\cos 2t - 4)\hat{j} + 8t^2\hat{k}.$$

Integrating again with respect to t, we get

$$\vec{r} = -3\cos 2t\,\hat{i} + (2\sin 2t - 4t)\hat{j} + \frac{8}{3}t^3\hat{k} + \vec{p},$$

where $\vec{p}$ is a constant of integration. But $\vec{r} = 0$ when $t = 0$. Therefore,

$$\vec{0} = -3\hat{i} + 0\hat{j} + 0\hat{k} + \vec{p}$$

and so, $\vec{p} = 3\hat{i}$. Hence,

$$\vec{r} = (3 - 3\cos 2t)\hat{i} + (2\sin 2t - 4t)\hat{j} + \frac{8}{3}t^3\hat{k}.$$

EXAMPLE 13.63

If $\vec{r} = \vec{0}$ when $t = 0$ and $\frac{d\vec{r}}{dt} = \vec{u}$ when $t = 0$, find the value of $\vec{r}$ satisfying the equation $\frac{d^2\vec{r}}{dt^2} = \vec{a}$, where $\vec{a}$ is a constant vector.

Solution. Integrating $\frac{d^2\vec{r}}{dt^2} = \vec{a}$ with respect to t, we get

$$\frac{d\vec{r}}{dt} = \vec{a}t + \vec{c},$$

where $\vec{c}$ is a constant vector of integration. When $t = 0$, $\frac{d\vec{r}}{dt} = \vec{u}$. Therefore,

$$\vec{u} = \vec{a}(0) + \vec{c} \text{ and so } \vec{c} = \vec{u}.$$

Therefore,

$$\frac{d\vec{r}}{dt} = \vec{a}t + \vec{u}$$

Integrating again with respect to t, we get

$$\vec{r} = \frac{1}{2}\vec{a}t^2 + \vec{u}t + \vec{p},$$

where $\vec{p}$ is the constant vector of integration. When $t = 0$, $\vec{r} = \vec{0}$. Therefore, $\vec{0} = \vec{p}$. Hence,

$$\vec{r} = \vec{u}t + \frac{1}{2}\vec{a}t^2.$$

13.14 LINE INTEGRAL

An integral which is evaluated along a curve is called a *line integral*. Note, however, that *a line integral is not represented by the area under the curve.*

Consider any arc of the curve C enclosed between two points A and B. Let a and b be the values of the parameter t for A and B, respectively. Partition the arc between A and B into n parts as given in the following equation:

$$A = P_{0,}\, P_1,\, P_n = B.$$

Let $\vec{r}_0, \vec{r}_1, \ldots, \vec{r}_n$ be the position vectors of the points $P_{0,}\, P_1, P_n$ respectively. Let ξ_i be any point on the subarc $P_{1-1}P_i$ and let $\delta\vec{r}_i = \vec{r}_i - \vec{r}_{i-1}$. Let $\vec{f}(\vec{r})$ be a continuous vector-point function. Then,

$$\lim_{\substack{n\to\infty \\ |\delta\vec{r}_i|\to 0}} \sum_{i=1}^{n} \vec{f}(\xi_i).\,\delta\vec{r}_i, \qquad (1)$$

if exists, is called a line integral of $\vec{f}$ along C and is denoted by $\int_C \vec{f}.d\vec{r}$ or $\int_C \vec{f}.\frac{d\vec{r}}{dt}dt$. Thus, the line integral is a scalar and is also called the *tangential line integral* of $\vec{f}$ along the curve C.

If $\vec{f} = f_1\hat{i} + f_2\hat{j} + f_3\hat{k}$ and $\vec{r} = x\hat{i} + y\hat{j} + z\hat{k}$, then $d\vec{r} = \hat{i}\,dx + \hat{j}\,dy + \hat{k}\,dz$ and so,

$$\int_C \vec{f}.d\vec{r} = \int_C (f_1dx + f_2dy + f_3dz)$$

$$\int_a^b \left(f_1\frac{dx}{dt} + f_2\frac{dy}{dt} + f_3\frac{dz}{dt} \right) dt,$$

where a and b are, respectively, values of the parameter t at the points A and B.

If we replace the dot product in (1) by a vector product, then the *vector line integral* is defined as $\int \vec{f} \times d\vec{r}$, which is a vector.

If C is a simple closed curve, then the tangent line integral of the vector function $\vec{f}$ around C is called the *circulation* of $\vec{f}$ around C and denoted by $\oint_C \vec{f}\,d\vec{r}$.

The vector function $\vec{f}$ is said to be *irrotational* in a region R if the circulation of $\vec{f}$ around any closed curve in R is zero.

EXAMPLE 13.64

If $\vec{f} = (3x^2 + 6y)\hat{i} - 14yz\,\hat{j} + 20xz^2\hat{k}$, evaluate $\int_C \vec{f}\,d\vec{r}$, where C is given by $x = t$, $y = t^2$, and $z = t^3$, and t varies from 0 to 1.

Solution. The parametric equation of C is

$$x = t, y = t^2, \text{ and } z = t^3,$$

where t varies from 0 to 1.

Now, $\vec{r} = x\hat{i} + y\hat{j} + z\hat{k} = t\hat{i} + t^2\hat{j} + t^3\hat{k}$. Therefore,

$$\frac{d\vec{r}}{dt} = \hat{i} + 2t\hat{j} + 3t^2\hat{k}.$$

Further,

$$\vec{f} = (3x^2 + 6y)\hat{i} - 14yz\,\hat{j} + 20xz^2\,\hat{k}$$

$$= (3t^2 + 6t^2)\hat{i} - 14t^2.t^3\hat{j} + 20t.t^6\hat{k}$$

$$= 9t^2\hat{i} - 14t^5\hat{j} + 20t^7\hat{k}.$$

Therefore,

$$\int_C \vec{f}.d\vec{r} = \int_C \left(\vec{f}.\frac{d\vec{r}}{dt} \right) dt = \left[\left(9t^2\hat{i} - 14t^5\hat{j} + 20t^7\hat{k} \right) \right.$$

$$\left. .\left(\hat{i} + 2t\,\hat{j} + 3t^2\hat{j} + 3t^2\hat{k} \right) \right] dt$$

$$= \int_0^1 (9t^2 - 28t^6 + 60t^9)dt$$

$$= \left[\frac{9t^3}{3} - \frac{28t^7}{7} + \frac{60t^{10}}{10} \right]_0^1 = 5.$$

EXAMPLE 13.65

If $\vec{f} = (\sin y)\hat{i} + x(1 + \cos y)\hat{j}$, evaluate the line integral $\int_C \vec{f}\,d\vec{r}$ along the circular path C given by $x^2 + y^2 = a^2$ and $z = 0$.

Solution. The parametric equation of the circular path C are $x = a \cos t$, $y = a \sin t$, and $z = 0$, where t varies from 0 to 2π. Now,

$$\vec{r} = x\hat{i} + y\hat{j} + z\,\hat{k} = (a \cos t)\hat{i} + (a \sin t)\hat{j}.$$

Therefore,

$$\frac{d\vec{r}}{dt} = (-a \sin t)\hat{i} + (a \cos t)\hat{j}.$$

Also, $\vec{f}$, in terms of parameter t, is given by

$$\vec{f} = \sin(a \sin t)\hat{i} + (a \cos t)(1 + \cos(a \sin t))\hat{j}.$$

Therefore,

$$\int_C \vec{f}.d\vec{r} = \int_C \left(\vec{f}.\frac{d\vec{r}}{dt} \right) dt$$

$$= \int_0^{2\pi} [\sin(a \sin t)\hat{i} + (a \cos t(1 + \cos(a \sin t))\hat{j})]$$

$$.\left[(-a \sin t)\hat{i} + (a \cos t)\hat{j} \right] dt$$

$$= \int_0^{2\pi} [-a \sin t \sin(a \sin t)$$

$$+ a^2\cos^2 t(1 + \cos(a \sin t))]dt$$

$$= \int_0^{2\pi} \left[-a \sin t \sin(a \sin t) + a^2\cos^2 t \right.$$

$$\left. + a^2 \cos^2 t \cos(a \sin t) \right] dt$$

$$= \int_0^{2\pi} \left\{ d\left[a \cos t \sin(a \sin t) \right] + a^2\cos^2 t \right\} dt$$

$$= [a \cos t \sin(a \sin t)]_0^{2\pi} + \int_0^{2\pi} a^2\cos^2 t\,dt$$

$$= a^2 \int_0^{2\pi} \frac{1 + \cos 2t}{2} dt = \frac{a^2}{2} \left[t + \frac{\sin 2t}{2} \right]_0^{2\pi}$$

$$= \frac{a^2}{2}(2\pi) = \pi a^2.$$

EXAMPLE 13.66

Calculate $\int_C \vec{f}.d\vec{r}$, where C is the part of the spiral $\vec{r} = (a\cos\theta, a\sin\theta, a\theta)$ corresponding to $0 \le \theta \le \frac{\pi}{2}$ and $\vec{f} = r^2\hat{i}$.

Solution. We have

$$\vec{r} = a\cos\theta\hat{i} + a\sin\theta\hat{j} + a\theta\hat{k} \text{ so that}$$

$$\frac{d\vec{r}}{d\theta} = -a\sin\theta\hat{i} + a\cos\theta\hat{j} + a\hat{k}.$$

Also,

$$\vec{f} = r^2\hat{i} = (a^2\cos^2\theta + a^2\sin^2\theta + a^2\theta^2)\hat{i}$$
$$= [a^2(1+\theta^2)]\hat{i}.$$

Therefore,

$$\int_C \vec{f}.d\vec{r} = \left(\vec{f}.\frac{d\vec{r}}{d\theta}\right)d\theta = \int_0^{\frac{\pi}{2}} \{[a^2(1+\theta^2)]\hat{i}$$
$$.[(-a\sin\theta)\hat{i} + (a\cos\theta)\hat{j} + a\hat{k}]\}d\theta$$
$$= -\int_0^{\frac{\pi}{2}} a^3(1+\theta^2)\sin\theta d\theta$$
$$= -a^3\int_0^{\frac{\pi}{2}} (\sin\theta + \theta^2\sin\theta)d\theta$$
$$= -a^3[\cos\theta + 2\theta\sin\theta - \theta^2\cos]_0^{\frac{\pi}{2}}$$
$$= -a^3(\pi - 1).$$

EXAMPLE 13.67

If $\vec{f} = xy\hat{i} - z\hat{j} + x^2\hat{k}$ and C is the curve $x = t^2$, $y = 2t$, and $z = t^3$ from $t = 0$ to $t = 1$, find the *vector line integral* $\int_C \vec{f} \times d\vec{r}$.

Solution. We have

$$\vec{r} = x\hat{i} + y\hat{j} + z\hat{k} = t^2\hat{i} + 2t\hat{j} + t^3\hat{k},$$

$$\frac{d\vec{r}}{dt} = 2t\hat{i} + 2\hat{j} + 3t^2\hat{k},$$

and

$$\vec{f} = xy\hat{i} - z\hat{j} + x^2\hat{k} = 2t^3\hat{i} - t^3\hat{j} + t^4\hat{k}.$$

Therefore,

$$\vec{f} \times \frac{d\vec{r}}{dt} = \begin{vmatrix} \hat{i} & \hat{j} & \hat{k} \\ 2t^2 & -t^3 & t^4 \\ 2t & 2 & 3t^2 \end{vmatrix}$$
$$= (-3t^5 - 2t^4)\hat{i} + (-4t^5)\hat{j} + (4t^3 + 2t^4)\hat{k}.$$

Hence,

$$\int_C \vec{f} \times d\vec{r} = \int_0^1 \left(\vec{f} \times \frac{d\vec{r}}{dt}\right)dt$$
$$= \int_0^1 \left[(-3t^5 - 2t^4)\hat{i} + (-4t^5)\hat{j} + (4t^3 + 2t^4)\hat{k}\right]dt$$
$$= \left[-\frac{3t^6}{6} - \frac{2t^5}{5}\right]_0^1 \hat{i} + \left[\left(-4\frac{t^6}{6}\right)\right]_0^1 \hat{j}$$
$$+ \left[4\frac{t^4}{4} + \frac{2t^5}{5}\right]_0^1 \hat{k}$$
$$= -\frac{9}{10}\hat{i} - \frac{2}{3}\hat{j} + \frac{7}{5}\hat{k}.$$

EXAMPLE 13.68

If $\vec{f} = (5xy - 6x^2)\hat{i} + (2y - 4x)\hat{j}$, evaluate $\int_C \vec{f}.d\vec{r}$ along the curve C in the xy plane $y = x^3$ from the point (1, 1) to (2, 8).

Solution. Substituting $x = t$, we get $y = t^3$. When $x = 1$, $t = 1$ and when $x = 2$, $t = 2$. Then,

$$\vec{r} = x\hat{i} + y\hat{j} + z\hat{k} = t\hat{i} + t^3\hat{j}, \frac{d\vec{r}}{dt} = \hat{i} + 3t^2\hat{j} \text{ and}$$

$$\vec{f} = (5xy - 6x^2)\hat{i} + (2y - 4x)\hat{j}$$
$$= (5t^4 - 6t^2)\hat{i} + (2t^3 - 4t)\hat{j}.$$

Hence,

$$\int_C \vec{f}.d\vec{r} = \int_1^2 \left(\vec{f}.\frac{d\vec{r}}{dt}\right)dt = \int_1^2 \{[(5t^4 - 6t^2)\hat{i}$$
$$+ (2t^3 - 4t)\hat{j}\big] \cdot \left(\hat{i} + 3t^2\hat{j}\right)\big\}dt$$
$$= \int_1^2 [(5t^4 - 6t^2) + 3t^2(2t^3 - 4t)]dt$$

$$= \int_1^2 [6t^5 + 5t^4 - 12t^3 - 6t^2]dt$$

$$= \left[6\frac{t^6}{6} + 5\frac{t^5}{5} - 12\frac{t^4}{4} - \frac{6t^3}{3}\right]_1^2$$

$$= [t^6 + t^5 - 3t^4 - 2t^3]_1^2 = 35.$$

EXAMPLE 13.69

Evaluate $\int_A^B [2xydx + (x^2 - y^2)dy]$ along the arc of the circle $x^2 + y^2 = 1$ in the first quadrant from A (1. 0) to B (0, 1).

Solution. On the circle, $y = \sqrt{1-x^2}$ so that $\frac{dy}{dx} = -x\left(1-x^2\right)^{-\frac{1}{2}}$ or $dy = -2x(1-x^2)^{-\frac{1}{2}}dx$. Therefore,

$$\int_A^B [2xydx + (x^2 - y^2)dy]$$

$$= \int_1^0 [2x(1-x^2)^{\frac{1}{2}}dx - (2x^2-1)x(1-x^2)^{-\frac{1}{2}}dx]$$

$$= \left[-\frac{2}{3}(1-x^2)^{\frac{3}{2}}\right]_1^0 - \int_1^0 \left(x^2 - \frac{1}{2}\right)2x(1-x^2)^{-\frac{1}{2}}dx$$

$$= -\frac{2}{3} + \left[2\left(x^2 - \frac{1}{2}\right)(1-x^2)^{\frac{1}{2}}\right]_1^0$$

$$-2\int_1^0 2x(1-x^2)^{\frac{1}{2}}dx$$

$$= -\frac{2}{3} - 1 + 2\left[\frac{(1-x^2)^{\frac{3}{2}}}{\frac{3}{2}}\right]_1^0$$

$$= -\frac{2}{3} - 1 + \frac{4}{3} = -\frac{1}{3}.$$

EXAMPLE 13.70

Evaluate $\int_C [(x^2+xy)dx + (x^2+y^2)dy]$, where C is the square formed by the lines $y = \pm 1$ and $x = \pm 1$.

Solution. The curve C is as shown in the following figure:

We note that

$$\int_C [(x^2+xy)dx + (x^2+y^2)dy] = \int_{AB} + \int_{BC} + \int_{CD} + \int_{DA}.$$

Along AB, we have $x = 1$ and so, $dx = 0$. Also along AB, y varies from −1 to 1. Thus,

$$\int_{AB} [(x^2+xy)dx + (x^2+y^2)dy]$$

$$= \int_{-1}^1 (1+y^2)dy = \left[y + \frac{y^3}{3}\right]_{-1}^1 = \frac{8}{3}.$$

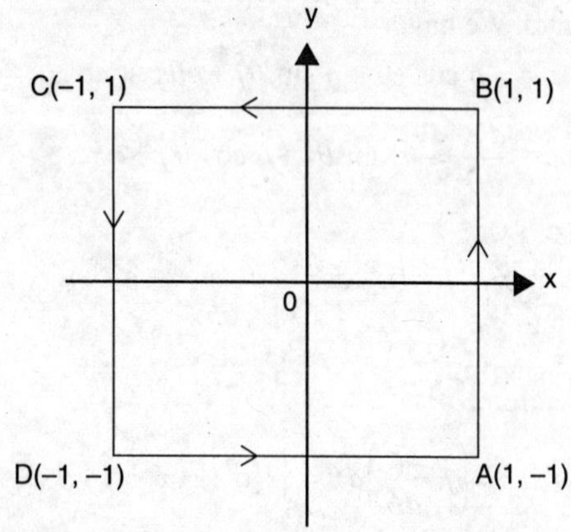

Along BC, we have y = 1 so that $dy = 0$. Also along BC, x varies from 1 to −1. Thus,

$$\int_{BC} [(x^2+xy)dx + (x^2+y^2)dy]$$

$$= \int_{-1}^1 \left(x^2 + x\right)dx = \left[\frac{x^3}{3} + \frac{x^2}{2}\right]_1^{-1} = -\frac{2}{3}.$$

Similarly,

$$\int_{CD} [(x^2+xy)dx + (x^2+y^2)dy]$$

$$= \int_1^{-1} (1+y^2)dy = -\frac{8}{3} \text{ and}$$

$$\int_{DA} [(x^2+xy)dx + (x^2+y^2)dy]$$

$$= \int_{-1}^1 (x^2+x)dx = \frac{2}{3}.$$

Hence,

$$\int_c [(x^2+xy)dx + (x^2+y^2)dy] = \frac{8}{3} - \frac{2}{3} - \frac{8}{3} + \frac{2}{3} = 0.$$

13.15 WORK DONE BY A FORCE

The work done as the point of application of a force $\vec{f}$ moves along a given path C can be expressed as a line integral. In fact, the work done, when the point of application moves from $P(\vec{r})$ to $Q(\vec{r}+\delta\vec{r})$, where $PQ = \delta\vec{r}$, is

$$\delta W = |\delta\vec{r}||\vec{f}|\cos\theta = \vec{f}.\delta\vec{r}.$$

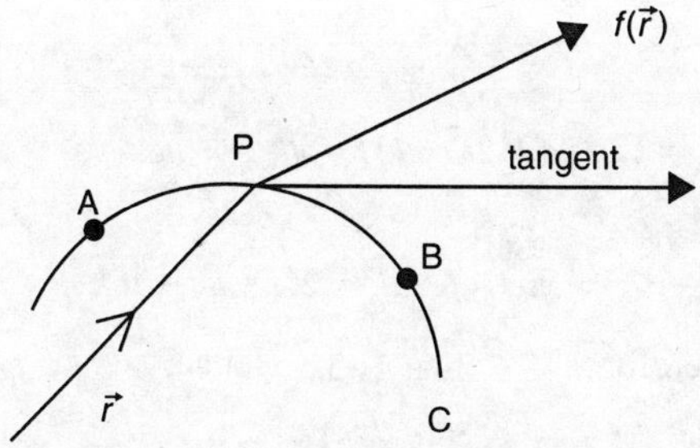

Therefore, the total work done as P moves from A to B is

$$W = \int_A^B \vec{f}.d\vec{r}.$$

Now, suppose that force $\vec{f}$ is conservative. Then, there exists a scalar function ϕ such that $\vec{f} = -\text{grad}\,\phi$, that is, $\vec{f} = -\nabla\phi = -\left[\hat{i}\frac{\partial\phi}{\partial x} + \hat{j}\frac{\partial\phi}{\partial y} + \hat{k}\frac{\partial\phi}{\partial z}\right]$

Therefore, the work done in this case is given by

$$W = \int_A^B -\left[\hat{i}\frac{\partial\phi}{\partial x} + \hat{j}\frac{\partial\phi}{\partial y} + \hat{k}\frac{\partial\phi}{\partial z}\right](\hat{i}dx + \hat{j}\,dy + \hat{k}dz)$$

$$= \int_B^A \left[\frac{\partial\phi}{\partial x}dx + \frac{\partial\phi}{\partial y}dy + \frac{\partial\phi}{\partial z}dz\right]$$

$$= \int_B^A d\phi = [\phi]_B^A = \phi_A - \phi_B.$$

Hence, in a conservative field, the work done depends on A and B and is the same for all paths joining A and B. Thus, in the case of conservative force $f(\vec{r}).d\vec{r}$ is an *exact differential* – $d\phi$. In such a case, ϕ is called the *potential energy*. The forces which do not have this property are said to be *dissipative* or *nonconservative*.

EXAMPLE 13.71

Find the total work done in moving a particle in a force field, given by $\vec{f} = 3xy\,\hat{i} - 5z\,\hat{j} + 10x\,\hat{k}$, along the curve $x = t^2+1$, $y = 2t^2$, and $z = t^3$, from $t=1$ to $t=2$.

Solution. The parametric equation of the curve is $x = t^2+1, y = 2t^2$, and $z = t^3, 1 \le t \le 2$.

We have

$$\vec{f} = 3xy\,\hat{i} - 5z\,\hat{j} + 10x\,\hat{k}.$$

$$= 3(t^2+1)(2t^2)\hat{i} - 5t^3\hat{j} + 10(t^2+1)\hat{k}$$

$$= 6(t^4+t^2)\hat{i} - 5t^3\hat{j} + 10(t^2+1)\hat{k}$$

and

$$\vec{r} = x\hat{i} + y\,\hat{j} + z\,\hat{k} = (t^2+1)\hat{i} + 2t^2\,\hat{j} + t^3\hat{k}.$$

Therefore, $\frac{d\vec{r}}{dt} = 2t\hat{i} + 4t\,\hat{j} + 3t^2\hat{k}$ and so, the total work done is given by

$$W = \int_C \vec{f}.d\vec{r} = \int_1^2 \vec{f}.\left(\frac{d\vec{r}}{dt}\right)dt$$

$$= \int_1^2 \left[6(t^4+t^2)\hat{i} - 5t^3\hat{j} + 10(t^2+1)\hat{k}\right] \cdot \left[2t\hat{i} + 4t\,\hat{j} + 3t^2\hat{k}\right]dt$$

$$= \int_1^2 [12(t^5+t^3) - 20t^4 + 30(t^4+t^2)]dt$$

$$= \int_1^2 (12t^2 + 10t^4 + 12t^3 + 30t^2)dt$$

$$= \left[12\frac{t^6}{6} + 10\frac{t^5}{5} - 12\frac{t^4}{4} + 30\frac{t^3}{3}\right]_1^2$$

$$= 320 - 17 = 303.$$

EXAMPLE 13.72

Find the work done by the force

$$\vec{f} = (2y+3)\hat{i} + xz\,\hat{j} + (yz - x)\hat{k},$$

when it moves a particle from the point (0,0,0) to the point (2, 1, 1) along the curve $x = 2t^2$, $y = t$, and $z = t^3$.

Solution. The parametric equations of the curve are $x = 2t^2$, $y = t$, and $z = t^3$. Further,

$$\vec{f} = (2y+3)\hat{i} + xz\,\hat{j} + (yz - x)\hat{k}$$
$$= (2t+3)\hat{i} + 2t^5\,\hat{j} + (t^4 - 2t^2)\hat{k}$$

and

$$\vec{r} = x\hat{i} + y\hat{j} + z\hat{k} = 2t^2\,\hat{i} + t\,\hat{j} + t^3\hat{k}.$$

Therefore,

$$\frac{d\vec{r}}{dt} = 4t\,\hat{i} + \hat{j} + 3t^2\hat{k}.$$

The given points (0,0,0) and (2, 1, 1) correspond to $t = 0$ and $t = 1$. Therefore, the work done by the force is given by

$$W = \int_C \vec{f}.d\vec{r}$$
$$= \int_0^1 \left[(2t+3)\hat{i} + 2t^5\,\hat{j} + (t^4 - 2t^2)\hat{k}\right] \cdot \left[4t\,\hat{i} + \hat{j} + 3t^2\hat{k}\right]dt$$
$$= \int_0^1 [(2t+3)4t + 2t^5 + 3(t^4 - 2t^2)t^2]dt$$
$$= \int_0^1 [8t^2 + 12t + 2t^5 + 3t^6 - 6t^4]dt$$
$$= \int_0^1 [3t^6 + 2t^5 - 6t^4 + 8t^2 + 12t]dt$$
$$= \left[3\frac{t^7}{7} + 2\frac{t^6}{6} - 6\frac{t^5}{5} + 8\frac{t^3}{3} + 12\frac{t^2}{2}\right]_0^1$$
$$= \frac{3}{7} + \frac{1}{3} - \frac{6}{5} + \frac{8}{3} + 6 = \frac{288}{35}.$$

EXAMPLE 13.73

Find the work done in moving a particle in the force field $\vec{f} = 3x^2\hat{i} + (2xz - y)\hat{j} + z\hat{k}$ along

(i) the straight line from (0,0,0) to (2, 1, 3) and

(ii) the curve defined by $x^2 = 4y$ and $3x^2 = 8z$ from $x = 0$ to $x = 2$.

Solution. (i) The curve C is the line joining (0,0,0) to (2, 1, 3) whose equation is

$$\frac{x-0}{2-0} = \frac{y-0}{1-0} = \frac{z-0}{3-0} \text{ or } \frac{x}{2} = \frac{y}{1} = \frac{Z}{3} = t, \text{ say.}$$

Thus, $x = 2t$, $y = t$, and $z = 3t$ are the parametric equations of the line. The point (0,0,0) corresponds to $t = 0$ and the point (2, 1, 3) corresponds to $t = 1$. Also,

$$\vec{f} = 3x^2\hat{i} + (2xz - y)\hat{j} + z\hat{k}$$
$$= 12t^2\hat{i} + (12t^2 - t)\hat{j} + 3t\hat{k}$$

and

$$\vec{r} = x\hat{i} + y\hat{j} + z\hat{k} = 2t\,\hat{i} + t\,\hat{j} + 3t\,\hat{k}.$$

Therefore, $\frac{d\vec{r}}{dt} = 2\hat{i} + \hat{j} + 3\hat{k}$. Hence,

$$W = \int_C \vec{f}.d\vec{r} = \int_0^1 \left(\vec{f}.\frac{d\vec{r}}{dr}\right)dt$$
$$= \int \{[12t^2\,\hat{i} + (12t^2 - t)\,\hat{j} + 3t\,\hat{k}] \cdot \left[2\hat{i} + \hat{j} + 3\hat{k}\right]\}dt$$
$$= \int_0^1 [24t^2 + (12t^2 - t) + 9t]dt$$
$$= \int_0^1 (36t^2 + 8t)dt = [12t^3 + 4t^2]_0^1 = 16.$$

(ii) Putting $x = t$ in the given curve, we get $y = \frac{t^2}{4}$ and $z = \frac{3t^3}{3}$, where $0 \le t \le 2$. Then,

$$\vec{f} = 3x^2\,\hat{i} + (2xz - y)\hat{j} + z\hat{k}$$
$$= 3t^2\hat{i} + \left(\frac{3}{4}t^4 - \frac{t^2}{4}\right)\hat{j} + \frac{3}{8}t^3\hat{k}$$

and

$$\vec{r} = x\hat{i} + y\,\hat{j} + z\,\hat{k} = t\,\hat{i} + \frac{t^2}{4}\hat{j} + \frac{3}{8}t^3\hat{k},\ 0 \le t \le 2.$$

Therefore,

$$\frac{d\vec{r}}{dt} = \hat{i} + \frac{1}{2}t\,\hat{j} + \frac{9}{8}t^2\hat{k}.$$

Hence,

$$W = \int_C \vec{f}.d\vec{r} = \int_0^2 \left(\vec{f}.\frac{d\vec{r}}{dt}\right)dt$$

$$= \int_0^2 (3t^2 + \frac{3}{8}t^5 - \frac{1}{8}t^3 + \frac{27}{64}t^5)dt$$

$$= \left[t^3 + \frac{t^6}{16} - \frac{t^4}{32} + \frac{9}{128}t^6\right]_0^2 = 16.$$

13.16 SURFACE INTEGRAL

An integral evaluated over a surface is called a *surface integral*. Two types of surface integral exist:

(i) $\iint_S f(x,y,z)dS$

and

(ii) $\iint_S \vec{f}(\vec{r}).\hat{n}\,dS = \iint_S \vec{f}(\vec{r}).d\vec{S}$.

In case (i), we have a scalar field f, whereas in case (ii), we have a vector field $\vec{f}(\vec{r})$, vector element of area $d\vec{S} = \hat{n}\,dS$, and $\hat{n}$ the outward-drawn unit normal vector to the element dS.

(i) Let $f(x, y, z)$ be a scalar-point function defined over a surface S of finite area. Partition the area S into n subareas $\delta S_1, \delta S_2, \ldots, \delta S_n$. In each area δS_i, choose an arbitrary point P_i (x_i, y_i, z_i). Define $f(P_i) = f_i\,(x_i, y_i, z_i)$ and form the sum $\sum_{i=1}^{n} f(x_i, y_i, z_i)\delta S_i$. Then, the limit of this sum as $n \to \infty$ in such a way that the largest of the subarea δS_i approaches zero is called the *surface integral of f (x,y,z)* over S and is denoted by $\iint_S f(x, y, z)$ dS.

(ii) Now, let $\vec{f}$ be a vector-point function defined and continuous over a surface S. Let P be any point on the surface S and let $\hat{n}$ be the unit vector at P in the direction of the outward-drawn normal to the surface S at P. Then, $\vec{f}.\hat{n}$ is the normal component of $\vec{f}$ at P. The integral of $\vec{f}.\hat{n}$ over S is called the *normal surface integral* of $\vec{f}$ over S and is denoted by $\iint_S \vec{f}.\hat{n}\,dS$. This integral is also known as *flux of* $\vec{f}$ over S. If we associate with the differential of surface area dS, a vector $d\vec{S}$, with magnitude dS, and whose direction is that of $\hat{n}$, then $d\vec{S} = \hat{n}dS$ and hence,

$$\iint_S \vec{f}.\hat{n}\,dS = \iint \vec{f}.d\vec{S}.$$

The surface integrals are easily evaluated by expressing them as double integrals, taken over an orthogonal projection of the surface S on any of the coordinate planes. But, the condition for this is that any line perpendicular to the coordinate plane chosen meets the surface S in not more than one point. However, if S does not satisfy this condition, then S can be subdivided into surfaces satisfying this condition.

Let S be the surface such that any line perpendicular to the xy-plane meets S in not more than one point. Then, the equation of the surface S can be written as $z = h(xy)$. Let R_1 be the projection of S on the xy-plane. Then, the projection of dS on the xy-plane is dS cos γ, where γ is the acute angle which the normal $\hat{n}$ at P to the surface S makes with z-axis. Therefore,

$$dS \cos \gamma = dxdy.$$

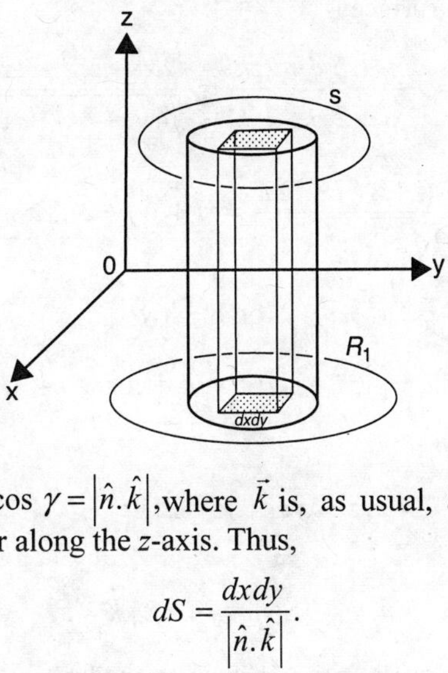

But $\cos \gamma = |\hat{n}.\hat{k}|$, where $\vec{k}$ is, as usual, a unit vector along the z-axis. Thus,

$$dS = \frac{dxdy}{|\hat{n}.\hat{k}|}.$$

Hence,

$$\iint_S \vec{f}.\hat{n}\,dS = \iint_{R_1} \vec{f}.\hat{n}\frac{dxdy}{|\hat{n}.\hat{k}|}.$$

Similarly, if R_2 and R_3 are projections of S on the yz, and zx-plane, respectively, then

$$\iint_S \vec{f}.\hat{n}\,dS = \iint_{R_2} \vec{f}.\hat{n}\frac{dxdy}{|\hat{n}.\hat{i}|},$$

and

$$\iint_S \vec{f}.\hat{n}\,dS = \iint_{R_3} \vec{f}.\hat{n}\frac{dxdy}{|\hat{n}.\hat{j}|}.$$

EXAMPLE 13.74

Evaluate $\iint_S \vec{f}.\hat{n}\,dS$, where $\vec{f} = yz\hat{i} + zx\hat{j} + xy\hat{k}$ and S is that part of the surface of the sphere $x^2 + y^2 + z^2 = 1$, which lies in the first octant.

Solution. A vector normal to the surface of the given sphere is

$$\nabla(x^2 + y^2 + z^2 - 1)$$

$$= \left(\hat{i}\frac{\partial}{\partial x} + \hat{j}\frac{\partial}{\partial y} + \hat{k}\frac{\partial}{\partial z}\right)(x^2 + y^2 + z^2 - 1)$$

$$= 2x\hat{i} + 2y\hat{j} + 2z\hat{k}.$$

Therefore, the unit normal to any point (x, y, z) of the surface is

$$\hat{n} = \frac{2x\hat{i} + 2y\hat{j} + 2z\hat{k}}{|2x\hat{i} + 2y\hat{j} + 2z\hat{k}|} = \frac{2x\hat{i} + 2y\hat{j} + 2z\hat{k}}{\sqrt{4x^2 + 4y^2 + 4z^2}}$$

$$= \frac{2x\hat{i} + 2y\hat{j} + 2z\hat{k}}{2\sqrt{x^2 + y^2 + z^2}} = x\hat{i} + y\hat{j} + z\hat{k},$$

since $x^2 + y^2 + z^2 = 1$ on S. Now,

$$\vec{f}.\hat{n} = \left(yz\hat{i} + zx\hat{j} + xy\hat{k}\right)\left(x\hat{i} + y\hat{j} + z\hat{k}\right)$$

$$= xyz + xyz + xyz = 3xyz$$

and

$$\vec{n}.\hat{k} = .\left(x\hat{i} + y\hat{j} + z\hat{k}\right).\hat{k} = z,$$

which gives $|\hat{n}.\hat{k}| = z$. Hence, in the first quadrant,

$$\iint_S \vec{f}.\hat{n}\,dS = \iint_S \vec{f}.\hat{n}\frac{dxdy}{|\hat{n}.\hat{k}|}$$

$$\iint_R \frac{3xyz}{z}\,dxdy$$

$$= 3\iint_R xy\,dx\,dy$$

$$= 3\int_0^1\left[\int_0^{\sqrt{1-x^2}} xy\,dy\right]dx$$

$$= \frac{3}{2}\int_0^1 x(1-x^2)\,dx$$

$$= \frac{3}{2}\left[\frac{x^2}{2} - \frac{x^4}{4}\right]_0^1 = \frac{3}{8}.$$

EXAMPLE 13.75

Evaluate $\iint_S \vec{f}.\hat{n}\,dS$, where $\vec{f} = 18z\hat{i} - 12\hat{j} + 3y\hat{k}$ and S is the surface $2x + 3y + 6z = 12$ in the first octant.

Solution. A vector normal to the surface S is

$$\nabla(2x + 3y + 6z = 12) = 2\hat{i} + 3\hat{j} + 6\hat{k}.$$

Therefore, the unit normal vector to the surface S is

$$\hat{n} = \frac{2\hat{i} + 3\hat{j} + 6\hat{k}}{\sqrt{4+9+36}} = \frac{2\hat{i} + 3\hat{j} + 6\hat{k}}{7}$$

and so,

$$\hat{n}.\hat{k} = \left(\frac{2\hat{i} + 3\hat{j} + 6\hat{k}}{77}\right).\hat{k} = \frac{6}{7}.$$

Also,

$$\vec{f}.\hat{n} = (18z\hat{i} - 12\hat{j} + 3y\hat{k})\cdot\left(\frac{2\hat{i} + 3\hat{j} + 6\hat{k}}{7}\right)$$

$$= \frac{36z}{7} - \frac{36}{7} + \frac{18}{7}y$$

$$= \frac{36}{7}\left[\frac{12 - 2x - 3y}{6}\right] - \frac{36}{7} + \frac{18}{7}y$$

$$= \frac{36}{7} - \frac{12x}{7}.$$

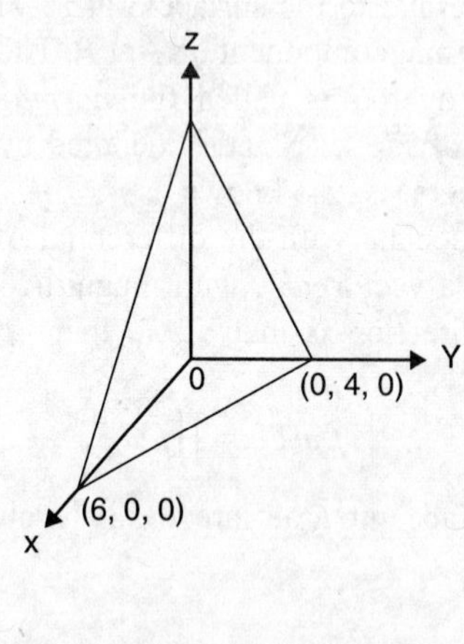

Hence,

$$\iint_S \vec{f}.\hat{n}\, dS = \iint_R \vec{f}.\hat{n}\frac{dxdy}{|\hat{n}.\hat{k}|}$$

$$= \frac{7}{6}\iint_R \left(\frac{36}{7} - \frac{12x}{7}\right)dx\, dy$$

$$= \iint_R (6-2x)dx\, dy.$$

But, R is the region of projection of S (triangle) on the xy plane. Thus, the projection is a triangle bounded by x-axis, y-axis, and the line $2x+3y=12$ and $z=0$. Hence, the limits of x are from 0 to 6 and that of y are from 0 to $\frac{12-2x}{3}$. Therefore,

$$\iint_S \vec{f}.\vec{n}\, dS = \int_0^6\left[\int_0^{\frac{12-2x}{3}} (6-2x)dy\right]dx$$

$$= \int_0^6 [6y - 2xy]_0^{\frac{12-2x}{3}}\, dx$$

$$= \frac{1}{3}\int_0^6 (72 - 36x + 4x^2)\, dx$$

$$= \frac{1}{3}\left[72x - 36\frac{x^2}{2} + \frac{4x^3}{3}\right]_0^6$$

$$= 144 - 216 + 96 = 24.$$

EXAMPLE 14.76

Evaluate $\vec{f}.d\vec{S}$, where $\vec{f} = z\hat{i} + x\hat{j} - 3y^2z\hat{k}$ and S is the surfaces of the cylinder $x^2+y^2=16$ in the first octant between $z=0$ and $z=5$.

Solution. The surface S is shown in the following figure:

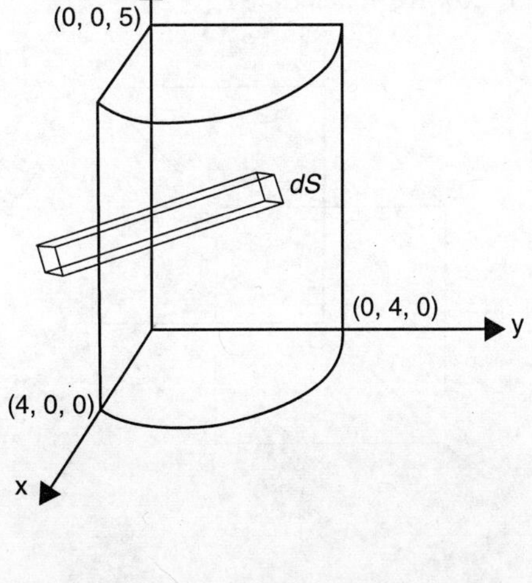

A vector normal to the surface S is given by

$$\nabla(x^2+y^2-16) = 2x\hat{i} + 2y\hat{j},$$

so that the unit normal vector $\hat{n}$ at any point of S is

$$\hat{n} = \frac{2x\hat{i}+2y\hat{j}}{\sqrt{4x^2+4y^2}} = \frac{2x\hat{i}+2y\hat{j}}{2\sqrt{x^2+y^2}}$$

$$= \frac{2x\hat{i}+2y\hat{j}}{8} = \frac{x\hat{i}+y\hat{j}}{4}.$$

Also,

$$\vec{f}.\hat{n} = (z\hat{i} + x\hat{j} - 3y^2z\hat{k}).\left(\frac{x\hat{i}+y\hat{j}}{4}\right) = \frac{1}{4}(xz+xy).$$

Let R be the projection of the surface S on xz-plane. Then

$$\hat{n} = \frac{1}{4}(x\hat{i}+y\hat{j}).(\hat{j}) = \frac{y}{4}.$$

Hence

$$\iint_S \vec{f}.dS = \iint_S \vec{f}.\hat{n}\, dS = \iint_R \frac{xz+xy}{4}\cdot\frac{dxdz}{|\hat{n}.\hat{j}|}$$

$$= \iint_R \frac{xz+xy}{4}.\frac{dxdz}{\frac{y}{4}} = \iint_R \frac{xz+xy}{y}.dxdz,$$

where R is the rectangular region in the xz-plane bounded by $0 \le x \le 4$, $0 \le z \le 5$. Since the integrand is still evaluated on the surface, we have $y = \sqrt{16-x^2}$ and so,

$$\vec{f}.d\vec{S} = \int_0^4\left[\int_0^5\left(x + \frac{XZ}{\sqrt{16-x^2}}\right)dz\right]dx$$

$$= \int_0^4\left[xz + \frac{xz^2}{2\sqrt{16-x^2}}\right]_0^5 dx$$

$$= \int_0^4\left(5x + \frac{25x}{2\sqrt{16-x^2}}\right)dx$$

$$= \left[5\frac{x^2}{2} - \frac{25}{2}\sqrt{16-x^2}\right]_0^4 = 90.$$

EXAMPLE 13.77

Evaluate $\iint_S \vec{f}.\hat{n}\, dS$, where $\vec{f} = 4x\hat{i} - 2y^2\hat{j} + z^2\hat{k}$ and S is the surface bounding the region

$x^2 + y^2 = 4,\ z = 0$, and $z = 3$.

Solution. The region is bounded by the cylinder $x^2 + y^2 = 4,\ z = 0$, and $z = 3$. Therefore, the surface S consists of three parts:

(i) S_1, the circular base of the cylinder in the plane $z = 0$,

(ii) S_2, the circular top in the plane $z = 3$, and

(iii) S_3, the curved surface of the cylinder given by $x^2 + y^2 = 4$.

Now, for the subsurface S_1, we have $z = 0, \hat{n} = -\hat{k}$, and $\vec{f} = 4x\hat{i} - 2y^2\hat{j}$.
Therefore,

$$\vec{f}.\hat{n} = (4x\hat{i} - 2y^2\hat{j}).\left(-\hat{k}\right) = 0.$$

Hence, $\iint_{S_1} \vec{f}.\hat{n}\, dS = 0.$

On S_2, we have $z = 3,\ \hat{n} = \hat{k}$, and $\vec{f} = 4x\hat{i} - 2y^2\hat{j} + 9\hat{k}$. Therefore,

$$\vec{f}.\hat{n} = \left(4x\hat{i} - 2y^2\hat{j} + 9\hat{k}\right).\left(\hat{k}\right) = 9.$$

Hence,

$$\iint_{S_2} \vec{f}.\hat{n}ds = \iint_{S_2} 9\, dxdy$$

$$= 9(4\pi) = 36\pi$$

For the surface S_3, which is the curved surface of the cylinder and is given by $x^2 + y^2 = 4$, the vector normal to the surface is

$$\nabla(x^2 + y^2 - 4) = 2x\hat{i} + 2y\hat{j}.$$

Therefore, the unit normal vector to the surface S3 is given by

$$\hat{n} = \frac{2x\hat{i} + 2y\hat{j}}{\sqrt{4x^2 + 4y^2}} = \frac{2x\hat{i} + 2y\hat{j}}{2\sqrt{x^2 + y^2}}$$

$$= \frac{2x\hat{i} + 2y\hat{j}}{2.2}$$

$$= \frac{x\hat{i} + y\hat{j}}{2}.$$

Therefore,

$$\vec{f}.\hat{n} = (4x\hat{i} - 2y^2\hat{j} + z^2\hat{k}).\left(\frac{x\hat{i} + y\hat{j}}{2}\right) = 2x^2 - y^3.$$

Now, on S_3, $x = 2\cos\theta, y = 2\sin\theta$, and $dS = 2d\theta dz$. For this surface, z varies from 0 to 3 and θ varies from 0 to 2π. Therefore,

$$\iint_{S_3} \vec{f}.\hat{n}\, dS = \int_0^{2\pi}\int_0^{3}\left[2(2\cos\theta)^2 - (2\sin\theta)^3\right]2d\theta\, dz$$

$$= \int_0^{2\pi} 16\left(\cos^2\theta - \sin^3\theta\right)[z]_0^3\, d\theta$$

$$= 48\int_0^{2\pi}\left(\cos^2\theta - \sin^3\theta\right)d\theta$$

$$= 48\int_0^{2\pi}\cos^2\theta\, d\theta - 48\int_0^{2\pi}\sin^3\theta\, d\theta$$

$$= (48)(4)\int_0^{\frac{\pi}{2}}\cos^2\theta\, d\theta - 0$$

$$= 192.\frac{1}{2}.\frac{\pi}{2}$$

$$= 48\pi.$$

Hence,

$$\iint_S \vec{f}.\hat{n}\, dS = \iint_{S_1} \vec{f}.\hat{n}dS + \iint_{S_2} \vec{f}.\hat{n}\, dS\ \iint_{S_3} \vec{f}.\hat{n}\, dS$$

$$= 0 + 36\,\pi + 48\,\pi$$

$$= 84\,\pi.$$

EXAMPLE 13.78

If $\vec{f} = 4xz\,\hat{i} - y^2\hat{j} + yz\,\hat{k}$, evaluate $\int \vec{f}.\hat{n}\, dS$, where S is the surface of the cube bounded by $x = 0, x = 1, y = 0, y = 1, z = 0$, and $z = 1$.

Solution. The surface of the cube is bounded by $x = 0, x = 1, y = 0, y = 1, z = 0$, and $z = 1$ and so, the surface can be subdivided into six parts in the following manner:

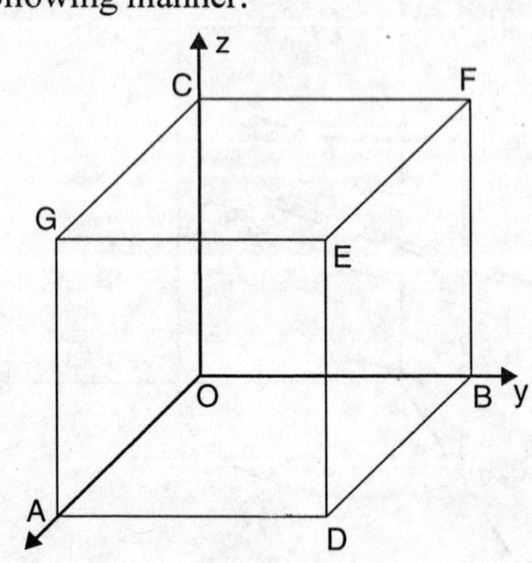

(i) S_1 is the surface formed by the face OADB, where
$z = 0,\ \hat{n} = -\hat{k}$, and $\vec{f} = -y^2\hat{j}$, so that

$$\vec{f}.\hat{n} = \left(-y^2\hat{j}\right).\left(-\hat{k}\right) = 0$$

and

$$\iint_{S_1} \vec{f}.\hat{n}\ dS = \int_0^1\int_0^1 0 dxdy = 0.$$

(ii) S_2 is the surface formed by the face GEFC, where $z = 1\ \hat{n} = \hat{k}$, and $dz = 0$. On this face, we have

$$\vec{f} = 4xz\ \hat{i} - y^2\hat{j} + yz\ \hat{k} = 4x\ \hat{i} - y^2\hat{j} + y\hat{k}$$

and so,

$$\vec{f}.\hat{n} = \left(4x\hat{i} - y^2\hat{j} + y\hat{k}\right).\left(\hat{k}\right) = y.$$

Hence,

$$\iint_{S_2} \vec{f}.\hat{n}\ dS = \int_0^1\int_0^1 ydxdy = \int_0^1 \left[\frac{y^2}{2}\right]_0^1 dx$$

$$= \frac{1}{2}\int_0^1 dx = \frac{1}{2}.$$

(iii) S_3 is the surface formed by the face ADEG, where $\hat{n} = \hat{i},\ x = 1$, and $dx = 0$. On this face,

$\vec{f} = 4xz\ \hat{i} - y^2\hat{j} + yz\hat{k} = 4z\hat{i} - y^2\hat{j} + yz\hat{k}$ and
$\vec{f}.\hat{n} = \left(4z\ \hat{i} - y^2\hat{j} + yz\ \hat{k}\ \right).\hat{i} = 4z.$

Hence,

$$\iint_{S_3} \vec{f}.\hat{n}\ ds = \int_0^1\int_0^1 4z\ dy\ dx$$

$$= 4\int_0^1 \left[\tfrac{z^2}{2}\right]_0^1 dy = 2\int_0^1 dy = 2.$$

(iv) S_4 is the surface formed by OBFC, where $\hat{n} = -\hat{i}\ \ x = 0$, and $dx = 0$. On this face,

$\vec{f} = 4xz\ \hat{i} - y^2\hat{j} + yz\ \hat{k} = -y^2\hat{j} + yz\ \hat{k}$ and
$\vec{f}.\hat{n} = \left(-y^2\hat{j} + yz\ \hat{k}\right)\cdot\left(-\hat{i}\right) = 0.$

Hence,

$$\iint_{S_4} \vec{f}.\hat{n}\ dS = \int_0^1\int_0^1 0\ dx\,dy = 0.$$

(v) S_5 is the surface formed by the face OCGA, where $\hat{n} = -\hat{j},\ y = 0$, and $dy = 0$. On this face,

$\vec{f} = 4xz\ \hat{i} - y^2\hat{j} + yz\hat{k} = 4xz\hat{i}$ and so,
$\vec{f}.\hat{n} = \left(4xz\ \hat{i}\right).\left(-\hat{j}\right) = 0.$

Hence,

$$\iint_{S_5} \vec{f}.\hat{n}\ dS = 0.$$

(vi) S_6 is the surface formed by the face DBFE, where $\hat{n} = \hat{j},\ y = 1$, and $dy = 0$. On this face,
$\vec{f} = 4xz\ \hat{i} - y^2\hat{j} + yz\ \hat{k} = 4xz\hat{i} - \hat{j} + z\ \hat{k}$ and
$\vec{f}.\hat{n} = \left(4xz\ \hat{i} - \hat{j} + z\ \hat{k}\right)\cdot\left(\hat{j}\right) = -1.$

Therefore,

$$\iint_{S_6} \vec{f}.\hat{n} \pm\pm\pm \int_0^1\int_0^1 (-1) \quad = -\int_0^1 [\ ]_0^1$$

$$= -1\int_0^1 dx = -1.$$

Hence,

$$\iint_S \vec{f}.\hat{n}\ dS = \iint_{S_1} + \iint_{S_2} + \iint_{S_3} + \iint_{S_4} + \iint_{S_5} + \iint_{S_6}$$

$$= 0 + \frac{1}{2} + 2 + 0 + 0 + -1 = \frac{3}{2}.$$

13.17 VOLUME INTEGRAL

Let ϕ be a scalar-point function defined throughout a given region of volume V. Partition the given region into n subregions of volumes $\delta V_1, \delta V_2, \ldots, \delta V_n$. Let $P(x_i\, y_i\, z_i)$ be any point inside or on the boundary of the subregion of volume δV_i. Then the limit

$$\lim_{\substack{n\to\infty \\ \delta V_i \to 0}} \sum_{i=1}^{n} \phi(P_i)\,\delta V_i,$$

if exists, for all mode of subdivision (partition), is called the *volume integral* of ϕ over the volume V, and this integral is denoted by

$$\iiint_V \phi\ dV = \iiint_V \phi\ dx\ dy\ dz.$$

Similarly, if $\vec{f} = f_1\,\hat{i} + f_2\,\hat{j} + f_3\,\hat{k}$ is a vector-point function, then

$$\iiint_V \vec{f}\ dV = \hat{i}\iiint_V f_1(x,y,z)\,dx\ dy\ dz$$

$$+\hat{j}\iiint_V f_2(x,y,z)\,dx\ dy\ dz$$

$$+\hat{k}\iiint_V f_3(x,y,z)\,dx\ dy\ dz.$$

EXAMPLE 13.79

If $\vec{f} = (2x^2 - 3z)\hat{i} - 2xy\ \hat{j} - 4x\ \hat{k}$, evaluate $\iiint_V \nabla\times\vec{f}\ dV$, where V is the region bounded by the coordinate planes and the plane $2x + 2y + z = 4$.

Solution. We have

$$\nabla\times\vec{f} = \begin{vmatrix} \hat{i} & \hat{j} & \hat{k} \\ \frac{\partial}{\partial x} & \frac{\partial}{\partial y} & \frac{\partial}{\partial z} \\ 2x^2-3z & -2xy & -4x \end{vmatrix}$$

$$= \hat{i}\left[\frac{\partial}{\partial y}(-4x) - \frac{\partial}{\partial z}(-2xy)\right]$$

$$+\hat{j}\left[\frac{\partial}{\partial z}(2x^2-3z) - \frac{\partial}{\partial x}(-4x)\right]$$

$$+\hat{k}\left[\frac{\partial}{\partial x}(-2xy) - \frac{\partial}{\partial y}(2x^2-3z)\right]$$

$$= \hat{j} - 2y\,\hat{k}.$$

The region V is bounded by the planes $x = 0$, $y = 0$, $z = 0$ and the plane $2x + 2y + z = 4$. Therefore, the limits of integration are:

z varies from 0 to $4 - 2x - 2y$,

y varies from 0 to $2 - x$, and

x varies from 0 to 2.

Hence,

$$\iiint_V \nabla\times\vec{f}\ dV$$

$$= \int_0^2\int_0^{2-x}\int_0^{4-2x-2y} (\hat{j} - 2y\,\hat{k})\,dx\ dy\ dz$$

$$= \int_0^2\int_0^{2-x} (\hat{j} - 2y\hat{k})[z]_0^{4-2x-2y}\ dx\ dy$$

$$= \int_x^2\int_0^{2-x}\left[(4-2x-2y)\hat{j} - (8y - 4xy + 4y^2)\hat{k}\right]dx\ dy$$

$$= \int_0^2\left[\left(4y - 2xy - 2\frac{y^2}{2}\right)\hat{j} - \left(8\frac{y^2}{2} - 4x\frac{y^2}{2} + 4\frac{y^3}{3}\right)\hat{k}\right]_0^{2-x}$$

$$= \int_0^2\left[(2-x)^2\,\hat{j} - \frac{2}{3}(2-x)^3\,\hat{k}\right]dx$$

$$= \left[\frac{(2-x)^3}{-3}\hat{j} - \frac{2}{3}\frac{(2-x)^4}{-4}\hat{k}\right]_0^2$$

$$= \frac{8}{3}\hat{j} - \frac{8}{3}\hat{k} = \frac{8}{3}(\hat{j} - \hat{k}).$$

EXAMPLE 13.80

If $\vec{f} = (2x^2 - 3z)\hat{i} - 2xy\ \hat{j} - 4x\,\hat{k}$, then evaluate $\iiint_V \nabla\cdot\vec{f}\ dV$, where V is bounded by the coordinate planes and the plane $2x + 2y + z = 4$.

Solution. We have

$$\nabla\cdot\vec{f} = \left(\frac{\partial}{\partial x}\hat{i} + \frac{\partial}{\partial y}\hat{j} + \frac{\partial}{\partial z}\hat{k}\right)$$

$$\cdot\left[(2x^2-3z)\hat{i} - 2xy\,\hat{j} - 4x\,\hat{k}\right]$$

$$= 4x - 2x = 2x.$$

The limits of integration are as mentioned in Example 13.79. Therefore,

$$\iiint_v \nabla.\vec{f}dv = \int_0^2\int_0^{2-x}\int_0^{4-2x-2y} 2x\,dx\,dy\,dz$$

$$= 2\int_0^2\int_0^{2-x} x(4-2x-2y)\,dx\,dy$$

$$= 4\int_0^2\int_0^{2-x} x(2-x-2)\,dx\ dy$$

$$= \frac{4}{2}\int_0^2 x(2-x)^2\ dx$$

$$= 2\left[\frac{x^2}{4} - \frac{4x^3}{3} + \frac{4x^2}{2}\right]_0^2 = \frac{8}{3}.$$

EXAMPLE 13.81

Evaluate $\iiint_V \nabla \cdot \vec{f}\, dV$ if $\vec{f} = 4xy\,\hat{i} + yz\,\hat{j} - xy\,\hat{k}$ and V is bounded by $x = 0, x = 2, y = 0, y = 2, z = 0$ and $z = 2$.

Solution. We have

$$\iiint_V \nabla \cdot \vec{f}$$

$$= \iiint_V \left(\frac{\partial}{\partial x}(4xy) + \frac{\partial}{\partial y}(yz) + \frac{\partial}{\partial z}(-xy)\right) dV$$

$$= \iiint_V (4y+z)dV = \int_0^2\int_0^2\int_0^2 (4y+z)dz\ dy\ dx$$

$$= \int_0^2\int_0^2 \left[4yz + \frac{z^2}{2}\right]_0^2 dy\ dx = 2\int_0^2\int_0^2 (4y+1)\,dy\ dx$$

$$= 2\int_0^2 \left[\frac{4y^2}{2} + y\right]_0^2 dx = 4\int_0^2 5dx = 20[x]_0^2 = 40.$$

13.18 GAUSS'S DIVERGENCE THEOREM

The following theorem of Gauss is useful in evaluating the surface integral over a closed surface by reducing it to a volume integral (triple integral) and vice versa.

Theorem 13.10. (Gauss's Divergence Theorem).

Let $\vec{f}$ be a vector-point function possessing continuous first-order partial derivatives at each point of a three-dimensional region V enclosed in a closed surface S. Then,

$$\iint_S \vec{f}.\hat{n}\ dS = \iiint_V \text{div}\ \vec{f}\ dV = \iiint_V \nabla \cdot \vec{f}\ dV,$$

where $\hat{n}$ is the outward-drawn unit normal vector to the surface S.

The divergence theorem can be expressed in the form of Cartesian coordinates as follows:

Let $\vec{f} = f_1\hat{i} + f_2\hat{j} + f_3\hat{k}$. Then

$$\text{div}\ \vec{f} = \nabla \cdot \vec{f} = \frac{\partial f_1}{\partial x} + \frac{\partial f_2}{\partial y} + \frac{\partial f_3}{\partial z}.$$

Let the outward-drawn unit normal vector $\hat{n}$ makes angles α, β, and γ, respectively, with positive directions of x-, y-, and z-axis. Thus, $\cos\alpha$, $\cos\beta$, and $\cos\gamma$ are the direction cosines of $\hat{n}$ and so,

$$\hat{n} = \cos\alpha\,\hat{i} + \cos\beta\,\hat{j} + \cos\gamma\,\hat{k}$$

and then,

$$\vec{f}.\hat{n} = \left(f_1\hat{i} + f_2\hat{j} + f_3\hat{k}\right).\left(\cos\alpha\,\hat{i} + \cos\beta\,\hat{j} + \cos\gamma\hat{k}\right)$$

$$= f_1\cos\alpha + f_2\cos\beta + f_3\cos\gamma.$$

Hence, the Gauss's Divergence Theorem takes the form

$$\iiint_V \left(\frac{\partial f_1}{\partial x} + \frac{\partial f_2}{\partial y} + \frac{\partial f_3}{\partial z}\right) dx\ dy\ dz$$

$$= \iint_S (f_1\,dy\,dz + f_2\,dz\,dx + f_3 dx\,dy),$$

since $\cos\alpha\ dS = dy\ dz$, $\cos\beta\ dS = dz\ dx$, and $\cos\gamma\ dS = dxdy$. This form of Gauss's Divergence Theorem is also known as *Green's Theorem in Space*.

Proof: Consider a closed surface S, which is such that it is possible to introduce a rectangular coordinate system, such that any line parallel to any coordinate axis cuts S in, at the most, two points. Let R be the projection of the surface S on the *xy*-plane. Then, in accordance to our assumption, a line through a point (x, y, 0) of R meets the boundary of S in two points. Suppose that the z coordinates of these points are $z = \phi_1(x, y)$ and $z = \phi_2(x, y)$, where $\phi_2(x,y) \ge \phi_1(x,y)$ Then,

$$\iiint_V \frac{\partial f_3}{\partial z} dV = \iiint_V \frac{\partial f_3}{\partial z} dz\ dy\ dx$$

$$= \iint_R \left[\int_{\phi_1(x,y)}^{\phi_2(x,y)} \frac{\partial f_3}{\partial z} dz\right] dy\,dx$$

$$= \iint_R [f_3(x,y,z)]_{\phi_1(xy)}^{\phi_2(xy)}\, dy\,dx$$

$$= \iint_R \left[f_3(x,y,\phi_2) - f_3(x,y,\phi)_1\right] dy\,dx. \quad (1)$$

Let S_1 and S_2 be the portion of the surface S corresponding to $z = \phi_1(x,y)$ and $z = \phi_2(x,y)$, respectively. Let $\hat{n}_2$ be the outward-drawn unit normal vector to S_2, making an acute angle γ_2 with the positive direction $(\hat{k})$ of z-axis. If dS_2

is projected on the xy-plane, then this projection $dy\,dx$ of dS_2 is

$$dy\,dx = \cos\gamma_2 dS_2 = \hat{k}.\hat{k}_2\, dS_2.$$

In the same fashion, let $\hat{n}_1$ be the outward-drawn unit normal vector to S_1, making an obtuse angle γ_1 with $\hat{k}$. Then,

$$dy\,dx = \cos(\pi-\gamma_1)dS_1 = -\cos\gamma_1 dS_1 = -\hat{k}.\hat{n}\, dS_1.$$

Therefore,

$$\iint_R f_3(x,y,\phi_2)\,dy\,dx = \iint_{S_2} f_3\hat{k}.\hat{n}_2 dS_2$$

and

$$\iint_R f_3(x,y,\phi_1)\,dy\,dx = -\iint_{S_1} f_3\hat{k}.\hat{n}_1\, dS_1.$$

Hence, (1) reduces to

$$\iiint_V \frac{\partial f_3}{\partial z}dV = \iint_{S_2} f_3\hat{k}.\hat{n}_2\, dS_2$$

$$+\iint_{S_1} f_3\hat{k}.\hat{n}_1 dS_1$$

$$= \iint_S f_3\hat{k}.\hat{n}\, dS. \quad (2)$$

Similarly, projecting S on the remaining two coordinate planes, we have

$$\iiint_V \frac{\partial f_1}{\partial x}dV = \iint_S f_1\hat{i}.\,\hat{n}\, dS \text{ and} \quad (3)$$

$$\iiint_V \frac{\partial f_2}{\partial y}dV = \iint_S f_2\hat{j}.\hat{n}\, dS. \quad (4)$$

Adding (2), (3), and (4), we obtain

$$\iiint_V \left(\frac{\partial f_1}{\partial x}+\frac{\partial f_2}{\partial y}+\frac{\partial f_3}{\partial z}\right)dV$$

$$= \iint_S \left(f_1\hat{i}+f_2\hat{j}+f_3\hat{k}\right).\hat{n}\, dS$$

$$= \iint_S \vec{f}.\hat{n}\, dS.$$

This proves the theorem.

The Gauss's Divergence Theorem can be extended to the surfaces which are such that lines parallel to the coordinate axes meet them in more than two points. For this, the region enclosed by S is partitioned into subregions whose surfaces satisfy the condition assumed in the above proof. Applying the theorem to each subregion and adding will yield the required result.

Deductions:

(i) If $\hat{n}$ is the outward-drawn unit normal vector to S, then

$$\iiint_V \vec{f}.\nabla\phi dV = \iint_S \phi\vec{f}.\hat{n}\, dS - \iiint_V \phi \operatorname{div}\vec{f}\, dV.$$

Proof: By Gauss's Divergence Theorem,

$$\iint_S \vec{F}.\hat{n}\, dS = \iiint_V \operatorname{div}\vec{F}\, dV.$$

Putting $\vec{F} = \phi\vec{f}$, we have

$$\iint_S \phi\vec{f}.\hat{n}\, dS = \iiint_V \operatorname{div}(\phi\vec{f})\,dV$$

$$= \iiint_V (\phi \operatorname{div}\vec{f} + \vec{f}.\nabla\phi)\,dV$$

$$= \iiint_V \phi \operatorname{div}\vec{f}\, dV + \iiint_V \vec{f}.\nabla\phi dV$$

and so,

$$\iiint_V \vec{f}.\nabla\phi dV = \iint_S \phi\vec{f}.\hat{n}\, dS - \iiint_V \phi\operatorname{div}\vec{f}\, dV.$$

(ii) $\iint_S \vec{f}\times\hat{n}\, dS = -\iiint_V \operatorname{curl}\vec{f}\, dV.$

Proof: By Gauss's Divergence Theorem,

$$\iint_S \vec{F}.\hat{n}\, dS = \iiint_V \operatorname{div}\vec{F}\, dV.$$

Putting $\vec{F} = \vec{a}\times\vec{f}$, whose $\vec{a}$ is an arbitrary constant vector, we have

$$\iint_S (\vec{a}\times\vec{f}).\hat{n}\, dS = \iiint_V \nabla\cdot(\vec{a}\times\vec{f})\,dV$$

or

$$\iint_S \vec{a}(\vec{f}\times\hat{n})\,dS = -\iiint_S \vec{a}.(\nabla\times\vec{f})\,dV$$

or

$$\vec{a}.\iint_S (\vec{f}\times\hat{n})\,dS = -\vec{a}.\iiint_V (\nabla\times\vec{f})\,dV$$

or

$$\vec{a}.\left[\iint_S (\vec{f}\times\hat{n})\,dS + \iiint_V (\nabla\times\vec{f})\,dV\right] = 0$$

or

$$\iint_S (\vec{f} \times \hat{n}) ds + \iiint_V (\nabla \times \vec{f}) dV = 0,$$

that is,

$$\iint_S (\vec{f} \times \hat{n}) dS = -\iiint_V \text{curl } \vec{f} \, dV.$$

(iii) $\iint_S \phi \hat{n} \, dS = \iiint_V \text{grad } \phi \, dV.$

Proof: By Gauss's Divergence Theorem

$$\iint_S \vec{f}.\hat{n} \, dS = \iiint_V \text{div } \vec{f} \, dV.$$

Putting $\vec{f} = \vec{a}\,\phi$,, where $\vec{a}$ is on arbitrary constant vector, we get

$$\iint_S (\vec{a}\,\phi).\hat{n} \, dS = \iiint_V \text{div}(\vec{a}\,\phi) dV$$

$$= \iiint_V \nabla \cdot (\vec{a}\,\phi) dV.$$

Thus,

$$\vec{a}.\iint_S (\phi \hat{n}) dS = \vec{a}.\iiint_V (\nabla \phi) dV$$

or

$$\vec{a}.\left[\iint_S (\phi \, \hat{n}) dS - \iiint_V (\nabla \phi) dV\right] = 0$$

or

$$\iint_S (\phi \, \hat{n}) dS - \iiint_V (\nabla \phi) dV = 0$$

or

$$\iint_S \phi \, \hat{n} \, dS = \iiint_V \nabla \phi \, dV.$$

EXAMPLE 13.82

If S is a closed surface, $\hat{n}$ is the outward-drawn normal to S and V is the volume enclosed by S, show that

(i) $\iint_S \vec{r}.\hat{n} \, dS = 3V,$

(ii) $\iiint_V \text{div } \hat{n} \, dV = S,$

(iii) $\iint_S \hat{n} \, dS = \vec{0},$ and

(iv) $\iint_S \vec{f}.\hat{n} \, dS = 6V,$

where $\vec{f} = x\hat{i} + 2y\hat{j} + 3z\hat{k}.$

Solution. (i) By the divergence theorem,

$$\iint_S \vec{r}.\hat{n} \, ds = \iiint_V \text{div } \vec{r} \, dV = \iiint_V \nabla \cdot \vec{r} \, dV$$

$$= \iiint_V \left(\hat{i}\frac{\partial}{\partial x} + \hat{j}\frac{\partial}{\partial y} + \hat{k}\frac{\partial}{\partial z}\right) . (x\,\hat{i} + y\,\hat{j} + z\,\hat{k}) dV$$

$$= 3\iiint_V dV = 3V.$$

(ii) By the divergence theorem,

$$\iiint_V \text{div } \hat{n} \, dV = \iint_S \hat{n}.\hat{n} \, dS$$

$$= \iint_S dS = S.$$

(iii) If $\hat{a}$ is any constant vector, then

$$\vec{a}.\iint_S \hat{n} \, dS = \iint_S \vec{a}.\hat{n} \, dS$$

$$= \iiint_V \text{div } \vec{a} \, dV,$$

by divergence theorem

$= \vec{0}$, because div $\vec{a} = \vec{0}$.

(iv) By the Gauss's Divergence Theorem

$$\iint_S \vec{f}.\hat{n} \, dS = \iiint_V \nabla \cdot \vec{f} \, dV$$

$$= \iiint_V \left(\hat{i}\frac{\partial}{\partial x} + \hat{j}\frac{\partial}{\partial y} + \hat{k}\frac{\partial}{\partial z}\right) . \left(x\hat{i} + 2y\,\hat{j} + 3z\,\hat{k}\right) dV$$

$$= \iiint_V (1 + 2 + 3) dV$$

$$= 6\iiint_V dV = 6V.$$

EXAMPLE 13.83

Verify the divergence theorem for $\vec{f} = 4x\,\hat{i} - 2y^2\,\hat{j} + z^2\hat{k}$, taken over the region bounded by the cylinder $x^2 + y^2 = 4$, $z = 0$ and $z = 3$.

Solution. In Example 13.77, we have shown that

$$\iint_S \vec{f}.\hat{n} \, dS = 84\pi.$$

On the other hand,

$$\iiint_V \text{div } \vec{f}\, dV = \iiint_V \nabla \cdot \vec{f}\, dV$$

$$= \iiint_V \left[\frac{\partial}{\partial x}(4x) + \frac{\partial}{\partial y}(-2y^2) + \frac{\partial}{\partial z}(z^2) \right] dV$$

$$= \iiint_V (4 - 4y + 2z)dx\, dy\, dz.$$

Since z varies from 0 to 3, y varies from $-\sqrt{4-x^2}$ to $\sqrt{4-x^2}$, and x varies from –2 to 2, we have

$$\iiint_V \text{div } \vec{f}\, dV$$

$$= \int_{-2}^{2} \int_{-\sqrt{4-x^2}}^{\sqrt{4-x^2}} \left[\int_0^3 (4 - 4y + 2z)dz \right] dy\, dx$$

$$= \int_{-2}^{2} \int_{-\sqrt{4-x^2}}^{\sqrt{4-x^2}} \left[\left(4z - 4yz + 2\frac{z^2}{2} \right) \right]_0^3 dy\, dx$$

$$= \int_{-2}^{2} \int_{-\sqrt{4-x^2}}^{\sqrt{4-x^2}} [12 - 12y + 9] dy\, dx$$

$$= 21 \int_{-2}^{2} \left[\int_{-\sqrt{4-x^2}}^{\sqrt{4-x^2}} dy \right] dx,$$

since $12y$ is an odd function of y

$$= 42 \int_{-2}^{2} \left[\int_0^{\sqrt{4-x^2}} dy \right] dx = 42 \int_{-2}^{2} \sqrt{4-x^2}\, dx$$

$$= 84 \int_0^2 \sqrt{4-x^2}\, dx, \text{ since integrand is even}$$

$$= 84 \left[\frac{x\sqrt{4-x^2}}{2} + \frac{4}{2}\sin^{-1}\frac{x}{2} \right]_0^2$$

$$= 84\left[0 + 2\sin^{-1}1\right] = 84\left[\frac{2\pi}{2}\right] = 84\pi.$$

Therefore,

$$\iint_S \vec{f}.\hat{n}\, dS = \iiint_V \text{div } \vec{f}\, dV$$

and thus, Gauss's Divergence Theorem is verified for the given function.

EXAMPLE 13.84

Verify Gauss's Divergence Theorem for the function $\vec{f} = 4xz\hat{i} - y^2\hat{j} + yz\hat{k}$ over the surface S of the cube bounded by $x = 0, x = 1, y = 0, y = 1, z = 0$, and $z = 1$.

Solution. In Example 13.78, we have shown that

$$\iint_S \vec{f}.\hat{n}\, dS = \frac{3}{2}.$$

On the other hand,

$$\iiint_V \text{div } \vec{f}\, dv$$

$$= \iiint_V \nabla \cdot \vec{f}\, dV$$

$$= \iiint_V \left[\frac{\partial}{\partial x}(4xz) + \frac{\partial}{\partial y}(-y^2) + \frac{\partial}{\partial z}(yz) \right] dV$$

$$= \iiint_V (4z - 2y + y)dV = \iiint_V (4z - y)dV$$

$$= \int_0^1\int_0^1 \left[\int_0^1 (4z - y)dz \right] dy\, dx = \int_0^1\int_0^1 \left[\frac{4z^2}{2} - yz \right]_0^1 dy\, dx$$

$$= \int_0^1 \left[\int_0^1 (2 - y)dy \right] dx = \int_0^1 \left[2y - \frac{y^2}{2} \right]_0^1 dx$$

$$= \int_0^1 \left(2 - \frac{1}{2} \right) dx = \frac{3}{2}\int_0^1 dx = \frac{3}{2}.$$

Therefore,

$$\iint_S \vec{f}.\hat{n}\, dS = \iiint_V \text{div } \vec{f}\, dV$$

and thus, the Gauss's Divergence Theorem is verified.

EXAMPLE 13.85

Evaluate $\iint_S \vec{f}.\hat{n}\, dS$, where $\vec{f} = ax\hat{i} + by\hat{j} + cz\hat{k}$ and S is the surface of the sphere $x^2 + y^2 + z^2 = 1$.

Solution. Since S is closed, by the divergence theorem, we have

$$\iint_S \vec{f}.\hat{n}\, dS = \iiint_V \text{div } \vec{f}\, dV$$

$$= \iiint_V \nabla \cdot \vec{f}\, dV$$

$$= \iiint_V \left[\frac{\partial}{\partial x}(ax) + \frac{\partial}{\partial y}(by) + \frac{\partial}{\partial z}(cz)\right] dV$$

$$= \iiint_V (a+b+c) dV$$

$$= (a+b+c) \iiint_V dV$$

$$= (a+b+c)\left(\frac{4}{3}\pi\right),$$

since $\iiint_V dV =$ volume of the sphere $x^2 + y^2 + z^2 = 1$, which is $\frac{4}{3}\pi.1^3$.

EXAMPLE 13.86

Verify Gauss's Divergence Theorem for $\vec{f} = (x^3 - yz)\hat{i} - 2x^2 y\,\hat{j} + 2\hat{k}$, taken over the cube bounded by the planes $x = 0, x = a, y = 0, y = a, z = 0$, and $z = a$.

Solution. The surface F is a cube with six faces as shown in the following figure:

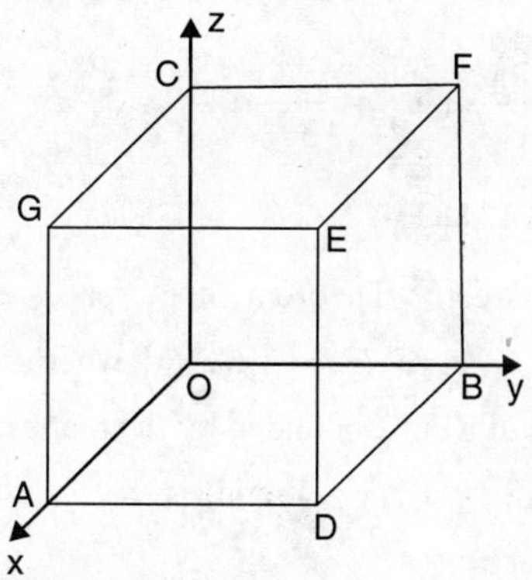

To calculate $\iint_S \vec{f}.\hat{n}\, dS$, we evaluate $\iint_S \vec{f}.\hat{n}\, dS$ over the six faces and then add those values.

(i) For the face OADB, we have $\hat{n} = -\hat{k}$ and $z = 0$. Therefore,

$$\iint_{OADB} \vec{f}.\hat{n}\, dS$$

$$= \iint_{OADB} \left[(x^3 - yz)\hat{i} - 2x^2 y\,\hat{j} + 2\hat{k}\right].(-\hat{k})\, dS$$

$$= \int_0^a\int_0^a (-2) dx dy = -2\int_0^a [y]_0^a\, dx$$

$$= -2a\int_0^a dx = -2a[x]_0^a$$

$$= -2a^2. \qquad (1)$$

(ii) For the face CGEF, we have $\hat{n} = \hat{k}$ and $z = a$. Therefore,

$$\iint_{CGEF} \vec{f}.\hat{n}\, dS = \int_0^a\int_0^a 2dx\, dy$$

$$= 2a\int_0^a dx$$

$$= 2a^2. \qquad (2)$$

(iii) For the face ADEG, we have $\hat{n} = \hat{i}, x = a$, and $dx = 0$. Therefore,

$$\iint_{ADEG} \vec{f}.\hat{n}\, dS = \int_0^a\int_0^a (a^3 - yz) dy dz$$

$$= \int_0^a \left[a^3 z - y\frac{z^2}{2}\right]_0^a dy$$

$$= \int_0^1 \left(a^4 - \frac{a^2 y}{2}\right) dy$$

$$= \left[a^4 y - \frac{a^2 y^2}{4}\right]_0^a$$

$$= a^5 - \frac{a^4}{4}. \qquad (3)$$

(iv) For the face OBFC, we have $\hat{n} = -\hat{i}$, $x = 0$, and $dx = 0$. Therefore,

$$\iint_{OBFC} \vec{f}.\hat{n}\, dS = \int_0^a\int_0^a (yz) dy dz$$

$$= \int_0^a \left[y\frac{z^2}{2}\right]_0^a$$

$$= \frac{1}{2}\int_0^a a^2 y dy$$

$$=\frac{a^2}{2}\left[\frac{y^2}{2}\right]_0^a$$

$$=\frac{a^2}{4}. \qquad (4)$$

(v) For the face OAGC, we have $\hat{n}=-\hat{j},\ y=0$, and $dy=0$. Therefore,

$$\int\int_{OAGC} \vec{f}.\hat{n}\ dS=\int_0^a\int_0^a 0\ dxdz=0. \qquad (5)$$

(vi) For the face DBFE, we have $\hat{n}=\hat{j},\ y=a$, and $dy=0$. Therefore,

$$\int\int_{DBFE} \vec{f}.\hat{n}\ dS$$

$$=\int_0^a\int_0^a -2x^2\ a\ dxdz=-2a\int_0^q\left[x^2z\right]_0^a\ dx$$

$$=-2a^2\int_0^a x^2dx=-2a^2\left[\frac{x^3}{3}\right]_0^a$$

$$=-\frac{2a^5}{3} \qquad (6)$$

Adding (1) – (6), we get

$$\iint_S \vec{f}.\hat{n}\ dS=-2a^2+2a^2+a^5$$

$$-\frac{a^4}{4}+\frac{a^2}{4}+0-\frac{2a^5}{3}=\frac{a^5}{3}.$$

On the other hand,

$$\iiint_V \text{div}\ \vec{f}\ dV=\iiint_V \nabla\cdot\vec{f}\ dV$$

$$=\iiint_V\left[\frac{\partial}{\partial x}(x^3-yz)+\frac{\partial}{\partial y}\left(-2x^2y\right)+\frac{\partial}{\partial z}(2)\right]dV$$

$$=\iiint_V\left(3x^2-2x^2\right)dV=\int_0^a\int_0^a\int_0^a x^2dx\ dy\ dz$$

$$=\int_0^a\int_0^a\left[x^2z\right]_0^a dy\,dx=\int_a\left[\int_0^a x^2\ a\ dy\right]dx$$

$$=a\int_0^a\left[x^2y\right]_0^a dx=a^2\int_0^a x^2dx=a^2\left[\frac{x^3}{3}\right]_0^a=\frac{a^5}{3}.$$

Thus,

$$\iint_S \vec{f}.\hat{n}\ dS=\iiint_V \text{div}\ \vec{f}\ dV$$

and thereby, Gauss's Divergence Theorem is verified.

EXAMPLE 13.87

Evaluate $\iint\left(x^3-yz\right)dy\,dz-2x^2ydzdx+zdxdy$ over the surface bounded by the coordinate planes and the planes $x=y=z=a$.

Solution. By giving divergence theorem in Cartesian form, we have

$$\iint_S\left(f_1dydz+f_2dzdx+f_3dxdy\right)$$

$$=\iiint_V\left(\frac{\partial f_1}{\partial x}+\frac{\partial f_2}{\partial y}+\frac{\partial f_3}{\partial z}\right)dV$$

$$=\iiint_V\left[\frac{\partial}{\partial x}\left(x^3-yz\right)+\frac{\partial}{\partial y}\left(-2x^2y\right)+\frac{\partial}{\partial z}(z)\right]dV$$

$$=\iiint_V\left(x^2+1\right)dV=\int_0^a\int_0^a\int_0^a\left(x^2+1\right)dx\ dy\ dz$$

$$=\int_0^a\int_0^a\left[x^2z+z\right]_0^a dydz=\int_0^a\left[\int_0^a a\left(x^2+1\right)dy\right]dx$$

$$=a\int_0^a\left[x^2y+y\right]_0^a dx=a^2\int_0^a\left(x^2+1\right)dx$$

$$=a^2\left[\frac{x^3}{3}+x\right]_0^a=\frac{a^5}{3}+a^3.$$

EXAMPLE 13.88

Using Green's Theorem of in space, evaluate $\iint_S\left(4xzdy\,dz-y^2dz\,dx+yzdx\ dy\right)$, where S is the surface of a cube bounded by the planes $x=0, y=0, z=0, x=1, y=1$, and $z=1$.

Solution. Let

$$f_1=4xz,\ f_2=-y^2,\text{ and } f_3=yz.$$

Then, by Green's Theorem in space, we have

$$\iint_S\left(f_1dydz+f_2dzdx+f_3dxdy\right)$$

$$=\iiint_V\left(\frac{\partial f_1}{\partial x}+\frac{\partial f_2}{\partial y}+\frac{\partial f_3}{\partial z}\right)dx\,dy\,dz$$

$$=\iiint_V \left(\frac{\partial}{\partial x}(4xz)+\frac{\partial}{\partial y}\left(-y^2\right)+\frac{\partial}{\partial z}(yz)\right)dx\,dy\,dz$$

$$=\iiint_V (4z-y)\,dx\,dy\,dz$$

$$=\int_0^1\int_0^1\left[4\frac{z^2}{2}-yz\right]_0^1 dy\,dx=\int_0^1\left[\int_0^1(2-y)dy\right]dx$$

$$=\int_0^1\left[2y-\frac{y^2}{2}\right]_0^1=\frac{3}{2}\int_0^1 dx=\frac{3}{2}.$$

13.19 GREEN'S THEOREM IN A PLANE

A domain D is said to be a *quadratic* with respect to y-axis, if it is bounded by the curves of the form

$$y=\phi(x),\ y=\psi(x): x=a,\ x=b,$$

where ϕ and ψ are continuous functions and $\phi(x)>\psi(x)$ for all $x\in[a,b]$. Thus, a domain which is quadratic with respect to *y*-axis is such that a line parallel to *y*-axis and lying between $x=a$ and $x=b$ meets the boundary of D in just two points. Similarly, we can define domains which are quadratic with respect to *x*-axis.

The Green's Theorem is useful in changing a line integral around a *closed curve C* into a double integral over the region R enclosed by C.

Theorem 13.11.

(Green's Theorem). Let $f, g, \frac{\partial f}{\partial y}$, and $\frac{\partial f}{\partial y}$ are continuous in a region *R*, which can be split up in finite number of regions quadratic with respect to either axis. Then,

$$\oint_C [f(x,y)dx+g(x,y)dy]=\iint_R\left(\frac{\partial g}{\partial x}-\frac{\partial f}{\partial y}\right)dx\,dy,$$

where the integral on the left is a line integral around the boundary *C* of the region, taken in such a way that the interior of the region remains on the left as the boundary is described.

Proof: Consider the region R bounded by the curves $x=a$, $x=b$, $y=\phi(x)$, and $y=\psi(x)$, such that $\phi(x)>\psi(x)$ for all $x\in[a,b]$. Let f be a real-valued continuous function defined in R, and let $\frac{\partial f}{\partial y}$ exists and is continuous in R. Then,

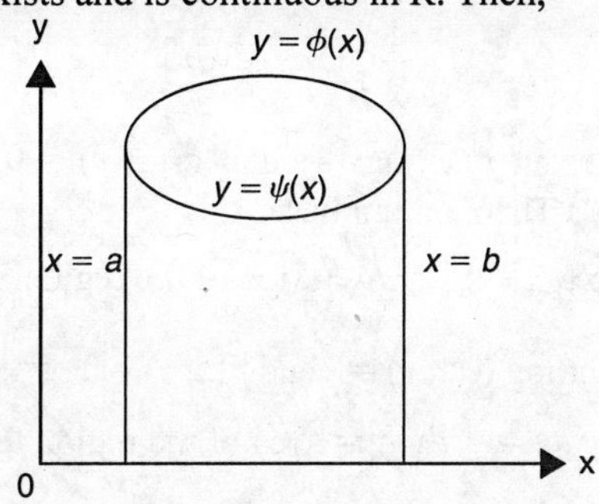

$$\iint_R\frac{\partial f}{\partial y}\,dxdy=\int_a^b\left[\int_{\psi(x)}^{\phi(x)}\frac{\partial f}{\partial y}dy\right]dx$$

$$=-\int_a^b f(x,.\phi(x))dx-\int_a^b f(x,\psi(x))\,dx$$

$$=-\int_b^a f(x,\phi(x))\,dx-\int_a^b f(x,\psi(x))\,dx$$

$$=-\left[\int_a^b f(x,\psi(x))\,dx+\int_b^a f(x,\phi(x))dx\right]$$

$$=-\oint_C f(x,y)dx.$$

Therefore,

$$\oint_C f(x,y)dx=-\iint_R\frac{\partial f}{\partial y}dx\,dy. \qquad (1)$$

Similarly, it can be shown that

$$\oint_C g(x,y)dy=\iint_R\frac{\partial g}{\partial x}dx\,dy. \qquad (2)$$

Adding (1) and (2), we obtain

$$\oint_C [f(x,y)dx+g(x,y)dy]=\iint_R\left[\frac{\partial g}{\partial x}-\frac{\partial f}{\partial y}\right]dx\,dy.$$

Deductions:

(i) If $f(x,y)=-y$ and $g(x,y)=x$, then by Green's Theorem, we have

$$\oint_C (xdy-ydx)=\iint_R(1+1)\,dxdy$$

$$= 2\iint_S dxdy = 2A,$$

where A denotes the area of the region R. Thus,

$$A = \frac{1}{2}\oint_C [xdy - ydx].$$

(ii) Putting $f(x, y) = -y$ and $g(x, y) = 0$, the Green's Theorem implies

$$-\oint ydx = \iint_R dxdy = \text{Area of the region R}.$$

(iii) Putting $g(x, y) = x$ and $f(x, y) = 0$, we get

$$\oint_C xdy = \iint_R dx\,dy = \text{Area of the region R}.$$

Hence, the area of a closed region R is given by any of the three formulae

$$\oint_C xdy,\ -\oint_C ydx,\ \text{or}\ \frac{1}{2}\oint_C (xdy - ydx),$$

where C denotes the boundary of the closed region R described in the positive sense.

EXAMPLE 13.89

Verify Green's theorem in the plane for $\oint_C [(xy + y^2)dx + x^2dy]$, where C is the closed curve of the region bounded by $y = x$ and $y = x^2$.

Solution. The region is bounded by the straight line $y = x$ and the parabola $y = x^2$. The point of intersection of $y = x$ and $y = x^2$ are (0, 0) and (1, 1). We note that

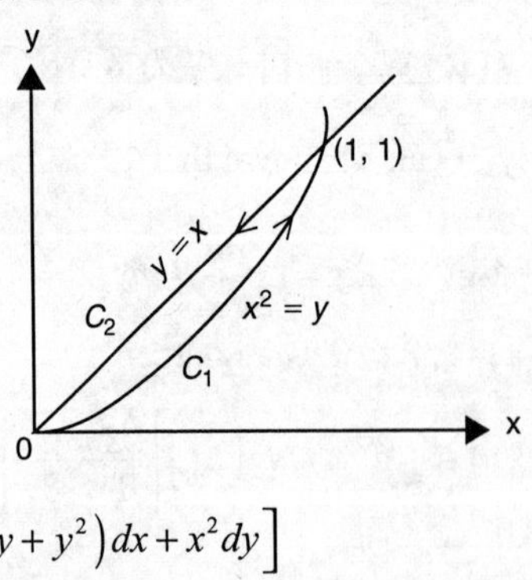

$$\oint_C [(xy + y^2)dx + x^2dy]$$

$$= \int_{C_1} [(xy + y^2)dx + x^2dy]$$

$$+ \int_{C_2} [(xy + y^2)dx + x^2dy] \qquad (1)$$

For the line integral along C_1, we have $y = x^2$ and so, $dy = 2xdx$ and x varies from 0 to 1. Thus,

$$\int_{C_1} [(xy + y^2)dx + x^2dy]$$

$$= \int_0^1 [(x^3 + x^4)dx + x^2(2x)dx]$$

$$= \int_0^1 (x^4 + 3x^3)dx = \left[\frac{x^5}{5} + \frac{3x^4}{4}\right]_0^1 = \frac{19}{20}.$$

For the line integral along (2), we have $y = x$ and so, $dy = dx$, and x varies from 1 to 0. Therefore,

$$\int_{C_2} [(xy + y^2)dx + x^2dy] \int_1^0 [(x^2 + x^2)dx + x^2dx]$$

$$= \int_1^0 3x^2dx = \left[\frac{3x^3}{3}\right]_1^0 = -1.$$

Hence, (1) yields

$$\int_C [(xy + y^2)dx + x^2dy] = \frac{19}{20} - 1 = -\frac{1}{20}.$$

On the other hand,

$$\iint_S \left(\frac{\partial f_2}{\partial x} - \frac{\partial f_1}{\partial y}\right) dx\,dy$$

$$= \iint_S \left[\frac{\partial}{\partial x}(x^2) - \frac{\partial}{\partial y}(xy + y^2)\right] dx\,dy$$

$$= \int_0^1 \int_{y=x^2}^{y=x} [2x - (x + 2y)]\,dy\,dx$$

$$= \int_0^1 \left[\int_{x^2}^{x} (x - 2y)\,dy\right] dx = \int_0^1 [xy - y^2]_{x^2}^{x}\,dx$$

$$= \int_0^1 (x^4 - x^3)dx = \left[\frac{x^5}{5} - \frac{x^4}{4}\right]_0^1 = \frac{1}{5} - \frac{1}{4} = -\frac{1}{20}.$$

Hence,

$$\oint_C [f_1(x, y)dx + f_2(x, y)dy] = \iint_S \left(\frac{\partial f_2}{\partial x} - \frac{\partial f_1}{\partial y}\right) dx\,dy,$$

and thus, Green's theorem is verified.

EXAMPLE 13.90

Apply Green's Theorem to show that the area bounded by a simple closed curve C is given by $\frac{1}{2}\oint_C (xdy - ydx)$. Hence, find the area of the ellipse $\frac{x^2}{a^2}+\frac{y^2}{b^2}=1$.

Solution. In Deduction (i) of Green's Theorem, we have shown that the area A bounded by a simple closed curve C is equal to $\frac{1}{2}(xdy - ydx)$. For the second parts, we know that the parametric equations of the ellipse $\frac{x^2}{a^2}+\frac{y^2}{b^2}=1$ are $x = a\cos\theta$ and $y = b\sin\theta$. Thus, $dx = -a\sin\theta d\theta$ and $dy = b\cos\theta d\theta$. Therefore, the area A of the ellipse is given by

$$A = \frac{1}{2}\oint_C (xdy - ydx)$$

$$= \frac{1}{2}\int_0^{2\pi}[(a\cos\theta)(b\cos\theta)d\theta$$

$$-(b\sin\theta)(-a\sin\theta)d\theta]$$

$$= \frac{1}{2}\int_0^{2\pi}[ab\cos^2\theta + absin^2\theta]d\theta$$

$$= \frac{ab}{2}\int_0^{2\pi}d\theta = \frac{ab}{2}[\theta]_\theta^{2\pi} = \pi ab.$$

EXAMPLE 13.91

Verify Green's theorem in the plane for $\oint_C[(3x^2 - 8y^2)dx + (4y - xy)dy]$, where C is the boundary of the region bounded by $x = 0$, $y = 0$, and $x + y = 1$.

Solution. We have

$f_1(x, y) = 3x^2 - 8y^2$ and $f_2(x, y) = 4y - 6xy$.

Therefore,

$$\iint_S\left(\frac{\partial f_2}{\partial x} - \frac{\partial f_1}{\partial y}\right)dx\,dy$$

$$= \int_0^1\left[\int_0^{1-x}(-6y + 16y)dy\right]dx = 10\int_0^1\left[\frac{y^2}{2}\right]_0^{1-x}dx$$

$$= 5\int_0^1(1-x)^2\,dx = -\frac{5}{3}[(1-x)^3]_0^1 = \frac{5}{3}. \quad (1)$$

Further, the line integral splits into three parts:

$$\oint_C[f_2dx + f_1dy] = \int_{OA} + \int_{AB} + \int_{BO}.$$

Along OA, we have $y = 0$ so that $dy = 0$ and x varies from 0 to 1. Hence,

$$\int_{OA} = \int_0^1 3x^2dx = 3\left[\frac{x^3}{3}\right]_0^1 = 1.$$

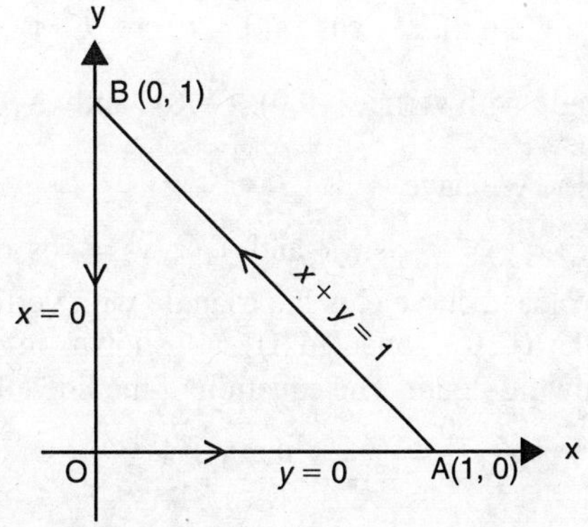

Along AB, we have $y = 1 - x$ and so $dy = -dx$ and x varies from 1 to 0. Therefore,

$$\int_{AB} = \int_1^0[3x^2 - 8(1-x)]^2\,dx$$

$$+[4(1-x) - 6x(1-x)](-1)dx$$

$$= \int_1^0(-11x^2 + 26x - 12)dx$$

$$= \left[-11\frac{x^3}{3} + 26\frac{x^2}{2} - 12x\right]_1^0$$

$$= \frac{11}{3} - 13 + 12 = \frac{8}{3}.$$

Along BO, we have $x = 0$ so that $dx = 0$ and y varies from 1 to 0. Therefore,

$$\int_{BO} = \int_1^0 4ydy = 4\left[\frac{y^2}{2}\right]_1^0 = -2.$$

Hence,

$$[f_2dx + f_1dy] = 1 + \frac{8}{3} - 2 = \frac{5}{3}. \quad (2)$$

From (1) and (2), it follows that

$$\oint_C [f_2 dx + f_1 dy] = \iint_S \left(\frac{\partial f_2}{\partial x} - \frac{\partial f_1}{\partial y}\right) dx\, dy,$$

and thus, Green's theorem is verified.

EXAMPLE 13.92

Using Green's theorem in a plane, evaluate $\oint_C [(y - \sin x)dx + \cos x dy]$, where C is the triangle with vertices (0,0), $(\frac{\pi}{2},0)$, and $(\frac{\pi}{2},1)$.

Solution. We have

$f_1(x,y) = y - \sin x$ and $f_2(x,y) = \cos x$.

The closed curve C is the triangle with vertices (0,0), $(\frac{\pi}{2},0)$, and $(\frac{\pi}{2},1)$ as shown in the following figure. The equation of the line OB is

$$\frac{y-0}{x-0} = \frac{0}{\frac{\pi}{2}-0}, \text{ that is, } y = \frac{2x}{\pi}.$$

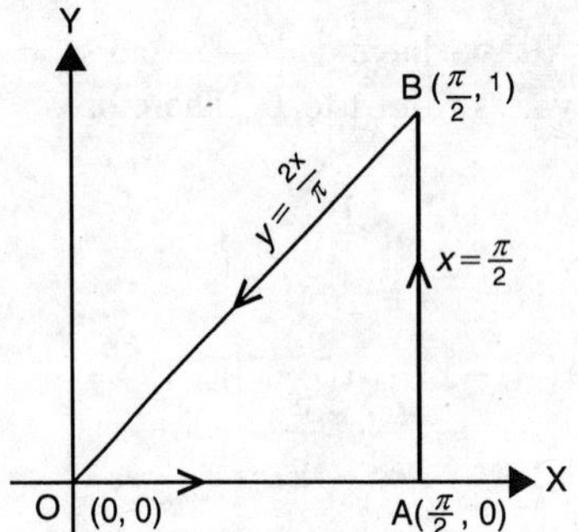

By Green's Theorem, we have

$$\oint_C [f_1 dx + f_2 dy] = \iint_S \left(\frac{\partial f_2}{\partial x} - \frac{\partial f_1}{\partial y}\right) dx\, dy$$

$$= \iint_S \left[\frac{\partial}{\partial x}(\cos x) - \frac{\partial}{\partial y}(y - \sin x)\right] dx\, dy$$

$$= \iint -(\sin x + 1) dx\, dy$$

$$= -\int_0^{\frac{\pi}{2}} \left[\int_0^{\frac{2x}{\pi}} (1 + \sin x) dy\right] dx$$

$$= -\int_0^{\frac{\pi}{2}} [y + y \sin x]_0^{\frac{2x}{\pi}} dx$$

$$= -\frac{2}{\pi}\int_0^{\frac{\pi}{2}} x(1 + \sin x) dx$$

$$= -\frac{2}{\pi}\left\{[x(x - \cos x)]_0^{\frac{\pi}{2}} - \int_0^{\frac{\pi}{2}} (x - \cos x) dx\right\}$$

$$= -\frac{\pi}{2} + \frac{2}{\pi}\int_0^{\frac{\pi}{2}} (x - \cos x) dx$$

$$= -\frac{\pi}{2} + \frac{2}{\pi}\left[\frac{x^2}{2} - \sin x\right]_0^{\frac{\pi}{2}}$$

$$= -\frac{\pi}{2} + \frac{2}{\pi}\left[\frac{\pi^2}{8} - 1\right]$$

$$= -\frac{\pi}{2} + \frac{\pi}{4} - \frac{2}{\pi} = -\frac{\pi}{4} - \frac{2}{\pi}.$$

EXAMPLE 13.93

Evaluate, by Green's Theorem, $\oint_C [(3x - y)dx + (2x + y)dy]$, where C is the curve $x + y^2 = a^2$

Solution. We have

$f_1(x,y) = 3x - y$ and $f_2(x,y) = 2x + y$.

By Green's Theorem, we have

$$\int_C (f_1 dx + f_2 dy)$$

$$= \iint_R \left(\frac{\partial f_2}{\partial x} - \frac{\partial f_1}{\partial y}\right) dx\, dy = \iint_R (2-1) dx\, dy$$

$$= \int_{-a}^{a} \int_{-\sqrt{a^2-x^2}}^{\sqrt{a^2-x^2}} dx\, dy = 4\int_0^a \int_0^{\sqrt{a^2-x^2}} dx\, dy$$

$$= 4\int_0^a \sqrt{a^2 - x^2}\, dx$$

$$= 4a^2 \int_0^{\frac{\pi}{2}} \cos^2\theta\, d\theta,\ x = \sin\theta,$$

$$= 4a^2 . \frac{1}{2} . \frac{\pi}{2} = \pi a^2.$$

EXAMPLE 13.94

Compute the area of the loop of Descartes's Folium, $x^3 + y^3 = 3axy$.

Solution. Putting $y = tx$, we get the parametric equations of the contour of the folium as

$$x = \frac{3at}{1+t^3} \text{ and } y = \frac{3at^2}{1+t^3}.$$

The loop is described as t varies from 0 to ∞, since $t = \frac{y}{x} = \tan\theta$, where θ varies from 0 to $\frac{\pi}{2}$. Thus, $dx = 3a\frac{1-2t^3}{(1+t^3)^2}dt$ and $dy = 3a\frac{2t-t^4}{(1+t^3)^2}dt$. Hence, by Green's Theorem,

$$\text{Area} = \frac{1}{2}\,(xd\,y - ydx) = \frac{9a^2}{2}\int_0^\infty \frac{t^2 dt}{(1+t^3)^2} = \frac{3}{2}a^2.$$

13.20 SIOKE'S THEOREM

The Stoke's Theorem provides a relation between a surface integral taken over a surface to a line integral along the boundary curve of the surface.

Theorem 13.12. (Stoke's Theorem). Let $\vec{f}$ be vector-point function possessing continuous first-order partial derivatives and S be a surface bounded by a closed curve C. Then,

$$\oint_C \vec{f}.d\vec{r} = \iint_S curl\,\vec{f}.\hat{n}dS$$

where $\hat{n}$ is a unit normal vector at any point of S, drawn in the sense in which a right-handed screw would move when rotated in the sense of description of the curve C.

Proof: Let the unit normal vector $\hat{n}$ makes angles α, β, and γ with the positive directions of coordinate axes x, y, and z, respectively. Then, $\hat{n} = \cos\alpha\hat{i} + \cos\beta\hat{j} + \cos\gamma\hat{k}$. Since $\vec{r} = x\hat{i} + y\hat{j} + z\hat{k}$, we have $d\vec{r} = \hat{i}dx + \hat{j}dy + \hat{k}dz$. Let $\vec{f} = f_1\hat{i} + f_2\hat{j} + f_3\hat{k}$. Then,

$$Curl\,\vec{f} = \nabla\times\vec{f} = \begin{vmatrix} \hat{i} & \hat{j} & \hat{k} \\ \frac{\partial}{\partial x} & \frac{\partial}{\partial y} & \frac{\partial}{\partial z} \\ f_1 & f_2 & f_3 \end{vmatrix}$$

$$= \hat{i}\left(\frac{\partial f_3}{\partial y} - \frac{\partial f_2}{\partial z}\right) + \hat{j}\left(\frac{\partial f_1}{\partial z} - \frac{\partial f_3}{\partial x}\right) + \hat{k}\left(\frac{\partial f_2}{\partial x} - \frac{\partial f_1}{\partial y}\right)$$

and so,

$$curl\,\vec{f}.\hat{n}\left(\frac{\partial f_3}{\partial y} - \frac{\partial f_2}{\partial z}\right)\cos\alpha + \left(\frac{\partial f_1}{\partial z} - \frac{\partial f_3}{\partial x}\right)\cos\beta + \left(\frac{\partial f_2}{\partial x} - \frac{\partial f_1}{\partial y}\right)\cos\gamma.$$

On the other hand,

$$\vec{f}.d\vec{r} = (f_1\hat{i} + f_2\hat{j} + f_3\hat{k}).(\hat{i}dx + \hat{j}dy + \hat{k}dz) = f_1dx + f_2dy + f_3dz.$$

Therefore, Stoke's Theorem takes the form

$$\oint_C (f_1dx + f_2dy + f_3dz)$$

$$= \iint_S \left[\left(\frac{\partial f_3}{\partial y} - \frac{\partial f_2}{\partial z}\right)\cos\alpha + \left(\frac{\partial f_1}{\partial z} - \frac{\partial f_3}{\partial x}\right)\cos\beta\right.$$
$$\left.\cos\beta + \left(\frac{\partial f_2}{\partial x} - \frac{\partial f_1}{\partial y}\right)\cos\gamma\right]dS.$$

We now prove the theorem in this form. Suppose $z = \phi(x, y)$ be the equation of the surface S and R be the projection of S on the xy-plane. Then, the projection of the curve C on the xy-plane shall be the curve C_1, which enclose the region R. Therefore,

$$\oint_C f_1(x,y,z)dx = \oint_{C_1} f_1(x,y,\phi(x,y))dx$$

$$= \int_{C_1} f_1(x,y,\phi)dx - 0dy]$$

$$= \iint_R\left[\frac{\partial}{\partial x}(0) - \frac{\partial}{\partial y}f_1(x,y,\phi)\right]dx\,dy$$

(by Green's theorem in plane)

$$= -\iint_R \frac{\partial}{\partial y}f_1(x,y,\phi)dx\,dy$$

$$= -\iint_R \left[\frac{\partial f_1}{\partial y} + \frac{\partial f_1}{\partial z}\frac{\partial \phi}{\partial y} \right] dx\, dy. \quad (1)$$

Since the direction ratios of the normal $\hat{n}$ to the surface S are $\frac{\partial \phi}{\partial x}, \frac{\partial \phi}{\partial y}$, and –l, we have

$$\frac{\cos\alpha}{\frac{\partial\phi}{\theta\tau}} = \frac{\cos\beta}{\frac{\partial\phi}{\partial y}} = \frac{\cos\gamma}{-1} \text{ and so } \frac{\partial\phi}{\partial y} = -\frac{\cos\beta}{\cos\gamma}.$$

Moreover, $dxdy$ being the projection of dS on the xy-plane, we have

$$dxdy = \cos\gamma dS.$$

Hence, (1) reduces to

$$\oint_C f_1(x,y,z)dx$$

$$= -\iint_S \left[\frac{\partial f_1}{\partial y} + \frac{\partial f_1}{\partial z}\left(-\frac{\cos\beta}{\cos\gamma}\right) \right] \cos\gamma dS$$

$$= \iint_S \left[\frac{\partial f_1}{\partial z}\cos\beta - \frac{\partial f_1}{\partial y}\cos\gamma \right] dS. \quad (2)$$

Similarly, it can be established that

$$f_2(x,y,z)dx = \iint \left[\frac{\partial f_2}{\partial x}\cos\gamma - \frac{\partial f_2}{\partial z}\cos\alpha \right] dS \; (3)$$

and

$$\oint_C f_3(x,y,z)dz = -\left[\frac{\partial f_3}{\partial y}\cos\alpha - \frac{\partial f_3}{\partial x}\cos\beta \right] dS. (4)$$

Adding (2), (3), and (4), we get

$$\oint_C (f_1 dx + f_2 dy + f_3 dz)$$

$$= \iint_S \left[\left(\frac{\partial f_3}{\partial y} - \frac{\partial f_2}{\partial z} \right)\cos\alpha \right.$$

$$\left. + \left(\frac{\partial f_1}{\partial z} - \frac{\partial f_3}{\partial x} \right)\cos\beta + \left(\frac{\partial f_2}{\partial x} - \frac{\partial f_1}{\partial y} \right)\cos\gamma \right] dS.$$

This completes the proof of the theorem.

Remark 13.1. The equivalent statement of Stoke's Theorem is that the line integral of the tangential component of a vector-point function $\vec{f}$ taken around a simple closed curve C is equal to the surface integral of the normal component of the curl of $\vec{f}$ taken over any surface S having C as its boundary.

EXAMPLE 13.95

Verify Stoke's Theorem for the ffinction $\vec{f} = x^2\hat{i} + xy\hat{j}$, integrated around the square in the plane $z = 0$, whose sides are along the lines $x = 0$, $x = a$, $y = 0$, and $y = a$.

Solution. Since $\vec{f} = x^2\hat{i} + xy\hat{j}$, we have

$$\vec{f}.d\vec{r} = (x^2\hat{i} + xy\hat{j}).(\hat{i}dx + \hat{j}dy) = x^2 dx \;\; xydy.$$

Therefore,

$$\oint_C \vec{f}.d\vec{r} = (x^2 dx + xydy),$$

where C is the square shown in the figure.

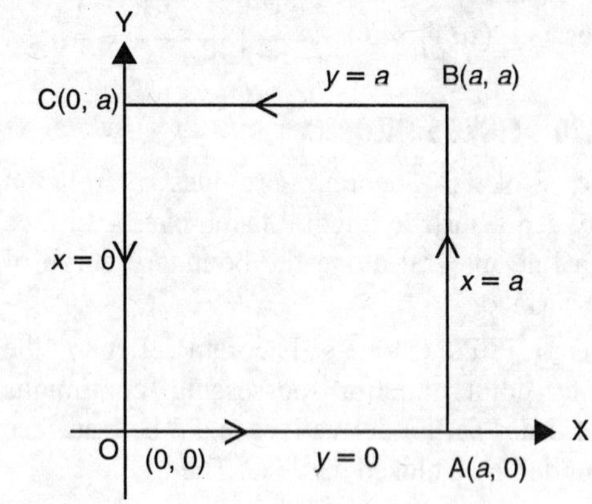

Thus,

$$\oint_C \vec{f}.d\vec{r} = \int_{OA} + \int_{AB} + \int_{BC} + \int_{CA}. \quad (1)$$

Along OA, we have $y = 0$ and so, $dy = 0$. Thus,

$$\int_{AB} \vec{f}.d\vec{r} = \int_0^a x^2 dx = \left[\frac{x^3}{3} \right]_0^a = \frac{a^3}{3}.$$

Along AB, $x = a$ and so, $dx = 0$. Thus,

$$\int_{AB} \vec{f}.d\vec{r} = \int_0^a aydy = \frac{a^2}{2}.$$

Along BC, we have $y = a$ and so, $dy = 0$. Thus,

$$\int_{BC} \vec{f}.d\vec{r} = \int_a^0 x^2 dx = -\frac{a^3}{3}.$$

Along CO, we have $x = 0$ and so, $dx = 0$. Thus,

$$\int \vec{f}.d\vec{r} = \int 0 dy = 0.$$

Hence, (1) yields

$$\int_{CO} \vec{f}.d\vec{r} = \frac{a^3}{3} + \frac{a^2}{2} - \frac{a^3}{3} = \frac{a^2}{2}.$$

On the other hand,

$$curl \vec{f} = \begin{vmatrix} \hat{i} & \hat{j} & \hat{k} \\ \frac{\partial}{\partial x} & \frac{\partial}{\partial y} & \frac{\partial}{\partial z} \\ x^2 & xy & 0 \end{vmatrix} = y\hat{k}$$

Since the square (surface) lies in the xy-plane, $\hat{n} = \hat{k}$. Therefore,

$$curl\ \vec{f}.\hat{n} = y\hat{k}.\hat{k} = y$$

and so,

$$\iint_S curl \vec{f}.\hat{n} dS = \int_0^a \int_0^a y dx dy = \int_0^a \left[\frac{y^2}{2}\right]_0^a dx$$

$$= \frac{a^2}{2} \int_0^a dx = \frac{a^3}{2}.$$

Hence,

$$\oint_C \vec{f}.d\vec{r} = \iint_S curl \vec{f}.\hat{n} ds.$$

EXAMPLE 13.96

Verifies Stoke's Theorem for $\vec{f} = y\hat{i} + z\hat{j} + x\hat{k}$, where S is the *upper-half* surface of the surface $x^2 + y^2 + z^2 = 1$ and C is its boundary.

Solution. Here, C is the boundary of the upper-half surface of $x^2 + y^2 + z^2 = 1$, that is, C is the boundary of the circle $x^2 + y^2 = 1$ in the xy-plane. Thus, the parametric equations of C are $x = \cos t$, $y = \sin t$, $z = 0$, and $0 \le t \le 2\pi$. Therefore,

$$\oint_C \vec{f} d\vec{r} = \oint_C (f_1 dx + f_2 dy + f_3 dz)$$

$$= \oint_C [y dx + z dy + x dz]$$

$$= \int_0^{2p} [\sin t(-\sin t)] dt$$

$$= -\int_0^{2\pi} \sin^2 t dt$$

$$= -4\int_0^{\frac{\pi}{2}} \sin^2 t\, dt = -4.\frac{1}{2}.\frac{\pi}{2} = -\pi.$$

On the other hand,

$$curl\ \vec{f} = \begin{vmatrix} \hat{i} & \hat{j} & \hat{k} \\ \frac{\partial}{\partial x} & \frac{\partial}{\partial y} & \frac{\partial}{\partial z} \\ y & z & x \end{vmatrix}$$

$$= \hat{i} = \left(\frac{\partial x}{\partial y} - \frac{\partial z}{\partial z}\right) = +\hat{j}\left(\frac{\partial y}{\partial z} - \frac{\partial x}{\partial x}\right) + \hat{k}\left(\frac{\partial z}{\partial x} - \frac{\partial y}{\partial y}\right)$$

$$= -(\hat{i} + \hat{j} + \hat{k})$$

and so,

$$curl\ \vec{f}.\hat{n} = -(\hat{i} + \hat{j} + \hat{k}).\hat{k} = -1.$$

Therefore,

$$\iint_S curl\ \vec{f}.\hat{n} dS = -\iint dx dy = -\pi(1)^2 = -\pi.$$

Hence,

$$\oint_C \vec{f}.d\vec{r} = \iint_S curl\ \vec{f}.\hat{n}\, ds$$

and thus, Stoke's theorem is verified:

EXAMPLE 13.97

Verify Stoke's theorem for the vector field $\vec{f} = (x^2 - y^2)\hat{i} + 2xy\hat{j}$, integrated around the rectangle $z = 0$, and bounded by the lines $x = 0$, $y = 0$, $x = a$, and $y = b$.

Solution. Since $\vec{f} = (x^2 - y^2)\hat{i} + 2xy\hat{j}$, we have

$$\vec{f}.d\vec{r} = [(x^2 - y^2)\hat{i} + 2xy\hat{j}]\cdot(\hat{i}dx + \hat{j}dy)$$
$$= (x^2 - y^2)dx + 2xydy.$$

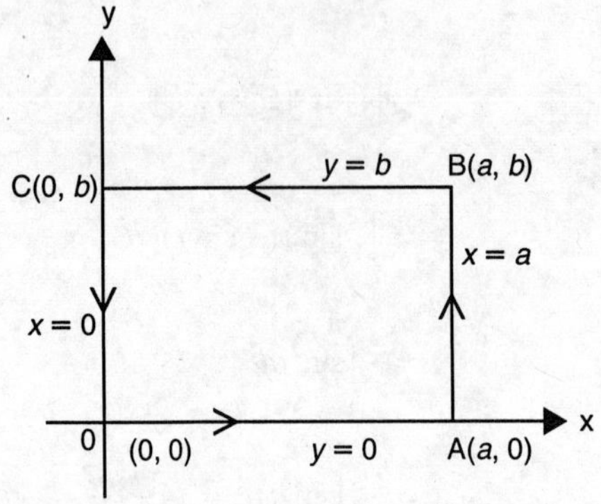

Therefore,

$$\oint_c \vec{f}d\vec{r} = \oint_c [(x^2 - y^2)dx + 2xydy]$$

$$= \int_{OA} + \int_{AB} + \int_{BC} + \int_{CO}$$

Along OA, we have $y = 0$ and dy $= 0$. Therefore,

$$\int_{OA} \vec{f}.d\vec{r} = \int_0^a x^2 dx = \left[\frac{x^3}{3}\right]_0^a = \frac{a^3}{3}$$

Along AB, we have $x = a$ and $dx = 0$. Therefore,

$$\int_{AB} \vec{f}d\vec{r} = \int_0^b 2aydy = ab^2$$

Along BC, we have $y = b$ and $dy = 0$. Therefore,

$$\int_{BC} \vec{f}.d\vec{r} = \int_a^0 \left(x^2 - b^2\right)dx$$
$$= \left[\frac{x^3}{3} - b^2 x\right]_a^0$$
$$= \frac{-a^3}{3} + ab^2$$

Along CO, we have x = 0 and dx = 0. Therefore

$$\int_{CO} \vec{f}.d\vec{r} = 0.$$

Hence,

$$\oint_C \vec{f}.d\vec{r} = \frac{a^3}{3} + ab^2 - \frac{a^3}{3} + ab^2 = 2ab^2.$$

On the other hand,

$$curl\,\vec{f} = \nabla\times\vec{f} = \begin{vmatrix} \hat{i} & \hat{j} & \hat{k} \\ \frac{\partial}{\partial x} & \frac{\partial}{\partial y} & \frac{\partial}{\partial z} \\ x^2 - y^2 & 2xy & 0 \end{vmatrix} k = 4y\hat{k}.$$

Therefore,

$$curl\,\vec{f}.\hat{n} = (4y\,\hat{k}).(\hat{k}) = 4y$$

And so,

$$\iint_S curl\,\vec{f}.\hat{n}\,ds = 4\int_0^a\left(\int_0^b ydy\right)dx = 4\int_0^a\left[\frac{y^2}{2}\right]_0^b dx$$

$$= 2d^2\int_0^a dx = 2ab^2$$

Hence

$$\oint_C \vec{f}.d\vec{r} = \iint_S curl\,\vec{f}.\hat{n}\,ds$$

And thus, Stoke's theorem is verified.

EXAMPLE 13.98

Apply Stoke's Theorem to evaluate

$$\int_C [(x+y)dx + (2x-z)dy + (y+z)dz],$$

where C is the boundary of the triangle with vertices (2,0,0), (0,3,0), and (0,0,6).

Solution. Taking projection on three planes, we note that the surface S consists of three triangles, OAB in xy-plane, OBC in yz-plane, and OAC in xz-plane. Using two-point formula (or intercept form), the equations of the lines AB, BC, and CA are respectively $3x + 2y = 6$, $2y + z = 6$ and $3x + z = 6$. We have

$$\vec{f} = (x+y)\hat{i} + (2x-z)\hat{j} + (y+z)\hat{k}.$$

Therefore,

$$curl\,\vec{f} = \begin{vmatrix} \hat{i} & \hat{j} & \hat{k} \\ \frac{\partial}{\partial x} & \frac{\partial}{\partial y} & \frac{\partial}{\partial z} \\ x+y & 2x-z & y+z \end{vmatrix}$$

$$= \hat{i}\left[\frac{\partial}{\partial y}(y+z) - \frac{\partial}{\partial z}(2x-z)\right]$$

$$+\hat{j}\left[\frac{\partial}{\partial z}(x+y) - \frac{\partial}{\partial x}(y+z)\right]$$

$$+\hat{k}\left[\frac{\partial}{\partial x}(2x-z) - \frac{\partial}{\partial y}(x+y)\right] = 2\hat{i} + \hat{k}.$$

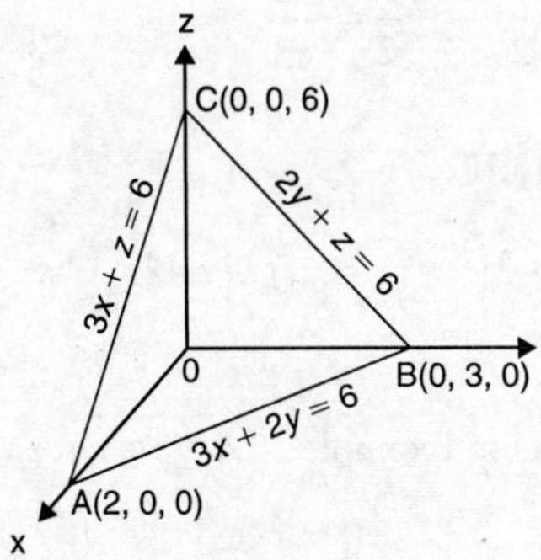

Now, by Stoke's Theorem,

$$\oint_C \vec{f}.d\vec{r} = \iint_S curl\ \vec{f}.\hat{n}\ ds$$

$$= \iint_{OAB} + \iint_{OBC} + \iint_{OAC}$$

$$= \iint_{OAB} \left(2\hat{i} + \hat{k}\right).\left(\hat{k}\right) ds$$

$$+ \iint_{OBC} \left(2\hat{i} + \hat{k}\right).\left(\hat{j}\right) ds$$

$$+ \iint_{OAC} (2\hat{i} + \hat{k}).\left(\hat{i}\right) ds$$

$$= \int_0^3 \left[\int_0^{\frac{6-2y}{3}} dx\right] dy + 0 + \int_0^6 \left[\int_0^{\frac{6-z}{2}} dy\right] dz$$

$$\int_0^3 \frac{6-2y}{3} dy + 2\int_0^6 \frac{6-z}{2} dz$$

$$= \frac{1}{3}\left[6y - 2\frac{y^2}{2}\right]_0^3 + \frac{2}{2}\left[6z - \frac{z^2}{2}\right]_0^6$$

$$= 3 + 18 = 21.$$

EXAMPLE 13.99

Evaluate $\oint \vec{f}.d\vec{r}$ by Stoke's theorem, where $\vec{f} = y^2\hat{i} + x^2\hat{j} - (x+z)\hat{k}$ and C is the boundary of the triangle with vertices at (0, 0, 0), (1, 0, 0), and (1, 1, 0).

Solution. We have

$$curl\ \vec{f} = \begin{vmatrix} \hat{i} & \hat{j} & \hat{k} \\ \frac{\partial}{\partial x} & \frac{\partial}{\partial y} & \frac{\partial}{\partial z} \\ y^2 & x^2 & -(x+z) \end{vmatrix} = \hat{j} + 2(x-y)\hat{k}.$$

Therefore,

$$curl\ \vec{f}.\hat{n} = \left[\hat{j} + 2(x-y)\hat{k}\right].\hat{n}.$$

We note that the z-coordinate of each vertex of the triangle is zero. Therefore, the triangle lies in the xy-plane. Hence, $\hat{n} = \hat{k}$. Thus,

$$curl\ \vec{f}.\hat{n} = \left[\hat{j} + 2(x-y)\hat{k}\right].\hat{k} = 2(x-y).$$

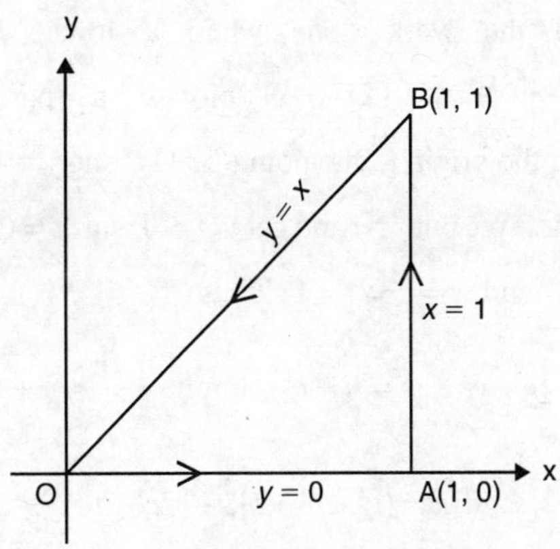

The equation of OB is $y = x$. Therefore, by Stoke's theorem, we have

$$\oint_C \vec{f}.d\vec{r} = \iint_S curl\ \vec{f}.\hat{n}\ dS = \int_0^1\int_0^x 2(x-y)dy\ dx$$

$$= 2\int_0^1 \left[xy - \frac{y^2}{2}\right]_0^x dx = 2\int_0^1 \left(x^2 - \frac{x^2}{2}\right) dx$$

$$= \int_0^1 x^2\,dx = \left[\frac{x^3}{3}\right]_0^1 = \frac{1}{3}.$$

13.21 MISCELLANEOUS EXAMPLES

EXAMPLE 13.100

Evaluate $\int_C (x^2 + yz)\,dz$, where C is the curve defined by $x = t$, $y = t^2$, $z = 3t$ for t lying in the interval $1 \le t \le 2$.

Solution. The parametric equation of the curve C are $x = t$, $y = t^2$ and $z = 3t$. Therefore

$$\int_C (x^2 + yz)\,dz = 3\int_1^2 (t^2 + 3t^3)\,dt$$

$$= 3\left[\frac{t^3}{3} + 3\frac{t^4}{4}\right]_1^2$$

$$= \frac{163}{4}$$

EXAMPLE 13.101

Find the work done when a force $\vec{F} = (x^2 - y^2 + x)i - (2xy + y)j$ moves a particle from the origin to the point (1, 1) along $y = x^2$.

Solution. We put $x = t$ and get $y = t^2$. Then $x = 0 \Rightarrow t = 0$ and $x = 1 \Rightarrow t = 1$. Thus

$$\vec{r} = x\hat{i} + y\hat{j} + z\hat{k} = t\hat{i} + t^2\hat{j} \text{ implies } \frac{d\vec{r}}{dt} = \hat{i} + 2t\,\hat{j}$$

$$\text{and } \vec{F} = (t^2 - t^4 + t)\hat{i} - (2t^3 + t^2)\hat{j}.$$

Therefore

$$W = \int_c \vec{F} \cdot d\vec{r} = \int_0^1 \left(\vec{F} \cdot \frac{d\vec{r}}{dt}\right) dt$$

$$= \int_0^1 \left\{\left[(t^2 - t^4 + t)\hat{i} - (2t^3 + t^2)\hat{j}\right] \cdot (\hat{i} + 2t\,\hat{j})\right\} dt$$

$$= \int_0^1 \left[(t^2 - t^4 + t) - 2t(2t^3 + t^2)\right] dt$$

$$= \int_0^1 \left[t^2 - t^4 + t - 4t^4 - 2t^3\right] dt$$

$$= \int_0^1 (-5t^4 - 2t^3 + t^2 + t)\,dt$$

$$= \left[\frac{-5t^5}{5} - \frac{2t^4}{4} + \frac{t^3}{3} + \frac{t^2}{2}\right]_0^1$$

$$= -1 - \frac{1}{2} + \frac{1}{3} + \frac{1}{2} = -\frac{2}{3}.$$

EXAMPLE 13.102

If $\vec{f} = (x + 3y)\hat{i} + (y - 2z)\hat{j} + (x + \lambda z)\hat{k}$ is Solenoidal, find λ

Solution. As in Example 13.44, we have

$$f_1 = x + 3y,$$

$$f_2 = y - 2z,$$

$$f_3 = x + \lambda z.$$

Then

$$div\,\vec{f} = \frac{\partial f_1}{\partial x} + \frac{\partial f_2}{\partial y} + \frac{\partial f_3}{\partial z} = 1 + 1 + \lambda = 2 + \lambda.$$

The vector will be solenoidal if $div\,\vec{f} = 0$, that is, if $2 + \lambda = 0$ or if $\lambda = -2$.

EXAMPLE 13.103

(a) Show that the vector field $\vec{F} = 2x(y^2 + z^3)\hat{i} + 2x^2 y\,\hat{j} + 3x^2 z^2\,\hat{k}$ is conservative.

(b) Prove that $(y^2 - z^2 + 3yz - 2x)\hat{i} + (3xz + 2xy)\hat{j} + (3xy - 2xz + 2z)\hat{k}$ is both solenoidal and irrotational.

Solution. (a) Since every irrotational field is conservative, it is sufficient to show that $curl\;\vec{F} = 0$. We note that

$$curl\ \vec{F} = \begin{vmatrix} \hat{i} & \hat{j} & \hat{k} \\ \frac{\partial}{\partial x} & \frac{\partial}{\partial y} & \frac{\partial}{\partial z} \\ 2x(y^2+z^3) & 2x^2y & 3x^2z^2 \end{vmatrix}$$

$$= \hat{i}\left[\frac{\partial}{\partial y}(3x^2z^2) - \frac{\partial}{\partial z}(2x^3y)\right]$$

$$-\hat{j}\left[\frac{\partial}{\partial x}(3x^2z^2) - \frac{\partial}{\partial z}\left(2x\left(y^2+z^3\right)\right)\right]$$

$$+\hat{k}\left[\frac{\partial}{\partial x}(2x^2y) - \frac{\partial}{\partial y}\left(2x\left(y^2+z^3\right)\right)\right]$$

$$= 0 - \left(6xz^2 - 6xz^2\right)\hat{j} + \hat{k}(4xy - 4xy) = 0.$$

Hence the force $\vec{F}$ is conservative.

(b) $\vec{f} = \left(y^2 - z^2 + 3yz - 2x\right)\hat{i}$

$$+(3xz + 2xy)\hat{j} + (3xy - 2xz + 2z)\hat{k}$$

Then

$$\nabla \cdot \vec{f} = \left(\hat{i}\frac{\partial}{\partial x} + \hat{j}\frac{\partial}{\partial y} + \hat{k}\frac{\partial}{\partial z}\right).\vec{f}$$

$$= -2 + 2x + 2 - 2x = 0.$$

Hence $\vec{f}$ is solenoidal. Further,

$curl\ \vec{f}$ =

$$\begin{vmatrix} \hat{i} & \hat{j} & \hat{k} \\ \frac{\partial}{\partial x} & \frac{\partial}{\partial y} & \frac{\partial}{\partial z} \\ y^2 - z^2 + 3yz - 2x & 3xy + 2xy & 3xy - 2xz + 2z \end{vmatrix}.$$

$$= \hat{i}[3x - 3x] - \hat{j}[3y - 2z + 2z - 3y]$$

$$+\hat{k}[3z + 2y - 2y - 3z] = \vec{0}$$

Hence $\vec{f}$ is irrotational.

EXAMPLE 13.104

Find the gradient of the sealer field $f(x, y) = y^2 - 4xy$ at (1, 2).

Solution.

$$\nabla f = \left(\hat{i}\frac{\partial}{\partial x} + \hat{j}\frac{\partial}{\partial y} + \hat{k}\frac{\partial}{\partial z}\right)\left(y^2 - 4xy\right)$$

$$= \hat{i}\frac{\partial}{\partial x}\left(y^2 - 4xy\right) + \hat{j}\frac{\partial}{\partial y}\left(y^2 - 4xy\right)$$

$$+\hat{k}\frac{\partial}{\partial z}\left(y^2 - 4xy\right)$$

$$= -4y\hat{i} + (2y - 4x)\hat{j} + 0$$

$$= -8\hat{i} + 0\hat{j} = -8\hat{i} \text{ at } (1,2).$$

EXAMPLE 13.105

A particle moves on the curve $x = 2t^2$, $y = t^2 - 4t$, $z = 3t - 5$, where t is time. Find the components of velocity and acceleration at time $t = 1$ in the direction of $I - 3j + 2k$.

Solution. Proceeding as in Example 13.27, we have

$$\vec{r} = x\hat{i} + y\hat{j} + z\hat{k} = 2t^2\hat{i} + \left(t^2 - 4t\right)\hat{j} + (3t - 5)\hat{k}.$$

Therefore

$$\vec{v} = \frac{d\vec{r}}{dt} = 4t\hat{i} + (2t - 4)\hat{j} + 3\hat{k}$$

$$= 4\hat{i} - 2\hat{j} + 3\hat{k} \quad \text{at } t = 1,$$

$$\vec{a} = \frac{d^2\vec{r}}{dt^2} = 4\hat{i} + 2\hat{j}$$

The unit vector in the direction of $\hat{i} - 3\hat{j} + 2\hat{k}$ is

$$\hat{n} = \frac{\hat{i} - 3\hat{j} + 2\hat{k}}{\left|\hat{i} - 3\hat{j} + 2\hat{k}\right|} = \frac{\hat{i} - 3\hat{j} + \hat{k}}{\sqrt{14}}.$$

Therefore the components of velocity and acceleration in the direction of $\hat{i} - 3\hat{j} + 2\hat{k}$ are

$$\vec{v}.\hat{n} = \left(4\hat{i} - 2\hat{j} + 3\hat{k}\right)\cdot\frac{\hat{i} - 3\hat{j} + 2\hat{k}}{\sqrt{14}} = \frac{4 + 6 + 6}{\sqrt{14}}$$

$$= \frac{16}{\sqrt{14}}$$

or

$$\vec{a}.\hat{n} = \left(4\hat{i} + 2\hat{j}\right)\cdot\frac{\hat{i} - 3\hat{j} + 2\hat{k}}{\sqrt{14}} = -\frac{2}{\sqrt{14}}.$$

EXAMPLE 13.106

Find the values of a and b so that the surfaces $ax^3 - by^2z = (a+3)x^2$ and $4x^2y - z^3 = 11$ may cut orthogonally at (2, –1, –3).

Solution. Following Example 13.40, we have

$$\phi = ax^3 - by^2z - (a+3)x^2$$

$$\psi = 4x^2y - z^3 - 11.$$

Then

$$\nabla\phi = \left(\hat{i}\frac{\partial}{\partial x} + \hat{j}\frac{\partial}{\partial y} + \hat{k}\frac{\partial}{\partial z}\right)\left(ax^3 - by^2z - a + z\right)x^2$$

$$= \hat{i}\left[3ax^2 - 2(a+3)x\right] + \hat{j}[-2byz] + \hat{k}\left[-by^2\right]$$

$$= \hat{i}(8a-12) - 6b\,\hat{j} - b\,\hat{k} \text{ at } (2,-1.-3)$$

$$\nabla\psi = \left(\hat{i}\frac{\partial}{\partial x} + \hat{j}\frac{\partial}{\partial y} + \hat{k}\frac{\partial}{\partial z}\right)\left(4x^3y - z^3 - 11\right)$$

$$= \hat{i}(8xy) + \hat{j}\left(4x^2\right) + \hat{k}\left(-3z^2\right)$$

$$= 16\hat{i} + 16\hat{j} - 27\hat{k} \text{ at } (2,-1,-3).$$

Then $\nabla\phi\cdot\nabla\psi = 0$ implies

$$\left[(8a-12)\hat{i} - 6b\,\hat{j} - b\hat{k}].[-16\hat{i} + 16\hat{j} - 27\,\hat{k}\right] = 0$$

$$\Rightarrow 128a + 69b = 192. \quad (1)$$

Also (2, –1, –3) lies on *f*. Therefore

$$8a + 3b - 4a - 12 = 0$$

or

$$4a + 3b = 12 \quad (2)$$

Solving (1) and (2), we get $a = -\frac{7}{3},\ b = \frac{64}{9}$.

EXAMPLE 13.107

(a) Find the directional derivative of $\varphi = x^2yz + 4xz^2$ at the point (1, –2, 1) in the direction of the vector $2\hat{i} - \hat{j} - 2\hat{k}$.

(b) Find a unit normal vector $\hat{n}$ of the cone of revolution $z^2 = 4\left(x^2 + y^2\right)$ at the point P(1, 0, 2).

(c) Find the directional derivative of *f* (*x*, *y*, *z*) = $2x^2 + 3y^2 + z^2$ at the point P (2, 1, 3) in the direction of the vector $\hat{a} = \hat{i} - 2\hat{k}$.

Solution. (a) We have

$$\nabla f = \left(\hat{i}\frac{\partial}{\partial x} + \hat{j}\frac{\partial}{\partial y} + \hat{k}\frac{\partial}{\partial z}\right)\left(x^2yz + 4xz^2\right)$$

$$= \hat{i}\left(2yz + 4z^2\right) + \hat{j}\left(x^2z\right) + \hat{k}\left(x^2y + 8xz\right)$$

$$= 8\hat{i} - \hat{j} - 10\hat{k} \text{ at } (1,-2,-1).$$

The unit vector in the direction of the given vector $2\hat{i} - \hat{j} - 2\hat{k}$ is

$$\hat{n} = \frac{2\hat{i} - \hat{j} - 2\hat{k}}{\sqrt{4+4+1}} = \frac{2\hat{i} - \hat{j} - 2\hat{k}}{3}.$$

Therefore the directional derivative of *f* at (1, –2, –1) in the direction of $2\hat{i} - \hat{j} - 2\hat{k}$ is

$$\nabla f.\hat{n} = \left(8\hat{i} - \hat{j} - 10\hat{k}\right).\left(\frac{2\hat{i} - \hat{j} - 2\hat{k}}{3}\right) = \frac{37}{3}.$$

(b) Similar to Exercise 24 of Chapter 13.

Let $\phi = z^2 - 4x^2 - 4y^2$. Then $\nabla\phi$ is along the normal vector.

But

$$\nabla\phi = \left(\hat{i}\frac{\partial}{\partial x} + \hat{j}\frac{\partial}{\partial y} + \hat{k}\frac{\partial}{\partial z}\right)\left(z^2 - 4x^2 - 4y^2\right)$$

$$= -8x\hat{i} - 8y\,\hat{j} + 2z\,\hat{k}$$

$$= -8\hat{i} + 4\,\hat{k} \text{ at the point } (1, 0, 2).$$

Therefore unit normal vector $\hat{n}$ to the given cone at (1,0,2) is

$$\hat{n} = \frac{-8\hat{i} + 4\hat{k}}{\sqrt{64+16}} = \frac{-8\hat{i} + 4\hat{k}}{\sqrt{80}} = \frac{-2\hat{i} + \hat{k}}{\sqrt{5}}$$

(c) Similar to Example 13.35.

We have

$$\nabla f = \left(\hat{i}\frac{\partial}{\partial x} + \hat{j}\frac{\partial}{\partial y} + \hat{k}\frac{\partial}{\partial z}\right)\left(2x^2 + 3y^2 + z^2\right)$$

$$= 4x\hat{i} + 6y\,\hat{j} + 2z\,\hat{k}$$

$$= 8\hat{i} + 6\,\hat{j} + 6\hat{k} \text{ at the point } (2, 13).$$

Now unit vector in the direction of $\hat{i} - 2\hat{k}$ is

$$\hat{u}=\frac{\hat{i}-2\hat{k}}{\sqrt{1+4}}=\frac{1}{\sqrt{5}}\left(\hat{i}-2\hat{k}\right).$$

Therefore, the directional derivative at (2, 1, 3) in the direction of $\hat{i}-2\hat{k}$ is

$$\nabla f.\hat{u}=\frac{1}{\sqrt{5}}\left(8\hat{i}+6\hat{j}+6\hat{k}\right)\left(\hat{i}-2\hat{k}\right)$$

$$=\frac{1}{\sqrt{5}}[8+0-12]=-\frac{4}{\sqrt{5}}$$

EXAMPLE 13.108

If $r=|\vec{r}|$ and $\vec{a}$ is a constant vector, prove that

$$\nabla\times\left[\frac{\vec{a}\times\vec{r}}{r^n}\right]=\frac{2-n}{r^n}\vec{a}+\frac{n}{r^{n+2}}(\vec{a}\cdot\vec{r})\vec{r}.$$

Solution. If $r=|\vec{r}|$ and $\vec{a}$ is a constant vector, then using 13.12 (B) (ii), we have

$$\nabla\times\left[\frac{\vec{a}\times\vec{r}}{r^n}\right]=\nabla\times\left[r^{-n}\left(\vec{a}\times\vec{r}\right)\right]$$

$$=\left(\nabla r^{-n}\right)\times\left(\vec{a}\times\vec{r}\right)+r^{-n}[\nabla\times\left(\vec{a}\times\vec{r}\right)$$

$$=\left[-\frac{n\vec{r}}{r^{n+2}}\right]\times\left(\vec{a}\times\vec{r}\right)+r^{-n}\left[\nabla\times\left(\vec{a}\times\vec{r}\right)\right]$$

$$=-\frac{n}{r^{n+2}}\left[\vec{r}\times\left(\vec{a}\times\vec{r}\right)\right]+r^{-n}\left[\nabla\times\left(\vec{a}\times\vec{r}\right)\right]$$

$$=-\frac{n}{r^{n+2}}\left[(\pm\pm)\mp\pm(^{-}.^{-})^{-}\right]+{}^{-n}\left[\nabla\times(^{-}\times^{-})\right]$$

$$=-\frac{n}{r^{n+2}}\left[r^2\vec{a}-(\vec{a}.\vec{r})\vec{r}\right]+r^{-n}\left[\nabla\times\left(\vec{a}\times\vec{r}\right)\right]$$

$$=-\frac{n}{r^n}\vec{a}+\frac{n}{r^{n+2}}(\vec{a}.\vec{r})\vec{r}]+r^{-n}\left[\nabla\times\left(\vec{a}\times\vec{r}\right)\right]$$

Also by Example 13.52, (ii)

$$\nabla\times\left(\vec{a}\times\vec{r}\right)=curl\left(\vec{a}\times\vec{r}\right)=2\vec{a}.$$

Therefore

$$\nabla\times\left[\frac{\vec{a}\times\vec{r}}{r^n}\right]=\frac{-n}{r^n}\vec{a}+\frac{n}{r^{n+2}}\left(\vec{a}.\vec{r}\right)\vec{r}+\frac{2\vec{a}}{r^n}$$

$$=\frac{2-n}{r^n}\vec{a}+\frac{n}{r^{n+2}}\left(\vec{a}.\vec{r}\right)\vec{r},$$

which is the required result.

EXAMPLE 13.109

Show that

$$grad\left(\vec{f}.\vec{g}\right)=\vec{f}\times curl\ \vec{g}+\vec{g}\times curl\ \vec{f}+\left(\vec{f}.\nabla\right)\vec{g}+\left(\vec{g}.\nabla\right)\vec{f}.$$

Solution. We have

$$grad\left(\vec{f}\cdot\vec{g}\right)=\sum\hat{i}\frac{\partial}{\partial x}\left(\vec{f}\cdot\vec{g}\right)$$

$$=\sum\hat{i}\frac{\partial}{\partial x}\left(\vec{f}\cdot\frac{\partial\vec{g}}{\partial x}+\frac{\partial\vec{f}}{\partial x}.\vec{g}\right)$$

$$=\sum\hat{i}\left(\vec{f}\cdot\frac{\partial\vec{g}}{\partial x}\right)+\sum\hat{i}\left(\frac{\partial\vec{f}}{\partial x}.\vec{g}\right).\quad(1)$$

On the other hand

$$\vec{f}\times\left(\hat{i}\times\frac{\partial\vec{g}}{\partial x}\right)=\left(\vec{f}.\frac{\partial\vec{g}}{\partial x}\right)\hat{i}-\left(\vec{f}.\hat{i}\right)\frac{\partial\vec{g}}{\partial x}$$

or

$$\left(\vec{f}\cdot\frac{\partial\vec{g}}{\partial x}\right)\hat{i}=\vec{f}\times\left(\hat{i}\times\frac{\partial\vec{g}}{\partial x}\right)+\left(\vec{f}\cdot\hat{i}\right)\frac{\partial\vec{g}}{\partial x}$$

Therefore

$$\sum\left(\vec{f}.\frac{\partial\vec{g}}{\partial x}\right)=\vec{f}\times\left(\hat{i}\times\frac{\partial\vec{g}}{\partial x}\right)+\left(\vec{f}.\hat{i}\right)\frac{\partial\vec{g}}{\partial x}.$$

$$=\vec{f}\times curl\,\vec{g}+\left(\vec{f}\cdot\nabla\right)\vec{g}.\quad(2)$$

Interchanging $\vec{f}$ and $\vec{g}$ in (2), we get

$$\sum\left(\vec{g}.\frac{\partial\vec{f}}{\partial x}\right)\hat{i}=\vec{g}\times curl\ \vec{f}+(\vec{g}.\nabla)\vec{f} \qquad (3)$$

From (1), (2) and (3), it follows that

$$grad\left(\vec{f}\cdot\vec{g}\right)=\vec{f}\times curl\ \vec{g}+\vec{g}\times curl\ \vec{f}$$

$$+\left(\vec{f}\cdot\nabla\right)\vec{g}+(\vec{g}.\nabla)\vec{f}.$$

EXAMPLE 13.110

Verify divergence theorem for

$$\vec{F}=\left(x^2-yz\right)\hat{i}+(f-zx)\hat{j}+\left(z^2-xy\right)\hat{k}$$

taken over the rectangular parallelopiped $0\le x\le a$, $0\le y\le b$, $0\le z\le c$.

Solution. To verify Gauss divergence theorem, we have to show that

$$\iiint_V div\,\vec{F}\,dv=\iint_S \vec{F}.\hat{n}\ ds.$$

Firstly,

$$\iiint_V div\ \vec{F}\ dv$$

$$=\int_0^c\int_0^b\int_0^a\left[\frac{\partial}{\partial x}\left(x^2-yz\right)+\frac{\partial}{\partial y}\left(y^2-zx\right)\right.$$

$$\left.+\frac{\partial}{\partial z}\left(z^2-xy\right)\right]dxdy$$

$$=\int_0^c\int_0^b\int_0^a 2(x+y+z)dxdydz$$

$$=a^2bc+ab^2c+abc^2=abc(a+b+c)$$

Now to calculate $\vec{F}\cdot\hat{n}\ ds$, we divide the surface s of the parallelepiped $0\le x\le a$, $0\le y\le b$, $0\le z\le c$ into six parts.

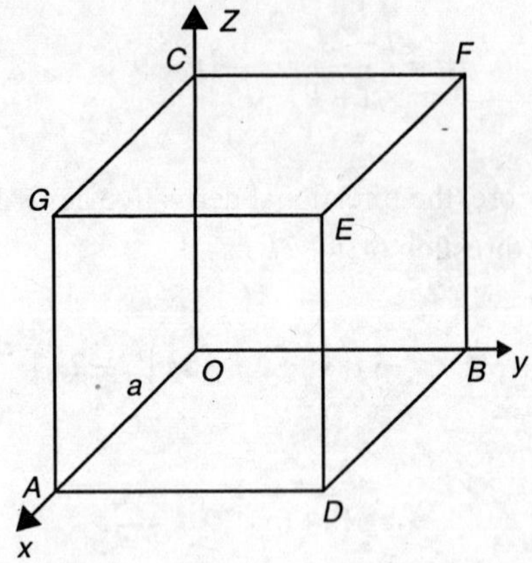

(i) For the face $OADB$, we have $\hat{n}=-\hat{k}$, $z=0$. Therefore

$$\iint_{OADB}\vec{F}.\hat{n}.\ ds=ds=\iint_{OADB}\left(x^2\hat{i}+y^2\hat{j}-xy\hat{k}\right).\left(-\hat{k}\right)ds$$

$$=\int_0^b\int_0^a xydxdy=\frac{a^2b^2}{4}.$$

(ii) For the face $CGEF$, we have $z=c,\hat{n}=\hat{k}$. Therefore

$$\iint_{CGEF}\vec{F}.\hat{n}.ds$$

$$=\iint_{CGEF}\left[\left(x^2-cy\right)\hat{i}+\left(y^2-cx\right)\hat{j}\right.$$

$$\left.+\left(c^2-xy\right)\hat{k}\right)\Big]\cdot\hat{k}\ ds$$

$$=\int_0^b\int_0^a\left(c^2-xy\right)dxdy=abc^2-\frac{a^2b^2}{4}.$$

(iii) For the face $ADEG$, we have $\hat{n}=\hat{i}, x=a$ and $dx=0$. Therefore

$$\iint_{ADEG}\vec{F}.\hat{n}.ds=\int_0^c\int_0^b\left(a^2-yz\right)dy\ dz.$$

$$=a^2bc-\frac{b^2c^2}{4}$$

(iv) For the face *OBFC*, we have $\hat{n} = -\hat{i}, x = 0, dx = 0$. Therefore

$$\int\int_{OBFC} \vec{F}.\hat{n}.\, ds = \int_0^a\int_0^b yz dy dz = \frac{b^2c^2}{4}.$$

(v) For the face *OAGC*, we have $\hat{n} = -\hat{j},\ y = 0,$ $dy = 0$ Therefore

$$\int\int_{OAGC} \vec{F}.\hat{n}.ds = \int_0^a\int_0^c zx\, dz\, dx = \frac{a^2c^2}{4}.$$

(vi) For the face *DBFE*, we have $\hat{n} = \hat{j},\ y = b,$ $dy = 0$ Therefore

$$\int\int_{DBFE} \vec{F}.\hat{n}.ds = \int_0^a\int_0^c \left(b^2 - zx\right) dz dx = ab^2c - \frac{a^2c^2}{4}.$$

Hence adding the values of the above integrals, we get

$$\int\int_S \vec{F}.\hat{n}.ds = abc(a+b+c).$$

Hence

$$\int\int\int_V div\,\vec{F}\, dv = \int\int_S \vec{F}.\hat{n}\, ds,$$

which verifies the Gauss's divergence theorem.

EXAMPLE 13.111

Evaluate using divergence theorem $\iint_S (\vec{v}\cdot\hat{n})dA,$ where $\vec{v} = x^2z\hat{i} + y\hat{j} - xz^2\hat{k}$ and S is the boundary of the region bounded by the paraboloid $z = x^2 + y^2$ and the plane $z = 4y$.

Solution. We have

$$\vec{v} = x^2z\hat{i} + y\hat{j} - xz^2\hat{k}\ .$$

Therefore

$$div\ \vec{v} = \nabla, \vec{v}$$

$$= \left(\hat{i}\tfrac{\partial}{\partial x} + \hat{j}\tfrac{\partial}{\partial y} + \hat{k}\tfrac{\partial}{\partial z}\right).\left(x^2z\hat{i} + y\hat{j} - x^2z\hat{k}\right)$$

$$= 2xz + 1 - 2xz = 1.$$

Using Divergence Theorem, we have

$$\iint(\vec{v}\cdot\hat{n})dA = \int\int\int_v div\,\vec{v}\, dv$$

$$= \iiint_v dv, \text{ since } div\,\vec{v} = 1.$$

$$= \int_0^4\int_{-\sqrt{4y-y^2}}^{\sqrt{4y-y^2}}\int_{x^2+y^2}^{4y} dz\ dx\ dy$$

$$= \int_0^4\int_{-\sqrt{4y-y^2}}^{\sqrt{4y-y^2}} [4y - x^2 - y^2]dx\ dy$$

$$= 2\int_0^4\int_0^{\sqrt{4y-y^2}} [4y - x^2 - y^2]dx\, dy$$

(even integrand in x)

$$= 2\int_0^4\left[(4y-y^2)x - \frac{x^3}{3}\right]_0^{\sqrt{4y-J^2}} dy$$

$$= \frac{4}{3}\int_0^4 (4y-y^2)^{\frac{3}{2}} dy = \frac{4}{3}\int_0^4 [4-(y-2)^2]^{\frac{3}{2}} dy$$

Substituting $y - 2 = 2\sin t$, we have $dy = 2\cos t dt$ and so

$$\iiint_v div\vec{v}dv = \frac{8}{3}\int_{-\frac{\pi}{2}}^{\frac{\pi}{2}} (4 - \sin^2 t)\cos t\, dt$$

$$= \frac{64}{3}\int_{-\frac{\pi}{2}}^{\frac{\pi}{2}} \cos^3 t \cos t dt$$

$$= \frac{128}{3}\int_0^{\frac{\pi}{2}} \cos^4 t dt = \frac{128}{3}.\frac{3}{8}.\frac{\pi}{2} = 8\pi.$$

EXAMPLE 13.112

(a) Using Green's theorem in the plane evaluate $\int_C [(2x^2 - y^2)dx + (x^2 + y^2)dy]$ where C is the boundary of the region bounded by $x = 0$, $y = 0$, $x + y = 1$.
(b) Using Green's Theorem find the area of the region in the first quadrant bounded by the curves $y = x$, $y = \frac{1}{x}$, $y = \frac{x}{4}$.

Solution. (a) We have

$$\iint_s \left(\frac{\partial f_2}{\partial x} - \frac{\partial f_1}{\partial y}\right) dx\, dy$$

$$= \int_0^1 \int_0^{1-x} 2(x-y)dy\, dx = 2\int_0^1 \left[xy - \frac{y^2}{2}\right]_0^{1-x} dx$$

$$= \int_0^1 [2xy - y^2]_0^{1-x} dx$$

$$= \int_0^1 [2x(1-x) - (1-x)^2]dx$$

$$= \int_0^1 [2x - 2x^2 - 1 - x^2 + 2x]dx$$

$$= \int_0^1 (4x - 3x^2 - 1)dx$$

$$= \left[\frac{4x^2}{2} - \frac{3x^3}{3} - x\right]_0^1 = 2 - 1 - 1 = 0.$$

(b) Using Green's Theorem,

$$A = \frac{1}{2}\oint_c (x\,dy - ydx),$$

$$= \frac{1}{2}\left[\oint_{C_1} + \oint_{C_2} + \oint_{C_3}\right],$$

$$= \frac{1}{2}\left[\int_{C_1} (xdy - ydx) + \int_{C_2} (xyd - ydx) + \int_{c_3} (xdy - ydx\right],$$

where c_1 is $y = \frac{x}{4}$, $c_2$is $y = \frac{1}{x}$, and c_3 is $y = x$. Along c_1, we have $y = \frac{x}{4}$ so that $dy = \frac{1}{4}dx$ and x varies from 0 to 2. Therefore

$$\oint_{c_1} (xdy - ydx) = \int_0^2 \left(\frac{x}{4}dx - \frac{x}{4}dx\right) = 0.$$

Along c_2 we have $y = \frac{1}{x}$ so that $dy = -\frac{1}{x^2}dx$ and x varies from 2 to 1. Therefore

$$\oint_{c_2} (xdy - ydx) = \int_2^1 \left(\frac{-1}{x}dx = \frac{1}{x}dx\right)$$

$$= -2\int_0^1 \frac{1}{x}dx = 2\log 2.$$

Along c_3, we have $y = x$ so that $dy = dx$ and x varies from 1 to 0. Therefore

$$\oint_{c_3} (xdy - ydx) = \oint_{c_3} (xdx - xdx) = 0.$$

Hence

$$\oint_{c_3} (xdy\text{-}ydx) = \frac{1}{2}[0 + 2\log 2 + 0] = \log 2.$$

EXAMPLE 13.113

Verify Gauss divergence theorem for the function $\bar{F} = yi + xj + z^2k$ over the cylindrical region bounded by $x^2 + y^2 = 9$, $z = 0$ and $z = 2$.

Solution. We have

$$\iiint div\, \bar{f}\, dv$$

$$= \int_{-3}^{3} \int_{-\sqrt{9-x^2}}^{\sqrt{9-x^2}} \left[\int_0^2 2zdz\right] dydx$$

$$4\int_{-3}^{3} \left[\int_{-\sqrt{9-x^2}}^{\sqrt{9-x^2}} dy\right] dx = 8\int_{-3}^{3} \left[\int_0^{\sqrt{9-x^2}} dy\right] dx$$

$$= 16\int_0^3 \sqrt{9-x^2}dx = 16\left[\frac{x\sqrt{9-x^2}}{2} + \frac{9}{2}\sin^{-1}\frac{x}{3}\right]_0^3$$

$$= 16\left[0 + \frac{9}{2}\sin^{-1}\frac{3}{3}\right] = \frac{16\times 9\times \pi}{2\times 2} = 36\pi.$$

Similarly (Proceeding as in Example 13.77), we have

$$\iint_S \vec{F}\cdot\hat{n}ds = 36\pi.$$

Hence the theorem is verified.

EXAMPLE 13.114

Evaluate and $\iint_S \vec{f}.d\vec{S}$ if $\vec{f} = yz\hat{i} + wy^2\hat{j} + xz^2\hat{k}$ S is the surface of the cylinder $x^2+y^2=9$ contained in the first octant between the planes $z=0$ and $z=2$.

Solution. By Gauss's Divergence Theorem

$$\iint_S \vec{f}.\hat{n}ds = \iiint_V div\vec{f}dv$$

$$= \iiint_V \left[\frac{\partial}{\partial x}(yz)+\frac{\partial}{\partial y}(2y^2)+\frac{\partial}{\partial z}(xz^2)\right]dv$$

$$= \iiint [4y+2xz]dz\,dy\,dx$$

$$= \int_0^3 \int_0^{\sqrt{9-x^2}} \int_0^2 (4y+2xz)dz\,dy\,dx$$

$$= \int_0^3 \int_0^{\sqrt{9-x^2}} (8y+4x)\,dy\,dx$$

$$= \int_0^3 [4(9-x)^2 + 4x(9-x^2)^{\frac{1}{2}}]dx = 108.$$

EXAMPLE 13.115

Verify Stoke's theorem for $\vec{F} = xy^2\hat{i} + y\hat{j} + z^2x\hat{k}$ for the surface of a rectangular lamina bounded by $x=0$, $y=0$, $x=1$, $y=2$, $z=0$.

Solution. Similar to Example 13.97.

We have

$$\vec{f} = xy^2 i + y\hat{j} + z^2x\hat{k}.$$

Therefore

$$\vec{f}\cdot d\vec{r} = (xy^2\hat{i} + y\hat{j} + z^2x\hat{k}).(\hat{i}dx + \hat{j}dy + \hat{k}dz)$$

$$= xy^2dx + ydy + z^2xdz\,.$$

Therefore

$$\oint_C \vec{f}.d\vec{r} = \oint_C (xy^2dx + ydy + z^2xdz)$$

$$= \int_{OA} + \int_{AB} + \int_{BC} + \int_{CO}$$

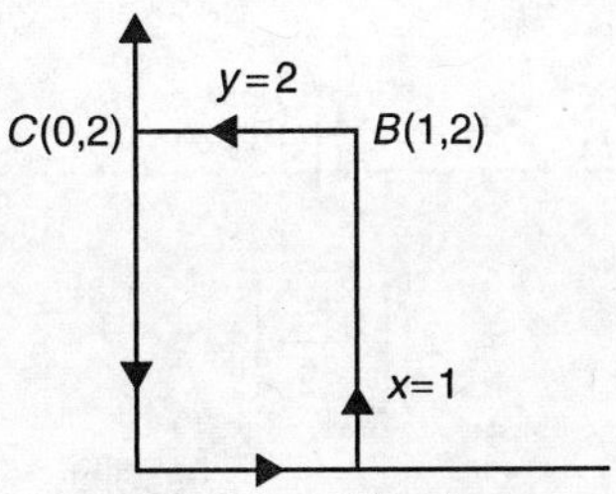

Along OA, we have $y=0$ and $dy=0$. Therefore

$$\int_{OA} \vec{f}.d\vec{r} = \int_0^1 0dx = 0.$$

Along AB, we have $x=1$ and $dx=0$. Therefore

$$\int_{AB} \vec{f}\cdot d\vec{r} = \int_0^2 ydy = \left[\frac{y^2}{2}\right]_0^2 = 2.$$

Along BC, we have $y=2$ and $dy=0$. Therefore

$$\int_{BC} \vec{f}\cdot d\vec{r} = \int_1^0 4xdx = 4\left[\frac{x^2}{2}\right]_1^0 = -2\,.$$

Along CO, we have $x=0$ and $dx=0$. Therefore

$$\int_{CO} \vec{f}\cdot d\vec{r} = \int_2^0 ydy = \left[\frac{y^2}{2}\right]_2^0 = -2\,.$$

Hence

$$\oint_C \vec{f}\cdot d\vec{r} = 0+2-2-2 = -2\,.$$

On the other hand,

$$curl\,\vec{f} = \nabla\times\vec{f} = \begin{vmatrix} \hat{i} & \hat{j} & \hat{k} \\ \frac{\partial}{\partial x} & \frac{\partial}{\partial y} & \frac{\partial}{\partial z} \\ xy^2 & y & z^2x \end{vmatrix}$$

$$= \hat{i}[0] + \hat{j}[0 - z^2] + \hat{k}[-2xy]$$

$$= -z^2\hat{j} + 2xy\hat{k}.$$

Therefore

$$curl\ \vec{f} \cdot \hat{n} \cdot ds = (-z^2\hat{j} + 2xy\hat{k})\hat{k} = -2xy$$

and so

$$\iint_S curl\ \vec{f} \cdot \hat{n} \cdot ds = -2\int_0^1 \left(\int_0^2 xy\,dy \right) dx$$

$$= -2\int_2^1 \left[x\frac{y^2}{2} \right]_0^2 dx$$

$$= -2\int_0^1 2x\,dx = -4\left[\frac{x^2}{2} \right]_0^1 = -2.$$

Hence

$$\oint_C \vec{f} \cdot d\vec{r} = \iint_S curl\vec{f} \cdot \hat{n}ds$$

and Stoke's Theorem is verified.

EXERCISES

Vector Product and Differentiation of Vectors

1. Find the length of the projection of the vector $\vec{a} = \hat{i} - 2\hat{j} + \hat{k}$ on the vector $\vec{b} = 4\hat{i} - 4\hat{j} + 7\hat{k}$.

   Ans. $\frac{19}{9}$

2. Find the value of λ so that $2\hat{i} + \lambda\hat{j} + \hat{k}$ and $4\hat{i} - 2\hat{j} - 2\hat{k}$ are perpendicular.

   Ans. $\lambda = 3$

3. Using dot product, find the angle between the lines AB and AC, where *A*, *B*, *C* are the points (1,2–1), (2,0,3) and(3,–1,2) respectively.

   Ans. $\theta = \cos^{-1}\frac{20}{\sqrt{21}\sqrt{22}}$

4. Dot products of a vector $\vec{a}$ with the vectors $\hat{i} - \hat{j} + \hat{k}$, $2\hat{i} + \hat{j} - 3\hat{k}$ and $\hat{i} + \hat{j} + \hat{k}$ are respectively 4, 0 and 2.Find the vector $\vec{a}$.

   Ans. $2\hat{i} - \hat{j} + \hat{k}$

5. Find the area of a triangle having vertices $A(2,3,-1)$, $B(1,-2,3)$ and $C(-2,1,4)$.

   Ans. $\frac{1}{2}\sqrt{734}$ sq.units

6. If $|\vec{a}| = 3, |\vec{b}| = 4, \vec{a}.\vec{b} = 9$, find $|\vec{a} \times \vec{b}|$.

   Ans. $\sqrt{63}$

7. Show that the vectors $2\hat{i} + \hat{k}$, $3\hat{j} + 4\hat{k}$ and $8\hat{i} - 3\hat{j}$ are coplanar. Also express the last vector as a linear combination of first two.

   Ans. $8\hat{i} - 3\hat{j} = 4(2\hat{i} + \hat{k}) - (3\hat{j} + 4\hat{k})$

8. Find the volume of the parallelepiped whose coterminus edges are $\hat{i} + \hat{j} + 2\hat{k}$, $\hat{i} + 2\hat{j} + 4\hat{k}$ and $2\hat{i} + \hat{j} + 3\hat{k}$.

   Ans. 1 cubic unit

9. If $\hat{r}$ is a unit vector in the direction of $\vec{r}$, show that $\hat{r} \times \frac{d\hat{r}}{dt} = \frac{1}{r^2}\vec{r} \times \frac{d\vec{r}}{dt}$, where $|\vec{r}| = r$.

10. If $\vec{a} = t^2\hat{i} - t\hat{j} + (2t+1)\hat{k}$ and $\vec{b} = (2t-3)\hat{i} + \hat{j} - t\hat{k}$, find (*i*) $\frac{d}{dt}(\vec{a}.\vec{b})$ and (*ii*) $\frac{d}{dt}(\vec{a} \times \vec{b})$, when $t = 1$.

    Ans. (*i*) -6, (*ii*) $7\hat{j} + 3\hat{k}$.

11. If the vector $\vec{a}$ has a constant magnitude, show that $\vec{a}$ and $\frac{d\vec{a}}{dt}$ are perpendicular, provided $|\frac{d\vec{a}}{dt}| \neq 0$.

    Hint: $\vec{a}.\vec{a} = |\vec{a}|^2 =$ constant implies $\frac{d}{dt}(\vec{a}.\vec{a}) = 0$ or $\vec{a}.\frac{d\vec{a}}{dt} + \vec{a}\frac{d\vec{a}}{dt} = 0$ or $2\vec{a}.\frac{d\vec{a}}{dt} = 0$ and so, $\vec{a}$ is orthogonal to $\frac{d\vec{a}}{dt}$, if $|\frac{d\vec{a}}{dt}| \neq 0$.

12. If $\vec{a}$, $\vec{b}$, and $\vec{c}$ are constant vectors, show that the vector $\vec{r} = \vec{a}t^2 + \vec{b}t + \vec{c}$ is the position vector of a point moving with a constant acceleration.

    Hint: $\frac{d^2\vec{r}}{dt} = 2\vec{a}$ (constant).

13. A particle moves along the curve $x = t^3+1$, $y = t^2$, and $z = 2t + 5$, where t is the time. Find the component of its velocity and acceleration at time $t = 1$ in the direction $\hat{i} + \hat{j} + 3\hat{k}$.

    Ans. $\sqrt{11}$, $\frac{8}{\sqrt{11}}$.

14. A particle moves so that its position vector is given by $\vec{r} = \cos \omega t\hat{i} + \sin \omega t\hat{j}$, where ω is constant. Show that (i) the velocity $\vec{v}$ of the particle is perpendicular to $\vec{r}$, (ii) the acceleration is directed toward the origin and has a magnitude proportional to the distance from the origin, and (iii) $\vec{r}\times\vec{v}$ is a constant vector.

Hint: $\vec{r}.\vec{v} = 0$ and $\vec{a} = \frac{d\vec{v}}{dt} = -\omega^2 \cos t\hat{i} - \omega^2 \sin \omega t\hat{j} = -\omega^2 .\vec{r}$.

15. Find the unit tangent vector at any point on the curve $x = t^2 + 2$, $y = 4t - 5$, and $z = 2t^2 - 6t$, where t is any variable.

Hint: $\hat{T} = \dfrac{d\vec{r}}{|\frac{dra}{dt}|}$, where $\vec{r} = x\hat{i} + y\hat{j} + z\hat{k}$.

Ans. $\frac{2\hat{i}+2\hat{j}+\hat{k}}{3}$.

16. Find the angle between the tangents to the curve $x = t$, $y = t^2$, and $z = t^3$ at $t = \pm 1$.

Hint: $\vec{r} = x\hat{i} + y\hat{j} + z\hat{k} = t\hat{i} + t^2\hat{j} + t^3\hat{k}$. Find $\frac{d\vec{r}}{dt}$ and put $t = 1$ and $t = -1$ to get $\vec{T}_1$ and $\vec{T}_2$. Then, the angle between $\vec{T}$ and $\vec{T}$ is given by $\cos\theta = \dfrac{\vec{T}_1\vec{T}_2}{|\vec{T}_1\vec{T}_2|}$.

Ans. $\cos^{-1}\frac{3}{7}$.

17. If $\vec{a} = x^2yz\hat{i} - 2xz^3\hat{j} + xz^2\hat{k}$, find the value of $\frac{\partial^2}{\partial x\partial y}(\vec{a}\times\vec{b})$ at the point $(1, 0, -2)$.

Ans. $-4\hat{i} - 8\hat{j}$.

18. The position vector of a point at any time t is given by $\vec{r} = e^t(\cos t\hat{j} + \sin t\hat{j})$. Show that (i) $\vec{a} = 2(\vec{v} - \vec{r})$, where $\vec{a}$ and $\vec{v}$ are respectively acceleration and velocity of the particle and (ii) the angle between the radius vector and the acceleration is constant.

Hint: $\vec{v} = \dfrac{d\vec{r}}{dt} = e^t(\cos t - \sin t)\hat{j} + e^t(\sin t + \cos t)\hat{j}$,

$\vec{a} = \sinh(t-1)\hat{i} + 2e^t \cos t\hat{j}$.

Clearly $\vec{a} = 2(\vec{v} - \vec{r})$.

19. The position vector of a particle at time t is $\vec{r} = \cos(t-1)\hat{i} + \sinh(t-1)\hat{j} + Kt^3\hat{k}$. Find the value of K such that at time $t = 1$, the acceleration is normal to the position vector $\vec{r}$.

Hint: $\frac{d^2\vec{r}}{dt^2}$ at $t = 1$ is $-\hat{i} + 6K\hat{k}$ and $\vec{r}$ at $t = 1$ is $\hat{i} + K\hat{k}$. Therefore,

$$\cos\theta = \frac{(-\hat{i} + 6K\hat{k})(i + Kk)}{\sqrt{7}(1+k^2)^{\frac{1}{2}}} = \frac{6K^2 - 1}{\sqrt{7}(1+K^2)^{\frac{1}{2}}}$$

normality, $6K^2 - 1 = 0$ and so, $K = \pm\frac{1}{\sqrt{6}}$.

20. A particle moves along the curve $x = t^3 + 1$, $y = t^2$, and $z = 2t + 5$, where t represents the time. Find the component of its velocity and acceleration at time $t = 1$ in the direction of $\hat{i} + \hat{j} + \hat{k}$.

Hint: $\vec{r} = (t^3 + 1)\hat{i} + t^2\hat{j} + (2t + 5)\hat{k}$ and $\frac{d\vec{r}}{dt} = 3t^2\hat{i} + 2t\hat{j} + 2\hat{k}$. At $t = 1, \vec{v} = 3\hat{i} + 2\hat{j} + 2\hat{k}$

The unit vector in the direction of $\hat{i} + \hat{j} + 3\hat{k}$ is $\frac{\hat{i}+\hat{j}+3k}{\sqrt{11}}$. Therefore, the component of $\vec{v}$ along $\hat{i} + \hat{j} + 3\hat{k}$ is $(3\hat{i} + 2\hat{j} + 2\hat{k})\left(\frac{\hat{i}+\hat{j}+3\hat{k}}{\sqrt{11}}\right) = \sqrt{11}$. Similarly, proceed for acceleration, which will be $\frac{8}{\sqrt{11}}$.

21. If $\vec{F} = xyz\hat{i} + xz^2\hat{j} - y^3\hat{k}$ and $\vec{g} = x^3\hat{i} - xyz\hat{j} + x^2z\hat{k}$, then determine $\dfrac{\partial^2\vec{f}}{\partial y^2}\times\dfrac{\partial^2\vec{g}}{\partial y^2}$ at the point (1, 1, 0).

Ans. $-36\hat{j}$.

Gradient and Fractional Derivatives

22. If $\vec{r}$ is the position vector of a point and $\vec{a}$ is any vector, show that $grad[\vec{r}\vec{a}\vec{b}] = \vec{a}\times\vec{b}$

Hint: $\vec{r} = x\hat{i} + y\hat{j} + z\hat{k}$ and $\vec{a} = a_1\hat{i} + a_2\hat{j} + a_3\hat{k}$. Then, $\vec{a}.\vec{r} = a_1x + a_2y + a_3z$ and $\nabla(\vec{a}.\vec{r}) = \vec{a}$.

Therefore,

$\nabla(\vec{r}\vec{a}\vec{b}) = \nabla[\vec{r}.(\vec{a}\times\vec{b})] = \nabla[(\vec{a}\times\vec{b}).\vec{r}] = \vec{a}\times\vec{b}$.

23. If $\phi(x,y,z)=3xy^2-y^3z^2$, find $\nabla\phi$ at the point $(-1,2,-1)$.

Ans. $12\,\hat{i}-24\hat{j}+16\hat{k}$.

24. Find a unit normal to the surface $x^2y + 2xy = 4$ at the point (2, –2,3). Hence, find the equation of the normal to the surface at (2, –2,3).

Hint: Let $\phi = x^2y + 2xz - 4$. Then, $\Delta\phi = \nabla\phi = 2(xy+z)\hat{i}+x^2\hat{j}+2x\hat{k}$ and $\nabla\phi$ at (2, –2, 3) $=-2\hat{i}+4\hat{j}+4\hat{k}$. Thus, the unit normal vector to the surface at (2, –2, 3) is $\frac{-2i+4j+4k}{\sqrt{36}}=\frac{1}{3}(-\hat{i}+2\hat{j}+2\hat{k})$. Thus, the equation of normal is $\frac{x-2}{-1}=\frac{y+2}{2}=\frac{z-3}{2}$.

25. Find a unit normal vector to the surface $x^3 + y^3 + 3xyz = 3$ at the point (1, 2 –1),

Ans. $\frac{-i+3j+2k}{\sqrt{14}}$.

26. Find the directional derivative of $\phi = (x,y,z) = x^2yz + 4xz^2$ at the point (1, 2 –1) in the direction of $2\,\hat{i}-\hat{j}-2\hat{k}$.

Hint: Proceed as in Example 13.16.

Ans. $\frac{37}{3}$.

27. Find the directional derivative of the function $\phi(x, y, z)=x^2-y^2+2z^2$ at $P(1,2,3)$, in the direction of the line PQ, where Q is the point (5, 0, 4). In what direction the directional derivative will be maximum?

Hint: $\nabla\phi=2x\hat{i}-2y\hat{j}+4z\hat{k}$. Therefore, $\nabla\phi$ at $(1,2,3)$ is $2\,\hat{i}-4\hat{j}+12\hat{k}$. Also, $\vec{PQ}=\vec{OQ}-\vec{OP}=(5\hat{i}+4\hat{j})-(\hat{i}+2\hat{j}+3\hat{k})=4\hat{i}-2\hat{j}+\hat{k}$.

Unit vector $\hat{a}$ in the direction of $\vec{PQ}$ is $\frac{4\hat{i}-2\hat{j}+\hat{k}}{\sqrt{21}}$. Then, the required directional derivative $=\nabla\phi(at(1,2,3)).\hat{a}=\frac{4}{3}\sqrt{21}$. It will be maximum in the direction of the normal to ϕ, that is, in the direction of $\nabla\phi$, which is equal to $2\hat{i}-4\hat{j}+12\hat{k}$. Its maximum value is $|\nabla\phi|=\sqrt{4+16+144}=\sqrt{164}=2\sqrt{41}$.

28. If $\phi(x,y,z)=2xy+z^2$, find the directional derivative of ϕ in the direction of $\hat{i}+2\hat{j}+2\hat{k}$.

Hint: $\nabla\phi=2y\hat{i}+2x\hat{j}+2z\hat{k}=-2\hat{i}+2\hat{j}+6\hat{k}$ at (1, –1, 3). Unit vector $\hat{a}$ in the direction of $\hat{i}+2\hat{j}+2\hat{k}$ is $\frac{\hat{i}+2\hat{j}+2\hat{k}}{3}$. Therefore, the required directional derivative is $\nabla\phi.\hat{a}=(-2\hat{i}+2\hat{j}+6\hat{k}).\left(\frac{\hat{i}+2\hat{j}+2\hat{k}}{3}\right)=\frac{14}{3}$.

29. Find the greatest rate of increase of $u=x^2+yz^2$ at the point $(1,-1,3)$.

Ans. $|\nabla u|=\sqrt{121}$.

30. Find the equation of the tangent plane to the surface $z=x^2+y^2$ at the point $(2,-1,5)$.

Hint: $\nabla\phi$ at (2, –1, 5) is $4\,\hat{i}-2\hat{j}-\hat{k}$. The unit normal vector at (2, –1, 5) is $\hat{a}=\frac{4i-2j-k}{\sqrt{21}}$. The equation of the line through (2, –1, 5) in the direction of normal vector $\hat{a}$ is $\frac{x-2}{4}=\frac{y+1}{-2}=\frac{z-5}{-1}$. Therefore, the equation of tangent plane to the surface at (2, –1, 5) is $4(x-2)-2(y+1)-(z-5)=0$ or $4x-2y-z=5$. We may also find a tangent plane using $(\vec{r}-\vec{a}).\nabla\phi=0$. Therefore, in the present case, we have $[(x\hat{i}+y\hat{j}+z\hat{k})-(2\hat{i}-\hat{j}+5\hat{k})].(4\hat{i}-2\hat{j}-\hat{k})=0$ or $4(x-2)-2(y+1)-2(y+1)-(z-5)=0$ or $4x-2y-z=5$.

Divergence and Curl of Vector-point function

31. Show that the vector $(-x^2+yz)\hat{i}+(4y-z^2x)\hat{j}+(2xz-4z)\hat{k}$ is solenoidal.

Hint: Show that $\nabla.\vec{f}=0$.

32. If $f=(x^2+y^2+z^2)^{-n}$, find div grand f and also n, so that div grand f = 0.

Ans. $\frac{2n(2n-1)}{(x^2+y^2+z^2)^{n+1}}$ and $n=\frac{1}{2}$.

33. Show that div $\left(\frac{\vec{r}}{r^3}\right)=0$, where $\vec{r}=x\hat{i}+y\hat{j}+z\hat{k}$.

Hint: Use div $(\phi\vec{f})=\phi$ div $\vec{f}+grad\phi.\vec{f}$. We get

$\text{div}\left(\frac{\vec{r}}{r^3}\right) = \text{div}(r^{-3}\vec{r}) = r^{-3}\text{div}\vec{r} + \vec{r}.grad r^{-3} =$

$3r^{-3} + \vec{r}.(-3r^{-4} grad r) = 3r^{-3} + \vec{r}$.

$\left[-3r^{-4}\frac{\vec{r}}{r}\right] =$

$3r^{-3} - 3r^{-5}(\vec{r}.\vec{r}) = 3r^{-3} - 3r^{-5}(r^2) = 0$.

Thus, it also follows that $\frac{\vec{r}}{r^3}$ is solenoidal.

34. Show that the function $\frac{1}{r}$, where $r = |\vec{r}| = \sqrt{x^2+y^2+z^2}$, is a harmonic function, if $r \neq 0$.

Hint: Show that $\nabla^2\left(\frac{1}{r}\right) = 0$ (see Example 13.36).

35. If $\vec{f} = \frac{1}{4}\nabla v$, where u and v are scalar fields and $\vec{f}$ is a vector field, show that $\vec{f}$ curl $\vec{f} = a$.

Hint: $curl\,\vec{f} = \nabla\times(\frac{1}{u}\nabla u).\nabla\frac{1}{u}\times\nabla v + \frac{1}{u}\nabla\times\nabla v = \nabla\frac{1}{u}\times\nabla v + 0$. Hence,

$\vec{f}$. Curl $\vec{f} = \frac{1}{u}\nabla v.(\nabla\frac{1}{u}\times\nabla v) = 0$.

36. Show that the vector $\nabla\phi\times\nabla\psi$ is solenoidal.

37. Find the value of a so that

$\vec{f} = (ax^2y + yz)\hat{i} + (xy^2 - xz^2)\hat{j}$

$+(2xyz - 2x^2y^2)\hat{k}$ is solenoidal. Also find the curl of this solenoidal vector.

Hint: $\text{div}\vec{f} = 2(a+2)xy$. Now, $\vec{f}$ will be solenoidal if div $\vec{f} = 0$, which yields $a = -2$. *curl* $\vec{f}$ can be found.

38. If $\vec{r} = x\hat{i} + y\hat{j} + z\hat{k}$ and $r = |\vec{r}| = \sqrt{x^2+y^2+z^2}$,

show that $div(\hat{r}) = \dfrac{2}{r}$.

Hint: $\hat{r} = \dfrac{x\hat{i}+y\hat{j}+z\hat{k}}{\sqrt{x^2+y^2+z^2}} = \dfrac{\vec{r}}{r}$ and

$$\text{div}(\tfrac{\vec{r}}{r}) = \frac{1}{r}div\vec{r} + grad\frac{1}{r}.\vec{r}$$

$$= \frac{3}{r} + [(-1)r^{-2} grad\, r].\vec{r}$$

$$= \frac{3}{r} - \frac{1}{r^2}\left(\frac{\vec{r}}{r}\right).\vec{r} = \frac{3}{r} - \frac{1}{r^3}(\vec{r}.\vec{r})$$

$$= \frac{3}{r} - \frac{r^2}{r^3} = \frac{3}{r} - \frac{1}{r} = \frac{2}{r}.$$

39. Show that $\nabla^2(r\vec{r}) = \left(\dfrac{4}{r}\right)\vec{r}$.

Hint:

$$\nabla^2(r\vec{r}) = \sum\frac{\partial^2}{\partial x^2}(r\vec{r})$$

$$= \sum\frac{\partial}{\partial x}\left(\frac{\partial}{\partial x}(r\vec{r})\right)$$

$$= \sum\frac{\partial}{\partial x}\left[\frac{\partial r}{\partial x}\vec{r} + r\frac{\partial\vec{r}}{\partial x}\right]$$

$$= \sum\frac{\partial}{\partial x}\left[\frac{x}{r}\vec{r} + ri\right]$$

$$= \sum\left(\left\{\frac{\vec{r}}{r} + \frac{x}{r}(i) - \frac{x}{r^2}\left(\frac{x}{r}\right)\vec{r}\right\} + i\frac{x}{r}\right)$$

$$= \left(\frac{4}{r}\right)\vec{r}.$$

40. Show that the vector field

$\vec{v} = (\sin y + z)\hat{i} + (x\cos y - z)\hat{j} + (x - y)\hat{k}$

is irrotational.

41. If $\vec{r} = xi + yj + z\hat{k}$, determine $\nabla\cdot\left(\dfrac{\vec{r}}{r}\right)$. (See Exercise 33).

Vector Integration and Line Integrals

42. If $\vec{r}(t) = 2\hat{i} - \hat{j} + 2\hat{k}$ for $t = 2$ and $\vec{r}(t) = 4\hat{i} - 2\hat{j} + 3\hat{k}$ for $t = 3$, show that $\int_2^3 [\vec{r}.\frac{d\vec{r}}{dt}]dt = 10$.

Hint: $\frac{d}{dt}[(\vec{r})^2] = 2\vec{r}\frac{d\vec{r}}{dt}$ implies $\vec{r}\frac{d\vec{r}}{dt} = \dfrac{1d}{2dt}[(\vec{r})^2]$

Therefore,

$\int_2^3 [\vec{r}\frac{d\vec{r}}{dt}]dt = \dfrac{1}{2}[(\vec{r})^2]_2^3 = \dfrac{1}{2}[29-9] = 10$, using

$\vec{r}(t) = 4\hat{i} - 2\hat{j} + 3\hat{k}$ for $t = 3$ and $2\hat{i} - \hat{j} + 2\hat{k}$ for $t = 2$.

43. Evaluate $\int_C \vec{f}\, d\vec{r}$, where $\vec{f} = (x^2 + y^2)\hat{i} - 2xy\hat{j}$ and the curve C is the rectangle bounded by $y = 0$, $x = a$, $y = b$, and $x = 0$.

Ans. $-2ab^2$.

44. If $\vec{f} = 2y\hat{i} - z\hat{j} + x\hat{k}$, find the vector line integral $\int_C \vec{f} \times d\vec{r}$ along the curve $x = \cos t$, $y = \sin t$, andc $z = 2\cos t$ and from $t = 0$ to $t = \frac{\pi}{2}$.

Ans. $(2 - \frac{\pi}{4})\hat{i} + \left(\pi - \frac{1}{2}\right)\hat{j}$.

45. Evaluate $\int_C \vec{f} . d\vec{r}$, where $\vec{f} = yz\hat{i} + zx\hat{j} + xy\hat{k}$ and C is the portion of the curve $\vec{r} = a\cos t\hat{i} + b\sin t\hat{j} + ct\hat{k}$ from $t = 0$ to $t = \dfrac{\pi}{2}$.

Hint: Parametric equations of the curve are $x = a\cos t$, $y = b\sin t$, and $z = ct$. Also $\frac{d\vec{r}}{dt} = -a\sin t\hat{i} + b\cos t\hat{j} + c\hat{k}$. Putting the values of x, y, z (in terms of t) in $\vec{f}$, we see that $\int_C \vec{f} . d\vec{r} = \int_C \vec{f} . (\frac{d\vec{r}}{dt}) dt = abc(0) = 0$.

46. Evaluate $\int_C [y^2 dx - x^2 dy]$ along the triangle whose vertices are (1, 0), (0,1) and (–1,0).

Hint: Find the equations of three sides by a two-point formula and evaluate the integral over those sides.

Ans. $-\dfrac{2}{3}$.

47. If $\vec{f} = (2x + y)\hat{i} + (3y - x)\hat{j}$, evaluate $\int_C \vec{f} . d\vec{r}$, where C is the curve in the xy-plane consisting of straight lines from (0,0) to (2,0) and (2, 0) to (3, 2).

Hint: If O (0,0), A(2,0), and B(3,2) are the points, then C consists of two lines OA and AB. On OA, we have $y = 0$ so that $dy = 0$ and x varies from 0 to 2. The equation of AB is $y = 2x - 4$ so that $dy = 2dx$ and on this line, x varies from 2 to 3. Therefore, $\int_C \vec{f} . d\vec{r} = \int_{OA} + \int_{AB}$, which will come out to be $4 + 7 = 11$.

48. Find the circulation of $\vec{f}$ around the curve C, where $\vec{f} = y\hat{i} + z\hat{j} + x\hat{k}$ and C is the circle $x^2 + y^2 = 1$ and $z = 0$.

Hint: Parametric equations of C are $x = \cos t$, $y = \sin t$, and $z = 0$, where t varies from 0 to 2π. Then, $\oint_C \vec{f} . d\vec{r} = -\int_0^{2\pi} \dfrac{1 - \cos 2t}{2} dt = -\pi$.

49. Find the work done when a force $\vec{f} = (x^2 - y^2 + x)\hat{i} - (2xy + y)\hat{j}$ moves a particle in xy-plane from (0,0) to (1, 1) along the parabola $y^2 = x$.

Hint: Proceed as in Example 13.49.

Ans. $-\frac{2}{3}$.

50. Compute the work done by a force $\vec{f} = x\hat{i} - z\hat{j} + 2y\hat{k}$ to displace a particle along a closed path C consisting of the segments C_1, C_2, and C_3, such that $0 \le x \le 1$, $y = x$, $z = 0$ *on* C_1, $0 \le z \le 1$, $x = 1$, $z = 1$ *on* C_2, and $0 \le x \le 1$, $y = z = x$ *on* C_3.

Ans. $\frac{3}{2}$.

51. Find the work done in moving a particle once around a circle C in the xy-plane, if the circle has its center at the origin with a radius 3, and if the force field is given by $\vec{f} = (2x - y + z)\hat{i} + (x + y - z^2)\hat{j} + (3x - 2y + 4z)\hat{k}$.

Hint: Parametric equations of C are $x = 3\cos t$, $y = 3\sin t$, and $0 \le t \le 2\pi$.

Ans. 18π.

Surface Integrals

52. Evaluate $\iint_S \vec{f}.\hat{n}\ ds$, where $\vec{f} = 12x^2 y\hat{j} - 3yz\hat{j} + 2z\hat{k}$ and S is the portion of the plane $x + y + z = 1$, included in the first octant.

Hint: $\hat{n} = \frac{\hat{i}+\hat{j}+\hat{k}}{\sqrt{3}}$ and $\vec{f}.\hat{n} = \frac{1}{\sqrt{3}}(12x^2y - 3yz + 2z)$

$= \frac{1}{\sqrt{3}}[12x^2y - 3y(1-x-y) + 2(1-x-y)\cdot$

$\hat{n}.\hat{k} = \frac{1}{\sqrt{3}}$. Evaluate $\iint \vec{f}.\hat{n}\ ds.$

Ans. $\frac{49}{120}$.

53. Evaluate $\iint_S \vec{f}.\hat{n}\,dS$, where $\vec{f} = (x+y^2)\hat{i} - 2x\hat{j} + 2yz\hat{k}$ and S is the surface of the plane $2x + y + 2z = 6$, in the first octant.

Hint: Proceed as in Example 13.56.

Ans. 81.

54. Evaluate $\iint_S \vec{f}.\hat{n}\,dS$, where $\hat{f} = 4xy\hat{i} + yz\hat{j} - xy\hat{k}$ and S is the surface bounded by the planes $x=0, x=2, y=0, y=2, z=0$, and $z=2$.

Hint: Proceed as in Example 13.59.

Ans. 40.

55. Evaluate $\iint_S \phi\,\hat{n}\,dS$, where $\phi = \frac{3}{8}xyz$ and S is the surface of the cylinder $x^2 + y^2 = 16$ included in the first octant between $z = 0$ and $z = 5$.

Hint: $\nabla(x^2+y^2-16) = 2x\hat{i} + 2y\hat{j}, \hat{n} = \frac{2xi + 2yj}{\sqrt{4x^2+4y^2}}$,

and $\hat{n}.\hat{j} = \frac{(x\hat{i}+y\hat{j})}{4}.\hat{j} = \frac{y}{4}$ Therefore;

$$\iint_S \phi\hat{n}ds = \iint_S \phi\hat{n}\frac{dxdz}{|\hat{n}.\hat{j}|}$$

$$= \iint_R \frac{3}{8}xyz\frac{(x\hat{i}+y\hat{j})}{4}.\frac{dxdz}{\frac{y}{4}}$$

$$= \frac{3}{8}\iint_R xz(x\hat{i}+y\hat{j})dxdz$$

$$= \frac{3}{8}\int_0^5\int_0^4 (x^2z\hat{i} + xz\sqrt{16-x^2}\,\hat{j})dxdz = 100(\hat{i}+\hat{j}).$$

56. Evaluate $\iint_S \vec{f}.\hat{n}dS$, where $\vec{f} = y\hat{i} + 2x\hat{j} - z\hat{k}$ And S is the surface of the plane $2x + y = 6$ in the first octant cut off by the plane $z = 4$.

Ans. 108.

Volume Integral

57. Evaluate $\iint_S \phi dV$, where $\phi = 45x^2y$ and V is the region bounded by the planes $4x + 2y + z = 8$, $x = 0$, $y = 0$, and $z = 0$.

Ans. 128.

58. Evaluate $\iiint_V (2x+y)dV$, where V is the closed Region bounded by the cylinder $z = 4 - x^2$ and the planes $x=0, y=0, y=2$, and $z=0$.

Hint: The limits of integration are $x = 0$ to $x = 2$, $y = 0$ to $y = 2$, and $z = 0$ to $z = 4 - x^2$.

Ans. $\frac{80}{3}$.

59. Evaluate $\iiint_V \text{div}\vec{f}dV$, where $\vec{f} = (x^2 - yz)\hat{i} - 2x^2y\hat{j} + 2\hat{k}$ and the region V is enclosed by the planes $x = 0$, $x = a$, $y = 0$, $y = a$. and $z = 0$, and $z = a$.

Hint: See Example 13.87.

Ans. $\frac{a^5}{3}$.

60. If $\vec{f} = 2xz\hat{i} - x\hat{j} + y^2\hat{k}$, evaluate $\iiint_V \vec{f}dV$, where V is the region bounded by the surface $x = 0$, $x = 2$, $y = 0$, $y = 6$, $z = x^2$ and $z = 4$.

Ans. $128\hat{i} - 24\hat{j} + 384\hat{k}$.

Gouss's Divergence Theorem

61. If $\vec{f} = x\hat{i} + 2y\hat{j} + 7z\hat{k}$, evaluate $\iint_S \vec{f}\hat{n}dS$,

Where S is the surface enclosing volumes V.

Hint: By Divergence Theorem,

$$\iint_S \vec{f}.\hat{n}dS = \iiint_V (\nabla.\vec{f})dV$$

$$= \iiint_V \left[\frac{\partial}{\partial x}(x) + \frac{\partial}{\partial y}(2y) + \frac{\partial}{\partial z}(7z)\right]dV$$

$$= \iiint_V (1+2+7)dV = 10V.$$

62. Verify divergence theorem for $\vec{f} = (x^2 - yz)\hat{i} + (y^2 - zx)\hat{j} + (z^2 - xy)\hat{k}$, taken over the rectangular parallel-opiped $0 \le x \le a$, $0 \le y \le b$, and $0 \le z \le c$.

Hint: $\text{div}\vec{f} = 2(x+y+z)$. Therefore,

$$\iiint_V \text{div}\,\vec{f}\,dV = \int_0^a\int_0^b\int_0^c 2(x+y+z)dxdydz$$

$$= abc(a+b+c).$$

Evaluate $\iint \vec{f}\cdot\hat{n}ds$ on all the six faces and add. We shall get $\iint_S \vec{f}.\hat{n}dS = abc(a+b+c)$

.Thus, the theorem is verified.

63. Evaluate $\iint_S \vec{f}.\hat{n}dS$, where $\vec{f} = (2x+3z)\hat{i} - (xz+y)\hat{j} + (y^2s+2z)\hat{k}$ and S is the surface of the sphere having a radius 3.

Hint: By divergence theorem,

$$\iint_S \vec{f}.\hat{n}dS = \iiint_V \text{div}\,\vec{f}\,dV$$

$$= \iiint_V \left[\frac{\partial}{\partial x}(2x+3z) + \frac{\partial}{\partial y}(-xz+y) + \frac{\partial}{\partial z}(y^2+2z)\right]dV$$

$$= \iiint_V 3dV$$

$$= 3V = 3\left(\frac{4}{3}\pi(3)^3\right) = 108\pi.$$

64. Verify Divergence Theorem for the function $\vec{f} = y\hat{i} + x\hat{j} + z^2\hat{k}$ over the cylindrical region bounded by $x^2+y^2=9$, $z=0$, and $z=2$.

Hint: Proceed as in Example 13.83.

65. Verify Divergence Theorem for $\vec{f} = y\hat{i} + x\hat{j} + z^2\hat{k}$ over the cylindrical region bounded by $x^2 + y^2 = a^2$, $z = 0$, and $z = h$. (Similar to Exercise 64.)

66. Evaluate $\iint_S \vec{f}.\hat{n}ds$, where $\vec{f} = x^3\hat{i} + y^3\hat{j} + z^3\hat{k}$ and S is the surface of the sphere $x^2+y^2+z^2=a^2$.

Hint: By divergence theorem,

$$\iint_S \vec{f}.\hat{n}dS = \iiint_V \text{div}\,\vec{f}\,dV$$

$$= \iiint_V (3x^2 + 3y^2 + 3z^2)dV$$

$$= 3a^2\iiint_V dV = 3a^2V$$

$$= 3a^2.\frac{4}{3}\pi a^3 = 4\pi a^5.$$

67. Evaluate $\iint \vec{f}\cdot\hat{n}ds$ for $\vec{f} = x\hat{i} - y\hat{j} + 2z\hat{k}$ over the sphere $x + y^2 + (z-1)^2 = 1$.

Ans. $\frac{8}{3}\pi$.

Green's Theorem

68. Verify Green's theorem in the xy-plane for $\oint_C [(xy^2 - 2xy)dx + (x^2y+3)dy]$ around the boundary C of the region enclosed by $y^2 = 8x$ and $x = 2$

Ans. $\oint_C (f_1dx + f_2dy) = \iint_S \left(\frac{\partial f_2}{\partial x} - \frac{\partial f_1}{\partial y}\right)dxdy = \frac{128}{5}$.

69. Evaluate by Green's Theorem $\oint_C e^{-x}(\sin y dx + \cos y dy)$, where C is the rectangle with vertices (0,0), $(\pi,0)$, $\left(\pi;\frac{\pi}{2}\right)$, and $\left(0,\frac{\pi}{2}\right)$.

Ans. $2e^{-\pi} - 2$.

70. Verify Green's theorem in the plane for $\oint_C [(2xy - x^2)dx + (x^2 + y^2)dy]$, where C is the boundary of the region enclosed by $y = x^2$ and $y^2 = x$.

Hint: The two parabolas intersect at (0,0) and (1,1).

$$\iint_S \left(\frac{\partial f_2}{\theta \tau} - \frac{\partial f_1}{\partial y}\right) dxdy = \int_0^1 \int_{y=x^2}^{\sqrt{x}} (2x - 2x)dxdy = 0.$$

Along the lower portion C_1, we have $x^2 = y$, so that $2xdx = dy$ and x varies from 0 to 1.

$$\int_{C_1} (f_1 dx + f_2 dy) = \int_0^1 [2x^3 - x^2 + (x^2 + x^4)2xdx$$

$$= \int_0^1 (4x^3 + 2x^5 - x^2)dx$$

$$= \left[\frac{4x^4}{5} + \frac{2x^6}{6} - \frac{x^3}{3}\right]_0^1$$

$$= 1 + \frac{1}{3} - \frac{1}{3} = 1.$$

Along the upper portion, we have $y^2 = x$ so that 2 $ydy = dx$ and y varies from 1 to 0. Thus,

$$\int_{C_2} (f_1 dx + f_2 dy)$$

$$= \int_1^0 [(2y^3 - y^4)2y + (y^4 + y^2)]dy$$

$$= \int_1^0 [4y^4 - 2y^5 + y^4 + y^2]dy$$

$$= \int_1^0 (5y^4 - 2y^5 + y^2)dy = \left[\frac{5y^5}{2} - \frac{2y^6}{6} + \frac{y^3}{3}\right]_1^0$$

$$= -1 - \frac{1}{3} + \frac{1}{3} = -1.$$

So, $\oint_C (f_1 dx + f_2 dy) = 1 - 1 = 0.$

71. Using Green's theorem in a plane, evaluate $\oint_C [(2x^2 - y^2)dx + (x^2 + y^2)]dy$, where C is the boundary in the xy-plane of the area enclosed by the x-axis and the semi-circle $x^2 + y^2 = 1$ in the upper half of the xy-plane.

Hint: $\oint_C (f_1 dx + f_2 dy) = \iint_S \left(\frac{\partial f_2}{\theta x} - \frac{\partial f_1}{\partial y}\right) dxdy = \frac{3}{2}.$

$$\int_{-1}^{1} \int_0^{\sqrt{1-x^2}} 2(x+y)dydx = \frac{4}{3}.$$

72. Verify Green's theorem in the plane for $\int_C (3x^2 - 8y^2)dx + (4y - 6xy)dy$, where C is the boundary of the region bounded by the parabolas $y = \sqrt{x}$ and $y = x^2$.

Ans. $\oint_C (f_1 dx + f_2 dy) = \iint_S \left(\frac{\partial f_2}{\theta \tau} - \frac{\partial f_1}{\partial y}\right) dxdy = \frac{3}{2}.$

Sloke's Theorem

73. Verify Stoke's Theorem for the function $\vec{f} = (x^2 + y^2)\hat{i} - 2xy\hat{j}$ $(x^2 + y^2)\hat{i} - 2xy\hat{j}$ taken around the rectangle bounded by $x = \pm$ a, $y = 0$, and $y = b$.

Hint: $curl\,\vec{f} = \begin{vmatrix} \hat{i} & \hat{j} & \hat{k} \\ \frac{\partial}{\partial x} & \frac{\partial}{\partial y} & \frac{\partial}{\partial z} \\ x^2 + y & -2xy & 0 \end{vmatrix} = -4y\hat{k}.$

For the given surface, $\hat{n} = \hat{k}$. Therefore,

$$\iint_S curl\,\vec{f}.\hat{n}dS = \int_0^b \int_{-a}^{a} -4yxdy = -4ab^2$$

It can beseen that the line integral $\oint_C \vec{f}.d\vec{r} = -4ab^2$.

74. Evaluate by Stoke's Theorem, the integral $(e^x dx + 2y dy - dz)$, where C is the curve $x^2 + y^2 = 4$ and $z = 2$.

Hint: $curl\ \vec{f} = 0$ and so, $curl\vec{f}.\hat{n} = 0$. Hence,

$$\iint_S curl\vec{f}.\hat{n}ds = 0.$$

75. Verify Stoke's Theorem for the function $\vec{f} = (2x - y)\hat{i} - yz^2\hat{j} - y^2z\hat{k}$, where S is the upper-half surface of the sphere $x^2 + y^2 + z^2 = 1$, bounded by its projection on the xy-plane.

Hint: Parametric equations of C are $x = \cos t, y = \sin t, z = 0$, and $0 \le t \le 2\pi$. Therefore,

$$\oint \vec{f}.d\vec{r} = \int (f_1 dx + f_2 dy + f_3 dz)$$

$$= \int_0^{2\pi} (-2 \sin t \cos t + \sin^2 t)dt = \pi.$$

Further, $curl\ \vec{f} = \hat{k}$. Therefore, $curl\vec{f}.\hat{n}dS =$

$\iint_R \hat{n}.\hat{k}\frac{dxdy}{|n.k|}$, where R is the projection of S on xy-plane. Then, $\iint_R \hat{n}.\hat{k}\frac{dxdy}{|n.k|} = dxdy = \iint_R dxdy =$ area of $R = \pi(1)^2 = \pi$.

76. Transform the integral $\iint_S curl\vec{f}.\hat{n}dS$ into a line integral, if S is a part the surface of the paraboloid $z = 1 - x^2 - y^2$ for which, $z \ge 0$ and $\vec{f} = y\hat{i} + z\hat{j} + x\hat{k}$.

Hint: Surface S is $x^2 + y^2 = 1$ and $z = 0$ with parametric equations $x = \cos\theta, y = \sin\theta$, $z = 0$, and $0 \le \theta < 2\pi$. Use Stoke's Theorem to transform the given integral into a line integral. The value of the line integral will come out to be $-\pi$.

14 Three-Dimensional Geometry

14.1 COORDINATE PLANES

Consider the right-handed three-dimensional coordinate system with origin O. The planes passing through the axes, taken two at a time, define the coordinate planes. Thus, the plane XOY passing through the axes OX and OY defines xy-plane, the plane YOZ passing through the axes OY and OZ defines yz-plane, and the plane ZOX passing through the axes OZ and OX defines zx-plane.

Since the axes OX, OY, and OZ are mutually perpendicular, the *coordinate planes are also mutually perpendicular*. Let $P(x,y,z)$ be any point in the space. Draw the perpendicular PM from P to the xy-plane. Then, by definition, $PM = z$. From M, draw perpendiculars MA and MB to the axes OX and OY, respectively, meeting the axes OX and OY at A and B, respectively. Then,

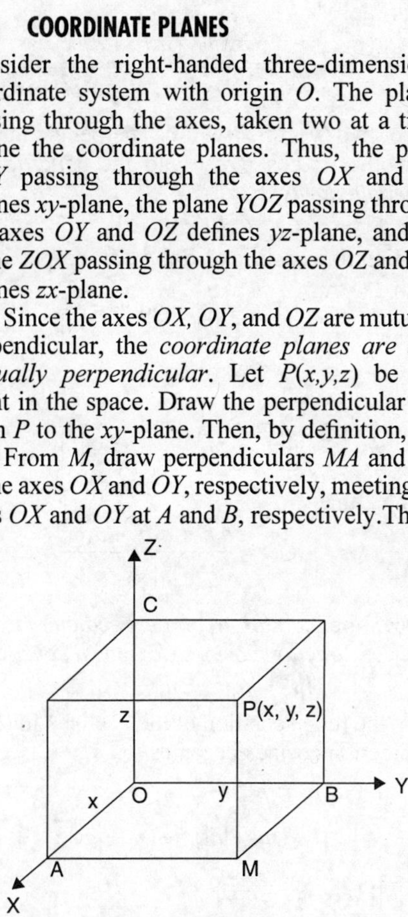

$OA = BM = x$, $OB = AM = y$, and $OC = MP = z$.

14.2 DISTANCE BETWEEN TWO POINTS

Let $P(x_1, y_1, z_1)$ be an arbitrary point in three dimensional space and let the position vector of P be

$$\vec{r} = \overrightarrow{OP} = x_1\hat{i} + y_1\hat{j} + z_1\hat{k}.$$

Then, the distance OP is defined as the magnitude of the vector $\vec{O}P$ and so,

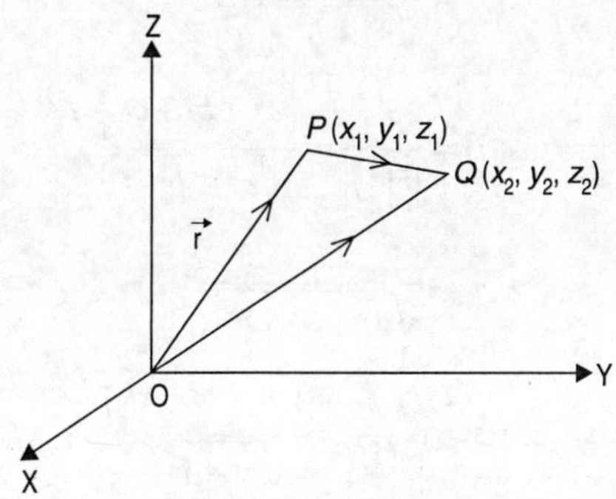

$$OP = \left|\overrightarrow{OP}\right| = |\vec{r}| = \sqrt{x^2 + y^2 + z^2}.$$

Let $Q(x_2, y_2, z_2)$ be another point in the space with position vector OQ. Then,

$$\overrightarrow{OP} + \overrightarrow{PQ} = \overrightarrow{OQ}$$

or

$$\overrightarrow{PQ} = \overrightarrow{OQ} - \overrightarrow{OP} = \left(x_2\hat{i} + y_2\hat{j} + z_2\hat{k}\right) - \left(x_1\hat{i} + y_1\hat{j} + z_1\hat{k}\right)$$

$$= (x_2 - x_1)\hat{i} + (y_2 - y_1)\hat{j} + (z_2 - z_1)\hat{k}.$$

Hence, the distance PQ between the points P and Q is given by

$$PQ = \left|\overrightarrow{PQ}\right| = \sqrt{(x_2 - x_1)^2 + (y_2 - y_1)^2 + (z_2 - z_1)^2},$$

which is known as the *distance formula*, giving the distance between two points in the space.

14.3 DIRECTION RATIOS AND DIRECTION COSINES OF A LINE

Let $\vec{r} = a\hat{i} + b\hat{j} + c\hat{k}$ be the position vector of a point $P\ (a,b,c)$ in space. Since a, b, and c determine the direction of the vector $\vec{r}$, the values *a*, *b*, and *c* are called the *direction ratios* of the vector $\vec{r}$.

If the vector $\vec{r}$ makes angles α, β, and γ with *x*–, *y*–, and *z*–axis, respectively, then α, β, and γ are called *direction angles* of the vector $\vec{r}$. Further, if $\cos\alpha = l$, $\cos\beta = m$, and $\cos\gamma = n$, then *l*, *m*, and *n*, are called the *direction cosines* of the vector $\vec{r}$. Thus, the cosines of the direction angles are called *direction cosines*.

We note that

$$l = \cos\alpha = \frac{\vec{r}.\hat{i}}{|\vec{r}|\,|\hat{i}|} = \frac{\left(a\hat{i}+b\hat{j}+c\hat{k}\right).\hat{i}}{\sqrt{a^2+b^2+c^2}}$$

$$= \frac{a}{\sqrt{a^2+b^2+c^2}},$$

$$m = \cos\beta = \frac{\vec{r}.\hat{j}}{|\vec{r}|\,|\hat{j}|} = \frac{\left(a\hat{i}+b\hat{j}+c\hat{k}\right).\hat{j}}{\sqrt{a^2+b^2+c^2}}$$

$$= \frac{b}{\sqrt{a^2+b^2+c^2}},$$

$$n = \cos\gamma = \frac{\vec{r}.\hat{k}}{|\vec{r}|\,|\hat{k}|} = \frac{\left(a\hat{i}+b\hat{j}+c\hat{k}\right).\hat{k}}{\sqrt{a^2+b^2+c^2}}$$

$$= \frac{c}{\sqrt{a^2+b^2+c^2}},$$

and

$$l^2+m^2+n^2 = \frac{a^2}{a^2+b^2+c^2} + \frac{b^2}{a^2+b^2+c^2} + \frac{c^2}{a^2+b^2+c^2} = 1.$$

It follows from the earlier derivations that the *direction ratios are constant multiples of the direction cosines*, that is, $a = \lambda l$, $b = \lambda m$, and $c = \lambda n$, where λ is a positive or negative constant. Thus,

$$a : b : c = l : m : n.$$

Let $P(x_1, y_1, z_1)$ and $Q(x_2, y_2, z_2)$ be any two points in space. Then,

$$\overrightarrow{PQ} = (x_2-x_1)\hat{i} + (y_2-y_1)\hat{j} + (z_2-z_1)\hat{k}.$$

Therefore, the direction ratios of $\overrightarrow{PQ}$ are $x_2 - x_1$, $y_2 - y_1$ and $z_2 - z_1$, whereas the *direction cosines of* $\overrightarrow{PQ}$ *are*

$$\frac{x_2-x_1}{\left|\overrightarrow{PQ}\right|}, \frac{y_2-y_1}{\left|\overrightarrow{PQ}\right|} \text{ and } \frac{z_2-z_1}{\left|\overrightarrow{PQ}\right|},$$

where

$$\left|\overrightarrow{PQ}\right| = \sqrt{(x_2-x_1)^2+(y_2-y_1)^2+(z_2-z_1)^2}.$$

If θ is the angle between two vectors $\vec{a} = a_1\hat{i}+a_2\hat{j}+a_3\hat{k}$ and $\vec{b} = b_1\hat{i}+b_2\hat{j}+b_3\hat{k}$, then

$$\cos\theta = \frac{\vec{a}.\vec{b}}{|\vec{a}||\vec{b}|} = \frac{\left(a_1\hat{i}+a_2\hat{j}+a_3\hat{k}\right).\left(b_1\hat{i}+b_2\hat{j}+b_3\hat{k}\right)}{\left|a_1\hat{i}+a_2\hat{j}+a_3\hat{k}\right|\left|b_1\hat{i}+b_2\hat{j}+b_3\hat{k}\right|}$$

$$= \frac{a_1b_1+a_2b_2+a_3b_3}{\sqrt{a_1^2+a_2^2+a_3^2}\sqrt{b_1^2+b_2^2+b_3^2}}.$$

These two vectors will be perpendicular to each other if $\cos\theta = 0$ and the *condition of perpendicularity* is

$$a_1b_1 + a_2b_2 + a_3b_3 = 0.$$

The direction cosines of $\vec{a}$ and $\vec{b}$ are respectively

$$l_1 = \frac{a_1}{|\vec{a}|}, m_1 = \frac{a_2}{|\vec{a}|}, \text{ and } n_1 = \frac{a_3}{|\vec{a}|}$$

and

$$l_2 = \frac{b_1}{|\vec{b}|}, m_1 = \frac{b_2}{|\vec{b}|}, \text{ and } n_1 = \frac{b_3}{|\vec{b}|}.$$

We note that

$$l_1l_2 + m_1m_2 + n_1n_2 = \frac{a_1b_1+a_2b_2+a_3b_3}{|\vec{a}|\,|\vec{b}|} = \cos\theta.$$

Hence, the *condition of perpendicularity, in terms of direction cosines* of $\vec{a}$ and $\vec{b}$, becomes

$$l_1l_2 + m_1m_2 + n_1n_2 = 0.$$

Also, the representation of the vector $\vec{a}$ in terms of direction cosines becomes

$$\vec{a} = a_1\hat{i} + a_2\hat{j} + a_3\hat{k}$$

$$= (l_1 + |\vec{a}|)\hat{i} + (m_1 + |\vec{a}|)\hat{j} + (n_1 + |\vec{a}|)\hat{k}$$

$$= |\vec{a}|\left(l_1\hat{i} + m_1\hat{j} + n_1\hat{k}\right).$$

14.4 SECTION FORMULAE—INIERNAL DIVISION OF A LINE BY A POINT ON THE LINE

Let *AB* be a straight line with position vectors of *A* (x_1, y_1, z_1) and *B* (x_2, y_2, z_2) as $\vec{r}_1$ and $\vec{r}_2$, respectively. Let *P* (*x*, *y*, *z*) be the point which divides internally, the straight line *AB* in the ratio *m*:*n*.

We have

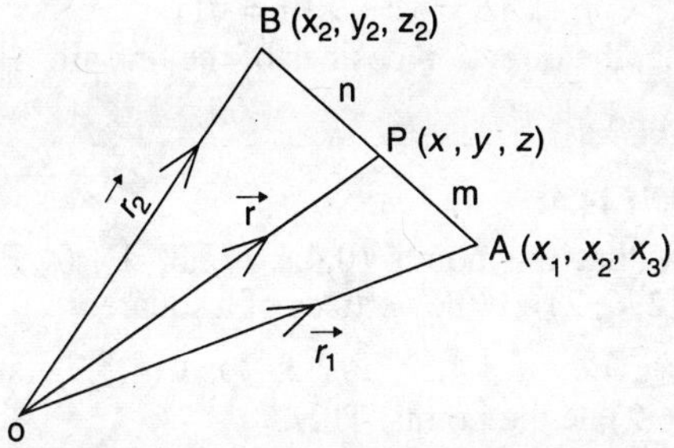

$$\frac{AP}{PB} = \frac{m}{n},$$

$\overrightarrow{AB} = \overrightarrow{OB} - \overrightarrow{OA} = \vec{r}_2 - \vec{r}_1$, and

$$\vec{AP} = \frac{m}{m+n}\vec{AB} = \frac{m}{m+n}(\vec{r}_2 - \vec{r}_1). \quad (1)$$

But,

$$\vec{AP} = \vec{OP} - \vec{OA},$$

which gives

$$\overrightarrow{OP} = \overrightarrow{AP} + \overrightarrow{OA}$$

or using (1),

$$\vec{r} = \vec{OP} = \frac{m}{m+n}(\vec{r}_2 - \vec{r}_1) + \overrightarrow{OA}$$

$$= \frac{m(\vec{r}_2 - \vec{r}_1) + (m+n)\vec{r}_1}{m+n}$$

$$= \frac{m\vec{r}_2 + n\vec{r}_1}{m+n} \text{ (Vector form).} \quad (2)$$

Since $\vec{r} = x\hat{i} + y\hat{j} + z\hat{k}, \vec{r}_1 = x_1\hat{i} + y_1\hat{j} + z_1\hat{k}$, and $\vec{r}_2 = x_2\hat{i} + y_2\hat{j} + z_2\hat{k}$, the vector formula (2) reduces to

$$x\hat{i} + y\hat{j} + z\hat{k} = \frac{1}{m+n}\Big[m\left(x_2\hat{i} + y_2\hat{j} + z_2\hat{k}\right) + n\left(x_1\hat{i} + y_1\hat{j} + z_1\hat{k}\right)\Big]$$

$$= \frac{1}{m+n}\Big[(mx_2 + nx_1)\hat{i} + (my_2 + ny_1)\hat{j} + (mz_2 + nz_1)\hat{k}\Big].$$

Hence,

$$x = \frac{mx_2 + nx_1}{m+n}, y = \frac{my_2 + ny_1}{m+n} \text{ and } z = \frac{mz_2 + nz_1}{m-n}, \quad (3)$$

which is called the *Cartesian form of the section formula*.

If $m = n = 1$, then the midpoint of AB is given by

$$x = \frac{1}{2}(x_1 + x_2), y = \frac{1}{2}(y_1 + y_2), \text{ and}$$

$$z = \frac{1}{2}(z_1 + z_2).$$

Thus, the midpoint is

$$\left(\frac{x_1 + x_2}{2}, \frac{y_2 + y_2}{2}, \frac{z_1 + z_2}{2}\right).$$

Dividing the numerator and denominator in (3) by n and setting $\frac{m}{n} = \lambda$, we get

$$x = \frac{\lambda x_2 + x_1}{\lambda + 1}, y = \frac{\lambda y_2 + y_1}{\lambda + 1} \text{ and } z = \frac{\lambda z_2 + z_1}{\lambda + 1},$$

where $\lambda \neq -1$. This result is known as the *parametric representation of any point P on AB*.

Remark 14.1. The section formula (2) yields

$$m\vec{r}_2 + n\vec{r}_2 - (m+n)\vec{r} = 0.$$

The sum of the coefficients in the above expression is zero. It follows from the expression that three points with position vectors $\vec{r}_1, \vec{r}_2$, and $\vec{r}$ are *collinear* if $\alpha\vec{r}_1 + \beta\vec{r}_2 + \gamma\vec{r} = 0$, where $\alpha + \beta + \gamma = 0$.

14.4.1 External Division of a Line by a Point on the Extended Line

If the point P divides AB externally in the ratio $m:n$, then

$$\frac{AP}{BP} = \frac{m}{n} \text{ or } m(BP) = n(AP).$$

Therefore,

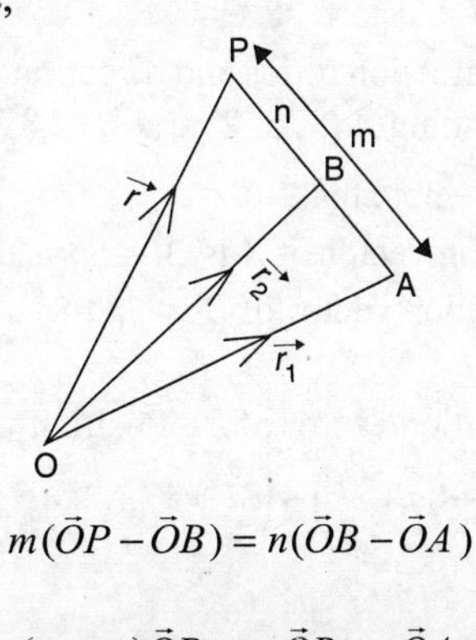

$$m(\vec{OP} - \vec{OB}) = n(\vec{OB} - \vec{OA})$$

or

$$(m-n)\vec{OP} = m\vec{OB} - n\vec{OA}$$

or

$$\vec{r}=\vec{O}P=\frac{m\vec{r}_2-n\vec{r}_1}{m-n} \text{ (Vector formula).}$$

In *Cartesian coordinates*, we get

$$x=\frac{mx_2-nx_1}{m-n}, y=\frac{my_2-ny_1}{m-n}, \text{and}$$

$$z=\frac{mz_2-nz_1}{m-n}$$

as the coordinates of the dividing point $P(x,y,z)$.

EXAMPLE 14. 1

If a line makes angles α, β, and γ with the axes, show that $\sin^2\alpha+\sin^2\beta+\sin^2\gamma=2$.

Solution. The direction angles are α , β, and γ, and so direction cosines of the line are $\cos\alpha$, $\cos\beta$, and $\cos\gamma$. Since the sum of the squares of direction cosines is equal to 1, we have

$$\cos^2\alpha+\cos^2\beta+\cos^2\gamma=1$$

or

$$(1-\sin^2\alpha)+(1-\sin^2\beta)+(1-\sin^2\gamma=2)=1$$

or

$$\sin^2\alpha+\sin^2\beta+\sin^2\gamma=2.$$

EXAMPLE 14.2

Find the direction cosines of axes.

Solution. The x-axis makes angles 0°, 90°, and 90° with the x-, y-, and z- axis, respectively. Therefore, the direction cosines of x-axis are

cos 0, cos 90, cos 90 or 1,0,0.

Similarly, direction cosines of y- and z-axis are

0,1,0 and 0,0,1, respectively.

EXAMPLE 14.3

Find the direction ratios and direction cosines of the line joining A (3, 5, 2) and B (2,8,3).

Solution. We note that

Position vector of A is $3\hat{i}+5\hat{j}+2\hat{k}$ and

Position vector of B is $2\hat{i}+8\hat{j}+3\hat{k}$.

Hence,

$$\vec{AB}=\text{Position vector of } B-\text{Position vector of } A$$

$$=\left(2\hat{i}+8\hat{j}+3\hat{k}\right)-\left(3\hat{i}+5\hat{j}+2\hat{k}\right)$$

$$=-\hat{i}+3\hat{j}+\hat{k}.$$

Thus, the direction ratios of the line are–1, 3, and 1. Further,

$$|\vec{AB}|=\sqrt{1+9+1}=\sqrt{11}.$$

Hence, the direction cosines of the line are $\frac{-1}{\sqrt{11}}$, $\frac{3}{\sqrt{11}}$, and $\frac{1}{\sqrt{11}}$.

EXAMPLE 14.4

Show that the points (0,4,1), (2,3,–1), (4,5,0) and (2, 6, 2) are the vertices of a square.

Solution. Let $A(0,4,1)$, $B(2,3,-1)$, $C(4,5,0)$, and $D(2,6,2)$ be the points. Then,

$$\text{distance } AB=\sqrt{(-2)^2+(4-3)^2+(1+1)^2}$$

$$=\sqrt{4+1+4}=3,$$

$$\text{distance } BC=\sqrt{(4-2)^2+(5-3)^2+(0+1)^2}$$

$$=\sqrt{4+4+1}=3,$$

$$\text{distance } CD=\sqrt{(4-2)^2+(5-6)^2+(0-2)^2}$$

$$=\sqrt{4+1+4}=3, \text{ and}$$

$$\text{distance } DA=\sqrt{(2-0)^2+(6-4)^2+(2-1)^2}$$

$$=\sqrt{4+4+1}=3.$$

Thus, $AB = BC = CD = DA = 3$. Further, the direction ratios of AB, BC, CD, and DA are respectively

AB : 2 – 0, 3 – 4, –1 –1 or 2, –1, –2,
BC : 4 – 2, 5 – 3, 0 + 1 or 2, 2, 1
CD: 2 – 4, 6 – 5, 2 – 0 or –2, 1, 2, and
DA : 0 – 2, 4 – 6, 1 – 2 or –2, –2, –1.

Since $2\times2+2\times-1+1\times-2=0$, it follows that $BC\perp AB$.

Since $2\times-2+2\times1+1\times2=0$, it follows that $CD\perp BC$.

Since $-2\times-2+1\times-2+2\times-1=0$, it follows that $AD\perp CD$.

Since $2\times-2+(-1)\times-2+(-2)\times-1=0$, it follows that $AB\perp AD$.

Hence, ABCD is a square with side equal to 3.

EXAMPLE 14.5

Show that the points $A(3,2,-4)$, $B(5,4,-6)$, and C (9,8,–10) are collinear and find the ratio in which B divides AC.

Solution. The position vector of A, B, and C are

$\vec{O}A=3i+2\hat{j}-4\hat{k}, \vec{O}B=5\hat{i}+4\hat{j}-6\hat{k}$, and

$\vec{O}C=9\hat{i}+8\hat{j}-10\hat{k}$.

We note that

$$\vec{OC} + 2\vec{OA} - 3\vec{OB} = 0,$$

which is of the form $\alpha r_1 + \beta r_2 + \gamma \vec{r} = 0$ with $\alpha = 2, \beta = 1$ and $\gamma = -3$, so that $\alpha + \beta + \gamma = 0$. Hence, A, B, and C are collinear.

Now, let B divide AC in the ratio k:1. Then, $5 = \frac{9k+3}{k+1}$, which yields $k = \frac{1}{2}$. Therefore, the ratio is $\frac{1}{2}:1$ or 1:2.

Second Method: We have

$$\vec{AB} = (5-3)\hat{i} + (4-2)\hat{j} + (-6+4)\hat{k}$$
$$= 2\hat{i} + 2\hat{j} - 2\hat{k} \text{ and}$$
$$\vec{BC} = (9-5)\hat{i} + (8-4)\hat{j} + (-10+6)\hat{k}$$
$$= 4\hat{i} + 4\hat{j} - 4\hat{k}.$$

We note that

$$\vec{AB} = 2\vec{BC} \qquad (1)$$

Therefore, $\vec{AB}$ and $\vec{BC}$ are parallel. Since B is common, A, B, and C are collinear. Also (1) shows that B divides AC in the ratio 1: 2.

EXAMPLE 14.6

Using section formula, show that the three points (–2,3,5), (1,2,3), and (7,0,– 1) are collinear.

Solution. Let $A(-2,3,5)$, $B(1,2,3)$, and $C(7,0,-1)$ be the given points. Let C divide AB in the ratio k:1.

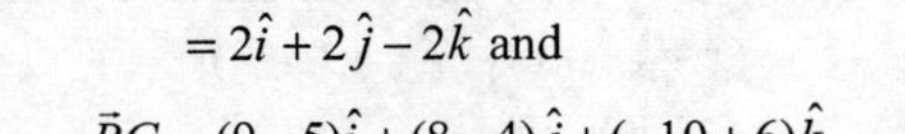

Then, using the section formula, we have

$$\frac{k-2}{k+1} = 7, \text{ which gives } k = -\tfrac{3}{2},$$

$$\frac{2k+3}{k+1} = 0, \text{ which gives } k = -\tfrac{3}{2},$$

$$\frac{3k+5}{k+1} = -1, \text{ which gives } k = -\tfrac{3}{2}.$$

Thus, we get the same value of k in each case. Hence, C divides AB externally in the ratio 3 : 2. Hence, A, B, and C are collinear.

EXAMPLE 14.7

Three vertices of a parallelogram ABCD are $A(3,-4,7)$, $B(5,3,-2)$, and $C(1,2,-3)$. Find the fourth vertex which is opposite to B.

Solution. The parallelogram ABCD is shown in the following figure:

Since the diagonals of a parallelogram bisect each other, the point O is the midpoint of AC. Therefore, the coordinates of O are $\left(\frac{3+1}{2}, \frac{-4+2}{2}, \frac{7-3}{2}\right)$ or (2, –1, 2). The point O divides the diagonal BD in the ratio 1:1. Therefore,

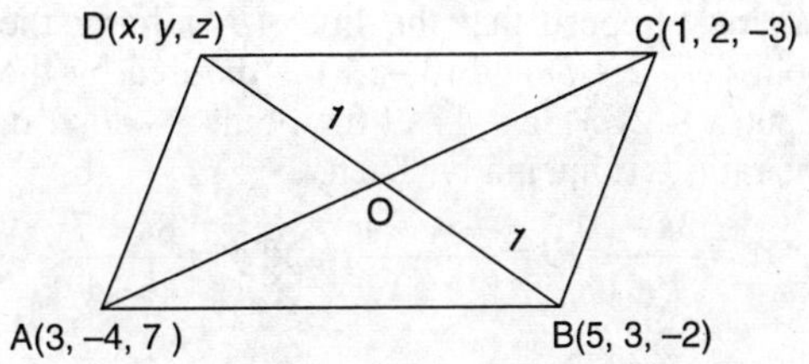

$$2 = \frac{x+5}{2}, -1 = \frac{y+3}{2}, \text{ and } 2 = \frac{z-2}{2},$$

which yields $x = 1, y = -5$, and $z = 6$. Hence, the coordinates of D are (1,–5,6).

EXAMPLE 14.8

Let $A(3,2,0)$, $B(5,3,2)$, and $C(-9,6,-3)$ be three points forming a triangle. If AD, the bisector of the angle BAC, meets BC in D, find the coordinates of D.

Solution. The line AD is the internal bisector of the angle BAC. By geometry,

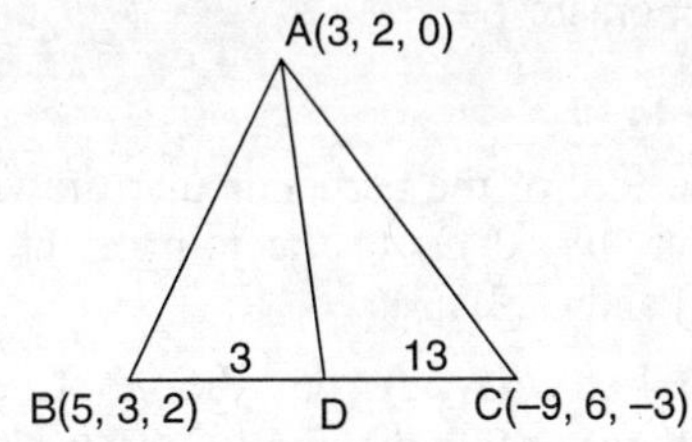

$$\frac{BD}{DC} = \frac{AB}{AC} = \frac{\sqrt{(5-3)^2 + (3-2)^2 + (2-0)^2}}{\sqrt{(3+9)^2 + (2-6)^2 + (3)^2}}$$

$$= \frac{\sqrt{9}}{\sqrt{169}} = \frac{3}{13}.$$

Thus, D divides BC in the ratio 3:13 internally. Hence, the coordinates of D are

$$\left(\frac{3(-9)+13(5)}{3+13}, \frac{3(6)+13(3)}{3+13}, \frac{3(-3)+13(2)}{3+13}\right)$$

$$\text{or} \left(\frac{19}{8}, \frac{57}{16}, \frac{17}{16}\right).$$

EXAMPLE 14.9

Find the ratio in which the plane $x-2y+3z=17$ divides the line joining the points (–2,4,7) and (3,–5,8). Also obtain the point of intersection.

Solution. Suppose that the line AB, joining the points $A(-2,4,7)$ and (3,–5,8) is divided by the plane $x - 2y + 3z = 17$ at the point $P(\alpha,\beta,\gamma)$ in the ratio k:1, intemally. Then,

$$\alpha = \frac{3k-2}{k+1}, \beta = \frac{-5k+4}{k+1}, \text{ and } \gamma = \frac{8k+7}{k+1}.$$

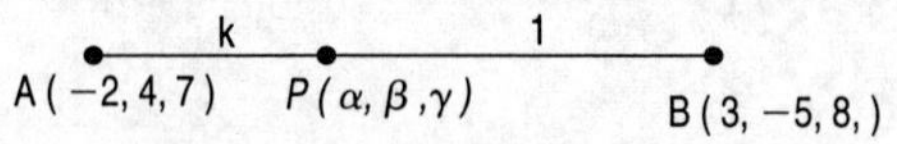

Since $P(\alpha,\beta,\gamma)$ lies on the plane $x-2y+3z=17$, we have $\frac{3k-2}{k+1} - 2\left(\frac{-5k+4}{k+1}\right) + 3\left(\frac{8k+7}{k+1}\right) = 17$ or $20\,k = 6$, which yields $k = \frac{3}{10}$. Therefore, the required ratio is 3:10, internally. Further, putting the above value of k in the expressions for α, β, and γ, we get

$$\alpha = \frac{3k-2}{k+1} = -\frac{11}{13}, \beta = \frac{-5k+4}{k+1} = \frac{25}{13}, \text{ and}$$

$$\gamma = \frac{8k+7}{k+1} = \frac{94}{13}.$$

Hence, the coordinates of the point of intersection are $\left(-\frac{11}{13}, \frac{15}{13}, \frac{94}{13}\right)$.

EXAMPLE 14.10

Find the foot of the perpendicular drawn from the point $A(1, 0, 3)$ to the join of the points $B(4,7,1)$ and C (3,5,3).

Solution. Let $P(x,y,z)$ be the foot of the perpendicular AP to the line BC. Then,

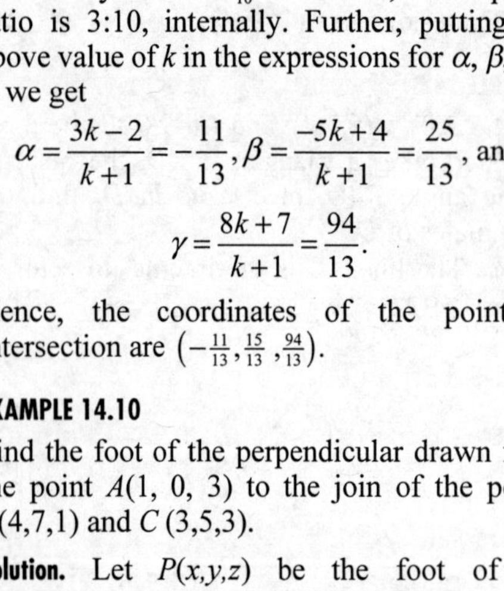

$$\vec{AP} = (x-1)\hat{i} + (y-0)\hat{j} + (z-3)\hat{k}$$

$$= (x-1)\hat{i} + y\hat{j} + (z-3)\hat{k} \text{ and}$$

$$\vec{BC} = (3-4)\hat{i} + (5-7)\hat{j} + (3-1)\hat{k}$$

$$= -\hat{i} + 2\hat{j} + 2\hat{k}.$$

Then, the direction ratios of AP and BC are, respectively, x–1, y, and z–3 and –1 –2, and 2. Since $AP \perp BC$, the condition of perpendicularity implies

$$(x-1)(-1) + y(-2) + (z-3)(2) = 0 \quad (1)$$

$$\text{or} \quad -x-2y+2z = -5.$$

Since P lies on BC, BP is parallel to BC. Therefore, $\vec{BP} = \lambda\vec{BC}$ for some scalar λ. Thus,

$$(x-4)\hat{i} + (y-7)\hat{j} + (z-1)\hat{k}$$

$$= \lambda\left(-\hat{i} - 2\hat{j} + 2\hat{k}\right),$$

which yields

$$x-4 = -\lambda, y-7 = -2\lambda, \text{ and } z-1 = 2\lambda \quad (2)$$

or

$$x-4 = -\lambda,\ y = -2\lambda + 7, \text{ and } z = 2\lambda + 1.$$

Putting these values of x, y, and z in (1), we get $\lambda = \frac{7}{3}$. Then (2) implies $x = \frac{5}{3}, y = \frac{7}{3}$, and $z = \frac{17}{3}$. Hence, the foot of the perpendicular is $P\left(\frac{5}{3}, \frac{7}{3}, \frac{17}{3}\right)$.

14.5 STRAIGHT LINE IN THREE DIMENSIONS

(A) Equation of Straight Line Through a Given Point and Parallel to a Given Vector

Let O be the origin of a rectangular coordinate system in the space and let $\vec{a}$ be the position vector of a fixed point $A(x_1, y_1, z_1)$. Let L be the straight line passing through A and parallel to a given vector $\vec{b}$. Let $\vec{r} = \vec{OP}$ be the position vector of an arbitrary point $P(x,y,z)$ on the straight line L.

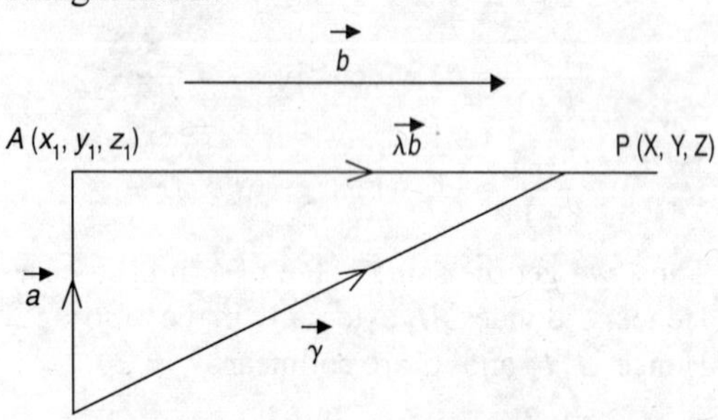

Since the vector $\vec{AP}$ is parallel to the given vector $\vec{b}$, we have $\vec{AP} = \lambda\vec{b}$, where λ is a scalar, which is positive for points on one side of A and

negative for points on the other side of A. Then,
$\vec{r} = \vec{OP} = \vec{OA} + \vec{AP} = \vec{a} + \lambda\vec{b}$ (Vector form). (1)

Thus, position of any point on the line L can be obtained by assigning positive or negative value to λ. Hence, equation (1) represents the equation of a straight line through a point A with a position vector $\vec{a}$ and parallel to vector $\vec{b}$. Equation (1) is also called *parametric representation of a straight line* in a vector form.

To obtain the *Cartesian form*, let the vector $\vec{b}$ be $\vec{b} = b_1\hat{i} + b_2\hat{j} + b_3\hat{k}$. Then $\vec{r} = \vec{OP} = x\hat{i} + y\hat{j} + z\hat{k}$ and $\vec{a} = \vec{OA} = x_1\hat{i} + y_1\hat{j} + z_1\hat{k}$.

Therefore, (1) reduces to

$$\begin{aligned} x\hat{i} + y\hat{j} + z\hat{k} &= x_1\hat{i} + y_1\hat{j} + z_1\hat{k} \\ &\quad + \lambda\left(b_1\hat{i} + b_2\hat{j} + b_3\hat{k}\right) \\ &= (x_1 + \lambda b_1)\hat{i} + (y_1 + \lambda b_2)\hat{j} \\ &\quad + (z_1 + \lambda b_3)\hat{k}. \end{aligned}$$

Comparing the coefficients of $\hat{i}, \hat{j}$, and $\hat{k}$ on both the sides, we get

$$x = x_1 + \lambda b_1, y = y_1 + \lambda b_2, \text{ and } z = z_1 + \lambda b_3.$$

Hence,

$$\frac{x - x_1}{b_1} = \frac{y - y_1}{b_2} = \frac{z - z_1}{b_3} \text{ (Cartesian form).} \quad (2)$$

The equation (2) is also called the *symmetrical form* of a straight line. The quantities b_1, b_2, and b_3 in the denominators of (2) are the *direction ratios* of the straight line.

If $\vec{a} = \vec{0}$, then the parametric form reduces to $\vec{r} = \lambda\vec{b}$ and the Cartesian form reduces to

$$\frac{x}{b_1} = \frac{y}{b_2} = \frac{z}{b_3},$$

which represents the equation of a straight line passing through the origin and parallel to a given line $\vec{b}$.

(B) Equation of a Straight line Passing Through Two Points

Let $A(x_1, y_1, z_1)$ and $B(x_2, y_2, z_2)$ be two given points with position vectors $\vec{a}$ and $\vec{b}$. Thus,
$\vec{a} = x_1\hat{i} + y_1\hat{j} + z_1\hat{k}$ and $\vec{b} = x_2\hat{i} + y_2\hat{j} + z_2\hat{k}$.

Let $P(x,y,z)$ an arbitrary point with a position vector $\vec{r} = x\hat{i} + y\hat{j} + z\hat{k}$ on the line L passing through A and B.

Then

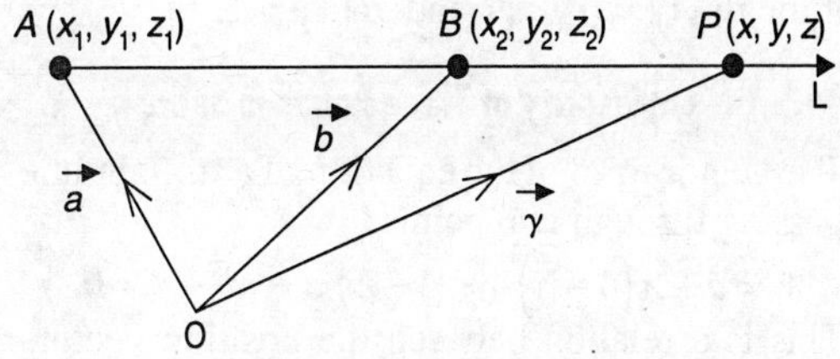

$$\vec{AB} = \vec{OB} - \vec{OA} = \vec{b} - \vec{a}.$$

Hence, the straight line passes through $A(x_1, y_1, z_1)$ and is parallel to vector $\vec{b} - \vec{a}$. Therefore, the *parametric form of the equation* is

$$\vec{r} = \vec{a} + \lambda\left(\vec{b} - \vec{a}\right) = (1 - \lambda)\vec{a} + \lambda\vec{b} \quad (1)$$

or

$$\vec{r} - \vec{a} = \lambda\left(\vec{b} - \vec{a}\right). \quad (2)$$

Similarly, we may also consider the line L passing through the point $B(x_2, y_2, z_2)$ and parallel to the vector $\vec{b} - \vec{a}$. In the above case, the required equation of the straight line becomes

$$\vec{r} = \vec{b} + \lambda\left(\vec{b} - \vec{a}\right).$$

To obtain the *Cartesian form* of (2), we have

$$\vec{r} - \vec{a} = \lambda(\vec{b} - \vec{a})$$

or

$$\left(x\hat{i} + y\hat{j} + z\hat{k}\right) - \left(x_1\hat{i} + y_1\hat{j} + z_1\hat{k}\right)$$
$$= \lambda\left[\left(x_2\hat{i} + y_2\hat{j} + z_2\hat{k}\right) - \left(x_1\hat{i} + y_1\hat{j} + z_1\hat{k}\right)\right]$$

or

$$\left(x - x_1\right)\hat{i} + \left(y - y_1\right)\hat{j} + \left(z - z_1\right)\hat{k}$$
$$= \lambda\left[\left(\pm\pm\pm\right)\hat{} + \left(_2 - _1\right)\hat{} + \left(_2 - _1\right)\hat{}\right].$$

Comparing the coefficients of $\hat{i}, \hat{j}$, and $\hat{k}$ on both sides, we get $x - x_1 = \lambda(x_2 - x_1)$, $y - y_1 = \lambda(y_2 - y_1)$, and $z - z_1 = \lambda(z_2 - z_1)$.

Hence,

$$\frac{x - x_1}{x_2 - x_1} = \frac{y - y_1}{y_2 - y_1} = \frac{z - z_1}{z_2 - z_1} = \lambda.$$

Since λ is a parameter, we have

$$\frac{x - x_1}{x_2 - x_1} = \frac{y - y_1}{y_2 - y_1} = \frac{z - z_1}{z_2 - z_1},$$

which is the required equation of the line passing through $A(x_1,y_1,z_1)$ and $B(x_2,y_2,z_2)$.

14.5.1 Collinearity of Three Points in Space

The vector form of the equation of a straight line passing through two points is

$$\vec{r}=\vec{a}+\lambda(\vec{b}-\vec{a}) \text{ or } (1-\lambda)\,\vec{a}+\lambda\vec{b}-\vec{r}=0.$$

This is a relation between the position vectors of three points $A(x_1,y_1,z_1)$, $B(x_2,y_2,z_2)$, and $P(x, y, z)$, which are collinear. We observe that the algebraic sum of the coefficients of $\vec{a},\vec{b}$ and $\vec{r}$ in the earlier relation is zero.

Conversely, suppose that the three vectors $\vec{a},\vec{b}$ and $\vec{r}$ satisfy the relation $c_1\vec{a}+c_2\vec{b}+c_3\vec{r}=0$ such that $c_1+c_2+c_3=0$. Dividing throughout by c_3, we get

$$\frac{c_1}{c_3}\vec{a}+\frac{c_2}{c_3}\vec{b}+\vec{r}=0 \text{ with } \frac{c_1}{c_3}+\frac{c_2}{c_3}+1=0.$$

Putting $\frac{c_2}{c_3}=-\lambda$, we get $\frac{c_1}{c_3}=-(1-\lambda)$.

Therefore,

$$-(1-\lambda)\vec{a}-\lambda\vec{b}+\vec{r}=0$$

or

$$\vec{r}=(1-\lambda)\vec{a}+\lambda\vec{b}$$
$$=\vec{a}+\lambda(\vec{b}-\vec{a}).$$

Thus, $P(x, y, z)$ is a point on the line joining $A(x_1,y_1,z_1)$ and $B(x_2,y_2,z_2)$ and so, A, B and P are collinear. Hence, the *three points A, B, and C are collinear if and only if there exists a linear relation between their position vectors such that the sum of the coefficients in the relation is zero.*

EXAMPLE 14.11

A straight line passes through the point with a position vector $2\hat{i}-\hat{j}+4\hat{k}$ and is in the direction of $\hat{i}+\hat{j}-2\hat{k}$. Find the equation of the line in both vector and Cartesian forms.

Solution. We have

$$\vec{a}=2\hat{i}-\hat{j}+4\hat{k} \text{ and } \vec{b}=\hat{i}+\hat{j}-2\hat{k}.$$

Therefore, the vector form of the required equation is

$$\vec{r}=\vec{a}+\lambda\vec{b}=(2\hat{i}-\hat{j}+4\hat{k})+\lambda(\hat{i}+\hat{j}-2\hat{k}).$$

Thus,

$$x\hat{i}+y\hat{j}+z\hat{k}=(2+\lambda)\hat{i}+(-1+\lambda)\hat{j}+(4-2\lambda)\hat{k}.$$

Comparing the coefficients of $\hat{i},\hat{j}$, and $\hat{k}$, we get

$$x=2+\lambda,\ y=-1+\lambda, \text{ and } z=4-2\lambda.$$

Hence, the equation in Cartesian form is

$$\frac{x-2}{1}=\frac{y+1}{1}=\frac{z-4}{-2}=\lambda$$

or

$$\frac{x-2}{1}=\frac{y+1}{1}=\frac{z-4}{-2}.$$

EXAMPLE 14.12

Determine the vector form of the following line:

$$\frac{x-5}{3}=\frac{y+4}{7}=\frac{z-6}{2}.$$

Solution. The equation of the given line in Cartesian form is

$$\frac{x-5}{3}=\frac{y+4}{7}=\frac{z-6}{2}.$$

The straight line passes through the point (5, –4, 6) and its direction ratios are 3, 7, and 2. Thus, the line is parallel to the vector $\vec{b}=3\hat{i}+7\hat{j}+2\hat{k}$. The position vector of the point (5, –4, 6) is $\vec{a}=5\hat{i}-4\hat{j}+6\hat{k}$. Therefore, if $\vec{r}$ is the position vector of the point (x, y, z) on the line, then the vector form of the equation of the line is

$$\vec{r}=\vec{a}+\lambda\vec{b}=5\hat{i}-4\hat{j}+6\hat{k}+\lambda(3\hat{i}+7\hat{j}+2\hat{k})$$
$$=(5+3\lambda)\hat{i}+(-4+7\lambda)\hat{j}+(6+2\lambda)\hat{k}.$$

EXAMPLE 14.13

Find the equation of the straight line passing through the points $A(-2, -1, 3)$ and $B(3, 2, -1)$. Find also the points on this line which are at a distance of $10\sqrt{2}$ from A.

Solution. The equation of the line passing through the given points is

$$\frac{x+2}{3+2}=\frac{y+1}{2+1}=\frac{z-3}{-4}=\lambda \text{ (parameter)}.$$

Any point on this line is

$$P(x,y,z)=(-2+5\lambda,-1+3\lambda,3-4\lambda).$$

Therefore, the distance AP is

$$(5\lambda-2+2)^2+(3\lambda-1+1)^2+(3-4\lambda-3)^2$$
$$=10\sqrt{2}$$

or

$$\sqrt{25\lambda^2+9\lambda^2+16\lambda^2}=10\sqrt{2}$$

or

$$\pm 5\sqrt{2}\lambda=10\sqrt{2} \text{ or } \lambda=\pm 2.$$

Hence, the required points are $(x, y, z) = (8, 5, -5)$ and $(-12, -7, 11)$.

14.6 ANGLE BETWEEN TWO LINES

Let θ be the angle between two given lines $\vec{r}=\vec{a}+\lambda\vec{b}$ and $\vec{r}=\vec{c}+\mu\vec{d}$. The given lines are respectively parallel to the vector $\vec{b}$ and $\vec{d}$. Therefore, the angle between the given lines is the angle between $\vec{b}$ and $\vec{d}$. Hence,

$$\cos\theta=\frac{\vec{b}.\vec{d}}{|\vec{b}||\vec{d}|}. \qquad (1)$$

If

$$\vec{r}=x\hat{i}+y\hat{j}+z\hat{k},\quad \vec{a}=x_1\hat{i}+y_1\hat{j}+z_1\hat{k},$$

$$\vec{b}=b_1\hat{i}+b_2\hat{j}+b_3\hat{k},$$

$$\vec{c}=x_2\hat{i}+y_2\hat{j}+z_2\hat{k}, \text{ and } \quad \vec{d}=d_1\hat{i}+d_2\hat{j}+d_3\hat{k},$$

then the Cartesian form of the given lines are

$$\frac{x-x_1}{b_1}=\frac{y-y_1}{b_2}=\frac{z-z_1}{b_3}$$

and

$$\frac{x-x_2}{d_1}=\frac{y-y_2}{d_2}=\frac{z-z_2}{d_3}.$$

Using (1), the angle between these lines is given by

$$\cos\theta=\frac{(b_1\hat{i}+b_2\hat{j}+b_3\hat{k}).(d_1\hat{i}+d_2\hat{j}+d_3\hat{k})}{\sqrt{b_1^2+b_2^2+b_3^2}\sqrt{d_1^2+d_2^2+d_3^2}}$$

$$=\frac{b_1d_1+b_2d_2+b_3d_3}{\sqrt{b_1^2+b_2^2+b_3^2}\sqrt{d_1^2+d_2^2+d_3^2}}. \qquad (2)$$

The direction cosines of $\vec{b}$ are

$$l_1=\frac{b_1}{|\vec{b}|},\ m_1=\frac{b_2}{|\vec{b}|},\ n_1=\frac{b_3}{|\vec{b}|},$$

whereas the direction cosines of $\vec{d}$ are

$$l_1=\frac{d_1}{|\vec{d}|},\ m_2=\frac{d_2}{|\vec{d}|},\ n_2=\frac{d_3}{|\vec{d}|}.$$

Therefore,

$$l_1l_2+m_1m_2+n_1n_2$$

$$=\frac{b_1d_1+b_2d_2+b_3d_3}{\sqrt{b_1^2+b_2^2+b_3^2}\sqrt{d_1^2+d_2^2+d_3^2}}. \qquad (3)$$

From (2) and (3), it follows that

$$\cos\theta=l_1l_2+m_1m_2+n_1n_2. \qquad (4)$$

Deductions:

(i) Two lines $\vec{r}=\vec{a}+\lambda\vec{b}$ and $\vec{r}=\vec{c}+\mu\vec{d}$ are *orthogonal* (perpendicular) if angle θ between them is 90°. Thus, the lines are orthogonal if

$$\frac{\vec{b}.\vec{d}}{|\vec{b}||\vec{d}|}=0,$$

that is, if $\vec{b}.\vec{d}=0$ or if $b_1d_1+b_2d_2+b_3d_3=0$ or, by (4), if $l_1l_2+m_1m_2+n_1n_2=0$.

(ii) Two lines $\vec{r}=\vec{a}+\lambda\vec{b}$ and $\vec{r}=\vec{c}+\mu\vec{d}$ are *parallel* if the angle between then is 0°. Thus, the lines are parallel if $\vec{b}=\lambda\vec{d}=0$ or if $b_1\hat{i}+b_2\hat{j}+b_3\hat{k}=\lambda(d_1\hat{i}+d_2\hat{j}+d_3\hat{k})$.

Comparing the coefficients of $\vec{i},\vec{j}$, and $\vec{k}$ on both sides, we get

$$b_1=\lambda d_1,\ b_2=\lambda d_2, \text{ and } b_3=\lambda d_3,$$

or

$$\frac{b_1}{d_1}=\frac{b_2}{d_2}=\frac{b_3}{d_3}=\lambda \text{ or } \frac{b_1}{d_1}=\frac{b_2}{d_2}=\frac{b_3}{d_3} \qquad (5)$$

or, in terms of direction cosines,

$$\frac{l_1}{l_2}=\frac{m_1}{m_2}=\frac{n_1}{n_2}. \qquad (6)$$

EXAMPLE 14.14

Find the coordinates of the foot of the perpendicular drawn from the point $A(1, 2, 1)$ to the line joining $B(1, 4, 6)$ and $C(5, 4, 4)$.

Solution. The equation of the straight line joining the points $B(1, 4, 6)$ and $C(5, 4, 4)$ is

$$\frac{x-1}{5-1}=\frac{y-4}{4-4}=\frac{z-6}{4-6}=\lambda$$

or

$$\frac{x-1}{4}=\frac{y-4}{0}=\frac{z-6}{-2}=\lambda.$$

Any point P on this line is $(4\lambda+1, 4, 6-2\lambda)$. Let this point be the foot of the perpendicular from A to BC. Direction ratios of the line AP are

$4\lambda + 1 - 1, 4 - 2$, and $6 - 2\lambda - 1$ or 4λ, 2, and $5 - 2\lambda$. Direction ratios of the line BC are 4, 0, and -2. Now, AP is perpendicular to BC if

$$4(4\lambda) + 0(2) - 2(5 - 2\lambda) = 0 \text{ or}$$

$$20\lambda - 10 = 0, \text{ which yields } \lambda = \frac{1}{2}.$$

Hence, the coordinates of the foot of the perpendicular are $(4\lambda + 1, 4, 6 - 2\lambda) = (3, 4, 5)$.

EXAMPLE 14.15

Find the angle between the lines

$$\vec{r} = 3\hat{i} + 2\hat{j} - 4\hat{k} + \lambda\left(\hat{i} + 2\hat{j} + 2\hat{k}\right) \text{ and}$$

$$\hat{r} = 5\hat{j} - 2\hat{k} + \mu(3\hat{i} + 2\hat{j} + 6\hat{k}).$$

Solution. The first line is parallel to the vector $\vec{b} = \hat{i} + 2\hat{j} + 2\hat{k}$ and the second line is parallel to the vector $\vec{d} = 3\hat{i} + 2\hat{j} + 6\hat{k}$. Therefore, the angle θ between these two lines is given by

$$\cos\theta = \frac{\vec{b}.\vec{d}}{|\vec{b}||\vec{d}|} = \frac{1(3) + 2(2) + 2(6)}{\sqrt{1^2 + 2^2 + 2^2}\sqrt{3^2 + 2^2 + 6^2}} = \frac{19}{21}.$$

Hence, $\theta = \cos^{-1}\left(\frac{19}{21}\right)$.

EXAMPLE 14.16

Find the angle between the lines

$$\frac{x+4}{3} = \frac{y-1}{5} = \frac{z+3}{4} \text{ and } \frac{x+1}{1} = \frac{y-4}{1} = \frac{z-5}{2}.$$

Solution. The first line is parallel to the vector $\vec{b} = 3\hat{i} + 5\hat{j} + 4\hat{k}$, whereas the second line is parallel to the vector $\vec{d} = \hat{i} + \hat{j} + 2\hat{k}$.

Hence, the angle between the two lines is given by

$$\cos\theta = \frac{3(1) + 5(1) + 4(2)}{\sqrt{3^2 + 5^2 + 4^2}\sqrt{1^2 + 1^2 + 2^2}} = \frac{16}{5\sqrt{2}\left(\sqrt{6}\right)}$$

$$= \frac{8\sqrt{3}}{15}$$

or

$$\theta = \cos^{-1}\left(\frac{8\sqrt{3}}{15}\right).$$

EXAMPLE 14.17

Find the perpendicular distance of the point (2,3,4) from the line

$$\frac{4-x}{2} = \frac{y}{6} = \frac{1-z}{3}.$$

Solution. Any point on the given line is $P(4 - 2\lambda, 6\lambda, 1 - 3\lambda)$. Let this point be the foot of the perpendicular from the point A (2, 3, 4). Then the direction ratios of the line AP are

$$4 - 2\lambda - 2,\ 6\lambda - 3, \text{ and } 1 - 3\lambda - 4$$
$$\text{or } 2 - 2\lambda,\ 6\lambda - 3, \text{ and } -3\lambda - 3.$$

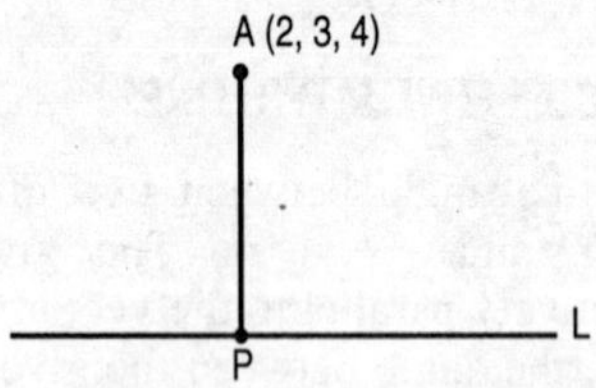

The direction ratios of the given line are 2, 6, and 3. Now, AP will be perpendicular to the given line if $(2 - 2\lambda)(2) + (6\lambda - 3)(6) + (-3\lambda - 3)(3) = 0$ or $49\lambda = 13$, which gives $\lambda = \frac{13}{49}$. Hence, the foot of the perpendicular is

$$(2 - 2\lambda, 6\lambda - 3, -3\lambda - 3) = \left(\frac{170}{49}, \frac{78}{49}, \frac{10}{49}\right).$$

The perpendicular distance AP from A(2, 3, 4) to $P\left(\frac{170}{49}, \frac{78}{49}, \frac{10}{49}\right)$ is

$$AP = \sqrt{\left(\frac{170}{49} - 2\right)^2 + \left(\frac{78}{49} - 3\right)^2 + \left(\frac{10}{49} - 4\right)^2}$$

$$= \frac{211.047}{49} = 4.307.$$

EXAMPLE 14.18

Find the image of the point (1,6,3) in the line

$$\frac{x}{1} = \frac{y-1}{2} = \frac{z-2}{3}.$$

Solution. The direction ratios of the given line are 1, 2, and 3. Let the image of the point A(1,6,3) in the given line be $B(x_1, y_1, z_1)$. Therefore, AB is perpendicular to the given line. The point M is the midpoint of AB. Since M lies on the given line, its coordinates are $(\lambda, 2\lambda + 1, 3\lambda + 2)$.

The direction ratios of AM are $\lambda - 1, 2\lambda + 1 - 6$, and $3\lambda + 2 - 3$

or

$$\lambda - 1, 2\lambda - 5, \text{ and } 3\lambda - 1.$$

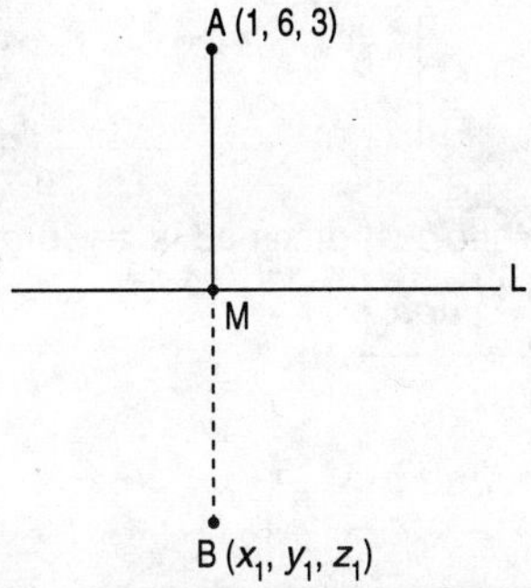

Since AM is perpendicular to the given line, we have $1(\lambda-1)+2(2\lambda-5)+3(3\lambda-1)=0$ or $14\lambda=14$ which gives $\lambda=1$. Therefore, the coordinates of M are (1, 3, 5). Since M is the midpoint of AB, we have

$$1=\frac{x_1+1}{1}, 3=\frac{y_1+6}{2}, \text{ and } 5=\frac{z_1+3}{2}.$$

Hence, the image of the point (1,6,3) is the point $(x_1, y_1, z_1)=(1,0,7)$.

EXAMPLE 14.19

Find the angle between the lines

$$x-2y+z=0, x+2y-2z=0 \text{ and}$$
$$x+2y+z=0, 3x+9y+5z=0$$

Solution. The first line is

$x-2y+z=0, x+2y-2z=0$,

which yields

$$\frac{x}{4-2}=\frac{y}{1-(-2)}=\frac{z}{2-(-2)} \text{ or } \frac{x}{2}=\frac{y}{3}=\frac{z}{4} \quad (1)$$

Similarly, the second line

$x+2y+z=0=3x+9y+5z$ yields

$$\frac{x}{1}=\frac{y}{-2}=\frac{z}{3} \quad (2)$$

The direction ratios of the line (1) are 2, 3, and 4 while the direction ratios of the line (2) are 1, −2, and 3. Therefore, the angle θ between the two lines is given by

$$\cos\theta=\frac{2(1)+3(-2)+4(3)}{\sqrt{2^2+3^2+4^2}\sqrt{1^2+(-2)^2+3^2}}$$

$$=\frac{8}{\sqrt{29}\sqrt{14}}=\frac{8}{\sqrt{406}}$$

and so,

$$\theta=\cos^{-1}\left(\frac{8}{\sqrt{406}}\right)$$

14.7 SHORTEST DISTANCE BETWEEN TWO SKEW LINES

Two lines l_1 and l_2 in space are said to be *skew lines* if they do not intersect and are not parallel. The shortest distance between the skew lines l_1 and l_2 is the length of the perpendicular to both of them. The direction of the shortest distance between the skew lines is the direction of the perpendicular to both the lines.

Let A and C be points with position vectors $\vec{a}$ and $\vec{c}$ on the lines l_1 and l_2, respectively. Join A and C. Let $\vec{PQ}$ be the vector representing the line of shortest distance. Then, the equation of the line through A and parallel to a vector $\vec{b}$ is

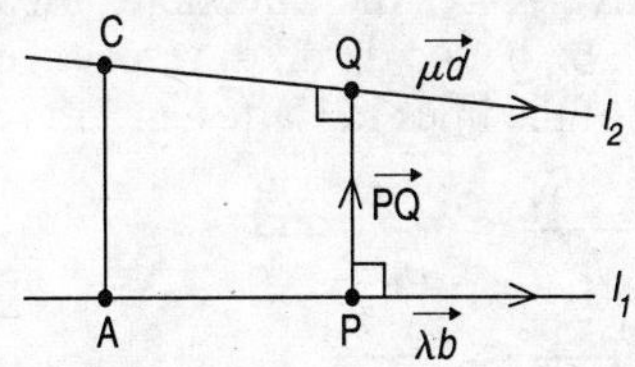

$$\vec{r}=\vec{a}+\lambda\vec{b} \quad (1)$$

Similarly, the equation of the line through C and parallel to the vector $\vec{d}$ is $\vec{r}=\vec{c}+\mu\vec{d}$ (2)

Since the line of shortest distance is perpendicular to both, line (1) and (2), it is perpendicular to both $\vec{b}$ and $\vec{d}$. Hence, the direction of the perpendicular is in the direction of $\vec{b}\times\vec{d}$, since $\vec{b}\times\vec{d}$ is perpendicular to both $\vec{b}$ and $\vec{d}$. The unit vector $\hat{n}$ in this direction is given by $\hat{n}=(\vec{b}\times\vec{d})/|\vec{b}\times\vec{d}|$. Therefore, $\vec{PQ}=s\hat{n}$, $s>0$. Thus, $|\vec{PQ}|=|s||\hat{n}|=s$ is the shortest distance between the lines. Let θ be the angle between AC and PQ. Since $\vec{PQ}$ is a projection of AC along the line of shortest distance, we have

$$PQ=AC\cos\theta=|\vec{c}-\vec{a}|\cos\theta$$

$$=|\vec{c}-\vec{a}|\frac{\vec{PQ}.\vec{AC}}{|\vec{PQ}||\vec{AC}|}$$

$$=|\vec{c}-\vec{a}|\frac{s\hat{n}.(\vec{c}-\vec{a})}{s|\vec{c}-\vec{a}|}=\hat{n}.(\vec{c}-\vec{a})$$

$$=\left|\frac{(\vec{b}\times\vec{d}).(\vec{c}-\vec{a})}{|\vec{b}\times\vec{d}|}\right|.$$

We have taken an absolute value because distance is always nonnegative. Thus, the shortest distance is given by

$$s = \left|\frac{(\vec{b}\times\vec{d}).(\vec{c}-\vec{a})}{|\vec{b}\times\vec{d}|}\right|$$

$$= \frac{[\vec{c}\vec{b}\vec{d}]-[\vec{a}\vec{b}\vec{d}]}{\vec{b}\times\vec{d}} \text{ (Vector form).}$$

To obtain the shortest distance between the skew lines in Cartesian form, let $A(x_1, y_1, z_1)$ and $C(x_2, y_2, z_2)$ be the points with position vectors $\vec{a}$ and $\vec{c}$, respectively on the lines l_1 and l_2, respectively. Let the direction ratios of the lines be b_1, b_2, b_3 and d_1, d_2, d_3 respectively. The equations of the lines in Cartesian form are

$$\frac{x-x_1}{b_1}=\frac{y-y_1}{b_2}=\frac{z-z_1}{b_3} \text{ and}$$

$$\frac{x-x_2}{d_1}=\frac{y-y_2}{d_2}=\frac{z-z_2}{d_3}.$$

In vector form, the lines are $\vec{r}=\vec{a}+\lambda\vec{b}$

$$= x_1\vec{i}+y_1\hat{j}+z_1\hat{k}+\lambda(b_1\hat{i}+b_2\hat{j}+b_3k)$$

and $\vec{r}=\vec{c}+\mu\vec{d}$

$$= x_2\vec{i}+y_2\hat{j}+z_2\hat{k}+\lambda(d_1\hat{i}+d_2\hat{j}+d_3k)$$

Therefore,

$$\vec{b}\times\vec{d} = \begin{vmatrix} \hat{i} & \hat{j} & \hat{k} \\ b_1 & b_2 & b_3 \\ d_1 & d_2 & d_3 \end{vmatrix}$$

$$= \hat{i}(b_2d_3-d_2b_3)+\hat{j}(d_1b_3-b_1d_3) +\hat{k}(b_1d_2-d_1b_2)$$

and so,

$$|\vec{b}\times\vec{d}| = \sqrt{(b_2d_3-d_2b_3)^2+(d_1b_3-b_1d_3)^2+(b_1d_2-d_1b_2)^2}.$$

Further,

$$[\vec{c}\vec{b}\vec{d}]-[\vec{a}\vec{b}\vec{d}] = \begin{vmatrix} x_2 & y_2 & z_2 \\ b_1 & b_2 & b_3 \\ d_1 & d_2 & d_3 \end{vmatrix} - \begin{vmatrix} x_1 & y_1 & z_1 \\ b_1 & b_2 & b_3 \\ d_1 & d_2 & d_3 \end{vmatrix}$$

$$= \begin{vmatrix} x_2-x_1 & y_2-y_1 & z_3-z_1 \\ b_1 & b_2 & b_3 \\ d_1 & d_2 & d_3 \end{vmatrix}$$

Hence, the shortest distance is given by

$$s = \frac{[\vec{c}\vec{b}\vec{d}]-[\vec{a}\vec{b}\vec{d}]}{\vec{b}\times\vec{d}}$$

$$= \frac{\begin{vmatrix} x_2-x_1 & y_2-y_1 & z_3-z_1 \\ b_1 & b_2 & b_3 \\ d_1 & d_2 & d_3 \end{vmatrix}}{\sqrt{(b_2d_3-b_3d_3)^2+(b_3d_1-b_1d_3)^2+(b_1d_2-b_2d_1)^2}}$$

$$= \frac{(x_2-x_1)(b_2d_3-b_3d_2)+(y_2-y_1)(b_3d_1-b_1d_3)+(z_2-z_1)(b_1d_2-b_2d_1)}{\sqrt{(b_2d_3-b_3d_2)^2+(b_3d_1-b_1d_3)^2+(b_1d_2-b_2d_1)^2}}$$

$$= l(x_2-x_1)+m(y_2-y_1)+n(z_2-z_1),$$

where l, m and n are direction cosines of the line perpendicular to both lines l_1 and l_2, that is, of the line $\vec{b}\times\vec{d}$.

Deductions: Intersection of Lines

The shortest distance between two lines $\vec{r}=\vec{a}+\lambda\vec{b}$ and $\vec{r}=\vec{c}+\mu\vec{d}$ is

$$s = \frac{[\vec{c}\vec{b}\vec{d}]-[\vec{a}\vec{b}\vec{d}]}{\vec{b}\times\vec{d}}$$

The lines will intersect if and only if the shortest distance between them is zero. Hence, the two lines l_1 and l_2 will intersect if

$$[\vec{c}\vec{b}\vec{d}]-[\vec{a}\vec{b}\vec{d}]=0 \text{ or } [\vec{c}-\vec{a}\ \vec{b}\ \vec{d}]=0$$

or, in Cartesian form, if

$$\begin{vmatrix} x_2-x_1 & y_2-y_1 & z_2-z_1 \\ b_1 & b_2 & b_3 \\ d_1 & d_2 & d_3 \end{vmatrix}=0$$

The above condition can be obtained in the following manner also:

Let $\vec{r}=\vec{a}+\lambda\vec{b}$ and $\vec{r}=\vec{c}+\mu\vec{d}$ be two lines. The first line passes through a point with a position vector $\vec{a}$ and is parallel to the vector $\vec{b}$, whereas the second line passes through $\vec{c}$ and is parallel to $\vec{d}$. Let these lines intersect at some point. Then, the lines will lie in a plane. Therefore, the vector $\vec{c}-\vec{a},\vec{b}$ and $\vec{d}$ are planar. Hence, $(\vec{c}-\vec{a}).(\vec{b}\times\vec{d})=0$ or $[\vec{c}-\vec{a},\vec{b}\ \vec{d}]=0$.

Remark 14.2.

$\left[\vec{c}-\vec{a}, \vec{b}\ \vec{d}\right]=0$ even if the lines are parallel because in that case $\vec{b}\times\vec{d}=0$ and so, $(\vec{c}-\vec{a}).(\vec{b}\times\vec{d})=0$. So, we must check whether $\vec{b}$ and $\vec{d}$ are parallel. In case they are not parallel $\left[\vec{c}-\vec{a},\vec{b}\ \vec{d}\right]=0$, then the given line $\vec{r}=\vec{a}+\lambda\vec{b}$ and $\vec{r}=\vec{c}+\mu\vec{d}$ intersect.

The *distance between the parallel lines* is given by $s=\dfrac{\vec{b}\times(\vec{c}-\vec{a})}{|\vec{b}|}$.

To find the *equation of the line of shortest distance* between the lines

$$\frac{x-x_1}{b_1}=\frac{y-y_1}{b_2}=\frac{z-z_1}{b_3} \quad (1)$$

and

$$\frac{x-x_2}{d_1}=\frac{y-y_2}{d_2}=\frac{z-z_2}{d_3}, \quad (2)$$

we observe that the line of shortest distance is coplanar with both the said lines. The equation of the plane containing the line (1) and the line of shortest distance,

$$\frac{x-\alpha}{l}=\frac{y-\beta}{m}=\frac{z-\gamma}{n}, \quad (3)$$

where

$$l=\frac{b_2d_3-b_3d_2}{\sqrt{(b_2d_3-b_3d_2)^2}},\ m=\frac{b_3d_1-b_1d_3}{\sqrt{(b_3d_1-b_1d_3)^2}},$$

$$n=\frac{b_1d_2-b_2d_1}{\sqrt{(b_1d_2-b_2d_1)^2}}$$

is

$$\begin{vmatrix} x-x_1 & y-y_1 & z-z_1 \\ b_1 & b_2 & b_3 \\ l & m & n \end{vmatrix}=0 \quad (4)$$

and that of the plane containing line (2) and the line (3) is

$$\begin{vmatrix} x-x_2 & y-y_2 & z-z_2 \\ d_1 & d_2 & d_3 \\ l & m & n \end{vmatrix}=0 \quad (5)$$

Hence (4) and (5) are the two equations of the line of shortest distance with l,m,n given in (3).

EXAMPLE 14.20

Find the magnitude and equation of the line of shortest distance between the lines

$$\frac{x-8}{3}=\frac{y+9}{-16}=\frac{z-10}{7} \text{ and}$$

$$\frac{x-15}{3}=\frac{y-29}{8}=\frac{z-5}{-5}.$$

Solution. Let $l_1, m_1,$ and n_1 be the direction ratios of the line of shortest distance. Since it is perpendicular to both the given lines, we have $3l_1-16m_1+7n_1=0$ and $3l_1+8m_1-5n_1=0$. Therefore,

$$\frac{l_1}{80-56}=\frac{m_1}{21+15}=\frac{n_1}{24+48} \text{ or } \frac{l_1}{2}=\frac{m_1}{3}=\frac{n_1}{6}.$$

Therefore, the direction cosines of the line of shortest distance between the given lines are

$$l=\frac{2}{\sqrt{2^2+3^2+6^2}},\ m=\frac{3}{\sqrt{2^2+3^2+6^2}}, \text{ and}$$

$$n=\frac{6}{\sqrt{2^2+3^2+6^2}}$$

or

$$l=\frac{2}{7},\ m=\frac{3}{7}, \text{ and } n=\frac{6}{7}$$

Hence, the shortest distance is

$$s=l(x_2-x_1)+m(y_2-y_1)+n(z_2-z_1)$$
$$=\frac{2}{7}(15-8)+\frac{3}{7}(29+9)+\frac{6}{7}(5-10)=14\cdot$$

Further, the equation of the line of shortest distance is

$$\begin{vmatrix} x-8 & y+9 & z-10 \\ 3 & -16 & 7 \\ \frac{2}{7} & \frac{3}{7} & \frac{6}{7} \end{vmatrix}=0 \text{ and}$$

$$\begin{vmatrix} x-15 & y-29 & z-5 \\ 3 & 8 & -5 \\ \frac{2}{7} & \frac{3}{7} & \frac{6}{7} \end{vmatrix}=0$$

or

$$117x+4y-41z-49=0 \text{ and}$$
$$9x-4y-z-14=0$$

Second Method: Let P and Q be points on the given lines, respectively. Then, the general coordinates of P and Q are

$$P(3\lambda+8,-16\lambda-9,7\lambda+10) \text{ and}$$
$$Q(3\mu+15,8\mu+29,-5\mu+5)$$

Therefore, the direction cosines of the line PQ are proportional to

$$3\lambda-3\mu-7,-16\lambda-8\mu-38, \text{ and } 7\lambda+5\mu+5$$

Now, PQ will be line of shortest distance if it is perpendicular to both the given lines. Therefore, by the condition of perpendicularity, we have and

$$3(3\lambda-3\mu-7)-16(-16\lambda-8\mu-38)+7(7\lambda+5\mu+5)=0$$

and

$$3(3\lambda-3\mu-7)+8(-16\lambda-8\mu-38)-5(7\lambda+5\mu+5)=0$$

or

$$157\lambda+77\mu+311=0 \text{ and } 11\lambda+7\mu+25=0$$

Solving these equations, we get $\lambda=-1$ and $\mu=-2$. Hence, the points are

$$P(5,7,3) \text{ and } Q(9,13,15).$$

The shortest distance PQ is

$$PQ=\sqrt{(9-5)^2+(13-7)^2+(15-3)^2}=14 \cdot$$

The equation of PQ, by the two-point formula, is

$$\frac{x-5}{9-5}=\frac{y-7}{13-7}=\frac{z-3}{15-3} \text{ or } \frac{x-5}{2}=\frac{y-7}{3}=\frac{z-3}{6}.$$

EXAMPLE 14.21

Show that the lines $\frac{x-4}{1}=\frac{y+3}{-4}=\frac{z+1}{7}$ and $\frac{x-1}{2}=\frac{y+1}{-3}=\frac{z+10}{8}$ intersect and find the coordinates of their point of intersection.

Solution. Any point on the first line is $(\lambda+4,\ -4\lambda-3,\ 7\lambda-1)$ whereas any point on the second line is $(2\mu+1,\ -3\mu-1,\ 8\mu-10)$

At the point of intersection, we must have

$$\lambda+4=2\mu-1,\ -4\lambda-3=-3\mu-1, \text{ and}$$
$$7\lambda-1=8\mu-10 \cdot$$

The first two members yield $\lambda=1$ and $\mu=2$. These values satisfy $7\lambda-1=8\mu-10$ Hence, the given lines intersect at the point $(\lambda+4,\ -4\lambda-3,\ 7\lambda-1)=(5,-7,6)$.

EXAMPLE 14.22

Find the equation of the straight line perpendicular to both the lines

$$\frac{x-1}{1}=\frac{y-1}{2}=\frac{z+2}{3} \text{ and } \frac{x+2}{2}=\frac{y-5}{-1}=\frac{z+3}{2}$$

and passing through their point of intersection.

Solution. Any point on the first lines is $(\lambda+1,\ 2\lambda+1,\ 3\lambda-2)$ and any point on the second line is $(2\mu-2,\ -\mu+5,\ 2\mu-3)$. At the point of intersection, we must have

$$\lambda+1=2\mu-2,\ 2\lambda+1=-\mu+5 \text{ and}$$
$$3\lambda-2=2\mu-3.$$

From the first two members, we get $\lambda=1$ and $\mu=3$. These values also satisfy $3\lambda-2=2\mu-3$. Hence, the given lines intersect at $(\lambda+1,\ 2\lambda+1,\ 3\lambda-2)=(2,3,1)$. Therefore, the required line is

$$\frac{x-2}{l}=\frac{y-3}{m}=\frac{z-1}{n}$$

This line will be perpendicular to the given lines if $l+2m+3n=0$ and $2l-m+2n=0$

Therefore,

$$\frac{l}{7}=\frac{m}{4}=\frac{n}{-5}$$

Hence, the required line is

$$\frac{x-2}{7}=\frac{y-3}{4}=\frac{z-1}{-5}$$

EXAMPLE 14.23

Find the magnitude and equation of the shortest distance between the lines

$$\frac{x-3}{1}=\frac{y-5}{-2}=\frac{z-7}{1} \text{ and } \frac{x+1}{7}=\frac{y+1}{-6}=\frac{z-1}{1}.$$

Also, find the points where the line of shortest distance intersects the given lines.

Solution. Any point P on the first line is $P(\lambda+3,\ -2\lambda+5,\ \lambda+7)$ and any point Q on the second line is $Q(7\mu-1,\ -6\mu-1,\ \mu-1)$ Therefore, direction cosines of PQ are proportional to $\lambda+3-7\mu+1,\ -2\lambda+5+6\mu+1$ and $\lambda+7-\mu+1$ or

$$\lambda-7\mu+4,\ -2\lambda+6\mu+6, \text{ and } \lambda-\mu+8.$$

The line PQ will be of shortest distance if it is perpendicular to both the given lines. Therefore, the condition for perpendicularity yields

$$\lambda-7\mu+4-2(-2\lambda+6\mu+6)+\lambda-\mu+8=0$$

and

$$7(\lambda-7\mu+4)-6(-2\lambda+6\mu+6)+\lambda-\mu+8=0$$

or

$$6\lambda-20\mu=0 \text{ and } 20\lambda-86\mu=0$$

Thus, we get trivial solution $\lambda=0$ and $\mu=0$ Thus, the points P and Q are (3, 5, 7) and (–1, –1, –1), respectively. The magnitude of PQ is

$$s=\sqrt{(3+1)^2+(5+1)^2+(7+1)^2}$$

$$=\sqrt{16+36+64}=10.7703\cdot$$

The equation of PQ is

$$\frac{x-3}{3+1}=\frac{y-5}{5+1}=\frac{z-7}{7+1} \text{ or } \frac{x-3}{2}=\frac{y-5}{3}=\frac{z-7}{4}$$

EXAMPLE 14.24

Find the shortest distance between the lines $\frac{x-1}{2}=\frac{y-2}{3}=\frac{z-3}{4}$ and $\frac{x-2}{3}=\frac{y-4}{4}=\frac{z-5}{5}$

Solution. The general points on the given lines are respectively

$$P(2\lambda+1,\ 3\lambda+2,\ 4\lambda+3) \text{ and}$$

$$Q(3\mu+2,\ 4\mu+4,\ 5\mu+5)$$

Therefore, the direction cosines of PQ are proportional to

$$2\lambda+1-3\mu-2,\ 3\lambda+2-4\mu-4, \text{ and}$$
$$4\lambda+3-5\mu-5$$

or

$$2\lambda-3\mu-1, 3\lambda-4\mu-2, \text{ and } 4\lambda-5\mu-2.$$

The line PQ will be perpendicular to both the lines if

$$2(2\lambda-3\mu-1)+3(3\lambda-4\mu-2)$$
$$+4(4\lambda-5\mu-2)=0$$

and

$$3(2\lambda-3\mu-1)+4(3\lambda-4\mu-2)$$
$$+5(4\lambda-5\mu-2)=0$$

or

$$29\lambda-38\mu-16=0 \text{ and } 38\lambda-50\mu-21=0.$$

Solving these equations for λ and μ, we get $\lambda=\frac{1}{3}$ and $\mu=-\frac{1}{6}$.

Thus, the points and are

$$P\left(\frac{5}{3},3,\frac{13}{3}\right) \text{ and } Q\left(\frac{3}{2},\frac{10}{3},\frac{25}{6}\right)$$

Hence, the shortest distance is

$$|PQ|=\sqrt{\left(\frac{3}{2}-\frac{5}{3}\right)^2+\left(\frac{10}{3}-3\right)^2+\left(\frac{25}{6}-\frac{13}{3}\right)^2}$$

$$=\sqrt{\frac{1}{36}+\frac{1}{9}+\frac{1}{36}}=\frac{1}{\sqrt{6}}$$

EXAMPLE 14.25

Show that the lines $\frac{x}{1}=\frac{y}{-1}=\frac{z+3}{1}$ and $\frac{x-6}{2}=\frac{y+8}{-4}=\frac{z+5}{3}$ do not intersect. Find the equation of a straight line parallel to the second line which does meet the first line.

Solution. The coordinates of the general points P and Q on the given lines, respectively, are

$$P(\lambda,-\lambda,\lambda-3) \text{ and}$$
$$Q(2\mu+6,-4\mu-8,3\mu-5).$$

The given lines will intersect only if there exist λ and μ such that

$$\lambda=2\mu+6,\ -\lambda=-4\mu-8, \text{ and } \lambda-3=3\mu-5.$$

Solving the first and the last equations, we get $\lambda=22$ and $\mu=8$. But the equation $-\lambda=-4\mu-8$ is not satisfied for these values. Hence, the given lines do not intersect.

Further, the equation of the line parallel to the second line shall be of the type

$$\frac{x-\alpha}{2}=\frac{y-\beta}{-4}=\frac{z-\gamma}{3}. \tag{1}$$

Taking, $\alpha=\beta=0$, the parametric equation of the line is $x=2\lambda$, $y=-4\lambda$, and $z=3\lambda+\gamma$. The parametric equation of the first line is $x=\mu$, $y=-\mu$ and $z=\mu-3$. Therefore, (1) with $\alpha=\beta=0$ will intersect the first given line if

$$2\lambda=\mu, -4\lambda=-\mu, \text{ and } 3\lambda+\gamma=\mu-3$$

The first two equations implies $\lambda=0$ and $\mu=0$ and then the third equation implies $\gamma=-3$. Hence, one of the required equation is

$$\frac{x}{2}=\frac{y}{-4}=\frac{z+3}{3}$$

EXAMPLE 14.26

Find the shortest distance between the lines $\vec{r}=(1-t)\hat{i}+(t-2)\hat{j}+(3-2t)\hat{k}$ and

$$\vec{r}=(s+1)\hat{i}+(2s-1)\hat{j}-(2s+1)\hat{k}.$$

Solution. The given lines can be expressed as $\vec{r}=(\hat{i}-2\hat{j}+3\hat{k})+t(-\hat{i}+\hat{j}-2\hat{k})=\vec{a}+t\vec{b}$, say
and
$\vec{r}=(\hat{i}-\hat{j}-\hat{k})+s(\hat{i}+2\hat{j}-2\hat{k})=\vec{c}+s\vec{d}$, say
Thus,

$$\vec{a}=\hat{i}-2\hat{j}+3\hat{k},\ \vec{b}=-\hat{i}+\hat{j}-2\hat{k},$$
$$\vec{c}=\hat{i}-\hat{j}-\hat{k},\ \vec{d}=\hat{i}+2\hat{j}-2\hat{k}.$$

Therefore, the shortest distance (S.D) between the given lines is

$$S.D=\frac{|(\vec{c}-\vec{a}).(\vec{b}\times\vec{d})|}{|\vec{b}\times\vec{d}|}. \qquad (1)$$

Now, $\vec{b}\times\vec{d}=\begin{vmatrix}\hat{i} & \hat{j} & \hat{k}\\ -1 & 1 & -2\\ 1 & 2 & 2\end{vmatrix}=2\hat{i}-4\hat{j}-3\hat{k},$

$$\vec{c}-\vec{a}=(\hat{i}-\hat{j}-\hat{k})-(\hat{i}-2\hat{j}+3\hat{k})=(\hat{j}-4\hat{k}),$$

and

$$(\vec{c}-\vec{a}).(\vec{b}\times\vec{d})=(\hat{j}-4\hat{k}).(2\hat{i}-4\hat{j}-3\hat{k})=8.$$

Hence, (1) implies

$$S.D=\frac{|8|}{|2\hat{i}-4\hat{j}-3\hat{k}|}=\frac{|8|}{\sqrt{4+16+9}}=\frac{|8|}{\sqrt{29}}.$$

EXAMPLE 14.27

Show that the line $\vec{r}=7\hat{i}-\hat{j}-\hat{k}+\lambda(2\hat{i}+3\hat{k})$ and $\vec{r}=\hat{i}+\hat{j}-\hat{k}+\mu(3\hat{i}-\hat{k})$ intersect and find their point of intersection.

Solution. The given lines are
$\vec{r}=7\hat{i}-\hat{j}-\hat{k}+\lambda(2\hat{i}+3\hat{k})=\vec{a}+\lambda\vec{b}$, say
and
$\vec{r}=\hat{i}+\hat{j}-\hat{k}+\mu(3\hat{i}-\hat{k})=\vec{c}+\mu\vec{d}$, say
Thus,
$\vec{a}=7\hat{i}-\hat{j}-\hat{k},\quad \vec{b}=2\hat{i}+3\hat{k},\quad \vec{c}=\hat{i}+\hat{j}-\hat{k}$, and
$\vec{d}=3\hat{i}-\hat{j}.$
The lines will intersect if

$$[\vec{c}\vec{b}\vec{d}]-[\vec{a}\vec{b}\vec{d}]=0. \qquad (1)$$

We have

$$[\vec{c}\vec{b}\vec{d}]=\begin{vmatrix}1 & 1 & -1\\ 2 & 0 & 3\\ 3 & -1 & 0\end{vmatrix}=3+9+2=14, \text{ and}$$

$$[\vec{a}\vec{b}\vec{d}]=\begin{vmatrix}7 & -1 & -1\\ 2 & 0 & 3\\ 3 & -1 & 0\end{vmatrix}=21-9+2=14.$$

Thus, (1) is satisfied and so, the given lines intersect. Now, at the point of intersection, we have $\vec{r}=\vec{a}+\lambda\vec{b}=c+\mu\vec{d}$,
which gives
$7\hat{i}-\hat{j}-\hat{k}+\lambda(2\hat{i}+3\hat{k})=\hat{i}+\hat{j}-\hat{k}+\mu(3\hat{i}-\hat{j}).$
Comparing the coefficients of $\hat{i}$, $\hat{j}$, and $\hat{k}$ on both sides, we have
$7+2\lambda=1+3\mu$, $-1=1-\mu$, and $-1+3\lambda=-1$.
These equations yield $\mu=2,\ \lambda=0$. Hence, the position vector of point of intersection is $\vec{r}=\vec{a}+0\vec{b}=7\hat{i}-\hat{j}-\hat{k}$. Therefore, the point of intersection is (7,–1, –1).

EXAMPLE 14.28

Find the shortest distance between the lines $\frac{x-3}{3}=\frac{y-8}{-1}=\frac{z-3}{1}$ and $\frac{x+3}{-3}=\frac{y+7}{2}=\frac{z-6}{4}$. Also, find the equation of the line of shortest distance.

Solution. Let PQ be the shortest distance between the given lines with point lying on the first line and lying on the second line. Then, we have

$$P(3\lambda+3,-\lambda+8,\ \lambda+3) \text{ and}$$
$$Q(-3\mu-3,\ 2\mu-7,\ 4\mu+6)$$

for some scalars λ and μ. The direction cosines of PQ are proportional to

$$-3\mu-3-3\lambda-3,\ 2\mu-7+\lambda-8, \text{ and}$$
$$4\mu+6-\lambda-3$$

or $-3\lambda-3\mu-6$, $\lambda+2\mu-15$, and $-\lambda+4\mu+3$.

The line PQ will be of shortest distance if it is perpendicular to both the given lines. So, we must have

$$3(-3\lambda-3\mu-6)-1(\lambda+2\mu-15)$$
$$+1(-\lambda+4\mu+3)=0$$

and

$$-3(-3\lambda-3\mu-6)+2(\lambda+2\mu-15)$$
$$+4(-\lambda+4\mu+3)=0$$

or

$-11\lambda - 7\mu = 0$ and $7\lambda + 29\mu = 0$.

These equations have a trivial solution $\lambda = \mu = 0$. Therefore, the points P and Q are P (3, 8, 3) and Q (−3, −7, 6).

Hence, the shortest distance is

$$PQ = \sqrt{(3+3)^2 + (8+7)^2 + (3-6)^2}$$

$$= \sqrt{270} = 3\sqrt{30}.$$

Also, using the two-point formula, the equation of PQ is

$$\frac{x-3}{-3-3} = \frac{y-8}{-7-8} = \frac{z-3}{6-3} \text{ or}$$

$$\frac{x-3}{2} = \frac{y-8}{5} = \frac{z-3}{-1}.$$

EXAMPLE 14.29

Find the direction cosines of the line bisecting the angles between the lines whose direction ratios are l_1, m_1, n_1 and l_2, m_2, n_2, respectively.

Solution. If $L = L(l_1, m_1, n_1)$ and $L' = L'(l_2, m_2, n_2)$ then OL and OL′ are the lines from the origin O with the given direction ratios. Also, $|OL| = 1$ and $|OL'| = 1$. The midpoint M of LL′ has the coordinates $\frac{l_1+l_2}{2}, \frac{m_1+m_2}{2}$ and $\frac{n_1+n_2}{2}$.

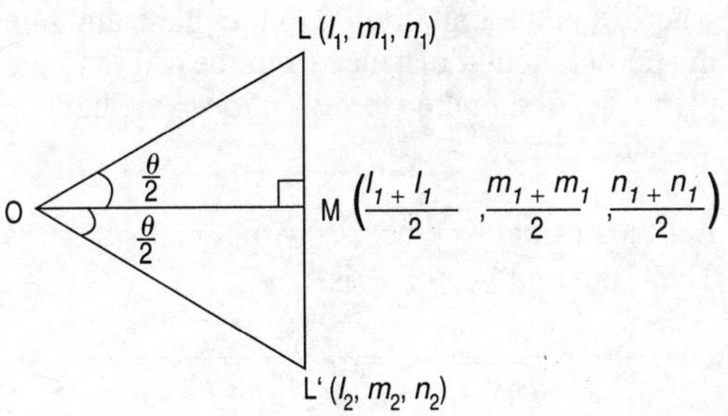

If $\angle LOL' = \theta$ then $\frac{OM}{OL} = \cos\frac{\theta}{2}$, that is $OM = OL\cos\frac{\theta}{2}$. Hence, the direction cosines of OM (the bisector of the angle θ) are

$$\frac{l_1+l_2}{2\cos\frac{\theta}{2}}, \frac{m_1+m_2}{2\cos\frac{\theta}{2}}, \text{ and } \frac{n_1+n_2}{2\cos\frac{\theta}{2}}.$$

Similarly, the direction cosines of the other bisector are

$$\frac{l_1-l_2}{2\cos\frac{\theta}{2}}, \frac{m_1-m_2}{2\cos\frac{\theta}{2}}, \text{ and } \frac{n_1-n_2}{2\cos\frac{\theta}{2}}.$$

Remark 14.3.

It follows from the previous example that the direction cosines of the bisectors are proportional to

$$l_1+l_2,\ m_1+m_2,\ n_1+n_2 \text{ and}$$
$$l_1-l_2,\ m_1-m_2,\ n_1-n_2.$$

EXAMPLE 14.30

Find the equations of the lines bisecting the angle between the lines

$$\frac{x-1}{2} = \frac{y+2}{-2} = \frac{z-3}{1} \text{ and } \frac{x-1}{12} = \frac{y+2}{2} = \frac{z-3}{-3}.$$

Solution. The direction ratios of the bisectors are proportional to

$$l_1+l_2,\ m_1+m_2,\ n_1+n_2 \text{ and}$$
$$l_1-l_2,\ m_1-m_2,\ n_1-n_2.$$

In the present case, the direction cosines of the given lines are respectively $l_1 = \frac{2}{3}, m_1 = -\frac{2}{3}$, and $n_1 = \frac{1}{3}$ and $l_2 = \frac{12}{\sqrt{157}}, m_2 = \frac{2}{\sqrt{157}}$, and $n_2 = \frac{-3}{\sqrt{157}}$. Therefore, the direction cosines of the bisectors are proportional to

$$l_1+l_2,\ m_1+m_2,\ n_1+n_2 \text{ and}$$
$$l_1-l_2,\ m_1-m_2,\ n_1-n_2.$$

or proportional to

$$61,\ -19,\ 3.5 \text{ and } -11,\ -31,\ 21.5.$$

Since the bisectors pass through (1, −2, 3) their equations are

$$\frac{x-1}{61} = \frac{y+2}{-19} = \frac{z-3}{3.5} \text{ and } \frac{x-1}{-11} = \frac{y+2}{-31} = \frac{z-3}{21.5}.$$

EXAMPLE 14.31

Find the equations of two straight lines that pass through the origin and intersect the line $\frac{x-3}{2} = \frac{y-3}{1} = \frac{z}{1}$ at an angle of 60°.

Solution. Any line that passes through the origin (0, 0, 0) is

$$\frac{x}{a} = \frac{y}{b} = \frac{z}{c}. \quad (1)$$

Any point on this line is given by

$$x = a\lambda,\ y = b\lambda, \text{ and } z = c\lambda.$$

The given line is

$$\frac{x-3}{2}=\frac{y-3}{1}=\frac{z}{1} \quad (2)$$

Any point on this line is

$$x = 2\mu+3,\ y = \mu+3, \text{ and } z = \mu.$$

For intersection, we have

$$a\lambda= 2\mu+3,\ b\lambda= \mu+3, \text{ and } c\lambda= \mu.$$

We note that

$$a\lambda-b\lambda=\mu=c\lambda$$

and so,

$$a=b+c \text{ or } a-b-c=0. \quad (3)$$

The angle between (1) and (2) is $60°$. Therefore,

$$\cos 60° = \frac{2a+b+c}{\sqrt{6}\sqrt{a^2+b^2+c^2}}.$$

or

$$\frac{1}{2}=\frac{3b+3c}{\sqrt{6}\sqrt{2b^2+2c^2+2b^2}}, \text{ using } a=b+c \text{ or}$$

$$6b+6c=\sqrt{6}\sqrt{2b^2+2c^2+2b^2}$$

Squaring both sides, we get

$$2b^2+2c^2+5bc=0 \text{ or } (2b+c)(b+2c)=0$$

Thus,

$$2b+c=0 \text{ or } b+2c=0$$

that is,

$$0a+2b+c=0 \text{ or } 0a+b+2c=0 \quad (4)$$

If (using (3) and (4))

$a-b-c=0$ and $0a+2b+c=0$,

then

$$\frac{a}{-1+2}=\frac{b}{0-1}=\frac{c}{2-0} \text{ or } \frac{a}{1}=\frac{b}{-1}=\frac{c}{2}.$$

If

$a-b-c=0$ and $0a+b+2c=0$,

then

$$\frac{a}{-2+1}=\frac{b}{0-2}=\frac{c}{1-0} \text{ or } \frac{a}{-1}=\frac{b}{-2}=\frac{c}{1}.$$

Hence, (1) gives the required equations as

$$\frac{x}{1}=\frac{y}{-1}=\frac{z}{2} \text{ and } \frac{x}{-1}=\frac{y}{-2}=\frac{z}{1},$$

that is,

$$\frac{x}{1}=\frac{y}{-1}=\frac{z}{2} \text{ and } \frac{x}{1}=\frac{y}{2}=\frac{z}{-1}.$$

14.8 EQUATION A PLANE

Let O be the origin of the three-dimensional coordinate system and $P(x, y, z)$ any point on the plane. Let ON = d be a perpendicular drawn from O to the plane. Let n denote the unit normal vector to the plane. Then $\vec{ON} = ON(\hat{n}) = d\hat{n}$.

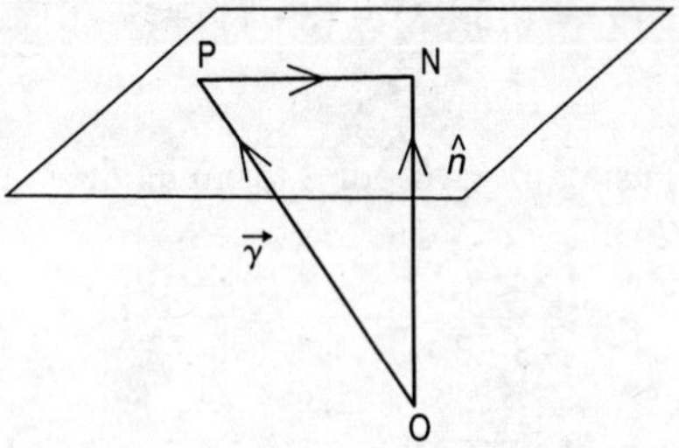

Since P is an arbitrary point on the plane, the vector $\vec{PN}$ is in the plane and is perpendicular to $\vec{ON}$. Therefore, $\vec{PN}.\vec{ON}=0$ or $\vec{PN}.d\hat{n}=0$

or

$$\vec{PN}.\hat{n}=0.$$

But $\vec{PN}=\vec{ON}-\vec{OP}=d\hat{n}-\vec{r}$. Therefore,

$$(d\hat{n}-\hat{r}).\hat{n}=0$$

or

$$\hat{r}.\hat{n}=d\hat{n}.\hat{n}=d.$$

Hence, the required equation of the plane is

$$\hat{r}.\hat{n}=d \text{ (Vector form)}, \quad (1)$$

where $\hat{n}$ is the unit normal to the plane and d is the perpendicular distance from the origin to the plane. To obtain the *Cartesian form*, we have $\vec{r}=x\hat{i}+y\hat{j}+z\hat{k}$.

Let the normal $\vec{ON}$ be $\vec{ON}=a\hat{i}+b\hat{j}+c\hat{k}$.
Then, the unit vector $\hat{n}$ is

$$\hat{n}=\frac{a\hat{i}}{\sqrt{a^2+b^2+c^2}}+\frac{b\hat{j}}{\sqrt{a^2+b^2+c^2}}+\frac{c\hat{k}}{\sqrt{a^2+b^2+c^2}}=l\hat{i}+m\hat{j}+n\hat{k},$$

where l, m, and n are direction cosines of the normal to the plane. Hence, the equation (1) of the plane reduces to

$$(x\hat{i}+y\hat{j}+z\hat{k}).(l\hat{i}+m\hat{j}+n\hat{k})=d$$

or

$$lx + my + nz = d \text{ (Normal form)}. \quad (2)$$

Multiplying (2) throughout by $\sqrt{a^2+b^2+c^2}$, we get

$$ax + by + cz = d\sqrt{a^2+b^2+c^2} = D \text{ say}, \quad (3)$$

as the equation of the plane. Hence, in a three-dimensional system, any linear equation of the form $ax+by+cz = D$ represents a plane.

A plane parallel to the plane $ax+by+cz = D$ has the same normal. Hence, the direction cosines of the normal to the two planes are same. However, the distance from the origin to the two planes is different. Hence the *equation of the plane parallel to the plane* $ax+by+cz = D$ is $ax+by+cz = p$, where $\frac{p}{\sqrt{a^2+b^2+c^2}}$ is the perpendicular distance from the origin to the parallel plane. If D and p are of opposite signs, then the two planes lie on the opposite sides with respect to the origin.

If the plane $ax+by+cz = D$ intersects the axes at, $(A,0,0)$, $(0,B,0)$ and $(0,0,C)$, then, since these points lie on the plane, we have $Aa = D, Bb = D$, and $Cc = D$, which yields $a = \frac{D}{A}, b = \frac{D}{B}$, and $c = \frac{D}{C}$. Hence, the equation of the plane reduces to

$$\frac{Dx}{A} + \frac{Dy}{B} + \frac{Dz}{C} = D \text{ or } \frac{x}{A} + \frac{y}{B} + \frac{z}{C} = 1, \quad (4)$$

which is called the *intercept form* of the equation of the plane.

14.9 EQUATION OF A PLANE PASSING THROUGH A GIVEN POINT AND PERPENDICULAR TO A GIVEN DIRECTION

Let the plane pass through a fixed point $A(a_1, a_2, a_3)$, whose position vector is $\vec{a}$. Let $\vec{N}$ be the given direction and let $\vec{r}$ be the position vector of a point $P(x, y, z)$ on the plane. Then, PA lies in the plane. Since the plane is perpendicular to $\vec{N}$, we have

$$\vec{PA}.\vec{N} = 0 \text{ or } (\vec{r} - \vec{a}).\vec{N} = 0 \text{ (Vector form)}, \quad (1)$$

which is the required equation of the plane.

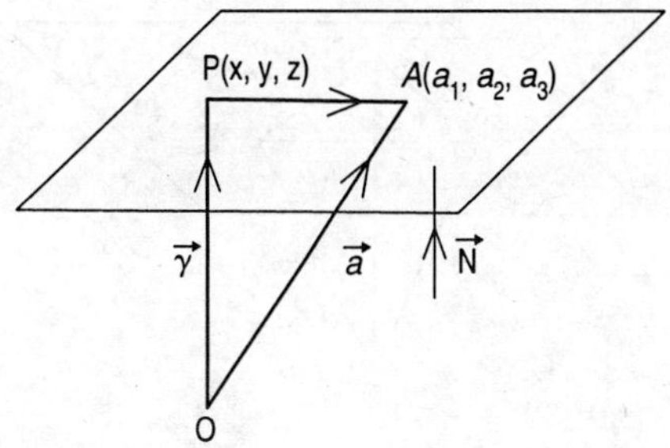

Substituting

$$\vec{r} = x\hat{i} + y\hat{j} + z\hat{k},\ \vec{a} = a_1\hat{i} + a_2\hat{j} + a_3\hat{k}, \text{ and}$$

$$\vec{N} = n_1\hat{i} + n_2\hat{j} + n_3\hat{k} \text{ in (1), we have}$$

$$(x-a_1)n_1 + (y-a_2)n_2 + (z-a_3)n_3 = 0, \quad (2)$$

which is the equation of the plane, in *Cartesian form*, passing through (a_1, a_2, a_3) and perpendicular to the direction $\vec{N}$.

14.10 EQUATION OF A PLANE PASSING THROUGH THREE POINTS

Let $P(x_1, y_1, z_1), Q(x_1, y_2, z_2)$, and $R(x_3, y_2, z_3)$ be the points and n_1, n_2, n_3 be the direction ratios of the normal to the plane. The equation of the plane passing through $P(x_1, y_1, z_1)$ and perpendicular to the line with direction ratios n_1, n_2, n_3 is

$$n_1(x-x_1) + n_2(y-y_1) + n_3(z-z_1) = 0 \quad (1)$$

The plane also passes through $Q(x_2, y_2, z_2)$ and $R(x_3, y_2, z_3)$ Therefore,

$$n_1(x_2-x_1) + n_2(y_2-y_1) + n_3(z_2-z_1) = 0 \quad (2)$$

and

$$n_1(x_3-x_1) + n_2(y_3-y_1) + n_3(z_3-z_1) = 0. \quad (3)$$

Eliminating n_1, n_2, n_3 from (1), (2), and (3), we get

$$\begin{vmatrix} x-x_1 & y-y_1 & z-z_1 \\ x_2-x_1 & y_2-y_1 & z_2-z_1 \\ x_3-x_1 & y_3-y_1 & z_3-z_1 \end{vmatrix} = 0, \quad (4)$$

which gives the required equation of the plane.

Second Method: We want to find the equation of the plane passing through three points $P(x_1, y_1, z_1)$, $Q(x_2, y_2, z_2)$ and $R(x_3, y_3, z_3)$.

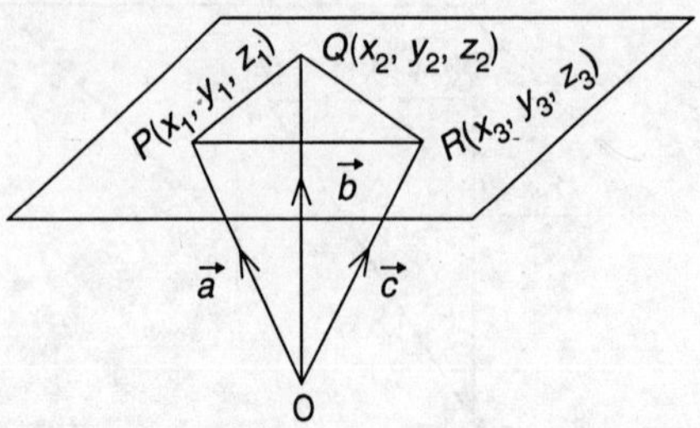

The vector $\vec{PQ}$ and $\vec{PR}$ lie in a given plane. Therefore, the cross product $\vec{PQ} \times \vec{PR}$ is in the direction of the normal to the plane. Let $\vec{a}, \vec{b}, \vec{c}$ be the position vectors of P, Q, and R respectively. The equation of the plane passing through $P(x_1, y_1, z_1)$ and normal to the vector $\vec{PQ} \times \vec{PR}$ is $(\vec{r}-\vec{a}).\left(\vec{PQ} \times \vec{PR}\right)=0$

or $\quad (\vec{r}-\vec{a}).\left[\left(\vec{b}-\vec{a}\right)\times\left(\vec{c}-\vec{a}\right)\right]=0 \qquad (5)$

or $\quad (\vec{r}-\vec{a}).\left[\vec{a}\times\vec{b}+\vec{b}\times\vec{c}+\vec{c}\times\vec{a}\right]=0. \qquad (6)$

14.11 EQUATION OF A PLANE PASSING THROUGH A POINT AND PARALLEL TO TWO GIVEN VECTORS

Let the plane pass through the point A with a position vector $\vec{a}$ and let the plane be parallel to the vectors $\vec{b}$ and $\vec{c}$. Now $\vec{b}\times\vec{c}$ is in the direction of the normal to the plane. Therefore, the equation of the plane passing through A and parallel to $\vec{b}$ and $\vec{c}$ is

$(\vec{r}-\vec{a}).\left(\vec{b}\times\vec{c}\right)=0$ (Scalar Product form). (1)

Second Method: Any vector parallel to the vectors $\vec{b}$ and $\vec{c}$ is $\lambda\vec{b}+\mu\vec{c}$. Since $\vec{r}-\vec{a}$ is the vector in the plane and hence, is parallel to $\vec{b}$ and $\vec{c}$. Therefore,

$$\vec{r}-\vec{a}=\lambda\vec{b}+\mu\vec{c} \qquad (2)$$

is the required equation of the plane.

14.12 EQUATION OF A PLANE PASSING THROUGH TWO POINT AND PARALLEL TO A LINE

Let A and B be two points on the plane with position vectors $\vec{a}$ and $\vec{b}$, respectively. If $\vec{c}$ is the direction of the given line and P is any point on the line with the position vector $\vec{P}$. Then the given line is $\vec{r}=\vec{p}+\lambda\vec{c}$.

The vector $\vec{AB}=\vec{b}-\vec{a}$ lies in the given plane and hence is parallel to $\vec{c}$.

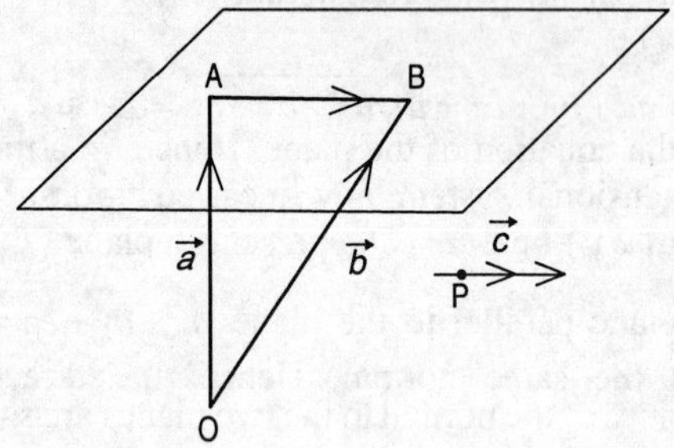

Therefore, the normal to the required plane is given by $(\vec{b}-\vec{a})\times\vec{c}$. Therefore, the equation of the plane passing through A and perpendicular to $(\vec{b}-\vec{a})\times\vec{c}$ is

$(\vec{r}-\vec{a}).\left[(\vec{b}-\vec{a})\times\vec{c}\right]=0$

or $\quad \left[\vec{r}-\vec{a} \quad \vec{b}-\vec{a} \quad \vec{c}\right]=0$

EXAMPLE 14.32

Find the direction cosines of the normal to the plane. What is the distance of the plane from the origin?

Solution. The equation of the plane is

$$3x+4y+5z=5$$

Dividing throughout by $\sqrt{3^2+4^2+5^2}=\sqrt{50}$, we get

$$\frac{3}{\sqrt{50}}x+\frac{4}{\sqrt{50}}y+\frac{5}{\sqrt{50}}z=\frac{5}{\sqrt{50}}$$

Comparing it with the normal form $lx+my+nz = d$, we get

$$l=\frac{3}{\sqrt{50}}, m=\frac{4}{\sqrt{50}}, \text{ and } n=\frac{5}{\sqrt{50}}$$

(direction cosines of the normal)

and

$$d=\frac{5}{\sqrt{50}}=\frac{1}{\sqrt{2}} \text{ (the distance from the origin).}$$

EXAMPLE 14.33

Find the equation of a plane which is at a distance of 7 units from the origin and is perpendicular to $4\hat{i}+2\hat{j}-3\hat{k}$, directed away from the origin.

Solution. Let $\hat{n}$ be the unit vector in the direction of the given vector $4\hat{i}+2\hat{j}-3\hat{k}$, directed away from the origin. Then

$$\hat{n}=\frac{4\hat{i}+2\hat{j}-3\hat{k}}{\sqrt{4^2+2^2+3^2}}=\frac{4\hat{i}}{\sqrt{29}}+\frac{2\hat{i}}{\sqrt{29}}-\frac{3\hat{k}}{\sqrt{29}}.$$

If $\vec{r}=x\hat{i}+y\hat{j}+z\hat{k}$ is the position vector of a general point (x, y, z) on the plane, then the equation of the plane is

$$\left(x\hat{i}+y\hat{j}-z\hat{k}\right).\left(\frac{4}{\sqrt{29}}\hat{i}+\frac{2}{29}\hat{j}-\frac{3}{\sqrt{29}}\hat{k}\right)=7$$

or

$$4x+2y-3z=7\sqrt{29}.$$

EXAMPLE 14.34

Find the equation of the plane passing through the point (1, 2, 3) and having the vector $\vec{N}=2\hat{i}-\hat{j}+3\hat{k}$ normal to it.

Solution. We know that if n_1, n_2, n_3 are direction ratios of the normal to the plane, then the equation of the plane passing through (a_1, a_2, a_3) is

$$(x-a_1)\,n_1+(y-a_2)n_2+(z-a_3)n_3=0.$$

In the present case, we have (a_1, a_2, a_3) = (1, 2, 3) and $n_1 = 2$, $n_2 = -1$, $n_3 = 3$. Therefore, the equation of the plane is

$$2(x-1)-(y-2)+3(z-3)=0$$

or

$$2x-y+3z=9.$$

EXAMPLE 14.35

Find the equation of the plane passing through the point (3, –3,1) and perpendicular (normal) to the line joining the points (3,4,–1) and (2, –1,5).

Solution. The direction of the line is

$$\vec{N}=\left(2\hat{i}-\hat{j}+5\hat{k}\right)-\left(3\hat{i}+4\hat{j}-\hat{k}\right)$$

$$=-\hat{i}-5\hat{j}+6\hat{k}.$$

We know that the equation of the plane passing through the point with the position vector $\vec{a}$ and perpendicular to the direction N is

$$(\vec{r}-\vec{a}).\vec{N}=0.$$

Therefore, the required equation is

$$\left[\left(x\hat{i}+y\hat{j}+z\hat{k}\right)-\left(3\hat{i}-3\hat{j}+\hat{k}\right)\right].\left(-\hat{i}-5\hat{j}+6\hat{k}\right)=0$$

or

$$\left[(x-3)\hat{i}+(y+3)\hat{j}+(z-1)\hat{k}\right].\left(-\hat{i}-5\hat{j}+6\hat{k}\right)=0$$

or

$$(x-3)(-1)+(y+3)(-5)+(z-1)(6)=0$$

or

$$x+5y-6z+18=0.$$

Second Method: Equation of the line joining (3,4,–1) and (2, –1, 5) is

$$\frac{x-3}{2-3}=\frac{y-2}{-1-4}=\frac{z+1}{5+1}$$

or

$$\frac{x-3}{-1}=\frac{y-2}{-5}=\frac{z+1}{6}.$$

The direction ratios of the normal to the above line are –1, –5, and 6. Therefore, the equation of the plane passing through (3, –3, 1) is

$$n_1(x-3)+n_2(y+3)+n_3(z-1)=0,$$

that is,

$$-1(x-3)-5(y+3)+6(z-1)=0 \text{ or}$$
$$x+5y-6z+18=0.$$

EXAMPLE 14.36

Find the equation of the plane which passes through the point (3, –3, 1) and is normal to the line joining the points (3, 2, –1) and (2, –1, 5).

Solution. As in Example 14.35, the required equation is

$$\left[\left(x\hat{i}+y\hat{j}+z\hat{k}\right)-\left(3\hat{i}-3\hat{j}+\hat{k}\right)\right].\left(-\hat{i}-3\hat{j}+6\hat{k}\right)=0$$

or

$$\left[(x-3)\hat{i}+(y+3)\hat{j}+(z-1)\hat{k}\right].\left(-\hat{i}-3\hat{j}+6\hat{k}\right)=0$$

or

$$(x-3)(-1)+(y+3)(-3)+(z-1)(6)=0$$

or

$$x+3y-6z+12=0.$$

EXAMPLE 14.37

Find the equation of the plane passing through the point (1, 2, 3) and parallel to the plane $4x+5y-3z=7$.

Solution. The equation of the plane parallel to the plane $4x+5y-3z=7$ is

$$4x+5y-3z+k=0.$$

Since it passes through (1,2,3), we have

$$4 + 10 - 9 + k = 0 \quad \text{or} \quad k = -5$$

Hence, the equation of the required plane is

$$4x + 5y - 3z = 5$$

EXAMPLE 14.38

Find the equation of the plane passing through the point (3, 4, –1), which is parallel to the plane

$$\vec{r}.\left(2\hat{i} - 3\hat{j} + 5\hat{k}\right) + 7 = 0.$$

Solution. The equation of the given plane is

$$\left(x\hat{i} + y\hat{j} + z\hat{h}\right).\left(2\hat{i} - 3\hat{j} + 5\hat{k}\right) + 7 = 0.$$

or

$$2x - 3y + 5z = 0.$$

The equation of the plane parallel to this plane is

$$2x - 3y + 5z + k = 0. \qquad (1)$$

Since the plane (1) passes through (3,4,–1), we have

$$6 - 12 - 5 + k = 0 \quad \text{or} \quad k = 11.$$

Hence, the equation of the required plane is

$$2x - 3y + 5z + 11 = 0.$$

EXAMPLE 14.39

The foot of the perpendicular from the origin to a plane is (4,–2, –5). Find the equation of the plane.

Solution. The position vector of the foot of the perpendicular A drawn from the origin to the plane is

$$\vec{a} = 4\hat{i} - 2\hat{j} - 5\hat{k}.$$

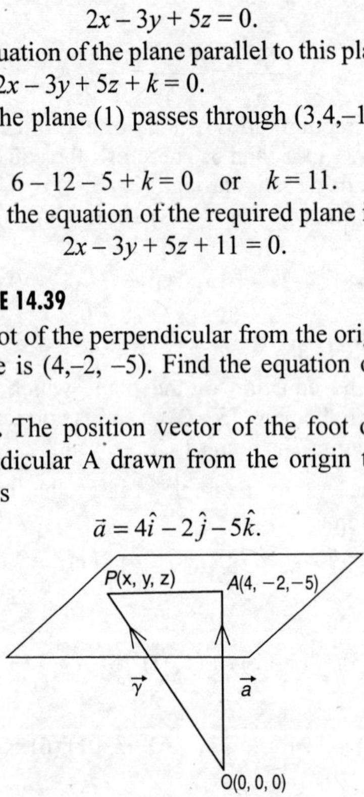

Also

$$\vec{OA} = \vec{N} = 4\hat{i} - 2\hat{j} - 5\hat{k}.$$

Thus, the plane passes through A and is perpendicular to the direction $\vec{OA}$. Therefore, its equation is $(\vec{r} - \vec{a}).\vec{N} = 0$

or

$$\left[\left(x\hat{i} + y\hat{j} + z\hat{k}\right) - \left(4\hat{i} - 2\hat{j} - 5\hat{k}\right)\right].\left(4\hat{i} - 2\hat{j} - 5\hat{k}\right) = 0$$

or

$$\left[(x-4)\hat{i} + (y+2)\hat{j} + (z+5)\hat{k}\right].\left(4\hat{i} - 2\hat{j} - 5\hat{k}\right) = 0$$

or

$$4(x - 4) - 2(y + 2) - 5(z + 5) = 0$$

or

$$4x - 2y - 5z = 45.$$

Second Method: The distance $OA = d = \sqrt{(4-0)^2 + (-2-0)^2 + (5-0)^2} = \sqrt{16+4+25} = \sqrt{45}.$

The unit normal vector along $\vec{OA}$ is

$$\hat{n} = \frac{4\hat{i} - 2\hat{j} - 5\hat{k}}{\sqrt{16+4+25}} = \frac{4\hat{i} - 2\hat{j} - 5\hat{k}}{\sqrt{45}}.$$

Therefore, the equation of the plane, in the normal form, is $\vec{r}.\hat{n} = d$ or

$$\left(x\hat{i} + y\hat{j} + z\hat{k}\right).\left(\frac{4\hat{i} - 2\hat{j} - 5\hat{k}}{\sqrt{45}}\right) = \sqrt{45}$$

or

$$4x - 2y - 5z = \sqrt{45}.\sqrt{45} = 45.$$

Third Method: Equation of the plane passing through (4,–2,–5) is

$$n_1(x - 4) + n_2(y + 2) + n_3(z + 5) = 0.$$

The direction ratios of the normal are 4,–2, and –5. Therefore, the equation of the required plane is

$$4(x - 4) - 2(y + 2) - 5(z + 5) = 0$$

or $\quad 4x - 2y - 5z = 45.$

EXAMPLE 14.40

The position vectors of two points A and B are $3\hat{i} + \hat{j} + 2\hat{k}$ and $\hat{i} - 2\hat{j} - 4\hat{k}$, respectively. Find the equation of the plane passing through B and perpendicular to AB.

Solution.

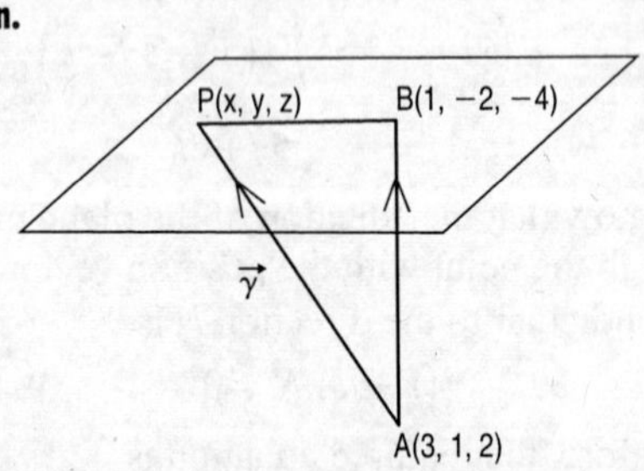

Let $\vec{r} = x\hat{i} + y\hat{j} + z\hat{k}$. Then

$$\vec{AB} = \vec{N} = \left(\hat{i} - 2\hat{j} - 4\hat{k}\right) - \left(3\hat{i} + \hat{j} + 2\hat{k}\right)$$
$$= -2\hat{i} - 3\hat{j} - 6\hat{k}.$$

The plane is to pass through B, whose position vector is $\vec{a} = \hat{i} - 2\hat{j} - 4\hat{k}$. Therefore, the required equation is $(\vec{r} - \vec{a}).\vec{N} = 0$

or

$$\left[\left(x\hat{i} + y\hat{j} + z\hat{k}\right) - \left(\hat{i} - 2\hat{j} - 4\hat{k}\right)\right].\left(-2\hat{i} - 3\hat{j} - 6\hat{k}\right) = 0$$

or

$$\left[(x-1)\hat{i} + (y+2)\hat{j} + (z+4)\hat{k}\right].\left(-2i - 3\hat{j} - 6\hat{k}\right) = 0$$

or

$$-2(x-1) - 3(y+2) - 6(z+4) = 0$$

or

$$2x + 3y + 6z + 28 = 0.$$

In Vector form, we can write

$$\vec{r}.\left(2\hat{i} + 3\hat{j} + 6\hat{k}\right) = -28.$$

EXAMPLE 14.41

Find the equation of the plane passing through the points (–2, 6, –6), (–3, 10, –9), and (–5, 0, –6).

Solution. The equation of the plane passing through (–2, 6, –6) is

$$a(x+2) + b(y-6) + c(z+6) = 0. \qquad (1)$$

Since (–3, 10, –9) and (–5, 0, –6) lie on it, we have

$$a(-3+2) + b(10-6) + c(-9+6) = 0$$

and

$$a(-5+2) + b(0-6) + c(-6+6) = 0,$$

that is

$$-a + 4b - 3c = 0. \qquad (2)$$

and

$$-3a - 6b + 0c = 0. \qquad (3)$$

From (2) and (3), we have

$$\frac{a}{-18} = \frac{b}{9} = \frac{c}{18} \text{ or } \frac{a}{-2} = \frac{b}{1} = \frac{c}{2}.$$

Hence, the equation of the plane (1) becomes

$$-2(x+2) + (y-6) + 2(z+6) = 0$$

or

$$2x - y - 2z - 2 = 0.$$

EXAMPLE 14.42

Show that the four points, (0, –1, 0), (2, 1, –1), (1, 1, 1), and (3, 3, 0) are coplanar. Find the equation of the plane passing through them.

Solution. The equation of the plane passing through the first three points is

$$\begin{vmatrix} x-0 & y+1 & z-0 \\ 2-0 & 1+1 & -1-0 \\ 1-0 & 1+1 & 1-0 \end{vmatrix} = 0 \text{ or } \begin{vmatrix} x & y+1 & z \\ 2 & 2 & -1 \\ 1 & 2 & 1 \end{vmatrix} = 0$$

or

$$x(2+2) + (y+1)(-1-2) + z(4-2) = 0$$

or

$$4x - 3y + 2z = 3. \qquad (1)$$

We note that (1) is satisfied by the fourth point (3, 3, 0) also. Hence, all the four points lie on the plane $4x - 3y + 2z = 3$ and so, are coplanar.

EXAMPLE 14.43

Find the equation of the plane which meets the coordinate axes at P, Q and R respectively and if the centroid of the triangle PQR is the point (a,b,c).

Solution. Suppose that the points P, Q and R are respectively

$$\text{P}(\text{A},0,0), \text{Q}(0,\text{B},0) \text{ and } \text{R}(0,0,\text{C}).$$

Since the centroid of the triangle PQR is (a,b,c) we have

$$a = \frac{A+0+0}{3} = \frac{A}{3},\ b = \frac{0+B+0}{3} = \frac{B}{3}, \text{ and}$$
$$c = \frac{0+0+C}{3} = \frac{C}{3}.$$

Hence $A = 3a$, $B = 3b$, and $C = 3c$. Also, the equation of the plane, in the intercept form, is

$$\frac{x}{A} + \frac{y}{B} + \frac{z}{C} = 1$$

Putting the values of A, B and C, we get

$$\frac{x}{3a} + \frac{y}{3b} + \frac{z}{3c} = 1 \text{ or } \frac{x}{a} + \frac{y}{b} + \frac{z}{c} = 3.$$

EXAMPLE 14.44

Find the image of the point $P(1,3,4)$ in the plane $2x - y + z + 3 = 0$.

Solution. Let $Q(x_1, y_1, z_1)$ be the image of the point $P(1,3,4)$ in the plane $2x - y + z + 3 = 0$. Then the direction ratios of PQ are

$$x_1 - 1,\ y_1 - 3, \text{ and } z_1 - 4.$$

Since PQ must be perpendicular to the given plane, we have

$$\frac{x_1 - 1}{2} = \frac{y_1 - 3}{-1} = \frac{z_1 - 4}{1} = \lambda, \text{ say}$$

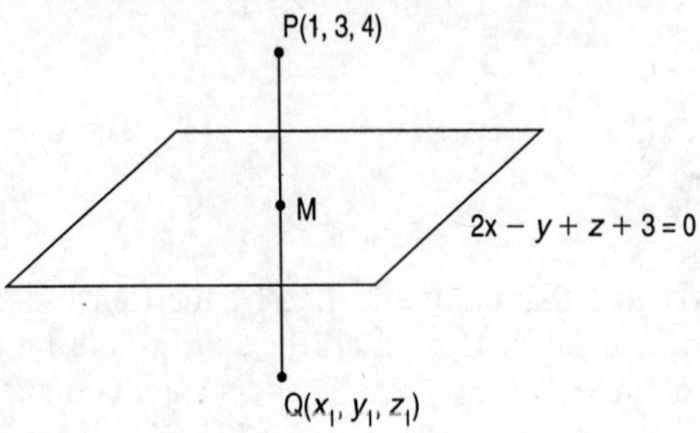

Thus, Q is $(2\lambda + 1, -\lambda + 3, \lambda + 4)$. Now the midpoint M of PQ is

$$\frac{2\lambda+1+1}{2}, \frac{-\lambda+3+3}{2}, \frac{\lambda+4+4}{2}.$$

But $M\left(\lambda+1, -\frac{\lambda}{2}+3, \frac{\lambda}{2}+4\right)$ lies on the given plane. Therefore,

$$2(\lambda+1) - \left(-\frac{\lambda}{2}+3\right) + \left(\frac{\lambda}{2}+4\right) + 3 = 0.$$

This yields $\lambda = -2$. Hence, the point Q is

$$(2\lambda+1, -\lambda+3, \lambda+4) = (-3, 5, 2).$$

EXAMPLE 14.45

Show that the line $\frac{x-2}{-1} = \frac{y-1}{6} = \frac{z-3}{-6}$ lies in the plane that passes through the points $A(1, -2, 3)$, $B(1, 1, 1)$, and $C(0, 1, -1)$.

Solution. The equation of the plane passing through the three given points is

$$\begin{vmatrix} x-1 & y+2 & z-3 \\ 0 & 3 & -2 \\ -1 & 3 & -4 \end{vmatrix} = 0$$

or

$$-6x + 2y + 3z + 1 = 0 \qquad (1)$$

Any point on the given line is $(-\lambda+2, 6\lambda+1, -6\lambda+3)$. This point clearly satisfies the equation (1). Hence, the given line lies in the plane $-6x + 2y + 3z + 1 = 0$.

14.13 ANGLE BETWEEN TWO PLANES

Let the equation of the two given planes be $a_1x + b_1y + c_1z = D_1$ and $a_2x + b_2y + c_2z = D_2$ Therefore, the direction ratios of the normal to these planes are respectively a_1, b_1, c_1, and a_2, b_2, c_2. The normals are $\vec{N}_1 = a_1\hat{i} + b_1\hat{j} + c_1\hat{k}$ and $\vec{N}_2 = a_2\hat{i} + b_2\hat{j} + c_2\hat{k}$. Therefore, the angle between the planes is the angle θ between these normals and so

$$\cos\theta = \frac{\vec{N}_1 . \vec{N}_2}{|\vec{N}_1 . \vec{N}_2|} = \frac{a_1a_2 + b_1b_2 + c_1c_2}{\sqrt{a_1^2 + b_1^2 + c_1^2}\sqrt{a_2^2 + b_2^2 + c_2^2}}$$

The two planes will be *perpendicular to each other* if

$$\frac{a_1a_2 + b_1b_2 + c_1 + c_2}{\sqrt{a_1^2 + b_1^2 + c_1^2}\sqrt{a_2^2 + b_2^2 + c_2^2}} = 0$$

or if

$$a_1a_2 + b_1b_2 + c_1c_2 = 0 \qquad (1)$$

The two planes will be *parallel to each other* if the vectors $a_1\hat{i} + b_1\hat{j} + c_1\hat{k}$ and $a_2\hat{i} + b_2\hat{j} + c_2\hat{k}$ are parallel. Therefore, for some scalar λ, we have

$$a_2 = \lambda a_1, b_2 = \lambda b_1, \text{ and } c_2 = \lambda c_1$$

or

$$\frac{a_2}{a_1} = \frac{b_2}{b_1} = \frac{c_2}{c_1} = \lambda$$

Therefore, the planes $a_1x + b_1y + c_1z = D_1$ and $a_2x + b_2y + c_2z = D_2$ are parallel if

$$\frac{a_2}{a_1} = \frac{b_2}{b_1} = \frac{c_2}{c_1}.$$

If we put $a_2 = \lambda a_1, b_2 = \lambda b_1$, and $c_2 = \lambda c_1$ in $a_2x + b_2y + c_2z = D_2$, we get

$$\lambda(a_1x + b_1y + c_1z) = D_2$$

or

$$a_1x + b_1y + c_1z = \frac{D^2}{\lambda} = k, \text{ say.}$$

Thus, the equation of any plane parallel to a given plane $x + by + cz = D$ is $ax + by + cz = k$.

EXAMPLE 14.46

Find the equation of the plane through the points (2, 2, 1) and (9, 3, 6) and perpendicular to the plane $2x+6y+6z=9$.

Solution. The equation of the plane passing through (2, 1, 0) is

$$a(x-2)+b(y-2)+c(z-1)=0 \quad (1)$$

It passes through the point (9, 3, 6) if

$$a(9-2)+b(3-2)+c(6-1)=0$$

or if

$$7a+b+5c=0 \quad (2)$$

Also (1) is perpendicular to the plane $2x+6y+6z=9$ if

$$2a+6b+6c=0 \quad (3)$$

Eliminating a,b and c from (1), (2), and (3), we get

$$\begin{vmatrix} x-2 & y-2 & z-1 \\ 7 & 1 & 5 \\ 2 & 6 & 6 \end{vmatrix}=0$$

or

$$(x-2)(6-30)+(y-2)(10-42)+(z-1)(42-2)=0$$

or

$$-24(x-2)-32(y-2)+40(z-1)=0$$

or

$$3x+4y-5z=9,$$

which is the equation of the required plane.

EXAMPLE 14.47

Find the equation of the plane through the point $(2,1,0)$ and perpendicular to the planes $2x-y-z=5$ and $x+2y-z=5$.

Solution. The equation of the plane passing through (2, 1, 0) is

$$a(x-2)+b(y-1)+c(z-0)=0. \quad (1)$$

The plane (1) will be perpendicular to the given planes if

$$2a-b-c=0 \quad (2)$$

and

$$a+2b-3c=0. \quad (3)$$

Eliminating a, b and c from (1), (2), and (3), we get

$$\begin{vmatrix} x-2 & y-1 & z \\ 2 & -1 & -1 \\ 2 & 2 & -3 \end{vmatrix}=0$$

or

$$(x-2)(3+2)+(y-1)(-1+6)+z(4+1)=0$$

or

$$5(x-2)+5(y-1)+5z=0$$

or

$$5x+5y+5z=15 \text{ or } x+y+z=3$$

EXAMPLE 14.48

Find the equation of the plane passing through the point (2, 3, 4) and parallel to the plane

$$5x-6y+7z=3.$$

Solution. The equation of the plane parallel to the given plane is $5x-6y+7z=k$.

If the plane passes through the point (2, 3, 4), we have

$$5(2)-6(3)+7(4)=k \text{ or } k=20$$

Hence, the required plane is

$$5x-6y+7z=20$$

EXAMPLE 14.49

Find the equation of the plane passing through the point (3, 4, –1) which is parallel to the plane

$$\vec{r}.\left(2\hat{i}-3\hat{j}+5\hat{k}\right)+7=0.$$

Solution. The given plane is

$$\left(x\hat{i}+y\hat{j}+z\hat{k}\right).\left(2\hat{i}-3\hat{j}+5\hat{k}\right)+7=0$$

or

$$2x - 3y + 5z + 7 = 0.$$

The plane parallel to the above plane is

$$2x - 3y + 5z = k.$$

It passes through (3, 4, –1), if

$$6 - 12 - 5 = k \text{ or } k = -11.$$

Hence, the equation of the required plane is

$$2x - 3y + 5z + 11 = 0$$

or

$$\left(\pm\pm\pm\hat{\pm}+\ \hat{}\right).\left(2\hat{}-3\hat{}+5\hat{}\right)+11=0$$

or

$$\vec{r}.\left(2\hat{i}-3\hat{j}+5\hat{k}\right)+11=0.$$

EXAMPLE 14.50

Find the angle between the planes

$$\vec{r}.\left(\hat{i}+\hat{j}\right)=1 \text{ and } \vec{r}.\left(\hat{i}+\hat{k}\right)=3.$$

Solution. Let

$$\vec{N}_1=\hat{i}+\hat{j} \textit{ and } \vec{N}_2=\hat{i}+\hat{k}.$$

Then, the angle θ between the given planes is given by

$$\cos\theta=\frac{\vec{N}_1.\vec{N}_2}{\left|\vec{N}_1.\vec{N}_2\right|}=\frac{\left(\hat{i}+\hat{j}\right).\left(\hat{i}+\hat{k}\right)}{\sqrt{1+1}\sqrt{1+1}}=\frac{1}{2}.$$

Hence, $\theta=60^{\circ}$.

14.14 ANGLE BETWEEN A LINE AND A PLANE

Let $\vec{r}=\vec{a}+\lambda\vec{b}$ be the given line and $\vec{r}.\vec{N}=d$ be the given plane, where $\vec{N}$ is normal to the plane. Let the line meet the plane at P.

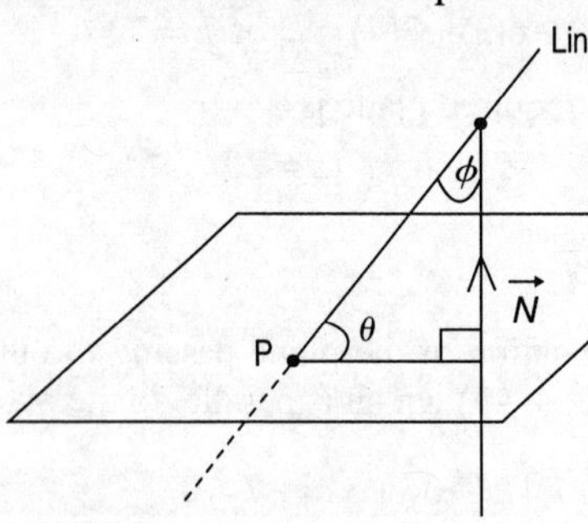

Let ϕ be the angle between the normal $\vec{N}$ and the line. Then

$$\cos\phi=\frac{\vec{b}.\vec{N}}{\left|\vec{b}.\vec{N}\right|}. \tag{1}$$

As we know, the angle between a line and a plane is defined as the *angle between the line and the projection of the line on the* plane. Let this angle be θ. Then, we note that $\phi=\frac{\pi}{2}-\theta$. Therefore, (1) yields

$$\cos\phi=\cos\left(\frac{\pi}{2}-\theta\right)=\sin\theta=\frac{\vec{b}.\vec{N}}{\left|\vec{b}.\vec{N}\right|}$$

or

$$\theta=\sin^{-1}\frac{\vec{b}.\vec{N}}{\left|\vec{b}.\vec{N}\right|} \text{ (Vector form).}$$

Cartesian Form.

Putting $\vec{r}=x\hat{i}+y\hat{j}+z\hat{k},\ \vec{a}=x_1\hat{i}+x_2\hat{j}+x_3\hat{k},$ and $\vec{b}=b_1\hat{i}+b_2\hat{j}+b_3\hat{k},$ the line becomes

$$\frac{x-x_1}{b_1}=\frac{y-y_1}{b_2}=\frac{z-z_2}{b_3}=\lambda.$$

On the other hand, the equation of the plane is

$$n_1x+n_2y+n_3z=d.$$

Thus, the direction ratios of the line are b_1, b_2, b_3 and the direction ratios of the normal to the plane are n_1, n_2, n_3. Thus, the angle ϕ between the line and the normal to the plane is given by

$$\cos\phi=\cos(90-\theta)=\sin\theta$$

$$=\frac{n_1b_1+n_2b_2+n_3b_3}{\sqrt{n_1^2+n_2^2+n_3^2}\sqrt{b_1^2+b_2^2+b_3^2}}$$

or

$$\theta=\sin^{-1}\frac{n_1b_1+n_2b_2+n_3b_3}{\sqrt{n_1^2+n_2^2+n_3^2}\sqrt{b_1^2+b_2^2+b_3^2}},$$

where θ is the angle between the given line and the plane.

The line will be parallel to the plane if and only if $\theta=0$, that is, if and only if $n_1b_1+n_2b_2+n_3b_3=0$.

14.15 PERPENDICULAR DISTANCE OF A POINT FROM A PLANE

Let $\vec{r}.\hat{n}=d$ be the given plane and let $A\left(x_1,y_1,z_1\right)$ be the point with the position vector $\vec{a}$. We have OB = d.

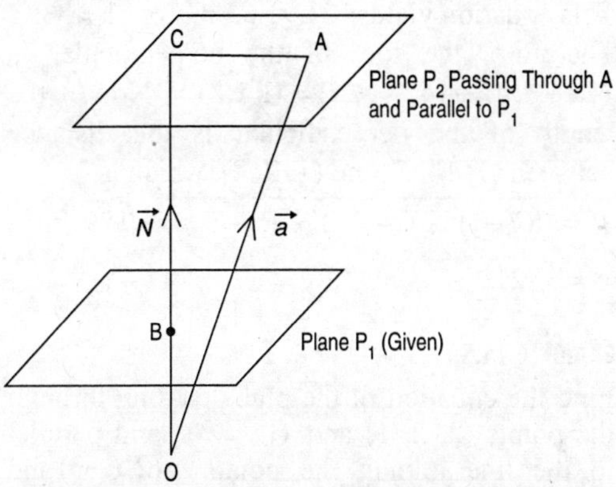

Let P_2 be a plane parallel to the given plane P_1 and passing through the point A. Extend the normal OB to meet the new plane P_2 at C. Then OC is the perpendicular distance from the origin to the plane P_2. Hence,

Distance BC between the planes = OC – OB. (1)

The equation of the plane P_2 passing through A and perpendicular to $\vec{N}$ is

$$(\vec{r}-\vec{a}).\hat{n}=0 \text{ or } \vec{r}.\hat{n}=.\vec{a}.\hat{n}, \qquad (2)$$

where $\hat{n}$ is the unit vector along $\vec{N}$. Thus, $OC=\vec{a}.\hat{n}$. Hence,

Perpendicular distance of A from Plane $P_1 = OC - OB = \vec{a}.\hat{n} - d$.

Since distance is always non-negative, the required perpendicular distance is $|\vec{a}.\hat{n}-d|$.

If the equation of the given plane is of the form, where $\vec{N}$ is normal to the plane, then the Perpendicular distance of A from

$$P_1 = \frac{|\vec{a}.\vec{N}-d|}{|\vec{N}|} \text{ (Vector form).}$$

To obtain the *Cartesian form*, let $ax + by + cz = D$ be the given plane and $A(x_1, y_1, z_1)$ be the given point. Then $\vec{N} = a\hat{i} + b\hat{j} + c\hat{k}$ and so Perpendicular distance of A form $P1$

$$= \left|\frac{(x_1\hat{i}+y_1\hat{j}+z_1\hat{k}).(a\hat{i}+b\hat{j}+z\hat{k})-D}{\sqrt{a^2+b^2+c^2}}\right|$$

$$= \left|\frac{ax_1+by_1+cz_1-D}{\sqrt{a^2+b^2+c^2}}\right|.$$

14.16 PLANES BISECTING THE ANGLES BETWEEN TWO PLANES

Let

$$a_1x+b_1y+c_1z=D_1 \qquad (1)$$

and

$$a_2x+b_2y+c_2z=D_2 \qquad (2)$$

be the given planes. Let P(*x*, *y*, *z*) be any point on either of the two planes bisecting the angle between the given planes (1) and (2). Then Perpendicular distance of P from (1) = perpendicular distance of P from (2) and so,

$$\frac{a_1x+b_1y+c_1z-D_1}{\sqrt{a_1^2+b_1^2+c_1^2}} = \pm\frac{a_2x+b_2y+c_2z-D_2}{\sqrt{a_2^2+b_2^2+c_2^2}},$$

which are the equations of the two bisector planes.

EXAMPLE 14.51

Find the angle between the line

$$\frac{x-2}{3}=\frac{y+1}{-1}=\frac{z-3}{2}$$

and the plane $3x + 4y + z + 5 = 0$.

Solution. The direction ratios of the given line are 3, –1, and 2, whereas the direction ratios of the given plane are 3, 4, and 1. Hence, the angle θ between the given line and the given plane is given by

$$\sin\theta = \frac{3(3)+(-1)(4)+2(1)}{\sqrt{3^2+(-1)^2+2^2}\sqrt{3^2+4^2+1^2}}$$

$$= \frac{7}{\sqrt{14}\sqrt{26}} = \frac{\sqrt{7}}{2\sqrt{13}}.$$

Hence, $\theta = \sin^{-1}\left(\frac{\sqrt{7}}{2\sqrt{13}}\right)$.

EXAMPLE 14.52

Find the distance between the parallel planes $\vec{r}.(\hat{i}+2\hat{j}-3\hat{k})=6$ and $\vec{r}.(3\hat{i}+6\hat{j}-9\hat{k})=-15$.

Solution. The standard forms of the given planes are

$$\frac{\vec{r}.(\hat{i}+2\hat{j}-3\hat{k})}{\sqrt{1^2+2^2+(-3)^2}} = \frac{6}{1^2+2^2+(-3)2} \text{ and}$$

$$\frac{\vec{r}.(3\hat{i}+6\hat{j}-9\hat{k})}{\sqrt{3^2+6^2+(-9)^2}}=\frac{-15}{3^2+6^2+(-9)2}$$

or

$$\vec{r}.\hat{n}=\frac{6}{\sqrt{14}} \text{ and } \vec{r}.\hat{n}=\frac{-15}{3\sqrt{14}}=\frac{-5}{\sqrt{14}}$$

The distances of the planes from the origin are $\frac{6}{\sqrt{14}}$ and $\frac{-5}{\sqrt{14}}$, which are the opposite sides of the origin. Hence, the distance between the planes is $|d_1|+|d_2|=\frac{6}{\sqrt{14}}+\frac{5}{\sqrt{14}}=\frac{11}{\sqrt{14}}$.

EXAMPLE 14.53

Find the distance between the parallel planes $2x-2y+z+3=0$ and $4x-4y+2z+5=0$.

Solution. The distance between the parallel planes is equal to the distance of any point lying on one plane from the second plane. We note that the point $(0,0,-3)$ satisfies the equation of the first plane. Thus, the required distance is the distance of $(0,0,-3)$ from the plane $4x-4y+2z+5=0$ Thus,

$$\text{Required distance }=\left|\frac{4x_1-4y_1+2z_1+5}{\sqrt{16+16+4}}\right|$$

$$=\left|\frac{4(0)-4(0)+2(-3)+5}{6}\right|$$

$$=\left|\frac{-1}{6}\right|=\frac{1}{6}$$

EXAMPLE 14.54

Find the length and foot of the perpendicular from the point (7, 14, 5) to the plane $2x+4y-z=2$.

Solution. Since the line passing through (7, 14, 5) is perpendicular to the plane $2x+4y-z=2$, its direction ratios are the direction ratios of the normal to the plane. Thus, the direction ratios of the perpendicular are 2, 4, and -1. So, the equation of the perpendicular is

$$\frac{x-7}{2}=\frac{y-14}{4}=\frac{z-5}{-1}$$

Any point on this line is

$$(2\lambda+7,\ 4\lambda+14,\ -\lambda+5).$$

Since the foot of the perpendicular lie on the plane, we have

$$2(2\lambda+7)+4(4\lambda+14)-(-\lambda+5)=2$$

This equation yields $21\lambda+63=0$ or $\lambda=-3$. Therefore, the foot of the perpendicular is $(2\lambda+7,4\lambda+14,-\lambda+5)=(1,2,8)$. Now, the length of the perpendicular is the distance between (7, 14, 5) and (1, 2, 8), which is

$$p=\sqrt{(7-1)^2+(14-2)^2+(5-8)^2}=\sqrt{189}$$

$$=3\sqrt{21}.$$

EXAMPLE 14.55

Find the equation of the plane passing through the points (2, 2, 1) and $(1,-2,3)$, and parallel to the line joining the points $(2,1,-3)$ and $(-1,5,-8)$.

Solution. The equation of the plane passing through the point (2, 2, 1) is

$$a(x-2)+b(y-2)+c(z-1)=0 \quad (1)$$

It passes through $(1,-2,3)$ and so,

$$a(1-2)+b(-2-2)+c(3-1)=0$$

or

$$a+4b-2c=0 \quad (2)$$

The direction ratios of the line joining $(2,1,-3)$ and $(-1,5,-8)$ are $3,-4,5$. The plane (1) will be parallel to the line joining $(2,1,-3)$ and $(-1,5,-8)$ if

$$3a-4b+5c=0 \quad (3)$$

From (2) and (3), we have

$$\frac{a}{20-8}=\frac{b}{-6-5}=\frac{c}{-16}.$$

Hence, the equation of the required plane is

$$12(x-2)-11(y-2)-16(z-1)=0$$

or

$$12x-11y-16z+14=0.$$

EXAMPLE 14.56

Find the equations of the planes bisecting the angle between the planes $6x+3y+2z=5$ and $4x+3y+12z+3=0$. Specify the plane which bisects the acute angle.

Solution. The equations of the bisector planes are

$$\frac{6x+3y+2z-5}{\sqrt{36+9+4}}=\pm\frac{4x+3y+12z+3}{\sqrt{16+9+144}}$$

or

$$13(6x+3y+2z-5) = \pm 7(4x+3y+12z+3)$$

which yields the equation of the bisectors as

$$25x+9y-29z-43=0 \text{ and}$$

$$53x+30y+55z-22=0.$$

Further, the angle between the plane $6x+3y+2z=5$ and $25x+9y-29z-43=0$ is given by

$$\cos\theta = \frac{6(25)+3(9)+2(-29)}{\sqrt{36+9+4}\sqrt{625+81+841}} = \frac{17}{\sqrt{1547}}$$

$$\approx 0.432 < \frac{1}{\sqrt{2}}.$$

Hence, $\theta > 45°$. Therefore, the second bisector plane $53x+30y+55z-22=0$ bisects the acute angle.

EXAMPLE 14.57

Show that the line $\vec{r} = (\hat{i}+\hat{j}) + \lambda(3\hat{i}-\hat{j}+2\hat{k})$ is parallel to the plane $\vec{r}.(2\hat{j}+\hat{k}) = 3$. Also find the distance between the line and the plane.

Solution. The equation of the plane is $\vec{r}.(2\hat{j}+\hat{k}) = 3$. Therefore, $2\hat{j}+\hat{k} = \vec{N}$ is normal to the plane. Further, the line is parallel to the vector $3\hat{i}-\hat{j}+2\hat{k}$ and we note that $(3\hat{i}-\hat{j}+2\hat{k}).(2\hat{j}+\hat{k}) = -2+2 = 0$

Therefore, the angle between the line and the normal to the plane is 90°. Hence, the line is parallel to the given plane.

The position vector of the point through which the line passes is $\hat{i}+\hat{j}$. Therefore, the distance between the line and the plane is

$$\left|\frac{(\hat{i}+\hat{j}).(2\hat{j}+\hat{k})-3}{|2\hat{j}+\hat{k}|}\right| = \left|\frac{2-3}{\sqrt{4+1}}\right| = \frac{1}{\sqrt{5}}.$$

14.17 INTERSECTION OF PLANES

Two planes intersect in a line. Therefore, the equations $a_1x+b_1y+c_1z=D_1$ and $a_2x_1+b_2y_1+c_2z=D_2$, *taken together, represent a line.* To find the equation of this line in a standard form, we take any one of the coordinates of a point on the line as 0. Let us take $z_1=0$. Then we have

$$a_1x_1+b_1y_1=D_1 \text{ and } a_2x_1+b_2y_1=D_2.$$

Solving by Cramer's Rule, we get

$$x_1 = \frac{b_2d_1-b_1d_2}{a_1b_2-a_2b_1},\ y_1 = \frac{a_1d_2-a_2d_1}{a_1b_2-a_2b_1}, \text{ and } z_1=0.$$

Further, if A, B, and C are the direction ratios of the line, then since the line is perpendicular to the normals of the planes, we have $a_1A+b_1B+c_1C=0$ and $a_2A+b_2B+c_2C=0$

Solving these equations, we get

$$\frac{A}{b_1c_2-b_2c_1} = \frac{B}{c_1a_2-a_1c_2} = \frac{C}{a_1b_2-a_2b_1}.$$

Therefore, the equation of the line in a standard form is

$$\frac{x-x_1}{A} = \frac{y-y_1}{B} = \frac{z-0}{C}.$$

14.18 PLANES PASSING THROUGH THE INTERSECTION OF TWO GIVEN PLANES

We know that the two planes $\vec{r}.\vec{N}_1 = d_1$ and $\vec{r}.\vec{N}_2 = d_2$ intersect along a line. The points on this line are common to both the planes. Further, the points that satisfy the equations of both the planes, also satisfy the equation

$$(\vec{r}.\vec{N}_1 - d_1) + \lambda(\vec{r}.\vec{N}_2 - d_2) = 0 \quad (1)$$

Hence, the equation of a plane passing through the line of intersection of the two given planes $\vec{r}.\vec{N}_1 = d_1$ and $\vec{r}.\vec{N}_2 = d_2$ is given by (1). Thus, if $P_1 = 0$ and $P_2 = 0$ represent the two given planes, then $P_1 + \lambda P_2 = 0$ represent the equation of the plane passing through the line of intersection of the given planes. The value of the parameter can be found by additional conditions like "plane passes through a point" or "the required plane is perpendicular to another plane," and so on.

EXAMPLE 14.58

Find the equation of the plane passing through the intersection of the planes $x+y+z=6$ and $2x+3y+4z+5=0$ and passing through the point (1, 1, 1).

Solution. The equation of the plane passing through the intersection of two given planes is

$$x+y+z-6+\lambda(2x+3y+4z+5)=0$$

Since this plane passes through (1, 1, 1), we have

$$1+1+1-6+\lambda(2+3+4+5)=0$$

and so, $\lambda=\frac{3}{14}$. Hence, the required plane is

$$x+y+z-6+\frac{3}{14}(2x+3y+4z+5)=0$$

or

$$20x+23y+26z-69=0.$$

EXAMPLE 14.59

Find the equation of the plane passing through the intersection of the planes $x+2y+3z-4=0$ and $2x+y-z+5=0$ perpendicular to the plane $5x+3y+6x+1=0$.

Solution. The equation of the plane passing through the intersection of the given planes is

$$x+2y+3z-4+\lambda(2x+y-z+5)=0$$

or

$$(1+2\lambda)x+(2+\lambda)y+(3-\lambda)z-4+5\lambda=0. \quad (1)$$

Since the plane (1) is perpendicular to the plane $5x+3y+6x+1=0$, we have

$$5(1+2\lambda)+3(2+\lambda)+6(3-\lambda)=0$$

This equation yields $\lambda=-\frac{29}{7}$. Hence, the required plane is

$$x+2y+3z-4-\frac{29}{7}(2x+y-z+5)=0$$

or

$$51x+15y-50z+173=0$$

EXAMPLE 14.60

Show that the planes $2x+5y+3z=0, x-y+4z-2=0$ and $x+6y-z+2=0$ have a common line of intersection.

Solution. The equation of the plane passing through the intersection of first two planes is

$$2x+5y+3z+\lambda(x-y+4z-2)=0 \quad (1)$$

The three given planes will have a common line of intersection if for some value of λ, the equation (1) yields the equation of the third plane. Comparing the coefficients of x, y and z and the constant terms in (1) and the equation of the third plane, we get

$$2+\lambda=1,\ 5-\lambda=6,\ 3+4\lambda=-1, \text{ and } -2\lambda=2.$$

We note that all these equations yields $\lambda=-1$ Hence, for $\lambda=-1$, the equation (1) reduces to $x+6y-z+2=0$. Therefore, the three given planes have a common line of intersection.

EXAMPLE 14.61

Find the equation of the plane passing through the line of intersection of the planes $x+y+z=1$ and $2x+3y-z+4=0$ and perpendicular to the plane $2y-3z=4$.

Solution. The equation of the plane passing through the line of intersection of the given planes is

$$x+y+z-1+\lambda(2x+3y-z+4)=0$$

or

$$(1+2\lambda)x+(1+3\lambda)y+(1-\lambda)z-1+4\lambda=0. \quad (1)$$

The plane (1) will be perpendicular to the plane $2y-3z-4=0$ if

$$0(1+2\lambda)+2(1+3\lambda)-3(1-\lambda)=0,$$

that is, if, $9\lambda=1$ or $\lambda=\frac{1}{9}$. Hence, the required plane is

$$x+y+z-1+\frac{1}{9}(2x+3y-z+4)=0$$

or

$$11x+12y+8z-5=0$$

14.19 SPHERE

A *sphere* is the locus of a point in a three-dimensional space, which remains at a constant distance from a fixed point. The fixed point is called the *Center of the sphere* and the constant distance is called the *radius of the sphere*.

Let $\vec{r}$ be the position vector of any point $P(x, y, z)$ on the surface of a sphere with respect to the origin O. Let the position vector of the center $C(c_1, c_2, c_3)$ be $\vec{c}$. Thus,

$$\vec{CP} = \vec{OP} - \vec{OC} = \vec{r} - \vec{c}.$$

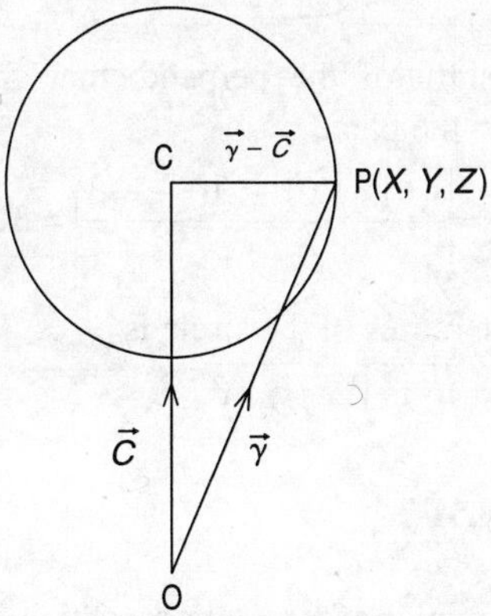

If $|\vec{CP}| = a$ be the radius of the sphere, then

$$|\vec{r} - \vec{c}| = a \quad (1)$$

or

$$(\vec{r} - \vec{c}).(\vec{r} - \vec{c}) = a^2 \text{ or } (\vec{r})^2 - 2\vec{r}.\vec{c} + (\vec{c}^2 - a^2) = 0. \quad (2)$$

Equations (1) and (2) are called the *vector equations* of the sphere with center C and radius *a*.

To obtain the *Cartesian form* of the equation of the sphere, we have

$$\vec{r} = x\hat{i} + y\hat{j} + z\hat{k} \text{ and } \vec{c} = c_1\hat{i} + c_2\hat{j} + c_3\hat{k}.$$

Therefore, (1) reduces to

$$\left|x\hat{i} + y\hat{j} + z\hat{k} - (c_1\hat{i} + c_2\hat{j} + c_3\hat{k})\right| = a$$

or $$\left|(x - c_1)\hat{i} + (y - c_2)\hat{j} + (z - c_3)\hat{k}\right| = a$$

or

$$\sqrt{(x - c_1)^2 + (y - c_2)^2 + (z - c_3)^2} = a$$

or

$$(x - c_1)^2 + (y - c_2)^2 + (z - c_3)^2 = a^2, \quad (3)$$

which is called the Cartesian equation of the sphere with center (c_1, c_2, c_3) and radius a.

The equation (3) can be expressed as

$$(x^2 + c_1^2 - 2c_1x) + (y^2 + c_2^2 - 2c_2y) + (z^2 + c_3^2 - 2c_3z) = a^2$$

or

$$x^2 + y^2 + z^2 - 2c_1x - 2c_2y - 2c_3z + c_1^2 + c_2^2 + c_3^2 - a^2 = 0. \quad (4)$$

If we put $g = -c_1$, $f = -c_2$, $h = -c_3$, and $c = c_1^2 + c_2^2 + c_3^2 - a^2$, then (4) reduces to

$$x^2 + y^2 + z^2 + 2gx + 2fy + 2hz + c = 0. \quad (5)$$

Equation (5) is called the *general form* of the equation of the sphere with center $(c_1, c_2, c_3), = (-g, -f, -h)$ and radius

$$a = \sqrt{c_1^2 + c_2^2 + c_3^2 - c} = \sqrt{g^2 + f^2 + h^2 - c}.$$

Hence, *the equation* (5) *represents a sphere with center* $(-g, -f, -h)$ *and radius* $\sqrt{g^2 + f^2 + h^2 - c}$.

We note that the equation of the sphere is a second-degree equation in *x*, *y*, and *z*, where the coefficients of x^2, y^2, and z^2 are equal, and there is no term involving *xy*, *yz*, and *zx*.

Remarks 14.4.

(i) If the center of the sphere is its origin, then (1) transforms to

$$|\vec{r}| = a \text{ or } \vec{r}.\vec{r} = a^2, \quad (6)$$

which, in the Cartesian form, is

$$x^2 + y^2 + z^2 = a^2.$$

(ii) Equation (6) can be rewritten as

$$(\vec{r} + \vec{a}).(\vec{r} - \vec{a}) = 0. \quad (7)$$

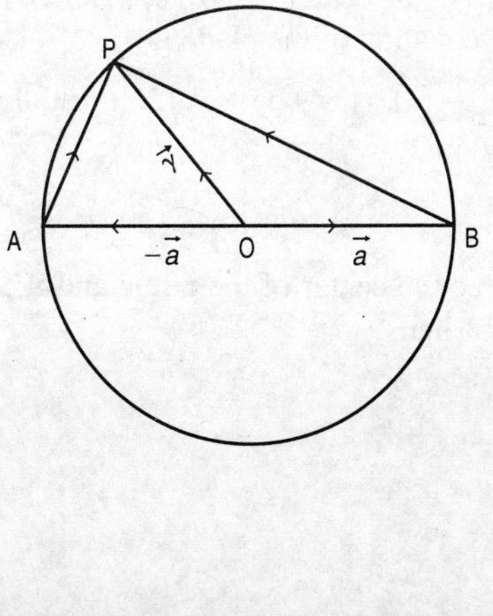

We have $\vec{A}P = \vec{r} + \vec{a}$ and $\vec{B}P = \vec{r} - \vec{a}$.

Therefore, (7) implies $\vec{A}P.\vec{B}P = 0$, which shows that $\vec{A}P$ is perpendicular to BP. Hence, the diameter of the sphere subtends a right angle at any point P on the surface of the sphere. Further, this result is true even if the center is not the origin.

(iii) If the origin lies on the sphere, then

$$|\vec{r} - \vec{c}| = |\vec{c}|$$

or

$$\vec{r} + \vec{c}^2 - 2\vec{r}.\vec{c} = \vec{c}^2 \text{ or } \vec{r}^2 - 2\vec{r}.\vec{c} = 0.$$

EXAMPLE 14.62

Find the center and radius of the sphere $x^2 + y^2 + z^2 + 4x - 8y + 6z + 4 = 0$.

Solution. Comparing the given equation with the general equation of the sphere $x^2 + y^2 + z^2 + 2gx + 2fy + 2hz + c = 0$, we have $g = 2, f = -4$, and $h = 3$. Hence, the center of the given sphere is $(-g, -f, -h) = (-2, 4, -3)$ and the radius of the sphere is $r = \sqrt{g^2 + f^2 + h^2 - c} = \sqrt{4 + 16 + 9 - 4} = 5$.

EXAMPLE 14.63

Find the radius of the circular section of the sphere $|\vec{r}| = 5$ by the plane $\vec{r}.(\hat{i} + \hat{j} + \hat{k}) = 3\sqrt{3}$.

Solution. The equation of the given sphere is

$$|\vec{r}| = 5 \text{ or } \sqrt{x^2 + y^2 + z^2} = 5 \text{ or}$$
$$x^2 + y^2 + z^2 = 25.$$

Therefore, the center C = (0, 0, 0) and radius $r = 5$. The equation of the plane is

$$\vec{r}.(\hat{i} + \hat{j} + \hat{k}) = (x\hat{i} + y\hat{j} + z\hat{k}).(\hat{i} + \hat{j} + \hat{k}) = 3\sqrt{3}$$

or

$$x + y + z = 3\sqrt{3}.$$

Let A be the center of the circle and AB be the radius. Then

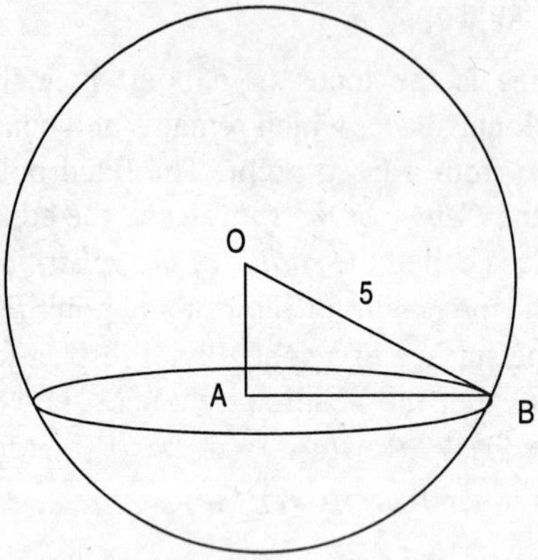

OA = Length of the perpendicular from the origin O to the plane

$$= \left|\frac{1(0) + 1(0) + 1(0) - 3\sqrt{3}}{\sqrt{1+1+1}}\right| = 3$$

and so, the radius of the circle is

$$AB = \sqrt{(OB)^2 - (OA)^2} = \sqrt{5^2 - 3^2} = 4.$$

EXAMPLE 14.64

Find the equation of the sphere for which the circle $x^2 + y^2 + z^2 + 10y - 4z - 8 = 0$, $x + y + z = 3$, is a great circle.

Solution. Section of a sphere by a plane through its center is called a *great circle*. Further, if $S = 0$ is a sphere and $U = 0$ is a plane, then the equation $S + \lambda U = 0$ represent a sphere; and the points of intersection of the sphere $S = 0$ and the plane $U = 0$ satisfy it. In view of this argument, the equation of any sphere through the given circle is

$$x^2 + y^2 + z^2 + 10y - 4z - 8 + \lambda(x + y + z - 3) = 0 \quad (1)$$

or

$$x^2 + y^2 + z^2 + \lambda x + (10 + \lambda)y - (4 - \lambda)z - (8 + 3\lambda) = 0.$$

The center of this sphere is $\left(-\frac{\lambda}{2}, -\frac{10+\lambda}{2}, \frac{4-\lambda}{2}\right)$. According to the definition of the great circle, this center must lie on the plane $x + y + z = 3$. Therefore,

$$-\frac{\lambda}{2}-\frac{10+\lambda}{2}+\frac{4-\lambda}{2}=3 \text{ and so, } \lambda=-4.$$

Hence, the equation (1) of the sphere reduces to $x^2+y^2+z^2-4x+6y-8z+4=0$.

EXAMPLE 14.65

Find the equation of the sphere whose center has the position vector $3\hat{i}+6\hat{j}-4\hat{k}$ and which touches the plane $\vec{r}.\left(2\hat{i}-2\hat{j}-\hat{k}\right)=10$.

Solution. The center of the sphere is (3, 6, –4). Since the sphere touches the plane $\vec{r}.\left(2\hat{i}-2\hat{j}-\hat{k}\right)=10$, the radius of the sphere is equal to the length of the perpendicular from the center of the sphere to the plane. The Cartesian form of the plane is

$$2x-2y-z-10=0. \quad (1)$$

Therefore, the length of the perpendicular from (3, 6, –4) to the plane (1) is its

$$\text{radius} = \left|\frac{2(3)+6(-2)-4(-1)-10}{\sqrt{2^2+2^2+1^2}}\right|$$

$$= \left|\frac{6-12+4-10}{3}\right| = 4.$$

Hence, the equation of the sphere with radius 4 and center (3, 6, –4) is

$$(x-3)^2+(y-6)^2+(z+4)^2=4^2$$

or

$$x^2+y^2+z^2+-6x-12y+8z+45=0\cdot$$

14.20 EQUATION OF A SPHERE WHOSE DIAMETER IS THE LINE JOINING TWO GIVEN POINTS

Let AB be the line joining two points A and B with position vectors $\vec{a}$ and $\vec{b}$ respectively. Thus, $\vec{OA}=\vec{a}$ and $\vec{OB}=\vec{b}$. Let AB be the diameter of the sphere. Further, let $\vec{r}$ be the position vector of an arbitrary point P on the sphere.

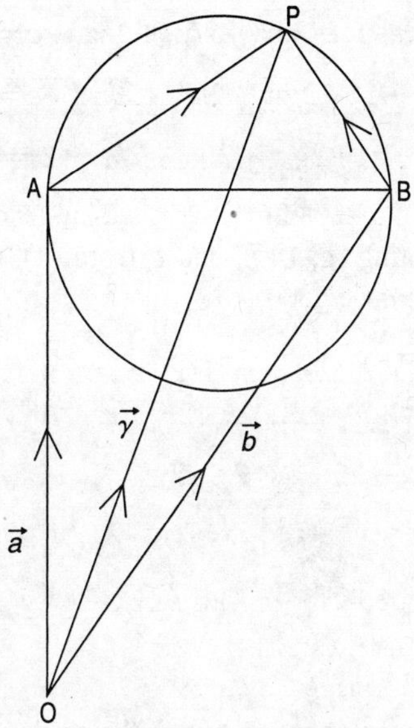

Then we have $\vec{AP}=\vec{r}-\vec{a}$ and $\vec{BP}=\vec{r}-\vec{b}$.

Since, by property of a sphere, $\vec{AP}$ and $\vec{BP}$ are perpendicular to each other,

$$\left(\vec{r}-\vec{a}\right).\left(\vec{r}-\vec{b}\right)=0, \quad (1)$$

which is the required equation of the sphere in *Vector form*.
Substituting

$$\vec{r}=x\hat{i}+y\hat{j}+z\hat{k},\ \vec{a}=a_1\hat{i}+a_2\hat{j}+a_3\hat{k}, \text{ and } \vec{b}=b_1\hat{i}+b_2\hat{j}+b_3\hat{k}$$

in (1), we get

$$\left(x-a_1\right)\left(x-b_1\right)+\left(y-a_2\right)\left(y-b_2\right)+\left(z-a_3\right)\left(z-b_3\right)=0, \quad (4)$$

which is the *Cartesian form* of the equation of a sphere with $\left(a_1,a_2,a_3\right)$ and $\left(b_1,b_2,b_3\right)$ as extremities of the diameter.

14.21 EQUATION OF A SPHERE PASSING THROUGH FOUR POINTS

Let the equation of the sphere be

$$x^2+y^2+z^2+2gx+2fy+2hz+c=0\cdot \quad (1)$$

Let this sphere pass through four points $\left(x_1,y_1,z_1\right)$, $\left(x_2,y_2,z_2\right)$, $\left(x_3,y_3,z_3\right)$, and $\left(x_4,y_4,z_4\right)$. Thus,

$$x_1^2+y_1^2+z_1^2+2gx_1+2fy_1+2hz_1+c=0, \quad (2)$$

$$x_2^2+y_2^2+z_2^2+2gx_2+2fy_2+2hz_2+c=0, \quad (3)$$

$$x_3^2+y_3^2+z_3^2+2gx_3+2fy_3+2hz_3+c=0, \quad (4)$$

and $x_4^2+y_4^2+z_4^2+2gx_4+2fy_4+2hz_4+c=0.$ (5)

Eliminating g, f, h, and c from (1), (2), (3), (4)and (5), we get

$$\begin{vmatrix} x^2+y^2+z^2 & 2x & 2y & 2z & 1 \\ x_1^2+y_1^2+z_1^2 & 2x_1 & 2y_1 & 2z & 1 \\ x_2^2+y_2^2+z_2^2 & 2x_2 & 2y_2 & 2z_2 & 1 \\ x_3^2+y_3^2+z_3^2 & 2x_3 & 2y_3 & 2z_3 & 1 \\ x_4^2+y_4^2+z_4^2 & 2x_4 & 2y_4 & 2z_4 & 1 \end{vmatrix}=0.$$

EXAMPLE 14.66

Find the center and radius of the circle $x^2+y^2+z^2-2y-4z=11, x+2y+2z=15.$

Solution. The center of the sphere is (0,1,2). The perpendicular distance from (0, 1, 2) to the plane $x+2y+2z=15$ is

$$\left|\frac{1(0)+2(1)+2(2)-15}{\sqrt{1^2+2^2+2^2}}\right|=\frac{9}{3}=3.$$

Radius of the sphere is

$$r=\sqrt{g^2+f^2+h^2-c}=\sqrt{1+4+11}=4$$

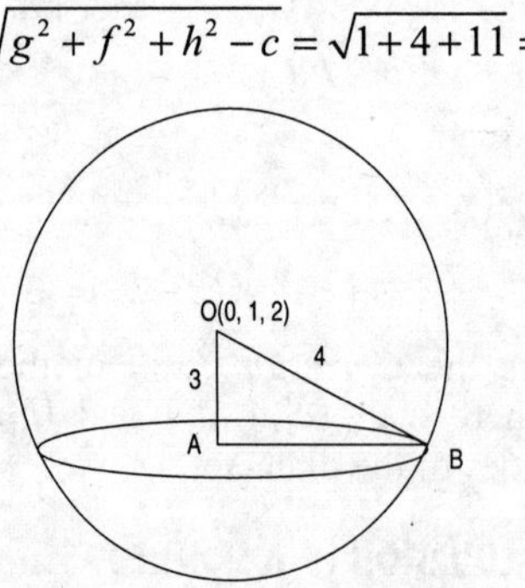

Therefore, the radius of the circle

$$(AB)=\sqrt{(OB)^2-(OA)^2}=\sqrt{16-9}=\sqrt{7}$$

Since OA is normal to the plane, its direction ratios are 1, 2, and 2. Therefore, the equation of OA is

$$\frac{x-0}{1}=\frac{y-1}{2}=\frac{z-2}{2}.$$

Any point on this line is $(\lambda, 2\lambda+1, 2\lambda+2)$. Since this point lies on the plane, we have

$$\lambda+2(2\lambda+1)+2(2\lambda+2)=15$$

and so, $\lambda=1$. Therefore, the point A is (1, 3, 4). Hence, the center of the circle is (1, 3, 4) and its radius is $\sqrt{7}$.

EXAMPLE 14.67

Find the equation of the sphere whose diameter is join of the points A and B, with position vectors $2\hat{i}+6\hat{j}-7\hat{k}$ and $-2\hat{i}+4\hat{j}-3\hat{k}$, respectively.

Solution. The equation of the sphere in a Vector form is

$$\left[\vec{r}-\left(2\hat{i}+6\hat{j}-7\hat{k}\right)\right]\cdot\left[\vec{r}-\left(-2\hat{i}+4\hat{j}-3\hat{k}\right)\right]=0,$$

whereas the Cartesian form of the equation of the sphere is

$$(x-2)(x+2)+(y-6)(y-4)+(z+7)(z+3)=0$$

or

$$x^2+y^2+z^2-10y+10z+41=0,$$

whose center is $(-g, -f -h)=(0, 5, -5)$ and radius is $\sqrt{g^2+f^2+h^2-c}=3$.

EXAMPLE 14.68

Find the equation of the sphere passing through the points (0, 0, 0), (–1, 2, 0), (0, 1, –1), and (1, 2, 5).

Solution. Let the equation of the sphere be

$$x^2+y^2+z^2+2gx+2fy+2hz+c=0$$

Since it passes through (0, 0, 0), (–1, 2, 0), (0, 1, –1), and (1, 2, 5), we have $c=0, -2g+4f=-5, 2f-2h=-2, 2g+4f+10h=-30.$

Solving these equations, we get

$$f=-\frac{5}{2},\ g=-\frac{5}{2},\ \text{and}\ h=-\frac{3}{2}.$$

Hence, the equation of the required sphere is

$$x^2+y^2+z^2-5x-5y-3z=0.$$

EXAMPLE 14.69

A plane passes through a fixed point (a, b, c) and cuts axes in A, B, and C. Show that the locus of the center of the sphere OABC is

$$\frac{a}{x}+\frac{b}{y}+\frac{c}{z}=2.$$

Solution. Let the equation of the plane be

$$px + qy + rz = 1. \quad (1)$$

Since it passes through (a, b, c), we have

$$ap + bq + cr = 1. \quad (2)$$

Putting $y = z = 0$ in (1), we get

$$px = 1 \text{ or } x = \frac{1}{p}.$$

Similarly, $y = \frac{1}{q}$ and $z = \frac{1}{r}$. Hence, the points A, B and C are

$$A\left(\frac{1}{p},0,0\right),\ B\left(0,\frac{1}{q},0\right), \text{ and } C\left(0,0,\frac{1}{r}\right).$$

Let the equation of the sphere OABC be

$$x^2 + y^2 + z^2 + 2gx + 2fy + 2hz + c = 0 \cdot \quad (3)$$

Since the origin O lies on the sphere, we have $c = 0$. Since $A\left(\frac{1}{p},0,0\right), B\left(0,\frac{1}{q},0\right)$, and $C\left(0,0,\frac{1}{r}\right)$ lie on the sphere, we have

$$\frac{1}{p^2} + \frac{2g}{p} = 0 \text{ or } 1 + 2gp = 0 \text{ or } g = -\frac{1}{2p},$$

$$\frac{1}{q^2} + \frac{2f}{q} = 0 \text{ or } f = -\frac{1}{2q}, \text{ and}$$

$$\frac{1}{r^2} + \frac{2h}{r} = 0 \text{ or } h = -\frac{1}{2r}.$$

Putting these values of $c, f, g,$ and h in (3), we get

$$x^2 + y^2 + z^2 - \frac{1}{p}x - \frac{1}{q}y - \frac{1}{r}z = 0$$

The center of the sphere is $\left(\frac{1}{2p}, \frac{1}{2q}, \frac{1}{2r}\right) = (\alpha, \beta, \gamma)$ say. Then

$$p = \frac{1}{2\alpha}, q = \frac{1}{2\beta}, \text{ and } r = \frac{1}{2\gamma}.$$

Thus, (2) reduces to

$$\frac{a}{\alpha} + \frac{b}{\beta} + \frac{c}{\gamma} = 2$$

Therefore, the locus of the center is

$$\frac{a}{x} + \frac{b}{y} + \frac{c}{z} = 2$$

EXAMPLE 14.70

Show that the two circles

$x^2 + y^2 + z^2 - 2x + 3y + 4z - 5 = 0, 5y + 6z + 1 = 0$ and $x^2 + y^2 + z^2 - 3x + 4y + 5z - 6 = 0, x + 2y - 7z = 0$

lie on the same sphere and find the equation for the same.

Solution. The equation of any sphere through the first circle is

$$x^2 + y^2 + z^2 - 2x + 3y + 4z - 5 + \lambda(5y + 6z + 1) = 0 \quad (1)$$

and the equation of any sphere through the second circle is

$$x^2 + y^2 + z^2 - 3x - 4y + 5z - 6 + \mu(x + 2y - 7z) = 0 \quad (2)$$

The equations (1) and (2) can be written respectively as

$$x^2 + y^2 + z^2 + -2x + (5\lambda + 3)y + (6\lambda + 4)z + (\lambda - 5) = 0 \quad (3)$$

and

$$x^2 + y^2 + z^2 + (\mu - 3)x + (2\mu - 4)y + (-7\mu + 5)z - 6 = 0 \quad (4)$$

The equations (3) and (4) will represent the same sphere if $-2 = \mu - 3, 5\lambda + 3 = 2\mu - 4,$

$$6\lambda + 4 = -7\mu + 5, \text{ and } \lambda - 5 = -6$$

From the last equation, we get $\lambda = -1$ and from the first equation, we get $\mu = 1$. These values of λ and μ satisfy the second- and third equation also. Hence, the given circles lie on the sphere

$$x^2 + y^2 + z^2 - 2x + 3y + 4z - 5 - 1(5y + 6z + 1) = 0$$

or

$$x^2 + y^2 + z^2 - 2x - 2y - 2z - 6 = 0$$

EXAMPLE 14.71

Find the equation of the spheres passing through the circle $x^2 + y^2 + z^2 - 6x - 2z + 5 = 0, y = 0$, and touching the plane $3y + 4z + 5 = 0$.

Solution. The equation of any sphere through the given circle is

$$x^2 + y^2 + z^2 - 6x - 2z + 5 + \lambda y = 0 \quad (1)$$

Its center is $(-g, -f, -h) = \left(3, \frac{-\lambda}{2}, 1\right)$. The sphere (1) will touch the plane $3y + 4z + 5 = 0$ if the perpendicular distance from the center to the plane is equal to the radius of the sphere. Thus, we must have

$$\frac{3\left(\frac{-\lambda}{2}\right) + 4(1) + 5}{\sqrt{3^2 + 4^2}} = \sqrt{g^2 + f^2 + h^2 - d}$$

$$= \sqrt{9 + \frac{\lambda^2}{4} + 1 - 5} = \sqrt{5 + \frac{\lambda^2}{4}}$$

Squaring yields

$4\lambda^2 + 27\lambda + 44 = 0$, so that $\lambda = -4, -\frac{11}{4}$.

Substituting these values of λ in (1), we get the required equations of the spheres as

$$x^2 + y^2 + z^2 - 6x - 4y - 2z + 5 = 0 \text{ and}$$

$$x^2 + y^2 + z^2 - 6x - \frac{11}{4}y - 2z + 5 = 0$$

EXAMPLE 14.72

Find the equation of the sphere through the circle $x^2 + y^2 + z^2 = 1, x + 2y + 3z = 4$ and the point $(1, -1, 2)$.

Solution. The equation of the sphere through the given circle is

$$x^2 + y^2 + z^2 - 1 + \lambda(x + 2y + 3z - 4) = 0$$

Since it passes through $(1, -1, 2)$, we have $1 + 1 + 4 - 1 + \lambda(1 - 2 + 6 - 4) = 0$, which yields $\lambda = -5$. . Hence, the equation of the sphere is

$$x^2 + y^2 + z^2 - 1 - 5(x + 2y + 3z - 4) = 0$$

or

$$x^2 + y^2 + z^2 - 5x - 10y - 15z + 19 = 0$$

14.22 EQUATION OF THE TANGENT PLANE TO A SPHERE

A line which meets a sphere in two coincident points is called the *tangent line* to the sphere. The locus of the tangent lines at a point P of a sphere is a plane known as *tangent plane* to the sphere at P.

Let $A\ (x, y, z)$ be any point on the tangent plane at $P\ (x_1, y_1, z_1)$ to the sphere

$$x^2 + y^2 + z^2 + 2gx + 2fy + 2hz + c = 0.$$

Then, the direction ratios of PA are $x - x_1, y - y_1$, and $z - z_1$. Since the center of the sphere is $(-g, -f, -h)$, the direction ratios of the radius OP are $x_1 + g, y_1 + f$, and $z_1 + h$., Since OP is normal to the tangent plane at P_1, we have $OP \perp PA$ and so

$$(x_1 + g)(x - x_1) + (y_1 + f)(y - y_1) + (z_1 + h)(z - z_1) = 0$$

or

$$xx_1 + yy_1 + zz_1 + gx + fy + hz$$
$$= x_1^2 + y_1^2 + z_1^2 + gx_1 + fy_1 + hz_1$$

or

$$xx_1 + yy_1 + zz_1 + g(x + x_1) + f(y + y_1) + h(z + z_1)$$
$$= x_1^2 + y_1^2 + z_1^2 + 2gx_1 + 2fy_1 + 2hz_1 = -c, \text{ since}$$

(x_1, y_1, z_1) lies on the sphere (1)

or

$$xx_1 + yy_1 + zz_1 + g(x + x_1) + f(y + y_1)$$
$$+ h(z + z_1) + c = 0, \quad (2)$$

which is the desired *equation of the tangent plane* at the point (x_1, y_1, z_1) of the sphere.

If the equation of the sphere is $x^2 + y^2 + z^2 = a^2$, then the center is O (0, 0, 0) that is, $g = f = h = 0$. Therefore, (1) reduces to

$$xx_1 + yy_1 + zz_1 = x_1^2 + y_1^2 + z_1^2 = a^2,$$

that is,

$$xx_1 + yy_1 + zz_1 = a^2, \quad (3)$$

which is the required equation of the tangent plane.

14.23 CONDITION OF TANGENCY

The plane $ax + by + cz = D$ will be a tangent plane to the given sphere $x^2 + y^2 + z^2 + 2gx + 2fy + 2hz + c = 0$, if the perpendicular distance of the plane from the center of the sphere is equal to the radius of the sphere, that is, if

$$\frac{a(-g) + b(-f) + c(-h) - D}{\sqrt{a^2 + b^2 + c^2}}$$
$$= \sqrt{g^2 + f^2 + h^2 - c},$$

or

$$(ag + bf + ch + D)^2 = (a^2 + b^2 + c^2)$$
$$\times (g^2 + f^2 + h^2 - c),$$

which is the required *condition of tangency*.

14.24 ANGLE OF INTERSECTION OF TWO SPHERES

The angle of intersection of two spheres is the angle between their tangent planes at a common point of intersection of the spheres.

Also we know that the angle between two planes is the angle between their normals. It follows, therefore, that the angle between two spheres at a point of intersection P is equal to the angle between their radii at P.

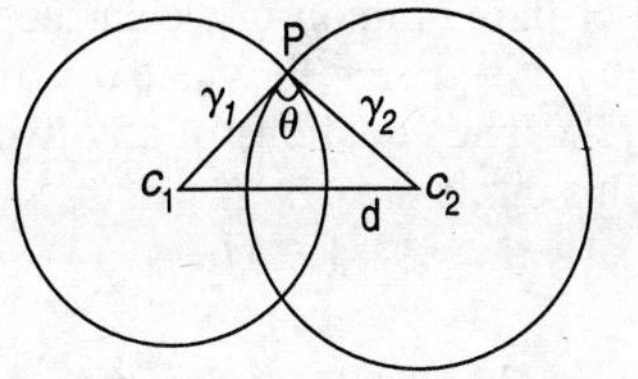

Thus, if c_1 and c_2 are the centers and r_1 and r_2 are the radii, respectively, of the given spheres, then the $\angle c_1Pc_2$ is the angle between the spheres at P. Let this angle be θ. Then, the cosine formula, when applied to the triangle Pc_1c_2, yields

$$\cos\theta = \frac{r_1^2 + r_2^2 - d^2}{2r_1r_2},$$

where d is the distance between their centers.

14.25 CONDITION OF ORTHOGONALITY OF TWO SPHERES

Let the spheres

$$x^2 + y^2 + z^2 + 2g_1x + 2f_1y + 2h_1z + c_1 = 0 \quad (1)$$

and

$$x^2 + y^2 + z^2 + 2g_2x + 2f_2y + 2h_2z + c_2 = 0 \quad (2)$$

cut orthogonally as shown in the following figure:

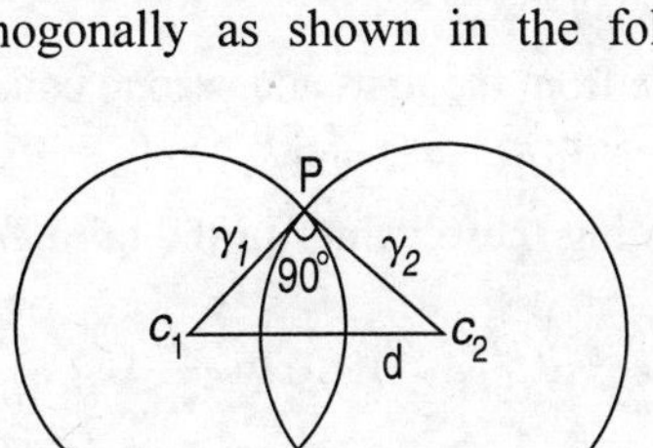

Then the angle c_1Pc_2 is 90°. The centers of the spheres are $c_1(-g_1, -f_1, -h_1)$ and $c_2(-g_2, -f_2, -h_2)$. The triangle Pc_1c_2 is a right-angled triangle. Therefore, $(c_1c_2)^2 = r_1^2 + r_2^2$ or

$$(g_1 - g_2)^2 + (f_1 - f_2)^2 + (h_1 - h_2)^2$$
$$= g_1^2 + f_1^2 + h_1^2 - c_1 + g_2^2 + f_2^2 + h_2^2 - c_2$$

or

$$2g_1g_2 + 2f_1f_2 + 2h_1h_2 = c_1 + c_2,$$

which is the required *condition of orthogonality*.

EXAMPLE 14.73

Find the equation of the sphere which is tangential to the plane $x - 2y - 2z = 7$ at $(3, -1, -1)$ and passes through $(1, 1, -3)$.

Solution. Let the equation of the sphere be

$$x^2 + y^2 + z^2 + 2gx + 2fy + 2hz + c = 0. \quad (1)$$

Since it passes through $(1, 1, -3)$, we have

$$2g + 2f - 6h + 11 + c = 0. \quad (2)$$

The equation of the tangent plane at $(3, -1, -1)$ is

$3x - y - z + g(x+3) + f(y - 1) + h(z - 1) + c = 0$

or

$$(g+3)\,x + (f-1)\,y + (h-1)\,z + 3g - f - h + c = 0. \quad (3)$$

But the given tangent plane is

$$x - 2y - 2z - 7 = 0. \quad (4)$$

The equations (3) and (4) will be the equations of the same plane if

$$\frac{g+3}{1} = \frac{f-1}{-2} = \frac{h-1}{-2} = \frac{3g - f - h + c}{-7} = k, \text{ say.}$$

Then

$$g = k - 3, f = -2k + 1, h = -2k + 1 \quad (5)$$

and

$$3g - f - h + c = -7k. \quad (6)$$

Substituting the values of g, f, and h from (5) in (2) and (6), we get

$$10k + 1 + c = 0 \quad (7)$$

and

$$14k - 11 + c = 0. \quad (8)$$

Subtracting (7) from (8), we get $4k - 12 = 0$ and so, $k = 3$. Putting this value in (7) and (5), we get

$$g = 0, f = -5, h = -5, \text{ and } c = -31.$$

Substituting these values in (1), the required equation of the sphere is

$$x^2 + y^2 + z^2 - 10y - 10z - 31 = 0.$$

EXAMPLE 14.74

Show that the spheres $x^2 + y^2 + z^2 + 6y + 2z + 8 = 0$ and $x^2 + y^2 + z^2 + 6x + 8y + 4z + 20 = 0$ intersect at right angles. Find their plane of intersection.

Solution. Comparing with the standard form, we have

$$g_1 = 0, f_1 = 3, h_1 = 1 \text{ and } c_1 = 8$$

and $\quad g_2 = 3, f_2 = 4, h_2 = 2$ and $c_2 = 20$.

Therefore,

$$2g_1g_2 + 2f_1f_2 + 2h_1h_2 = 0 + 24 + 4 = 28$$

and $\quad c_1 + c_2 = 8 + 20 = 28.$

Thus, the condition for orthogonality $2g_1 g_2 + 2f_1 f_2 + 2h_1 h_2 = c_1 + c_2$ is satisfied. The plane of intersection of these spheres is

$$x^2 + y^2 + z^2 + 6y + 2z + 8$$
$$-(x^2 + y^2 + z^2 + 6x + 8y + 4z + 20) = 0$$

or

$$3x + y + z + 6 = 0.$$

EXAMPLE 14.75

If any tangent plane to the sphere $x^2 + y^2 + z^2 = r^2$ makes intercepts a, b, and c on the coordinate axes, show that

$$\frac{1}{a^2}+\frac{1}{b^2}+\frac{1}{c^2}=\frac{1}{r^2}.$$

Solution. Let (x_1, y_1, z_1) be any point on the given sphere $x^2 + y^2 + z^2 = r^2$. The equation of the tangent plane to the sphere at the point (x_1, y_1, z_1) is

$$xx_1 + yy_1 + zz_1 = r^2 \text{ or } \frac{x}{\frac{r^2}{x_1}}+\frac{y}{\frac{r^2}{y_1}}+\frac{z}{\frac{r^2}{z_1}}=1. \qquad (1)$$

But the tangent plane to the given sphere makes intercepts a, b, and c, respectively, on the coordinate axes. Therefore, its equation is

$$\frac{x}{a}+\frac{y}{b}+\frac{z}{c}=1. \qquad (2)$$

Comparing (1) and (2), we get

$$\frac{r^2}{x_1}=a, \frac{r^2}{y_1}=b, \text{ and } \frac{r^2}{z_1}=c$$

or

$$x_1=\frac{r^2}{a}, y_1=\frac{r^2}{b}, \text{ and } z_1=\frac{r^2}{c}.$$

Since (x_1, y_1, z_1) lies on the given sphere, we have

$$x_1^2 + y_1^2 + z_1^2 = r^2$$

or

$$\frac{r^4}{a^2}+\frac{r^4}{b^2}+\frac{r^4}{c^2}=r^2 \text{ or } \frac{1}{a^2}+\frac{1}{b^2}+\frac{1}{c^2}=\frac{1}{r^2}.$$

EXAMPLE 14.76

Two spheres of radii r_1 and r_2 cut orthogonally. Show that the radius of the common circle is $\dfrac{r_1 r_2}{\sqrt{r_1^2+r_2^2}}$.

Solution. Let the common circle in the xy-plane be $x^2 + y^2 + z^2 = a^2$, $z = 0$, where a is the radius of the circle. The equations of any two spheres through this circle are

$$x^2 + y^2 + z^2 - a^2 + k_1 z = 0 \qquad (1)$$

and

$$x^2 + y^2 + z^2 - a^2 + k_2 z = 0. \qquad (2)$$

The spheres (1) and (2) will cut orthogonally if

$$2g_1 g_2 + 2f_1 f_2 + 2h_1 h_2 = c_1 + c_2,$$

that is, if

$$2\frac{k_1}{2}.\frac{k_2}{2}=-a^2-a^2 \text{ or if } k_1k_2=-4a^2. \qquad (3)$$

On the other hand, the radius r_1 and r_2 of the spheres (1) and (2), respectively, are

$$r_1=\sqrt{g_1^2+f_1^2+h_1^2-c_1}=\sqrt{\frac{k_1^2}{4}+a^2}.$$

and

$$r_2=\sqrt{g_2^2+f_2^2+h_2^2-c_2}=\sqrt{\frac{k_2^2}{4}+a^2}.$$

Thus,

$$r_1^2=\frac{k_1^2}{4}+a^2, r_2^2=\frac{k_2^2}{4}+a^2, \text{ and } k_1^2k_2^2=16a^4.$$

We have from the first- and second equation,

$$k_1^2=4(r_1^2-a^2) \text{ and } k_2^2=4(r_2^2-a^2).$$

Substituting these values in the third equation, we get

$$16(r_1^2-a^2)(r_2^2-a^2)=16a^4$$

or

$$r_1^2r_2^2-a^2(r_1^2+r_2^2)+a^4=a^4$$

or

$$a^2=\frac{r_1^2r_2^2}{r_1^2+r_2^2} \text{ or } a=\frac{r_1r_1}{\sqrt{r_1^2+r_2^2}}.$$

EXAMPLE 14.77

Find the equation of the sphere which touches the plane $3x + 2y - z + 2 = 0$ at the point $(1, -2, 1)$ and cuts orthogonally the sphere $x^2 + y^2 + z^2 - 4x + 6y + 4 = 0$.

Solution. Let the equation of the required sphere be

$$x^2 + y^2 + z^2 + 2gx + 2fy + 2hz + c = 0. \qquad (1)$$

The equation of the tangent plane at (1, –2,1) to the sphere (1) is

$$x-2y+z+g(x+1)+f(y-2)+h(z+1)+c=0$$

or

$$(g+1)x+(f-2)y+(h+1)z$$
$$+(g-2f+h+c)=0. \quad (2)$$

But the tangent plane is given to be

$$3x+2y-z+2=0. \quad (3)$$

Since (2) and (3) represent the same planes, we have

$$\frac{g+1}{3}=\frac{f-2}{2}=\frac{h+1}{-1}=\frac{g-2f+h+c}{2}=k, \text{ say.}$$

This yields

$g = 3k - 1, f = 2k + 2$, and $h = -k - 1$ (4)

and

$$g - 2f + h + c = 2k. \quad (5)$$

Since the sphere (1) cut orthogonally the sphere $x^2 + y^2 + z^2 - 4x + 6y + 4 = 0$, we have

$$2g(-2) + 2f(3) + 2h(0) = c + 4 \text{ or} \quad (6)$$
$$-4f + 6f = c + 4.$$

Substituting the values of g, f, and h from (4) in equations (5) and (6), we obtain $-4k + c = 6$ and $c = 12$, respectively. Therefore, $k=\frac{3}{2}$. Thus, (4) yields $g=\frac{7}{2}$, $f = 5$, and $h=-\frac{5}{2}$. Substituting these values of g, f, h, and c in (1), the required equation of the sphere is $x^2+y^2+z^2+7x+10y-5z+12=0$.

EXAMPLE 14.78

Find the equation of a sphere which passes through the circle $x^2 + y^2 + z^2 - 2x + 3y - 4z + 6 = 0$, $3x - 4y + 5z - 15 = 0$ and cuts orthogonally the sphere $x^2 + y^2 + z^2 + 2x + 4y - 6z + 11 = 0$.

Solution. The equation of the sphere through the given circle is

$$x^2+y^2+z^2-2x+3y-4z+6$$
$$+\lambda(3x-4y+5z-15)=0 \quad (1)$$

or

$$x^2+y^2+z^2+(3\lambda-2)x+(-4\lambda+3)y$$
$$+(5\lambda-4)z+6-15\lambda=0.$$

It cuts the sphere

$$x^2+y^2+z^2+2x+4y-6z+11=0$$

orthogonally if

$$2\left(\frac{3\lambda-2}{2}\right)(1)+2\left(\frac{-4\lambda+3}{2}\right)(2)+2\left(\frac{5\lambda-4}{2}\right)(-3)$$
$$=6-15\lambda+11$$

or if $\lambda=-\frac{1}{5}$. Putting this value of λ in (1), the equation of the required sphere is

$$5(x^2+y^2+z^2)-13x+19y-25z+45=0.$$

EXAMPLE 14.79

Show that the spheres $x^2 + y^2 + z^2 = 25$ and $x^2 + y^2 + z^2 - 24x - 40y - 18z + 225 = 0$ touch externally. Find their point of contact and the equation of the common tangent plane.

Solution. The centers of the given spheres are (0,0,0) and (12,20,9), respectively. Their radii are

$$r_1 = 5 \text{ and } r_2=\sqrt{144+400+81-225}=20.$$

If c_1 and c_2 are the centers of the spheres, then

$$c_1c_2=\sqrt{(12)^2+(20)^2+9^2}=25.$$

Also,

$$r_1 + r_2 = 5 + 20 = 25.$$

Thus, $c_1c_2 = r_1 + r_2$ and so, the given spheres touch externally.

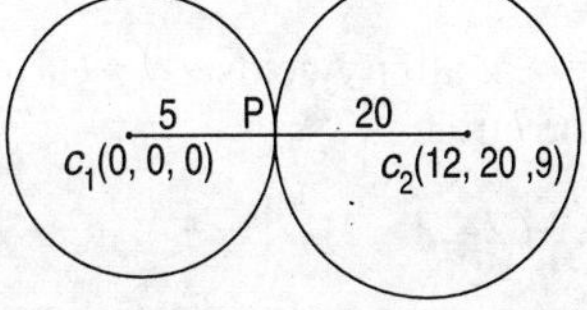

If P is the point of contact, then P divides c_1c_2 internally in the ratio 5:20. Therefore, the coordinates of P are

$$\left(\frac{60-0}{25},\frac{100-0}{25},\frac{45}{25}\right) \text{ or } \left(\frac{12}{5},4,\frac{9}{5}\right).$$

The equation of the tangent plane common to the given sphere is

$$x\left(\frac{12}{5}\right)+y(4)+z\left(\frac{9}{5}\right)=25$$

or $\quad 12x + 20y + 9z = 125.$

EXAMPLE 14.80

Find the center and the radius of the circle

$$x^2+y^2+z^2-8x+4y+8z-45=0,$$

$$x-2y+2z-3=0.$$

Solution. The center of the sphere is (–g, –f, –h) = (4,–2,–4). The radius of the sphere is

$$r=\sqrt{g^2+f^2+z^2-c}=\sqrt{16+4+16+45}$$
$$=9$$

The distance of the plane from the center of the sphere is

$$d=\left|\frac{4+4-8-3}{\sqrt{1+4+4}}\right|=1.$$

Therefore, the radius of the circle is

$$a=\sqrt{r^2-d^2}=\sqrt{81-1}=4\sqrt{5}.$$

Further, the center of the circle is the foot of the perpendicular drawn from the center of the sphere to the plane. The equations of the perpendicular from the center are

$$\frac{x-4}{1}=\frac{y+2}{-2}=\frac{z+4}{2}=\lambda, \text{ say.}$$

Any point on this perpendicular is $(\lambda+4, -2\lambda-2, 2\lambda-4)$. This point will be the foot of the perpendicular, that is, the center of the circle if it lies on the plane. Therefore,

$$(\lambda+4)-2(-2\lambda-2)+2(2\lambda-4)-3=0.$$

The above equation yields $\lambda=\frac{1}{3}$. Hence, the center of the circle is

$$(\lambda+4,-2\lambda-2,2\lambda-4)=\left(\frac{1}{3}+4,-\frac{2}{3}-2,\frac{2}{3}-4\right)$$
$$=\left(\frac{13}{3},-\frac{8}{3},\frac{10}{3}\right).$$

EXAMPLE 14.81

Find the equation of the tangent line to the circle $x^2+y^2+z^2=3$, $3x-2y+4z+3=0$ at the point (1,1,–1).

Solution. The equation of the sphere is $x^2+y^2+z^2=3$ and the equation of the plane is $3x-2y+4z+3=0$. The tangent plane to the sphere at (1,1,–1) is

$$x(1)+y(1)+z(1)=3 \text{ or } x+y+z=3.$$

The tangent line is the intersection of the given plane $3x-2y+4z+3=0$ and the tangent plane $x+y+z=3$. Taking $z=0$, the point of intersection of these planes is given by

$$x+y=3 \text{ and } 3x-2y=-3.$$

Solving the equations, we get $x=0.6$ and $y=2.4$. The direction ratios of the tangent line are given by $x+y+z=0$ and $3x-2y+z=0$. Thus,

$$\frac{x}{4-2}=\frac{y}{-3-4}=\frac{Z}{-2-3} \text{ or } \frac{x}{2}=\frac{y}{-7}=\frac{z}{-5}.$$

Hence, the tangent line through (0.6, 2.4, 0) with direction ratios 2, –7, and –5 is

$$\frac{x-0.6}{2}=\frac{y-2.4}{-7}=\frac{Z}{-5}.$$

14.26 CYLINDER

The surface generated by a variable line which remains parallel to a fixed line and intersects a given curve (or touches a given surface) is called a *cylinder*.

The variable line is called the *generator*, the fixed line is called the *axis*, and the given curve is called the *guiding curve*.

14.27 EQUATION OF A CYLINDER WITH GIVEN AXIS AND GUIDING CURVES

Let the equation of the axis be

$$\frac{x}{l}=\frac{y}{m}=\frac{z}{n} \quad (1)$$

and let the equation of the guiding curve be

$$ax^2+2xy+by^2+2gx+2fy+c=0, z=0. \quad (2)$$

Let (x_1, y_1, z_1) be any point on the cylinder. Then, the equation of the generator (being parallel to (1)) is

$$\frac{x-x_1}{l}=\frac{y-y_1}{m}=\frac{z-z_1}{n}. \quad (3)$$

The generator (3) meets the plane $z=0$ at the point given by $\left(x_1-\frac{lz_1}{n}, y_1-\frac{mz_1}{n}, 0\right)$. Since the generator intersect the conic (2), the point $\left(x_1-\frac{lz_1}{n}, y_1-\frac{mz_1}{n}, 0\right)$ lies on (2). Therefore,

$$a\left(x_1-\frac{lz_1}{n}\right)^2+2h\left(x_1-\frac{lz_1}{n}\right)\left(\frac{y_1-mz_1}{n}\right)$$
$$+b\left(y_1-\frac{mz_1}{n}\right)^2+2g\left(x_1-\frac{lz_1}{n}\right)$$

$$+2f\left(y_1-\frac{mz_1}{n}\right)+c=0$$

or

$$a(nx_1-lz_1)^2+2h(nx_1-lz_1)(ny_1-mz_1)$$
$$+b(ny_1-mz_1)^2+2gn(nx_1-lz_1)$$
$$+2fn(ny_1-mz_1)+cn^2=0.$$

Thus, the locus of (x_1,y_1,z_1) is

$$a(nx-lz)^2+2h(nx-lz)(ny-mz)+b(ny-mz)^2$$
$$+2gn(nx-lz)+2fn(ny-mz)+cn^2=0$$

which is the required equation of the cylinder.

14.28 RIGHT CIRCULAR CYLINDER

A surface generated by a line which intersects a fixed circle (called the *guiding curve*) and is perpendicular to the plane of the circle is called the *right circular cylinder*.

The normal to the plane of the circle through its center is called the *axis of the cylinder* and the radius of the circle is called the *radius* of the cylinder.

To find the equation of a right circular cylinder with radius r, let $A(x_1,y_1,z_1)$ be a fixed point on the axis of the cylinder. Then, the equations of the axis are

$$\frac{x-x_1}{l}=\frac{y-y_1}{m}=\frac{z-z_1}{n}.$$

The direction cosines of the axis are

$$\frac{l}{\sqrt{l^2+m^2+n^2}},\ \frac{m}{\sqrt{l^2+m^2+n^2}},\ \text{and}\ \frac{n}{\sqrt{l^2+m^2+n^2}}.$$

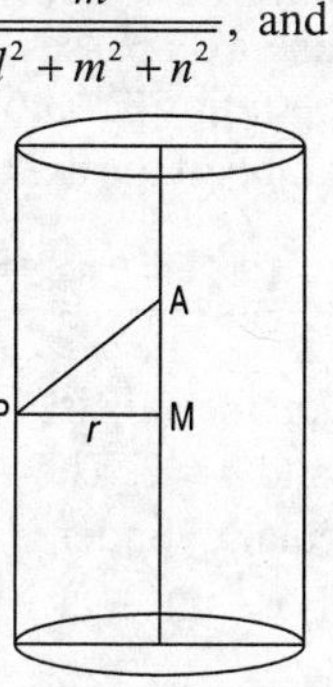

Let $P(x,y,z)$ be any point on the cylinder and let M be the foot of the perpendicular from P on the axis. Then, $PM = r$ and $(AP)^2-(AM)^2=(PM)^2$

or

$$(x-x_1)^2+(y-y_1)^2+(z-z_1)^2-(AM)^2=r^2. \quad (1)$$

But

AM = length of projection of AP on the axis of the cylinder

$$=\left|(x-x_1)\frac{l}{\sqrt{l^2+m^2+n^2}}+(y-y_1)\frac{m}{\sqrt{l^2+m^2+n^2}}+(z-z_1)\frac{n}{l^2+m^2+n^2}\right|$$

$$=\left|\frac{1}{\sqrt{l^2+m^2+n^2}}\times[l(x-x_1)+m(y-y_1)+n(z-z_1)]\right|.$$

Therefore, (1) reduce to

$$(x-x_1)^2+(y-y_1)^2+(z-z_1)^2-\frac{[l(x-x_1)+m(y-y_1)+n(z-z_1)]^2}{l^2+m^2+n^2}=r^2$$

or

$$(l^2+m^2+n^2)[(x-x_1)^2+(y-y_1)^2+(z-z_1)^2]$$
$$-[l(x-x_1)+m(y-y_1)+n(z-z_1)]^2$$
$$=r^2(l^2+m^2+n^2),$$

which is the required *equation of the right circular cylinder*.

If the axis of the cylinder is z-axis, then A is $(0, 0, z)$, $l = 0$, $m = 0$, and $n = 1$ and so, the equation of the right circular cylinder reduces to

$$x^2+y^2=r^2.$$

Similarly, the equation of a right circular cylinder whose axis is $\frac{x}{l}=\frac{y}{m}=\frac{z}{n}$ is

$$(l^2+m^2+n^2)(x^2+y^2+z^2-r^2)=(lx+my+nz)^2.$$

EXAMPLE 14.82

Find the equation of the right circular cylinder whose guiding curve is the circle passing through the points (1,0,0), (0,1,0), and (0,0,1).

Solution. The equation of the plane passing through $A(1,0,0)$, $B(0,1,0)$, and $C(0,0,1)$ is

$$\begin{vmatrix} x-1 & y & z \\ -1 & 1 & 0 \\ -1 & 0 & 1 \end{vmatrix}=0$$

or

$$x+y+z=1 \text{ or } \frac{x}{1}+\frac{y}{1}+\frac{z}{1}=1.$$

Let O be the origin. Then OA = OB = OC = 1. Therefore, a sphere with center O (0, 0, 0) and radius unity passes through A, B, and C. Thus, the guiding curve is the circle

$$x^2+y^2+z^2=1, x+y+z=1.$$

The equation of the axis of the cylinder (which passes through (0, 0, 0) and is perpendicular to the plane $x+y+z=1$) is

$$\frac{x-0}{1}=\frac{y-0}{1}=\frac{z-0}{1} \text{ or } \frac{x}{1}=\frac{y}{1}=\frac{z}{1}.$$

The perpendicular distance from the center of the sphere $x^2+y^2+z^2=1$ to the plane $x+y+z=1$ is

$$p=\frac{1}{\sqrt{1+1+1}}=\frac{1}{\sqrt{3}}.$$

Therefore, the radius of the circle is

$$r=\sqrt{(\text{radius of the sphere})^2-p^2}=\sqrt{1-\frac{1}{3}}=\sqrt{\frac{2}{3}}.$$

Therefore, the equation of the right circular cylinder is

$(l^2+m^2+n^2)(x^2+y^2+z^2-r^2)=(lx+my+nz)^2$

or

$$(1+1+1)\left(x^2+y^2+z^2-\frac{2}{3}\right)=(x+y+z)^2$$

or

$3x^2+3y^2+3z^2-2=x^2+y^2+z^2+2xy+2yz+2xy$

or

$$x^2+y^2+z^2-xy-yz-xy-1=0.$$

EXAMPLE 14.83

Find the equation of the right circular cylinder whose guiding curve is

$$x^2+y^2+z^2=9,\ x-y+z=3.$$

Solution. Proceeding as in the above example, the required equation of the right circular cylinder is $x^2+y^2+z^2+xy+yz+zx+9=0$.

Second Method: The direction ratios of the plane $x-y+z=3$ are 1, –1, and 1. Since the axis of the cylinder is perpendicular to the plane, its direction cosines are proportional to 1, –1, and 1. Let (x_1, y_1, z_1) be any point on the cylinder. The equation of the generators through this point is

$$\frac{x-x_1}{1}=\frac{y-y_1}{-1}=\frac{z-z_1}{1}=\lambda, \text{ say.}$$

Any point on the generator is $(\lambda+x_1, -\lambda+y_1, \lambda+z_1)$. This point lies on the guiding curve if

$$(\lambda+x_1)^2+(-\lambda+y_1)^2+(\lambda+z_1)^2=9,$$
$$(\lambda+x_1)-(-\lambda+y_1)+(\lambda+z_1)=3$$

or

$$x_1^2+y_1^2+z_1^2+2\lambda(x_1-y_1+z_1)+3\lambda^2=9,$$
$$x_1-y_1+z_1+3\lambda=3.$$

From second member, we have $\lambda=\frac{x_1-y_1+z_1-3}{-3}$. Substituting this value of λ in the first member, we get

$$x_1^2+y_1^2+z_1^2-\frac{2(x_1-y_1+z_1-3)(x_1-y_1+z_1)}{3}$$
$$+\frac{3(x_1-y_1+z_1-3)^2}{9}=9$$

or

$$x_1^2+y_1^2+z_1^2+y_1z_1-z_1x_1+x_1y_1-9=0.$$

Hence, the locus of (x_1, y_1, z_1) is

$$x^2+y^2+z^2+yz-zx+xy-9=0,$$

which is the required equation of the right circular cylinder.

EXAMPLE 14.84

Find the equation of the right circular cylinder of radius 2 whose axis passes through (1, 2, 3) and has direction ratios (2,–3,6).

Solution. The equation of the axis of the cylinder is

$$\frac{x-1}{2}=\frac{y-2}{-3}=\frac{z-3}{6}.$$

Thus, the point on the axis is $A(1, 2, 3)$ and $l=2$, $m=-3$, $n=6$, and $r=2$. The standard equation of the right circular cylinder is

$$(l^2+m^2+n^2)[(x-x_1)^2+(y-y_1)^2+(z-z_1)^2]$$
$$-[l(x-x_1)+m(y-y_1)+n(z-z_1)]^2$$
$$=r^2(l^2+m^2+n^2).$$

Putting the value of l, m, n, r, x_1, y_1, and z_1, we get

$$49[(x-1)^2+(y-2)^2+(z-3)^2]$$

$$-[2(x-1)-3(y-2)+6(z-3)]^2 = 4(49)$$

or

$$45x^2+40y^2+13z^2+12xy+36yz$$
$$-24zx-42x-112y-126z-392=0,$$

which is the required equation of the right circular cylinder.

EXAMPLE 14.85

Find the equation of the cylinder whose generators are parallel to the line $\frac{x}{1}=\frac{y}{-2}=\frac{z}{3}$ and whose guiding curve is the ellipse $x^2+2y^2=1$, $z=3$.

Solution. Let (x_1, y_1, z_1) be any point on the cylinder. The generator is parallel to the line $\frac{x}{1}=\frac{y}{-2}=\frac{z}{3}$. Therefore, the direction ratios of the generator are 1, –2, and 3. Thus, the equation of the generator is

$$\frac{x-x_1}{1}=\frac{y-y_1}{-2}=\frac{z-z_1}{3}. \quad (1)$$

The generator (1) meets the plane $z = 3$ in the point $x = x_1 + \frac{3-z_1}{3}, y = y_1 - \frac{2}{3}(3-z_1), z = 3$. This point will lie on the guiding curve if

$$\left(x_1+\frac{3-z_1}{3}\right)^2+2\left(y_1-\frac{2}{3}(3-z_1)\right)^2=1$$

or

$$3(x_1^2+2y_1^2+z_1^2)+8y_1z_1-2z_1x_1$$
$$+6x_1-24y_1-18z_1+24=0.$$

Hence, the locus of (x_1, y_1, z_1) is

$$3(x^2+2y^2+z^2)+8yz-2zx+6x-24y-18z+24=0.$$

14.29 CONE

A *cone* is a surface generated by a variable line which passes through a fixed point and intersects a given curve (or touches a given surface). The fixed point is called the *vertex*, the variable line is called the *generator*, and the given curve is called the *guiding curve*.

14.30 EQUATION OF A CONE WITH ITS VERTEX AT THE ORIGIN

Let $f(x,y,z) = 0$ be the equation of a cone with its vertex at the origin and let $P(x_1, y_1, z_1)$ be any point on the cone. Therefore, $f(x_1, y_1, z_1) = 0$. Let the vertex of the cone be O (0,0,0). Then, the equation of the generator OP is

$$\frac{x-0}{x_1-0}=\frac{y-0}{y_1-0}=\frac{z-0}{z_1-0} \text{ or } \frac{x}{x_1}=\frac{y}{y_1}=\frac{z}{z_1} \quad (1)$$

Any point on the generator (1) is $(\lambda x_1, \lambda y_1, \lambda z_1)$. Since the generator lies on the cone, the point $(\lambda x_1, \lambda y_1, \lambda z_1)$ lies on the cone and so, $f(\lambda x_1, \lambda y_1, \lambda z_1) = 0$ for all values of λ. But this is possible only if $f(x, y, z) = 0$ is homogeneous. It follows, therefore, that the equation of the cone $f(x, y, z) = 0$, with vertex at the origin, is a homogeneous equation in x, y, and z.

Conversely, suppose that $f(x, y, z) = 0$ is homogeneous in x, y and z, and let $P(x_1, y_1, z_1)$ be a point on this surface. Then, $f(x_1, y_1, z_1) = 0$ and due to homogeneity $f(tx_1, ty_1, tz_1) = 0$ for any real t. In particular, for $t = 0$, we have $f(0,0,0) = 0$. Therefore, $O(0,0,0)$ is a point on the locus of $f(x, y, z) = 0$. The equation of OP is

$$\frac{x-0}{x_1-0}=\frac{y-0}{y_1-0}=\frac{z-0}{z_1-0} \text{ or } \frac{x}{x_1}=\frac{y}{y_1}=\frac{z}{z_1}. \quad (1)$$

Any point on this line is (tx_1, tx_2, tx_3). Since $f(tx_1, tx_2, tx_3) = 0$, the point (tx_1, tx_2, tx_3) lies on the surface of $f(x_1, x_2, x_3) = 0$. Therefore, the line (1) lies wholly on $f(x, y, z) = 0$. Thus, the surface $f(x, y, z) = 0$ is generated by a straight line through the fixed point O (0,0,0). Hence the homogeneous equation $f(x, y, z) = 0$ represents a cone with vertex at the origin.

14.31 EQUATION OF A CONE WITH GIVEN VERTEX AND GUIDING CURVE

Let $A(\alpha, \beta, \gamma)$ be the vertex and let the equation of the basic conic be

$$ax^2+2hxy+by^2+2gx+2fy+c=0, z=0. \quad (1)$$

The equation of the generator through the vertex (α, β, γ) is

$$\frac{x-\alpha}{l}=\frac{y-\beta}{m}=\frac{z-\gamma}{n}. \quad (2)$$

This line meets the plane $z = 0$ at the point

$$P\left(\alpha-\frac{l\gamma}{n}, \beta-\frac{m\gamma}{n}, 0\right).$$

Since this point lies on the conic (1), we have

$$a\left(\alpha-\frac{l\gamma}{n}\right)^2+2h\left(\alpha-\frac{l\gamma}{n}\right)\left(\beta-\frac{m\gamma}{n}\right)+b\left(\beta-\frac{m\gamma}{n}\right)^2$$

$$+2g\left(\alpha-\frac{l\gamma}{n}\right)+2f\left(\beta-\frac{m\gamma}{n}\right)+c=0. \quad (3)$$

To eliminate l, m, and n from (2) and (3), we note that (2) gives

$$\frac{l}{n}=\frac{x-\alpha}{z-\gamma} \text{ and } \frac{m}{n}=\frac{y-\beta}{z-\gamma}.$$

Substituting these values of $\frac{l}{n}$ and $\frac{m}{n}$ in (3), we get

$$a\left(\alpha-\gamma\frac{x-\alpha}{z-\gamma}\right)^2+2h\left(\alpha-\gamma\frac{x-\alpha}{z-\gamma}\right)\left(\beta-\gamma\frac{y-\beta}{z-\gamma}\right)$$
$$+b\left(\beta-\gamma\frac{y-\beta}{z-\gamma}\right)^2+2g\left(\alpha-\gamma\frac{x-\alpha}{z-\gamma}\right)$$
$$+2f\left(\beta-\gamma\frac{y-\beta}{z-\gamma}\right)+c=0$$

or

$$a(\alpha z-\gamma x)^2+2h(\alpha z-\gamma x)(\beta z-\gamma y)+b(\beta z-\gamma y)^2$$
$$+2g(\alpha z-\gamma x)(z-\gamma)+2f(\beta z-\gamma y)(z-\gamma)$$
$$+c(z-\gamma)^2=0,$$

which is the required equation of the cone.

EXAMPLE 14.86

Find the equation of the cone whose vertex is the point (1,2,3) and guiding curve is the circle $x^2+y^2+z^2=4$, $x+y+z=1$.

Solution. The equation of the guiding curve is

$$x^2+y^2+z^2=4,\ x+y+z=1. \quad (1)$$

The vertex is (1,2,3). The equation of the generators is

$$\frac{x-1}{l}=\frac{y-2}{m}=\frac{z-3}{n}=t, \text{ say.} \quad (2)$$

Any point on the line (2) is $(lt+1, mt+2, nt+3)$. The line (2) will be a generator if it intersects the circle (1). Thus, the point $(lt+1, mt+2, nt+3)$ should satisfy (1). Therefore,

$$(lt+1)^2+(mt+2)^2+(nt+3)^2=4 \text{ and}$$
$$lt+1+mt+2+nt+3=1. \quad (3)$$

From the second member of (3), we have

$$t=\frac{-5}{l+m+n}.$$

Putting this value of t in the first member of (3), we have

$$5l^2+3m^2+n^2-6mn-4nl-2lm=0. \quad (4)$$

Eliminating l, m, and n from (2) and (4), we obtain

$$5(x-1)^2+3(y-2)^2+(z-3)^2-6(y-2)(z-3)$$
$$-4(z-3)(x-1)-2(x-1)(y-2)=0$$

or

$$5x^2+3y^2+z^2-6yz-4zx-2xy$$
$$+6x+8y+10z-26=0.$$

EXAMPLE 14.87

The plane $\frac{x}{a}+\frac{y}{b}+\frac{z}{c}=1$ meets the coordinate axes in A, B, and C. Show that the equation of the cone generated by the lines drawn from O to meet the circle ABC is

$$yz\left(\frac{b}{c}+\frac{c}{b}\right)+zx\left(\frac{c}{a}+\frac{a}{c}\right)+xy\left(\frac{a}{b}+\frac{b}{a}\right)=0.$$

Solution. The given plane meets the axes in $A(a,0,0)$, $B(0,b,0)$, and $C(0,0,c)$. The equation of the sphere OABC is

$$x^2+y^2+z^2-ax-by-cz=0. \quad (1)$$

Therefore, the circle ABC is

$$x^2+y^2+z^2-ax-by-cz=0, \frac{x}{a}+\frac{y}{b}+\frac{z}{c}=1. \quad (2)$$

Since the vertex is the origin (0,0,0), the equation of the required cone is obtained by making equation (1) homogeneous with the help of $\frac{x}{a}+\frac{y}{b}+\frac{z}{c}=1$. Towards this task, we have

$$x^2+y^2+z^2-(ax+by+cz)\left(\frac{x}{a}+\frac{y}{b}+\frac{Z}{c}\right)=0$$

or

$$yz\left(\frac{b}{c}+\frac{c}{b}\right)+zx\left(\frac{c}{a}+\frac{a}{c}\right)+xy\left(\frac{a}{b}+\frac{b}{a}\right)=0,$$

as the required equation of the cone.

EXAMPLE 14.88

Find the equation of the cone whose vertex is the point (–1,1,2) and whose guiding curve is $3x^2-y^2=1$, $z=0$.

Solution. The equation of the guiding curve is

$$3x^2-y^2=1, z=0. \quad (1)$$

The equation of the generator through (–1,1,2) is

$$\frac{x+1}{l}=\frac{y-1}{m}=\frac{z-2}{n}. \quad (2)$$

This line meets the plane $z=0$ in the point

$$\left(-\frac{n+2l}{n},\frac{n-2m}{n},0\right) \text{ or } \left(-1-2\frac{l}{n},1-2\frac{m}{n},0\right).$$

This point lies on (1), and so,

$$3\left(-1-2\frac{l}{n}\right)^2-\left(1-2\frac{m}{n}\right)^2=1. \qquad (3)$$

But from (2), we have

$$\frac{l}{n}=\frac{x+1}{z-2} \text{ and } \frac{m}{n}=\frac{y-1}{z-2}.$$

Hence, (3) reduces to

$$3\left(-1-2\frac{x-1}{z-2}\right)^2-\left(1-2\frac{y-1}{z-2}\right)^2=1$$

or

$$12x^2-4y^2+z^2+4yz+12zx+4z-4=0.$$

EXAMPLE 14.89

The section of a cone whose vertex is P and guiding curve is the ellipse $\frac{x^2}{a^2}+\frac{y^2}{b^2}=1, z=0$ by the plane $x=0$ is a rectangular hyperbola. Show that the locus of the vertex P is $\frac{x^2}{a^2}+\frac{y^2+z^2}{b^2}=1$.

Solution. Let $P(\alpha, \beta, \gamma)$ be the vertex of the cone. The equation of the generator through (α, β, γ) is

$$\frac{x-\alpha}{l}=\frac{y-\beta}{m}=\frac{z-\gamma}{n}. \qquad (1)$$

This line meets the plane $z = 0$ in the point $\left(\alpha-\frac{l\gamma}{n},\beta-\frac{m\gamma}{n},0\right)$. This point will lie on the ellipse if

$$\frac{1}{a^2}\left(\alpha-\frac{l\gamma}{n}\right)^2+\frac{1}{b^2}\left(\beta-\frac{m\gamma}{n}\right)^2=1. \qquad (2)$$

From (1), we have

$$\frac{l}{n}=\frac{x-\alpha}{z-\gamma} \text{ and } \frac{m}{n}=\frac{y-\beta}{z-\gamma}.$$

Putting these values of $\frac{l}{n}$ and $\frac{m}{n}$ in (2), we get

$$\frac{1}{a^2}\left(\alpha-\gamma\frac{x-\alpha}{z-\gamma}\right)^2+\frac{1}{b^2}\left(\beta-\gamma\frac{y-\beta}{z-\gamma}\right)^2=1$$

or

$$\frac{1}{a^2}(\alpha z-\gamma x)^2+\frac{1}{b^2}(\beta z-\gamma y)^2=(z-\gamma)^2,$$

which is the equation of the cone with vertex (α, β, γ). The section cone (1) by the plane $x = 0$ is given by

$$\frac{\alpha^2 z^2}{b^2}+\frac{1}{b^2}(\beta z-\gamma y)^2=(z-\gamma)^2$$

or

$$\frac{\gamma^2 y^2}{b^2}+\left(\frac{\alpha^2}{a^2}+\frac{\beta^2}{b^2}-1\right)z^2-\frac{2\beta\gamma yz}{b^2}+2\gamma z-\gamma^2=0. \qquad (3)$$

The equation (3) will represent a rectangular hyperbola if

coeff. of y^2 + *coeff. of* $z^2 = 0$,

that is, if

$$\frac{\gamma^2}{b^2}+\left(\frac{\alpha^2}{a^2}+\frac{\beta^2}{b^2}-1\right)=0.$$

Hence, the locus of (α, β, γ) is

$$\frac{x^2}{a^2}+\frac{y^2}{b^2}+\frac{z^2}{b^2}=1 \text{ or } \frac{x^2}{a^2}+\frac{y^2+z^2}{b^2}=1.$$

14.32 RIGHT CIRCULAR CONE

A *right circular cone* is a surface generated by a straight line which passes through a fixed point and makes a constant angle with a fixed line through the fixed point. The fixed point is called the *vertex of the cone*, the fixed line is called the *axis of the cone*, and the constant angle is called the *semivertical angle* of the cone.

Let $P(x, y, z)$ be any point on the cone so that the semivertical angle $ZOP = \theta$.

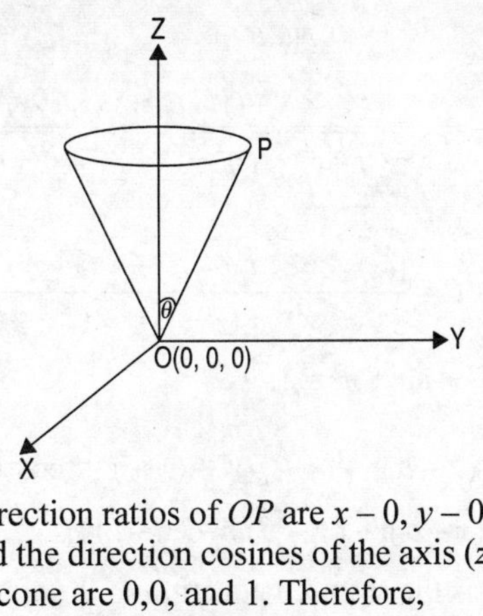

The direction ratios of OP are $x-0$, $y-0$, and $z-0$ and the direction cosines of the axis (z–axis) of the cone are 0,0, and 1. Therefore,

$$\cos\theta=\frac{x(0)+y(0)+z(1)}{\sqrt{0^2+0^2+1^2}\sqrt{x^2+y^2+z^2}}$$

$$= \frac{z}{\sqrt{x^2 + y^2 + z^2}}.$$

Thus,

$$\cos^2\theta = \frac{z^2}{x^2 + y^2 + z^2} \text{ or } \sec^2\theta = \frac{x^2 + y^2 + z^2}{z^2}.$$

This yields

$$(\sec^2\theta - 1)z^2 = x^2 + y^2 \text{ or } x^2 + y^2 = z^2\tan^2\theta,$$

which is called the *standard form of the right circular cone.*

14.33 RIGHT CIRCULAR CONE WITH VERTEX (α, β, γ), SEMI-VERTICAL ANGLE θ, AND $\langle l, m, n\rangle$ THE DIRECTION COSINES OF THE AXIS.

Let $P(x, y)$ be any point on the right circular cone whose vertex is $A(\alpha, \beta, \gamma)$ and the direction cosines of its axis AL are $<l,m,n>$. Let θ be the semi vertical angle of the cone. The direction ratios of AP are $<x-\alpha, y-\beta, z-r>$. Therefore,

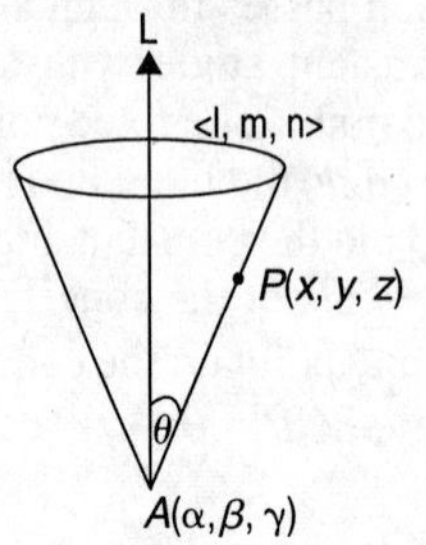

$$\cos\theta = \frac{(x-\alpha)l + (y-\beta)m + (z-\gamma)n}{\sqrt{(x-\alpha)^2 + (y-\beta)^2 + (z-\gamma)^2}.\sqrt{l^2 + m^2 + n^2}}$$

or

$$\cos^2\theta = \frac{[(x-\alpha)l + (y-\beta)m + (z-\gamma)n]^2}{(x-\alpha)^2 + (y-\beta)^2 + (z-\gamma)^2},$$

since $l^2 + m^2 + n^2 = 1$,

or

$$[(x-\alpha)^2 + (y-\beta)^2 + (z-\gamma)^2]\cos^2\theta$$

$$= [(x-\alpha)l + (y-\beta)m + (z-\gamma)n]^2, \quad (1)$$

which is the required equation of the cone.

Corollary 14.1: If the vertex is the origin $A(0,0,0)$, then taking $\alpha = \beta = \gamma = 0$ in the equation (1), we have

$$(x^2 + y^2 + z^2)\cos^2\theta = (lx + my + nz)^2,$$

which the *equation a right circular cone with vertex* (0,0,0).

EXAMPLE 14.90

Find the equation of the right circular cone which passes through the point (1,1,2) and has vertex at the origin and axis $\frac{x}{2} = \frac{y}{-4} = \frac{z}{3}$.

Solution. The direction cosines of the axis are

$$< l, m, n > = < \frac{2}{\sqrt{29}}, \frac{-4}{\sqrt{29}}, \frac{3}{\sqrt{29}} >.$$

We know that the equation of the cone through the origin, with semivertical angle θ and direction cosines of axis l, m, and n, is

$$(x^2 + y^2 + z^2)\cos^2\theta = (lx + my + nz)^2. \quad (1)$$

Since the cone passes through (1,1,2), we have

$$6\cos^2\theta = (l + m + 2n)^2 = \frac{1}{29}(2 - 4 + 6)^2 = \frac{16}{29}.$$

Thus, $\cos^2\theta = \frac{8}{87}$. Putting the values of l, m, n, and $\cos^2\theta$ in (1), we get

$$8x^2 + 8y^2 + 8z^2 - 3(2x - 4y + 3z)^2 = 0$$

or

$$8x^2 + 8y^2 + 8z^2 - 3[4x^2 + 16y^2 + 9z^2$$

$$-16xy - 24yz + 12xz] = 0$$

or

$$4x^2 + 40y^2 + 19z^2 - 48xy - 72yz + 36xz = 0,$$

which is the required equation of the cone.

EXAMPLE 14.91

If a right circular cone has three mutually perpendicular generators, Show that the semivertical angle is $\tan^{-1}\sqrt{2}$.

Solution. The equation, in a standard form, of a right circular cone with vertex at the origin, semivertical angle θ, and axis along z-axis, is

$$x^2 + y^2 - z^2\tan^2\theta = 0. \quad (1)$$

The cone (1) will have three mutually perpendicular generators if coefficient of x^2 + coefficient of y^2+ coefficient of $z^2 = 0$, that is, if $1 + 1 - \tan^2\theta = 0$ or $\tan\theta = \sqrt{2}$ or $\theta = \tan^{-1}\sqrt{2}$.

EXAMPLE 14.92

Find the equation of the right circular cone, whose vertex is the origin, axis is the line $\frac{x}{1}=\frac{y}{2}=\frac{z}{3}$, and which makes semivertical angle of 30°.

Solution. We know that the equation of the right circular cone with vertex (0,0,0), is

$$(x^2+y^2+z^2)\cos^2\theta=(lx+my+nz)^2, \qquad (1)$$

where θ is the semivertical angle and $< l, m, n >$ are direction cosines of the axis. The direction cosines of the axis are $<\frac{1}{\sqrt{14}},\frac{2}{\sqrt{14}},\frac{3}{\sqrt{14}}>$ and $\theta =$ 30°. Therefore, (1) reduces to

$$(x^2+y^2+z^2)\frac{3}{4}=\frac{1}{14}(x+2y+3z)^2$$

or

$$42(x^2+y^2+z^2)=4(x+2y+3z)^2$$

or

$$19x^2+13y^2+3z^2-24yz-12zx-8xy=0,$$

which is the required equation of the cone.

EXAMPLE 14.93

If θ is the semivertical angle of the right circular cone passing through the lines OX, OY, and $x = y = z$, show that

$$\cos\theta=\left[9-4\sqrt{3}\right]^{-\frac{1}{2}}.$$

Solution. Since the circular cone passes through OX, OY, and $\frac{x}{1}=\frac{y}{1}=\frac{z}{1}$, and all the above lines pass through (0,0,0), the origin is the vertex of the cone. Let $<l, m, n>$ be the direction cosines of the axis of the cone. The direction cosines of OX, OY, and the given line are respectively, <1, 0, 0>, <0, 1, 0>, and $<\frac{1}{\sqrt{3}},\frac{1}{\sqrt{3}},\frac{1}{\sqrt{3}}>$. Since the axis makes the same angle θ with each of OX, OY, and $\frac{x}{1}=\frac{y}{1}=\frac{z}{1}$, we have

$$\cos\theta=l(1)+m(0)+n(0)$$

$$=l(0)+m(1)+n(0)=\frac{l(1)+m(1)+n(1)}{\sqrt{3}}.$$

Therefore, $l=\cos\theta, m=\cos\theta$, and The last member yields $n=\sqrt{3}\cos\theta-l-n=\left(\sqrt{3}-2\right)\cos\theta$. Since l, m, and n are direction cosines, $l^2+m^2+n^2=1$, that is

$$\cos^2\theta+\cos^2\theta+(\sqrt{3}-2)^2\cos^2\theta=1$$

or

$$(9-4\sqrt{3})\cos^2\theta=1 \text{ or } \cos\theta=(9-4\sqrt{3})^{-\frac{1}{2}}.$$

EXAMPLE 14.94

Find the equation of the right circular cone generated by rotating the line $\frac{x}{1}=\frac{y}{2}=\frac{z}{3}$, about the line $\frac{x}{-1}=\frac{y}{1}=\frac{z}{2}$.

Solution. The equation of the axis is $\frac{x}{-1}=\frac{y}{1}=\frac{z}{2}$, whose direction cosines are $<\frac{-1}{\sqrt{6}},\frac{1}{\sqrt{6}},\frac{2}{\sqrt{6}}>$. The point of intersection of the rotating line $\frac{x}{1}=\frac{y}{2}=\frac{z}{3}$ with the axis is clearly (0,0,0). Therefore, the vertex of the cone is the origin (0,0,0). The angle θ (semivertical angle of the cone) between the rotating line and the axis is given by

$$\cos\theta=\frac{1(-1)+2(1)+3(2)}{\sqrt{1^2+2^2+3^2}\sqrt{(-1)^2+1^2+2^2}}$$

$$=\frac{7}{\sqrt{6}\sqrt{14}}$$

and so,

$$\cos^2\theta=\frac{49}{84}.$$

The equation of the right circular cone with vertex as the origin, semivertical angle θ, and direction cosines of the axis as $< l, m, n >$ is

$$\left(x^2+y^2+z^2\right)\cos^2\theta=(lx+my+nz)^2. \qquad (1)$$

Putting the values of $\cos^2\theta$ and $< l, m, n > = <\frac{-1}{\sqrt{6}},\frac{1}{\sqrt{6}},\frac{2}{\sqrt{6}}>$, in (1), we get

$$\left(x^2+y^2+z^2\right)\frac{49}{84}=\frac{1}{6}(-x+y+2z)^2$$

or

$$49\left(x^2+y^2+z^2\right)=14(-x+y+2z)^2$$

or

$$5x^2+5y^2-z^2+4xy-8yz+8zx=0.$$

14.34 CONICOIDS

The locus of the general equation of the second degree

$$ax^2+by^2+cz^2+2fyz+2gzx+2hxy+2ux+2vy+2wz+d=0 \qquad (1)$$

in x, y, and z is called a *quadric* or *conicoid*.

By the transformation of coordinate axes, the equation (1) can be transformed to any of following important surfaces:

1. $\frac{x^2}{a^2}+\frac{y^2}{b^2}+\frac{\neq}{d}=1$, surface is called the *ellipsoid.*
2. $\frac{x^2}{a^2}+\frac{y^2}{b^2}-\frac{z^2}{c^2}=1$, surface is called the *hyperboloid of one sheet.*
3. $\frac{x^2}{a^2}-\frac{y^2}{b^2}-\frac{z^2}{c^2}=1$, surface is called the *hyperboloid of two sheets.*
4. $\frac{x^2}{a^2}+\frac{y^2}{b^2}=\frac{2z}{c}$, surface is called the *elliptic paraboloid.*
5. $\frac{x^2}{a^2}-\frac{y^2}{b^2}=\frac{2z}{c}$, surface is called the *hyperbolic paraboloid.*
6. $\frac{x^2}{a^2}+\frac{y^2}{b^2}=1$, surface is called the *elliptic cylinder.*
7. $\frac{x^2}{a^2}-\frac{y^2}{b^2}=1$, surface is called the *hyperbolic cylinder.*
8. $\frac{x^2}{a^2}+\frac{y^2}{b^2}=-1$, surface is called the *imaginary cylinder.*

A conicoid whose center is the origin, that is, whose equation remains unchanged on changing x to $-x$, y to $-y$, and z to $-z$, is called a *central conicoid.* For example, the ellipsoid, the hyperboloid of one sheet, and the hyperboloid of two sheets are central conicoids.

14.35 SHAPE OF AN ELLIPSOID

The equation of the ellipsoid is

$$\frac{x^2}{a^2}+\frac{y^2}{b^2}+\frac{z^2}{c^2}=1. \qquad (1)$$

We note that

(i) The equation (1) is symmetrical with respect to all the three coordinate axes since the equation remains unaltered by changing x to $-x$, y to $-y$, and z to $-z$.

(ii) It meets the x-axis at $(a, 0, 0)$, $(-a, 0, 0)$, the y-axis at $(b, 0, 0)$, $(-b, 0, 0)$, and the z-axis at $(0, 0, c)$, $(0, 0, -c)$. These six points are called the *vertices* of the ellipsoid. The three-line segments joining the pair of vertices on each coordinate axis are called the *axes of the ellipsoid.* The three axes intersect at the point (0,0,0), called the *center of the ellipsoid.* The line segments joining the center to the vertices have lengths a, b, and c. These lengths are called the *semiaxes* of the ellipsoid. The largest of lengths, $2a$, $2b$, $2c$ is called the *major axis,* the next to largest is called the *mean axis,* and the smallest is called the *minor axis* of the ellipsoid.

(iii) Its sections by the coordinate planes xy-, yz-, and zx-plane, respectively, are the ellipses

$$\frac{x^2}{a^2}+\frac{y^2}{b^2}=1, \frac{y^2}{b^2}+\frac{z^2}{c^2}=1 \text{ and } \frac{x^2}{a^2}+\frac{z^2}{c^2}=1.$$

(iv) The plane section by the plane $z = k$ (arbitrary fixed constant) are the ellipses

$$\frac{x^2}{a^2}+\frac{y^2}{b^2}=\frac{a^2-k^2}{c^2},\ z=k, \text{ and } |k|<c.$$

The size of the ellipses decreases as $|k|$ increases. Similarly, the plane sections by the planes $x = k$ and $y = k$ are ellipses.

In view of the above facts, the sketch of the surface is as shown below:

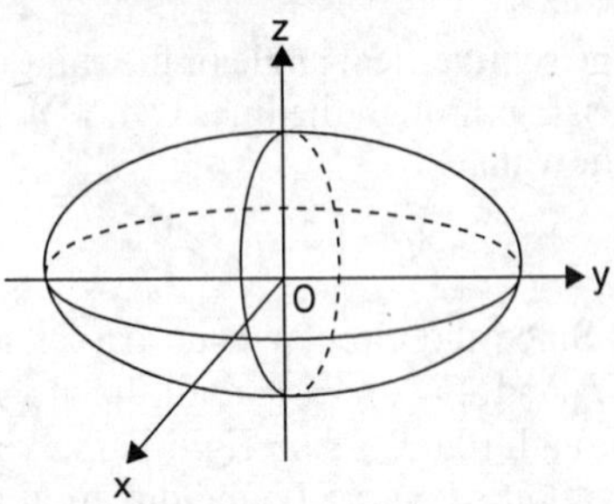

14.36 SHAPE OF THE HYPERBOLOID OF ONE SHEET

The equation of the hyperboloid of one sheet is

$$\frac{x^2}{a^2}+\frac{y^2}{b^2}-\frac{z^2}{c^2}=1. \qquad (1)$$

We observe that

(i) The equation (1) is symmetrical with respect to all the three coordinate axes.

(ii) The intersection of the surface with x- and y-axis are real, $(\pm a, O, O)$ and $(0, \pm b, O)$ respectively; but the intersection with z-axis is imaginary.

(iii) The plane section by the yz-plane is the

hyperbola $\frac{y^2}{b^2}-\frac{z^2}{c^2}=1$. Similarly, the plane sections by the *zx*- and xy-plane are respectively $\frac{x^2}{a^2}-\frac{z^2}{c^2}=1$ (hyperbola) and $\frac{x^2}{a^2}+\frac{y^2}{b^2}=1$ (ellipse).

(iv) The plane sections by the plane $z=k$ (plane parallel to xy-plane) are the ellipses

$$\frac{x^2}{a^2}+\frac{y^2}{b^2}=1+\frac{k^2}{c^2},\ z=k$$

The sizes of these ellipses increase as $|k|$ increases. The size becomes infinite as $|k|\to\infty$ For $k=0$, we get the smallest ellipse.

The shape of the surface is as shown below:

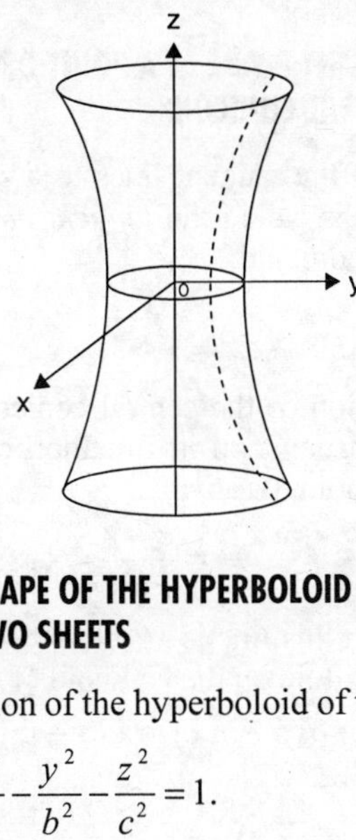

14.37 SHAPE OF THE HYPERBOLOID OF TWO SHEETS

The equation of the hyperboloid of two sheets is

$$\frac{x^2}{a^2}-\frac{y^2}{b^2}-\frac{z^2}{c^2}=1. \tag{1}$$

We note that

(i) The surface is symmetrical about all the three coordinate axes.

(ii) The surface meets the *x*-axis at the point $(a, 0, 0)$ and $(-a, 0, 0)$, whereas the *y*- and *z*-axis do not meet the surface at all.

(iii) The section by the planes $z = k$ and $y = k$ are the hyperbolas

$$\frac{x^2}{a^2}-\frac{y^2}{b^2}=1+\frac{k^2}{c^2}, z=k \text{ and}$$

$$\frac{x^2}{a^2}-\frac{z^2}{c^2}=1+\frac{k^2}{b^2}, y=k,$$

respectively. The plane $x = k$ does not meet the surface if $-a<k<a$. Hence, the surface does not exist between the planes $x=-a$ and $x=a$. But when $k^2>a^2$, that is, when $k\ge a$ or $k\ge -a$, the plane $x=k$ cuts the surface in the ellipse

$$\frac{y^2}{b^2}+\frac{z^2}{c^2}=\frac{k^2}{a^2}-1, x=k.$$

The sizes of these ellipses increase as k^2 increases. Hence, the sketch of the surface is as shown below.

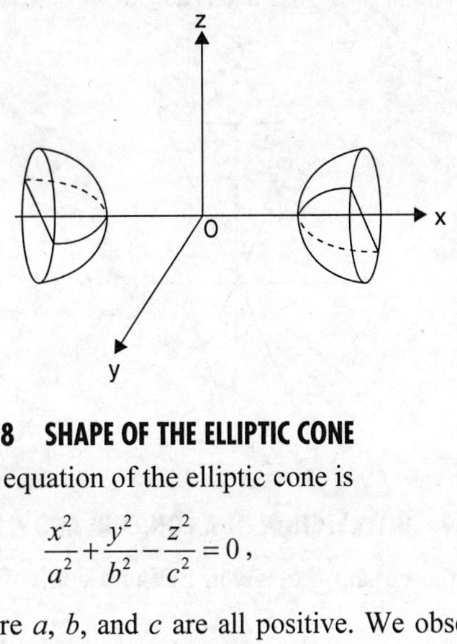

14.38 SHAPE OF THE ELLIPTIC CONE

The equation of the elliptic cone is

$$\frac{x^2}{a^2}+\frac{y^2}{b^2}-\frac{z^2}{c^2}=0, \tag{1}$$

where *a*, *b*, and *c* are all positive. We observe that

(i) The surface passes through the origin.

(ii) The surface is symmetrical about all the three coordinate axes.

(iii) The surface meets each of the three coordinate axes only at the origin.

(iv) The surface (1) meets *yz*-plane in $\frac{y^2}{b^2}-\frac{z^2}{c^2}=0$ or $\frac{y}{b}=\pm\frac{z}{c}$, which is a pair of straight lines through the origin. Similarly, it meets *zx*- plane in a pair of lines $\frac{x}{a}=\pm\frac{z}{c}$ and the *xy*- plane in $\frac{x^2}{a^2}+\frac{y^2}{b^2}=0$, which is the point ellipse $x = 0$ and $y = 0$ in the xy-plane. Thus, it meets the xy-plane in (0,0,0).

(v) The surface (1) is generated by a variable curve. The plane $z = k$ meets the surface (1) in the curve

$$\frac{x^2}{a^2}+\frac{y^2}{b^2}=\frac{k^2}{c^2},\ z=k.$$

Thus, the surface is generated by a variable ellipse $\frac{x^2}{a^2}+\frac{y^2}{b^2}=\frac{k^2}{c^2}$.

(vi) The surface (1) extends to infinity both above and below the xy-plane. Hence, the shape of the surface is as shown below:

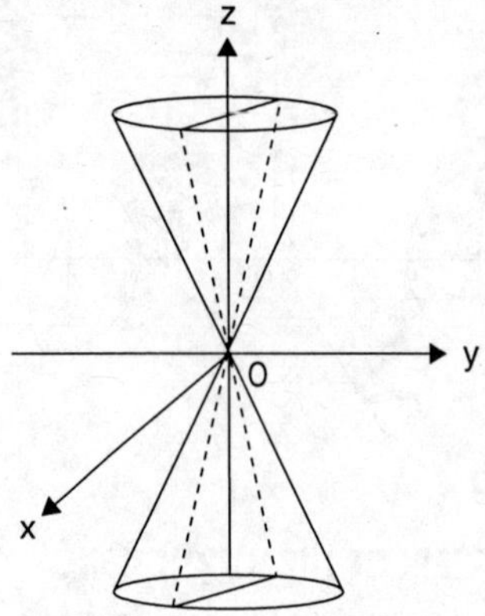

14.39 INTERSECTION OF A CONICOID AND A LINE

Let the equation of given line and conicoid be respectively

$$\frac{x-\alpha}{l}=\frac{y-\beta}{m}=\frac{z-\gamma}{n}=\lambda,\text{ say} \quad (1)$$

and

$$ax^2+by^2+cz^2=1. \quad (2)$$

Any point on the line (1) is $(\alpha+l\lambda,\beta+m\lambda,\gamma+n\lambda)$. This point will be on the conicoid (2) if

$$a(\alpha+l\lambda)^2+b(\beta+m\lambda)^2+c(\gamma+n\lambda)^2=1$$

or

$$\lambda^2(al^2+bm^2+cn^2)+2\lambda(a\alpha l+b\beta m+c\gamma n)$$
$$+a\alpha^2+b\beta^2+c\gamma^2-1=0. \quad (3)$$

The line (3) is a quadratic in λ and so, yields two values λ_1 and λ_2 of λ. Hence, a *line intersects a conicoid in two points* which may be real and distinct, imaginary or coincident.

If $\lambda_1=\lambda_2$, then these two points are coincident and so, the line becomes a *tangent line* to the conicoid. Thus, the *condition that the line* (1) *is a tangent line* to conicoid (2) is that the discriminant of (3) is equal to zero, that is

$$4(al\alpha+bm\beta+cn\gamma)^2-4(al^2+bm^2+cn^2)$$
$$\times(a\alpha^2+b\beta^2+c\gamma^2-1)=0$$

or

$$(al\alpha+bm\beta+cn\gamma)^2$$
$$=(al^2+bm^2+cn^2)(a\alpha^2+b\beta^2+c\gamma^2-1)\cdot$$

14.40 TANGENT PLANE AT A POINT OF CENTRAL CONICOID

The locus of the tangent lines to a conicoid at a point on it is called the *tangent plane* at that point of the conicoid.

Let

$$ax^2+by^2+cz^2=1 \quad (1)$$

be the equation of the central conicoid and let (α, β, γ)be a point on it. Equation of any line through the point (α, β, γ) is

$$\frac{x-\alpha}{l}=\frac{y-\beta}{m}=\frac{z-\gamma}{n}=\lambda,\text{ say.}$$

Any point on this line is $(\alpha+l\lambda,\beta+,m\lambda,\gamma+n\lambda)$. This point will lie on the conicoid (1) if

$$a(\alpha+l\lambda)^2+b(\beta+m\lambda)^2+c(\gamma+n\lambda)^2-1=0$$

or

$$\lambda^2\left(al^2+bm^2+cn^2\right)+2\lambda(a\alpha l+b\beta m+c\gamma n)+$$
$$a\alpha^2+b\beta^2+c\gamma^2-1=0. \quad (3)$$

Since (α, β, γ) is a point on the conicoid, $a\alpha^2+b\beta^2+\ c\gamma^2=1$ and so, (3) reduces to

$$\lambda^2\left(al^2+bm^2+cn^2\right)+2\lambda(a\alpha l+b\beta m+c\gamma n)=0.$$

Thus, $\lambda=0$ is one of the root, and the reduced equation is

$$\lambda\left(al^2+bm^2+cn^2\right)+2(a\alpha l+b\beta m+c\gamma n)=0$$

and so,

$$\lambda = -\frac{2(a\alpha l + b\beta m + c\gamma n)}{al^2 + bm^2 + cn^2}.$$

The line (2) will be the tangent line if both roots are equal. But as one of the root is zero, the second root should also be zero only. For this, we must have

$$a\alpha + b\beta m + c\gamma n = 0. \quad (4)$$

Eliminating l, m, and n from (2) and (4), we get the locus of the tangent line as

$$a\alpha(x-\alpha) + b\beta(y-\beta) + c\gamma(z-\gamma) = 0$$

or

$$a\alpha x + b\beta y + c\gamma z = a\alpha^2 + b\beta^2 + c\gamma^2$$

or

$$a\alpha x + b\beta y + c\gamma z = 1. \quad (5)$$

The equation (5) represents a plane, called the *tangent plane* to the conicoid $ax^2 + by^2 + cz^2 = 1$.

14.41 CONDITION OF TANGENCY

Let

$$lx + my + nz = p \quad (1)$$

be a plane. The equation of the tangent plane to the conicoid

$$ax^2 + by^2 + cz^2 = 1 \quad (2)$$

at any point (α, β, γ) is

$$a\alpha x + b\beta y + c\gamma z = 1. \quad (3)$$

The plane (1) will be a tangent plane to (2) if

$$lx + my + nz - p \equiv a\alpha x + b\beta y + c\lambda z - 1,$$

that is, if

$$\frac{a\alpha}{l} = \frac{b\beta}{m} = \frac{c\gamma}{n} = \frac{-1}{-p}$$

or if

$$\alpha = \frac{l}{ap}, \ \beta = \frac{m}{bp}, \text{ and } \gamma = \frac{n}{cp}.$$

But (α, β, γ) is a point on the conicoid (2). Therefore,

$$a\left(\frac{l}{ap}\right)^2 + b\left(\frac{m}{bp}\right)^2 + c\left(\frac{n}{cp}\right)^2 = 1$$

or

$$\frac{l^2}{a} + \frac{m^2}{b} + \frac{n^2}{c} = p^2, \quad (4)$$

which is the required *condition of tangency*.

From equation (4), we have

$$p = \pm\sqrt{\frac{l^2}{a} + \frac{m^2}{b} + \frac{n^2}{c}}.$$

Thus, $lx + my + nz = \sqrt{\frac{l^2}{a} + \frac{m^2}{b} + \frac{n^2}{c}}$ is always a tangent plane to the conicoid $ax^2 + by^2 + cz^2 = 1$ whatever be the values of l, m, and n. This shows that it is a *family of tangent planes* to the central conicoid $ax^2 + by^2 + cz^2 = 1$.

14.42 EQUATION OF NORMAL TO THE CENTRAL CONICOID AT ANY POINT (α, β, γ) ON IT

A line passing through a point P on a surface, which is perpendicular to the tangent plane, to the surface at P is called the *normal* to the surface at P.

Let $ax^2 + by^2 + cz^2 = 1$ be a central conicoid. Then, the equation of the tangent plane at (α, β, γ) to this conicoid is

$$a\alpha x + b\beta y + c\gamma z = 1.$$

The direction cosines of the normal to this plane are proportional to $a\alpha$, $b\beta$, and $c\gamma$. Hence, the *equation of the normal to the conicoid* at (α, β, γ) is

$$\frac{x-\alpha}{a\alpha} = \frac{y-\beta}{b\beta} = \frac{z-\gamma}{c\gamma}.$$

EXAMPLE 14.95

Show that from any point (f, g, h), six normals can be drawn to a conicoid.

Solution. The equation of the normal at any point (α, β, γ) on the conicoid $ax^2 + by^2 + cz^2 = 1$ is

$$\frac{x-\alpha}{a\alpha} = \frac{y-\beta}{b\beta} = \frac{z-\gamma}{c\gamma}.$$

It passes through (f, g, h) if

$$\frac{f-\alpha}{a\alpha} = \frac{g-\beta}{b\beta} = \frac{h-\gamma}{c\gamma} = \lambda, \text{ say.}$$

This yields

$$\alpha = \frac{f}{1+a\lambda}, \ \beta = \frac{g}{1+b\lambda}, \text{ and } \gamma = \frac{h}{1+c\lambda}.$$

Since (α, β, γ) lies on $ax^2 + by^2 + cz^2 = 1$, we get

$$\frac{af^2}{(1+a\lambda)^2} + \frac{bg^2}{(1+b\lambda)^2} + \frac{ch^2}{(1+c\lambda)^2} = 1,$$

which is of sixth degree in λ and so, yields six values of λ. Hence, there are six normals that can be drawn from a point (f, g, h) to the conicoid.

EXAMPLE 14.96

What surfaces are represented by the following equations?

(i) $z^2 = 4(1+x^2+y^2)$ and (ii) $\dfrac{x^2}{2}-\dfrac{y^2}{3}=z$.

Solution. (i) The given equation can be written as

$$4x^2+4y^2-z^2=-4$$

or $\dfrac{x^2}{1}+\dfrac{y^2}{1}-\dfrac{z^2}{4}=-1$ or $\dfrac{x^2}{-1}+\dfrac{y^2}{-1}+\dfrac{z^2}{4}=1,$

which is the hyperboloid of two sheets.

(ii) Taking $a=\sqrt{2}$, $b=\sqrt{3}$, and $c=2$, the given equation can be written as

$$\frac{x^2}{a^2}-\frac{y^2}{b^2}=\frac{2z}{c},$$

which is the hyperbolic paraboloid.

EXAMPLE 14.97

Show that the equation of the tangent plane to the ellipsoid $\dfrac{x^2}{a^2}+\dfrac{y^2}{b^2}+\dfrac{z^2}{c^2}=1$ at the point (α, β, γ) on it is $\dfrac{\alpha x}{a^2}+\dfrac{\beta y}{b^2}+\dfrac{\gamma z}{c^2}=1$.

Solution. We know that the equation of the tangent plane to the conicoid $Ax^2+By^2+Cz^2=1$ at the point (α, β, γ) is

$$A\alpha x+B\beta y+C\gamma z=1. \qquad (1)$$

But for the ellipsoid $\frac{x^2}{a^2}+\frac{y^2}{b^2}+\frac{z^2}{c^2}=1$, we have

$A=\dfrac{1}{a^2}$, $B=\dfrac{1}{b^2}$, and $C=\dfrac{1}{c^2}$.

Substituting the values of A, B, and C in (1), we get the equation of the tangent plane as

$$\frac{\alpha x}{a^2}+\frac{\beta y}{b^2}+\frac{\gamma z}{c^2}=1.$$

EXAMPLE 14.98

Find the equation of two tangent plane passing through the line $7x + 10y - 30 = 0$, $5y - 3z = 0$ and which touch the conicoid $7x^2+5y^2+3z^2-60=0$.

Solution. The equation of the conicoid is $\frac{7}{60}x^2+\frac{5}{60}y^2+\frac{3}{60}z^2=1$. Comparing it with the standard equation of the central conicoid $ax^2+by^2+cz^2=1$, we have $a=\frac{7}{60}$, $b=\frac{1}{12}$, and $c=\frac{1}{20}$.

Now, the equation of any plane through the given line is

$$7x+10y-30+\lambda(5y-3z)=0$$

or

$$7x+(10+5\lambda)y-3\lambda z-30=0. \qquad (1)$$

Comparing this with $lx+my+nz=p$, we get

$$l=7,\ m=10+5\lambda,\ n=-3\lambda, \text{ and } p=30.$$

The condition for tangency is

$$\frac{l^2}{a}+\frac{m^2}{b}+\frac{n^2}{c}=p^2$$

and so,

$$\frac{49}{\frac{7}{60}}+\frac{(10+5\lambda)^2}{\frac{1}{12}}+\frac{(-3\lambda)^2}{\frac{1}{20}}=(30)^2$$

or

$$2\lambda^2+5\lambda+3=0.$$

Solving this quadratic, we get $\lambda=-1$ and $-\frac{3}{2}$. Substituting these values of λ in (1), we get

$$7x+5y+3z-30=0 \text{ and}$$
$$14x+5y+9z-60=0$$

as the equations of the two tangent planes.

EXAMPLE 14.99

Show that the plane $lx + my + nz = p$ touches the ellipsoid $\dfrac{x^2}{a^2}+\dfrac{y^2}{b^2}+\dfrac{z^2}{c^2}=1$ if $p^2=a^2l^2+b^2m^2+c^2n^2$.

Solution. The plane $lx + my + nz = p$ touches the conicoid

$$Ax^2+By^2+Cz^2=1 \text{ if } \frac{l^2}{A}+\frac{m^2}{B}+\frac{n^2}{C}=p^2. \quad (1)$$

Here,

$$A=\frac{1}{a^2},\ B=\frac{1}{b^2}, \text{ and } C=\frac{1}{c^2}.$$

Therefore, the condition of tangency (1) becomes

$$a^2l^2+b^2m^2+c^2n^2=p^2.$$

EXAMPLE 14.100

Tangent planes are drawn to the ellipsoid $\frac{x^2}{a^2}+\frac{y^2}{b^2}+\frac{z^2}{c^2}=1$ through the point (α, β, γ). Show that the perpendiculars to these planes through the origin generate the cone $(ax+\beta y+\gamma z)^2 = a^2x^2+b^2y^2+c^2$

Solution. The equation of the plane through the point (α, β, γ) is

$$l(x-\alpha)+m(y-\beta)+n(z-\gamma)=0$$

or

$$lx+my+nz=\alpha l+\beta m+\gamma n. \qquad (1)$$

By the condition of tangency, this plane will touch ellipsoid if (see Example 14.99)

$$(\alpha l+\beta m+\gamma n)^2 = a^2l^2+b^2m^2+c^2n^2. \qquad (2)$$

On the other hand, the equation of the perpendicular to the plane (1) passing through the origin is

$$\frac{x}{l}=\frac{y}{m}=\frac{z}{n}. \qquad (3)$$

Eliminating l, m, and n from (2) and (3), we get

$$(\alpha x+\beta y+\gamma z)^2 = a^2x^2+b^2y^2+c^2z^2,$$

which is the required equation of the cone.

EXAMPLE 14.101

Show that the plane $x + 2y + 3z = 2$ touches the conicoid $x^2 - 2y^2 + 3z^2 = 2$. Find also the point of contact.

Solution. Equation of the given plane is $x + 2y + 3z = 2$. Comparing it with $lx + my + nz = p$, we have $l = 1, m = 2, n = 3$, and $p = 2$. The equation of the given conicoid is $x^2 - 2y^2 + 3z^2 = 2$ or $\frac{x^2}{2}-y^2+\frac{3}{2}z^2=1$. Comparing with the standard equation of the central conicoid $ax^2 + by^2 + cz^2 = 1$, we have $a=\frac{1}{2}, b=-1$, and $c=\frac{3}{2}$. The given plane will touch the given conicoid if

$$\frac{l^2}{a}+\frac{m^2}{b}+\frac{n^2}{c}=p^2 \text{ (Condition of tangency)}$$

or if

$$\frac{1}{\frac{1}{2}}+\frac{4}{-1}+\frac{9}{\frac{3}{2}}=(2)^2$$

or if $\quad 2-4+6=4$, which is true.

Hence, the given plane touches the given conicoid.

Now let $P(x_1, y_1, z_1)$ be the point of contact. The equation of tangent plane to the conicoid at (x_1, y_1, z_1) is

$$xx_1-2yy_1+3zz_1=2.$$

Comparing it with $x + 2y + 3z = 2$, we have

$$\frac{x_1}{1}=\frac{-2y_1}{2}=\frac{3z_1}{3}=\frac{2}{2} \text{ or } \frac{x_1}{1}=\frac{y_1}{-1}=\frac{z_1}{1}=1.$$

Therefore, $x_1 = 1, y_1 = -1$, and $z_1 = 1$. Hence, the point of contact is $(1, -1, 1)$.

EXAMPLE 14.102

Show that the locus of the foot of the perpendicular drawn from the center of the ellipsoid $\frac{x^2}{a^2}+\frac{y^2}{b^2}+\frac{z^2}{c^2}=1$ to any of its tangent planes is

$$a^2x^2+b^2y^2+c^2z^2=(x^2+y^2+z^2)^2.$$

Solution. Let $P(x_1,y_1,z_1)$ be the foot of the perpendicular from the center (0,0,0) of the given ellipsoid to any of its tangent plane. The equation of the plane to the ellipsoid at (x_1,y_1,z_1) is

$$l(x-x_1)+m(y-y_1)+n(z-z_1)=0. \qquad (1)$$

The direction ratios of the perpendicular line OP are $< x_1-0,\ y_1-0,\ z_1-0 >$. Since OP is perpendicular to the plane (1), we have $l=x_1$, $m=y_1$, and $n=z_1$. Therefore, (1) transforms to

$$x_1(x-x_1)+y_1(y-y_1)+z_1(z-z_1)=0$$

or

$$xx_1+yy_1+zz_1=x_1^2+y_1^2+z_1^2. \qquad (2)$$

Since the plane (2) touches the ellipsoid, the tangency condition $a^2l^2+b^2m^2+c^2n^2=p^2$ implies

$$a^2x_1^2+b^2y_1^2+c^2z_1^2=(x_1^2+y^2+z^2)^2.$$

Therefore, the locus of the foot of the perpendicular is

$$a^2x^2 + b^2y^2 + c^2z^2 = \left(x^2 + y^2 + z^2\right)^2.$$

14.43 MISCELLANEOUS EXAMPLES

EXAMPLE 14.103

Find the equation of the plane parallel to $2x - 3y + z = 6$ and which passes through the point (1,–2,1).

Solution. The equation of the plane parallel to $2x - 3y + z = 6$ is $2x - 3y + z + k = 0$. Since it passes through (1, –2,1), we have

$$2 + 6 + 1 + k = 0 \text{ or } k = -7.$$

Hence the equation of the required plane is

$$2x - 3y + z - 7 = 0.$$

EXAMPLE 14.104

Find the equation of the plane which passes through (2, 1, 3) and which contains the line $\frac{x}{1} = \frac{y-1}{2} = \frac{z+1}{-2}$.

Solution. The equation of the plane containing the given line is

$$a(x - 0) + b(y - 1) + c(z + 1) = 0.$$

Since the plane passes through (2, 1, 3), we have

$$2a + 0b + 4c = 0. \qquad (1)$$

Also, direction cosines of the line are proportional to 1, 2 and –2 and so

$$a + 2b - 2c = 0. \qquad (2)$$

From (1) and (2), we have

$$\frac{a}{8} = \frac{b}{-8} = \frac{c}{-4}$$

or

$$\frac{a}{2} = \frac{b}{-2} = \frac{c}{-1}.$$

Hence the equation of the required plane is

$$2(x - 0) + 2(y - 1) - 1(z + 1) = 0$$

or

$$2x - 2y - z + 1 = 0.$$

EXAMPLE 14.105

Find the image of the point (4,5,–2) in the plane $x - y + 3z - 4 = 0$.

Solution. Let (x_1, y_1, z_1) be the image of the point $P(4,5,-2)$ in the plane $x - y + 3z - 4 = 0$. Then the direction ratios of PQ are $x_1 - 4, y_1 - 5, z_1 + 2$. Since PQ must be perpendicular to the given plane, we have

$$\frac{x_1 - 4}{1} = \frac{y_1 - 5}{-1} = \frac{z_1 + 2}{3} = \lambda, \text{ say.}$$

Therefore $Q = (\lambda + 4, -\lambda + 5, 3\lambda - 2)$. Then midpoint M of PQ is

$$\left(\frac{\lambda+4+4}{2}, \frac{-\lambda+5+5}{2}, \frac{3\lambda-2-2}{2}\right) = \left(\frac{\lambda+8}{2}, \frac{\lambda+10}{2}, \frac{3\lambda-4}{2}\right).$$

But M lies on the given plane. Therefore,

$$\frac{\lambda+8}{2} - \left(\frac{-\lambda+10}{2}\right) + 3\left(\frac{3\lambda-4}{2}\right) - 4 = 0$$

or

$$\lambda + 8 + \lambda - 10 + 9\lambda - 12 - 8 = 0$$

or

$$11\lambda - 22 = 0 \text{ and so } \lambda = 2.$$

Hence the image Q of P is $Q = (6,3,4)$

EXAMPLE 14.106

Find the equation to the plane parallel to $x + y - z + 1 = 0$ and is $\sqrt{3}$ units from the origin.

Solution. The equation of the plane parallel to the plane $x + y - z + 1 = 0$ is $x + y - z = p$, where $\frac{p}{\sqrt{1+1+1}} =$ perpendicular distance from the origin to the parallel plane $= \sqrt{3}$ (given).

Thus $p = \sqrt{3}\left(\sqrt{3}\right) = 3$. Hence the equation of the parallel plane is

$$x + y - z = 3.$$

EXAMPLE 14.107

Find the image of the line $x + 5 = \frac{y+7}{6} = z$ in the plane $2x - y + z + 3 = 0$.

Solution. Let P(p,q,r) be any point on the given line $\frac{x+5}{1} = \frac{y+7}{6} = \frac{z}{1}$. The equation of the line through P and perpendicular to the given plane $2x - y + z + 3 = 0$ is

$$\frac{x-p}{2} = \frac{y-q}{-1} = \frac{z-r}{1} = \lambda, \text{ say.}$$

Any point on this line is $Q(2\lambda + p, -\lambda + q, \lambda + r)$. If this point is image of P, then the midpoint M of PQ is $M\left(\frac{2\lambda+2p}{2}, \frac{-\lambda+2q}{2}, \frac{\lambda+2r}{2}\right)$. Then M lies on the plane $2x - y + z + 3 = 0$ and so

$$2\lambda + 2p - \left(\frac{-\lambda+2q}{2}\right) + \left(\frac{\lambda+2r}{2}\right) + 3 = 0$$

or $\qquad 3\lambda + 2p - q + r + 3 = 0,$

which yields

$$\lambda = \frac{-2p+q-r-3}{3}.$$

Putting this value of λ in the coordinates of Q, we get

$$Q\left(\frac{-p+2q-2r-6}{3}, \frac{2p+2q+r+3}{3}, \frac{-2p+q+2r-3}{3}\right).$$

Two points on the given line are

$P_1(-5,-7,0)$ and $P_2(-4,-1,1)$.

Their images are

$Q_1(-5,-7,0)$ and $Q_2(-2,-2,2)$

Hence equation of the image line (using two point formula) is

$$\frac{x+5}{-2+5} = \frac{y+7}{-2+7} = \frac{z}{2}$$

or

$$\frac{x+5}{3} = \frac{y+7}{5} = \frac{z}{2}.$$

EXAMPLE 14.108

Find the length and equations of the shortest distance between

$$\frac{x+1}{2} = \frac{y+1}{3} = \frac{z+1}{4} \text{ and } \frac{x+1}{3} = \frac{y}{4} = \frac{z}{5}.$$

Solution. Let PQ be the shortest distance with l, m, n as its direction cosine. Then PQ is perpendicular to both lines. Therefore $2l + 3m + 4n = 0$ and $3l + 4m + 5n = 0$. Solving these equations, use get

$$\frac{l}{-1} = \frac{m}{2} = \frac{n}{-1} = \frac{\sqrt{l^2+m^2+n^2}}{\sqrt{1+4+1}} = \frac{1}{\sqrt{6}}.$$

The given lines pass through $(-1,-1,-1)$ and $(-1,0,0)$. Therefore the shortest distance between them is

$$PQ = (-1+1)\frac{1}{\sqrt{6}} + (0+1)\frac{1}{\sqrt{6}} + (0+1)\frac{1}{\sqrt{6}} = \frac{2}{\sqrt{6}}.$$

The equations of the line of shortest distance are

$$\begin{vmatrix} x & y & z \\ 2 & 3 & 4 \\ -1 & 2 & -1 \end{vmatrix} = 0 \text{ and } \begin{vmatrix} x & y & Z \\ 3 & 4 & 5 \\ -1 & 2 & -1 \end{vmatrix} = 0$$

or $\quad 11x + 2y + 7z = 0$ and $7x + y - 5z = 0$

EXAMPLE 14.109

Find the image of the point $(1, -1, 2)$ in the plane $2x + 2y + z = 1$.

Solution. Let $Q(x_1, y_1, z_1)$ be the image of the point $P(1,-1, 2)$ in the plane $2x + 2y + z - 1 = 0$. Then the direction rations of PQ are $x_1 - 1$, $y_1 + 1$, $z_1 - 2$. Since PQ is perpendicular to the given plane, we have

$$\frac{x_1-1}{2} = \frac{y_1+1}{2} = \frac{z_1-2}{1} = \lambda, \text{ say.}$$

Thus Q is $(2\lambda + 1, 2\lambda - 1, \lambda + 2)$. Now midpoint M of PQ is $\left(\frac{2\lambda+1+1}{2}, \frac{2\lambda-1-1}{2}, \frac{\lambda+2+2}{2}\right)$. But $M(\lambda + 1, \lambda - 1, \frac{\lambda}{2}+2)$ lies on the given plane. Therefore

$$2(\lambda+1) + 2(\lambda-1) + \left(\frac{\lambda}{2}+2\right) - 1 = 0,$$

which yields $\lambda = -\frac{2}{9}$. Hence the point Q is $\left(\frac{5}{9}, \frac{-13}{9}, \frac{14}{9}\right)$.

EXAMPLE 14.110

Find the magnitude and equations of the shortest distance between the lines $\frac{x-1}{2} = \frac{y-2}{3} = \frac{z-3}{4}$ and $\frac{x-2}{3} = \frac{y-4}{4} = \frac{z-5}{5}$.

Solution. The shortest distance is $\frac{1}{\sqrt{6}}$. Also, using two point formula, the equation of the line is

$$\frac{x-\frac{5}{3}}{\frac{3}{2}-\frac{5}{3}} = \frac{y-3}{\frac{10}{3}-3} = \frac{z-\frac{13}{3}}{\frac{25}{6}-\frac{13}{3}}$$

or

$$\frac{6x-10}{-1} = \frac{3y-9}{1} = \frac{6z-26}{-1}.$$

EXAMPLE 14.111

Find the equation of the sphere whose centre is the same as that of the sphere $x^2 + y^2 + z^2 - 2x - 4y - 6z + 7 = 0$ and which passes through $(1,-1,1)$.

Solution. The center of the given sphere is $(-g, -f, -h) = (1,2,3)$. The equation of the sphere with centre $(1, 2, 3)$ is $x^2 + y^2 + z^2 - 2x - 4y - 6z + k = 0$. Since it passes through $(1,-1,1)$, we have

$1+1+1-2+4-6+k=0$ or $k=1$.

Hence the required equation of the sphere is

$$x^2+y^2+z^2-2x-4y-6z+1=0.$$

EXAMPLE 14.112

Find the equation of the tangent planes to the sphere $x^2+y^2+z^2-2x-4y-6z-2=0$ which passes through the line

$$9x-3y+25=0=3x+4z+9.$$

Solution. Taking $z=0$, we have

$$9x-3y+25=0,\ 3x+9=0.$$

These equations give $x=-3,\ y=-\frac{2}{3},\ z=0$.

The equation of the sphere is

$$x^2+y^2+z^2-2x-4y-6z-2=0.$$

Comparing it with $x^2+y^2+z^2-2gx+2fy+2hz+c=0$, we get

$$g=-1, f=-2, h=-3, c=-2.$$

The equation of the tangent plane at (x_1, y_1, z_1) is

$$xx_1+yy_1+zz_1+g(x+x_1)+f(y+y_1)$$
$$+h(z+z_1)+c=0.$$

Therefore the equation of the required plane is

$$x(-3)+y\left(-\frac{2}{3}\right)+z(0)-(x-3)-2\left(y-\frac{2}{3}\right)-3z-2=0$$

or $-9x-2y-3(x-3)-2(3y-2)-9z-6=0$

or $12x+8y+9z-7=0$.

EXAMPLE 14.113

Find the centre, radius and area of the circle which is the intersection of the sphere $x^2+y^2+z^2-8x+4y+8z=45$ and the plane $x-2y+2z=3$.

Solution. The given sphere is

$$x^2+y^2+z^2-8x+4y+8z-45=0.$$

Its centre is $(-g,-f,-h)=(4,-2,-4)$. Further, the length of the perpendicular from $(4,-2,-4)$ to the plane $x-2y+2z=3$ is

$$p=\left|\frac{4(1)-2(-2)-4(2)-3}{\sqrt{1+4+4}}\right|=1$$

The radius of the sphere is

$$a=\sqrt{g^2+f^2+h^2-c}=\sqrt{16+4+16+45}=9.$$

Therefore the radius r of the circle is

$$r=\sqrt{a^2-p^2}=\sqrt{81-1}=\sqrt{80}=4\sqrt{5}.$$

The area of the circle is $A=\pi r^2=80\pi$.

Further, the perpendicular from $(4,-2,-4)$, being perpendicular to the plane, has direction ratios $1,-2,2$. Therefore the equation of the perpendicular is

$$\frac{x-4}{1}=\frac{y+2}{-2}=\frac{z+4}{2}=\lambda,\text{ say.}$$

Any point on this line is $(\lambda+4,-2\lambda-2,2\lambda-4)$. Since this point lies on the plane, we have $\lambda+4-2(-2\lambda-2)+2(2\lambda-4)=3$, and so $\lambda=\frac{1}{3}$. Hence centre of the circle is $(\lambda+4,-2\lambda-2,2\lambda-4)=\left(\frac{13}{3},\frac{-8}{3},\frac{-10}{3}\right)$.

EXAMPLE 14.114

The point (2, 3, 4) is one end of the diameter of a sphere $x^2+y^2+z^2-2x-2y+4z-1=0$, find the other end.

Solution. The equation of the sphere is

$$x^2+y^2+z^2-2x-2y+4z-1=0 \quad (1)$$

and (2, 3, 4) is one of the ends of the diameter of (1). Let the other end of the diameter be (x, y, z). The centre of the sphere (1) is $(-g,-f,-h)=(1,1,-2)$. Since centre is the midpoint of the diameter, we have $\frac{x+2}{2}=1, \frac{y+3}{2}=1$ and $\frac{z+4}{2}=-2$. Solving these equations, we get the other end of the diameter as $(x,y,z)=(0,-1,-8)$.

EXERCISES

Direction Ratios and Direction Cosines

1. The vertices of a triangle are $A(1,3,2), B(2,-1,3)$ and $C(-1,4,2)$. Find the direction cosines of the vectors $\vec{A}B$ and $\vec{A}C$ and find the angle A.

   Ans. $<\frac{1}{\sqrt{18}},\frac{-4}{\sqrt{18}},\frac{1}{\sqrt{18}}>,<\frac{-2}{\sqrt{5}},\frac{1}{\sqrt{5}},0>, A=\cos^{-1}\left(-\frac{2}{\sqrt{5}}\right)$

2. Determine whether the points $P(2,-1,3), Q(-3,2,4)$ and $R(-8,5,5)$ are collinear. Find the ratio in which R divides PQ.

   Ans. Collinear, R divides PQ externally in the ratio 2:1.

3. A line makes angles α,β,γ and δ with four diagonals of a cube. Show that

   Ans. $\cos^2\alpha+\cos^2\beta+\cos^2\gamma+\cos^2\delta=\frac{4}{3}$

4. Find the ratio in which the line joining $(2,4,16)$ and $(3,5,-4)$ is divided by the plane $2x-3y+z+6=0$

Hint: See Example 14.9.

Ans. 2:1.

5. Show that the points $\hat{i}-2\hat{j}+3\hat{k}, 2\hat{i}+3\hat{j}-4\hat{k}$ and $-7\hat{j}+10\hat{k}$ are collinear.

6. Show that the vectors $\vec{a}=3\hat{i}+\hat{j}-2\hat{k}$, $\vec{b}=\hat{i}+3\hat{j}+4\hat{k}$, and $\vec{c}=4\hat{i}-2\hat{j}-6\hat{k}$ form the sides of a triangle.

7. Find the coordinates of the foot of the perpendicular drawn from the point $A(1,8,4)$ to the line joining $B(0,-1,3)$ and $C(2,-3,-1)$.

Ans. $\left(-\frac{5}{3},\frac{2}{3},\frac{19}{3}\right)$

8. Find the angle between the line of intersection of the planes $3x+2y+z=5$ and $x+y-2z=3$ and the line of intersection of the planes $2x=y+z$ and $7x+10y=8z$.

Hint: Planes intersect in a line (see Example 14.19).

Ans. 90°

9. Find the image of the point (p,q,r) in the plane $2x+y+2=6$.

Ans. $\left(\frac{1}{3}[12-p-2q-2r],\frac{1}{3}[6-2p+2q-r],\frac{1}{3}[6-2p-q+2r]\right)$

10. Find the distance of the point $(3,-4,5)$ from the plane $2x+5y-6z=16$, measured parallel to the line $\frac{x}{2}=\frac{y}{1}=\frac{z}{-2}$.

Hint: Line through $(3,-4,5)$ having direction ratios 2, 1, − 2 is $\frac{x-3}{2}=\frac{y+4}{1}=\frac{z-5}{-2}=\lambda$.. Any point on this line is $(2\lambda+3,\lambda-4,-2\lambda+5)$ This point lies on the given plane if $2(2\lambda+3)+5(\lambda-4)-6(-2\lambda+5)=16$ that is, if $\lambda=\frac{20}{7}$. Thus, the point is $\left(\frac{61}{7},\frac{-8}{7},\frac{-5}{7}\right)$ The distance between $(3,-4,5)$ and this point is

$$\sqrt{\left(\tfrac{61}{7}-3\right)^2+\left(\tfrac{-8}{7}+4\right)^2+\left(\tfrac{-5}{7}-5\right)^2}=\tfrac{60}{7}=8.571.$$

11. Find the angle between the lines, whose direction cosines are given by $3l+m+5n=0$ and $6mn-2nl+5lm=0$

Hint: Solving the given equations, we get two sets of the direction cosines $<1,2,-1>$ and $<-2,1,1>$. Therefore, $\cos\theta=\frac{1(-2)+2(1)+(-1)(1)}{\sqrt{1+4+1}\sqrt{4+1+1}}=\frac{1}{6}$. or $\cos^{-1}\left(\frac{1}{6}\right)$.

12. Find the perpendicular distance of the point $(2,4,-1)$ from the line $\frac{x+5}{1}=\frac{y+3}{4}=\frac{z-6}{-9}$

Hint: See Example 14.17.

Ans. 7.

13. Find the foot of the perpendicular from the point $(0,2,3)$ to the line $\frac{x+3}{5}=\frac{y-1}{2}=\frac{z+4}{3}$. Also find the length of the perpendicular.

Ans. $(2,3,-1)$, Length $\sqrt{21}$.

14. Find the angle between the lines $\frac{x-2}{3}=\frac{y+1}{-2}=\frac{z}{2}$ and $\frac{x-1}{1}=\frac{y+3}{3}=\frac{z+5}{2}$.

Ans. $\theta=\cos^{-1}\left(\frac{1}{\sqrt{238}}\right)$.

15. The points A(4,5,10), B(2,3,4) and C(1,2,−1) are three vertices of a parallelogram ABCD. Find vector equation for the side AB and the coordinates of D.

Ans. $\vec{r}=(4\vec{i}+5\vec{j}+10\vec{k})+\lambda(-2\vec{i}+2\vec{j}-6\vec{k})$, $D(3,4,5)$

16. Determine the value of α for which the lines $\frac{x-\alpha}{1}=\frac{y-\beta}{-2}=\frac{z-\gamma}{0}$ and $\frac{x}{1}=\frac{y-3}{-1}=\frac{z+2}{2}$ intersect.

Ans. $\frac{5}{4}$.

17. Find the shortest distance between the lines $\vec{r}=3\vec{i}+8\vec{j}+3\vec{k}+\lambda(3\vec{i}-\vec{j}+\vec{k})$ and $\vec{r}=-3\vec{i}-7\vec{j}+6\vec{k}+\mu(-3\vec{i}+2\vec{j}+4\vec{k})$

Ans. $3\sqrt{30}$.

18. Find the shortest distance between the lines $\frac{x-3}{3}=\frac{y-8}{-1}=\frac{z-3}{1}$ and $\frac{x+3}{-3}=\frac{y+7}{2}=\frac{z-6}{4}$ Also find the equation of the line of shortest distance.

Ans. $3\sqrt{30}$, Equation: $\frac{x-3}{2}=\frac{y-8}{5}=\frac{z-3}{-1}$.

19. Find the shortest distance between the lines $\vec{r}=(3-t)\hat{i}+(4+2t)\hat{j}+(t-2)\hat{k}$ and $\vec{r}=(1+s)\hat{i}+(3s-7)\hat{j}+(2s-2)\hat{k}$, where t and s are scalars.

Ans. $\sqrt{35}$.

20. Find the shortest distance between the lines $\frac{x-5}{1}=\frac{y-4}{-2}=\frac{z-4}{1}$ and $\frac{x-1}{7}=\frac{y+2}{-6}=\frac{z+4}{1}$
Ans. $2\sqrt{29}$.

21. Determine whether or not the following pair of lines intersect:
$\vec{r}=\hat{i}-2\hat{j}+3\hat{k}+\lambda(-\hat{i}+\hat{j}-2\hat{k})$ and
$\vec{r}=\hat{i}-\hat{j}-\hat{k}+\mu(\hat{i}+2\hat{j}-2\hat{k})$
Ans. Do not intersect.

22. Show that the lines, $\frac{x+5}{3}=\frac{y+4}{1}=\frac{z-7}{-2}, 3x+2y+z=2$, and $x-3y+2z=13$ are coplanar.
Hint: Show that these lines intersect, that is, the shortest distance between them is zero.

23. Find the condition that the lines and $\frac{x-\alpha}{l}=\frac{y-\beta}{m}=\frac{z-\gamma}{n}$ and $ax+by+cz+d=0=a'x+b'y+c'z+d'$ are coplanar.
Ans. $\frac{a\alpha+b\beta+c\gamma+d}{al+bm+cn}=\frac{a'\alpha+b'\beta+c'\gamma+d'}{a'l+b'm+c'n}$.

24. Find the magnitude and the equation of the shortest distance between the lines
$\frac{x}{2}=\frac{y}{-3}=\frac{z}{1}$ and $\frac{x-2}{3}=\frac{y-1}{-5}=\frac{z+2}{2}$.
Hint: Proceed as in Example 14.23.
Ans. $\frac{1}{\sqrt{3}}, 4x+y-5z=0=7x+y-8z-31$

25. Show that the lines $\frac{x-1}{2}=\frac{y-2}{3}=\frac{z-3}{4}$ and $\frac{x-2}{3}=\frac{y-3}{4}=\frac{z+4}{5}$ are coplanar, and also find the equation of the plane containing them.
Hint: We note that

$$\begin{vmatrix} x_2-x_1 & y_2-y_1 & z_2-z_1 \\ l_1 & m_1 & n_1 \\ l_2 & m_2 & n_2 \end{vmatrix}$$

$$=\begin{vmatrix} 2-1 & 3-2 & 4-3 \\ 2 & 3 & 4 \\ 3 & 4 & 5 \end{vmatrix}=\begin{vmatrix} 1 & 1 & 1 \\ 2 & 3 & 4 \\ 3 & 4 & 5 \end{vmatrix}=0$$

Therefore, the given lines intersect and so, are coplanar. Further,

$$\begin{vmatrix} x-x_1 & y-y_1 & z-z_1 \\ l_1 & m_1 & n_1 \\ l_2 & m_2 & n_2 \end{vmatrix}=\begin{vmatrix} x-1 & y-2 & z-3 \\ 2 & 3 & 4 \\ 3 & 4 & 5 \end{vmatrix}$$

=0 yields
$(x-1)(-1)+(y-2)(2)+(z-3)(-1)=0$ or $x-2y+z=0$, which is the equation of the plane containing the given lines.

Plane

26. Find the equation of the plane passing through (1, 2, 2) and perpendicular to the planes $22x+3y-z=3y-z=2$ and $5x+3y-4z=5$
Ans. $3x+y+3z=7$.

27. Show that the points, (0, –1, –1), (4, 5, 1), (3, 9, 4) and $(-4,4,4)$ lie on a plane.
Hint: See Example 14.42.

28. Find the vector equation of the plane $\vec{r}=\hat{i}+\hat{j}+\lambda(\hat{i}+\hat{j}+\hat{k})+\mu(\hat{i}-22\hat{j}+3\hat{k})$ in the scalar-product form:
Hint: Setting $\vec{a}=\hat{i}-\hat{j}, \vec{b}=\hat{i}+\hat{j}+\hat{k}$, and $\vec{c}=\hat{i}-2\hat{j}+3\hat{k}$, we have $\vec{r}=\vec{a}+\lambda\vec{b}+\mu\vec{c}$
Therefore, the given plane represents a plane which passes through A(1,–1, 0) and is parallel to the vectors $\vec{b}$ and $\vec{c}$. Thus, the vectors $\vec{r}-\vec{a}, \vec{b}$ and $\vec{c}$ are coplanar. Therefore $\left[\vec{r}-\vec{a}\,\vec{b}\,\vec{c}\right]=0$, that is,

$$\begin{vmatrix} x-1 & y-2 & z-3 \\ 1 & 1 & 1 \\ 1 & -2 & 3 \end{vmatrix}=0 \text{ or}$$

$5x-2y-3z=7$

or $\left(x\hat{i}+y\hat{j}+z\hat{k}\right).\left(5\hat{i}-2\hat{j}-3\hat{k}\right)=7$ or
$\vec{r}.\left(5\hat{i}-2\hat{j}-3\hat{k}\right)=0$, which is the required form.

29. Find the equation of the plane passing through the points, $(-3,-2,-4),(2,4,1)$ and (4, 2, 4).
Hint: The equation of the plane is

$$\begin{vmatrix} x-x_1 & y-y_1 & z-z_1 \\ x_2-x_1 & y_2-y_1 & z_2-z_1 \\ x_3-x_1 & y_3-y_1 & z_3-z_1 \end{vmatrix} = 0 \text{ or}$$

$$\begin{vmatrix} x+3 & y+2 & z+4 \\ 5 & 6 & 5 \\ 7 & 4 & 8 \end{vmatrix} = 0$$

or $28x-5y-22z-14=0$.

30. A variable plane which remains at a constant distance $3p$ from the origin cuts the coordinate axes at A, B, and C. Show that the locus of the centroid of the triangle ABC is $\frac{1}{x^2}+\frac{1}{y^2}+\frac{1}{z^2}=\frac{1}{p^2}$.

Hint: If $< l, m, n >$ are direction cosines of the normal to the plane, then since distance remains $3p$, the equation of the plane is $lx+my+nz=3p$. Since it cuts the axes at A, B, and C, respectively, we have $A\left(\frac{3p}{l},0,0\right)$, $B\left(0,\frac{3p}{m},0\right)$, and $C=\left(0,0,\frac{3p}{n}\right)$. If (x_1,y_1,z_1) is centroid of ΔABC, then

$x_1=\frac{3p}{3l}$, $y_1=\frac{3p}{3m}$, $z_1=\frac{3p}{3n}$

or $l=\frac{p}{x_1}$, $m=\frac{p}{y_1}$, $n=\frac{p}{z_1}$.

Since, $l^2+m^2+n^2=1$, we have

$$\left(\frac{p}{x_1}\right)^2+\left(\frac{p}{y_1}\right)^2+\left(\frac{p}{z_1}\right)^2=1 \text{ or}$$

$$\frac{1}{x_1^2}+\frac{1}{y_1^2}+\frac{1}{z_1^2}=\frac{1}{p^2}.$$

Therefore, the locus of the centroid $(x_1,y_1;z_1)$ is $\frac{1}{x^2}+\frac{1}{y^2}+\frac{1}{z^2}=\frac{1}{p^2}$.

31. Show that the plane through the points $(1,1,1),(1,-1,1)$, and $(-7,3,-5)$ is perpendicular to the xz-plane.

Hint: Equation of the plane through the given points is $3x-4z+1=0$. The given plane is $y=0$. Use orthogonality condition $a_1a_2+b_1b_2+c_1c_2=0$ to get $3(0)+0(1)+(-4)(0)=0$.

32. Find the equation of the plane passing through $(3,4,2)$ and $(7,0,6)$ perpendicular to the plane $2x-5y=15$.

Hint: Plane passing through (3, 4, 2) is $A(x-3)+B(y-4)+C(z-2)=0$ As it passes through (7,0,6), we have $A-B+C=0$. Since it is perpendicular to the given plane, we have $2A-5B=0$. Eliminating A, B, and C from these three equations, we get $5x+2y-3z=17$.

33. Find the angle between the lines $3x+2y+z-5=0, x+y-2z-3=0$, and $8x-4y-2z-2=0, 7x+10y-8z=0$

Hint: The lines are perpendicular to the normal of the planes.

Ans. $\theta=\cos^{-1}\left(\frac{99}{5\sqrt{12651}}\right)$.

34. Find the angle between the planes $\vec{r}.(2\hat{i}-3\hat{j}+4\hat{k})=1$ and $\vec{r}.(-\hat{i}+\hat{j})=4$

Hint: See Example 14.50.

Ans. $\theta=\cos^{-1}\left(\frac{-5}{\sqrt{58}}\right)$.

35. Find the equation of the plane through the point (1, 1, 1) and perpendicular to $x+2y+3z-7=0$ and $2x-3y+4z=0$.

Hint: The plane passing through (1,1,1) is $A(x-1)+B(y-1)+C(z-1)=0$. It is perpendicular to the given Planes. Therefore, $A+2B+3C=0$ and $2A+3B+4C=0$. Eliminating A, B, and C between these equations we get $17x+2y-7z-12=0$.

36. Find the angle between the line $\frac{x-1}{2}=\frac{y+2}{3}=\frac{z-5}{4}$ and the plane $3x+y-z+3=0$.

Hint: See Example 14.51.

Ans. $\theta=\sin^{-1}\left(\frac{5}{\sqrt{319}}\right)$.

37. Show that the lines $\frac{x+4}{2}=\frac{y+6}{5}=\frac{z-1}{-2}$ and $3x-2y+z+5=0=2x+3y+4z-4$ intersect (are coplanar). Find their point of

intersection and the plane in which they lie.

Hint: Any point on the first line is $(3\lambda-4,\ 5\lambda-6,\ -2\lambda+1)$. This point will lie on the planes $3x-2y+z+5=0$ and $2x+3y+4z-4=0$, if $3(3\lambda-4)-2(5\lambda-6)+(-2\lambda+1)+5=0$,

$2(3\lambda-4)-2(5\lambda-6)+4(-2\lambda+1)-4=0$

Both these equations yield $\lambda=2$. Therefore, this given lines intersect, that is, they are coplanar. The point of intersection is $(3\lambda-4,\ 5\lambda-6,\ -2\lambda+1)=(2,\ 4,\ -3)$. The equation of plane through the second line is $3x-2y+z+k(2x+3y+4z-4)=0$. (1)

This plane will contain the first line if the point (2,4,–3) lies on it. Putting $x=2$, $y=4$, and $z=-3$ in the above equation of the plane (1), we get $k=\frac{3}{13}$. Substituting this value in (1), we get the plane as $45x-17y+25z+53=0$.

38. Show that the line

$\vec{r}=2\hat{i}-3\hat{j}+5\hat{k}+\lambda(\hat{i}-\hat{j}+2\hat{k}s)$

lies in the plane $\vec{r}.\left(3\hat{i}+\hat{j}-\hat{k}\right)=-2$.

Hint: Comparing the equation of the line with $\vec{r}=\vec{a}+\lambda\vec{b}$, we have $\vec{b}=\hat{i}-\hat{j}+2\hat{k}$. Comparing the equation of the plane with $\vec{r}.\vec{N}=d$, we have $\vec{N}=3\hat{i}+\hat{j}-\hat{k}$. The angle θ between the line and the plane is given by

$$\sin\theta=\frac{\vec{b}.\vec{N}}{|\vec{b}||\vec{N}|}=\frac{\left(\hat{i}-\hat{j}+2\hat{k}\right).\left(3\hat{i}+\hat{j}-\hat{k}\right)}{\sqrt{1+1+4}\sqrt{9+1+1}}$$

$$=\frac{3-1-2}{\sqrt{6}\sqrt{11}}=0.$$

Hence, the line is parallel to the plane. The perpendicular distance from (2,–3,5) to the plane is

$$\frac{\vec{a}.\vec{N}-D}{|\vec{N}|}=\frac{\left(2\hat{i}-3\hat{j}+5\hat{k}\right).\left(3\hat{i}+\hat{j}-\hat{k}\right)+2}{\sqrt{9+1+1}}=0.$$

Therefore, the line lies in the plane.

39. Show that the line

$\vec{r}=2\hat{i}-2\hat{j}+3\hat{k}+\lambda\left(\hat{i}-\hat{j}+4\hat{k}\right)$

is parallel to the plane $\vec{r}.\left(\hat{i}+5\hat{j}+\hat{k}\right)=5.$. Find the distance between them.

Hint: The angle between the line and the plane should be zero degree. Therefore, we should have $\vec{b}.\vec{N}=0$. We note that $\vec{b}.\vec{N}=\left(\hat{i}-\hat{j}+4\hat{k}\right).\left(\hat{i}+5\hat{j}+\hat{k}\right)=0.$

Hence, they are parallel. Distance between them

$$\frac{\vec{a}.\vec{N}-D}{|\vec{N}|}=\frac{\left(2\hat{i}-2\hat{j}+3\hat{k}\right).\left(\hat{i}+5\hat{j}+\hat{k}\right)+2}{\sqrt{1^2+5^2+1^2}}$$

$$\frac{|2-10+3-5|}{\sqrt{27}}=\frac{10}{3\sqrt{3}}.$$

40. Find the length and the coordinates of the foot of the perpendicular from the point (1, 1, 2) to the plane $2x-2y+4z+5=0$

Hint: Proceed as in Example 14.54.

Ans. Length $\frac{13}{\sqrt{24}}$, foot $\left(\frac{1}{12},\frac{25}{12},\frac{-1}{6}\right)$.

41. Find the equation of the plane passing through the intersection of the planes $x+2y+3x+5=0$ and $2x-4y+z-3=0$ and the point (0,1, 0).

Hint: Proceed as in Example 14.58.

Ans. $3x-2y+4z+2=0$.

42. Find the equation of the plane passing through the intersection of the planes $2x+3y+z=5$ and $3x-y+4z+3=0$ and perpendicular to the plane $x+y-3z=6$.

Hint: Proceed as in Example 14.59.

Ans. $13x+14y+9z-22=0$.

43. Find the equation of the planes bisecting the angle between the planes $3x-4y+5z=3$ and $5x+3y-4z=9$. Specify which of the planes bisect the acute angle.

Hint: Proceed as in Example 14.56.

Ans. $2x+7y-9z-6=0$ and $8x-y+z-12=0$, the plane $2x+7y-9z-6=0$ bisects the acute angle.

44. Find the symmetrical form of the equation of the line represented by the planes $3x-4y+5z=3$ and $5x+3y-4z=9$.

Hint: Putting $z = 0$ in these equations, we get $3x-4y-3=0$ and $5x+3y-9=0$. Solving, we get $\frac{x}{45}=\frac{y}{12}=\frac{1}{29}$ and so, a point on the line is $\left(\frac{45}{29},\frac{12}{29},0\right)$. The line lies on both the planes. Therefore, it is perpendicular to their normal. Hence, $\langle l, m, n\rangle$ are direction cosines of the line, then $3l-4m+5n=0$ and $5l+3m-4n=0$. Therefore, $\frac{l}{1}=\frac{m}{27}=\frac{n}{29}$, Thus, the line is

$$\frac{x-\frac{45}{29}}{1}=\frac{y-\frac{12}{29}}{27}=\frac{z}{29}.$$

45. Find the equation of the line passing through the point (–2, 3, 4) and parallel to the planes $2x+3y+4z=5$ and $4x+3y+5z=6$.

Hint: The line passing through the point (–2, 3, 4) is $\frac{x+2}{l}=\frac{y-3}{m}=\frac{z-4}{n}$. Since the above line is parallel to the planes, it is perpendicular to the normals to the given planes. Therefore, $2l+3m+4n=0$ and $4l+3m+5n=0$. Solving these, we have $\frac{l}{3}=\frac{m}{6}=\frac{n}{-6}$. Hence, the equation of the line is

$$\frac{x+2}{3}=\frac{y-3}{6}=\frac{z-4}{-6} \text{ or}$$

$$\frac{x+2}{1}=\frac{y-3}{2}=\frac{z-4}{-2}.$$

46. Find the volume of a tetrahedron in terms of the coordinates of its vertices.

Solution. Let $A(x_1,y_1,z_1)$, $B(x_2,y_2,z_2)$, $C(x_3,y_3,z_3)$, and $D(x_4,y_4,z_4)$, be the vertices of the tetrahedron. Then, D does not lie on the plane determined by the points A, B, and C. Thus, the three points may be assumed as lying on the plane of the paper and the point D as lying above this plane. The triangles ABC, BCD, CAD, and ABD are the four faces and the lines AB, CD, BC, AD, CA, and BD are the six edges. The edges AB and CD are called opposite edges. Similarly, BC, AD and CA, BD are opposite edges. Thus, the graph of the tetrahedron is as shown in the following figure:

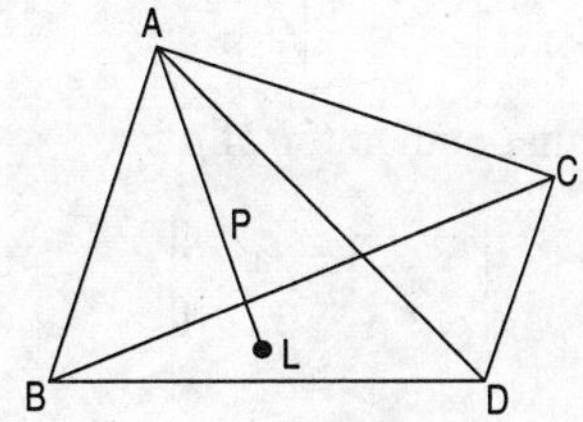

or

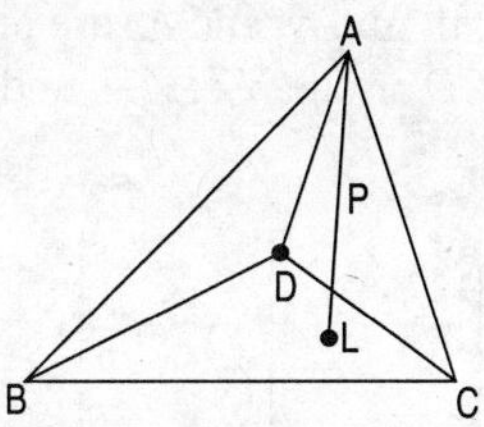

Let V be the volume of the tetrahedron ABCD. If p is the length of the perpendicular AL from vertex A on the opposite face BCD and Δ is the area of the triangle BCD, the volume (V) $=\frac{1}{3}p\Delta$.

Now, the equation of the plane BCD passing through three points $B(x_2,y_2,z_2)$, $C(x_3,y_3,z_3)$, and $D(x_4,y_4,z_4)$ is

$$\begin{vmatrix} x & y & z & 1 \\ x_2 & y_2 & z_2 & 1 \\ x_3 & y_3 & z_3 & 1 \\ x_4 & y_4 & z_4 & 1 \end{vmatrix}=0$$

or

$$x\begin{vmatrix} y_2 & z_2 & 1 \\ y_3 & z_3 & 1 \\ y_4 & z_4 & 1 \end{vmatrix}-y\begin{vmatrix} x_2 & z_2 & 1 \\ x_3 & z_3 & 1 \\ x_4 & z_4 & 1 \end{vmatrix}+z\begin{vmatrix} x_2 & y_2 & 1 \\ x_3 & y_3 & 1 \\ x_4 & y_4 & 1 \end{vmatrix}$$

$$-\begin{vmatrix} x_2 & y_2 & z_2 \\ x_3 & y_3 & z_3 \\ x_4 & y_4 & z_4 \end{vmatrix}=0. \qquad (1)$$

Now the length of the perpendicular from A to the plane (1) is equal to

$$\frac{x_1\begin{vmatrix}y_2 & z_2 & 1\\ y_3 & z_3 & 1\\ y_4 & z_4 & 1\end{vmatrix} - y_1\begin{vmatrix}x_2 & z_2 & 1\\ x_3 & z_3 & 1\\ x_4 & z_4 & 1\end{vmatrix} + z_1\begin{vmatrix}x_2 & y_2 & 1\\ x_3 & y_3 & 1\\ x_4 & y_4 & 1\end{vmatrix} - \begin{vmatrix}x_2 & y_2 & z_2\\ x_3 & y_3 & z_3\\ x_4 & y_4 & z_4\end{vmatrix}}{\left[\begin{vmatrix}y_2 & z_2 & 1\\ y_3 & z_3 & 1\\ y_4 & z_4 & 1\end{vmatrix}^2 + \begin{vmatrix}x_2 & z_2 & 1\\ x_3 & z_3 & 1\\ x_4 & z_4 & 1\end{vmatrix}^2 + \begin{vmatrix}x_2 & y_2 & 1\\ x_3 & y_3 & 1\\ x_4 & y_4 & 1\end{vmatrix}^2\right]^{\frac{1}{2}}}$$

We note that numerator of p is

$$\begin{vmatrix}x_1 & y_1 & z_1 & 1\\ x_2 & y_2 & z_2 & 1\\ x_3 & y_3 & z_3 & 1\\ x_4 & y_4 & z_4 & 1\end{vmatrix}.$$

Further, if $\Delta x, \Delta y$, and Δz are projections of the ΔBCD on the YZ, ZX, and XY planes, then

$$\Delta_x = \frac{1}{2}\begin{vmatrix}y_2 & z_2 & 1\\ y_3 & z_3 & 1\\ y_4 & z_4 & 1\end{vmatrix}, \ \Delta_y = \frac{1}{2}\begin{vmatrix}x_2 & z_2 & 1\\ x_3 & z_3 & 1\\ x_4 & z_4 & 1\end{vmatrix},$$

and

$$\Delta_z = \frac{1}{2}\begin{vmatrix}x_2 & y_2 & 1\\ x_3 & y_3 & 1\\ x_4 & y_4 & 1\end{vmatrix}.$$

Therefore, the denominator of p is equal to

$$\left[4\Delta_x^2 + 4\Delta_y^2 + 4\Delta_z^2\right]^{\frac{1}{2}} = 2\sqrt{\Delta_x^2 + \Delta_y^2 + \Delta_z^2} = 2\Delta.$$

Hence, $$p = \left|\frac{\begin{vmatrix}x_1 & y_1 & z_1 & 1\\ x_2 & y_2 & z_2 & 1\\ x_3 & y_3 & z_3 & 1\\ x_4 & y_4 & z_4 & 1\end{vmatrix}}{2\Delta}\right|.$$

Hence, the volume V of the tetrahedron is

$$V = \tfrac{1}{3}p\Delta = \tfrac{1}{6}\begin{vmatrix}x_1 & y_1 & z_1 & 1\\ x_2 & y_2 & z_2 & 1\\ x_3 & y_3 & z_3 & 1\\ x_4 & y_4 & z_4 & 1\end{vmatrix}.$$

47. Show that the three planes $2x-3y-7z=0$, $3x-14y-13z=0$ and $8x-31y-33z=0$ pass through one line.

Hint: Any plane that passes through the line of intersection of first two planes is

$2x-3y-7z=0+\lambda(3x-14y-13z)=0$

or

$(2+3\lambda)x-(3+14\lambda)y-(7+13\lambda)z=0.$

The given three planes will intersect in a common line if (1) and the third plane represent the same plane. Therefore,

$$\frac{2+3\lambda}{8}=\frac{3+14\lambda}{31}=\frac{7+13\lambda}{33}. \quad (2)$$

From the first two members of equation (2), we have $19\lambda = 38$ or $\lambda = 2$. This value of λ satisfies all equations in (2). Hence, the three given planes intersect in a line.

Sphere

48. Find the center and the radius of the sphere $x^2+y^2+z^2+4x-8y+6z+4=0$

Ans. $(-2,4,-3), 5$

49. Find the equation of the sphere concentric with the sphere $x^2+y^2+z^2-6x+2y-4z=2$ and double its radius.

Hint: The radius of the given sphere $= \sqrt{g^2+f^2+h^2-c} = \sqrt{9+1+4+2} = 4$ and its center is $(-g,-f,-h)=(3,-1,2)$.

Therefore, the center of the new sphere shall be $(3,-1,2)$ and its radius be 8. Thus, the equation of the required sphere is

$(x-3)^2+(y+1)^2+(z-2)^2=8^2$

or

$x^2+y^2+z^2-6x-2y-4z=50.$

50. Find the equation of the sphere for which the circle $x^2+y^2+z^2-3x+4y-2z-5=0$, $5x+2y+4z+7=0$ is a great circle.

Hint: Proceed as in Example 14.64

Ans. $x^2+y^2+z^2+2x+2y+2z+2=0.$

51. Find the equation of the sphere having its center on the plane $4x-5y-z=3$ and passing through the circle
$x^2+y^2+z^2-2x-3y+4z+8=0,$
$x-2y+z=8.$

Hint: Here the circle becomes great circle

Ans. $13x^2+13y^2+13z^2+-35x-21y+43z+176=0.$

52. Show that the plane $2x+2y-z+10=0$ touches the sphere $x^2+y^2+z^2-4x+2y-6z+5=0$. Also find its point of contact.

Ans. $(0,-3,4)$

53. Show that the two circles

$x^2+y^2+z^2-y+2z=0, x-y+z-2=0$

and

$x^2+y^2+z^2+x-3y+z=5, 2x-y+4z-1=0$

lie on the same sphere and find its equation.

Ans. $x^2+y^2+z^2+3x-4y+5z-6=0$

54. A sphere of constant radius k passes through the origin and meets the axes in A, B and C. Show that the centroid of the triangle ABC lies on the sphere $9(x^2+y^2+z^2)=4k^2$.

Hint: Let the coordinates of the points O, A, B and C be (0,0,0), $A(\alpha 0,0)$, $B(0,\beta,0)$, and $C(0,0,\gamma)$.
Using the four-point formula, the equation of the sphere through these points is $x^2+y^2+z^2-\alpha x-\beta y-\gamma z=0.$

The radius of this sphere is $\sqrt{\frac{\alpha^2}{4}+\frac{\beta^2}{4}+\frac{\gamma^2}{4}}=k$ (given). Thus $\alpha^2+\beta^2+\gamma^2=4k^2$. If (x_1,y_1,z_1) is the centroid of the triangle ABC, then

$$x_1=\frac{\alpha+0+0}{3}=\frac{\alpha}{3},\quad y_1=\frac{0+\beta+0}{3}=\frac{\beta}{3},$$

and $z_1=\dfrac{0+0+\gamma}{3}=\dfrac{\gamma}{3}.$

Thus, $\alpha=3x_1$, $\beta=3y_1$, and $\gamma=3z_1$. Putting these values in $\alpha^2+\beta^2+\gamma^2=4k^2$, we get $9x_1^2+9y_1^2+9z_1^2=4k^2$ and hence, the locus of (x_1,y_1,z_1) is

$9(x^2,y^2,z^2)=4k^2.$

55. Find the equation of the spheres which passes through the circle $x^2+y^2=4, z=0$ and is cut by the plane $x+2y+2z=0$ in a circle of radius 3.

Ans. $x^2+y^2+z^2\pm 6z-4=0$

56. A sphere of constant radius k passes through the origin) and meets the axes in A, B, and C. Prove that the locus of the foot of the perpendicular from O to the plane ABC is given by

$$(x^2+y^2+z^2)(x^{-2}+y^{-2}+z^{-2})=4k^2. \quad (1)$$

Hint: As in Exercise 54, the equation of the sphere through O, A, B, and C is $x^2+y^2+z^2-\alpha x-\beta y-\gamma z=0$. Since the radius is k, we also have $\alpha^2+\beta^2+\gamma^2=4k^2$. The intercept form of the plane ABC is $\frac{x}{\alpha}+\frac{y}{\beta}+\frac{z}{\gamma}=1$. The direction cosines of the normal to this plane are proportional to $\frac{1}{\alpha},\frac{1}{\beta},\frac{1}{\gamma}$. Therefore, the equation of the perpendicular from O to this plane is $\frac{x}{\frac{1}{\alpha}}+\frac{y}{\frac{1}{\beta}}+\frac{z}{\frac{1}{\gamma}}=\lambda$, say. Any point on this perpendicular is $\left(\frac{\lambda}{\alpha}+\frac{\lambda}{\beta}+\frac{\lambda}{\gamma}\right)$. This point lies on the plane if $\frac{\lambda}{\alpha^2}+\frac{\lambda}{\beta^2}+\frac{\lambda}{\gamma^2}=1$ and so $\lambda=\frac{1}{\alpha^{-2}+\beta^{-2}+\gamma^{-2}}$. Therefore, the foot of the perpendicular is

$$(x_1, y_1, z_1) = \left(\frac{\lambda}{\alpha}, \frac{\lambda}{\beta}, \frac{\lambda}{\gamma}\right) = \left(\frac{\alpha^{-1}}{\alpha^{-2}+\beta^{-2}+\gamma^{-2}}, \frac{\beta^{-1}}{\alpha^{-2}+\beta^{-2}+\gamma^{-2}}, \frac{\gamma^{-1}}{\alpha^{-2}+\beta^{-2}+\gamma^{-2}}\right).$$

Then

$$x_1^2 + y_1^2 + z_1^2 = \frac{1}{\alpha^{-2} + \beta^{-2} + \gamma^{-2}}$$

and

$$x_1^{-2} + y_1^{-2} + z_1^{-2} = \frac{\alpha^2 + \beta^2 + \gamma^2}{\left(\alpha^{-2} + \beta^{-2} + \gamma^{-2}\right)^{-2}}.$$

Putting these values in (1) and using $\alpha^2 + \beta^2 + \gamma^2 = 4k^2$, we get

$$\left(x^2 + y^2 + z^2\right)\left(x^{-2} + y^{-2} + z^{-2}\right) = 4k^2.$$

Hence, the locus of (x_1, y_1, z_1) is

$$\left(x^2 + y^2 + z^2\right)\left(x^{-2} + y^{-2} + z^{-2}\right) = 4k^2.$$

57. A sphere of constant radius k passes through the origin and meets the axes in A, B, and C. Show that the controid of the tetrahedron OABC lies on the sphere

$$x^2 + y^2 + z^2 = \frac{k^2}{4}.$$

Hint: Proceed as in Exercise 54. The only difference is that the controid of the tetrahedron is

$$\left(\frac{x_1 + x_2 + x_3 + x_4}{4}, \frac{y_1 + y_2 + y_3 + y_4}{4}, \frac{z_3 + z_2 + z_3 + z_4}{4}\right)$$

58. The plane ABC, whose equation is $\frac{x}{a} + \frac{y}{b} + \frac{z}{c} = 1$ meets the axes in A, B, and C. Find the equation of the circumcircle of the triangle ABC.

Hint: The circumcircle of the triangle ABC is the intersection of the plane ABC and any sphere through A, B, and C. We consider the sphere OABC, where O is the origin. As in Exercise 54, the equation of the sphere is $x^2 + y^2 + z^2 - ax - by - cz = 0$. Hence, the equation of the circumcircle of ΔABC is

$$x^2 + y^2 + z^2 - ax - by - cz = 0, \quad \frac{x}{a} + \frac{y}{b} + \frac{z}{c} = 1$$

59. Find the equations of the spheres which pass through the circle

$x^2 + y^2 + z^2 - 2x + 2y + 4z - 3 = 0$, $2x + y + z = 4$ and touch the plane $3x + 4y = 14$.

Hint: Proceed as in Example 14.71.

Ans. $x^2 + y^2 + z^2 - 2x + 2y + 4z - 3 = 0$ and $x^2 + y^2 + z^2 - 2x + 4y + 6z - 11 = 0$.

60. Find the equation of the sphere that passes through the circle

$x^2 + y^2 + z^2 - 2x + 3y - 4z + 6 = 0$,

$3x - 4y + 5z = 15$ and cuts orthogonally the sphere $x^2 + y^2 + z^2 + 2x + 4y - 6z + 11 = 0$.

Hint: First find the sphere through the circle and then use the condition of orthogonality to get the value of λ.

Ans. $5\left(x^2 + y^2 + z^2\right) - 13x + 19y - 25z + 45 = 0$.

61. Find the center and radius of the circle $x^2 + y^2 + z^2 = 49,\ 2x + 3y + 6z = 14$.

Hint: See Example 14.66.

Ans. $\left(\frac{4}{7}, \frac{6}{7}, \frac{12}{7}\right), \sqrt{45}$

62. Prove that the spheres

$x^2 + y^2 + z^2 + 2ax + c = 0$ and

$x^2 + y^2 + z^2 + 2by + c = 0 \quad (a^2, b^2 > c > 0)$

touch if and only if and only if $\frac{1}{a^2} + \frac{1}{b^2} = \frac{1}{c}$

Cylinder

63. Find the equation of the right circular cylinder of radius 2 and whose axis is the line $\frac{x-1}{2} = \frac{y-2}{1} = \frac{z-3}{2}$

Ans. $5x^2 + 8y^2 + 5z^2 - 4xy - 4yz - 8zx + 22x - 16y - 14z - 10 = 0$.

64. Obtain the equation of the right circular cylinder whose generating circle is $x^2+y^2+z^2=4,\ x+y+z-3=0.$

Hint: Proceed as in Example 14.83.

65. Find the equation of the cylinder whose generators are parallel to the line $\frac{x}{1}=\frac{y}{2}=\frac{z}{3}$ and which passes through the curve $x^2+y^2=16,\ z=0.$

Hint: Direction ratios of the generator are <1, 2, 3>. If (x_1, y_1, z_1) is any point on the cylinder, then the equation of generator through (x_1, y_1, z_1) is $\frac{x-x_1}{1}=\frac{y-y_1}{2}=\frac{z-z_1}{3}$. Equation of the guiding curve is $x^2+y^2=16,\ z=0.$

The line must intersect the guiding curve. Taking $z=0$, the line becomes $\frac{x-x_1}{1}=\frac{y-y_1}{2}=\frac{z-z_1}{3}$. The result yields $x=x_1-\frac{z_1}{3}$ and $y=y_1-\frac{2z_1}{3}$. Putting these values in $x^2+y^2=16,$ we get $9x_1^2+z_1^2-6z_1x_1+9y_1^2+4z_1^2-12y_1z_1=144$

Therefore, the locus of (x_1, y_1, z_1) is $9x^2+z^2-6zx+9y^2+4z^2-12yz=144.$

66. Find the equation of the cylinder whose generators are parallel to the line $\frac{x}{3}=\frac{y-4}{5}=\frac{z+1}{-4}$ and whose guiding curve is the hyperbola $3x^2-4y^2=5,\ z=2.$

Ans. $48x^2-64y^2-73z^2-160yz+72zx-144x+320y+292z-372=0.$

67. Find the equation of the right circular cylinder whose radius is 2 and axis is $\frac{x-1}{2}=\frac{y}{3}=\frac{z-3}{1}.$

Ans. $10x^2+5y^2+13z^2-12xy-6yz-4zx-8x+30y-74z+59=0.$

Cone

68. Find the equation of the cone with vertex at the origin and which passes through the curve $ax^2+by^2+cz^2=1,\ lx+my+nz=p.$

Ans. $p^2(ax^2+by^2+cz^2)=(lx+my+nz)^2.$

69. Find the equation of the cone with vertex (1, 1, 1) and passing through the curve of intersection of $x^2+y^2+z^2=1$ and $x+y+z=1.$

Ans. $x^2+y^2+z^2-2xy-2yz-2zx+2x+2y+2z-3=0.$

70. Find the equation of the cone with vertex at (2, 3, 1) and passing through the curve of intersection of $x^2+y^2+z^2-2x+4y-6x+7=0$ and $x+2y+2z=5.$

Ans. $38x^2+17y^2+157z^2+22xy+124yz+92zx-310z-270y-870z+1150=0.$

71. Find the equation of the right circular cone whose vertex is at the origin, axis is $\frac{x}{1}=\frac{y}{2}=\frac{z}{3}$, and vertical angle is $60°$.

Ans. $19x^2+13y^2+3z^2-24yz-12zx-8xy=0\cdot$

(see Example 14.92)

72. Find the equation of the right circular cone whose vertex is $(2,-3,5)$, axis makes equal angles with the axes, and the cone passes through $(1,-2,3)$.

Ans. $x^2+y^2+z^2+6(yz+zx+xy)-16x-36y-4z-28=0.$

73. Find the equation of the right circular cone generated when the straight line

$2y+3z=6,\ x=0$ revolves about z-axis.

Hint: Vertex is the point of intersection of the given line with z-axis. Thus, putting x=0 and y = 0 in the line, we get vertex as (0, 0, 2). One of the generator is $\frac{x-0}{0}=\frac{y-0}{3}=\frac{z-2}{-2}$.

Ans. $4\left(x^2+y^2\right)-9z^2+36z-36=0.$

Conicoids

74. Name the conicoid represented by (i) $x^2+y^2=9z^2$ and (ii) $x^2+y^2=5-2y$.

Ans. (i) Cone, (ii) Right circular cylinder

75. Show that the equation

$2y^2-8yz-4zx-8xy$

$+6x-4y-2z+5=0$ represents a cone. Find the coordinates of its vertex.

Hint: Make the given equation homogeneous by introducing another variable t. Thus, we have $F(x,y,z,t)=2y^2-8yz-4zx-8xy$ $+6xt-4yt-2zt+5t^2$.

This will represent a cone if and only if $F_x=F_y=F_z=F_t=0$ has a unique solution (means these equations are consistent). The unique solution to these equations will yield the vertex. These equations will come out to be

$F_x=-4z-8y+6t=0,$

$F_y=4y-8z-8x-4t=0,$

$F_z=-8y-4x-2t=0,$ and

$F_t=6x-4y-2z+10t=0.$

Taking t=1 and solving first three equations, we get $(x,y,z)=\left(\frac{-7}{6},\frac{1}{3},\frac{5}{6}\right)$, which satisfies the last equation $6x-4y-2z=0$.

76. Show that the equation

$x^2-2y^2+5z^2-4xy-5yz-6zx$

$+8x-19y-2z-20=0$ represents a cone.

Hint: Proceed as in Exercise 75.

77. Find the points of intersection of the line joining (4, 3, –1) and (1, 0, 2) with the hyperboloid $x^2-y^2+2z^2=5$.

Hint: Equation of the line joining the given points is $\frac{x-4}{3}=\frac{y-3}{3}=\frac{z+1}{-3}$ or $\frac{x-4}{1}=\frac{y-3}{1}=\frac{z+1}{-1}=\lambda$, say.

Any point on this line is $(\lambda+4,\lambda+3,-\lambda-1)$. That point lies on the hyperboloid and so, $(\lambda+4)^2-(\lambda+3)^2+2(-\lambda-1)^2=5$, which yields $\lambda=-1,-20$. Hence, the points of intersection are (3,2,0) and (2,1,1).

78. Find the points of intersection of the line $\frac{x+5}{-3}=\frac{y-4}{1}=\frac{z-11}{7}$ with the central conicoid $12x^2-17y^2-+8z^2-7=0$

Hint: Proceed as in Exercise 77.

Ans. $(-2,3,4),(1,2,-3)$.

79. Find the equation to the tangent planes to the surface $4x^2-5y^2+7z^2+13=0$ parallel to the plane $4x+20y-21z=0$.

Hint: We know that the tangent planes to the conicoid $ax^2+by^2+cz^2=1$ are given as $x+my+nz=\pm\sqrt{\frac{l^2}{a}+\frac{m^2}{b}+\frac{n^2}{c}}$. Here, l = 4, m = 20, n = –21 and the conicoid is $-\frac{4}{13}x^2+\frac{5}{13}y^2-\frac{7}{13}z^2=1$. Thus, $a=-\frac{4}{13}$, $b=\frac{5}{13}$, and $c=\frac{-7}{13}$. Hence, the required tangent planes are

$$4x+20y-21z=\pm\sqrt{\frac{16}{-\frac{4}{13}}+\frac{400}{\frac{5}{13}}+\frac{441}{-\frac{7}{13}}}$$
$$=\pm\sqrt{169}=\pm 13.$$

80. Show that the plane $3x+12y-6z-17=0$ touches the conicoid

$3x^2-6y^2+9z^2+17=0$.

81. Prove that the lines drawn from the origin parallel to the normals to $\frac{x^2}{a^2}+\frac{y^2}{b^2}+\frac{z^2}{c^2}=1$ at the point of intersection with the plane $lx+my+nz=p$, generate the cone $P^2(a^2x^2+b^2y^2+c^2z^2)=(a^2lx+b^2my+c^2nz)^2$.

15 Logic

Logic is the study of reasoning and is specifically concerned with whether a particular reasoning is valid. It is a science of the necessary laws of thought, without which no employment of the understanding and the reason takes place.

15.1 PROPOSITIONS

Definition 15.1 A declarative sentence that is either true or false, but not both is called a **Proposition** or a **Statement.**

EXAMPLE 15.1

Which of the following are propositions?

(i) London is the capital of France
(ii) Open the door
(iii) Take two tablets of medicine
(iv) $x + y > 0, x, y \in \mathbf{Z}$
(v) The only positive integers that divide a prime number are 1 and the number itself.
(vi) The sun is hot

Solution.

(i) The sentence is declarative and false. Hence, it is a proposition.
(ii) The sentence "Open the door" is not declarative. It is rather a command. Therefore, it is not a proposition.
(iii) The given sentence is not declarative and so it is not a proposition.
(iv) The sentence is not a statement because it is true for some values of x and y whereas for other values of x and y it is false. For example, if $x = 2$, $y = 1$, then it is true but if $x = -2, y = 1$, then it is false.
(v) The given sentence is declarative as well as true. Hence it is a statement.
(vi) The sentence "The sun is hot" is both a declarative sentence and true. Hence it is a proposition.

Notations. The propositions are represented by lower case letters such as p, q and r. Thus, the notation p: $3 + 7 = 10$ means that p is a proposition $3 + 7 = 10$.

Many propositions are composite, that is, composed of sub-propositions and various connectives. Thus, we have

Definition 15.2 Composite propositions are called **compound propositions.**

A proposition which is not compound is said to be **primitive.**

Thus, a primitive proposition cannot be broken into simpler propositions.

EXAMPLE 15.2

Consider the sentence: The sun is shining and it is cold. This is a compound proposition composed of two propositions:

The sun is shining

and

It is cold

connected by the connective "and". On the other hand, the proposition

London is in Denmark

is a **primitive statement**.

Definition 15.3 The truth values of a compound statement in terms of its component parts is called a **truth table.**

15.2 BASIC LOGICAL OPERATIONS

The three basic logical operations are

1. **Conjunction**
2. **Disjunction**
3. **Negation**

which correspond, respectively, to "and", "or" and "not".

Definition 15.4

The **conjunction** of two propositions p and q is the proposition p and q.

It is denoted by $p \wedge q$.

EXAMPLE 15.3

Let

p: This child is a boy

q: This child is intelligent.

Then

$p \wedge q$: This child is a boy and intelligent.

Thus, $p \wedge q$ is true, if the child is a boy and intelligent both.

Even if one of the component is false, $p \wedge q$ is false.

Thus,

"the proposition $p \wedge q$ is true if and only if the propositions p and q are both true".

The truth value of the compound proposition $p \wedge q$ is defined by the truth table given below:

p	q	$p \wedge q$
T	T	T
T	F	F
F	T	F
F	F	F

EXAMPLE 15.4

If

p: London is capital of India

q: A decade is 10 years

then p is false, q is true and the conjunction

$p \wedge q$: London is capital of India and a decade is 10 years

is **false**.

Definition 15.5 The **disjunction** of two proposition p and q is the proposition p or q and is denoted by $p \vee q$.

The compound statement $p \vee q$ is true if at least one of p or q is true. **It is false when both p and q are false.**

The truth value of the compound proposition $p \vee q$ is defined by the following truth table:

p	q	$p \vee q$
T	T	T
T	F	T
F	T	T
F	F	F

For example, if

p: London is capital of India

q: A decade is 10 years,

then p is false, q is true. The disjunction

$p \vee q$: London is capital of India or a decade is 10 years

is **true**.

EXAMPLE 15.5

Form the disjunction of p and q for each of the following:

(a) p: 2 is a positive integer, q: $\sqrt{2}$ is a rational number

(b) p: $2 + 3 \neq 5$, q: London is the capital of France.

Solution.

(a) $p \vee q$: 2 is a positive integer or $\sqrt{2}$ is a rational number. Since p is true, the disjunction $p \vee q$ is true, even though q is false.

(b) $p \vee q$: $2 + 3 \neq 5$ or London is the capital of France. Since both p and q are false, $p \vee q$ is false.

Remark 15.1.

It is clear from the above example that in logic, unlike in ordinary English, **we may join totally unrelated statements by the connective "or"**.

EXAMPLE 15.6

Consider the following four statements:

(i) Taj is in Agra and 7 is a prime number

(ii) Paris is in France and $\sqrt{2}$ is a rational number

(iii) Paris is in England and $2 + 2 = 4$

(iv) Paris is in England and $2 + 2 = 5$

Solution. Here the propositions are combined by connective "and". Therefore, the compound proposition shall be conjunction. We observe that in (i) both sub-propositions are true. In the rest three statements, at least one sub-proposition is false. Hence only (i) is true. The truth table is

p	q	$p \wedge q$
T	T	T
T	F	F
F	T	F
F	F	F

Definition 15.6 If p is a statement, the **negation** of p is the statement "not p", denoted by $\sim p$.

Thus, $\sim p$ is the statement "it is not the case that p".

Hence, if p is true, then $\sim p$ is false and if p is false, then $\sim p$ is true.

The truth table for negation is

p	$\sim p$
T	F
F	T

EXAMPLE 15.7

Give the negation of the following statements:

(a) p: $2 + 3 > 1$

(b) q: It is cold.

Solution.

(a) $\sim p$: $2 + 3$ is not greater than 1, that is, $\sim p$: $2 + 3 \leq 1$.

Since p is true in this case, $\sim p$ is false.

(b) $\sim q$: It is not the case that it is cold. More simply, $\sim q$: It is not cold.

Remark 15.2.

(i) In expressions that include the symbol $\sim$ as well as $\wedge$ or $\vee$, the order of operation is that **$\sim$ is performed first**. For example,

$$\sim p \wedge q = (\sim p) \wedge q.$$

(ii) An expression such as $p \wedge q \vee r$ is considered ambiguous. This expression must be written as either $(p \wedge q) \vee r$ or $p \wedge (q \vee r)$ to have meaning.

15.2.1 Translating from English to Symbols

We consider

EXAMPLE 15.8

Write each of the following sentences symbolically, letting p: "It is hot" and q: "It is sunny":

(a) It is not hot but it is sunny

(b) It is neither hot nor sunny

Solution.

(a) In logic, the words "but" and "and" mean the same thing. Generally, "but" is used in place of "and" when the part of the sentence that follows is in some way unexpected.

The given sentence is equivalent to " It is not hot and it is sunny" which can be written symbolically as $\sim p \wedge q$.

(b) The phrase neither A nor B means the same as not A and not B. Thus, to say "It is neither hot nor sunny" means that it is not hot and it is not sunny. Therefore, the given sentence can be written symbolically as $\sim p \wedge \sim q$.

Remark 15.3.

The notation for inequalities involves "and" and "or" statements. For example, if x, a and b are particular real numbers, then

$$x \leq a \quad \text{means} \quad x < a \quad \text{or} \quad x = a,$$
$$a \leq x \leq b \quad \text{means} \quad a \leq x \quad \text{and} \quad x \leq b.$$

We note that $2 \leq x \leq 1$ means $2 \leq x$ and $x \leq 1$, which is false, no matter what x happens to be. We have taken x, a and b as particular real numbers to ensure that sentences such as $x < a$ and $x \geq b$ are either true or false and hence that such sentences are statements.

Definition 15.7

A "Statement form" or "Propositional form" is an expression made up of **statement variables** (such as ~ , ∧, ∨) that becomes a statement when actual statements are substituted for the component statement variable. The **truth table** for a given statement form displays the truth values that correspond to the different combinations of truth values for the variables.

15.2.2 Truth Table for Exclusive OR

When OR is used in its exclusive sense, the statement "p or q" means "p or q but not both" or "p or q and not both p and q" which translates into symbols as

$$(p \vee q) \wedge \sim(p \wedge q).$$

This is sometimes abbreviated as

$$p \oplus q \quad \text{or} \quad p \text{ XOR } q.$$

The truth table for $p \oplus q$ is

p	q	$p \vee q$	$p \wedge q$	$\sim(p \wedge q)$	$(p \vee q) \wedge \sim(p \wedge q)$
T	T	T	T	F	F
T	F	T	F	T	T
F	T	T	F	T	T
F	F	F	F	T	F

It can be shown that $(p \wedge \sim q) \vee (q \wedge \sim p)$ have same truth table as $(p \vee q) \wedge \sim (p \wedge q)$. Hence these are logically equivalent (see Definition 15.8).

EXAMPLE 15.9

Construct a truth table for the statement form:

$$(p \wedge q) \vee \sim r.$$

Solution. The truth table for the given statement form is

p	q	r	$p \wedge q$	$\sim r$	$(p \wedge q) \vee \sim r$
T	T	T	T	F	T
T	T	F	T	T	T
T	F	T	F	F	F
T	F	F	F	T	T
F	T	T	F	F	F
F	T	F	F	T	T
F	F	T	F	F	F
F	F	F	F	T	T

Definition 15.8

Two different compound propositions (or statement forms) are said to be **logically equivalent** if they have the same truth value no matter what truth values their constituent propositions have. Thus,

"Two different compound propositions (or statement forms) are said to be **logically equivalent** if they have **identical truth tables**".

We use the symbol ≡ for logical equivalence.

EXAMPLE 15.10

Negation of the negation of a statement is equal to the statement. Symbolically, $\sim(\sim p) \equiv p$.

Solution. The truth table of $\sim(\sim p)$ is

p	$\sim p$	$\sim(\sim p)$
T	F	T
F	T	F

Thus, truth values for p and $\sim(\sim p)$ are same and hence p and $\sim(\sim p)$ are logically equivalent. The logical equivalence $\sim(\sim p) \equiv p$ is called **Involution Law.**

EXAMPLE 15.11

Show that the statement forms $\sim(p \wedge q)$ and $\sim p \wedge \sim q$ are not logically equivalent.

Solution. Construct the truth table for both statement forms:

p	q	$\sim p$	$\sim q$	$p \wedge q$	$\sim(p \wedge q)$		$\sim p \wedge \sim q$
T	T	F	F	T	F		F
T	F	F	T	F	T	$\neq$	F
F	T	T	F	F	T	$\neq$	F
F	F	T	T	F	T		T

Thus we have different truth values in rows 2 and 3 and so $\sim(p \wedge q)$ and $\sim p \wedge \sim q$ are not logically equivalent.

Remark 15.4.

If we consider $\sim p \vee \sim q$, then its truth values shall be

F
T
T
T

and hence $\sim(p \wedge q)$ and $\sim p \vee \sim q$ are logically equivalent. Symbolically,

$$\sim(p \wedge q) \equiv \sim p \vee \sim q.$$

Analogously,

$$\sim(p \vee q) \equiv \sim p \wedge \sim q.$$

The above two logical equivalence are known as **De Morgan's Laws of Logic**.

EXAMPLE 15.12

Use De Morgan's Law to write the negation of $-4 < x \leq 2$.

Solution. The given statement is equivalent to

$$-4 < x \quad \text{and} \quad x \leq 2.$$

By De Morgan's Law, the negation is

$$-4 \nless x \quad \text{or} \quad x \nleq 2,$$

which is equivalent to

$$-4 \geq x \quad \text{or} \quad x > 2.$$

EXAMPLE 15.13

Use De Morgan's Laws to write the negation of

p: John is a boy and John is handsome

Solution. The negation of p is

$\sim p$: John is not a boy or John is not handsome

Remark 15.5.

In logic, the connectives "and" and "or" are used between complete statements. Thus, we cannot write "John is not a boy and handsome."

Definition 15.9 A compound proposition which is **always true** regardless of truth values assigned to its component propositions is called a **Tautology**.

Definition 15.10 A compound proposition which is **always false** regardless of truth values assigned to its component propositions is called a **Contradiction.**

Definition 15.11 A compound proposition which can be either true or false depending on the truth values of its component propositions is called a **Contingency**.

EXAMPLE 15.14

Consider the statement form

$$p \vee \sim p.$$

The truth table for this statement form is

p	$\sim p$	$p \vee \sim p$
T	F	T
F	T	T
		↑ all T's

Hence $p \vee \sim p$ is a tautology.

EXAMPLE 15.15

Consider the statement form

$$p \wedge \sim p.$$

The truth table for this statement form is

p	$\sim p$	$p \wedge \sim p$
T	F	F
F	T	F

↑ all F's

Hence the statement form $p \wedge \sim p$ is a **Contradiction**.

Remark 15.6.

If τ and c denote tautology and contradictions, respectively, then we notice that

$$\sim \tau \equiv c \quad (1)$$

and

$$\sim c \equiv \tau. \quad (2)$$

Also from the above two examples

$$p \vee \sim p \equiv \tau \quad (3)$$

and

$$p \wedge \sim p \equiv c. \quad (4)$$

The logical equivalence (1), (2), (3) and (4) are known as **Complement Laws**.

15.3 LOGICAL EQUIVALENCE INVOLVING TAUTOLOGIES AND CONTRADICTIONS

If τ is a tautology and c is a contradiction, then the truth tables for $p \wedge \tau$ and $p \wedge c$ are:

p	τ	$p \wedge \tau$
T	T	T
F	T	F

Same truth values

p	c	$p \wedge c$
T	F	F
F	F	F

Same truth values

Hence, we have

$$p \wedge \tau \equiv p \quad \text{and} \quad p \wedge c \equiv c.$$

Similarly, the truth tables for $p \vee \tau$ and $p \vee c$ are

p	τ	$p \vee \tau$
T	T	T
F	T	T

Same truth values

p	c	$p \vee c$
T	F	T
F	F	F

Same truth values

Hence, we have

$$p \vee \tau \equiv \tau \quad \text{and} \quad p \vee c \equiv p.$$

Thus we have the following logical equivalence:

$$p \wedge \tau \equiv p,\ p \wedge c \equiv c$$

$p \vee \tau \equiv \tau$, $p \vee c \equiv p$ (**universal bound laws**).

These four logical equivalence are known as **Identity Laws**.

EXAMPLE 15.16 (Idempotent Laws)

Consider the truth tables for $p \wedge p$ and $p \vee p$ given below:

p	p	$p \wedge p$	$p \vee p$
T	T	T	T
F	F	F	F

We note that
(i) $p \wedge p$ and p have same truth values
(ii) $p \vee p$ and p have same truth values

Hence,

$$p \wedge p \equiv p \quad \text{and} \quad p \vee p \equiv p.$$

These two logical equivalence are known as **Idempotent Laws**.

EXAMPLE 15.17 (Commutative Laws)

Consider the truth tables of $p \wedge q$, $q \wedge p$; $p \vee q$ and $q \vee p$. We have

p	q	$p \wedge q$	$q \wedge p$
T	T	T	T
T	F	F	F
F	T	F	F
F	F	F	F

(Same truth values and so $p \wedge q \equiv q \wedge p$)

and

p	q	$p \vee q$	$q \vee p$
T	T	T	T
T	F	T	T
F	T	T	T
F	F	F	F

(Same truth values and so $p \vee q \equiv q \vee p$).

It follows therefore that $p \wedge q \equiv q \wedge p$ and $p \vee q \equiv q \vee p$.

These logical equivalences are known as **Commutative Laws.**

EXAMPLE 15.18 (Absorption Laws)

Consider the truth tables of $p \wedge (p \vee q)$ and $p \vee (p \wedge q)$. We have

p	q	$p \wedge q$	$p \vee (p \wedge q)$
T	T	T	T
T	F	F	T
F	T	F	F
F	F	F	F

(Same truth values and so $p \vee (p \wedge q) \equiv p$)

and

p	q	$p \vee q$	$p \wedge (p \vee q)$
T	T	T	T
T	F	T	T
F	T	T	F
F	F	F	F

(Same truth values and so $p \wedge (p \vee q) \equiv p$)

Hence,

$$p \wedge (p \vee q) \equiv p$$

and

$$p \vee (p \wedge q) \equiv p.$$

These logical equivalence are known as **Absorption Laws.**

EXAMPLE 15.19 (Associative Laws and Distributive Laws)

If p, q and r are propositions, then:

$$(p \wedge q) \wedge r \equiv p \wedge (q \wedge r),$$

$$(p \vee q) \vee r \equiv p \vee (q \vee r) \text{ (\textbf{Associative Laws})}$$

and

$$p \wedge (q \vee r) = (p \wedge q) \vee (p \wedge r),$$

$$p \vee (q \wedge r) = (p \vee q) \wedge (p \vee r) \text{ (\textbf{Distributive Laws})}.$$

These laws can be established by the readers themselves.

EXAMPLE 15.20

Simplify the logical statement

$$\sim(\sim p \wedge q) \wedge (p \vee q).$$

Solution. We have

$$\sim(\sim p \wedge q) \wedge (p \vee q) \equiv (\sim(\sim p) \vee \sim q) \wedge (p \vee q) \quad \text{(De Morgan Law)}$$

$$\equiv (p \vee \sim q) \wedge (p \vee q) \quad \text{(Involution Law)}$$

$$\equiv p \vee (\sim q \wedge q) \quad \text{(Distributive Law)}$$

$$\equiv p \vee (q \wedge \sim q) \quad \text{(Commutative Law)}$$

$$\equiv p \vee c \quad \text{(Complement laws or Negation Law)}$$

$$\equiv p \quad \text{(Identity law).}$$

15.4 CONDITIONAL PROPOSITIONS

Definition 15.12 If p and q are propositions, the compound proposition

if p then q or p implies q

is called a **conditional proposition** or **implication** and is denoted by

$$p \rightarrow q.$$

The proposition p is called the **hypothesis** or **antecedent** whereas the proposition q is called the **conclusion** or **consequent.**

The connective if……then is denoted by the symbol $\rightarrow$.

It is false when p is true and q is false, otherwise it is true. In particular, if p is false, then $p \rightarrow q$ is true for any q.

Definition 15.13 A conditional statement that is true by virtue of the fact that its hypothesis is false is called **true by default** or **vacuously true.**

For example, the conditional statement "If 4 is a prime number, then I am President of America" is **true** simply because p: 4 is a prime number is false. So it is not the case that p is true and q is false simultaneously.

Thus the truth values of the conditional proposition $p \rightarrow q$ are defined by the truth table:

p	q	$p \rightarrow q$
T	T	T
T	F	F
F	T	T
F	F	T

Each of the following expressions is an equivalent form of the conditional statement $p \rightarrow q$:

p implies q

q if p

p only if q

p is sufficient condition for q

q is necessary condition for p.

EXAMPLE 15.21

Restate each proposition in the form of a conditional proposition:

(a) I will eat if I am hungry.
(b) 3 + 5 = 8 if it is snowing.
(c) When you sing, my ears hurt.
(d) Ram will be a good teacher if he teaches well.
(e) A sufficient condition for Sohan to visit Calcutta is that he goes to Disney land.

Solution.

(a) If I am hungry, then I will eat
(b) If it is snowing, then 3 + 5 = 8.
(c) If you sing, then my ears hurt.
(d) If Ram teaches well, then he will be a good teacher.
(e) If Sohan visits Calcutta, then he goes to Disney land.

15.4.1 Representation of "if ... then" as "or"

Lemma 15.1.

Show that for propositions p and q,

$$p \rightarrow q \equiv \sim p \vee q.$$

Proof: The truth values for $p \rightarrow q$ and $\sim p \vee q$ are given below:

p	q	$p \rightarrow q$	$\sim p$	$\sim p \vee q$
T	T	T	F	T
T	F	F	F	F
F	T	T	T	T
F	F	T	T	T

↑ Same truth values ↑ (columns $p \rightarrow q$ and $\sim p \vee q$)

Hence,

$$p \rightarrow q \equiv \sim p \vee q.$$

EXAMPLE 15.22

Rewrite the statement in "If … then" form:

Either you teach well or you are terminated.

Solution. Let

$\sim p$: you teach well

and

q: you are terminated.

Then the given statement is $\sim p \vee q$. But,

p: you do not teach well.

Hence, the equivalent "If….then" version of the given statement is

If you do not teach well, then you are terminated.

Negation of a conditional statement. We

know that $p \to q$ is false if and only if p is true and its conclusion q is false. Also, we have shown above that

$$p \to q \equiv \sim p \vee q.$$

Taking negation of both sides, we have

$$\sim(p \to q) \equiv \sim(\sim p \vee q)$$

$$\equiv \sim(\sim p) \wedge (\sim q) \text{ (De Morgan's Law)}$$

$$\equiv p \wedge \sim q \text{ (Double negative Law or Involution Law).}$$

This can also be obtained by constructing the truth tables for $\sim(p \to q)$ and $p \wedge \sim q$; the truth tables would have the same truth values proving the logical equivalence.

Thus,

The negation of "If p then q" is **logically equivalent to** "p and not q".

EXAMPLE 15.23

Write negations for each of the following statements:

(a) If I am ill, then I cannot go to university.
(b) If my car is in the repair shop, then I cannot attend the class.

Solution. We know that negation of "If p then q" is logically equivalent to "p and not q". Using this fact, the negations of (a) and (b) are respectively

1. I am ill and I can go to university.
2. My car is in the repair shop and I can attend the class.

Remark 15.7.

The negation of "if ... then" proposition does not start with the word "if".

Definition 15.14 If $p \to q$ is an implication, then the **converse** of $p \to q$ is the implication $q \to p$.

Definition 15.15 The **contrapositive** of a conditional statement "If p then q" is "If $\sim q$ then $\sim p$".

In symbols,

The contrapositive of $p \to q$ is $\sim q \to \sim p$.

Lemma 15.2.

A conditional statement is logically equivalent to its contrapositive.

Proof: The truth tables of $\boldsymbol{p \to q}$ and $\boldsymbol{\sim q \to \sim p}$ are:

$p \to q$

p	q	$p \to q$
T	T	T
T	F	F
F	T	T
F	F	T

$\sim q \to \sim p$

p	q	$\sim p$	$\sim q$	$\sim q \to \sim p$
T	T	F	F	T
T	F	F	T	F
F	T	T	F	T
F	F	T	T	T

Same truth values

Hence,

$$p \to q \equiv \sim q \to \sim p.$$

EXAMPLE 15.24

Give the converse and contrapositive of the implications:

(a) If it is hot, then I take cold drinks.
(b) If today is Monday, then tomorrow is Tuesday.

Solution. (a) We have

p: It is hot

q: I take cold drinks

The converse is $q \to p$: If I take cold drinks, then it is hot.

The contrapositive is $\sim q \to \sim p$: If I do not take cold drinks, then it is not hot.

(b) We have

p: Today is Monday

q: Tomorrow is Tuesday

The converse is $q \to p$: If Tomorrow is Tuesday, then today is Monday.

The contrapositive is $\sim q \to \sim p$: If tomorrow is not Tuesday, then today is not Monday.

Definition 15.16 The **inverse** of the conditional statement $p \to q$ is $\sim p \to \sim q$.

For example, the inverse of "If today is Sunday, then tomorrow is Monday" is

"If today is not Sunday, then tomorrow is not Monday".

Remark 15.8.

If a conditional statement is true, then its converse and inverse may or may not be true.

Only if: "p only if q" means that p can take place only if q also takes place. That is, if q does not take place, then p cannot take place, i.e., $\sim q \to \sim p$. Therefore, equivalence between a statement and its contrapositive imply that " if p occurs, then q must also occur". Hence,

If p and q are statements, "p only if q" means "if not q, then not p" or equivalently "if p then q".

Remark 15.9.

"p only if q" does not mean "p if q".

EXAMPLE 15.25

Use contrapositive to rewrite the following statement in "if then" form:

"Ram will stand first in the class only if he works 12 hours a day."

Solution. Version 1: We have

p: Ram will stand first in the class,

q: He works 12 hours a day.

The contrapositive is $\sim q \to \sim p$: If Ram does not works 12 hours a day, then he will not stand first in the class.

Version 2: If Ram stands first in the class, then he will work 12 hours a day.

Definition 15.17 If p and q are statements, the compound statement "p if and only if q" is called a **Biconditional statement** or an **equivalence**. It is denoted by $p \leftrightarrow q$.

Observe that $p \leftrightarrow q$ is true only when both p and q are true or when both p and q are false (i.e., if both p and q have same truth values) and is false if p and q have opposite truth values. Thus, the biconditional statement has the following truth table:

$p \leftrightarrow q$

p	q	$p \leftrightarrow q$
T	T	T
T	F	F
F	T	F
F	F	T

Lemma 15.3.

If p and q are propositions, then

$$p \leftrightarrow q \equiv (p \to q) \wedge (q \to p).$$

Hence, $p \leftrightarrow q \equiv (\sim p \vee q) \wedge (\sim q \vee p)$.

Proof: We know that "$\boldsymbol{p}$ if and only if $\boldsymbol{q}$" means that both "$\boldsymbol{p}$ if $\boldsymbol{q}$" and "$\boldsymbol{p}$ only if $\boldsymbol{q}$" hold. This means $\boldsymbol{p \leftrightarrow q}$ should be logically equivalent to $\boldsymbol{(p \to q) \wedge (q \to p)}$. We verify it using the truth table:

p	q	$p \to q$	$q \to p$	$p \leftrightarrow q$	$(p \to q) \wedge (q \to p)$
T	T	T	T	T	T
T	F	F	T	F	F
F	T	T	F	F	F
F	F	T	T	T	T

Same truth values.

Hence,

$$P \leftrightarrow q \equiv (p \to q) \wedge (q \to p).$$

It follows therefore that biconditional statement can be written as the **conjunction** of two "if then" statements namely $p \to q$ and $q \to p$. Also we know that

$$p \to q \equiv \sim p \vee q$$

and so

$$q \to p \equiv \sim q \vee p.$$

Hence,

$$p \leftrightarrow q \equiv (p \to q) \wedge (q \to p)$$
$$\equiv (\sim p \vee q) \wedge (\sim q \vee p).$$

Thus, the statements having $\rightarrow$ or $\leftrightarrow$ symbol are logically equivalent to statement having $\sim$, $\wedge$ and $\vee$.

Definition 15.18 Let p and q be statements. Then, p is a sufficient condition for q means "if p then q" whereas p is a necessary condition for q means "if not p then not q".

15.4.2 Hierarchy of Operations of Logical Connectives

The order of operations of connectives are

$$\sim, \ \wedge, \ \vee, \ \rightarrow, \ \leftrightarrow.$$

Definition 15.19 An **argument** is a sequence of statements. **All statements but the final one** are called **premises** (or **assumptions** or **hypothesis**). The **final statement** is called the **conclusion.**

The symbol $\therefore$, read "therefore", is generally placed just before the conclusion.

15.4.3 Logical Form of an Argument

The logical form of an argument can be obtained from the contents of the given argument. For example, consider the argument:

If a man is a bachelor, he is unhappy.
If a man is unhappy, he dies young.
$\therefore$ Bachelors die young.

The logical form of the argument is:

If p then q.
If q then r.
$\therefore p \rightarrow r$,

where

p: He is bachelor
q: He is unhappy
r: He dies young

Definition 15.20 An argument is said to be valid if the conclusion is true whenever all the premises are true.

Definition 15.21 An **argument** which is **not true** is called a **fallacy**.

15.4.4 Method to Test Validity of an Argument

1. Identify the premises and conclusion of the argument.
2. Construct a truth table showing the truth values of all the premises and conclusion.
3. Find the rows (called **critical rows**) in which all the **premises are true.**
4. In each critical row, determine whether the conclusion of the argument is also true.
   (a) If in each critical row the conclusion is also true, then the **argument form is valid.**
   (b) If **there is at least one critical row** in which conclusion is **false**, the argument form is **fallacy** (invalid).

EXAMPLE 15.26

Show that the argument

$$p$$
$$p \rightarrow q$$
$$\therefore q$$

is valid.

Solution. The premises are $\boldsymbol{p}$ and $\boldsymbol{p \rightarrow q}$. The conclusion is $\boldsymbol{q}$. The truth table is

		Premises		Conclusion	
p	q	p	$p \rightarrow q$	q	
T	T	T	T	T	$\rightarrow$ Critical row
T	F	T	F	F	
F	T	F	T	T	
F	F	F	T	F	

In the first row, all the premises are true. Therefore, the first row is critical row. The conclusion in this critical row is also true. Hence the argument is valid.

The argument (discussed above)

$$p$$
$$p \rightarrow q$$
$$\therefore q$$

is known as **Law of Detachment.** The validity of this law may also be established by showing that $p \wedge (p \rightarrow q) \rightarrow q$ is a tautology.

EXAMPLE 15.27

Verify the validity of the following argument form:

$$\begin{array}{c} p \to q \\ p \\ \therefore q. \end{array}$$

Solution. The truth table for the premises and conclusion is

		Premises		Conclusion	
p	q	$p \to q$	p	q	
T	T	T	T	T	→ Critical row
T	F	F	T	F	
F	T	T	F	T	
F	F	T	F	F	

The first row is critical row and the conclusion in the critical row is true. Hence, the given argument form is **Valid**. Thus $(p \to q) \wedge p \to q$ is a tautology.

The fact that this argument form is valid is called **Modus ponens**. This Latin term means **"Method of affirming"** (since the conclusion is an affirmation).

EXAMPLE 15.28

Verify the validity of the argument form

$$\begin{array}{c} p \to q \\ \sim q \\ \therefore \sim p. \end{array}$$

Solution. The truth table for the premises and conclusion is

		Premises		Conclusion	
p	q	$p \to q$	$\sim q$	$\sim p$	
T	T	T	F	F	
T	F	F	T	F	
F	T	T	F	T	
F	F	T	T	T	→ Critical row

The last row is critical row and conclusion in this row is also true. Hence, the argument form is valid.

The fact that this argument is valid is called **Modus Tollens** which means **Method of denying**, since the conclusion is a denial.

Theorem 15.1

(Rule of Inference, Law of Syllogism or Hypothetical Syllogism)
The argument

$$\begin{array}{c} p \to q \\ q \to r \\ \therefore p \to r \end{array}$$

is universally valid and so is a **rule of inference.**

In other words,

$$(p \to q) \wedge (q \to r) \to (p \to r)$$

is a Tautology.

Proof: The truth table for the argument and the conclusion is given below:

p	q	r	$p \to q$	$q \to r$	$p \to r$	
T	T	T	T	T	T	→ Critical row
T	T	F	T	F	F	
T	F	T	F	T	T	
T	F	F	F	T	F	
F	T	T	T	T	T	→ Critical row
F	T	F	T	F	T	
F	F	T	T	T	T	→ Critical row
F	F	F	T	T	T	→ Critical row

We observe that the critical rows for the premises $p \to q$, $q \to r$ are first row, fifth row, seventh row and eighth row. The conclusion $(p \to r)$ in these rows is also true. Hence the argument is valid. Also we note that $(p \to q) \wedge (q \to r) \to (p \to r)$ is a tautology because all the truth values in this case are true.

EXAMPLE 15.29

Consider the argument

If you invest in the stock market, then you will get rich.

If you get rich, then you will be happy.

∴ If you invest in the stock market, then you will be happy.

Solution. By rule of inference, the argument is valid.

EXAMPLE 15.30

Verify the validity of the argument:

Drinking is healthy.

If drinking is healthy, then wine is prescribed by the physicians.

∴ Wine is prescribed by the physicians.

Solution. In symbols, the argument is

$$p$$
$$p \to q$$
$$\therefore q.$$

The argument is of the form **Modus Ponens** (or Law of Detachment) and so it is valid. However, the **conclusion is false**. Observe that the first premises, p: "Drinking is healthy", is **false**. The second premises, $p \to q$ is then true and conjunction of the two premises $(p \wedge (p \to q))$ is false.

EXAMPLE 15.31

The following arguments are valid:

(a)	(b)
p	q
$\therefore p \vee q$	$\therefore p \vee q.$

These arguments are called "**Disjunctive Addition**" and are used for **making generalizations**.

Solution. (a) The truth table is

Premise		Conclusion	
p	q	$p \vee q$	
T	T	T	→ Critical row
T	F	T	→ Critical row
F	T	T	
F	F	F	

The conclusion in the critical rows is also true. Hence, the argument is valid.

(b) The truth table is

Premise		Conclusion	
p	q	$p \vee q$	
T	T	T	→ Critical row
T	F	T	
F	T	T	→ Critical row
F	F	F	

The conclusion in the critical rows is true. Hence the argument is valid.

EXAMPLE 15.32

The following arguments are valid:

(a)	(b)
$p \wedge q$	$p \wedge q$
$\therefore p$	$\therefore q.$

(These arguments are called **Conjunctive Simplification** and are used for particularizing.)

For example (a) says that if both p and q are true, then in particular, p is true. The validity of these arguments can be proved using truth tables.

EXAMPLE 15.33

The arguments

(a)		(b)
$p \vee q$	and	$p \vee q$
$\sim q$		$\sim p$
$\therefore p$		$\therefore q$

are valid

For example, (a) says that there are two possibilities p or q but q is not there. Hence p is there. So (a) is valid. Similarly, (b) is valid. These arguments are called "**Disjunctive Syllogism**".

EXAMPLE 15.34 Dilemma: (Proof by Division into Cases)

The argument

$$\begin{array}{l} p \vee q \\ p \to r \\ q \to r \\ \therefore \; r \end{array}$$

is valid because $(p \vee q) \wedge (p \to r) \wedge (q \to r) \to r$ is a tautology.

EXAMPLE 15.35

Using the following true statements, locate the treasure hidden in the estate:

(i) If the house is next to a lake, then the treasure is not in the kitchen.
(ii) If the tree in front yard is a mango tree, then the treasure is in the kitchen.
(iii) The house is next to a lake.
(iv) The tree in the front yard is a mango tree or the treasure is placed behind the mirror in the wall of the bed room.
(v) If the tree in the backyard is an oak tree, then the treasure is in the drawing room.

Solution. Let

p: the house is next to a lake
q: the treasure is in the kitchen
r: the tree in the front yard is a mango tree
s: the treasure is placed behind the mirror in the wall of the bed room
t: the tree in the backyard is an oak tree
u: the treasure is in the drawing room

Then the logical forms of the clues are:

(i) $p \to \sim q$
(ii) $r \to q$
(iii) p
(iv) $r \vee s$
(v) $t \to u$

The following deductions can be made:

(1) $p \to \sim q$	by (i)
p	by (iii)
$\therefore \; \sim q$	by Modus Ponens
(2) $r \to q$	by (ii)
$\sim q$	by the conclusion of (1)
$\therefore \; \sim r$	by Modus Tollen
(3) $r \vee s$	by (iv)
$\sim r$	by conclusion of (2)
$\therefore \; s$	by Disjunctive Syllogism

Hence, the treasure is placed behind the mirror in the wall of the bed room. The fifth clue has not been used to get the conclusion.

EXAMPLE 15.36

If there is a gas in the car, then John will go to a store. If John goes to the store, then he will get a soda. There is gas in the car. Will John get a soda?

Solution. Let

p: There is a gas in the car
q: John will go to a store
r: John will get a soda.

Then, logical forms of the given statements are

(a) $p \to q$ (b) $q \to r$ (c) p

The following deductions can be made:

(1) $p \to q$	by (a)
p	by (c)
$\therefore q$	by Modus Ponens
(2) $q \to r$	by (b)
q	by the conclusion of (1)
$\therefore r$	by Modus Ponens

Hence, John will get a soda.

EXAMPLE 15.37

A software engineer makes the following observations in a computer programming:

(i) There is an undeclared variable or there is a syntax error in the first five lines.
(ii) If there is a syntax error in the first five lines, then there is a missing semicolon

or a variable name is misspelled.

(iii) There is not a missing semicolon.

(iv) There is not a misspelled variable name.

Find the mistake in the program.

Solution. Let

p: There is undeclared variable

q: There is a syntax error in the first five lines

r: There is a missing semicolon

s: Variable name is misspelled.

The logical forms of the given statements are

(i) $p \vee q$

(ii) $q \to r \vee s$

(iii) $\sim r$

(iv) $\sim s$.

Deductions:

(1) $\sim r$ by (iii)

$\sim s$ by (iv)

$\therefore \sim r \wedge \sim s$ by conjunction

But $\sim r \wedge \sim s = \sim(r \vee s)$. Thus the first conclusion is $\sim(r \vee s)$.

(2) $q \to r \vee s$ by (ii)

$\sim(r \vee s)$ by the conclusion of (1)

$\therefore \sim q$ by Modus Tollen

(3) $p \vee q$ by (i)

$\sim q$ by the conclusion of (2)

$\therefore\ p$ by Disjunctive Syllogism

Hence, there is an undeclared variable.

EXAMPLE 15.38

Show that the following argument is invalid.

If taxes are lowered, then income rises

Income rises

∴ Taxes are lowered.

Solution. Let

p: Taxes are lowered, q: Income rises

Then the argument form is

$$p \to q$$
$$q$$
$$\therefore p.$$

The truth table for the premises and conclusion is

	Premises		Conclusion	
p	q	$p \to q$	p	
T	T	T	T	← Critical row
T	F	F	T	
F	T	T	F	← Critical row
F	F	T	F	

In the third critical row, the conclusion is not true. Hence, the argument is **not valid**.

EXAMPLE 15.39

Is the following argument valid?

If two sides of a triangle are equal, then the opposite angles are equal

Two sides of a triangle are **not** equal

∴ The opposite angles are not equal.

Solution. In symbolic form the argument is

$$p \to q$$
$$\sim p$$
$$\therefore \sim q.$$

The truth table is

	Premises			Conclusion	
p	q	$p \to q$	$\sim p$	$\sim q$	
T	T	T	F	F	
T	F	F	F	T	
F	T	T	T	F	← Critical row
F	F	T	T	T	← Critical row

In this case, the third and fourth rows are critical rows. But in the third row, the conclusion is false. Hence the argument is a **fallacy**.

EXAMPLE 15.40

Consider the following argument for validity:

If I study, then I will not fail in mathematics.

If I do not play basketball, then I will study.

But I failed in mathematics.

∴ I must have played basketball.

Solution. The symbolic form of the argument is

(a) $p \to q$

(b) $r \to p$

(c) $\sim q$

$\therefore \sim r.$

The validity can be verified by the following two methods:

First Method: The following deductions can be made:

(1)	$p \to q$	by (a)
	$\sim q$	by (c)
	$\therefore \sim p$	by Modus Tollen
(2)	$r \to p$	by (b)
	$\sim p$	by (1)
	$\therefore \sim r$	by Modus Tollen

which is true (conclusion). Hence the argument is **valid.**

Second Method: The truth table for the argument form is:

p	q	r	$p \to q$ (Premises)	$r \to p$ (Premises)	$\sim q$ (Premises)	$\sim r$ (Conclusion)	
T	T	T	T	T	F	F	
T	T	F	T	T	F	T	
T	F	T	F	T	T	F	
T	F	F	F	T	T	T	
F	T	T	T	F	F	F	
F	T	F	T	T	F	T	
F	F	T	T	F	T	F	
F	F	F	T	T	T	T	← Critical row

We see that there is only one critical row (eighth row) in the truth table. The conclusion is also true in the critical row. Hence the argument is valid.

Remark 15.10.

It is possible for a valid argument to have a false conclusion and for an invalid argument to have a true conclusion. For example:

1. If Sohan was an artist, then Sohan had green hair

   Sohan was an artist.

   ∴ Sohan had green hair.

In symbol, we have

$$p \to q$$
$$p$$
$$\therefore q.$$

The argument is valid by Modus Ponens. But its major premises is false and so is its conclusion.

2. If Delhi is a big city, then Delhi has tall buildings.

   Delhi has tall buildings.

   ∴ Delhi is a big city.

The symbolic form of the argument is

$$p \to q$$
$$q$$
$$\therefore p.$$

Using truth table, it can be shown that the **argument is fallacy**. But the conclusion is true.

If we replace first premise by $q \to p$, then argument becomes

$$q \to p$$

$$q$$

$\therefore p$ which is valid.

That is why, such type of fallacy is called **Converse Error.**

Remark 15.11.

Consider the argument form

$$p \to q$$

$$\sim p$$

$$\therefore \sim q.$$

We can see that the argument form is **invalid**. But if we replace the premises by $\sim p \to \sim q$, then it becomes valid.

Such type of fallacy is called **Inverse Error**.

15.4.5 Contradiction Rule

If the supposition that the statement p is false leads logically to a contradiction, then you can conclude that p is true.

In symbols,

$\sim p \to c$, where c is a contradiction

$$\therefore p.$$

The truth table for the premise and the conclusion of this argument are given below:

p	$\sim p$	c	$\sim p \to c$	p	
T	F	F	T	T	→ Critical row
F	T	F	F	F	

The premises and conclusion are both true in the critical row and hence the argument is valid.

EXAMPLE 15.41

Knights and Knaves (Raymond Smullyan's description of an island containing two types of people):

This island contains two types of people: knights who always tell the truth and Knaves who always lie. A visitor visits the island and approached two natives who spoke to the visitor as follows:

A says: B is a knight

B says: A and I are of opposite type.

What are A and B?

Solution. Suppose A is a knight. Because A always tells the truth, it follows that B is a knight. Therefore what B says is true (by the definition of Knight). Therefore A and B are of opposite type. Thus we arrive at a contradiction: A and B are both Knights and A and B are of opposite type. Therefore supposition is wrong. Hence A is not a Knight. So A is a Knave. Therefore what A says is false. Hence B is not a Knight and so is a Knave. Hence A and B **are both Knaves**.

15.4.5.1 Summary of Rules of Inference

The following are the valid argument forms:

1. **Modus Ponens:**

$$p \to q$$
$$p$$
$$\therefore q.$$

2. **Modus Tollens:**

$$p \to q$$

$$\sim q$$

$$\therefore \sim p.$$

3. **Disjunctive Addition:**

(a) p

$\therefore p \vee q$ for any q

(b) q

$\therefore p \vee q$ for any p.

4. Conjunctive Simplification:

(a) $p \wedge q$

$\therefore p$.

(b) $p \wedge q$

$\therefore q$.

5. Conjunctive Addition:

p

q

$\therefore p \wedge q$.

6. Disjunctive Syllogism:

(a) $p \vee q$

$\sim q$

$\therefore p$.

(b) $p \vee q$

$\sim p$

$\therefore q$.

7. Hypothetical Syllogism:

$p \to q$

$q \to r$

$\therefore p \to r$.

8. Dilemma: proof by division into cases:

$p \vee q$

$p \to r$

$q \to r$

$\therefore r$.

9. Rule of contradiction:

$\sim p \to c$, where c is a contradiction

$\therefore p$.

EXERCISES

1. Show that the logical expression
$$\{[p \to (q \vee r)] \wedge (\sim q)\} \to (p \to r)$$
is a tautology.

2. Restate the following as implications "If…, then …" :
   (i) A sufficient condition that a figure be rectangle is that it be a square.
   (ii) I am citizen of India if I am living in Bangalore.
   (iii) A necessary condition for Pakistanis to win a cricket match is that they have at least two left-handed batsmen.

   Ans. (i) If a figure is a square, then it is a rectangle

   (ii) If I live in Bangalore, then I am a citizen of india

   (iii) If Pakistanis win cricket match, then they have at least two left-handed batsmen

3. Write negations of the following propositions:
   (i) No rectangle is a square.
   (ii) He takes his bath if and only if the water is cold.
   (iii) If it rains, then they do not go for a walk.

   Ans. (i) Some rectangles are squares

   (ii) He takes his bath if and only if the water is not cold

(iii) If it rains and they go for a walk

4. Verify the validity of the following arguments:
   (i) (a) If there are more pigeons than there are pigeonholes, then at least two pigeons fly into the same pigeon-hole.
   (b) There are more pigeons than there are pigeonholes.
   (c) Therefore, at least two pigeons fly into the same pigeonhole.

   (ii) (a) If this number is divisible by 9, then it is divisible by 3.
   (b) This number is not divisible by 3.
   (c) Therefore, this number is not divisible by 9.

   Ans. (a) The logical expression for the argument is

   $p \to q$

   p

   $\therefore q.$

   which is valid by Modus Ponens

   (b) In logical symbols, the premises and conclusion are

   $p \to q$

   $\sim q$

   $\therefore \sim p.$

   which is valid by Modus Tollen

5. State the converse, inverse and contrapositive to the following statement:

   "If triangle ABC is a right triangle, then $|AB|^2 + |BC|^2 = |AC|^2$".

   Ans. **Converse:** If $|AB|^2 + |BC|^2 = |AC|^2$, then triangle ABC is a right triangle

   Inverse: If ABC is not a right triangle, then $|AB|^2 + |BC|^2 \neq |AC|^2$

   Contrapositive: If $|AB|^2 + |BC|^2 \neq |AC|^2$, then triangle ABC ia not a right triangle

6. Show that the following argument is a fallacy:
   If today is Smith's birthday, then today is May 18.

   Today is May 18

   Therefore, today is Smith's birthday.

7. Let x be a real number. Is the following argument valid?

   x is positive or x is negative

   If x is positive, then $x^2 > 0$

   If x is negative, then $x^2 > 0$

   $\therefore x^2 > 0.$

   Ans. The argument is valid. The logical expression of the argument is

   $p \vee q$

   $p \to r$

   $q \to r$

   $\therefore r.$

   This is because $(p \vee q) \wedge (q \to r) \wedge (p \to r) \to r$ is a tautology

8. Verify the validity of the following argument by using rules of inference:

   If Julia does not live in Italy, then she does not speak Italian.
   Julia does not drive a car
   If Julia lives in Italy, then she travels by train.
   Either Julia speaks Italian or she drives a car.
   Therefore, Julia travels by train.

   Ans. Let

   p: Julia live in Italy

   q: Julia speaks Italian

   r: Julia drives a car

   s: Julia travels by a train

   Then the logical form of the given statement is:

   (i) $\sim p \to \sim q$ (ii) $\sim r$ (iii) $p \to s$ (iv) $\sim q \to r$

   The following deduction can be made:

(i) $\sim q \rightarrow r$ by (iv) (ii) $\sim p \rightarrow \sim q$ by (i)

$\sim r$ by (ii) q by conclusion of (i)

$\because q$ $\therefore p$

(iii) $p \rightarrow s$ by (iii)

p by conclusion (ii)

$\therefore s$

"Hence Julia travels by a train."

9. Verify the validity of the following argument:

$$\begin{array}{l} \sim r \\ p \rightarrow q \\ q \rightarrow r \\ \therefore \sim p. \end{array}$$

Ans. We note that

(a) $q \rightarrow r$ and (b) $p \rightarrow q$

$\sim r$ $\sim p$ by (a)

$\therefore \sim q$ $\therefore \sim p$

Hence the argument is valid

10. Fill in the blanks:

(i) If the graphs are isomorphic, then their degree spectrum will be the same

Their degree spectrums are different

Hence ____________________

(ii) If it rains, Bill will be happy

Bill is not happy

Therefore ____________________

Ans. (i) The graphs are not isomorphic

(ii) It did not rain

16 Elements of Fuzzy Logic

According to Georg Cantor, a *set* may be viewed as a well-defined collection of objects, called the *element* or *member* of the set. The term *'well defined'* means that it is possible to decide if a given object belongs to the collection or not. Thus, if S is a set, then an element s is either a member of S or it is not.

Let χ_S be a *characteristic* (indicator) *function* of S. Then, by definition, χ_S is a function from universe of discourse to the set $\{0, 1\}$ such that

$$\chi_S(s) = \begin{cases} 1 \text{ if } & s \in S \\ 0 \text{ if } & s \notin S \end{cases}$$

Hence, a set S can be represented in term of ordered pairs as

$$\{(s, \chi_S): \chi_S(s) = 1\}.$$

If we call χ_S as a *membership function,* then, in standard set theory, the membership has only two values 1 and 0. The membership of an elements s is 1 if $s \in S$ and 0 if $s \notin S$.

16.1 FUZZY SET

The fuzzy set is a generalization of the standard set in which membership has values in the closed interval [0, 1]. Let X be universe of discourse. Then a *fuzzy subset* S of X is a set of ordered pairs $(s, \mu_S(s))$, $s \in X$, where membership function μ_S assumes the values in [0, 1] called the *degree of membership* or *level of membership* in S.

For example, let $X = \{x, y, z\}$. Then

$$S = \{(x, 0.2\}, (y, 0.5), (z, 0.7)\}$$

is a fuzzy subset of X in which

0.2 is the degree of membership of x in S,
0.5 is the degree of membership of y in S,
0.7 is the degree of membership of z in S.

Let S and T be two fuzzy subsets of a universal set X. If $\mu_S(x) \leq \mu_T(x)$ for every $x \in X$, then S is called a *subset* of T and we write $S \subseteq T$.

For example, if $X = \{x, y, z\}$ and

$$S = \{x, 0.2\}, (y, 0.5), (z, 0.7)\},$$

$$T = \{x, 04\}, (y, 0.7), (z, 0.8)\}$$

then, we note that

$$\mu_S(x) < \mu_T(x),$$
$$\mu_S(y) < \mu_T(y), \text{ and}$$
$$\mu_S(z) < \mu_T(z)$$

Hence S is a subset of T.

Obviously, every fuzzy set is a subset of itself. The fuzzy set S will be called *proper subset* of a fuzzy set T if $\mu_S(x) \leq \mu_T(x)$ for every $x \in X$ and $\mu_S(x) < \mu_T(x)$, for at least one $x \in X$. In this case, we write $S \subset T$.

For example, if $X = \{x, y, z\}$ is universal set and

$$S = \{(x, 0.4), (y, 0.7), (z, 0.9)\}$$
$$T = \{(x, 0.4), (y, 0.8), (z, 0.9)\}$$

then we note that

$$\mu_S(x) = \mu_T(x),$$
$$\mu_S(y) < \mu_T(y),$$
$$\mu_S(z) = \mu_T(z).$$

Hence $\mu_S(u) \leq \mu_T(u)$, for every $u \in X$ and $\mu_S(y) < \mu_T(y)$, $y \in X$. Hence S is a proper subset of T.

Two fuzzy subset S and T of a universal set X are called *equal* if $S \subseteq T$ and $T \subseteq S$, that is, if $\mu_S(x) = \mu_T(x)$ for every $x \in X$.

Let S be a fuzzy subset of a *finite* universal set X. Then

$$|S| = \sum_{x \in X} \mu_S(x)$$

is called the *scalar cardinality* of S.

Thus the scalar cardinality of a fuzzy set S is the sum of degree of membership of its members.

For example, if $X = (x, y, z)$ is universal set and

$$S = \{(x, 0.2), (y, 0.5), (z, 0.7)\}$$

is a fuzzy subset of X, then scalar cardinality of S is

$$|S| = \mu_S(x) + \mu_S(y) + \mu_S(z)$$
$$= 0.2 + 0.5 + 0.7 = 1.4.$$

Let S be a fuzzy subset of universal set X. Then the *support* of S in X is the set of all elements of X that have a non-zero degree of membership. Thus

$$\text{supp}(s) = \{x \in X: \mu_S(x) > 0\}.$$

For example, let $X = \{x, y, z, t\}$ be universal set and let

$$S = \{(x, 0), (y, 0.7), (z, 0), (t, 0.9)\}$$

be a fuzzy subset of X. Then

$$\text{supp}(S) = \{u \in X: \mu_S(u) > 0\}$$
$$= \{y, t\}.$$

The *height* of a fuzzy set S is the largest degree of membership attained by any element in S.

For example, the height of the fuzzy set $S = \{(x, 0.2), (y, 0.4), (z, 0.9)\}$ is 0.9.

A fuzzy set S is called *normal* if at least one of its member attains the membership value1. Thus a fuzzy set of height 1 is normal. A given fuzzy set S can be normalized by dividing each degree of membership by the height of that fuzzy set. The normalized version of a given fuzzy set S is denoted by S_N. Obviously, S_N is also a fuzzy set.

EXAMPLE 16.1

Normalize the fuzzy set

$$S = \{(x, 0.2), (y, 0.5), (z, 0.7)\}.$$

Solution. The height of the fuzzy set S is 0.7. Dividing each degree of membership of the members of S by 0.7, we get the normalized set

$$S_N = \left\{\left(x, \frac{0.2}{0.7}\right), \left(y, \frac{0.5}{0.7}\right), \left(z, \frac{0.7}{0.7}\right)\right\}$$
$$= \{(x, 0.287), (y, 0.714), (z, 1)\}.$$

Let S be fuzzy subset of a universal set X and $\alpha \in [0, 1]$. Then the set

$$S_\alpha = \{x \in X: \mu_S(x)) \geq \alpha\}$$

is called the α-cut of the fuzzy set S.

EXAMPLE 16.2

Show that a fuzzy set A is normal if and only if $A_{1.0} \neq \phi$.

Solution. We know that α-cut of a fuzzy set A is defined by

$$A_\alpha = \{x: \mu_A(x)) \geq \alpha\}.$$

Thus

$$A_{1.0} = \{x: \mu_A(x)) \geq 1\}.$$

If A is normal, then by definition, there is at least one element x in the universal set such that $\mu_A(x)) = 1$. Hence $A_{1.0}$ is not an empty set, that is, $A_{1.0} \neq \phi$.

Conversely, if, $A_{1.0} \neq \phi$, then there is at least one element x in $A_{1.0}$ such that $\mu_A(x) \geq 1$. But, by the definition of membership function $\mu_A(x)$ cannot be greater than 1. Hence $\mu_A(x) = 1$. Thus $A_{1.0}$ contains at least one element x such that $\mu_A(x) = 1$. Hence A is normal.

EXAMPLE 16.3

If $X = \{x, y, z\}$ is universal set and $S = \{(x, 0. 2), (y, 0.5), (z, 0.7)\}$ be fuzzy subset of X, determine α-cut of S for $\alpha = 0.5$ and $\alpha = 0.1$.

Solution.

We have

$$S = \{(x, 0, 2), (y, 0.5), (z, 0.7)\}$$

and $\alpha = 0.5$. Therefore

$$S_{0.5} = \{u \in X: \mu_S(u) \geq 0.5\}$$
$$= \{y, z\}.$$

Similarly,

$$S_{0.1} = \{u \in X: \mu_S(u) \geq 0.1\}$$
$$= \{x, y, z\}.$$

If S is a fuzzy subset of a universal set X and $\alpha \in [0, 1]$. Then the set

$$S_{\alpha+} = \{x \in X: \mu_S(x) > \alpha\}$$

is called the strong α-cut of the fuzzy set S.
In Example 16.3, we have

$$S_{0.5+} = \{x \in X: \mu_S(x) > 0.5\} \\ = \{z\}.$$

The set of all levels $\alpha \in [0, 1]$ that represent α-cuts of a given fuzzy set S is called a *level set* of the fuzzy set S and is denoted by $\wedge_S$. Thus

$$\wedge_S = \{\alpha: \mu_S(x) = \alpha \text{ for some } x \in X\}.$$

An ordinary (Crisp) set X is said to be *convex* if $\lambda x + (1 - \lambda)y \in X$ for all $x, y \in X$ and $\lambda \in [0, 1]$.

Let S be a fuzzy subset of a universal set X. Then S is called *convex* if each of its α-cuts is a convex set.

Theorem 16.1 A fuzzy set S on a universal set X is convex if and only if

$$\mu_S[\lambda x_1 + (1-\lambda)x_2] \geq \min[\mu_S(x_1), \mu_S(x_2)]$$
$$\text{for } x_1, x_2 \in X \text{ and } 0 \leq \lambda \leq 1.$$

Proof: Suppose first that S is convex. Therefore each α-cut S_α is convex. Let $\alpha = \mu_S(x_1) < \mu_S(x_2)$. Then $x_1, x_2 \in S_\alpha$. Since S_α is convex, we have

$$\lambda x_1 + (1 - \lambda)x_2 \in S_\alpha$$

Therefore, by definition of α-cut

$$\mu_S[\lambda x_1 + (1-\lambda)x_2] \geq \alpha = \mu_S(x_1) \\ = \min[\mu_S(x_1), \mu_S(x_2)].$$

Conversely, suppose that

$$\mu_S(\lambda x_1 + (1-\lambda)x_2] \geq \min[\mu_S(x_1), \mu_S(x_2)] \quad (1)$$

for $x_1, x_2 \in X$ and $0 \leq \lambda \leq 1$. We wish to show that each α-cut of S is convex.

Let $x_1, x_2 \in S_\alpha$. Therefore $\mu_S(x_1) \geq \alpha$ and $\mu_S(x_2) \geq \alpha$. Now S_α will be convex if $\lambda x_1 + (1-\lambda)x_2 \in S_\alpha$, that is, if

$$\mu_S[\lambda x_1 + (1-\lambda)x_2] \geq \alpha. \quad (2)$$

Since (1) holds, we have

$$\mu_S(\lambda x_1 + (1-\lambda)x_2] \geq \min[\mu_S(x_1), \mu_S(x_2)] \\ \geq \min[\alpha, \alpha] = \alpha.$$

Hence (2) holds and so S_α is convex.

Remark 16.1 Since $\lambda \in [0, 1]$, $x_1, x_2 \in X$ implies that $\lambda x_1 + (1 - \lambda)x_2 \in [x_1, x_2]$. Thus the condition (1) implies that degree of membership of any element in $[x_1, x_2]$ is never less than the degree of membership of x_1 or x_2.

For example, let R be universal set of real numbers and $S = \left\{-1, -\frac{1}{2}, 0, \frac{1}{2}, 1\right\}$. If $\mu_S(x) = \dfrac{1}{1 + 5x^2}$ is the membership function, then

$$S = \left\{\left(-1, \tfrac{1}{6}\right), \left(-\tfrac{1}{2}, \tfrac{4}{9}\right), (0.1), \left(\tfrac{1}{2}, \tfrac{4}{9}\right), \left(1, \tfrac{1}{6}\right)\right\}$$
$$(-1, 0.16), \left(-\tfrac{1}{2}, 0.44\right), (0, 1),$$
$$\left(\tfrac{1}{2}, 0.44\right) (1, 0.16)\Big\}$$

is a fuzzy subset of R. We note that for any $x, y \in R$, the membership value of any element in $[x, y]$ is greater than or equal to $\mu_S(x)$ and $\mu_S(y)$. Hence S is convex in this case. Graph of the function μ_S is shown below:

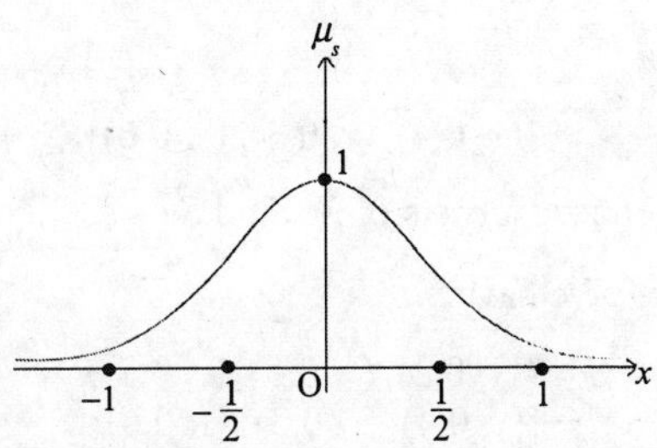

A normalized convex fuzzy set S on the set $\mathfrak{R}$ of real numbers is called a *fuzzy number* if its membership function μ_S is continuous.

Let $S = \{(u, \mu_S(u), u \in X)\}$ be a fuzzy subset of the universal set X. Then the set $\{x \in X, \mu_S(x) = 1\}$ set is called the *Core* of the fuzzy set S and is denoted by core (S). For example, core of the fuzzy set $S = \{(x, 1), (y, 0.5), (z, 1)\}$is core $(S) = \{x, z\}$.

16.2 STANDARD OPERATIONS ON A FUZZY SET

(a) Complement of a Fuzzy set: Let $S = \{(S, (\mu_S(S)), s \in X\}$be a fuzzy set on the universal set X. Then the complement of S, denoted by S', is the set

$$S' = \{(s, 1 - \mu_S(s), s \in X\}.$$

EXAMPLE 16.4

Let $X = \{x, y, z\}$ be universal set and

$$S = \{(x, 0.2), (y, 0.6), (z, 0.7)\}$$

be a fuzzy set on X. Find the complement of S.

Solution. We have

$$S = \{(x, 0.2), (y, 0.5), (z, 0.7)\}.$$

Therefore complement of S is the set

$$S' = \{(x, 1 - 0.2), (y, 1 - 0.5), (z, 1 - 0.7)\}$$
$$= \{(x, 0.8), (y, 0.5), (z, 0.3)\}$$

(b) Union of Fuzzy Sets: Let $S = \{(s, \mu_S(s)), s \in X\}$ and $T = \{t, \mu_T(t), t \in X\}$ be two fuzzy sets defined on a universal set X. Then union of S and T, denoted by $S \cup T$, is the set defined by

$$S \cup T = \{(u, \max(\mu_S(u), \mu_T(u)), u \in X\}.$$

EXAMPLE 16.5

Let $X = \{x, y, z\}$ be universal set and

$$S = \{(x, 0.2), (y, 0.5), (z, 0.7)\}$$

and

$$T = \{(x, 0.4), (y, 0.3), (z, 0.6)\}$$

be two fuzzy subsets of X. Find $S \cup T$.

Solution. We have

$$S = \{(x, 0.2), (y, 0.5), (z, 0.7)\}$$
$$T = \{(x, 0.4), (y, 0.3), (z, 0.6)\}.$$

Since

$$\max(0.2, 0.4) = 0.4,$$
$$\max(0.5, 0.3) = 0.5,$$
$$\max(0.7, 0.6) = 0.7,$$

it follows that

$$S \cup T = \{(x, 0.4), (y, 0.5), (z, 0.7)\}.$$

EXAMPLE 16.6

If $X = \{x, y, z, t\}$ be universal set and

$$A = \{(x, 0.4), (y, 0.6), (z, 0.2), (t, 0)\}$$

and

$$B = \{(x, 0.7), (y, 0), (z, 0.4), (t, 0.3)\}$$

are two fuzzy subsets of X, find $A \cup B$.

Solution. The given fuzzy sets are

$$A = \{(x, 0.4), (y, 0.6), (z, 0.2), (t, 0)\}$$

and

$$B = \{(x, 0.7), (y, 0), (z, 0.4), (t, 0.3)\}$$

Therefore

$$A \cup B = \{u, \max(X_A(u), X_B(u)), u \in X\}$$
$$= \{(x, 0.7), (y, 0.6), (z, 0.4), (t, 0.3)\}.$$

(c) Intersection of Fuzzy sets: Let X be universal set and let

$$S = \{(s, \mu_S(s), s \in X\} \text{ and}$$
$$T = \{(t, \mu_t(t), t \in X\}$$

be two fuzzy subsets of X. Then the *intersection* of S and T, denoted by $S \cap T$, is the set defined by

$$S \cap T = \{(u, \min(\mu_S(u), \mu_T(u)), u \in X\}.$$

EXAMPLE 16.7

Let $X = \{x, y, z\}$ be universal set and let

$$A = \{(x, 0.2), (y, 0.5), (z, 0.7)\}$$

and

$$B = \{(x, 0.4), (y, 0.3), (z, 0.6)\}$$

be fuzzy subsets of X, Find $A \cap B$.

Solution. We have

$$A = \{(x, 0.2), (y, 0.5), (z, 0.7)\}$$
$$B = \{(x, 0.4), (y, 0.3), (z, 0.6)\}.$$

Since

$$\min(0.2, 0.4) = 0.2,$$
$$\min(0.5, 0.3) = 0.3,$$
$$\min(0.7, 0.6) = 0.6,$$

the intersection of A and B is

$$A \cap B = \{(x, 0.2), (y, 0.3), (z, 0.6)\}.$$

EXAMPLE 16.8

Let X be a universal set and let A and B be two fuzzy subset of X? Show that

(i) $\text{supp}(X) = X$

(ii) supp ($A \cup B =$ supp $(A) \cup$ supp (B)

(iii) supp ($A \cap B =$ supp $(A) \cap$ supp (B).

Solution.

(i) Let X be a universal set, then X is a fuzzy set with all its membership values equal to 1. Hence

$$\text{Supp}(X) = \{x: x_S(x) > 0\}$$
$$= X, \text{ since } \chi_X(x) = 1 \text{ for all } x \in X.$$

(ii) If A and B are two fuzzy subsets of a universal set X, then

$$A \cup B = \{(x, \max(\mu_A(x), \mu_B(x)), x \in X\}$$

Thus

$$\text{supp}(A \cup B) = \{(x: \max(\mu_A(x), \mu_B(x)) > 0\} \quad (1)$$

On the other hand,

$$\text{supp}(A) = \{x: \mu_A(x) > 0\}$$
$$\text{supp}(B) = \{x: \mu_B(x) > 0\},$$

and so

$$\text{supp}(A) \cup \text{supp}(B) = \{x, \max(\mu_A(x), \mu_B(x)) > 0)\} \quad (2)$$

Now (1) and (2) imply that

$$\text{supp}(A \cup B) = \text{supp}(A) \cup \text{supp}(B).$$

(iii) We have

$$A \cap B = \{x, \min(\mu_A(x), \mu_B(x)), x \in X\}.$$

Therefore

$$\text{supp}(A \cap B) = \{x: \min(\mu_A(x), \mu_B(x)) > 0\} \quad (3)$$

On the other hand,

$$\text{supp}(A) = \{x: \mu_A(x) > 0\}$$
$$\text{supp}(B) = \{x: \mu_B(x) > 0\}$$

and so

$$\text{supp}(A) \cap \text{supp}(B) = \{x: \min(\mu_A(x), \mu_B(x)) > 0\} \quad (4)$$

Form (3) and (4), it follows that

$$\text{supp}(A \cap B) = \text{supp}(A) \cap \text{supp}(B).$$

16.3 MANY VALUED LOGIC

The classical logic deals with the propositions which are either true or false. Thus classical logic is *two valued logic.* The concept of classical logic has been extended to three value logic by allowing a third truth value called *indeterminate*. Thus in three value logic a proposition may be *true, false or indeterminate*. Five best known three value logics are:

1. Lukasiewicz three value logic
2. Bochvar three value logic
3. Kleene three value logic
4. Heyting three value logic
5. Reichenback three value logic.

In all these "*three value logic*", we denote truth, falsity and indeterminacy by 1, 0 and $\frac{1}{2}$ respectively. If p is a proposition, then negation $\bar{p}$ of p is defined by $\bar{p} = 1 - p$. Thus

$$\bar{1} = 0, \bar{0} = 1 \text{ and } \overline{\left(\frac{1}{2}\right)} = \frac{1}{2}.$$

However operations of conjunctivity ($\wedge$) and disjunctivity ($\vee$) differ from one logic to another in the above mentioned "three value logics".

In Lukasiewiez logic, if p and q are two propositions, then the logic operations are defined by

$$\bar{p} = 1 - p$$
$$p \wedge q = \min(p, q)$$
$$p \vee q = \max(p, q)$$
$$p \Rightarrow q = \min(1, 1 + q - p)$$
$$p \Leftrightarrow q = 1 - |1 - q|.$$

Thus, for this particular 'three value logic', the truth tables for $p \wedge q$, $p \vee q$, $p \Rightarrow q$ and $p \Leftrightarrow q$ turn out to be as shown in the below tables:

p	q	$p\ q$
0	0	0
0	$\frac{1}{2}$	0
0	1	0
$\frac{1}{2}$	0	0
$\frac{1}{2}$	$\frac{1}{2}$	$\frac{1}{2}$
$\frac{1}{2}$	1	$\frac{1}{2}$
1	0	0
1	$\frac{1}{2}$	$\frac{1}{2}$
1	1	1

p	q	$p \vee q$
0	0	0
0	$\frac{1}{2}$	$\frac{1}{2}$
0	1	1
$\frac{1}{2}$	0	$\frac{1}{2}$
$\frac{1}{2}$	$\frac{1}{2}$	$\frac{1}{2}$
$\frac{1}{2}$	1	1
1	0	1
1	$\frac{1}{2}$	1
1	1	1

p	q	$p\ q$
0	0	1
0	$\frac{1}{2}$	1
0	1	1
$\frac{1}{2}$	0	$\frac{1}{2}$
$\frac{1}{2}$	$\frac{1}{2}$	1
$\frac{1}{2}$	1	1
1	0	0
1	$\frac{1}{2}$	$\frac{1}{2}$
1	1	1

p	q	$p \Leftrightarrow q$
0	0	1
0	$\frac{1}{2}$	$\frac{1}{2}$
0	1	0
$\frac{1}{2}$	0	$\frac{1}{2}$
$\frac{1}{2}$	$\frac{1}{2}$	1
$\frac{1}{2}$	1	$\frac{1}{2}$
1	0	0
1	$\frac{1}{2}$	$\frac{1}{2}$
1	1	1

The concept "three-valued logic" has further been generalized into "*n-valued logic*" for which the truth values are:

$$\frac{0}{n-1}, \frac{1}{n-1}, \frac{2}{n-1}, \ldots, \frac{n-2}{n-1}, \frac{n-1}{n-1}$$

or

$$0, \frac{1}{n-1}, \frac{2}{n-1}, \ldots, \frac{n-2}{n-1}, 1.$$

These values are called *degree of truth.*

16.4 FUZZY LOGIC

Logic based on fuzzy set theory is called *fuzzy logic*. It is an extension of *n*-valued logics. In fact *fuzzy logic is any logic having the closed interval* [0, 1] *as its truth value set*. If p and q are fuzzy propositions, then logical connectives are defined by

$$T(p \wedge q) = \min [T(p), T(q)]$$
$$T(p \vee q) = \max [T(p), T(q)]$$
$$T(\bar{p}) = 1 - T(p),$$

where $T(p), T(q) \in [0, 1]$.

However, the conditional proposition is defined in different ways. For example,

(i) $T(p \to q) = \min[1 - T(p) + T(q)]$

and

(ii) $T(p \to q) = \max[\min\{T(p), T(q)\}, 1 - T(p)]$.

The definition (ii) is due to L. A. Zadeh.

16.5 FUZZY PROPOSITIONS

Linguistic variables. Linguistic variables are those variables, where the meaning remains constant but form (values) varies. For example,

1. *Pressure* is a linguistic variable with linguistic values *low pressure* and *high pressure* etc.
2. *Temperature* is a linguistic variable with linguistic values *cold, hot, cool*, *warm* etc.
3. *Age* is a linguistic variable with linguistic values *infant, young, middle-aged, old* etc.

Fuzzy Predicate. The linguistic values of a linguistic variable are called *fuzzy predicates.*

For example, consider the propositions

(1) Today is hot.
(2) Ram is young.
(3) Sohan is tall.

In (1), hot is a linguistic value of the linguistic variable *temperature* and so it is a fuzzy predicate. In (2), young is a linguistic value of the linguistic variable age and so it is a fuzzy predicate. In (3), tall is a linguistic value of the linguistic variable *height* and so it is a fuzzy predicate.

Fuzzy proposition. If P is a fuzzy predicate, then a proposition of the form 'x is P' is called a *fuzzy proposition.*

Using the logical connectives $\wedge$, $\vee$ and-, compound fuzzy preposition can be obtained.

EXERCISES

1. Let $X = \{x, y, z, t\}$ be universal set and

$$S = \{(x, 0), (y, 0.4), (z, 0.8), (t, 1)\}$$

be a fuzzy subset of X. Find (*i*) S', (ii) height (S), (iii) supp (S) and (iv) core (S).

Ans. (i) $S' = \{(x, 1), (y, 0.6), (z, 0.2), (t, 0)\}$
(ii) height $(S) = 1$
(iii) supp $(S) = \{y, z, t\}$
(iv) core $(S) = \{t\}$.

2. Let $X = \{a, b, c\}$ be universe of discourse and

$$A = \{(a, 0.5), (b, 1), (c, 0.4)\}$$

and

$$B = \{(a, 0.6), (b, 0.4), (c, 0.7)\}$$

be fuzzy subsets of X. Find $A \cup B, A \cap B, A_\alpha$ for $\alpha = 0.4$ and B_N (normalized).

Ans. $A \cup B = \{(a, 0.6), (b, 1), (c, 0.7)\}$

$$A \cap B = \{(a, 0.5), (b, 0.4), (c, 0.4)\}$$

$$B_{0.4} = \{a, b, c\}$$

$$B_N = \left\{\left(a, \frac{0.6}{0.7}\right), \left(b, \frac{0.4}{0.7}\right), (c, 1)\right\}$$

$$= \{(a, 0.86), (b, 0.56), (c, 1)\}$$

3. Let $X = \{x, y, z\}$ be universal set and let

$$A = \{(x, 0.4), (y, 0.2), (z, 0.6)\}$$

and

$$B = \{(x, 0.3), (y, 0.5), (z, 0.2)\}$$

be two fuzzy subsets of X. Show that

$$\text{supp } (A \cup B) = \text{supp } (A) \cup \text{supp } (B).$$

Hint: $A \cup B = \{(x, 0.4), (y, 0.5), (z, 0.6)\}$

Therefore

$$\text{supp } (A \cup B) = \{x, y, z\}$$

Also

$$\text{supp (A)} = \{x, y, z\}$$
$$\text{supp (B)} = \{x, y, z\}$$

Therefore

$$\text{supp (A)} \cup \text{supp (B)} = \{x, y, z\}.$$

Hence

$$\text{supp } (A \cup B) = \text{supp } (A) \cup \text{supp } (B).$$

4. Normalize the fuzzy set

$$A = \{(x, 0), (y, 0.8), (3, 0.4)\}.$$

Hint: Height $(A) = 0.8$. So dividing the memberships by 0.8, we get

$$A_N = \{(x, 0), (y, 1), (3, 0.5)\}$$

as the normalized fuzzy set.

5. Let $X = \{x, y, z\}$ be universal set and let

$$A = \{(x, 0.2), (y, 0.4), (z, 0.7)\}$$

and

$$B = \{(x, 0.3), (y, 0.2), (z, 0.6)\}$$

be two fuzzy subsets of X. Show that

$$(A \cup B)' = A' \cap B' \text{ (De-Morgan Law)}$$

Hint:

$$(A \cup B) = \{(x, 0.3), (y, 0.4), (z, 0.7)\}$$
$$(A \cup B)' = \{(x, 0.7), (y, 0.6), (z, 0.3)\}$$
$$A' = \{(x, 0.8), (y, 0.6), (z, 0.3)\}$$
$$B' = \{(x, 0.7), (y, 0.8), (z, 0.4)\}$$
$$A' \cap B' = \{(x, 0.7), (y, 0.6), (z, 0.3)\}$$

Hence the result.

6. For the fuzzy sets in exercise 5, verify

$$A \cup (A \cap B) = A \text{ (absorption law)}$$

Hint:

$$A \cap B = \{(x, 0.2), (y, 0.2), (z, 0.6)\}$$

and so

$$A \cup (A \cap B) = \{(x, 0.2), (y, 0.4), (z, 0.7)\}$$
$$= A.$$

17 Graphs

The interest in the study of graph theory has increased due to its applicability in so many fields like artificial intelligence, electrical engineering, transportation system, scheduling problems, economics, chemistry and operations research.

17.1 DEFINITIONS AND BASIC CONCEPTS

Definition 17.1 A **graph** $G = (V, E)$ is a mathematical structure consisting of two finite sets V and E. The elements of V are called **vertices (or nodes)** and the elements of E are called edges. Each edge is associated with a set consisting of **either one** or **two vertices** called its **endpoints**.

The correspondence from edges to endpoints is called edge-endpoint function. This function is generally denoted by γ. Due to this function, some authors denote graph by $G = (V, E, \gamma)$.

Definition 17.2 A graph consisting of one vertex and no edges is called a **trivial graph**.

Definition 17.3 A graph whose vertex and edge sets are empty is called a **null graph**.

Definition 17.4 An edge with just one endpoint is called a **loop** or a **self-loop**.

Thus, a loop is an edge that joins a single endpoint to itself.

Definition 17.5 An edge that is not a self-loop is called a **proper edge**.

Definition 17.6 If two or more edges of a graph G have the same vertices, then these edges are said to be **parallel** or **multi-edges**.

Definition 17.7 Two vertices that are connected by an edge are called **adjacent**.

Definition 17.8 An endpoint of a loop is said to be **adjacent to itself**.

Definition 17.9 An edge is said to be **incident** on each of its endpoints.

Definition 17.10 Two edges incident on the same endpoint are called **adjacent edges**.

Definition 17.11 The number of edges in a graph G which are incident on a vertex is called the **degree of that vertex**.

Definition 17.12 A vertex of degree zero is called an **isolated vertex**.

Thus, a vertex on which no edges are incident is called isolated.

Definition 17.13 A graph without multiple edges **(parallel edges)** and loops is called **simple graph.**

Notations: In pictorial representations of a graph, the vertices will be denoted by dots and edges by line segments.

EXAMPLE 17.1

(a) Let

$V = \{1, 2, 3, 4\}$ and $E = \{e_1, e_2, e_3, e_4, e_5\}$.

Let γ be defined by

$$\gamma(e_1) = \gamma(e_5) = \{1, 2\}, \quad \gamma(e_2) = \{4, 3\},$$
$$\gamma(e_3) = \{1, 3\}, \quad \gamma(e_4) = \{2, 4\}.$$

We note that both edges e_1 and e_5 have same endpoints $\{1, 2\}$. The endpoints of e_2 are $\{4, 3\}$, the endpoints of e_3 are $\{1, 3\}$ and endpoints of e_4 are $\{2, 4\}$. Thus the graph is as shown in the Figure 17.1.

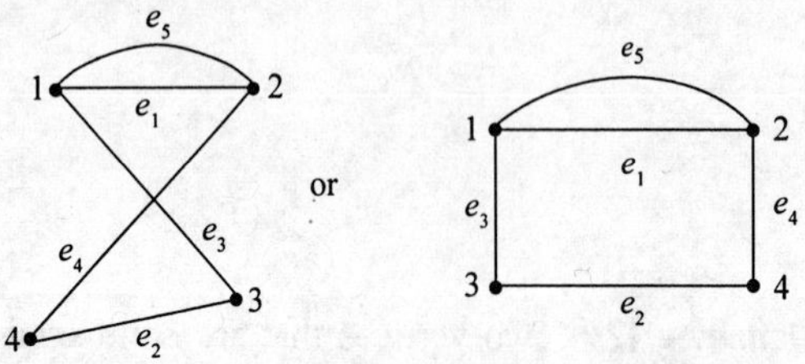

Figure 17.1

(b) For the graph pictured in the Figure 17.2,
  (i) Write down the degree of the vertices A, B, D.
  (ii) Which of the edges are parallel?
  (iii) Which vertex is isolated?
  (iv) Point out whether it is a simple graph or a multi-graph.
  (v) Point out one pair of adjacent vertices and one pair of adjacent edges.

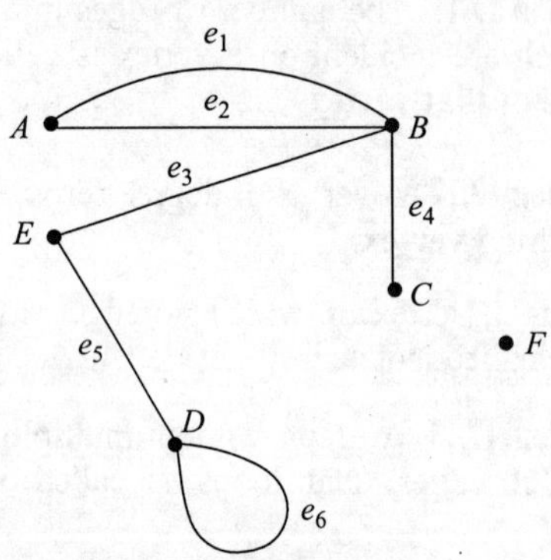

Figure 17.2

In this graph,
(i) We notice that
Degree of the vertex $A = 2$
Degree of the vertex $B = 4$
Degree of the vertex $D = 3$
(because loop e_6 has degree 2).
The vertex D has degree 3 (because e_6 is a loop and so has degree 2)
(ii) The edges e_1 and e_2 are parallel
(iii) The vertex F is isolated because no edge is incident on F
(iv) It is a multi-graph because it has multi-edges and a loop.
(v) A and B are adjacent vertices because they are connected by the edge e_2 or e_1, whereas A and E are not adjacent.

The edges e_2 and e_3 are adjacent edges because they are incident on the same vertex B.

(c) Consider the graph with the vertices A, B, C, D and E pictured in Figure 17.3.

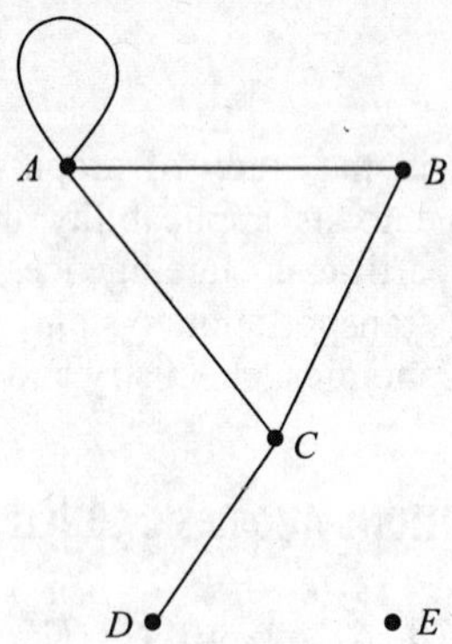

Figure 17.3

In this graph, we note that
Number of edges $= 5$
Degree of vertex $A = 4$
Degree of vertex $B = 2$
Degree of vertex $C = 3$
Degree of vertex $D = 1$
Degree of vertex $E = 0$
Sum of the degree of vertices
$= 4 + 2 + 3 + 1 + 0 = 10.$
Thus, we observe that

$$\sum_{i=1}^{5} \deg(v_i) = 2e,$$

where $\deg(v_i)$ denotes the degree of vertex v_i and e denotes the number of edges.

Euler's Theorem 17.1 (The First Theorem of Graph Theory)

The sum of the degrees of the vertices of a graph G is equal to twice the number of edges in G. **Thus, total degree of a graph is even.**

Proof: Each edge in a graph contributes a count of 1 to the degree of two vertices (endpoints of the edge), that is, each edge contributes 2 to the degree sum. Therefore, the sum of degrees of the vertices is equal to twice the number of edges.

Corollary 17.1 There can be only an even number of vertices of odd degree in a given graph G.

Proof: We know, by the fundamental theorem, that

$$\sum_{i=1}^{n} \deg(v_i) = 2 \times \text{number of edges.}$$

Thus the right-hand side is an even number. Hence to make the left-hand side an even number there can be only even number of vertices of odd degree.

Remarks 17.1

(i) A vertex of degree d is also called a **d-valent vertex.**

(ii) The degree (or valence) of a vertex v in a graph G is the number of proper edges incident on v plus twice the number of self-loops.

Theorem 17.2 A nontrivial simple graph G must have at least one pair of vertices whose degrees are equal.

Proof: Let the graph G has n vertices. Then there appear to be n possible degree values, namely $0, 1, \ldots, n-1$. But there cannot be both a vertex of degree 0 and a vertex of degree $n-1$ because if there is a vertex of degree 0 then each of the remaining $n-1$ vertices is adjacent to at most $n-2$ other vertices. Hence the n vertices of G can realize at most $n-1$ possible values for their degrees. Hence, the pigeonhole principle implies that at least two of the vertices have equal degree.

EXAMPLE 17.2

Is there a graph with eight vertices of degree 2, 2, 3, 6, 5, 7, 8, 4?

Solution. The answer is No. In such a graph, there will be three vertices of odd degree which is impossible.

We can also argue as follows: Total degree of the graph (if possible) is equal to $2+2+3+6+5+7+8+4=37$, which is odd. This contradicts the first theorem of graph theory, according to which, total degree of a graph is always even.

17.2 SPECIAL GRAPHS

Definition 17.14 A graph G is said to **simple** if it has no parallel edges or loops. In a simple graph, an edge with endpoints v and w is denoted by $\{v, w\}$.

Definition 17.15 For each integer $n \geq 1$, let D_n denote the graph with n **vertices** and **no edges.** Then D_n is called the **discrete graph on n vertices.**

For example, we have

• • • and • • • • •

D_3 D_5

Definition 17.16 Let $n \geq 1$ be an integer. Then a simple graph with n vertices in which there is an edge between each pair of distinct vertices is called the **complete graph** on n vertices. It is denoted by K_n.

For example, the complete graphs K_2, K_3 and K_4 are shown in Figure 17.4.

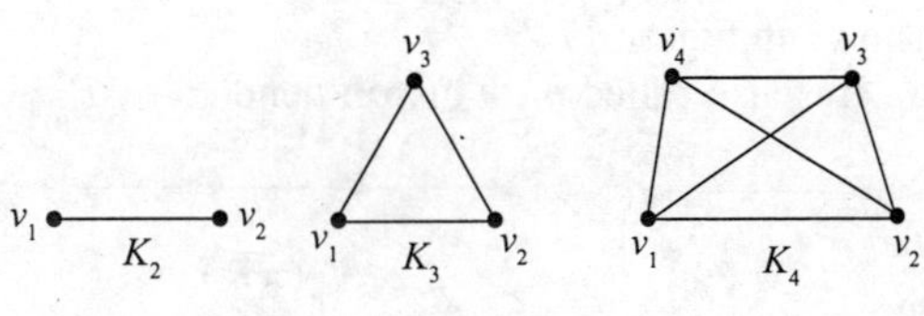

Figure 17.4

Definition 17.17 If each vertex of a graph G has the same degree as every other vertex, then G is called a **regular graph**.

A **k-regular graph** is a regular graph whose common degree is k.

For example, consider K_3. The degree of each vertex in K_3 is 2. Hence K_3 is regular. Similarly, K_4 is regular. Also the graph shown in Figure 17.5 is regular because degree of each vertex here is 2.

But this graph is not complete because v_2 and v_4 have not been connected through an edge. Similarly, v_1 and v_3 are not connected by any edge.

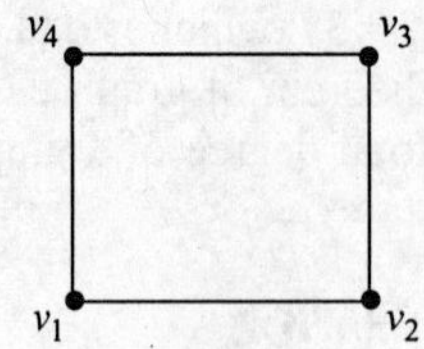

2-regular graph

Figure 17.5

Thus, **a complete graph is always regular** but **a regular graph need not be complete.**

EXAMPLE 17.3

The oxygen molecule O_2, made up of two oxygen atoms linked by a double bond can be represented by the regular graph shown in Figure 17.6.

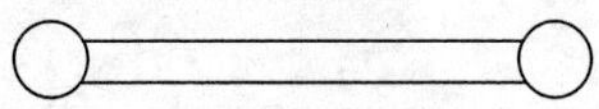

Figure 17.6

Definition 17.18 Let $n \geq 1$ be an integer. Then a graph L_n with n vertices $\{v_1, v_2, \ldots, v_n\}$ and with edges $\{v_i, v_{i+1}\}$ for $1 \leq i < n$ is called a **linear graph on *n* vertices.**

For example, the linear graphs L_2 and L_4 are shown in Figure 17.7.

It is also called **path graph** denoted by P_n.

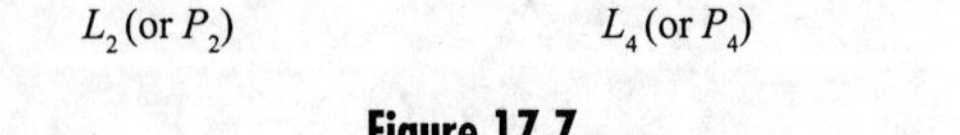

L_2 (or P_2) L_4 (or P_4)

Figure 17.7

Definition 17.19 A **bipartite graph** G is a graph whose vertex set V can be partitioned into two subsets U and W, such that each edge of G has one endpoint in U and one endpoint in W.

The pair (U, W) is called a **vertex bipartition of** G where as U and W are called the bipartition subsets. Obviously, a bipartite graph cannot have any self-loop.

EXAMPLE 17.4

If vertices in U are solid vertices and vertices in W are hollow vertices, then the graphs shown in Figure 17.8 are bipartite graphs.

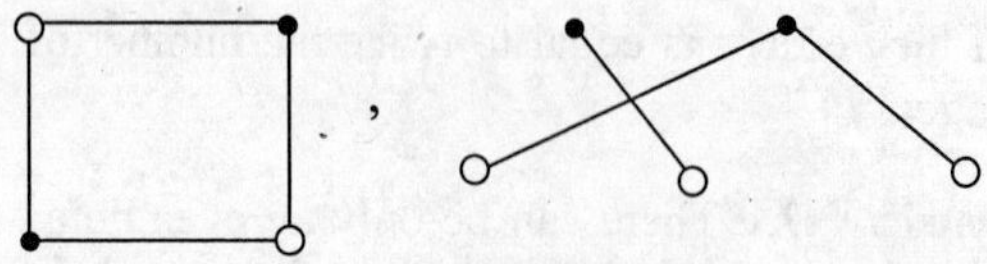

Figure 17.8

EXAMPLE 17.5

The smallest possible simple graph that is not bipartite is the complete graph K_3 shown in Figure 17.9.

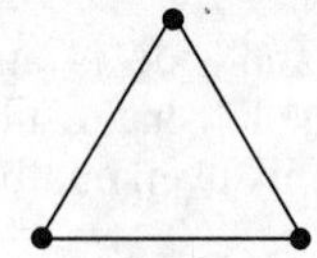

K_3

Figure 17.9

Definition 17.20 A **complete bipartite graph** G is a simple graph whose vertex set V can be partitioned into two subsets $U = \{v_1, v_2, \ldots, v_m)$ and $W = \{w_1, w_2, \ldots, w_n\}$ such that for all i, k in $\{1, 2, \ldots, m\}$ and j, l in $\{1, 2, \ldots, n\}$

(i) There is an edge from each vertex v_i to each vertex w_j.
(ii) There is not an edge from any vertex v_i to any other vertex v_k.
(iii) There is not an edge from any vertex w_j to any other vertex w_l.

A complete bipartite graph on (m, n) vertices is denoted by $K_{m,n}$.

EXAMPLE 17.6

The complete bipartite graphs $K_{3,2}$ and $K_{3,4}$ are shown in Figure 17.10.

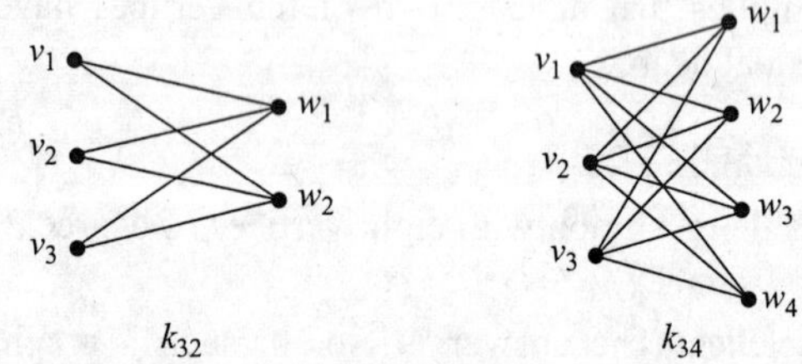

k_{32} k_{34}

Figure 17.10

Definition 17.21 A **cycle graph** is a single vertex with a self-loop or a simple connected graph C

with $|V_c| = |E_c|$, that can be drawn so that all of its vertices and edges lie on a circle.

An n-vertex cycle graph is denoted by C_n. The cycle graphs C_1, C_2, and C_4 are shown in Figure 17.11.

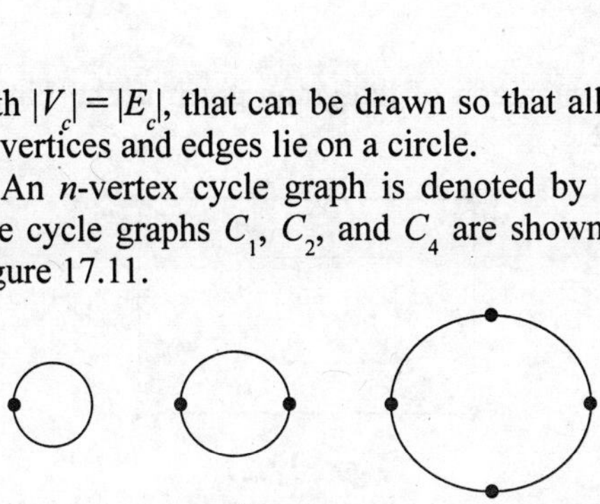

C_1 C_2 C_4

Figure 17.11

Definition 17.22 The **Petersen graph** is the 3-regular graph shown in Figure 17.12.

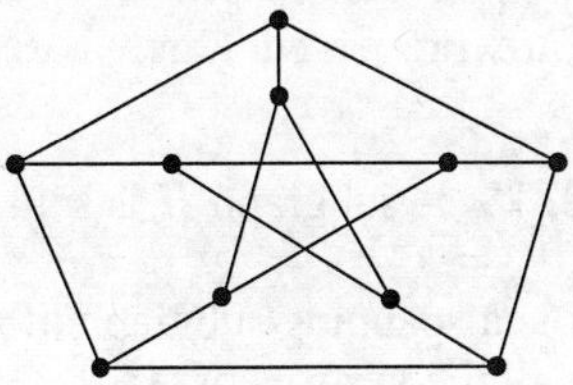

(Petersen graph)

Figure 17.12

This graph is frequently used to establish theorems and to test conjectures.

Definition 17.23 The **hypercube graph** Q_n is the n-regular graph whose vertex set is the set of bit string of length n and such that there is an edge between two vertices if and only if they differ in exactly one bit.

For example, 8-vertex cube graph is a hypercube graph Q_3 shown in Figure 17.13.

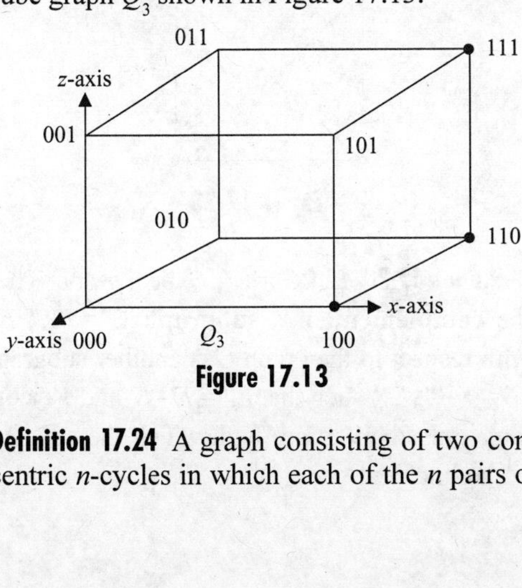

Figure 17.13

Definition 17.24 A graph consisting of two concentric n-cycles in which each of the n pairs of the corresponding vertices is joined by an edge is called the **circular ladder graph** CL_n.

For example, the graphs CL_2 and CL_4 are shown in the Figure 17.14.

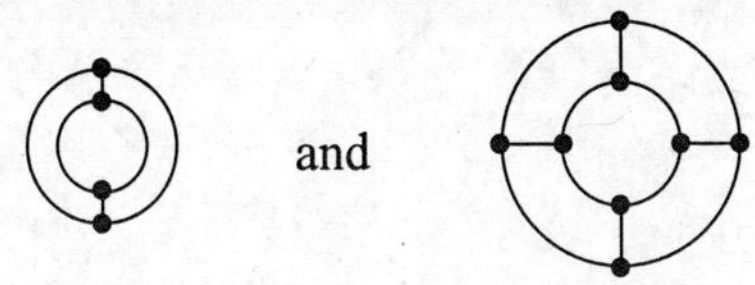

CL_2 CL_4

Figure 17.14

Definition 17.25 A graph consisting of a single vertex having n loops is called a **bouquet** and is denoted by B_n.

For example, the graphs shown in Figure 17.15 represent bouquet B_2 and bouquet B_4.

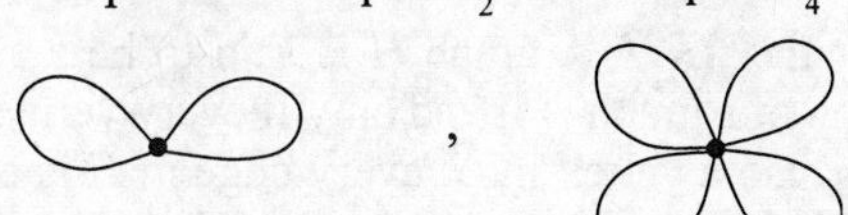

B_2 B_4

Figure 17.15

Definition 17.26 A graph consisting of two vertices and n edges joining them is called a **dipole**. It is denoted by D_n.

For example, Figure 17.16 represents D_2, D_3 and D_4.

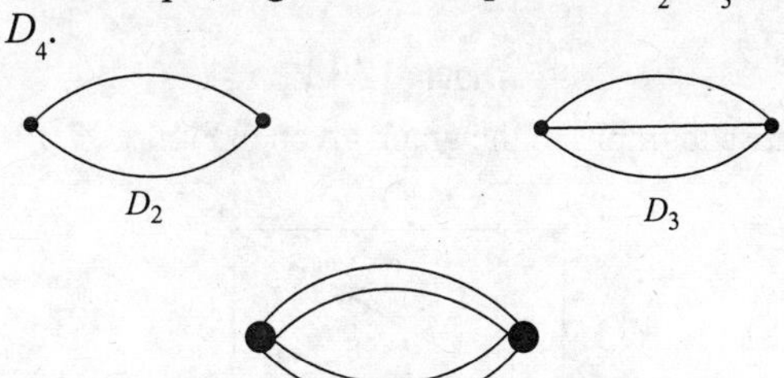

Figure 17.16

EXAMPLE 17.7

Draw all simple graphs with vertices $\{u, v, x, y\}$ and two edges, one of which is $\{u, v\}$.

Solution. In a simple graph, an edge corresponds to a subset of two vertices. Since there are four vertices, we can have $4c_2 = 6$ such subsets in all and they are $\{u, v\}$, $\{u, x\}$, $\{u, y\}$, $\{v, x\}$,

$\{v, y\}$ and $\{x, y\}$. But one edge of the graph is specified to be $\{u, v\}$. Hence the possible simple graphs are as shown in Figure 17.17.

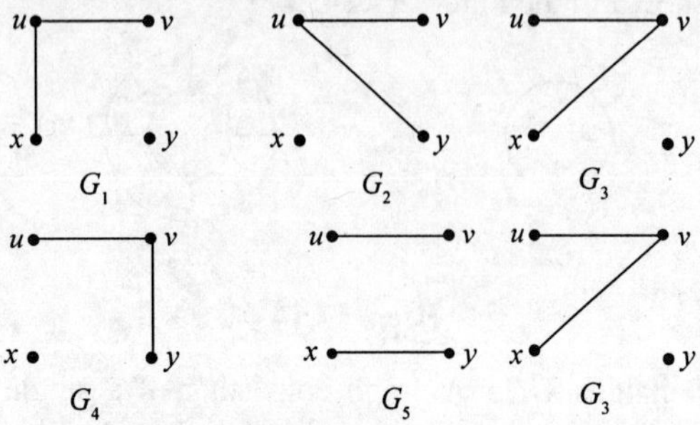

Figure 17.17

17.3 SUBGRAPHS

Definition 17.27 A graph H is said to be a subgraph of a graph G if and only if every vertex in H is also a vertex in G, every edge in H is also an edge in G and every edge in H has the same endpoints as in G.

We may also say that G **is a supergraph of** H. For example, the graphs (Figure 17.18).

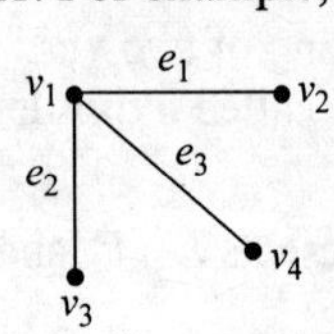

and

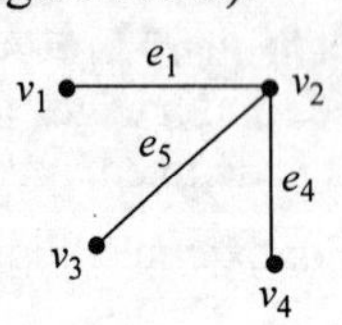

Figure 17.18

are subgraphs of the graph given in Figure 17.19.

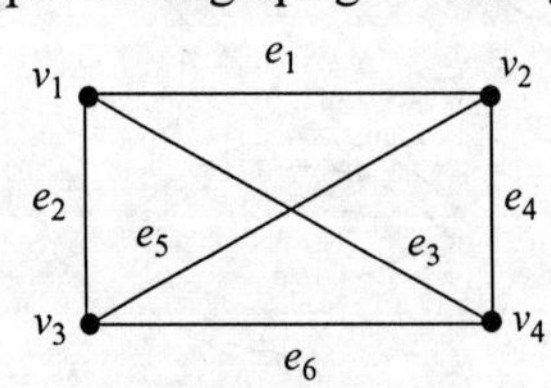

Figure 17.19

Similarly, the graph (Figure 17.20)

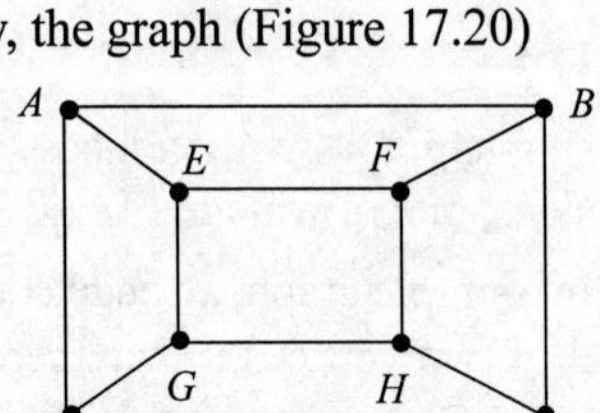

Figure 17.20

is a subgraph of the graph given in the Figure 17.21.

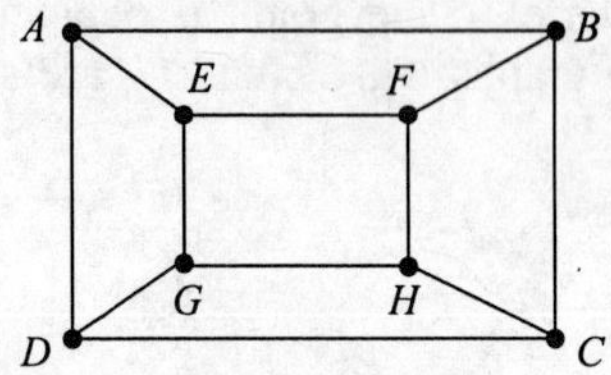

Figure 17.21

Definition 17.28 A subgraph H is said to be a **proper subgraph** of a graph G if vertex set V_H of H is a proper subset of the vertex set V_G of G or edge set E_H is a proper subset of the edge set E_G.

For example, the subgraphs in the above examples are proper subgraphs of the given graphs.

Definition 17.29 A subgraph H is said to **span** a graph G if $V_H = V_G$.

Thus H is a spanning subgraph of graph G if it contains all the vertices of G.

For example, the subgraph (Figure 17.22)

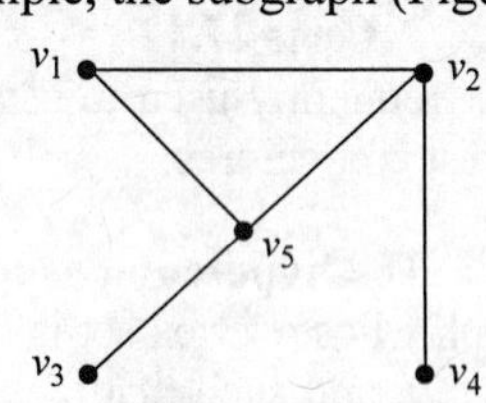

Figure 17.22

spans the graph give in Figure 17.23.

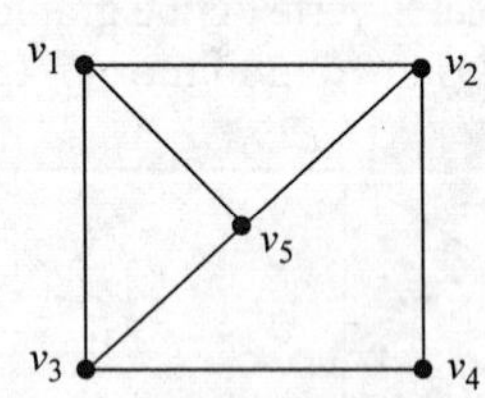

Figure 17.23

Definition 17.30 Let $G = (V, E)$ be a graph. Then the **complement of a subgraph** $G' = (V', E')$ with respect to the graph G is another subgraph $G'' = (V'', E'')$ such that $E'' = E - E'$ and V'' contains only the vertices with which the edges in E'' are incident.

For example, the subgraph (Figure 17.24)

Figure 17.24

is the complement of the subgraph (Figure 17.25).

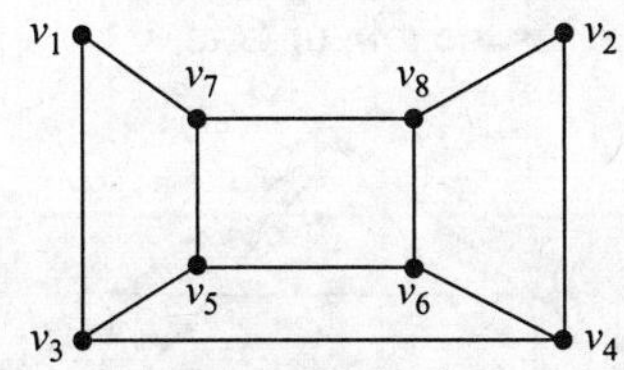

Figure 17.25

with respect to the graph G shown in Figure 17.26.

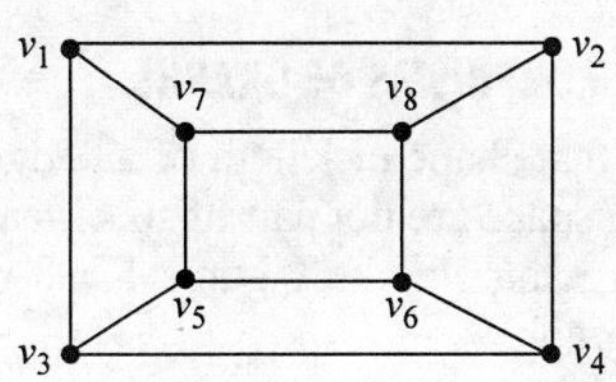

Figure 17.26

Definition 17.31 If G is a simple graph, the **complement of** G (**edge complement**), denoted by G' or G^c is a graph such that

(i) The vertex set of G' is identical to the vertex set of G, that is, $V_{G'} = V_G$.
(ii) Two distinct vertices v and w of G' **are connected by** an edge if and only if v and w **are not connected** by an edge in G.

For example, consider the graph G shown in Figure 17.27.

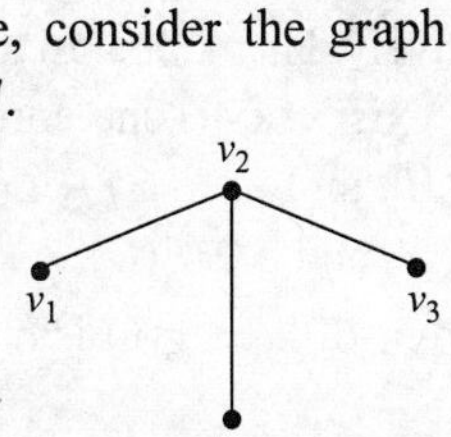

Figure 17.27

Then complement G' of G is the graph shown in Figure 17.28.

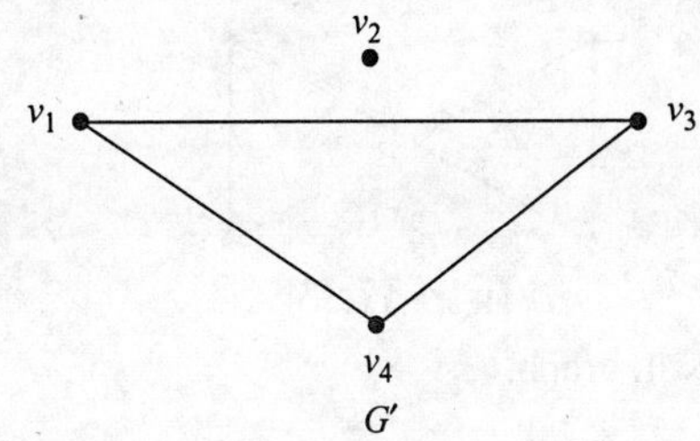

Figure 17.28

EXAMPLE 17.8

Find the complement of the graphs:

(a)

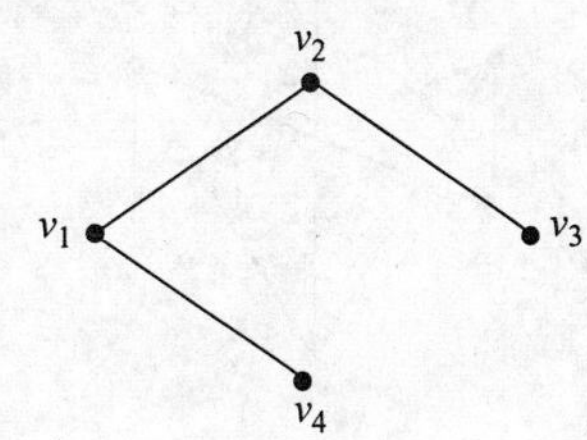

Figure 17.29

(b)

Figure 17.30

(c) Complete graph K_4:

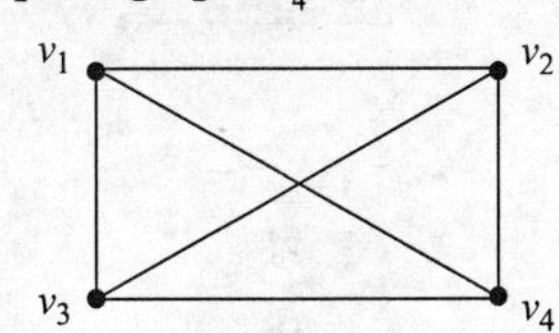

Figure 17.31

Solution.

(a)

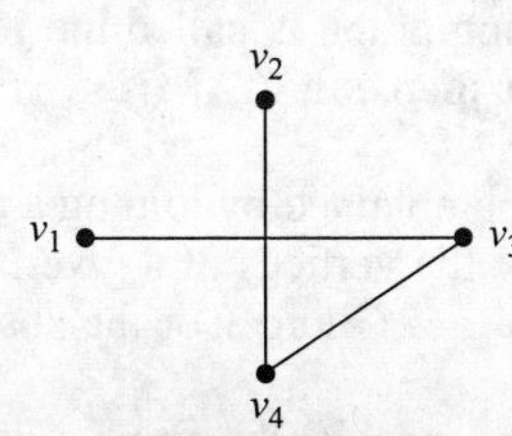

Figure 17.32

(b)

Figure 17.33

(c) Null graph.

EXAMPLE 17.9

Find the edge complement of the graph G shown in Figure 17.34.

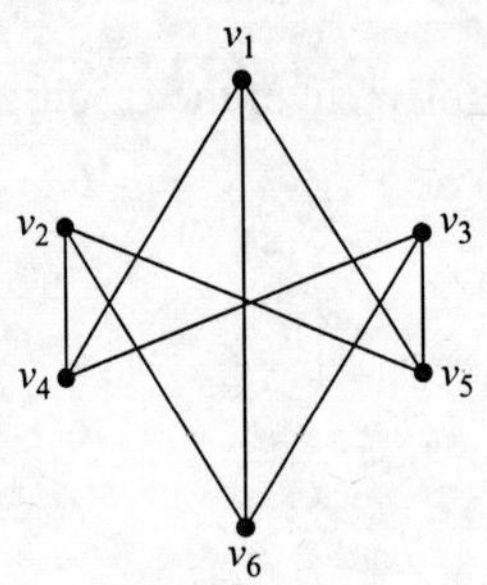

G

Figure 17.34

Solution. The edge complement of G is the following graph G^c (Figure 17.35).

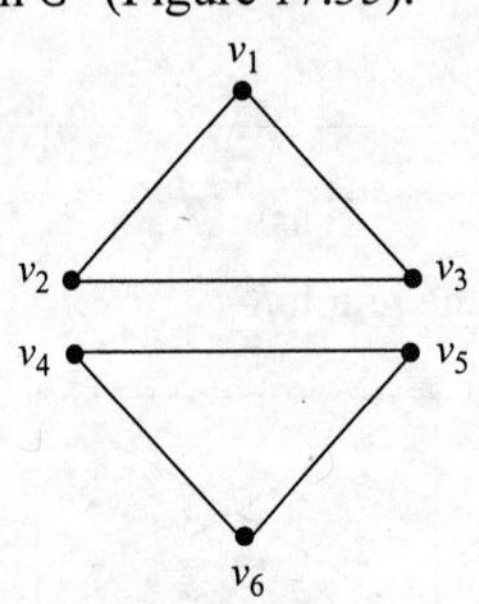

G^C

Figure 17.35

Definition 17.32 If a new vertex v is joined to each of the pre-existing vertices of a graph G, then the resulting graph is called the **join of** G **and** v or the **suspension of** G **from** v. It is denoted by $G + v$.

Thus, a graph obtained by joining a new vertex v to each of the vertices of a given graph G is called the **join of** G **and** v or the **suspension of** G **from** v.

For example, consider the graph given below (Figure 17.36).

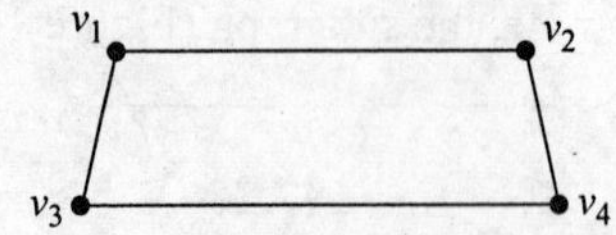

Figure 17.36

Let v be a vertex. Then the graph shown in the Figure 17.37 is the join of G to v.

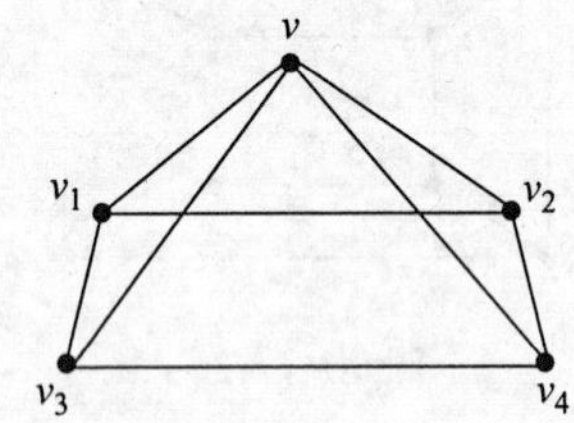

$G + v$

Figure 17.37

17.4 ISOMORPHISMS OF GRAPHS

We know that shape or length of an edge and its position in space are not part of specification of a graph. For example, the Figures 17.38 represent the same graph.

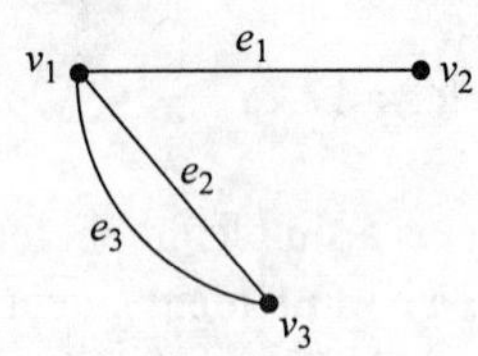

and

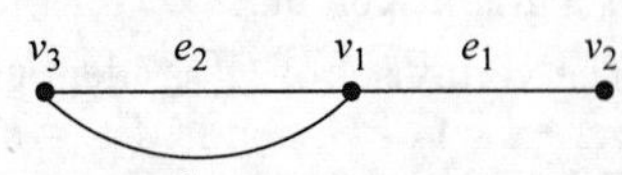

Figure 17.38

Definition 17.33 Let G and H be graphs with vertex sets $V(G)$ and $V(H)$ and edge sets $E(G)$ and $E(H)$, respectively. Then G **is said to isomorphic to** H if there exist one-to-one correspondences $g: V(G) \to V(H)$ and $h: E(G) \to E(H)$ such that for all $v \in V(G)$ and $e \in E(G)$,

v is an endpoint of $e \Leftrightarrow g(v)$ is an endpoint of $h(e)$.

Definition 17.34 The property of mapping endpoints to endpoints is called **preserving incidence** or **the continuity rule** for graph mappings.

As a consequence of this property, a self-loop must map to a self-loop.

Thus, two isomorphic graphs are same except for the labelling of their vertices and edges. For example, the graphs shown in the Figure 17.39 are isomorphic.

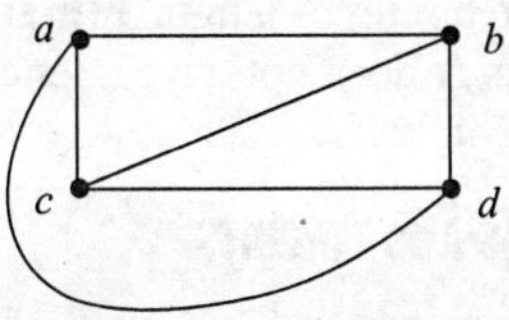

and

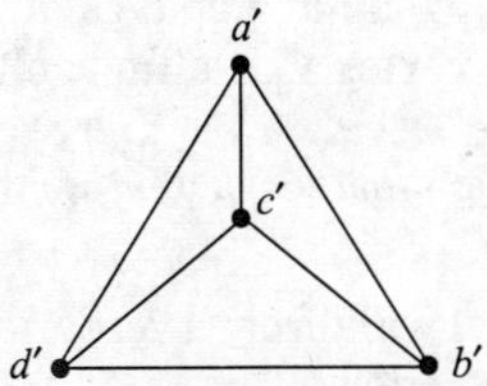

Figure 17.39

Similarly, the hypercube graph Q_3 and circular ladder CL_4 shown in Figure 17.40 are isomorphic.

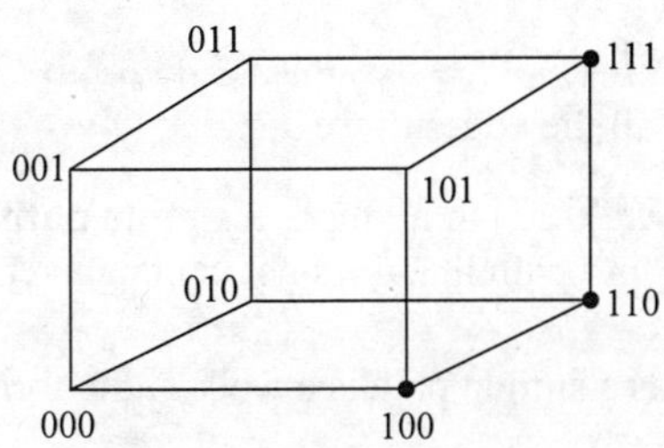

Hypercube Q_3

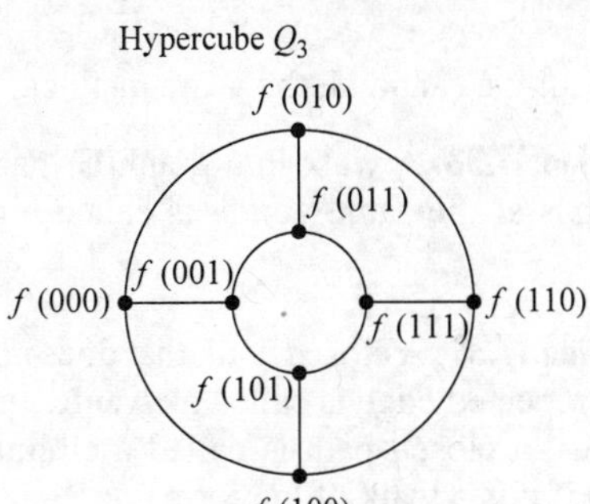

Circular ladder CL_4

Figure 17.40

EXAMPLE 17.10

Show that the graphs shown in Figure 17.41 are isomorphic.

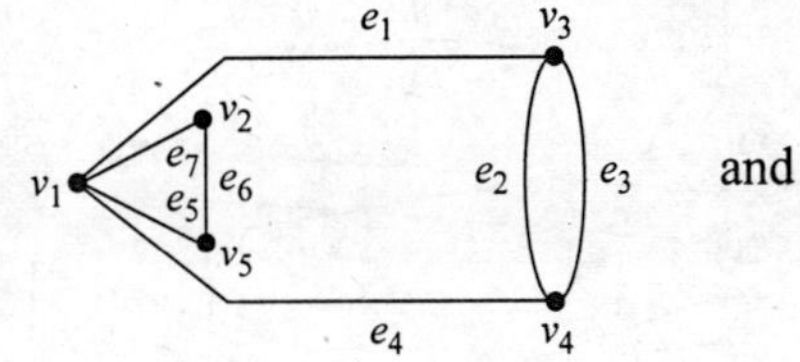

and

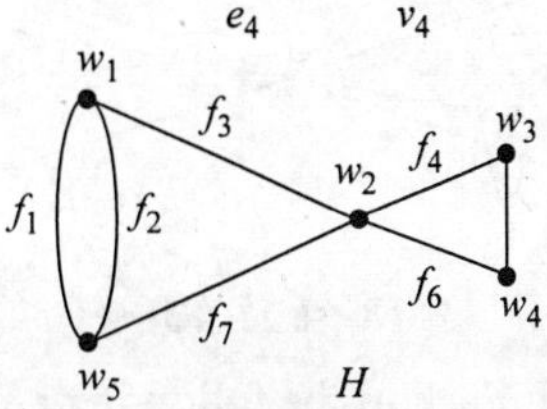

Figure 17.41

Solution. To solve this problem, we have to find $g: V(G) \to V(H)$ and $h: E(G) \to E(H)$ such that for all $v \in V(G)$ and $e \in E(G)$,

v is an endpoint of $e \Leftrightarrow g(v)$ is an endpoint of $h(e)$.

Since e_2 and e_3 are parallel (have the same endpoints), $h(e_2)$ and $h(e_3)$ must also be parallel. Thus we have

$h(e_2) = f_1$ and $h(e_3) = f_2$ or $h(e_2) = f_2$ and $h(e_3) = f_1$.

Also the endpoints of e_2 and e_3 must correspond to the endpoints of f_1 and f_2 and so

$g(v_3) = w_1$ and $g(v_4) = w_5$ or $g(v_3) = w_5$ and $g(v_4) = w_1$.

Further, we note that v_1 is the endpoint of four distinct edges e_1, e_7, e_5 and e_4 and so $g(v_1)$ should be the endpoint of four distinct edges. We observe that w_2 is the vertex having four edges and so $g(v_1) = w_2$. If $g(v_3) = w_1$, then since v_1 and v_3 are endpoints of e_1 in G, $g(v_1) = w_2$ and $g(v_3) = w_1$ must be endpoints of $h(e_1)$ in H. This implies that $h(e_1) = f_3$.

Continuing this way, we can find g and h to define the isomorphism between G and H.

One such pair of functions (of course there exist several) is shown in Figure 17.42.

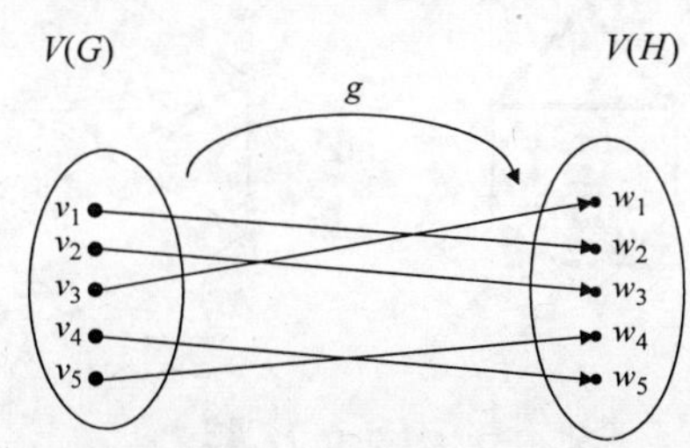

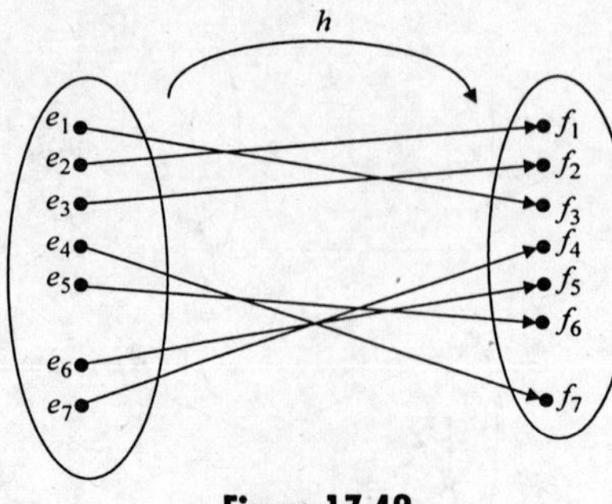

Figure 17.42

Remark 17.2 Each of the following properties of a graph is invariant under graph isomorphism, where n, m and k are all non-negative integers:

1. Has n vertices
2. Has m edges
3. Has a vertex of degree k
4. Has m vertices of degree k
5. Has a circuit of length k
6. Has a simple circuit of length k
7. Has m simple circuits of length k,
8. Is connected
9. Has an Euler circuit
10. Has a Hamiltonian circuit

(Items 4–9: to be studied later on in this chapter)

EXAMPLE 17.11

Examine for isomorphism

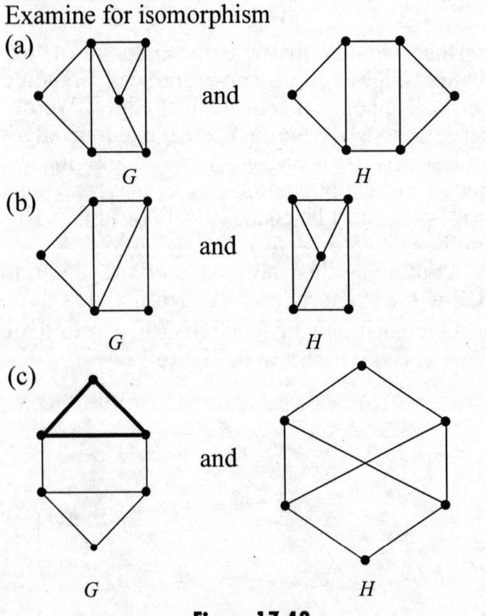

Figure 17.43

Solution. We note in Figure 17.43 that

(a) G has nine edges whereas H has only eight. Hence, G is not isomorphic to H.

(b) G has a vertex v of degree 4, whereas H has no vertex of degree 4. Hence, G is not isomorphic to H.

(c) G has a simple circuit of length 3 **(marked dark)** whereas H does not have. Hence, G is not isomorphic to H.

17.5 WALKS, PATHS AND CIRCUITS

Definition 17.35 In a graph G, **a walk from vertex v_0 to vertex v_n** is a finite alternating sequence $\{v_0, e_1, v_1, e_2, \ldots, v_{n-1}, e_n, v_n\}$ of vertices and edges such that v_{i-1} and v_i are the endpoints of e_i.

The **trivial walk** from a vertex v to v consists of the single vertex v.

Definition 17.36 In a graph G, a **path** from the vertex v_0 to the vertex v_n is a walk from v_0 to v_n that does not contain a repeated edge.

Thus a **path** from v_0 to v_n is a walk of the form

$$\{v_0, e_1, v_1, e_2, v_2, \ldots, v_{n-1}, e_n, v_n\},$$

where all the edges e_i are distinct.

Definition 17.37 In a graph, a **simple path** from v_0 to v_n is a path that does not contain a repeated vertex.

Thus a simple path is a walk of the form

$$\{v_0, e_1, v_1, e_2, v_2, \ldots, v_{i-1}, e_n, v_n\},$$

where all the e_i and all the v_i are distinct.

Definition 17.38 A walk in a graph G that starts and ends at the same vertex is called a **closed walk.**

Definition 17.39 A closed walk that does not contain a repeated edge is called a **circuit**.

Thus, a closed path is called a circuit (or a **cycle**) if it is a walk of the form

$$\{v_0, e_1, v_1, e_2, v_2, \ldots, v_{n-1}, e_n, v_n\},$$

where $v_0 = v_n$ and all the e_i are distinct.

Definition 17.40 A **simple circuit** is a circuit that does not have any other repeated vertex except the first and the last.

Thus, a simple circuit is a walk of the form

$$\{v_0, e_1, v_1, e_2, v_2, \ldots, v_{n-1}, e_n, v_n\},$$

where all the e_i are distinct and all the v_j are distinct except that $v_0 = v_n$.

EXAMPLE 17.12

In the graph given in the Figure 17.44, determine whether the following walks are paths, simple paths, closed walks, circuits, simple circuits or are just walks.

1. $v_1\, e_2\, v_2\, e_3\, v_3\, e_4\, v_4\, e_5\, v_2\, e_2\, v_1\, e_1\, v_0$,
2. $v_5\, v_4\, v_2\, v_1$,
3. $v_2\, v_3\, v_4\, v_5\, v_2$,
4. $v_4\, v_2\, v_3\, v_4\, v_5\, v_2\, v_4$,
5. $v_2\, v_1\, v_5\, v_2\, v_3\, v_4\, v_2$,
6. $v_0\, v_5\, v_2\, v_3\, v_4\, v_2\, v_1$.

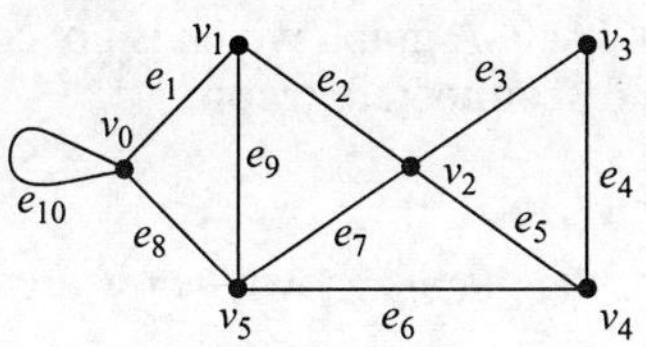

Figure 17.44

Solution.

1. It is **just a walk** because it has repeated edges and repeated vertices.
2. It is a **simple path**, because here no vertex is repeated and no edge is repeated.
3. It is **closed walk** in which no edge is repeated and hence it is a **circuit**. It starts and ends at the same vertex, and therefore is a **simple circuit**.
4. It is a closed walk in which no edge is repeated but vertices v_2 and v_4 are repeated. Hence it is a **circuit**.
5. It is a closed walk in which no edge is repeated. Hence it is a circuit. Only one vertex is repeated twice, hence it is **not a simple circuit.**
6. It is a walk in which no edge is repeated but the vertex v_2 is repeated. Hence it is a **path.**

EXAMPLE 17.13

Consider the graph shown in Figure 17.45.

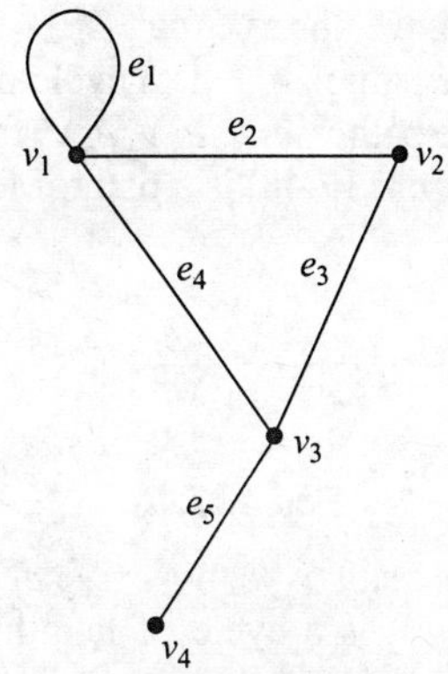

Figure 17.45

We note that e_3, e_5 is a path. The walk e_1, e_2, e_3, e_5 is a path but it is not a simple path because the vertex v_1 is repeated (e_1 being a self-loop). The walk e_2, e_3, e_4 is a circuit. The walk e_2, e_3, e_4, e_1 is a circuit but it is not simple circuit because vertex v_1 repeats twice (or we may write that v_1 is met twice).

Definition 17.41 In a graph the number of edges in the path $\{v_0, e_1, v_1, e_2, \ldots, e_n, v_n\}$ from v_0 to v_n is called the **length of the path.**

Definition 17.42 A cycle with *k*-edges is called a ***k*-cycle** or **cycle of length *k*.**

For example, loop is a cycle of length 1. On the other hand, a pair of parallel edges e_1 and e_2, shown in the Figure 17.46, is a cycle of length 2.

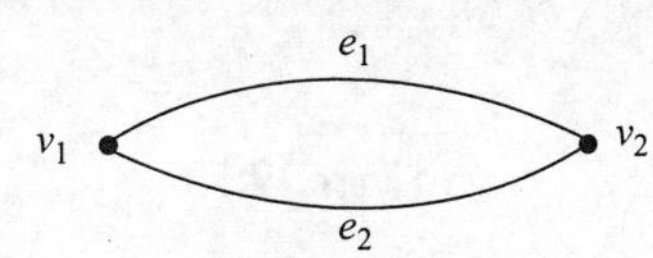

Figure 17.46

EXAMPLE 17.14

Consider the graph shown in Figure 17.47.

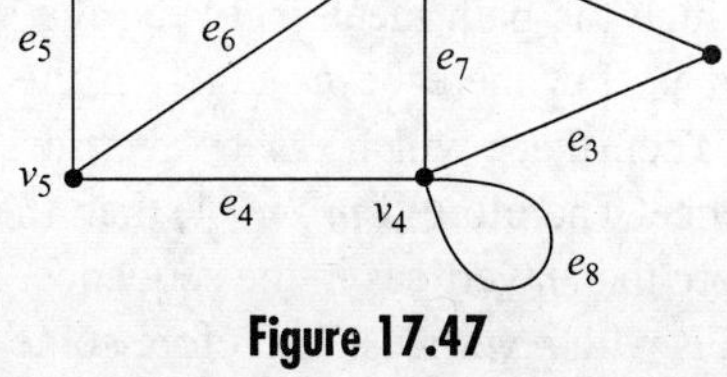

Figure 17.47

In this graph, we observe that

(i) The path $v_4 e_8 v_4$ is a cycle of length 1.

(ii) The path $v_5 e_4 v_4 e_8 v_4 e_3 v_3$ from v_5 to v_3 (Figure 17.48) is a path of length 3.

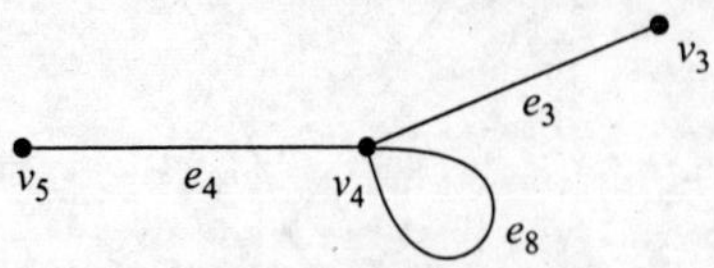

Figure 17.48

(iii) The path $v_1 e_1 v_2 e_6 v_5 e_5 v_1$ (Figure 17.49) is a cycle of length 3, since it consists of three edges.

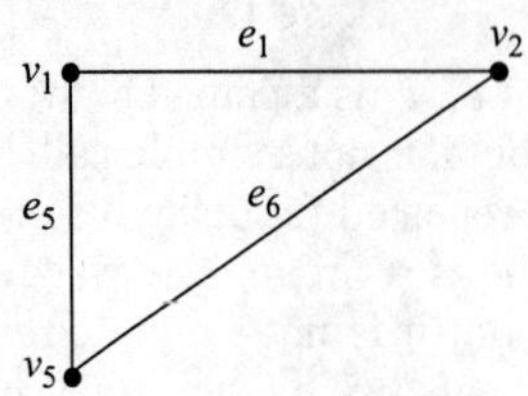

Figure 17.49

Definition 17.43 A graph is said to be **acyclic** if it contains no cycle.

For example, the graphs shown in Figure 17.50 are acyclic.

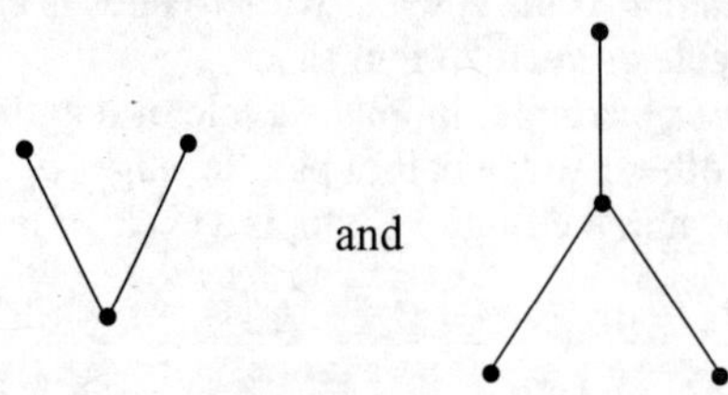

Figure 17.50

Theorem 17.3 If there is a path from vertex v_1 to v_2 in a graph with n vertices, then there does not exist a simple path of more than $n-1$ edges from vertex v_1 to v_2.

Proof: Suppose there is a path from v_1 to v_2. Let $v_1, \ldots, v_i, \ldots, v_2$ be the sequence of vertices which the path meets between the vertices v_1 and v_2. Let there be m edges in the simple path. Then there will be $m+1$ vertices in the sequence. Therefore if $m > n-1$, then there will be more than n vertices in the sequence. But the graph is with n vertices. Therefore some vertex, say v_k, appears more than once in the sequence. So the sequence of vertices shall be $v_1, \ldots, v_i, \ldots, v_k, \ldots, v_k, \ldots, v_2$. Deleting the edges in the path that lead v_k back to v_k we have a path from v_1 to v_2 that has fewer edges than the original one. This argument is repeated until we get a path that has $n-1$ or fewer edges.

Definition 17.44 Two vertices v_1 and v_2 of a graph G are said to be **connected** if and only if there is a walk from v_1 to v_2.

Definition 17.45 A graph G is said to be **connected** if and only if given any two vertices v_1 and v_2 in G, there is a walk from v_1 to v_2.

Thus, a graph G is connected if there exists a walk between every two vertices in the graph.

Definition 17.46 A graph which is not connected is called **disconnected graph**.

EXAMPLE 17.15

Which of the graphs shown below are connected?

(a)

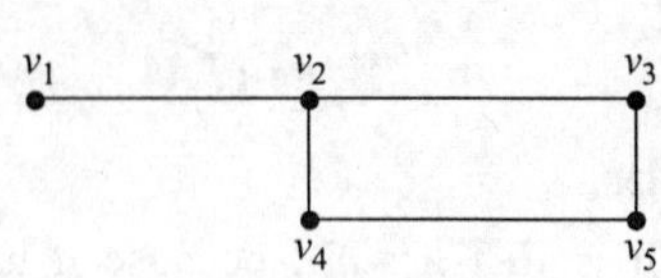

(b)

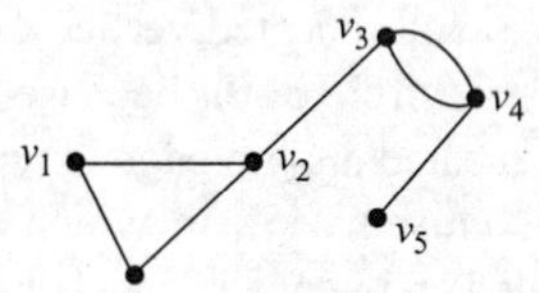

(c)

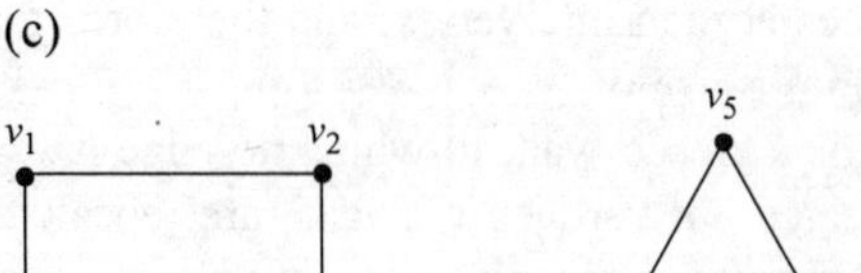

(d)

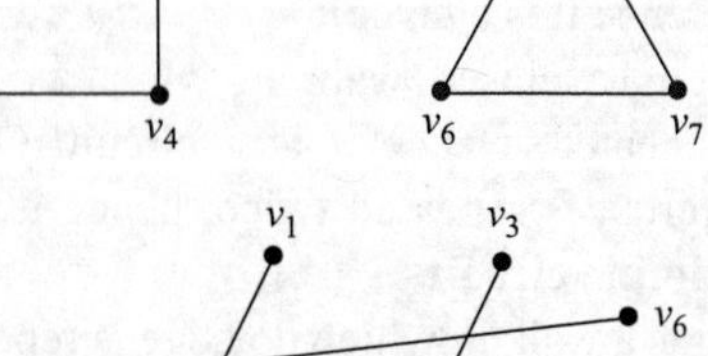

Figure 17.51

Solution. In Figure 17.51, we observe that

(a) Since every two vertices in the given graph are connected by a path, therefore the given graph is connected.
(b) This graph is also connected.
(c) The graph in (c) is disconnected. It has two connected components.
(d) In this graph, the edge (v_5, v_6) cross the edges (v_1, v_2) and (v_3, v_4) at points which are not vertices. Therefore the graph can be redrawn as shown in Figure 17.52.

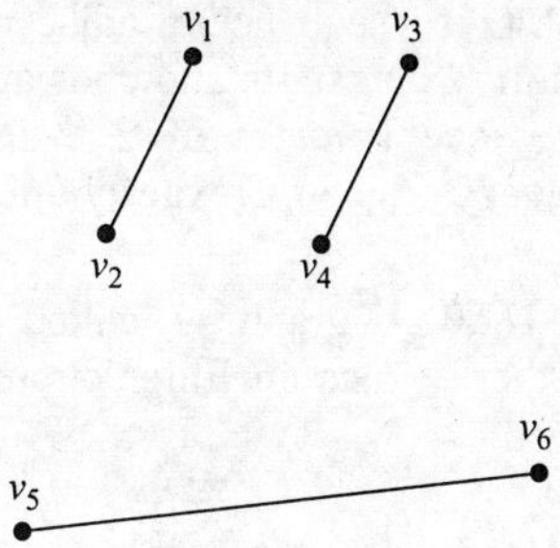

Figure 17.52

There is no path from v_2 to v_4, etc. Hence the given graph is disconnected and has three connected components.

EXAMPLE 17.16

Which of the graphs shown in Figure 17.53 are connected?

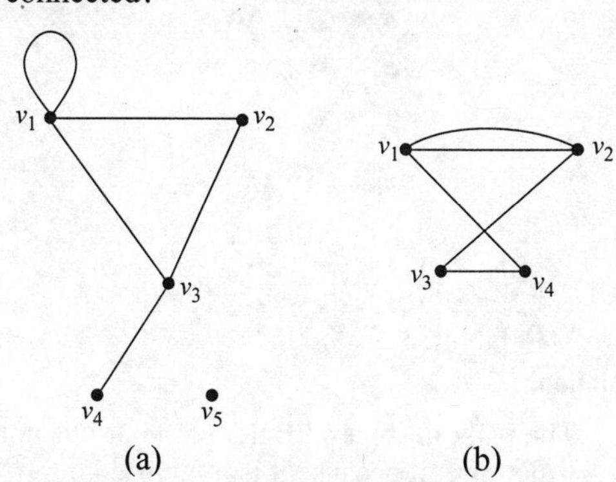

Figure 17.53

Solution. Graph (a) is not connected as there is no walk from any of v_1, v_2, v_3, v_4 to the vertex v_5.

The graph (b) is clearly connected.

Definition 17.47 If a graph G is disconnected, then the various connected pieces of G are called the **connected components of the graph.**

EXAMPLE 17.17

Find the connected components of the graphs given in the following (Figure 17.54).

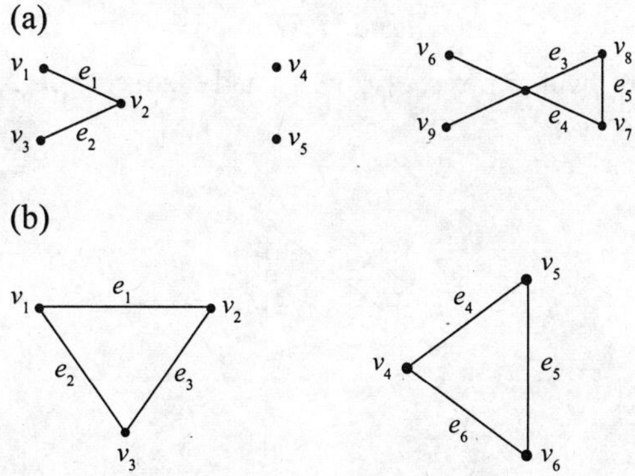

Figure 17.54

Solution.

(a) The graph in (a) has four connected components (Figures 17.55 and 17.56)

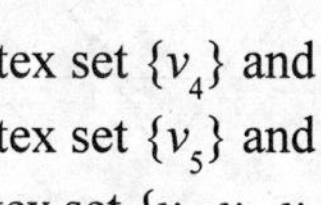

Figure 17.55

H_1: with vertex set $\{v_1, v_2, v_3\}$ and edge set $\{e_1, e_2\}$
H_2: with vertex set $\{v_4\}$ and edge set ϕ
H_3: with vertex set $\{v_5\}$ and edge set ϕ
H_4: with vertex set $\{v_6, v_7, v_8, v_9\}$ and edge set $\{e_3, e_4, e_5\}$

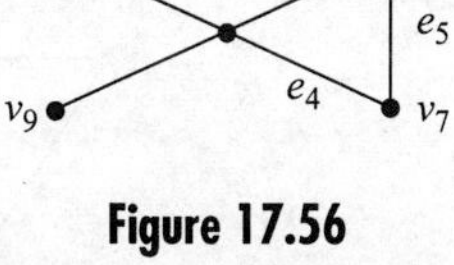

Figure 17.56

(b) The graph in (b) is disconnected and have two connected components (Figures 17.57(a), 17.57(b))

H_1:

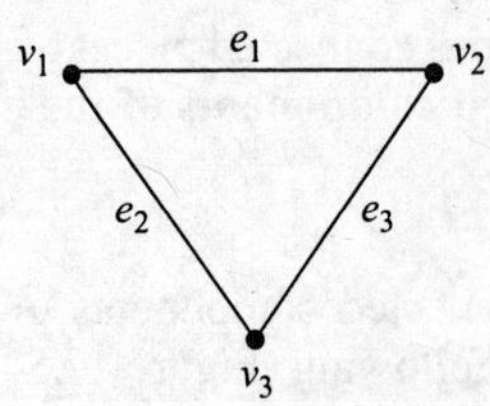

Figure 17.57(a)

with vertex set $\{v_1, v_2, v_3\}$ and edge set $\{e_1, e_2, e_3\}$

H_2:

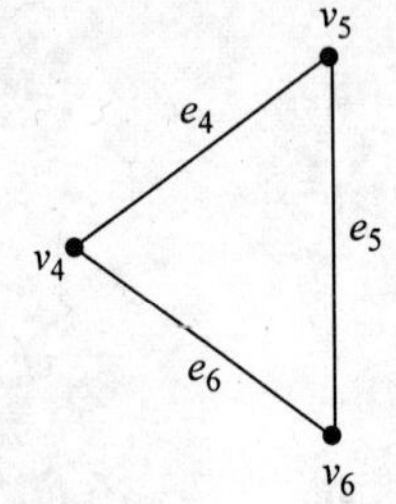

Figure 17.57(b)

with vertex set $\{v_4, v_5, v_6\}$ and edge set $\{e_4, e_5, e_6\}$.

EXAMPLE 17.18

Find the number of connected components in the graph shown below (Figure 17.58).

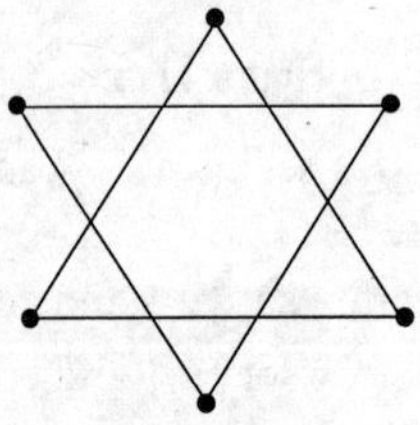

Figure 17.58

Solution. The connected components are shown in Figure 17.59.

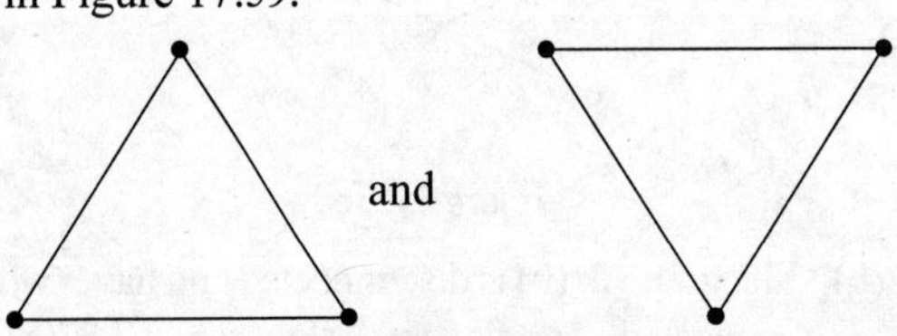

Figure 17.59

Remark 17.3 If a connected component has n vertices, then degree of any vertex cannot exceed $n-1$.

17.6 EULERIAN PATHS AND CIRCUITS

Definition 17.48 A path in a graph G is called an **Euler path** if it includes **every edge exactly once.**

Definition 17.49 A circuit in a graph G is called an **Euler circuit** if it includes every edge exactly once. Thus, an Euler circuit (Eulerian trail) for a graph G is a sequence of adjacent vertices and edges in G that starts and ends at the same vertex, uses **every vertex of** G **at least once,** and uses **every edge of** G **exactly once.**

Definition 17.50 A graph is called **Eulerian graph** if there exists an Euler circuit for that graph.

EXAMPLE 17.19

Find which of the following are Euler paths and Euler circuits in the graph given below (Figure 17.60)

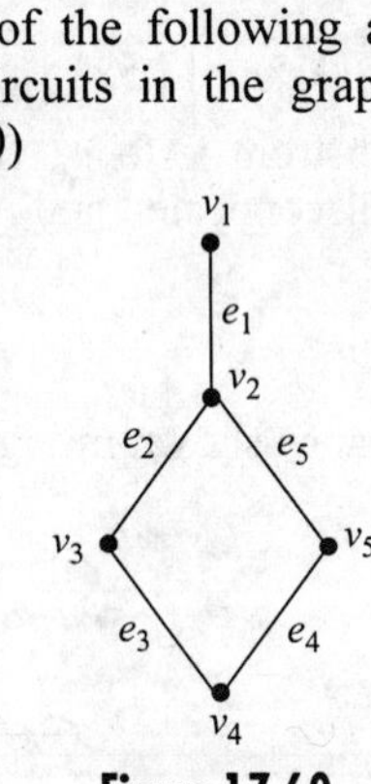

Figure 17.60

(a) $v_1\, e_1\, v_2\, e_2\, v_3\, e_3\, v_4\, e_4\, v_5\, e_5\, v_2$
(b) $v_1\, e_1\, v_2\, e_2\, v_3\, e_3\, v_4\, e_4\, v_5\, e_5\, v_2\, e_1\, v_1$
(c) $v_2\, e_2\, v_3\, e_3\, v_4\, e_4\, v_5\, e_5\, v_2$

Solution.

(a) The walk in (a) is an Euler path but it is not Euler circuit because it is not closed.
(b) It is not an Euler circuit because the edge e_1 is covered twice.
(c) It is neither an Euler path nor an Euler circuit because the vertex v_1 of G has not been used.

Theorem 17.4 If a graph has an Euler circuit, then every vertex of the graph has even degree.

Proof: Let G be a graph which has an Euler circuit. Let v be a vertex of G. We shall show that degree of v is even. By definition, Euler circuit contains every edge of graph G. Therefore the Euler circuit contains all edges incident on v. We start a journey beginning in the middle of one of the edges adjacent to the start of Euler circuit and continue around the Euler circuit to end in the middle of the starting edge. Since Euler circuit uses every edge exactly once, the edges incident on v occur in entry/exist pair and hence the degree of v is a multiple of 2. Therefore, the degree of v is even. This completes the proof of the theorem.

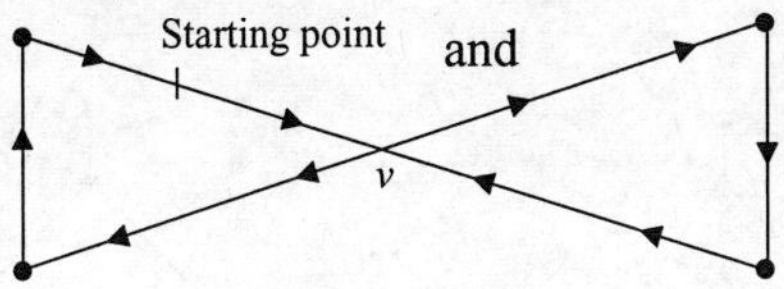

Figure 17.61

We know that ***contrapositive of a conditional statement is logically equivalent to the statement.*** Thus, Theorem 17.4 is equivalent to the following:

Theorem 17.5 If a vertex of a graph is not of even degree, then it does not have an Euler circuit.
Thus,
If some vertex of a graph has odd degree, then that graph does not have an Euler circuit.

EXAMPLE 17.20

Show that the graphs shown in Figure 17.62 do not have Euler circuits.

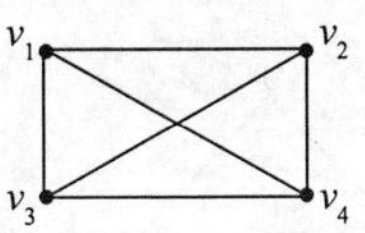

(a)

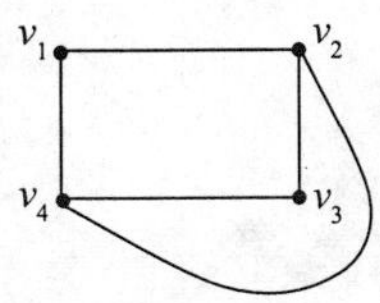

(b)

Figure 17.62

Solution. In graph (a), degree of each vertex is 3. Hence this **does not** have an Euler circuit.

In graph (b), we have

$$\deg(v_2) = 3 \quad \text{and} \quad \deg(v_4) = 3.$$

Since there are vertices of odd degree in the given graph, therefore it **does not** have an Euler circuit.

Remark 17.4 The converse of Theorem 17.4 is not true. There exist graphs in which every vertex has even degree but the Euler circuits do not exist.

For example,

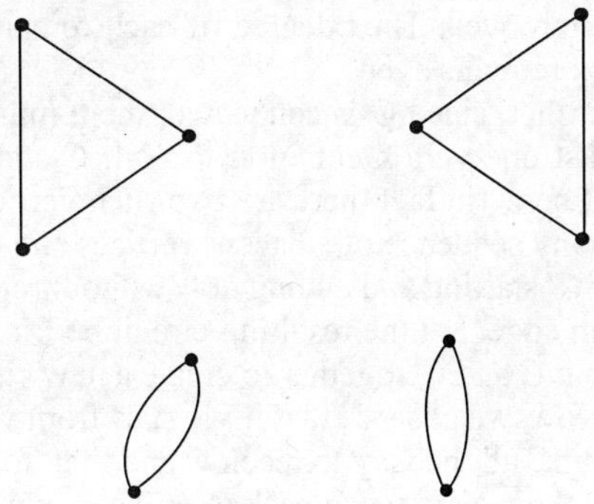

Figure 17.63

are graphs (Figure 17.63) in which each vertex has degree 2 but these graphs do not have Euler circuits since there is no path which uses each vertex at least once.

Theorem 17.6 If G is a connected graph and every vertex of G has even degree, then G has an Euler circuit.

Proof: Let every vertex of a connected graph G has even degree. If G consists of a single vertex v, the trivial walk from v to v is an Euler circuit. So suppose that G consists of more than one vertices. We start from any vertex v of G. Since the degree of each vertex of G is even, if we reach each vertex other than v by travelling on one edge, the same vertex can be reached by travelling on another previously unused edge. Thus a sequence of distinct adjacent edges can be produced indefinitely as long as v is not reached. Since the number of edges of the graph is finite (by definition of graph), the sequence of distinct edges will terminate. Thus the sequence must return to the starting vertex. We thus obtain a sequence of adjacent vertices and edges starting and ending at v without repeating any edge. Thus we get a circuit C.

If C contains every edge and vertex of G, then C is an Euler circuit.

If C does not contain every edge and vertex of G, remove all edges of C from G and also any vertices that become isolated when the edges of C are removed. Let the resulting subgraph be G'. We note that when we removed edges of C, an even number of edges from each vertex have been removed. Thus degree of each remaining vertex remains even.

Further, since G is connected, there must be at least one vertex common to both C and G'. Let it be w (in fact there are two such vertices). Pick any sequence of adjacent vertices and edges of G' starting and ending at w without repeating an edge. Let the resulting circuit be C'.

Join C and C' together to create a new circuit C''. Now, we observe that if we start from v and follow C all the way to reach w and then follow C' all the way to reach back to w. Then continuing travelling along the untravelled edges of C, we reach v.

If C'' contains every edge and vertex of C, then C'' is an Euler circuit. If not, then we again repeat our process. Since the graph is finite, the process must terminate.

The process followed has been described in the graph G shown below (Figure 17.64).

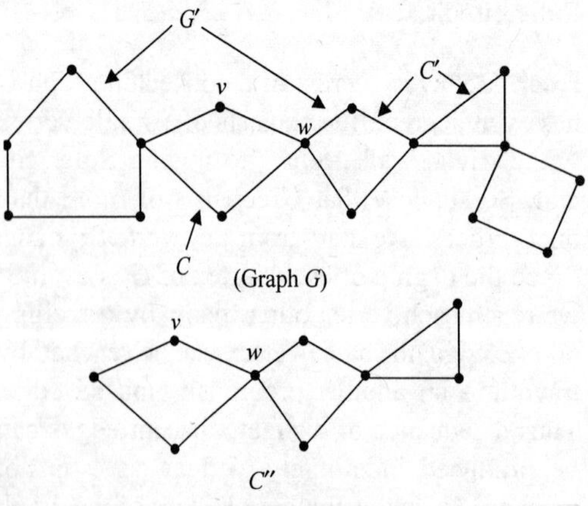

Figure 17.64

Theorems 17.4 and 17.6 taken together imply:

Theorem 17.7 (Euler's Theorem) A finite connected graph G has an Euler circuit if and only if every vertex of of G has even degree.

Thus, finite connected graph is Eulerian if and only if each vertex has even degree.

Theorem 17.8 If a graph G has more than two vertices of odd degree, then there can be no Euler path in G.

Proof: Let v_1, v_2 and v_3 be vertices of odd degree. Since each of these vertices had odd degree, any possible Euler path must leave (arrive at) each of v_1, v_2, v_3 with no way to return (or leave). One vertex of these three vertices may be the beginning of Euler path and another the end, but this leaves the third vertex at one end of an untravelled edge. Thus, there is no Euler path.

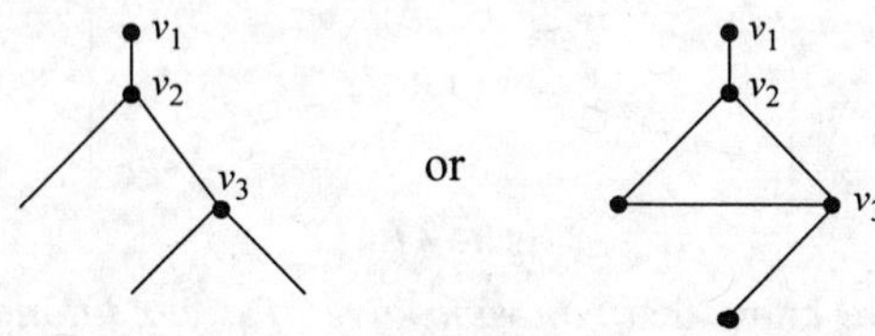

(Graphs having more than two vertices of odd degree).

Figure 17.65

Theorem 17.9 If G is a connected graph and has exactly two vertices of odd degree, then there is an Euler path in G. Further, any Euler path in G must begin at one vertex of odd degree and end at the other.

Proof: Let u and v be two vertices of odd degree in the given connected graph G (see Figure 17.66)

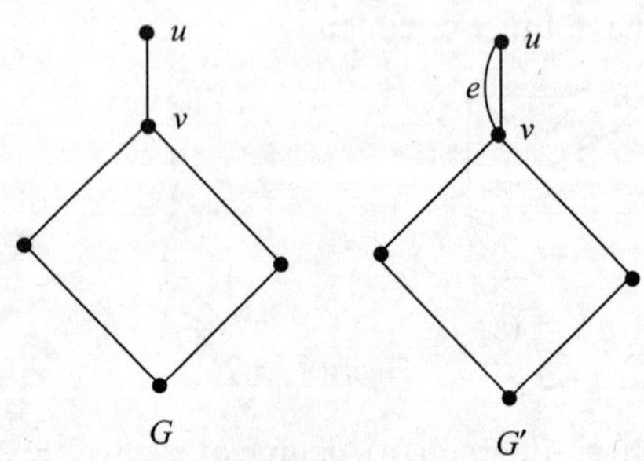

Figure 17.66

If we add the edge e to G, we get a connected graph G' all of whose vertices have even degree. Hence there will be an Euler circuit in G'. If we omit e from Euler circuit, we get an Euler path beginning at u (or v) and ending at v (or u).

EXAMPLE 17.21

Has the graph given in Figure 17.67 an Eulerian path?

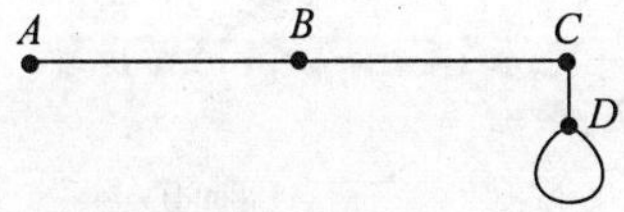

Figure 17.67

Solution. In the given graph,

$$\deg(A) = 1, \quad \deg(B) = 2,$$
$$\deg(C) = 2, \quad \deg(D) = 3.$$

Thus the given connected graph has exactly two vertices of odd degree. Hence, it has an Eulerian path.

If it starts from A (vertex of odd degree), then it ends at D (vertex of odd degree). If it starts from D (vertex of odd degree), then it ends at A (vertex of odd degree).

But, on the other hand, let the graph be as shown below (Figure 17.68)

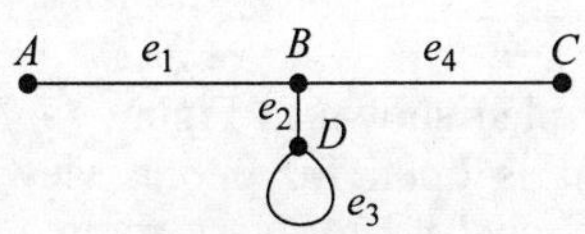

Figure 17.68

Then $\deg(A) = 1$, $\deg(B) = 3$, $\deg(C) = 1$, degree of $D = 3$ and so we have four vertices of odd degree. Hence it does not have an Euler path.

EXAMPLE 17.22

Does the graph given in Figure 17.69 possess an Euler circuit?

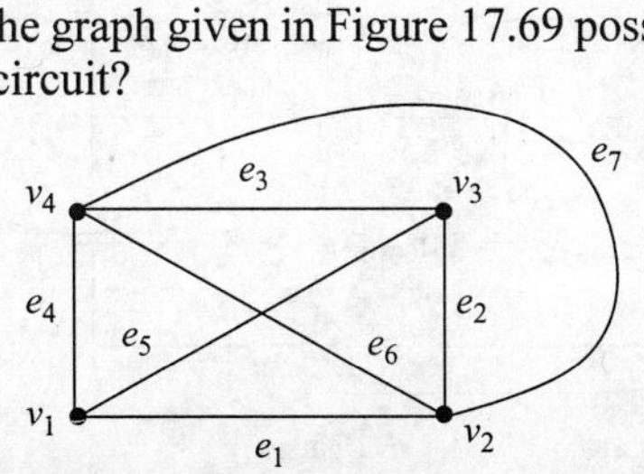

Figure 17.69

Solution. The given graph is connected. Further

$$\deg(v_1) = 3, \quad \deg(v_2) = 4,$$
$$\deg(v_3) = 3, \quad \deg(v_4) = 4.$$

Since this connected graph has vertices with odd degree, it cannot have Euler circuit. But this graph has Euler path, since it has exactly two vertices of odd degree. For example, $v_3\, e_2\, v_2\, e_7\, v_4\, e_6\, v_2\, e_1\, v_1\, e_4\, v_4\, e_3\, v_3\, e_5\, v_1$ is an Euler path.

EXAMPLE 17.23

Consider the graph given below (Figure 17.70):

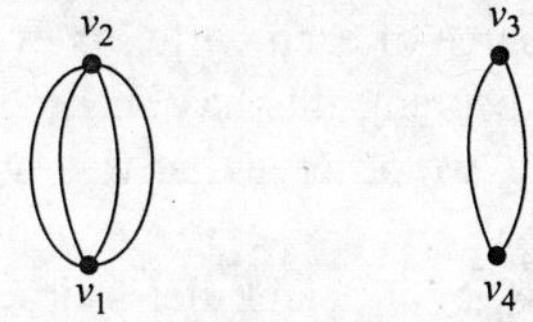

Figure 17.70

Here, $\deg(v_1) = 4$, $\deg(v_2) = 4$, $\deg(v_3) = 2$, $\deg(v_4) = 2$. Thus degree of each vertex is even. But the graph is not Eulerian since it is **not connected.**

EXAMPLE 17.24

Is it possible to trace the graph in Figure 17.71 without lifting the pencil?

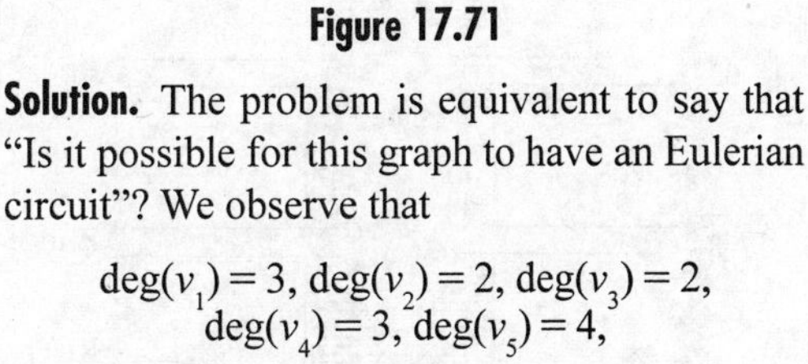

Figure 17.71

Solution. The problem is equivalent to say that "Is it possible for this graph to have an Eulerian circuit"? We observe that

$$\deg(v_1) = 3, \deg(v_2) = 2, \deg(v_3) = 2,$$
$$\deg(v_4) = 3, \deg(v_5) = 4,$$

$$\deg(v_6) = 2, \deg(v_7) = 2, \deg(v_8) = 4,$$
$$\deg(v_9) = 2.$$

Since the graph contains vertex of odd degree, therefore it cannot have Euler's circuit and therefore can not be traced without lifting the pencil.

EXAMPLE 17.25

Find the Euler's circuit for the graph given below (Figure 17.72):

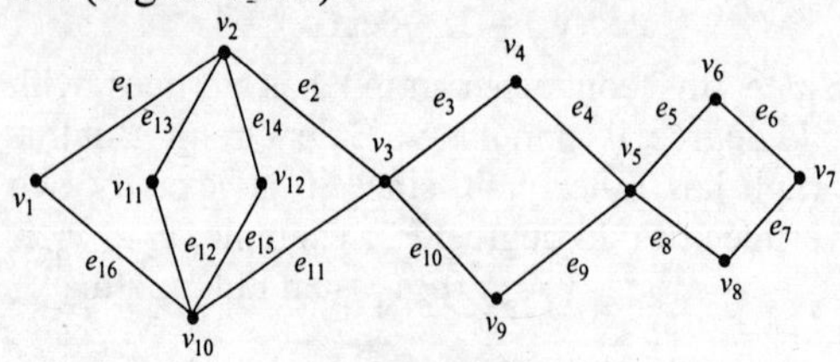

Figure 17.72

Solution. The given connected graph has 12 vertices and degree of each vertex is even. Hence, by Euler's theorem, this has an Euler's circuit. For example, we observe that $v_1\, e_1\, v_2\, e_{13}\, v_{11}\, e_{12}\, v_{10}\, e_{15}\, v_{12}\, e_{14}\, v_2\, e_2\, v_3\, e_3\, v_4\, e_4\, v_5\, e_5\, v_6\, e_6\, v_7\, e_7\, v_8\, e_8\, e_9\, v_9\, e_{10}\, v_3\, e_{11}\, v_{10}\, e_{16}\, v_1$ is an Euler's circuit. In short we can represent it by $e_1, e_{13}, e_{12}, e_{15}, e_{14}, e_2, e_3, e_4, e_5, e_6, e_7, e_8, e_9, e_{10}, e_{11}, e_{16}$.

EXAMPLE 17.26 (The bridges of Konigsberg)

The graph theory began in 1736 when Leonhard Euler solved the problem of seven bridges on Pregel River in the town of Konigsberg in Prussia (now Kaliningrad in Russia). The two islands and seven bridges are shown in Figure 17.73.

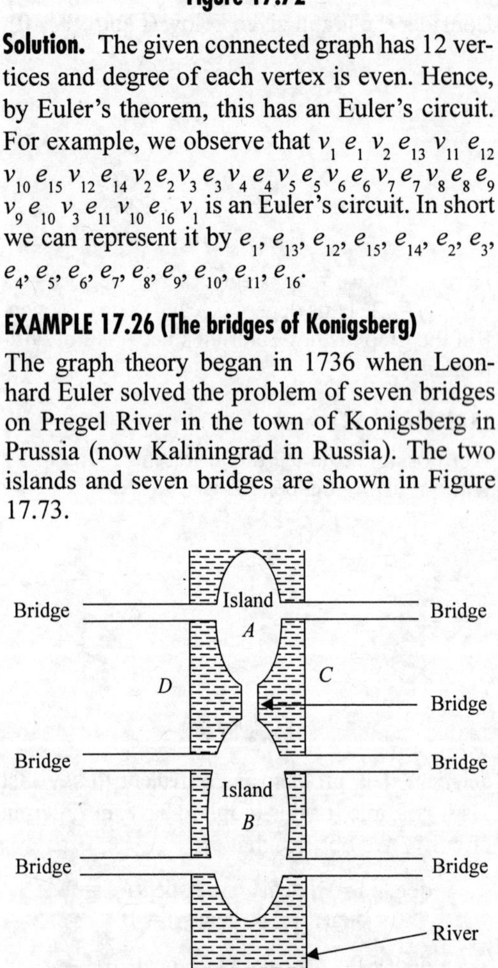

Figure 17.73

The people of Konigsberg posed the following question to famous Swiss Mathematician Leonhard Euler:

"Beginning anywhere and ending any where, can a person walk through the town of Konigsberg crossing all the seven bridges exactly once?"

Euler showed that such a walk is impossible. He replaced the islands A, B and the two sides (banks) C and D of the river by vertices and the bridges as edges of a graph. We note then that

$$\deg(A) = 3, \quad \deg(B) = 5,$$
$$\deg(C) = 3, \quad \deg(D) = 3.$$

Thus the graph of the problem is as shown in Figure 17.74.

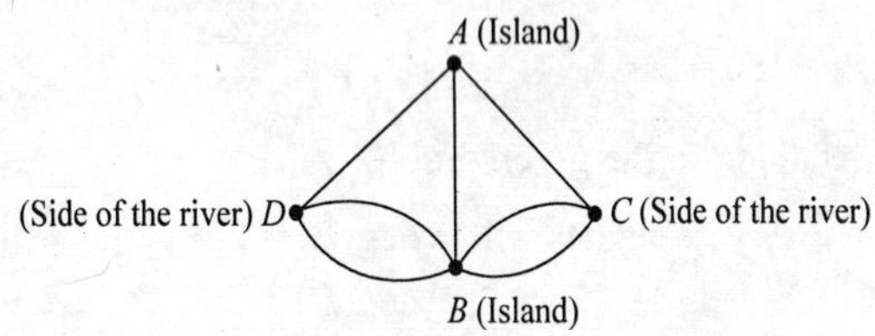

(Euler's graphical representation of seven bridges problem)

Figure 17.74

The problem then re1duces to

"Is there any Euler's path in the above diagram?".

To find the answer, we note that there are more than two vertices having odd degree. Hence there exists no Euler path for this graph.

EXAMPLE 17.27

The floor plan shown in Figure 17.75 is for a house that is open for public viewing. Each room is connected to every room with which it has a common wall and to the outside along each wall. Is it possible to begin in a room or outside and take a walk which goes through each door exactly once?

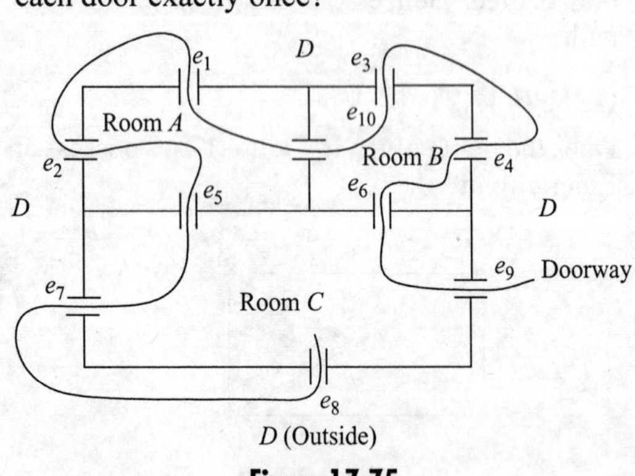

Figure 17.75

If each room and out side constitute a vertex and each door corresponds to an edge, then the floor plan converts into the graph. There are two edges e_1 and e_2 from A to D, two edges e_3 and e_4 from B to D, three edges e_7, e_8, e_9 from C to D, one edge e_{10} from A to B, one edge e_5 from A to C, one edge e_6 from B to C. Hence the graph is as shown below (Figure 17.76):

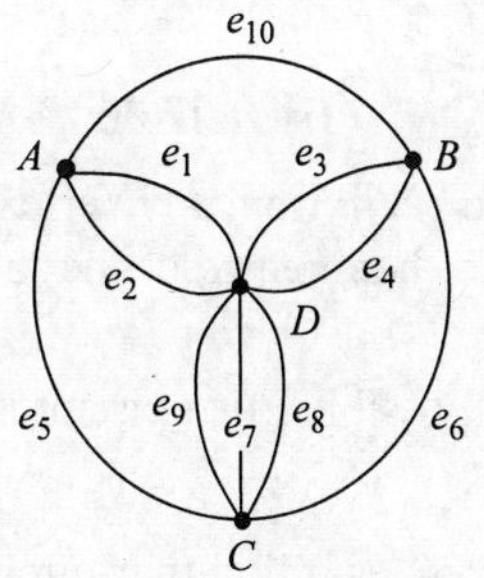

Figure 17.76

We note that

$$\deg(A) = \deg(B) = 4, \quad \deg(C) = 5, \quad \deg(D) = 7.$$

Since the graph is connected and degree of exactly two vertices C and D is odd, there exists an Euler's path and that path should start from one vertex of odd degree and end to the other vertex of odd degree. Therefore, either the path will begin from C and end at D or it will begin from D and end at C. An Euler's path is shown (which begins from C).

Definition 17.51 An edge in a connected graph is called a **bridge** or a **cut edge** if deleting that edge creates a disconnected graph.

For example, consider the graph shown below (Figure 17.77):

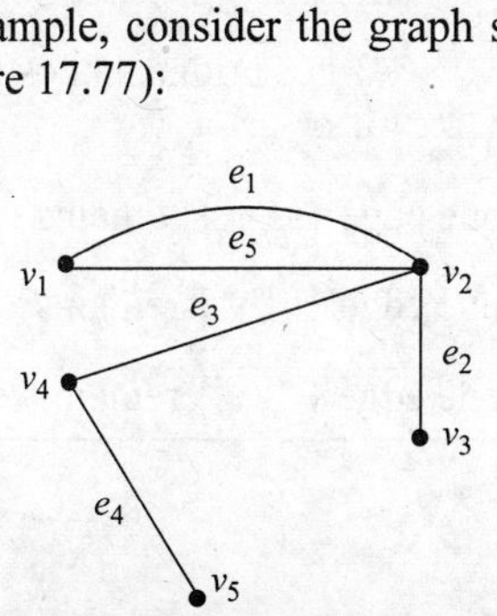

Figure 17.77

In this graph, if we remove the edge e_3, then the graph breaks into two connected components given below (Figure 17.78):

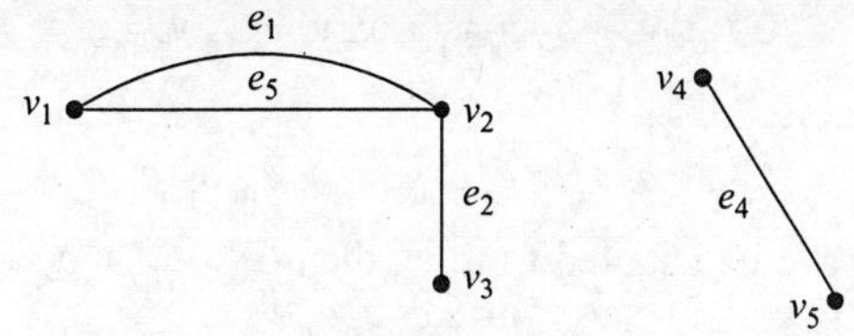

Figure 17.78

Hence the edge e_3 is a bridge in the given graph.

17.6.1 METHODS FOR FINDING EULER CIRCUIT

Method 1: We know that if every vertex of a non-empty connected graph has even degree, then the graph has an Euler circuit. We shall make use of this result to find an Euler path in a given graph.

Consider the graph shown in Figure 17.79.

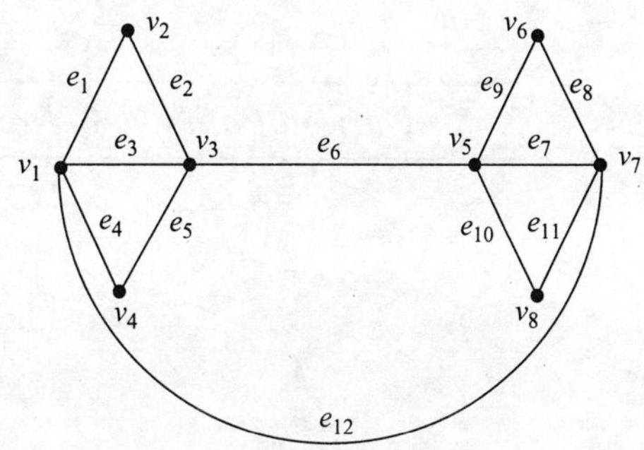

Figure 17.79

We note that

$$\deg(v_2) = \deg(v_4) = \deg(v_6) = \deg(v_8) = 2,$$
$$\deg(v_1) = \deg(v_3) = \deg(v_5) = \deg(v_7) = 4.$$

Hence all vertices have even degree. Also the given graph is connected. Hence the given graph has an Euler circuit. We start from the vertex v_1 and let C be

$$C: v_1\, v_2\, v_3\, v_1.$$

Then C is not an Euler circuit for the given graph but C intersects the rest of the graph at v_1 and v_3.
Let C' be

$$C': v_1 v_4\, v_3\, v_5\, v_7\, v_6\, v_5\, v_8\, v_7\, v_1.$$

(In case we start from v_3, then C' will be $v_3\, v_4\, v_1\, v_7\, v_6\, v_5\, v_7\, v_8\, v_5$).
Path C' into C and obtain

$$C'': v_1 v_2 v_3 v_1 v_4 v_3 v_5 v_7 v_6 v_5 v_8 v_7 v_1,$$

that is,

$$C'': e_1 e_2 e_3 e_4 e_5 e_6 e_7 e_8 e_9 e_{10} e_{11} e_{12}.$$

(If we had started from v_2, then C'': $v_1 v_2 v_3 v_4 v_1 v_7 v_6 v_5 v_7 v_8 v_5 v_3 v_1$ **or** $e_1 e_2 e_5 e_4 e_{12} e_8 e_9 e_7 e_{11} e_{10} e_6 e_3$). In C'' all edges are covered exactly once. Also every vertex has been covered at least once. Hence C'' is an Euler circuit.

Method 2: Fleury's Algorithm: This algorithm is used to find Euler's circuit for a connected graph with no vertices of odd degree.

We shall illustrate the method with the help of the following example:

Let G be the connected graph as shown in the Figure 17.80.

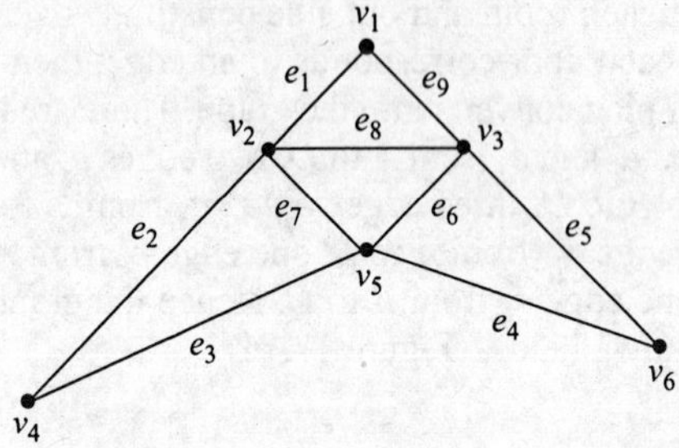

Figure 17.80

Step 1. We begin from any vertex, say v_1

Step 2. We then construct the table given below.

Hence one possible Euler circuit is

$$C: v_1 v_2 v_4 v_5 v_6 v_3 v_5 v_2 v_3 v_1.$$

In term of edges, this Euler circuit can be expressed as

$$e_1 e_2 e_3 e_4 e_5 e_6 e_7 e_8 e_9.$$

Current Path	*Next Edge*	*Reason*
v_1	$\{v_1, v_2\}$	No Edge from v_1 is a bridge, so choose any edge, say $\{v_1, v_2\}$
$v_1 v_2$	$\{v_2, v_4\}$	(v_2, v_5) a bridge, so choose $\{v_2, v_4\}$
$v_1 v_2 v_4$	$\{v_4, v_5\}$	only one edge $\{v_4, v_5\}$ from v_4 remains
$v_1 v_2 v_4 v_5$	$\{v_5, v_6\}$	Since $\{v_2, v_5\}$ and $\{v_5, v_3\}$ are bridges, choose $\{v_5, v_6\}$
$v_1 v_2 v_4 v_5 v_6$	$\{v_6, v_3\}$	only one edge $\{v_6, v_3\}$ remains
$v_1 v_2 v_4 v_5 v_6 v_3$	$\{v_3, v_5\}$	Since $\{v_1, v_3\}$ is a bridge, choose either $\{v_3, v_2\}$ or $\{v_3, v_5\}$
$v_1 v_2 v_4 v_5 v_6 v_3 v_5$	$\{v_5, v_2\}$	only one edge $\{v_5, v_3\}$ remains
$v_1 v_2 v_4 v_5 v_6 v_3 v_5 v_2$	$\{v_2, v_3\}$	only one edge $\{v_2, v_3\}$ remains
$v_1 v_2 v_4 v_5 v_6 v_3 v_5 v_2 v_3$	$\{v_3, v_1\}$	only one edge $\{v_3, v_1\}$ remains

17.7 HAMILTONIAN CIRCUITS

Definition 17.52 A **Hamiltonian path** for a graph G is a sequence of adjacent vertices and distinct edges in which every vertex of G appears exactly once.

Definition 17.53 A **Hamiltonian circuit** for a graph G is a sequence of adjacent vertices and distinct edges in which every vertex of G appears exactly once, except for the first and the last which are the same.

Definition 17.54 A graph is called **Hamiltonian** if it admits a Hamiltonian circuit.

EXAMPLE 17.28

The wooden graph shown in Fig. 17.81 and constructed by William Hamilton in the shape of a regular dodecahedron is a Hamiltonian circuit.

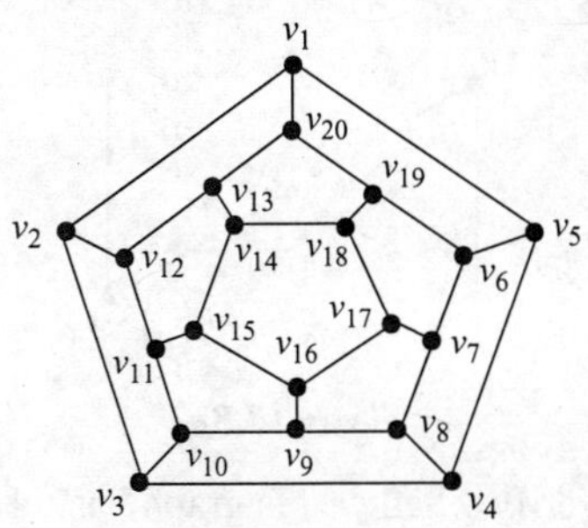

Figure 17.81

The Hamilton circuit is $v_1\, v_2\, v_3 \ldots v_{18}\, v_{19}\, v_{20}\, v_1$.

EXAMPLE 17.29

A complete graph K_n has a Hamiltonian circuit. In particular, the graphs shown in Figure 17.82 are Hamiltonian.

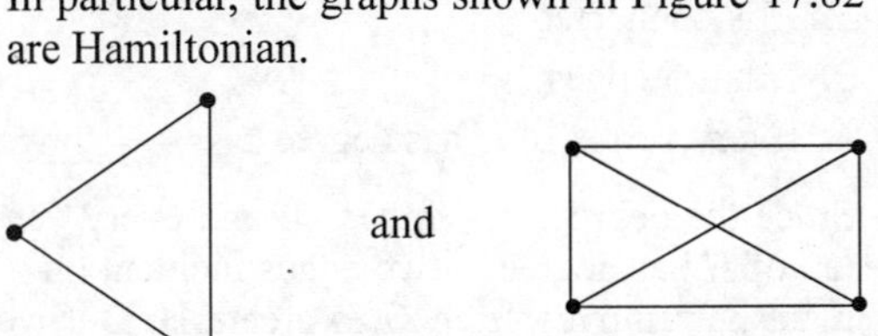

Figure 17.82

EXAMPLE 17.30

The graph shown below does not have a Hamiltonian circuit.

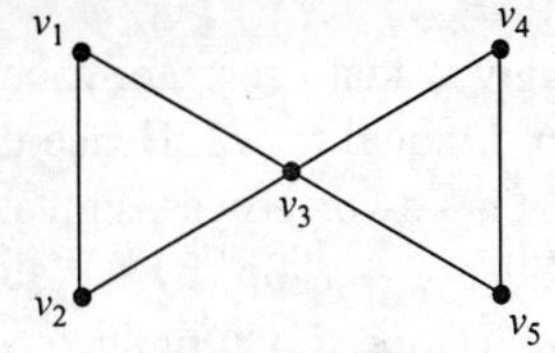

Figure 17.83

EXAMPLE 17.31

The graph shown in Figure 17.84 does not have a Hamiltonian circuit.

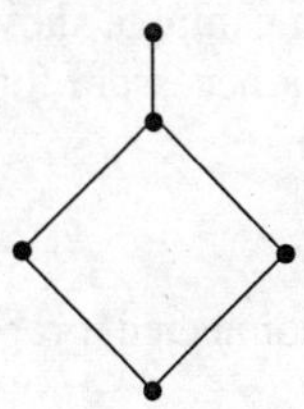

Figure 17.84

Remark 17.5 It is clear that **only connected graphs can have Hamiltonian circuit**. However, there is no simple criterion to tell us whether or not a given graph has Hamiltonian circuit. The following results give us some sufficient conditions for the existence of Hamiltonian circuit/path.

Theorem 17.10 Let G be a linear graph of n vertices. If the sum of the degrees for each pair of vertices in G is greater than or equal to $n-1$, then there exists a Hamiltonian path in G.

Theorem 17.11 Let G be a connected graph with n vertices. If $n \geq 3$ and $\deg(v) \geq n$ for each vertex v in G, then G has a Hamiltonian circuit.

Theorem 17.12 Let G be a connected graph with n vertices and let u and v be two vertices of G that are not adjacent. If $\deg(u) + \deg(v) \geq n$, then G has a Hamiltonian circuit.

Corollary 17.2 Let G be a connected graph with n vertices. If each vertex has degree greater than or equal to $n/2$, then G has a Hamiltonian circuit.

Proof: It is given that degree of each vertex is greater than or equal to $n/2$. Hence the sum of the degree of any two vertices is greater than or equal to $n/2 + n/2 = n$. So, by the above theorem, the graph G has a Hamiltonian circuit.

Theorem 17.13 Let n be the number of vertices and m be the number of edges in a connected graph G. If $m \geq \frac{1}{2}(n^2-3n+6)$, then G has a Hamiltonian circuit.

The following example shows that the above conditions are not necessary for the existence of Hamiltonian path.

EXAMPLE 17.32

Let G be the connected graph shown in the Figure 17.85.

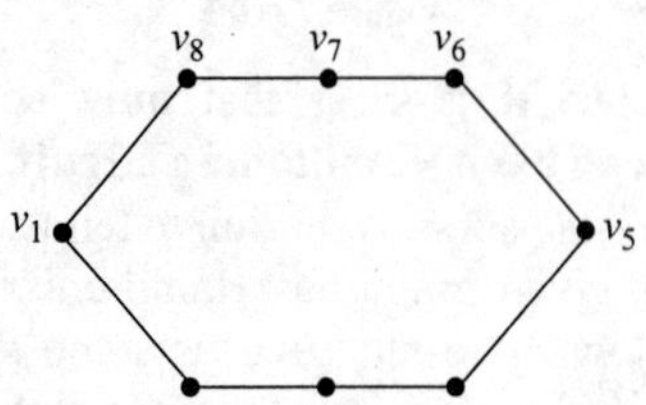

Figure 17.85

We note that

$$n = \text{Number of vertices in } G = 8,$$
$$m = \text{Number of edges in } G = 8,$$
$$\text{Degree of each vertex} = 2.$$

Thus, if u and v are non-adjacent vertices, then

$$\deg(u) + \deg(v) = 2 + 2 = 4 \ngeq 8.$$

Also,

$$\frac{1}{2}(n^2-3n+6) = \frac{1}{2}(64-24+6) = 23.$$

Clearly $m \ngeq \frac{1}{2}(n^2-3n+6)$. Therefore the above two theorems fail. But the given graph has Hamiltonian circuit. For example, $v_1, v_2, v_3, v_4, v_5, v_6, v_7, v_8, v_1$ is an Hamiltonian circuit for the graph.

Proposition 17.1 Let G be a graph with at least two vertices. If G has a Hamiltonian circuit, then G has a subgraph H with the following properties:

1. H contains every vertex of G
2. H is connected
3. H has the same number of edges as vertices
4. Every vertex of H has degree 2

The contrapositive of this proposition is

"If a graph G with at least two vertices does not have a subgraph H satisfying (1) – (4), then G does not have a Hamiltonian circuit".

Also we know that contrapositive of a statement is logically equivalent to the statement. Therefore the above result can be used to show non-existence of a Hamiltonian circuit.

EXAMPLE 17.33

Consider the graph G shown below (Figure 17.86):

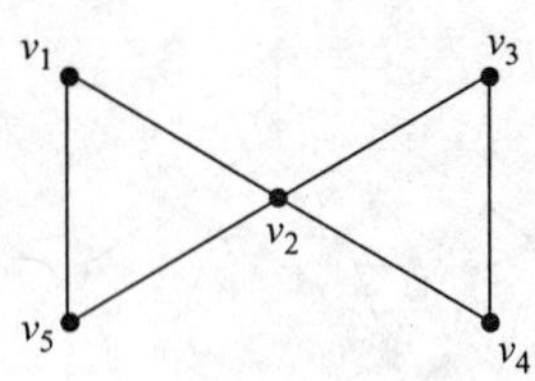

Figure 17.86

Let us verify whether this graph has a Hamiltonian circuit.

Suppose that G has a Hamiltonian circuit. Then it has a subgraph H such that

1. H contains every vertex of G, i.e., H has 5 vertices v_1, v_2, v_3, v_4, v_5
2. H is connected
3. H has 5 edges
4. Every vertex of H has degree 2

Since the degree of v_2 in G is 4 and every vertex of H has degree 2, two edges incident on v_2 must be removed from G to create H. But we note that the edge $\{v_1, v_2\}$ cannot be removed, because the removal of $\{v_1, v_2\}$ will decrease the degree of v_1 to 1, which is less than 2. Similarly the edge $\{v_2, v_3\}$, $\{v_2, v_4\}$ and $\{v_2, v_5\}$ cannot be removed. Hence the degree of v_2 in H must stand at 4 which contradicts the condition that every vertex in H has degree 2 in H. Therefore, no such subgraph H exists. Hence G does not have a Hamiltonian circuit.

EXAMPLE 17.34

Does the graph G given below (Figure 17.87) have Hamiltonian circuit?

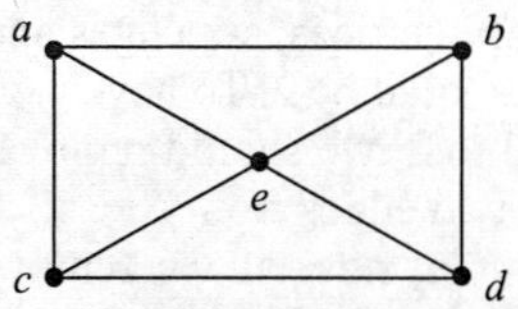

Figure 17.87

Solution. The given graph has

n = Number of vertices = 5

m = Number of edges = 8

$\deg(a) = \deg(b) = \deg(c) = \deg(d) = 3$

$\deg(e) = 4$

We observe that

(i) Degree of each vertex is greater than $n/2$

(ii) The sum of degrees of any non-adjacent pair of vertices is greater than n

(iii) $\frac{1}{2}(n^2 - 3n + 6) = \frac{1}{2}(25 - 15 + 6) = 8.$

Thus the condition $m \geq \frac{1}{2}(n^2 - 3n + 6)$ is satisfied.

(iv) The sum of degrees of each pair of vertices in the given graph is greater than $n - 1 = 5 - 1 = 4$.

Thus four sufficiency conditions are satisfied (whereas one condition out of these four conditions is sufficient for the existence of Hamiltonian path/circuit). Hence the graph has a Hamiltonian circuit.

For example, the following circuits in G are Hamiltonian (Figure 17.88).

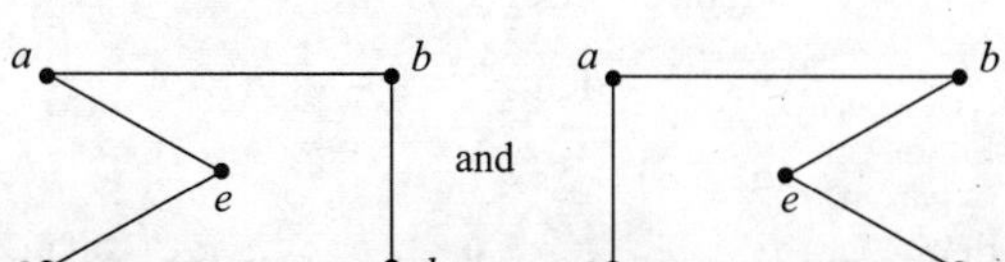

Figure 17.88

EXAMPLE 17.35

Does the graph shown below (Figure 17.89) has Hamiltonian circuit?

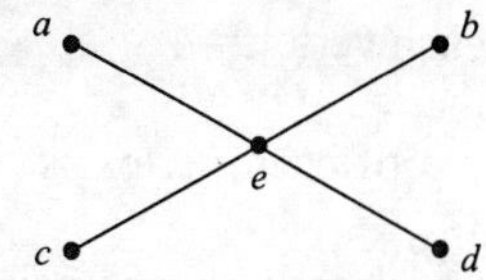

Figure 17.89

Solution. Here

Number of vertices $(n) = 5$

Number of edges $(m) = 4$

$\deg(a) = \deg(b) = \deg(c) = \deg(d) = 1$

$\deg(e) = 4$

We note that

(i) $\deg(a) = \deg(b) = \deg(c) = \deg(d) \ngeq \frac{5}{2}$

(ii) $\deg(a) + \deg(b) = 2 \ngeq 5$, that is sum of any non-adjacent pair of vertices is not greater than 5

(iii) $\frac{1}{2}(n^2 - 3n + 6) = \frac{1}{2}(25 - 15 + 6) = 8$

Therefore the condition $m \geq \frac{1}{2}(n^2 - 3n + 6)$ is **not satisfied,**

(iv) $\deg(a) + \deg(b) = 2 \ngeq 4$, i.e., the condition that sum of degrees of each pair of vertices in the graph is not greater than or equal to $n - 1$.

Hence no sufficiency condition is satisfied. So we try the Proposition 17.1. Suppose that G has a Hamiltonian circuit. Then G should have a subgraph which contains every vertex of G, and number of vertices and number of edges in H should be same. Thus, H should have five vertices a, b, c, d, e and five edges. Since G has only four edges, H cannot have more than four edges. Hence no such subgraph is possible. Hence, the given graph does not have Hamiltonian circuit.

EXAMPLE 17.36

Does the graph shown below (Figure 17.90) possess a Hamiltonian circuit?

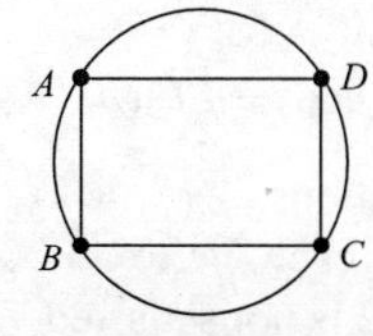

Figure 17.90

Solution. In the given graph

Number of vertices $(n) = 4$
Number of edges $(m) = 8$
Degree of each vertex $= 4$

Thus we see that

(1) Degree of each vertex is greater than $n/2$

(ii) Sum of degree of each pair of vertices is greater than $n-1$,

(iii) $\frac{1}{2}(n^2-3n+6)=\frac{1}{2}(16-12+6)=5$ and so $m \geq \frac{1}{2}(n^2-3n+6)$.

Hence the given graph has a Hamiltonian path. For example, *ABCDA* is a Hamiltonian path in the given graph (see Figure 17.91).

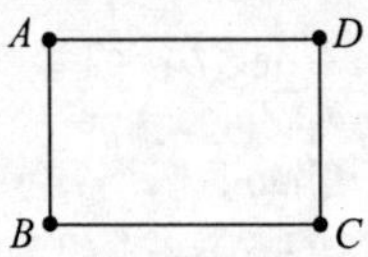

Figure 17.91

EXAMPLE 17.37

Is the graph shown in Figure 17.92 a Hamiltonian?

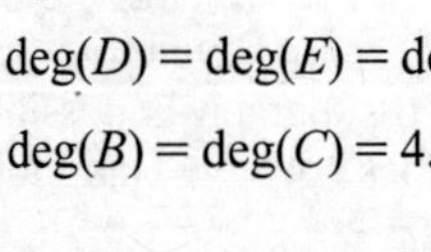

Figure 17.92

Solution. We note that

$$\deg(A) = \deg(D) = \deg(E) = \deg(F) = 2,$$
$$\deg(B) = \deg(C) = 4.$$

Further,

Number of vertices $(n) = 6$
Number of edges $(m) = 8$.

So,

$$\frac{1}{2}(n^2-3n+6)=\frac{1}{2}(36-18+6)=12.$$

Thus the conditions

(i) Degree of each vertex is greater than or equal to $n/2$ is not satisfied

(ii) $m \geq \frac{1}{2}(n^2-3n+6)$ is not satisfied

(iii) $\deg(A)+\deg(E)=2+2=4$ and so the condition that "the sum of degrees of non-adjacent vertices is greater than or equal to n" is not satisfied

(iv) "The sum of degrees of any pair of vertices is greater than or equal to $n-1$" is not satisfied.

So suppose that G has a Hamiltonian circuit. Then it should have a connected subgraph H containing six vertices, six edges and degree of each vertex should be 2. To have degree of each vertex equal to 2, we should remove two edges from C and two edges from B.

For example, now, if we remove CE from C, then $\deg(E)\neq 2$, If we remove CF, then $\deg(F)\neq 2$. So, we cannot remove CE and CF. If we remove CD, then $\deg(D)\neq 2$. If we remove CA, then $\deg(A)\neq 2$. Hence no such subgraph H exists. So, G cannot have a Hamiltonian circuit.

Remark 17.6 Since the degree of each vertex in the above graph is even, it has Eulerian circuit. Thus the graph in EXAMPLE 17.37 is Eulerian but not Hamiltonian.

EXAMPLE 17.38

Is the graph given in Figure 17.93 Hamiltonian?

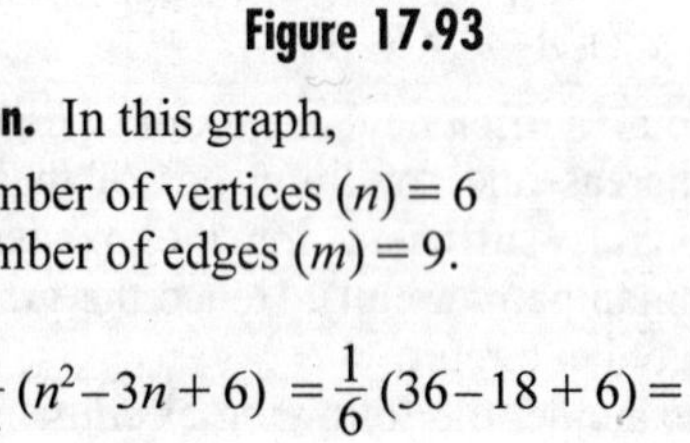

Figure 17.93

Solution. In this graph,

Number of vertices $(n) = 6$
Number of edges $(m) = 9$.

So,

$$\frac{1}{2}(n^2-3n+6) = \frac{1}{6}(36-18+6) = 12,$$

$$\deg(A) = \deg(C) = 2,$$
$$\deg(D) = \deg(F) = 3 = \deg(E),$$
$$\deg(B) = 5.$$

Thus, no sufficient condition is satisfied in this case. **But, the graph has Hamiltonian circuit**

ADEFCBA shown in the Figure 17.94.

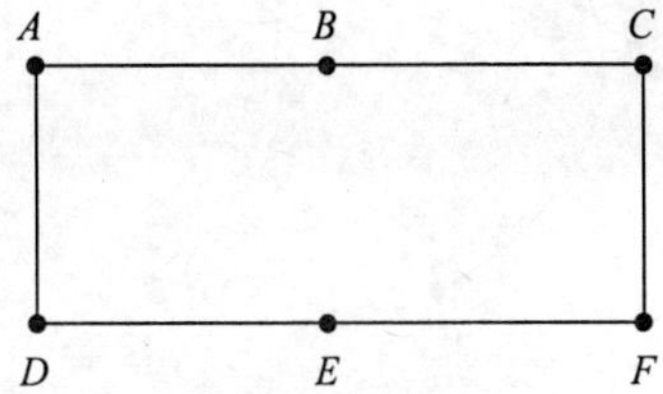

Figure 17.94

EXAMPLE 17.39

Show that the graph *G* shown in Figure 17.95 is not Hamiltonian.

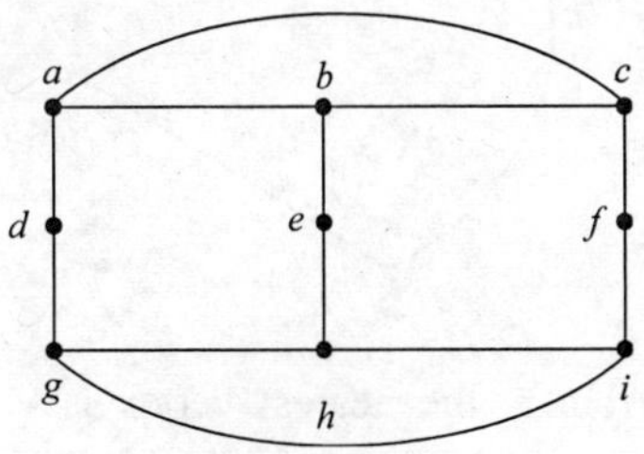

Figure 17.95

Solution. If *G* is Hamiltonian, then it has a subgraph *H* that

1. Contains every vertex of *G*
2. Is connected
3. Has the same number of edges as vertices
4. Is such that every vertex has degree 2

Thus, if such a graph exists, then it will have vertices ($a, b, c, d, e, f, g, h, i$), will be connected, will have six edges and the degree of each vertex shall be 2. We note that $\deg(d) = \deg(e) = \deg(f) = 2$. The degree of the vertex *h* in G is 3, one edge incident on *h* must be deleted from *G* to create *H*. The edge $\{h, e\}$ is required since otherwise $\deg(e) \neq 2$. So we have to delete either $\{g, h\}$ or $\{h, i\}$. If we delete $\{h, i\}$ and retain $\{g, h\}$, then $\{g, i\}$ has to be deleted to keep g of degree 2. In such a case, there is only one edge $\{i, f\}$ on *i* and so there is a contradiction to (4). So we cannot remove $\{h, i\}$. Similarly, we see that $\{g, h\}$ cannot be deleted. Hence no such subgraph *H* exists and so *G* does not have a Hamiltonian circuit.

Definition 17.55 *A* **weighted graph** is a graph for which each edge or each vertex or both is (are) labelled with a numerical value, called its **weight.**

For example, if vertices in a graph denote recreational sites of a town and weights of edges denote the distances in kilometers between the sites, then the graph shown in Figure 17.96 is a weighted graph.

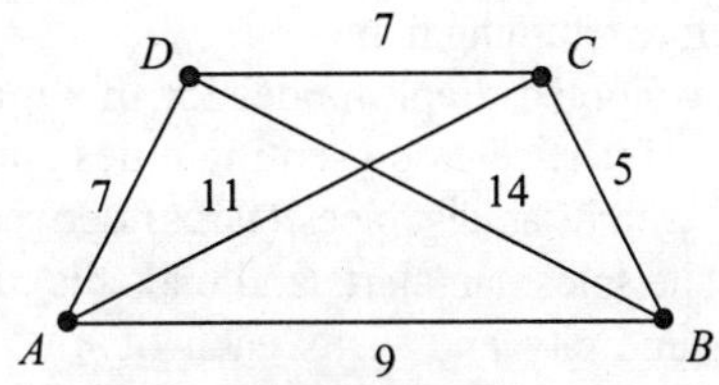

Figure 17.96

Definition 17.56 The **weight of an edge** (v_i, v_j) is called **distance between the vertices** v_i **and** v_j.

Definition 17.57 A vertex *u* is a **nearest neighbour** of vertex *v* in a graph if *u* and *v* are adjacent and no other vertex is joined to *v* by an edge of lesser weight than (u, v).

For example, in the above example, *B* is the nearest neighbour of *C*, whereas *A* and *C* are both nearest neighbours of the vertex *D*. **Thus, nearest neighbour of a set of vertices is not unique.**

Definition 17.58 A vertex *u* is a nearest neighbour of a set of vertices $\{v_1, v_2, \ldots, v_n\}$ in a graph if *u* is adjacent to some member v_i of the set and no other vertex adjacent to member of the set is joined by an edge of lesser weight than (u, v_i).

In the above example, if we have set of vertices as $\{B, D\}$, *C* is the neatest neighbour of $\{B, D\}$ because the edge (C, B) has weight 5 and no other vertex adjacent to $\{B, D\}$ is linked by an edge of lesser weight than (C, B).

Definition 17.59 The **length of a path** in a graph is the sum of lengths of edges in the path.

Definition 17.60 Let $G = (V, E)$ be a graph and let l_{ij} denote the length of edge (v_i, v_j) in *G*. Then a **shortest path** from v_i to v_k is a path such that the sum $l_{12} + l_{23} + + l_{k-1,k}$ of lengths of its edges is **minimum**, that is, total edge weight is minimum.

17.7.1 Travelling Salesperson Problem

This problem requires the determination of a **shortest Hamiltonian circuit** in a given graph of cities and lines of transportation to minimize the total fare for a travelling person who wants to make a tour of n cities visiting each city exactly once before returning home.

The weighted graph model for this problem consists of vertices representing cities and edges with weight as distances (fares) between the cities. The salesman starts and ends his journey at the same city and visits each of $n-1$ cities once and only once. We want to find minimum total distance.

We discuss the case of five cities and so consider the weighted graph shown in the Figure 17.97.

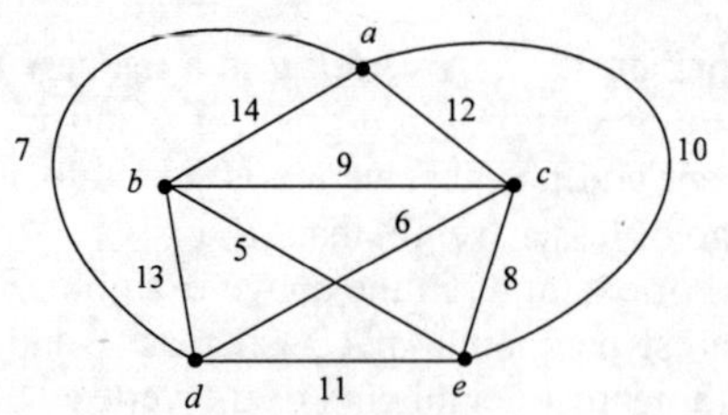

Figure 17.97

We shall use **nearest neighbour** algorithm to solve the problem:

Algorithm: Nearest neighbour (closest insertion)

Input: A weighted complete graph G.

Output: A sequence of labelled vertices that forms a Hamiltonian cycle.

Start at any vertex v.
Initialize $l(v) = 0$.
Initialize $i = 0$.

While there are unlabelled vertices

$i: = i + 1$

Traverse the cheapest edge that join v to an unlabelled vertex, say w

Set $l(w) = i$.
$v: = w$.

For the present example,

(i) Let us choose a as the starting vertex. Then d is the nearest vertex and then (a, d) is the corresponding edge. Thus we have the figure

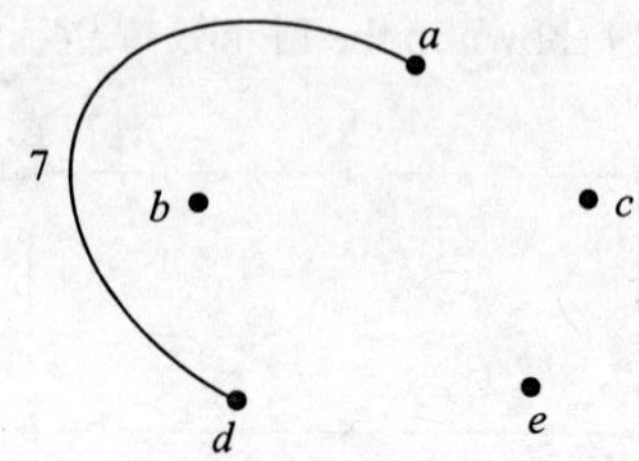

(ii) From d, the nearest vertex is c, so we have a path shown below:

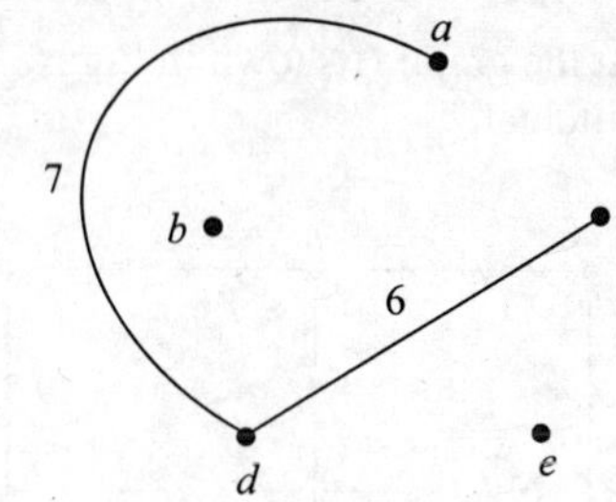

(iii) From c, the nearest vertex is e. So we have the path as shown below:

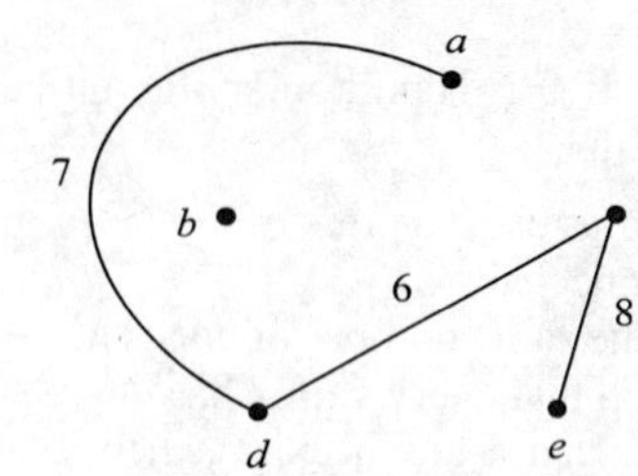

(iv) From e, the nearest vertex is b and so we have the path

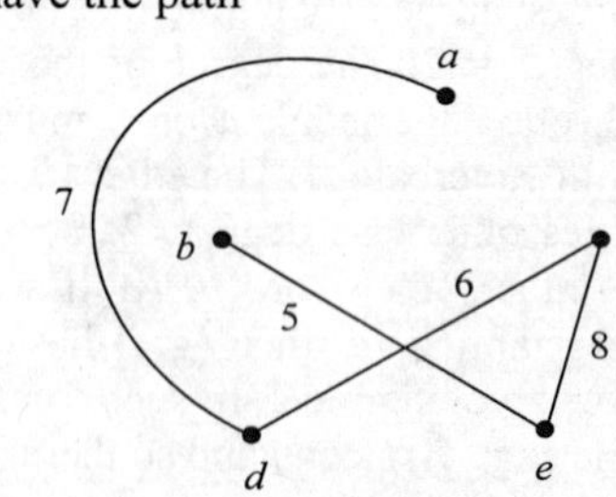

(v) Now, from b, the only vertex to be covered is a to **form Hamiltonian circuit**. Thus we have a Hamiltonian circuit as given below. The length of this Hamiltonian circuit is

$$7 + 6 + 8 + 5 + 14 = 40.$$

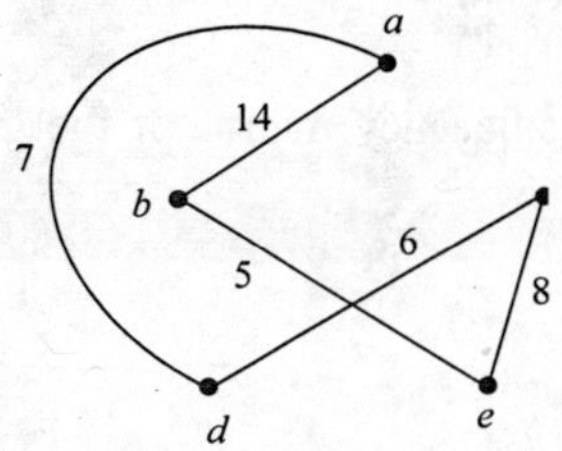

However, **this is not Hamiltonian circuit of minimal length.**

The **total distance** of a **minimum Hamiltonian** circuit (Figure 17.98) **is 37.**

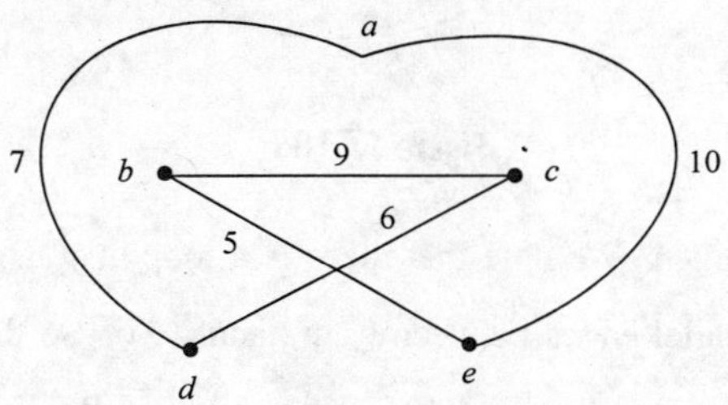

Figure 17.98

Total length $= 7 + 6 + 9 + 5 + 10 = 37$.

Remark 17.7 Unless otherwise stated, try to start from a vertex of largest weight.

EXAMPLE 17.40

Find a Hamiltonian circuit of minimal weight for the graph given below (Figure 17.99)

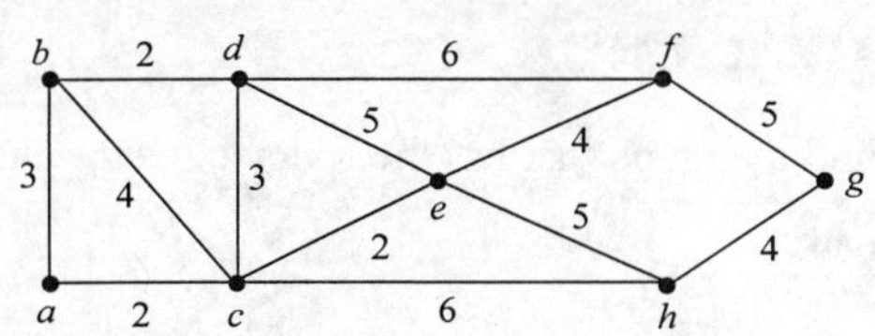

Figure 17.99

Solution. Starting from c and applying nearest neighbour method, we have the required Hamiltonian circuit as $c\ a\ b\ d\ e\ f\ g\ h\ c$ with total length as

$$2 + 3 + 2 + 5 + 4 + 5 + 4 + 6 = 31.$$

EXAMPLE 17.41

Find a Hamiltonian circuit of minimal weight for the graph shown below (Figure 17.100)

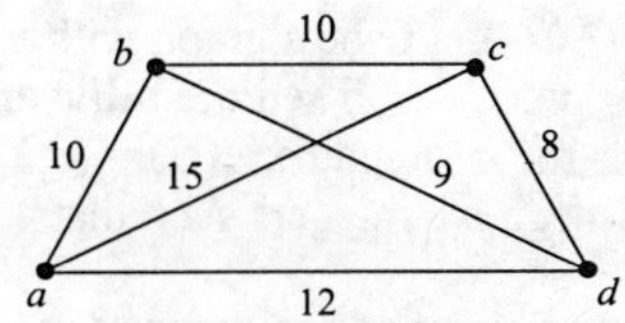

Figure 17.100

Solution. Starting from the point a and using nearest neighbour method, we have the required Hamiltonian circuit as $a\ b\ c\ d\ a$ with total length as

$$10 + 10 + 8 + 12 = 40.$$

Definition 17.61 A **k-factor of a graph** is a spanning subgraph of the graph **with the degree of its vertices being k.**

Consider the graph shown in Figure 17.101.

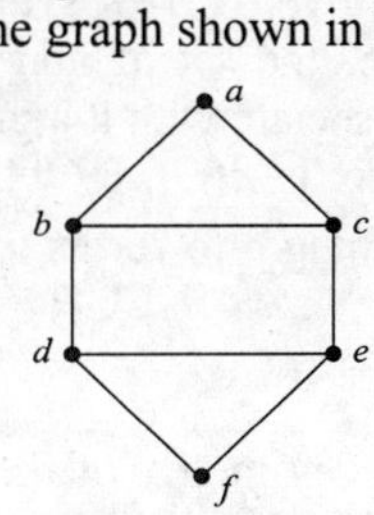

Figure 17.101

Then the graph shown in Figure 17.102 shows a 1-factor of the given graph.

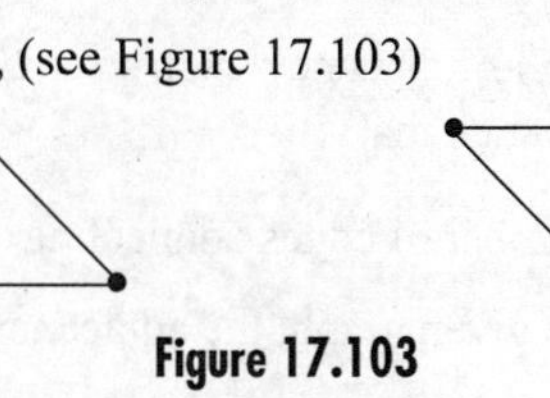

Figure 17.102

Also then, (see Figure 17.103)

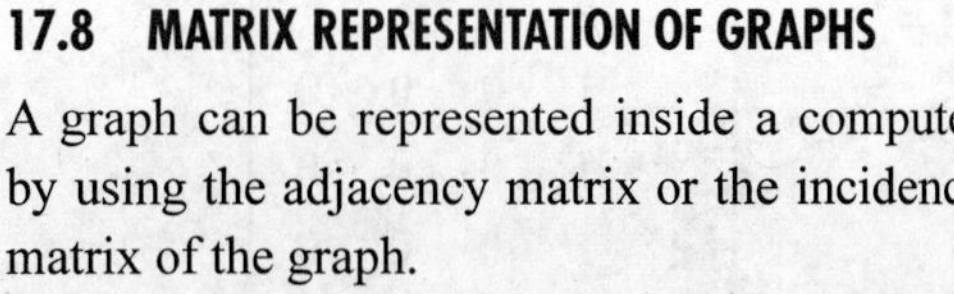

Figure 17.103

is a 2-factor of the given graph.

17.8 MATRIX REPRESENTATION OF GRAPHS

A graph can be represented inside a computer by using the adjacency matrix or the incidence matrix of the graph.

Definition 17.62 Let G be a graph with n ordered vertices $v_1, v_2, \ldots, v_n$. Then the **adjacency matrix** of G is the $n\times n$ matrix $A(G) = (a_{ij})$ over the set of non-negative integers such that

$$a_{ij} = \text{the number of edges connecting } v_i \text{ and } v_j \text{ for all } i, j = 1, 2, \ldots, n.$$

We note that if G has no loop, then there is no edge joining v_i to v_i, $i = 1, 2, \ldots, n$. Therefore, in this case, all the entries on the main diagonal will be 0.

Further, if G has no parallel edge, then the entries of $A(G)$ are either 0 or 1.

It may be noted that adjacent matrix of a graph is symmetric.

Conversely, given a $n\times n$ symmetric matrix $A(G) = (a_{ij})$ over the set of non-negative integers, we can associate with it a graph G, whose adjacency matrix is $A(G)$, by letting G have n vertices and joining v_i to vertex v_j by a_{ij} edges.

EXAMPLE 17.42

Find the adjacency matrix of the graph shown below (Figure 17.104):

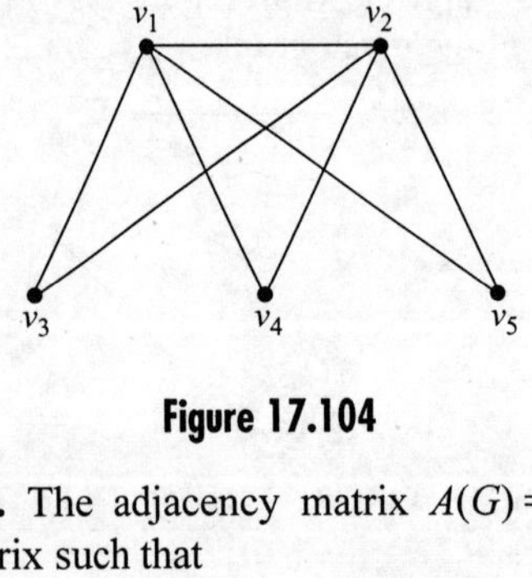

Figure 17.104

Solution. The adjacency matrix $A(G) = (a_{ij})$ is the matrix such that

$$a_{ij} = \text{Number of edges connecting } v_i \text{ and } v_j.$$

So, for the given graph, the adjacency matrix is

$$A(G) = \begin{bmatrix} 0 & 1 & 1 & 1 & 1 \\ 1 & 0 & 1 & 1 & 1 \\ 1 & 1 & 0 & 0 & 0 \\ 1 & 1 & 0 & 0 & 0 \\ 1 & 1 & 0 & 0 & 0 \end{bmatrix}$$

EXAMPLE 17.43

Find the adjacency matrix of the graph given below:

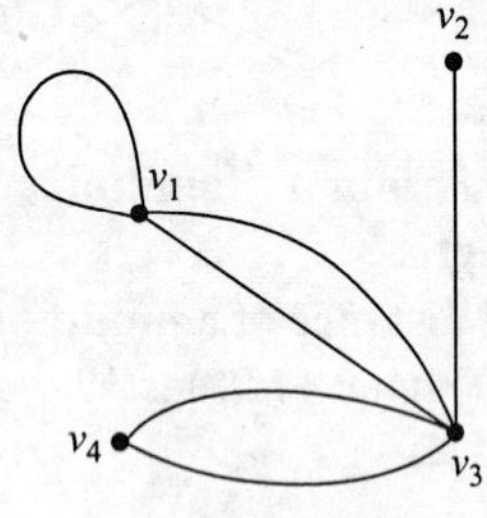

Figure 17.105

Solution. We note that there is a loop at v_1 and parallel edges between v_1, v_3, and v_3, v_4. So the adjacency matrix $A(G)$ is given by the following 4×4 matrix:

$$A(G) = \begin{bmatrix} 1 & 0 & 2 & 0 \\ 0 & 0 & 1 & 0 \\ 2 & 1 & 0 & 2 \\ 0 & 0 & 2 & 0 \end{bmatrix}$$

EXAMPLE 17.44

Find the graph that has the following adjacency matrix:

$$\begin{bmatrix} 1 & 2 & 1 & 2 \\ 2 & 0 & 2 & 1 \\ 1 & 2 & 1 & 0 \\ 2 & 1 & 0 & 0 \end{bmatrix}$$

Solution. We note that there is a loop at v_1 and a loop at v_3. There are parallel edges between v_1, v_2; v_1, v_4; v_2, v_1; v_2, v_3; v_3, v_2; v_4, v_1. Thus the graph is as shown in Figure 17.106.

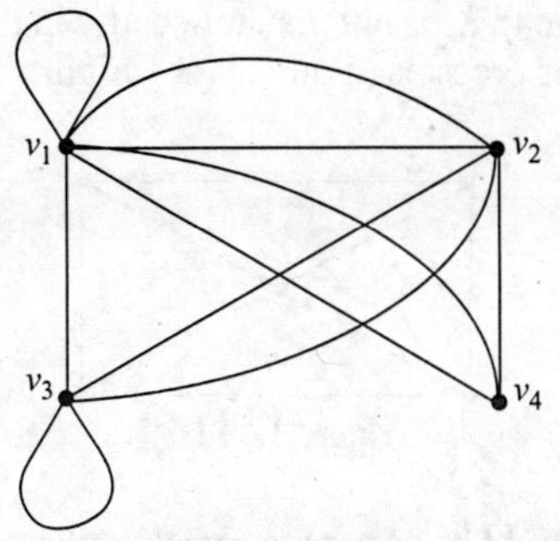

Figure 17.106

The following theorem is stated without proof.

Theorem 17.14 Let G be a graph with n vertices $v_1, v_2, \ldots, v_n$ and let $A(G)$ denote the matrix of G. If $B = (b_{ij})$ is the matrix

$$B = A + A^2 + \cdots + A^{n-1},$$

then the graph G is a connected graph if and only if $b_{ij} \neq 0$ for $i \neq j$, that is, B has no zero entry off the main diagonal.

EXAMPLE 17.45

A graph G has the following adjacency matrix. Verify whether it is connected.

$$A = \begin{bmatrix} 0 & 0 & 1 & 0 & 0 \\ 0 & 0 & 0 & 1 & 0 \\ 1 & 0 & 0 & 0 & 1 \\ 0 & 1 & 0 & 0 & 1 \\ 0 & 0 & 1 & 1 & 0 \end{bmatrix}.$$

Solution. We have

$$A^2 = \begin{bmatrix} 1 & 0 & 0 & 0 & 1 \\ 0 & 1 & 0 & 0 & 1 \\ 0 & 0 & 2 & 1 & 0 \\ 0 & 0 & 1 & 2 & 0 \\ 1 & 1 & 0 & 0 & 2 \end{bmatrix},$$

$$A^3 = \begin{bmatrix} 0 & 0 & 2 & 1 & 0 \\ 0 & 0 & 1 & 2 & 0 \\ 2 & 1 & 0 & 0 & 3 \\ 1 & 2 & 0 & 0 & 3 \\ 0 & 0 & 3 & 3 & 0 \end{bmatrix},$$

$$A^4 = \begin{bmatrix} 2 & 1 & 0 & 0 & 3 \\ 1 & 2 & 0 & 0 & 3 \\ 0 & 0 & 5 & 4 & 0 \\ 0 & 0 & 4 & 5 & 0 \\ 3 & 3 & 0 & 0 & 6 \end{bmatrix},$$

Therefore,

$$B = A + A^2 + A^3 + A^4 = \begin{bmatrix} 3 & 1 & 3 & 1 & 4 \\ 1 & 3 & 1 & 3 & 4 \\ 3 & 1 & 7 & 5 & 4 \\ 1 & 3 & 5 & 7 & 4 \\ 4 & 4 & 4 & 4 & 8 \end{bmatrix},$$

Since B has no zero entry off the main diagonal, the graph is connected.

Definition 17.63 Suppose a graph G has n vertices $v_1, v_2, \ldots, v_n$ and t edges $e_1, e_2, \ldots, e_t$. The **incidence matrix** $B(G)$ of G is the $n \times t$ matrix $B(G) = (b_{ij})$, where

b_{ij} = the number of times that the vertex v_i is incident with the edge e_j,
that is,

$$b_{ij} = \begin{cases} 0 & \text{if } v_i \text{ is not end of } e_j \\ 1 & \text{if } v_i \text{ is an end of the non-loop } e_j \\ 2 & \text{if } v_i \text{ is an end of the loop } e_j \end{cases}$$

17.9 PLANAR GRAPHS

Definition 17.64 A graph which can be drawn in the plane so that its edges do not cross is said to be **planar.**

For example, the graph shown in Figure 17.107 is planar:

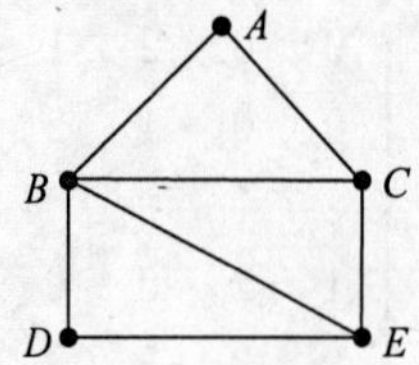

Figure 17.107

Also the complete graph K_4 shown below (Figure 17.108) is planar.

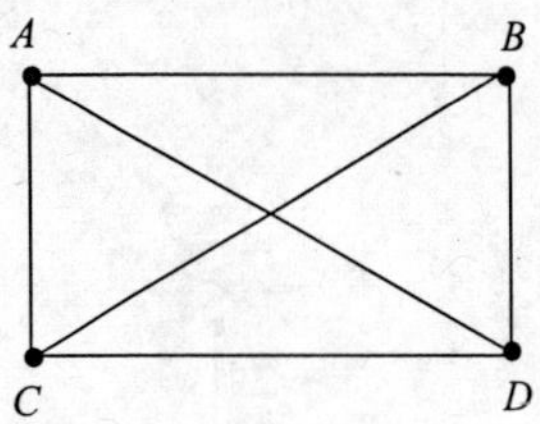

Figure 17.108

In fact, it can be redrawn as shown below in Figure 17.109 so that no edges cross.

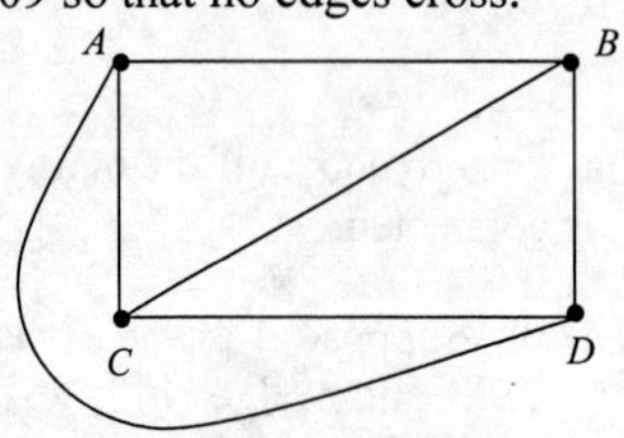

Figure 17.109

But the map K_5 is not planar because in this case, the edges cross each others (see Figure 17.110).

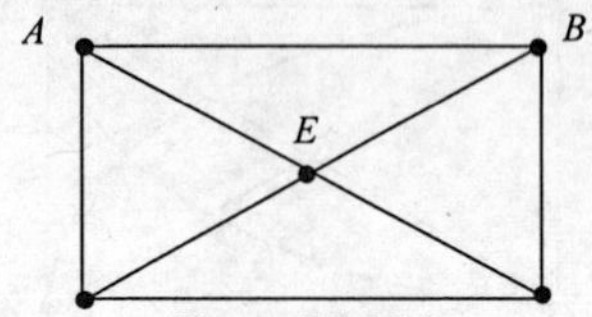

Figure 17.110

Definition 17.65 An area of the plane that is bounded by edges of the planar graph and is not further subdivided into subareas is called a **region** or **face** of the planar graph.

A face is characterized by the cycle that forms its boundary.

Definition 17.66 A region is said to be **finite** if its area is finite and **infinite** if its area is infinite. Clearly a planar graph has exactly one infinite region.

For example, consider the graphs shown in Figure 17.111.

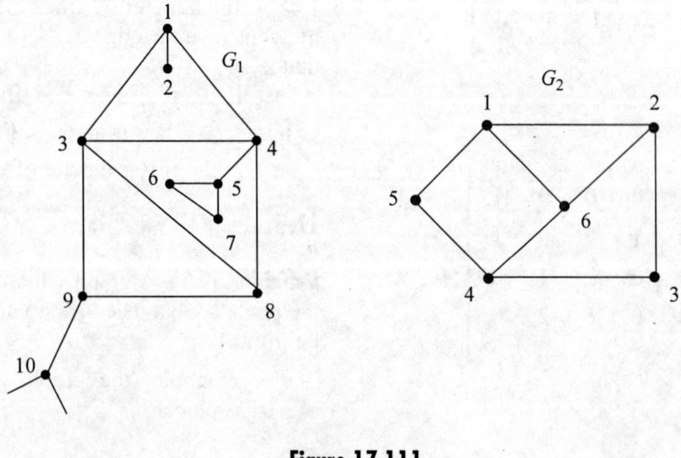

Figure 17.111

We note that in the graph G_1, there are five regions **A, B, C, D, E** as shown below:

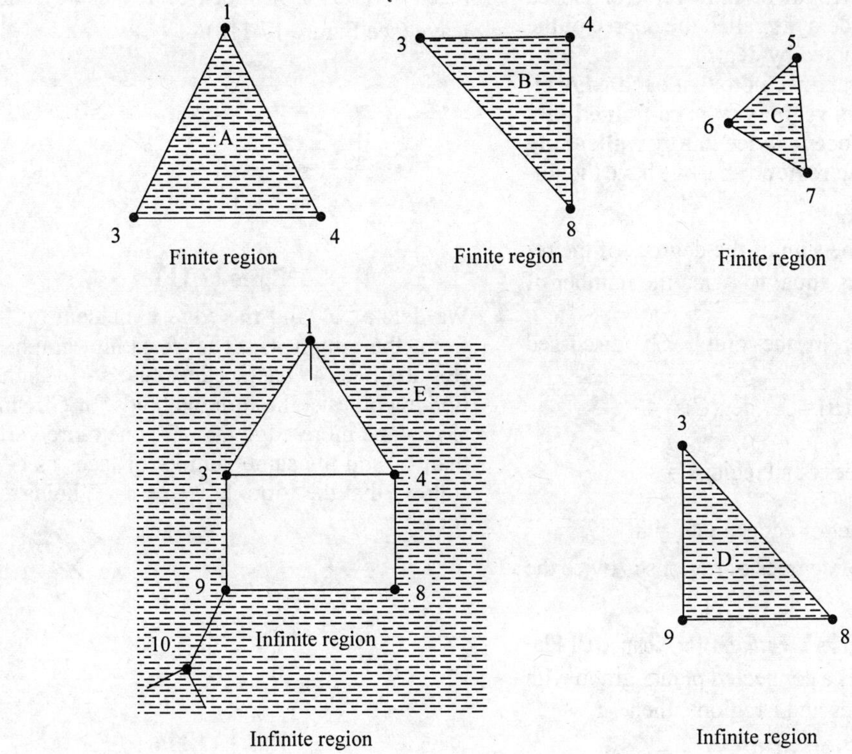

In graph G_2, there are four region **A, B, C, D** (Figure 20.112)

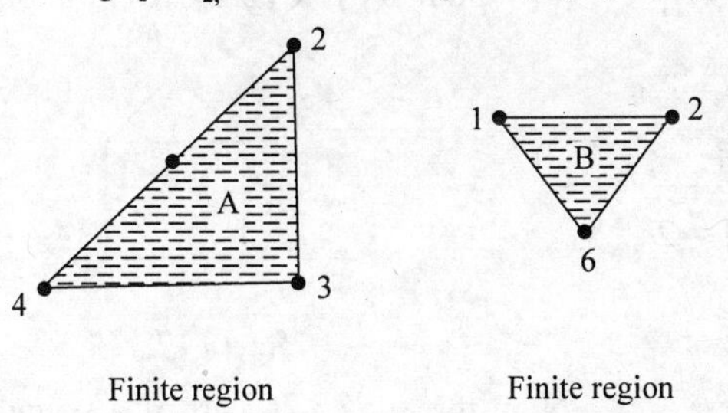

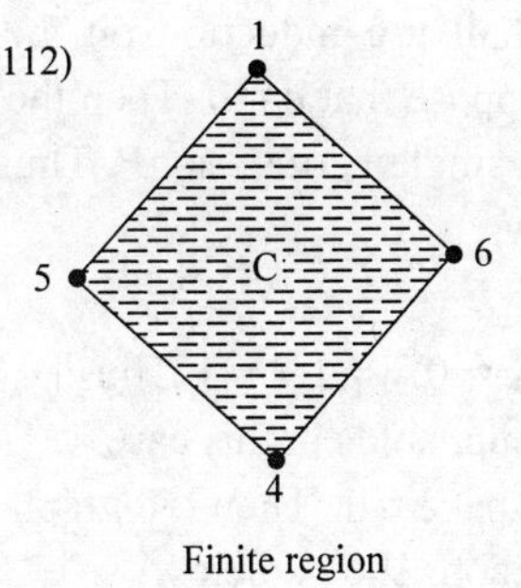

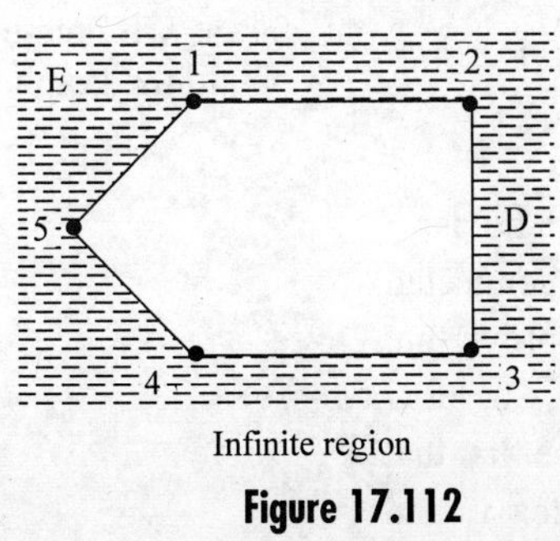

Figure 17.112

Definition 17.67 Let f be a face (region) in a planar graph. The length of the cycle (or closed walk) which borders f is called the degree of the region f. It is denoted by $\deg(f)$.

In a planar graph we note that **each edge either borders two regions or is contained in a region and will occur twice in any walk along the border of the region.** Thus we have the following:

Theorem 17.15 The sum of the degrees of the regions of a map is equal to twice the number of edges.

For example, in the graph G_2, discussed above, we have

$\deg(A) = 4$, $\deg(B) = 3$, $\deg(C) = 4$, $\deg(D) = 5$.

The sum of degrees of all regions $= 4 + 3 + 4 + 5 = 16$.

Number of edges in $G_2 = 8$.

Hence the sum of degrees of regions is twice the number of edges.

Theorem 17.16 (Euler's Formula for Connected Planar Graphs) If G is a connected planar graph with e edges, v vertices and r regions, then

$$v - e + r = 2.$$

Proof: We shall use induction on the number of edges. Suppose that $e = 0$. Then the graph G consists of a single vertex, say P. Thus, G is as shown below:

$$\bullet\, P$$

and we have $e = 0, v = 1, r = 1$. Thus $1 - 0 + 1 = 2$ and the formula holds in this case.

Suppose that $e = 1$. Then the graph G is one of the two graphs shown below:

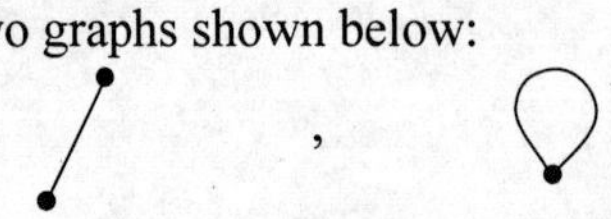

$e = 1, v = 2, r = 1$ $\qquad$ $e = 1, v = 1, r = 2$

We see that, in either case, the formula holds.

Suppose that the formula holds for connected planar graph with n edges. We shall prove that this holds for graph with $n + 1$ edges. So, let G be the graph with $n + 1$ edges. Suppose first that G contains no cycles. Choose a vertex v_1 and trace a path starting at v_1. Ultimately, we will reach a vertex a with degree 1, that we cannot leave (see Figure 17.113).

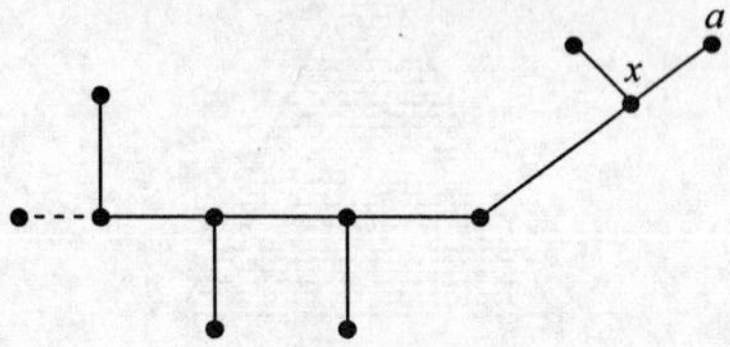

G

Figure 17.113

We delete "a" and the edge x incident on "a" from the graph G. The resulting graph G' (Figure 17.114) has n edges and so by induction hypothesis, the formula holds for G'. Since G has one more edge than G', one more vertex than G' and the same number of faces as G', it follows that the formula $v - e + r = 2$ holds also for G.

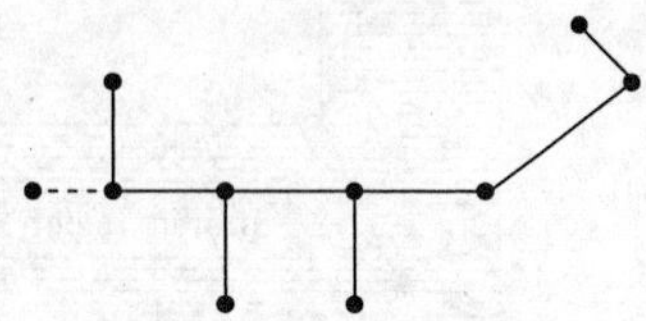

G'

Figure 17.114

Now suppose that G contains a cycle. Let x be an edge in a cycle as shown in Figure 17.115.

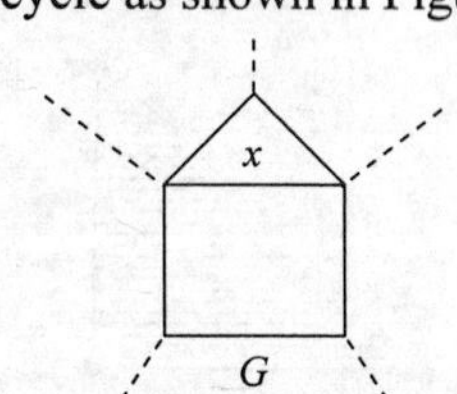

Figure 17.115

Now the edge x is part of a boundary for two faces. We delete the edge x but no vertices to obtain the graph G' as shown in Figure 17.116.

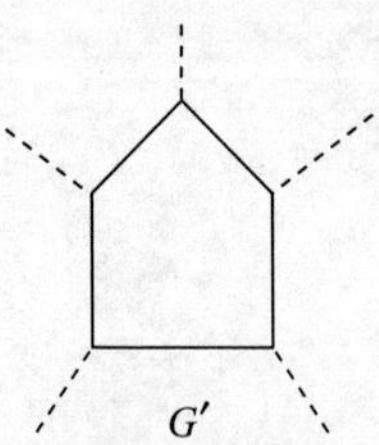

Figure 17.116

Thus G' has n edges and so by induction hypothesis the formula holds. Since G has one more face (region) than G', one more edge than G' and the same number of vertices as G', it follows that the formula $v-e+r=2$ also holds for G. Hence, by mathematical induction, the theorem is true.

Remark 17.8 Planarity of a graph is not affected if

(i) An edge is divided into two edges by the insertion of new vertex of degree 2 (Figure 17.117).

Figure 17.117

(ii) Two edges that are incident with a vertex of degree 2 are combined as a single edge by the removal of that vertex (Figure 17.118).

Figure 17.118

Definition 17.68 Two graphs G_1 and G_2 are said to be **isomorphic to within vertices of degree 2 (or homeomorphic)** if they are isomorphic or if they can be transformed into isomorphic graphs by repeated insertion and/or removal of vertices of degree 2.

Definition 17.69 The repeated insertion/removal of vertices of degree 2 is called **sequence of series reduction**.

For example, the graphs shown in Figure 17.119 are isomorphic to within vertices of degree 2.

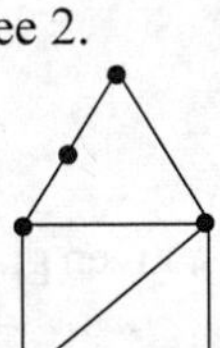

and

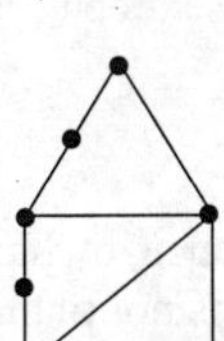

Figure 17.119

If we define a relation R on the set of graphs by $G_1 \, R \, G_2$ if G_1 and G_2 are homeomorphic, then R is an equivalence relation. Each equivalence class consists of a set of mutually homeomorphic graphs.

EXAMPLE 17.46

Show that the graph $K_{3,3}$, given in Figure 17.117.below, is not planar.

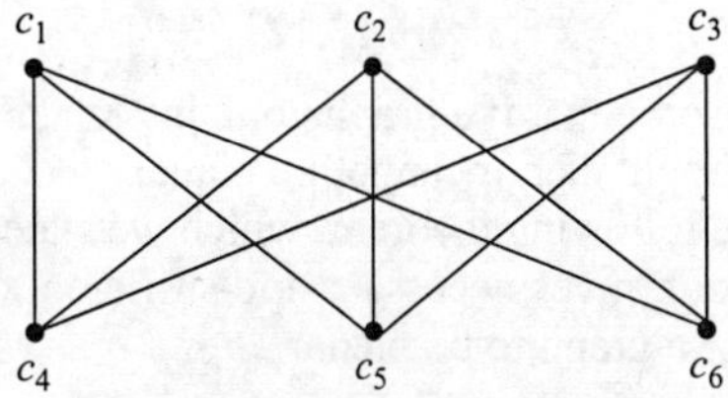

$K_{3,3}$

Figure 17.117.

A problem based on this example can be stated as "Three cities c_1, c_2 and c_3 are to be directly connected by express ways to each of three cities c_4, c_5 and c_6. Can this road system be designed so that the express ways do not cross?" This example shows that it cannot be done.

Solution. Suppose that $K_{3,3}$ is planar. Since every cycle in $K_{3,3}$ has at least four edges, each face (region) is bounded by at least four edges. Thus the number of edges that bound regions is at least $4r$. Also, in a planar graph each edge belongs to at most two bounding cycles. Therefore,

$2e \geq 4r$ (sums of degrees of region is equal to twice the number of edges)

But, by Euler's formula for planar graph,

$$r = e - v + 2.$$

Hence,

$$2e \geq 4(e - v + 2) \qquad (1)$$

In case of $K_{3,3}$ we have $e = 9$, $v = 6$ and so (1) yields

$$18 \geq 4\,(9 - 6 + 2) = 20$$

which is a contradiction. Therefore $K_{3,3}$ is not planar.

Remark 17.9 By a argument similar to the above example, we can show that the graph K_5 (Figure 17.121) is not planar.

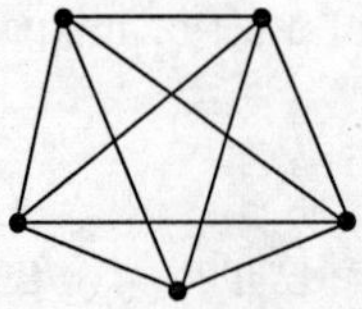

(Non-planar graph K_5)

Figure 17.121

We observe that if a graph contains $K_{3,3}$ or K_5 as a subgraph, then it cannot be planar.

The following theorem, which we state without proof, gives necessary and sufficient condition for a graph to be planar.

Kuratowski's Theorem 17.17 A graph G is planar if and only if G does not contain a subgraph homeomorphic to $K_{3,3}$ or K_5.

The complete graph K_5 and the complete bipartite graph $K_{3,3}$ are called the **Kuratowski graphs.**

EXAMPLE 17.47

Using Kuratowski's theorem show that the graph G, shown in Figure 17.122, is not planar.

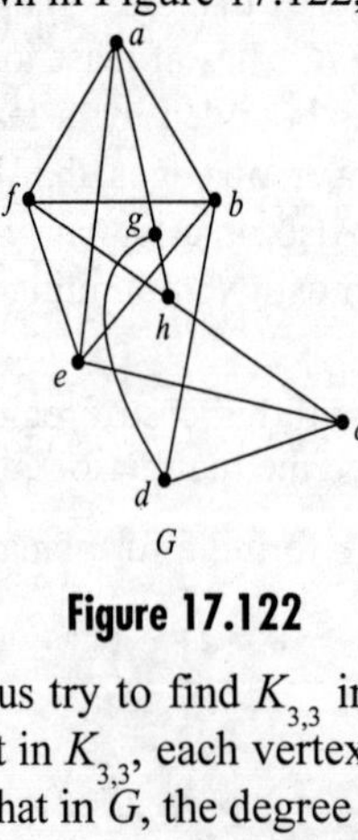

G

Figure 17.122

Solution. Let us try to find $K_{3,3}$ in the graph G. We know that in $K_{3,3}$, each vertex has degree 3. But we note that in G, the degree of a, b, f and e each is 4. So we eliminate the edges (a, b) and (f, e) so that all vertices have degree 3. If we eliminate one more edge, we will obtain two vertices of degree 2 and we can then carry out series reduction. The resulting graph will have nine edges.

Also we know that $K_{3,3}$ **has nine edges**. So this approach seems promising. Using trial and error, we find that the edge (g, h) should be removed. Then g and h have degree 2.

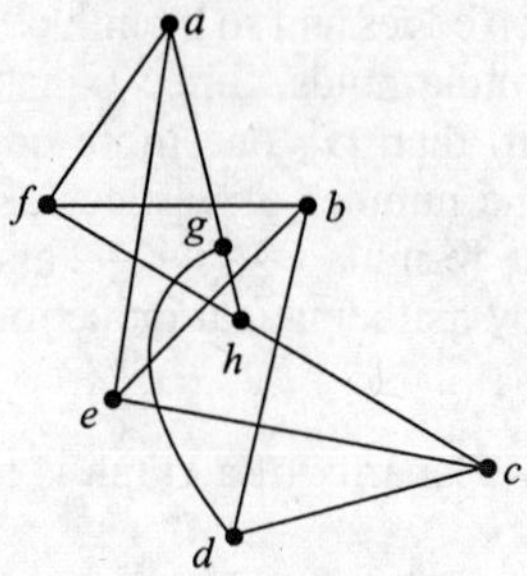

(Graph obtained by deleting edges (a, b) and (f, e)).

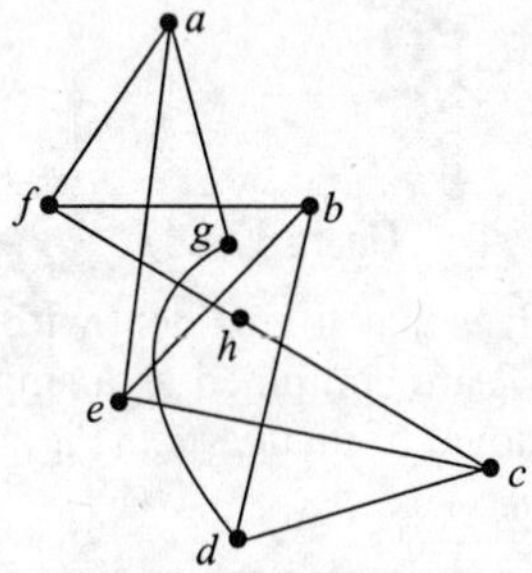

(Graph obtained by eliminating the edge (g, h).

Performing series reduction now, we obtain an isomorphic copy of $K_{3,3}$ (Figure 17.123).

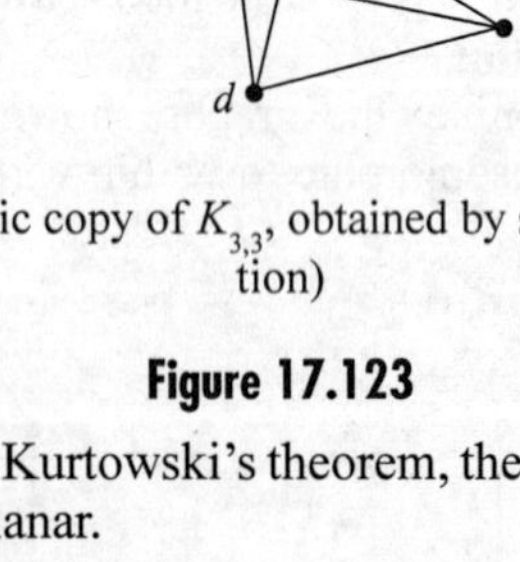

(Isomrophic copy of $K_{3,3}$, obtained by series reduction)

Figure 17.123

Hence, by Kurtowski's theorem, the given graph G is not planar.

EXAMPLE 17.48

If the graph given in Figure 17.124 is planar, redraw it so that no edges cross, otherwise find a subgraph homeomorphic to either K_5 or $K_{3,3}$.

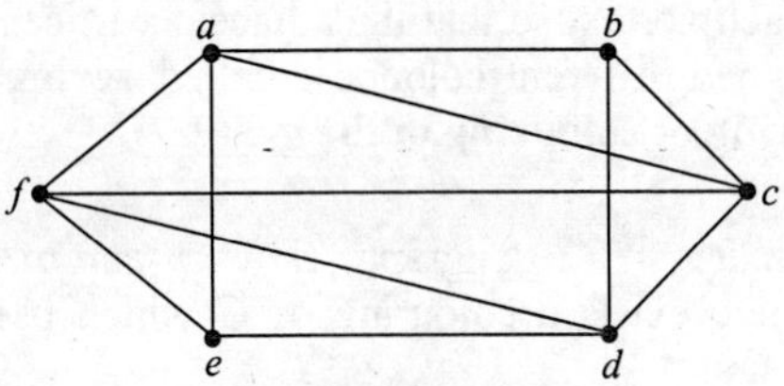

Figure 17.124

Solution. The given graph can be redrawn as shown in the Figure 17.125.

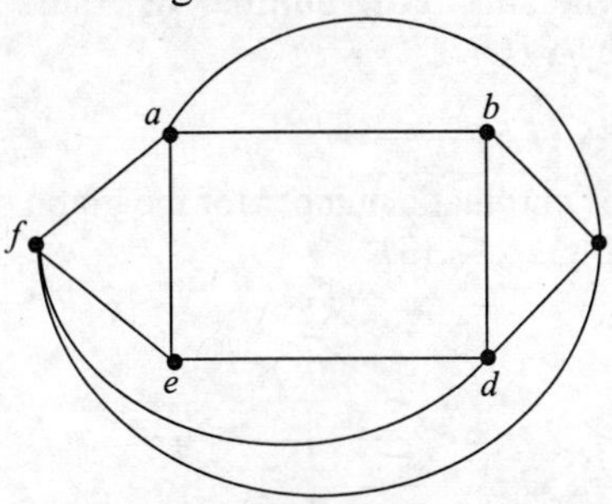

Figure 17.125

We then observe that no edges cross. Hence the given graph is planar.

Theorem 17.18 Let G be a connected planar graph with v vertices, e edges, where $v \geq 3$. Then $e \leq 3v - 6$ or $6 \leq 3v - e$.

(Note that the theorem is not true for K_1, where $v = 1$ and $e = 0$ and is not true for K_2, where $v = 2$ and $e = 1$).

Proof: Let r be the number of regions in a planar representation of the graph G. The sum of the degrees of the regions is equal to $2e$. But each region has degree greater than or equal to 3. Hence $2e \geq 3r$, that is, $r \leq \frac{2e}{3}$. But, by Euler's formula,

$$v - e + r = 2.$$

So,

$$2 = v - e + r \leq v - e + \frac{2e}{3} = v - \frac{e}{3}.$$

Hence,

$$6 \leq 3v - e \quad \text{or} \quad e \leq 3v - 6$$

Remark 17.10 Using above theorem, let us consider planarity of K_5. Here $v = 5$, $e = 10$. Suppose K_5 is planar, then by the above theorem,

$$6 \leq 3(5) - 10 = 5,$$

which is impossible. Hence K_5 is non-planar.

Definition 17.70 Any graph homeomorphic either to K_5 or to $K_{3,3}$ is called **Kuratowski subgraph**.

EXAMPLE 17.49

Find Kuratowski subgraph of the following graph (Figure 17.126):

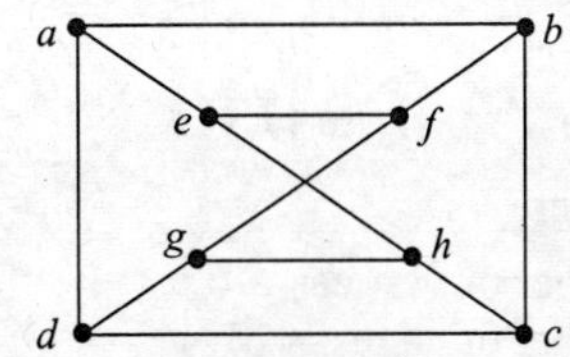

Figure 17.126

Solution. In the given graph,

Number of vertices = 8

Number of edges = 12.

Firstly, we remove the edge (g, h). Then $\deg(g) = 2$, $\deg(h) = 2$ and so the vertices g and h can be removed. Then, we have

Number of vertices = 6

Number of edges = 9

and the graph reduces to

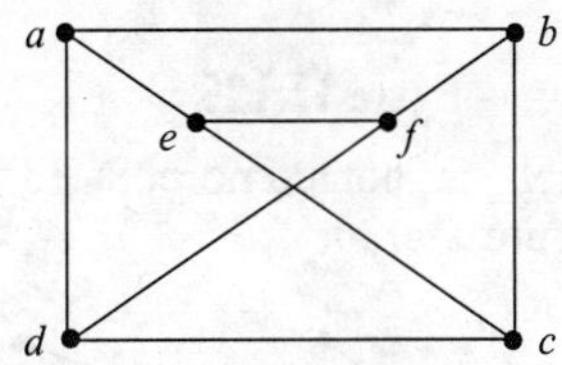

which is homeomorphic to $K_{3,3}$ (Kuratowski graph) shown in Figure 17.127.

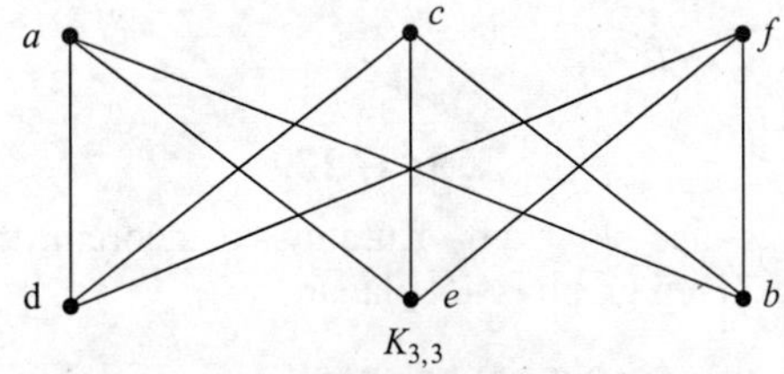

$K_{3,3}$

Figure 17.127

Thus the given graph is non-planar by Kuratowski theorem.

EXAMPLE 17.50

Find a subgraph homeomorphic to either K_5 or $K_{3,3}$ in the graph given below (Figure 17.128):

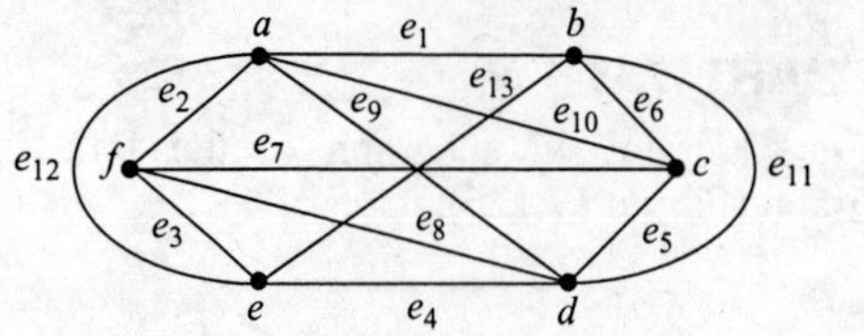

Figure 17.128

Solution. Here,

Number of vertices = 6
Number of edges = 13

$$\deg(a)=5,\quad \deg(f)=4,\quad \deg(c)=4,\quad \deg(d)=5$$

But, in $K_{3,3}$, the degree of each vertex is 3. We delete the edges (a, c), (f, d), (a, e), (b, d), then degree of each vertex a, f, c and d becomes 3. As such, the subgraph becomes as given below (Figure 17.129), which has six vertices each with degree 3 and has 9 edges.

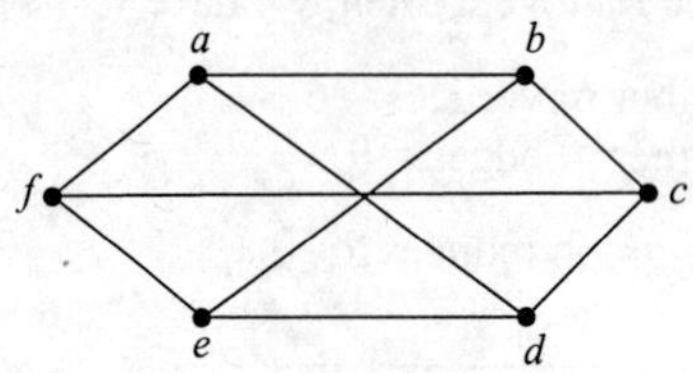

Figure 17.129

This subgraph is homeomorphic to $K_{3,3}$ as shown in Figure 17.130.

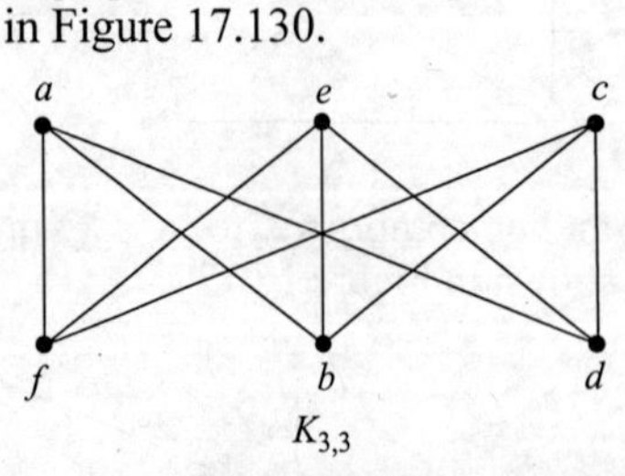

Figure 17.130

This also shows, by Kuratowski theorem, that the given graph is not planar.

17.10 COLOURING OF GRAPH

Definition 17.71 Let G be a graph. The assignment of colours to the vertices of G, one colour to each vertex, so that the adjacent vertices are assigned different colours is called **vertex colouring** or **colouring of the graph** G.

Definition 17.72 A graph G is n-**colourable** if there exists a colouring of G which uses n colours.

Definition 17.73 The minimum number of colours required to paint (colour) a graph G is called the **chromatic number of** G and is denoted by $\chi(G)$.

EXAMPLE 17.51

Find the chromatic number for the graph shown in the Figure 17.131.

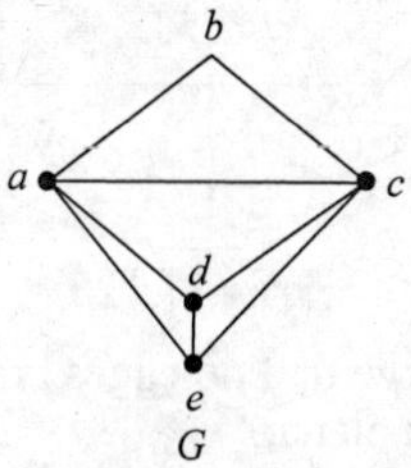

Figure 17.131

Solution. The triangle $a\ b\ c$ needs three colours. Suppose that we assign colours c_1, c_2, c_3 to a, b and c respectively. Since d is adjacent to a and c, d will have different colour than c_1 and c_3. So we paint d by c_2. Then e must be painted with a colour different from those of a, d and c, that is, we cannot colour e with c_1, c_2 or c_3. Hence, we have to give e a fourth colour c_4. Hence

$$\chi(G)=4.$$

Welsh–Powell algorithm to determine upper bound to the chromatic number of a given graph.

The input is a given graph G.

1. Order the vertices of G according to decreasing degree.
2. Assign the first colour, say c_1, to the first vertex and then, in sequential order, assign c_1 to each vertex, which is not adjacent to a previous vertex assigned c_1.
3. Repeat Step 2 with a second colour c_2 and the subsequence of the remaining non-painted vertices.

4. Repeat Step 3 with a third colour c_3, then a fourth colour c_4 and so on until all vertices are coloured.
5. Exit.

EXAMPLE 17.52

Use Welsh-Powell algorithm to determine an upper bound to the chromatic number of the "Wheel" graph shown in Figure 17.132.

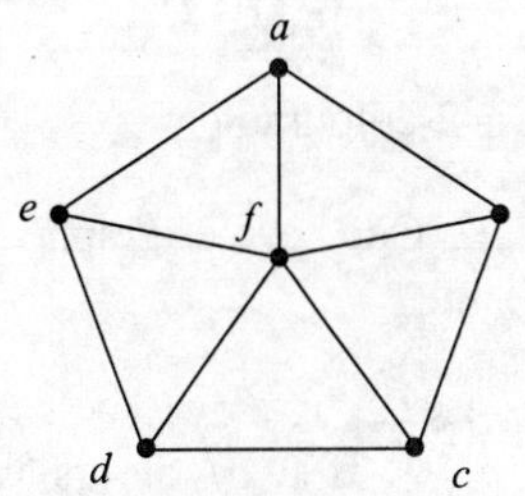

Figure 17.132

Solution. We note that

$$\deg(f) = 5,$$

$$\deg(a) = \deg(b) = \deg(c) = \deg(d) = \deg(e) = 3.$$

Step 1. Ordering the vertices according to decreasing degree yields

$$f, a, b, c, d, e$$

Step 2. Paint f with colour c_1.
Step 3. Paint a, d with colour c_2.
Step 4. Paint b, e with colour c_3.
Step 5. Paint c with colour c_4.

Hence $\chi(G) \leq 4$. Also since there is a triangle in the given graph so we have $\chi(G) \geq 3$. Thus $3 \leq \chi(G) \leq 4$. But we do not yet know exactly what the chromatic number is. We try to build a 3-colouring of G.

Let us start colouring the triangle abf with the colours c_1, c_2, c_3, respectively. Since c is adjacent to the vertices b and f of colour c_2 and c_3, respectively, c is forced to be coloured c_1 and then d is forced to be c_2. However, now the adjacent vertices a and e cannot both have colour c_1. Thus the graph cannot be 3-coloured. But using a fourth colour c_4 for e gives us 4-colouring of G. Hence, $X(G) = 4$.

EXAMPLE 17.53

Use Welsh-Powell algorithm to colour the graph shown in Figure 17.133.

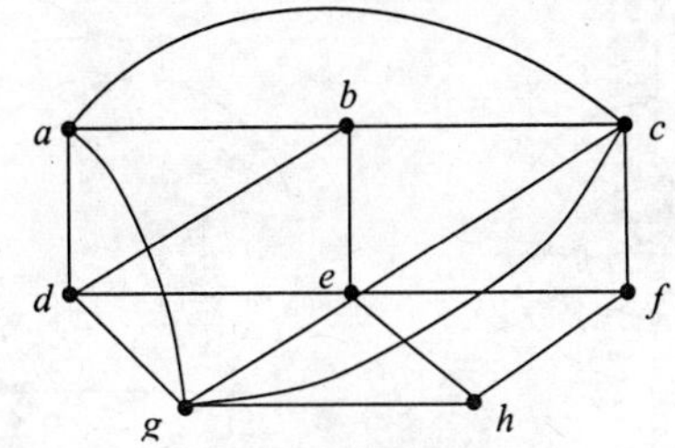

17.133

Solution. Ordering the vertices according to decreasing degrees, we get the sequence e, c, g, a, b, d, f, h.

Use the colour c_1 to colour (paint) e and a.
Use the colour c_2 to paint c, d and h.
Use third colour c_3 to paint vertices g, b and f.

Thus all the vertices are painted such that no adjacent vertices get the same colour. Hence $\chi(G) = 3$.

Some rules for colouring a graph

1. $\chi(G) \leq |V|$, where $|V|$ is the number of vertices of G.
2. A triangle always requires three colours, that is, $\chi(K_3) = 3$. Similarly $\chi(K_n) = n$, where K_n is the complete graph of n vertices.
3. If some subgraph of G requires k-colours, then $\chi(G) \geq k$.
4. If $\deg(v) = n$, then at most n colours are required to colour the vertices adjacent to v.
5. $\chi(G) = \max\ \{\chi(C)$: C is a connected component of $G\}$.
6. Every n-colourable graph has at least n vertices v such that $\deg(v) \geq n-1$.
7. $\chi(G) \leq 1 + \Delta(G)$, where $\Delta(G)$ is the largest degree of any vertex of G.
8. The following statements are equivalent:
   (i) A graph G is 2-colourable
   (ii) G is bipartite
   (iii) Every cycle of G has even length
9. If $\delta(G)$ is the smallest degree of any vertex of G, then

$$\chi(G) \geqslant \frac{|v|}{|v| - \delta(G)}$$

EXAMPLE 17.54

Find chromatic number of the graphs shown in Figure 17.134.

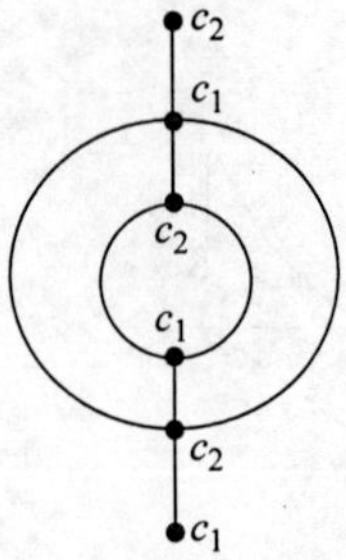

(i)

and

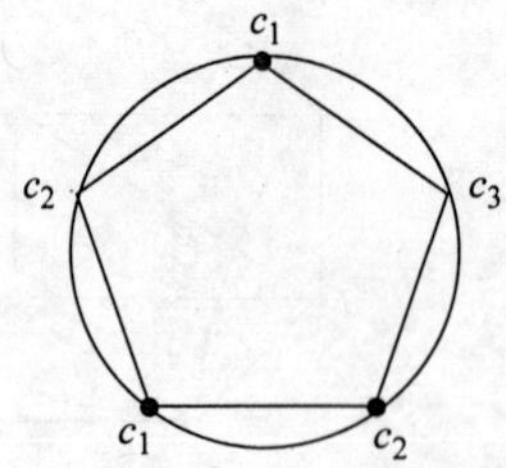

(ii)

Figure 17.134.

Solution. The graph (i) is 2-colourable whereas the graph (ii) is 3-colourable.

Theorem 17.19 Bipartite graph is 2-colourable unless G is edgeless.

Proof: A two colouring is obtained by assigning one colour to every vertex in one of the bipartition parts and another colour to every vertex in the other partition part as shown in Fig. 17.135.

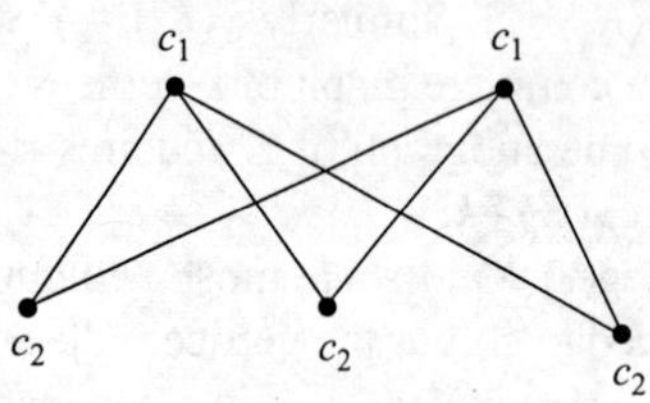

Figure 17.135

Corollary 17.3 Even cycle graphs C_{2n} have $\chi(C_{2n}) = 2$.

Proof: An even cycle graph is bipartite and therefore, by Theorem 17.19, its chromatic number is 2.

For example consider the graph shown in Figure 17.136.

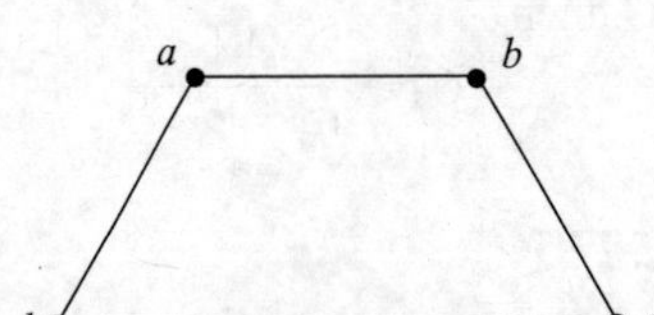

Figure 17.136

It is equivalent to the following bipartite graph shown in Figure 17.137.

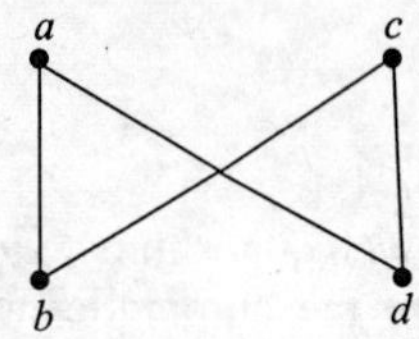

Figure 17.137

Therefore, it is 2-colourable.

Proposition 17.2 Odd cycle graph C_{2n+1} has $\chi(C_{2n+1}) = 3$.

Proof: Let $v_1, v_2, \ldots, v_{2n}, v_{2n+1}$ be vertices of a cycle graph c_{2n+1}. If two colours were sufficient, then they would have to alternate around the cycle. Thus, the odd subscripted would have to be one colour and the even subscripted vertices have to be second colour but vertex v_{2n+1} is adjacent to v_1, and so according to this scheme two adjacent vertices v_1 and v_{2n+1} have the same colour. This is a contradiction and so c_{2n+1} is not 2-colourable but 3-colourable.

It follows, therefore, that "The chromatic number of a cycle is either two or three, depending on whether its length is even or odd."

Definition 17.74 The **join** $G + H$ of the graphs G and H is obtained from the graph $G \cup H$ by adding an edge between each vertex of H.

For example, if we have a graph as shown below (Figure 17.138):

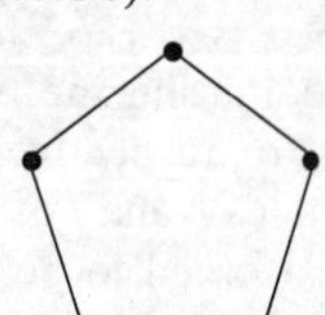

(G)

Figure 17.138

and a graph

$\cdot K_1$,

then the join $G + K_1$ of the graphs G and K_1 is the graph as shown below (Figure 17.139):

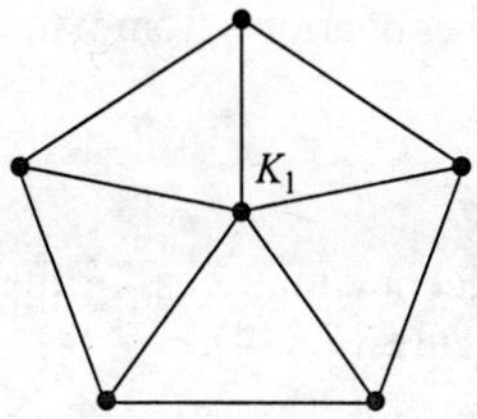

Figure 17.139

Regarding the chromatic number of the join $G+H$ of the graphs G and H we have the following:

Proposition 17.3 The join of a graph G and H has chromatic number

$$\chi(G+H)=\chi(G)+\chi(H).$$

Definition 17.75 The n vertex wheel graph W_n is called an odd-order wheel if n is odd and an even-order wheel if n is even.

Proposition 17.4 Odd-order wheel graph has

$$\chi(W_{2m+1})=3, \quad \text{for all } m\geq 1.$$

Proof: Using the fact that the wheel graph W_{2m+1} is the join $C_{2m}+K_1$, it follows that

$$\chi(W_{2m+1})=\chi(C_{2m})+\chi(K_1)=2+1=3.$$

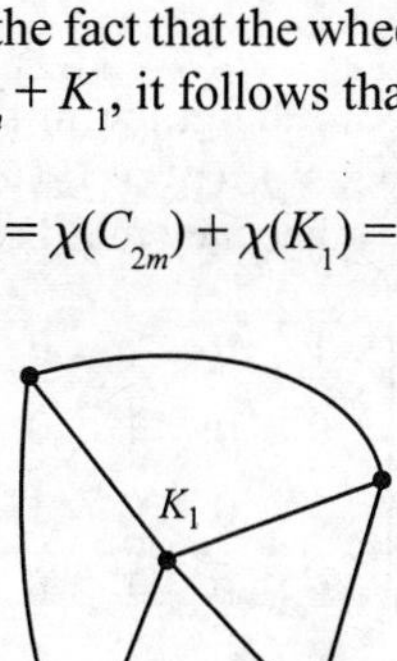

Figure 17.140 (W_5)

Proposition 17.5 Even-order wheel graph has $\chi(W_{2m})=4$.

Proof: Using the fact that $W_{2m}=c_{2m-1}+K_1$, it follows that

$$\chi(W_{2m})=\chi(c_{2m-1})+\chi(K_1)=3+1=4.$$

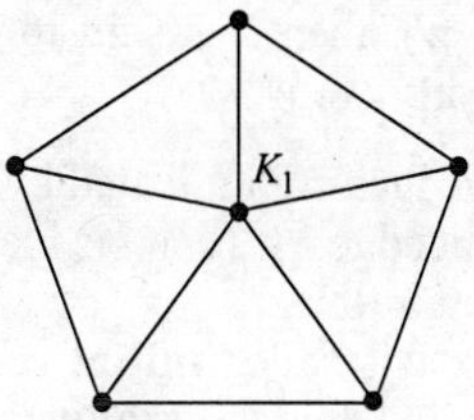

Figure 17.141 (W_6)

EXAMPLE 17.55

Find chromatic number of the graph shown in Figure 17.142:

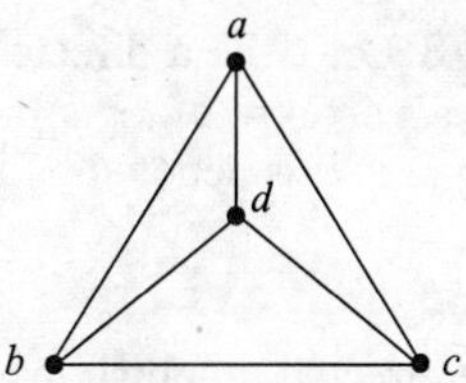

Figure 17.142

Solution. Here degree of each vertex is 3. Therefore, $\chi(G)\leq 1+3=4$. Since it has a triangle, $\chi(G)\geq 3$. Thus, $3\leq \chi(G)\leq 4$.

We first colour the triangle abd with colours c_1, c_2, c_3, respectively. Since c is adjacent to each of a, b and d, therefore a different colour c_4 have to be given to it. Hence the graph is 4-colourable and so $\chi(G)=4$.

Note: If we apply rule 9, then $\chi \geqslant \dfrac{4}{(4-3)}=4$.

Hence $\chi(G)=4$.

Theorem 17.20.(Four-Colour Theorem of Appel and Haken)

Any planar graph is 4-colourable.

(Proof of the theorem is out of the scope of this book).

17.11 DIRECTED GRAPHS

Definition 17.76 A **directed graph** or **digraph** consists of two finite sets:

(i) A set V of vertices (or nodes or points).
(ii) A set E of directed edges (or arcs), where each edge is associated with an ordered pair (v, w) of vertices called its endpoints. If edge e is associated with the ordered

pair (v, w), then e is said to be **directed edge from v to w.**

The directed edges are indicated by arrows.

We say that edge $e = (v, w)$ is incident from v and is incident into w.

The vertex v is called **initial vertex** and the vertex w is called the **terminal vertex** of the directed edge (v, w).

Definition 17.77 Let G be a directed graph. The **outdegree of a vertex v of G** is the number of edges beginning at v. It is denoted by outdeg(v).

Definition 17.78 Let G be a directed graph. The **indegree of a vertex v of G** is the number of edges ending at v. It is denoted by indeg(v).

EXAMPLE 17.56

Consider the directed graph shown below (Figure 17.143):

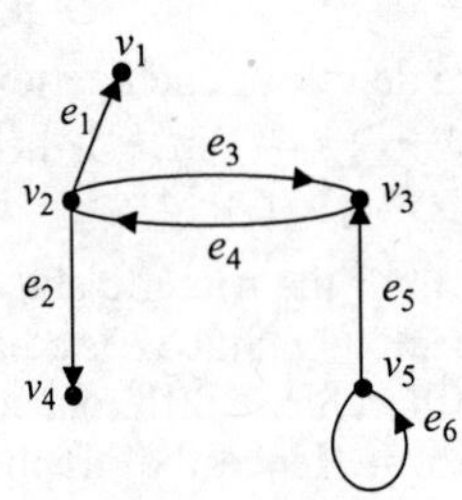

Figure 17.143

Here edge e_1 is (v_2, v_1) whereas e_6 is denoted by (v_5, v_5) and is called a loop. The indegree of v_2 is 1, outdegree of v_2 is 3.

Definition 17.79 A vertex with 0 indegree is called a **source,** whereas a vertex with 0 outdegree is called a **sink.**

For instance, in the above example, v_1 is a sink.

Definition 17.80 If the edges and/or vertices of a directed graph G are labelled with some type of data, then G is called a **labelled directed graph**.

Definition 17.81 Let G be a **directed graph** with ordered vertices $v_1, v_2, \ldots, v_n$. **The adjacency matrix of G is the matrix** $A = (a_{ij})$ over the set of non-negative integers such that

a_{ij} = the number of arrows from v_i to v_j, $i, j = 1, 2, \ldots, n$.

EXAMPLE 17.57

Find the adjacency matrices for the graphs given below (Figure 17.144):

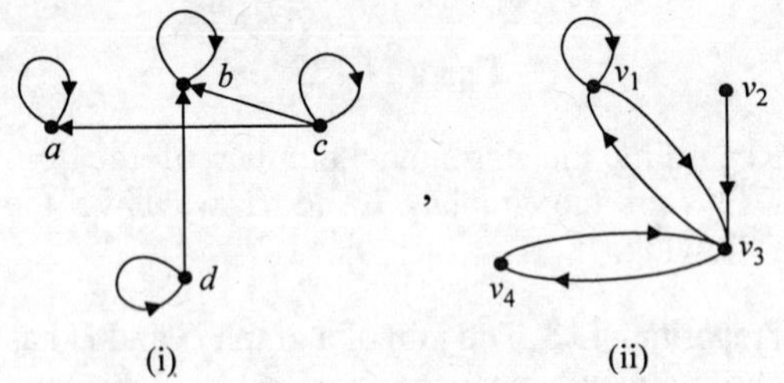

Figure 17.144

Solution. The edges in the directed graph are (a, a), (b, b), (c, c), (d, d), (c, a), (c, b) and (d, b). Therefore the adjacency matrix $A = (a_{ij})$ is

$$\begin{bmatrix} 1 & 0 & 0 & 0 \\ 0 & 1 & 0 & 0 \\ 1 & 1 & 1 & 0 \\ 0 & 1 & 0 & 1 \end{bmatrix}$$

(iii) The edges in the graph in (ii) are (v_2, v_3), (v_1, v_1), (v_1, v_3), (v_3, v_1), (v_3, v_4), (v_4, v_3). Hence the adjacency matrix is

$$\begin{bmatrix} 1 & 0 & 1 & 0 \\ 0 & 0 & 1 & 0 \\ 1 & 0 & 0 & 1 \\ 0 & 0 & 1 & 0 \end{bmatrix}$$

EXAMPLE 17.58

Find the directed graph represented by the adjacency matrix:

$$\begin{bmatrix} 0 & 1 & 0 & 0 & 0 \\ 0 & 0 & 1 & 0 & 0 \\ 0 & 0 & 0 & 1 & 1 \\ 1 & 1 & 0 & 0 & 0 \\ 0 & 0 & 0 & 0 & 0 \end{bmatrix}$$

Solution. We observe that $a_{12} = 1$, $a_{23} = 1$, $a_{34} = 1$, $a_{35} = 1$, $a_{41} = 1$, $a_{42} = 1$. Hence the digraph is as shown in Figure 17.145.

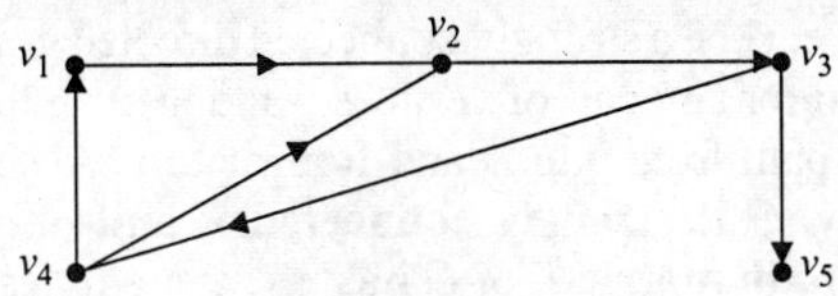

Figure 17.145

Definition 17.82 In a directed graph, if there is no more than one directed edge in a particular direction between a pair of vertices, then it is called **simple directed graph.**

For example, the graph shown in Figure 17.146 is a simple directed graph.

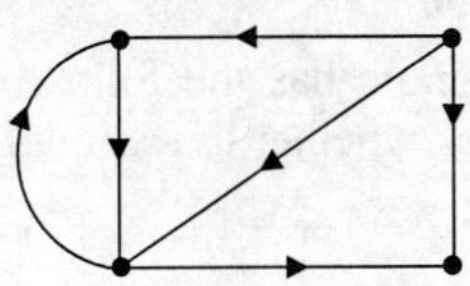

Figure 17.146

A directed graph which is not simple is called **directed multi-graph**.

Definition 17.83 A simple digraph is said to be **strongly connected** if for each pair of vertices *v, w,* there are path, from *v* to *w* and *w* to *v*.

For example, the graphs shown in Figure 17.147 are strongly connected.

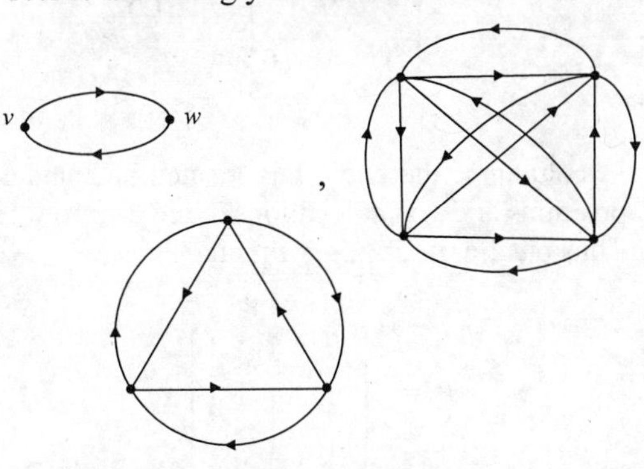

Figure 17.147

Definition 17.84 Let G be a simple digraph with n vertices. Then a $n \times n$ matrix $P = (p_{ij})$ such that

$$p_{ij} = \begin{cases} 1 \text{ if there is a path from } v_i \text{ to } v_j \\ 0 \text{ otherwise} \end{cases}$$

is called **path matrix** or **reachability matrix** of G.

The path matrix can be given in term of adjacency matrix:

$$P = A \vee A^{(2)} \vee A^{(3)} \vee \ldots \vee A^{(n)}$$

and

$$A^{(2)} = A \odot A, \quad A^{(r)} = A^{(r-1)} \odot A,$$

where $\vee$ represents Boolean addition and $\odot$ denotes the Boolean matrix multiplication.

Warshall's algorithm for finding path matrix from adjacency matrix.

We discuss this algorithm with the help of the following example.

EXAMPLE 17.59

Find the adjacency matrix and the path matrix for the digraph shown in Figure 17.148.

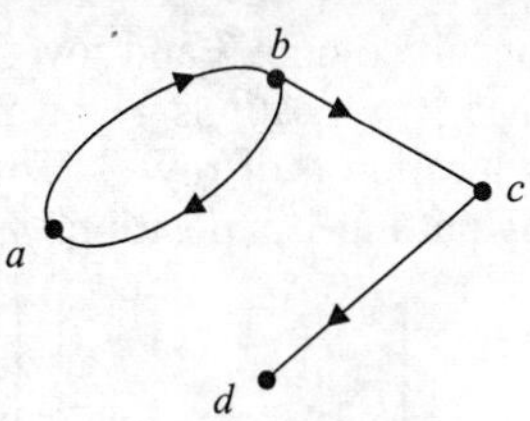

Figure 17.148

Solution. The adjacency matrix of this digraph is:

$$A = \begin{bmatrix} 0 & 1 & 0 & 0 \\ 1 & 0 & 1 & 0 \\ 0 & 0 & 0 & 1 \\ 0 & 0 & 0 & 0 \end{bmatrix}.$$

Then the path matrix $P = (p_{ij})$ is found as follows: We take

$$P_0 = \begin{bmatrix} 0 & 1 & 0 & 0 \\ 1 & 0 & 1 & 0 \\ 0 & 0 & 0 & 1 \\ 0 & 0 & 0 & 0 \end{bmatrix} = A \text{ itself.}$$

Now we find P_1 by consulting column 1 and row 1. We note that P_0 has 1 in location **2** of column 1 and location **2** of row 1. Thus the element p_{22} is replaced by 1 in P_0. Thus we have

$$P_1 = \begin{bmatrix} 0 & 1 & 0 & 0 \\ 1 & 1 & 1 & 0 \\ 0 & 0 & 0 & 1 \\ 0 & 0 & 0 & 0 \end{bmatrix}$$

To find P_2 we consult column 2 and row 2 of P_1. We note that P_1 has 1 in locations 1 and 2 of column 2 and locations 1, 2 and 3 of row 2. Thus, to obtain P_2 we should put 1 in positions p_{11}, p_{12}, p_{13}, p_{21}, p_{22} and p_{23} of matrix P_1. We thus have

$$P_2 = \begin{bmatrix} 1 & 1 & 1 & 0 \\ 1 & 1 & 1 & 0 \\ 0 & 0 & 0 & 1 \\ 0 & 0 & 0 & 0 \end{bmatrix}.$$

We now consult column 3 and row 3 of P_2. The matrix P_2 has 1 in locations 1 and 2 of column 3 and 1 in location 4 of row 3. Thus we shall obtain P_3 by putting 1 in positions p_{14}, p_{24}. Thus

$$P_3 = \begin{bmatrix} 1 & 1 & 1 & 1 \\ 1 & 1 & 1 & 1 \\ 0 & 0 & 0 & 1 \\ 0 & 0 & 0 & 0 \end{bmatrix}.$$

Finally, we consult column 4 and row 4 of P_3. We have 1 in location 1, 2, 3 of column 4 and there is no 1 in row 4. Hence $P_4 = P_3$. Thus the path matrix is

$$P = (p_{ij}) = \begin{bmatrix} 1 & 1 & 1 & 1 \\ 1 & 1 & 1 & 1 \\ 0 & 0 & 0 & 1 \\ 0 & 0 & 0 & 0 \end{bmatrix}$$

Remark 17.11 Given an adjacency matrix of a graph G, we first draw digraph of G and then apply definition of path matrix. We will obtain path matrix.

Now, we state a result without proof.

Theorem 17.21 Let A be the adjacency matrix of a graph G with n vertices and let $B_n = A + A^2 + A^3 + \ldots + A^n$. Then the path matrix P and B_n have the same non-zero entries.

Let G be a strongly connected directed graph. Then for any pair of vertices v and w in G, there is a path from v to w and from w to v. Accordingly, G is **strongly connected if and only if the path matrix P of G has no zero entries.** Thus, in view of the above theorem, we have:

Theorem 17.22 Let A be adjacency matrix of a graph G with n vertices and let $B_n = A + A^2 + \ldots + A^n$. Then G is strongly connected if and only if B_n has no zero entries.

EXAMPLE 17.60

A directed graph has the following adjacency matrix. Check whether it is strongly connected.

$$A(G) = \begin{bmatrix} 0 & 1 & 1 \\ 1 & 0 & 1 \\ 1 & 1 & 0 \end{bmatrix}.$$

Solution. To find the path matrix, we first consider column 1 and row 1 of

$$P_0 = \begin{bmatrix} 0 & 1 & 1 \\ 1 & 0 & 1 \\ 1 & 1 & 0 \end{bmatrix} = A(G).$$

In column 1, the entry 1 is located at 2 and 3 positions and 1 is at locations 2 and 3 in row 1. Thus put 1 at p_{22}, p_{23}, p_{32}, p_{33} and get

$$P_1 = \begin{bmatrix} 0 & 1 & 1 \\ 1 & 1 & 1 \\ 1 & 1 & 1 \end{bmatrix}.$$

We note that 1 is located in column 2 of P_1 at position 1, 2, 3, and 1 is located at position 1, 2, 3 of row 2. Thus put 1 at p_{11}, p_{12}, p_{13}, p_{21}, p_{22}, p_{23}, p_{31}, p_{32}, p_{33}, and get

$$P_2 = \begin{bmatrix} 1 & 1 & 1 \\ 1 & 1 & 1 \\ 1 & 1 & 1 \end{bmatrix}.$$

The next step will not change the position. Hence,

$$P = \begin{bmatrix} 1 & 1 & 1 \\ 1 & 1 & 1 \\ 1 & 1 & 1 \end{bmatrix} = \text{(all non-zero entries)}.$$

Therefore, G is strongly connected.
The graph of G is shown in Figure 17.149.

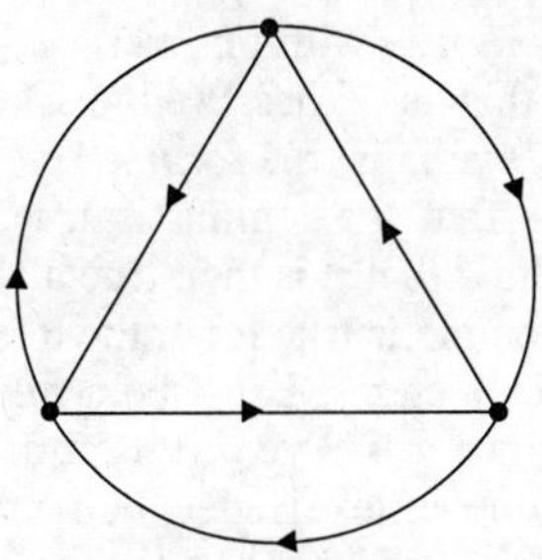

Figure 17.149

Remark 17.12 In the above example, we also note that

$$B_3 = A + A^2 + A^3$$

$$= \begin{bmatrix} 1 & 1 & 1 \\ 1 & 0 & 1 \\ 1 & 1 & 0 \end{bmatrix} + \begin{bmatrix} 2 & 1 & 1 \\ 1 & 2 & 1 \\ 1 & 1 & 2 \end{bmatrix} + \begin{bmatrix} 2 & 3 & 3 \\ 3 & 2 & 3 \\ 3 & 3 & 2 \end{bmatrix}$$

$$= \begin{bmatrix} 4 & 5 & 5 \\ 5 & 4 & 5 \\ 5 & 5 & 4 \end{bmatrix} \text{ (all non-zero entries)}$$

showing that G is strongly connected.

17.12 TREES

Definition 17.85 A graph is said to be a **tree** if it is a connected acyclic graph.

A **trivial tree** is a graph that consists of a single vertex. An **empty tree** is a tree that does not have any vertices or edges.

For example, the graphs shown in Figure 17.150 are all trees.

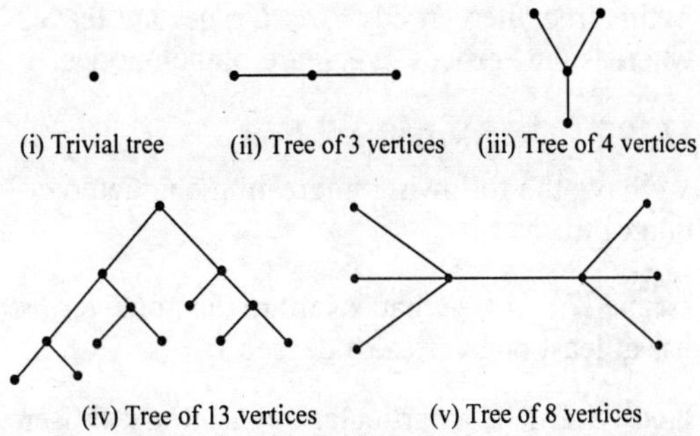

Figure 17.150

But the graphs shown in Figure 17.151 are not trees:

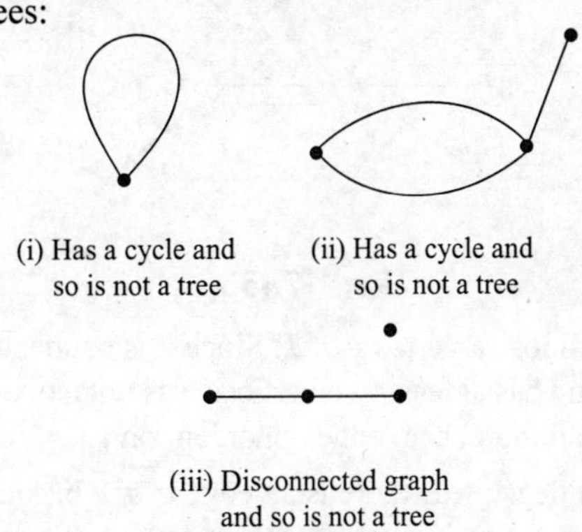

Figure 17.151

Definition 17.86 A collection of disjoint trees is called a **forest**.

Thus a graph is a forest if and only if it is circuit free.

Definition 17.87 A vertex of degree 1 in a tree is called a **leaf** or a **terminal node** or a **terminal vertex**.

Definition 17.88 A vertex of degree greater than 1 in a tree is called a **branch node** or **internal node** or **internal vertex**.
Consider the tree shown in Figure 17.152.

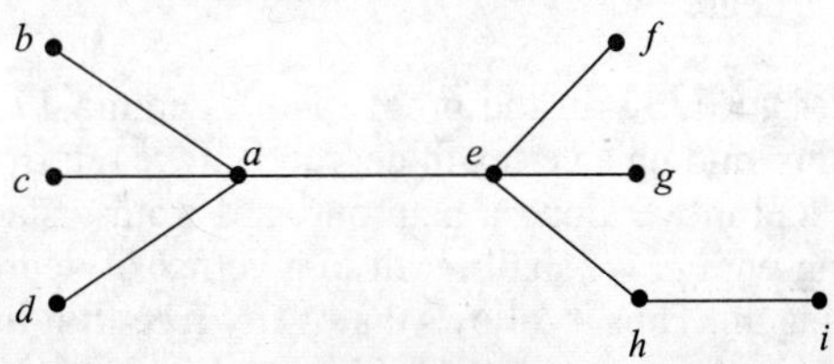

Figure 17.152

In this tree, the vertices b, c, d, f, g, i are leaves whereas the vertices a, e, h are branch nodes.

17.12.1 Characterization of Trees

We have the following interesting characterization of trees:

Lemma 17.1 A tree that has more than one vertex has at least one vertex of degree 1.

Proof: Let T be a particular but arbitrary chosen tree having more than one vertex (Figure 17.153).

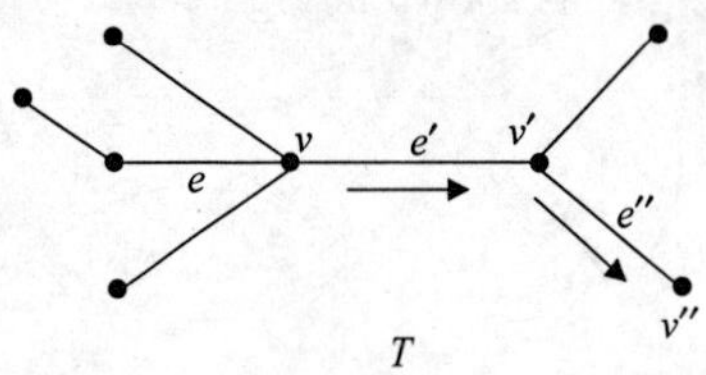

Figure 17.153

1. Choose a vertex v of T. Since T is connected and has at least two vertices, v is not isolated and there is an edge e incident on v.
2. If $\deg(v) > 1$, there is an edge $e' \neq e$ because there are, in such a case, at least two edges incident on v. Let v' be the vertex at the other end of e'. This is possible because e' is not a loop by the definition of a tree.
3. If $\deg(v') > 1$, then there are at least two edges incident on v'. Let e'' be the other edge different from e' and v'' be the vertex at other end of e''. This is again possible because T is acyclic.
4. If $\deg(v'') > 1$, repeat the above process. Since the number of vertices of a tree is finite and T is circuit free, the process must terminate and we shall arrive at a vertex of degree 1.

Remark 17.13 In the proof of the Lemma 17.1, after finding a vertex of degree 1, if we return to v and move along a path outward from v starting with e, we shall reach to a vertex of degree 1 again. Thus it follows that **"Any tree that has more than one vertex has at least two vertices of degree 1"**.

Lemma 17.2 There is a unique path between every two vertices in a tree.

Proof: Suppose on the contrary that there are more than one path between any two vertices in a given tree T. Then T has a cycle which contradicts the definition of a tree because T is acyclic. Hence the lemma is proved.

Lemma 17.3 The number of vertices is one more than the number of edges in a tree.

Equivalently, we may state this lemma as "For any positive integer n, a tree with n vertices has $n-1$ edges".

Proof: We shall prove the lemma by mathematical induction.

Let T be a tree with **one** vertex. Then T has no edges, that is, T has 0 edge. But $0 = 1-1$. Hence the lemma is true for $n = 1$.

Suppose that the lemma is true for $k > 1$. We shall show that it is then true also for $k+1$. Since the lemma is true for k, the tree has k vertices and $k-1$ edges. Let T be a tree with$k+1$ vertices. Since k is +ve, $k+1 \geq 2$ and so T has more than one vertex. Hence, by Lemma 17.1, T has a vertex v of degree 1. Also there is another vertex w and so there is an edge e connecting v and w. Define a subgraph T' of T so that

$$V(T') = V(T) - \{v\},$$
$$E(T') = E(T) - \{e\}.$$

Then number of vertices in $T' = (k+1) - 1 = k$ and since T is circuit free and T' has been obtained on removing one edge and one vertex, it follows that T' is acyclic. Also T' is connected. Hence T' is a tree having k vertices and therefore by induction hypothesis, the number of edges in T' is $k-1$.

But then

Number of edges in T = number of edges in $T' + 1$

$$= k - 1 + 1 = k.$$

Thus the Lemma is true for tree having $k+1$ vertices. Hence the lemma is true by mathematical induction.

Corollary 17.4 Let $C(G)$ denote the number of components of a graph. Then a forest G on n vertices has $n - C(G)$ edges.

Proof: Apply Lemma 17.3 to each component of the forest G.

Corollary 17.5 Any graph G on n vertices has at least $n - C(G)$ edges.

Proof: If G has cycle-edges, remove them one at a time until the resulting graph G^* is acyclic. Then G^* has $n - C(G^*)$ edges by Corollary 17.4. Since we have removed only circuit, $C(G^*) = C(G)$. Thus G^* has $n - C(G)$ edges. Hence G has at least $n - C(G)$ edges.

Lemma 17.4 A graph in which there is a unique path between every pair of vertices is a tree

(This lemma is converse of Lemma 17.2).

Proof: Since there is a path between every pair of points, therefore the graph is connected. Since a path between every pair of points is unique, there does not exist any circuit because existence of circuit implies existence of distinct paths between pair of vertices. Thus the graph is connected and acyclic and so is a tree.

Lemma 17.5 (Converse of Lemma 17.3)

A connected graph G with $e = v - 1$ is a tree

First Proof: The given graph is connected and $e = v - 1$. To prove that G is a tree, it is sufficient to show that G is acyclic. Suppose on the contrary that G has a cycle. Let m be the number of vertices in this cycle. Also, we know that **number of edges in a cycle is equal to number of vertices in that cycle**. Therefore number of edges in the present case is m. Since the graph is connected, every vertex of the graph which is not in cycle must be connected to the vertices in the cycle (see Figure 17.154).

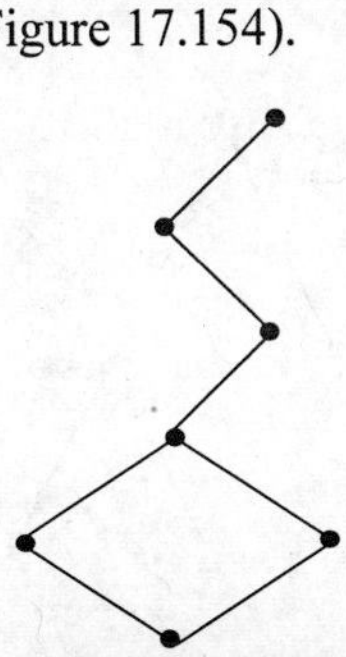

Figure 17.154

Now each edge of the graph that is not in the cycle can connect only one vertex to the vertices in the cycle. There are $v - m$ vertices that are not in the cycle. So the graph must contain at least $v - m$ edges that are not in the cycle. Thus we have $e \geq v - m + m = v$, which is a contradiction to our hypothesis. Hence there is no cycle and so the graph is a tree.

Second Proof: We shall show that a connected graph with v vertices and $v - 1$ edges is a tree. It is sufficient to show that G is acyclic. Suppose on the contrary that G is not circuit free and has a nontrivial circuit C. If we remove one edge of C from the graph G, we obtain a graph G' which is connected (see Figure 17.155).

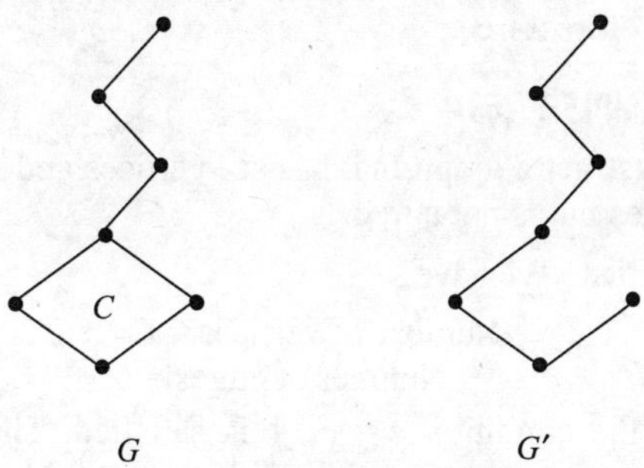

Figure 17.155

If G' still has a nontrivial circuit, we repeat the above process and remove one edge of that circuit obtaining a new connected graph. Continuing this process, we obtain a connected graph G^* which is circuit free. Hence G^* is a tree. Since no vertex has been removed, the tree G^* has v vertices. Therefore, by Lemma 17.3, G^* has $v - 1$ edges. But at least one edge of G has been removed to form G^*. This means that G^* has not more than $v - 1 - 1 = v - 2$ edges. Thus, we arrive at a contradiction. Hence our supposition is wrong and G has no cycle. Therefore G is connected and cycle free and so is a tree.

Lemma 17.6 A graph G with $e = v - 1$, that has no circuit is a tree.

Proof: It is sufficient to show that G is connected. Suppose G is not connected and let G', G'', ... be the connected components of G. Since each of G', G'', ... is connected and has no cycle, they all are tree. Therefore, by Lemma 17.3,

$$e' = v' - 1,$$
$$e'' = v'' - 1,$$
$$\overline{}$$
$$\overline{},$$

where $e', e'', \ldots$ are the number of edges and v', $v'', \ldots$ are the number of vertices in $G', G'', \ldots$, respectively. We have, on adding

$$e' + e'' + \ldots = (v' - 1) + (v'' - 1) + \ldots$$

Since

$$e = e' + e'' + \ldots,$$
$$v = v' + v'' + \ldots,$$

we have

$$e < v - 1,$$

which contradicts our hypotheses. Hence G is connected. So G is connected and acyclic and is therefore a tree.

EXAMPLE 17.61

Construct a graph that has six vertices and five edges but is not a tree.

Solution. We have

Number of vertices = 6
Number of edges = 5

So the condition $e = v - 1$ is satisfied. Therefore, to construct graph with six vertices and five edges that is not a tree, we should keep in mind that the graph should not be connected. The graph shown in Figure 17.156 has six vertices and five edges but is not connected. Hence it is not a tree.

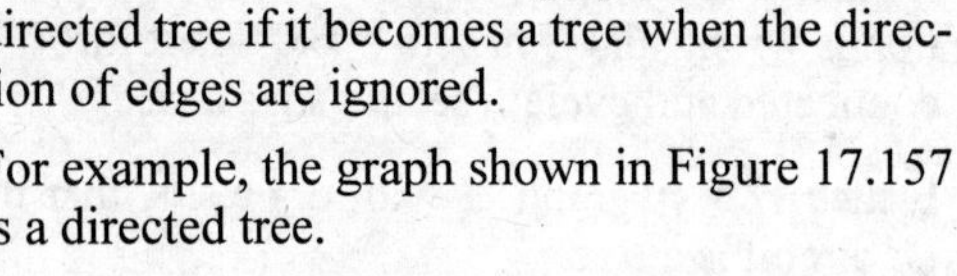

Figure 17.156

Definition 17.89 A directed graph is said to be a directed tree if it becomes a tree when the direction of edges are ignored.

For example, the graph shown in Figure 17.157 is a directed tree.

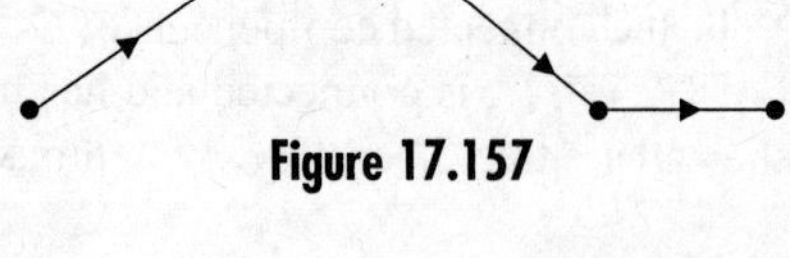

Figure 17.157

Definition 17.90 A directed tree is called a **rooted tree** if there is exactly one vertex whose incoming degree is 0 and the incoming degrees of all other vertices are 1.

The vertex with incoming degree 0 is called the **root** of the rooted tree.

A tree T with root v_0 will be denoted by (T, v_0).

Definition 17.91 In a rooted tree, a vertex, whose outgoing degree is 0 is called a **leaf** or **terminal node**, whereas a vertex whose outgoing degree is non-zero is called a **branch node** or an **internal node**.

Definition 17.92 Let u be a branch node in a rooted tree. Then a vertex v is said to be **child (son** or **offspring)** of u if there is an edge from u to v. In this case, u is called **parent (father)** of v.

Definition 17.93 Two vertices in a rooted tree are said to be **siblings (brothers)** if they are both children of same parent.

Definition 17.94 A vertex **v** is said to be a **descendent** of a vertex u if there is a unique directed path from u to v.

In this case, u is called the **ancestor** of v.

Definition 17.95 The **level** (or **path length**) of a vertex u in a rooted tree is the number of edges along the unique path between u and the root.

Definition 17.96 The **height** of a rooted tree is the maximum level to any vertex of the tree.

As an example of these terms, consider the rooted tree shown in Figure 17.158:

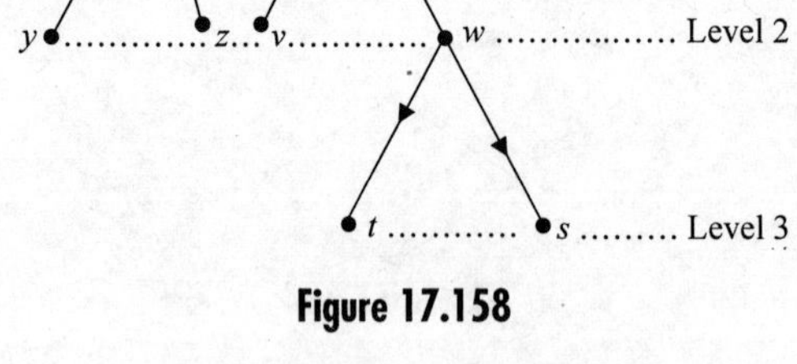

Figure 17.158

Here y is a child of x; x is the parent of y and z. Thus y and z are siblings. The descendents of u are v, w, t and s. Levels of vertices are shown in the figure. The height of this rooted tree is 3.

Definition 17.97 Let u be a branch node in the tree $T = (V, E)$. Then the subgraph $T' = (V', E')$ of T such that the vertices set V' contains u and all of its descendents andE' contains all the edges in all directed paths emerging from u is called a **subtree** with u as the root.

Definition 17.98 Let u be a branch node. By a subtree of u, we mean a subtree that has child of u as root.

In the above example, we note that the Figure 17.159 is a subtree of T,

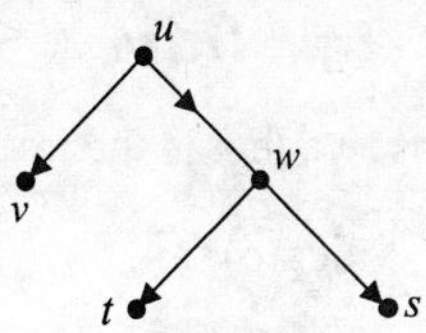

,

Figure 17.159

whereas the Figure 17.160 is a subtree of the branch node u.

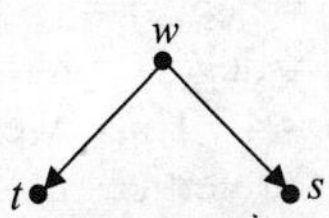

Figure 17.160

EXAMPLE 17.62

Let

$$V = \{v_1, v_2, v_3, v_4, v_5, v_6, v_7, v_8\}$$

and let

$$E = \{(v_2, v_1), (v_2, v_3), (v_4, v_2), (v_4, v_5), (v_4, v_6), (v_6, v_7), (v_5, v_8)\}.$$

Show that (V, E) is rooted tree. Identify the root of this tree.

Solution. We note that

Incoming degree of $v_1 = 1$,
Incoming degree of $v_2 = 1$,
Incoming degree of $v_3 = 1$,
Incoming degree of $v_4 = 0$,
Incoming degree of $v_5 = 1$,
Incoming degree of $v_6 = 1$,
Incoming degree of $v_7 = 1$,
Incoming degree of $v_8 = 1$.

Since incoming degree of the vertex v_4 is 0, it follows that v_4 is root.

Further,

Outgoing degree of $v_1 = 0$,
Outgoing degree of $v_3 = 0$,
Outgoing degree of $v_7 = 0$,
Outgoing degree of $v_8 = 0$.

Therefore, v_1, v_3, v_7, v_8 are leaves. Also,

Outgoing degree of $v_2 = 2$,
Outgoing degree of $v_4 = 3$,
Outgoing degree of $v_5 = 1$,
Outgoing degree of $v_6 = 1$.

Now the root v_4 is connected to v_2, v_5 and v_6. So, we have the following diagram:

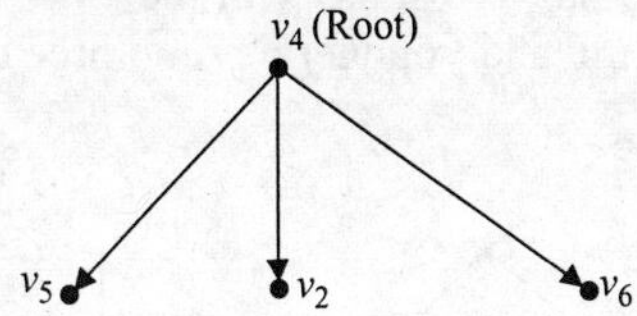

Now v_2 is connected to v_1 and v_3, v_5 is connected to v_8, v_6 is connected to v_7. Thus, we have the graph shown in Figure 17.161.

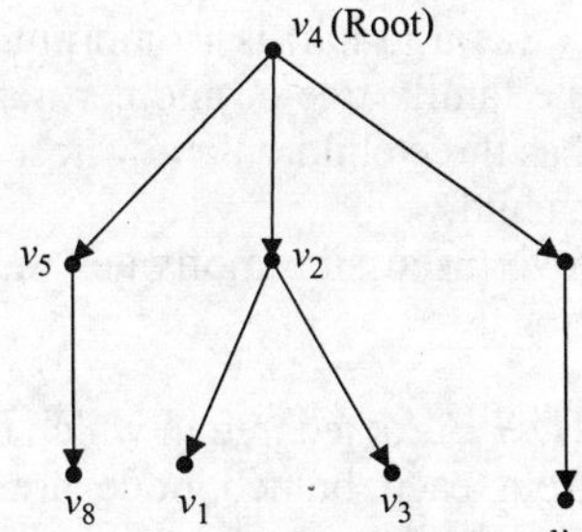

Figure 17.161

We thus have a connected acyclic graph and so (V, E) is a rooted tree with root v_4.

EXAMPLE 17.63

Let F be the set of all female descendent of a lady v_0. We define a relation T on F as follows: if $v_1, v_2, \in F$, then $v_1 \; T \; v_2$ if v_1 is mother of v_2. Show that the relation T on F is a rooted tree with root v_0. Is this relation an equivalence relation?

Solution. The graph of T is cycle free, because descendents cannot be parents of their parents. The graph is also connected. Hence T is rooted tree. We further note that

(i) Since v_1 is not mother of itself, $(v_1, v_1) \notin T$.
(ii) If $(v_1, v_2) \in T$, then v_1 is mother of v_2. But then v_2 cannot be mother of v_1. Hence $(v_1, v_2) \in T$ does not imply $(v_2, v_1) \in T$.
(iii) If $(v_1, v_2) \in T$, $(v_2, v_3) \in T$, then

v_1 is mother of v_2 and v_2 is mother of v_3.

This does not imply that v_1 is mother of v_3, i.e., $(v_1, v_3) \notin T$. Thus T is irreflexive, asymmetric and is not transitive. Hence T is not an equivalence relation.

Consider the rooted trees T and T' shown in the Figure 17.162, where T is family tree of a man who has two sons, with the elder son having no child and younger having three children.

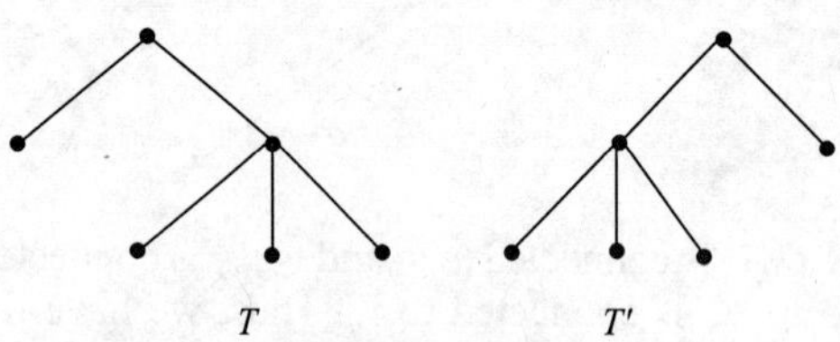

Figure 17.162

Although T' (as a graph) is isomorphic to T, it could be the family tree of another man whose elder son has three children and whose younger son has no child.

The above discussion motivates the following definitions:

Definition 17.99 A rooted tree in which the edges incident from each branch node are labelled with integers 1, 2, 3, … is called an **ordered tree.**

17.13 ISOMORPHISM OF TREES

Definition 17.100 Two ordered trees are said to be **isomorphic** if (i) there exists a one-to-one correspondence between their vertices and edges and that preserves the incident relation (ii) labels of the corresponding edges match.

In view of this definition, the ordered trees shown in Figure 17.163 are not isomorphic.

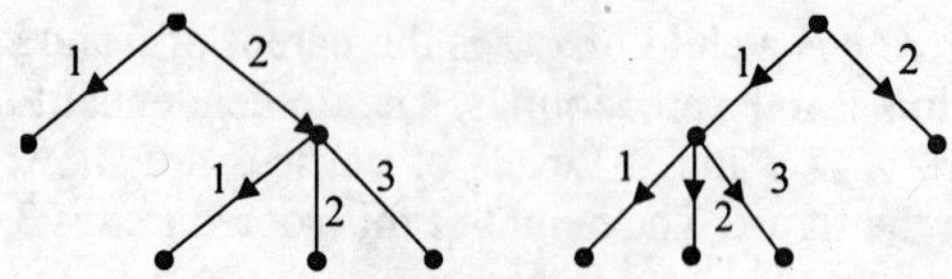

Figure 17.163

EXAMPLE 17.64

Show that the tree T_1 and T_2 shown in the diagram below are isomorphic.

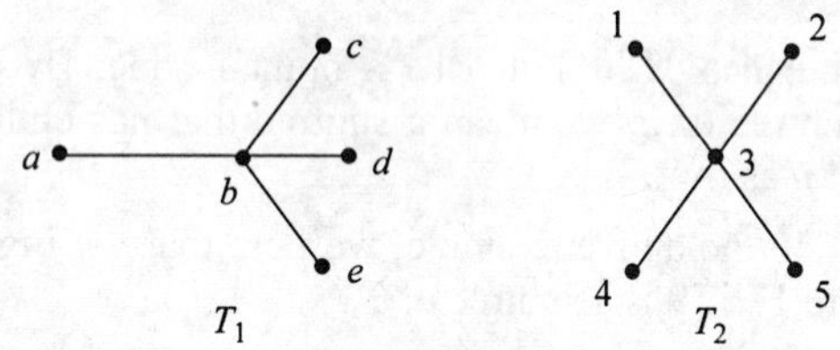

Figure 17.164

Solution. We observe that in the tree T_1,

$$\deg(b) = 4.$$

In the tree T_2,

$$\deg(3) = 4.$$

Further, $\deg(a) = \deg(1) = 1$, $\deg(c) = \deg(2)$, $\deg(d) = \deg(4) = \deg(e) = 1 = \deg(5)$ and $\deg(b) = \deg(3) = 4$. Thus we may define a function f from the vertices of T_1 to the vertices of T_2 by
$f(a) = 1$, $f(b) = 3$, $f(c) = 2$, $f(d) = 4$, $f(e) = 5$.

This is a one-to-one and onto function. Also, adjacency relation is preserved because if v_i and v_j are adjacent vertices in T_1, then $f(v_i)$ and $f(v_j)$ are adjacent vertices in T_2. Hence T_1 is isomorphic to T_2.

EXAMPLE 17.65

Show that the tree T_1 and T_2, shown in the Figure 17.165 are isomorphic

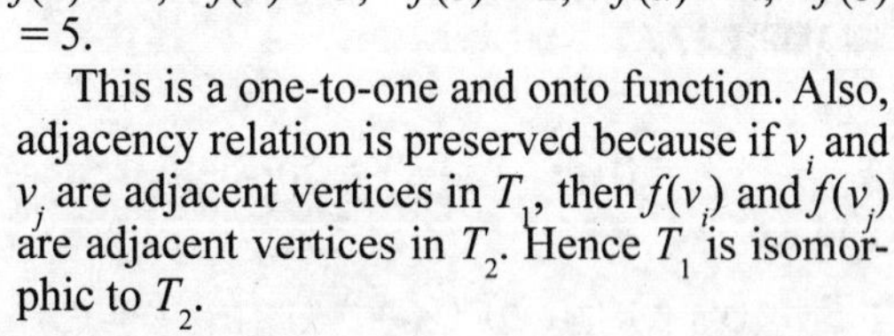

Figure 17.165

Solution. Let f be a function defined by

$$f(z)=v,\quad f(y)=w,\quad f(x)=m,$$
$$f(s)=u,\quad f(t)=l.$$

Then f is an one-one, onto mapping which preserves adjacency. Hence, T_1 and T_2 are isomorphic.

Remark 17.14 There are three non-isomorphic trees (Figure 17.166) with five vertices:

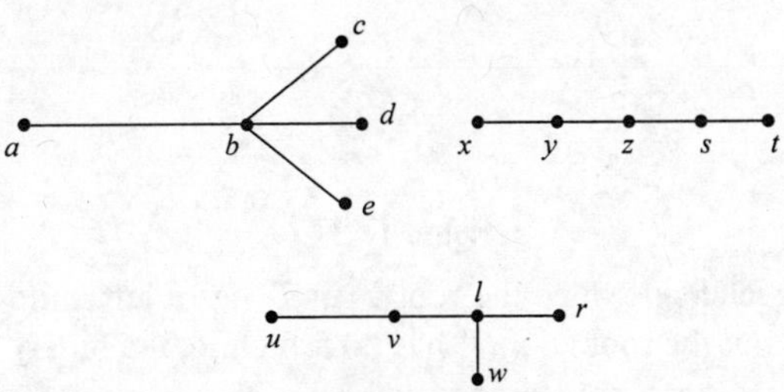

Figure 17.166

Definition 17.101 Let T_1 and T_2 be rooted trees with roots r_1 and r_2, respectively. Then T_1 and T_2 are **isomorphic** if there exists a one-one, onto function f from the vertex set of T_1 to the vertex set of T_2 such that

(i) Vertices v_i and v_j are adjacent in T_1 if and only if the vertices $f(v_i)$ and $f(v_j)$ are adjacent in T_2,

(ii) $f(r_1)=r_2$.

The function f is then called an **isomorphism**.

EXAMPLE 17.66

Show that the trees T_1 and T_2 shown in Figure 17.167 are isomorphic.

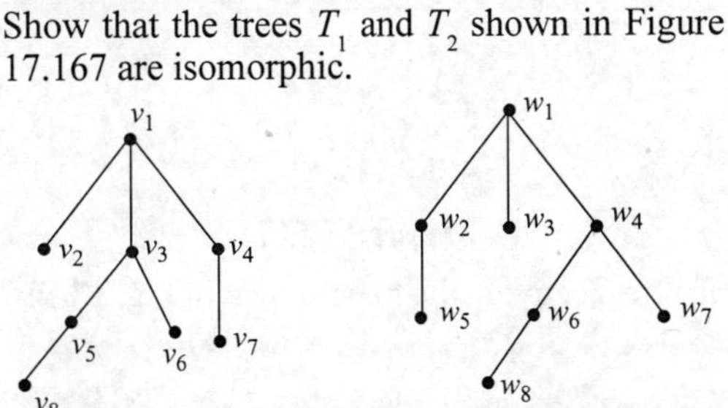

Figure 17.167

Solution. We observe that T_1 and T_2 are rooted trees.

Define f: (vertex set of T_1)$\rightarrow$(vertex set of T_2) by

$$f(v_1)=w_1,\quad f(v_2)=w_3,\quad f(v_3)=w_4$$
$$f(v_4)=w_2,\quad f(v_5)=w_6,\quad f(v_6)=w_7$$
$$f(v_7)=w_5\quad f(v_8)=w_8.$$

Then f is one-to-one and adjacency relation is preserved. Hence f is an isomorphism and so the rooted trees T_1 and T_2 are isomorphic

EXAMPLE 17.67

Show that the rooted trees shown in Figure 17.168 are not isomorphic:

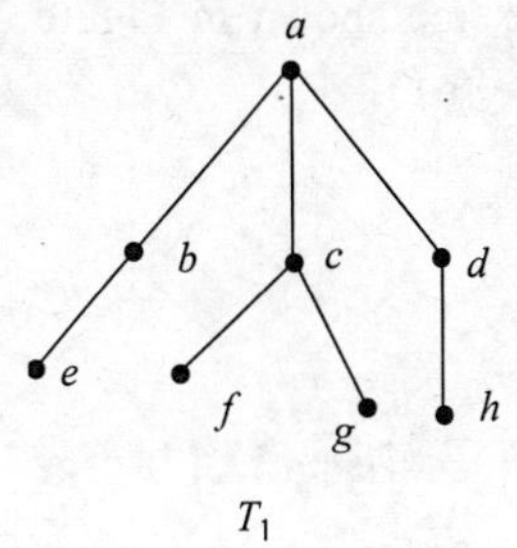

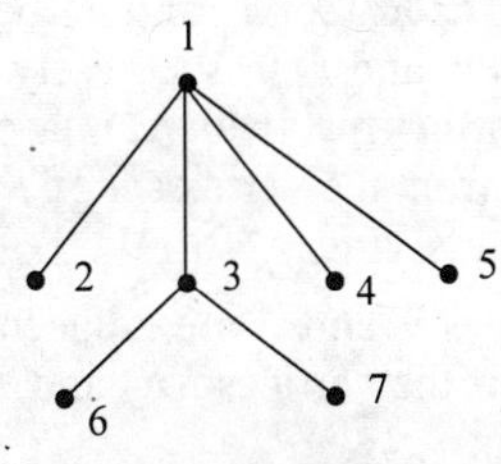

Figure 17.168

Solution. We observe that the degree of root in T_1 is 3, whereas the degree of root in T_2 is 4. Hence T_1 is not isomorphic to T_2.

Definition 17.102 An ordered tree in which every branch node has at most n offspring is called a ***n*-ary tree** (or ***n*-tree**).

Definition 17.103 An n-ary tree is said to be **fully *n*-ary tree (complete *n*-ary tree or regular *n*-ary tree)** if every branch node has exactly n offspring.

Definition 17.104 An ordered tree in which every branch node has at most two offspring is called a **binary tree** (or **2-tree**).

Definition 17.105 A binary tree in which every branch node (internal vertex) has exactly two offspring is called a **fully binary tree.**

For example, the tree given in Figure 17.169 is a binary tree,

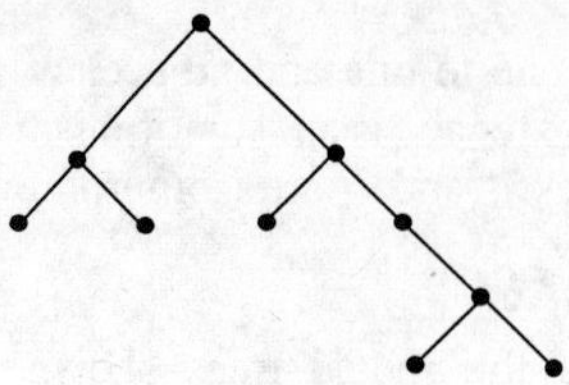

Figure 17.169

whereas the tree shown in Figure 17.170 is a fully binary tree.

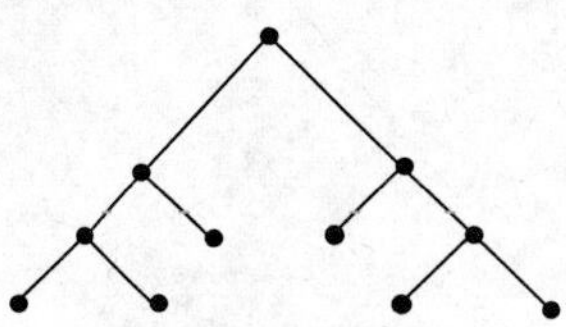

Figure 17.170

Definition 17.106 Let T_1 and T_2 be binary trees with roots r_1 and r_2, respectively. Then T_1 and T_2 are **isomorphic** if there is a one-one, onto function f from the vertex set of T_1 to the vertex set of T_2 satisfying

(i) Vertices v_i and v_j are adjacent in T_1 if and only if the vertices $f(v_i)$ and $f(v_j)$ are adjacent in T_2

(ii) $f(r_1) = r_2$

(iii) v is a left child of w in T_1 if and only if $f(v)$ is a left child of $f(w)$ in T_2

(iv) v is a right child of w in T_1 if and only if $f(v)$ is a right child of $f(w)$ in T_2

The function f is then called an **isomorphism** between binary trees T_1 and T_2.

EXAMPLE 17.68

Show that the trees given in Figure 17.171 are isomorphic.

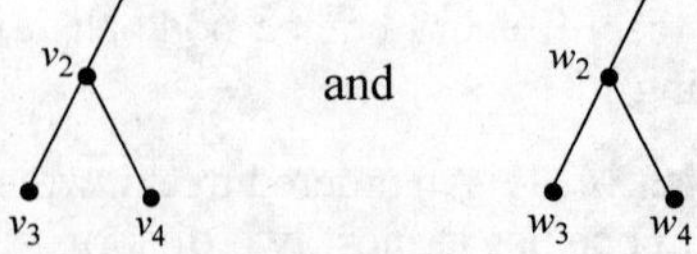

Figure 17.171

Solution. Define f by $f(v_i) = w_i$, $i = 1, 2, 3, 4$. Then f satisfies all the properties for isomorphism. Hence T_1 and T_2 are isomorphic.

EXAMPLE 17.69

Show that the trees given in Figure 17.172 are not isomorphic.

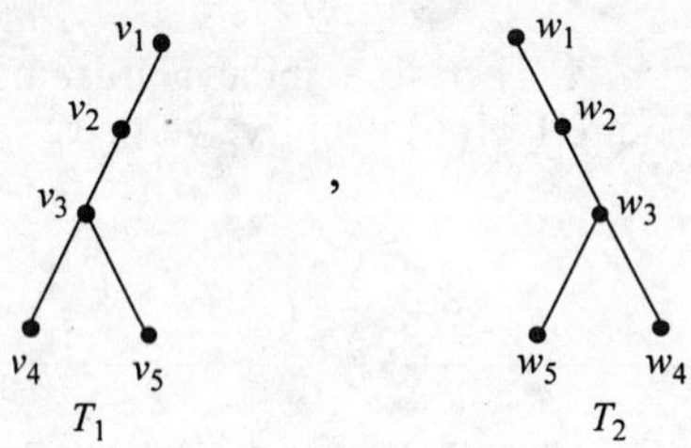

Figure 17.172

Solution. Since the root v_1 in T_1 has a left child but the root w_1 in T_2 has no left child, the binary trees are not isomorphic.

Definition 17.107 Let v be a branch node of a binary tree T. The **left subtree** of v is the binary tree whose root is the left child of v, whose vertices consists of the left child of v and all its descendents and whose edges consists of all those edges of T that connects the vertices of the left subtree together.

The **right subtree** can be defined analogously.

For example, the left subtree and the right subtree of v in the tree (Figure 17.173)

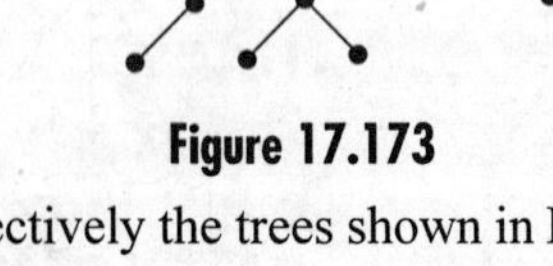

Figure 17.173

are, respectively the trees shown in Fig. 17.174.

Figure 17.174

17.14 REPRESENTATION OF ALGEBRAIC EXPRESSIONS BY BINARY TREES

Binary trees are used in computer science to represent algebraic expressions involving parentheses. For example, the binary trees (Figure 17.175) represent the expressions, $a+b$, a/b and $b * (c * d)$, respectively.

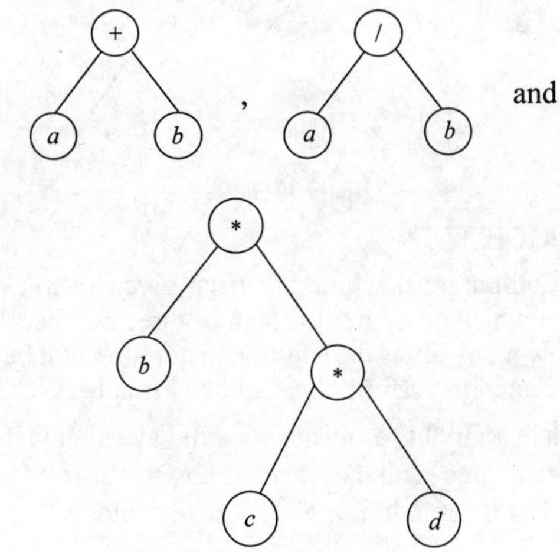

Figure 17.175

Thus, the **central operator acts as root of the tree.**

EXAMPLE 17.70

Draw a binary tree to represent

(i) $(2-(3\times x))+((x-3)-(2+x))$
(ii) $ab-(c/(d+e))$.

Solution.

(i) In this expression + is the central operator. Therefore the root of tree is + . The binary tree is shown in Figure 17.176.

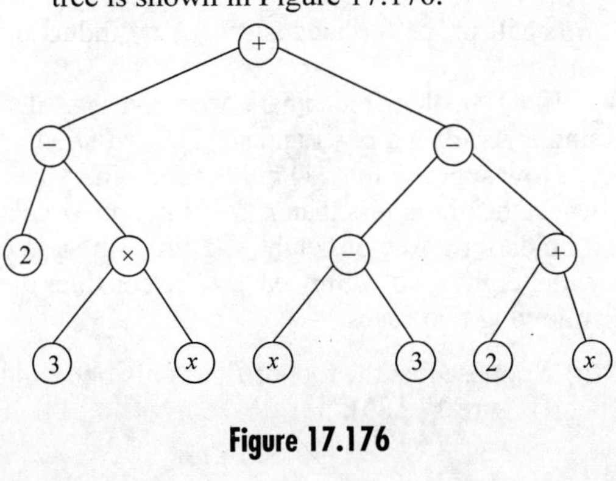

Figure 17.176

(ii) Here the central operator is −. Therefore it is the root of the tree. We have the following binary tree (Figure 17.177) to represent this expression:

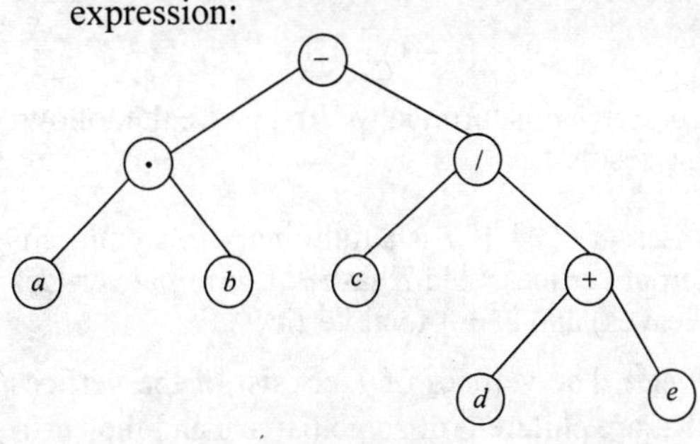

Figure 17.177

EXAMPLE 17.71

Consider a single elimination tournament with eight players in a wrestling game. How many rounds will be there to declare the champion and how many games are to be played?

Solution. In this example, the leaves of the tree shown in Figure 17.178 will represent player, the branch nodes will represent winners in the first round and the second round. Then the tree showing the tournament shall be as given below: The graph of single elimination tournament is a full binary tree. The champion shall be at the root.

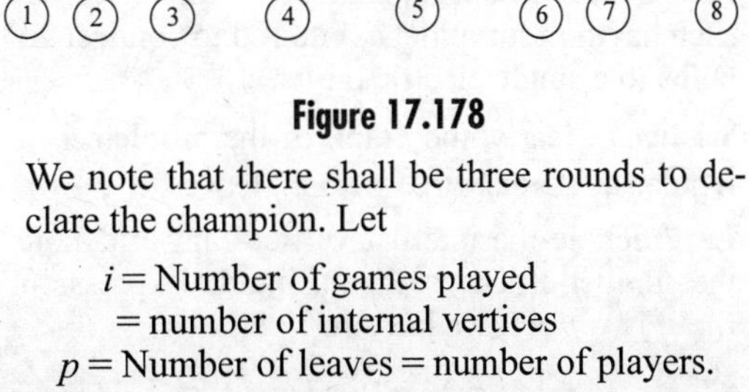

Figure 17.178

We note that there shall be three rounds to declare the champion. Let

i = Number of games played
= number of internal vertices
p = Number of leaves = number of players.

Then, we see that

$$i = p - 1 = 7.$$

In general, for complete n-ary tree, we will have

$$(n-1)\,i = p - 1.$$

To derive this formula we first prove the following result:

Theorem 17.23 If T is a full binary tree with i internal vertices, then T has $i + 1$ terminal vertices (leaves) and $2i + 1$ total vertices.

Proof: The vertices of T consist of the vertices that are children (of some parent) and the vertices that are not children (of any parent). There is a non-child (the root). Since there are i internal vertices, each having two children, there are $2i$ children. Thus the total number of vertices of T is $2i + 1$ and the number of terminal vertices is

$$(2i+1) - i = i + 1.$$

This completes the Proof.
In the context of above example, we have

$$\text{Number of leaves} = p = i + 1$$
$$\Rightarrow i = p - 1.$$

Remark 17.15 In case of full n-ary tree, if i denotes the number of branch nodes, then total number of vertices of T is $ni + 1$ and the number of terminal vertices is

$$ni + 1 - i = i(n-1) + 1.$$

If p is the number of terminal vertices, then

$$p = i(n-1) + 1,$$

which yields

$$(n-1)\,i = p - 1.$$

EXAMPLE 17.72

Find the minimum number of extension cords, each having four outlets, required to connect 22 bulbs to a single electric outlet.

Solution. Clearly, the graph of the problem is a regular quaternary tree with 22 leaves.

Let i denote the internal vertices and p denote the number of leaves, then using the expression $(n-1)\,i = p - 1$, we have

$$(4-1)\,i = 22 - 1 \text{ or } i = \frac{21}{3} = 7.$$

Thus seven extension cords, as shown in Figure 17.179, are required.

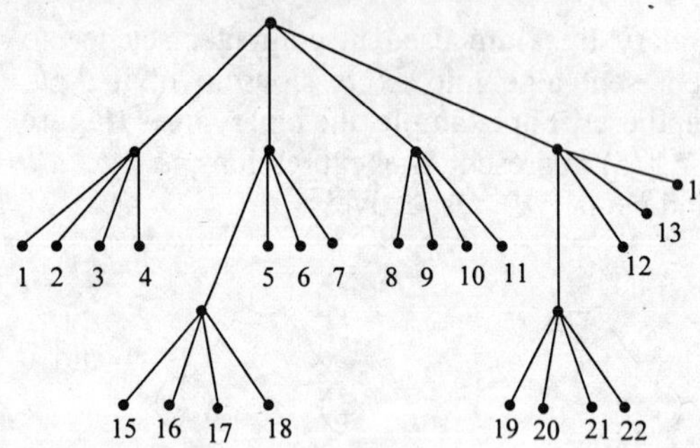

Figure 17.179

EXAMPLE 17.73

A computing machine has been given instruction which computes the sum of three numbers. How many times the addition instruction will be executed to perform the sum of 11 numbers?

Solution. In this problem, we will have a regular ternary tree with 11 leaves. Thus $n = 3$, $p = 11$ and so the relation $(n-1)\,i = p - 1$ yields

$$i = \frac{p-1}{n-1} = \frac{10}{2} = 5.$$

Theorem 17.24 If a binary tree of height h has p leaves, then

$$\log_2 p \le h \quad \text{or} \quad p \le 2^h.$$

Proof: It is sufficient to prove that

$$p \le 2^h, \ (1)$$

because then taking logarithm to the base 2 of both sides, we will get

$$\log_2 p \le h.$$

We shall prove the inequality (1) by induction on h.

If $h = 0$, then the binary tree consists of a single vertex and $p = 1$ and so (1) is satisfied.

Now suppose that (1) holds for a binary tree whose height is less than h. We shall prove that it holds for a tree of height h. So let T be a tree with height $h > 0$ with p leaves. We consider the following two cases:

(a) Suppose that the root of T has only one child (Figure 17.180).

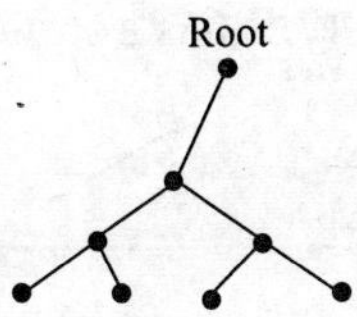

Figure 17.180

Eliminating the root and the edge incident on the root, we get a tree T' of height $h-1$ in which number of leaves is same as in T. Using induction hypothesis on T', we get $p \leq 2^{h-1}$. Since $2^{h-1} < 2^h$, we have$p \leq 2^h$ and the inequality holds in this case.

(b) Suppose the root of T has offspring's v and w (Figure 17.181).

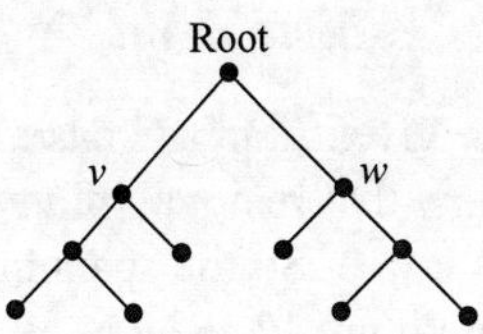

Figure 17.181

Let T_1 be the subtree rooted at v whose height is h_1 and has p_1 leaves. Similarly, let T_2 be the subtree rooted at w whose height is h_2 and has p_2 leaves. Using mathematical induction, to these trees, we have $p_1 \leq 2^{h_1}$ and $p_2 \leq 2^{h_2}$. But the leaves of T consist of the leaves of T_1 and T_2. Hence,

$$p = p_1 + p_2 \leq 2^{h_1} + 2^{h_2}$$

$$\leq 2^{h-1} + 2^{h-1} = 2^{h-1}(1+1) = 2^{\text{h}}.$$

Hence, by mathematical induction, the result holds.

It can be proved on the same lines that an *n*-**ary tree of height** h **has atmost** n^h **leaves**.

EXAMPLE 17.74

Find the number of terminal vertices in a regular binary tree of height 4.

Solution. Let p be the number of terminal vertices. We are given that height (h) of the tree is 4. So, we must have $\text{p} \leq 2^4 = 16$. Since the tree is regular, we have $p = 16$.

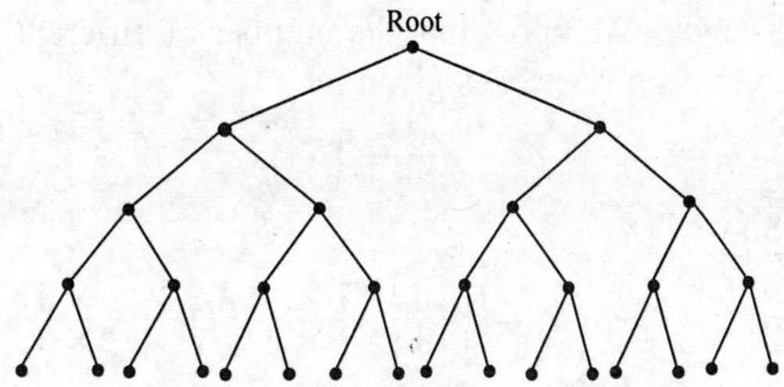

(A regular binary tree of height 4 with 16 leaves)

Figure 17.182

EXAMPLE 17.75

Does there exist a full binary tree with 12 internal vertices and 15 leaves?

Solution. We know that if i is the number of branch nodes in a full binary tree, then the number of leaves is $i+1$. Therefore for a tree with 12 branch nodes, the number of leaves should be 13 and not 15. Hence such tree does not exist.

EXAMPLE 17.76

Is there a binary tree with height 6 and 65 leaves?

Solution. No, since

$$p \leq 2^h = 26 = 64.$$

We now state some results without proof on *n*-ary tree.

Theorem 17.25 The number b_n of different binary trees on n vertices is given by

$$b_n = \frac{1}{n+1}\binom{2n}{n}.$$

Theorem 17.26 A regular *n*-ary tree of height h has **at least** $n + (n-1)(h-1)$ leaves.

Theorem 17.27 If in a rooted regular *n*-ary tree,

I = Sum of the path length of all branch nodes
E = Sum of the path length of all leaves
i = Number of branch nodes

then

$$E = (n-1)I + n\,i$$

Thus, in particular, for a full binary tree,

$$E = I + 2i.$$

EXAMPLE 17.77

How many binary trees are possible on three vertices?

Solution. We know that the number of different trees on n vertices is given by

$$b_n = \frac{1}{n+1}\binom{2n}{n}.$$

So we have

$$b_n = \frac{1}{3+1}\binom{6}{3} = \frac{1}{4}\left(\frac{6\times5\times4\times3\times2\times1}{3\times2\times1\times3\times2\times1}\right)$$
$$= 5 \text{ trees}$$

and those are shown in Figure 17.183.

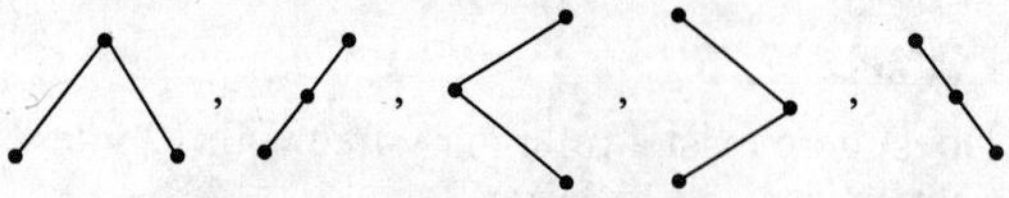

Figure 17.183

17.15 SPANNING TREE OF A GRAPH

Definition 17.108 Let G be a graph, then a subgraph of G which is a tree is called **tree of the graph**.

Definition 17.109 A **spanning tree** for a graph G is a subgraph of G that contains every vertex of G and is a tree.

Thus,

"A **spanning tree** for a graph G is a spanning subgraph of G which is a tree".

EXAMPLE 17.78

Determine a tree and a spanning tree for the connected graph given below (Figure 17.184):

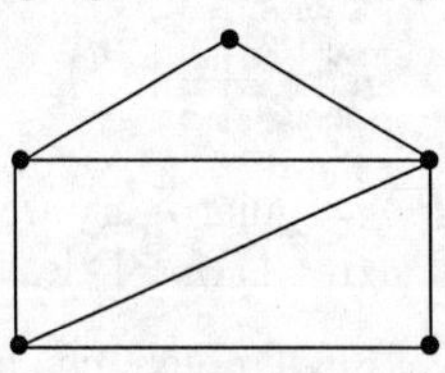

G

Figure 17.184

Solution. The given graph G contains circuits and we know that removal of the circuits gives a tree. So, we note that the Figure 17.185 is a tree.

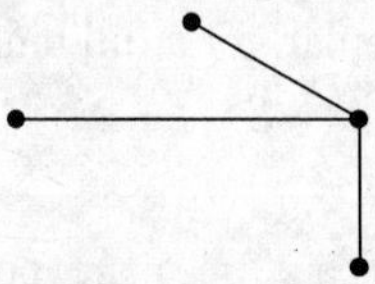

Figure 17.185

And the Figure 17.186 is a spanning tree of the graph G.

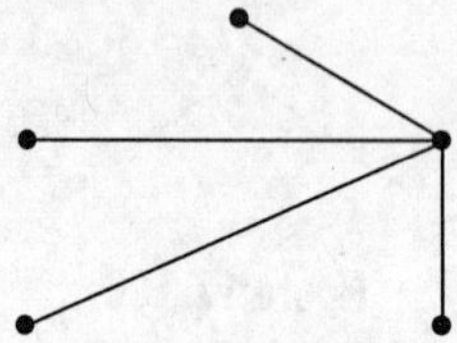

Figure 17.186

EXAMPLE 17.79

Find all spanning trees for the graph G shown below (Figure 17.187).

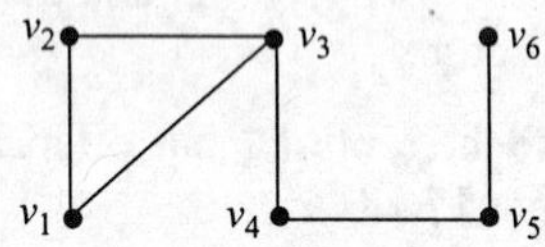

Figure 17.187

Solution. The given graph G has a circuit $v_1\, v_2\, v_3\, v_1$. We know that removal of any edge of the circuit gives a tree. So the spanning trees of G are shown in Figure 17.188.

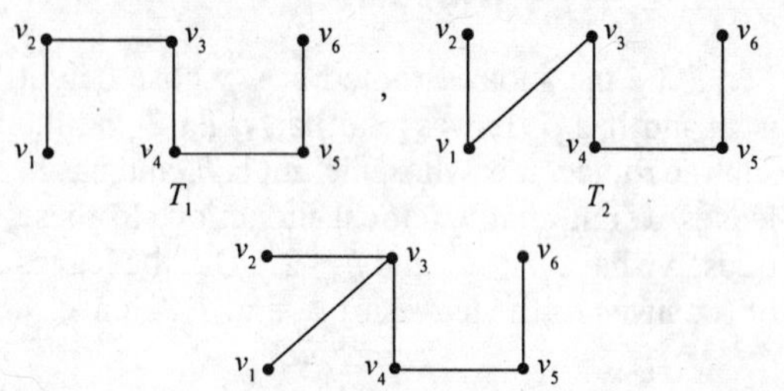

Figure 17.187

Remark 17.16 We know that a tree with n vertices has exactly $n-1$ edges. Therefore if G is a connected graph with n vertices and m edges, a spanning tree of G must have $n-1$ edges. Hence the number of edges that must be removed before a spanning tree is obtained must be

$$m-(n-1) = m-n+1.$$

For illustration, in EXAMPLE 17.79, $n=6$, $m=6$, so, we had to remove one edge to obtain a spanning tree.

Definition 17.110 A **branch** of a tree is an edge of the graph that is in the tree.

Definition 17.111 A **chord** (or a **link**) of a tree is an edge of the graph that is not in the tree.

It follows from the above remark that the number of chords in a tree is equal to $m-n+1$, where n is the number of vertices and m is the number of edges in the graph related to the tree.

Definition 17.112 The set of the chords of a tree is called the **complement of the tree.**

EXAMPLE 17.80

Consider the graph discussed in the above example. We note that the edge (v_2, v_3) is a branch of the tree T_1, whereas (v_1, v_3) is a chord of the tree T_1.

Theorem 17.28 A graph G has a spanning tree if and only if G is connected.

Proof: Suppose first that a graph G has a spanning tree T. If v and w are vertices of G, then they are also vertices in T and since T is a tree there is a path from v to w in T. This path is also a path in G. Thus every two vertices are connected in G. Hence G is connected.

Conversely, suppose that G is connected. If G is acyclic, then G is its own spanning tree and we are done. So suppose that G contains a cycle C_1. If we remove an edge from the cycle, the subgraph of G so obtained is also connected. If it is acyclic, then it is a spanning tree and we are done. If not, it has at least one circuit, say C_2. Removing one edge from C_2, we get a subgraph of G which is connected. Continuing in this way, we obtain a connected circuit free subgraph T of G. Since T contains all vertices of G, it is a spanning tree of G.

Theorem 17.29 (Cayley'sFormula) The number of spanning trees of the complete graph K_n, $n \geq 2$ is n^{n-1}.

(Proof of this formula is out of scope of this book)

EXAMPLE 17.81

Find all the spanning trees of K_4 (Figure 17.189).

Solution. According to Cayley's formula, K_4 has $4^{4-2} = 4^2 = 16$ different spanning trees.

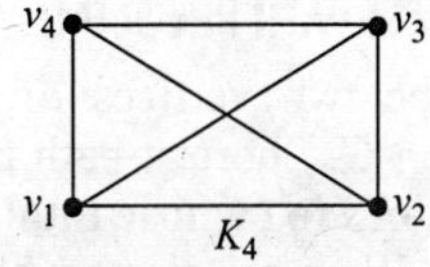

Figure 17.189

Here $n=4$, so the number of edges in any tree should be $n-1=4-1=3$. But here number of edges is equal to 6. So to get a tree, we have to remove three edges of K_4. The 16 spanning trees so obtained are shown in Figure 17.190.

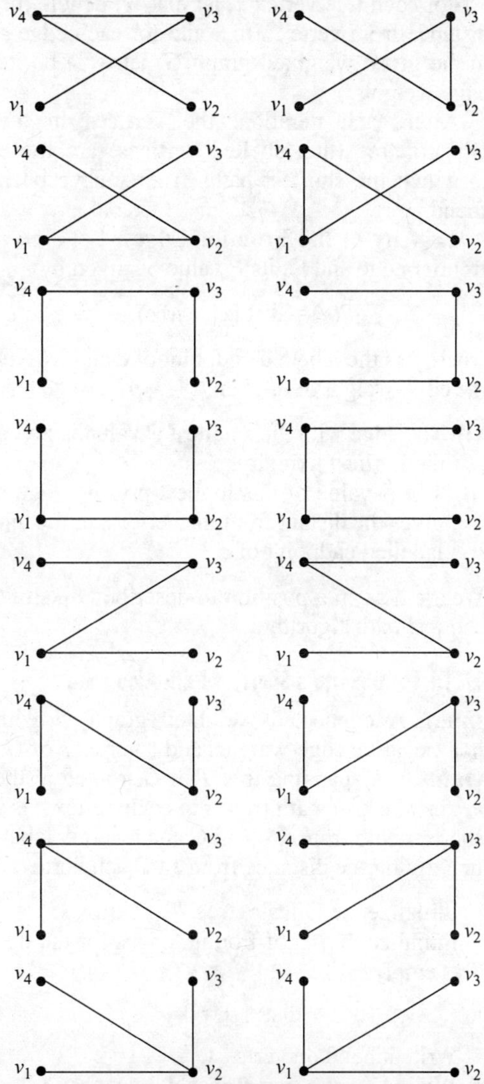

Figure 17.190

17.16 SHORTEST PATH PROBLEM

Let s and t be two vertices of a connected weighted graph G. Shortest path problem is to find a **path from s to t whose total edge weight is minimum**. We now discuss algorithm due to $E \cdot W$. Dijkstra which efficiently solves the shortest path problem. The idea is to grow a Disjkstra tree, starting at the vertex s, by adding, at each iteration, a frontier edge, whose non-tree endpoint is as close as possible to s. The algorithm involves assigning labels to vertices.

For each tree vertex x, let dist $[x]$ denote the distance from vertex s to x and for each edge e in the given weighted graph G, let $w(e)$ be its edge-weight.

After each iteration, the vertices in the Dijkstra tree (the labelled vertices) are those to which the shortest paths from s have been found.

Priority of the Frontier Edges: Let e be a frontier edge and let its P value be given by

$$P(e) = \text{dist } [x] + w(e),$$

where x is the labelled endpoint of e and $w(e)$ is the edge-weight of e. Then,

(i) The edge with the smallest P value is given the **highest priority**.
(ii) The P value of this highest priority edge e gives the distant from the vertex s to the unlabelled endpoint of e.

We are now in a position to describe Dijkstra's shortest path algorithm.

17.16.1 Dijkstra's Shortest Path Algorithm

Input: A connected weighted graph G with non-negative edge-weights and a vertex s of G.
Output: A spanning tree T of G, rooted at the vertex s, whose path from s to each vertex v is a shortest path from s to v in G and a vertex labelling giving the distance from s to each vertex.

Initialize the Dijkstra tree T as vertex s.
Initialize the set of frontier edges for the tree T as empty.

$$\text{dist: } [s] = 0.$$

Write label 0 on vertex s.
While Dijkstra tree T does not yet span G.
For each frontier edge e for T,
Let x be the labelled endpoint of edge e.
Let y be the unlabelled endpoint of edge e.
Set $P(e) = \text{dist } [x] + w(e)$
Let e be a frontier edge for T that has smallest P value
Let x be the labelled endpoint of edge e
Let y be the unlabelled endpoint of edge e
Add edge e (and vertex y) to tree T

$$\text{dist } [y]\text{: } P(e)$$

Write label dist $[y]$ on vertex y.

Return Dijkstra tree T and its vertex labels.

EXAMPLE 17.82

Apply Dijkstra algorithm to find shortest path from s to each other vertex in the graph given in Figure 17.191.

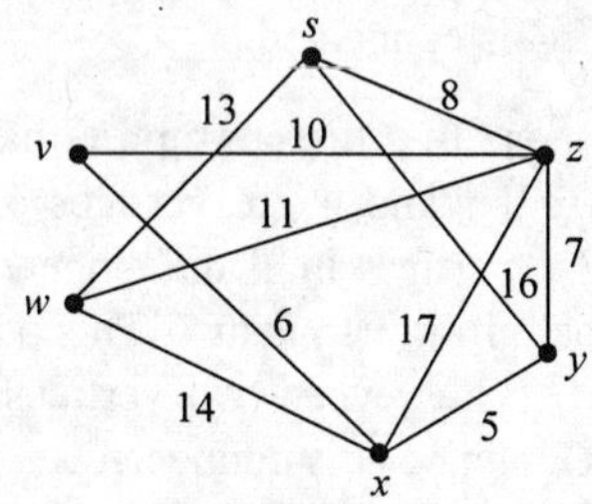

Figure 17.191

If t is the labelled endpoint of edge e, then P values are given by

$$P(e) = \text{dist } [t] + w(e),$$

where dist $[t]$ = distance from s to t and $w(e)$ is the edge weight of edge e. For each vertex v, dist $[v]$ appears in the parenthesis. Iteration tree at the end of each iteration is **drawn in dark line**.

Iteration 1

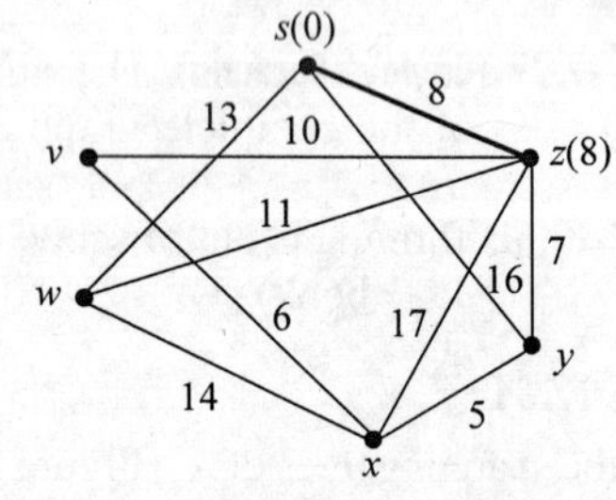

dist $[s] = 0$ $\quad$ $P(sw) = 13$ (minimum)
dist $[z] = 8$ $\quad$ $P(zy) = 8 + 7 = 15$

$P(sy) = 16$
$P(zv) = 8 + 10 = 18$
$P(zw) = 8 + 11 = 19$
$P(zx) = 8 + 17 = 25$

Iteration 2

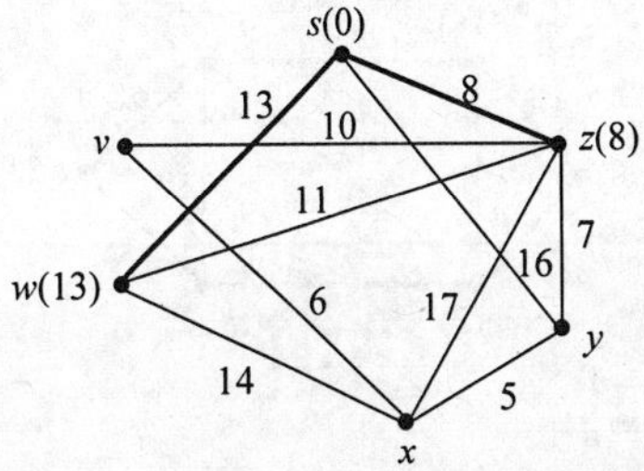

dist [s] = 0
dist [z] = 8
dist [w] = 13

$P(zy) = 8 + 7 = 15$ (minimum)
$P(zx) = 8 + 17 = 25$
$P(zv) = 8 + 10 = 18$
$P(sy) = 16$
$P(wx) = 13 + 14 = 27$

Iteration 3

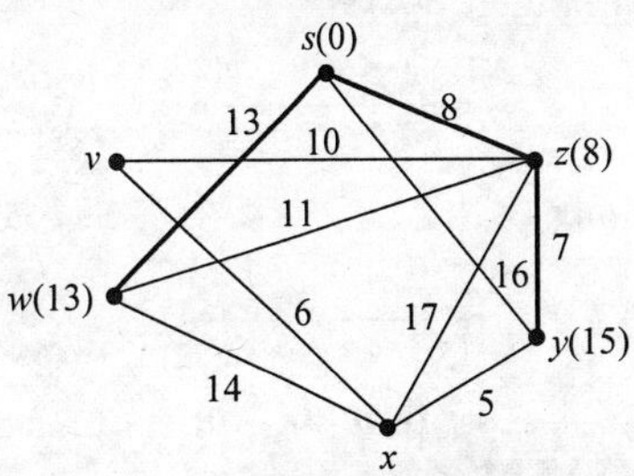

dist [s] = 0
dist [z] = 8
dist [w] = 13
dist [y] = 15

$P(zv) = 18$ (minimum)
$P(zx) = 8 + 17 = 25$
$P(wx) = 13 + 14 = 27$
$P(yx) = 15 + 5 = 20$

Iteration 4

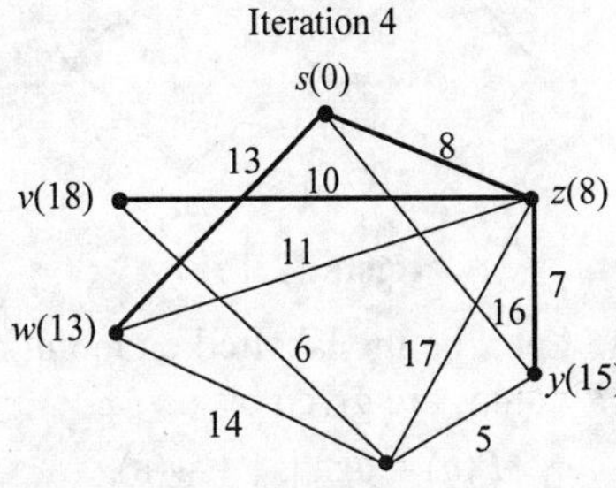

dist [s] = 0
dist [z] = 8
dist [w] = 13
dist [y] = 15
dist [v] = 18

$P(yx) = 20$ (minimum)
$P(zx) = 8 + 17 = 25$
$P(vx) = 18 + 6 = 24$
$P(wx) = 13 + 14 = 27$

Iteration 5

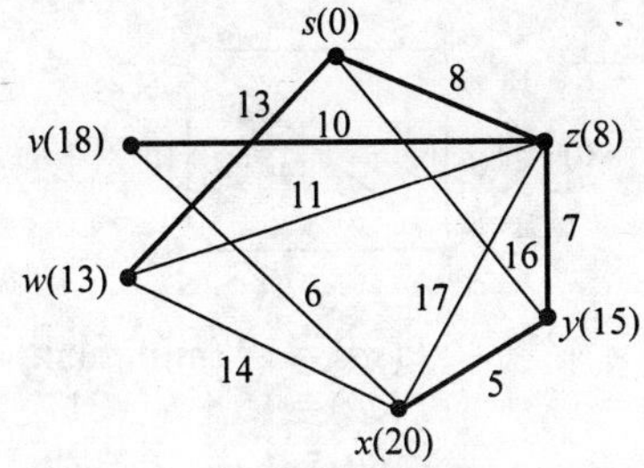

dist [s] = 0
dist [z] = 8
dist [w] = 13
dist [y] = 15
dist [v] = 18
dist [x] = 20,

which are the required shortest paths from s to any other point. The Dijkstra tree is shown in dark lines in Figure 17.192.

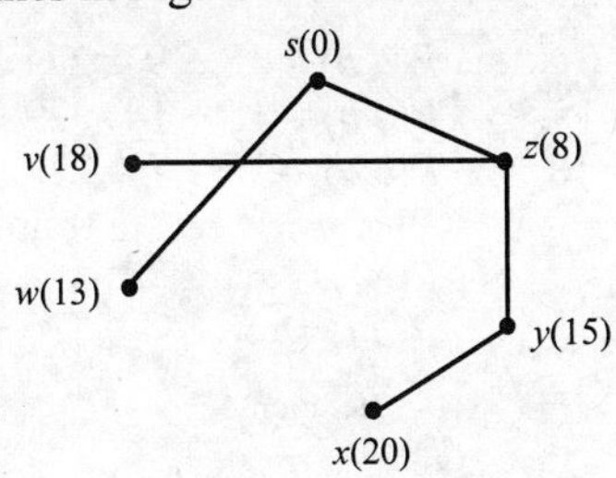

Figure 17.192

EXAMPLE 17.83

Apply Dijkstra algorithm to find the shortest path from s to t in the graph given below (Fig. 17.193).

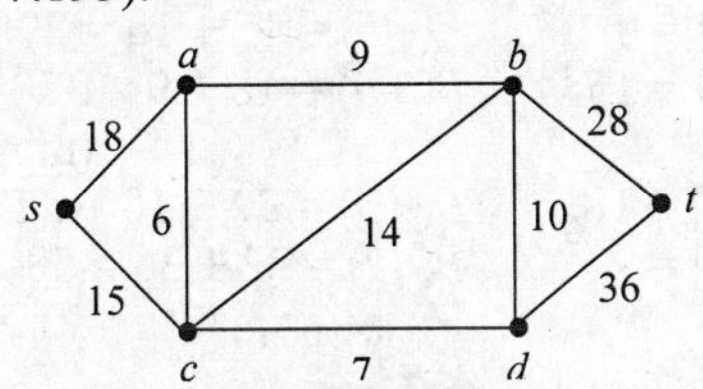

Figure 17.193

Solution. Let x be the labelled endpoint of the edge e, then P values are given by $P(e) = \text{dist}[x] + w(e)$, where dist $[x]$ denotes the distance of x from s and $w(e)$ is the edge weight of e.

For each vertex v, dist $[v]$ appears in the bracket. Iteration tree at the end of each iteration is shown by dark lines:

Iteration 1

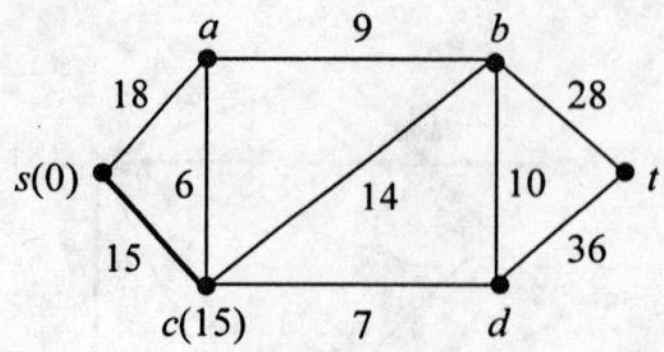

dist $[s] = 0$ $\quad$ $P(sa) = 18$ (minimum)

dist $[c] = 15$ $\quad$ $P(ca) = 15 + 6 = 21$

$P(cb) = 15 + 14 = 29$

$P(cd) = 15 + 7 = 22$

$P(c\,t) = \infty$

Iteration 2

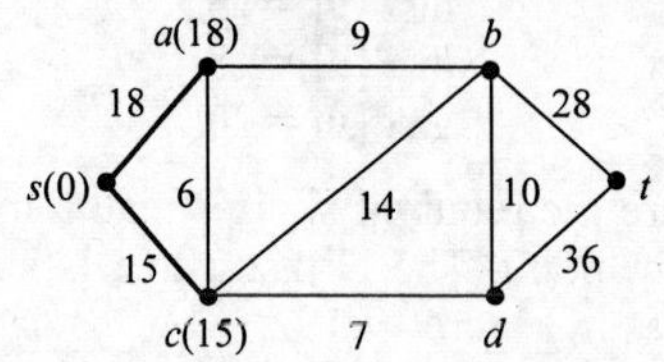

dist $[s] = 0$ $\quad$ $P(a\,b) = 18 + 9 = 27$

dist $[c] = 15$ $\quad$ $P(c\,b) = 15 + 14 = 29$

dist $[a] = 18$ $\quad$ $P(c\,d) = 15 + 7 = 22$ (minimum)

$P(c\,t) = \infty$

$P(a\,t) = \infty$

Iteration 3

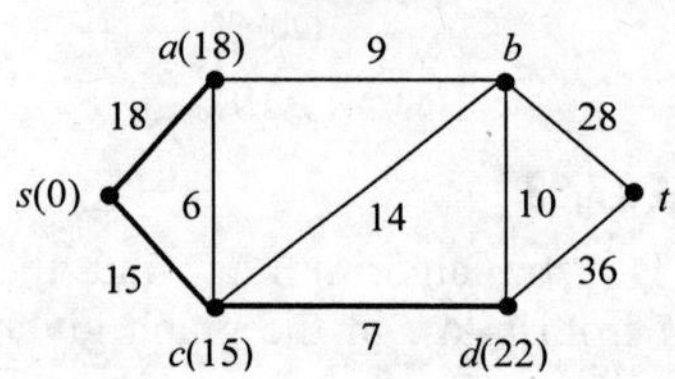

dist $[s] = 0$ $\quad$ $P(c\,b) = 15 + 14 = 29$

dist $[c] = 15$ $\quad$ $P(a\,b) = 18 + 9 = 27$ (minimum)

dist $[a] = 18$ $\quad$ $P(d\,b) = 22 + 10 = 32$

dist $[d] = 22$ $\quad$ $P(d\,t) = 22 + 36 = 58$

Iteration 4

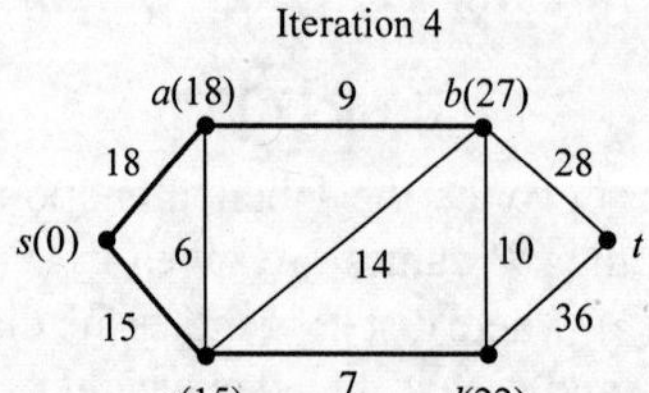

dist $[s] = 0$ $\quad$ $P(d\,t) = 22 + 36 = 58$

dist $[c] = 15$ $\quad$ $P(b\,t) = 27 + 28 = 55$ (minimum)

dist $[a] = 18$

dist $[a] = 18$

dist $[d] = 22$

dist $[b] = 27$

Iteration 5

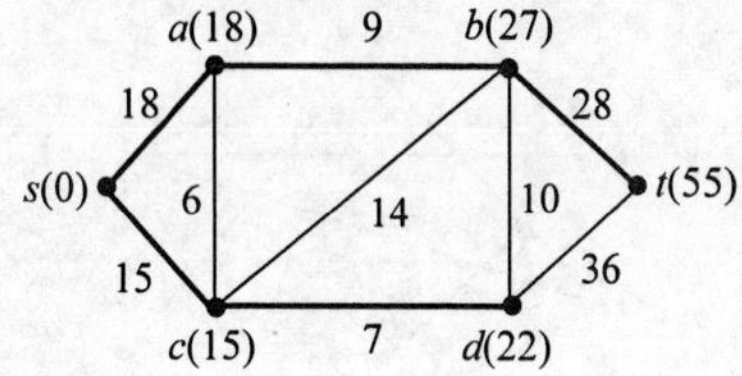

dist $[s] = 0$

dist $[c] = 15$

dist $[a] = 18$

dist $[d] = 22$

dist $[b] = 27$

dist $[t] = 55$

Hence, **the length of the shortest path** (s, a, b, t) **from** s **to** t *is 55* and the Dijkstra's tree is shown in Figure 17.194.

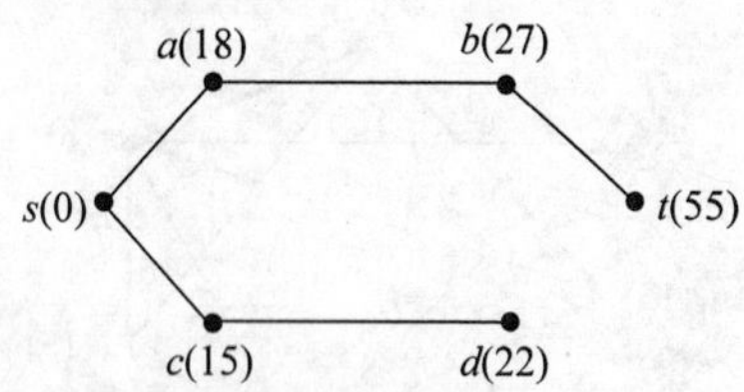

Figure 17.194

EXAMPLE 17.84

Find a shortest path from s to t and its length for the graph given below:

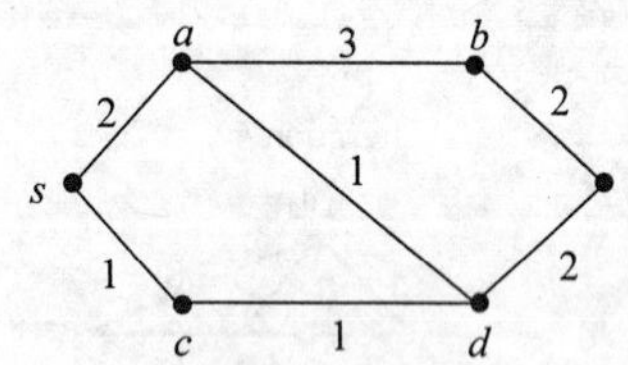

Figure 17.195

Solution. Let x be the labelled endpoint of edge e, then P values are given by

$$P(c) = \text{dist}\,[x] + w(e),$$

where dist $[x]$ denotes the distance from s to x and $w(e)$ is the weight of the edge e.

For each vertex v, dist $[v]$ appears in the bracket. Iteration tree at the end of each iteration is shown in dark lines.

Iteration 1

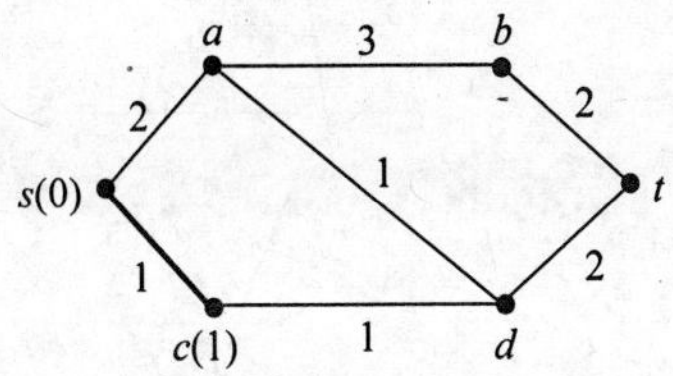

dist $[s] = 0$ $\quad P(c\ d) = 2$
dist $[c] = 1$ $\quad P(s\ a) = 2$ (minimum)
$P(s\ b) = \infty$
$P(s\ t) = \infty$
$P(d\ t) = \infty$
$P(b\ t) = \infty$

Iteration 2

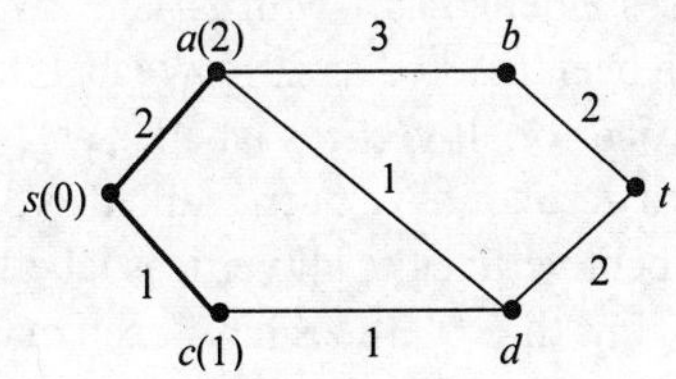

dist $[s] = 0$ $\quad P(a\ b) = 2 + 3 = 5$
dist $[c] = 1$ $\quad P(c\ d) = 1 + 1 = 2$ (minimum)
dist $[a] = 2$ $\quad P(d\ t) = \infty$
$P(b\ t) = \infty$
$P(a\ d) = 3$

Iteration 3

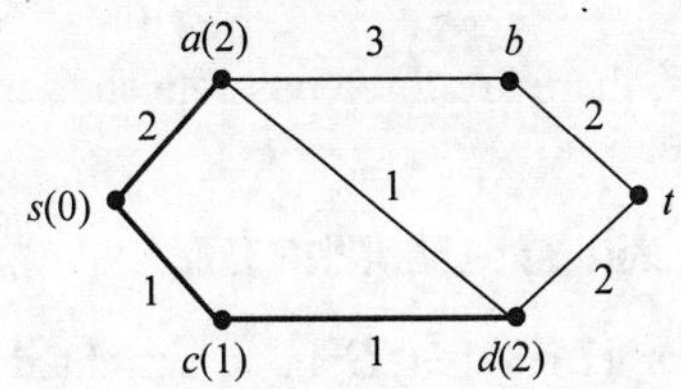

dist $[s] = 0$ $\quad P(d\ t) = 2$ (minimum)
dist $[c] = 1$ $\quad P(a\ b) = 5$
dist $[a] = 2$ $\quad P(b\ t) = \infty$
dist $[d] = 2$

Iteration 4

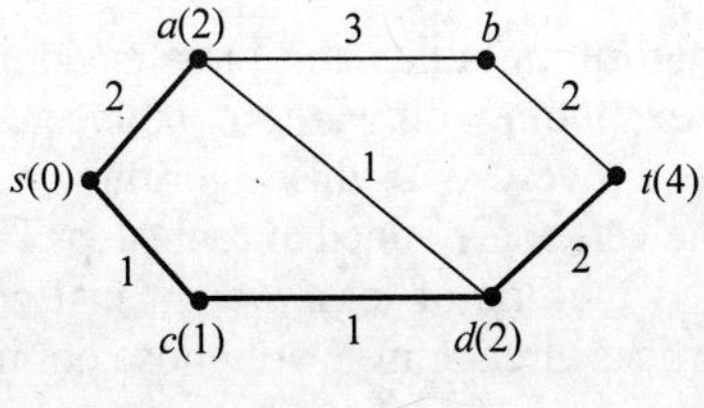

dist $[s] = 0$ $\quad P(a\ b) = 5$ (minimum)
dist $[c] = 1$ $\quad P(b\ t) = \infty$
dist $[c] = 1$
dist $[a] = 2$
dist $[d] = 2$
dist $[t] = 4$

Iteration 5

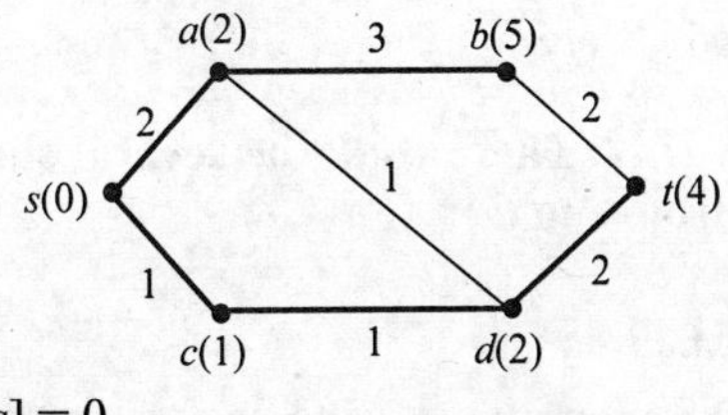

dist $[s] = 0$
dist $[c] = 1$
dist $[a] = 2$
dist $[d] = 2$
dist $[t] = 4$
dist $[b] = 5$

Thus, the Dijkstra tree is

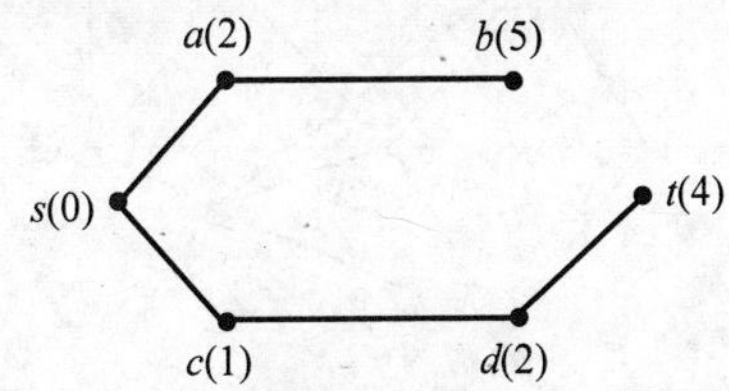

Figure 17.196

Thus the shortest path is *scdt* and its length is 4.

17.16.2 Shortest Path if All Edges Have Length 1

If all edges in a connected graph G have length 1, then a **shortest path** $v_1 \to v_k$ is the path that has the smallest number of edges among all paths $v_1 \to v_k$ in the given graph G.

Moore's Breadth First Search Algorithm

This method of finding shortest path in a connected graph G from a vertex s to a vertex t is used when all edges have length 1.

Input: Connected graph $G = (V, E)$, in which one vertex is denoted by s and one by t and each edge (v_i, v_j) has length 1.
Initially all vertices are unlabelled.

Output: A shortest path $s \to t$ in $G = (V, E)$.

1. Label s with 0.
2. Set $v_i = 0$.
3. Find all unlabelled vertices **adjacent to a vertex labelled** v_i.

4. Label the vertices just found with v_{i+1}.
5. If vertex t is labelled, then **"back tracking"** gives the shortest path. If k is label of t(i.e., $t = v_k$), then

Output: $v_k, v_{k-1}, \ldots, v_1, 0$.

Else increase i by 1. Go to Step 3.
End Moore.

Remark 17.17 There could be several shortest paths from *s* to *t*.

EXAMPLE 17.85

Use BFS algorithm to find shortest path from s to t in the connected graph G given in Figure 17.197.

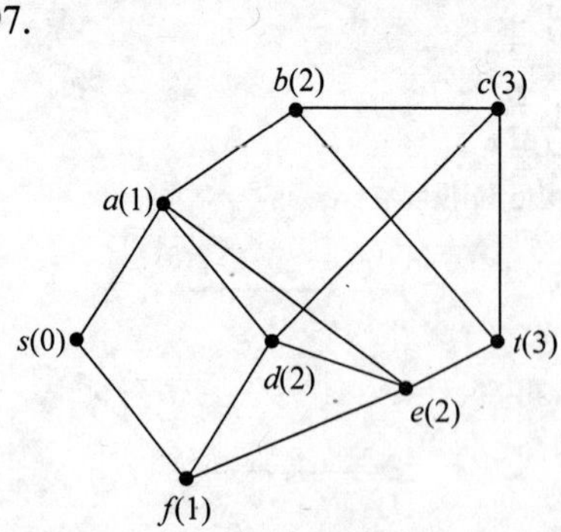

Figure 17.197

Solution. Label s with 0 and then label the adjacent vertices with 1. Thus two vertices have been labelled by 1. Now label the adjacent vertices of all vertices labelled by 1 with label 2. Thus three vertices have been labelled with 2. Label the vertices adjacent to these vertices (labelled by 2) with 3. Thus two vertices have been labelled with 3. We have reached t. Now backtracking yields the following three shortest paths:

(i) $t(3), e(2), f(1), s(0)$, that is, $s\,f\,e\,t$,
(ii) $t(3), b(2), a(1), s(0)$, that is $s\,a\,b\,t$,
(iii) $t(3), e(2), a(1), s(0)$, that is, $s\,a\,e\,t$.

Thus there are three possible shortest paths of length 3.

EXAMPLE 17.86

Find a shortest path $s \rightarrow t$ in the graph given in Figure 17.198 (using Moore's BFS algorithm).

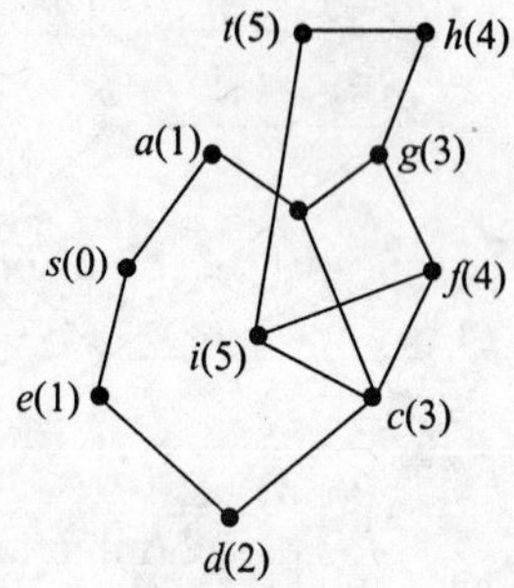

Figure 17.198

Solution. Label s with 0 and then label the adjacent vertices with 1. Thus two vertices have been marked as 1. Label the adjacent vertices of al vertices labelled as 1 with 2. Thus two vertices have been marked as 2. Now label the adjacent vertices of all vertices labelled as 2 with 3. Thus two vertices have been marked as 3. Label the adjacent vertices of all vertices labelled as 3 with 4. Thus three vertices have been marked as 4. Now there is only one vertex t to be labelled. Label it with 5. We have reached at t.

Now backtracking gives the following shortest paths:

$t\,h\,g\,b\,a\,s$ or $s\,a\,b\,g\,h\,t$ with length 5,
$t\,i\,c\,d\,e\,s$ or $s\,e\,b\,c\,i\,t$ of length 5,
$t\,i\,c\,b\,a\,s$ or $s\,a\,b\,c\,i\,t$ of length 5.

Thus, there are three possible shortest paths of length 5.

17.17 MINIMAL SPANNING TREE

Definition 17.113 Let G be a weighted graph. A spanning tree of G with minimum weight is called **minimal spanning tree of** G.

We discuss two algorithms to find a minimal spanning tree for a weighted graph G.

17.17.1 Prim Algorithm

Prim algorithm builds a minimal spanning tree T by expanding outward in connected links from some vertex. In this algorithm, one edge and one vertex are added at each step. The edge added is the one of least weight that connects the vertices already in T with those not in T.

Input: A connected weighted graph G with n vertices.
Output: The set of edges E in a minimal spanning tree.

1. Choose a vertex v_1 of G. Let $V = \{v_1\}$ and $E = \{\ \}$.
2. Choose a nearest neighbour v_i of V that is adjacent to v_j, $v_j \in V$ and for which the edge (v_i, v_j) does not form a cycle with members of E. Add v_i to V and add (v_i, v_j) to E.
3. Repeat Step 2 till number of edges in T is $n-1$. Then V contains all n vertices of G and E contains the edges of a minimal spanning tree for G.

Definition 17.114 A greedy algorithm is an algorithm that optimizes the choice at each iteration without regard to previous choices.

For example, Prim algorithm is a greedy algorithm.

EXAMPLE 17.87

Find a minimal spanning tree for the graph shown in Figure 17.199.

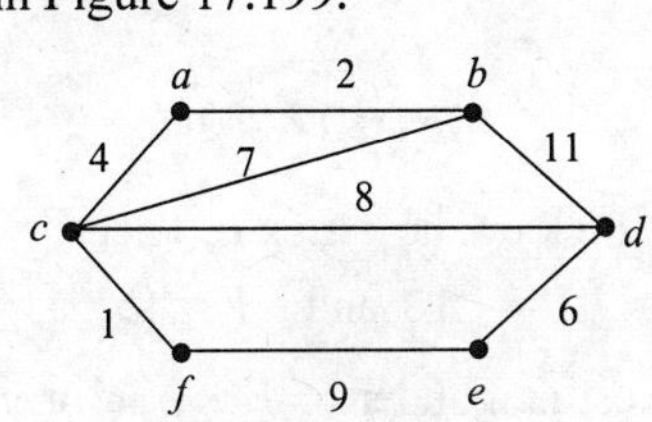

Figure 17.199

Solution. We shall use **Prim algorithm to find the required minimal spanning tree**. We note that number of vertices in this connected weighted graph is 6. Therefore the tree will have five edges.

We start with any vertex, say c. The nearest neighbour of c is f and (c, f) does not form a cycle. Therefore (c, f) is the first edge selected.

Now we consider the set of vertices $V = \{c, f\}$. The vertex a is nearest neighbour to $V = \{c, f\}$ and the edge (c, a) does not form a cycle with the member of set of edges selected so far. Thus,

$$E = \{(c, f), (c, a)\} \quad \text{and} \quad V = \{c, f, a\}.$$

The vertex b is now nearest neighbour to $V = \{c, f, a\}$ and the edge (a, b) do not form a cycle with the member of $E = \{(c, f), (c, a)\}$. Thus,

$$E = \{(c, f), (c, a), (a, b)\} \quad \text{and} \quad V = \{c, f, a, b\}.$$

Now the edge (b, c) cannot be selected because it forms a cycle with the members of E. We note that d is the nearest point to $V = \{c, f, a, b\}$ and (c, d) is the edge which does not form a cycle with members of $E = \{(c, f), (c, a), (a, b)\}$. Thus we get

$$E = \{(c, f), (c, a), (a, b)\}, (c, d)\},$$
$$V = \{c, f, a, b, d\}.$$

The nearest vertex to V is now e and (d, e) is the corresponding edge. Thus,

$$E = \{(c, f), (c, a), (a, b), (c, d), (d, e)\},$$
$$V = \{c, f, a, b, d, e\}$$

Since number of edges in the Prim tree is 5, the process is complete. The minimal spanning tree is shown in Figure 17.200:

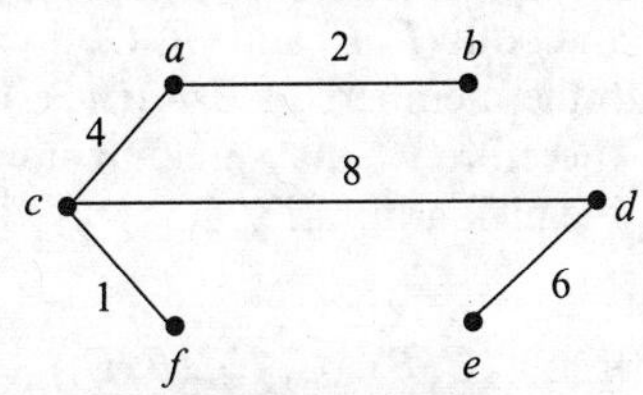

Figure 17.200

The length of the tree is $1 + 4 + 2 + 8 + 6 = 21$

EXAMPLE 17.88

Build the lowest-cost road system that will connect the cities in the graph given in Figure 17.201.

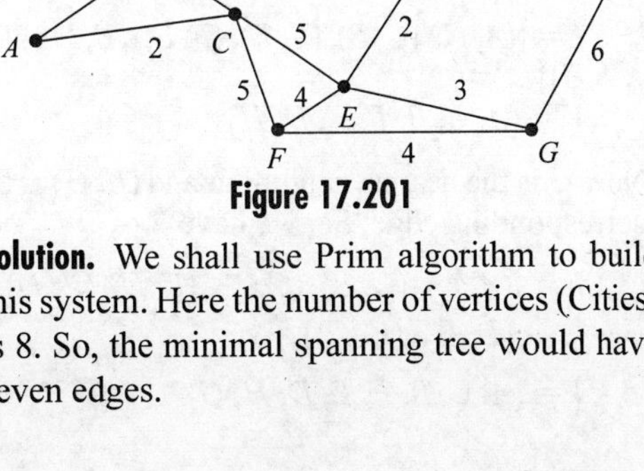

Figure 17.201

Solution. We shall use Prim algorithm to build this system. Here the number of vertices (Cities) is 8. So, the minimal spanning tree would have seven edges.

We start from any vertex, say A and have at the initial stage:

$$V=\{A\},\quad E=\{\ \}.$$

The nearest neighbour of $V=\{A\}$ is C and the corresponding edge is (A, C). So we have

$$E=\{(A, C)\},\quad V=\{A, C\}.$$

Now B is the nearest neighbour of $V=\{A, C\}$ and edge (C, B) does not form cycle with the edges in $E=\{(A, C)\}$. Here the edge (A, B) can also be chosen. Thus, choosing arbitrarily one out of (A, B) and (C, B), we have

$$E=\{(A, C), (C, B)\},\quad V=\{A, C, B\}.$$

Now, E and F are both nearest neighbour to V and (C, E) or (C, F) do not form cycle with the members of $E=\{(A, C), (C, B)\}$. So we may choose any of E or F. Let us choose F and the corresponding edge (C, F). Then,

$$E=\{(A, C), (C, B), (C, F)\},\quad V=\{A, C, B, F\}.$$

Now E and G are nearest neighbours of $V=\{A, C, B, F\}$ and both (F, E) and (F, G) do not form cycle with the members of $E=\{(A, C), (C, B), (C, F)\}$. Therefore we may pick up any of these vertices. We pick arbitrarily E.
Then,

$$E=\{(A, C), (C, B), (C, F), (F, E)\},$$
$$V=\{A, C, B, F, E\}.$$

Now D is the nearest neighbour of $V=\{A, C, B, F, E\}$ and (E, D) is the corresponding edge to be added to the tree. So, we have

$$E=\{(A, C), (C, B), (C, F), (F, E), (E, D)\},$$
$$V=\{A, C, B, F, E, D\}.$$

Now H is the nearest neighbour and (D, H) is the corresponding edge and we have

$$E=\{(A, C), (C, B), (C, F), (F, E), (E, D), (D, H)\},$$
$$V=\{A, C, B, F, E, D, H\}.$$

Now G is the nearest neighbour and (E, G) is the corresponding edge. So, we have

$$E=\{(A, C), (C, B), (C, F), (F, E), (E, D), (D, H), (E, G)\},$$
$$V=\{A, C, B, F, E, D, H, G\}$$

Since the number of edges in the Prim tree is 7, the process is complete. The minimal spanning tree is shown in Figure 17.202:

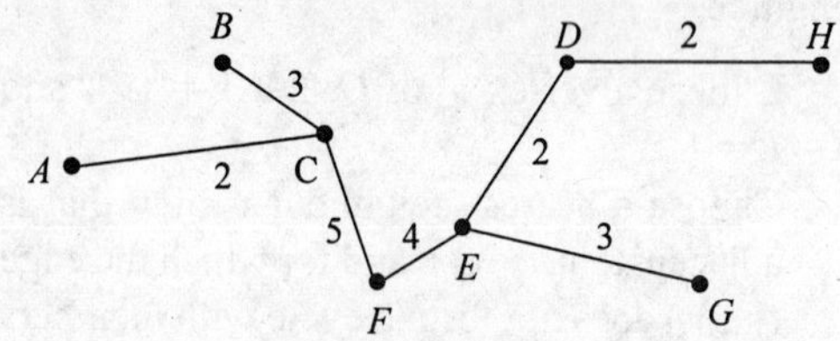

Figure 17.202

The total length of the roads is

$$2+3+5+4+2+2+3=21.$$

EXAMPLE 17.89

Using Prim algorithm, find the minimal spanning tree of the following graph (Figure 17.203):

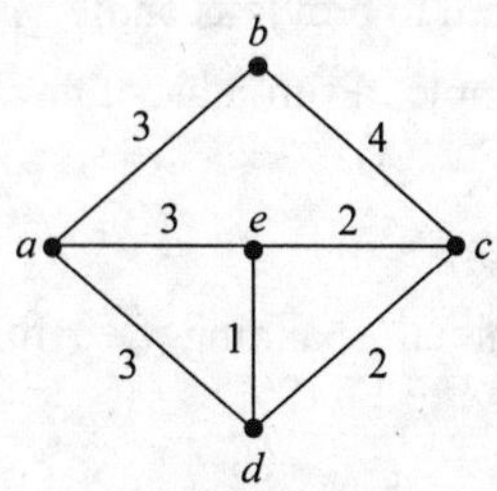

Figure 17.203

Solution. Pick up the vertex a. Then

$$E=\{\ \}\quad\text{and}\quad V=\{a\}.$$

The nearest neighbour of V is b, d or e and the corresponding edges are (a, b) or (a, d) or (a, e). We choose arbitrarily (a, b) and have

$$E=\{(a, b)\},\quad V=\{a, b\}.$$

Now d is the nearest neighbour of $V=\{a, b\}$ and the corresponding edge (a, d) does not form cycle with (a, b). Thus we get

$$E=\{(a, b), (a, d)\},\quad V=\{a, b, d\}.$$

Now e is the nearest neighbour of $\{a, b, d\}$ and (d, e) does not form cycle with $\{(a, b), (a, d)\}$.
Hence,

$$E=\{(a, b), (a, d), (d, e)\},\quad V=\{a, b, d, e\}.$$

Now e is the nearest neighbour of $V=\{a, b, d, e\}$ and the corresponding edges are (e, c), (d, c).

Thus we have, choosing (e, c),

$$E = \{(a, b), (a, d), (d, e), (e, c)\}, \quad V = \{a, b, d, e, c\}$$

Total weight $= 3 + 3 + 1 + 2 = 9$.

(If we choose (d, c), then total weight is $3 + 3 + 1 + 2 = 9$).

The minimal tree is shown in Figure 17.204.

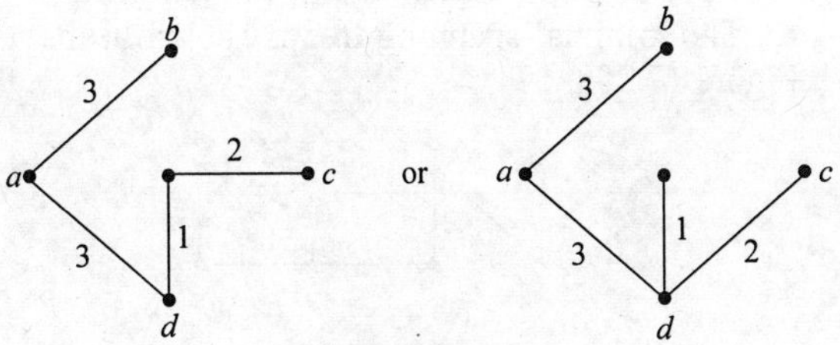

Figure 17.204

17.17.2 Kruskal's Algorithm

In Kruskal's algorithm, the edges of a weighted graph are examined one by one in order of increasing weight. At each stage, an edge with least weight out of edge-set remaining at that stage is added provided this additional edge does not create a circuit with the members of existing edge set at that stage. After $n-1$ edges have been added, these edges together with the n vertices of the connected weighted graph form a minimal tree.

Algorithm

Input: A connected weighted graph G with n vertices and the set $E = \{e_1, e_2, \ldots, e_k\}$ of weighted edges of G.

Output: The set of edges in a minimal spanning tree T for G.

Step 1. Initialize T to have all vertices of G and no edges.

Step 2. Choose an edge e_1 in E of least weight. Let

$$E^* = \{e_1\}, \quad E = E - \{e_1\}.$$

Step 3. Select an edge e_i in E of least weight that does not form circuit with members of E^*. Replace

$$E^* \text{ by } E^* \cup \{e_i\} \quad \text{and} \quad E \text{ with } E - \{e_i\}.$$

Step 4. Repeat Step 3 until number of edges in E^* is equal to $n-1$.

EXAMPLE 17.90

Use Kruskal's algorithm to determine a minimal spanning tree for the connected weighted graph G shown in Figure 17.205.

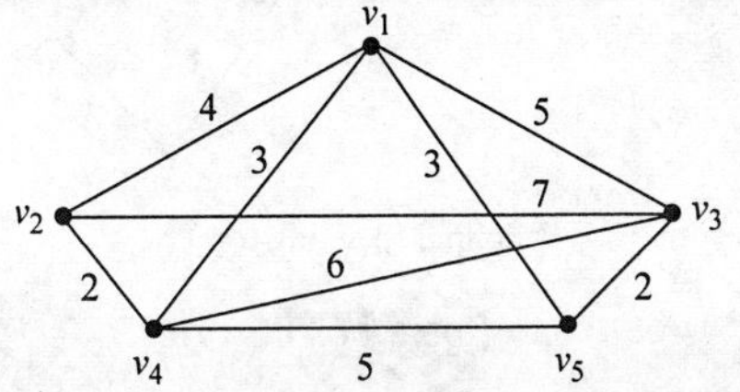

Figure 17.205

Solution. The given weighted graph has five vertices. The minimal spanning tree would, therefore, have four edges.

Let

$$E = \{(v_1, v_2), (v_1, v_4), (v_1, v_5), (v_2, v_3), (v_1, v_3), (v_2, v_4), (v_4, v_5), (v_5, v_3), (v_3, v_4)\}.$$

The edges (v_2, v_4) and (v_3, v_5) have minimum weight. We choose arbitrarily one of these, say (v_2, v_4).

Thus

$$E^* = \{(v_2, v_4)\},$$
$$E = E - \{(v_2, v_4)\}.$$

The edge (v_3, v_5) has minimum weight, so we pick it up. We have thus

$$E^* = \{(v_2, v_4), (v_3, v_5)\},$$
$$E = E - \{(v_2, v_4), (v_3, v_5)\}.$$

The edges (v_1, v_4) and (v_1, v_5) have minimum weight in the remaining edge set. We pick (v_1, v_4) say, as it does not form a cycle with E^*. Thus

$$E^* = \{(v_2, v_4), (v_3, v_5), (v_1, v_4),$$
$$E = E - \{(v_2, v_4), (v_3, v_5), (v_1, v_4)\}.$$

Now the edge (v_1, v_5) has minimum weight in $E - \{(v_2, v_4), (v_3, v_5), (v_1, v_4)\}$ and it does not form a cycle with E^*. So, we have

$$E^* = \{(v_2, v_4), (v_3, v_5), (v_1, v_4), (v_1, v_5)\}$$

and

$$E = E - \{(v_2, v_4), (v_3, v_5), (v_1, v_4), (v_1, v_5)\}.$$

Thus all the four edges have been selected. The minimal tree has the edges.

$$(v_2, v_4), (v_3, v_5), (v_1, v_4), (v_1, v_5)$$

and is shown in the Figure 17.206.

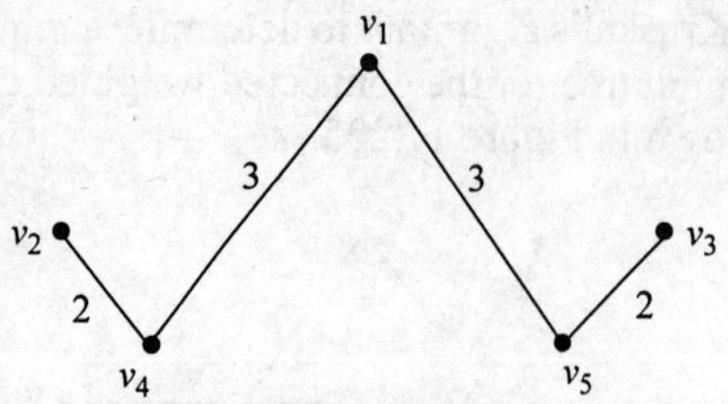

(Minimal spanning tree)

Figure 17.206

EXAMPLE 17.91

Using Kruskal's algorithm, find the minimal spanning tree for the graph given in the Figure 17.207.

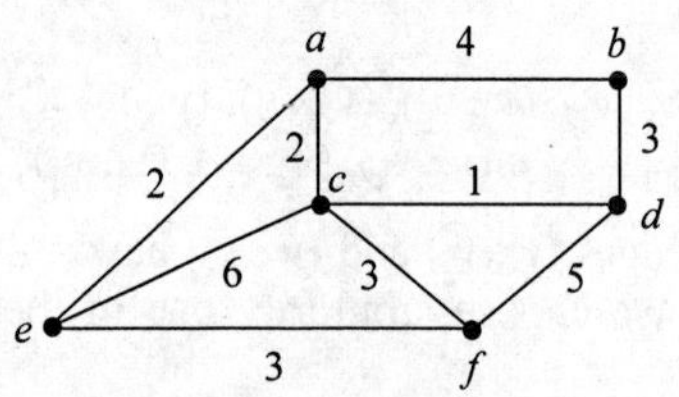

Figure 17.207

Solution. The given graph has six vertices, therefore the minimal spanning tree would have $6-1=5$ edges

Let

$$E=\{(a,b),(a,c),(b,d),(c,f),(d,f),(c,d),\\ (a,e),(e,f),(e,c)\}.$$

Choose the edge (c, d) first because it has minimum weight. Therefore,

$$E^*=\{(c,d)\},\quad E=E-\{(c,d)\}.$$

Now choose the edge (a, c) because it has minimum weight. So,

$$E^*=\{(c,d),(a,c),\quad E=E-\{(c,d),(a,c)\}.$$

Now choose the edge (a, e), since it has also minimum weight in $E-\{(c,d),(a,c)\}$ and it does not form cycle with E^*. So

$$E^*=\{(c,d),(a,c),(a,e)\},\\ E=E-\{(c,d),(a,c),(a,e)\}.$$

Now choose (b, d) since it has minimum weight and it does not form circuit with E^*. Thus

$$E^*=\{(c,d),(a,c),(a,e),(b,d)\},\\ E=E-\{(c,d),(a,c),(a,e),(b,d)\}.$$

Now choose (c,f) or (e,f) because they have equal and minimal weight. Let us choose (c,f). Then

$$E^*=\{(c,d),(a,c),(a,e),(b,d),(c,f)\},\\ E=E-\{(c,d),(a,c),(a,e),(b,d),(c,f)\}.$$

We have thus obtained all the five required edges. The minimal spanning tree is shown in the Figure 17.208.

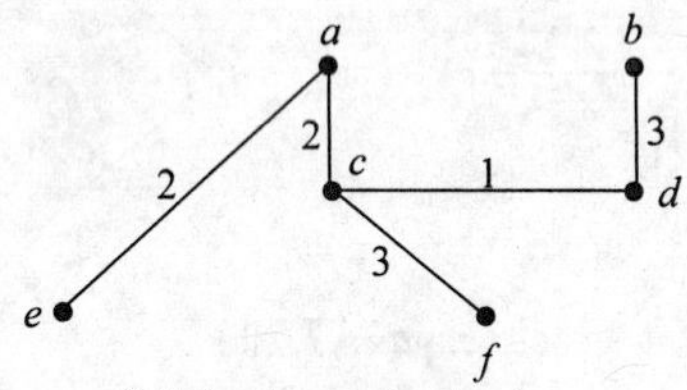

(Minimal spanning tree)

Figure 17.208

Remark 17.18 In the above example, if we had chosen (e,f) in place of (c,f) in the last step, then the minimal spanning tree would have been as shown in the Figure 17.209.

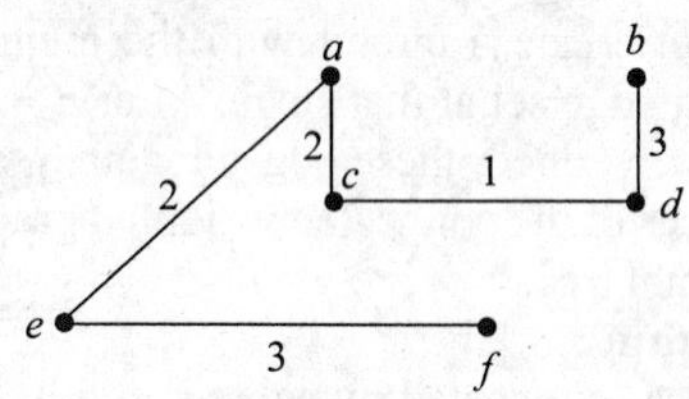

(Minimal spanning tree)

Figure 17.209

EXAMPLE 17.92

Use Kruskal's algorithm to find a shortest spanning tree for the graph G given in the Figure 17.210.

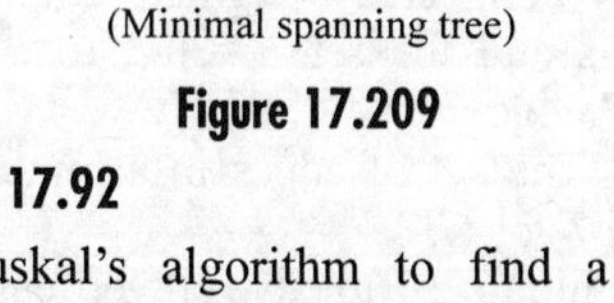

Figure 17.210

Solution. The given graph has five vertices. So the required spanning tree would have $5-1=4$ edges. Let

$E = \{(a_1, a_2), (a_1, a_3), (a_3, a_4), (a_2, a_4), (a_2, a_5), (a_4, a_5), (a_2, a_3), (a_1, a_4)\}$.

Choose the edge (a_2, a_3) since it has minimum weight. Then

$$E^* = \{(a_2, a_3)\},$$
$$E = E - \{a_2, a_3\} = \{(a_1, a_2), (a_1, a_3), (a_3, a_4), (a_2, a_4), (a_2, a_5), (a_4, a_5), (a_1, a_4)\}.$$

Now choose the edge (a_2, a_4) or (a_3, a_4) having minimum weight in E. Let us take (a_2, a_4), say. Then

$$E^* = \{(a_2, a_3), (a_2, a_4)\}$$

and

$$E = E - \{(a_2, a_3), (a_2, a_4)\} = \{(a_1, a_2), (a_1, a_3), (a_3, a_4), (a_2, a_5), (a_4, a_5), (a_1, a_4)\}.$$

We cannot choose (a_3, a_4) now because it will form circuit with already chosen edges (a_2, a_3), (a_2, a_4).
Therefore we can choose (a_1, a_3) or (a_4, a_5). Suppose we choose (a_1, a_3). Then

$$E^* = \{(a_2, a_3), (a_2, a_4), (a_1, a_3)\},$$
$$E = E - \{(a_2, a_3), (a_2, a_4), (a_1, a_3)\} = \{(a_1, a_2), (a_3, a_4), (a_2, a_5), (a_4, a_5), (a_1, a_4)\}.$$

Now choose (a_4, a_5) as it has minimal weight in E. Then

$$E^* = \{(a_2, a_3), (a_2, a_4), (a_1, a_3), (a_4, a_5)\}$$

and

$$E = E - \{(a_2, a_3), (a_2, a_4), (a_1, a_3), (a_4, a_5)\} = \{(a_1, a_2), (a_3, a_4), (a_2, a_5), (a_1, a_4)\}.$$

Thus we have obtained four edges and therefore the process is complete.

The Kruskal minimal spanning tree is shown in the Figure 17.211.

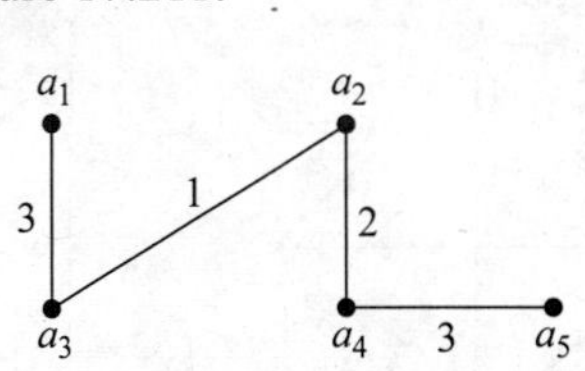

Figure 17.211

Its length is $3 + 1 + 2 + 3 = 9$.

17.18 CUT SETS

Let G be a connected graph. We know that the distance between two vertices v_1 and v_2, denoted by $d(v_1,v_2)$, is the length of the shortest path.

Definition 17.115 The diameter of a connected graph G, denoted by diam (G), is the maximum distance between any two vertices in G.

For example, in graph G shown in the Figure 17.212, we have

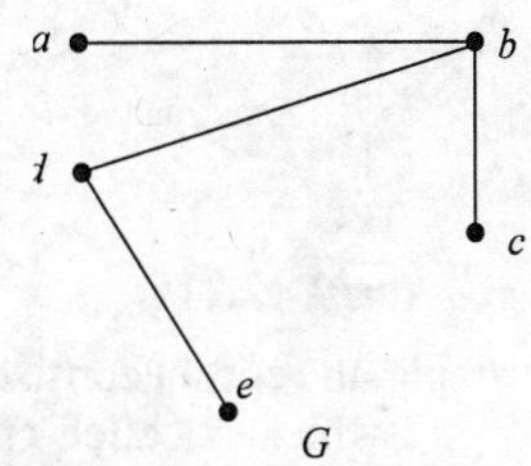

Figure 17.212

$d(a,e) = 3$, $d(a,c) = 2$, $d(b,e) = 2$ and diam $(G) = 3$.

Definition 17.116 A vertex v in a connected graph G is called a **cut point** if $G - v$ is disconnected, where $G - v$ is the graph obtained from G by deleting v and all edges containing v.

For example, in the above graph, d is a cut point.

Definition 17.117 An edge e of a connected graph G is called a **bridge** (or **cut edge**) if $G - e$ is disconnected, where $G - e$ is the graph obtained by deleting the edge e.

For example, consider the graph G shown in the Figure 17.213.

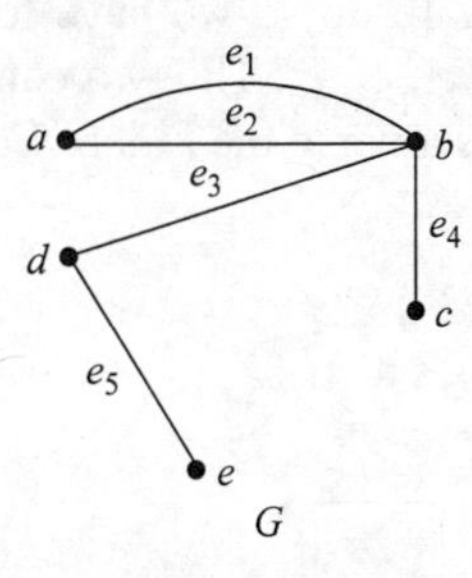

Figure 17.213

We observe that $G - e_3$ is disconnected. Hence the edge e_3 is a bridge.

Definition 17.118 A minimal set C of edges in a connected graph G is said to be a **cut set** (or **minimal edge-cut**) if the subgraph $G-C$ has more connected components than G has.

For example, in the above graph, if we delete the edge $(b,d) = e_3$, the resulting subgraph $G - e_3$ has two connected components shown in the Figure 17.214.

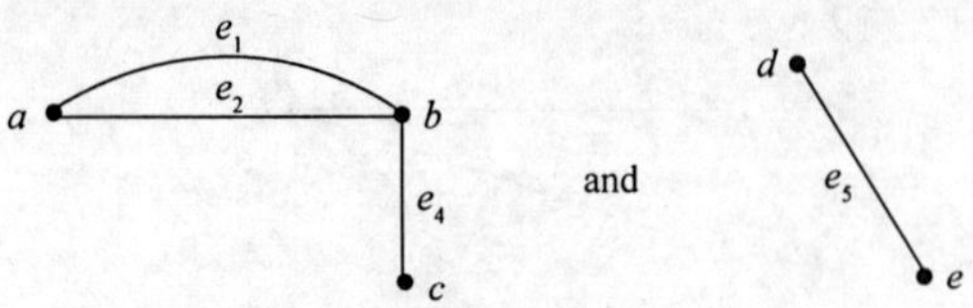

Figure 17.214

So, in this example, the cut set consists of single edge $(b, d) = e_3$, which is called **cut edge** or **bridge**.

EXAMPLE 17.93

Find a cut set for the graph given below in the Figure 17.215.

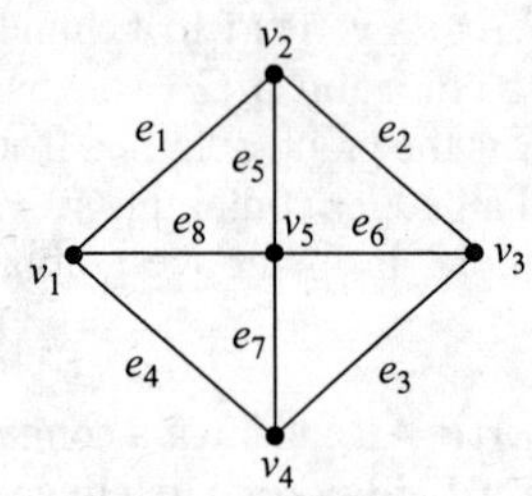

Figure 17.215

Solution. The given graph is connected. It is sufficient to reduce the graph into two connected components. To do so we have to remove the edges e_1, e_4, e_5, e_6, e_7. The two connected components are shown in the Figure 17.216.

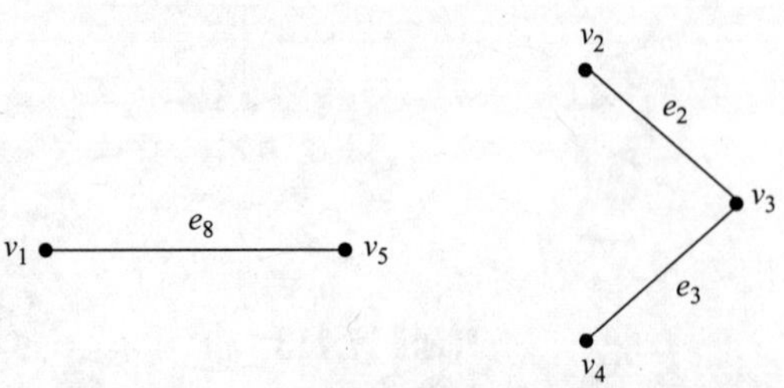

Figure 17.216

But, if we remove any proper subset of $\{e_1, e_4, e_5, e_6, e_7\}$, then there is no increase in connected components of G. Hence $\{e_1, e_4, e_5, e_6, e_7\}$ is a cut set.

EXAMPLE 17.94

Find a cut set for the graph given in the Figure 17.217.

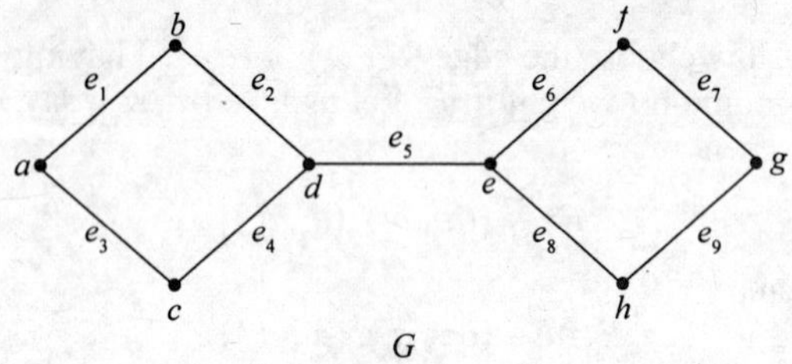

Figure 17.217

The given graph G is connected. Clearly $G - e_5$ has two connected components given in the Figure 17.218.

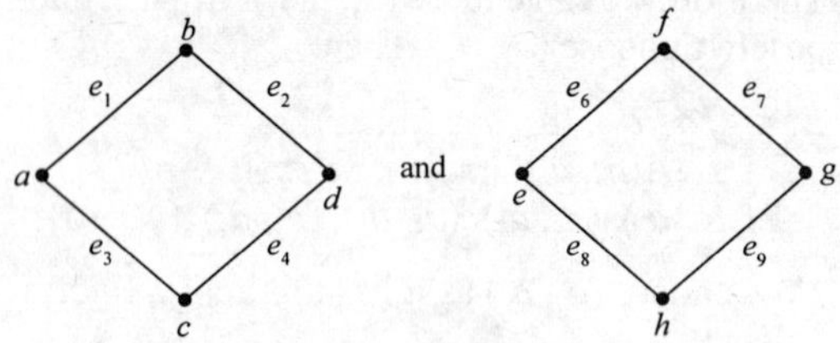

Figure 17.218

Hence a cut set for this graph is $\{e_5\}$. Some other cut sets for G are $\{e_1, e_3\}$, $\{e_3, e_2\}$, $\{e_6, e_9\}$, $\{e_6, e_8\}$.

EXAMPLE 17.95

Find a cut set for the graph represented by the Figure 17.219.

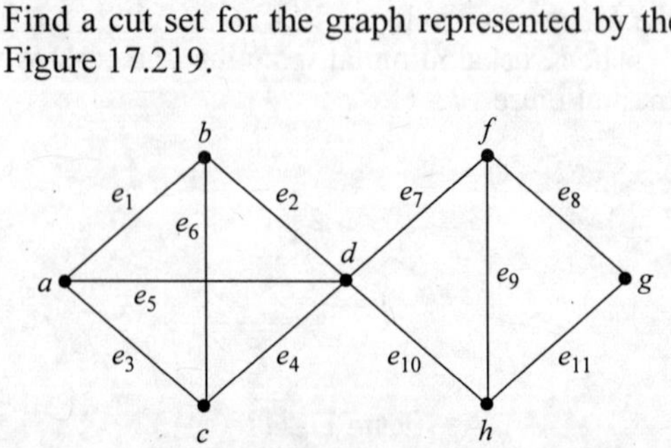

Figure 17.219

Solution. The given graph is a connected graph. We note that removal of the edges e_7 and e_{10} creates two connected components of G shown in Figure 17.220.

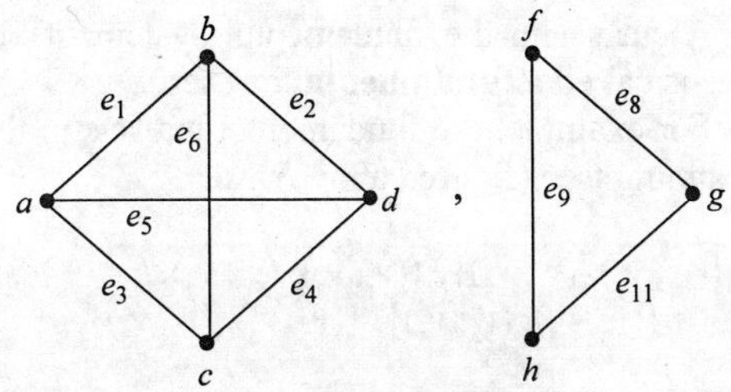

Figure 17.220

Hence the set $\{e_7, e_{10}\}$ is a cut set for the given graph G.

EXAMPLE 17.96

Find a cut set (minimal edge cut) for the following graph (Figure 17.221).

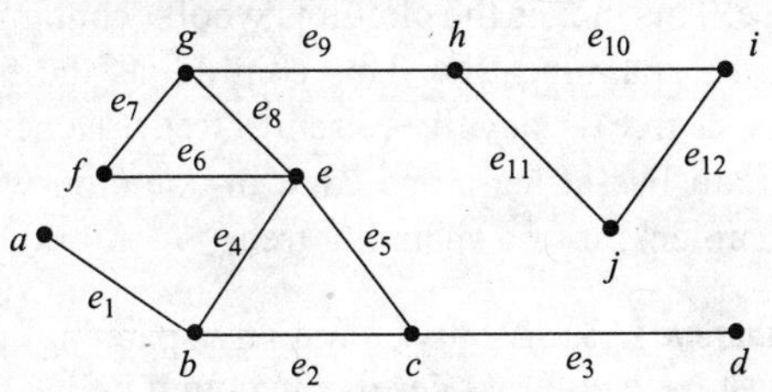

Figure 17.221

Solution. We note that $\{e_4, e_5\}$ is a cut set (minimal edge cut) for the given graph because removal of e_4 and e_5 gives rise to two components shown in the Figure 17.222.

and

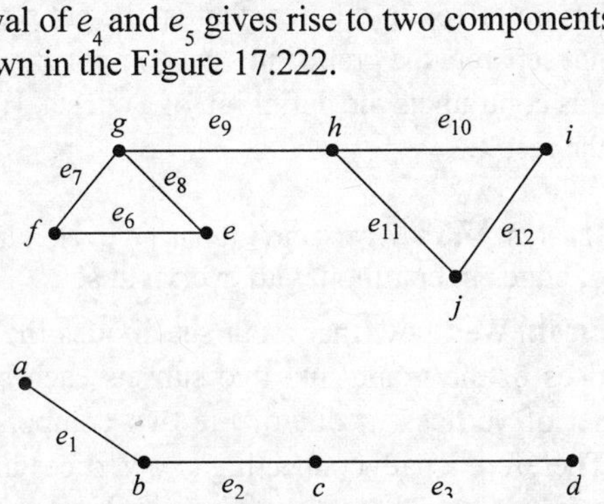

Figure 17.222

Similarly, $\{e_9\}$ forms a cut set (bridge) for this graph.

Theorem 17.30 Let G be a connected graph with n vertices. Then G is a tree if and only if every edge of G is a bridge (cut edge).

(This theorem asserts that every edge in a tree is a bridge).

Proof: Let G be a tree. Then it is connected and has $n - 1$ edges (proved already). Let e be an arbitrary edge of G. Since $G-e$ has $n-2$ edges, and also we know that a graph G with n vertices has at least $n-\text{C}(G)$ edges, it follows that $n-2 \geq n-\text{C}(G-e)$. Thus, $G-e$ has at least two components. Thus removal of the edge e created more components than in the graph G. Hence e is a cut edge. This proves that every edge in a tree is a bridge.

Conversely, suppose that G is connected and every edge of G is a bridge. We have to show that G is a tree. To prove it, we have only to show that G is circuit-free. Suppose on the contrary that there exists a cycle between two points x and y in G (Figure 17.223). Then any edge on this cycle is not a cut edge which contradicts the fact that every edge of G is a cut edge. Hence G has no cycle. Thus G is connected and acyclic and so is a tree.

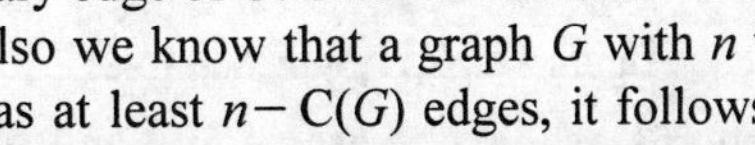

Figure 17.223

17.18.1 Relation Between Spanning Trees, Circuits and Cut Sets

A spanning tree contains a unique path between any two vertices in the graph. Therefore, addition of a chord to the spanning tree yields a subgraph that contains exactly one circuit. For example, consider the graph G shown below in the Figure 17.224.

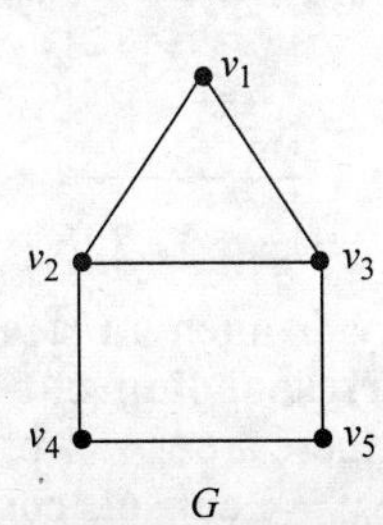

Figure 17.224

For this graph, the Figure 17.225 is a spanning tree.

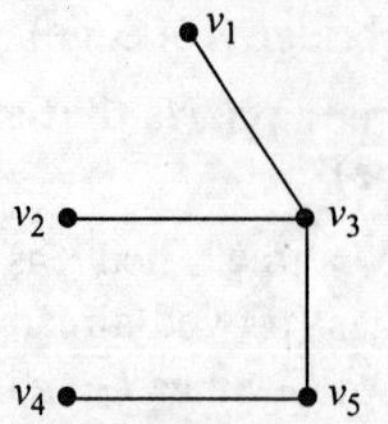

Figure 17.225 (Spanning tree)

The chords of this tree are (v_1, v_2) and (v_2, v_4). If we add (v_1, v_2) to this spanning tree, we get a circuit $v_1\, v_2\, v_3\, v_1$. Similarly addition of (v_2, v_4) gives one more circuit $v_2\, v_3\, v_5\, v_4\, v_2$. If there are v vertices and e edges in a graph, then there are $e-v+1$ chords in a spanning tree. Therefore, if we add all the chords to the spanning tree, there will be $e-v+1$ circuits in the graph.

Definition 17.119 Let v be the number of vertices and e be the number of edges in a graph G.

Then the set of $e-v+1$ circuits obtained by adding $e-v+1$ chords to a spanning tree of G is called the **fundamental system of circuits relative to the spanning tree**.

A circuit in the fundamental system is called a **fundamental circuit.**

For example, $\{v_1, v_2, v_3, v_1\}$ is the fundamental circuit corresponding to the chord (v_1, v_2).

On the other hand, since each branch of a tree is cut edge, removal of any branch from a spanning tree breaks the spanning tree into two trees. For example, if we remove (v_1, v_3) from the above figured spanning tree, the resulting components are shown in the Figure 17.226.

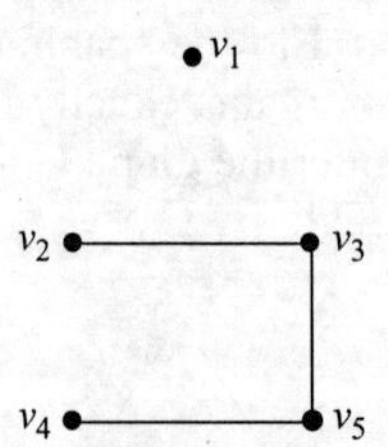

Figure 17.226

Thus, **to every branch in a spanning tree, there is a corresponding cut set**. But, in a spanning tree, there are $v - 1$ branches. Therefore, **there are $v - 1$ cut sets corresponding to $v - 1$ branches.**

Definition 17.120 The set of $v - 1$ cut sets corresponding to $v - 1$ branches in a spanning tree of a graph with v vertices is called the **fundamental system of cut sets relative to the spanning tree.**

A cut set in the fundamental system of cut sets is called a **fundamental cut set.**

For example, the fundamental cut sets in the spanning tree (figured above) are

$$\{(v_1, v_2), (v_1, v_3)\}, \{(v_1, v_3), (v_2, v_3), (v_3, v_4)\}, \{(v_3, v_5), (v_4, v_5)\}, \{(v_2, v_4), (v_4, v_5)\}.$$

Theorem 17.31 A circuit and the complement of any spanning tree must have at least one edge in common.

Proof: We recall that the set of all chords of a tree is called the complement of the tree. Suppose on the contrary that a circuit has no common edge with the complement of a spanning tree. This means the circuit is wholly contained in the spanning tree. This contradicts the fact that a tree is acyclic (circuit- free). Hence a circuit has at least one edge in common with complement of a spanning tree.

Theorem 17.32 A cut set and any spanning tree must have at least one edge in common.

Proof: Suppose on the contrary that there is a cut set which does not have a common edge with a spanning tree. Then removal of cut set has no effect on the tree, that is, the cut set will not separate the graph into two components. But this contradicts the definition of a cut set. Hence the result.

Theorem 17.33 Every circuit has an even number of edges in common with every cut set.

Proof: We know that a cut set divides the vertices of the graph into two subsets each being set of vertices in one of the two components. Therefore a path connecting two vertices in one subset must traverse the edges in the cut set an even number of times. Since a circuit is a path from some vertex to itself, it has an even number of edges in common with every cut set.

17.19 TREE SEARCHING

Let T be a binary tree of height $h \geq 1$ and root v. Since $h \geq 1$, v has at least one child : v_L and/or v_R. Now v_L and v_R are the roots of the left and right subtrees of v called T_L and T_R, respectively (see figure 17.227).

Definition 17.121 Performing appropriate tasks at a vertex is called **visiting the vertex.**

Definition 17.122 The process of visiting each vertex of a tree in some specified order is called **searching the tree** or **walking or traversing the tree.**

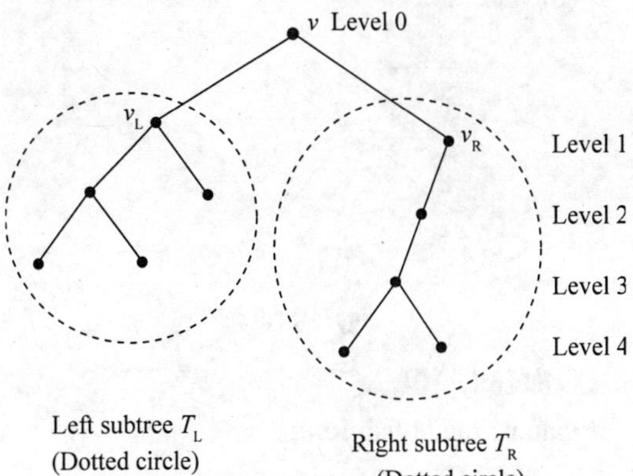

Figure 17.227

We now discuss methods of searching a tree.

1. Pre-order Search Method

Input: The root v of a binary tree.

Output: Vertices of a binary tree using pre-order traversal

1. **Visit** v
2. If v_L (left child of v) exists, then apply the algorithm to $(T(v_L), v_L)$
3. If v_R (right child of v) exists, then apply this algorithm to $(T(v_R), v_R)$

End of algorithm pre-order.

In other words, pre-order search of a tree consists of the following steps:

Step 1. Visit the root.
Step 2. Search the left subtree if it exists.
Step 3. Search the right subtree if it exists.

EXAMPLE 17.97

Find the order of the vertices of the following tree T processed using pre-order traversal.

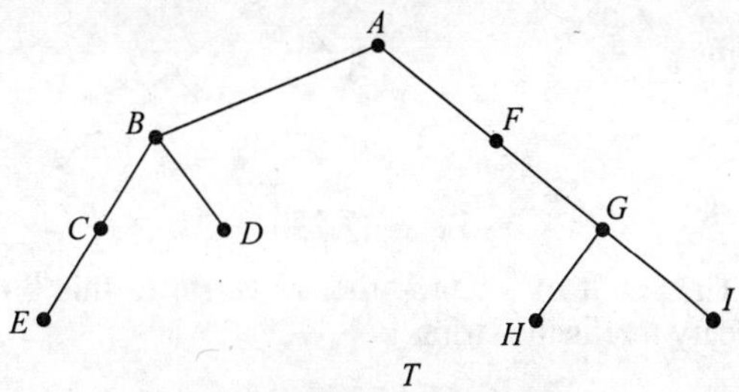

Figure 17.228

Solution. The root of the given tree is A. It has two children B and F. The left subtree and right subtree are shown in Figure 17.229.

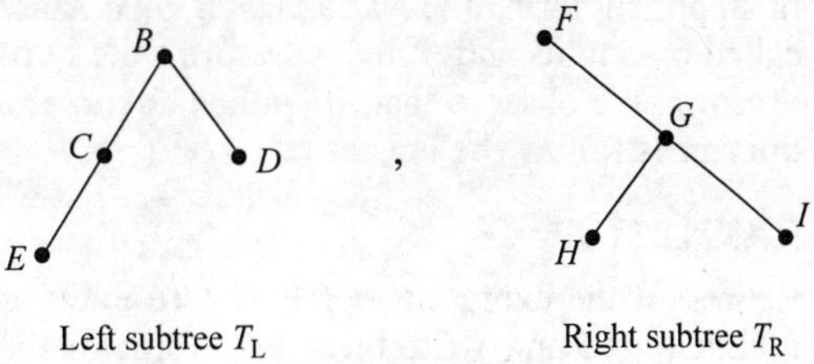

Figure 17.229

If we apply pre-order algorithm to the given tree T, we visit the root and print A. Using pre-order to left subtree, we visit the root B and print B and then search T_L printing C and E and then D. Up to this point, the search has yielded the string $A\ B\ C\ E\ D$. Thus the search of T_L is complete. We now go to subtree T_R. Using the same procedure, we get the string $F\ G\ H\ I$. Thus the complete search of the tree T is

A B C E D F G H I.

EXAMPLE 17.98

Find binary tree representation of the expression

$$(a - b) \times (c + (d \div e))$$

and represent the expression in string form using pre-order traversal.

Solution. In the given expression, $\times$ is the central operator and therefore shall be the root of the binary tree. Then the operator $-$ acts as v_L and the operator $+$ acts as v_R. Thus the tree representation of the given expression is as shown in the Figure 17.230.

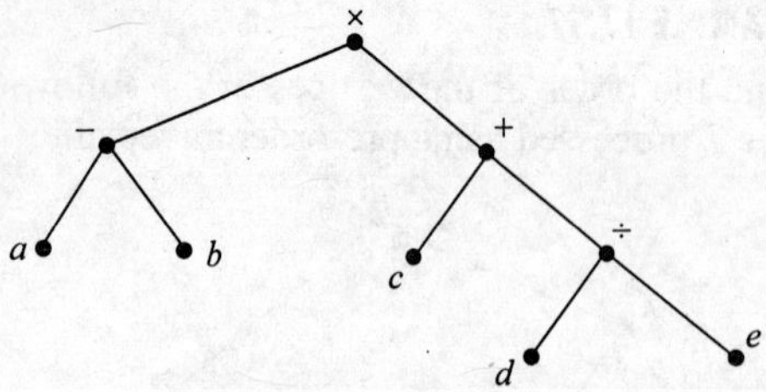

Figure 17.230

The result of the pre-order traversal to this binary tree is the string

$$\times - a\, b + c \div d\, e.$$

This form of the expression is called **prefix form** or **polish form** of the expression

$$(a-b)\times(c+(d\div e))$$

In a polish form, the variables a, b, c,...are called **operands** and $-, +, \times, \div$ are called **operators**. We observe that, **in polish form, the operands follow the operator.**

EXAMPLE 17.99

Represent the expression $(A+B) * (C-D)$ as a binary tree and write the prefix form of the expression.

Solution. Here * is the central operator. Further + and − operators are v_L and v_R. Hence the binary tree for the given experssion is as shown in the Figure 17.231.

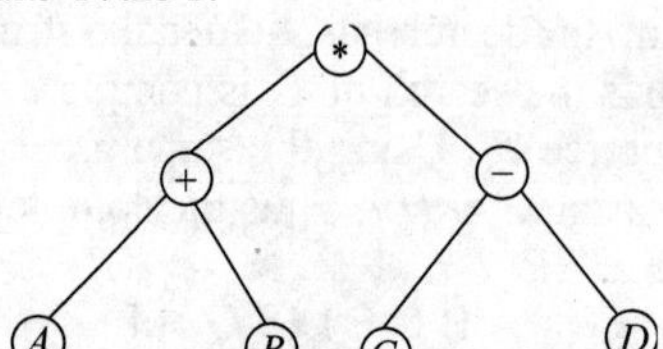

Figure 17.231

Using pre-order traversal, the prefix expression for it is

$$* + A\, B - C\, D.$$

17.19.1 Procedure to Evaluate an Expression Given in Polish Form

To find the value of a polish form, we proceed as follows:

Move from left to right until we find a string of the form K x y, where K is operator and x, y are operands.

Evaluate x K y and substitute the answer for the string K x y. Continue this procedure until only one number remains.

EXAMPLE 17.100

Find parenthesized form of the polish expression

$$-+A\,B\,C.$$

Solution. The parenthesized form of the given polish expression is derived as follows:

$$-(A+B)\,C,$$
$$(A+B)-C.$$

The corresponding binary tree is shown in the Figure 17.232.

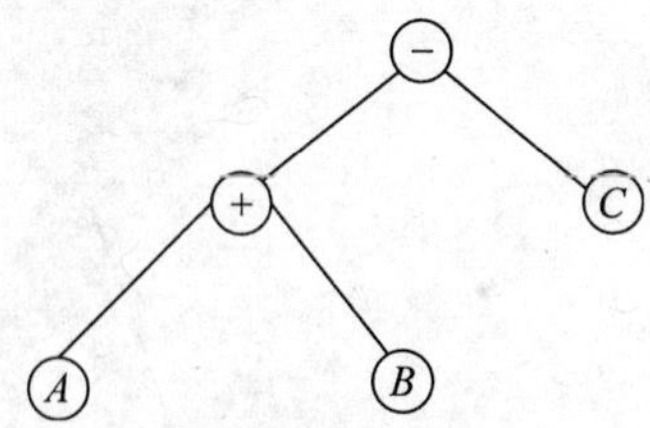

Figure 17.232

EXAMPLE 17.101

Evaluate the polish form

$$\times - 64 + 5 \div 22$$

Solution. We have the following steps in this regard:

1. $\times(6-4)+5\div 22$
2. $\times 2+5\div 22$
3. $\times 2+5\,(2\div 2)$
4. $\times 2+51$
5. $\times 2\,(5+1)$
6. $\times 26$
7. 2×6
8. 12, which is the required value of the expression.

2. Post-order Search Method

Algorithm

Step 1. Search the left subtree if it exists
Step 2. Search the right subtree if it exists
Step 3. Visit the root

End of algorithm

EXAMPLE 17.102

Represent the expression

$$(A+B)*(C-D)$$

as a binary tree and write the result of post-order search for that tree.

Solution. The binary tree expression (as shown earlier) of the given algebraic expression is given in Figure 17.233.

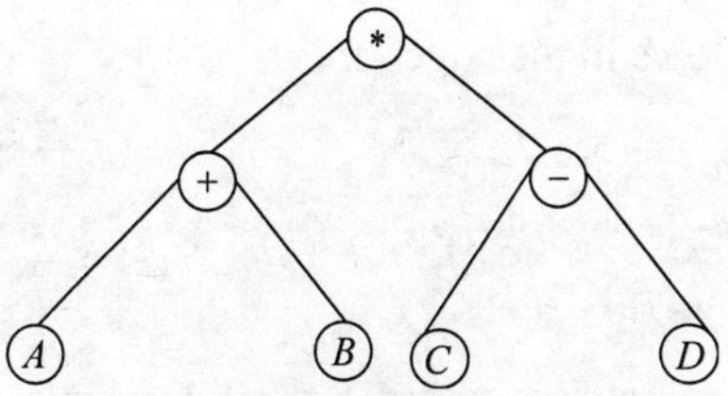

Figure 17.233

The result of post-order search of this tree is

$$AB+CD-*$$

This form of the expression is called **postfix form** of the expression or **reverse polish form** of the expression.

In postfix form, the operator follows its operands.

EXAMPLE 17.103

Find the parenthesized form of the postfix form

$$ABC**CDE+/-.$$

Solution. We have

1. $ABC**CDE+/-$
2. $A(B*C)*C(D+E)/-$
3. $(A*(B*C))(C/(D+E))-$
4. $(A*(B*C))-(C/(D+E))$.

The corresponding binary tree is shown in the Figure 17.234.

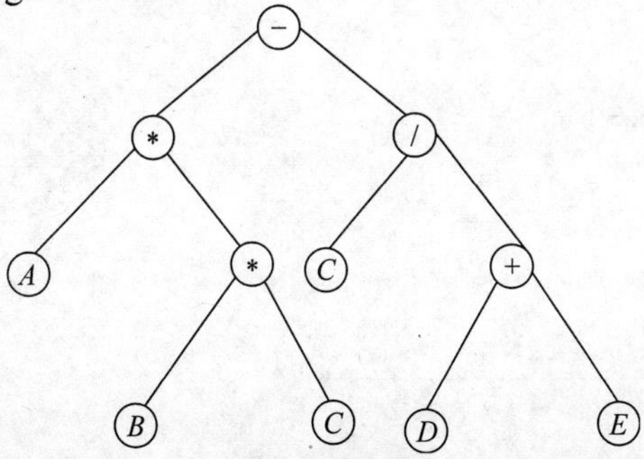

Figure 17.234

EXAMPLE 17.104

Evaluate the postfix form

$$21-342\div+\times$$

Solution. We have

$$\begin{aligned}21-342\div+\times &= (2-1)\,342\div+\times\\ &= 13\,(4\div 2)+\times\\ &= 132+\times\\ &= 1\,(3+2)\times\\ &= 15\times\\ &= 1\times 5 = 5.\end{aligned}$$

17.20 TRANSPORT NETWORKS

An important application of weighted directed graph is to model transport networks like oil pipeline, water supply pipeline, communication network, electric power distribution system and highway system. The purpose of a network is to implement the flow of oil, water, electricity, messages, traffic, etc and we desire that the flow through the network should be largest possible value. Such a flow will be called maximum flow.

Definition 17.123 A ***transport network*** or simply a network is a simple weighted directed graph with the following properties:

(i) There is a designated vertex (node), called the source, that has no incoming edge.
(ii) There is a designated vertex, called the ***sink***, that has no outgoing edge.
(iii) There is a non-negative weight c_{ij} on the directed edge (i, j), called the ***capacity*** of the edge (i, j).

Since the graph is simple, it follows that if the edge (i, j) is in the network, then (j, i) is not there.

Definition 17.124 A flow in a network is a function that assigns to each (i, j) of the network a non-negative number f_{ij} such that

(i) $0 \le f_{ij} \le c_{ij}$, where c_{ij} is the capacity of the edge (i, j)
(ii) For each vertex j, which is neither the source nor the sink,

$$\sum_i f_{ij} = \sum_k f_{jk}$$

The non-negative number f_{ij} is called the ***flow*** in the edge (i, j). For any vertex (distinct from source and sink) j, the sum $\sum_i f_{ij}$ is called the ***flow into*** j and the sum $\sum_k f_{jk}$ is called the ***flow out of*** j. Thus the condition (ii) implies that material cannot accumulate, be created, dissipate or lost at any vertex other than the source or the sink. Hence the equation is called ***conservation of flow***. As a consequence of conservation, it follows that the ***sum of flows leaving the source must be equal to the sum of the flows entering the sink***. This sum is called the ***value of the flow*** and is denoted by value (F). A flow F in a network is represented by labelling each edge (i, j) with the pair (c_{ij}, f_{ij}).

For example consider the oil pipeline network shown in Figure 17.235.

In this network, the crude oil unloaded at the dock O (source) is being pumped to the refinery S (sink). The capacities of the flow are

$$c_{OA} = 5, \quad c_{AB} = 4, \quad c_{BS} = 5, \quad c_{OC} = 3$$

$$c_{CB} = 2, \quad c_{CD} = 4, \quad c_{DB} = 2, \quad c_{DS} = 5$$

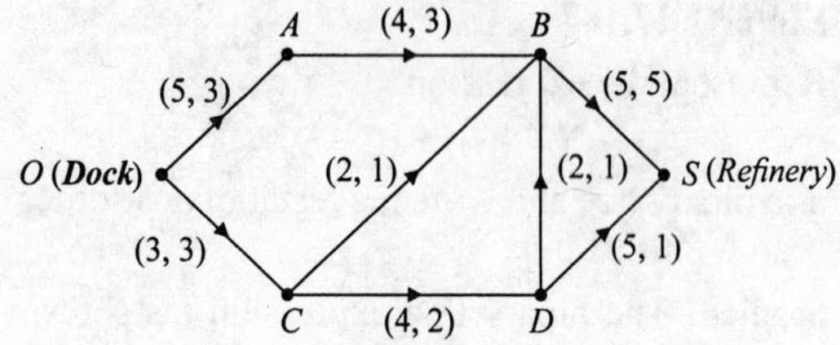

Figure 17.235

The flows in the edges are

$$f_{OA} = 3, \quad f_{AB} = 3, \quad f_{BS} = 5, \quad f_{OC} = 3$$

$$f_{CB} = 1, \quad f_{CD} = 2, \quad f_{DB} = 1, \quad f_{DS} = 1$$

We note that

Flow into the vertex $B = 3 + 1 + 1 = 5$
Flow out of the vertex $B = 5$.

Therefore, ***conservation of flow*** is satisfied. Further, value $(F) = 3 + 3 = 5 + 1 = 6$

EXAMPLE 17.105

Consider the water pipeline network shown in Figure 17.236 for two towns A and B in which water is supplied from four borewells w_1, w_2, w_3, w_4 and a, b, c, d represent intermediate pumping stations. Form an equivalent network with unique source and unique sink and determine the value of the flow.

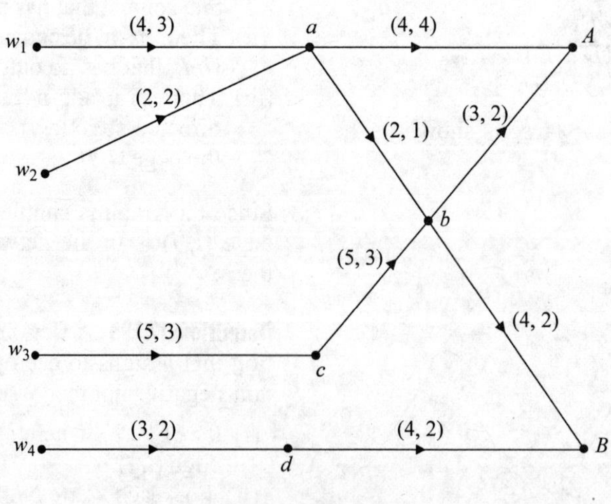

Figure 17.236

Solution. To obtain the required equivalent network with designated source and designated sink, we tie together the given sources w_1, w_2, w_3, and w_4 into a ***super source*** O and the towns A and B into a ***super sink*** S as shown in Figure 17.237.

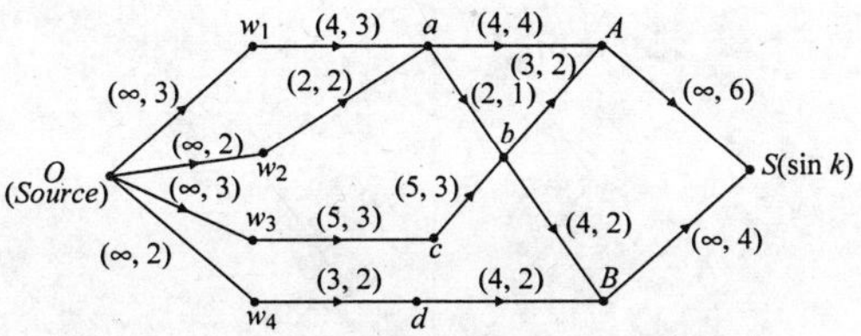

Figure 17.237

The value of the flow is

$$\text{Value } (F) = 3 + 2 + 3 + 2 = 6 + 4 = 10.$$

Definition 17.125 A ***maximal flow*** in a network N is a flow with maximum value.

In general, there will be several flows having the same maximum value. For example, consider the network shown in Figure 17.238.

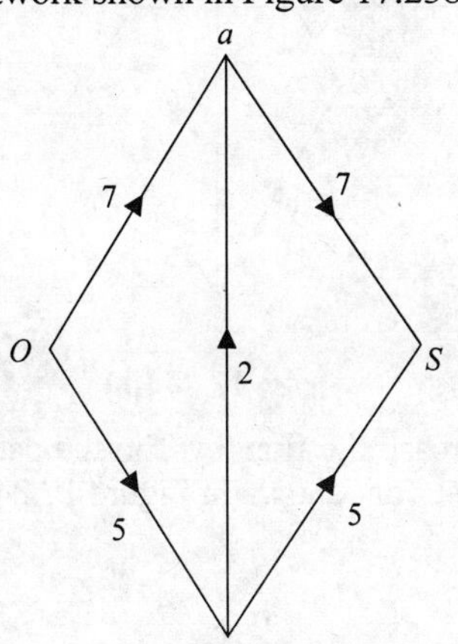

Figure 17.238

Consider the following two flows for this network:
The value of the flow in Fig. 17.239(a) is 10 and three of the five edges in this network attain their maximum capacity. But for the same network, the value of the flow in Fig. 17.239(b) is 12 and four out of five edges are carrying maximum capacity. Therefore, the flow function in the second case is better than the first one.

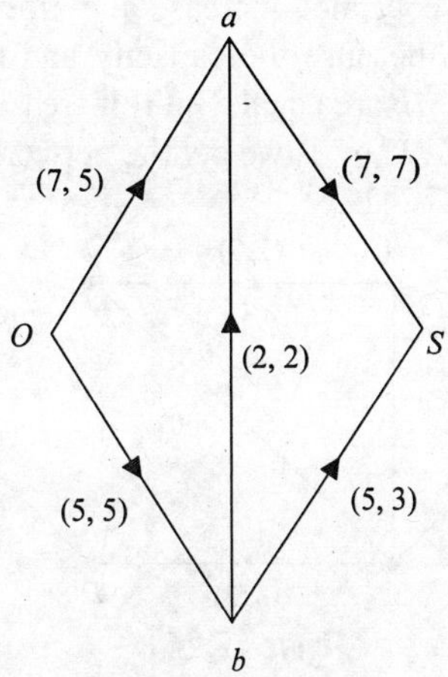

Figure 17.239(a)

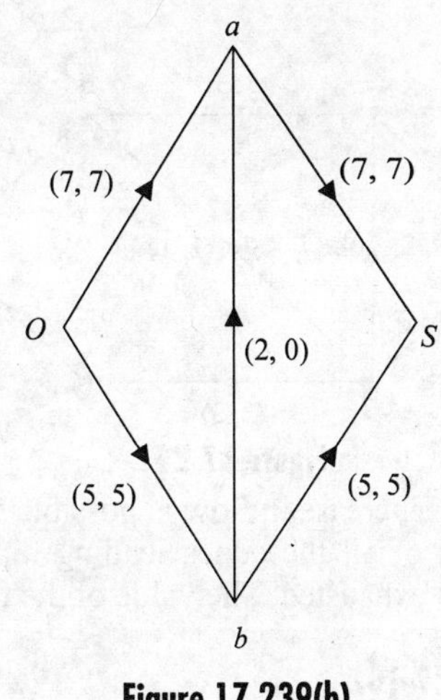

Figure 17.239(b)

EXAMPLE 17.106

Find the maximum possible increase in the following network (Figure 17.240).

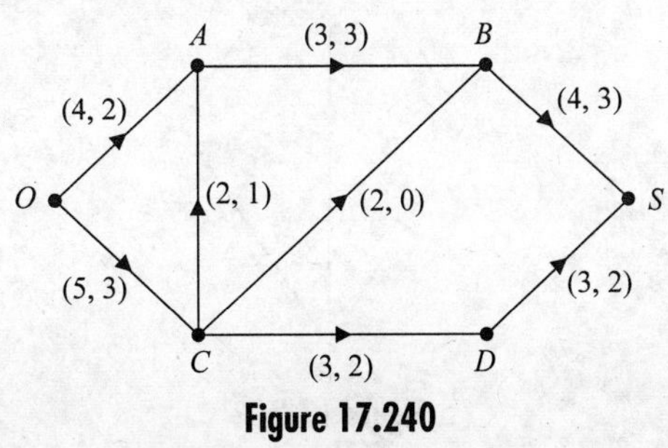

Figure 17.240

Solution. We cannot increase the flow in the path *OABS* because the capacity and the flow of the edge *AB* are equal. But if increase flow in the path *OCDS* by 1, we get the network shown in Figure 17.241.

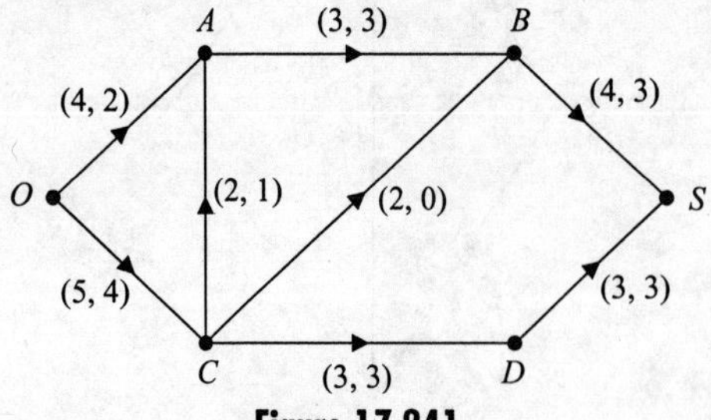

Figure 17.241

Further, if we increase the flow in the path *OCBS* by 1, we get the network shown in Figure 17.242.

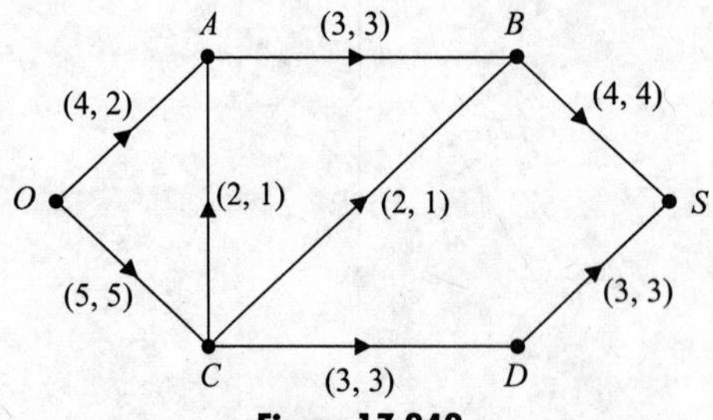

Figure 17.242

No further increase in flow is possible since the capacities of all the edges leading to the sink have been exhausted. The value of the flow is 7.

EXAMPLE 17.107

Find a maximal flow in the network shown in Figure 17.243.

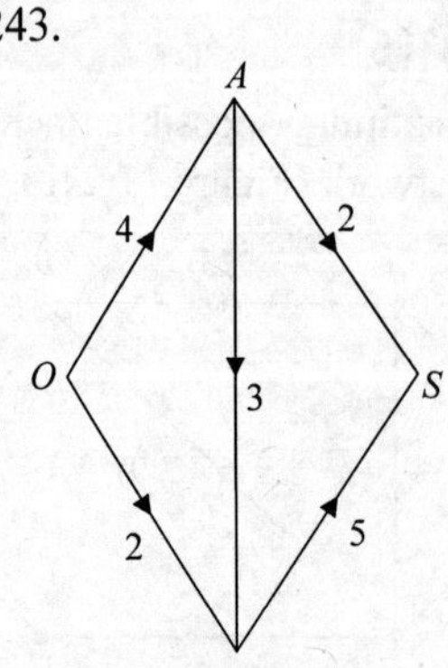

Figure 17.243

Solution. The capacity of each edge is shown in Figure 17.243. Let the initial flow be 0 in each edge. Thus we have the network as shown in Figure 17.244(a).

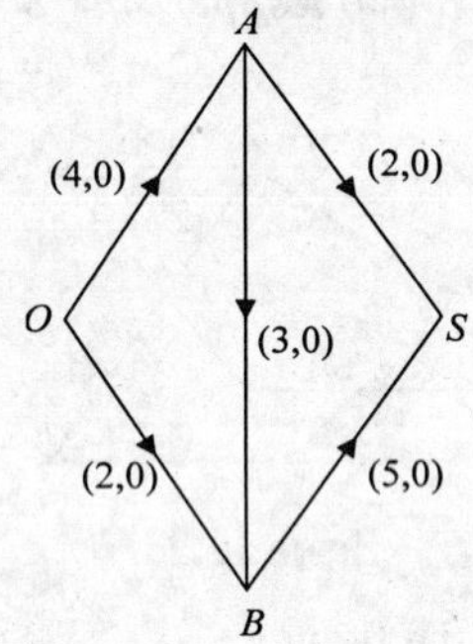

Figure 17.244(a)

Increase the flow by 2 in the path *OAS* to get the network shown in the Figure 17.244(b).

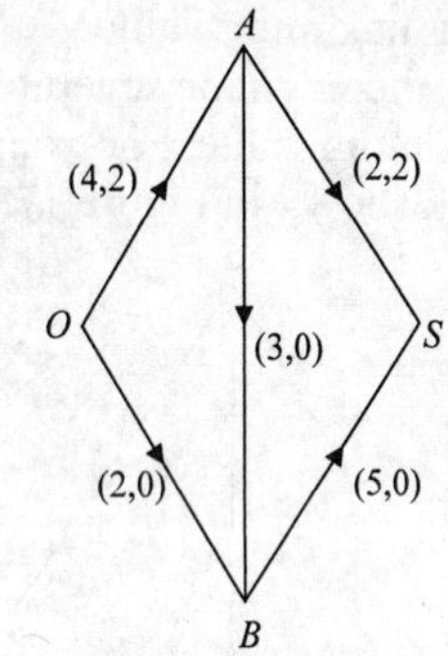

Figure 17.244(b)

Now increase the flow by 2 in the path *OBS* to get the network shown in Figure 17.244(c).

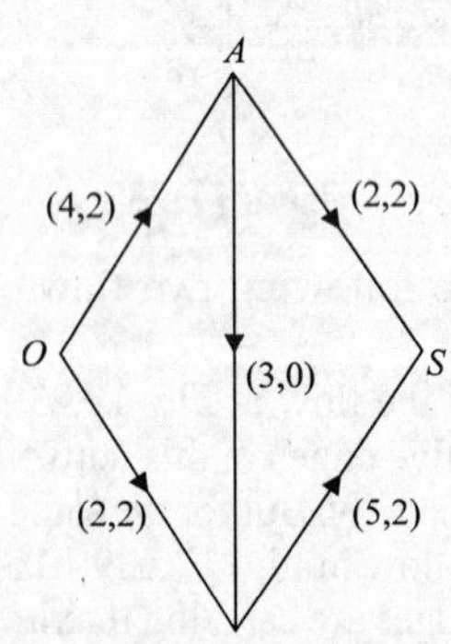

Figure 17.244(c)

Now increase the flow by 2 in the path *OABS* to get the network shown in Figure 17.244(d).

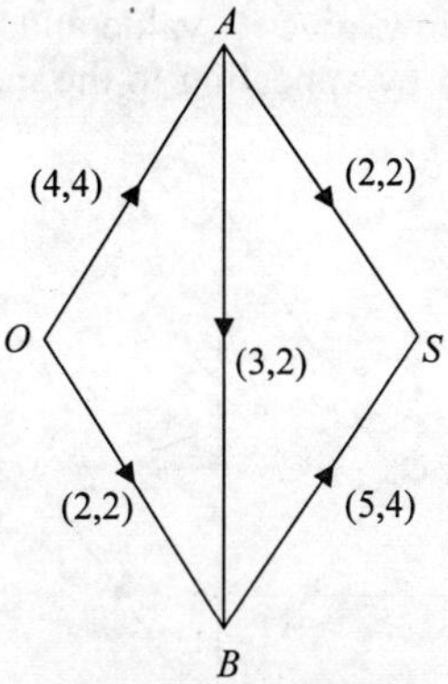

Figure 17.244(d)

Any further increase in the flow is not possible. Thus maximal flow in the network has been reached. Thus, value $(F) = 2 + 4 = 6$.

EXAMPLE 17.108

Find a maximal flow in the network shown in Figure 17.245.

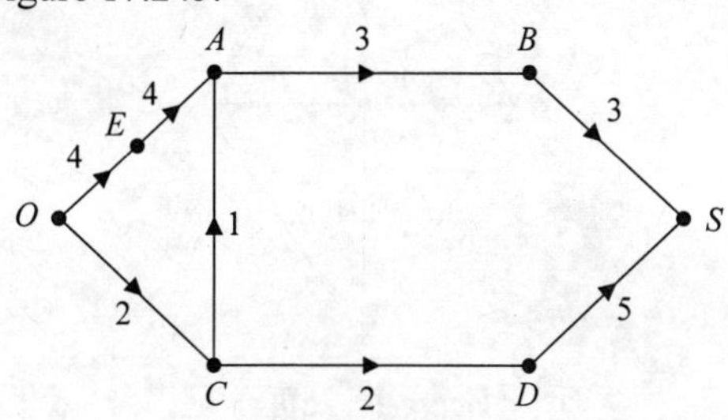

Figure 17.245

Solution. The initial labelling yields the network shown in Figure 17.246(a).

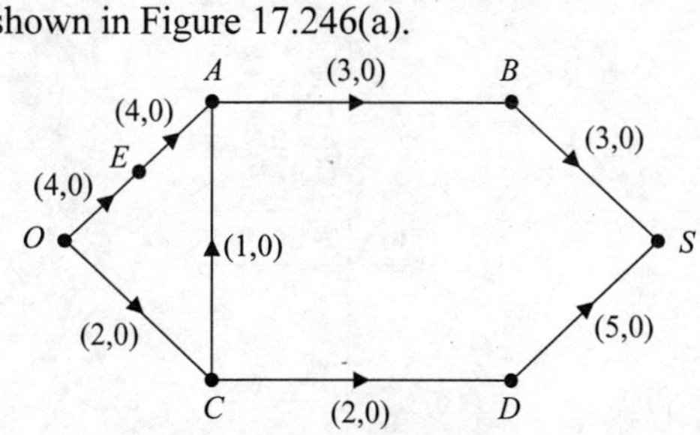

Figure 17.246(a)

Increase the flow by 2 in the path *OCDS* by to get the network shown in Figure 17.246(b).

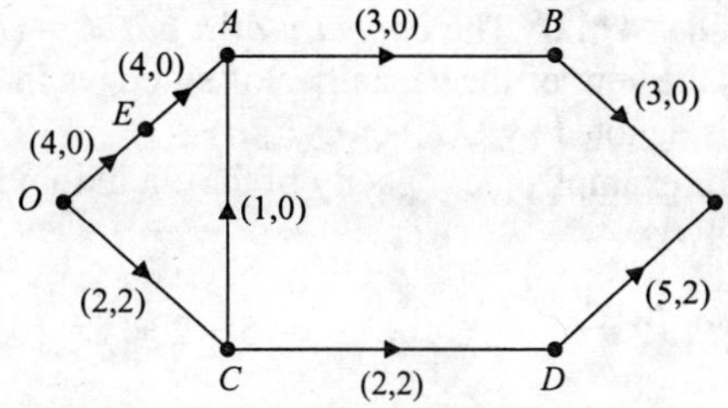

Figure 17.246(b)

Now increase the flow by 3 in the path *OEABS* to get the network shown in Figure 17.246(c).

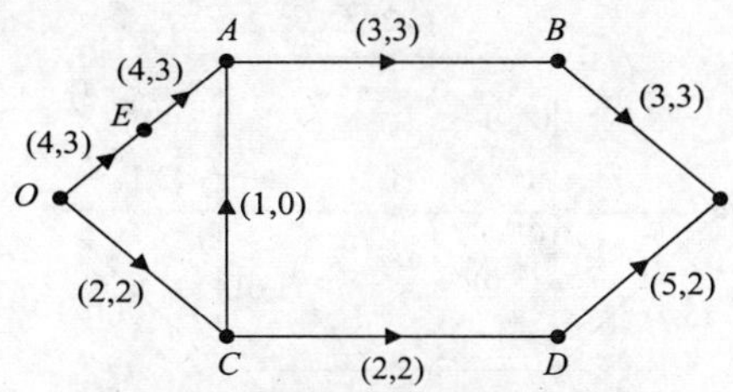

Figure 17.246(c)

No further increase in flow in possible since the capacities in the edges (A, B) and (C, D) has been exhausted. Hence value $(F) = 5$.

Definition 17.126 Let N be a network. A ***cut*** in N is a partition $(P, \overline{P})$ of the set of vertices in N such that the source $O \in P$ and the sink $S \in \overline{P}$.

Thus, a cut $(P, \overline{P})$ in a network N is a set K of edges (v, w), $v \in P$, $w \in P$ such that every path from source to sink contains at least one edge from K. In fact, a cut does cut a directed graph into two pieces, one containing the source and the other containing the sink. ***Nothing can flow from source to sink if edges of a cut are removed.*** We indicate a cut by drawing a dashed line to partition the vertex set. As an illustration, consider the network of Example 17.106. The dashed line divides the vertex set into the sets $P = \{O, A\}$ and $\overline{P} = \{C, B, D, S\}$.

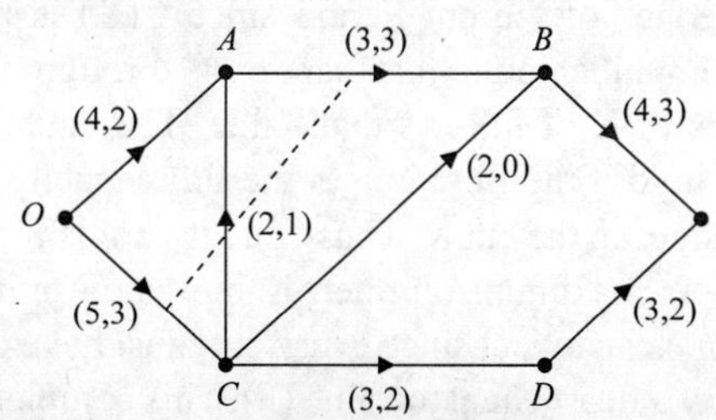

Figure 17.247

Definition 17.127 The ***capacity of a cut*** $K = (P, \overline{P})$ is the sum of the capacities of all edges in K and is denoted by $C(K)$ or $C(P, \overline{P})$.

For example, the capacity of the cut in Figure 17.247 is

$$C(P, \overline{P}) = C_{OC} + C_{AC} + C_{AB} = 5 + 2 + 3 = 10$$

EXAMPLE 17.109

Find the capacity of the cut, shown by dashed line, in the network shown in Figure 17.248.

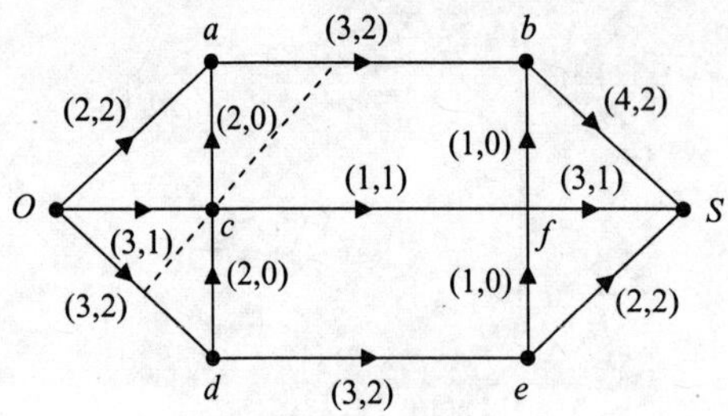

Figure 17.248

Solution. The cut $(P, \overline{P})$ is given by

$$P = \{o, b, c\}, \overline{P} = \{d, e, f, b, S\}.$$

Therefore the capacity of the given cut is

$$C(P, \overline{P}) = C_{Od} + C_{cf} + C_{ab} = 3 + 1 + 3 = 7.$$

Definition 17.128 A minimal cut is a cut having minimum capacity.

Theorem 17.34 (The Max Flow Min Cut Theorem) A maximum flow F in a network has value equal to the capacity of a minimum cut of the network.

Proof: Let F be any flow and K be any cut in a network. Since all parts of F must pass through the edges of the cut K and since C(K) is the maximum amount that can pass through the edges of K, it follows that value. The equality will hold if the flow F uses the full capacity of all edges in the cut K. Thus equality holds if the flow is maximum. Further, K must be a minimum capacity cut since every cut must have capacity at least equal to value (F). This completes the proof of the theorem.

EXAMPLE 17.110

In the network shown in Figure 17.249, find a maximum flow, give its value and prove that it is maximum by appealing to the max flow min cut theorem.

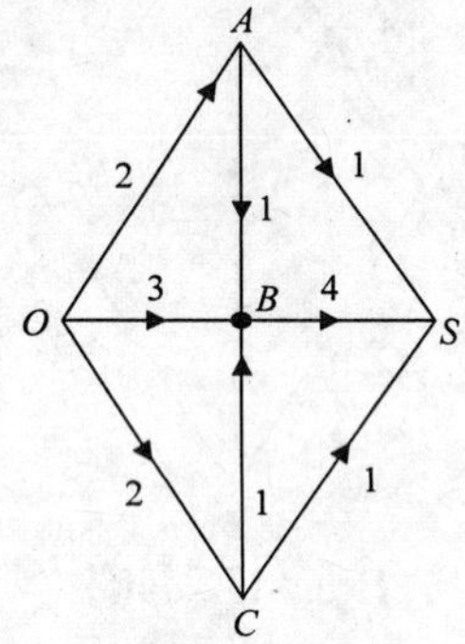

Figure 17.249

Solution. The initial labelling yields the following network shown in Figure 17.250(a):

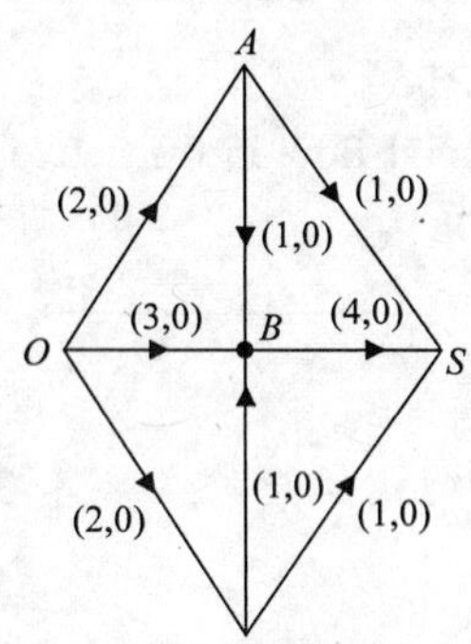

Figure 17.250(a)

Increase the flow by 1 in the path *OAS* to get the network shown in Figure 17.250(b).

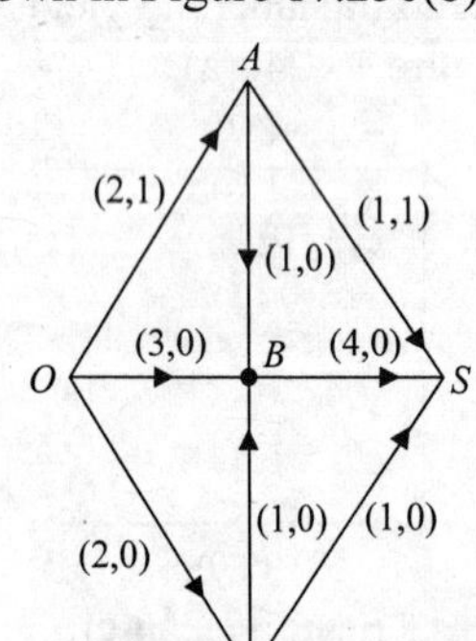

Figure 17.250(b)

Now increase the flow by 1 in the path *OCS* to get the network shown in Figure 17.250(c).

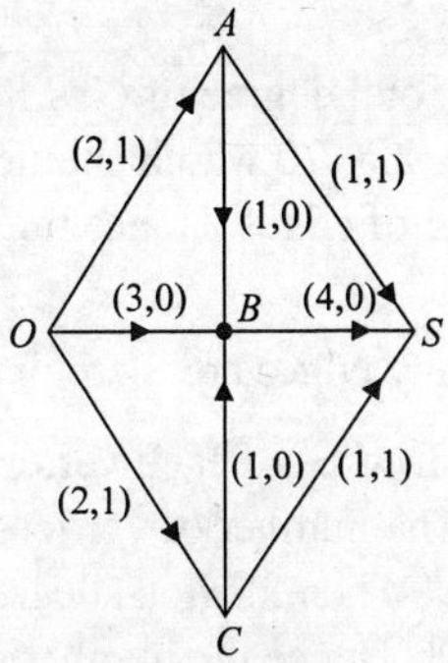

Figure 17.250(c)

Further, increase in flow by 1 in the path *OABS* to get the network of Figure 17.250(d).

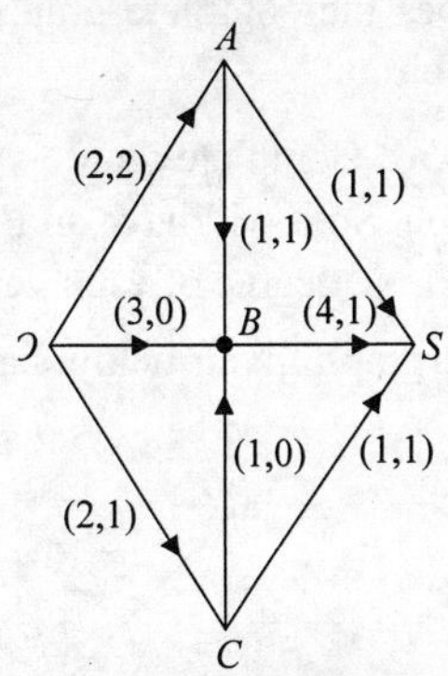

Figure 17.250(d)

Now increase the flow by 1 in the path *OCBS* and get the network shown in Figure 17.250(e).

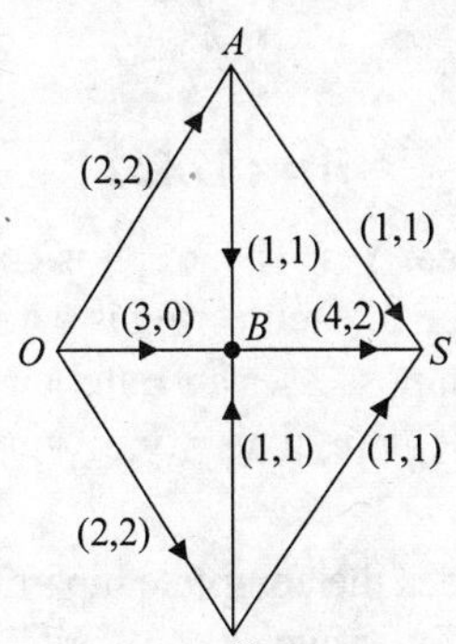

Figure 17.250(e)

Lastly, we increase the flow by 2 in the path OBS and get the network with maximum flow as shown in Figure 17.250(f). In fact the capacities in the edges leading to the sink have been completely exhausted.

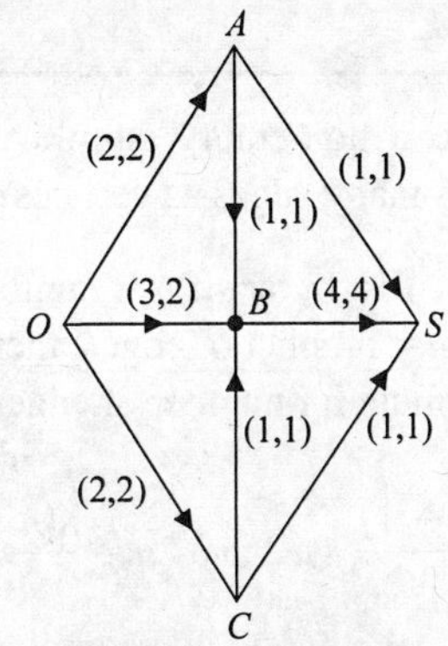

Figure 17.250(f)

We note that value ($F = 2 + 2 + 2 = 1 + 4 + 1 = 6$). To check whether it is maximum flow, we appeal to max flow min cut theorem. We consider the following cuts:

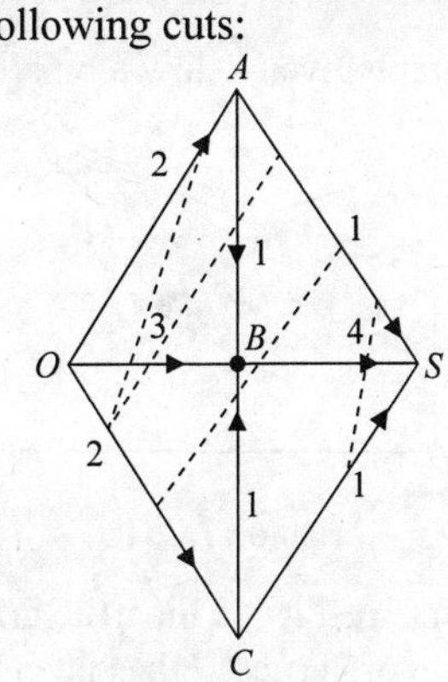

Figure 17.250(g)

(i) $K_1 = \{OC, OB, OA\}$
with
cap $(K_1) = C_{OC} + C_{OB} + C_{OA} = 2 + 3 + 2 = 7$,
with cap

(ii) $K_2 = \{OC, BC, BS, AS\}$
with
cap $(K_2) = C_{OC'} + C_{BC} + C_{BS'} + C_{AS}\}$
$= 2 + 1 + 4 + 1 = 8$

(iii) $K_3 = \{CS, BS, AS\}$
with cap $(K_3) = C_{CS} + C_{BS} + C_{AS} = 1 + 4 + 1 = 6$,

(iv) $K_4 = \{OC, OB, AS, AS\}$
with cap $(K_4) = C_{OC} + C_{OB} + C_{AS} + C_{AB}$
$= 2 + 3 + 1 + 1 = 7$,

We note that capacity of the minimum cut in 6. Hence, the maximum flow is 6.

EXERCISES

1. Is there a non-empty simple graph with twice as many edges as vertices?

Ans. Yes. In fact if n = number of vertices, $2n$ = number of edges, then using the formula for number of edges in a complete graph:, $\frac{n(n-1)}{2}$, we have $2n = \frac{n(n-1)}{2}$ which yields $n = 5$. But in a complete graph K_5 the number of edges is 10. Thus the complete graph K_5, which is simple by definition, has five vertices and ten edges.

2. Is the graph given below a bipartite graph?

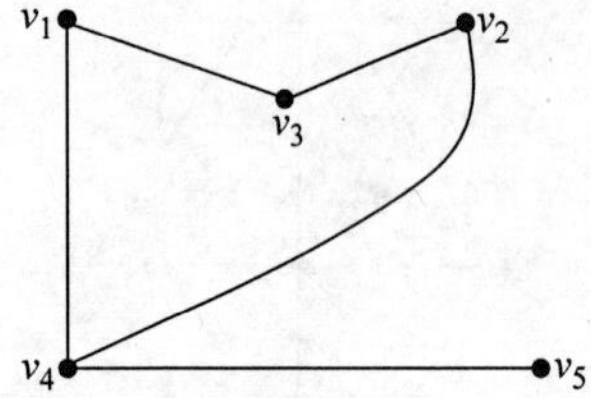

Figure 17.251

Ans. Yes, it is a bipartite graph, where the set of vertices V has been partitioned into two subsets, $V_1 = \{v_3, v_4\}$, $V_2 = \{v_1, v_2, v_5\}$. But it is not a complete bipartite graph.

3. Find a formula for the number of edges in complete bipartite graph K_{mn}.

Ans. The $m + n$ vertices having been divided into two subsets V_1 and V_2 containing m vertices and n vertices, respectively. Every verticex in V_1 is joined to all the n vertices in V_2. Thus the task of joining each vertex of V1 to each vertex in V_2 can be performed in mn ways. Hence the number of edges in K_{mn} is mn.

4. A graph G has vertices of degrees 1, 4, 3, 7, 3 and 2. Find the number of edges in the graph.

Ans. Total degree of G is $1 + 4 + 3 + 7 + 3 + 2 = 20$ which should be twice the number of edges. Hence number of edges in G is 10.

5. How many edges are there in an n-cube?

Ans. Degree of each vertex in an n-cube in n. The number of vertices in an n cube is 2^n. Therefore total degree of n-cube is $n\,2^n$. Hence the number of edges is an n-cube is

$$\frac{n \cdot 2^n}{2} = n \cdot 2^{n-1}.$$

6. When does the complete graph K_n possess an Euler circuit?

Ans. Complete graph is a connected graph. So it will have an Euler cycle if degree of each vertex is even.

7. Does the graph given below have an Euler circuit?

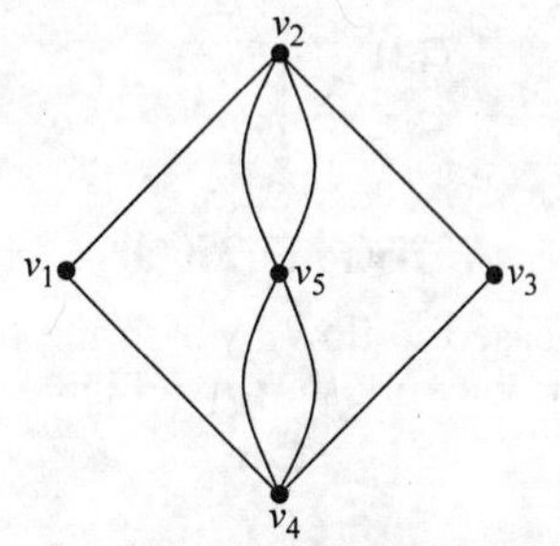

Figure 17.252

Ans. Yes. The graph is connected and degree of each vertex is even. Hence the graph possesses an Euler circuit. For example, $v_1\, v_2\, v_5\, v_4\, v_5\, v_2\, v_3\, v_4\, v_1$ is an Euler cycle.

8. When does the complete bipartite graph K_{mn} contain Euler cycle?

Ans. When both m and n are even. In such a case the degree of each vertex will be even.

9. In seven bridges problem, was it possible for a citizen of Konigsberg to make a tour of the city and cross each bridge exactly twice? Give reasons.

Ans. Yes, because in such a case degree of each vertex of the graph of seven bridges problem shall be even.

10. Show that the graph given below does not contain a Hamiltonian cycle.

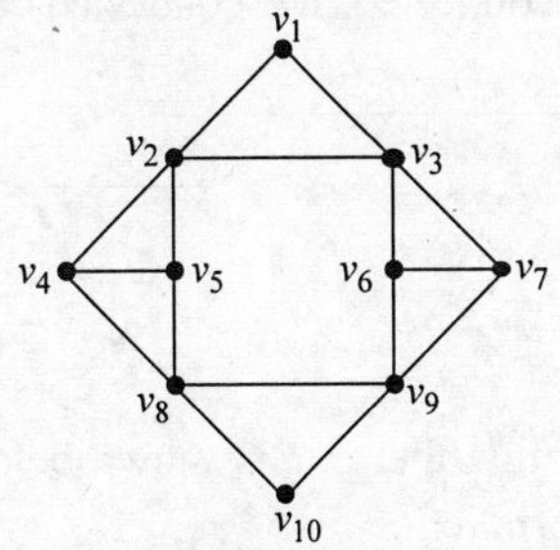

Figure 17.253

11. Give example of a graph that has Euler circuit but not Hamiltonian circuit.

Ans.

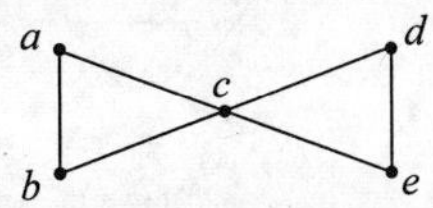

12. Give example of a graph that has an Hamiltonian circuit but not an Euler circuit.

Ans.

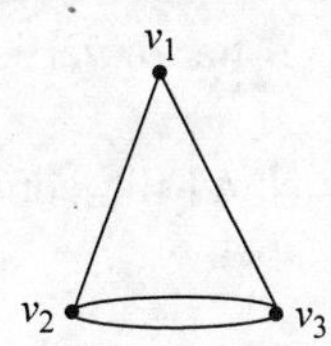

13. Give example of a graph that has both an Euler circuit and an Hamiltonian circuit.

Ans.

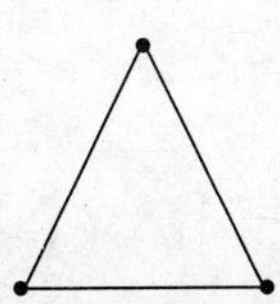

K_3

14. Use Dijkstra's shortest path algorithm to find the shortest path from a to f in the graph given below:

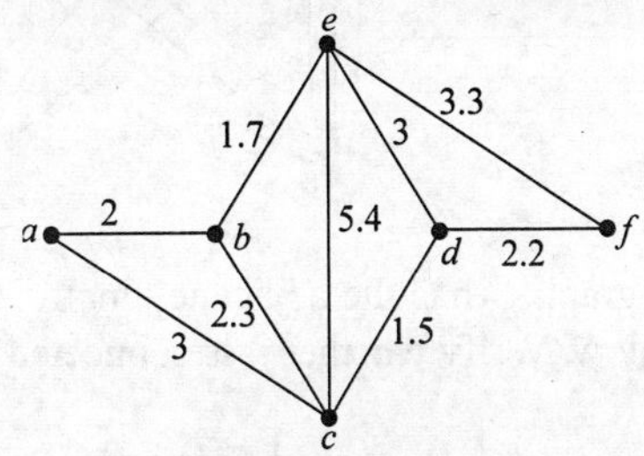

Figure 17.254

Ans. 6.7

15. Find the adjacency matrix of the graph.

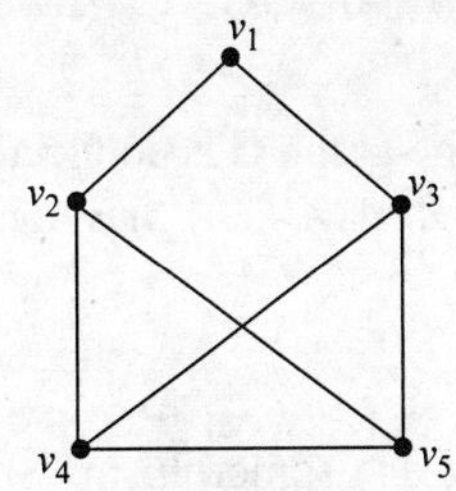

Figure 17.255

Ans. $\begin{bmatrix} 0 & 1 & 1 & 0 & 0 \\ 1 & 0 & 0 & 1 & 1 \\ 1 & 0 & 0 & 1 & 1 \\ 0 & 1 & 1 & 0 & 1 \\ 0 & 1 & 1 & 1 & 0 \end{bmatrix}$

16. Find directed graph that have the following adjacency matrix.

$$\begin{bmatrix} 0 & 0 & 1 & 1 \\ 1 & 0 & 2 & 0 \\ 2 & 1 & 0 & 0 \\ 1 & 1 & 0 & 0 \end{bmatrix}$$

Ans.

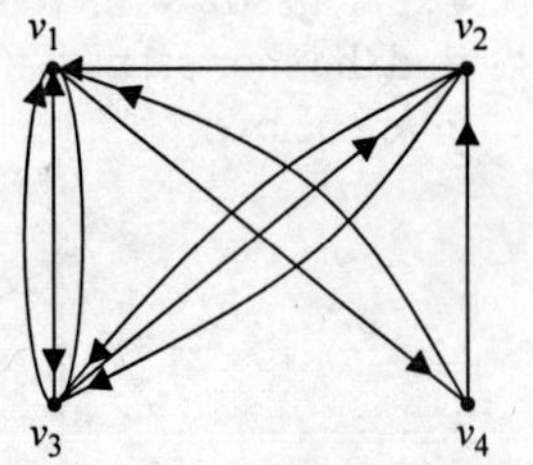

17. A graph G has the adjacency matrix given below. Verify whether it is connected.

$$A = \begin{bmatrix} 0 & 0 & 1 & 0 & 0 \\ 0 & 0 & 0 & 1 & 0 \\ 0 & 0 & 0 & 0 & 1 \\ 0 & 1 & 0 & 0 & 1 \\ 0 & 0 & 1 & 1 & 0 \end{bmatrix}$$

Ans. The graph G is connected because $B = A + A^2 + A^3 + A^4$ has non-zero entries off the main diagonal.

18. Defining an isomorphism. Show that the graphs shown below are isomorphic.

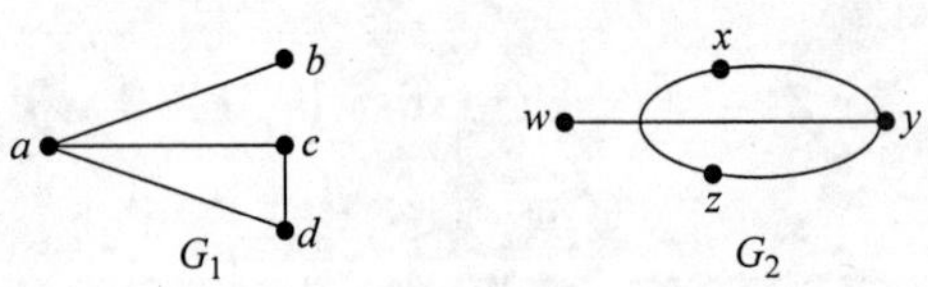

(Clearly, the graph G_2 can be drawn as)

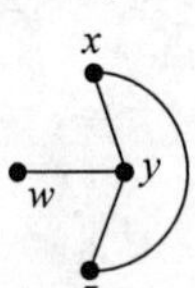

Figure 17.256

Ans. Vertex set of G_1 is $V(G_1) = \{a, b, c, d\}$ vertex set of G_2 is $V(G_2) = \{x, y, z, w\}$
Define $f: V(G_1) \to V(G_2)$ by
$f(a) = y, f(b) = w, f(c) = x, f(d) = z.$
Then f is bijective.

19. Show that the graphs shown below are not isomorphic. Give reasons.

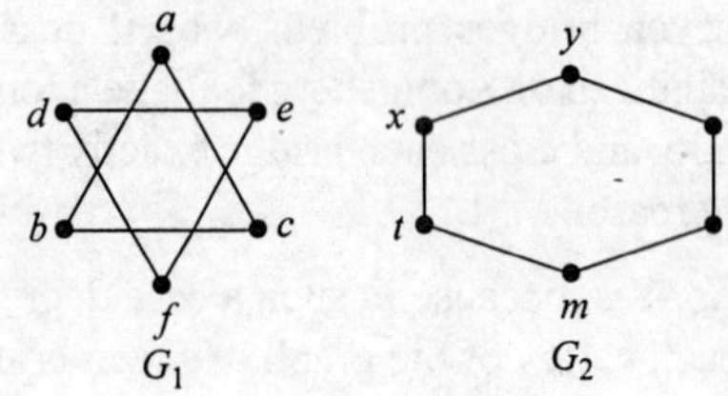

Figure 17.257

Ans. Graph G_2 is connected whereas G_1 is not connected. The connected components of G_1 are

20. Show that the graphs shown below are not isomorphic.

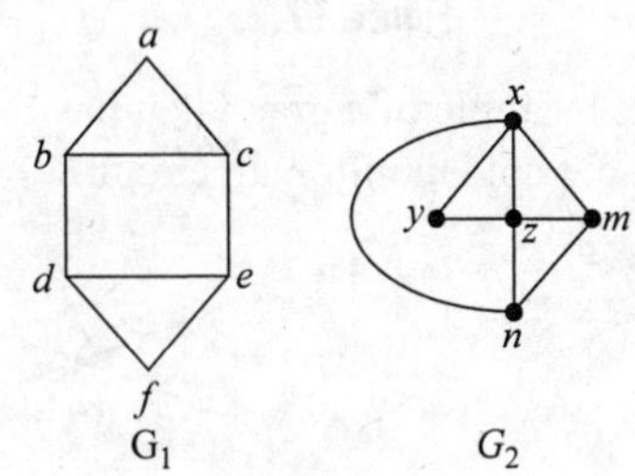

Figure 17.258

Ans. The graphs do not have same number of vertices and so are not isomorphic.

21. Find the complement of the graph shown below.

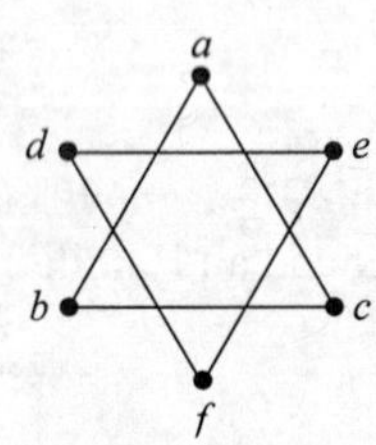

Figure 17.259

Ans. The complement of the given graph is

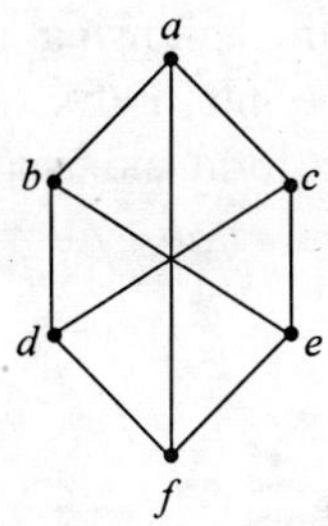

22. Let G be a graph and V be its vertex set. Then

(i) Eccentricity of $v \in V$ is defined as
$e(v) = \max \{d(u, v): u \in V, u \neq v\}$.

(ii) The radius of the graph is defined by
$\text{rad}(G) = \min \{e(v): v \in V\}$.

(iii) The diameter of G is defined by
$\text{diam}(G) = \max \{e(v): v \in V\}$.
$= \max [d(u, v): u \in V, v \in V, u \neq v\}$.

(iv) v is a central point if
$e(v) = \text{rad}(G)$.

(v) The centre of G is the set of all central points.

Find the centre of the graph given below:

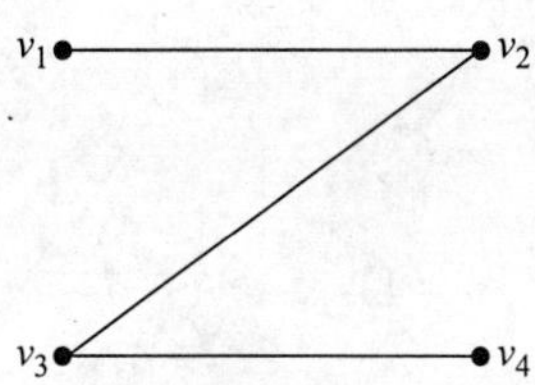

Figure 17.260

Ans. We note that

(i) $d(v_1, v_2) = 1, d(v_1, v_3) = 2, d(v_1, v_4) = 3$ and so $e(v_1) = 3$

(ii) $d(v_2, v_1) = 1, d(v_2, v_3) = 1, d(v_2, v_4) = 2$ and so $e(v_2) = 2$

(iii) $d(v_3, v_1) = 2, d(v_3, v_2) = 1, d(v_3, v_4) = 1$ and so $e(v_3) = 2$

(iv) $d(v_4, v_1) = 3, d(v_4, v_2) = 2, d(v_4, v_3) = 1$ and so $e(v_4) = 3$

Hence $\text{rad}(G) = \min \{3, 2, 2, 3\} = 2$. The centre is $\{v_2, v_3\}$.

23. Show that any graph having n vertices, $n = 1, 2, 3, 4$ is planar.

24. Show that the colouring of map given below requires at least three colours

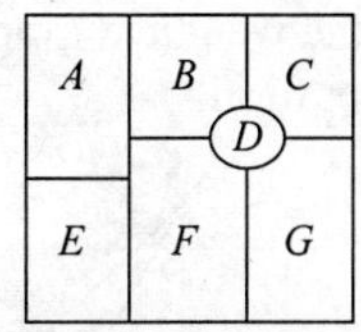

Figure 17.261

Ans. Give one colour to A and D, give second colour to B, E and G, and third colour to C and F. Thus all the adjoining boundaries have different colours.

25. How many colours are required to paint the graph given below?

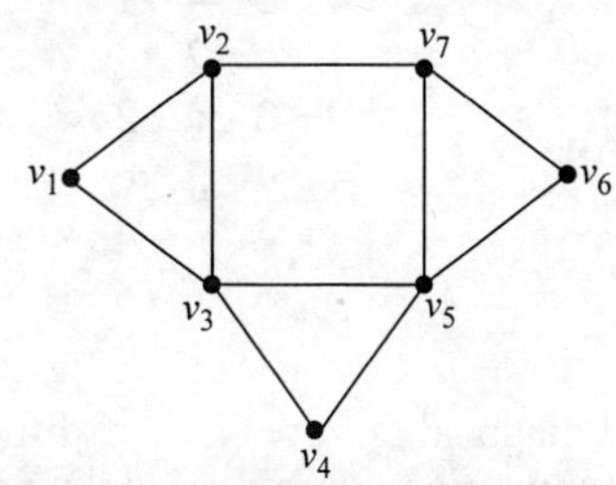

Figure 17.262

Ans. Three. We can paint v_1, v_2, v_3 with colours C_1, C_2, C_3. Then paint v_4 by C_1, v_5 by C_2, v_6 by C_3 and v_7 by C_1.

26. For which values of m and n is the complete bipartite graph K_{mn} a tree?

Ans. Number of edges in $K_{mn} = mn$. Total number of vertices in $K_{mn} = m + n$. If K_{mn} is a tree, then number of edges in K_{mn} sp-5which is satisfied if either $m = 1$ or $n = 1$.

27. Let G be a graph with n vertices and $n-2$ or fewer edges. Show that G is not connected.

Ans. Suppose G is connected. The graph itself will be a tree or it can be made a tree by eliminating some edges. In either case it must have $n-1$ edges. But in the questions, number of edges is less than or equal to $n-2$. Hence G cannot be connected.

28. Represent the weighted graph given below in the matrix form.

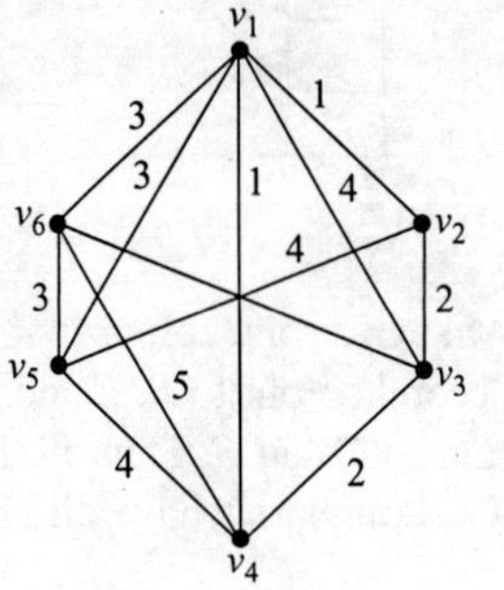

Figure 17.263

Ans.
$$\begin{bmatrix} \infty & 1 & 4 & 1 & 3 & 3 \\ 1 & \infty & 2 & \infty & 4 & \infty \\ 4 & 2 & \infty & 2 & \infty & \infty \\ 1 & \infty & 2 & \infty & 4 & 5 \\ 3 & 4 & \infty & 4 & \infty & 3 \\ 3 & \infty & \infty & 5 & 3 & \infty \end{bmatrix}$$

The diagonal is $\infty\ \infty \ldots \infty$ and the matrix is symmetric with respect to the diagonal.

29. Represent the expression given below by a binary tree and obtain the prefix form and postfix form of the expression.

$$a{\cdot}b-(c/(d+e)).$$

Ans.

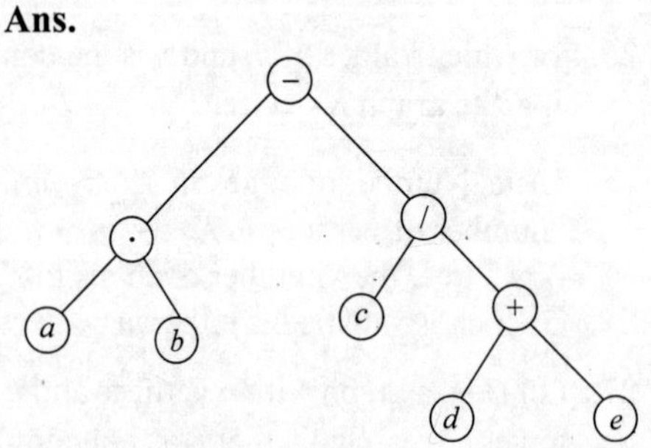

Prefix form: $- \cdot ab - c + de$
Postfix form: $ab \cdot de + c/-$

30. Convert the following postfix form of a binary tree into prefix form and parenthesized infix form and usual infix form

$$ab + c - .$$

Ans.

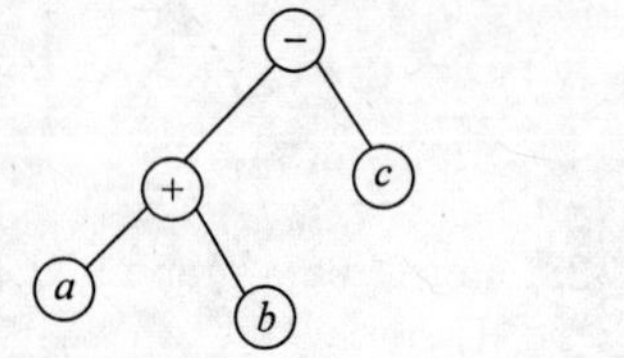

Prefix form: $- + abc$
Parenthesized form: $(a+b)-c$
Usual infix form: $a+b-c$

31. Draw a tree for the algebraic expression

$$(3x+y)\,(4m+5n)^2$$

and find its prefix polish form.

Ans. Using ↑ for exponentiation and * for multiplication, the binary tree representing the algebraic expression is

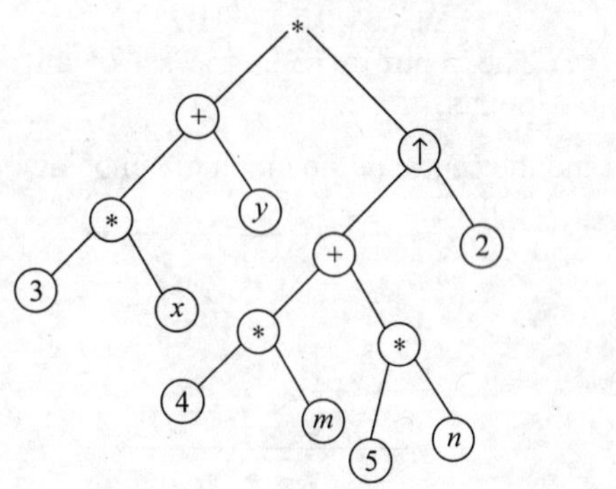

The prefix form is

$$* + * 3xy \uparrow + * 4m * 5n$$

$$232* \uparrow + 233 - 24 = * \uparrow (2+3)^3 (2-4)$$
$$= * (2+3)^3 (2-4)$$
$$= (2+3)^3 * (2-4)$$
$$= 5^3 * (-2)$$
$$= -250$$

32. If ↑ denotes exponentiation, evaluate the polish form

$$* \uparrow + 233 - 24.$$

33. Find all cut sets for the graph G given below.

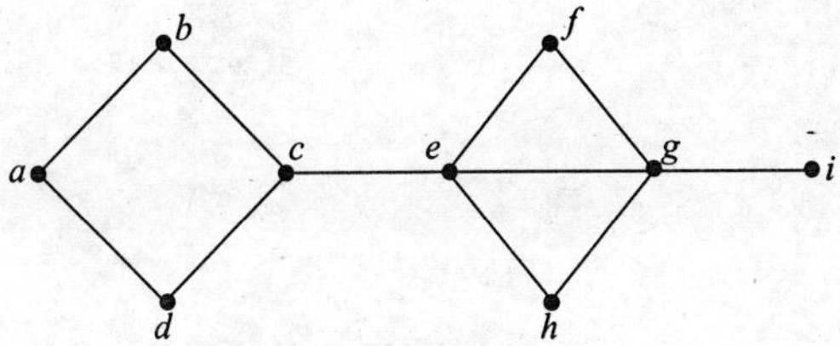

Figure 17.264

Ans. $\{c, e\}$; $\{\{a, b\}, \{a, d\}\}$; $\{\{a, d\}, \{b, c\}\}$; $\{\{e, f\}, \{e, g\}\}$ $\{h, g\}\}$; $\{g, i\}$; $\{\{e, h\}, \{h, g\}\}$. Try for other cut edges.

34. Find a minimal spanning tree for the graph shown below:

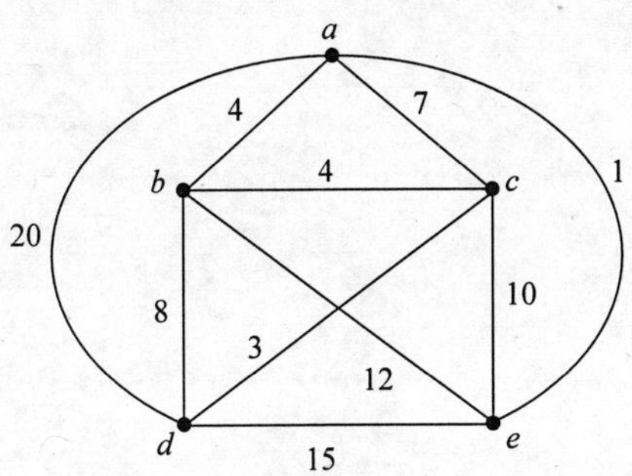

Figure 17.265

Ans.

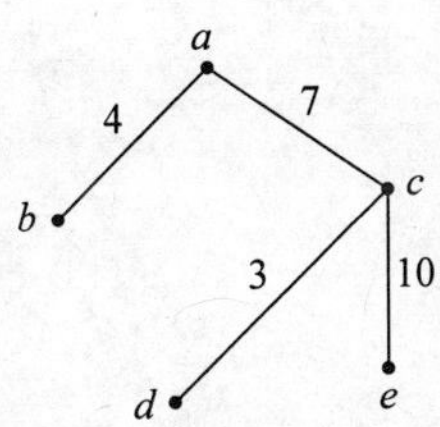

35. Find a minimal tree for the graph of Exercise 34 by deleting one by one those costliest edges whose deletion does not disconnect the graph.

Ans. First remove $\{a, \mathrm{d}\}$, then $\{d, c\}$, $\{a, e\}$, $\{b, e\}$. We cannot remove $\{c, e\}$ now because the graph will be disconnected by doing so. But we can delete $\{b, d\}$. Now delete any of $\{a, c\}$ or $\{b, c\}$. We will get two minimal trees.

36. Find all spanning trees of the graph shown below:

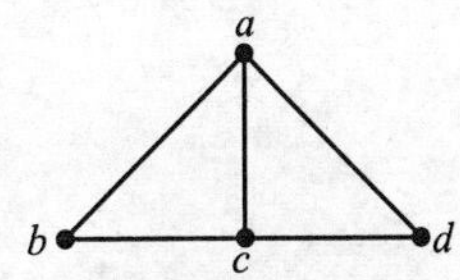

Figure 17.266

Ans. Each spanning tree must have $n-1 = 4-1 = 3$ edges. There are eight such trees.

37. Find the value of the maximum flow in the network of Example 17.109 and verify the value obtained by max flow min cut theorem.

Ans. Value (F) for max flow = 7, minimum cut (k) is shown in Figure 17.248.

38. Find the value of the maximum flow in the network shown in Figure 17.267 and verify your answer using max flow min cut theorem.

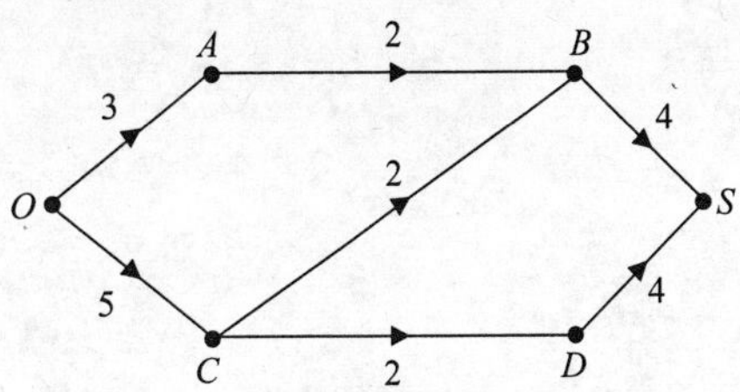

Figure 17.267

Ans. Max flow = 6

Index